Development of Unconventional Reservoirs 2020

Development of Unconventional Reservoirs 2020

Editor

Reza Rezaee

MDPI • Basel • Beijing • Wuhan • Barcelona • Belgrade • Manchester • Tokyo • Cluj • Tianjin

Editor
Reza Rezaee
Curtin University
Australia

Editorial Office
MDPI
St. Alban-Anlage 66
4052 Basel, Switzerland

This is a reprint of articles from the Special Issue published online in the open access journal *Energies* (ISSN 1996-1073) (available at: https://www.mdpi.com/journal/energies/special_issues/development_unconventional_reservoirs_2020).

For citation purposes, cite each article independently as indicated on the article page online and as indicated below:

LastName, A.A.; LastName, B.B.; LastName, C.C. Article Title. *Journal Name* **Year**, *Volume Number*, Page Range.

ISBN 978-3-0365-1753-7 (Hbk)
ISBN 978-3-0365-1754-4 (PDF)

Contents

About the Editor

Reza Rezaee (Professor) of Curtin's Department of Petroleum Engineering has a PhD in Reservoir Characterization. His research has been mostly on integrated solutions for reservoir characterization, formation evaluation, and petrophysics. He has also worked on the application of artificial intelligence in the oil and gas industry for many years. Currently, he is focused on unconventional gas including shale gas and tight gas sand studies. He has over 27 years' experience in academia being responsible for both teaching and research. During his career, he has been engaged in several research projects supported by major national and international oil and gas companies and these commissions, together with his supervisory work at various universities, have involved a wide range of achievements. During his research career, he has led several major research projects funded by various oil and gas companies. He has received a total of more than $3.2M funds through his collaborative research projects. He has supervised over 75 M.Sc. and PhD students during his university career to date. He has published more than 180 peer-reviewed journal and conference papers and is the author of 5 books on petroleum geology, logging and log interpretation, and gas shale reservoirs. As a founder of the "Unconventional Gas Research Group" of Australia, he has established a unique and highly sophisticated research lab at the Department of Petroleum Engineering, Curtin University. This lab was established to research petrophysical evaluation of tight gas sands and shale gas formations.

He is the Editor-in-Chief of Improved Oil and Gas Recovery journal, Associate Editor of Marine and Petroleum Geology, and Associate Editor of Geofluids.

Preface to "Development of Unconventional Reservoirs 2020"

After the successful publication of the Development of Unconventional Reservoirs book a subsequent special issue with the same theme "Development of Unconventional Reservoirs 2020" was commenced and a total of 41 papers were published in this edited book.

The topics covered by this special issue are very diverse but they can be grouped into several categories based on the type of unconventional plays covered by each paper.

The first set of the papers covers some aspects of **shale gas plays** that include the following works:

Review of Formation and Gas Characteristics in Shale Gas Reservoirs; by Boning Zhang, Baochao Shan, Yulong Zhao, Liehui Zhang

Effect of Particle Size on Pore Characteristics of Organic-Rich Shales: Investigations from Small-Angle Neutron Scattering (SANS) and Fluid Intrusion Techniques; by Yi Shu, Yanran Xu, Shu Jiang, Linhao Zhang, Xiang Zhao, Zhejun Pan, Tomasz P. Blach, Liangwei Sun, Liangfei Bai, Qinhong Hu, Mengdi Sun

Porosity and Water Saturation Estimation for Shale Reservoirs: An Example from Goldwyer Formation Shale, Canning Basin, Western Australia; by Atif Iqbal, Reza Rezaee

Factors Affecting Shale Gas Chemistry and Stable Isotope and Noble Gas Isotope Composition and Distribution: A Case Study of Lower Silurian Longmaxi Shale Gas, Sichuan Basin; by Chunhui Cao, Liwu Li, Yuhu Liu, Li Du, Zhongping Li, Jian He

Quantitative Analysis of Amorphous Silica and Its Influence on Reservoir Properties: A Case Study on the Shale Strata of the Lucaogou Formation in the Jimsar Depression, Junggar Basin, China; by Ke Sun, Qinghua Chen, Guohui Chen, Yin Liu, Changchao Chen

Molecular-Scale Considerations of Enhanced Oil Recovery in Shale; by Mohamed Mehana, Qinjun Kang, Hari Viswanathan

Dynamic Pore-Scale Network Modeling of Spontaneous Water Imbibition in Shale and Tight Reservoirs; by Xiukun Wang, James J. Sheng

Patent Analysis on the Development of the Shale Petroleum Industry Based on a Network of Technological Indices; by Jong-Hyun Kim, Yong-Gil Lee

An Analytical Model for Production Analysis of Hydraulically Fractured Shale Gas Reservoirs Considering Irregular Stimulated Regions; by Kaixuan Qiu, Heng Li

Development and Calibration of a Semianalytic Model for Shale Wells with Nonuniform Distribution of Induced Fractures Based on ES-MDA Method; by Qi Zhang, Shu Jiang, Xinyue Wu, Yan Wang, Qingbang Meng

Experimental Investigation of the Impacts of Fracturing Fluid on the Evolution of Fluid Composition and Shale Characteristics: A Case Study of the Niutitang Shale in Hunan Province, South China; by Jingqiang Tan, Guolai Li, Ruining Hu, Lei Li, Qiao Lyu, Jeffrey Dick

Reservoir Characteristics of the Lower Jurassic Lacustrine Shale in the Eastern Sichuan Basin and Its Effect on Gas Properties: An Integrated Approach; by Jianhua He, Hucheng Deng, Ruolong Ma, Ruyue Wang, Yuanyuan Wang, Ang Li

The second set of articles are about **tight gas sands** that include the following works:

Controls on Pore Structures and Permeability of Tight Gas Reservoirs in the Xujiaweizi Rift, Northern Songliao Basin; by Luchuan Zhang, Shu Jiang, Dianshi Xiao, Shuangfang Lu, Ren Zhang, Guohui Chen, Yinglun Qin, Yonghe Sun

Investigation of the Pore Structure of Tight Sandstone Based on Multifractal Analysis from NMR Measurement: A Case from the Lower Permian Taiyuan Formation in the Southern North China Basin; by Kaixuan Qu, Shaobin Guo

Finite Element Simulation of Multi-Scale Bedding Fractures in Tight Sandstone Oil Reservoir; by Qianyou Wang, Yaohua Li, Wei Yang, Zhenxue Jiang, Yan Song, Shu Jiang, Qun Luo, Dan Liu Occurrence, Classification and Formation Mechanisms of the Organic-Rich Clasts in the Upper Paleozoic Coal-Bearing Tight Sandstone, Northeastern Margin of the Ordos Basin, China; by Guanqun Yang, Wenhui Huang, Jianhua Zhong, Ningliang Sun

An Analytical Solution for Transient Productivity Prediction of Multi-Fractured Horizontal Wells in Tight Gas Reservoirs Considering Nonlinear Porous Flow Mechanisms; by Qiang Wang, Jifang Wan, Langfeng Mu, Ruichen Shen, Maria Jose Jurado, Yufeng Ye

The third group of papers deals with some aspects of **tight carbonate reservoirs** that include the following works:

Pore-Structural Characteristics of Tight Fractured-Vuggy Carbonates and Its Effects on the P- and S-Wave Velocity: A Micro-CT Study on Full-Diameter Cores; by Wei Li, Xiangjun Liu, Lixi Liang, Yinan Zhang, Xiansheng Li, Jian Xiong

Reservoir Properties of Low-Permeable Carbonate Rocks: Experimental Features; by Aliya Mukhametdinova, Andrey Kazak, Tagir Karamov, Natalia Bogdanovich, Maksim Serkin, Sergey Melekhin, Alexey Cheremisin

Natural Fractures in Carbonate Basement Reservoirs of the Jizhong Sub-Basin, Bohai Bay Basin,

China: Key Aspects Favoring Oil Production; by Guoping Liu, Lianbo Zeng, Chunyuan Han, Mehdi Ostadhassan, Wenya Lyu, Qiqi Wang, Jiangwei Zhu, Fengxiang Hou

The fourth group of papers focuses on **coalbeds** that include the following works:

Multi-Phase Tectonic Movements and Their Controls on Coalbed Methane: A Case Study of No. 9 Coal Seam from Eastern Yunnan, SW China; by Ming Li, Bo Jiang, Qi Miao, Geoff Wang, Zhenjiang You, Fengjuan Lan

A Novel Data-Driven Method to Estimate Methane Adsorption Isotherm on Coals Using the Gradient Boosting Decision Tree: A Case Study in the Qinshui Basin, China; by Jiyuan Zhang, Qihong Feng, Xianmin Zhang, Qiujia Hu, Jiaosheng Yang, Ning Wang

Development and Performance Evaluation of Solid-Free Drilling Fluid for CBM Reservoir Drilling in Central Hunan; by Pinghe Sun, Meng Han, Han Cao, Weisheng Liu, Shaohe Zhang, Junyi Zhu

Two papers investigate **gas hydrates** that include the following works:

Research on the Estimate of Gas Hydrate Saturation Based on LSTM Recurrent Neural Network; by Chuanhui Li, Xuewei Liu

Fractional Time Derivative Seismic Wave Equation Modeling for Natural Gas Hydrate; by Yanfei Wang, Yaxin Ning, Yibo Wang

The last group of papers is classified as **miscellaneous subjects** that cover a variety of works dealing with some technical aspects of different unconventional plays:

Reservoir Formation Model and Main Controlling Factors of the Carboniferous Volcanic Reservoir in the Hong-Che Fault Zone, Junggar Basin; by Danping Zhu, Xuewei Liu, Shaobin Guo

Productivity-Index Behavior for a Horizontal Well Intercepted by Multiple Finite-Conductivity Fractures Considering Nonlinear Flow Mechanisms under Steady-State Condition; by Maojun Cao, Hong Xiao, Caizhi Wang

Numerical Investigation of Injection-Induced Fracture Propagation in Brittle Rocks with Two Injection Wells by a Modified Fluid-Mechanical Coupling Model; by Song Wang, Jian Zhou, Luqing Zhang, Zhenhua Han

Proppant Transportation in Cross Fractures: Some Findings and Suggestions for Field Engineering; by Yan Zhang, Xiaobing Lu, Xuhui Zhang, Peng Li

A Novel Equivalent Continuum Approach for Modelling Hydraulic Fractures; by Eziz Atdayev, Ron C. K. Wong, David W. Eaton

Keywords: Unconventional reservoirs; shale gas; shale oil; tight gas sand; coalbed methane; tight carbonates, gas hydrate, hydraulic fracturing, machine learning

Reza Rezaee
Editor

Review

Review of Formation and Gas Characteristics in Shale Gas Reservoirs

Boning Zhang [1], Baochao Shan [2,*], Yulong Zhao [1] and Liehui Zhang [1]

[1] State Key Laboratory of Oil and Gas Reservoir Geology and Exploitation, Southwest Petroleum University, Chengdu 610500, China; zhangboning@zhenhuaoil.com (B.Z.); swpuzhao@swpu.edu.cn (Y.Z.); zhangliehui@vip.163.com (L.Z.)

[2] State Key Laboratory of Coal Combustion, Huazhong University of Science and Technology, Wuhan 430074, China

* Correspondence: shanbaochao@hust.edu.cn or rzshan@foxmail.com

Received: 3 June 2020; Accepted: 14 October 2020; Published: 17 October 2020

Abstract: An accurate understanding of formation and gas properties is crucial to the efficient development of shale gas resources. As one kind of unconventional energy, shale gas shows significant differences from conventional energy ones in terms of gas accumulation processes, pore structure characteristics, gas storage forms, physical parameters, and reservoir production modes. Traditional experimental techniques could not satisfy the need to capture the microscopic characteristics of pores and throats in shale plays. In this review, the uniqueness of shale gas reservoirs is elaborated from the perspective of: (1) geological and pore structural characteristics, (2) adsorption/desorption laws, and (3) differences in properties between the adsorbed gas and free gas. As to the first aspect, the mineral composition and organic geochemical characteristics of shale samples from the Longmaxi Formation, Sichuan Basin, China were measured and analyzed based on the experimental results. Principles of different methods to test pore size distribution in shale formations are introduced, after which the results of pore size distribution of samples from the Longmaxi shale are given. Based on the geological understanding of shale formations, three different types of shale gas and respective modeling methods are reviewed. Afterwards, the conventional adsorption models, Gibbs excess adsorption behaviors, and supercritical adsorption characteristics, as well as their applicability to engineering problems, are introduced. Finally, six methods of calculating virtual saturated vapor pressure, seven methods of giving adsorbed gas density, and 12 methods of calculating gas viscosity in different pressure and temperature conditions are collected and compared, with the recommended methods given after a comparison.

Keywords: shale gas reservoir; geology; Gibbs excess adsorption; supercritical adsorption; gas viscosity

---

1. Introduction

Shale gas has received great attention from governments all over the world, especially after the successful shale gas revolution in North America. As a type of clean energy with abundant reserves, shale gas is believed to be one of the most promising replacements for conventional energy in the future. According to the Energy Information Administration (EIA) (2016), shale gas production is expected to drive world natural gas production growth in the coming decades and will account for approximately 30% of world natural gas production by 2040 (Figure 1a). The United States and China are predicted to be the two largest shale gas producers in the world by the end of the forecast period, with shale gas production making up 70% and 40%, respectively, of each country's total natural gas production (Figure 1b).

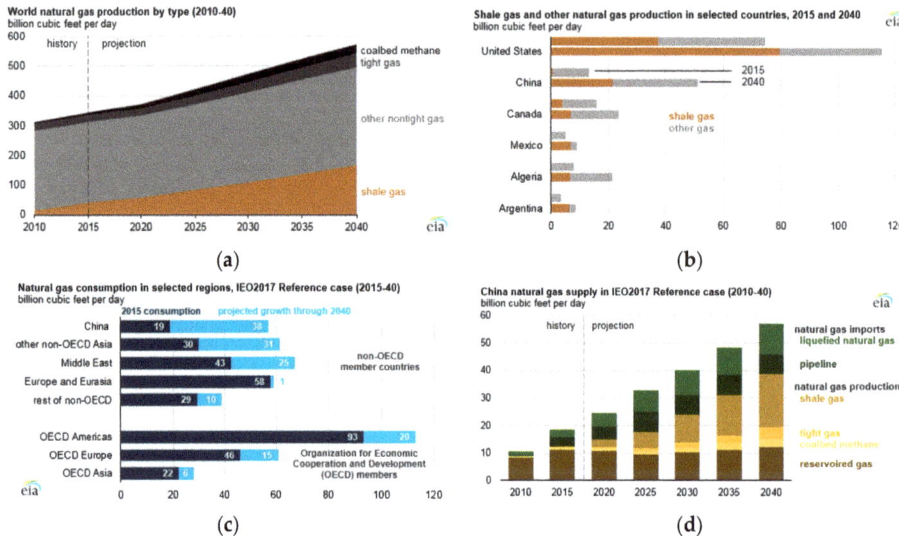

Figure 1. Current and predicted situations of shale gas resources in different countries or regions (from EIA). (**a**) Natural gas production by type. (**b**) Shale gas production in different countries. (**c**) Natural gas consumption in different regions. (**d**) Natural gas supply by type in China.

Taking China as an example, due to the relatively high economic growth and increasing attention to environmental protection, natural gas consumption is expected to increase from 19 Bcf/d in 2015 to 57 Bcf/d in 2040 (Figure 1c), accounting for a quarter of all global natural gas consumption growth between 2015 and 2040 (EIA, 2017). Driven by the development of shale gas resources, China's domestic natural gas supply will grow from 13 Bcf/d in 2016 to 39 Bcf/d by 2040, with shale gas production increasing from 0.7 Bcf/d in 2016 to 10 Bcf/d by 2030 and 19 Bcf/d by 2040. As we can see from Figure 1d, shale gas is expected to increase the fastest and will account for more than 30% of the total natural gas supply in China by 2040.

As one of the unconventional energy resources, shale gas reservoirs have uniqueness and complexity in terms of the gas storage type, transporting mechanism, and reservoir development mode, which makes the commercial production of shale plays a very challenging task for petroleum engineers. The United States, Canada, and China are the only three countries that produced commercial volumes of natural gas from shale formations by 2015. Although horizontal well drilling and hydraulic fracturing have been applied to produce shale gas in Australia and Russia, no commercial gas volumes were obtained from low-permeability shale formations. Currently, the commercial development of shale gas resources in North America and China mainly benefits from advanced engineering technology. The theoretical understanding of shale gas storage capacity, gas transporting mechanism in nanopores or micropores, and pore structure characterization is still not clear, being far behind engineering practice [1].

In this review, our attention will mainly be paid to three aspects: (1) petrological, organic geochemical characteristics and micropore structures of shale formations; (2) different types of adsorption models as well as their principles and application range, including Gibbs excess sorption, supercritical adsorption phenomenon, and adsorption/absorption models; (3) different methods of calculating gas physical properties, such as virtual saturated vapor pressure, adsorbed gas density, free gas density, free gas viscosity, etc. Different models on each subject will be compared and evaluated based on their physical meaning, reliability, accuracy, and applicability, which are significant for accurate numerical simulation and enhancing hydrocarbon recovery in shale gas reservoirs.

2. An Overview of Pore Structures and Gas Types in Shale Formations

The complex micropore structure of shale plays is determined by its special accumulation processes. An overall and deep understanding of pore structures and gas storage types is key to proper reservoir assessment and precise numerical simulation.

The pore structure of shale formations can be detected by the observation description method and the physical test method [2,3]. The observation description method adopts radiation techniques, mainly referring to the means of optical microscopy, scanning electron microscopy (SEM), scanning transmission electron microscopy (STEM), transmission electron microscopy (TEM), nuclear magnetic resonance (NMR), and small angle X-ray scattering (SAXS) to directly describe pore geometrical shapes, connectivity, and pore filling of shale formations [4,5]. The physical test method mainly refers to fluid penetration experiments, which utilize tests of fluid mass, volume, and pressure to obtain the pore sizes volumes indirectly, including mercury injection capillary pressure (MICP), helium (He) porosity, liquid N_2 adsorption, low-temperature CO_2 adsorption, etc. [6–9]. The resolution of the different methods is shown in Figure 2.

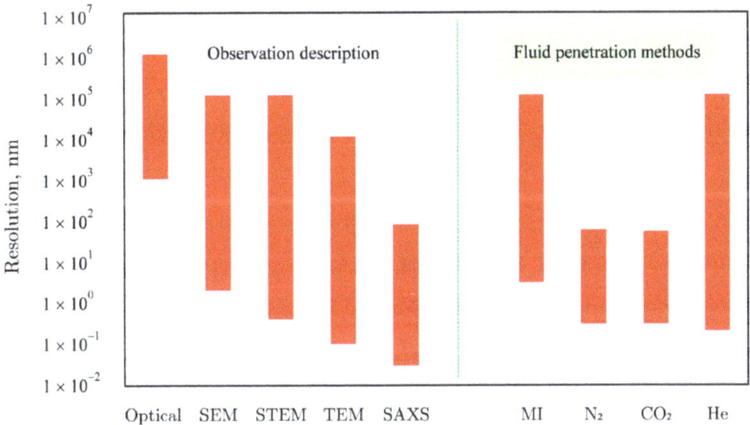

Figure 2. The resolution of different methods to characterize micropore structures [2–4].

In this section, we summarize the petrological and organic-geochemical characteristics of shale gas reservoirs in Sichuan basin in China compared to the shale gas development in North America. The chosen samples are marine shale of the Lower Silurian Longmaxi Formation. The porosity and permeability of the main shale gas reservoirs in North America and China are collected and tested, respectively, based on which the storage space types and pore size distributions are analyzed. Finally, different kinds of shale gas as well as their modeling methods are identified and compared.

2.1. Petrological and Geochemical Characteristics

In this part, our previous work related to shale gas reservoir characterization and assessment is introduced. Samples of the Longmaxi Formation in south Sichuan basin are collected as experimental objects.

The highly mature Longmaxi marine shale is one of the most important candidates for the commercial development of shale gas resources in China [10]. The total organic carbon (TOC) content ranges from 0.4% to 18.4%, with the organic matter (OM) mainly composed of type I and II1 kerogen [10,11]. Its vitrinite reflectance (R_0) values range from 1.8% to 4.2% [11]. The Longmaxi Shale is found to be porous and permeable [11,12], with porosity ranging from 1.2% to 10.8% and permeability ranging from 0.25 μD to 1.737 mD. Other geological and petrophysical characteristics of the Longmaxi Shale Formation can be found in previous publications [13,14].

2.1.1. Mineral Composition

The rock mechanics, adsorption capacity, and well productivity of shale gas reservoirs are directly determined by the relative content of different minerals due to the property differences among the minerals. Taking 18 samples from X1 well, eight samples from X2 well, and 60 samples from X3 well, we tested the mineral compositions using an X'Pert Pro type X-ray scattering diffractometer produced by PANalytical B.V. (Almelo, The Netherlands) and following the Chinese Oil and Gas Industry Standard SY/T 5163 1995 and SY/T 5983 94. The laboratory temperature is 24 °C and the humidity is 30%. The test results are shown in Figure 3. As shown, the main mineral compositions of the Longmaxi Formation are quartz (11–70%, average 31.06%) and clay minerals (7–64%, average 33.87%). Comparing the results with those in the Mississippian Barnett Shale of the Fort Worth Basin in North America [15,16], the content of brittle minerals (such as quartz, feldspar, and calcite) in the Longmaxi Formation is relatively lower, while the content of clay minerals is higher (Figure 4).

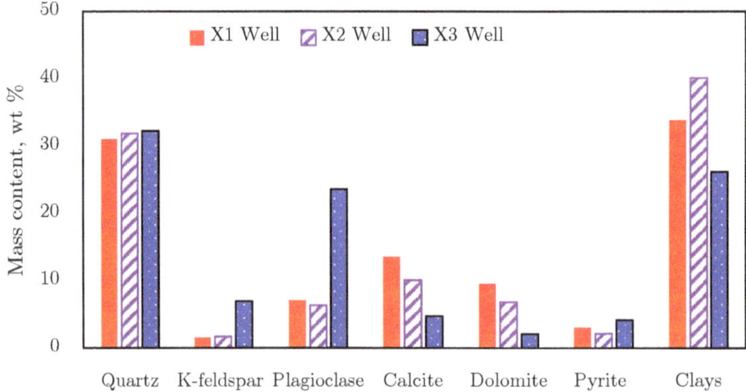

Figure 3. Relative content of minerals of samples from the Longmaxi Formation in the south Sichuan Basin.

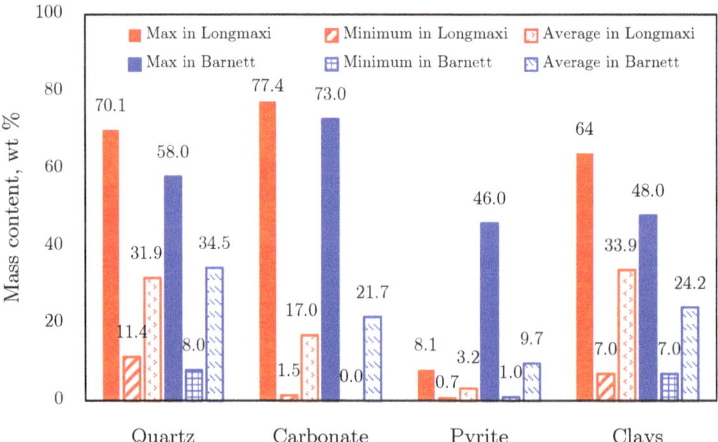

Figure 4. Comparison of mineral content between shale samples from the Longmaxi Formation and the Barnett Shale [15,16].

With less clay mineral and more brittle minerals, natural or induced fractures more easily develop under external forces. On the contrary, the higher clay mineral content has a negative effect on volume stimulation, since most of the energy is absorbed by shale formations. Under such circumstances, plane fractures are more likely to be generated, rather than tree-like or reticular structural fractures. Generally, the brittle minerals need to be higher than 40% and the clay minerals need to be less than 30% for a potential shale gas reservoir to be commercially developed [16,17].

The content of different minerals also exhibits different trends in terms of formation depth, as shown in Figure 5. For the formation depth increasing from 2120 m to 2250 m, the content of calcite decreases from 12% to 5%, and the content of brittle minerals increases from 28% to 60%. The content of clay minerals increases from 44% to 48% for the formation depth, increasing from 2120 m to 2205 m, while it decreases rapidly from 48% to 8% for formation depth increasing from 2205 m to 2250 m.

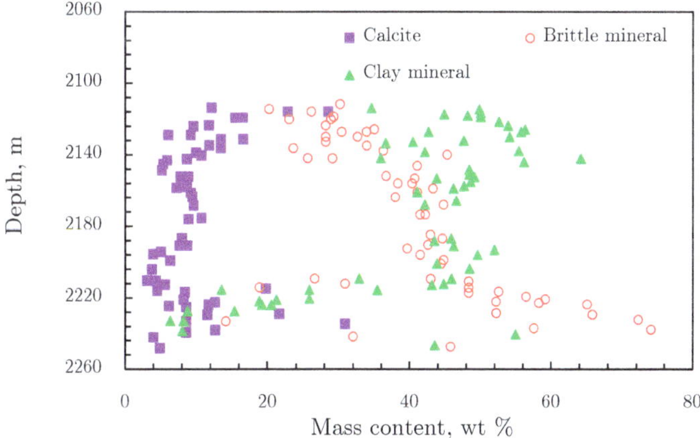

Figure 5. The relationship between mineral mass content and depth.

The TOC content is determined by a CS230 carbon/sulfur analyzer (LECO, St. Joseph, USA) with samples crushed into powder less than 100-mesh. The powder is pyrolyzed up to 600 °C and the inorganic carbon is removed by hydrochloric acid. The relationship between quartz and TOC content can help to explain the origin of quartz, i.e., detrital or biogenic.

There is no linear relationship between detrital quartz and TOC content, while a positive correlation can be found for biogenic quartz and TOC. As shown in Figure 6, the correlation coefficients between quartz and TOC content in X1, X2, and X3 wells are larger than 0.53, indicating that the quartz in the targeted formation is biogenic.

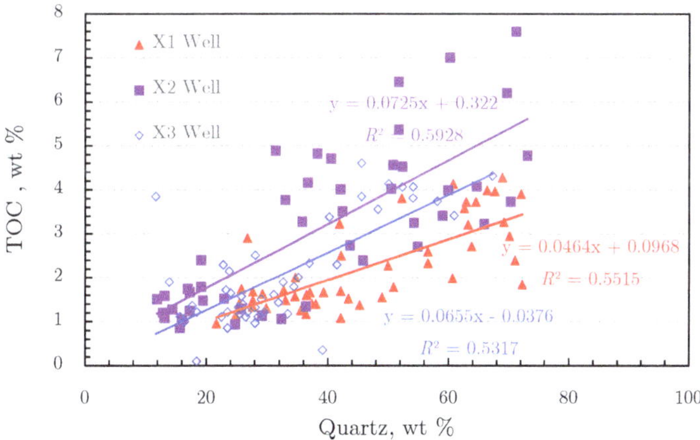

Figure 6. The relationship between quartz and TOC contents in Longmaxi Formation.

2.1.2. Organic Geochemical Characteristics of Shale

The organic matter richness, thermal maturity, and kerogen types are three key parameters for the accurate assessment of hydrocarbon-forming conditions. The organic matter richness not only affects the hydrocarbon generating strength, but also the development of organic pores and the adsorbed gas content. The lower limit value of the TOC content for economic exploitation of shale gas reservoirs is approximately at 2.5–3 wt % [18]. However, with the development of technology, this value could become even lower.

Measuring 122 samples of TOC content, we find that the TOC content ranges from 0.43% to 8.39%, with an average value of 2.20% in Longmaxi Formation of Sichuan Basin. The samples with a TOC content less than 2.00% account for 57.38%, while 42.62% of samples have a TOC content larger than 2.00%. This reflects the fact that TOC is abundant in Longmaxi Formation, which is advantageous for shale gas generation and storage. However, comparing with shale gas reservoirs in North America, the TOC content in Sichuan Basin is smaller. The TOC content of Antrim shale and New Albany shale is between 1% and 25%, while it is between 0.45% and 4.5% in the Barnett Shale and Lewis Shale [19].

Analyzing the TOC content data of three wells in the Longmaxi Formation longitudinally, we find a positive relationship between TOC content and depth, as shown in Figure 7. The TOC content at the bottom of the formation is much larger than that at the top. The TOC content increases with depth at 2.3–10.0% per 100 m in the targeted formation.

The kerogen types can be classified into sapropelic type (type I), mixed -type (type II), and humic type (type III) [20]. All three types of kerogen can generate natural gas. For type I and II1 kerogen, oil is generated first and then cracked into gas. The type III kerogen is not advantageous for oil generation and gas is formed directly from organic matter [17]. The abundance of organic matter is the material basis for hydrocarbon generation, while the type of organic matter determines the hydrocarbon generating potential and hydrocarbon characteristics.

The thermal evaluation extent of organic matter can be characterized by thermal maturity, reflected by the vitrinite reflectance R_o, which is the basis of assessing the hydrocarbon generating potential of source rocks (Table 1). The organic matter maturity range of $1.1\% < R_o < 3.5\%$ is advantageous for the generation of shale gas [3,21]. Single well production in more mature shale formations is larger than in less mature ones, because more gas is generated by kerogen or the thermal cracking of crude oil in more mature areas. The average thermal maturity of the targeted Barnett Shale is $R_o = 1.7\%$, with the maximum value larger than 2.0% [22]. Oil and gas are generated from the kerogen with initial $R_o < 1.1\%$ in the Barnett Shale, but gas is produced within the formation in the Newark East and surrounding areas at higher thermal maturity, i.e., $R_o > 1.1\%$ [23]. The effect of organic matter maturity R_o on shale gas reservoirs is very complicated and needs further study.

Table 1. Characterization of thermal evaluation of organic matter [24].

Maturity of OM	Maturity Class	Hydrocarbon Generating Stage
$R_o < 0.5\%$	Immaturity	Biochemical gas-genous stage
$0.5\% < R_o < 1.3\%$	Maturity	Thermal catalytic oil and gas genous stage
$1.3\% < R_o < 2.0\%$	High maturity	Thermal cracking condensate gas-genous stage
$R_o > 2.0\%$	Overmaturity	Deep high temperature gas-geneous stage

The R_o of the Longmaxi Shale Formation in the Sichuan Basin ranges from 2.4 wt % to 4.0 wt %, mainly in the stage of high maturity and overmaturity.

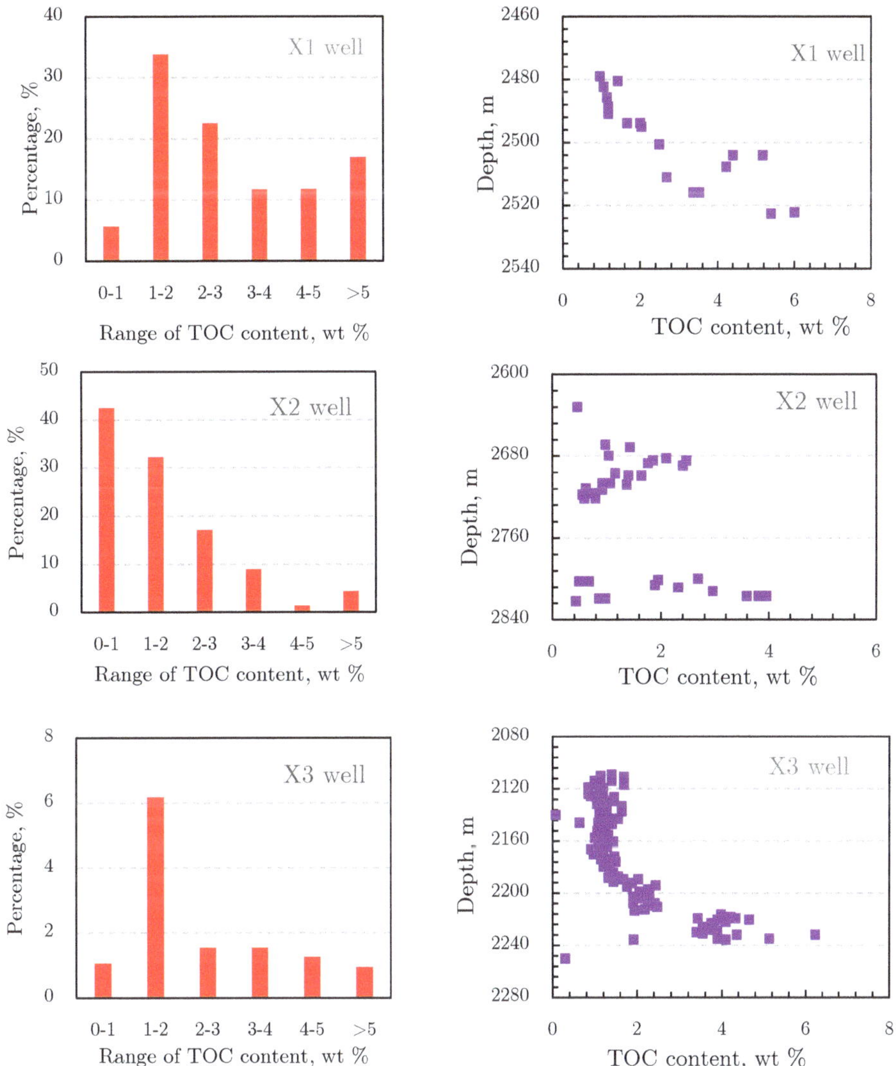

Figure 7. The frequency distribution histogram of TOC content as well as its relationship with depth in the Longmaxi Formation.

2.2. Porosity and Permeability Characterization

Porosity and permeability are the two most important parameters to characterize gas storage and seepage capacities in shale gas reservoirs. Compared to conventional reservoirs, shale gas reservoirs are ultra-tight formations with extremely low porosity and permeability. Corresponding formation physical properties of main shale gas reservoirs in North America are attributed in Table 2 [3,15,18,19,25,26].

Table 2. Statistical physical properties of the main shale gas reservoirs in North America [3,15,18,19,25,26].

Properties	Woodford	Marcellus	Fayetteville	Haynesville	Barnett	Antrim	New Albany	Lewis	Ohio
Permeability, mD	-	-	-	-	0.01	<0.1	<0.1	<0.1	<0.1
Total porosity, %	3-9	10	2-8	8-9	4-5	9	10-14	3-5.5	4.7
Logging porosity, %	3-6.5	5.5-7.5	4-12	8-10	6.5-8.5	-	-	-	-
Gas porosity, %	-	-	-	6-7.5	2.5	4	5	1-3.5	2
Water porosity, %	-	-	-	-	1.9	4	4-8	1-2	2.5-3
Water-filled porosity, %	10	12-35	15-35	15-20	25	-	-	-	-

As the earliest country to commercially develop shale gas resources, the USA has formulated a standard system to evaluate the physical properties of shale formations. The Gas Research Institute (GRI) of America proposed a test method for shale cores to determine both the total porosity and the gas-bearing porosity of the shale matrix. Generally, the porosity range of shale formations is between 2% and 15%. From the statistical data in Table 2, we can see that the total porosity of shale formation in North America is between 2% and 14%, and the average value is between 4.22% and 6.51%. Following the procedure of GRI, the statistical results of measured gas-filled porosity is between 1% and 7.5%, and water-filled porosity is between 1% and 8% in the Longmaxi Formation. The permeability of measured samples in the Longmaxi Formation is less than 0.1 mD, and the average pore-throat radius is smaller than 0.002 μm.

2.3. Pore Structure Division

Pore structures in shale formations can roughly be divided into two types: matrix pores and fractures (Table 3). Matrix pores are the main storage space of shale gas, directly determining the reserve of a shale gas reservoirs. The development of fractures as well as the connectivity of pores determines the gas-transporting and -producing capabilities [27].

Loucks et al. [28] studied the pore structures of Barnett Shale and concluded that micropores ($d \geq 0.75$ μm) and nanopores ($d < 0.75$ μm) are the two main pore types. Meanwhile, nanopores were divided into organic pores, intergranular pores, intragranular pores, and mixed pores. Slatt and O'Brien [29] analyzed pore types in the Barnett and Woodford Shale and the main pore types were intergranular pores of clay minerals, micropores in organic matter, pores in fecal spherulites, pores in bioclastic, and intergranular micropores. According to pore sizes, Loucks et al. [30] divided mudrock pores into: picopores ($r < 1$ nm), nanopores (1 nm $< r < 1$ μm), micropores (1 μm $< r < 62.5$ μm), mesopores (62.5 μm $< r < 4$ mm), and macropores (4 mm $< r < 256$ mm).

Table 3. Classification of storage space in shale formation.

Major Classes	Subgroups		Types
Fractures	Fractures		Tectonic extensional fractures (Figure 8a) [31] Tectonic shear fractures (Figure 8b) [31] Interlayer bedding fractures (Figure 8c) Rock convergent fractures (Figure 8d) Abnormal pressure fractures (Figure 8e)
Matrix pores	Inorganic matrix pores	Intergranular pores	Intergranular pores among grains (Figure 8f) Clay interslice pores (Figure 8g) Intercrystalline pores (Figure 8h) Marginal pores (Figure 8i)
		Intragranular pores	Dissolved pores (Figure 8j) Cleavage crack (Figure 8k) Biological pores (Figure 8l)
	Organic pores		

Based on previous studies [16,28–31], we came up with a new classification method of gas storage space in shale formations in this paper by comprehensively utilizing core observation, thin-section analysis, SEM analysis, and emission scanning electron microscopy after argon ion polishing technologies. The main storage and seepage space in shale formations can be classified into fractures and matrix pores (Table 3). According to the origin mode, matrix pores are further divided into inorganic pores and organic pores, among which inorganic pores include intergranular pores and intragranular pores. Fractures can be classified into tectonic extensional fractures, tectonic shear fractures, interlayer bedding fractures, rock convergent fractures, and abnormal pressure fractures (Table 3).

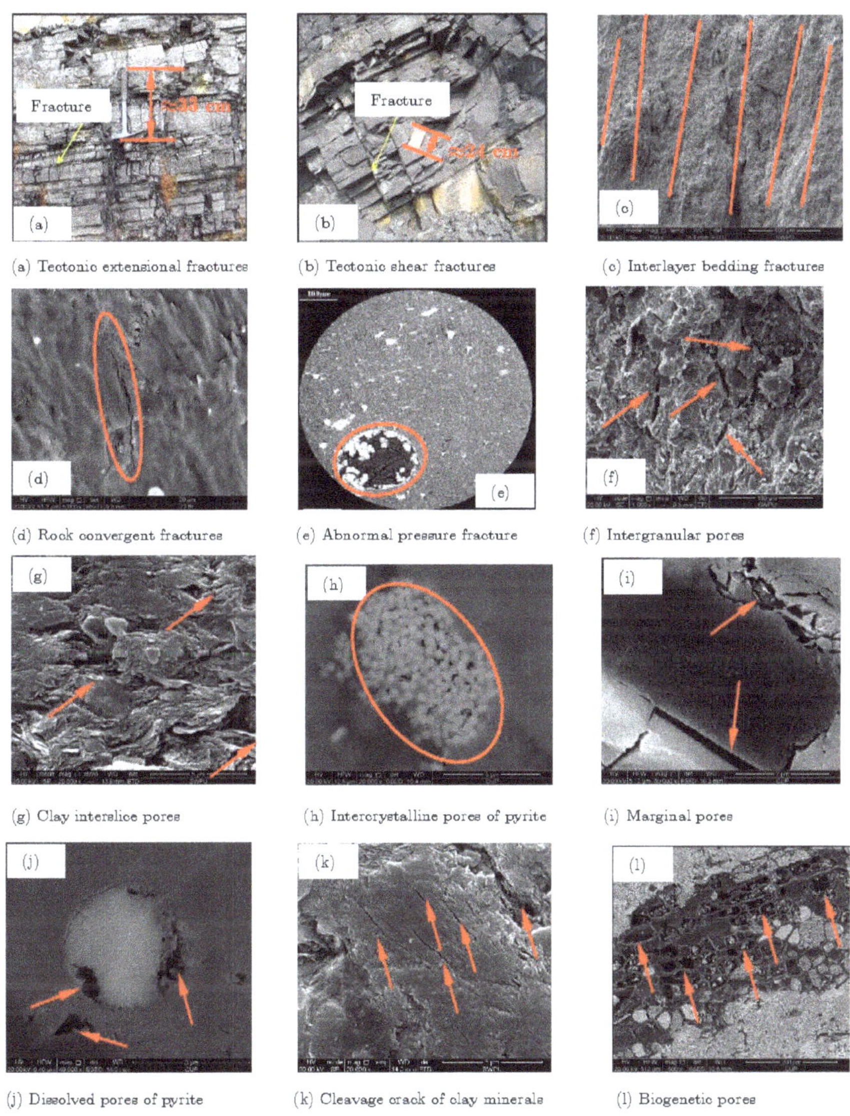

Figure 8. Different types of pores or fractures in the Longmaxi Shale Formation. (**a**) Tectonic extensional fractures. (**b**) Tectonic shear fractures. (**c**) Interlayer bedding fractures. (**d**) Rock convergent fractures. (**e**) Abnormal pressure fracture. (**f**) Intergranular pores. (**g**) Clay interslice pores. (**h**) Intercrystalline pores of pyrite. (**i**) Marginal pores. (**j**) Dissolved pores of pyrite. (**k**) Cleavage crack of clay minerals. (**l**) Biogenetic pores.

2.4. Pore Size Distribution and Influential Factors

Gas storage and seepage mechanisms vary significantly due to the difference of pore sizes. Clear knowledge of pore size distribution (PSD) in shale formations is essential for shale gas exploitation and development. MICP, gas adsorption, and NMR are three commonly used methods to determine PSD. In this section, the principles and results of different methods will be introduced and analyzed, based on the measurements of samples from the Longmaxi Formation in Sichuan Basin, China.

2.4.1. Test Methods and Principles

N_2 Adsorption Measurement

The Brunauer-Emmett-Teller (BET) adsorption model [32] is adopted to determine the specific surface of shale samples when $0.05 < p/p_s < 0.35$. The pressure is too small to achieve multilayer adsorption when $p/p_s < 0.05$, and capillary condensation may happen when $p/p_s > 0.40$. Some studies (e.g., [33]) pointed out that capillary condensation could not happen in shale gas reservoirs, since the common shale gas reservoir temperature is much higher than the critical temperature of shale gas (mainly methane). The two-parameter BET equation can be expressed as follows [32]:

$$\frac{p/p_s}{V(1-p/p_s)} = \frac{1}{V_m b} + \frac{b-1}{V_m b}\frac{p}{p_s}, \tag{1}$$

where V is the adsorbed gas volume, mL; V_m is the saturated adsorption volume of monolayer, mL; p is pressure, Pa; p_s is saturated vapor pressure, Pa; and b is a dimensionless constant related to the adsorption capacity.

After measuring the adsorbed gas amount G, a linear relationship between $p/[V(p_s - p)]$ and p/p_s $(0.05 < p/p_s < 0.35)$ can be found. According to the slope and intercept of the straight line, the saturated adsorption amount V_m can be calculated, by which the specific surface of samples can be obtained:

$$S_g = \frac{V_m A_m N_A}{22400 W} \times 10^{-18}, \tag{2}$$

where N_A is the Avogadro constant; A_m is the cross-section area of N_2 ($0.162\ \text{nm}^2$); W is the weight of the samples, g; and S_g is the specific surface of samples, m^2/g.

The Barrett-Joyner-Halenda (BJH) equation [34] is used to calculate PSD when $p/p_s > 0.40$:

$$r = -2\gamma V_m/[RT \ln(p/p_s)] + 0.354[-5/\ln(p/p_s)]^{1/3}, \tag{3}$$

where γ is the surface tension, N/m; R is the mole heat capacity, J/(mol·K); T is the environmental temperature, K; and r is the pore radius, m.

Mercury Injection Capillary Pressure

A mercury intrusion porosimeter is widely adopted to determine PSD in conventional sandstone reservoirs, where the pressure of mercury and pore radius r satisfy the Washburn equation [35]:

$$r = \frac{2\sigma \cos \xi}{p}, \tag{4}$$

where ξ is the contact angle between mercury and shale surface; σ is surface tension of mercury, 10^{-3} N/m; and p is the injection pressure, Pa.

The smallest pore radius that can be tested is determined by the highest pressure that mercury porosimetry can hold. In our study, a PoreMaster 60 mercury porosimeter is employed and its measurement range of pore size lies between 3.6 nm and 950 µm, but values near the lower limit can hardly be detected. This is because it is difficult to inject mercury into micro/nanopores due to the high capillary pressure. Meanwhile, high pressure may create artificial crack and stress sensitivity, which reduces the credibility of the measurement. Therefore, MICP is mainly used to analyze mesopores and macropores in shale samples.

Nuclear Magnetic Resonance (NMR)

The relaxation characteristic of a hydrogen nucleus under an external magnetic field is used to obtain the PSD by the NMR method, which causes no harm to the shale samples. The speed of relaxation is characterized by the longitudinal relaxation time T_1 and transverse relaxation time T_2. The relaxation characteristics of fluid in different-sized pores are different, based on which PSD can be calculated. The transverse relaxation time T_2 is composed of bulk phase relaxation T_{2B}, surface relaxation T_{2S}, and diffusion relaxation T_{2D}, which is expressed as follows:

$$\frac{1}{T_2} = \frac{1}{T_{2B}} + \frac{1}{T_{2D}} + \frac{1}{T_{2S}}. \tag{5}$$

The diffusion relaxation speed can be ignored compared to the surface relaxation speed in a uniform magnetic field, and the reciprocal of diffusion relaxation time T_{2D} is almost 0. Meanwhile, the bulk phase relaxation time T_{2B} is much bigger than the surface relaxation time T_{2S}. Therefore, $1/T_{2D}$ and $1/T_{2B}$ in Equation (5) can be ignored, and we have

$$\frac{1}{T_2} \approx \frac{1}{T_{2S}} = \rho_2 \frac{F_s}{r}. \tag{6}$$

Letting $C = F_s \cdot \rho_2$, we obtain the relationship between relaxation time T_2 and pore radius r via

$$r = CT_2. \tag{7}$$

Note that the transformation coefficient C in Equation (7) is an empirical parameter varying from one area to another, which can be determined by experiments [36]. In order to obtain the value of C, the T_2 spectrum need to be measured for specific shale samples first, and a N_2 adsorption test needs to be conducted on exactly the same samples (or samples from the same formation) afterwards [36]. Since core plugs are used in NMR and crushed rock samples are needed for the N_2 adsorption test, it is essential to perform the NMR test prior to N_2 adsorption. Comparing the two PSD results, the transformation coefficient C can be fitted. Accurate determination of C is a key part of determining the PSD of shale samples by NMR.

2.4.2. Pore Size Distribution

N_2 Adsorption Results

According to the principle of N_2 adsorption on measuring PSD, the measurable pore-throat size range is on the magnitude of nanometers, mainly micropores (<2 nm) and mesopores (2–50 nm). N_2 adsorption results (Figure 9) show that PSD displays a high single peak at the pore size range of 2 to 5 nm, implying that nanopores ranging from 2 nm to 5 nm are very developed in shale formations, which is advantageous for the storage of adsorbed gas. Meanwhile, we see that pores larger than 10 nm are not very developed. In a PSD frequency histogram (Figure 10), mesopores account for the largest percentage of all pores, followed by micropores and macropores. Micropores and mesopores accounted for more than 90% of the total pore volume.

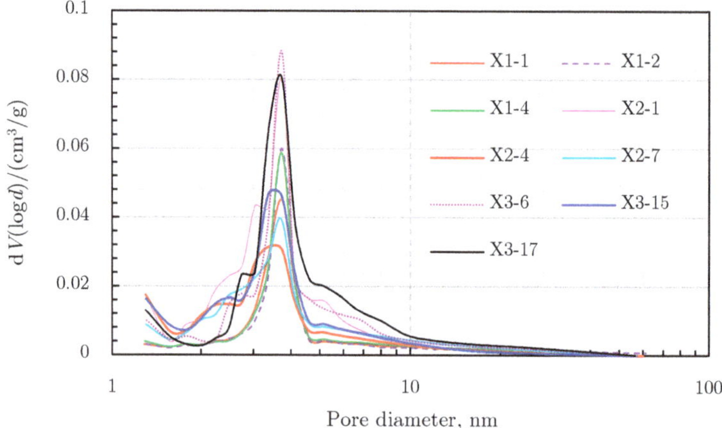

Figure 9. PSD of samples from Longmaxi formation based on low-temperature N_2 adsorption.

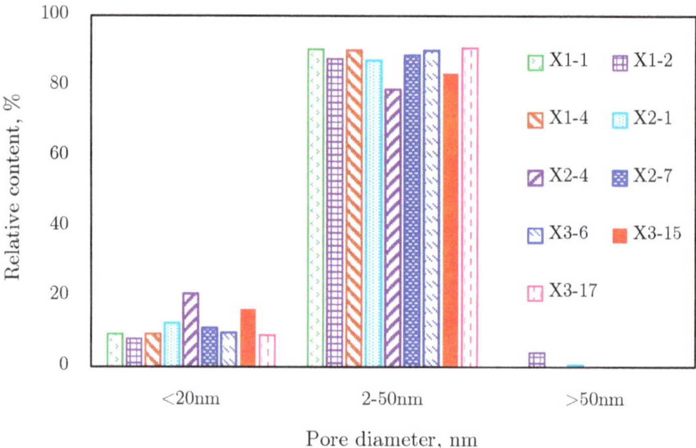

Figure 10. PSD histogram of samples from the Longmaxi Formation in Sichuan Basin based on N_2 adsorption.

Mercury Intrusion Results

Two peaks can be found on the MICP measurement results of PSD, as shown in Figure 11. The left peak is relatively small and smooth, corresponding to macropores of 10 nm to 1000 nm in organic matter and clay minerals. The right peak is very high, corresponding to a pore size of 40–200 μm. The highly developed lamellation in shale formations generates micro fractures, which may happen during sample preparation or mercury injection tests. Therefore, there is a high probability that the right peak corresponds to artificial fractures. Considering the fact that large pores (>5 μm) correspond to mercury injection pressure less than 0.14 MPa, these artificial fractures are more likely to be induced during sample preparation. Figure 12 shows that macropores account for the largest percentage (73.17%), followed by mesopores (26.83%). Due to the limitations of the instrument, no micropores are detected by MICP. Whether the tested macropores are primitive or induced needs to be determined by combining with other techniques, such as NMR.

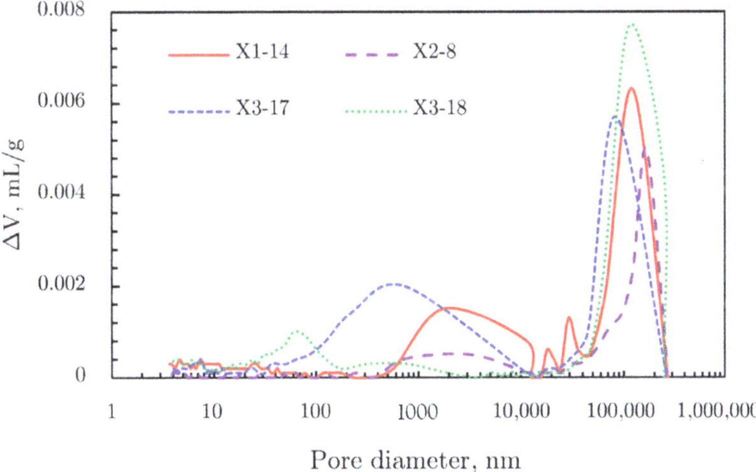

Figure 11. PSD of samples from the Longmaxi Formation based on mercury intrusion.

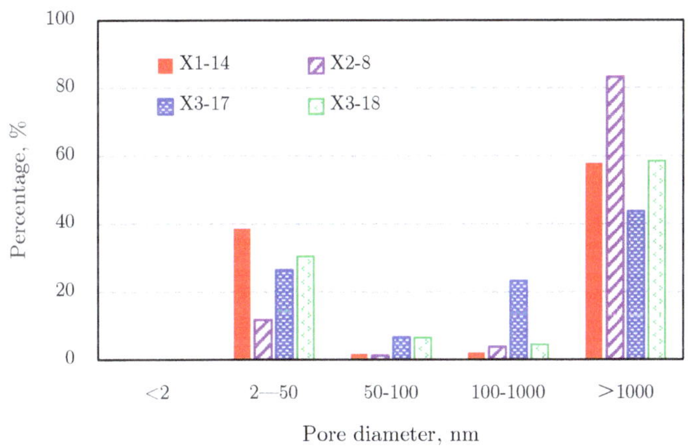

Figure 12. PSD histogram of samples from the Longmaxi Formation in Sichuan Basin based on mercury intrusion.

Nuclear Magnetic Resonance Results

Two or three peaks can be found on PSD, measured by NMR (Figure 13). The left peak, corresponding to pores smaller than 10 nm, has the largest percentage, while the other two peaks correspond to larger pore sizes of 800 nm and 7000 nm, respectively. The NMR results indicate that small pores are very developed in shale formations, while large pores account for a small but non-negligible percentage (Figure 14). Generally, the left two peaks correspond to micro-, meso-, and macro-pores in matrix, while the third peak corresponds to micro fractures. The PSD histogram measured by the NMR is similar to that of N_2 adsorption in terms of the small pore size range, while large pores could not be detected by N_2 adsorption, but by NMR.

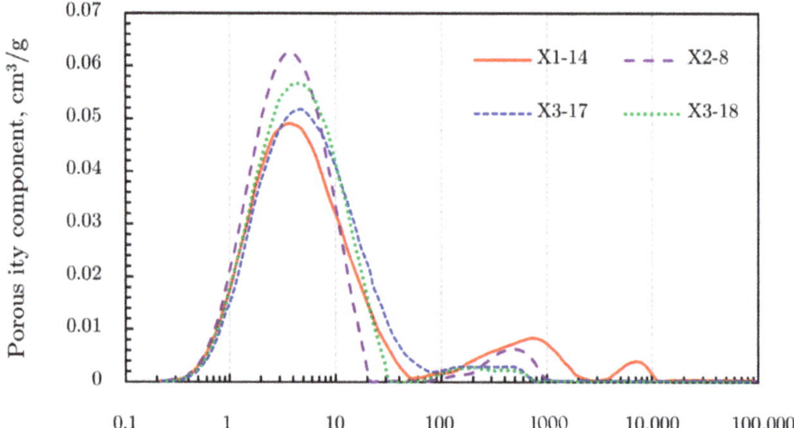

Figure 13. PSD of samples from the Longmaxi Formation based on NMR.

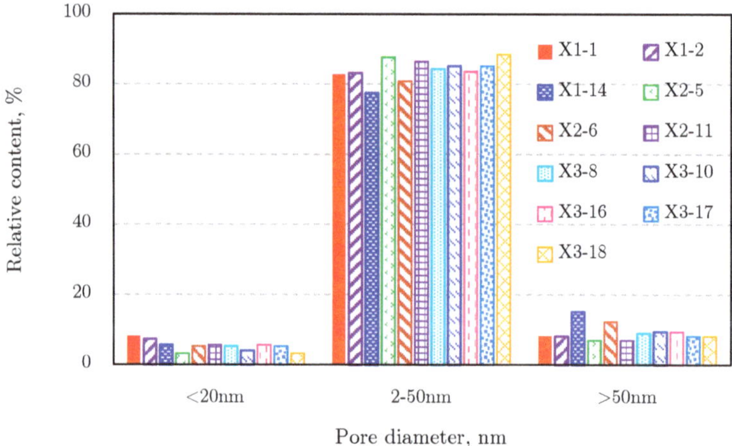

Figure 14. PSD histogram of samples from the Longmaxi Formation in the Sichuan Basin based on NMR.

2.4.3. Comprehensive Analysis of Pore Size Distribution

N_2 adsorption, MICP, and NMR can all measure PSD and reflect the heterogeneity of shale samples, with different test ranges. The lowest limit of N_2 adsorption is 0.35 nm, while the MICP test range is 3.6 nm–950 μm, and the NMR test range is 1 nm–5 mm. Comparing the results from the three methods, we find that the N_2 adsorption results mainly reflect the micropore and mesopore size distribution, while the NMR results reflect all pore size range and MICP results mainly test the development of macropores and micro fractures. Although the N_2 adsorption and NMR results display similar PSD trends of small pore size ranges, the peak of N_2 adsorption results (3–4 nm) is slightly smaller than that of NMR (4–5 nm). This is because samples are saturated by plant oil in the NMR test, and they struggle to enter micropores due to their large diameter compared to nitrogen molecules. Therefore, the micropore size determined by NMR is larger than that measured by N_2 adsorption. The nitrogen molecule is much smaller, so it enters micropores more easily than oil molecules. Therefore, the N_2 adsorption results are closer to the real data compared to NMR results.

N_2 adsorption can measure micropores and mesopores accurately, while MICP mainly tests macropores. Consequently, combining the two methods can better characterize PSD in shale formations.

Figures 15 and 16 show the PSD of the Longmaxi shale samples, tested comprehensively by N_2 adsorption and MICP, where pores smaller than 50 nm are measured by N_2 adsorption and pores larger than 50 nm are measured by MICP. Pore volume in the Longmaxi Shale Formation is mainly mesopores and macropores (including artificial fractures).

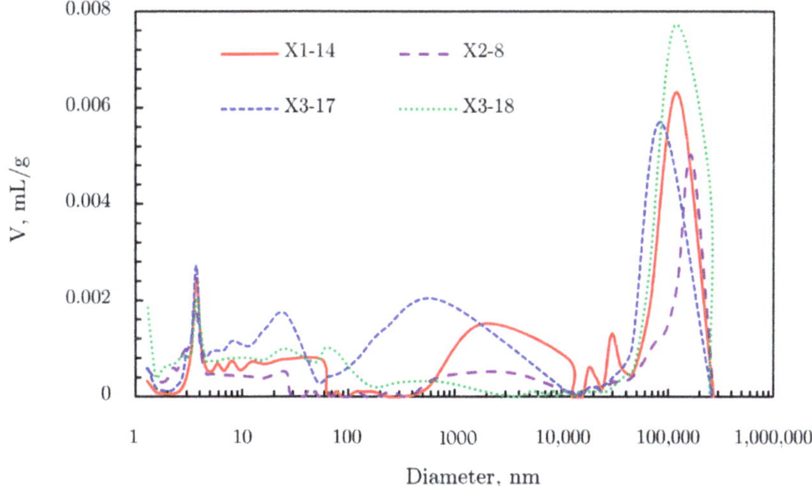

Figure 15. PSD of samples from the Longmaxi Formation based on N_2 adsorption and MICP.

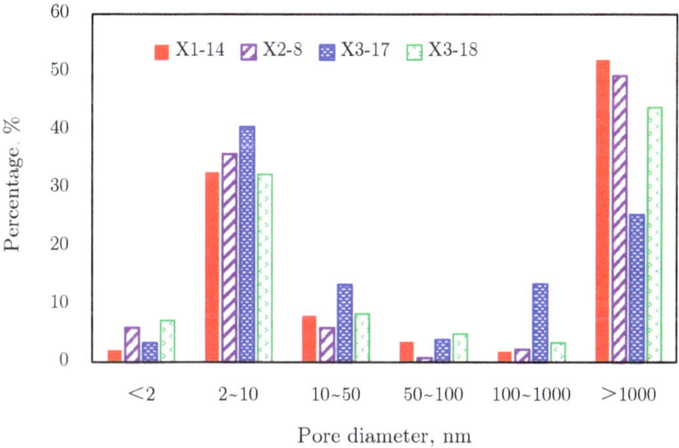

Figure 16. PSD histogram of samples from the Longmaxi Formation in the Sichuan Basin based on N_2 adsorption and MICP.

2.5. Gas Composition and Origin

Milkov et al. [37] studied gas composition and origins based on around 2600 shale gas samples from 76 geological formations in 38 sedimentary basins located in eleven countries. It is found that methane is the predominated hydrocarbon component, with more than 80% in volume concentration, followed by ethane with around 6% in volume concentration, and propane with around 2% in volume concentration. Nitrogen and carbon dioxide are two main non-hydrocarbon components in shale gas samples, with average volume concentration around 6% and 2%, respectively. For most shale plays in the USA, China and Argentina, it is found that the most productive and commercially successful shale plays have pure thermogenic origin. This is very different from the study of Curtis [19], where it is found that shale gas has predominantly microbial origin.

2.6. Shale Gas Occurrence Types

The types of natural gas in shale formations are determined by diverse formation physics and pore characteristics. In accordance with the classification of pore structures in shale formations (Section 2.3), free gas, adsorbed gas, and dissolved gas are three possible gas occurrence states underground [38,39]. Generally, free gas is stored not only in fractures, but also in pore systems, including organic pores and inorganic pores. Adsorbed gas is mostly stored on the surface of organic matter in equilibrium state with free gas. Dissolved gas is usually stored in liquid hydrocarbons, formation water, but most importantly in solid kerogen. Organic kerogen serves as the source rock and generates shale gas continuously [40].

The percentage of different gas types varies from one reservoir to another, since it is significantly influenced by pressure, temperature, organic matter types, organic matter content, organic matter maturity, the development of micro fractures, and liquid hydrocarbon content. The different gas types and gas flow mechanisms in organic shale nanopores can be seen in Figure 17.

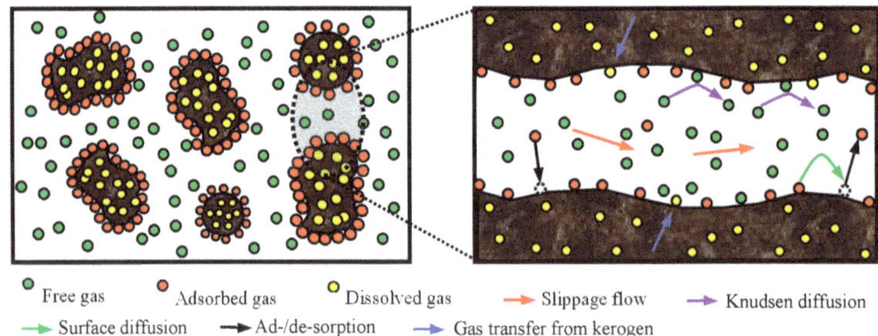

Figure 17. The storage form and flow mechanism of shale gas in organic nanopores [38,41].

2.6.1. Free Gas Characterization

Free gas is stored in organic or inorganic pores, micro fractures, and hydraulic fractures. The content of free gas is determined by adsorbed gas and dissolved gas. Only when the total gas amount is larger than the sum of adsorbed and dissolved gas amount does a free gas state exist. Under high-pressure or high-temperature reservoir conditions, gas behaviors do not satisfy the ideal gas equation of state (EOS), so the real gas EOS needs to be adopted to describe its behaviors [42,43].

Semi-Empirical Formula

EOS is one of the most important models to calculate the thermodynamic properties of real gas. More than 150 types of EOS have been proposed to describe the pressure-volume-temperature relationships of real gas, but none is able to include all properties of any gas in the engineering application range [44]. Typical and widely applied EOS, as well as their pros and cons, are displayed in Table 4.

Empirical Formula

According to the corresponding state principle, gas deviation factor Z could be introduced to describe real gas behaviors:

$$pV = ZnRT, \tag{8}$$

where p is the pressure, Pa; V is the gas volume, m^3; Z is the gas deviation factor, dimensionless; n is the molar mass, kg/mol; R is the universal gas constant, J/(mol K); and T is temperature, K.

In Equation (8), gas deviation factor Z could be obtained by experiments, referring to Z-plate [44], or calculating from an empirical formula [27]. The Z-plate and empirical formula were mainly obtained by assuming Z_c as a constant in the range of 0.23–0.29. That is to say, gas deviation factor Z is a function of the reduced pressure p_r and reduced temperature T_r. Therefore, we also name the Z-plate as the two-parameter generalized compressibility chart. The critical gas deviation factor Z_c of most materials varies in the range of 0.23–0.29. Therefore, a more precise expression of Z is expected to be obtained by regarding Z as a function of p_r, T_r, and another parameter (Z_c or acentric factor ω)—a three-parameter relationship [44].

Table 4. Typical EOS of describing real gas behaviors [44].

Types	Z_c	Pros	Cons
vdW	0.375	Simple; basis of other EOS	Low accuracy; seldom practical for application
RK	0.333	Practical; accurate to calculate gas phase volume	Failed to calculate liquid volume accurately
SRK	0.333	Calculate gas and liquid phase equilibrium; widely used	The error is large when calculating liquid volume
PR	0.307	Higher accuracy than SRK to calculate liquid volume	Z_c is slightly bigger than the practical value
Virial		Able to describe viscosity, sound velocity, and heat capacity of gases	Failed to calculate liquid volume; inaccurate at high pressure

Note: a_e and b_e are energy and volume parameters, respectively, and their expressions vary in different EOS; Z_c is the critical gas deviation factor.

2.6.2. Adsorbed Gas

Adsorbed gas is mainly stored on the surface of matrix particles, kerogen, and clay minerals, and can account for 20–85% of total gas reserves [19,45–47]. Gas adsorption on shale matrix particles belongs to physical adsorption [48]. Although adsorbed gas contributes to the total gas production, its exact contribution is not clear. Compared to the contribution of adsorbed gas in total gas production, the percentage of adsorbed gas in original gas in place (OGIP) is much clearer. Recent studies have found that adsorbed gas accounts for 50–80% of OGIP when the pressure is lower than 13.79 MPa, while it accounts for 30–50% when the pressure is higher than 13.79 MPa [49,50].

Organic nanopores of 5–750 nm, are significantly developed in shale gas reservoirs when the maturity of organic matter is larger than 0.6% [28,30]. Due to the small pore radius, organic matter has a large surface area. For example, the specific surface area can be up to 300 m^2/g in nanoporous kerogen [51]. The enormous surface area provides favorable places for gas adsorption. The gas adsorption capacity is mainly affected by pore structures, mineral compositions, metamorphism degree, gas components, pressure, temperature, water vapor content, etc. Since adsorbed gas is mainly stored in organic matter, the TOC content significantly affects the adsorbed gas content in shale gas reservoirs. As we can see from Figure 18, there is a positive relationship between the adsorbed gas content and TOC content in different shale gas reservoirs. This is because large TOC content means more organic matter in the shale matrix, which can provide sufficient space for gas storage due to its large surface area.

Taking the Barnett Shale as an example, as shown in Figure 18, we can distinguish different gas types from the relationship between adsorbed gas and the total gas amount in shale matrix. Adsorbed gas and free gas in the organic matrix increase as the TOC content increases, while the amount of free gas in the inorganic matrix is not affected by the TOC content. In shale gas plays, adsorbed gas is a non-ignorable component in shale gas reserve calculation [27,39,53–55], and gas ad-/desorption is important in the study of gas flow behaviors [56–59]. If we analyze the case further, we could ask how much adsorbed gas could be produced during shale gas production, and how significantly gas ad-/desorption affects gas transient flow behaviors in shale gas reservoirs.

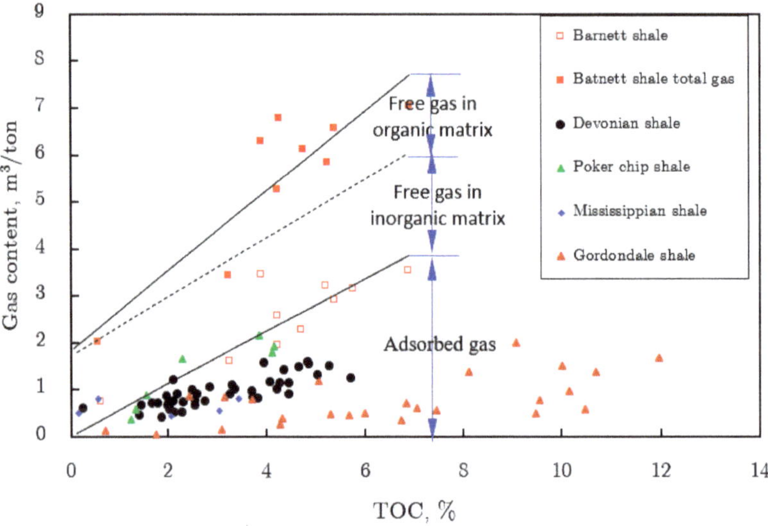

Figure 18. Adsorbed gas in different shales and its relationship with TOC content (revised from [50,52]).

The gas transporting process of CH_4 and He in organic shale samples was compared by an experimental study at 3.4 MPa and 308 K [60], where CH_4 serves as the adsorptive gas and He is non-adoptive. The production dynamics of CH_4 and He can be seen in Figure 19a. Assuming the free gas amount of CH_4 and He is equal in shale samples, the adsorbed volume of CH_4 can be obtained by the difference between the total CH_4 production volume and the total He production volume, which are 2.60 cm³/g and 1.33 cm³/g respectively. Therefore, the produced volume of free gas for unit mass shale particles under standard conditions is 1.33 cm³/g, while the adsorbed gas amount is 1.27 cm³/g. Similarly, simulation results from dynamic adsorption diffusion model show that the production of free gas dominates at an early production period (before point A) and drops very fast, while adsorbed gas dominates the later production after point A for a relatively long time. Experimental study and model simulation signified that both free gas and adsorbed gas played an important role in gas production.

The above research is conducted at low pressure (3.4 MPa) and temperature (308 K) compared to practical shale gas reservoir conditions. Gas desorption pressure in shale is usually below 12 MPa, which is close to the abandonment pressure of shale gas reservoirs. Meanwhile, the formation pressure is mainly depleted in a small area near the wellbore, i.e., the average pressure of shale formation is much higher than the abandon pressure. Consequently, gas desorption may only occur in a small area near the wellbore or hydraulic fractures, meaning a limited amount of adsorbed gas is produced during the life cycle of the shale gas reservoir. The significance of adsorbed gas, as well as the corresponding seepage mechanisms, need further investigation.

Assuming adsorbed gas volume is a function of pressure, Tang et al. [61] obtained the absolute adsorbed gas amount from excess adsorption and studied the adsorbed gas proportion to total gas at different shale depths (Figure 20). The conventional absolute adsorption refers to results obtained by fitting low and intermediate pressure sorption data using the Langmuir model of Equation (10), while the new absolute adsorption refers to a dual-site Langmuir model considering the adsorbed layer variation and excess adsorption. The conventional model severely underestimates the absolute adsorption amount when the pressure is higher than 6 MPa, as shown in Figure 20a. The percentage of adsorbed gas to total gas in place (GIP) is a function of shale formation depth, where it increases fast in shallow areas and slows down after 2000 m, as shown in Figure 20b. The adsorbed gas accounts for approximately 40–80% at different depths of formation. Meanwhile, the excess adsorption amount needs to be corrected to the absolute adsorption amount when considering the adsorbed gas percentage in GIP. Otherwise, it will massively underestimate the adsorbed gas amount and overestimate the free gas amount.

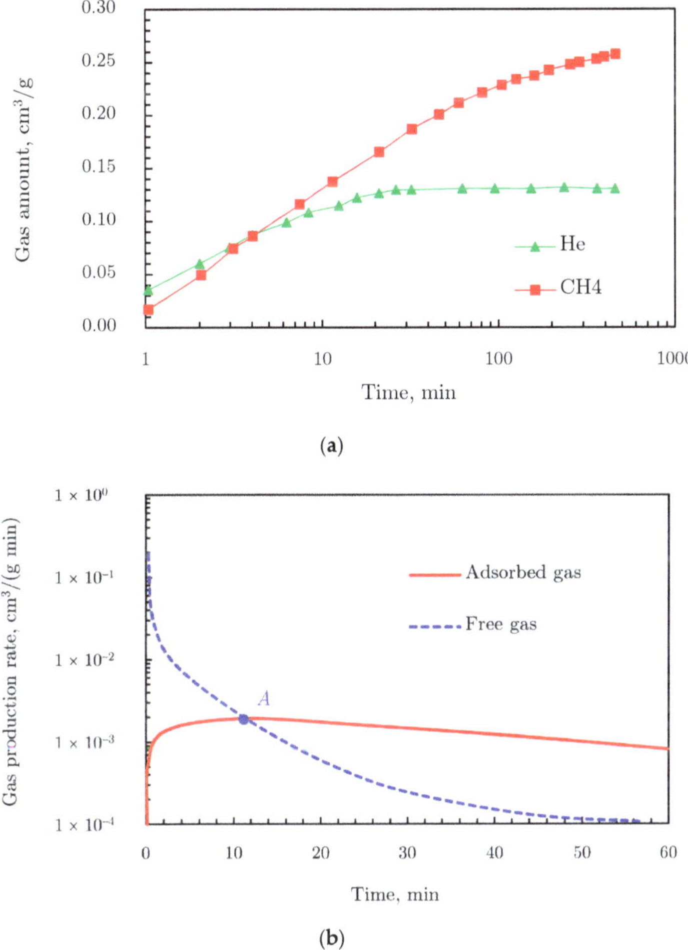

(**a**)

(**b**)

Figure 19. Experimental data (**a**) and mathematical simulation results (**b**) for gas transport process at 3.4 MPa and 308 K [60].

2.6.3. Dissolved Gas

After the equilibrium between adsorption and desorption is found, shale gas could dissolve into the liquid hydrocarbon or formation water during the hydrocarbon accumulation process. Meanwhile, organic kerogen continuously generates shale gas and contains a certain amount of gas molecules [62]. The gas in liquid hydrocarbon, formation water, and kerogen is called the dissolved gas, which has been overlooked, but may play a significant role in shale gas reservoir development [40,59]. Gap-filling and hydration are the two main storage mechanism of dissolved gas, and can be described theoretically by Henry's law [63]:

$$C_b = \frac{p}{K_c},\tag{9}$$

where C_b is the mole concentration of dissolved gas, mol/m^3 and K_c is the Henry constant, m^3 Pa/mol.

Since it is hard to differentiate dissolved gas from adsorbed gas, both gas types are usually attributed to one type, namely adsorbed gas. Moreover, adsorbed gas and dissolved gas can be transformed to the other under proper circumstances. Therefore, it can be roughly seen as one type in some cases [48,64].

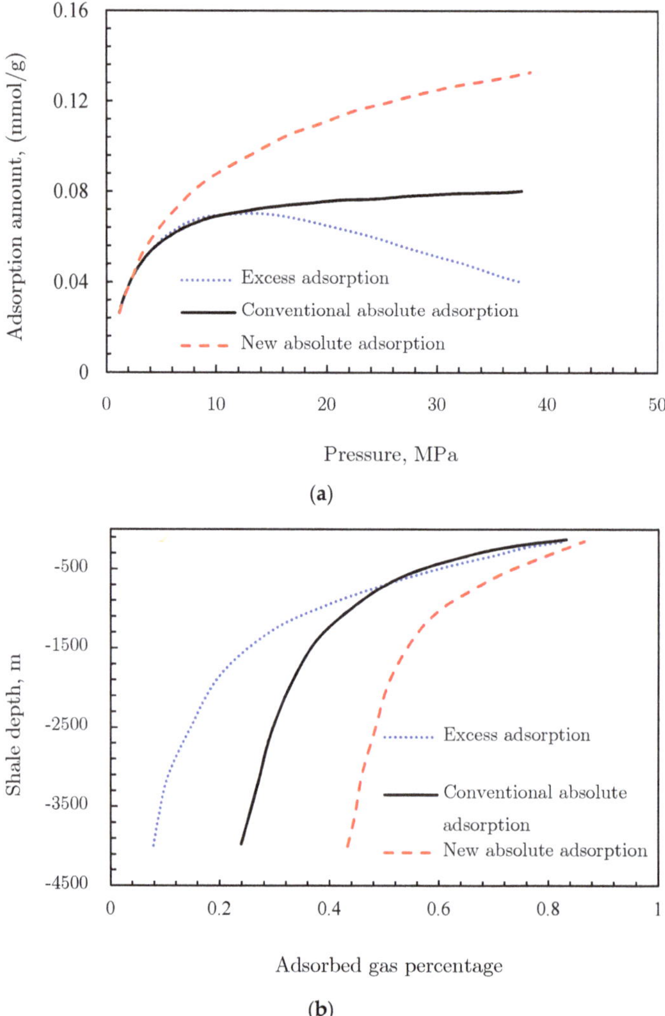

Figure 20. Adsorption amount versus pressure (**a**) and adsorbed gas percentage versus shale formation depth (**b**) in different models [61].

3. Gas Adsorption and Desorption

Shale gas can be stored on pore surfaces of organic matter and clays by gas adsorption. Organic matter in the shale matrix is a key parameter that influences gas adsorption characteristics in shale gas reservoirs. On the one hand, a large amount of nanopores are developed in organic shales, which provide enormous surface area for the gas to be adsorbed on. On the other hand, the adsorption potential is significant in organic nanopores compared to in inorganic nanopores or large organic pores. The adsorption-desorption law in organic shale nanopores is a key scientific problem in the practice of shale gas development, affecting the accuracy of evaluating shale adsorption capacity, studying the seepage flow behaviors, and developing transient seepage mathematical models [65].

3.1. Different Sorption Types and Models

Methane is the main component of shale gas underground, with a critical pressure of 4.59 MPa and a critical temperature of 190.53 K. Therefore, shale gas is in a supercritical state under in situ formation conditions (3000–6000 m, with high pressure up to 60 MPa) [66]. The study of supercritical gas sorption is essential for an accurate understanding of adsorption and desorption mechanisms in shale gas reservoirs. Gas sorption mechanisms are quite confusing, and no unified conclusions have been reached. Monolayer adsorption, multilayer adsorption, and micropore filling are three common assumptions in shale gas sorption research. Based on these assumptions, the Langmuir model, BET model, the Dubinin–Radushkevich (D-R) model, and the Dubinin-Astakhov (D-A) model have been established to fit the sorption data, and have obtained good results. However, good fitting results do not guarantee the validity of the assumption in the adsorption models. For example, even if the Langmuir model fits the experimental data very well, we cannot say that gas adsorption belongs to monolayer adsorption.

3.1.1. Monolayer Adsorption Type

Monolayer adsorption means that gas molecules adsorb on the pore surface in one layer, so the thickness of adsorbed gas equals the molecular diameter. Due to the huge surface area in a shale matrix, a considerable amount of adsorbed gas exists in shale gas reservoirs. Assuming 80% gas saturation in shale samples, the adsorbed gas ratio can be 22.65% of the total gas amount according to the Ono-Kondo lattice model established by Zhou et al. [67].

(1) Langmuir model [68]: The Langmuir model was established by monolayer adsorption assumption. Due to its simplicity and accurate fitting to the experimental data, it is widely used to describe monolayer gas adsorption, which can be written as:

$$G_a = G_m \frac{bp}{1 + bp}, \tag{10}$$

where G_a is the absolute adsorbed gas amount; G_m is the maximum adsorbed capacity, p is the pressure, and b is the Langmuir sorption constant, which can be obtained by fitting the experimental or field test data.

(2) Freundlich model [69]: With pressure decreasing, the Langmuir equation of Equation (10) approaches the Henry's law of Equation (9). Therefore, the Henry's law can describe low-pressure sorption behaviors, since any sorption isotherm satisfies the linear relationship between adsorption amount and pressure at low pressure. To broaden the application range of the Henry's law into high-pressure areas, an exponential empirical formulation, namely the Freundlich model, was used:

$$G_a = k_f p^{1/n_f}, \tag{11}$$

where k_f is related to the adsorption interaction and adsorption amount; n_f is a constant usually between 2 and 3, reflecting the intensity of adsorption. The values of both k_f and n_f depend on the type of adsorbent and adsorbate as well as the temperature.

With temperature increasing, constant n_f approaches to unity and the Freundlich model of Equation (11) becomes the Henry model of Equation (9). The Freundlich model can properly describe monolayer adsorption, especially for low-concentration gases and in the meso pressure range. However, there is no explicit physical meaning of the constants k_f and n_f, and it cannot explain the mechanisms of adsorption.

(3) Langmuir-Freundlich model: To modify the assumption of uniform adsorption sites in the Langmuir model, the Freundlich equation and the Langmuir equation were coupled to form a new adsorption model, namely the Langmuir-Freundlich model, which is:

$$G_a = G_m \frac{bp^l}{1 + bp^l}, \tag{12}$$

where l reflects the heterogeneity of adsorbents, $l \leq 1$. The smaller the value of l, the stronger the heterogeneity of the adsorbent. If an adsorbent possesses an ideal surface, l tends to be 1 and the Langmuir-Freundlich equation is equivalent to the Langmuir equation.

(4) Toth model [69]: To improve the fitting capacity of the Langmuir model, the Toth model was proposed:

$$G_a = G_m \frac{bp}{\left[1 + (bp)^t\right]^{1/t}}, \tag{13}$$

where t is a constant related to the adsorbent properties.

Note that the Toth model of Equation (13) solves two problems: (1) the Freundlich model of Equation (11) and the DR model of Equation (16) do not satisfy Henry's law at low pressure; and (2) no maximum adsorption amount appears in the Freundlich model of Equation (11) with increasing pressure. Bae et al. [69] found that the Toth equation fitted the experimental data better than the extended three-parameter and Langmuir equation, and yielded realistic values of pore volumes of coal samples and adsorbed gas density.

3.1.2. Multilayer Sorption Type

(1) Two-parameter BET model [32]: Multilayer adsorption can be modeled by the BET sorption theory, which assumes that gas molecules can adsorb on a solid surface by infinite layers and no interaction exists between contiguous layers. In other words, any monolayer obeys the Langmuir adsorption theory in the BET model, which can be expressed as follows:

$$\frac{G_a}{G_m} = \frac{C_b(p/p_s)}{(1 - p/p_s)[1 + (C - 1)(p/p_s)]}, \tag{14}$$

where G_m is the maximum monolayer adsorption amount, p_s is the saturated vapor pressure, and C_b is the dimensionless constant controlling the time of multilayer adsorption.

In applying Equation (14), its equivalent expression needs to be adopted, which is Equation (1). A plot of $(p/p_s)/[G_a(1 - p/p_s)]$ versus p/p_s is employed to figure out whether the adsorption follows the BET theory. If the plot satisfies the linear relationship in the range of $0.005 < p/p_s < 0.35$, we can use the scope and intercept of the straight line to obtain the values of G_m and C.

(2) Three-parameter BET model [44,69]: The above two-parameter BET model of Equation (14) assumes infinite layers of adsorption. If the adsorption layers are finite, the three-parameter BET model can be employed:

$$\frac{G_a}{G_m} = \frac{C_b(p/p_s)}{(1 - p/p_s)} \frac{\left[1 - (n + 1)(p/p_s)^{n_b} + n_b(p/p_s)^{n_b+1}\right]}{\left[1 + (C_b - 1)(p/p_s) - C_b(p/p_s)^{n_b+1}\right]}. \tag{15}$$

If $n_b = 1$, Equation (15) simplifies into the Langmuir model of Equation (10); if $n \to \infty$, Equation (15) transforms into the two-parameter BET model of Equation (14). Note that the three-parameter BET model of Equation (15) is applicable to describe adsorption behaviors for p/p_s in the range of 0.35–0.60.

Note that, although in theory the multilayer adsorption assumption may produce a wider application scope with the BET model than the Langmuir model, it may not be suitable to adopt the

BET theory in a shale gas adsorption study, since sorption behaviors in shale gas reservoirs belong to supercritical adsorption and the saturated vapor pressure p_s of shale gas (mainly methane) does not exist in practice. Besides, as reported, the BET model may have a poorer performance than the Langmuir model at fitting absolute sorption data, as reported by [70].

3.1.3. Micropore Volume Filling

Dubinin–Radushkevich and Dubinin-Astakhov Models

Gas molecule behavior in nanopores is significantly different from that in mesopores or macropores, since there is a superposition of adsorption potential from both pore sides. Consequently, the adsorption force of micropore walls on gas molecules is much greater than in mesopores or macropores, leading to large adsorption. Dubinin named the gas adsorption in these small-scale pores micropore volume filling [71]. Compared to the Langmuir adsorption theory, micropore volume filling is more helpful for understanding the gas adsorption mechanism and gas true storage forms, and for evaluating gas adsorption properties. Meanwhile, it has been reported that the D-A model provides a better fit to sorption data of coal than the Langmuir model [72].

The condensed adsorbate looks like microemulsion droplets when adsorption occurs in micropores, which is greatly affected by interfaces. Based on the Polanyi adsorption potential theory [73], the D-R model and D-A model are commonly used in shale gas adsorption studies, and can be expressed as follows:

$$W = W_0 \exp\left\{-D[\ln(p_s/p)]^m\right\}, \tag{16}$$

where W is the pore volume filled with gas molecules at relative pressure p/p_s; W_0 is the total volume of micropores; D is a parameter related to the adsorbate-adsorbent system; and m is a parameter ranging from 2 to 6, reflecting the heterogeneity of potential energy on adsorbent surfaces. If m constantly equals 2, Equation (16) is the D-R model. If m is a random parameter between 2 and 6, then Equation (16) is the D-A model.

In applying Equation (16), W is equivalent to the absolute adsorbed gas amount G_a, and W_0 is equivalent to the maximum adsorbed gas amount G_m, Equation (16) can be re-expressed as follows:

$$G_a = G_m \exp\left\{-D[\ln(p_s/p)]^m\right\}. \tag{17}$$

Compared to the D-R model, the D-A model performs better fitting with experimental data, according to the study of Wang et al. [70]. This is because the chosen range of structure heterogeneity parameter m in the D-A model is broader than that in the D-R model, which is related to pore size distribution in shale formations. The chosen parameter m in the D-A model brings the micropore structure information into adsorption prediction and modeling, while it has a constant value of 2 in the D-R model, without considering the structural heterogeneity in shale samples.

Calculation of Virtual Saturated Vapor Pressure

Note that the above D-R and D-A models were not initially proposed for supercritical adsorption, but for subcritical sorption. Therefore, the concept of saturated vapor pressure in the D-R and D-A model was replaced by virtual saturated vapor pressure or supercritical adsorption limited pressure [72]. Generally, the virtual saturated vapor pressure can be calculated by the following empirical formulations or approaches:

(1)   The first is the Dubinin method [71,74,75]:

$$p_s = p_c \left(\frac{T}{T_c}\right)^2, \tag{18}$$

where p_c is the critical pressure and T_c is the critical temperature.

(2) The second is the Reid method [72]:

$$p_s = p_c \exp\left[\frac{T_b}{T_c} \times \frac{\ln p_c}{1 - \frac{T_b}{T_c}} \times (1 - \frac{T_c}{T})\right], \tag{19}$$

where T_b is the boiling point of gas at atmospheric pressure.

(3) The third is the Antoine method [74]:

$$p_s = 0.1 \times \exp(B_A - \frac{C_A}{D_A + T}). \tag{20}$$

Equation (20) is a three-parameter vapor pressure equation, where the extrapolation technique and saturated vapor pressure data under subcritical conditions are needed. For methane, three parameters can be obtained: $B_A = 8.784$, $C_A = 933.51$, and $D_A = -5.37$, respectively [74]. Then, saturated vapor pressure could be calculated by Equation (20).

(4) The fourth is the Astakhov method [72], the calculation results of which fit the experimental data well in the interval of p_s from 0.1 MPa up to critical pressure p_c [71]. Meanwhile, it should be noted that this method gives satisfactory results for temperatures exceeding the critical temperature by 50–100 K.

$$p_s = \exp(\frac{c_A}{T} + d_A), \tag{21}$$

where parameters c_A and d_A are determined by the gas critical point (T_c, p_c) and boiling point (T_b, 101,325 Pa). For methane, $c_A = -1032.693$ and $d_A = 6.945$.

(5) The Amankwah method [76] is an improved calculation method for the Dubinin method, which involves a parameter k_A to account for interactions in an adsorbate-adsorbent system:

$$p_s = p_c(\frac{T}{T_c})^{k_A}, \tag{22}$$

where k_A is a parameter accounting for interactions in the adsorbate-adsorbent systems.

(6) The linearization of isotherm adsorption data is another processing method [77] to extend the D-R and D-A models into supercritical area. By transforming isotherms from G_{ex} versus p space to $\ln[\ln G_{ex}] - 1$ versus $\ln p$ space, where G_{ex} is the excess adsorption amount, a bunch of fitting straight lines could be obtained; they converge to a single point B, as shown in Figure 21. This merge point B is defined as the limiting state of the adsorbate, corresponding to the extreme condition of the adsorption potential field, where no more adsorptive molecules can enter the adsorbent micropores [77]. Therefore, the limiting pressure and limiting adsorption amount of the merge point correspond to the saturated vapor pressure and the saturated adsorption amount in the D-R and D-A models, respectively.

In order to compare these six different methods, the experimental data in Zhang's study for organic-rich Woodford shale [78] are collected for analysis at temperatures of 308.53 K, 323.53 K, and 338.53 K, respectively, as shown in Figure 22a. The linearization processing method of adsorption data in Zhou's study is adopted, which transforms the sorption data into three straight lines and defines a limiting state at the intersection point A in Figure 22b. The other five methods are also employed to calculate the virtual saturated vapor pressure for the same shale sample and sorption data. The calculated results are shown in Figure 23, which exhibits great differences to the results from different methods. The results from Antoine and Astakhov have obviously larger values than the others, while the linearization processing results, which are independent of temperature, have much smaller values than the others. Dubinin and Reid's results are basically the same, with slightly lower values than in the Amankwah results. Dubinin's method only considers the properties of the individual adsorbates, while the Amankwah method also takes adsorbent properties into consideration. Moreover, the value of parameter k is obtained by nonlinear fitting for isotherm sorption data, which is more practical than the constant value in the Dubinin method. As a result, the Amankwah method is recommended to calculate virtual saturated vapor pressure.

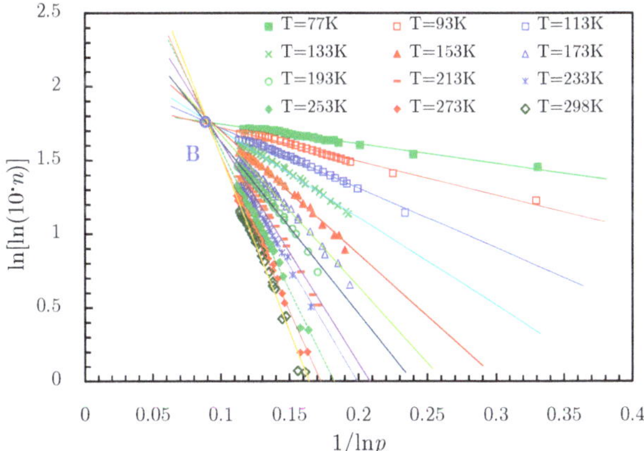

Figure 21. Linearized adsorption isotherms of hydrogen [77], where p is in KPa and G_{ex} is in mmol/g.

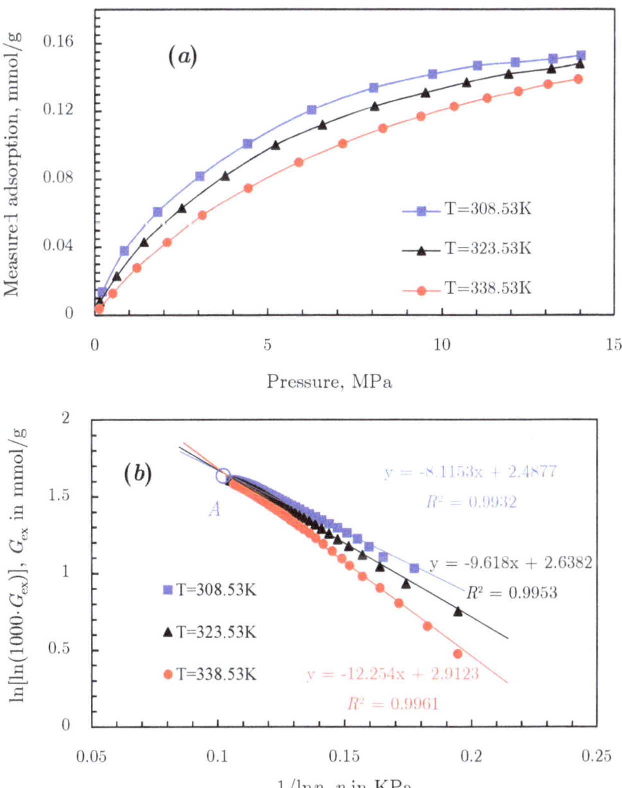

Figure 22. The sorption experimental data of Woodford shale (**a**) [78] as well as their linearized isotherms (**b**).

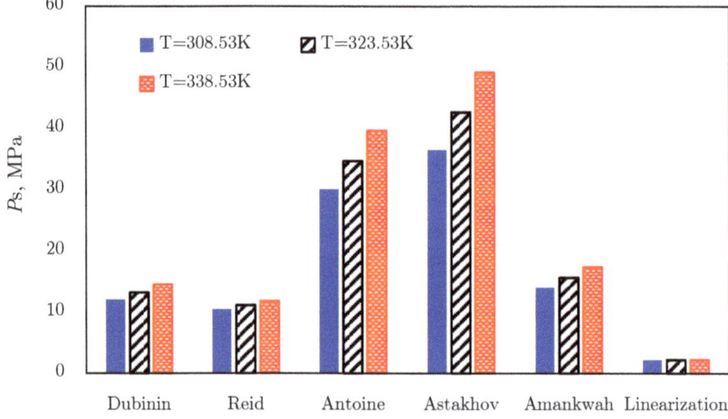

Figure 23. The calculated virtual saturated vapor pressure for Woodford shale at different temperatures.

Note that the linearization processing method was proposed to tackle sorption problems in the ranges of 77–298 K and 0–7 MPa [77]. The pressure and temperatures in shale gas reservoirs are generally beyond this range, so this method is not recommended in shale gas sorption studies.

3.2. Sorption Study in Supercritical Area

3.2.1. Gibbs Excess Adsorption

For high-pressure and -temperature sorption, Gibbs excess adsorption is adopted to describe its unique behaviors [79]. Generally, adsorbed gas and bulk gas both exist for an adsorbate-adsorbent system, where adsorbed gas is distributed on the pore surface as a layer and bulk gas is far from the surface. Bulk phase gas is also distributed in the adsorption layer, which is irrelevant to gas-solid molecular interactions and can be ignored at low pressure. However, it needs to be considered in shale gas sorption research, since the in situ pressure is high in shale gas reservoirs (>30 MPa) [65]. Therefore, the excess adsorption amount corresponds to the part that is larger than the bulk phase density in the adsorption layer. The difference between the absolute adsorption and the excess adsorption can be seen in Figure 25, where the absolute adsorption (Figure 25e) consists of excess adsorption (Figure 25c) and bulk phase gas in the adsorbed layer (Figure 25d). The relationship can be expressed as follows:

$$G_{ex} = G_a - \rho_g v_{ad},\tag{23}$$

where G_{ex} is the Gibbs excess adsorption amount, G_a is the absolute adsorption amount, and v_{ad} is the adsorbed gas volume, as can be seen in Figure 25.

The measured adsorption amount in high-pressure sorption experiments is the Gibbs excess adsorption amount G_{ex} in Equation (23). The relationship has another explanation: namely, the adsorbed gas is under the effect of bulk phase gas buoyancy. Thus, the measured adsorbed gas weight equals the difference between absolute adsorbed gas weight and the buoyancy it received in bulk-phase gas.

For adsorbed gas, the following relationship exists:

$$v_{ad} = \frac{G_a}{\rho_{ad}},\tag{24}$$

where ρ_{ad} is the adsorbed gas density.

Then, incorporating Equations (23) and (24), we can obtain:

$$G_{ex} = G_a \left(1 - \frac{\rho_g}{\rho_{ad}}\right). \tag{25}$$

Associating Equation (25) with the calculation methods for bulk and adsorbed gas density in Sections 3.4.1 and 3.4.2, the simulation results of absolute adsorption amount G_a can be transformed into measured excess adsorption amounts G_{ex}. In low-pressure sorption studies, the bulk gas density ρ_g is much lower than the adsorbed gas density ρ_{ad}, and the excess adsorption amount G_{ex} is approximately the same as the absolute adsorption amount G_a. However, gas density ρ_g becomes comparable to adsorbed gas density ρ_{ad} with pressure increasing, as to be introduced in Section 3.4.2. Thus, the difference between the absolute adsorption amount G_a and the excess adsorption amount G_{ex} cannot be ignored and there is a maximum on the plot of measured adsorption amount versus pressure or bulk phase gas density, as can be seen from Figure 24a. The location of the maximum depends on the interaction between adsorbate and adsorbent, as well as the thermodynamic state of the adsorptive [80].

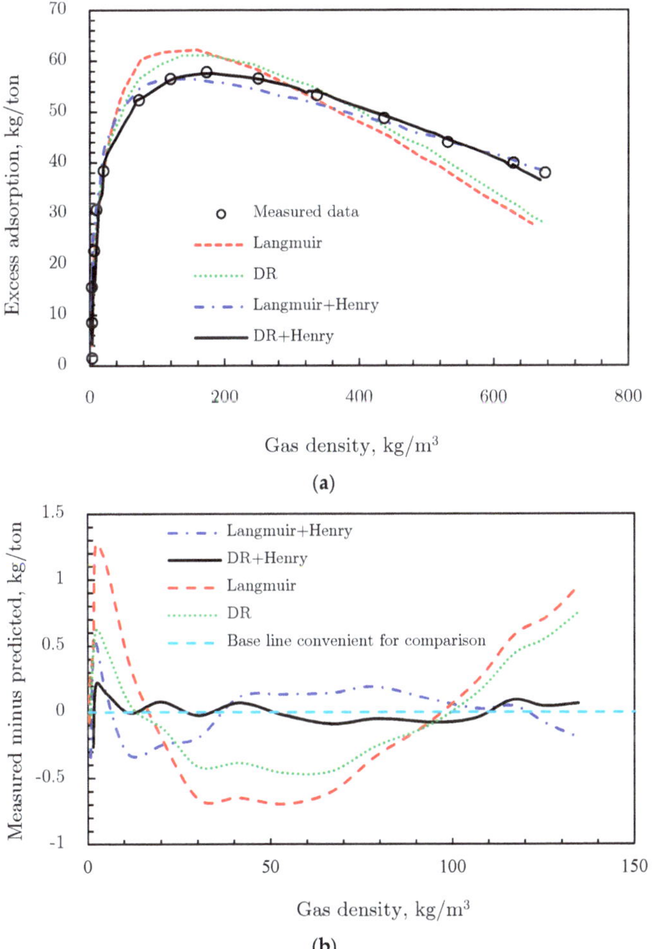

Figure 24. Fitting of models to CO_2 adsorption data (**a**) and residuals of curves from different models to CH_4 adsorption data, and the base line means the measured equals to the predicted (**b**) [81].

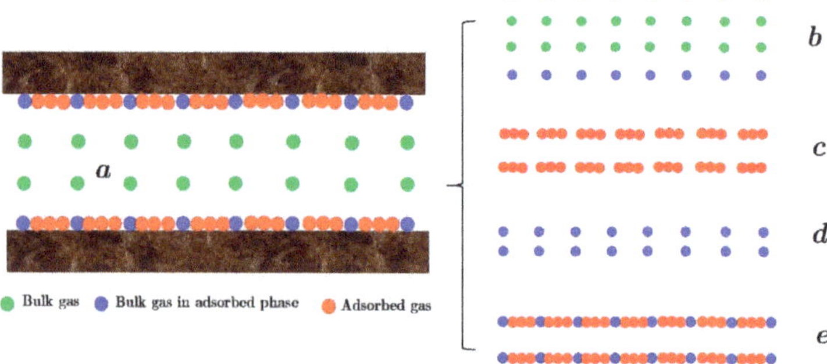

a: Adsorbed gas and free gas distribution in shale nanopores

b: Total bulk phase gas in the system

c: Excess adsorption amount

d: Bulk phase gas in adsorbed layer

e: Absolute adsorption amount

Figure 25. Scheme of absolute adsorption amount and excess adsorption amount.

3.2.2. Supercritical Adsorption Models

The introduction of the virtual saturated vapor pressure in Section 3.1.3 extends the micropore filling models from subcritical area into supercritical range. However, Sakurovs et al. [81] noted that this method cannot easily accommodate adsorption at conditions where both the pressure and temperature are above the critical values. Since the adsorbed gas density is greater than the free gas density in supercritical conditions, another method, which replaced the saturated vapor pressure p_s by adsorbed phase gas density ρ_{ad}, and gas pressure p by gas density ρ_g, was proposed to extend the volume-filling models to a wider pressure and temperature application range [72]. Based on this idea and associating Equation (17) with Equation (25), we obtain:

$$G_{ex} = G_m(1 - \rho_g/\rho_{ad})\exp\left\{-D[\ln(\rho_{ad}/\rho_g)]^m\right\}. \tag{26}$$

Similarly, other models in Section 3.1 can also be transformed into this form, i.e., replacing gas pressure p and saturated vapor pressure p_s with gas density ρ_g and adsorbed gas density ρ_{ad}, respectively, and employing $(1 - \rho_g/\rho_{ad})$ to correct for the true adsorbed gas amount. For example, the Langmuir equation can be transformed into the following form:

$$G_{ex} = G_m\left(1 - \frac{\rho_g}{\rho_{ad}}\right)\frac{b_r\rho_g}{1 + b_r\rho_g}, \tag{27}$$

where b_r is a constant similar to the Langmuir constant, which has the relationship $\rho_L = 1/b_r$. Langmuir density ρ_L refers to the gas density at which the adsorption amount is half of the maximum.

The transformation of other models occurs in the same way. A previous study [82] pointed out that gas density is the most meaningful variable in high-pressure sorption areas, so it is recommended that they be used in high-pressure studies instead of pressure.

3.3. Adsorption/Absorption Models

Reucroft [83] reported that CO_2 dissolved in coal and caused it to swell in addition to being adsorbed on the coal surface. When studying gas sorption behaviors on polymers, Sato et al. [84] found that gas can not only adsorb on a solid surface, but also can be absorbed into the interior of solid material. Larsen [85] proposed a similar two-component sorption on coal samples. Adsorbed gas and dissolved gas exist on the kerogen surface and in the kerogen interior, respectively, which restrict and connect with each other in shale gas reservoirs [38,41]. To model the two types of sorption (adsorption and absorption), different methods were proposed in previous studies.

The first category is the hybrid type, namely the superposition of gas adsorption law and gas absorption law. Gas adsorption can be described by the abovementioned adsorption models, such as the monolayer adsorption models or the micropore filling models, while gas absorption is described by Henry's law. Meanwhile, supercritical sorption characteristics need to be considered in high-pressure and high-temperature sorption studies. Here, we take the Langmuir adsorption and D-A models as examples to introduce a hybrid method, and other adsorption models, such as the Freundlich, Toth, and D-R models, can be handled by the same procedure.

(1) Langmuir/Henry combination: In this model, gas adsorption is modeled by the Langmuir equation, and gas absorption is described by a term proportional to pressure, following Henry's law. Here, the subcritical adsorption and absorption are described in terms of pressure, as in Equation (28), while supercritical adsorption and absorption are in terms of gas density [81] as in Equation (29):

$$G_a = G_m \frac{bp}{1 + bp} + kp, \tag{28}$$

$$G_{ex} = G_m(1 - \frac{\rho_g}{\rho_{ad}}) \frac{b_r \rho_g}{1 + b_r \rho_g} + k\rho_g. \tag{29}$$

(2) Volume filling/Henry combination: Gas adsorption is described by the D-R or D-A model, and gas absorption is described by Henry's law. For subcritical adsorption and absorption, this can be described in terms of pressure:

$$G_a = G_m \exp\left\{-D[\ln(p_s/p)]^m\right\} + kp. \tag{30}$$

For supercritical adsorption and absorption description in terms of gas density [81], this is:

$$G_{ex} = G_m(1 - \rho_g/\rho_{ad}) \exp\left\{-D[\ln(\rho_{ad}/\rho_g)]^m\right\} + k\rho_g. \tag{31}$$

If gas pressure is adopted in a supercritical adsorption model [86], then the virtual saturated vapor pressure concept introduced in Section 3.1.3 needs to be employed, i.e.:

$$G_{ex} = G_m(1 - \rho_g/\rho_{ad}) \exp\left\{-D[\ln(p_s/p)]^m\right\} + kp. \tag{32}$$

(3) Swelling contribution: Dissolved gas usually swells the solid materials after absorption [83,85]. If the swelling contribution was equal to the condensed gas volume in adsorbents, Equations (28)–(32) need to be modified, because the swelling occupies space that would otherwise be taken up by gases [81]. The term $(1 - \rho_g/\rho_{ad})$ needs to be multiplied by the absorption term. Taking the supercritical volume filling/Henry combination as an example, the model considering swelling contribution can be expressed as follows:

$$G_{ex} = G_m(1 - \rho_g/\rho_{ad}) \exp\left\{-D[\ln(\rho_{ad}/\rho_g)]^m\right\} + k\rho_g(1 - \rho_g/\rho_{ad}), \tag{33}$$

$$G_{ex} = G_m(1 - \rho_g/\rho_{ad}) \exp\left\{-D[\ln(p_s/p)]^m\right\} + kp(1 - \rho_g/\rho_{ad}). \tag{34}$$

Sakurovs et al. [81] compared the calculation results from the Langmuir model of Equation (27), the D-R model of Equation (26), the Langmuir/Henry combination model of Equation (29), and the D-R/Henry combination model of Equation (31) using CO_2 and CH_4 adsorption data, as shown in Figure 24a,b. The fitting effects of the Langmuir and the D-R models are both poor, while the Langmuir/Henry or D-R/Henry combination model has a much better effect. This means that the added term (kp) in adsorption models improves the fitting effect and reduces the calculated surface sorption capacity. The improvement of this term is greater in the D-R model than in the Langmuir model, since the D-R/Henry combination fits the measured sorption data better than the Langmuir/Henry combination. This suggest that the gas sorption mechanism in coal is more likely the volume filling, rather than monolayer coverage.

(4) Bi-Langmuir adsorption model: Assuming absorption and swelling are related, Pini et al. [82] applied the Bi-Langmuir model [87] to describe the combination of adsorption and absorption, where linear superposition is adopted for each Langmuir adsorption term, i.e.:

$$G_a = G_m^{ad} \frac{b_{ad}\rho_g}{1 + b_{ad}\rho_g} + G_m^{ab} \frac{b_{ab}\rho_g}{1 + b_{ab}\rho_g},$$

(35)

where the first term on the right side of the equation is the adsorption term, and the second term is the absorption term.

Assuming the Langmuir equilibrium constants for adsorption and absorption are equal [82], i.e., $b_{ad} = b_{ab}$, the excess adsorption amount can be expressed as follows:

$$G_a = G_t \frac{b\rho_g}{1 + b\rho_g} - \rho_g v_{ad},$$

(36)

where G_t is the sum of maximum adsorption amount and maximum absorption amount.

This Bi-Langmuir model was also compared to the experimental data and the D-R/Henry combination model, as shown in Figure 26. Both models fitted the excess sorption data well, but the fitted curve for absolute sorption data from the D-R/Henry combination model is much higher than the experimental data, while the Bi-Langmuir model fitted the absolute sorption data excellently. This is caused by the neglect of the swelling effect and the assumption of unlimited sorption capacity in the D-R/Henry combination model. Therefore, it should be seen as an empirical approach to describe excess sorption isotherms, and cannot be used for gas storage capacity estimation. Contrarily, the Bi-Langmuir model has a solid physical basis from experimental observations of a saturation-limited equilibrium between the gas phase and the condensed phase [82]. From this point, the Bi-Langmuir model is more suitable for adsorption and absorption modeling of shale gas.

3.4. Physical Properties Calculation of Shale Gas

3.4.1. Bulk Gas Properties

Gas Density

(1) Calculated by EOS: Generally, the bulk phase gas density can be calculated by the real gas EOS, as mentioned in Section 2.6.1. After calculation, we can also obtain the relationship between gas density and pressure by the nonlinear fitting technique, which expresses density in terms of pressure:

$$\rho_g = c_0 + c_1 p + c_2 p^2 + c_3 p^3 + \cdots,$$

(37)

where c_0, c_1, c_2, and c_3 are fitting parameters.

As we can see from Figure 27a, free gas density decreases with increasing temperature when $0.1\,\text{MPa} < p < 30\,\text{MPa}$ and increases with increasing pressure when $273.13\,\text{K} < T < 373.13\,\text{K}$. Compared to temperature, the influence of pressure on free gas density is more obvious. Since the change in pressure is more marked than that of temperature during shale gas reservoir development, more attention needs to be paid to the change in free gas density with pressure. The influence of temperature on free gas density is the most severe at a pressure of around 15 MPa, while it is weaker at lower or higher temperatures.

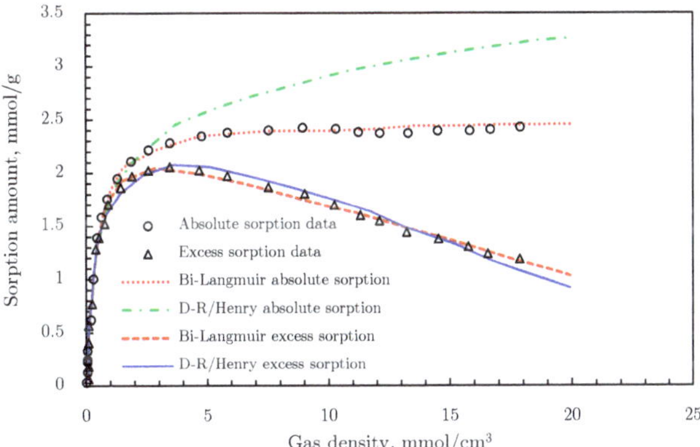

Figure 26. The comparison of Bi-Langmuir model with CO_2 experimental sorption data on coal and D-R/Henry combination model.

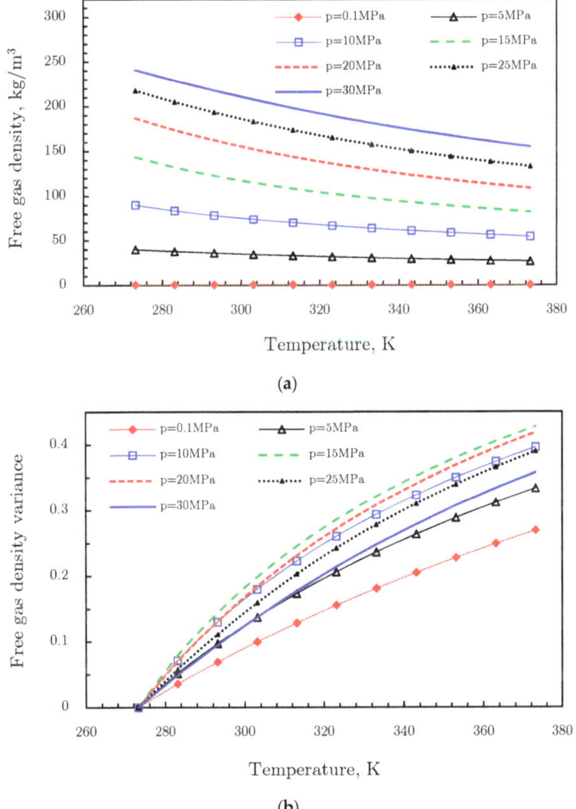

(**a**)

(**b**)

Figure 27. Free gas density (**a**) as well as its variance (**b**) in different pressure and temperature conditions.

(2) Measured by experiments: Bulk phase gas density can also by measured by experiments. Since analytical modeling is the main method introduced in this article, experimental measurement apparatus and procedures are not introduced.

Gas Viscosity

Accurate determination of natural gas viscosity plays a key role in its management as it is one of the most important parameters in calculations. It is a pressure- and temperature-dependent parameter that can be calculated by different empirical methods. In this section, different calculation formulations are compared. First, we will introduce some density-based models, where the calculation accuracy of viscosity is based on the prediction of gas density. We note that the following formula in the original work may use different units, and we have transformed all parameters into the SI unit for convenience, i.e., viscosity in Pa s, density in kg/m^3, molar mass in kg/mol, and temperature in K. With different units, the coefficients in the equation have different values.

(1)   Lee method [88,89]

Lee et al. reported a viscosity calculation formula of light hydrocarbons based on accurate density data of pure or mixed gas components. Using a linear molecular weight mixing rule, gases' viscosity is expressed as a function of the molecular weight and the gas density, which is:

$$\mu = K \exp\left[X\left(0.001\rho_g\right)^Y\right] \times 10^{-7},\tag{38}$$

where:

$$K = \frac{(7.77 + 6.3M)(1.8T)^{1.5}}{122.4 + 12900M + 1.8T},\tag{39}$$

$$X = 2.57 + \frac{1914.5}{1.8T} + 9.5M,\tag{40}$$

$$Y = 1.11 + 0.04X,\tag{41}$$

where μ is the viscosity, Pa s; M is the molar mass, mol/kg; T is the temperature, K; and K, X, and Y are intermediate parameters.

The results of Equation (38) provided satisfactory fitting capability to the data on methane, ethane, propane, and n-butane simultaneously, with a standard deviation of 1.89% [88,89]. The coefficients in Equations (39)–(41) were determined from the experimental data for pressures ranging from 0.69 MPa to 55.16 MPa and temperatures ranging from 310.93 K to 444.26 K.

(2)   Improved Lee method [90,91]

If experimental data on gas density are not available, it can be predicted by an empirical method. In this situation, the coefficients of K, X, Y can be obtained as follows:

$$K = \frac{(9.379 + 16.07M)(1.8T)^{1.5}}{209.2 + 19260M + 1.8T},\tag{42}$$

$$X = 3.448 + \frac{986.4}{1.8T} + 10.09M,\tag{43}$$

$$Y = 2.447 - 0.2224X.\tag{44}$$

The calculated viscosity of this method agrees with parts of the published viscosity data within 2% at low pressure and within 4% at high pressure when the specific gravity of the gas is smaller than 1.0. The method is less accurate for gases of higher specific gravities, usually giving lower estimates by up to 20% for retrograde gases with specific gravities over 1.5. Different from the original study, we take SI units here in Equations (43)–(45), i.e., μ in Pa·s, ρ_g in kg/m^3, M in kg/mol and T in K.

(3)   Londono method [92]

Londono et al. determined the coefficients of K, X, and Y in Equation (38) by applying pure-component and light natural gas mixture data to improve the accuracy of the model, with:

$$K = \frac{(16.7175 + 41.9188M)(1.8T)^{1.40256}}{212.209 + 18134.9M + 1.8T}, \tag{45}$$

$$X = 2.12574 + \frac{2063.71}{1.8T} + 11.926M, \tag{46}$$

$$Y = 1.09809 + 0.0392851X. \tag{47}$$

Another polynomial gas viscosity model was proposed in the study of Londono et al. [92] based on nonlinear regression techniques, which can be expressed as:

$$\mu_g = \mu_{1atm} + 10^{-3}f(\rho), \tag{48}$$

$$f(\rho) = \frac{a_l + b_l(0.001\rho) + c_l(0.001\rho)^2 + d_l(0.001\rho)^3}{e_l + f_l(0.001\rho) + g_l(0.001\rho)^2 + h_l(0.001\rho)^3}, \tag{49}$$

$$a_l = a_{l0} + a_{l1}(1.8T) + a_{l2}(1.8T)^2, \tag{50}$$

$$b_l = b_{l0} + b_{l1}(1.8T) + b_{l2}(1.8T)^2, \tag{51}$$

$$c_l = c_{l0} + c_{l1}(1.8T) + c_{l2}(1.8T)^2, \tag{52}$$

$$d_l = d_{l0} + d_{l1}(1.8T) + d_{l2}(1.8T)^2, \tag{53}$$

$$e_l = e_{l0} + e_{l1}(1.8T) + e_{l2}(1.8T)^2, \tag{54}$$

$$f_l = f_{l0} + f_{l1}(1.8T) + f_{l2}(1.8T)^2, \tag{55}$$

$$g_l = g_{l0} + g_{l1}(1.8T) + g_{l2}(1.8T)^2, \tag{56}$$

$$h_l = h_{l0} + h_{l1}(1.8T) + h_{l2}(1.8T)^2. \tag{57}$$

The values of the parameters in Equations (50)–(57) are given in Table 5.

Table 5. The values of coefficients in Equations (50)–(57) for viscosity calculation.

Coefficients	$i = 0$	$i = 1$	$i = 2$
a_{li}	0.953363	−1.07384	0.00131729
b_{li}	−0.971028	11.2077	0.09013
c_{li}	1.01803	4.98986	0.302737
d_{li}	−0.990531	4.17585	−0.63662
e_{li}	1	−3.19646	3.90961
f_{li}	−1.00364	−0.181633	−7.79089
g_{li}	0.99808	−1.62108	0.000634836
h_{li}	−1.00103	0.676875	4.62481

(4) Sutton method [93]

Sutton correlated the effect of intermolecular forces as a function of apparent molar weight, pseudocritical pressure, and pseudocritical temperature and used the following coefficients:

$$K = \frac{1}{\xi_L}\left[0.807T_{pr}^{0.618} - 0.357\exp\left(-0.449T_{pr}\right) + 0.34\exp\left(-4.058T_{pr}\right) + 0.018\right], \tag{58}$$

$$\xi_L = 0.949\left[\frac{1.8T_{pc}}{(1000M)^3(p_{pc}/6894.8)^4}\right]^{1/6}, \tag{59}$$

$$X = 3.47 + \frac{1588}{1.8T} + 9M, \tag{60}$$

$$Y = 1.66378 - 0.04679X. \tag{61}$$

(5) Heidaryan and Jarrahian (H-J) method [94].

The coefficients were determined empirically through multiple regression analyses by Heidaryan and Jarrahian, as follows:

$$K = H_1 + H_2\frac{1.8T}{\ln(1.8T)} + \frac{H_3}{(1000M)^{1.5}} + \frac{H_4}{(1000M)^2} + H_5\exp(-1000M), \tag{62}$$

$$X = H_6 + H_7\ln(1000M)\sqrt{1000M} + H_8\ln(1000M) + \frac{H_9}{1.8T}, \tag{63}$$

$$Y = H_{10} + H_{11}\frac{1000M}{\ln(1000M)} + H_{12}\ln(1000M) + \frac{H_{13}}{1.8T}. \tag{64}$$

where the coefficients of H_1 to H_{13} are shown in Table 6.

Table 6. The coefficients of H_1–H_{13} in Equations (62)–(64).

Coefficient	Tuned Coefficient
H_1	−336.6309996192
H_2	1.432588837697
H_3	113,908.9211105
H_4	−380,054.8489939
H_5	384,369,898.3053
H_6	4.098692867002
H_7	−0.003307304447043
H_8	−0.004252150356903
H_9	718.2941490688
H_{10}	1.658818877773
H_{11}	−0.00580479565535
H_{12}	−0.001129165058823
H_{13}	−204.6192651917

(6) Abooali-Khamehchi (A-K) method [95]

Abooali and Khamehchi developed a natural gas viscosity calculation method covering 1938 data points with the following expression:

$$\mu = 10^{-3}\left\{ \begin{array}{c} 0.007393 + 0.2738481408\left(\frac{0.001\rho_g}{T_{pr}}\right)^2 + 0.594577152\left[\frac{(0.001\rho_g)^2 p_{pr}}{0.0624\rho_g + p_{pr}}\right] \\ -1.5620581417 \times 10^{-3}(0.001\rho_g)^3(1000M + 0.0624\rho_g) + 9.59 \times 10^{-5}(1000MT_{pr}^2) \end{array} \right\}. \tag{65}$$

The above density-based models provide reliable ways to calculate natural gas viscosity [96]. However, the accuracy of the gas density calculation in the models depends on the prediction of gas deviation factor Z at elevated pressure and temperature [97]. To overcome this problem, correlations based on the corresponding states theory were also established.

(7) Heidaryan-Moghadasi-Salarabadi (H-M-S) method [98]

Based on the falling body viscometer experiment, a correlation for methane gas viscosity was presented for temperatures up to 400 K and pressures up to 140 MPa, with an average absolute percent relative error of 0.794:

$$\mu = 10^{-3} \times \frac{A_{h1} + A_{h2}p_r + A_{h3}p_r^2 + A_{h4}p_r^3 + A_{h5}T_r + A_{h6}T_r^2}{1 + A_{h7}p_r + A_{h8}T_r + A_{h9}T_r^2 + A_{h10}T_r^3}, \tag{66}$$

where the coefficients A_{h1}–A_{h10} have values of $A_{h1} = -2.25711259 \times 10^{-2}$, $A_{h2} = -1.31338399 \times 10^{-4}$, $A_{h3} = 3.44353097 \times 10^{-6}$, $A_{h4} = -4.69476607 \times 10^{-8}$, $A_{h5} = 2.2303086 \times 10^{-2}$, $A_{h6} = -5.56421194 \times 10^{-3}$, $A_{h7} = 2.90880717 \times 10^{-5}$, $A_{h8} = -1.90511457$, $A_{h9} = 1.14082882$, and $A_{h10} = -0.225890087$. Note that the method is proposed according to the experimental data of methane. Therefore, this equation is strictly proposed for pure methane. That is also why the reduced pressure and reduced temperature, rather than the pseudoreduced pressure and pseudoreduced temperature, are employed in Equation (66).

(8) Sanjari et al. method [99]

Sanjari et al. presented a rapid method to calculate natural gas viscosity with high accuracy compared with other empirical methods:

$$\mu = 10^{-7} \frac{-0.141645 + 0.018076p_{pr} + 0.00214p_{pr}^2 - 0.004192\ln p_{pr} - 0.000386\ln^2 p_{pr} + \frac{0.187138}{T_{pr}} + 0.569211\ln^2 T_{pr}}{1 + 0.000387p_{pr}^2 - \frac{2.857176}{T_{pr}} + \frac{2.925776}{T_{pr}^2} - \frac{1.062425}{T_{pr}^3}}. \tag{67}$$

(9) Heidaryan-Esmaeilzadeh-Moghadasi (H-E-M) method [96]

In this method, gas viscosity is expressed as follows:

$$\mu = \mu_{1atm} \exp\left[\frac{1 + A_{e1}T_{pr} + A_{e2}p_{pr} + A_{e3}\left(2T_{pr}^2 - 1\right) + A_{e4}\left(2p_{pr}^2 - 1\right)}{A_{e5}T_{pr} + A_{e6}p_{pr} + A_{e7}\left(2T_{pr}^2 - 1\right) + A_{e8}\left(2p_{pr}^2 - 1\right) + A_{e9}\left(4T_{pr}^3 - 3T_{pr}\right) + A_{e10}\left(4p_{pr}^3 - 3p_{pr}\right)}\right], \tag{68}$$

where the values of coefficients A_{e1}–A_{e10} can be seen in Table 7.

Table 7. Values of coefficients in Equation (68).

Coefficient	For $p_{\mathrm{pr}} \leq 3$	For $p_{\mathrm{pr}} > 3$
A_{e1}	−0.574429785927299	−1.61486373676777
A_{e2}	−0.161401455390735	0.393317983084269
A_{e3}	0.111211457326131	0.273470537671412
A_{e4}	0.0300553354088043	0.00142823962661707
A_{e5}	5.56157730361211	−0.779279490434404
A_{e6}	−4.08580223632285	0.142553527534496
A_{e7}	−1.08843921447191	0.581332491577921
A_{e8}	0.694563966721831	0.00329912709369652
A_{e9}	0.13871744921249	0.0243974777653492
A_{e10}	−0.0420997501646278	−0.0000632194669476397

(10) Jarrahian and Heidaryan Method [97]

A generalized correlation method was proposed based on 3231 data points of 29 multicomponent mixtures at wide ranges of pressure (0.1–137.8 MPa), temperature (241–473 K), and specific gravity (0.573–1.337) by Jarrahian and Heidaryan [97], which can be expressed as follows:

$$\mu = \mu_{1atm}\left[1 + \frac{J_1}{T_{pr}^5}\left(\frac{p_{pr}^4}{T_{pr}^{20} + p_{pr}^4}\right) + J_2\left(\frac{p_{pr}}{T_{pr}}\right)^2 + J_3\left(\frac{p_{pr}}{T_{pr}}\right)\right], \tag{69}$$

where μ_{1atm} is the gas viscosity at 0.1 MPa; p_{pr} and T_{pr} are pseudoreduced pressure and pseudoreduced temperature; and J_1, J_2, and J_3 are fitting parameters with the values J_1 = 7.86338004624174, J_2 = −9.00157084101445 × 10^{−6}, and J_3 = 0.278138950019508.

(11) Izadmehr method [100]

Two models were developed for pure natural gas and impure natural gas viscosity calculation based on genetic programming techniques, covering 6484 data points and suitable for temperatures ranging from 109.6 K to 600 K, pressures ranging from 0.01 MPa to 199.95 MPa, and gas specific gravity ranging from 0.553 to 1.5741.

For the prediction of sour or sweet natural gases, the following formula can be employed:

$$\mu = 10^{-3} \times \left(a_{iz} + b_{iz} \times p_{pr} + \frac{c_{iz}}{T_{pr}} + d_{iz} \times p_{pr}^2 + \frac{e_{iz}}{T_{pr}^2} + f_{iz} \times \frac{p_{pr}}{T_{pr}} \right). \tag{70}$$

To improve the precision of pure natural gas viscosity prediction, the viscosity is calculated as follows:

$$\mu = 10^{-3} \times \left(a_{iz} \times T_{pr} + b_{iz} \times p_{pr} + c_{iz} \times \sqrt{p_{pr}} + d_{iz} \times T_{pr}^2 + e_{iz} \times \frac{p_{pr}}{T_{pr}} + f_{iz} \right). \tag{71}$$

where the coefficients of a_{iz} to f_{iz} are shown in Table 8.

Table 8. Values of coefficients in Equations (70) and (71).

Coefficient	Sour/Sweet Natural Gas Equation (70)	Sweet Natural Gas Equation (71)
a_{iz}	0.033359716350877	0.0085050748654501
b_{iz}	−0.00303297726852698	−0.00104065426590739
c_{iz}	−0.0514170415427857	−0.00217777225933512
d_{iz}	0.0000187555872613017	−0.000510724061609292
e_{iz}	0.0233088265671431	0.00595154429253907
f_{iz}	0.00880957960915389	−0.000548942531453252

(12) Low-pressure gas viscosity calculation

As can be seen, gas viscosity at low-pressure conditions, namely 1 atmospheric pressure, is required in some models. Dempsey used graphical correlations and gave the following expression [96]:

$$\mu_{1atm} = 10^{-3} \begin{bmatrix} 1.11231913 \times 10^{-2} + 1.67726604 \times 10^{-5}T + 2.11360496 \times 10^{-9}T^2 \\ -1.0948505 \times 10^{-4}M - 6.40316395 \times 10^{-8}MT - 8.99374533 \times 10^{-11}MT^2 \\ +4.57735189 \times 10^{-7}M^2 + 2.1290339 \times 10^{-7}M^2T + 3.97732249 \times 10^{-13}M^2T^2 \end{bmatrix}. \tag{72}$$

Standing improved the procedure of Dempsey for atmospheric viscosity calculation; the result is known as the Dempsey-Standing method [96]:

$$\mu_{1atm} = 10^{-3}\left[\left(1.709 \times 10^{-5} - 2.062 \times 10^{-6}\gamma_g \right)(1.8T - 459.67) + 8.188 \times 10^{-3} - 6.15 \times 10^{-3}\lg\gamma_g \right]. \tag{73}$$

A correlation method for this calculation was proposed by Londono et al. [92] as follows:

$$\mu_{1atm} = 10^{-3} \exp\left[\frac{-6.39821 - 0.6045922 \ln\left(\gamma_g\right) + 0.749768 \ln(1.8T) + 0.1261051 \ln\left(\gamma_g\right)\ln(1.8T)}{1 + 0.069718 \ln\left(\gamma_g\right) - 0.1013889 \ln(1.8T) - 0.0215294 \ln\left(\gamma_g\right)\ln(1.8T)} \right]. \tag{74}$$

Meanwhile, the coefficient K in the Sutton method above is also equivalent to the low-pressure gas viscosity, the value of which is 10^4 cp in the original work [93].

Figure 28a shows the relationship between gas viscosity at low pressure ($p = 0.1$ MPa). The Dempsey method [96] leads to much higher values than the other three models, while the Londono method [92] values are slightly higher than those from the Standing method [96] and Sutton method [93]. The Standing method [96] and Sutton method [93] produce basically the same viscosity results. Figure 28b displays gas viscosity variance with temperature at different pressures. For pressure lower than 14 MPa, the gas viscosity increases as temperature increases, which is the opposite situation to that of liquids. With temperature increasing, the gas characteristics become more similar to those of liquids, and gas viscosity decreases as temperature increases.

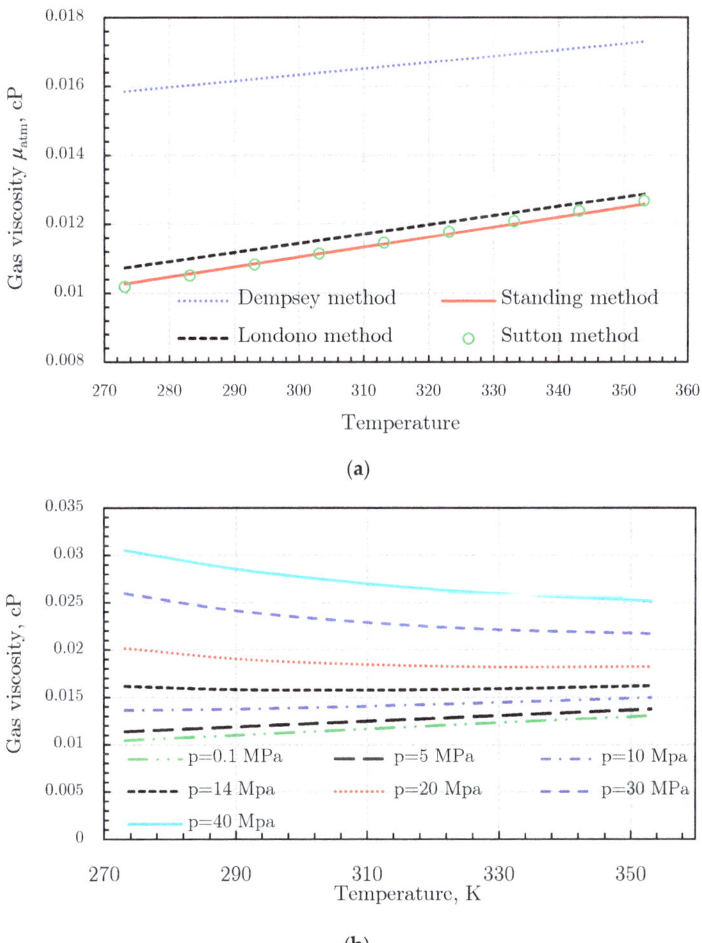

Figure 28. Gas viscosity variance with temperature at different pressure conditions. (**a**) Gas viscosity calculated from different methods at atmospheric pressure. (**b**) Gas viscosity calculated from the Standing method at different pressures.

In Figure 29, the gas viscosity from different models is compared with the experimental data for pressures ranging from 0.1 MPa to 3.3 MPa at a temperature of 293.15 K. The experimental data were chosen from the study of Hurly et al. [101]. As we can see, the calculation results vary: some match the experimental data well, while others deviate from the experimental data significantly, such as the H-J [94] and H-E-M methods [96]. To quantitatively evaluate the different models, the relative deviation and average absolute relative deviation (AARD), in terms of the experimental data, are given in Figures 30 and 31, respectively. Figure 30 illustrates that the H-J [94] and H-E-M methods [94] have poor performance for gas viscosity prediction for pressures ranging from 0.1 MPa to 3.3 MPa, while the Izadmehr method [100] performs badly for both sour and sweet gas viscosity prediction when the pressure is smaller than 1.4 MPa. This is because this method is obtained based on the regression of different kinds of gases, while the experimental data are for the gas viscosity of pure methane. Although the H-J method [97] predicts gas viscosity accurately when $p < 1$ MPa, its deviation from the experimental data becomes more and more obvious as pressure increases. The Lee method, Improved Lee method, Londono method, Sutton method, A-K method, Sanjari method, and Izadmehr sweet gas method can predict gas viscosity for 0.1 MPa $< p <$ 3.3 MPa with a relative deviation smaller than 0.3%. Among these methods, the Izadmehr sweet gas, Sanjari, A-K, and Sutton methods perform the best, with AARD equaling to 0.57%, 0.68%, 0.88%, and 0.95%, respectively, as shown in Figure 31.

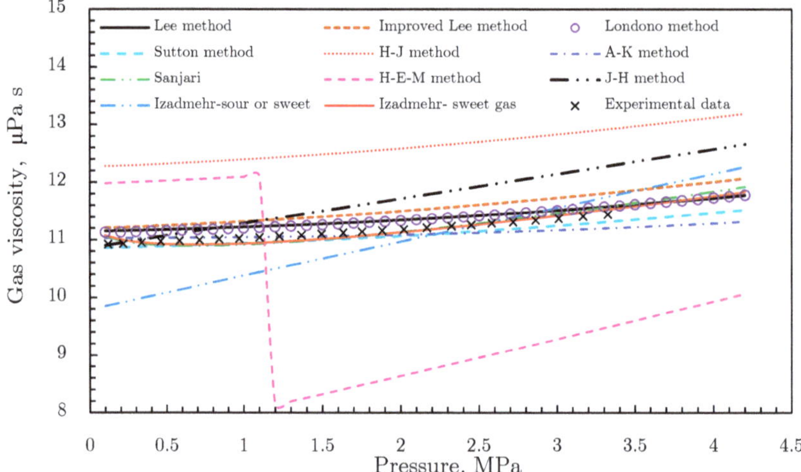

Figure 29. Model comparison for gas viscosity calculations as well as compared with experimental data at 293.15 K.

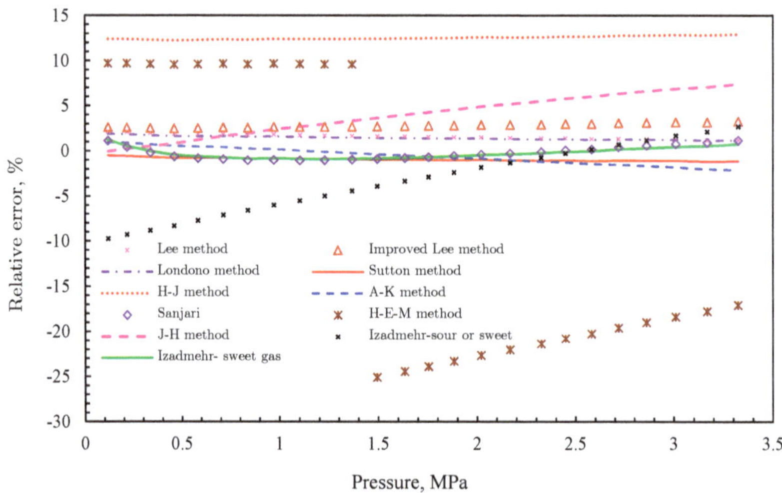

Figure 30. Relative deviation of different models for gas viscosity calculation at 293.15 K.

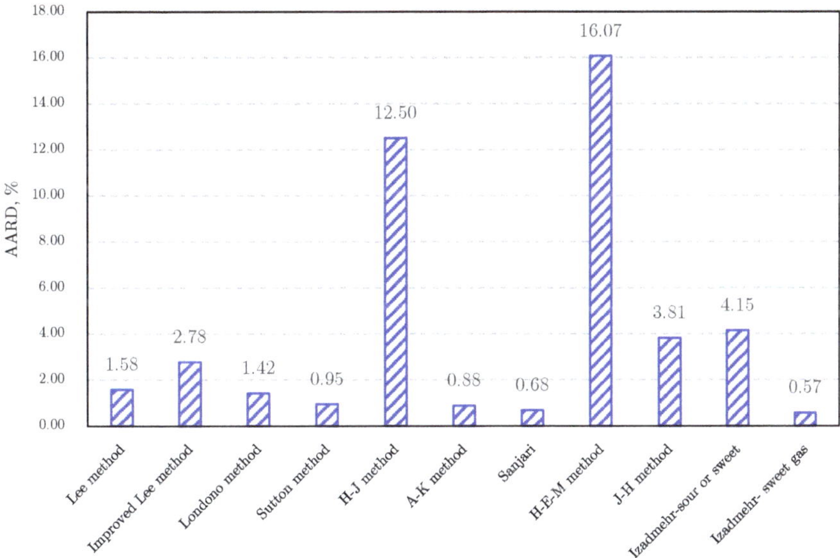

Figure 31. The AARD of different models for gas viscosity calculation at 293.15 K.

Gas viscosity calculation results at different temperatures (273.13 K, 293.13 K, 313.13 K, and 333.13 K) for 0.1 MPa $< p <$ 73 MPa are compared in Figure 32. The results of the H-M-S method vary significantly as the temperature changes. The discrepancies between the different models become more prominent as pressure increases. However, the discrepancies become less obvious as temperature increases.

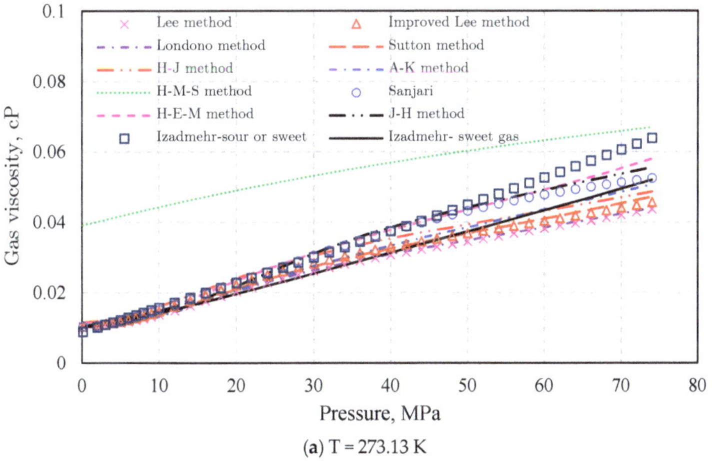

(**a**) T = 273.13 K

Figure 32. *cont.*

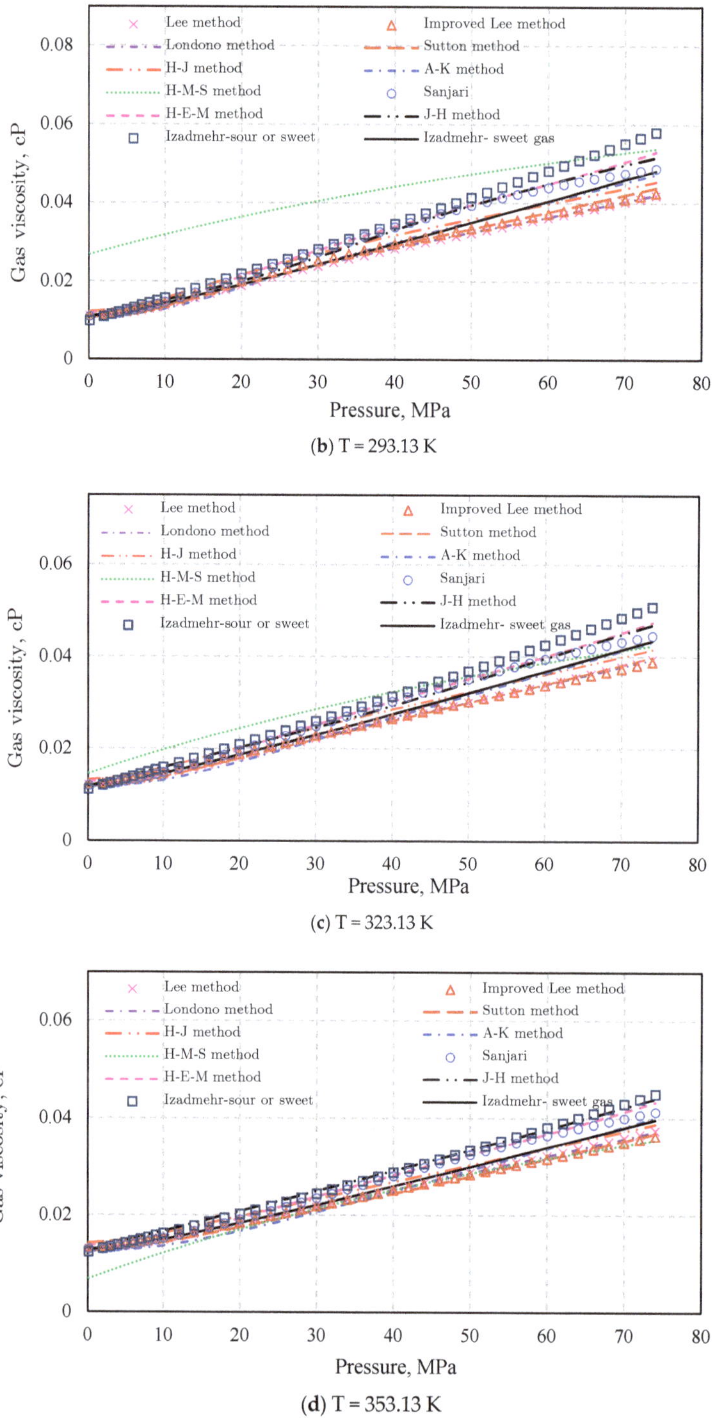

(**b**) T = 293.13 K

(**c**) T = 323.13 K

(**d**) T = 353.13 K

Figure 32. Comparison of gas viscosity results from different models at different temperature conditions. (**a**) 273.13 K. (**b**) 293.13 K. (**c**) 323.13 K. (**d**) 353.13 K.

3.4.2. Adsorbed Gas Density

Empirical Formula

(1) Dubinin method: Adsorbed gas density cannot be directly measured by experiments as the free gas density. As a result, many empirical formulations were proposed to approximately calculate its value. A temperature-independent formula was proposed by Dubinin [75,102] to approximate adsorbed gas density as follows:

$$\rho_{ad} = \frac{M}{b},$$ (75)

where b is the van der Waals covolume, representing the volume occupied by a single gas molecule.

(2) Gas density at critical point (p_c, T_c): The second way to calculate adsorbed phase gas density is from the critical pressure and critical temperature [72,103]:

$$\rho_{ad} = \frac{8Mp_c}{RT_c}.$$ (76)

Note that the above two methods are equivalent for calculating adsorbed gas density, which can be linked by the van der Waals equation:

$$(p + \frac{a}{V^2})(V - b) = RT,$$ (77)

where V is the molar volume of real gas; a and b are van der Waals constants, where a is related to intermolecular forces and b reflects the effects of the molecular volume of a real gas. According to an experimental study, the first and second derivative of the pressure of pure gases towards molar volume at the critical point (T_c, p_c) equal 0, i.e.:

$$\left(\frac{\partial p}{\partial V}\right)_{T_c} = 0,$$ (78)

$$\left(\frac{\partial^2 p}{\partial V^2}\right)_{T_c} - 0.$$ (79)

Combining Equations (77)–(79), the values of the constants a and b are solved as follows:

$$a = \frac{27R^2T_c^2}{64p_c},$$ (80)

$$b = \frac{RT_c}{8p_c}.$$ (81)

Substituting Equation (81) into Equation (75), Equation (76) could be obtained. Therefore, the methods of Equations (75) and (76) for calculating adsorbed gas density are the same.

(3) Ozawa method: Assuming adsorbed phase gas in a sort of superheated liquid state, Ozawa [75] calculated the density of adsorbed gas by the following equation [75]:

$$\rho_{ad} = \rho_b \exp[-0.0025(T - T_b)],$$ (82)

where ρ_{ad} is the adsorbed gas density, ρ_b is the liquid density at normal boiling point, T is the temperature, and T_b is the normal boiling point.

Although ρ_b and T_b of gases are functions of pressure, their values are assigned at the atmosphere pressure, when the pressure dependences of these quantities are negligibly small in the pressure range of the study. If the effects of pressure on ρ_b and T_b values cannot be neglected, they as well as the coefficient -0.0025 should be assigned other values.

According to the relationship between absolute adsorption and excess adsorption, the isotherm linearization method can also be used to obtain adsorbed phase gas density. Since it is not recommended to calculate the virtual saturated vapor pressure, using this method to obtain adsorbed gas density will not be introduced here. One of the conclusions from this paper [77] is that the value of adsorbed gas density is between the critical density (162.66 kg/m^3, at 4.59 MPa and 190.53 K) and normal boiling density of liquid phase (422.36 kg/m^3, at 0.101 MPa and 111.67 K), which can be confirmed from the calculation results of different models, as can be seen in Figure 33a. Note that the data of the Ono-Kondo method and ZGR EOS are collected from a previous study of Sudibandriyo et al. [104].

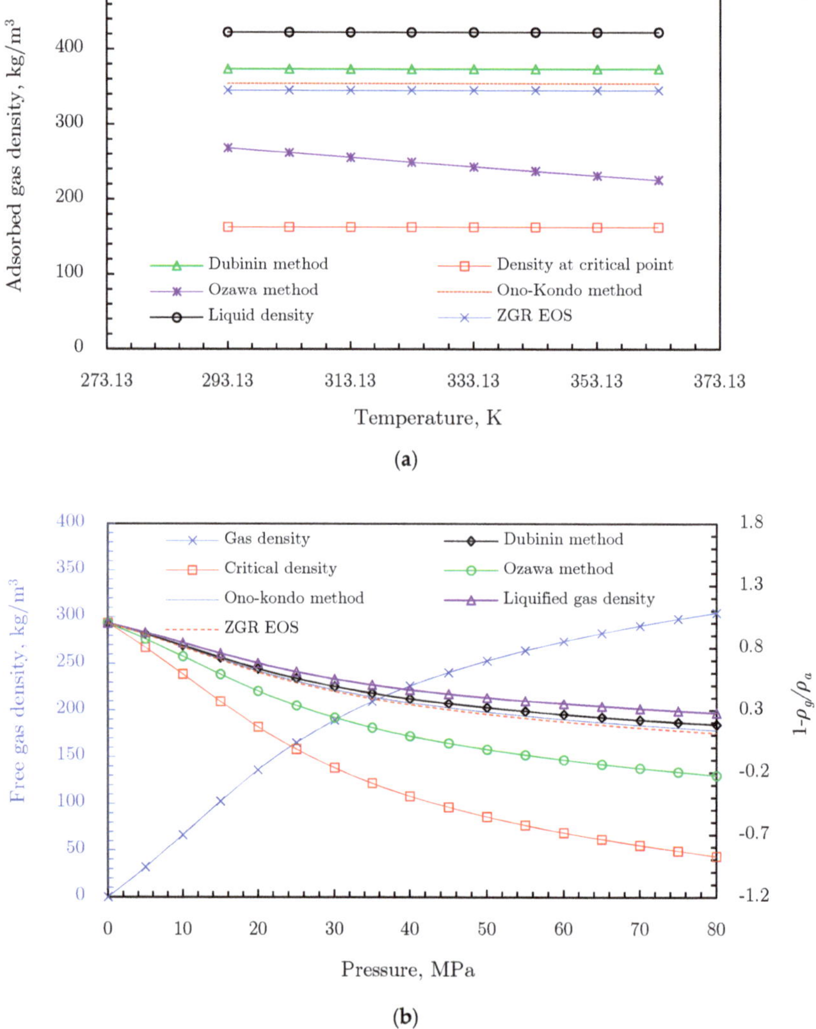

Figure 33. Comparison of adsorbed gas density calculated from different models (**a**) and the change of $(1 - \rho_g/\rho_a)$ with increasing pressure (**b**).

The adsorbed gas density is independent of temperature in all methods except the Ozawa model [75] of Equation (82), which assumed adsorbed gas as a superheated liquid. It seems that the Ozawa model is more suitable to predict adsorbed gas density since superheated liquid can expand with increasing temperature. However, the excess adsorption, which is calculated by Equation (25), may have a negative value, as can be inferred from Figure 33b. This is not physically right. As we mentioned, ρ_b and T_b of gases are functions of pressure for superheated liquid. These pressure-dependent characteristics are ignored in the original work [75], which cannot be neglected in a sorption study of shale gas reservoirs. Therefore, corresponding research needs to be done on this subject to develop a more practical model for shale gas sorption study.

Experimental Method

The experimental surface sorption amount can be a starting point for the calculation of absolute adsorption and adsorbed gas density. From Equations (24) and (25), we can obtain another expression of excess sorption:

$$G_{ex} = (\rho_{ad} - \rho_g)v_{ad}. \tag{83}$$

For high-pressure adsorption, bulk gas is more compressible than adsorbed gas and the excess sorption isotherm is mainly influenced by bulk gas EOS, where the adsorbed gas volume can be seen as a constant with increasing pressure. From this standpoint, the line segment after the inflection point on plot of excess sorption amount versus bulk gas density can be used for calculating the volume and density of adsorbed gas, as we can see from Figure 34. The density and volume of adsorbed gas after the maximum sorption amount can be speculated by the slope and intercept of line, according to Equation (83), where the absolute value of the slope is adsorbed gas volume and the intersection point of linear segment and x-axis is adsorbed gas density [80,104]. The calculated adsorbed gas density of methane is between 413.78 kg/m^3 and 433.41 kg/m^3, which is in the range of liquid density at its boiling point $\rho_b = 422.36$ kg/m^3. The calculated adsorbed gas volume for methane is between 0.423 cm^3/g and 0.467 cm^3/g for 303 K < T < 333 K. Note that, due to data error and the difference in data processing methods, there is a slight variation in the results between our calculation (see in Figure 35) and that of Moellmer et al. [80].

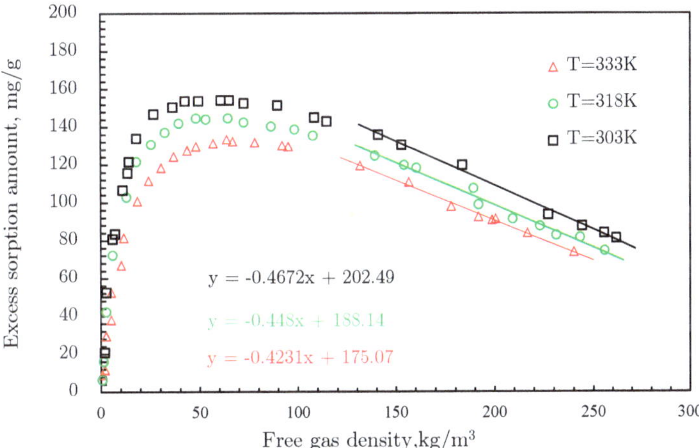

Figure 34. Excess sorption amount of methane in three different temperatures and the line segment after inflection point of maximum [80].

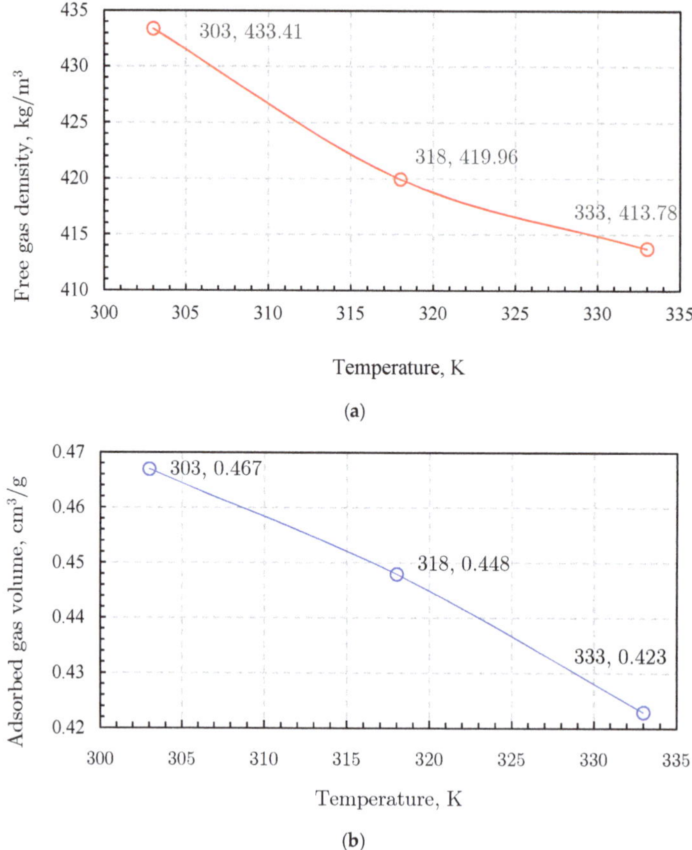

(a)

(b)

Figure 35. The change in adsorbed gas density (**a**) and adsorbed gas volume (**b**) with temperature, calculated by an experimental method.

As we can see from Figure 35a, the adsorbed gas density of methane decreases as temperature increases, showing a similar tendency to the Ozawa prediction [75], but with a higher value. This may provide a way to modify the Ozawa model of Equation (82), as we have mentioned that constant values of ρ_b and T_b for gases as well as the coefficient −0.0025 may not be suitable for methane sorption studies at high pressure. The calculated adsorbed gas volume of methane also shows a decreasing tendency with increasing temperature, as shown in Figure 35b, which corresponds to a decreasing amount adsorbed at increasing adsorption temperatures [80].

4. Concluding Remarks

Due to the massive development of shale gas reservoirs in recent years, the understanding of shale formation characteristics and shale gas storage forms has become a research hotspot. Our increased understanding has helped to extend the application of the classical seepage theory to new fields, where nanoscale flow spaces, gas sorption behaviors, and real gas properties are taken into account. However, much disagreement and confusion on this subject still exist, so it requires comprehensive investigation in the future.

This review includes a summary, discussion, and comparison of shale formation characteristics, shale gas occurrence types, and property calculation methods of adsorbed gas and free gas, providing fundamental support to the deep understanding of shale gas reservoirs. (1) The typical mineral composition and organic geochemical characteristics, as well as pore size distribution, of shale formations are given based on our measurements and analysis of samples from the Longmaxi Formation. We found that mesopores are the mainly developed pore types in the Longmaxi Shale Formation, with the gas-filled porosity of shale samples ranging from 1% to 7.5%, and permeability usually smaller than 0.1 mD. Three shale gas types are usually classified, free gas, adsorbed gas, and dissolved gas, where adsorbed gas and dissolved gas are often considered as one type due to the equilibrium state between them. Therefore, it is claimed that there are two gas types, free gas and adsorbed gas, in shale gas plays in some research. (2) Methane in shale gas reservoirs is in a supercritical state, so Gibbs excess sorption models and supercritical state sorption models are employed to capture the gas adsorption and desorption behaviors. Meanwhile, different models considering not only the gas adsorption but also the absorption are introduced. Great discrepancies could occur if the supercritical state and Gibbs excess adsorption characteristics are ignored. Different mechanisms of adsorption in micropores and macropores may explain the hysteresis between adsorption and desorption, rather than capillary condensation. (3) Different methods of calculating gas properties, such gas free gas density, free gas viscosity, and adsorbed gas density considering high-pressure and high-temperature conditions, are summarized, with recommended approaches given after the comparison. From our review, we can see that the geological characteristics of shale formations are quite different from those of conventional ones, and need further assessment using a high-resolution apparatus. Gas adsorption mechanisms are still not clear, although numerous models have been developed to account for this phenomenon. For example, current supercritical adsorption models based on the potential theory are modifications of a previous adsorption theory, which are dependent on empirical parameters and lack of a universal theoretical basis. Applicable multicomponent gas adsorption models considering high-pressure and high-temperature conditions and pore size distribution are in demand to describe gas behavior in shale gas reservoirs. They are a rough way to clarify adsorbed gas and dissolved as one type, since dissolved gas may play an important role in shale gas production. Therefore, current adsorption and absorption models should be improved, based on the practical relationship between adsorbed gas and dissolved gas under in situ conditions.

Author Contributions: Conceptualization, Y.Z. and L.Z.; methodology, B.Z. and B.S.; validation, B.S. and Y.Z.; formal analysis, B.Z. and B.S.; investigation, B.Z.; resources, L.Z..; writing—original draft preparation, B.Z., B.S. and Y.Z.; writing—review and editing, L.Z.; supervision, B.S. and L.Z.; All authors have read and agreed to the published version of the manuscript.

Funding: This work was supported by the National Natural Science Foundation of China (grant nos. 51874251 and 51704247) and the Science and Technology Cooperation Project of the CNPC-SWPU Innovation Alliance.

Acknowledgments: The authors would like to thank the reviewers and editors, whose critical comments were very helpful in preparing this article.

Conflicts of Interest: The authors declare no conflict of interest.

References

1. Shen, W.; Li, X.; Cihan, A.; Lu, X.; Liu, X. Experimental and numerical simulation of water adsorption and diffusion in shale gas reservoir rocks. *Adv. Geo-Energy Res.* **2019**, *3*, 165–174. [CrossRef]
2. Bustin, R.M.; Bustin, A.M.M.; Cui, A.; Ross, D.; Pathi, V.M. Impact of Shale Properties on Pore Structure and Storage Characteristics. In Proceedings of the SPE Shale Gas Production Conference, Fort Worth, TX, USA, 16–18 November 2008; p. 28.
3. Zhang, L.; Guo, J.; Tang, H. *The Fundamental of Shale Gas Reservoir Development*; Petroleum Industry Press: Beijing, China, 2014; (In Chinese). [CrossRef]
4. Clarkson, C.R.; Jensen, J.L.; Chipperfield, S. Unconventional gas reservoir evaluation: What do we have to consider? *J. Nat. Gas Sci. Eng.* **2012**, *8*, 9–33. [CrossRef]
5. Li, J.; Zhang, P.; Lu, S.; Chen, C.; Xue, H.; Wang, S.; Li, W. Scale-Dependent Nature of Porosity and Pore Size Distribution in Lacustrine Shales: An Investigation by BIB-SEM and X-Ray CT Methods. *J. Earth Sci.* **2018**, *30*, 823–833. [CrossRef]

6. Daigle, H.; Johnson, A. Combining Mercury Intrusion and Nuclear Magnetic Resonance Measurements Using Percolation Theory. *Transp. Porous Media* **2015**, *111*, 669–679. [CrossRef]
7. Deng, H.; Hu, X.; Li, H.A.; Luo, B.; Wang, W. Improved pore-structure characterization in shale formations with FESEM technique. *J. Nat. Gas Sci. Eng.* **2016**, *35*, 309–319. [CrossRef]
8. Zhao, P.; Wang, Z.; Sun, Z.; Cai, J.; Wang, L. Investigation on the pore structure and multifractal characteristics of tight oil reservoirs using NMR measurements: Permian Lucaogou Formation in Jimusaer Sag, Junggar Basin. *Mar. Pet. Geol.* **2017**, *86*, 1067–1081. [CrossRef]
9. Karimi, S.; Saidian, M.; Prasad, M.; Kazemi, H. Reservoir Rock Characterization Using Centrifuge and Nuclear Magnetic Resonance: A Laboratory Study of Middle Bakken Cores. In Proceedings of the SPE Annual Technical Conference and Exhibition, Houston, TX, USA, 28–30 September 2015; p. 18.
10. Shao, D.; Zhang, T.; Ko, L.T.; Luo, H.; Zhang, D. Empirical plot of gas generation from oil-prone marine shales at different maturity stages and its application to assess gas preservation in organic-rich shale system. *Mar. Pet. Geol.* **2019**, *102*, 258–270. [CrossRef]
11. Dai, J.; Zou, C.; Liao, S.; Dong, D.; Ni, Y.; Huang, J.; Wu, W.; Gong, D.; Huang, S.; Hu, G. Geochemistry of the extremely high thermal maturity Longmaxi shale gas, southern Sichuan Basin. *Org. Geochem.* **2014**, *74*, 3–12. [CrossRef]
12. Yang, F.; Ning, Z.; Liu, H. Fractal characteristics of shales from a shale gas reservoir in the Sichuan Basin, China. *Fuel* **2014**, *115*, 378–384. [CrossRef]
13. Zou, C.; Zhu, R.; Chen, Z.Q.; Ogg, J.G.; Wu, S.; Dong, D.; Qin, Z.; Wang, Y.; Wang, L.; Lin, S.; et al. Organic-matter-rich shales of China. *Earth-Sci. Rev.* **2019**, *189*, 51–78. [CrossRef]
14. Fan, C.; Li, H.; Qin, Q.; He, S.; Zhong, C. Geological conditions and exploration potential of shale gas reservoir in Wufeng and Longmaxi Formation of southeastern Sichuan Basin, China. *J. Pet. Sci. Eng.* **2020**, *191*, 107138. [CrossRef]
15. Montgomery, S.L.; Jarvie, D.M.; Bowker, K.A.; Pollastro, R.M. Mississippian Barnett Shale, Fort Worth basin, north-central Texas: Gas-shale play with multi–trillion cubic foot potential: Reply. *AAPG Bull.* **2006**, *90*, 967–969. [CrossRef]
16. Loucks, R.G.; Ruppel, S.C. Mississippian Barnett Shale: Lithofacies and depositional setting of a deep-water shale-gas succession in the Fort Worth Basin, Texas. *AAPG Bull.* **2007**, *91*, 579–601. [CrossRef]
17. Zou, C.; Dong, D.; Wang, S.; Li, J.; Li, X.; Wang, Y.; Li, D.; Cheng, K. Geological characteristics, formation mechanism and resource potential of shale gas in China (in Chinese with English abstract). *Pet. Explor. Dev.* **2010**, *37*, 641–653. [CrossRef]
18. Bowker, K.A. Barnett Shale gas production, Fort Worth Basin: Issues and discussion. *AAPG Bull.* **2007**, *91*, 523–533. [CrossRef]
19. Curtis, J.B. Fractured shale-gas systems. *AAPG Bull.* **2002**, *86*, 1921–1938.
20. Jarvie, D.M.; Hill, R.J.; Ruble, T.E.; Pollastro, R.M. Unconventional shale-gas systems: The Mississippian Barnett Shale of north-central Texas as one model for thermogenic shale-gas assessment. *AAPG Bull.* **2007**, *91*, 475–499. [CrossRef]
21. Jarvie, D.M.; Hill, R.; Pollastro, R.; Wavrek, D.; Claxton, B.; Tobey, M. Evaluation of unconventional natural gas prospects: The Barnett Shale fractured shale gas model. In Proceedings of the European Association of International Organic Geochemists Meeting, Kraków, Poland, 8–12 September 2003.
22. Martineau, D.F. History of the Newark East field and the Barnett Shale as a gas reservoir. *AAPG Bull.* **2007**, *91*, 399–403. [CrossRef]
23. Pollastro, R.M.; Jarvie, D.M.; Hill, R.J.; Adams, C.W. Geologic framework of the Mississippian Barnett Shale, Barnett-Paleozoic total petroleum system, Bend arch-Fort Worth Basin, Texas. *AAPG Bull.* **2007**, *91*, 405–436. [CrossRef]
24. Welte, D.H.; Tissot, B.P. *Petroleum Formation and Occurrence*; Springer: Berlin/Heidelberg, Germany, 1984.
25. David, G.; Charles, R.N. Gas productive fractured shales: An overview and update. *Gas Tips* **2000**, *6*, 4–13. [CrossRef]
26. Warlick, D. Gas shale and CBM development in North America. *Oil Gas Financ. J.* **2006**, *3*, 1–5.
27. Zhang, L.H.; Shan, B.C.; Zhao, Y.L.; Guo, Z.L. Review of micro seepage mechanisms in shale gas reservoirs. *Int. J. Heat Mass Transfer* **2019**, *139*, 144–179. [CrossRef]

28. Loucks, R.G.; Reed, R.M.; Ruppel, S.C.; Jarvie, D.M. Morphology, Genesis, and Distribution of Nanometer-Scale Pores in Siliceous Mudstones of the Mississippian Barnett Shale. *J. Sediment. Res.* **2009**, *79*, 848–861. [CrossRef]

29. Slatt, R.M.; O'Brien, N.R. Pore types in the Barnett and Woodford gas shales: Contribution to understanding gas storage and migration pathways in fine-grained rocks. *AAPG Bull.* **2011**, *95*, 2017–2030. [CrossRef]

30. Loucks, R.G.; Reed, R.M.; Ruppel, S.C.; Hammes, U. Spectrum of pore types and networks in mudrocks and a descriptive classification for matrix-related mudrock pores. *AAPG Bull.* **2012**, *96*, 1071–1098. [CrossRef]

31. Long, P.; Zhang, J.; Tang, X.; Nie, H.; Liu, Z.; Han, S.; Zhu, L. Feature of muddy shale fissure and its effect for shale gas exploration and development (in Chinese with English abstract). *Nat. Gas Geosci.* **2011**, *22*, 525–532.

32. Brunauer, S.; Emmett, P.H.; Teller, E. Adsorption of gases in multimolecular layers. *J. Am. Chem. Soc.* **1938**, *60*, 309–319. [CrossRef]

33. Chen, J.; Wang, F.C.; Liu, H.; Wu, H.A. Molecular mechanism of adsorption/desorption hysteresis:dynamics of shale gas in nanopores. *Sci. China (Phys. Mech. Astron.)* **2017**, *60*, 014611. [CrossRef]

34. Barrett, E.P.; Joyner, L.G.; Halenda, P.P. The determination of pore volume and area distributions in porous substances. I. Computations from nitrogen isotherms. *J. Am. Chem. Soc.* **1951**, *73*, 373–380. [CrossRef]

35. Washburn, E.W. The Dynamics of Capillary Flow. *Phys. Rev.* **1921**, *17*, 273–283. [CrossRef]

36. Chen, Y.; Zhang, L.; Li, J. Study of Pore Structure of Gas Shale with Low-Field NMR: Examples from the Longmaxi Formation, Southern Sichuan Basin, China. *Soc. Pet. Eng.* **2015**. [CrossRef]

37. Milkov, A.V.; Faiz, M.; Etiope, G. Geochemistry of shale gases from around the world: Composition, origins, isotope reversals and rollovers, and implications for the exploration of shale plays. *Org. Geochem.* **2020**, *143*, 103997. [CrossRef]

38. Zhang, L.; Shan, B.; Zhao, Y.; Tang, H. Comprehensive Seepage Simulation of Fluid Flow in Multi-scaled Shale Gas Reservoirs. *Transp. Porous Media* **2017**, *121*, 263–288. [CrossRef]

39. Cai, J.; Lin, D.; Singh, H.; Zhou, S.; Meng, Q.; Zhang, Q. A simple permeability model for shale gas and key insights on relative importance of various transport mechanisms. *Fuel* **2019**, *252*, 210–219. [CrossRef]

40. Javadpour, F. Nanopores and Apparent Permeability of Gas Flow in Mudrocks (Shales and Siltstone). *J. Can. Petrol. Technol.* **2009**, *48*, 16–21. [CrossRef]

41. Zhang, L.; Shan, B.; Zhao, Y.; Du, J.; Chen, J.; Tao, X. Gas Transport Model in Organic Shale Nanopores Considering Langmuir Slip Conditions and Diffusion: Pore Confinement, Real Gas, and Geomechanical Effects. *Energies* **2018**, *11*, 223. [CrossRef]

42. Zhang, T.; Li, Y.; Sun, S. Phase equilibrium calculations in shale gas reservoirs. *Capillarity* **2019**, *2*, 8–16. [CrossRef]

43. Shen, W.; Song, F.; Hu, X.; Zhu, G.; Zhu, W. Experimental study on flow characteristics of gas transport in micro- and nanoscale pores. *Sci. Rep.* **2019**, *9*, 10196. [CrossRef]

44. Li, S. *Natural Gas Engineering*; Petroleum Industry Press: Beijing, China, 2008.

45. Zhang, R.H.; Zhang, L.H.; Tang, H.Y.; Chen, S.N.; Zhao, Y.L.; Wu, J.F.; Wang, K.R. A Simulator for Production Prediction of Multistage Fractured Horizontal Well in Shale Gas Reservoir Considering Complex Fracture Geometry. *J. Nat. Gas Sci. Eng.* **2019**, *67*, 14–29. [CrossRef]

46. Zhang, R.H.; Zhang, L.H.; Wang, R.H.; Zhao, Y.L.; Zhang, D.L. Research on transient flow theory of a multiple fractured horizontal well in a composite shale gas reservoir based on the finite-element method. *J. Nat. Gas Sci. Eng.* **2016**, *33*, 587–598. [CrossRef]

47. Wang, Y.; Shahvali, M. Discrete fracture modeling using Centroidal Voronoi grid for simulation of shale gas plays with coupled nonlinear physics. *Fuel* **2016**, *163*, 65–73. [CrossRef]

48. Deliang, Z.; Liehui, Z.; Jingjing, G.; Yuhui, Z.; Yulong, Z. Research on the production performance of multistage fractured horizontal well in shale gas reservoir. *J. Nat. Gas Sci. Eng.* **2015**, *26*, 279–289. [CrossRef]

49. Mengal, S.A.; Wattenbarger, R.A. Accounting For Adsorbed Gas in Shale Gas Reservoirs. In Proceedings of the SPE Middle East Oil and Gas Show and Conference, Manama, Bahrain, 25–28 September 2011; p. 15.

50. Wang, L.; Wang, S.; Zhang, R.; Wang, C.; Xiong, Y.; Zheng, X.; Li, S.; Jin, K.; Rui, Z. Review of multi-scale and multi-physical simulation technologies for shale and tight gas reservoirs. *J. Nat. Gas Sci. Eng.* **2017**, *37*, 560–578. [CrossRef]

51. Taotao, C.; Zhiguang, S.; Sibo, W.; Jia, X. A comparative study of the specific surface area and pore structure of different shales and their kerogens. *Sci. China Earth Sci.* **2015**, *58*, 510–522.

52. Wang, F.P.; Reed, R.M. Pore Networks and Fluid Flow in Gas Shales. In Proceedings of the SPE Annual Technical Conference and Exhibition, New Orleans, LA, USA, 4–7 October 2009; p. 8.

53. Zhao, Y.; Shan, B.; Zhang, L.; Liu, Q. Seepage flow behaviors of multi-stage fractured horizontal wells in arbitrary shaped shale gas reservoirs. *J. Geophys. Eng.* **2016**, *13*, 674–689. [CrossRef]

54. Fathi, E.; Akkutlu, I.Y. Multi-component gas transport and adsorption effects during CO_2 injection and enhanced shale gas recovery. *Int. J. Coal Geol.* **2014**, *123*, 52–61. [CrossRef]

55. Cai, J.; Lin, D.; Singh, H.; Wei, W.; Zhou, S. Shale gas transport model in 3D fractal porous media with variable pore sizes. *Mar. Petrol. Geol.* **2018**, *98*, 437–447. [CrossRef]

56. Kim, T.H.; Cho, J.; Lee, K.S. Evaluation of CO_2 injection in shale gas reservoirs with multi-component transport and geomechanical effects. *Appl. Energy* **2017**, *190*, 1195–1206. [CrossRef]

57. Yin, Y.; Qu, Z.G.; Zhang, J.F. Multiple diffusion mechanisms of shale gas in nanoporous organic matter predicted by the local diffusivity lattice Boltzmann model. *Int. J. Heat Mass Transfer* **2019**, *143*, 118571. [CrossRef]

58. Zhang, T.; Sun, S. A coupled Lattice Boltzmann approach to simulate gas flow and transport in shale reservoirs with dynamic sorption. *Fuel* **2019**, *246*, 196–203. [CrossRef]

59. Javadpour, F.; Fisher, D.; Unsworth, M. Nanoscale Gas Flow in Shale Gas Sediments. *J. Can. Petrol. Technol.* **2007**, *46*. [CrossRef]

60. Wang, J.; Yang, Z.; Dong, M.; Gong, H.; Sang, Q.; Li, Y. Experimental and Numerical Investigation of Dynamic Gas Adsorption/Desorption Diffusion Process in Shale. *Energy Fuels* **2016**, *30*, 10080–10091. [CrossRef]

61. Tang, X.; Ripepi, N.; Stadie, N.P.; Yu, L.; Hall, M.R. A dual-site Langmuir equation for accurate estimation of high pressure deep shale gas resources. *Fuel* **2016**, *185*, 10–17. [CrossRef]

62. Lopez, B.; Aguilera, R. Sorption-Dependent Permeability of Shales. In Proceedings of the Spe/csur Unconventional Resources Conference, Calgary, AB, Canada, 20–22 October 2015.

63. Ross, D.J.K.; Marc Bustin, R. The importance of shale composition and pore structure upon gas storage potential of shale gas reservoirs. *Mar. Pet. Geol.* **2009**, *26*, 916–927. [CrossRef]

64. Zhang, L.; Li, J.; Jia, D.U.; Zhao, Y.; Xie, C.; Tao, Z. Study on the Adsorption Phenomenon in Shale with the Combination of Molecular Dynamic Simulation and Fractal Analysis. *Fractals* **2018**, *26*, 1840004. [CrossRef]

65. Zhang, L.; Chen, Z.; Zhao, Y. *Well Production Performance Analysis for Shale Gas Reservoirs*; Candice Janco: Amsterdam, The Netherlands, 2019; Volume 66.

66. Yin, Y.; Qu, Z.G.; Zhang, J.F. An analytical model for shale gas transport in kerogen nanopores coupled with real gas effect and surface diffusion. *Fuel* **2017**, *210*, 569–577. [CrossRef]

67. Zhou, S.; Wang, H.; Xue, H.; Guo, W.; Li, X. Discussion on the Supercritical Adsorption Mechanism of Shale Gas Based on Ono-Kondo Lattice Model (in Chinese with English abstract). *Earth Sci.* **2017**, *42*, 1421–1430.

68. Langmuir, I. The Adsorption of Gases on Plane Surfaces of Glass, Mica and Platinum. *J. Am. Chem. Soc.* **1918**, *40*, 1361–1403. [CrossRef]

69. Bae, J.-S.; Bhatia, S.K. High-pressure adsorption of methane and carbon dioxide on coal. *Energy Fuels* **2006**, *20*, 2599–2607. [CrossRef]

70. Wang, Y.; Zhu, Y.; Liu, S.; Zhang, R. Methane adsorption measurements and modeling for organic-rich marine shale samples. *Fuel* **2016**, *172*, 301–309. [CrossRef]

71. Dubinin, M.M. *Progress in Membrane and Surface Science*; Cadenhead, D.A., Ed.; Academic Press: New York, NY, USA, 1975; Volume 9, pp. 1–70. Available online: https://scholar.google.com/scholar?q=M.+M.+Dubinin,+Progress+in+Surface+and+Membrane+Science,+Academic+Press,+New+York,+1975&hl=zh-CN&as_sdt=0,5 (accessed on 4 April 2020).

72. Yang, F.; Xie, C.; Xu, S.; Ning, Z.; Krooss, B.M. Supercritical Methane Sorption on Organic-Rich Shales over a Wide Temperature Range. *Energy Fuels* **2017**, *31*, 13427–13438. [CrossRef]

73. Manes, M.; Hofer, L.J.E. Application of the Polanyi adsorption potential theory to adsorption from solution on activated carbon. *J. Phys. Chem.* **1969**, *73*, 584–590. [CrossRef]

74. Agarwal, R.K.; Amankwah, K.A.G.; Schwarz, J.A. Analysis of adsorption entropies of high pressure gas adsorption data on activated carbon. *Carbon* **1990**, *28*, 169–174. [CrossRef]

75. Wakasugi, Y.; Ozawa, S.; Ogino, Y. Physical Adsorption of Gases at High Pressure (IV). *J. Colloid Interface Sci.* **1981**, *79*, 399–409. [CrossRef]

76. Amankwah, K.A.G.; Schwarz, J.A. A Modified Approach for Estimating Pseudo Vapor Pressures in the Application of the Dubinin-Astakhov Equation. *Carbon* **1995**, *33*, 1313–1319. [CrossRef]

77. Zhou, L.; Zhou, Y.P. Linearization of adsorption isotherms for high-pressure applications—Argon and methane onto graphitized carbon black. *Chem. Eng. Sci.* **1998**, *53*, 2531–2536. [CrossRef]

78. Zhang, T.; Ellis, G.S.; Ruppel, S.C.; Milliken, K.; Yang, R. Effect of organic-matter type and thermal maturity on methane adsorption in shale-gas systems. *Org. Geochem.* **2012**, *47*, 120–131. [CrossRef]

79. Zhou, S.; Ning, Y.; Wang, H.; Liu, H.; Xue, H. Investigation of methane adsorption mechanism on Longmaxi shale by combining the micropore filling and monolayer coverage theories. *Adv.Geo-Energy Res.* **2018**, *2*, 269–281. [CrossRef]

80. Moellmer, J.; Moeller, A.; Dreisbach, F.; Glaeser, R.; Staudt, R. High pressure adsorption of hydrogen, nitrogen, carbon dioxide and methane on the metal–organic framework HKUST-1. *Microporous Mesoporous Mater.* **2011**, *138*, 140–148. [CrossRef]

81. Sakurovs, R.; Day, S.; Weir, S.; Duffy, G. Application of a Modified Dubinin—Radushkevich Equation to Adsorption of Gases by Coals under Supercritical Conditions. *Energy Fuels* **2007**, *21*, 992–997. [CrossRef]

82. Pini, R.; Ottiger, S.; Burlini, L.; Storti, G.; Mazzotti, M. Sorption of carbon dioxide, methane and nitrogen in dry coals at high pressure and moderate temperature. *Int. J. Greenh. Gas Control* **2010**, *4*, 90–101. [CrossRef]

83. Reucroft, P.J.; Patel, K.B. Surface area and swellability of coal. *Fuel* **1983**, *62*, 279–284. [CrossRef]

84. Sato, Y.; Takikawa, T.; Yamane, M.; Takishima, S.; Masuoka, H. Solubility of carbon dioxide in PPO and PPO/PS blends. *Fluid Phase Equilib.* **2002**, *194*, 847–858. [CrossRef]

85. Larsen, J.W. The effects of dissolved CO_2 on coal structure and properties. *Int. J. Coal Geol.* **2004**, *57*, 63–70. [CrossRef]

86. Ozdemir, E.; Morsi, B.I.; Schroeder, K. CO_2 adsorption capacity of argonne premium coals. *Fuel* **2004**, *83*, 1085–1094. [CrossRef]

87. Fornstedt, T.; Zhong, G.; Bensetiti, Z.; Guiochon, G. Experimental and theoretical study of the adsorption behavior and mass transfer kinetics of propranolol enantiomers on cellulase protein as the selector. *Anal. Chem.* **1996**, *68*, 2370–2378. [CrossRef] [PubMed]

88. Lee, A.L.; Gonzalez, M.H.; Eakin, B.E. The Viscosity of Natural Gases. *J. Pet. Technol.* **1966**, *18*, 997–1000. [CrossRef]

89. Lee, A.; Starling, K.; Dolan, J.; Ellington, R. Viscosity correlation for light hydrocarbon systems. *AIChE J.* **1964**, *10*, 694–697. [CrossRef]

90. McCain, W.D., Jr. Reservoir-Fluid Property Correlations-State of the Art. *SPE Reserv. Eng.* **1991**, *6*, 266–277. [CrossRef]

91. Naraghi, M.E.; Javadpour, F. A stochastic permeability model for the shale-gas systems. *Int. J. Coal Geol.* **2015**, *140*, 111–124. [CrossRef]

92. Londono, F.E.; Archer, R.A.; Blasingame, T.A. Correlations for Hydrocarbon Gas Viscosity and Gas Density—Validation and Correlation of Behavior Using a Large-Scale Database. *SPE Reserv. Eval. Eng.* **2005**, *8*, 561–572. [CrossRef]

93. Sutton, R.P. Fundamental PVT Calculations for Associated and Gas/Condensate Natural-Gas Systems. *SPE Reserv. Eval. Eng.* **2007**, *10*, 270–284. [CrossRef]

94. Heidaryan, E.; Jarrahian, A. Natural gas viscosity estimation using density based models. *Can. J. Chem. Eng.* **2013**, *91*. [CrossRef]

95. Abooali, D.; Khamehchi, E. Estimation of dynamic viscosity of natural gas based on genetic programming methodology. *J. Nat. Gas Sci. Eng.* **2014**, *21*, 1025–1031. [CrossRef]

96. Heidaryan, E.; Esmaeilzadeh, F.; Moghadasi, J. Natural gas viscosity estimation through corresponding states based models. *Fluid Phase Equilib.* **2013**, *354*, 80–88. [CrossRef]

97. Jarrahian, A.; Heidaryan, E. A simple correlation to estimate natural gas viscosity. *J. Nat. Gas Sci. Eng.* **2014**, *20*, 50–57. [CrossRef]

98. Heidaryan, E.; Moghadasi, J.; Salarabadi, A. A new and reliable model for predicting methane viscosity at high pressures and high temperatures. *J. Nat. Gas Chem.* **2010**, *19*, 552–556. [CrossRef]

99. Sanjari, E.; Lay, E.N. An accurate empirical correlation for predicting natural gas viscosity. *J. Nat. Gas Chem.* **2011**, *20*, 654–658. [CrossRef]

100. Izadmehr, M.; Shams, R.; Ghazanfari, M.H. New correlations for predicting pure and impure natural gas viscosity. *J. Nat. Gas Sci. Eng.* **2016**, *30*, 364–378. [CrossRef]

101. Hurly, J.J.; Gillis, K.A.; Mehl, J.B.; Moldover, M.R. The Viscosity of Seven Gases Measured with a Greenspan Viscometer. *Int. J. Thermophys.* **2003**, *24*, 1441–1474. [CrossRef]

102. Dubinin, M. The potential theory of adsorption of gases and vapors for adsorbents with energetically nonuniform surfaces. *Chem. Rev.* **1960**, *60*, 235–241. [CrossRef]

103. Li, J.; Chen, Z.; Wu, K.; Wang, K.; Luo, J.; Feng, D.; Qu, S.; Li, X. A multi-site model to determine supercritical methane adsorption in energetically heterogeneous shales. *Chem. Eng. J.* **2018**, *349*, 438–455. [CrossRef]
104. Sudibandriyo, M.; Pan, Z.; Fitzgerald, J.E.; Robinson, R.L.; Gasem, K.A.M. Adsorption of Methane, Nitrogen, Carbon Dioxide, and Their Binary Mixtures on Dry Activated Carbon at 318.2 K and Pressures up to 13.6 MPa. *Langmuir* **2003**, *19*, 5323–5331. [CrossRef]

Article

Effect of Particle Size on Pore Characteristics of Organic-Rich Shales: Investigations from Small-Angle Neutron Scattering (SANS) and Fluid Intrusion Techniques

Yi Shu [1], Yanran Xu [1], Shu Jiang [1,*], Linhao Zhang [1], Xiang Zhao [1], Zhejun Pan [1,2], Tomasz P. Blach [3], Liangwei Sun [4], Liangfei Bai [4], Qinhong Hu [5] and Mengdi Sun [1,*]

[1] Key Laboratory of Tectonics and Petroleum Resources, China University of Geosciences, Wuhan 430074, China; 1201620026@cug.edu.cn (Y.S.); yxu26@art.edu (Y.X.); linhaozhang@cug.edu.cn (L.Z.); 1504010328@hhu.edu.cn (X.Z.); Zhejun.Pan@csiro.au (Z.P.)

[2] CSIRO Energy, Private Bag 10, Clayton South, VIC 3169, Australia

[3] School of Minerals and Energy Resources Engineering, University of New South Wales, Sydney, NSW 2052, Australia; t.blach@unsw.edu.au

[4] Key Laboratory of Neutron Physics and Institute of Nuclear Physics and Chemistry, China Academy of Engineering Physics (CAEP), Mianyang 621999, China; Sunlw209@caep.cn (L.S.); bai1985@mail.ustc.edu.cn (L.B.)

[5] Department of Earth and Environment Sciences, University of Texas at Arlington, Arlington, TX 76019, USA; maxhu@uta.edu

* Correspondence: jiangsu@cug.edu.cn (S.J.); sunmd@cug.edu.cn (M.S.)

Received: 20 October 2020; Accepted: 17 November 2020; Published: 19 November 2020

Abstract: The sample size or particle size of shale plays a significant role in the characterization of pores by various techniques. To systematically investigate the influence of particle size on pore characteristics and the optimum sample size for different methods, we conducted complementary tests on two overmature marine shale samples with different sample sizes. The tests included small-angle neutron scattering (SANS), gas (N_2, CO_2, and H_2O) adsorption, mercury injection capillary pressure (MICP), and field emission-scanning electron microscopy (FE-SEM) imaging. The results indicate that artificial pores and fractures may occur on the surface or interior of the particles during the pulverization process, and some isolated pores may be exposed to the particle surface or connected by new fractures, thus improving the pore connectivity of the shale. By comparing the results of different approaches, we established a hypothetical model to analyze how the crushing process affects the pore structure of overmature shales. Our results imply that intact wafers with a thickness of 0.15–0.5 mm and cubic samples (~1 cm^3) are optimal for performing SANS and MICP analyses. Meanwhile, the 35–80 mesh particle size fraction provides reliable data for various gas physisorption tests in overmature shale. Due to the intrinsic heterogeneity of shale, future research on pore characteristics in shales needs a multidisciplinary approach to obtain a more comprehensive, larger scale, and more reliable understanding.

Keywords: sample size; neutron scattering; mercury injection capillary pressure; adsorption; shale

---

1. Introduction

With the commercial development of shale gas in North America and China, the pore characteristics of shale reservoirs have been extensively studied [1–3]. Pore characteristics not only control the hydrocarbon gas storage capacity [4,5], but also have an important influence on the gas flow mechanism and producibility [6,7], which can provide basic information for the evaluation of shale reservoirs and shale gas accumulation mechanisms. Therefore, a series of quantitative and visual techniques are used to characterize the pore structure of shales and the characterization scales are summarized in Figure 1.

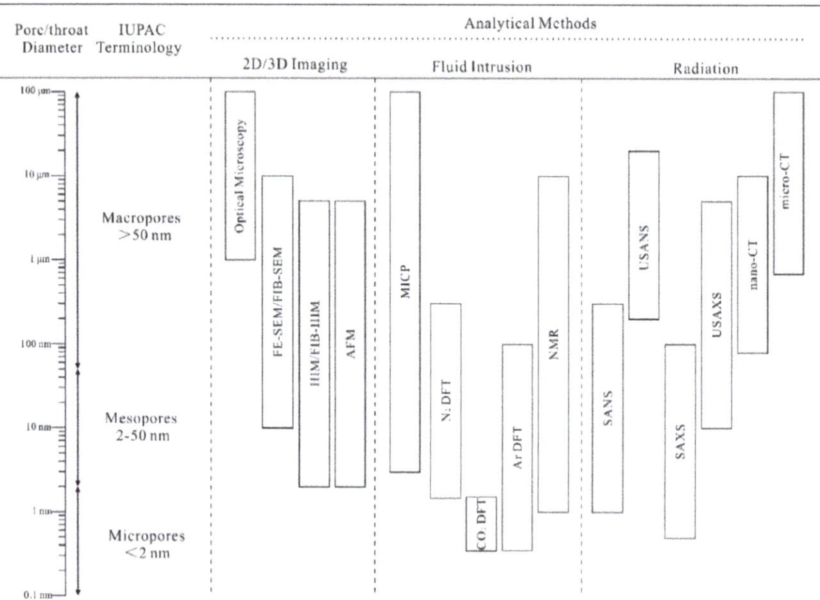

Figure 1. Methods used to investigate pore characteristics in shale (modified from [3]).

Among the above methods, fluid intrusion analytical methods are the most widely applied to characterize the pore structure of shale [8–10]. However, for each test, there is no uniform standard as to whether the test sample size should be an intact or crushed sample, as well as the particle size of the crushed sample. Chen et al. [11] carried out gas (N_2 and CO_2) physisorption measurements on New Albany Shales with different particle sizes (4 mesh, 20 mesh, and 60 mesh), and the results indicated that the mesopore volume increased with a decrease in particle size, whereas the micropore volume changed irregularly. Subsequently, Wei et al. [12] and Han et al. [13] evaluated the effect of particle size (5–250 mesh) on the change in pore structure through gas (N_2 and CO_2) adsorption experiments for Longmaxi Shale samples. Their conclusions were consistent with those of Chen et al. [11], who found that the decrease in particle size primarily affected the pores larger than 10 nm, and suggested that 60–140 mesh is the most suitable particle size for gas adsorption tests. In addition, Mastalerz et al. [14] suggested that gas adsorption tests on low-maturity (R_o~0.57%) and high-maturity (R_o~1.30%) shale samples with smaller particle size (200 mesh) could eliminate the equilibration problems and attain accurate results. However, Hazra et al. [15] proposed that shale particle sizes that were too fine would lead to destruction or alteration of the mesopore structures. Similarly, the results of the mercury injection capillary pressure (MICP) analysis for Barnett and Haynesville shale samples demonstrate that permeability and accessible porosity increase with decreasing sample particle size, indicating that the shale matrix has a higher connectivity on a small scale [16]. Moreover, the water vapor adsorption results for shale show that the total adsorption at 95% relative humidity (RH) is smaller in larger particle size samples, which is associated with fewer accessible pores [17,18]. Nevertheless, there is still a lack of a systematic analysis explaining the above results in the various fluid intrusion experiments.

In addition to fluid intrusion techniques, the nondestructive small-angle neutron scattering (SANS) technique has been used to evaluate the pore characteristics of shale reservoirs in recent years [19–21]. The main advantage of SANS in characterizing pore structure compared with fluid intrusion methods is that it contains information on closed pores (inaccessible to fluids) [3]. Thin sections of thickness from 0.15 mm to 0.5 mm are commonly used in the SANS test for shale for neutron transmission and avoidance of multiple scattering [22,23]. In addition, shale grain samples can also be used for the SANS test to avoid the anisotropy of the SANS images for wafer samples [24–26]. However, few studies have been performed on the effect of particle size on pore structure using SANS. Previous studies using fluid intrusion techniques considered the enhancement of the pore connectivity and ignored the artificial pores and fractures generated in the process of particle size reduction [14,16,27]. Therefore, the revelation of pore structure changes in the shale samples with different particle sizes by SANS can compensate for the deficiencies of previous studies. Moreover, the results of the crushed shale pressure-decay test for different particle sizes show that the helium permeability decreases with decreasing particle size [28]. However, the adsorption capacity of methane increased with decreasing particle size [29]. The reasons for the above phenomena can be explained by the mechanism of pore structure changes during the process of shale particle size reduction.

This study aims to reveal the influence of particle size on the pore characteristics of overmature organic-rich shales. Two shale samples were prepared as 1 cm cubes and particles of 20–35 mesh, 35–80 mesh, and 80–200 mesh. With a combination of SANS, low-pressure gas (N_2, CO_2, and H_2O) physisorption, and MICP, pore structure changes in shale with different sample sizes were first analyzed quantitatively. Then, the grinding positions of the shale were observed and characterized using field emission-scanning electron microscopy (FE-SEM). Finally, the effect of the pulverization process on the original pore characteristics of the shale was revealed. Thus, this study attempts to provide a reasonable suggestion on the size of shale samples that should be selected for the characterization techniques of different principles.

2. Materials and Methods

2.1. Sample Preparation

In this work, two fresh, overmature marine shale samples were collected from the Upper Ordovician Wufeng Formation of Well TY1 and Lower Cambrian Niutitang Formation of Well RY2, northwest of Guizhou Province, respectively (Figure 2). Information regarding the composition and maturity of selected samples is listed in Table 1. The raw shale samples were cut into cubes with a side length of 1 cm for the FE-SEM observation and MICP test. Then, the cubes were carefully hand-crushed and sieved into three particle size subsamples: 20–35 mesh (size A), 35–80 mesh (size B), and 80–200 mesh (size C). The shale samples used for analysis were well preserved, and with no sign of oxidation or weathering. Each subsample was dried in a vacuum oven at 60 °C for more than 48 h (until mass constancy) to remove the initial moisture content and subsequently analyzed via SANS, low-pressure N_2 and CO_2 physisorption, water vapor adsorption, and MICP measurements to determine various parameters of pore characteristics.

Table 1. Basic properties of shale samples used in this work.

Sample	Depth (m)	TOC [1] (wt.%)	R_o [2] (%)	Quantitative Analysis of Whole-Rock Minerals (wt.%)							Relative Content of Clay Minerals (wt.%)				
				Quartz	K-Feldspar	Plagioclase	Calcite	Dolomite	Pyrite	Clays	K [3]	C [4]	I [5]	I/S [6]	%S [7]
TY1-20	677.5	2.93	2.61	45	3	6	1	0	2	43	11	14	48	27	15
RY2-18	926.7	11.6	3.56	74	1	2	1	2	4	16	0	2	79	19	5

[1] TOC = total organic carbon; [2] R_o = equivalent vitrinite reflectance converted from the reflectance of bitumen; [3] K = kaolinite; [4] C = chlorite; [5] I = illite; [6] I/S = illite-smectite mixed-layer mineral; [7] %S = percentage of smectite in mixed-layer mineral.

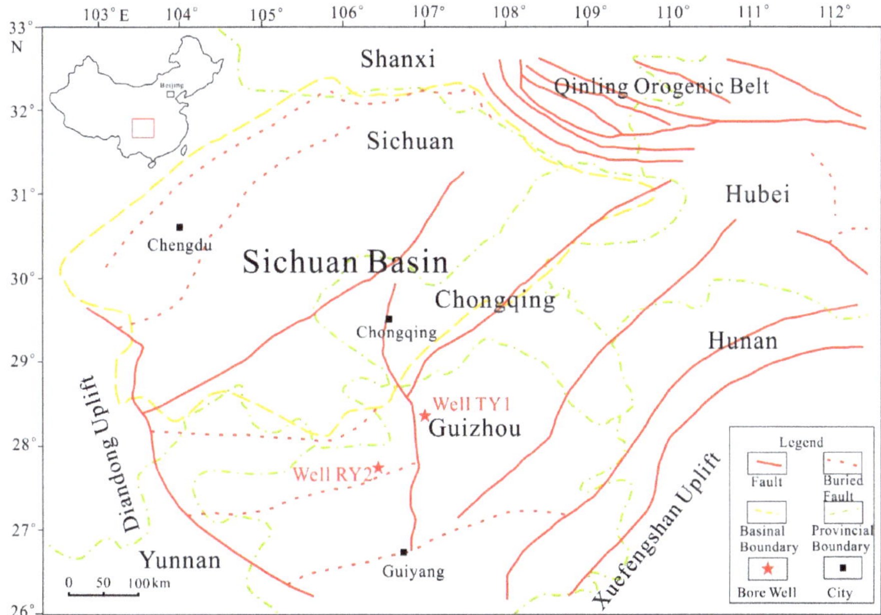

Figure 2. Location map of sampling wells in southern China (modified from [10]).

2.2. SANS Experiment

SANS was performed at the Suanni SANS instrument at the China Mianyang Research Reactor using three sample-to-detector distances (10 m, 4 m, and 1 m) and two neutron wavelengths λ = 5.3 Å (4 m, 1 m) and λ = 8 Å (10 m). The scattering vector (Q) range of this test was 0.0039–0.3 Å^{-1}, which corresponds to the pore diameters (D) from 128 nm to 1.7 nm, according to an approximate relation D = 5/Q [30]. Shale samples with different particle sizes were placed into Hellma cells with a 1 mm path length for the SANS measurement. The raw scattering data were corrected for scattering from the background and space between sample particles by acid-washed quartz sand with the same mesh and empty-cell [21]. The corrected SANS data were analyzed using the polydisperse size-distribution model (PDSM) in IRENA macros of the IGOR Pro software, which assumes that the pores are in a spherical shape and have a random size distribution [31]. Additional background information on the application of the SANS technique for pore characterization of shales can be found in a review article of Sun [3].

2.3. Low-Pressure N_2 and CO_2 Physisorption

Shale samples with different particle sizes were analyzed via low-pressure N_2 and CO_2 physisorption on a Quantachrome Autosorb-iQ apparatus after the SANS tests. The samples were degassed at 105 °C for 12 h to remove any adsorbed moisture and volatile matter. The relative pressures (P/P_0, where P_0 is the vapor pressure of the adsorbing gas, and P is the actual gas pressure) of the N_2 and CO_2 adsorption ranged from 0.0009 to 0.995 and 0.0006 to 0.03, respectively. The surface area and pore size distribution (PSD) of the samples were calculated from N_2 adsorption data. N_2-based and CO_2-based adsorption data were interpreted using the density functional theory (DFT) [32].

2.4. Water Vapor Adsorption Experiment

Water vapor adsorption (WVA) tests were carried out on the shale samples, which completed low-pressure gas physisorption using the dynamic vapor sorption (DVS) method at 25 °C. The DVS apparatus (Quantachrome Aquadyne) accurately measures the mass change (resolution of 0.1 µg ± 1%) of shale samples from 2% to 95% RH to obtain the water vapor ad-/desorption isotherm. The RH usually corresponds to the ratio of the pressure of water vapor (P) to the pressure of saturated vapor (P_0). Based on the Kelvin equation, the relation between RH and pore radius (r_p) of the water-filled capillary can be described [33]:

$$\ln\left(\frac{1}{RH}\right) = \frac{2\gamma V_m \cos\theta / RT}{r_p} \tag{1}$$

where γ is the surface tension, V_m is the molar volume of the water vapor, θ is the contact angle, R is the universal gas constant, and T is the temperature. In addition to the water-filled in the pores by capillary condensation, the water adsorbed on the pore surfaces also occupies part of the pore volume. Before capillary condensation, it is assumed that the pore surface of the shale is covered by multiple layers of water vapor with the same interfacial forces to form a water film of a specified thickness. When van der Waals force is the main controlling factor of water film thickness, the thickness (t) of the water-adsorbed layer on the pore surface could be calculated using Hasley's equation [17]:

$$t = 0.354\left[-5 / \ln\left(\frac{P}{P_0}\right)\right]^{\frac{1}{3}} \tag{2}$$

Since the bound water in clay minerals is not removed during sample treatment, only water film and capillary water are considered when calculating the PSD of the shale sample by WVA isotherm. The mass change could then be converted to the pore volume. The actual pore radius (r) measured by WVA could be expressed as follows:

$$r = r_p + t \tag{3}$$

2.5. MICP Measurement

MICP measurements were conducted on intact cube samples and crushed samples (size A, size B, and size C) using a porosimeter (Autopore IV 9520, Micromeritics) located at the China University of Geosciences (Wuhan) at pressures up to 60,000 psia (~413 MPa). In addition, to determine the influence of the space between the particles on the intrusion curves, acid-washed quartz sand with different particle sizes was used as a reference for MICP. The porosities and distribution of pore-throat sizes ranging from 3 nm to 36 µm were calculated from the mercury intrusion data.

2.6. FE-SEM Imaging

The intact TY1-20 shale sample was first cut into a 10 mm × 10 mm × 5 mm slice, and then ion-milled on a 10 mm × 10 mm surface using an argon-ion-beam polisher (LEICA EM XTP) to obtain a smooth surface for FE-SEM observation. After imaging, the sample was carefully hand-crushed into particles of approximately 0.5 mm in diameter (~35 mesh). Then, the crushed subsamples were further observed using FE-SEM to identify the influence of crushing on the shale sample.

3. Experimental Results

3.1. Characteristics of SANS Results

Figure 3a,b display the neutron scattering curves of the two shale samples with different particle sizes. As shown in Figure 3a,b, the scattering intensity of the shale samples with size C in the low Q region is higher than that of shale samples with size A. The neutron SLD values of the shale samples were calculated by averaging the SLD value of each component in the shale, recorded in Table 2. The detailed calculation method can be found in our previous study [3,20]. Shale is treated as a pseudo-two-phase system of pores and solids when the pore structure parameters are determined from SANS tests [34]. Table 2 shows the results of the porosity and specific surface area (SSA) derived from SANS. For the two shale samples, with decreasing particle size, the PDSM porosity increased

(Table 2). Because the porosity measured by SANS represents the total porosity (open and closed pores), the increase in porosity with the decrease in particle size is primarily due to the artificially increased pore and fracture space induced during the sample crushing. However, the SSA of the shale samples derived from SANS did not show the same trend as porosity.

The relationship between the scattering vector (Q) and pore radius (R) can be transformed by the empirical equation $R = 2.5/Q$ [30]. The PSD of the samples with different particle sizes are illustrated in Figure 3c,d for comparison. For the TY1-20 sample (Figure 3c), the pore volumes of both size B and size C were significantly higher than those of size A within the pore size range tested by SANS. For the RY2-18 sample (Figure 3d), as the particle size decreased, the pore volume of the pores with a diameter greater than 20 nm increased considerably. In addition, the pore volumes of size B and size C within the pore size range of 2~5 nm were also significantly higher than that of size A.

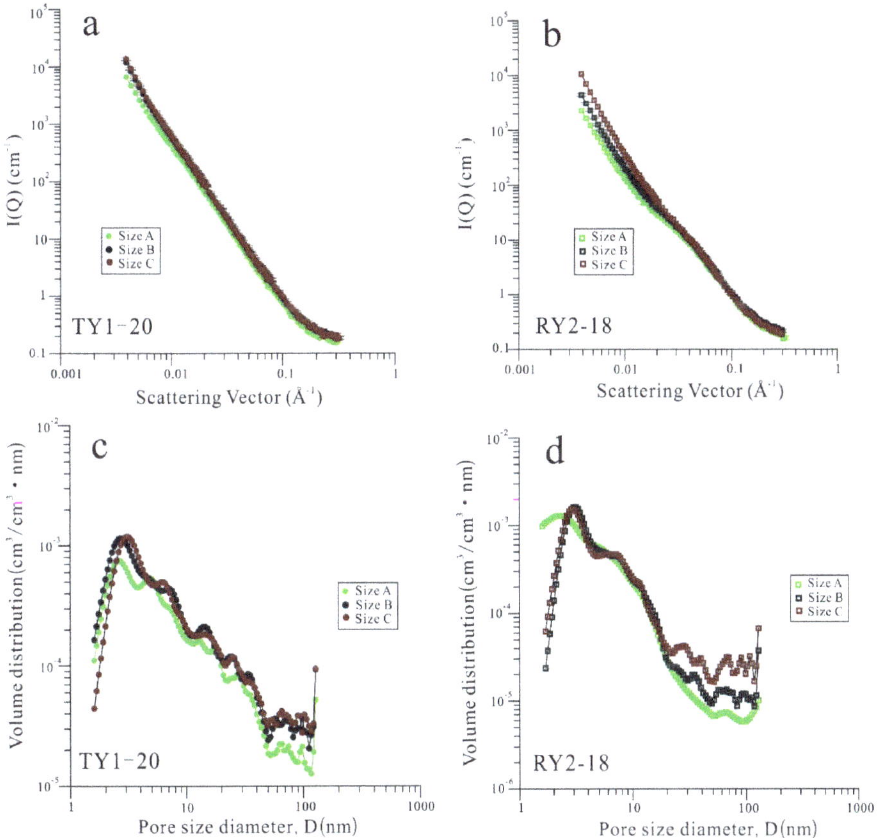

Figure 3. (**a**,**b**) Small-angle neutron scattering (SANS) profiles and (**c**,**d**) pore size distribution (PSD) for the TY1-20 and RY2-18 samples with three different particle sizes.

Table 2. Pore parameters of samples from SANS and gas (N_2 and CO_2) adsorption.

Sample ID	SANS SLD [1] ($\times 10^{10}$ cm^{-2})	PDSM [2] Porosity (%) (1.6–127 nm)	PDSM SSA [3] (m^2/g)	N_2 Pore Volume (10^{-2} cm^3/g)	N_2 [4] Porosity (%) (1.6–127 nm)	N_2 BET SSA (m^2/g)	CO_2 Pore Volume (10^{-2} cm^3/g)	CO_2 DFT SSA (m^2/g)
TY1-20 (size A)		7.83	23.87	2.22	5.25	27.79	0.34	5.42
TY1-20 (size B)	3.94	10.80	32.36	2.36	5.37	27.41	0.33	5.29
TY1-20 (size C)		11.05	30.39	2.50	5.70	26.65	0.32	5.03
RY2-18 (size A)		6.91	36.22	1.01	2.42	10.77	0.45	7.19
RY2-18 (size B)	4.03	7.44	30.40	1.30	3.10	15.75	0.44	7.03
RY2-18 (size C)		8.75	30.72	1.18	2.66	14.13	0.43	6.84

[1] The SLDs ($\times 10^{10}$ cm^{-2}) of organic phase used in the samples (TY1-20 = 3.3, RY2-18 = 3.7); [2] PDSM = Polydisperse sphere model; [3] SSA = Specific surface area; [4] N_2 Porosity is calculated by N_2 pore volume and bulk density of shale cubes obtained from mercury injection capillary pressure (MICP).

3.2. Low-Pressure N_2 and CO_2 Physisorption

Figure 4a,b show the N_2 adsorption–desorption isotherms of the two shale samples for the three different particle sizes. All samples display distinct hysteresis loops, and the hysteresis influence of desorption is mostly due to the pore morphology (ink-bottle shape). A comparison of the adsorption branches shows that the shale samples with the minimum particle size exhibit the maximum adsorption capacity at the maximum equivalent pressure. The observation of the desorption branches demonstrates that the desorption rate increases as the particle size decreases from size A to size C. In other words, the decrease in sample particle size shortens the distance required for desorption, thus enhancing the pore connectivity and gas transport capacity.

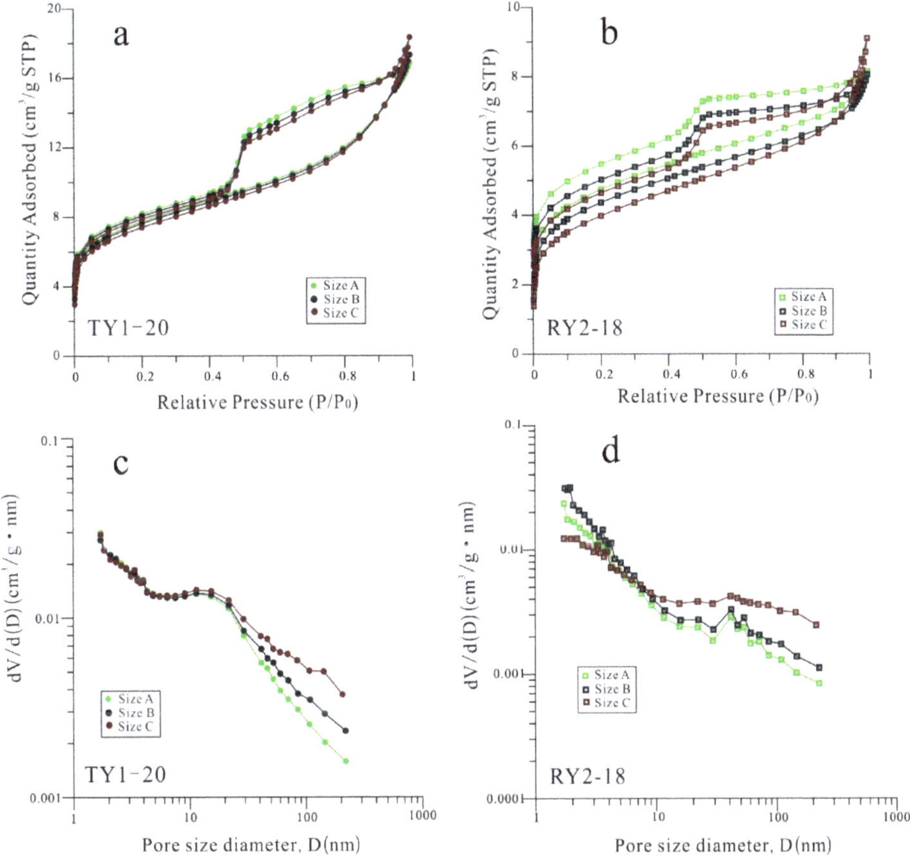

Figure 4. (**a,b**) Low-pressure N_2 adsorption/desorption isotherms and (**c,d**) PSD for the TY1-20 and RY2-18 samples with three different particle sizes.

The N_2 pore volume and surface area of the two samples are listed in Table 2. For sample TY1-20, the N_2 pore volume increased with decreasing particle size, and the increased pore volume is predominantly concentrated on the range of pore sizes larger than 10 nm (Figure 4c). Similarly, the BET SSA of sample TY1-20 does not show the same trend as the pore volume changes with particle size. For sample RY2-18, the N_2 pore volume and SSA increased with the particle size reduction from size A to size B, but then decreased for size C. Similar to sample TY1-20, the pore volume with a pore size larger than 10 nm increased with the decrease in particle size (Figure 4d). However, sample RY2-18 at particle size C exhibited a significant reduction in pore volume within the pore size range of less than 10 nm.

The CO_2 adsorption capacity of the two samples decreased slightly with a decrease in particle size (Figure 5a,b). Determined by CO_2 adsorption, the pore volume and SSA of the shale samples were consistent (Table 2). The consistency of the CO_2 PSD curves of shale samples with different particle sizes (Figure 5c,d) indicates that the effect of particle size reduction on the micropore volume of the overmature shale is limited.

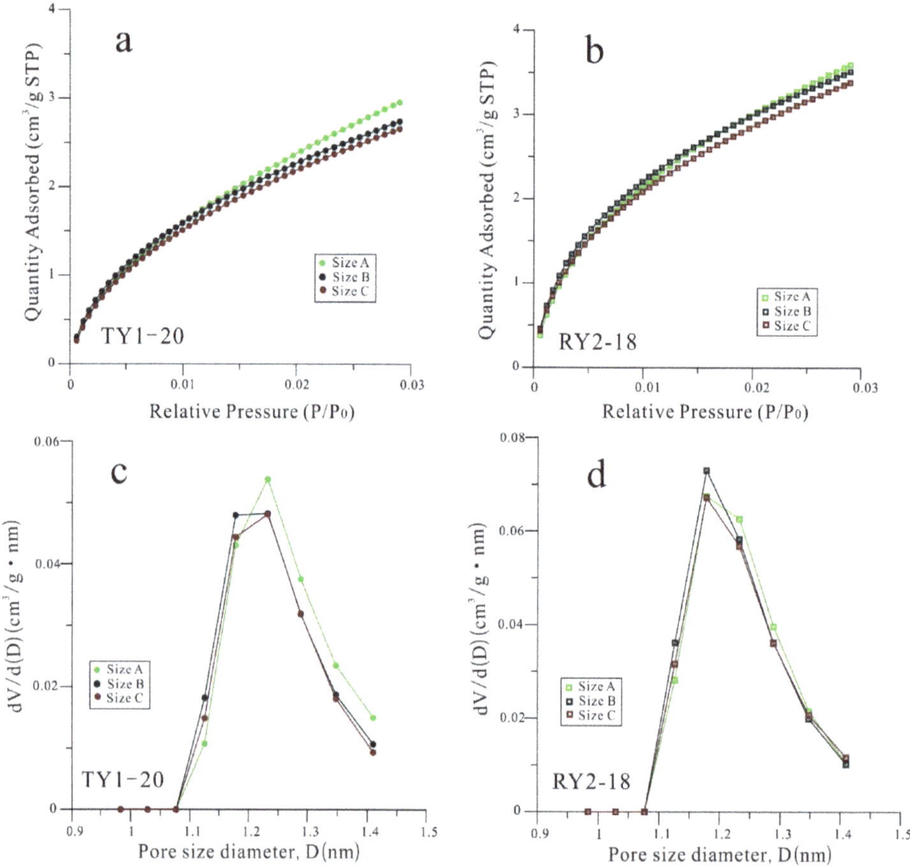

Figure 5. (**a,b**) CO_2 adsorption isotherms and (**c,d**) PSD for the TY1-20 and RY2-18 samples with three different particle sizes.

3.3. WVA Analysis

The water vapor adsorption–desorption isotherms of the shale samples are presented in Figure 6a,b. As the particle size decreased, the total water adsorption of the shale samples under 95% RH increased continuously (Table 3). When the RH is higher than 70%, the WVA curves of samples with different particle sizes are the most distinct. According to Equations (1)–(3), the results of the distribution relationship between pore size diameter and water incremental intrusion are shown in Figure 6c,d. Within the pore size range, less than 6.2 nm (corresponding to approximately 70% RH), the water absorption capacity of the samples with different particle sizes did not differ significantly. However, with pore sizes larger than 6.2 nm, the water vapor uptake increased dramatically with decreasing particle size.

In general, the adsorption of water vapor in shale occurs in three stages: monomolecular-layer coverage, multimolecular-layer adsorption, and capillary condensation with an increase in humidity [18,35]. Therefore, the influence of particle size on the WVA of the overmature shale is primarily reflected in the stage of capillary condensation.

Moreover, pronounced hysteresis loops were observed in all the water vapor adsorption/desorption isotherms (Figure 6a,b). Based on the IUPAC classification, the hysteresis loops of sample TY1-20 and sample RY2-18 can be classified as type H3 and type H2, indicating slit-like and ink-bottle shape pore networks, respectively [36]. In this work, the Areal Hysteresis Index (AHI) was used to quantitatively describe the characteristics of the hysteresis loop. The AHI is expressed as follows [37]:

$$AHI = \frac{A_{de} - A_{ad}}{A_{ad}} \times 100\% \tag{4}$$

where A_{ad} and A_{de} are the areas under the adsorption and desorption isotherms, respectively. The AHI values of sample RY2-18 were significantly higher than those of sample TY1-20 (Table 3). In addition, the values of AHI tended to increase with the decreasing particle size.

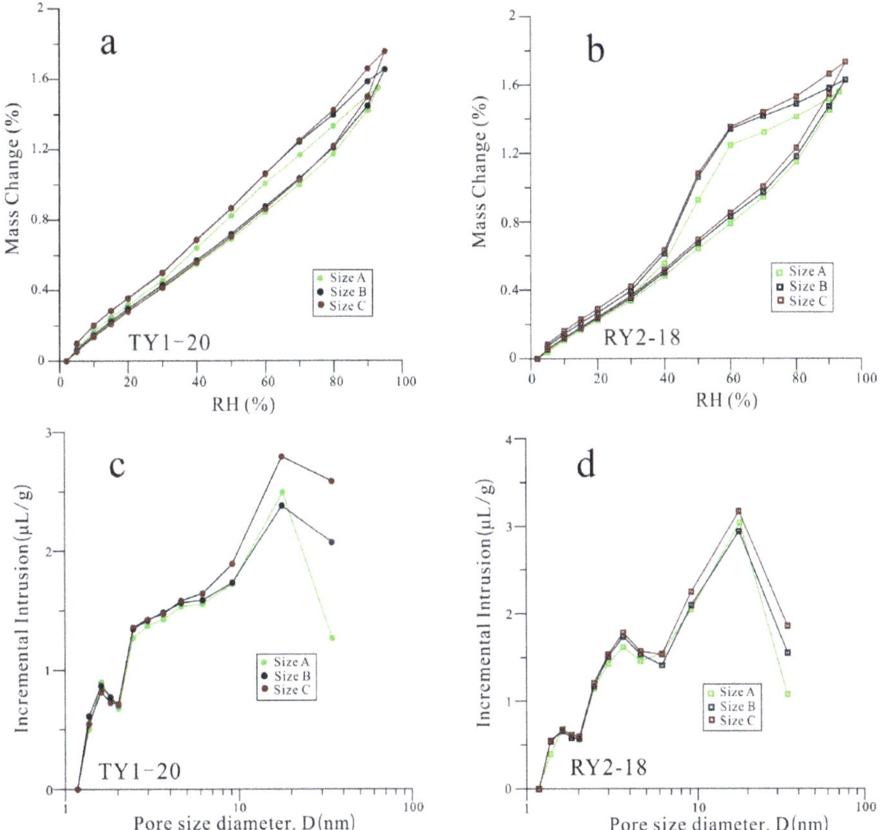

Figure 6. (**a**,**b**) Water vapor adsorption/desorption isotherms and (**c**,**d**) PSD for the TY1-20 and RY2-18 samples with three different particle sizes.

Table 3. Parameters obtained from water vapor adsorption (WVA) and MICP analyses.

Sample ID	WVA		MICP			
	Total Adsorption at 95% RH [1] (µL/g)	AHI [2] (%)	Particle Density (g/cm³)	Porosity (%)	Total Pore Area (cm²/g)	APtS [3] (nm)
TY1-20 cube	-		2.55	2.83	10.9	5.9
TY1-20 (size A)	15.51	13.96	2.4	3.09	6.3	9.6
TY1-20 (size B)	16.54	18	2.62	9.29	9.1	19.1
TY1-20 (size C)	17.56	20.23	2.41	13.94	0.3	908.1
RY2-18 cube	-		2.5	0.79	1.6	7.8
RY2-18 (size A)	15.62	25.17	2.55	1.79	0.3	107.5
RY2-18 (size B)	16.31	30.62	2.48	4.92	0.2	663.9
RY2-18 (size C)	17.34	30.44	2.33	11.65	0.3	1537.7
Quartz (size A)	-		2.57	0.39	-	
Quartz (size B)	-		2.57	0.85	-	
Quartz (size C)	-		2.55	9.96	-	

[1] RH = relative humidity; [2] AHI = areal hysteresis index; [3] APtS = average pore-throat size.

3.4. MICP Analysis

The cumulative mercury intrusion curves and pore-throat size distribution curves for all sample sizes of TY1-20 and RY2-18 are illustrated in Figure 7. As shown in Figure 7, as the sample size decreased, the cumulative mercury intrusion volume increased. The cumulative intrusions of sample TY1-20 from the cube to size B increased even at the maximum pressure (413 MPa), indicating that mercury will continue to enter the pore space if the pressure is increased. However, for TY1-20 with size C, the cumulative intrusion would not increase when the pressure was higher than 4000 psia. For sample RY2-18, the cumulative intrusion volume of all the samples except the cube will be constant after a certain pressure.

The pore-throat size distribution curves (Figure 7) indicate that the accessible pore volume connected with the pore throat less than 10 nm in the cube samples will be greatly reduced in the particle samples. For the samples with particle size C, the pore network space connected with a pore throat of less than 100 nm almost disappeared. Contrarily, the connected pore volume of the samples with pore-throat diameters larger than 100 nm dramatically increased with the decrease in sample size. Furthermore, the MICP results of acid-washed quartz sand with different particle sizes demonstrated that the influence between particles was mostly manifested on the micron scale (Figure 8).

Table 3 summarizes the results of the pore structure parameters corrected by the acid-washed quartz sand using MICP data and conformance volume calculations [16,38] for the different sample sizes of TY1-20 and RY2-18 samples. The results demonstrate that porosity is strongly related to the sample size, which increased with decreasing sample size. With the decrease in the sample size, the pore throat with a small diameter gradually disappeared, which led to an increase in the average pore-throat size (APtS) and a decrease in the total pore area. The particle density of the same samples with different sizes also exhibited slight differences, indicating that there will inevitably be subtle differences in composition in the sorting process.

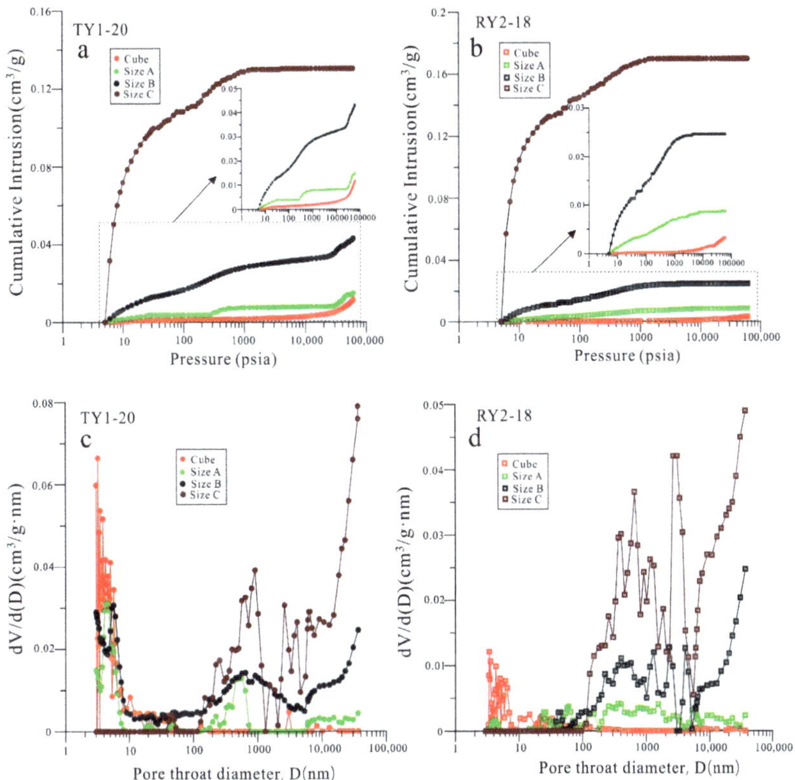

Figure 7. (**a**,**b**) MICP intrusion curves and (**c**,**d**) pore volume distribution vs. pore-throat diameters for the TY1-20 and RY2-18 samples with four different sample sizes.

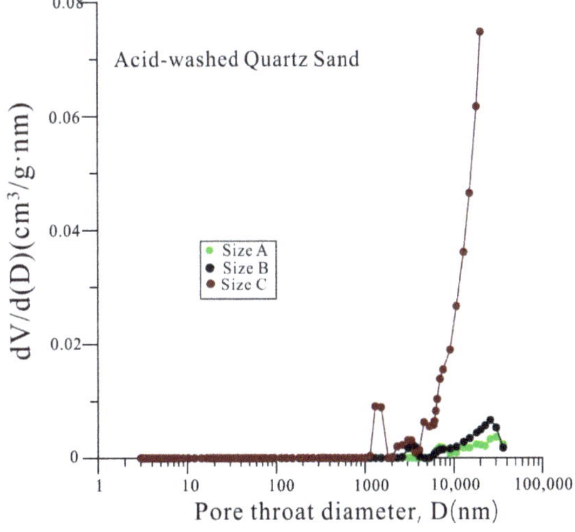

Figure 8. Pore volume distribution vs. pore-throat diameters of quartz samples with three different particle sizes.

3.5. Observation of FE-SEM

Various mineral components and organic matter can be clearly observed under FE-SEM on the smooth surface of shale obtained by argon-ion-beam polishing (Figure 9a). The crushed subsamples were then fixed with epoxy resin and further observed using FE-SEM, revealing that the smooth surface became rough (Figure 9b). Artificial microcracks at the nanometer scale formed by the fracture of rigid minerals could be observed on the rough surface of crushed subsamples (Figure 9c). In addition, in the process of particle size reduction, the structure of the clay minerals was prone to collapse, leading to the generation of smaller fragments and new artificial pore space (Figure 9c).

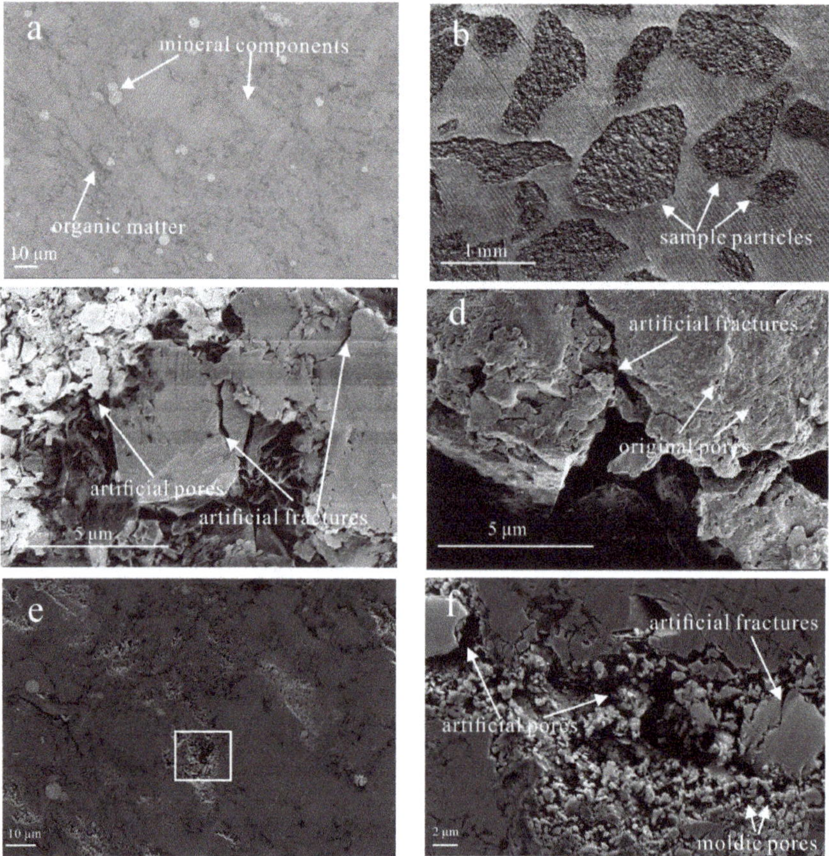

Figure 9. Field emission-scanning electron microscopy (FE-SEM) images of the TY1-20 sample. (**a**) Smooth surface of shale obtained by argon-ion-beam polishing; (**b**) Crushed subsamples were then fixed with epoxy resin; (**c**) Artificial pores and microcracks; (**d**) Artificial microfractures perpendicular to the bedding direction; (**e**) Marks of the crushing process on the smooth surface; (**f**) An enlargement of the crush mark in the rectangle area in Figure 9e.

Among the artificial microcracks, artificial microfractures perpendicular to the bedding direction are more conducive to enhancing pore connectivity. As illustrated in Figure 9d, the hydrocarbon fluids in the original pores could flow out through artificial microfractures, whereas the small molecular fluid in fluid-invasion porosimetry can enter the original pores through the artificial microfractures. The marks of the crushing process on the smooth surface by argon-ion-beam polishing could be observed in the local area of the particle sample (Figure 9e). By magnifying the observation of the crush area (Figure 9f), it was found that the artificial pore-fracture is manifested in three forms: (1) the separation of rigid minerals (quartz, calcite, etc.) and plastic components (organic matter and clay) to form artificial pore-fractures; (2) the fragmentation of mineral grains to form artificial fractures; and (3) the spalling of mineral grains, such as pyrite, to form artificial moldic pores.

4. Discussion

4.1. Effect of Particle Size on Pore Structure Characteristics

The results of the SANS show that the total porosities (including both open and closed pores) increased with the decrease in particle size (Table 2). This characteristic indicates that the increased porosity is due to the artificial pores or fractures created by the pulverization process. Previous studies [11,15] have shown that shear and compression forces in the crushing process induce the formation of new fractures and smaller fragments in the shale, generating new porosity, which is consistent with the results of our FE-EM observations. In addition, during the crushing process, the shale samples commonly fracture along the inherent weak parts, which will also lead to the disappearance of some microfractures [28,39]. Based on the analysis of the above results, a hypothetical model of the effect of the crushing process on the pore structure in shale was established (Figure 10). As illustrated in Figure 10, when an intact shale sample is crushed into particles, the pore connectivity in the shale is significantly enhanced. For a particle, artificial fractures or pores may form on the surface or interior of the particle. Meanwhile, some isolated pore networks may also be exposed to the particle surface or connected by new fractures.

The results of the MICP data indicate that MICP porosity increases with the decrease in sample size (Table 3). For the MICP test, shale can be divided into three constituents: accessible pores, inaccessible pores, and solid matrix [40,41]. The primary pores in the overmature shale have been greatly reduced after a long period of compaction and cementation [42,43]. In addition, the closure of minerals to the organic matter pore system results in low overall pore connectivity in the overmature shale [10,44,45]. Therefore, with the increase in sample size, the pore connectivity of shale decreases further, which results in a lower cumulative mercury intrusion, consistent with the reduction in accessible porosity. Moreover, as can be seen from the schematic representation of particle size reduction (Figure 10c,d), the disappearance of some pore throats results in a lower driving pressure for mercury entering the pores. When pore throats less than 3 nm (corresponding to a maximum pressure of 60,000 psia) are damaged, the pores that are inaccessible to mercury become accessible, thus increasing the accessible porosity. Gas injection porosimetry experiments from Sun et al. [46] also indicated that the crushing process decreases the required total diffusion time and minimum gas injection pressure. The results of MICP show that the destruction of the smaller pore throat is the primary cause of the increase in APtS and the decrease in total pore area with an increase in the crushing level.

The studies of Davudov et al. [47] and Hu et al. [7] on Barnett shale show that permeability and diffusivity increase as the sample size decreases. The permeability measured by MICP in shale is a strongly correlated function of accessible porosity and APtS [47]. Therefore, the increase in these two parameters is bound to increase shale permeability. However, the permeability values obtained from the helium shale matrix permeameter (Core Laboratories SMP 200) in the Woodford Shale decrease with a decrease in the sample particle size [28]. Compared with the permeability measured by MICP, the permeability of the shale matrix measured by the helium pressure-decay method is primarily at the nano-Darcy scale [28], which is inconsistent with the micro-Darcy scale permeability reported by Davudov and Moghanloo [16]. This also indicates that the shale matrix permeability is controlled by pore-throat distribution. With the reduction in particle size, the disappearance of a smaller pore throat will cause some transport pores to become dead-end pores, which will lead to a loss of seepage capacity and a decrease in matrix permeability.

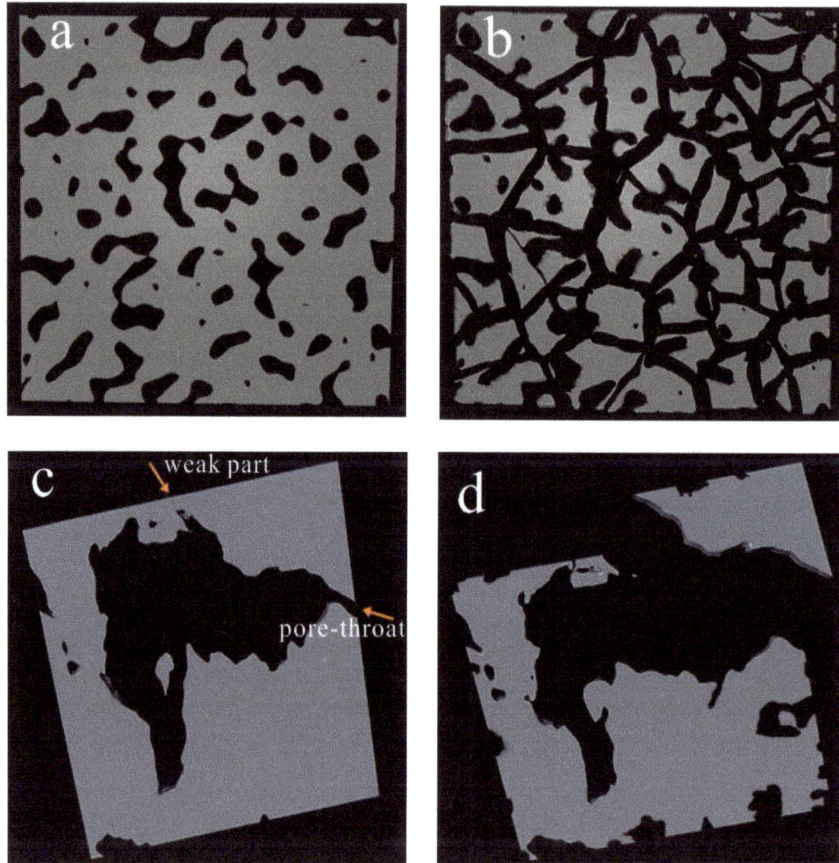

Figure 10. Model of an intact shale and a particle shale (**a,c**) before and (**b,d**) after crushing (the gray areas denote the shale solid matrix, and the black areas denote pores or fractures space).

The effect of the particle size on the pore structure parameters obtained by low-pressure gas (N_2, CO_2, and H_2O) physisorption are compared with those generated by MICP and SANS (Tables 2 and 3). The pore structure characterization in the shales using the gas adsorption techniques range from micropores to macropores with the increase in pressure, which is opposite to the order of Hg entering the pores in MICP [48,49]. For N_2 adsorption, the connectivity of the pore network, especially for sizes larger than 10 nm, is significantly enhanced (Figure 4), which is consistent with previous studies [11,13]. However, according to the results of sample RY2-18 (Figure 4), the particle sizes of 80~200 mesh will cause some small pores to be damaged. A similar phenomenon of pore destruction was observed in the results of CO_2 adsorption, but the overall effect on the pore system parameters was not significant. For WVA, an increasing trend was found for the total adsorption at 95% RH with a decreasing particle size, which could be dependent on the enhanced connectivity of the hydrophilic pore network [50]. The pore structure changes associated with clay minerals observed by FE-SEM may be responsible for the increased water adsorption sites. Similarly, previous studies also found that the methane uptake rate and excess sorption capacity of shale increased with a decrease in particle size [29,51]. This also indicates that the physical disruption of shale fabric will improve the pore connectivity and, thus, more methane adsorption sites. In addition, the increased artificial space can improve the swelling ability of shale after methane adsorption [52,53].

Moreover, evidence for the weaker hysteresis effect of N_2 in the pores associated with smaller particle sizes (Figure 4) is related to the increased pore connectivity due to shorter flow paths. However, the hysteresis loops of water vapor sorption did not tighten with a decreasing particle size (Figure 6). In contrast, AHI increased with decreasing particle size, suggesting that wettability, in addition to pore structure, also caused fluid retention in the pores [35,36,50]. The closed porosities of various shales calculated by comparing the results of N_2 adsorption and SANS have been reported in some studies [20,54,55]. However, in this work, the results of SANS were found to contain information on artificial pores rather than those of gas adsorption. Therefore, for the same particle size, the closed porosity measured by comparing the results of N_2 adsorption and SANS is generally overestimated.

4.2. Implications for Sample Size Selection of Different Methods

The size of the sample has a great influence on the determination of the petrophysical parameters of shale using different measurement methods. The heterogeneity of shale is characterized by rich microstructural features (microfractures and laminae), complex components (organic matter and minerals), and multi-scale pore structure (from nanometers to micrometers) [56–58]. As shown in Figure 11, for the same intact shale sample, sampling at different locations for testing may produce different results. In addition, based on the analysis of Section 4.1, the difference in the results of particle size variation obtained by several methods is primarily related to the different theoretical bases.

For the neutron scattering technique, the most significant advantage is the nondestructive determination of the total porosity, including open and closed pores. Gu et al. [59] studied Marcellus shale with different sampling directions (parallel and perpendicular to the lamination) and found that the scattering pattern of the parallel layered samples is isotropic, while that of the vertically layered samples is anisotropic. The study of pore anisotropy in oriented shale wafers using SANS provides new insights for shale gas storage, migration, and preservation [23,60,61]. The SANS test on particle shale samples results in isotropic scattering patterns due to the reduced effects of bedding and microfractures. For the same shale sample, the scattering intensity for particles is higher overall in the low Q range than in the intact wafer [22]. The results of this work also indicate that the downsizing of particles will increase the SANS porosity due to the creation of more artificial pore space. Therefore, it is imperative to select an intact shale wafer in the SANS test to limit the analytical error caused by artificial space on the shale surface. Shale wafers with a thickness of 0.15~0.5 mm seem to be optimal for performing SANS analysis, considering the scattering time and multiple scattering of the shale.

The MICP measurement provides multi-scale (nm-μm-scale) information regarding the pore throats of shale [62–64]. Compared with the intact shale sample, the real pore throat distribution characteristics could not be provided with the destruction of small pore throat with particle reduction (Figure 7). Meanwhile, the artificially increased pore space is also reflected in the results of MICP. According to Yu et al. [27], the MICP results for 20~35 mesh particle shale samples can be corrected by fractal theory. However, the study only eliminated the error of mercury intrusion between particles without mentioning the effect of pore-throat damage. In addition, the smaller the particle is, the larger the modified exterior surface will be, causing the edge-accessible pores in shale to constitute a high percentage of the interior volume, which cannot truly reflect the pore-throat distribution of the shale [7,65]. As shown in Figure 11, the selection of representative positions at the centimeter scale can reflect the pore characteristics of shale more comprehensively. Therefore, the cubic samples (~1 cm^3) with polished surfaces reduced the effect of cutting pockmarks and pore-throat damage and were more suitable for the MICP test. Moreover, previous studies have confirmed that MICP with cubes from overmature shale has minimal compression effects [40,66].

Many studies have demonstrated that more pores can be accessible to the adsorption gas (e.g., N_2, CO_2, and H_2O) as the particle size of pulverized shale decreased [11,14,50]. Meanwhile, samples with smaller particle sizes can shorten the equilibration time and increase the desorption rate [41]. Therefore, Mastalerz et al. [14] suggested that a 200 mesh size of shale is more suitable for N_2 and CO_2 analysis to determine the total pore volume. However, the N_2 and CO_2 adsorption results in this work indicate that an extremely fine particle size (80–200 mesh) may result in the destruction of finer mesopores (i.e., less than 10 nm) and micropores, which is consistent with the study of Hazra et al. [15]. Gas adsorption experiments typically measure pore size ranges of less than 100 nm and rarely involve artificial pores and fractures on the shale surface during the grinding process [12,13]. Considering the equilibration time and original pore connectivity characteristics of shale, 35~80 mesh particle samples provide good quality and reliable data for characterizing the PSD and calculating the pore volume of the overmature shale. For WVA of overmature shale, it is recommended to use consistent particle sizes for comparison with other gas adsorption results.

Figure 11. Sedimentary structure characteristics of the TY1-20 sample.

Due to the intrinsic heterogeneity of shale, the information to be included should be selected when characterizing the pore characteristics of organic-rich shales. Pore structure characterization is a significant part of a petrophysical evaluation of shale reservoirs and for assessing shale gas productivity. However, even for samples with the same particle size, discrepant results will be produced owing to the different principles and error sources of each test method. Therefore, future research on pore characteristics will need a multidisciplinary approach to obtain a more comprehensive, larger scale, and more reliable results.

5. Conclusions

Multiple tests, including SANS, MICP, gas (N_2, CO_2, and H_2O) physisorption, and FE-SEM, were conducted to analyze the influence of particle size on the pore characteristics of the two overmature shales. The following main conclusions can be summarized:

(1) The results consistently show that artificial pores and fractures are created on the surface or interior of the particles during the pulverization process. The pore connectivity of the shale is enhanced as some isolated pore networks may be exposed to the particle surface or connected by new fractures.

(2) For the SANS analysis of the shale, intact wafers with a thickness of 0.15~0.5 mm and an area of approximately 1 cm^2 are the optimal sample size. Fine particles less than 80 mesh are not recommended because the result will include information regarding artificial pores and fractures.

(3) For the MICP test, the cubic samples (~1 cm^3) appear to be optimal. The downsizing of particles will lead to the destruction of the pore-throat, which cannot reflect the true pore-throat distribution of shale using the MICP test.

(4) Gas adsorption is usually conducted on the same shale samples for comparison. Considering the equilibration time and original pore connectivity characteristics of shale, 35~80 mesh is the most practical for overmature shale.

Highlights

(1) The analytical sample size influences small-angle neutron scattering and fluid intrusion results.
(2) A model of the effect of the crushing process on the pore structure was established by visual and quantitative techniques.
(3) An intact wafer with a thickness of 0.15~0.5 mm is recommended for SANS measurements.
(4) Cubic samples (~1 cm^3) appear to be optimal for MICP analysis of shale.
(5) The 35–80 mesh particle size range is optimum for overmature shale samples used in gas adsorption.

Author Contributions: As corresponding authors, S.J. and M.S. have made substantial contributions to the conception of the work. Y.S. analyzed the experimental results and drafted the manuscript. Y.X., L.Z., X.Z., L.S., L.B., Q.H. and T.P.B. performed the various tests and helped with the synthesis results. Z.P. contributed to the preparation and revision of the manuscript. All authors have read and agreed to the published version of the manuscript.

Funding: This research was funded by National Natural Science Foundation (Grant Nos. 41802146, 41830431and 41690134), the Strategic Priority Research Program of the Chinese Academy of Sciences (No. XDA14010302), Key Laboratory of Tectonics and Petroleum Resources (TPR-2019-09), and the Fundamental Research Funds for the Central Universities (No. CUG180608) of China, for their financial support.

Conflicts of Interest: The authors declare no conflict of interest.

References

1. Clarkson, C.R.; Solano, N.; Bustin, R.; Bustin, A.; Chalmers, G.R.L.; He, L.; Melnichenko, Y.; Radlinski, A.P.; Blach, T. Pore Structure Characterization of North American Shale Gas Reservoirs Using Usans/Sans, Gas Adsorption, and Mercury Intrusion. *Fuel* **2013**, *103*, 606–616. [CrossRef]
2. Guo, T. The Fuling Shale Gas Field—A Highly Productive Silurian Gas Shale With High Thermal Maturity and Complex Evolution History, Southeastern Sichuan Basin, China. *Interpretation* **2015**, *3*, SJ25–SJ34. [CrossRef]
3. Sun, M.; Zhao, J.; Pan, Z.; Hu, Q.; Yu, B.; Tan, Y.; Sun, L.; Bai, L.; Wu, C.; Blach, T.P.; et al. Pore Characterization of Shales: A Review of Small Angle Scattering Technique. *J. Nat. Gas Sci. Eng.* **2020**, *78*, 103294. [CrossRef]
4. Chalmers, G.R.; Bustin, R.M.; Power, I.M. Characterization of Gas Shale Pore Systems by Porosimetry, Pycnometry, Surface Area, and Field Emission Scanning Electron Microscopy/Transmission Electron Microscopy Image Analyses: Examples from the Barnett, Woodford, Haynesville, Marcellus, and Doig Units. *AAPG Bull.* **2012**, *96*, 1099–1119. [CrossRef]

5. Ko, L.T.; Ruppel, S.C.; Loucks, R.G.; Hackley, P.C.; Zhang, T.; Shao, D. Pore-Types and Pore-Network Evolution in Upper Devonian-Lower Mississippian Woodford and Mississippian Barnett Mudstones: Insights From Laboratory Thermal Maturation and Organic Petrology. *Int. J. Coal Geol.* **2018**, *190*, 3–28. [CrossRef]

6. Javadpour, F.; Fisher, D.; Unsworth, M. Nanoscale Gas Flow in Shale Gas Sediments. *J. Can. Pet. Technol.* **2007**, *46*, 10. [CrossRef]

7. Hu, Q.; Ewing, R.P.; Rowe, H.D. Low Nanopore Connectivity Limits Gas Production in Barnett Formation. *J. Geophys. Res. Solid Earth* **2015**, *120*, 8073–8087. [CrossRef]

8. Zhang, L.; Xiong, Y.; Li, Y.; Wei, M.; Jiang, W.; Lei, R.; Wu, Z. DFT Modeling of CO_2 and Ar Low-Pressure Adsorption for Accurate Nanopore Structure Characterization in Organic-Rich Shales. *Fuel* **2017**, *204*, 1–11. [CrossRef]

9. Rijfkogel, L.S.; GhanbarianiD, B.; Hu, Q.; Liu, H.-H. Clarifying Pore Diameter, Pore Width, and Their Relationship Through Pressure Measurements: A Critical Study. *Mar. Pet. Geol.* **2019**, *107*, 142–148. [CrossRef]

10. Sun, M.; Zhang, L.; Hu, Q.; Pan, Z.; Yu, B.; Sun, L.; Bai, L.; Fu, H.; Zhang, Y.; Zhang, C.; et al. Multiscale Connectivity Characterization of Marine Shales in Southern China by Fluid Intrusion, Small-Angle Neutron Scattering (SANS), and FIB-SEM. *Mar. Pet. Geol.* **2020**, *112*, 104101. [CrossRef]

11. Chen, Y.; Wei, L.; Mastalerz, M.; Schimmelmann, A. The Effect of Analytical Particle Size on Gas Adsorption Porosimetry of Shale. *Int. J. Coal Geol.* **2015**, *138*, 103–112. [CrossRef]

12. Wei, M.; Xiong, Y.; Zhang, L.; Li, J.; Peng, P. The Effect of Sample Particle Size on the Determination of Pore Structure Parameters in Shales. *Int. J. Coal Geol.* **2016**, *163*, 177–185. [CrossRef]

13. Han, H.; Cao, Y.; Chen, S.-J.; Lu, J.-G.; Huang, C.-X.; Zhu, H.-H.; Zhan, P.; Gao, Y. Influence of Particle Size on Gas-Adsorption Experiments of Shales: An Example From a Longmaxi Shale Sample from the Sichuan Basin, China. *Fuel* **2016**, *186*, 750–757. [CrossRef]

14. Mastalerz, M.; Hampton, L.; Drobniak, A.; Loope, H. Significance of Analytical Particle Size in Low-Pressure N_2 and CO_2 Adsorption of Coal and Shale. *Int. J. Coal Geol.* **2017**, *178*, 122–131. [CrossRef]

15. Hazra, B.; Zhang, Z.; Vishal, V.; Singh, A.K. Pore Characteristics of Distinct Thermally Mature Shales: Influence of Particle Size on Low-Pressure CO_2 and N_2 Adsorption. *Energy Fuels* **2018**, *32*, 8175–8186. [CrossRef]

16. Davudov, D.; Moghanloo, R.G. Scale-Dependent Pore and Hydraulic Connectivity of Shale Matrix. *Energy Fuels* **2018**, *32*, 99–106. [CrossRef]

17. Yang, R.; Jia, A.; He, S.; Hu, Q.; Dong, T.; Hou, Y.; Yan, J. Water Adsorption Characteristics of Organic-Rich Wufeng and Longmaxi Shales, Sichuan Basin (China). *J. Pet. Sci. Eng.* **2020**, *193*, 107387. [CrossRef]

18. Yang, R.; Jia, A.; He, S.; Hu, Q.; Sun, M.; Dong, T.; Hou, Y.; Zhou, S. Experimental Investigation of Water Vapor Adsorption Isotherm on Gas-Producing Longmaxi Shale: Mathematical Modeling and Implication for Water Distribution in Shale Reservoirs. *Chem. Eng. J.* **2021**, *406*, 125982. [CrossRef]

19. Xu, H. Probing Nanopore Structure and Confined Fluid Behavior in Shale Matrix: A Review on Small-Angle Neutron Scattering Studies. *Int. J. Coal Geol.* **2020**, *217*, 103325. [CrossRef]

20. Sun, M.; Yu, B.; Hu, Q.; Zhang, Y.; Li, B.; Yang, R.; Melnichenko, Y.B.; Cheng, G. Pore Characteristics of Longmaxi Shale Gas Reservoir in the Northwest of Guizhou, China: Investigations Using Small-Angle Neutron Scattering (SANS), Helium Pycnometry, and Gas Sorption Isotherm. *Int. J. Coal Geol.* **2017**, *171*, 61–68. [CrossRef]

21. Sun, M.; Zhang, L.; Hu, Q.; Pan, Z.; Yu, B.; Sun, L.; Bai, L.; Connell, L.D.; Zhang, Y.; Cheng, G. Pore Connectivity and Water Accessibility in Upper Permian Transitional Shales, Southern China. *Mar. Pet. Geol.* **2019**, *107*, 407–422. [CrossRef]

22. Zhang, Y.; Hu, Q.; Long, S.; Zhao, J.; Peng, N.; Wang, H.; Lin, X.; Sun, M. Mineral-Controlled NM-μM-Scale Pore Structure of Saline Lacustrine Shale in Qianjiang Depression, Jianghan Basin, China. *Mar. Pet. Geol.* **2019**, *99*, 347–354. [CrossRef]

23. Zhang, R.; Liu, S.; He, L.; Blach, T.P.; Wang, Y. Characterizing Anisotropic Pore Structure and Its Impact on Gas Storage and Transport in Coalbed Methane and Shale Gas Reservoirs. *Energy Fuels* **2020**, *34*, 3161–3172. [CrossRef]

24. Zhang, Y.; Barber, T.J.; Hu, Q.; Bleuel, M.; El-Sobky, H.F. Complementary Neutron Scattering, Mercury Intrusion and SEM Imaging Approaches to Micro- and Nano-Pore Structure Characterization of Tight Rocks: A Case Study of the Bakken Shale. *Int. J. Coal Geol.* **2019**, *212*, 103252. [CrossRef]

25. Sun, M.; Yu, B.; Hu, Q.; Yang, R.; Zhang, Y.; Li, B.; Melnichenko, Y.B.; Cheng, G. Pore Structure Characterization of Organic-Rich Niutitang Shale From ChinA: Small Angle Neutron Scattering (SANS) Study. *Int. J. Coal Geol.* **2018**, *186*, 115–125. [CrossRef]

26. Bahadur, J.; Radlinski, A.P.; Melnichenko, Y.B.; Mastalerz, M.; Schimmelmann, A. Small-Angle and Ultrasmall-Angle Neutron Scattering (SANS/USANS) Study of New Albany Shale: A Treatise on Microporosity. *Energy Fuels* **2015**, *29*, 567–576. [CrossRef]

27. Yu, Y.; Luo, X.; Wang, Z.; Cheng, M.; Lei, Y.; Zhang, L.; Yin, J. A New Correction Method for Mercury Injection Capillary Pressure (MICP) to Characterize the Pore Structure of Shale. *J. Nat. Gas Sci. Eng.* **2019**, *68*, 102896. [CrossRef]

28. Achang, M.; Cui, X.; Pashin, J.C. The Influence of Particle Size, Microfractures, and Pressure Decay on Measuring the Permeability of Crushed Shale Samples. *Int. J. Coal Geol.* **2017**, *183*, 174–187. [CrossRef]

29. Gaus, G.; Kalmykov, A.; Krooss, B.M.; Fink, R. Experimental Investigation of the Dependence of Accessible Porosity and Methane Sorption Capacity of Carbonaceous Shales on Particle Size. *Geofluids* **2020**, *2020*, 2382153. [CrossRef]

30. Radliński, A.; Boreham, C.; Lindner, P.; Randl, O.; Wignall, G.; Hinde, A.; Hope, J. Small Angle Neutron Scattering Signature of Oil Generation in Artificially and Naturally Matured Hydrocarbon Source Rocks. *Org. Geochem.* **2000**, *31*, 1–14. [CrossRef]

31. Zhang, Y.; Hu, Q.; Barber, T.J.; Bleuel, M.; Anovitz, L.M.; Littrell, K. Quantifying Fluid-Wettable Effective Pore Space in the Utica and Bakken Oil Shale Formations. *Geophys. Res. Lett.* **2020**, *47*. [CrossRef]

32. Mastalerz, M.; Schimmelmann, A.; Drobniak, A.; Chen, Y. Porosity of Devonian and Mississippian New Albany Shale Across a Maturation Gradient: Insights From Organic Petrology, Gas Adsorption, and Mercury Intrusion. *AAPG Bull.* **2013**, *97*, 1621–1643. [CrossRef]

33. Zolfaghari, A.; Dehghanpour, H.; Xu, M. Water Sorption Behaviour of Gas Shales: II. Pore Size Distribution. *Int. J. Coal Geol.* **2017**, *179*, 187–195. [CrossRef]

34. Anovitz, L.A.; Cole, D.R. Characterization and Analysis of Porosity and Pore Structures. *Rev. Miner. Geochem.* **2015**, *80*, 61–164. [CrossRef]

35. Zolfaghari, A.; Dehghanpour, H.; Holyk, J. Water Sorption Behaviour of Gas Shales: I. Role of Clays. *Int. J. Coal Geol.* **2017**, *179*, 130–138. [CrossRef]

36. Cheng, P.; Tian, H.; Xiao, X.; Gai, H.; Li, T.; Wang, X. Water Distribution in Overmature Organic-Rich Shales: Implications from Water Adsorption Experiments. *Energy Fuels* **2017**, *31*, 13120–13132. [CrossRef]

37. Tang, X., Ripepi, N.; Valentine, K.A.; Keles, C.; Long, T.; Gonciaruk, A. Water Vapor Sorption on Marcellus Shale: Measurement, Modeling and Thermodynamic Analysis. *Fuel* **2017**, *209*, 606–614. [CrossRef]

38. Wang, S.; Javadpour, F.; Feng, Q. Confinement Correction to Mercury Intrusion Capillary Pressure of Shale Nanopores. *Sci. Rep.* **2016**, *6*, 20160. [CrossRef]

39. Chen, Y.; Qin, Y.; Wei, C.; Huang, L.; Shi, Q.; Wu, C.; Zhang, X. Porosity Changes in Progressively Pulverized Anthracite Subsamples: Implications for the Study of Closed Pore Distribution in Coals. *Fuel* **2018**, *225*, 612–622. [CrossRef]

40. Davudov, D.; Moghanloo, R.G. Impact of Pore Compressibility and Connectivity Loss on Shale Permeability. *Int. J. Coal Geol.* **2018**, *187*, 98–113. [CrossRef]

41. Zheng, D.; Su, Y.; Reza, Z. Integrated Pore-Scale Characterization of Mercury Injection/Imbibition and Isothermal Adsorption/Desorption Experiments Using Dendroidal Model for Shales. *J. Pet. Sci. Eng.* **2019**, *178*, 751–765. [CrossRef]

42. Zhao, J.; Jin, Z.; Hu, Q.; Liu, K.; Liu, G.; Gao, B.; Liu, Z.; Zhang, Y.; Wang, R. Geological Controls on the Accumulation of Shale Gas: A Case Study of the Early Cambrian Shale in the Upper Yangtze Area. *Mar. Pet. Geol.* **2019**, *107*, 423–437. [CrossRef]

43. Zhao, J.; Jin, Z.; Jin, Z.; Hu, Q.; Hu, Z.; Du, W.; Yan, C.; Geng, Y. Mineral Types and Organic Matters of the Ordovician-Silurian Wufeng and Longmaxi Shale in the Sichuan Basin, China: Implications for Pore Systems, Diagenetic Pathways, and Reservoir Quality in Fine-Grained Sedimentary Rocks. *Mar. Pet. Geol.* **2017**, *86*, 655–674. [CrossRef]

44. Tang, X.; Jiang, Z.; Li, Z.; Gao, Z.; Bai, Y.; Zhao, S.; Feng, J. The Effect of the Variation in Material Composition on the Heterogeneous Pore Structure of High-Maturity Shale of the Silurian Longmaxi Formation in the Southeastern Sichuan Basin, China. *J. Nat. Gas Sci. Eng.* **2015**, *23*, 464–473. [CrossRef]

45. Tang, X.; Jiang, Z.; Jiang, S.; Wang, P.; Xiang, C. Effect of Organic Matter and Maturity on Pore Size Distribution and Gas Storage Capacity in High-Mature to Post-Mature Shales. *Energy Fuels* **2016**, *30*, 8985–8996. [CrossRef]
46. Sun, J.; Dong, X.; Wang, J.; Schmitt, D.R.; Xu, C.; Mohammed, T.; Chen, D. Measurement of Total Porosity for Gas Shales by Gas Injection Porosimetry (GIP) Method. *Fuel* **2016**, *186*, 694–707. [CrossRef]
47. Davudov, D.; Moghanloo, R.G.; Lan, Y. Evaluation of Accessible Porosity Using Mercury Injection Capillary Pressure Data in Shale Samples. *Energy Fuels* **2018**, *32*, 4682–4694. [CrossRef]
48. Sun, M.; Yu, B.; Hu, Q.; Chen, S.; Xia, W.; Ye, R. Nanoscale pore characteristics of the Lower Cambrian Niutitang Formation Shale: A Case Study From Well Yuke #1 in the Southeast of Chongqing, China. *Int. J. Coal Geol.* **2016**, 16–29. [CrossRef]
49. Sun, M.; Yu, B.; Hu, Q.; Yang, R.; Zhang, Y.; Li, B. Pore Connectivity and Tracer Migration of Typical Shales in South China. *Fuel* **2017**, *203*, 32–46. [CrossRef]
50. Yang, R.; Jia, A.; Hu, Q.; Guo, X.; Sun, M. Particle Size Effect on Water Vapor Sorption Measurement of Organic Shale: One Example From Dongyuemiao Member of Lower Jurassic Ziliujing Formation in Jiannan Area of China. *Adv. Geo-Energy Res.* **2020**, *4*, 207–218. [CrossRef]
51. Zou, J.; Rezaee, R. Effect of Particle Size on High-Pressure Methane Adsorption of Coal. *Pet. Res.* **2016**, *1*, 53–58. [CrossRef]
52. Wang, Y.; Liu, L.; Sheng, Y.; Wang, X.; Zheng, S.; Luo, Z. Investigation of Supercritical Methane Adsorption of Overmature Shale in Wufeng-Longmaxi Formation, Southern Sichuan Basin, China. *Energy Fuels* **2019**, *33*, 2078–2089. [CrossRef]
53. Ju, Y.; He, J.; Chang, E.; Zheng, L. Quantification of CH4 Adsorption Capacity in Kerogen-Rich Reservoir Shales: An Experimental Investigation and Molecular Dynamic Simulation. *Energy* **2019**, *170*, 411–422. [CrossRef]
54. Mastalerz, M.; He, L.; Melnichenko, Y.B.; Rupp, J.A. Porosity of Coal and Shale: Insights from Gas Adsorption and SANS/USANS Techniques. *Energy Fuels* **2012**, *26*, 5109–5120. [CrossRef]
55. Liu, K.; Ostadhassan, M.; Sun, L.; Zou, J.; Yuan, Y.; Gentzis, T.; Zhang, Y.; Carvajal-Ortiz, H.; Rezaee, R. A Comprehensive Pore Structure Study of the Bakken Shale with SANS, N_2 Adsorption and Mercury Intrusion. *Fuel* **2019**, *245*, 274–285. [CrossRef]
56. Jiang, S.; Xu, Z.; Feng, Y.; Zhang, J.; Cai, D.; Chen, L.; Wu, Y.; Zhou, D.; Bao, S.; Long, S. Geologic Characteristics of Hydrocarbon-Bearing Marine, Transitional and Lacustrine Shales in China. *J. Asian Earth Sci.* **2016**, *115*, 404–418. [CrossRef]
57. Milliken, K.L.; Esch, W.L.; Reed, R.M.; Zhang, T. Grain Assemblages and Strong Diagenetic Overprinting in Siliceous Mudrocks, Barnett Shale (Mississippian), Fort Worth Basin, Texas. *AAPG Bull.* **2012**, *96*, 1553–1578. [CrossRef]
58. Cao, X.; Gao, Y.; Cui, J.; Han, S.; Kang, L.; Song, S.; Wang, C. Pore Characteristics of Lacustrine Shale Oil Reservoir in the Cretaceous Qingshankou Formation of the Songliao Basin, NE China. *Energies* **2020**, *13*, 2027. [CrossRef]
59. Gu, X.; Cole, D.R.; Rother, G.; Mildner, D.F.R.; Brantley, S.L. Pores in Marcellus Shale: A Neutron Scattering and FIB-SEM Study. *Energy Fuels* **2015**, *29*, 1295–1308. [CrossRef]
60. Liu, S.; Zhang, R. Anisotropic Pore Structure of Shale and Gas Injection-Induced Nanopore Alteration: A Small-Angle Neutron Scattering Study. *Int. J. Coal Geol.* **2020**, *219*, 103384. [CrossRef]
61. Blach, T.; Radliński, A.P.; Edwards, D.S.; Boreham, C.J.; Gilbert, E.P. Pore Anisotropy in Unconventional Hydrocarbon Source Rocks: A Small-Angle Neutron Scattering (SANS) Study on the Arthur Creek Formation, Georgina Basin, Australia. *Int. J. Coal Geol.* **2020**, *225*, 103495. [CrossRef]
62. Huang, H.; Chen, L.; Dang, W.; Luo, T.; Sun, W.; Jiang, Z.; Tang, X.; Zhang, S.; Ji, W.; Shao, S.; et al. Discussion on the Rising Segment of the Mercury Extrusion Curve in the High Pressure Mercury Intrusion Experiment on Shales. *Mar. Pet. Geol.* **2019**, *102*, 615–624. [CrossRef]
63. Moghadam, A.; Vaisblat, N.; Harris, N.B.; Chalaturnyk, R. On the Magnitude of Capillary Pressure (Suction Potential) in Tight Rocks. *J. Pet. Sci. Eng.* **2020**, *190*, 107133. [CrossRef]
64. El Sharawy, M.S.; Gaafar, G.R. Pore-Throat Size Distribution Indices and Their Relationships with the Petrophysical Properties of Conventional and Unconventional Clastic Reservoirs. *Mar. Pet. Geol.* **2019**, *99*, 122–134. [CrossRef]

65. Hu, Q.; Liu, H.; Yang, R.; Zhang, Y.X.; Kibria, G.; Sahi, S.; Alatrash, N.; MacDonnell, F.M.; Chen, W. Applying Molecular and Nanoparticle Tracers to Study Wettability and Connectivity of Longmaxi Formation in Southern China. *J. Nanosci. Nanotechnol.* **2017**, *17*, 6284–6295. [CrossRef]
66. Zhao, J.; Jin, Z.; Jianhua, Z.; Jin, Z.; Barber, T.J.; Zhang, Y.; Bleuel, M. Integrating SANS And Fluid-Invasion Methods to Characterize Pore Structure of Typical American Shale Oil Reservoirs. *Sci. Rep.* **2017**, *7*, 15413. [CrossRef]

Publisher's Note: MDPI stays neutral with regard to jurisdictional claims in published maps and institutional affiliations.

Article

Porosity and Water Saturation Estimation for Shale Reservoirs: An Example from Goldwyer Formation Shale, Canning Basin, Western Australia

Muhammad Atif Iqbal * and Reza Rezaee

Department of Petroleum Engineering, Western Australia School of Mines: Minerals,
Energy and Chemical Engineering, Curtin University, 26 Dick Perry Avenue, Kensington, WA 6151, Australia;
r.rezaee@curtin.edu.au
* Correspondence: m.iqbal14@postgrad.curtin.edu.au

Received: 12 November 2020; Accepted: 27 November 2020; Published: 29 November 2020

Abstract: Porosity and water saturation are the most critical and fundamental parameters for accurate estimation of gas content in the shale reservoirs. However, their determination is very challenging due to the direct influence of kerogen and clay content on the logging tools. The porosity and water saturation over or underestimate the reserves if the corrections for kerogen and clay content are not applied. Moreover, it is very difficult to determine the formation water resistivity (R_w) and Archie parameters for shale reservoirs. In this study, the current equations for porosity and water saturation are modified based on kerogen and clay content calibrations. The porosity in shale is composed of kerogen and matrix porosities. The kerogen response for the density porosity log is calibrated based on core-based derived kerogen volume. The kerogen porosity is computed by a mass-balance relation between the original total organic carbon (TOC_o) and kerogen maturity derived by the percentage of convertible organic carbon (Cc) and the transformation ratio (TR). Whereas, the water saturation is determined by applying kerogen and shale volume corrections on the Rt. The modified Archie equation is derived to compute the water saturation of the shale reservoir. This equation is independent of R_w and Archie parameters. The introduced porosity and water saturation equations are successfully applied for the Ordovician Goldwyer formation shale from Canning Basin, Western Australia. The results indicate that based on the proposed equations, the total porosity ranges from 5% to 10% and the water saturation ranges from 35% to 80%. Whereas, the porosity and water saturation were overestimated by the conventional equations. The results were well-correlated with the core-based porosity and water saturation. Moreover, it is also revealed that the porosity and water saturation of Goldwyer Formation shale are subjected to the specific rock type with heterogeneity in total organic carbon total clay contents. The introduced porosity and water saturation can be helpful for accurate reserve estimations for shale reservoirs.

Keywords: shale reservoirs; matrix porosity; kerogen porosity; water saturation; well logs

---

1. Introduction

The organic-rich shale reservoirs have gained increasing attention in the last decades due to the depletion of conventional reservoirs [1,2]. For reliable volumetric calculation of the reserve, the porosity and water saturation are the most critical parameters to estimate [3–6]. The shale reservoirs contain free and adsorbed gases. The free gas associates within the pore spaces whereas the adsorbed gas is usually linked with the clay minerals and organic matter [2,4,7–10]. However, the complex pore system and organic matter together with inorganic mineral constituents affect the well logging tool responses needing to take them into account during petrophysical evaluation. Previous studies demonstrate that the porosity can be overestimated by using empirical equations without applying kerogen corrections. Therefore, the conventional approaches for porosity estimation are not feasible for organic-rich shale reservoirs. Many authors selected petrophysical models based on wireline logs to generate a set of simultaneous equations to estimate the kerogen content, mineral volume, and pore volume [7,11–13]. The introduced methods are most suitable for composition computation; however, it is hard to accurately determine all the required coefficients. Similarly, few authors standardised the well logs by multiplying the log data with defined coefficients to match the results with the core-derived porosity [12]. However, such equations were limited to a specific area and data set due to heterogeneity of shale in terms of thermal maturity, mineral composition, and organic matter content. Moreover, the organic-rich shales consist of the organic as well as matrix porosities [5,10,14–16]. In this study, the porosity for the shale reservoir is estimated by using a kerogen corrected density log, and the kerogen porosity is calculated by using a mass balance method based on original total organic carbon (TOCo) and kerogen maturity. The core-based total organic carbon (TOC) and porosity were used to validate the results.

Similarly, the accurate estimation of water saturation also plays a key role in economic evaluations of shale reservoirs. However, the investigations of the water saturation determination methods did not get much attention in the literature. Already available water saturation equations, e.g., Archie and Simandoux work better for conventional reservoirs (e.g., sandstone and shaly sands) [17,18]. However, the accurate determination of the unknown parameters such as formation water resistivity (Rw), cementation exponent (m), and saturation exponent (n) is very challenging for shale reservoirs [2,4,19–21]. The shale reservoir is a mixture of inorganic material (e.g., clays and detrital grains), kerogen, clay bound water, free and capillary held water, free and adsorbed gas [2,4]. However, the resistivity tool measures a reflection of constituent minerals and fluids of shales. Therefore, it is very critical to correct the resistivity log for shale and kerogen effects. In this research, a water saturation equation independent of water resistivity and Archie's parameters is introduced. Based on core derived water saturation validation, this equation worked very well as compared to other equations. However, it is always hard to take and interpret pressurised core samples from shale reservoirs. Therefore, sometimes it is impractical to measure water saturation through core samples in shale.

A case study from organic-rich Ordivician Goldwyer Formation (Goldwyer-III shale unit), Canning Basin, Western Australia is presented to verify both techniques for porosity and water saturation estimations. The Goldwyer Formation of Lower to Middle Ordovician age has an average thickness of almost 400 m, whereas, it's the thickest encounter (740 m) is recorded in Blackstone 1, a Lennard Shelf Sub-basin well. The Goldwyer shale is deposited in an open marine setting [22] having thin laminations of quartz silt and carbonates bands with alternating black shale layers. The mineral composition of Goldwyer shale includes quartz, carbonates, clay minerals, and pyrite [14]. The illite is a more abundant clay mineral in this shale. The Goldwyer shale is thermally mature having kerogen types-II and III and the total organic carbon content (TOC) varies from 0.35 to 4.5 wt% [23,24]. The results indicate a good match between core-based and corrected well logs-based estimations. Archie equation overestimated the water saturation, however, the proposed modified equation provided us better results. [25,26].

2. Materials and Methods

As illustrated in the simple shale reservoirs petrophysical model (Figure 1), the organic-rich shales are composed of kerogen and non-kerogen parts. A systematic workflow is developed to estimate the porosity and water saturations by considering the organic matter and matrix of the shale.

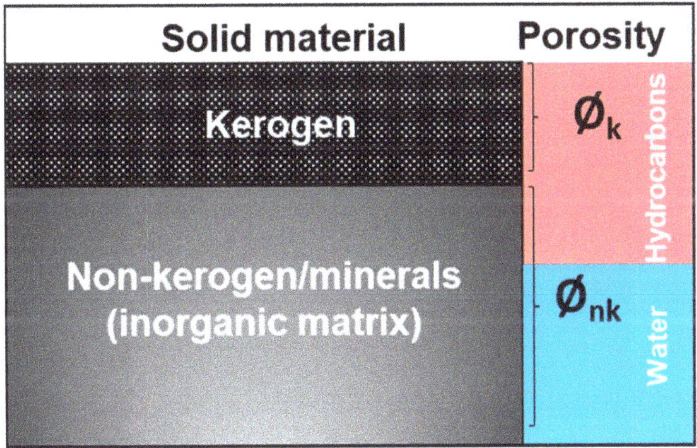

Figure 1. A typical conceptual petrophysical model for shale reservoirs showing kerogen porosity Ø$_k$ and non-kerogen Ø$_{nk}$ (inorganic matrix) porosity, modified from Yu et al. [5].

2.1. Porosity Estimation

The conventional density-based porosity equation is described in Equation (1):

$$\emptyset D = \frac{\rho_{ma} - \rho_b}{\rho_{ma} - \rho_f} \tag{1}$$

where $\emptyset D$ = density porosity (%), ρ_{ma} = matrix density (g/cc), ρ_b = bulk density (g/cc), ρ_f = fluid density (g/cc). Unlike in conventional reservoirs (sandstone or limestone), the bulk density acquired through density log in organic-rich shale usually overestimates the porosity. Therefore, the kerogen correction is applied to avoid porosity overestimation. The kerogen volume is determined by using Equation (2) [25]:

$$V_k = \frac{\gamma \times TOC \times \rho_b}{100 \times \rho_k} \tag{2}$$

where, V_k is the kerogen volume (fractions); TOC is total organic carbon content (wt %); ρ_b is the bulk density from the density log (g/cc); γ is the kerogen conversion factor; and ρ_k is the kerogen density (g/cc). TOC is determined by the rock eval pyrolysis method on powdered shale samples, and the continuous TOC for the whole interval is estimated by Passey method [27]; γ is proposed by [25], and the selected values are shown in Table 1.

Table 1. Conversion factors for total organic carbon (TOC) to kerogen, adapted from Tissot and Welte [25].

Stage	Type of Kerogen		
	I	II	III
Diagenesis	1.25	1.34	1.48
End of Catagenesis	1.20	1.19	1.18

For this study, based on rock eval pyrolysis results, the kerogen types are 30% type-II and 70% type-III. Therefore, the kerogen conversion factor for the studied formation is calculated as 1.18; and ρ_k is determined by the relationship of lab-based TOC and reciprocal of lab-based derived grain density on shale samples by the Equation (3). A good relationship between TOC and reciprocal of grain density

(ρ_g read as RHOG) is observed in Figure 2. The Equation (3) is derived based on the relationship between TOC and reciprocal of grain density (Figure 2).

$$\frac{1}{\rho_g} = A \times TOC + B \tag{3}$$

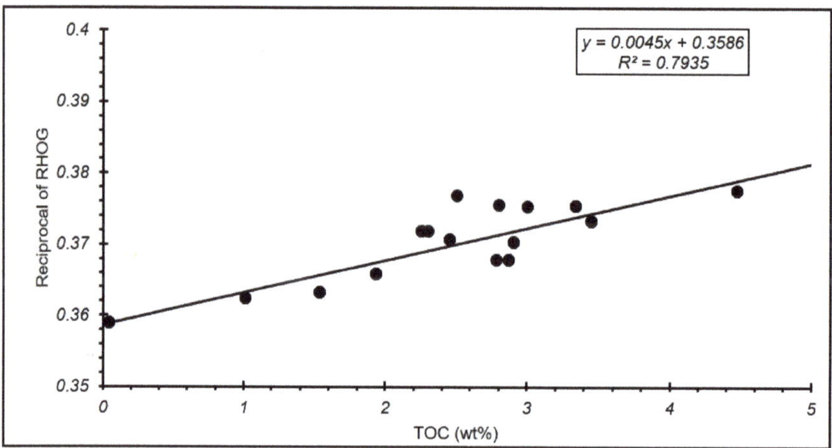

Figure 2. The direct relationship between core-based derived total organic carbon and reciprocal of grain density providing helpful information for estimation of kerogen and matrix densities.

ρ_g is the matrix density if *TOC* is zero and ρ_{g_k} is kerogen density if *TOC* is 100%. *A* and *B* are based on the linear relationship seen in Figure 2. From the relation found in Figure 2, the matrix density for the samples of this study is 2.79 g/cc, and kerogen density is 1.24 g/cc. The well logs are calibrated by eliminating the kerogen effect, and the following equations Equations (4) and (5) are applied for matrix porosity estimation through density log:

$$\rho_{bk_c} = \frac{\rho_b - \rho_k \times V_k}{1 - V_k} \tag{4}$$

$$\emptyset kc = \frac{\rho_{ma} - \rho_{bk}}{\rho_{ma} - \rho_f} \tag{5}$$

where, ρ_{bk_c} is kerogen corrected bulk density (g/cc); ρ_k is kerogen density (g/cc); V_k is kerogen volume (fractions) and $\emptyset kc$ is kerogen corrected density porosity (%). As the porosity in organic-rich shale is associated with organic matter and inorganic minerals, so it is crucial to estimate the porosity within organic matter (kerogen). An equation for kerogen porosity was proposed by [28] using mass-balance relation Equation (6).

$$\emptyset_k = ([TOC_o \times C_c] \times \gamma)TR\frac{\rho_b}{\rho_k} \tag{6}$$

where, $\emptyset_k$ = kerogen porosity (%), *TOCo* = original total organic carbon, *Cc* = convertible carbon fraction and *TR* = transformation ratio.

$$TOC_o = \frac{TOC}{1 - TR \times C_c} \tag{7}$$

$$TR = 1 - \frac{HI_p[1200 - HI_o(1 - PI_o)]}{HI_o[1200 - HI_p(1 - PI_p)]} \tag{8}$$

where: HIp = present hydrogen index (mg/g), HIo = original hydrogen index (mg/g), PIp = present production index and PIo = original production index. The following equations were used to estimate the original hydrogen index and present hydrogen index proposed by [28]:

$$HI_o = \frac{TypeII}{100} \times 450 + \frac{TypeIII}{100} \times 125 \tag{9}$$

For this study:

$$HI_o = 225 \text{ mg/g}$$

$$HI_p = 170 \text{ mg/g}$$

S1/S1 + S2 = PI_p = 0.35

The convertible carbon fraction is determined by using the relationship proposed by [29], such as $C_c = 0.085 \times HI_o$=18.91%.

Although, the transformation ratio (TR) can be determined by Claypool equation as explained in Equation (8) [28]. However, for this study, the TR value is taken as 88% that is adapted from [24,30] based on organic geochemistry and basin modelling of Goldwyer shale. So, the equation for kerogen porosity will be as Equation (10). By eliminating the kerogen effect and adding the kerogen porosity Equation (11), the final Equation (12) is applied to compute total density porosity for shale reservoirs.

$$\emptyset_k = 0.2 \times TOC \times \rho_b \tag{10}$$

$$\emptyset D_{Total} = \left[\left(\frac{\rho_{ma} - \rho_{bk_c}}{\rho_{ma} - \rho_f}\right) + \emptyset_k\right] \tag{11}$$

$$\emptyset D_{Total} = \left[\left(\frac{\rho_{ma} - \left(\frac{\rho_b - \rho_{k} \times V_k}{1 - V_k}\right)}{\rho_{ma} - \rho_f}\right) + (0.2 \times TOC \times \rho_b)\right] \tag{12}$$

2.2. Calculation of Water Saturation

The water saturation estimation in shale is mainly dependent on its organic (kerogen) and inorganic components (minerals). Archie equation [17] is mainly popular for water saturation calculation in clean reservoirs. The equation was developed based on a function between formation conductivity and the conductivity of fluids in the pore spaces of a reservoir, such as:

$$C_t = \frac{S_w^n \times C_w}{F} \tag{13}$$

where C_t = total conductivity (ohm^{-1} m^{-1}), C_w = formation water conductivity (ohm^{-1} m^{-1}), n = saturation exponent usually equals to 2, S_w = water saturation (%). The equation can be written in terms of resistivity as follows:

$$\frac{1}{R_t} = \frac{\emptyset^m \times S_w^n}{a \times R_w} \tag{14}$$

where R_t = true resistivity measured by logging tool (ohm-m), $\emptyset$ = porosity (%), m = cementation exponent, n = saturation exponent usually equals to 2, a = tortuosity factor usually considered as 1 and R_w = formation water resistivity (ohm-m). The Equation (14) is known as the Archie equation for clean formations. Later, this equation did not provide acceptable and accurate results for the shaly formations. Therefore, other approaches such as Simandoux considered the shale effect on water saturation and developed an equation Equation (15) by considering the volume of shale in the equation that was further modified by Schlumberger, 1972 and the modified Simandoux equation is [18]:

$$\frac{1}{R_t} = \frac{\varnothing^m \times S_w^n}{a.R_w \times (1 - V_{sh})} + \frac{V_{sh} \times S_w}{R_{sh}} \tag{15}$$

where R_{sh} is the resistivity of shale (ohm-m) and V_{sh} is the volume of shale (fraction). The conventional water saturation models, e.g., Simandoux equation, modified Simandoux, total shale, and modified total shale equations provided better results for shaly formations as these equations are derived based on the conductivities of clays and non-clay matrix. However, these models overestimate the water saturation for organic-rich shales. Therefore, a modified water saturation equation is applied in this study. An equation was proposed by [2,4] for water saturation calculation for shale reservoirs. The derivation details of the equation are explained by [17] simplified equation for water saturation:

$$S_w = \sqrt{\frac{R_o}{R_t}} \tag{16}$$

where, R_o is the rock resistivity in lean shale interval where water saturation is deemed 100% (ohm-m) and R_t is the rock resistivity in the organic-rich shale reservoir with some degree of oil/gas saturation (ohm-m). Therefore, Ro and Rt are the key parameters for water saturation calculations.

As the organic-rich shale reservoirs have a higher content of total clay and organic matter it is necessary to conduct corrections (total organic carbon and total clay) for the true formation resistivity (Rt). The clay minerals decrease the formation resistivity and the kerogen increases the resistivity. So, the TOC and shale corrections are used for Rt. First, the correlation is developed between true resistivity log and TOC measurements (on powdered shale samples through rock eval pyrolysis) (Equation (17), Figure 3).

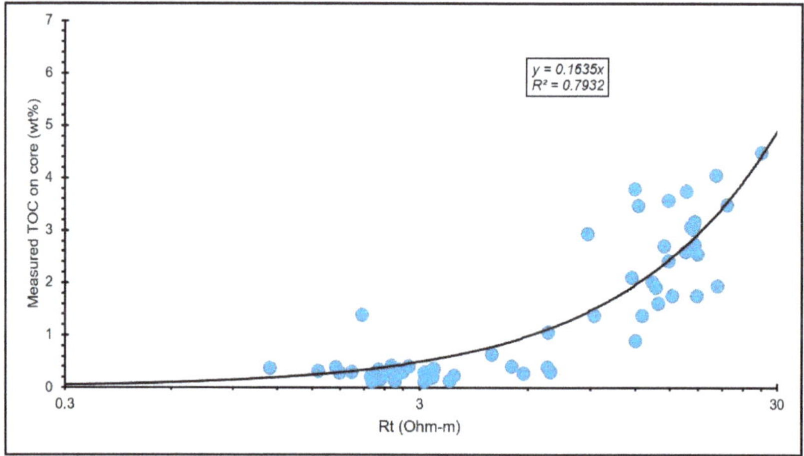

Figure 3. Direct relationship between true resistivity and measured total organic carbon showing influence of organic matter on resistivity tool.

A negative correlation Equation (18) is found between laboratory-based water saturation measured on shale samples and rock eval pyrolysis-based TOC. This relationship shows that with the increase in TOC, the water saturation reduces that provides an indication of hydrocarbon saturation in the shale interval (Figure 4).

$$TOC = 0.1635 \times R_t \tag{17}$$

$$Sw_{core} = -0.0981 \times TOC_{core} + 0.825 \tag{18}$$

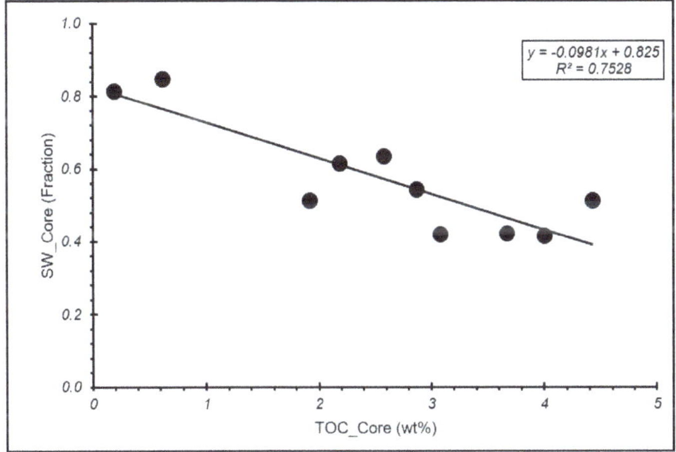

Figure 4. An inverse relationship between core-based total organic carbon and water saturation showing the fact that the organic matter increases gas saturation.

The true resistivity is corrected in terms of subtracting a factor A Equation (19) due to TOC that can be evaluated by making arrangements, such as:

$$A = V_k^2 \times R_k \tag{19}$$

If TOC is 100% then Rt will be considered as kerogen resistivity R_k (based on Equation (17)) so for this study based on Figure 3 R_k = 613 ohm-m and Figure 5 R_{sh} = 1.97 ohm-m are used.

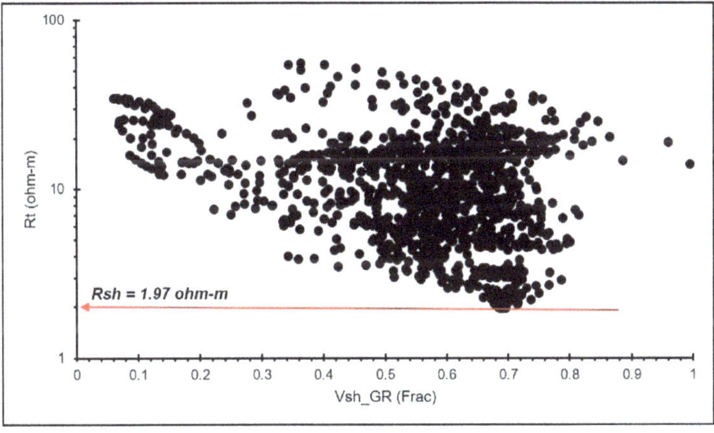

Figure 5. The shale resistivity estimation based on shale volume and true resistivity relationship.

Based on the correlation, the TOC_{max} is found as 4.91 wt %. Another factor B Equation (20) because of clay minerals effect on resistivity is defined by many authors [18,31,32], such as:

$$B = V_{sh}^2 \times R_o \tag{20}$$

The squared form of the shale volume will be more convincing in the calculation of reduced resistivity as a result of shale volume. It can be due to the nonlinear relationship between R_o and R_w in shales [18,31]. For this study, the R_o is taken as 1.97 ohm-m (Figure 5).

By compensating the shale and organic matter effects on the true resistivity, the modified equation is introduced as:

$$S_w = \sqrt{\frac{R_o}{R_t - (V_{kr}^2 \times R_k) + (V_{sh}^2 \times R_{sh})}} \tag{21}$$

3. Results and Discussion

In this section, the applications of proposed porosity and water saturation equations are implemented for the Ordovician Goldwyer shale formation drilled in Theia-1, Pictor East-1, and Canopus-1 wells in Canning Basin, Western Australia.

The kerogen corrected total porosity (matrix porosity plus kerogen porosity) was estimated by using Equation (12). The total porosity on crushed shale samples (core porosity) ranges from 2 to 13%, measured through the difference between the bulk volume of shale samples and the grain volume of the crushed, cleaned, and dried samples. The Goldwyer shale porosity shows the same range of porosity as most of the organic-rich shales [5,7,14,33–36]. The Goldwyer shale consists three types of pores such as organic pores, interparticle and intraparticle pores as shown in Figure 6. The results show that the conventional porosity estimation through density log overestimates the porosity that may affect the accurate reserve estimation in shale. Such as, the porosity based on Equation (1) provided the porosity range from 8 to 15% for Goldwyer shale (Figure 7). However, after applying the kerogen corrections, the corrected porosity ranging from 5 to 10% gives more accurate results that can be well-compared with core porosity (Table 2 and Figure 7). Moreover, the clay minerals also affect the pore structure of shale that directly affects the water saturation [37,38]. The Goldwyer shale also consists interparticle pores influenced by illite that may change the water saturation (Figure 6). The core derived TOC varies from 0.35 to 4.5 wt % in this study. The log derived TOC matches well with core-based *TOC* and the equivalent kerogen volume also validates the results (Figure 7). It can also be observed in Table 2 and Figure 7 that the clusters (e.g., siliceous and argillaceous shales) with higher TOC value have higher porosity (about 8–10%) due to the addition of organic pores (kerogen porosity) in the matrix porosity.

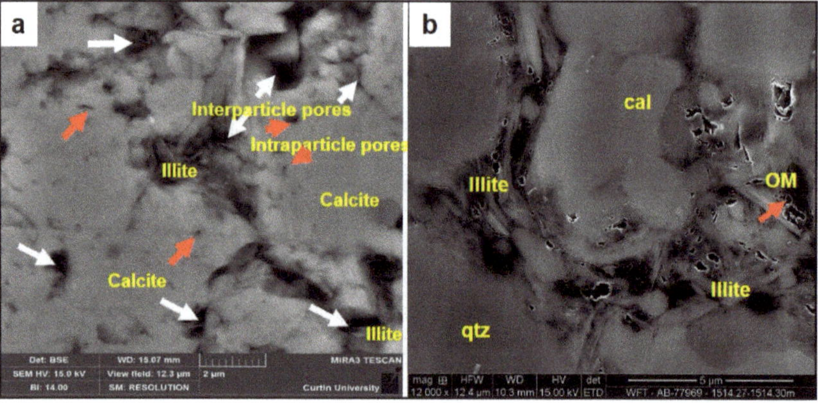

Figure 6. Different pore types observed in Goldwyer shale based on scanning electron microscope images, such as (**a**) interparticle pores indicated by white arrows and intraparticle pores indicated by red arrows; (**b**) organic matter pores (OM), mineral components include calcite (cal), quartz (qtz) and illite.

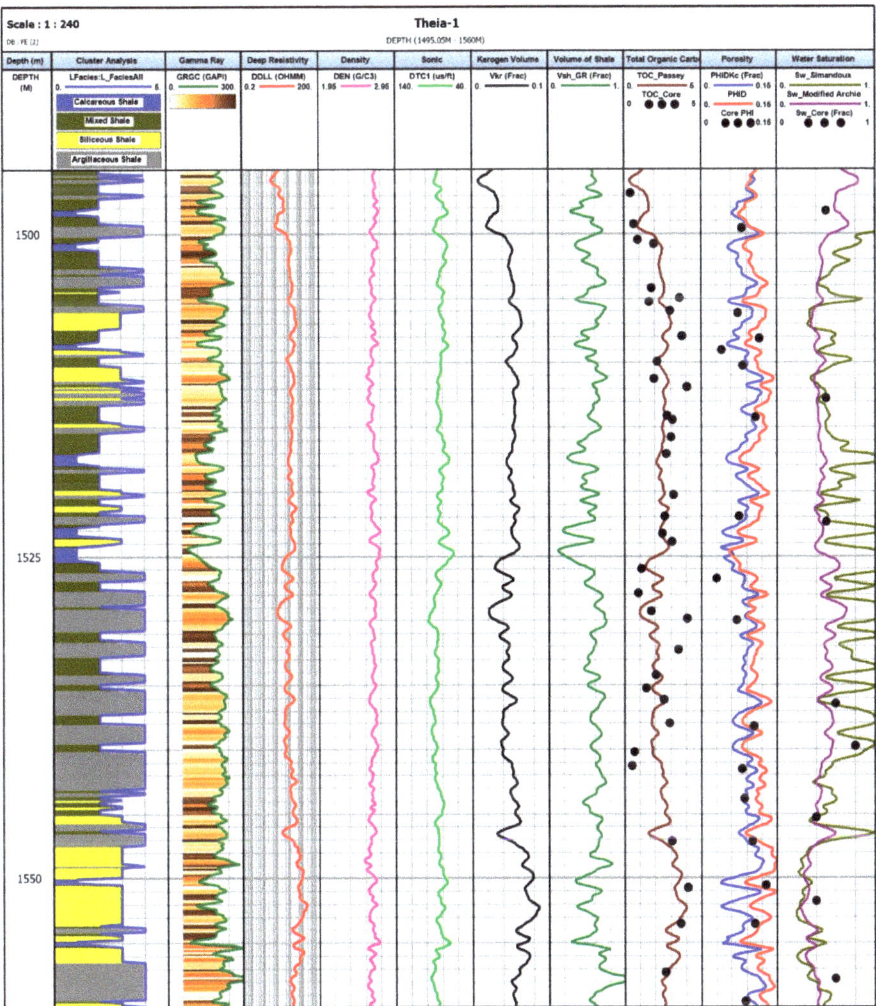

Figure 7. Petrophysical evaluation of Goldwyer shale providing accurate estimation of porosity and water saturation through proposed equations as validated by core-based measurements. Track-1: Depth in meters; Track-2: Cluster analysis to identify cluster based facies; Track-3: Gamma ray log; Track-4: Deep resistivity log; Track-5: Density log; Track-4: Sonic (DT) log; Track-4: Kerogen volume; Track-4: Shale volume based on Gamma ray log; Track-4: TOC based on Passey's method and core measurements; Track-4: Kerogen corrected total density porosity (PHIDKc) based on proposed equation in this study, density based porosity (PHID) & Total porosity based on core samples; Track-4: Water saturation (Sw) based on Simandoux equation (overestimated) and modified Archie's equation (by this study) and core derived Sw.

The water saturation was estimated by Equation (21) by considering the kerogen and shale effects on the resistivity. The required kerogen volume and kerogen resistivity were computed by using the data set (well logs) and core information from three wells (Theia-1, Pictor East-1 and Canopus-1) drilled in Canning Basin. The results for Theia-1 well are illustrated in Figure 7. Similarly, the shale resistivity was taken based on the data set for these three wells. It can be observed in Figure 7 that with the increase in shale volume (e.g., at depth 1546.5 m), the deep resistivity is decreased that enhances the water saturation. In conventional reservoirs, shale resistivity is usually determined from the averaged deep resistivity log reading against shale interval having higher gamma-ray log reading. However, in shale reservoirs, the shale resistivity is obtained from the average reading of the deep resistivity log against an organic lean interval. In this study, the shale resistivity in the organic lean interval is determined as 1.97 ohm-m based on the relationship between shale volume and true resistivity developed by this study (Figure 5). It is impractical to determine the fluid-water contact in heterogeneous shale reservoirs; therefore, an organic lean shale is treated to be fully brine saturated rock, $S_w = 1$ [4].

Table 2. Comparison of averaged total porosity and water saturation determined by conventional equations (PHID and Sw_Simandoux) and introduced by this study (PHIDKc and Sw_modified Archie). The conventional equations overestimated the porosity and water saturation in shale.

Cluster	Lithofacies	TOC	PHIDKc	Sw_Modified Archie	PHID	Sw_Simandoux
		(wt. %)	%	%	%	%
Cluster-1 (Blue)	Calcareous shale	0.7	5	55	6	90
Cluster-2 (Olive)	Mixed shale	1.4	8.5	45	10	80
Cluster-3 (Yellow)	Siliceous shale	2.5	8	35	12	45
Cluster-4 (Grey)	Argillaceous shale	3.5	9	80	13	>100

In the same way, the zones with higher TOC value and kerogen volume (such as organic-rich siliceous shale–cluster 3 (siliceous shale) at depth 1550 m) have the lowest water saturation. The inverse relationship between core-based TOC and Sw is also confirmed in this study (Figure 4). So, the kerogen resistivity ($R_{kr} = 613$ ohm-m) is determined by Equation (17) by putting TOC value as 100%. Therefore, the modified Archie equation applied in this study provides much better results (well correlated with core derived Sw) than the Simandoux equation (Table 2 and Figure 7). It can be observed that the Simandoux method overestimated water saturation as it is impossible to have more than 100% Sw. Another key factor of this overestimation is inaccurate determination of water resistivity and cementation exponent (m) values. Therefore, the modified Archie equation applied in this study is simple and accurate subject to the resistivity corrections for shale and kerogen.

4. Conclusions

In this research, effective equations for two critical petrophysical parameters of shale reservoirs (total porosity and water saturation) have been introduced. These equations are compensated based on kerogen effects for density logs to estimate more accurate total porosity. Similarly, the resistivity log was corrected based on kerogen and shale effects to compute the accurate water saturation for shale reservoirs. This study shows that the density log overestimates the total porosity (8–15%). Whereas the total porosity based on kerogen corrected density log and kerogen porosity matches perfectly with the core-based porosity having porosity ranged from 5 to 10%. In the same way, the Simandoux equation overestimated the water saturation with more than 100% Sw in most of the intervals. However, the proposed water saturation equation (modified Archie's equation) provided better results and correlation with core-based water saturation ranged from 35 to 80%. Moreover, the introduced modified Archie equation is independent of water resistivity and Archie parameters as these inputs are very difficult to obtain for shale reservoirs. It is also revealed that the porosity and water saturation in shale reservoirs are mainly dependent on the specific rock type. Such as the cluster-2 (mixed shale lithofacies with mixed lithologies and moderate TOC value) and cluster-3 (siliceous shale lithofacies with higher silica, less clay content and moderate to high TOC) have more shale gas potential in Goldwyer shale due to higher porosity and water saturation. This study has proposed a step to step workflow for accurate estimation of porosity and water saturation based on well logs for organic-rich shale. This workflow will be helpful for accurate reserve estimations in the shale reservoirs.

Author Contributions: Conceptualisation, M.A.I.; methodology, M.A.I, and R.R.; software, M.A.I.; data curation, M.A.I.; validation, M.A.I. and R.R.; writing—original draft, M.A.I.; writing—review and editing, R.R.; supervision, R.R. All authors have read and agreed to the published version of the manuscript.

Funding: This research received no external funding.

Acknowledgments: The authors would like to acknowledge the contributions of Chief Minister Merit Scholarship (CMMS), Pakistan and the Unconventional Gas Research group at the department of Petroleum Engineering, Western Australia School of Mines: Minerals, Energy and Chemical Engineering, Curtin University and Senergy Interactive Petrophysics v4.5 software for supporting this research. Special thanks to Department of Mines Industry Regulation and Safety (DMIRS) Western Australia and Finder Exploration Pty Ltd. (Perth, Western Australia). for providing data and reports about core-based measurements for water saturation.

Conflicts of Interest: The authors declare no conflict of interest.

Nomenclature

$\varnothing D$	density porosity
ρ_{ma}	matrix density
ρ_b	bulk density
ρ_f	fluid density
ρ_b	bulk density (g/cc)
γ	kerogen conversion factor
ρ_k	kerogen density (g/cc).
ρ_g	grain density
ρ_{bk}	kerogen corrected bulk density
$\varnothing$	porosity
$\varnothing k$	kerogen porosity
$\varnothing D_{Total}$	total density porosity
a	tortuosity factor
Cc	convertible carbon fraction
C_t	total conductivity
C_w	formation water conductivity
HIp	present hydrogen index
HIo	original hydrogen index
m	cementation exponent
n	saturation exponent
PIp	present production index
PIo	original production index
R_w	formation water resistivity
R_{sh}	resistivity of shale
R_t	true resistivity in ohm-m
R_o	the rock resistivity in lean shale interval where water saturation is deemed 100%
R_k	Kerogen resistivity
S_w	water saturation
TOC	total organic carbon content
$TOCo$	original total organic carbon
TR	transformation ratio
V_k	kerogen volume in fractions
V_{sh}	volume of shale

References

1. Jenner, S.; Lamadrid, A.J. Shale gas vs. coal: Policy implications from environmental impact comparisons of shale gas, conventional gas, and coal on air, water, and land in the United States. *Energy Policy* **2013**, *53*, 442–453. [CrossRef]
2. Rezaee, R. *Fundamentals of Gas Shale Reservoirs*; John Wiley & Sons: Hoboken, NJ, USA, 2015.
3. Ross, D.J.K.; Bustin, R.M. The importance of shale composition and pore structure upon gas storage potential of shale gas reservoirs. *Mar. Pet. Geol.* **2009**, *26*, 916–927. [CrossRef]
4. Kadkhodaie, A.; Rezaee, R. A new correlation for water saturation calculation in gas shale reservoirs based on compensation of kerogen-clay conductivity. *J. Pet. Sci. Eng.* **2016**, *146*, 932–939. [CrossRef]
5. Yu, H.; Wang, Z.; Rezaee, R.; Zhang, Y.; Han, T.; Arif, M.; Johnson, L. Porosity estimation in kerogen-bearing shale gas reservoirs. *J. Nat. Gas Sci. Eng.* **2018**, *52*, 575–581. [CrossRef]
6. Walls, J.D.; Sinclair, S.W. Eagle Ford shale reservoir properties from digital rock physics. *First Break* **2011**, *29*. [CrossRef]
7. Sondergeld, C.H.; Newsham, K.E.; Comisky, J.T.; Rice, M.C.; Rai, C.S. Petrophysical Considerations in Evaluating and Producing Shale Gas Resources. In Proceedings of the SPE Unconventional Gas Conference, Society of Petroleum Engineers, Pittsburgh, PA, USA, 23–25 February 2010; p. 34.
8. Kale, S.; Rai, C.; Sondergeld, C. Rock Typing in Gas Shales. In Proceedings of the SPE Annual Technical Conference and Exhibition, Society of Petroleum Engineers, Florence, Italy, 19–22 September 2010; p. 20.
9. Ambrose, R.J.; Hartman, R.C.; Diaz-Campos, M.; Akkutlu, I.Y.; Sondergeld, C.H. Shale gas-in-place calculations part I: New pore-scale considerations. *Spe J.* **2012**, *17*, 219–229. [CrossRef]
10. Yu, H.; Rezaee, R.; Wang, Z.; Han, T.; Zhang, Y.; Arif, M.; Johnson, L. A new method for TOC estimation in tight shale gas reservoirs. *Int. J. Coal Geol.* **2017**, *179*, 269–277. [CrossRef]
11. Jacobi, D.J.; Breig, J.J.; LeCompte, B.; Kopal, M.; Hursan, G.; Mendez, F.E.; Bliven, S.; Longo, J. Effective geochemical and geomechanical characterization of shale gas reservoirs from the wellbore environment: Caney and the Woodford shale. In Proceedings of the SPE Annual Technical Conference and Exhibition, Society of Petroleum Engineers, New Orleans, LA, USA, 4–7 October 2009.
12. Fu, Q.; Horvath, S.C.; Potter, E.C.; Roberts, F.; Tinker, S.W.; Ikonnikova, S.; Fisher, W.L.; Yan, J. Log-derived thickness and porosity of the Barnett Shale, Fort Worth basin, Texas: Implications for assessment of gas shale resources. *Aapg Bull.* **2015**, *99*, 119–141. [CrossRef]
13. Arredondo-Ramírez, K.; Ponce-Ortega, J.M.; El-Halwagi, M.M. Optimal planning and infrastructure development for shale gas production. *Energy Convers. Manag.* **2016**, *119*, 91–100. [CrossRef]
14. Yuan, Y.; Rezaee, R.; Al-Khdheeawi, E.A.; Hu, S.-Y.; Verrall, M.; Zou, J.; Liu, K. Impact of Composition on Pore Structure Properties in Shale: Implications for Micro-/Mesopore Volume and Surface Area Prediction. *Energy Fuels* **2019**, *33*, 9619–9628. [CrossRef]
15. Yuan, Y.; Rezaee, R.; Verrall, M.; Hu, S.-Y.; Zou, J.; Testmanti, N. Pore characterization and clay bound water assessment in shale with a combination of NMR and low-pressure nitrogen gas adsorption. *Int. J. Coal Geol.* **2018**, *194*, 11–21. [CrossRef]
16. Labani, M.M.; Rezaee, R.; Saeedi, A.; Al Hinai, A. Evaluation of pore size spectrum of gas shale reservoirs using low pressure nitrogen adsorption, gas expansion and mercury porosimetry: A case study from the Perth and Canning Basins, Western Australia. *J. Pet. Sci. Eng.* **2013**, *112*, 7–16. [CrossRef]
17. Archie, G.E. The electrical resistivity log as an aid in determining some reservoir characteristics. *Trans. AIME* **1942**, *146*, 54–62. [CrossRef]
18. Simandoux, P. Dielectric measurements in porous media and application to shaly formation: Revue del'Institut Francais du Petrole. *Suppl. Issue* **1963**, *18*, 193–215.
19. Wang, F.P.; Gale, J.F. Screening Criteria for Shale-Gas Systems. Gulf Coast Association of Geological Societies Transactions: Tulsa, OK, USA, 2009; Volume 59, pp. 779–793.
20. Bust, V.K.; Majid, A.A.; Oletu, J.U.; Worthington, P.F. The petrophysics of shale gas reservoirs: Technical challenges and pragmatic solutions. *Pet. Geosci.* **2013**, *19*, 91–103. [CrossRef]
21. Akbar, M.N.A.; Musu, J.T.; Milad, B. Water Saturation Interpretation Model for Organic-Rich Shale Reservoir: A Case Study of North Sumatra Basin. In Proceedings of the Unconventional Resources Technology Conference (URTEC), Houston, TX, USA, 23–25 July 2018.

22. Haines, P. *Depositional Facies and Regional Correlations of the Ordovician Goldwyer and Nita Formations, Canning Basin, Western Australia, with Implications for Petroleum Exploration*; Geological Survey of Western Australia, Record: East Perth, WA, Australia, 2004; p. 7.
23. Johnson, L.M.; Rezaee, R.; Kadkhodaie, A.; Smith, G.; Yu, H. Geochemical property modelling of a potential shale reservoir in the Canning Basin (Western Australia), using Artificial Neural Networks and geostatistical tools. *Comput. Geosci.* **2018**, *120*, 73–81. [CrossRef]
24. Johnson, L.M. Integrated Reservoir Characterization of the Goldwyer Formation, Canning Basin; Curtin University, Perth, Western Australia, 2019.
25. Tissot, B.P.; Welte, D.H. *Diagenesis, Catagenesis and Metagenesis of Organic Matter, in Petroleum Formation and Occurrence*; Springer: Berlin/Heidelberg, Germany, 1984; pp. 69–73.
26. Espitalie, J.; Madec, M.; Tissot, B.; Mennig, J.; Leplat, P. Source rock characterization method for petroleum exploration. In Proceedings of the Offshore Technology Conference, Houston, TX, USA, 2–5 May 1977.
27. Passey, Q.; Creaney, S.; Kulla, J.; Moretti, F.; Stroud, J. A practical model for organic richness from porosity and resistivity logs. *Aapg Bull.* **1990**, *74*, 1777–1794.
28. Peters, K.E.; Walters, C.; Moldowan, J.M. *Biomarkers and Isotopes in the Environment and Human History*; Cambridge University Press: Cambridge, UK, 2005.
29. Kilgore, E.; Land, A.; Schmidt, A.; Yunker, J. Applications of the Coriband Technique to Complex Lithologies. *Log Anal.* **1972**, *13*, 24.
30. Johnson, L.M.; Rezaee, R.; Smith, G.C.; Mahlstedt, N.; Edwards, D.S.; Kadkhodaie, A.; Yu, H. Kinetics of hydrocarbon generation from the marine Ordovician Goldwyer Formation, Canning Basin, Western Australia. *Int. J. Coal Geol.* **2020**, *232*, 103623. [CrossRef]
31. Leveaux, J.; Poupon, A. Evaluation of water saturation in shaly formations. *Log Anal.* **1971**, *12*, 6.
32. Clavier, C.; Coates, G.; Dumanoir, J. Theoretical and Experimental Bases forthe Dual-Water Model for the Interpretation of Shaly Sands. *Soc. Pet. Eng. J.* **1984**, *24*, 153–167. [CrossRef]
33. Chalmers, G.R.; Bustin, R.M.; Power, I.M. Characterization of gas shale pore systems by porosimetry, pycnometry, surface area, and field emission scanning electron microscopy/transmission electron microscopy image analyses: Examples from the Barnett, Woodford, Haynesville, Marcellus, and Doig units. *Aapg Bull.* **2012**, *96*, 1099–1119.
34. Wu, T.; Li, X.; Zhao, J.; Zhang, D. Multiscale pore structure and its effect on gas transport in organic-rich shale. *Water Resour. Res.* **2017**, *53*, 5438–5450. [CrossRef]
35. Wei, W.; Zhu, X.; Meng, Y.; Xiao, L.; Xue, M.; Wang, J. Porosity model and its application in tight gas sandstone reservoir in the southern part of West Depression, Liaohe Basin, China. *J. Pet. Sci. Eng.* **2016**, *141*, 24–37. [CrossRef]
36. Mastalerz, M.; Schimmelmann, A.; Drobniak, A.; Chen, Y. Porosity of Devonian and Mississippian New Albany Shale across a maturation gradient: Insights from organic petrology, gas adsorption, and mercury intrusion. *Aapg Bull.* **2013**, *97*, 1621–1643. [CrossRef]
37. Tian, H.; Wang, M.; Liu, S.; Zhang, S.; Zou, C. Influence of Pore Water on the Gas Storage of Organic-Rich Shale. *Energy Fuels* **2020**, *34*, 5293–5306. [CrossRef]
38. Cao, T.; Xu, H.; Liu, G.; Deng, M.; Cao, Q.; Yu, Y. Factors influencing microstructure and porosity in shales of the Wufeng-Longmaxi formations in northwestern Guizhou, China. *J. Pet. Sci. Eng.* **2020**, *191*, 107181. [CrossRef]

Publisher's Note: MDPI stays neutral with regard to jurisdictional claims in published maps and institutional affiliations.

Article

Factors Affecting Shale Gas Chemistry and Stable Isotope and Noble Gas Isotope Composition and Distribution: A Case Study of Lower Silurian Longmaxi Shale Gas, Sichuan Basin

Chunhui Cao [1,2,*], Liwu Li [1,2], Yuhu Liu [3], Li Du [1,2], Zhongping Li [1,2] and Jian He [1,2]

[1] Northwest Institute of Eco-Environment and Resources, Chinese Academy of Sciences, Lanzhou 730000, China; llwu@lzb.ac.cn (L.L.); duli@lzb.ac.cn (L.D.); lzp7748@163.com (Z.L.); hejian@lzb.ac.cn (J.H.)
[2] Key Laboratory of Petroleum Resources, Gansu Province, Lanzhou 730000, China
[3] Northeast Oil and Gas Branch, Sinopec, Changchun 130062, China; liuyuhu2008@163.com
* Correspondence: caochunhui@lzb.ac.cn; Tel.: +86-093-1496-0995

Received: 11 October 2020; Accepted: 5 November 2020; Published: 16 November 2020

Abstract: The Weiyuan (WY) and Changning (CN) fields are the largest shale gas fields in the Sichuan Basin. Though the shale gases in both fields are sourced from the Longmaxi Formation, this study found notable differences between them in molecular composition, carbon isotopic composition, and noble gas abundance and isotopic composition. CO_2 (av. 0.52%) and N_2 (av. 0.94%) were higher in Weiyuan than in Changning by an average of 0.45% and 0.70%, respectively. The $\delta^{13}C_1$ (−26.9% to −29.7%) and $\delta^{13}C_2$ (−32.0% to −34.9%) ratios in the Changning shale gases were about 8% and 6% heavier than those in Weiyuan, respectively. Both shale gases had similar $^3He/^4He$ ratios but different $^{40}Ar/^{36}Ar$ ratios. These geochemical differences indicated complex geological conditions and shed light on the evolution of the Lonmaxi shale gas in the Sichuan Basin. In this study, we highlight the possible impacts on the geochemical characteristics of gas due to tectonic activity, thermal evolution, and migration. By combining previous gas geochemical data and the geological background of these natural gas fields, we concluded that four factors account for the differences in the Longmaxi Formation shale gas in the Sichuan Basin: a) A different ratio of oil cracking gas and kerogen cracking gas mixed in the closed system at the high over-mature stage. b) The Longmaxi shales in WY and CN have had differential geothermal histories, especially in terms of the effects from the Emeishan Large Igneous Province (LIP), which have led to the discrepancy in evolution of the shales in the two areas. c) The heterogeneity of the Lower Silurian Longmaxi shales is another important factor, according to the noble gas data. d) Although shale gas is generated in closed systems, natural gas loss throughout geological history cannot be avoided, which also accounts for gas geochemical differences. This research offers some useful information regarding the theory of shale gas generation and evolution.

Keywords: gas geochemical characteristics; noble gas; shale gas evolution; Large Igneous Province (LIP); heterogeneity; gas loss

1. Introduction

Shale gas is produced from organic-rich black shale and self-generation and self-storage natural gas, and is continuously accumulated in nano-scale micropores in shale [1]. Shale gas is a thermogenic natural gas generated by organic matter pyrolysis. The shale gas relative elemental abundance patterns and isotopic compositions fluctuate continuously throughout the pyrolysis process due to the fractionation effect [2,3].

Shale gas is produced in a closed system and gas does not easily migrate. As a result, compared with conventional natural gas it has a greater genetic accumulation impact [4–6]. Hence, shale gas maintains more of the original information regarding the means of oil and gas generation from source rocks than conventional natural gas does, and its geochemical characteristics could be a reflection of the evolutionary process of closed-system fossil energy production. Natural gas formation theory has focused on the generation and evolution of shale gas, since differences the in geochemical characteristics between conventional natural gas and shale gas were discovered [7–11]. Shale gas geochemical irregularities include (1) the rollover of iso-alkane/normal alkane ratios [12]; (2) the rollover of ethane and propane isotopic compositions [13]; and (3) abnormally light ethane and propane $\delta^{13}C$ values and isotope reversals among methane, ethane, and propane [11,12,14–16]. Together, these irregularities reflect the complicated history of shale gas generation and the isotopic fractionation associated with it, as well as the in situ "mixing and accumulation" of gases that are generated from different precursors at various thermal maturities [4]. In addition, shale gases from different areas around the world also have many different geochemical characteristics [7,12,17–20]. Even if these shale gases come from the same area and same strata, variations in molecular composition, carbon isotopic composition, and noble gas abundance and isotopic composition can be found [15,16]. Recently, we found that the gas geochemical characteristics of shale gases from the Longmaxi Formation, Sichuan Basin, China, show several apparent differences between the Weiyuan (WY) and Changning (CN) areas [14–17,21–23]. For instance, there is more CH_4 in CN shale gas, and its carbon isotope composition is heavier than that of WY shale gas [14,16,17,21–23]. Meanwhile, the He and Ar abundance and isotope composition are higher in WY shale gas than in CN shale gas [15]. Although previous studies have found differential gas geochemical characteristics between WY and CN shale gas, few studies have explained the potential reason for these differences.

We collected and compared the geochemical data from our previous works [14–16] and other studies [16,21,22] and combined the geological background and oil/gas generation theory to clarify the causes and mechanics of variations in the geochemical characteristics of shale gas from the Longmaxi Formation, Sichuan Basin, China. These results should increase our understanding of the generation and evolution of shale gas.

2. Geological Background

The Sichuan Basin is a structurally complex, superimposed basin that comprises an area of over 18×10^4 km^2. The Sichuan Basin and the surrounding areas contain many gas fields (Figure 1). The two largest shale gas fields in the Sichuan Basin, Weiyuan and Changning, are located on the south and east borders of the basin, respectively. The organic-rich shales of the Wufeng–Longmaxi Formation (Ordovician–Silurian), one of the most important hydrocarbon source rocks in China, are the source of the shale gas generated ftom these two shale fields. The Silurian shales have equivalent vitrinite reflectance (EqVRo, %) values ranging from 2.4% to 3.8% [24–26], indicating that they are largely thermally over-mature and in a dry gas generation stage [24,27]. One possible external heat source for the maturation of organic-rich shale is the Emei Large Igneous Province (LIP), located in the southwestern part of the basin [28,29].

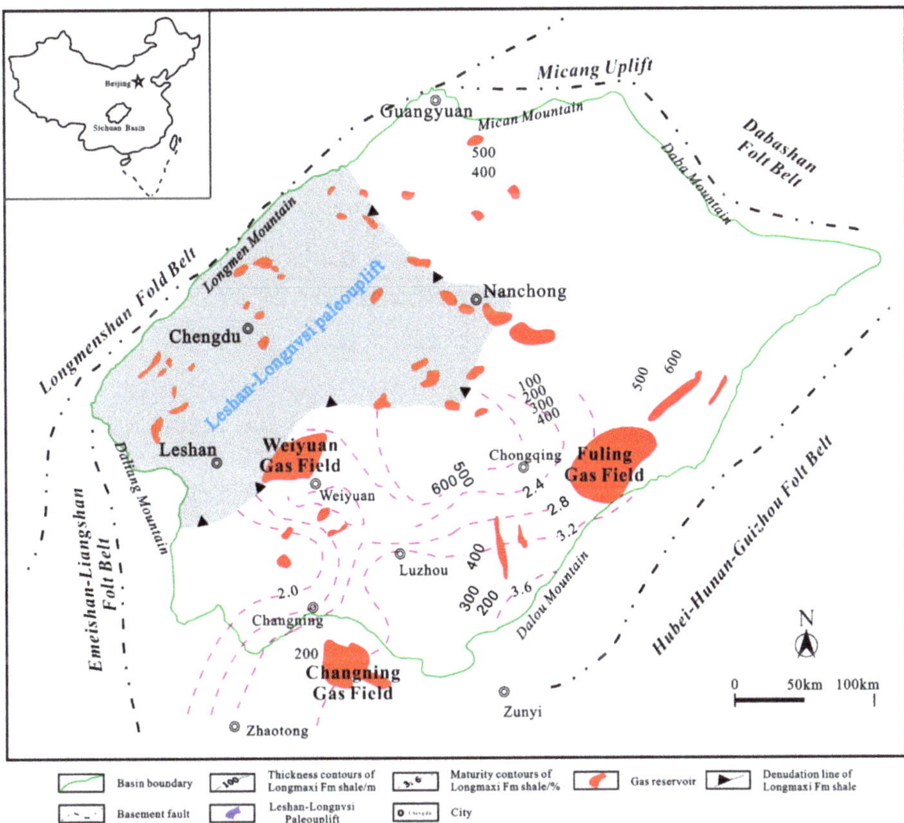

Figure 1. Geological sketch map of the Sichuan Basin, showing the locations of the main gas sampling sites, isolines of Ro values, and shale thickness of the Longmaxi Formation.

The Sichuan Basin is in the transition zone between the Palaeo–Pacific tectonic area and the Tethys–Himalayan tectonic area [30]. The Caledonian, Hercynian, Indosinian, and Yanshanian orogenies and Himalayan movement have all been recorded in the Sichuan Basin (Figure 2). There were two major tectonic evolution stages in the history of the Sichuan Basin: an early cratonic depression during the Palaeozoic era, followed by a foreland basin stage in the Triassic era [30–33]. This could have generated a large number of faults and unconformity surfaces, leading to diverse hydrocarbon migration and gas preservation [34]. The Lower Silurian Longmaxi Formation experienced deep burial in the Yanshan period, causing gas generation to reach its peak. Subsequent repeated uplift and erosion caused hydrocarbons to migrate out of the formation [35,36].

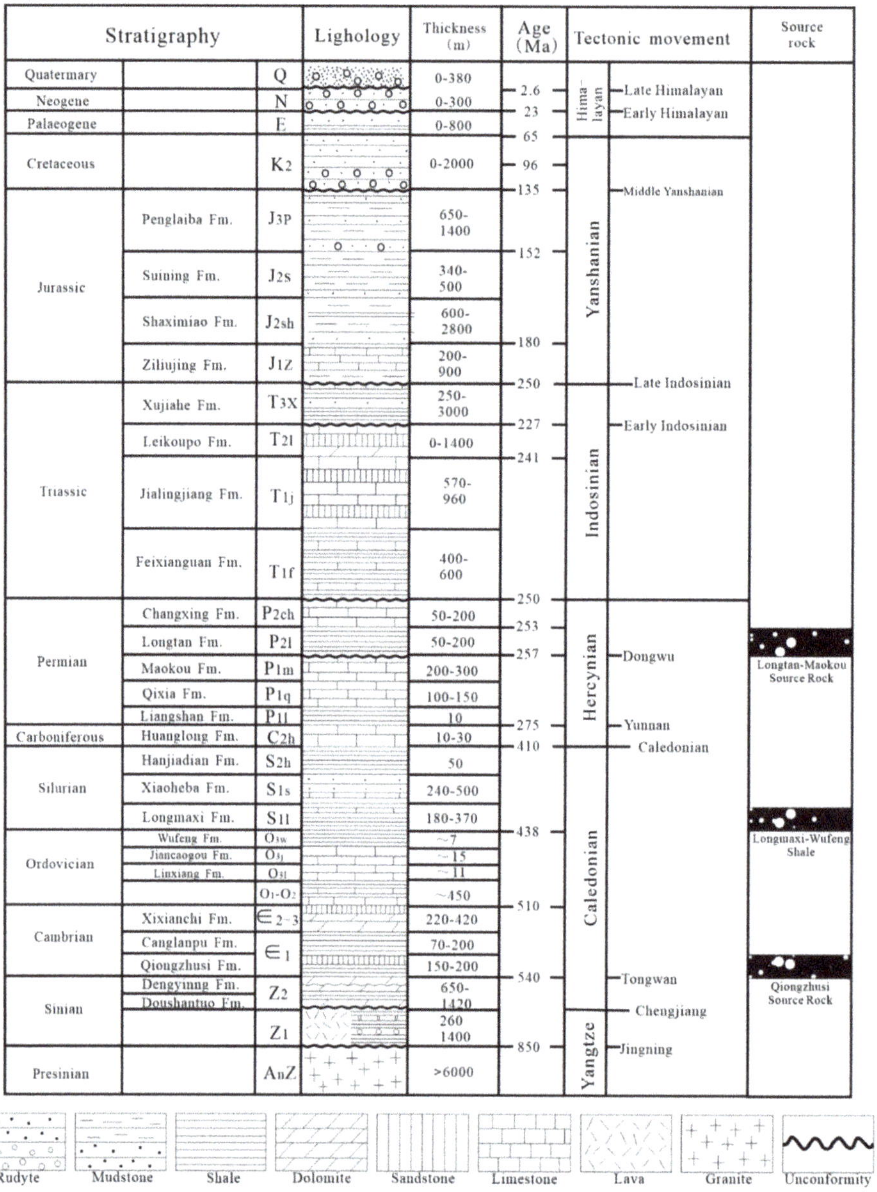

Stratigraphy	Formation	Symbol	Lighology	Thickness (m)	Age (Ma)	Tectonic movement		Source rock
Quatermary		Q		0–380	2.6	Himalayan	Late Himalayan	
Neogene		N		0–300	23	Himalayan	Early Himalayan	
Palaeogene		E		0–800	65			
Cretaceous		K_2		0–2000	96 / 135	Yanshanian	Middle Yanshanian	
Jurassic	Penglaiba Fm.	J_3P		650–1400	152	Yanshanian		
Jurassic	Suining Fm.	J_2s		340–500		Yanshanian		
Jurassic	Shaximiao Fm.	J_2sh		600–2800	180	Yanshanian		
Jurassic	Ziliujing Fm.	J_1Z		200–900	250	Yanshanian		
Triassic	Xujiahe Fm.	T_3X		250–3000	227	Indosinian	Late Indosinian	
Triassic	Leikoupo Fm.	T_2l		0–1400	241	Indosinian	Early Indosinian	
Triassic	Jialingjiang Fm.	T_1j		570–960		Indosinian		
Triassic	Feixianguan Fm.	T_1f		400–600	250	Indosinian		
Permian	Changxing Fm.	P_2ch		50–200	253	Hercynian		
Permian	Longtan Fm.	P_2l		50–200	257	Hercynian	Dongwu	Longtan-Maokou Source Rock
Permian	Maokou Fm.	P_1m		200–300		Hercynian		Longtan-Maokou Source Rock
Permian	Qixia Fm.	P_1q		100–150		Hercynian		
Permian	Liangshan Fm.	P_1l		10	275	Hercynian	Yunnan	
Carboniferous	Huanglong Fm.	C_2h		10–30	410	Hercynian	Caledonian	
Silurian	Hanjiadian Fm.	S_2h		50		Caledonian		
Silurian	Xiaoheba Fm.	S_1s		240–500		Caledonian		
Silurian	Longmaxi Fm.	S_1l		180–370	438	Caledonian		Longmaxi-Wufeng Shale
Ordovician	Wufeng Fm.	O_3w		~7		Caledonian		Longmaxi-Wufeng Shale
Ordovician	Jiancaogou Fm.	O_3j		~15		Caledonian		
Ordovician	Linxiang Fm.	O_3l		~11		Caledonian		
Ordovician		$O_1\text{-}O_2$		~450	510	Caledonian		
Cambrian	Xixianchi Fm.	$\epsilon_{2\text{-}3}$		220–420		Caledonian		
Cambrian	Canglanpu Fm.	ϵ_1		70–200		Caledonian		
Cambrian	Qiongzhusi Fm.	ϵ_1		150–200	540	Caledonian	Tongwan	Qiongzhusi Source Rock
Sinian	Dengyinng Fm.	Z_2		650–1420		Caledonian	Chengjiang	
Sinian	Doushantuo Fm.	Z_2		650–1420				
Sinian		Z_1		260 / 1400	850	Yangtze	Jingning	
Presinian		AnZ		>6000		Yangtze		

Legend: Rudyte · Mudstone · Shale · Dolomite · Sandstone · Limestone · Lava · Granite · Unconformity

Figure 2. Schematic diagram showing the stratigraphy system of the Sichuan Basin, as well as the main tectonic events.

The Weiyuan (WY) area extends over 2700 km^2 on the southeastern edge of the Leshan–Longnvsi paleo uplift in the southwestern part of the Sichuan Basin (Figure 1) [37]. Silurian strata are present in the southeastern part of the WY area, but are missing from its northwestern part. Silurian Longmaxi Formation shale is characterized as graptolite shale, mainly organic type I with an EqRo range from 1.80% to 2.24% [38,39]. It is primarily found on the southeastern flank of the Weiyuan anticline at a current burial depth of 1600–3200 m. The Changning (CN) area is located southwest of the Changning anticline in a gentle flank zone. The CN has an overall area of nearly 4000 km^2 (Figure 1). The CN area is closer to the core zone of Emei Large Igneous Province (LIP) than the WY area (Figure 3). In this area, the Longmaxi shale mainly belongs to type I-II1 organic matter, with a thermal maturity Ro of 2.8% to 3.3% [40,41].

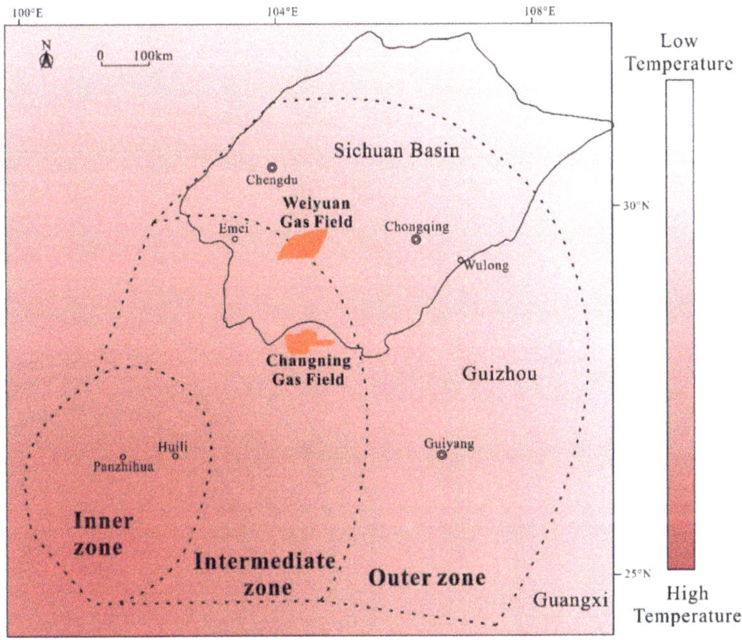

Figure 3. Schematic map of the Emeishan large igneous province (ELIP) and its thermal effect (Jiang, 2017, modified from Sun et al., 2010). The dashed lines show the boundaries of the inner, intermediate, and outer zones of the ELIP, as defined by He et al. (2003). For the interpretation of the references for color in this figure legend, the reader is referred to the web version of this article. The red column represents the Permian heat flow and the green column represents the present heat flow.

3. Differences in Gas Geochemical Characteristics

3.1. Molecular Composition

With an average content of 98.2%, CH_4 dominates the Longmaxi shale gases. The average contents of C_2H_6 and C_2H_6 are 0.51% and 0.02%, respectively. Non-hydrocarbon gases are mainly composed of N_2 and CO_2, with a small amount of He and Ar. H_2S has not been detected. The content of methane in Weiyuan (WY) shale gas (av. 97.9%) is slightly lower than that in Changning (CN) shale gas (av. 98.6%), while the contents of CO_2 (av. 0.47%) and N_2 (av. 0.98%) in the WY area are higher than those in CN area (av. 0.45% and av. 0.70%, respectively).

3.2. Carbon Isotope Composition

As shown in the carbon isotope correlation diagrams in Figures 4 and 5, there were distinct differences in the carbon isotopic distribution characteristics between the WY and CN shale gases. The $\delta^{13}C_1$ values of WY Longmaxi shale gas ranged from −34.1‰ to −37.3‰, the $\delta^{13}C_2$ values ranged from −37.6‰ ~ −43.4‰, and the $\delta^{13}C_3$ values ranged from −33.6‰ ~ −43.5‰. The $\delta^{13}C_1$ (−26.8‰ to −31.3‰) and $\delta^{13}C_2$ (−32.3‰ to −34.9‰) values of the CN Longmaxi shale gas were heavier by about 8‰ and 6‰, respectively, than those in the WY area (Figure 4). The $\delta^{13}C_3$ values ranged from −34.8‰ to −37.2‰ in the CN area.

Reversed distribution patterns of carbon isotopic compositions for CH_4 to C_3H_8 were found in the Longmaxi shale gas in both the WY and CN areas. Full reversal distribution patterns of the carbon isotopic composition according to the carbon number were found in most Longmaxi Formation shale gases in these areas—that is, $\delta^{13}C_1 > \delta^{13}C_2 > \delta^{13}C_3$. However, shale gases from three wells (W202, W201, and W201-H1) in the WY area showed a partial reversal distribution pattern of carbon isotopic composition according to the carbon number—that is, $\delta^{13}C_2 > \delta^{13}C_1$ and $\delta^{13}C_3 < \delta^{13}C_2$ (Figure 5).

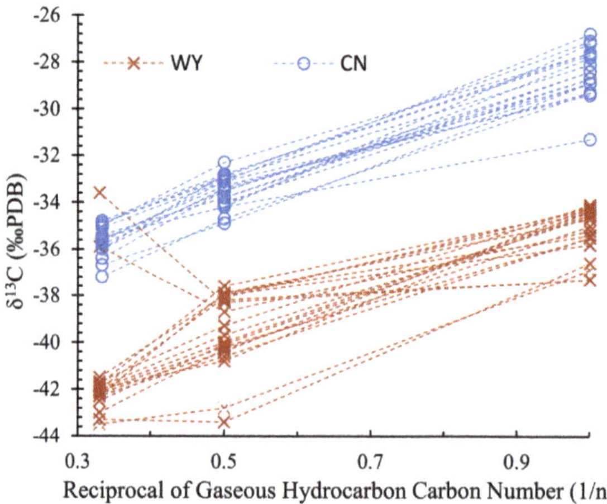

Figure 4. Carbon isotopic composition of shale gas from the Longmaxi Formation in the Weiyuan and Changning areas, Sichuan Basin, China.

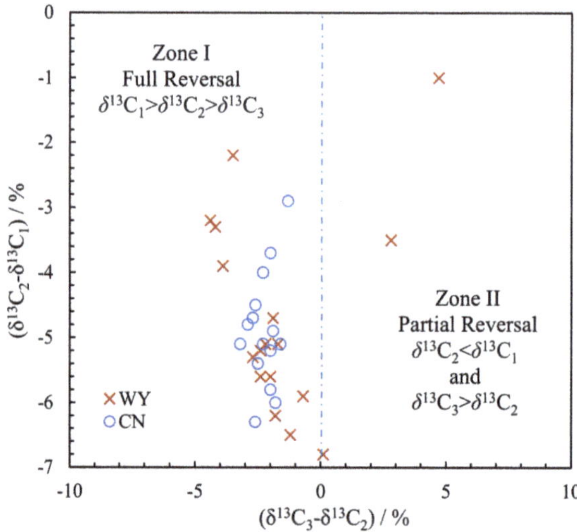

Figure 5. Variation in $\delta^{13}C_2-\delta^{13}C_1$ as a function of $\delta^{13}C_3-\delta^{13}C_2$ for gases from the Weiyuan (WY) and Changning (CN) areas, showing the isotope distribution patterns among methane, ethane, and propane.

3.3. Noble Gases

The Longmaxi formation shale gases in the Sichuan Basin showed regional differences in the abundance and isotopic compositions of noble gases [15]. The abundance and isotopic ratios of He and Ar in the WY shale gas are slightly higher than those in the CN shale gas (Figure 6). The concentrations of ^{4}He and ^{40}Ar ranged from 304.5 to 1286.3 ppm and 473.7 to 734.7 ppm, respectively, in the WY shale gas. The ^{3}He/^{4}He ratio was mainly around 0.02Ra, and the ^{40}Ar/^{36}Ar ratios ranged from 1276.2 to 6640.3 in the WY shale gas, while the concentration of ^{4}He and ^{40}Ar in the CN area varied in a small range from 386.1 to 445.9 ppm and 32.0 to 176.4 ppm, respectively. The ^{3}He/^{4}He ratios were around 0.01Ra, and the ^{40}Ar/^{36}Ar ratios were clustered around 1700 in the CN area. The ratios of ^{20}Ne/^{22}Ne and ^{21}Ne/^{22}Ne of the Longmaxi shale gases showed similar values to those of atmospheric Ne.

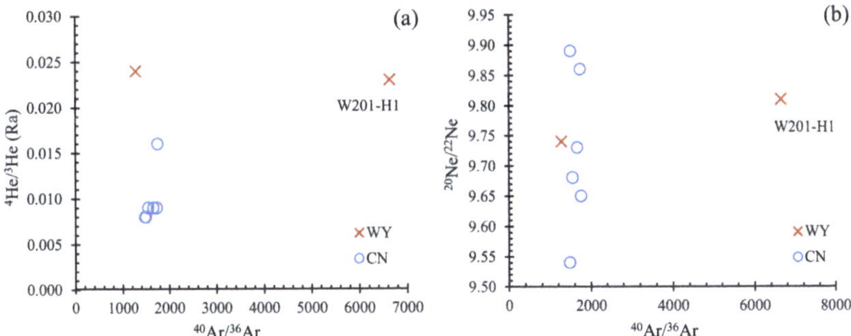

Figure 6. Plots of (**a**) ^{3}He/^{4}He vs. ^{40}Ar/^{36}Ar and (**b**) ^{20}Ne/^{22}Ne vs. ^{40}Ar/^{36}Ar of Longmaxi shale gases in the Weiyuan (WY) and Changning (CN) areas, China.

4. Causes of Gas Geochemical Variation

Hydrocarbon gases are ubiquitous products of organic maturation at all stages of burial. During the burial history, complex geological events may occur that could influence their maturity and lead to secondary alteration processes (migration, preservation, and water–rock interactions) that may result in changes in gas geochemical characteristics. The carbon isotope compositions of shale gas are closely related to the thermal alteration of organic matter. The different thermal histories of source rocks could bring about various patterns of carbon isotope composition [42–44]. Apart from the effect of temperature, the loss of natural gas during tectonic processes also affects the distribution of the molecular and isotope compositions of shale gas [45–50]. Water–organic matter redox reactions are another factor which could reform the gas geochemical characteristics of shale gases [51–54]. Lastly, the heterogeneity of Longmaxi Formation shale can lead to different concentrations of some molecules/elements in shale gases.

4.1. Mixing of Secondary Cracking Gas

Most shale gases are generated by the thermal degradation of sedimentary organic matter. The origin of this sedimentary organic matter is tightly linked to organic matter diagenetic and thermal alteration [42–44]. Although differences in the thermal maturity and/or organic type (marine and terrestrial shale gas) could bring about various δ^{13}C values and carbon isotopic distribution patterns [24], this seems not to be the reason for the differences in gas geochemistry between Weiyuan (WY) and Changning (CN) shale gas, as they have the same conditions in terms of these two factors [21,22,24]. Tissot and Welte [55] found that, in the early thermal evolution stage, gaseous hydrocarbons are formed concurrently with oil from kerogen in source rocks, whereas in the late thermal evolution stage gaseous hydrocarbons are generated by the thermal cracking of both residual kerogen and oil. The produced natural gas becomes progressively drier and isotopically more positive with the improvement in the thermal evolution degree [43,56]. The source rocks of the Longmaxi formation in the WY and CN areas are all in the high to over-maturity stage [24–26]. However, according to the tectonic activity history, Longmaxi shale in the WY and CN areas has experienced different processes of temperature change. Therefore, the differences in the molecular and carbon isotopic compositions of shale gas from the

Longmaxi Formation between the WY and CN areas could be caused by the different proportions of secondary cracking gas generated by residual kerogen and liquid hydrocarbons.

The different effects of tectonic movement in the WY and CN areas could have led to the different burial and thermal histories of Longmaxi shale. During the Triassic to Early Cretaceous era, the WY area underwent strong subsistence and then experienced extensive uplifting and erosion after the Late Cretaceous era. These events resulted in large fluctuations in temperature in the shales [57]. As shown in Figure 7, from the Middle Triassic (70 °C, started to generate oil) to the Late Cretaceous era (210 °C, maximum gas generation), a complete evolution of the hydrocarbon generation stages occurred in the WY Silurian Longmaxi shale, including oil generation, oil cracking to gas, and residual kerogen cracking to gas [57,58]. High temperature ranges from 172 °C to 205 °C were revealed by the homogenous temperature of the fluid inclusions taken from N202 in the CN area [59,60], providing evidence that the CN Silurian Longmaxi shale also went through the complete evolution of the hydrocarbon generation stages. The complex tectonic activities in the Weiyuan and Changning areas will cause source rocks to undergo different evolutionary processes, leading to differences in the shale gas geochemical characteristics. However, this needs to be proven by accurate source rock burial history and other information in each region.

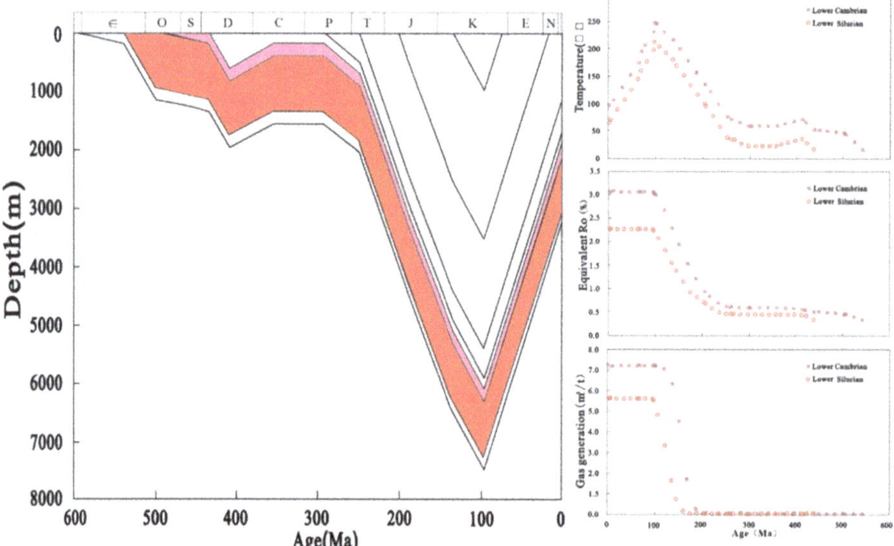

Figure 7. Plots of the burial history and thermal evolution of the Silurian Longmaxi shales in the Weiyuan area (Well W117). The thermal gradient evolution history was established by the reflectance inversion method: 32 °C/km (>96 Ma), 30 °C/km (96–65 Ma), 27 °C/km (<65 Ma). The thermal evolution histories of Lower Silurian shale were reconstructed by combining the thermal gradient model and their burial histories [57,58].

The Emeishan large igneous province (ELIP) covers an area of about 2.5×10^5 km^2 in southwest China. The heat flow in the inner and intermediate zones is abnormally high compared with that in the outer zone, where a decrease in the average heat flow from 76 to 51 mW/m^2 has been observed [59]. This provides a differentiated heat source for overlying and underlying strata in different areas. The appearance of pyrobitumen in the Sinian–Cambrian reservoirs is clear evidence of an abrupt hydrothermal fluid event, which might correspond to the Emei mantle plume in the late Permian era [60,61]. WY and CN are in the Emeishan large igneous province region (Figure 3). The thermal evolution of source rocks in the WY and CN areas was strongly affected by the ELIP [62–64].

From Figure 3, we can see that the CN area is in the intermediate zone of ELIP, while the WY area is in the outer zone. We can conclude that the CN area received relatively more heat energy than the WY area during the Emeishan mantle plume activity.

The shale gases from the Longmaxi formation in the WY and CN areas are thermogenic gases, which are formed at higher temperatures by the thermal decomposition of higher molecular weight organic matter (kerogen or oil) [27]. It is known that ^{12}C forms slightly weaker chemical bonds in the process of thermal decomposition than ^{13}C, resulting in a "kinetic isotope fractionation" in which the reaction product (gas in this case) is enriched in ^{12}C (isotopically "lighter") and the rest of the source material (kerogen or oil) becomes similarly enriched in ^{13}C (isotopically "heavier") in a process known as the Rayleigh fractionation effect [27]. As the maturity degree increases, the $\delta^{13}C_1$ ratio decreases until it reaches the lightest point, after which it increases [9]. Closed-system kerogen pyrolysis experiments and the study of geologic systems have determined that the secondary cracking of heavier hydrocarbons is a crucial pathway for gas generation [47,65–68]. Primary gases generated from kerogen and secondary gases cracked by oil and/or gaseous hydrocarbons are the main components of thermogenic shale gases. Shale gas (e.g., CH_4, C_2H_6, and C_3H_8) generated at different temperatures will have different isotopic compositions due to the Rayleigh fractionation effect. This may be one of the primary causes of the differences in the carbon isotopic composition of Longmaxi shale gases between the WY and CN areas, which experienced different temperature changes throughout their thermal evolution.

4.2. The Loss of Shale Gas

Shale gas aggregates continuously in gas reservoirs and is characterized by relatively short hydrocarbon migration distances [69]. Tectonic movement and preservation conditions are the main drivers of the accumulation and migration of shale gas [33,36]. The Sichuan Basin experienced complex tectonic movements during the evolution from the Craton basin (Palaeozoic) to the foreland basin (Triassic) [33,36]. Silurian Longmaxi shale was affected by the Yanshan, Indo-China, Dongwu, and Yunnan movements after deposition (Figure 2), which generated a large number of faults and unconformity surfaces and resulted in various pathways of hydrocarbon migration and gas loss [33,70,71]. Repeated uplift and erosion and numerous faults destroyed the preservation conditions of Longmaxi shale gas in the WY and CN areas [36,70] and consequently caused shale gas loss. The formation of the Leshan–Longnvsi paleo uplift involved several periods of tectonic movements, from the Tongwan movement to Yanshanian movement [35]; its tectonic evolution has had a greater influence on the formation and distribution of the WY shale gas reservoirs than that of the CN shale gas reservoirs [72–74].

During shale gas loss in the geological history, diffusive leakage from reservoirs and source rocks could induce carbon isotope fractionation ranges from 1% to 30% [75], and this loss of fractionation is universal in sedimentary basins [75,76]. Further, the smaller the volume of gas in the accumulation, the more likely any type of secondary fractionation will be significant [77]. Cao et al. [15] and Zhang et al. [16] discovered changes in noble gas abundance and isotopic composition and molecular and carbon isotope variation in the Longmaxi Formation shale gas in the WY and CN areas over the course of 3.5 years. Shale gas production is a kind of artificial diffusion process; the methane carbon isotope composition become slightly heavier, with its content decreasing in WY shale gases, while there are no changes in CN shale gases [16]. The differences between the gas geochemical characteristics in these two areas is due to the lower gas pressure (which means smaller volume) of the Longmaxi reservoir in the WY area [15,16,75–77].

Therefore, we can conclude that the Longmaxi shale in the WY area has been more affected by the intense tectonic activities, resulting in more shale gas loss over geological history. According to the diffusive fractionation theory [75,77], there should be a heavier $\delta^{13}C_1$ ratio in WY shale gases, but just the opposite is true [16,21,22,27]. Some other secondary reactions may have occurred in the Longmaxi shale gas reservoirs, either in WY or CN, leading to the present carbon isotope composition characteristics.

4.3. Water–Rock Interaction

The $\delta^{13}C_1$ ratios of shale gases in the CN area range from −26.8% to −31.3%, which is heavier than that of the thermogenic methane from type I and II kerogen (−50% to −30%, [78,79]). Its carbon isotope composition and distribution pattern are similar to those of abiogenic gas (>−30%, [80]). Abiogenic hydrocarbons could be generated by Fischer–Tropsch-type reactions in granitic rocks, whose $\delta^{13}C_1$ ratios range from −32% to −20% and $\delta^{13}C_1 > \delta^{13}C_{2+}$ [81,82]. Tang et al. [83] recognized a new mechanism of shale gas generation: a Fischer–Tropsch-type synthesis of hydrocarbon from CO_2 and H_2, resulting from the water reforming of residual organic matter in shale.

$$CH_x \text{ (organic matter)} + 2H_2O \rightarrow CO_2 + (2 + x/2)\ H_2, \tag{1}$$

$$CO_2 + mH_2 \rightarrow xCH_4 + yC_2H_6 + \ldots + zH_2O. \tag{2}$$

The Longmaxi shale in the CN area contains a large amount of formation water, providing the base materials for this methane generation mechanism, which may account for 50% or as much as 80% of the gas in shale, especially in particularly high-producing wells [83]. This also could increase the porosity and permeability of the shales [83]. Therefore, high $\delta^{13}C_1$ ratios, the full reversal distribution pattern of the carbon isotopic composition, and high gas pressure may be related to this new mechanism of shale gas generation. However, much more detailed work, including on the temperature and catalyst of the reaction and the matrix pore features, should be undertaken to properly understand this Fischer–Tropsch-type reaction.

4.4. Heterogeneity of Longmaxi Shale

Longmaxi shale has obvious lateral and vertical heterogeneity in its mineral composition, Total Organic Carbon (TOC), porosity, and trace elements [84–89]. Figure 6 shows the differences in ^{4}He and ^{40}Ar content between the WY and CN shale gases. The ^{4}He production in the crust is dominated by the α-decay of the ^{235,238}U and ^{232}Th decay chains, and is therefore directly proportional to the concentration of these radioelements in the crust, while the decay of ^{40}K dominates the ^{40}Ar production in the crust, which is thus directly proportional to the K concentrations. The contents of U, Th, and K are varied in Silurian Longmaxi shale [88,89], which may be one reason for the differences in noble gas isotope abundance (^{4}He and ^{40}Ar).

In addition, the ^{40}Ar/^{36}Ar ratio of W201-H1 is extremely high (6640.3), and close to that (7000) of conventional natural gas from the Sinian Dengying Formation reservoirs [15,90]. The two sets of gas reservoirs are in the same area (WY), and this combined with the intense tectonic activities in this area makes it very likely that some deeper conventional natural gas has leaked into the W201-H1 well from the fractures. Due to the low permeability and connectivity of shale rock, these deeper gases remained contained and did not spread to other shale gas wells.

5. Conclusions

There are differences in the molecular and carbon isotopic composition and noble gas abundance and isotopic composition in the Longmaxi shale gas sourced from the Weiyuan (WY) and Changning (CN) areas. This could be accounted for by the intense tectonic activities of the Sichuan Basin and secondary reactions in the shale gas reservoir. Additionally, the differential effect on WY and CN of the complex burial history and Emeishan super mantle plume activities in the Sichuan Basin has resulted in different thermal histories and hydrocarbon generation processes of Longmaxi shale in these two areas. The carbon isotope fractionation effect during the processes of shale gas generation results in different molecular and carbon isotopic compositions in the primary gases generated from kerogen compared with the secondary gases cracked by oil and/or gaseous hydrocarbons. The different partial mixing of primary and secondary gases is a governing factor for the differences in the gas geochemical characteristics between WY and CN. In addition, the different geotectonic movements have caused different fault and fracture systems in the WY and CN Longmaxi shale strata, which have led to various amounts of shale gas loss and consequently carbon isotope fractionation in the gas loss processes. Furthermore, water–rock interactions could enhance the molecular and carbon isotope composition differences, and the heterogeneity of the shale could bring about the different noble gas abundances and isotope compositions.

Author Contributions: Conceptualization, C.C.; funding acquisition, C.C. and Z.L.; investigation, C.C., L.L., Y.L.; methodology, C.C. and L.L.; project administration, C.C.; resources, C.C., L.L., L.D., and J.H.; supervision, C.C., and L.L.; writing—original draft, C.C.; writing—review and editing, C.C. and L.L. All authors have read and agreed to the published version of the manuscript.

Funding: This research was funded by National Natural Science Foundation of China, grant number 41502143, and Key Laboratory Project of Gansu Province, grant number 1309RTSA041.

Conflicts of Interest: The authors declare no conflict of interest.

Data Availability: The shale gas geochemical data used to support the findings of this study are included within the supplementary information file(s). Previously reported shale gas geochemical data were used to support this study, and are available at: https://doi.org/10.1016/j.marpetgeo.2017.01.023, https://doi.org/10.1016/j.marpetgeo.2017.01.022, https://doi.org/10.1016/S1876-3804(16)30092-1, https://doi.org/10.1016/j.petrol.2018.04.030. These prior studies are cited at relevant places within the text as references: [15,16,21,22,27].

References

1. Zou, C.; Dong, D.; Wang, S.; Li, J.; Li, X.; Wang, Y.; Li, D.; Cheng, K. Geological characteristics and resource potential of shale gas in China. *Pet. Explor. Dev.* **2010**, *37*, 641–653. [CrossRef]
2. Campbell, J.H.; Koskinas, G.J.; Gallegos, G.; Gregg, M. Gas evolution during oil shale pyrolysis. 1. Nonisothermal rate measurements. *Fuel* **1980**, *59*, 718–726. [CrossRef]
3. Campbell, J.H.; Gallegos, G.; Gregg, M. Gas evolution during oil shale pyrolysis. 2. Kinetic and stoichiometric analysis. *Fuel* **1980**, *59*, 727–732. [CrossRef]
4. Hao, F.; Zou, H. Cause of shale gas geochemical anomalies and mechanisms for gas enrichment and depletion in high-maturity shales. *Mar. Pet. Geol.* **2013**, *44*, 1–12. [CrossRef]
5. Lorant, F.; Prinzhofer, A.; Behar, F.; Huc, A.-Y. Carbon isotopic and molecular constraints on the formation and the expulsion of thermogenic hydrocarbon gases. *Chem. Geol.* **1998**, *147*, 249–264. [CrossRef]
6. Behar, F.; Vandenbroucke, M.; Teermann, S.; Hatcher, P.; Leblond, C.; Lerat, O. Experimental simulation of gas generation from coals and a marine kerogen. *Chem. Geol.* **1995**, *126*, 247–260. [CrossRef]
7. Tilley, B.; Muehlenbachs, K. Isotope reversals and universal stages and trends of gas maturation in sealed, self-contained petroleum systems. *Chem. Geol.* **2013**, *339*, 194–204. [CrossRef]
8. Xia, X.; Chen, J.; Braun, R.D.; Tang, Y. Isotopic reversals with respect to maturity trends due to mixing of primary and secondary products in source rocks. *Chem. Geol.* **2013**, *339*, 205–212. [CrossRef]
9. Qu, Z.; Sun, J.; Shi, J.; Zhan, Z.; Zou, Y.; Peng, P. Characteristics of stable carbon isotopic composition of shale gas. *J. Nat. Gas Geosci.* **2016**, *1*, 147–155. [CrossRef]
10. Tang, Y.C.; Xia, X.Y. Kinetics and mechanism of shale-gas formation: A quantitative interpretation of gas isotope 'rollover' for shale gas formation. In Proceedings of the AAPG Hedberg Conference, Austin, TX, USA, 5–10 December 2010; pp. 5–10.
11. Gao, L.; Schimmelmann, A.; Tang, Y.; Mastalerz, M. Isotope rollover in shale gas observed in laboratory pyrolysis experiments: Insight to the role of water in thermogenesis of mature gas. *Org. Geochem.* **2014**, *68*, 95–106. [CrossRef]
12. Zumberge, J.; Ferworn, K.; Brown, S. Isotopic reversal ('rollover') in shale gases produced from the Mississippian Barnett and Fayetteville formations. *Mar. Pet. Geol.* **2012**, *31*, 43–52. [CrossRef]
13. Ferworn, K.J.; Zumberge, J.; Reed, J.; Brown, S. Gas Character Anomalies Found in Highly Productive Shale Gas Wells. 2008. Available online: http://www.zenzebra.net/quebec/Ferworn_et_al_2008.pdf (accessed on 16 October 2008).
14. Cao, C.; Lv, Z.; Li, L.; Du, L. Geochemical characteristics and implications of shale gas from the Longmaxi Formation, Sichuan Basin, China. *J. Nat. Gas Geosci.* **2016**, *1*, 131–138. [CrossRef]

15. Cao, C.; Zhang, M.; Tang, Q.; Yang, Y.; Lv, Z.; Zhang, T.; Chen, C.; Yang, H.; Li, L. Noble gas isotopic variations and geological implication of Longmaxi shale gas in Sichuan Basin, China. *Mar. Pet. Geol.* **2018**, *89*, 38–46. [CrossRef]

16. Zhang, M.; Tang, Q.; Cao, C.; Lv, Z.; Zhang, T.; Zhang, D.; Li, Z.; Du, L. Molecular and carbon isotopic variation in 3.5 years shale gas production from Longmaxi Formation in Sichuan Basin, China. *Mar. Pet. Geol.* **2018**, *89*, 27–37. [CrossRef]

17. Yang, R.; He, S.; Hu, Q.H.; Hu, D.F.; Yi, J.Z. Geochemical characteristics and origin of natural gas from Wufeng-Longmaxi shales of the Fuling gas field, Sichuan Basin (China). *Int. J. Coal Geol.* **2017**, *171*, 1–11. [CrossRef]

18. Jiang, K.; Lin, C.; Zhang, X.; He, W.; Xiao, F. Geochemical characteristics and possible origin of shale gas in the Toolebuc Formation in the northeastern part of the Eromanga Basin, Australia. *J. Nat. Gas Sci. Eng.* **2018**, *57*, 68–76. [CrossRef]

19. Rodriguez, N.D.; Philp, R.P. Geochemical characterization of gases from the Mississippian Barnett Shale, Fort Worth Basin, Texas. *AAPG Bull.* **2010**, *94*, 1641–1656. [CrossRef]

20. Tilley, B.; McLellan, S.; Hiebert, S.; Quartero, B.; Veilleux, B.; Muehlenbachs, K. Gas isotope reversals in fractured gas reservoirs of the western Canadian Foothills: Mature shale gases in disguise. *AAPG Bull.* **2011**, *95*, 1399–1422. [CrossRef]

21. Feng, Z.; Liu, D.; Huang, S.; Wu, W.; Dong, D.; Peng, W.; Han, E. Carbon isotopic composition of shale gas in the Silurian Longmaxi Formation of the Changning area, Sichuan Basin. *Pet. Explor. Dev.* **2016**, *43*, 769–777. [CrossRef]

22. Feng, Z.; Dong, D.; Tian, J.; Qiu, Z.; Wu, W.; Zhang, C. Geochemical characteristics of Longmaxi Formation shale gas in the Weiyuan area, Sichuan Basin, China. *J. Pet. Sci. Eng.* **2018**, *167*, 538–548. [CrossRef]

23. Wei, X.; Guo, T.; Liu, R. Geochemical features and genesis of shale gas in the Jiaoshiba Block of Fuling Shale Gas Field, Chongqing, China. *J. Nat. Gas Geosci.* **2016**, *1*, 361–371. [CrossRef]

24. Dai, J.; Zou, C.; Dong, D.; Ni, Y.; Wu, W.; Gong, D.; Wang, Y.; Huang, S.; Huang, J.; Fang, C.; et al. Geochemical characteristics of marine and terrestrial shale gas in China. *Mar. Pet. Geol.* **2016**, *76*, 444–463. [CrossRef]

25. Zou, C.; Dong, D.; Wang, Y.; Li, X.; Huang, J.; Wang, S.; Guan, Q.; Zhang, C.; Wang, H.; Liu, H.; et al. Shale gas in China: Characteristics, challenges and prospects (I). *Pet. Explor. Dev.* **2015**, *42*, 753–767. [CrossRef]

26. Guo, T. Key geological issues and main controls on accumulation and enrichment of Chinese shale gas. *Pet. Explor. Dev.* **2016**, *43*, 349–359. [CrossRef]

27. Dai, J.; Zou, C.; Liao, S.; Dong, D.; Ni, Y.; Huang, J.; Wu, W.; Gong, D.; Huang, S.; Hu, G. Geochemistry of the extremely high thermal maturity Longmaxi shale gas, southern Sichuan Basin. *Org. Geochem.* **2014**, *74*, 3–12. [CrossRef]

28. Zhu, C.; Xu, M.; Yuan, Y.; Zhao, Y.; Shan, J.; He, Z.; Tian, Y.; Hu, S. Palaeogeothermal response and record of the effusing of Emeishan basalts in the Sichuan basin. *Chin. Sci. Bull.* **2009**, *55*, 949–956. [CrossRef]

29. Xu, Y.G.; Chung, S.-L.; Jahn, B.-M.; Wu, G.Y. Petrologic and geochemical constraints on the petrogenesis of Permian–Triassic Emeishan flood basalts in southwestern China. *Lithos* **2001**, *58*, 145–168. [CrossRef]

30. Liu, S.G.; Deng, B.; Zhong, Y.S.; Ran, B.; Yong, Z.; Sun, W.; Yang, D.Z.; Jiang, L.; Ye, Y. Unique geological features of burial and superimposition of the Lower Paleozoic shale gas across the Sichuan Basin and its periphery. *Earth Sci. Front.* **2016**, *23*, 11–28. [CrossRef]

31. Ma, L.; Chen, H.J.; Gan, K.W. *Tectonics and Marine Petroleum Geology in South China*; Geology Press: Beijing, China, 2004; pp. 37–41.

32. Hao, F.; Zou, H.; Lu, Y. Mechanisms of shale gas storage: Implications for shale gas exploration in China. *AAPG Bull.* **2013**, *97*, 1325–1346. [CrossRef]

33. Xu, H.; Zhou, W.; Cao, Q.; Xiao, C.; Zhou, Q.; Zhang, H.; Zhang, Y. Differential fluid migration behaviour and tectonic movement in Lower Silurian and Lower Cambrian shale gas systems in China using isotope geochemistry. *Mar. Pet. Geol.* **2018**, *89*, 47–57. [CrossRef]

34. Liu, S.G.; Qin, C.; Sun, W.; Wang, G.Z.; Xu, G.S.; Zhang, Z.J.; Zhang, Z.J. The coupling formation process of four centers of hydrocarbon in Sinian Dengying Formation of Sichuan Basin. *Acta Petrol. Sin.* **2012**, *28*, 879–888.

35. Xu, H.; Wei, G.; Jia, C.; Yang, W.; Zhou, T.; Xie, W.; Li, C.; Luo, B. Tectonic evolution of the Leshan-Longnüsi paleo-uplift and its control on gas accumulation in the Sinian strata. *Pet. Explor. Dev.* **2012**, *39*, 436–446. [CrossRef]

36. Feng, W.P.; Wang, F.Y.; Guan, J.; Zhou, J.X.; Wei, F.B.; Dong, W.J.; Xu, Y.F. Geologic structure controls on initial productions of lower Silurian Longmaxi shale in south China. *Mar. Pet. Geol.* **2018**, *91*, 163–178. [CrossRef]

37. Zhai, G.M.; Gao, W.L.; Song, J.G.; An, Z.X.; Cheng, K.M.; Dai, J.X.; Guan, D.S.; Liu, F.H.; Li, J.C.; Qiu, Y.N.; et al. *Petroleum Geology of China—Volume 10, Oil and Gas Zone in Sichuan Basin*; Petroleum Industry Press: Beijing, China, 1989; pp. 389–424.

38. Wang, S.Q.; Chen, G.S.; Dong, D.Z.; Yang, G.; Lv, Z.G.; Xu, H.Y.; Huang, Y.B. Accumulation conditions and exploitation prospect of shale gas in the Lower Paleozoic Sichaun Basin. *Nat. Gas Ind.* **2009**, *29*, 51–58. [CrossRef]

39. Huang, J.L.; Zou, C.N.; Li, J.Z.; Dong, D.Z.; Wang, S.J.; Wang, S.Q.; Wang, Y.M.; Li, D.H. Shale gas accumulation conditions and favorable zones of Silurian Longmaxi Formation in south Sichuan Basin. *China J. China Coal Soc.* **2012**, *37*, 782–787.

40. Dong, D.; Gao, S.; Huang, J.; Guan, Q.; Wang, S.; Wang, Y. Discussion on the exploration & development prospect of shale gas in the Sichuan Basin. *Nat. Gas Ind. B* **2015**, *2*, 9–23. [CrossRef]

41. Gao, B. Geochemical characteristics of shale gas from lower Silurian Longmaxi formation in the Sichuan Basin and its geological significance. *J. Nat. Gas Geosci.* **2015**, *26*, 1173–1182. [CrossRef]

42. Schoell, M. The hydrogen and carbon isotopic composition of methane from natural gases of various origins. *Geochim. Cosmochim. Acta* **1980**, *44*, 649–661. [CrossRef]

43. Schoell, M. Genetic Characterization of Natural Gases. *AAPG Bull.* **1983**, *67*, 2225–2238. [CrossRef]

44. Tissot, B.P.; Pelet, R.; Ungerer, P. Thermal history of sedimentary basins, maturation indices, and kinetics of oil and gas generation. *AAPG Bull.* **1987**, *71*, 1445–1466. [CrossRef]

45. Wang, X.F.; Li, X.F.; Wang, X.Z.; Shi, B.G.; Luo, X.R.; Zhang, L.X.; Lei, Y.H.; Jiang, C.F.; Meng, Q. Carbon isotopic fractionation by desorption of shale gases. *Mar. Pet. Geol.* **2015**, *60*, 79–86. [CrossRef]

46. Hill, R.J.; Zhang, E.; Katz, B.J.; Tang, Y. Modeling of gas generation from the Barnett Shale, Fort Worth Basin, Texas. *AAPG Bull.* **2007**, *91*, 501–521. [CrossRef]

47. Prinzhofer, A.A.; Huc, A.Y. Genetic and post-genetic molecular and isotopic fractionations in natural gases. *Chem. Geol.* **1995**, *126*, 281–290. [CrossRef]

48. Tang, Y.; Perry, J.; Jenden, P.; Schoell, M. Mathematical modeling of stable carbon isotope ratios in natural gases. *Geochim. Cosmochim. Acta* **2000**, *64*, 2673–2687. [CrossRef]

49. Cramer, B.; Faber, E.; Gerling, P.; Krooss, B.M. Reaction Kinetics of Stable Carbon Isotopes in Natural GasInsights from Dry, Open System Pyrolysis Experiments. *Energy Fuels* **2001**, *15*, 517–532. [CrossRef]

50. Zou, Y.-R.; Wang, L.; Shuai, Y.; Peng, P. EasyDelta: A spreadsheet for kinetic modeling of the stable carbon isotope composition of natural gases. *Comput. Geosci.* **2005**, *31*, 811–819. [CrossRef]

51. Lewan, M. Experiments on the role of water in petroleum formation. *Geochim. Cosmochim. Acta* **1997**, *61*, 3691–3723. [CrossRef]

52. Burruss, R.; Laughrey, C. Carbon and hydrogen isotopic reversals in deep basin gas: Evidence for limits to the stability of hydrocarbons. *Org. Geochem.* **2010**, *41*, 1285–1296. [CrossRef]

53. Cheng, B.; Xu, J.B.; Qian, D.; Liao, Z.W.; Wang, Y.P.; Faboya, O.L.; Li, S.D.; Liu, J.Z.; Peng, P.A. Methane cracking within shale rocks: A new explanation for carbon isotope reversal of shale. *Mar. Pet. Geol.* **2020**, *121*, 104591. [CrossRef]

54. Price, L.C. Chapter H—A Possible Deep-Basin High-Rank Gas Machine via Water Organic-Matter Redox Reactions. In *U.S. Geological Survey Digital Data Series DDS-67*; Version 1.00.; Dyman, T.S., Kuuskraa, V.A., Eds.; U.S. Department of the Interior: Washington, DC, USA; U.S. Geological Survey: Reston, VA, USA, 2001; pp. 1–29. Available online: http://pubs.usgs.gov/dds/dds-067/CHH.pdf (accessed on 16 January 2001).

55. Tissot, P.B.P.; Welte, P.D.H. *Petroleum Formation and Occurrence*, 2nd ed.; Springer: Berlin, Germany, 1984.

56. Rooney, M.A.; Claypool, G.E.; Chung, H.M. Modeling thermogenic gas generation using carbon isotope ratios of natural gas hydrocarbons. *Chem. Geol.* **1995**, *126*, 219–232. [CrossRef]

57. Zhou, Q.; Xiao, X.; Tian, H.; Pan, L. Modeling free gas content of the Lower Paleozoic shales in the Weiyuan area of the Sichuan Basin, China. *Mar. Pet. Geol.* **2014**, *56*, 87–96. [CrossRef]

58. Xiao, X.M.; Tian, H.; Liu, D.H.; Liu, Z.F. Evaluation of Deep Burial Source Rocks and Their Natural Gas Potentials in Sichuan Basin. In *Annual Report of State Major Research Program of China (Project No. 2011ZX05008-002-40), Internal Report*; Guangzhou Institute of Geochemistry of Chinese Academy of Sciences: Guangzhou, China, 2012; pp. 1–68.

59. Jiang, Q.; Qiu, N.; Zhu, C. Heat flow study of the Emeishan large igneous province region: Implications for the geodynamics of the Emeishan mantle plume. *Tectonophysics* **2018**, 11–27. [CrossRef]

60. Xu, Y.G.; He, B.; Huang, X.L.; Luo, Z.Y.; Zhu, D.; Ma, J.L.; Shao, H. Late Permian Emeishan Flood Basalts in Southwestern China. *Earth Sci. Front.* **2007**, *14*, 1–9. [CrossRef]

61. Yang, C.; Ni, Z.-Y.; Li, M.; Wang, T.; Chen, Z.; Hong, H.; Tian, X. Pyrobitumen in South China: Organic petrology, chemical composition and geological significance. *Int. J. Coal Geol.* **2018**, *188*, 51–63. [CrossRef]

62. Zhu, C.-Q.; Tian, Y.-T.; Xu, M.; Rao, S.; Yuan, Y.-S.; Zhao, Y.-Q.; Hu, S.-B. The Effect of Emeishan Supper Mantle Plume on the Thermal Evolution of Hydrocarbon Source Rocks in the Sichuan Basin. *Chin. J. Geophys.* **2010**, *53*, 83–91. [CrossRef]

63. Wu, X.; Dai, J.; Liao, F.; Huang, S. Origin and source of CO_2 in natural gas from the eastern Sichuan Basin. *Sci. China Earth Sci.* **2013**, *56*, 1308–1317. [CrossRef]

64. Feng, Z.Q.; Hao, F.; Zhou, S.; Wu, W.; Tian, J.; Xie, C.; Cai, Y. Pore characteristics and methane adsorption capacity of different lithofacies of the Wufeng formation-Longmaxi formation shales, Southern Sichuan Basin. *Energ. Fuel.* **2020**, *34*, 8046–8062. [CrossRef]

65. Arneth, J.-D.; Matzigkeit, U. Variations in the carbon isotope composition and production yield of various pyrolysis products under open and closed system conditions. *Org. Geochem.* **1986**, *10*, 1067–1071. [CrossRef]

66. Horsfield, B.; Schenk, H.; Mills, N.; Welte, D. An investigation of the in-reservoir conversion of oil to gas: Compositional and kinetic findings from closed-system programmed-temperature pyrolysis. *Org. Geochem.* **1992**, *19*, 191–204. [CrossRef]

67. Berner, U.; Faber, E.; Scheeder, G.; Panten, D. Primary cracking of algal and landplant kerogens: Kinetic models of isotope variations in methane, ethane and propane. *Chem. Geol.* **1995**, *126*, 233–245. [CrossRef]

68. Chen, Y.X.; Tian, C.T.; Li, K.N.; Cui, X.Y.; Wu, Y.Q.; Xia, Y.Q. Influence of thermal maturity on carbon isotopic composition of individual aromatic hyrocarbons during anhydrous closed-system pyrolysis. *Fuel* **2016**, *186*, 466–475. [CrossRef]

69. Curtis, J.B. Fractured shale-gas systems. *AAPG Bull.* **2002**, *86*, 1921–1938. [CrossRef]

70. Wei, S.L.; He, S.; Pan, Z.J.; Guo, X.W.; Yang, R.; Dong, T.; Yang, W.; Gao, J. Models of shale gas storage capacity during burial and uplift: Application to Wufeng-Longmaxi shales in the Fuling shale gas field. *Mar. Pet. Geol.* **2019**, *109*, 233–244. [CrossRef]

71. Huang, H.Y.; He, D.F.; Li, Y.Q.; Li, J.; Zhang, L. Silurian tectonic-sedimentary setting and basin evolution in the Sichuan area, southwest China: Implications for palaeogeographic reconstructions. *Mar. Pet. Geol.* **2018**, *92*, 403–423. [CrossRef]

72. Jiang, H.; Wang, Z.C.; Du, H.Y.; Zhang, C.M.; Wang, R.J.; Zou, N.N.; Wang, T.S.; Gu, Z.D.; Li, Y.X. Tectonic evolution of the Leshan-Longnvsi paleo-uplift and reservoir formation of Neoproterozoic Sinian gas. *Nat. Gas Geosci.* **2014**, *25*, 192–200. [CrossRef]

73. Zhang, B.; Xiao, D.; Wang, X.; Zhao, L.; Luo, S.; Yang, X. Sedimentary characteristics and distribution patterns of grain shoals in the Lower Cambrian Longwangmiao Formation, southern Sichuan Basin, SW China. *Arab. J. Geosci.* **2018**, *11*, 135. [CrossRef]

74. Wei, L.; Haiyong, Y.; Wangshui, H.; Geng, Y.; Xuan, X.; Li, W.; Yi, H.; Hu, W.; Xiong, X. Tectonic evolution of Caledonian Palaeohigh in the Sichuan Basin and its relationship with hydrocarbon accumulation. *Nat. Gas Ind. B* **2014**, *1*, 58–65. [CrossRef]

75. Prinzhofer, A. Pernaton, Éric Isotopically light methane in natural gas: Bacterial imprint or diffusive fractionation? *Chem. Geol.* **1997**, *142*, 193–200. [CrossRef]

76. Ballentine, C.J.; O'Nions, R.; Coleman, M. A Magnus opus: Helium, neon, and argon isotopes in a North Sea oilfield. *Geochim. Cosmochim. Acta* **1996**, *60*, 831–849. [CrossRef]

77. Fuex, A. Experimental evidence against an appreciable isotopic fractionation of methane during migration. *Phys. Chem. Earth* **1980**, *12*, 725–732. [CrossRef]

78. Bernard, B.B.; Brooks, J.M.; Sackett, W.M. Light hydrocarbons in recent Texas continental shelf and slope sediments. *J. Geophys. Res. Space Phys.* **1978**, *83*, 4053–4061. [CrossRef]

79. Whiticar, M.J. A geochemial perspective of natural gas and atmospheric methane. *Org. Geochem.* **1990**, *16*, 531–547. [CrossRef]

80. Zhang, X.; Guan, R.-F.; Wu, D.-Q.; Chan, K.Y. Enzyme immobilization on amino-functionalized mesostructured cellular foam surfaces, characterization and catalytic properties. *J. Mol. Catal. B Enzym.* **2005**, *33*, 43–50. [CrossRef]

81. Salvi, S.; Williams-Jones, A.E. Fischer-Tropsch synthesis of hydrocarbons during sub-solidus alteration of the Strange Lake peralkaline granite, Quebec/Labrador, Canada. *Geochim. Cosmochim. Acta* **1997**, *61*, 83–99. [CrossRef]
82. Potter, J.; Salvi, S.; Longstaffe, F.J. Abiogenic hydrocarbon isotopic signatures in granitic rocks: Identifying pathways of formation. *Lithos* **2013**, 114–124. [CrossRef]
83. Tang, Y.C.; Xia, X.Y. Quantitative Assessment of Shale Gas Potential Based on Its Special Generation and Accumulation Processes. In Proceedings of the AAPG Annual Convention and Exhibition, Houston, TX, USA, 10–13 April 2011. Available online: http://www.searchanddiscovery.com/pdfz/documents/2011/40819tang/ndx_tang.pdf.html (accessed on 31 October 2011).
84. Chen, S.; Zhu, Y.; Wang, H.; Liu, H.; Wei, W.; Fang, J. Shale gas reservoir characterisation: A typical case in the southern Sichuan Basin of China. *Energy* **2011**, *36*, 6609–6616. [CrossRef]
85. Chen, L.; Lu, Y.; Jiang, S.; Li, J.; Guo, T.; Luo, C. Heterogeneity of the Lower Silurian Longmaxi marine shale in the southeast Sichuan Basin of China. *Mar. Pet. Geol.* **2015**, *65*, 232–246. [CrossRef]
86. Tang, X.; Jiang, Z.; Li, Z.; Gao, Z.; Bai, Y.; Zhao, S.; Feng, J. The effect of the variation in material composition on the heterogeneous pore structure of high-maturity shale of the Silurian Longmaxi formation in the southeastern Sichuan Basin, China. *J. Nat. Gas Sci. Eng.* **2015**, *23*, 464–473. [CrossRef]
87. Tuo, J.; Wu, C.; Zhang, M. Organic matter properties and shale gas potential of Paleozoic shales in Sichuan Basin, China. *J. Nat. Gas Sci. Eng.* **2016**, *28*, 434–446. [CrossRef]
88. Yan, D.; Wang, H.; Fu, Q.; Chen, Z.; He, J.; Gao, Z. Geochemical characteristics in the Longmaxi Formation (Early Silurian) of South China: Implications for organic matter accumulation. *Mar. Pet. Geol.* **2015**, *65*, 290–301. [CrossRef]
89. Li, Y.F.; Shao, D.Y.; Lv, H.G.; Zhang, Y.; Zhang, X.L.; Zhang, T.W. A relationship between elemental geochemical characteristics and organic matter enrichment in marine shale of Wufeng Formation-Longmaxi Formation, Sichuan Basin. *Acta Pet. Sin.* **2015**, *36*, 1470–1483. [CrossRef]
90. Yang, M.-X.; Qu, X.; Liu, F.-L.; Zheng, G.-J. HCA520, a novel tumor associated antigen, involved in cell proliferation and apoptosis. *Chin. J. Cancer Res.* **2003**, *15*, 282–285. [CrossRef]

Publisher's Note: MDPI stays neutral with regard to jurisdictional claims in published maps and institutional affiliations.

Article

Quantitative Analysis of Amorphous Silica and Its Influence on Reservoir Properties: A Case Study on the Shale Strata of the Lucaogou Formation in the Jimsar Depression, Junggar Basin, China

Ke Sun [1,*], Qinghua Chen [1], Guohui Chen [2,3], Yin Liu [1] and Changchao Chen [4]

[1]   School of Geosciences, China University of Petroleum (East China), Qingdao 266580, China; chenqhua@upc.edu.cn (Q.C.); liuyin@upc.edu.cn (Y.L.)

[2]   School of Earth Resources, China University of Geosciences, Wuhan 430074, China; chenguohui@cug.edu.cn

[3]   Key Laboratory of Theory and Technology of Petroleum Exploration and Development in Hubei Province, China University of Geosciences, Wuhan 430074, China

[4]   Exploration and Development Research Institite, PetroChina Tarim Oilfield Company, Korla 841000, China; ccccup1989@sina.com

*   Correspondence: sunke@s.upc.edu.cn; Tel.: +86-185-6065-3166

Received: 30 October 2020; Accepted: 23 November 2020; Published: 24 November 2020

Abstract: To establish a new quantitative analysis method for amorphous silica content and understand its effect on reservoir properties, the amorphous silica (SiO_2) in the shale strata of the Lucaogou Formation in the Jimsar Depression was studied by scanning electron microscopy (SEM) observation, X-ray diffraction (XRD), and X-ray fluorescence spectrometry (XRF). Amorphous silica shows no specific morphology, sometimes exhibits the spherical or ellipsoid shapes, and usually disorderly mounds among other mineral grains. A new quantitative analysis method for observing amorphous SiO_2 was established by combining XRD and XRF. On this basis, while the higher content of amorphous SiO_2 lowers the porosity of the reservoir, the permeability shows no obvious changes. The higher the content of amorphous SiO_2, the lower the compressive strength and Young's modulus and the lower the oil saturation. Thus, amorphous SiO_2 can reduce the physical properties of reservoir rocks and increase the reservoir plasticity, which is not only conducive to the enrichment of shale oil but also increases the difficulty of fracturing in later reservoir development.

Keywords: amorphous SiO_2; X-ray diffraction; X-ray fluorescence spectrometry; scanning electron microscope; quantitative analysis; reservoir properties

---

1. Introduction

The success of shale gas exploration and development in North America has promoted the development of the shale gas industry around the world. At present, successful exploration and development of shale gas in China is mainly concentrated in the Sichuan Basin and surrounding areas, such as the Weiyuan, Zhaotong, Zhengan, and Jiaoshiba areas [1–3]. The shale sections containing commercial scale gas in these areas are located at the top of the Wufeng and the bottom of the Longmaxi Formations, corresponding to the 2-3 graphitic biozones of the Wufeng Formation and the 1-4 graphitic biozones of the Longmaxi Formation [4,5]. These high-quality shale sections contain high content of silica: as much as 60% [6–11]. Although there are different opinions about the evidence of biogenesis, most researchers consider that the silica in these high-quality shale sections has biogenic sources [12–17]. Shale oil sources are mainly concentrated in basins in China, where lacustrine shale is widely developed, such as the Ordos, Songliao, and Bohai Bay Basins. Shale oil exploration has been particularly successful in the second member of the Kongdian Formation in

Cangdong Depression of Bohai Bay Basin, where commercial-scale oil has been obtained in several wells [18]. This quartz-feldspathic shale exhibits good quality, high TOC content, and high hydrocarbon potential [18]. The tuffaceous shale sections of the Lucaogou Formation shale strata in the oil reservoir of the Malang Depression contain high total organic carbon and exhibit high hydrocarbon generation potential. These tuffaceous shales are also mainly composed of quartz and feldspar [19]. In both marine shale gas and or lacustrine shale oil reservoirs, silica is an important component, having a significant impact on shale reservoir properties, organic matter enrichment, shale oil and gas accumulation, and fracturing potential [14,15,20–25]. Hence, silica is a hot spot in shale reservoir research at present.

Studies on silica diagenesis have shown that the end members are amorphous SiO_2 and crystalline quartz. The quartz can be further divided into authigenic quartz formed during diagenesis and detrital quartz from deposition. During diagenesis, amorphous SiO_2 will gradually change from the amorphous state (opal-A) to the cryptocrystalline state and finally to the fully crystalline state (α-quartz), which is authigenic quartz. Amorphous SiO_2 can be from biological organisms or an abioticearly diagenesis stage. Some studies suggest that amorphous SiO_2 has already transformed into crystalline quartz during early diagenesis (Ro is 0.35%~0.5%) [23,24]. Others suggest that the conversion of amorphous SiO_2 to crystalline quartz in shale reservoirs may be much later, because amorphous SiO_2 has been seen in the middle diagenetic stage A (Ro is 0.5%~1.3%) [21]. During clay mineral conversion, a large amount of silica is generated, and its content is closely related to mineral composition, crystallinity, and thermal conditions; it also affects the physical properties and brittleness of the reservoir [26–28]. The influence of amorphous SiO_2 on reservoir properties can make a large difference in different evolution stages. From the beginning of diagenesis to the cryptocrystalline state, formation porosity has been shown to be reduced from about 45% to less than 25%, and the permeability declines to be difficult to be measured [29]. In the authigenic quartz stage, reservoir physical properties and brittleness increases instead, which improves reservoir fracturability [29]. Thus, it can be seen that amorphous SiO_2 also plays a great impact on reservoir properties. If the influence of amorphous SiO_2 on reservoirs can be clarified, it will be of great significance for evaluating shale oil reservoirs and fracturing potential, especially for immature lacustrine shale oil reservoirs.

Accurate calculation of amorphous SiO_2 content is the key problem to understand the influence of amorphous SiO_2 on reservoir properties. There are currently four methods for the quantitative analysis of amorphous SiO_2 in heterogeneous systems. The first is chemical dissolution, which means removing minerals other than amorphous SiO_2. However, chemical dissolution incudes crystalline, which affects the accuracy of quantitative analysis. The second is quantitative analysis using XRD as proposed by Lin (1997) [30]. Although the method is correct in theory, human error enters into in the quantification [31]. Thirdly, Chu (1998) proposed a new quantitative XRD method based on the increment method proposed by Popović et al. (1983) [32,33], but this method required preparation of a standard sample having a known mineral composition and proportions; the error was relatively large in the actual experiment. Fourth, Huang et al. (2015) established a calculation method for amorphous SiO_2 in the Yanchang Formation shale of the Ordos Basin by using XRD combined with QEMSCAN analysis [34]. However, this method has two disadvantages. Firstly, it is too expensive to conduct large-scale tests. Secondly, the mineral composition obtained by QEMSCAN analysis can be understood as a volume percentage. Hence, it needs to be converted into a mass percentage, but the density of minerals was not determined in Huang et al. (2015) [34].

In view of the shortcomings of previous methods for calculating the content of amorphous silica [30–34], a new quantitative analysis method for amorphous silica content was established in this research, based on XRD and XRF analysis of core samples from the Lucaogou Formation in the Jimsar Depression. Through the analysis of the relationship between physical parameters, rock mechanical parameters, oil saturation, and amorphous silica content in shale strata, the effect of amorphous SiO_2 on reservoir properties and its geological significance was determined.

2. Geological Settings

The Junggar Basin is located in the northwestern part of China with an area of about 1.30×10^5 km^2 (Figure 1A); it is geotectonically located at the intersection of Kazakhstan, Siberian, and Tarim plates. The Jimusar Depression is in the southeast of Junggar Basin, covering an area of 1.278×10^3 km^2; it is surrounded by the Shaqi Uplift to the north, the Guxi Uplift to the east, the Fukang faults zone to the south, and the Santai Uplift to the west (Figure 1B). The periphery of the Jimsar Depression is bounded by six faults (Figure 1B). The Permian Lusaogou Formation has a thickness of 200~350 m and is in conformable contact with the lower Jingjingzigou Formation and in unconformable contact with the upper Wutonggou Formation (Figure 1C). The Lucaogou Formation is mainly composed of deep and semideep lake facies formed of fine-grained, mixed sedimentary rocks [35,36]. It was formed in an intracontinental rifted saline lake basin environment, accompanied by volcanic eruptions and hydrothermal activity [37,38]. Since September 2011, J25, J23, J28, J30, and other exploration and evaluation wells have been successively drilled in the Jimusar Depression, oil testing shows industrial potential, and shale oil was discovered in the Lucaogou Formation. After nine years of development, the calculated reserves of shale reservoir have reached 11.12×10^8 t [39].

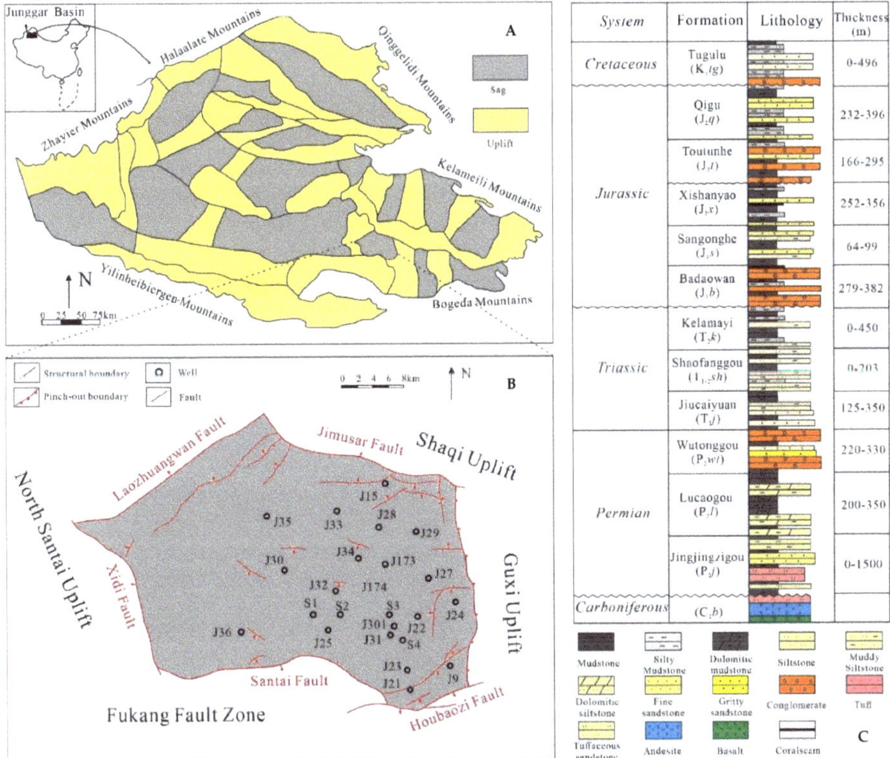

System	Formation	Lithology	Thickness (m)
Cretaceous	Tugulu (K$_1$tg)		0-496
Jurassic	Qigu (J$_3$q)		232-396
	Toutunhe (J$_2$t)		166-295
	Xishanyao (J$_1$x)		252-356
	Sangonghe (J$_1$s)		64-99
	Badaowan (J$_1$b)		279-382
Triassic	Kelamayi (T$_3$k)		0-450
	Shaofanggou (T$_{1-2}$sh)		0-203
	Jiucaiyuan (T$_1$j)		125-350
Permian	Wutonggou (P$_3$wt)		220-330
	Lucaogou (P$_2$l)		200-350
	Jingjingzigou (P$_2$j)		0-1500
Carboniferous	(C$_2$b)		

Figure 1. Diagrams showing (**A**) Junggar tectonic units and location of the Jimusar Depression, (**B**) structure and well location map of the Jimsar Depression, and (**C**) the stratigraphic sequence from Upper Caboniferous to Lower Cretaceous in the Jimsar Depression (modified from [40]).

3. Materials and Methods

3.1. Materials

The samples of the Lucaogou Formation in this study are from four cored wells (S1–S4) in the Jimsar Depression (Figure 1B). We selected 42 samples that met experimental needs. Their lithology includes tuffaceous shale (also called siliceous shale), shale, dolomite, and dolomitic mudstone. Generally, the lithology can be divided into three lithofacies: tuffaceous shale lithofacies, transitional lithofacies (also called mixed lithofacies), and carbonate lithofacies [41,42].

3.2. Experimental Method

XRD analysis was completed at Sichuan Keyuan Engineering Technology Testing Center. XRF analysis, rock mechanics experiments, and reservoir physical properties analysis were completed at the Experimental Research Center of East China Oil and Gas Branch of Sinopec.

3.2.1. XRD and XRF Analysis

The mineral composition of samples was obtained by XRD, which was determined on the premise of deducting background values through the Jade 5.0 software package. The principle of XRD analysis is that different minerals show different XRD diffraction effects. Data calculated by the XRD accurately represents the relative content of each mineral. However, XRD cannot measure the content of amorphous silica because it shows no diffraction peaks.

The secondary X-rays were emitted when the X-ray irradiated on the material. Different elements show their specific secondary X-ray with certain features or wavelength characteristics. XRF analysis uses secondary X-rays to convert the data into specific elements and their abundance. Elemental Si occurs in quartz, plagioclase, k-feldspar, clay minerals, and amorphous silica.

3.2.2. Rock Mechanics Experiment

Samples were tested using a TAW-2000 computer-controlled electrohydraulic servo testing machine under constant confining pressure conditions. The size of test samples is 25 mm (diameter) × 50 mm (length). In the process of testing, strain rate was controlled by the DUOLI microcomputer control system, mostly 0.01–0.03, which was convenient to obtain smooth stress–strain curves. The compressive strength, Young's modulus, and Poisson's ratio can be calculated by the stress–strain curves.

3.2.3. Reservoir Physical Properties

The total porosity was obtained by calculating the difference between the bulk density and the skeleton density. Permeability was obtained by calculating the expansion of He with increasing pressure (5 MPa–30 Mpa) at a constant temperature. Oil saturation was measured by nuclear magnetic resonance (NMR).

3.3. A New Method for Calculating the Content of Amorphous SiO_2

In this study, a new method for quantitative analysis of amorphous SiO_2 in the Lucaogou Formation of the Jimusar Depression was established by using a combination of XRD and XRF. Through XRD analysis, the shale strata mainly consist of quartz, plagioclase, potash feldspar, dolomite, calcite, pyrite, and clay minerals (Figure 2A). Elemental Si is in quartz, plagioclase, potash feldspar, and clay minerals.

The combination of XRD and XRF can calculate amorphous silica as follows. Suppose the sample mass is M, where the mass of amorphous SiO_2, quartz, plagioclase, K-feldspar, and clay minerals are respectively represented by m_{SiO_2}, m_{quartz}, $m_{plagioclase}$, $m_{K-feldspar}$, and m_{clay}.

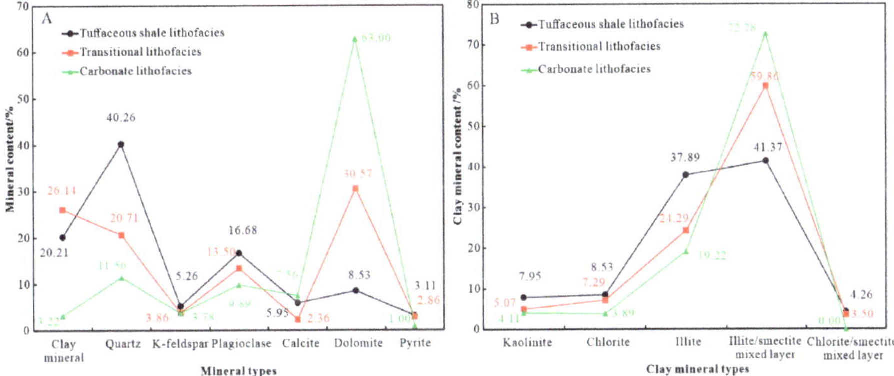

Figure 2. Mineral composition of different lithofacies samples in the Lucaogou Formation. (**A**) The mineral content of the different lithofacies; (**B**) clay mineral composition in the different lithofacies.

According to XRD analysis:

$$\frac{m_{quartz}}{M - m_{SiO_2}} = W_{quartz}. \tag{1}$$

$$\frac{m_{plagioclase}}{M - m_{SiO_2}} = W_{plagioclase} \tag{2}$$

$$\frac{m_{K-feldspar}}{M - m_{SiO_2}} = W_{K-feldspar} \tag{3}$$

$$\frac{m_{clay}}{M - m_{SiO_2}} = W_{clay} \tag{4}$$

The W_{quartz}, $W_{plagioclase}$, $W_{K-feldspar}$, and W_{clay} represent the percentage of quartz, plagioclase, k-feldspar, and clay minerals measured by XRD analysis.

According to XRF analysis:

$$\frac{m_{SiO_2} \times P_{Si-SiO_2} + m_{quartz} \times P_{Si-quartz} + m_{plagioclase} \times P_{Si-plagioclase}}{M} + $$
$$\frac{m_{K-feldspar} \times P_{Si-K-feldspar} + m_{Clay} \times P_{Si-clay}}{M} = W_{Si} \tag{5}$$

The mass percentages of Si in amorphous SiO_2, quartz, plagioclase, k-feldspar, clay minerals, and the sample are represented by P_{Si-SiO_2}, $P_{Si-quartz}$, $P_{Si-plagioclase}$, $P_{Si-K-feldspar}$, $P_{Si-clay}$, and W_{Si}, respectively.

Placing Formulas (1)–(4) into Formula (5), thus creating Formula (6)

$$\frac{m_{SiO_2} \times P_{Si-SiO_2} + W_{quartz} \times (M - m_{SiO_2}) \times P_{Si-quartz} + W_{plagioclase} \times (M - m_{SiO_2}) \times P_{Si-plagioclase}}{M} + $$
$$\frac{W_{K-feldspar} \times (M - m_{SiO_2}) \times P_{Si-K-feldspar} + W_{clay} \times (M - m_{SiO_2}) \times P_{Si-clay}}{M} = W_{Si} \tag{6}$$

Formula (6) can be changed to Formula (7):

$$W_{SiO_2} = \frac{m_{SiO_2}}{M} = $$
$$\frac{W_{Si} - W_{quartz} \times P_{Si-quartz} - W_{plagioclase} \times P_{Si-plagioclase} - W_{K-feldspar} \times P_{Si-K-feldspar} - W_{clay} \times P_{Si-clay}}{P_{Si-SiO_2} - W_{quartz} \times P_{Si-quartz} - W_{plagioclase} \times P_{Si-plagioclase} - W_{K-feldspar} \times P_{Si-K-feldspar} - W_{clay} \times P_{Si-clay}} \tag{7}$$

In Formula (7), only the mass percentage of element Si in clay minerals is difficult to determine, because the molecular formulas of other minerals are known. The molecular formulas of clay minerals are variable. Therefore, the ideal molecular formulas of different types of clay minerals are applied in this research. For the mass percentage of Si in mixed clay minerals, it is calculated according to the mixed layer ratio based on XRD measurements. Molecular formulas used for kaolinite, montmorillonite, chlorite, and illite are respectively $Al_4(Si_4O_{10})(OH)_8$, $Al_4Si_8O_2(OH)_2$, $Al_6Si_4O_{10}(OH)_8$, and $Al_4(Si_8O_{20})(OH)_4$. The mass percentages of element Si in these are 21.7%, 56.3%, 19.6%, and 31.1%, respectively. The P_{clay} of the tuffaceous shale lithofacies, transitional lithofacies, and carbonate lithofacies samples can be calculated. Then, the contents of amorphous SiO_2 in these samples can be calculated by Formula (7).

4. Results

4.1. Occurrence and Characteristics of Amorphous SiO_2

The shale strata of the Lucaogou Formation in the Jimusar Depression can be divided into tuffaceous shale lithofacies, transitional lithofacies, and carbonate lithofacies [41,42]. The tuffaceous shale lithofacies is mainly composed of feldspathic minerals including quartz and feldspar. The carbonate lithofacies mainly consists of dolomite and includes dolomite and argillaceous dolomite. The mineral composition and lithology of the transitional lithofacies is primarily a hybrid of the other two lithofacies. It can be seen by SEM that in addition to the development of authigenic quartz in the shale strata (Figure 3A,B), amorphous SiO_2 is also present (Figure 3C–H). Amorphous SiO_2 shows no fixed form and usually fills randomly between mineral grains (Figure 3C–E). Some of the amorphous SiO_2 was wrapped in tuffaceous components (Figure 3F), and other forms were spherical or ellipsoid shapes having varying sizes (Figure 3G,H).

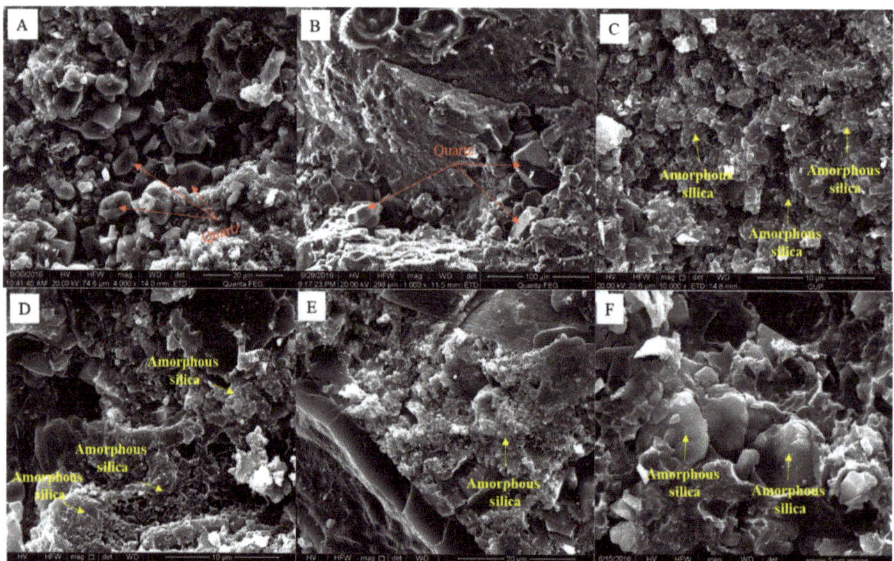

Figure 3. *Cont.*

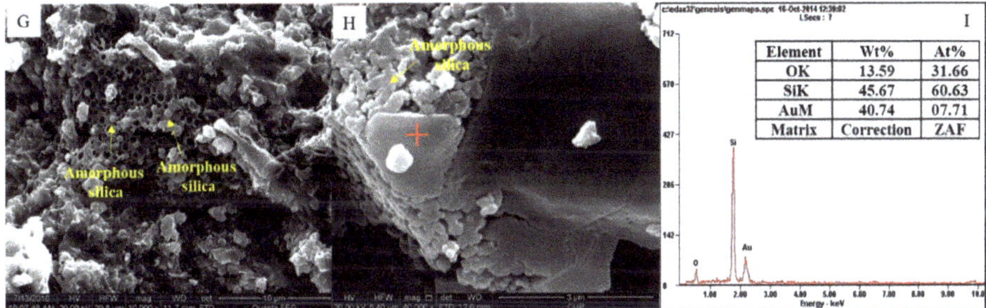

Figure 3. SEM images of quartz and amorphous silica in the Lucaogou Formation, Jimsar Depression. (**A**) Transitional lithofacies, S1 well, 3147.64 m; (**B**) tuffaceous shale lithofacies, S2 well, 3348.08 m; (**C**) transitional lithofacies, S1 well, 3147.64 m; (**D**) tuffaceous shale lithofacies, S2 well, 3343.00 m; (**E**) tuffaceous shale lithofacies, S2 well, 3348.08 m; (**F**) transitional lithofacies, S2 well, 3359.95 m; (**G**) tuffaceous shale lithofacies, S3 well, 2815.21 m; (**H**) tuffaceous shale lithofacies, S4 well, 2601.81 m; (**I**) energy spectrum analysis of point "+" in image H.

4.2. Composition Characteristics of Crystalline Minerals

Analysis of the XRD test results (Table 1) shows that the tuffaceous shale lithofacies samples exhibit the highest content of quartz-feldspathic minerals. The average content of quartz is as much as 40.26%; the average content of plagioclase and k-feldspar are as much as 16.68% and 5.26% respectively (Figure 2A). The carbonate lithofacies samples show the highest content of dolomite, reaching 63% on average. The transitional lithofacies samples present the highest content of clay minerals, which is as much as 26.14% (Figure 2A). In clay minerals, the content of the illite/smectite mixed layer is the highest, followed by illite. The average contents of the illite/smectite mixed layer in tuffaceous shale lithofacies, transitional lithofacies, and carbonate lithofacies are 41.37%, 59.86%, and 72.78%, respectively (Figure 2B). The tuffaceous lithofacies show the highest content of illite (average 37.89%), followed by transitional lithofacies (average 24.29%). The content of kaolinite, chlorite, and chlorite/smectite mixed layer is relatively low (Figure 2B).

4.3. Content of Amorphous SiO$_2$

Analysis of the XRF test results (Table 2) shows that the tuffaceous shale lithofacies samples have the highest content of Si, reaching 34.21% on average. As expected, the carbonate lithofacies samples exhibit the lowest content of Si, only 11.51% on average (Figure 4A). Moreover, the tuffaceous shale lithofacies samples also exhibit the highest values of Si in crystalline minerals calculated by the above method, reaching 33.18% on average (Figure 4A). According to the calculations, the shale strata of the Lucaogou Formation thereby contains a small amount of amorphous SiO$_2$. The tuffaceous shale lithofacies samples show the highest content of amorphous SiO$_2$, reaching an average of 7.07%, and the carbonate lithofacies samples show the lowest, only 1.52% (Figure 4A). Amorphous SiO$_2$ has a certain negative correlation with crystalline quartz (Figure 4B). During burial diagenesis, amorphous silica will gradually convert to crystalline quartz. The silica in the Lucaogou Formation is mainly derived from tuffaceous materials alteration in previous studies [17,21]. Therefore, the content of amorphous SiO$_2$ in the tuffaceous shale lithofacies sample is the highest among the three lithofacies. The content of silica in a sample is generally definite. Hence, the higher the content of crystalline quartz, the lower the content of amorphous SiO$_2$.

Table 1. Test data table of mineral composition of different lithofacies samples in the Luchaogou Formation.

Well	Depth (m)	Lithofacies	Clay Mineral/%	Quartz/%	K-Feldspar /%	Plagioclase/%	Calcite/%	Dolomite/%	Pyrite/%	Kaolinite/%	Chlorite/%	Illite/%	Illite/Smectite Mixed Layers/%	Chlorite/Smectite Mixed Layers/%	Illite/Smectite mixed Layer RATIO/%	Chlorite/Smectite Mixed Layers Ratio/%
S1	3063.80	Carbonate lithofacies	5	19	4	3	8	61	0	5	6	15	74	0	55	0
S1	3086.50	Carbonate lithofacies	2	10	4	5	0	79	0	0	0	50	50	0	56	0
S1	3136.85	Transitional lithofacies	28	22	0	13	3	34	0	8	9	16	67	0	70	0
S1	3147.64	Transitional lithofacies	31	21	4	8	0	33	3	25	29	14	32	0	44	0
S1	3149.00	Carbonate lithofacies	2	5	6	3	0	81	3	3	3	14	80	0	57	0
S2	3343.00	Tuffaceous shale lithofacies	22	23	5	22	0	25	3	0	0	64	36	0	40	0
S2	3315.00	Carbonate lithofacies	3	12	0	18	13	53	1	5	4	10	81	0	50	0
S2	3323.00	Transitional lithofacies	23	28	0	17	3	27	2	0	6	10	35	49	67	60
S2	3332.00	Carbonate lithofacies	2	13	0	18	5	60	2	9	6	9	76	0	49	0
S2	3344.00	Transitional lithofacies	17	17	9	12	7	35	3	0	11	24	65	0	56	0
S2	3347.00	Transitional lithofacies	39	26	0	10	10	12	3	0	7	15	78	0	55	0
S2	3348.08	Tuffaceous shale lithofacies	22	31	8	12	0	23	4	0	0	85	15	0	40	0
S2	3351.66	Carbonate lithofacies	2	19	0	10	24	45	0	4	4	32	60	0	52	0

Table 1. *Cont.*

Well	Depth (m)	Lithofacies	Clay Mineral/%	Quartz/%	K-Feldspar /%	Plagioclase/%	Calcite/%	Dolomite/%	Pyrite/%	Kaolinite/%	Chlorite/%	Illite/%	Illite/Smectite Mixed Layers/%	Chlorite/Smectite Mixed Layers/%	Illite/Smectite mixed Layer RATIO/%	Chlorite/Smectite Mixed Layers Ratio/%
S2	3353.14	Carbonate lithofacies	2	10	1	14	2	71	0	8	9	28	55	0	56	0
S2	3355.13	Transitional lithofacies	32	21	0	11	0	33	3	0	0	43	57	0	54	0
S2	3359.95	Transitional lithofacies	22	14	8	14	6	36	0	2	2	81	15	0	67	0
S2	3379.55	Transitional lithofacies	23	20	7	16	4	25	5	0	8	9	83	0	61	0
S2	3380.01	Tuffaceous shale lithofacies	16	48	11	20	0	0	5	12	15	0	73	0	90	0
S2	3463.00	Transitional lithofacies	27	15	6	14	0	36	2	0	0	38	62	0	51	0
S2	3477.00	Transitional lithofacies	26	20	9	14	0	28	3	0	7	13	80	0	65	0
S2	3600.34	Carbonate lithofacies	4	10	9	8	6	60	3	3	3	15	79	0	49	0
S3	2794.71	Tuffaceous shale lithofacies	18	40	4	21	12	4	1	13	15	9	63	0	87	0
S3	2795.80	Tuffaceous shale lithofacies	14	59	3	4	8	12	0	0	0	80	20	0	40	0
S3	2805.21	Transitional lithofacies	12	15	0	27	0	43	3	5	6	0	89	0	70	0
S3	2808.52	Carbonate lithofacies	7	6	10	10	10	57	0	0	0	0	100	0	52	0

Table 1. *Cont.*

Well	Depth (m)	Lithofacies	Clay Mineral/%	Quartz/%	K-Feldspar /%	Plagioclase/%	Calcite/%	Dolomite/%	Pyrite/%	Kaolinite/%	Chlorite/%	Illite/%	Illite/Smectite Mixed Layers/%	Chlorite/Smectite Mixed Layers/%	Illite/Smectite mixed Layer RATIO/%	Chlorite/Smectite Mixed Layers Ratio/%
S3	2815.21	Tuffaceous shale lithofacies	27	30	9	18	6	5	5	0	0	94	6	0	40	0
S3	2817.81	Tuffaceous shale lithofacies	18	54	0	17	4	4	3	11	12	8	69	0	89	0
S3	2832.89	Tuffaceous shale lithofacies	22	38	7	10	19	1	3	0	0	98	2	0	40	0
S3	2857.65	Tuffaceous shale lithofacies	22	40	13	15	3	2	5	0	0	95	5	0	40	0
S3	2869.14	Tuffaceous shale lithofacies	23	38	8	5	0	26	0	0	0	86	14	0	40	0
S3	2870.01	Tuffaceous shale lithofacies	19	37	10	15	5	11	3	23	18	0	59	0	85	0
S3	2871.01	Tuffaceous shale lithofacies	18	35	13	23	3	4	4	14	17	0	69	0	88	0
S3	2873.01	Tuffaceous shale lithofacies	18	36	0	26	6	11	3	15	14	6	65	0	83	0
S3	2874.01	Tuffaceous shale lithofacies	23	38	0	26	4	5	4	14	16	0	70	0	86	0

Table 1. *Cont.*

Well	Depth (m)	Lithofacies	Clay Mineral/%	Quartz/%	K-Feldspar/%	Plagioclase/%	Calcite/%	Dolomite/%	Pyrite/%	Kaolinite/%	Chlorite/%	Illite/%	Illite/Smectite Mixed Layers/%	Chlorite/Smectite Mixed Layers/%	Illite/Smectite mixed Layer RATIO/%	Chlorite/Smectite Mixed Layers Ratio/%
S3	2876.81	Tuffaceous shale lithofacies	22	41	0	19	6	9	3	8	7	4	0	81	0	50
S4	2580.81	Transitional lithofacies	25	20	0	16	0	37	2	12	11	49	28	0	40	0
S4	2585.01	Transitional lithofacies	21	15	8	14	0	37	5	4	6	28	62	0	59	0
S4	2591.91	Tuffaceous shale lithofacies	21	51	0	13	8	5	2	13	16	4	67	0	86	0
S4	2592.21	Tuffaceous shale lithofacies	26	36	0	26	0	9	3	16	18	6	60	0	86	0
S4	2601.21	Tuffaceous shale lithofacies	17	50	3	15	8	3	4	12	14	7	67	0	86	0
S4	2601.81	Tuffaceous shale lithofacies	16	40	6	10	21	3	4	0	0	74	26	0	40	0
S4	2607.60	Transitional lithofacies	40	36	3	3	0	12	6	15	0	0	85	0	60	0

Table 2. Statistical table of calculated silica content, porosity, permeability, mechanical properties, and oil saturation in different lithofacies samples of the Lucaogou Formation.

Well	Depth (m)	Lithofacies	Oil Saturation/%	Young's Modulus/N*mm^{-2}	Poisson's Ratio	Compressive Strength/Kg*cm^{-2}	Porosity/%	Permeability/mD	Si Content Test by XRF/%	Amorphous Silica Content Calculated Through the New Method/%	Calculated Si Content in Crystalline Minerals/%	Calculated Si Content in Clay Minerals/%
S1	3063.80	Carbonate lithofacies	17.6	–	–	–	–	–	13.535	1.017	13.184	40.196
S1	3086.50	Carbonate lithofacies	10.9	–	–	–	0.7041	0.0013	9.197	2.410	8.246	38.156
S1	3136.85	Transitional lithofacies	–	–	–	–	–	–	26.937	4.652	25.924	41.131
S1	3147.64	Transitional lithofacies	9.70	–	–	–	0.2367	0.0000923	24.348	6.921	22.612	28.963
S1	3149.00	Carbonate lithofacies	–	–	–	–	0.6041	0.00215	7.013	2.540	5.939	41.964
S2	3343.00	Tuffaceous shale lithofacies	5.80	–	–	–	0.5851	0.000043	29.234	11.885	26.743	34.728
S2	3315.00	Carbonate lithofacies	14.4	6.396	0.211	306.527	–	–	12.545	0.542	12.353	40.376
S2	3323.00	Transitional lithofacies	15.40	36.212	0.246	196.722	–	–	25.334	2.878	24.671	26.948
S2	3332.00	Carbonate lithofacies	1	16.381	0.353	368.663	1.1986	0.01	12.978	1.643	12.398	38.948
S2	3344.00	Transitional lithofacies	12.00	–	–	–	–	–	21.979	3.412	21.070	39.007
S2	3347.00	Transitional lithofacies	26.10	10.707	0.308	94.94	–	–	31.536	0.570	31.443	41.105
S2	3348.08	Tuffaceous shale lithofacies	–	40.386	0.356	324.954	–	–	30.056	10.484	27.990	32.612

Table 2. *Cont.*

Well	Depth (m)	Lithofacies	Oil Saturation/%	Young's Modulus/N*mm^{-2}	Poisson's Ratio	Compressive Strength/Kg*cm^{-2}	Porosity/%	Permeability/mD	Si Content Test by XRF/%	Amorphous Silica Content Calculated Through the New Method/%	Calculated Si Content in Crystalline Minerals/%	Calculated Si Content in Clay Minerals/%
S2	3351.66	Carbonate lithofacies	–	12.181	0.225	234.695	0.9727	0.00236	13.028	0.472	12.836	38.126
S2	3353.14	Carbonate lithofacies	22.7	–	–	–	–	–	10.224	0.522	10.027	37.074
S2	3355.13	Transitional lithofacies	9.90	35.739	0.333	173.612	–	–	26.256	2.251	25.762	38.856
S2	3359.95	Transitional lithofacies	–	–	–	–	–	–	22.943	8.594	20.615	33.214
S2	3379.55	Transitional lithofacies	8.60	14.625	0.202	184.049			26.952	2.838	26.346	42.938
S2	3380.01	Tuffaceous shale lithofacies	–	–	–	–	–	–	39.916	6.143	39.407	44.803
S2	3463.00	Transitional lithofacies	–	–	–	–	0.4650	0.0109	25.431	7.099	23.729	39.068
S2	3477.00	Transitional lithofacies	–	18.328	0.269	114.190	1.1957	0.135	28.659	4.529	27.756	43.399
S2	3600.34	Carbonate lithofacies	2.10	26.013	0.317	373.744	–	–	12.147	1.773	11.505	40.227
S3	2794.71	Tuffaceous shale lithofacies	40.90	35.675	0.244	178.849	1.2807	0.012	34.525	2.676	34.163	41.965
S3	2795.80	Tuffaceous shale lithofacies	32.10	26.702	0.241	139.743	1.1727	0.0162	35.295	3.167	34.889	33.116
S3	2805.21	Transitional lithofacies	–	–	–	–	0.6000	0.0244	22.358	5.936	20.759	45.639
S3	2808.52	Carbonate lithofacies	–	12.089	0.271	319.996	0.9194	0.00105	12.957	2.718	11.986	44.204

Table 2. *Cont.*

Well	Depth (m)	Lithofacies	Oil Saturation/%	Young's Modulus/N*mm^{-2}	Poisson's Ratio	Compressive Strength/Kg*cm^{-2}	Porosity/%	Permeability/mD	Si Content Test by XRF/%	Amorphous Silica Content Calculated Through the New Method/%	Calculated Si Content in Crystalline Minerals/%	Calculated Si Content in Clay Minerals/%
S3	2815.21	Tuffaceous shale lithofacies	8.90	–	–	–	0.9194	0.00244	32.932	11.535	31.006	31.704
S3	2817.81	Tuffaceous shale lithofacies	70.30	13.176	0.278	67.705	1.8524	0.0301	38.87632	0.589	38.824	44.161
S3	2832.89	Tuffaceous shale lithofacies	–	–	–	–	–	–	31.874	9.893	30.136	31.301
S3	2857.65	Tuffaceous shale lithofacies	36.60	–	–	–	–	–	35.511	7.841	34.474	31.604
S3	2869.14	Tuffaceous shale lithofacies	13.20	39.502	0.293	181.272	1.4511	0.076	31.016	8.205	29.524	32.511
S3	2870.01	Tuffaceous shale lithofacies	31.80	–	–	–	–	–	33.429	4.922	32.690	39.505
S3	2871.01	Tuffaceous shale lithofacies	–	18.927	0.327	128.162	–	–	36.128	6.621	35.307	43.130
S3	2873.01	Tuffaceous shale lithofacies	–	–	–	–	–	–	33.013	3.377	32.499	41.675
S3	2874.01	Tuffaceous shale lithofacies	–	35.684	0.355	132.212	–	–	36.425	4.705	35.868	43.114
S3	2876.81	Tuffaceous shale lithofacies	31.50	33.164	0.192	276.39	–	–	30.243	9.486	28.413	14.260
S4	2580.81	Transitional lithofacies	1.70	39.594	0.287	435.174	0.5100	0.000089	23.923	6.616	22.238	31.529
S4	2585.01	Transitional lithofacies	–	–	–	–	0.5600	0.043	22.973	3.681	22.027	39.252

Table 2. *Cont.*

Well	Depth (m)	Lithofacies	Oil Saturation/%	Young's Modulus/N*mm^{-2}	Poisson's Ratio	Compressive Strength/Kg*cm^{-2}	Porosity/%	Permeability/mD	Si Content Test by XRF/%	Amorphous Silica Content Calculated Through the New Method/%	Calculated Si Content in Crystalline Minerals/%	Calculated Si Content in Clay Minerals/%
S4	2591.91	Tuffaceous shale lithofacies	29.00	—	—	—	—	—	37.636	4.359	37.177	42.558
S4	2592.21	Tuffaceous shale lithofacies	—	—	—	—	—	—	35.947	3.382	35.535	40.529
S4	2601.21	Tuffaceous shale lithofacies	56.20	31.654	0.396	148.228	—	—	36.932	3.331	36.561	42.882
S4	2601.81	Tuffaceous shale lithofacies	—	—	—	—	—	—	31.012	9.318	29.297	33.721
S4	2607.60	Transitional lithofacies	9.10	—	—	—	1.1364	0.05	36.134	1.164	35.998	42.542

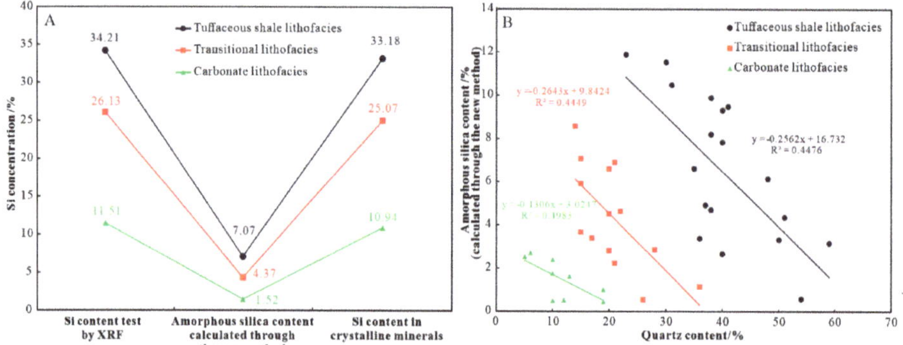

Figure 4. Diagrams showing (**A**) Si content tested by XRF, amorphous silica content calculated through the new method, and Si content in crystalline minerals. (**B**) cross plot of the amorphous silica content with crystalline quartz content in samples of the different lithofacies in the Lucaogou Formation.

5. Discussion

5.1. Advantages and Disadvantages of the New Method

Compared with the previous quantitative analysis methods for amorphous SiO_2, the new method does not require chemical dissolution. The most important is that the cost of this method is much lower. The equipment required has already been widely used for a large-scale sample testing. This method also has some shortcomings: the ideal formula of clay mineral is used to calculate the mass percentage of elemental Si in clay minerals. Using illite as an example, its ideal structural molecular formula is $Al_4(Si_8O_{20})(OH)_4$, and the mass percentage of Si is 31.1%. However, due to the fact that the illite in the actual sample contains impurities, its molecular formula is diverse, which introduces small errors into the calculated value.

5.2. The Influence of Amorphous SiO_2 on Reservoir Properties

The silica content is mainly derived from the alteration of tuffaceous material in the shale strata. It was found through the cross plot between the calculated amorphous SiO_2 content and the reservoir physical property data that amorphous SiO_2 content was negatively correlated with reservoir porosity and permeability (Figure 5). The content of amorphous SiO_2 is negatively correlated with the content of crystalline quartz (Figure 4B). Hence, it indicates that the higher the content of crystalline quartz, the higher the porosity and permeability of the reservoir. Alteration is an important cause of pore formation in the Lucaogou Formation because it is a process of volume reduction for the total material [43,44]. From the perspective of density, it is easy to understand this process of volume reduction. The density of volcanic ash is only 2.3 g/cm^3, while the mineral density after its alteration is much higher than 2.3 g/cm^3, such as quartz 2.6–2.7 g/cm^3. According to the law of conservation of mass, the overall volume must decrease. In other words, a large amount of silica was released during the alteration of tuffaceous components. Some silica crystallized to authigenic quartz, which increases the physical properties of the reservoir, while some silica did not crystallize and occurs between the grains in the form of amorphous SiO_2 cement, which reduces the storage space of the reservoir.

The rock mechanical parameters of the Lucaogou Formation were measured by triaxial stress experiment under given confining pressure (Table 2). The calculated content of amorphous SiO_2 was positively correlated with Young's modulus and compressive strength (Figure 6A,B). It indicates that the higher the content of amorphous SiO_2 was, the harder the samples were to be deformed and fractured. Amorphous SiO_2 cements various grains together, making the reservoir more compacted. Amorphous SiO_2 is negatively correlated with oil saturation (Figure 6D). It indicates that the existence of amorphous SiO_2 is unfavorable for hydrocarbon enrichment. Previous studies suggested that volcanic ash would lead to algal blooms, and the alteration of volcanic ash would also generate a large number of pore spaces, which provided storage space for hydrocarbon enrichment. During volcanic eruptions, a large amount of volcanic ash was deposited with particulate organic matter and well preserved in a strong reduction environment. At last, they further condensed into kerogen and became source rocks with high organic matter. The organic matter type of Lucaogou Formation shale is mainly I~II_1 type, which suggests an origin of bacteria, algae, and other aquatic organisms [19]. However, the presence of amorphous SiO_2 makes the tuffaceous shale lithofacies lack sufficient storage space. Furthermore, part of hydrocarbon migrated to the adjacent carbonate lithofacies. On the whole, amorphous SiO_2 in Lucaogou Formation in Jimsar Depression is not high in content (Figure 4A and Table 2), which is merely the same to that of K-feldspar. Therefore, the changes in reservoir properties are likely to be caused by other factors, such as the development of laminae, the direction of stress in triaxial stress experiments, and so on. In the early diagenetic stage (Ro is 0.35%~0.5%), amorphous SiO_2 has already started to crystallize to quartz in large quantities [23,24]. It can be inferred that the amorphous SiO_2 should have a greater physical influence on shale samples in the earlier diagenetic stage.

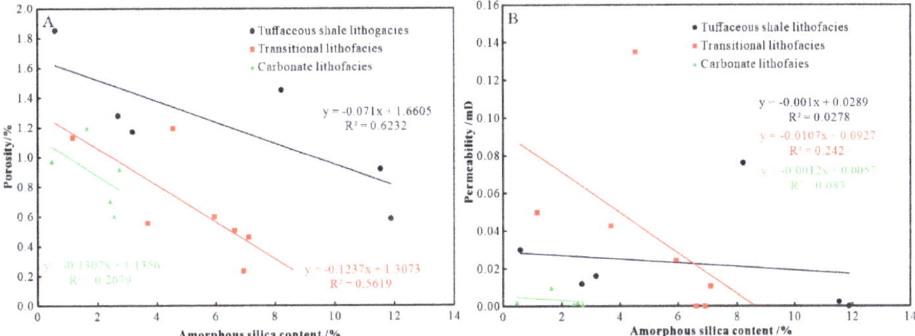

Figure 5. Cross plot of amorphous silica content with (**A**) porosity, (**B**) permeability of different lithofacies in Lucaogou Formation.

5.3. Factors Controlling the Conversion of Amorphous SiO_2 into Quartz

The conversion of amorphous SiO_2 into quartz in diagenesis was affected by many factors, including temperature, properties of fluid medium, burial, and formation pressure, etc. [45–48]. It was proposed that the hydrocarbon injection and formation overpressure can inhibit the formation of authigenic quartz [46–48]. However, in the same one sample, both authigenic quartz and amorphous SiO_2 occur (Figures 3 and 7), the contents of amorphous silica in the four samples (Figure 7A–D) are 6.921%, 10.484%, 11.535%, and 9.318% (Table 2). It means temperature, fluid properties, and formation pressure was not the key factor. It was found that authigenic quartz tended to develop in pores, holes, or fractures through a large number of scanning electron microscope observations (Figure 7). It was a reasonable presumption that the authigenic quartz can only grow when there was space. Without growth space, it can only be amorphous SiO_2 without crystal morphological characteristics. The silica in shale strata of Lucaogou Formation mainly came from the tuffaceous material alteration. A large amount of silica was released. When these pores were filled with a large amount of amorphous SiO_2, there was no room left for the growth of the authigenic quartz. Hence, the amorphous SiO_2 merely existed in the amorphous state. Only when the silica-rich fluid entered one of those large pores, holes, or cracks was there enough space for silica to grow to authigenic quartz.

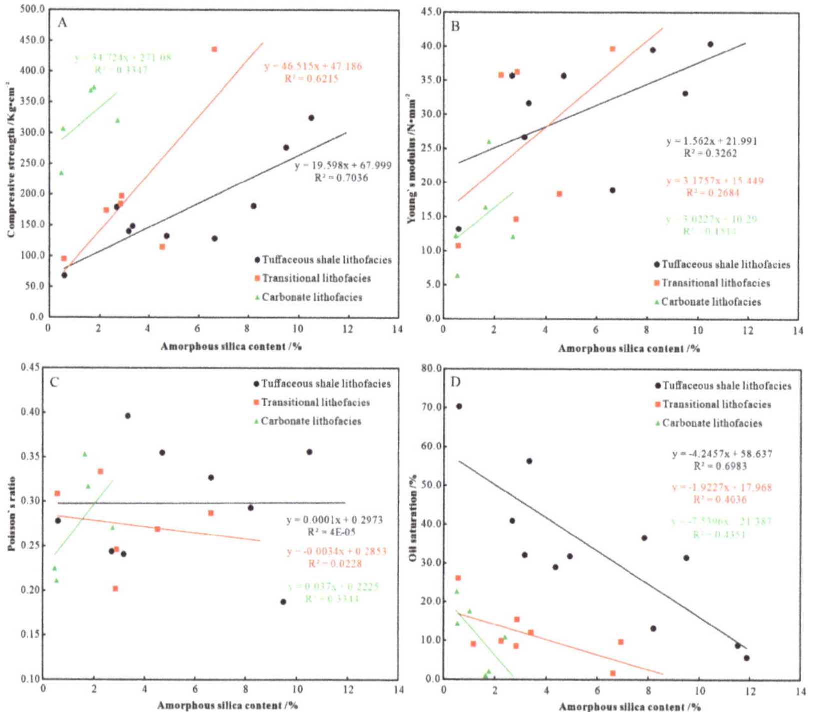

Figure 6. Cross plot of amorphous silica content with (**A**) Young's modulus, (**B**) Poisson's ratio, (**C**) compressive strength and (**D**) oil saturation of different lithofacies in Lucaogou Formation.

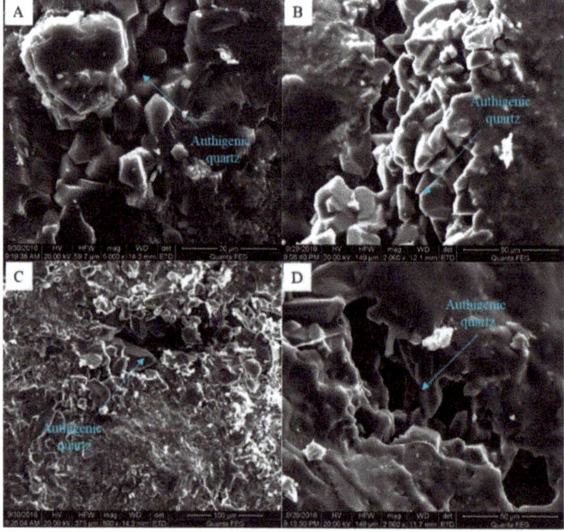

Figure 7. Scanning electron microscope of authigenic quartz in the pores, cavities, and cracks of Lucaogou Formation. (**A**) Transitional lithofacies, S1 well, 3147.64 m; (**B**) tuffaceous shale lithofacies, S2 well, 3348.08 m; (**C**) tuffaceous shale lithofacies, S3 well, 2815.21 m; (**D**) tuffaceous shale lithofacies, S4 well, 2601.81 m.

6. Conclusions

The amorphous SiO_2 in the shale strata of the Lucaogou Formation of the Jimusar Depression had no specific form and was usually mounded among mineral grains. XRD analysis measured the percentage of crystalline minerals, while XRF measured the percentage of elemental Si. Therefore, a new quantitative analysis method for calculating the percentage of amorphous SiO_2 was established by combining the two methods. The content of amorphous SiO_2 in the tuffaceous shale lithofacies of the Lucaogou Formation was the highest, with an average of 7.07%.

The calculation confirmed that the higher the content of amorphous SiO_2, the lower the porosity of the reservoir. Moreover, amorphous SiO_2 was found to be inversely proportional to the compressive strength, Young's modulus, and oil saturation of the reservoir. It indicates that amorphous SiO_2 reduces the physical properties of the reservoir, increases the plasticity, and increases the difficulty of fracturing during development for hydrocarbon extraction. The lack of growing space is the key factor affecting the conversion of amorphous SiO_2 into crystalline quartz. Thus, the existence of amorphous SiO_2 is harmful to shale reservoirs in many ways and has economic impact deleterious to oil and gas exploration and development.

Author Contributions: Conceptualization, Q.C.; methodology, K.S.; software, K.S. and C.C.; investigation, K.S.; resources, Q.C. and G.C.; writing—original draft preparation, K.S.; writing—review and editing, K.S., Y.L., G.C. and C.C.; supervision, Q.C. and G.C.; project administration, Q.C. and G.C. All authors have read and agreed to the published version of the manuscript.

Funding: This research was funded by National Science and Technology Major Project Foundation of China (2016ZX05006-003-002; 2016ZX05014-002); National Nature Science Foundation of China (41802157); Fundamental Research Funds for the Central Universities, China University of Geoscience (Wuhan) (102-162301202627); China Postdoctoral Science Foundation (2016M592265).

Conflicts of Interest: The authors declare no conflict of interest.

References

1. Zhang, J.C.; Xu, B.; Nie, H.K.; Wang, Z.Y.; Lin, T. Exploration potential of shale gas resources in China. *Natl. Gas Ind.* **2008**, *28*, 136–140.
2. Guo, T.I..; Zhang, H.R. Formation and enrichment mode of Jiaoshiba shale gas field, Sichuan Basin. *Pet. Explor. Dev.* **2014**, *41*, 31–40. [CrossRef]
3. Zou, C.; Zhao, Q.; Dong, D.; Yang, Z.; Qiu, Z.; Liang, F.; Wang, N.; Huang, Y.; Duan, A.; Zhang, Q.; et al. Geological characteristics, main challenges and future prospect of shale gas. *J. Natl. Gas Geosci.* **2017**, *2*, 273–288. [CrossRef]
4. Jin, Z.J.; Nie, H.K.; Liu, Q.Y.; Zhao, J.H.; Jiang, T. Source and seal coupling mechanism for shale gas enrichment in upper Ordovician Wufeng Formation—Lower Silurian Longmaxi Formation in Sichuan Basin and its periphery. *Mar. Pet. Geol.* **2018**, *97*, 78–93. [CrossRef]
5. Chen, X.; Fan, J.X.; Wang, W.H.; Wang, H.Y.; Nie, H.K.; Shi, X.W.; Wen, Z.D.; Chen, D.Y.; Li, W.J. Stage-progressive distribution pattern of the Lungmachi black graptolitic shales from Guizhou to Chongqing, Central China. *Sci. China Earth Sci.* **2017**, *60*, 1133–1146. [CrossRef]
6. Bowker, K.A. Barnett Shale gas production, Fort Worth Basin: Issues and discussion. *AAPG Bull.* **2007**, *91*, 523–533. [CrossRef]
7. Jarvie, D.M.; Hill, R.J.; Ruble, T.E.; Pollastro, R.M. Unconventional shale-gas systems: The Mississippian Barnett shale of north-central Texas as one model for thermo-genic shale-gas assessment. *AAPG Bull.* **2007**, *91*, 475–499. [CrossRef]
8. Ruyue, W.; Wenlong, D.; Dajian, G.; Jigao, L.; Xinghua, W.; Shuai, Y. Gas preservation conditions of marine shale in northern Guizhou area: A case study of the Lower Cambrian Niutitang Formation in the Cen'gong block, Guizhou Province. *Oil Gas Geol.* **2016**, *37*, 45–55.

9.   Wang, R.; Hu, Z.; Sun, C.; Liu, Z.; Zhang, C.; Gao, B.; Du, W.; Zhao, J.; Tang, W. Comparative analysis and discussion of shale reservoir characteristics in the Wufeng-Longmaxi and Niutitang formations. *Pet. Geol. Exp.* **2018**, *40*, 639–649.

10.   Dong, T.; Harris, N.B.; Ayranci, K.; Yang, S. The impact of rock composition on geomechanical properties of a shale formation: Middle and Upper Devonian Horn River Group shale, Northeast British Columbia, Canada. *AAPG Bull.* **2017**, *101*, 177–204. [CrossRef]

11.   Guo, X.S.; Hu, D.F.; Wen, Z.D.; Liu, R.B. Major factors controlling the accumulation and high productivity in marine shale gas in the Lower Paleozoic of Sichuan Basin and its periphery: A case study of the Wufeng-Longmaxi Formation of Jiaoshiba area. *Geol. China* **2014**, *41*, 893–901.

12.   Lu, L.; Qin, J.; Shen, B.; Teng, G.; Liu, W.; Zhang, Q. Biogenic origin and hydrocarbon significance of siliceous shale from the Wufeng-Longmaxi formations in Fuling area, southeastern Sichuan Basin. *Pet. Geol. Exp.* **2016**, *38*, 460–472.

13.   Lu, L.; Qin, J.; Shen, B.; Teng, G.; Liu, W.; Zhang, Q. The origin of biogenic silica in siliceous shale from Wufeng-Longmaxi Formation in the Middle and Upper Yangtze region and its relationship with shale gas enrichment. *Earth Sci. Front.* **2018**, *25*, 226–236.

14.   Zhao, J.; Jin, Z.; Jin, Z.; Wen, X.; Geng, Y.K.; Yan, C.N. The genesis of quartz in Wufeng-Longmaxi gas shales, Sichuan Basin. *Natl. Gas Geosci.* **2016**, *27*, 377–386.

15.   Wang, S.; Zou, C.; Dong, D.; Wang, Y.; Huang, J.; Guo, Z. Biogenic silica of organic-rich shale in Sichuan Basin and its significance for shale gas. *Acta Sci. Nat. Univ. Pekin.* **2014**, *50*, 476–486.

16.   Liu, G.; Zhai, G.; Zou, C.; Cheng, L.; Guo, X.; Xia, X.; Shi, D.; Yang, Y.; Zhang, C.; Zhou, Z. A comparative discussion of the evidence for biogenic silica in Wufeng-Longmaxi siliceous shale reservoir in the Sichuan basin, China. *Mar. Pet. Geol.* **2019**, *109*, 70–87. [CrossRef]

17.   Liu, G.; Zhai, G.; Huang, Z.; Zou, C.; Xia, X.; Shi, D.; Zhou, Z.; Zhang, C.; Chen, R.; Yu, S.; et al. The effect of tuffaceous material on characteristics of different lithofacies: A case study on Lucaogou Formation fine-grained sedimentary rocks in Santanghu Basin. *J. Pet. Sci. Eng.* **2019**, *179*, 355–377. [CrossRef]

18.   Xianzheng, Z.H.; Lihong, Z.H.; Xiugang, P.U.; Fengming, J.; Wenzhong, H.; Dunqing, X.; Shiyue, C.H.; Zhannan, S.H.; Zhang, W.; Fei, Y.A. Geological characteristics of shale rock system and shale oil exploration breakthrough in a lacustrine basin: A case study from the Paleogene 1st sub-member of Kong 2 Member in Cangdong sag, Bohai Bay Basin, China. *Pet. Explor. Dev.* **2018**, *45*, 377–388.

19.   Liu, G.; Liu, B.; Huang, Z.; Chen, Z.; Jiang, Z.; Guo, X.; Li, T.; Chen, L. Hydrocarbon distribution pattern and logging identification in lacustrine fine-grained sedimentary rocks of the Permian Lucaogou Formation from the Santanghu basin. *Fuel* **2018**, *222*, 207–231. [CrossRef]

20.   Liu, G.H.; Huang, Z.L.; Guo, X.B.; Liu, Z.Z.; Gao, X.Y.; Chen, C.C.; Zhang, C.L. The research of SiO_2 occurrence in mud shale reservoir of the Yanchang formation in Ordos Basin. *Acta Geol. Sin.* **2016**, *90*, 1016–1029.

21.   Liu, G.H.; Huang, Z.L.; Guo, X.B.; Liu, Z.Z.; Gao, X.Y.; Chen, C.C.; Zhang, C.L. The SiO_2 occurrence and origin in the shale system of middle Permian series Lucaogou Formation in Malang Sag, Santanghu Basin, Xinjiang. *Acta Geol. Sin.* **2016**, *90*, 1220–1235.

22.   Huang, Z.; Liu, G.; Li, T.; Li, Y.; Yin, Y.; Wang, L. Characterization and control of mesopore structural heterogeneity for low thermal maturity shale: A case study of Yanchang Formation shale, Ordos Basin. *Energy Fuels* **2017**, *31*, 11569–11586. [CrossRef]

23.   Zhao, J.; Jin, Z.; Jin, Z.; Hu, Q.; Hu, Z.; Du, W.; Yan, C.; Geng, Y. Mineral types and organic matters of the Ordovician-Silurian Wufeng and Longmaxi Shale in the Sichuan Basin, China: Implications for pore systems, diagenetic pathways, and reservoir quality in fine-grained sedimentary rocks. *Mar. Pet. Geol.* **2017**, *86*, 655–674. [CrossRef]

24.   Zhao, J.; Jin, Z.; Jin, Z.; Wen, X.; Geng, Y. Origin of authigenic quartz in organic-rich shales of the Wufeng and Longmaxi Formations in the Sichuan Basin, South China: Implications for pore evolution. *J. Natl. Gas Sci. Eng.* **2017**, *38*, 21–38. [CrossRef]

25.   Zhang, L.; Li, B.; Jiang, S.; Xiao, D.; Lu, S.; Zhang, Y.; Gong, C.; Chen, L. Heterogeneity characterization of the lower Silurian Longmaxi marine shale in the Pengshui area, South China. *Int. J. Coal Geol.* **2018**, *195*, 250–266. [CrossRef]

26. Peltonen, C.; Marcussen, Ø.; Bjørlykke, K.; Jahren, J. Clay mineral diagenesis and quartz cementation in mudstones: The effects of smectite to illite reaction on rock properties. *Mar. Pet. Geol.* **2009**, *26*, 887–898. [CrossRef]

27. Thyberg, B.; Jahren, J.; Winje, T.; Bjørlykke, K.; Faleide, J.I.; Marcussen, Ø. Quartz cementation in Late Cretaceous mudstones, northern North Sea: Changes in rock properties due to dissolution of smectite and precipitation of microquartz crystals. *Mar. Pet. Geol.* **2010**, *27*, 1752–1764. [CrossRef]

28. Thyberg, B.; Jahren, J. Quartz cementation in mudstones: Sheet-like quartz cement from clay mineral reactions during burial. *Pet. Geosci.* **2011**, *17*, 53–63. [CrossRef]

29. Chaika, C.; Williams, L.A. Density and mineralogy variations as a function of porosity in Miocene Monterey Formation oil and gas reservoirs in California. *AAPG Bull.* **2001**, *85*, 149–167.

30. Lin, J.H. Quantitative analysis of amorphous silica in multicomponent system by X-ray diffraction. *J. Mineral. Petrol.* **1997**, *17*, 22–25.

31. Liu, G.H. The Formation Mechanism of Authigenic Quartz in Lacustrine Shale and Influence to Reservoir Property [D]. Ph.D. Thesis, China University of Petroleum, Beijing, China, 2017.

32. Chu, G. A Doping Method for Quantitative X-Ray Diffraction Phase Analysis of Samples Containing Amorphous Material. *Acta Phys. Sin.* **1998**, *47*, 1143–1148.

33. Popović, S.; Gržeta-plenković, B.; Balić-žunić, T. The doping method in quantitative x-ray diffraction pahase analysis addendum. *J. Appl. Crystallogr.* **1983**, *16*, 505–507. [CrossRef]

34. Huang, Z.; Liu, G.; Ma, J.; Xue, D.; Han, W.; Chen, J.; Chen, Z. A new method for semiquantitative analysis of amorphous SiO_2 in terrestrial shale: A case study about Yanchang Formation shale, Ordos Basin. *Natl. Gas Geosci.* **2015**, *26*, 2017–2028.

35. Cao, Z.; Liu, G.; Kong, Y.; Wang, C.; Niu, Z.; Zhang, J.; Wei, Z. Lacustrine tight oil accumulation characteristics: Permian Lucaogou Formation in jimusaer sag, Junggar Basin. *Int. J. Coal Geol.* **2016**, *153*, 37–51. [CrossRef]

36. Qu, C.S.; Qiu, L.W.; Cao, Y.C. Organic petrology characteristics and occurrence of source rocks in Permian Lucaogou Formation, Jimsar sag. *J. China Univ. Pet. (Ed. Natl. Sci.)* **2017**, *41*, 30–38.

37. Zhou, D.W.; Liu, Y.Q.; Xing, X.J.; Hao, J.R.; Dong, Y.P.; Ouyang, Z.J. Formation of the Permian basalts and implications of geochemical tracing for paleo-tectonic setting and regional tectonic background in the Turpan-Hami and Santanghu basins, Xinjiang. *Sci. China Ser. D* **2006**, *49*, 584–596. [CrossRef]

38. Jiang, Y.Q.; Liu, Y.Q.; Zhao, Y.; Nan, Y.; Wang, R.; Zhou, P.; Yang, Y.J.; Kou, J.Y.; Zhou, N.C. Characteristics and origin of tuff-type tight oil in Jimusar sag, Junggar Basin, NW China. *Pet. Explor. Dev.* **2015**, *42*, 810–818. [CrossRef]

39. Wang, J.; Zhou, L.; Liu, J.; Zhang, X.J.; Zhang, F.; Zhang, B. Acid-base alternation diagenesis and its influence on shale reservoirs in the Permian Lucaogou Formation, Jimusar Sag, Junggar Basin, NW China. *Pet. Explor. Dev.* **2020**, *47*, 962–976. [CrossRef]

40. Liu, C.; Liu, K.; Wang, X.; Wu, L.; Fan, Y. Chemostratigraphy and sedimentary facies analysis of the Permian Lucaogou Formation in the Jimusaer Sag, Junggar Basin, NW China: Implications for tight oil exploration. *J. Asian Earth Sci.* **2019**, *178*, 96–111. [CrossRef]

41. Wang, C.Y.; Kuang, L.C.; Gao, G.; Cui, W.; Kong, Y.H.; Xiang, B.L.; Liu, G.D. Difference in hydrocarbon generation potential of the shaly source rocks in Jimusar sag' Permian Lucaogou formation. *Acta Sedimentol. Sin.* **2014**, *32*, 385–390.

42. Zhi, Y.; Lianhua, H.; Senhu, L.; Xia, L.; Lijun, Z.; Songtao, W.; Jingwei, C. Geologic characteristics and exploration potential of tight oil and shale oil in Lucaogou formation in Jimsar sag. *China Pet. Explor.* **2018**, *23*, 76–85.

43. Cai, D.M.; Sun, L.D.; Qi, J.S.; Dong, J.H.; Zhu, Y.K. Reservoir characteristics and evolution of volcanic rocks in Xujiaweizi fault depression. *Acta Pet. Sin.* **2010**, *31*, 400–407.

44. Hu, W.T.; Liu, C.Z.; Zhao, H.; Luan, K.; Yao, H.P. The Diagenesis of Volcanic Rocks and Its Effects on the Reservior Quality in Xujiaweizi Fault Depression, Soliao Basin. *Sci. Technol. Eng.* **2011**, *11*, 1176–1181.

45. Yang, J.S.; Meng, Y.L.; Zhang, H. Kinetic model of quartz cementation and its application. *Pet. Geol. Exp.* **2002**, *24*, 372–376.

46. Yuan, D.S.; Zhang, Z.H.; Zhu, L.; Liu, H.J. Effects of oil charge on quartz cement. *Acta Petrol. Sin.* **2007**, *23*, 2315–2320.

47. Li, R.; Ge, Y.J.; Chen, Y.; Zhou, Y.Q.; Zhou, H.J. Experimental evidence of influential factor of quartz growth in reservoir. *J. China Univ. Pet.* **2010**, *34*, 13–18.
48. Yuanlin, M.; Cheng, X.; Hongyu, X.; Weizhi, T.; Chuanxin, T.; Jinghuan, L.; Youchun, W.A.; Yuting, G.A. A new kinetic model for authigenic quartz formation under overpressure. *Pet. Explor. Dev.* **2013**, *40*, 701–707.

Publisher's Note: MDPI stays neutral with regard to jurisdictional claims in published maps and institutional affiliations.

 *energies*

Article

Molecular-Scale Considerations of Enhanced Oil Recovery in Shale

Mohamed Mehana *, Qinjun Kang and Hari Viswanathan

Los Alamos National Lab, Computational Earth Science Group, Earth and Environmental Science Division, Los Alamos, NM 87545, USA; qkang@lanl.gov (Q.K.); Viswana@lanl.gov (H.V.)
* Correspondence: mzm@lanl.gov

Received: 1 November 2020; Accepted: 11 December 2020; Published: 15 December 2020

Abstract: With only less than 10% recovery, the primary production of hydrocarbon from shale reservoirs has redefined the energy equation in the world. Similar to conventional reservoirs, Enhanced Oil Recovery (EOR) techniques could be devised to enhance the current recovery factors. However, shale reservoirs possess unique characteristics that significantly affect the fluid properties. Therefore, we are adopting a molecular simulation approach that is well-suited to account for these effects to evaluate the performance of three different gases, methane, carbon dioxide and nitrogen, to recover the hydrocarbons from rough pore surfaces. Our hydrocarbon systems consists of either a single component (decane) or more than one component (decane and pentane). We simulated cases where concurrent and countercurrent displacement is studied. For concurrent displacement (injected fluids displace hydrocarbons towards the production region), we found that nitrogen and methane yielded similar recovery; however nitrogen exhibited a faster breakthrough. On the other hand, carbon dioxide was more effective in extracting the hydrocarbons when sufficient pressure was maintained. For countercurrent displacement (gases are injected and hydrocarbons are produced from the same direction), methane was found to be more effective, followed by carbon dioxide and nitrogen. In all cases, confinement reduced the recovery factor of all gases. This work provides insights to devise strategies to improve the current recovery factors observed in shale reservoirs.

Keywords: molecular simulation; enhanced oil recovery; methane; shale

---

1. Introduction

While abundant, shale formations are unique and serve as unconventional hosts for hydrocarbons [1]. In recent times, a shale revolution has redefined the energy equation in the world [2]. Although the hydrocarbon content of these reservoirs was known for long, the economic developments of these reservoirs became possible only by the coupling of multi-stage hydraulic fracturing with horizontal drilling [3–5]. This coupling overcomes the ultra-low permeability nature of the shale by providing highly conductive pathways connecting the natural fractures. However, the current recovery factors are less than 10%, even with the most efficient completion schemes [6,7].

Enhanced oil recovery techniques have been widely-used in conventional reservoirs [8,9]. Some provide pressure maintenance and others improve the hydrocarbon mobility by tuning the interfacial and physical properties of the reservoir fluids [10]. In contrary to conventional reservoirs, the interconnectivity between shale wells does not always warrant a field-wide design of EOR operations [11,12]. Alternatively, single-well approaches have been proposed [13,14]. Zhang et al. [15] numerically evaluated the efficiency of cyclic methane injection considering both the molecular diffusion and nano-confinement effects. Their results suggest implementing a cyclic injection strategy during the early stages of production. On the other hand, Meng et al. [16] evaluated the efficiency of carbon dioxide cyclic injection, and they experimentally observed more than 30% increase in the recovery.

Assef and Pereira Almao [17] further economically optimized the cyclic gas injection operations. Currently, the Huff and Puff technique is a commonly-used technique that involves injecting the EOR fluid into the well, shutting down the well for a soaking time and resuming the production from the well [18,19]. While the EOR fluid could be gas, water or surfactant, Sheng and Chen showed that gas flooding yields better performance compared to water flooding [20].

The optimization of the gas injection has been the focus of several studies [21–23]. For instance, Hoffman [24] studied the feasibility of various gases for injection in Bakken and reported better performance for the miscible cases. In addition, they encouraged the implementation of EOR since significant oil could be recovered regardless of the gas type. Sheng [25] favored lean gases to enhance the liquid hydrocarbon from shale condensate reservoirs. In addition, Fragoso et al. [26] proposed a holistic approach to develop fields with multiple fluid-type windows, like Eagle Ford and Duvernay, where the gas production is used to enhance the liquid production from the liquid and condensate windows.

Experiments and lab studies have been providing numerous insights about the potential of shale EOR [27,28]. Tovar et al. [29] showed promising results for CO_2-EOR (18–55% recovery factors). In addition, Gamadi et al. [30] observed up to 85% improvement in the oil recovery from Eagle Ford samples using cyclic CO_2 injection. On the other hand, Yu and Sheng [31] studied the N_2 performance in Eagle Ford core and found better results for a longer flooding time and a higher injection pressure. Nguyen et al. [32] used microfluidic experiments to probe the CO_2 and N_2 performance and attributed the superior CO_2 performance to its miscibility characteristics. Alharthy et al. [23] found that both CO_2 and Natural gas liquids, C1-C4+, have a similar efficiency when extracting oil through countercurrent flow from the matrix instead of displacing oil. Hawthorne et al. [33] identified the molecular diffusion of CO_2 as the main mechanism for oil recovery in Bakken samples.

Following the promising lab results, field pilots of CO_2-EOR have been conducted in Bakken and Eagle Ford [34]. Liu et al. [35] designed a case study to evaluate the potential of CO_2-EOR in the Bakken formation where promising results were observed. Pankaj et al. [36] devised an Eagle Ford case study to investigate the CO_2-EOR potential where they refuted the need for infill wells and reported an extra 9% increase in the recovery factor. Kerr et al. [37] developed an Eagle Ford case study to engineer single- and multi-well CO2-EOR techniques and reasonable agreement with the field results was reported. Although promising results were observed for Eagle Ford, fruitless results were reported for Bakken [28]. Rassenfoss [38] attributed these contradictory results to the reservoir containment of each field. For CO_2-EOR to work, sufficient and sustainable contact between CO_2 and the matrix should be achievable. While Eagle Ford pilots provided the containment required for CO_2 to work, the fractured nature of the Bakken formation obstructed the containment. However, the fractured nature of the Bakken formation allowed multi-well EOR techniques to be implemented. Todd and Evans analyzed Huff and Puff and continuous injection pilots in Bakken, and found that continuous injection was favored for CO_2-EOR operations [34].

Molecular simulations have been widely used to probe the gas injection in shale at the molecular scale [39–41]. For instance, Wu et al., studied the displacement mechanisms of CO_2, N_2 and CH_4 [42]. They attributed the slow breakthrough of CO_2 to the superior adsorption characteristics. By contrast, N_2 exhibits a fast breakthrough and a wide front. Wang et al., confirmed the adsorption selectivity of kerogen pores to CO_2 and CH_4 to range from 2.53 up to 7.25 [43]. Sun et al., studied the diffusion of methane and CO_2 in kerogen and observed that the diffusion of dissolved molecules was smaller than that of those adsorbed, which were smaller than the bulk [44]. Liu et al., studied the oil flow displacement by CO_2 in silica nano channels and recommended a small injection rate to assure that the miscible front is developed [45].

Zhou et al. found that the pressure drawdown is efficient in extracting the lighter hydrocarbons and the CO_2 is more efficient is stripping the heavier hydrocarbons from the middle of the pore [46]. Zhang et al., reported the CO_2 behavior in organic and inorganic pores [47]. They observed that C2 and C3 remain adsorbed during the primary production regardless of the pore type compared to C1.

On the other hand, CO_2 injection disrobes the hydrocarbon from the pore surface, which enhances the extraction of heavier components.

Fluid behavior and phase properties deviate from the bulk behavior under confinement [48–50]. Consequently, confinement affects the efficiency of the gases to extract liquid hydrocarbons and requires revisiting both the design and the operation to account for the confinement effects. We used molecular dynamics to evaluate the performance of various gases for enhancing the hydrocarbon recovery from shale formations. We organized the rest of this article as follows: the methodology section briefly presents the modeling approach and the simulation details, and the results section discusses the impact of confinement and operational parameters on the performance of different gases. The main findings are summarized in the conclusions section.

2. Methods

In this section, we briefly present the modeling approach and the simulation details.

2.1. Modeling Approach

Molecular dynamics rely on decoupling between the nucleus and electron motions. While the nucleus motion is classically treated through Newton's mechanics, the effects of the electron motions are considered through partial charges placed on the molecular structure. The force field is a set of equations that describe the molecular interactions and includes intermolecular and intramolecular interactions. Intermolecular interactions describe the interactions among atoms from different molecular entities and include both the Van der Waals (VDW) and electrostatics interactions. While the Lenard Jones potential (Equation (1)) is one of the widely-used models to model the VDW interactions, electrostatic interactions are modeled by Coulomb's law (Equation (2)). On the other hand, the intramolecular interactions maintain the molecular entity and include the bond, angle, dihedral and improper interactions. Given the initial configuration of the system and the molecular interactions,

$$u(r)_{LJ} = 4\varepsilon\left[\left(\frac{\sigma}{r}\right)^{12} - \left(\frac{\sigma}{r}\right)^{6}\right] \tag{1}$$

$$u(r)_{coulomb} = \frac{Q_1 Q_2}{4\pi\epsilon_0 r} \tag{2}$$

where ε is the depth of the potential well (dimensions: ML^2T^{-2}), σ is the distance at which the potential energy is zero (dimension: L), Q is the atomic charge (dimension: AT) and ϵ_0 is the permittivity of free space (dimensions: $M^{-1}L^{-3}T^4A^2$).

2.2. Simulation Details

We used graphite to represent the pore material, decane and pentane to represent the hydrocarbons and carbon dioxide, respectively, and nitrogen and methane to represent the EOR's gases. We elected to simulate rough solid surfaces, which realistically capture the complexity of the porous media. As shown in Figure 1, our system consists of a silt pore connected to a reservoir containing the EOR's gas. This reservoir is constrained by a wall that acts as a fixed boundary or constant-rate boundary. In addition, a production region is defined on the other side of the pore. The whole simulation matrix is shown in Table 1. The dimensions of the simulation cell are lx = 504 Å, ly = 42.6 Å and lz = 90/70 Å. The solid substrate comprise three sheets of graphite. We used equal molar gas density for all gases. Note that equal gas densities do not correspond to equal fluid pressures, as shown in Figure 2.

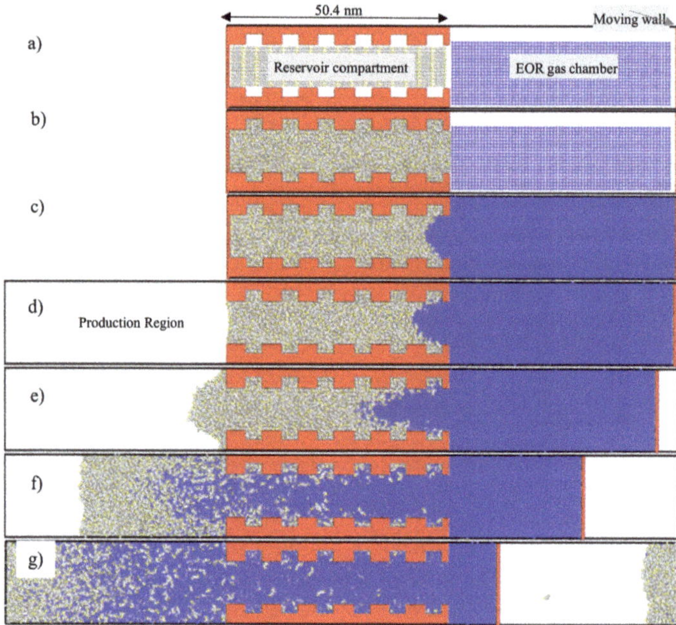

Figure 1. System setup and the simulation evolution: (**a**) the initial setup of the system where the hydrocarbons are placed inside the pore and the Enhanced Oil Recovery (EOR) gas (in this case CO_2) is placed outside; (**b**) the hydrocarbon equilibration; (**c**) the EOR gas equilibration; (**d**) the system extended to include a production region, where the pore was opened for production and a constant-velocity wall is allowed; (**e**) the system after 0.1 ns; (**f**) the system after 0.5 ns; (**g**) the system after 1 ns. Color code: red is graphite, blue is gas and yellow is hydrocarbon.

Table 1. Force field parameters of the system components.

LJ	$\mathcal{E}$ (Kcal/Mole)	σ (Å)
C	0.068443	3.407
CO_2	0.717017	3.72
CH_4	0.294	3.73
N_2	0.18918	3.75
CH_3	0.175	3.905
CH_2	0.118	3.905
Bond	**Kb (Kcal/mole)**	**b0 (Å)**
CH_3-CH_2 CH_2-CH_2	260	1.526
Angle	**Ka (Kcal/mole)**	**θ (°)**
CH_3-CH_2-CH_2 CH_2-CH_2-CH_2	63	112.4

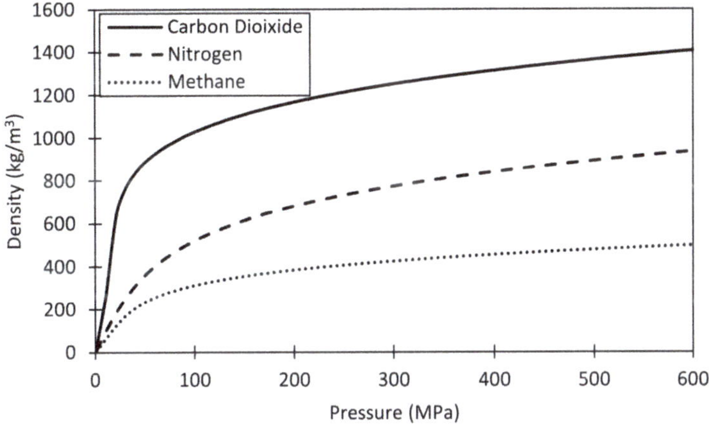

Figure 2. The impact of pressure on the bulk density of carbon dioxide, nitrogen and methane at temperature = 350 K (National Institute of Standards and Technology (NIST) data [51]).

We used the OPLS_UA force field [52] to model the hydrocarbon interactions and united-atom force fields for the EOR's gases as detailed in Table 2. We used Moltemplate software to set up the initial system [53] and LAMMPS to run all the simulations [54]. We started by annealing the hydrocarbons inside the pore for 0.1 ns using canonical ensemble and then equilibrating the EOR's gas for 0.1 ns. After that, the simulation was extended to include a production region (an extra 250 Å in the x-direction). In this scenario, we simulated concurrent displacement, where both the injected fluid and hydrocarbons are moving in the same directions. For constant-rate boundary cases, we applied a constant velocity of 0.0002 Kcal/mole-Angstrom on the wall. We ran the simulation for 1 ns and recorded the hydrocarbon production. We created rough surfaces by carving out grooves inside the solid substrate. We used the Lorentz–Berthelot mixing rules to model the cross interactions. We used the Nosé–Hoover thermostat with a damping parameter of 100 time steps.

Table 2. Simulation matrix of the scenarios studied.

#	Pore Width (nm)	Hydrocarbon Mixture	Boundary Condition	EOR Gas
1				CO_2
2			Moving	CH_4
3		Decane		N_2
4				CO_2
5			Fixed	CH_4
6	7			N_2
7				CO_2
8			Moving	CH_4
9		Decane and Pentane		N_2
10				CO_2
11			Fixed	CH_4
12				N_2
13				CO_2
14	5	Decane	Fixed	CH_4
15				N_2

We also simulated the Huff and Puff process, where the EOR's gas is kept in contact with the hydrocarbon system for 10 ns. After that, the region of the EOR's gas is evacuated to allow the hydrocarbon recovery. This scenario depicts countercurrent displacement. We recorded the production for another 5 ns.

3. Results

In the section, we discuss the results of our single, binary and confined systems.

3.1. Concurrent Displacement

We used the recovery factor as a quantitative measure of the gas efficiency to extract the hydrocarbons from the pore. We estimated the recovery factor by normalizing the number of hydrocarbon molecules leaving the systems by the total number of hydrocarbon molecules in the system. Figures 3 and 4 present the performance of different gases to extract single-component hydrocarbon systems with and without constant-rate injection. We observed that both nitrogen and methane yielded similar recovery factors with and without injection. However, nitrogen exhibited faster breakthrough and displacement compared to carbon dioxide and methane, respectively. We could attribute this behavior to the miscibility of each gas in the hydrocarbons [55]. Given that the solubility of nitrogen in decane is only 15% of that of carbon dioxide at 5 MPa and 50 °C with more than six times minimum miscibility pressure at a temperature of 343.2 K, nitrogen had a stable displacement front and in turn a faster breakthrough [56–60]. In addition, we observed lower recovery rates for all gases without injection. However, the reduction in the case of CO_2 was the most significant, where the breakthrough was not achieved.

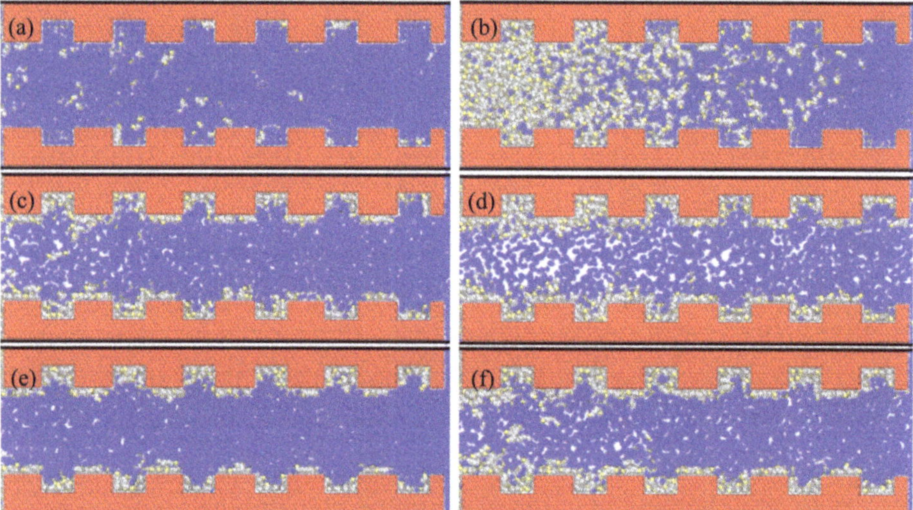

Figure 3. Final snapshots of single-component systems: (**a**) CO_2 with constant-rate injection; (**b**) CO_2 without constant-rate injection; (**c**) CH_4 with constant-rate injection; (**d**) CH_4 without constant-rate injection; (**e**) N_2 with constant-rate injection; (**f**) N_2 without constant-rate injection. The color code is the same as in Figure 1.

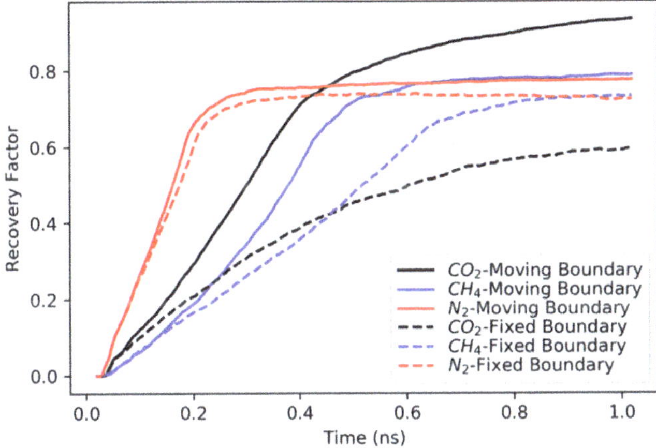

Figure 4. Recovery factors for single-component hydrocarbon systems.

On the other hand, carbon dioxide had better results than the rest with continuous injection and worse than the rest without injection. This could be attributed to the superior adsorption and diffusion characteristics of supercritical CO_2, which allow the extraction of the trapped hydrocarbons in the pore grooves [61,62]. While carbon dioxide could not achieve the breakthrough without the injection, it did not induce a phase separation. Li et al. [63] experimentally and numerically compared the performance of miscible and immiscible CO_2 displacement using the recovery factor. In our case, we believe that the miscibility was initially achieved; however there was no pressure to maintain the miscible front.

Herein, our hydrocarbon system is a multi-component system with a 50:50 mixture of pentane and decane. Figure 5 presents the results of the multi-component systems. Regardless of the boundary conditions, more pentane was extracted compared to decane. Similar to the single-component systems, both nitrogen and methane had similar recovery factors. However, methane and nitrogen had the same displacement speed. In addition, CO_2 still yielded a relatively slower and better performance with constant-rate injection. However, it failed to reach the breakthrough without the injection. It is worth noting that the methane performance was quite different than in the single-component system, especially regarding the displacement speed. This behavior could be attributed to a stable displacement front caused by diluting the decane with pentane. Therefore, there is limited room for methane solubility before phase separation. On the other hand, the pentane presence did not affect the carbon dioxide performance.

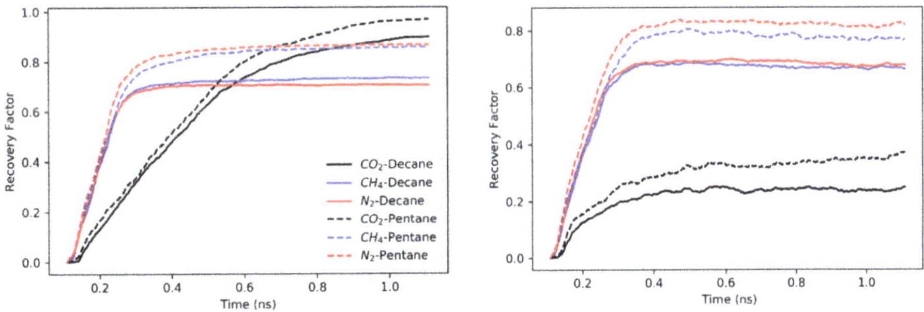

Figure 5. Recovery Factor for multi-component systems with constant-rate injection (**Left**) and without (**Right**).

As the pore width decreased, we observed that the displacement of hydrocarbons became slower. Figure 6 presents the gas performance to extract single-component systems from confined pores (2 nm). For constant-rate injection scenarios, nitrogen started to displace the hydrocarbons faster than the rest. However, both methane and carbon dioxide eventually yielded better recovery rates. We observed a no-production period for all gases at the start, which was shorter for nitrogen. This period might be attributed to the stronger adsorption hydrocarbons experienced as the pore size decreases. Theoretically, higher capillary pressure is required for gases to enter smaller pores. However, our results suggest that both cases, with and without injection, start to displace hydrocarbons around the same time. This behavior could be attributed to the compressibility of the injected fluid. Even though we did not observe significant impacts on the boundary conditions in the early stages, significant effects were observed later on.

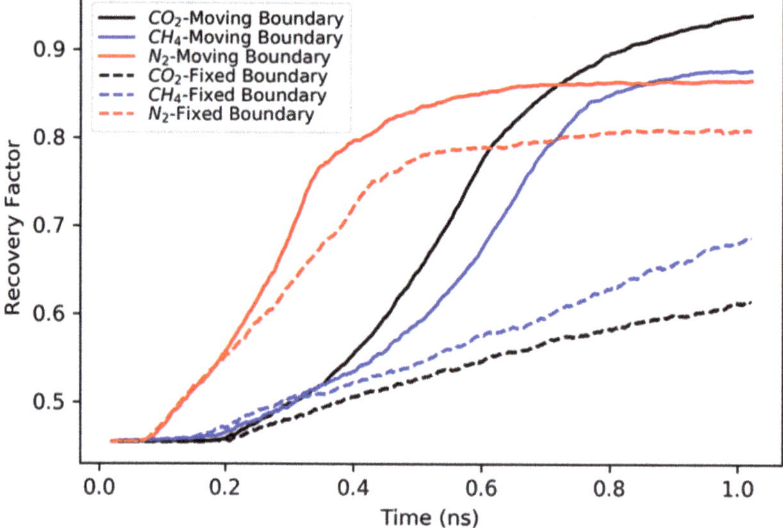

Figure 6. Recovery factors for single-component systems in confined pores.

It is worth noting that the curve shape of the recovery factor of confined pores differs from the one observed from large pores. In confined pores, we observed more of a convex shape compared to the linear response observed in large pores. In addition, the breakthrough is more transitional instead of the abrupt change observed in large pores. Both observations suggest a stronger adsorption of hydrocarbons on the pore surface under confinement. On the other hand, less recovery is observed without injection. Both methane and carbon dioxide did not reach the breakthrough. However, methane had a slightly better recovery than carbon dioxide.

3.2. Countercurrent Displacement

In this section, we simulate the performance of the gases to extract single-component hydrocarbon systems. The gases were soaked in contact with the hydrocarbon system for 10 ns, while the hydrocarbons were extracted from the pore to the EOR's gas region. Figure 7 presents the results of the Huff and Puff simulations. During the soaking time, CH$_4$ was the most effective in extracting the hydrocarbon, followed by CO$_2$ and then N$_2$. As the puff process started, a large influx of hydrocarbons left the pore. We observed that some of the hydrocarbons returned back to the pore for 5 nm cases, especially with nitrogen or methane. While similar behavior was observed for 2 nm pores, less recovery was observed for all gases. In addition, the CO$_2$'s performance was relatively improved compared to the methane. While reduced compared to the concurrent displacement, the countercurrent displacement's recovery factors were in line with field observations and recent molecular simulation studies [64,65]. The superiority of CO_2 over N_2 has been previously experimentally and numerically reported. However, we found that CH_4 outperformed all of them in this scenario. This behavior could be attributed to the better vaporization characteristics of methane [66,67].

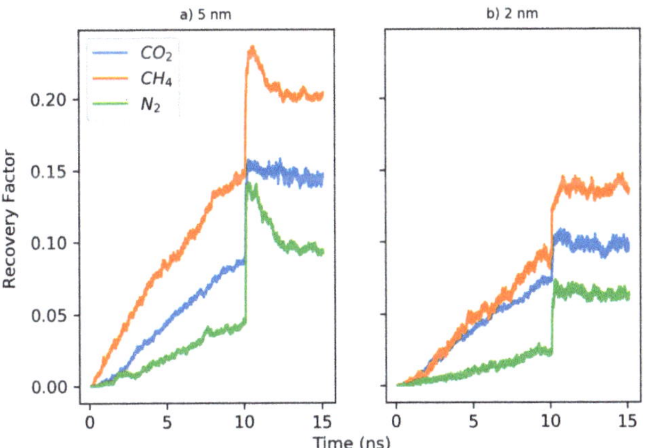

Figure 7. Recovery factors observed for Huff and Puff simulation: (**a**) pore width is 5 nm and (**b**) pore width is 2 nm.

Field recommendations include using CO_2 for multi-well EOR operations and CH_4 for single-well EOR operations. In addition, the injection pressure significantly affects the CO_2 performance. Consequently, pressure support should be maintained throughout the operations. While valid, these recommendations were derived based on single-pore simulations with single or binary component hydrocarbons. However, the heterogeneity of the porous media and the complexity of the crude oil mixture might dramatically affect the EOR operations. Therefore, further research is required to quantify the impact of these factors along with more integrated lab and field pilots.

4. Conclusions

In this study, we have conducted a comprehensive molecular simulation study to examine the efficiency of carbon dioxide, nitrogen and methane to extract hydrocarbons from organic rough pores. We found that:

- Confinement enhances the adsorption of hydrocarbons to the pore surface, which hinders both concurrent and counter-current displacements.
- All gases are more efficient in displacing hydrocarbons in concurrent displacement relative to counter-current displacement. While the recovery factors observed in counter-current displacements are usually less than 20%, the concurrent displacement could reach up to 90%.
- Nitrogen usually exhibited faster breakthrough regardless of the hydrocarbons' type, pore size and the boundary conditions for the concurrent displacement. Interestingly, the limited diffusion and miscibility of nitrogen in hydrocarbons led to faster recovery in the case of concurrent displacement, while the opposite was observed for counter-current displacement. On the other hand, methane yielded better recovery for counter-current displacement.
- Carbon dioxide proved more efficient in extracting the hydrocarbons from rough pores (from the grooves) if enough pressure was maintained. Having favorable adsorption characteristics and capability to improve the hydrocarbon mobility, carbon dioxide provides the best candidate. However, constant pressure support is needed to overcome the unstable displacement front.

Author Contributions: Conceptualization, methodology and analysis, M.M. and Q.K.; resources, Q.K. and H.V.; writing—original draft preparation, M.M.; writing—review and editing, Q.K. and H.V.; supervision, Q.K. and H.V.; funding acquisition, Q.K. All authors have read and agreed to the published version of the manuscript.

Funding: This work was supported by the Laboratory Directed Research and Development (LDRD) program of Los Alamos National Laboratory (LANL).

Conflicts of Interest: The authors declare no conflict of interest.

References

1. Mehana, M.; El-monier, I. Shale characteristics impact on Nuclear Magnetic Resonance (NMR) fluid typing methods and correlations. *Petroleum* **2016**, *2*, 138–147. [CrossRef]
2. Hughes, J.D. Energy: A reality check on the shale revolution. *Nature* **2013**, *494*, 307. [CrossRef]
3. Middleton, R.; Carey, B.; Currier, R.P.; Hyman, J.D.; Kang, Q.; Karra, S.; Jiménez-Martínez, J.; Porter, M.L.; Viswanathan, H.S. Shale gas and non-aqueous fracturing fluids: Opportunities and challenges for supercritical CO_2. *Appl. Energy* **2015**, *147*, 500–509. [CrossRef]
4. King, G.E. Thirty years of gas shale fracturing: What have we learned? In Proceedings of the SPE Annual Technical Conference and Exhibition, Florence, Italy, 19–22 September 2010; Society of Petroleum Engineers: Richardson, TX, USA, 2010.
5. King, G.E. Hydraulic fracturing 101: What every representative, environmentalist, regulator, reporter, investor, university researcher, neighbor and engineer should know about estimating frac risk and improving frac performance in unconventional gas and oil wells. In Proceedings of the SPE Hydraulic Fracturing Technology Conference, The Woodlands, TX, USA, 6–8 February 2012; Society of Petroleum Engineers: Richardson, TX, USA, 2012.
6. Clark, A.J. Determination of recovery factor in the Bakken formation, Mountrail County, ND. In Proceedings of the SPE Annual Technical Conference and Exhibition, New Orleans, LA, USA, 4–7 October 2009; Society of Petroleum Engineers: Richardson, TX, USA, 2009.
7. LeFever, J.A.; Helms, L.D. *Bakken Formation Reserve Estimates*; North Dakota Geological Survey: Bismarck, ND, USA, 2006.
8. Green, D.W.; Willhite, G.P. Enhanced oil recovery. In *Doherty Memorial Fund of AIME*; Henry, L., Ed.; Society of Petroleum Engineers: Richardson, TX, USA, 1998; Volume 6.
9. Lake, L.W. Enhanced Oil Recovery. 1989. Available online: https://store.spe.org/Enhanced-Oil-Recovery--P436.aspx (accessed on 22 September 2020).
10. Lake, L.W.; Johns, R.; Rossen, W.R.; Pope, G.A. Fundamentals of Enhanced Oil Recovery. 2014. Available online: https://store.spe.org/Fundamentals-ofEnhanced-Oil-Recovery-P921.aspx (accessed on 22 September 2020).
11. Chalmers, G.R.; Ross, D.J.; Bustin, R.M. Geological controls on matrix permeability of Devonian Gas Shales in the Horn River and Liard basins, northeastern British Columbia, Canada. *Int. J. Coal Geol.* **2012**, *103*, 120–131. [CrossRef]
12. Yu, W.; Wu, K.; Zuo, L.; Tan, X.; Weijermars, R. Physical models for inter-well interference in shale reservoirs: Relative impacts of fracture hits and matrix permeability. In Proceedings of the Unconventional Resources Technology Conference, San Antonio, TX, USA, 1–3 August 2016.
13. Alfarge, D.; Wei, M.; Bai, B. IOR methods in unconventional reservoirs of North America: Comprehensive review. In Proceedings of the SPE Western Regional Meeting, Bakersfield, CA, USA, 23–27 April 2017; Society of Petroleum Engineers: Richardson, TX, USA, 2017.
14. Du, F.; Nojabaei, B. A review of gas injection in shale reservoirs: Enhanced oil/gas recovery approaches and greenhouse gas control. *Energies* **2019**, *12*, 2355. [CrossRef]
15. Zhang, Y.; Di, Y.; Shi, Y.; Hu, J. Cyclic CH4 injection for enhanced oil recovery in the Eagle Ford shale reservoirs. *Energies* **2018**, *11*, 3094. [CrossRef]
16. Meng, X.; Meng, Z.; Ma, J.; Wang, T. Performance Evaluation of CO_2 Huff-n-Puff Gas Injection in Shale Gas Condensate Reservoirs. *Energies* **2018**, *12*, 42. [CrossRef]
17. Assef, Y.; Almao, P.P. Evaluation of Cyclic Gas Injection in Enhanced Recovery from Unconventional Light Oil Reservoirs: Effect of Gas Type and Fracture Spacing. *Energies* **2019**, *12*, 1370. [CrossRef]

18. Yu, W.; Lashgari, H.; Sepehrnoori, K. Simulation study of CO_2 huff-n-puff process in Bakken tight oil reservoirs. In Proceedings of the SPE Western North American and Rocky Mountain Joint Meeting, Denver, CO, USA, 17–18 April 2014; Society of Petroleum Engineers: Richardson, TX, USA, 2014.

19. Yu, W.; Lashgari, H.R.; Wu, K.; Sepehrnoori, K. CO_2 injection for enhanced oil recovery in Bakken tight oil reservoirs. *Fuel* **2015**, *159*, 354–363. [CrossRef]

20. Sheng, J.J.; Chen, K. Evaluation of the EOR potential of gas and water injection in shale oil reservoirs. *J. Unconv. Oil Gas. Resour.* **2014**, *5*, 1–9. [CrossRef]

21. Sheng, J.J. Enhanced oil recovery in shale reservoirs by gas injection. *J. Nat. Gas. Sci. Eng.* **2015**, *22*, 252–259. [CrossRef]

22. Wang, L.; Tian, Y.; Yu, X.; Wang, C.; Yao, B.; Wang, S.; Winterfeld, P.H.; Wang, X.; Yang, Z.; Wang, Y. Advances in improved/enhanced oil recovery technologies for tight and shale reservoirs. *Fuel* **2017**, *210*, 425–445. [CrossRef]

23. Alharthy, N.; Teklu, T.W.; Kazemi, H.; Graves, R.M.; Hawthorne, S.B.; Braunberger, J.; Kurtoglu, B. Enhanced oil recovery in liquid-rich shale reservoirs: Laboratory to field. *SPE Reserv. Eval. Eng.* **2018**, *21*, 137–159. [CrossRef]

24. Hoffman, B.T. Comparison of various gases for enhanced recovery from shale oil reservoirs. In Proceedings of the SPE Improved Oil Recovery Symposium, Tulsa, OK, USA, 14–18 April 2012; Society of Petroleum Engineers: Richardson, TX, USA, 2012.

25. Sheng, J.J. Increase liquid oil production by huff-n-puff of produced gas in shale gas condensate reservoirs. *J. Unconv. Oil Gas. Resour.* **2015**, *11*, 19–26. [CrossRef]

26. Fragoso, A.; Wang, Y.; Jing, G.; Aguilera, R. Improving recovery of liquids from shales through gas recycling and dry gas injection. In Proceedings of the SPE Latin American and Caribbean Petroleum Engineering Conference, Quito, Ecuador, 18–20 November 2015; Society of Petroleum Engineers: Richardson, TX, USA, 2015.

27. Jin, L.; Hawthorne, S.; Sorensen, J.; Pekot, L.; Kurz, B.; Smith, S.; Heebink, L.; Herdegen, V.; Bosshart, N.; Torres, J. Advancing CO_2 enhanced oil recovery and storage in unconventional oil play—Experimental studies on Bakken shales. *Appl. Energy* **2017**, *208*, 171–183. [CrossRef]

28. Jia, B.; Tsau, J.S.; Barati, R. A review of the current progress of CO2 injection EOR and carbon storage in shale oil reservoirs. *Fuel* **2019**, *236*, 404–427. [CrossRef]

29. Tovar, F.D.; Eide, O.; Graue, A.; Schechter, D.S. Experimental investigation of enhanced recovery in unconventional liquid reservoirs using CO_2: A look ahead to the future of unconventional EOR. In Proceedings of the SPE Unconventional Resources Conference, The Woodlands, TX, USA, 1–3 April 2014; Society of Petroleum Engineers: Richardson, TX, USA, 2014.

30. Gamadi, T.; Sheng, J.; Soliman, M.; Menouar, H.; Watson, M.; Emadibaladehi, H. An experimental study of cyclic CO_2 injection to improve shale oil recovery. In Proceedings of the SPE Improved Oil Recovery Symposium, New Orleans, LA, USA, 30 September–2 October 2014; Society of Petroleum Engineers: Richardson, TX, USA, 2014.

31. Yu, Y.; Sheng, J.J. Experimental evaluation of shale oil recovery from Eagle Ford core samples by nitrogen gas flooding. In Proceedings of the SPE Improved Oil Recovery Conference, Tulsa, OK, USA, 11–13 April 2016; Society of Petroleum Engineers: Richardson, TX, USA, 2016.

32. Nguyen, P.; Carey, J.W.; Viswanathan, H.S.; Porter, M. Effectiveness of supercritical-CO_2 and N_2 huff-and-puff methods of enhanced oil recovery in shale fracture networks using microfluidic experiments. *Appl. Energy* **2018**, *230*, 160–174. [CrossRef]

33. Hawthorne, S.B.; Jin, L.; Kurz, B.A.; Miller, D.J.; Grabanski, C.B.; Sorensen, J.A.; Pekot, L.J.; Bosshart, N.W.; Smith, S.A.; Burton-Kelly, M.E. Integrating petrographic and petrophysical analyses with CO_2 permeation and oil extraction and recovery in the Bakken Tight oil formation. In Proceedings of the SPE Unconventional Resources Conference, Calgary, AB, Canada, 15–16 February 2017; Society of Petroleum Engineers: Richardson, TX, USA, 2017.

34. Todd, H.B.; Evans, J.G. Improved oil recovery IOR pilot projects in the Bakken formation. In Proceedings of the SPE Low Perm Symposium, Denver, CO, USA, 5–6 May 2016; Society of Petroleum Engineers: Richardson, TX, USA, 2016.

35. Liu, G.; Sorensen, J.; Braunberger, J.; Klenner, R.; Ge, J.; Gorecki, C.; Steadman, E.; Harju, J. CO_2-based enhanced oil recovery from unconventional reservoirs: A case study of the Bakken formation. In Proceedings

of the SPE Unconventional Resources Conference, The Woodlands, TX, USA, 1–3 April 2014; Society of Petroleum Engineers: Richardson, TX, USA, 2014.

36. Pankaj, P.; Mukisa, H.; Solovyeva, I.; Xue, H. Enhanced oil recovery in eagle ford: Opportunities using huff-n-puff technique in unconventional reservoirs. In Proceedings of the SPE Liquids-Rich Basins Conference-North America, Midland, TX, USA, 5–6 September 2018; Society of Petroleum Engineers: Richardson, TX, USA, 2018.

37. Kerr, E.; Venepalli, K.K.; Patel, K.; Ambrose, R.; Erdle, J. Use of Reservoir Simulation to Forecast Field EOR Response-An Eagle Ford Gas Injection Huff-N-Puff Application. In Proceedings of the SPE Hydraulic Fracturing Technology Conference and Exhibition, The Woodlands, TX, USA, 4–6 February 2020; Society of Petroleum Engineers: Richardson, TX, USA, 2020.

38. Rassenfoss, S. Shale EOR Works, But Will It Make a Difference? *J. Pet. Technol.* **2017**, *69*, 34–40. [CrossRef]

39. Lan, Y.; Yang, Z.; Wang, P.; Yan, Y.; Zhang, L.; Ran, J. A review of microscopic seepage mechanism for shale gas extracted by supercritical CO_2 flooding. *Fuel* **2019**, *238*, 412–424. [CrossRef]

40. Mehana, M.; Fahes, M.; Huang, L. The Density of Oil/Gas Mixtures: Insights from Molecular Simulations. *SPE J.* **2018**, *23*, 1798–1808. [CrossRef]

41. Mehana, M.; Fahes, M.; Huang, L. Asphaltene Aggregation in Oil and Gas Mixtures: Insights from Molecular Simulation. *Energy Fuels* **2019**, *33*, 4721–4730. [CrossRef]

42. Wu, H.; Chen, J.; Liu, H. Molecular dynamics simulations about adsorption and displacement of methane in carbon nanochannels. *J. Phys. Chem. C* **2015**, *119*, 13652–13657. [CrossRef]

43. Wang, T.; Tian, S.; Li, G.; Sheng, M.; Ren, W.; Liu, Q.; Zhang, S. Molecular simulation of CO_2/CH_4 competitive adsorption on shale kerogen for CO_2 sequestration and enhanced gas recovery. *J. Phys. Chem. C* **2018**, *122*, 17009–17018. [CrossRef]

44. Sun, H.; Zhao, H.; Qi, N.; Li, Y. Molecular insights into the enhanced shale gas recovery by carbon dioxide in kerogen slit nanopores. *J. Phys. Chem. C* **2017**, *121*, 10233–10241. [CrossRef]

45. Liu, B.; Wang, C.; Zhang, J.; Xiao, S.; Zhang, Z.; Shen, Y.; Sun, B.; He, J. Displacement mechanism of oil in shale inorganic nanopores by supercritical carbon dioxide from molecular dynamics simulations. *Energy Fuels* **2016**, *31*, 738–746. [CrossRef]

46. Zhou, J.; Jin, Z.; Luo, K.H. Insights into recovery of multi-component shale gas by CO_2 injection: A molecular perspective. *Fuel* **2020**, *267*, 117247. [CrossRef]

47. Zhang, M.; Zhan, S.; Jin, Z. Recovery mechanisms of hydrocarbon mixtures in organic and inorganic nanopores during pressure drawdown and CO_2 injection from molecular perspectives. *Chem. Eng. J.* **2020**, *382*, 122808. [CrossRef]

48. Neil, C.W.; Mehana, M.; Hjelm, R.P.; Hawley, M.E.; Watkins, E.B.; Mao, Y.; Viswanathan, H.; Kang, Q.; Xu, H. Reduced methane recovery at high pressure due to methane trapping in shale nanopores. *Commun. Earth Environ.* **2020**, *1*, 1–10. [CrossRef]

49. Santos, J.E.; Mehana, M.; Wu, H.; Prodanovic, M.; Kang, Q.; Lubbers, N.; Viswanathan, H.; Pyrcz, M.J. Modeling nanoconfinement effects using active learning. *J. Phys. Chem. C* **2020**, *124*, 22200–22211. [CrossRef]

50. Lubbers, N.; Agarwal, A.; Chen, Y.; Son, S.; Mehana, M.; Kang, Q.; Karra, S.; Junghans, C.; Germann, T.C.; Viswanathan, H.S. Modeling and scale-bridging using machine learning: Nanoconfinement effects in porous media. *Sci. Rep.* **2020**, *10*, 1–13. [CrossRef]

51. NIST. National Institute of Standards and Technology (NIST) Chemistry WebBook. 2009. Available online: https://webbook.nist.gov/chemistry/fluid/ (accessed on 5 October 2020).

52. Jorgensen, W.L.; Maxwell, D.S.; Tirado-Rives, J. Development and testing of the OPLS all-atom force field on conformational energetics and properties of organic liquids. *J. Am. Chem. Soc.* **1996**, *118*, 11225–11236. [CrossRef]

53. Jewett, A.I.; Zhuang, Z.; Shea, J.E. Moltemplate a coarse-grained model assembly tool. *Biophys. J.* **2013**, *104*, 169. [CrossRef]

54. Plimpton, S. Fast parallel algorithms for short-range molecular dynamics. *J. Comput. Phys.* **1995**, *117*, 1–19. [CrossRef]

55. Fang, T.; Wang, M.; Gao, Y.; Zhang, Y.; Yan, Y.; Zhang, J. Enhanced oil recovery with CO_2/N_2 slug in low permeability reservoir: Molecular dynamics simulation. *Chem. Eng. Sci.* **2019**, *197*, 204–211. [CrossRef]

56. Pereira, L.M.; Chapoy, A.; Burgass, R.; Tohidi, B. Measurement and modelling of high pressure density and interfacial tension of (gas+ n-alkane) binary mixtures. *J. Chem.* **2016**, *97*, 55–69. [CrossRef]

57. Tong, J.; Gao, W.; Robinson, R.L.; Gasem, K.A. Solubilities of nitrogen in heavy normal paraffins from 323 to 423 K at pressures to 18.0 MPa. *J. Chem. Eng. Data* **1999**, *44*, 784–787. [CrossRef]

58. Georgiadis, A.; Llovell, F.; Bismarck, A.; Blas, F.J.; Galindo, A.; Maitland, G.C.; Trusler, J.M.; Jackson, G. Interfacial tension measurements and modelling of (carbon dioxide+ n-alkane) and (carbon dioxide+ water) binary mixtures at elevated pressures and temperatures. *J. Supercrit. Fluids* **2010**, *55*, 743–754. [CrossRef]

59. Zamudio, M.; Schwarz, C.; Knoetze, J. Phase equilibria of branched isomers of C10-alcohols and C10-alkanes in supercritical carbon dioxide. *J. Supercrit. Fluids* **2011**, *59*, 14–26. [CrossRef]

60. Nourozieh, H.; Bayestehparvin, B.; Kariznovi, M.; Abedi, J. Equilibrium properties of (carbon dioxide+ n-decane+ n-octadecane) systems: Experiments and thermodynamic modeling. *J. Chem. Eng. Data* **2013**, *58*, 1236–1243. [CrossRef]

61. Huang, L.; Ning, Z.; Li, H.; Wang, Q.; Ye, H.; Qin, H. Molecular simulation of CO_2 sequestration and enhanced gas recovery in gas rich shale: An insight based on realistic kerogen model. In Proceedings of the Abu Dhabi International Petroleum Exhibition & Conference, Abu Dhabi, United Arab Emirates, 13–16 November 2017; Society of Petroleum Engineers: Richardson, TX, USA, 2017.

62. Fang, T.; Zhang, Y.; Liu, J.; Ding, B.; Yan, Y.; Zhang, J. Molecular insight into the miscible mechanism of CO_2/C_{10} in bulk phase and nanoslits. *Int. J. Heat Mass Transf.* **2019**, *141*, 643–650. [CrossRef]

63. Li, L.; Zhang, Y.; Sheng, J.J. Effect of the injection pressure on enhancing oil recovery in shale cores during the CO_2 huff-n-puff process when it is above and below the minimum miscibility pressure. *Energy Fuels* **2017**, *31*, 3856–3867. [CrossRef]

64. Li, L.; Su, Y.; Hao, Y.; Zhan, S.; Lv, Y.; Zhao, Q.; Wang, H. A comparative study of CO_2 and N_2 huff-n-puff EOR performance in shale oil production. *J. Pet. Sci. Eng.* **2019**, *181*, 106174. [CrossRef]

65. Tovar, F.D.; Barrufet, M.A.; Schechter, D.S. Enhanced Oil Recovery in the Wolfcamp Shale by Carbon Dioxide or Nitrogen Injection: An Experimental Investigation. *SPE J.* **2020**. [CrossRef]

66. Choudhary, N.; Nair, A.K.N.; Ruslan, M.F.A.C.; Sun, S. Bulk and interfacial properties of decane in the presence of carbon dioxide, methane, and their mixture. *Sci. Rep.* **2019**, *9*, 1–10. [CrossRef]

67. Potoff, J.J.; Siepmann, J.I. Vapor–liquid equilibria of mixtures containing alkanes, carbon dioxide, and nitrogen. *AIChE J.* **2001**, *47*, 1676–1682. [CrossRef]

Publisher's Note: MDPI stays neutral with regard to jurisdictional claims in published maps and institutional affiliations.

Article

Dynamic Pore-Scale Network Modeling of Spontaneous Water Imbibition in Shale and Tight Reservoirs

Xiukun Wang [1,2,*] **and James J. Sheng** [1,3]

[1] State Key Laboratory of Petroleum Resources and Prospecting, China University of Petroleum (Beijing), Beijing 102249, China; james.sheng@ttu.edu
[2] Key Laboratory of Unconventional Oil & Gas Development, China University of Petroleum (East China), Ministry of Education, Qingdao 266580, China
[3] Bob L. Herd Department of Petroleum Engineering, Texas Tech University, Lubbock, TX 43111, USA
* Correspondence: xiukunwang@cup.edu.cn

Received: 10 July 2020; Accepted: 8 September 2020; Published: 10 September 2020

Abstract: Spontaneous water imbibition plays an imperative role in the development of shale or tight oil reservoirs. Spontaneous water imbibition is helpful in the extraction of crude oil from the matrix, although it decreases the relative permeability of the hydrocarbon phase dramatically. The dynamic pore-scale network modeling of water imbibition in shale and tight reservoirs is presented in this work; pore network generation, local capillary pressure function, conductance calculation and boundary conditions for imbibition are all presented in detail in this paper. The pore network is generated based on the characteristics of Barnett shale formations, and the corresponding laboratory imbibition experiments are matched using this established dynamic pore network model. The effects of the wettability, throat aspect ratio, viscosity, shape factor, micro-fractures, etc. are all investigated in this work. Attempts are made to investigate the water imbibition mechanisms from a micro-scale perspective. According to the simulated results, wettability dominates the imbibition characteristics. Besides this, the viscous effects including viscosity, initial capillary pressure and micro-fractures increase the imbibition rate, while the final recovery factor is more controlled by the capillarity effect including the cross-area shape factor, contact angle and the average pore-throat aspect ratio.

Keywords: dynamic pore network modeling; shale reservoirs; water imbibition

---

1. Introduction

The spontaneous water imbibition phenomenon is commonly encountered in the development of shale or tight reservoirs. Water imbibition is helpful for the replacement of oil from the matrix and increases the oil recovery, while it decreases the relative permeability of the hydrocarbon phase dramatically, which is detrimental to the production rate and recovery factor [1,2]. Generally, the capillarity is believed to be the controlling factor for the imbibition of water in shale and tight formations. Analogous to waterflooded carbonate reservoirs [3–5], taking full advantage of water imbibition to yield the highest oil recovery is a major concern for the development of shale and tight reservoirs. Moreover, surfactant additives are proposed to alter the formation wettability and improve the performance of water imbibition [6–8]. Large numbers of experimental studies of spontaneous imbibition have been reported in the literature for both conventional [9–13] and unconventional reservoirs [14–16]. Large numbers of analytical models based on a bundle of capillary tubes have been proposed following the early works by Lucas [17]. Viscosity ratios, the tortuosity of capillary tubes, the shapes of cross-sectional areas and the assumption of fractal porous media are involved in these models [18,19]. An attempt has been made to find a universal upscaling equation since the imbibition rate and its recovery factor is sensitive to the sample size; i.e., the characteristic length. Unfortunately, these upscaling techniques are still not applicable for some other factors including wettability, complex rock structures and a wide range of fluid viscosities, etc.

Recently, scholars have become more focused on the micro-scale studies of water imbibition [20], and pore-scale network models have therefore been proposed to simulate this process. There are two types of pore-scale network fluid flow simulation models: quasi-static and dynamic models. For quasi-static models, the fluid flow is controlled by capillarity effects and the viscous forces are neglected. In contrast, inspired by the continuum reservoir simulation method, the dynamic pore network models consider both capillarity and viscous effects. For every increase or decrease of capillary pressures, the two-phase displacement mechanisms—including the piston-like entry of the non-wetting phase, the piston-like throat filling of the wetting phase, the snap-off of the non-wetting phase and the cooperative pore filling of the wetting phase [21,22]—are considered by quasi-static models. Based on the calculated capillary pressure and relative permeability from quasi-static models, the spontaneous imbibition process can then be simulated using conventional reservoir simulators or some simple analytical solutions [23–26]. However, we cannot simulate the process of imbibition directly from the quasi-static model, as not all calculated macro-properties have a relationship with time. Meanwhile, the quasi-static models assume capillary-dominated flow, and the capillary number is assumed to be less than 10^{-5} [27,28].

For the dynamic pore network models, Sheng and Thompson [29] reviewed and divided the dynamic models into three categories: semi-dynamic models (perturbative models), Washburn equation-based dynamic models and coupled dynamic models. The semi-dynamic models separately calculate each phase pressure and update the saturation profiles by ranking the pore elements using a defined global potential which is a combination of the local capillary pressure, viscous pressure drop and gravity. This idea was first derived by Blunt and Scher [30] and then further developed by Hughes and Blunt [31], Idowu and Blunt [32] and Aghaei and Piri [33]. The second type of dynamic model uses the Washburn equation to combine the viscous and capillary forces. Aker et al. [34] did pioneering work and established a framework in this area. Accounting for the wetting film flow, this model is further improved [35,36]. The third type of dynamic model is based on the mass conservation of multiphase flow in every pore element, which is similar to the conventional macroscale reservoir simulation. The Implicit Pressure and Explicit Saturation (IMPES) algorithm is applied to solve the dynamic model. As noted by Khayrat [37], there are two key difficulties resulting in the instability of numerical solutions: (1) local rules in dynamic network models cause a strong non-linearity in the non-wetting conductance because of the occurrence of non-continuity; (2) the capillary pressure is fully coupled within the displacement process, which is a sensitive function of phase saturation. Therefore, an appropriate algorithm is needed to ensure numerical stability, and several related studies have been conducted [38,39].

In this work, we propose a dynamic pore-scale network fluid flow simulation model that simulates the process of spontaneous water imbibition in shale or tight formations. The newly derived mathematical model and corresponding solution algorithm are presented. The proposed model is first applied and validated for Barnett shale formations for the single-phase flow, and then the spontaneous imbibition process is simulated and corresponding systematical sensitivity studies are conducted. Attempts are made to propose an approach to investigate the spontaneous water imbibition mechanisms from a micro-scale perspective.

2. Dynamic Pore Network Modelling of Water Imbibition

2.1. Pore Network Generation

Capillary tube models are suitable for single-phase flow; however, they fail to simulate the multiphase flow scenarios in a porous medium. Pore-scale network models bridge the gap between the single capillary tube model and the porous media system. Firstly, we need to generate the pore network for the studied porous media before simulating two-phase flow processes within them. A structured network consisting of nodes and connecting bonds is used, where nodes are assigned as pore bodies and bonds as pore-throats. A truncated lognormal density distribution function (c.f. Equation (1)) commonly yields a good fit for the profiles of the pore-throat distribution of reservoir rocks [38]. Since the larger pores tend to be connected with larger throats, we cannot randomly assign the pore bodies connected to the pore-throats; here, we sort the pore bodies by the mean values of the connected throats' radii, and then the corresponding sorted pore body radii are assigned.

$$f(r,\sigma) = \frac{\sqrt{2}\exp\left[-\frac{1}{2}\left(\ln\frac{r}{r_m}\right)^2\right]}{\sqrt{\pi\sigma^2}r_i\left[\mathrm{erf}\left(\frac{\ln\frac{r_{max}}{r_m}}{\sqrt{2\sigma^2}}\right) - \mathrm{erf}\left(\frac{\ln\frac{r_{min}}{r_m}}{\sqrt{2\sigma^2}}\right)\right]} \tag{1}$$

where, r refers to the radius of pores or throats, and r_m, r_{max}, r_{min} refer to the mean, maximum and minimum radii of pores or throats, respectively; σ is the standard derivation of the radius distribution.

Following the work by Mason and Morrow [40], in which they studied the capillarity behavior of drainage and imbibition in irregular perfect wetting triangular micro-tubes, the definition of the shape factor is used as shown in Equation (2).

$$G = \frac{A}{P^2} \tag{2}$$

where A and P are the cross-sectional area and the perimeter of a pore or throat.

For a cross-sectional shape, if we define the corner half angles of the triangle as $\beta_1 < \beta_2 < \beta_3$, two constraints need to be met, given a shape factor value:

$$\beta_1 + \beta_2 + \beta_3 = \frac{\pi}{2} \tag{3}$$

$$G = \frac{A}{P^2} = \frac{1}{4\sum_{i=1}^{3}\cot\beta_i} = \frac{1}{4}\tan\beta_1\tan\beta_2\cot(\beta_1+\beta_2) \tag{4}$$

where β_2 is randomly chosen to range from the lower and upper boundaries in Equations (5) and (6).

$$\beta_2^{max} = \mathrm{atan}\left\{\frac{2}{\sqrt{3}}\cos\left[\frac{\mathrm{acos}\left(-12\sqrt{3}G\right)}{3} + \frac{4\pi}{3}\right]\right\} \tag{5}$$

$$\beta_2^{min} = \mathrm{atan}\left\{\frac{2}{\sqrt{3}}\cos\left[\frac{\mathrm{acos}\left(-12\sqrt{3}G\right)}{3}\right]\right\} \tag{6}$$

The mean values of Equations (5) and (6) are used; thus, the smallest half corner angle can be calculated by

$$\beta_1 = -\frac{1}{2}\beta_2 + \frac{1}{2}\mathrm{asin}\left(\frac{\tan\beta_2 + 4G}{\tan\beta_2 - 4G}\sin\beta_2\right) \tag{7}$$

Finally, β_3 is calculated by applying Equation (3). In this way, we can use a single shape factor for the whole pore network. Therefore, one single shape factor will help us to use a single dimensionless local capillary function for every pore or throat with variable wetting phase saturation, which will be presented later. The schematic of the pore network and the triangular cross-section of every pore element are presented in Figure 1.

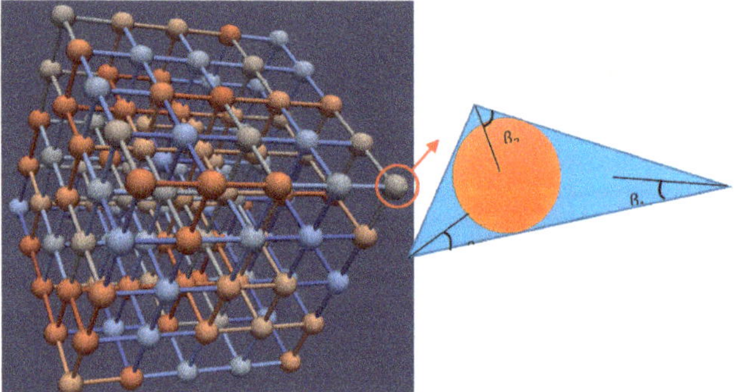

Figure 1. Schematic of the pore network and the triangular cross-section of every pore element.

2.2. Control Equations and Conductance Calculation

The proposed dynamic pore-scale network model is based on the following assumptions:

(1) The volume of throats is negligible compared with that of pore bodies, while the throats control the conductances between pore bodies; (2) the pressure field is continuous, and an upwind scheme is used for the numerical treatment, while the saturation could be discontinuous, especially for the connecting throats; (3) fluids are assumed to be incompressible, and the pore network is rigid.

Applying the mass conservation for every pore body, the governing equations are obtained as

$$V_i \frac{\partial S_w}{\partial t} = \sum_j \frac{K_{ij} K_{ij}^{rw}}{\mu_w} \left(p_j^w - p_i^w \right) \tag{8a}$$

$$V_i \frac{\partial S_n}{\partial t} = \sum_j \frac{K_{ij} K_{ij}^{rn}}{\mu_n} \left(p_j^n - p_i^n \right) \tag{8a}$$

where V, S, and p refer to the volume, phase saturation and pressure for every pore body, respectively, and K_{ij} and K_{ij}^r are the absolute and relative conductance between the ith and j^{th} pores.

The geometry of pores or throats controls the conductance of multiphase flow. For single-phase flow in an irregular triangle, the absolute conductance is obtained by [28]

$$K = 0.6 \frac{GA^2}{L} \tag{9}$$

where L is the length of a pore or throat.

For the two-phase flow scenario of the triangular cross-section, the non-wetting phase occupies the center, while the wetting phase resides in corners, which leads to extra resistances for wetting phase flow [41]. Following the study by Sheng and Thompson [29], the relative conductances for two-phase flow in a throat during the imbibition process are further derived in Equation (10a,b):

$$K^{rn} = \begin{cases} S_n^2, & S_w < \left[F(G,\theta) \frac{\sigma}{r} \frac{1}{p_c} \right]^2 \\ 0, & S_w \geq \left[F(G,\theta) \frac{\sigma}{r} \frac{1}{p_c} \right]^2 \end{cases} \tag{10a}$$

$$K^{rw} = \begin{cases} C(G,\theta) S_w^2, & S_w < \left[F(G,\theta) \frac{\sigma}{r} \frac{1}{p_c} \right]^2 \\ 1, & S_w \geq \left[F(G,\theta) \frac{\sigma}{r} \frac{1}{p_c} \right]^2 \end{cases} \tag{10b}$$

where $C(G, \theta)$ is the function of the shape factor and contact angle, accounting for the extra resistance in corners of the wetting phase, and S_w and S_n refer to the saturations of wetting and non-wetting phases, respectively; $F(G, \theta)$ will be introduced later in Equation (13).

Following the analytical derivation of Øren et al. [21] and Valvatne and Blunt [22], the expression of $C(G, \theta)$ is rearranged in Equation (11a–c). Note that $C(G, \theta)$ does not depend on the local capillary pressure and saturation, which is only related to the cross-sectional shape factor and contact angle. Given a shape factor and a contact angle, a constant value of $C(G, \theta)$ is calculated and will be used in Equation (10b).

$$C(G, \theta) = \frac{\sum_{i=1}^{3} \frac{\left(0.364 G_{ci} + 0.28 G_i^*\right)}{0.6G} S_{wi}^2}{S_w^2} = \frac{\sum_{i=1}^{3} \frac{\left(0.364 G_{ci} + 0.28 G_i^*\right)}{0.6G} \left[\frac{\cos\theta \cos(\theta+\beta_i)}{\sin\beta_i} + \theta + \beta_i - \frac{\pi}{2}\right]^2}{\left\{\sum_{i=1}^{3}\left[\frac{\cos\theta \cos(\theta+\beta_i)}{\sin\beta_i} + \theta + \beta_i - \frac{\pi}{2}\right]\right\}^2} \tag{11a}$$

$$G_i^* = \frac{\sin\beta_i \cos\beta_i}{4(1 + \sin\beta_i)^2} \tag{11b}$$

$$G_{ci} = \frac{\left[\frac{\cos\theta \cos(\theta+\beta_i)}{\sin\beta_i} + \theta + \beta_i - \frac{\pi}{2}\right]}{4\left[\frac{\cos(\theta+\beta_i)}{\sin\beta_i} + \frac{\pi}{2} - \theta - \beta_i\right]^2} \tag{11c}$$

2.3. Local Capillary Pressure Function

For every pore element (pore body or throat), the fluid distribution is controlled by the local capillary pressure function, which is dependent on the element radius, shape of the cross area and contact angle. Based on the previous studies [27,42,43], the local capillary pressure function is reformulated as shown in Equation (12):

$$p_c = B(G, \theta) \frac{\sigma}{r} \frac{1}{\sqrt{S_w}} \tag{12}$$

where $B(G, \theta) = 2\sqrt{G \sum_{i=1}^{3}\left[\frac{\cos\theta \cos(\theta+\beta_i)}{\sin\beta_i} + \theta + \beta_i - \frac{\pi}{2}\right]}$.

However, Equation (12) is only partially applicable, as either snap-off or piston-like-filling happens before the wetting phase saturation reaches unity. The snap-off occurs when the upstream pore element is not filled by the wetting phase, while piston-like-filling occurs when the upstream element is filled by the wetting phase. The corresponding criteria [21,22] are shown in Equations (13) and (14) for snap-off and piston-like-filling, respectively.

$$F(G, \theta) = \cos\theta - \frac{2\sin\theta}{\cot\beta_1 + \cot\beta_2} \tag{13}$$

$$E(G, \theta) = \cos\theta + \sqrt{\cos^2\theta + 4G\left(\pi - \frac{2}{3}\theta + 3\sin\theta\cos\theta - \frac{\cos^2\theta}{4G}\right)} \tag{14}$$

If using the dimensionless capillary pressure as $\frac{p_c}{\sigma/r}$, the local capillary function is shown in Figure 2. Before snap-off or piston-like-filling occurs, the local capillary pressure function follows Equation (12). After the pore filling occurs, the local capillary function becomes a horizontal line (Equation (13) or (14)) until the wetting phase saturation reaches unity. A smooth technique is used in this work to make the local capillary function monotonic. The general capillary fitting correlation by Andersen et al. [44] is used. Thus, given either saturation or capillary pressure, the other situation can be reached.

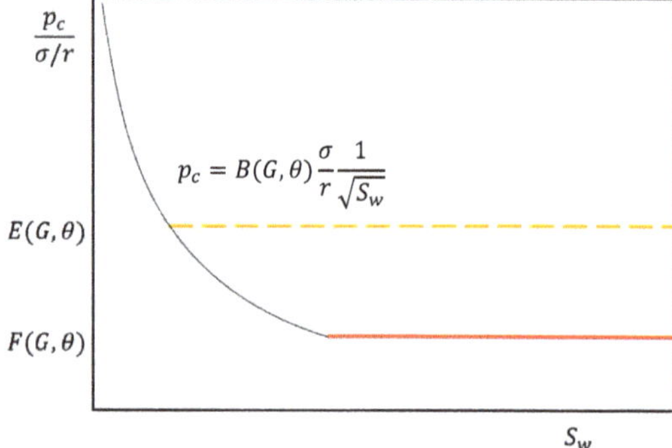

Figure 2. Schematic of local capillary pressure function for variable wetting phase saturation.

2.4. Solution Scheme and Boundary Conditions

The solution scheme follows the IMPES algorithm. Firstly, Equation (8a,b) are combined and discretized to yield the implicit pressure equation as

$$\sum_j \left\{ \left[\frac{K_{ij}K_{ij}^{rw}}{\mu_w} \right]^l \left(p_j^{w,l+1} - p_i^{w,\,l+1} \right) + \left[\frac{K_{ij}K_{ij}^{rn}}{\mu_n} \right]^l \left(p_j^{n,l+1} - p_i^{n,\,l+1} \right) \right\} = 0 \tag{15}$$

where the local capillary pressure is treated in an explicit form (using the value from current time step l); i.e., $p^{n,l+1} = p^{w,l+1} - p_c\!\left(S_w^l\right)$.

After solving the pressure equations, the next step is to update the saturation explicitly using Equation (8a) a in discretized form:

$$S_w^{l+1} = S_w^l + \frac{\Delta t}{V_i} \sum_j \left\{ \left[\frac{K_{ij}K_{ij}^{rw}}{\mu_w} \right]^l \left(p_j^{w,l+1} - p_i^{w,l+1} \right) \right\} \tag{16}$$

Setting the maximum time step size for every time step interval is necessary to ensure that the solution scheme is stable. Here, we only allow a single pore body to be filled by the wetting phase for every step. The maximum time step size estimation from the current step l cannot ensure the stability of the calculation of the next step $l + 1$ due to the severe nonlinearity of this problem. This instability issue can be prevented by rechecking the maximum water saturation changing at the $l + 1$ step. If this is less than unity, the calculation proceeds to the next step $l + 1$; otherwise, the water saturation at the $l + 1$ step is recalculated using a smaller time step size. The details can be found in the work by Aziz and Settari [45]. After obtaining the saturation field, the capillary pressure is updated; then, the calculation proceeds until the preset maximum time. Moreover, the throats are treated explicitly by applying the Equations (8) and (15), which significantly reduces the nonlinearity of the conductivity of non-wetting phase.

The boundary conditions for spontaneous imbibition are set as follows:

$$p^w(@\text{Inlet}) = 0 \tag{17a}$$

$$p_c(@\text{Inlet}) = 0 \tag{17b}$$

$$p_c(@\text{Outlet}) = p_c^{max} \tag{17c}$$

$$p^w(@\text{Outlet}) = -p_c^{max} \tag{17d}$$

where the initially capillary pressure of every pore element is set as the maximum pressure p_c^{max}.

3. Dynamic Pore-Scale Network Modeling of Water Imbibition in Shale and Tight Formations

Firstly, the pore network of Barnett shale is generated using the truncated lognormal distribution of Equation (1), and the single-phase flow simulation is validated. Moghaddam and Jamiolahmady [46] presented the experimental study of the apparent gas permeability of Barnett shale samples; the corresponding pore-throat distributions were obtained from high-pressure mercury injection capillary pressure (HPMICP) experiments. We use Equation (1) to fit the experimental throat radius distribution data of HPMICP, and the mean, minimum and maximum of the throat radii are 6.2, 1 and 50 nanometers, respectively. Since the pore throat aspect ratio is non-trivial to be obtained for shale, the value is set as 1.7 following the works of Valvatne and Blunt [22]. For gas flow in shale and tight formations, the smaller the pore size, the more important the effect of a gas–solid collision; therefore, the corresponding non-Darcy effect needs to be considered. After considering the gas non-Darcy flow (the details are provided in our previous work [47]) for every pore element, the calculated apparent gas permeability and experimentally measured data show a good match, as shown in Figure 3, which demonstrates that the generated network is representative for Barnett Shale, which forms the foundation for the following two-phase flow simulation.

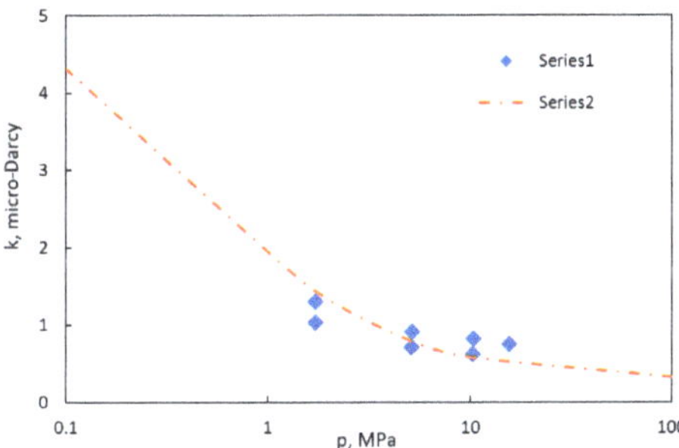

Figure 3. Apparent gas permeability of a Barnett shale sample: experiment and pore network modeling.

Using the obtained pore network for Barnett shale, the initialization is conducted by setting every pore and throat element to have the local capillary pressure $p_c^{max} = 10$ MPa. Then, boundary conditions (Equation (17a) through (17d)) are applied in our dynamic pore network model, and the upper and lower boundaries are set as periodic. The water imbibition process can then be simulated directly within this proposed dynamic pore network model representing Barnett shale. Figure 4 shows the effect of the network size on the simulated spontaneous imbibition recovery with respect to time. We can see that a size of 20×60 is sufficient to capture the accuracy and efficiency for the following studies. The water saturation profiles of the water imbibition process are shown in Figure 5, where the average water saturation equals 0.2239. Note that the contact angle is set as 60° as the advancing contact angle is larger than the intrinsic contact angle due to the hysteresis effect [48]. All the mentioned conditions are set as the base case.

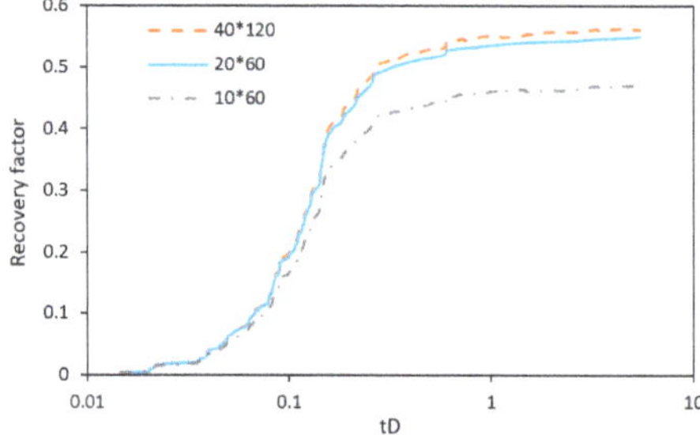

Figure 4. Effect of pore network size on the simulated imbibition recovery.

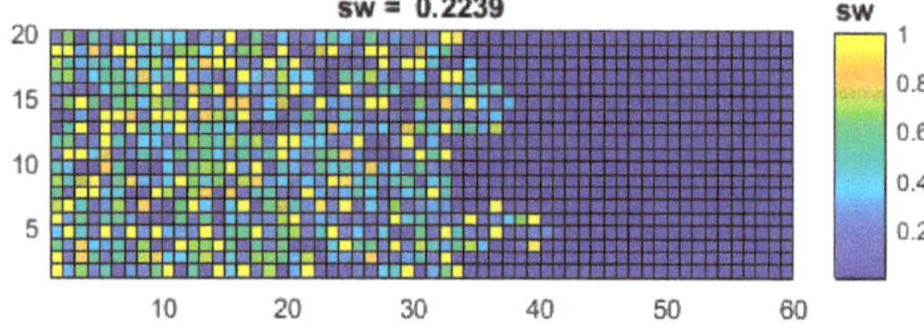

Figure 5. Water saturation profiles of the water imbibition process using the proposed dynamic pore network model with an average water saturation equal to 0.2239.

Generally, a mixed wettability is assumed for shale and tight formations. Here, we considered mixed wet conditions for Barnett shale. We assigned three types of mixed wet conditions—30% oil-wet, 50% oil-wet and 70% oil-wet—as shown in Figure 6. The corresponding spontaneous water imbibition-induced oil recoveries are shown in Figure 7 for dimensionless time $t_D = Ct\sqrt{\frac{k}{\phi}\frac{\sigma}{\mu_w}\frac{1}{L^2}}$, where C is the unit conversion factor, which is equal to 0.018849 if t is in minutes, k in md, ϕ in decimal, σ in mN/m, μ_w in cp, and L is a characteristic length in cm [49]; we follow the same units here. According to the figure, with the increasing proportion of oil-wet pores, the imbibition rate and recovery factor decrease dramatically. The results indicate that wettability dominates the water imbibition characteristics.

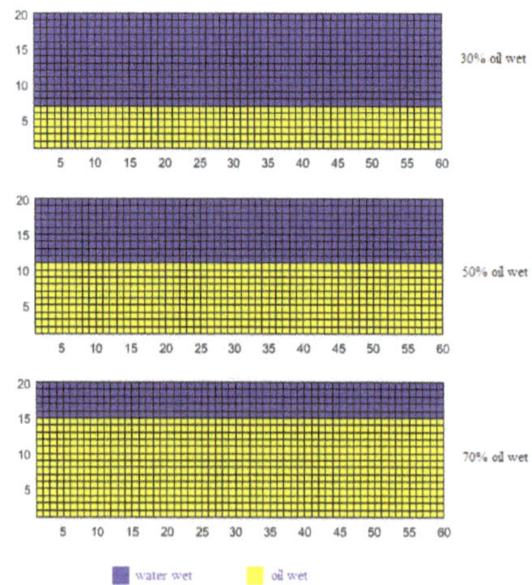

Figure 6. Three types of mixed wet conditions.

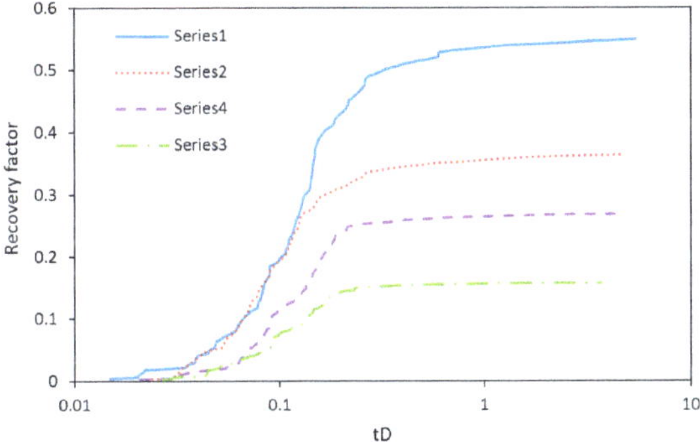

Figure 7. Spontaneous imbibition recovery with respect to dimensionless time for different wettability distributions.

The existence of micro-fractures is considered in this work to investigate their effect on water imbibition, as natural and hydraulic fractures are presented within shale and tight formations. Four lines of micro-fractures are assigned within the Barnett shale pore network, and the pore radius is set as 20 nanometers (c.f. Figure 8), which is 3.23 times the mean radius of the Barnett shale matrix; therefore, the micro-fracture permeability is approximately 10 times that of the matrix permeability. The corresponding imbibition recovery with respect to dimensionless time is shown in Figure 9. With the existence of micro-fractures, the imbibition rate increases significantly while the final recovery remains similar. The micro-factures contribute to the conductance of the pore network instead of the pore volume, and the increase of conductance increases the imbibition rate but not the final recovery, as the final recovery is controlled by the capillarity of the matrix pores and throats.

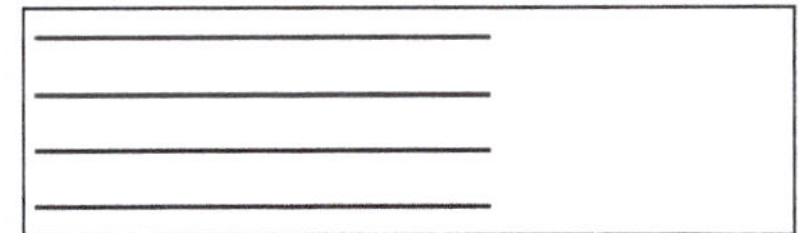

Figure 8. Schematic of micro-fracture distributions in the pore network.

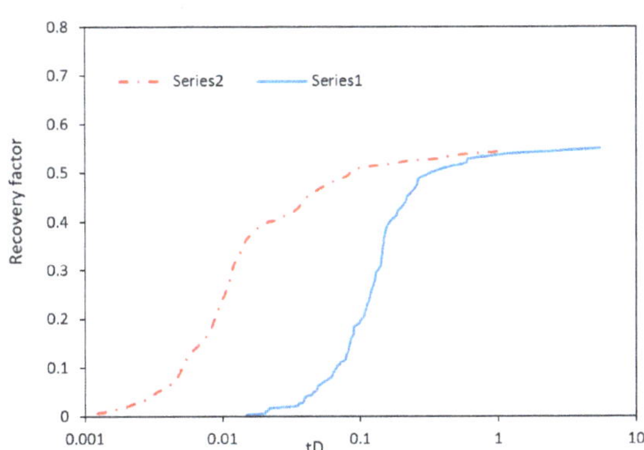

Figure 9. Water imbibition recovery factor with and without micro-fractures.

Based on the above analysis, the simulated imbibition recovery factor with respect to dimensionless time is compared with the laboratory experimental work of Barnett shale samples by Morsy and Sheng [16]. The corresponding results are shown in Figure 10. Both the effects of wettability and micro-fractures are considered to fit the experimental data. Mixed wettability for Barnett shale is found within the fitting process. Since micro-fractures are observed in the imbibition experiments, the characteristic length is reduced significantly. In the fitting results, the effective characteristic length of Barnett shale samples is approximately several millimeters, and the fraction of oil-wet pores ranges from 70% to 50%.

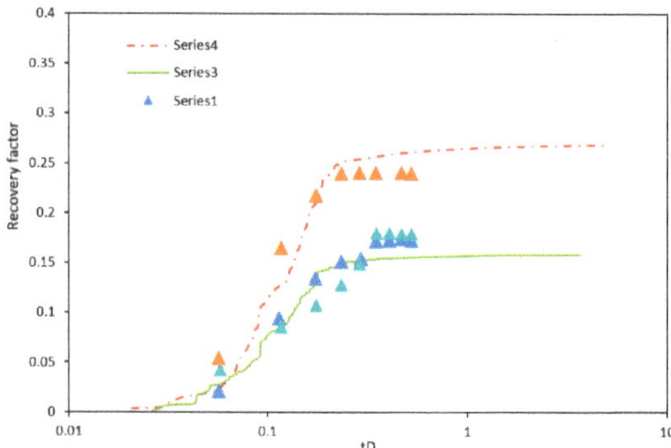

Figure 10. Imbibition recovery factors of Barnett shale: pore network modeling and experiments [16].

Moreover, the effects of the initial maximum capillary pressure and the viscosity of the non-wetting phase on water imbibition are studied in this work, and the corresponding results are shown in Figures 11 and 12. According to the figures, when the initial capillary pressure increases and the non-wetting phase viscosity decreases, the corresponding imbibition rates increase. However, the final imbibition recovery factor tends to be similar, which is because the water imbibition process is a capillary-dominated flow and the recovery factor is controlled by capillarity, which is affected by pore shape, aspect ratio, contact angle, etc. Therefore, the viscous forces in the reservoir conditions increase the imbibition rates but not the final recovery factor.

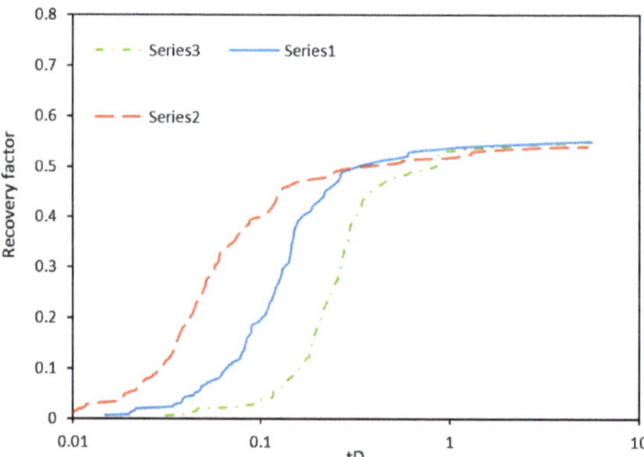

Figure 11. Spontaneous imbibition with different initial maximum capillary pressures.

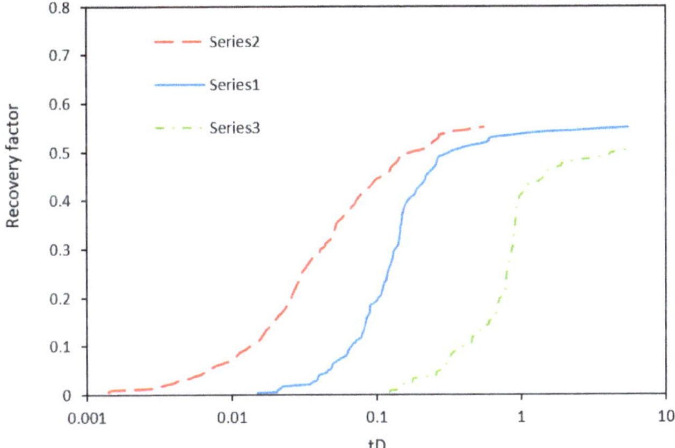

Figure 12. Spontaneous imbibition with different non-wetting phase viscosities.

To further investigate the capillarity effect on the water imbibition of Barnett shale, the following sensitivity studies are conducted: the determination of the pore throat aspect ratio, contact angle and shape factor of the cross-area. The corresponding results for the water imbibition recovery factor with respect to dimensionless time are shown in Figures 13–15, respectively. A higher aspect ratio tends to increase the percentage of the non-wetting phase trapped by the snap-off effect, which increases the residual non-wetting phase saturation and hence the final recovery factor (c.f. Figure 13). When the contact angle and cross-area shape factor decrease, leading to an increase of capillary pressure, the imbibition rate increases at the beginning. However, the final recovery factor decreases slightly (c.f. Figures 13 and 14) because snap-off tends to occur more frequently with a lower contact angle and shape factor, as shown in Equation (13).

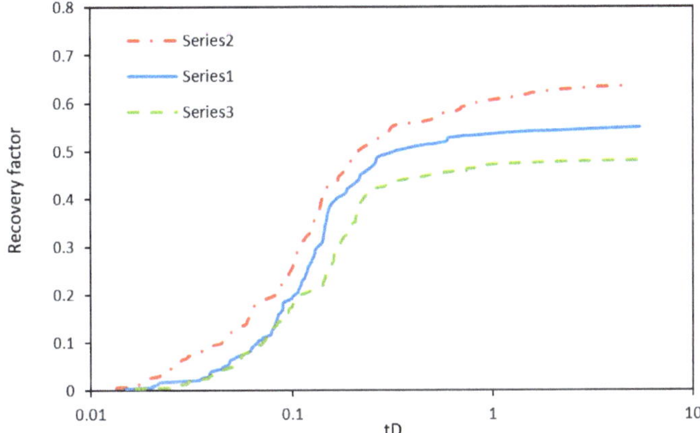

Figure 13. Water imbibition recovery factor with different pore throat aspect ratios.

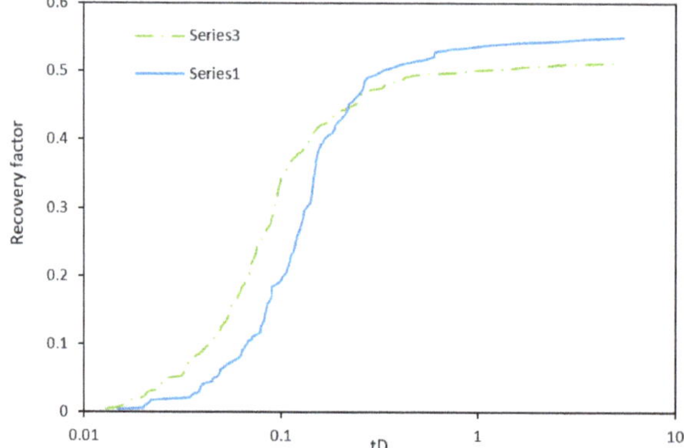

Figure 14. Water imbibition recovery factor with different contact angles.

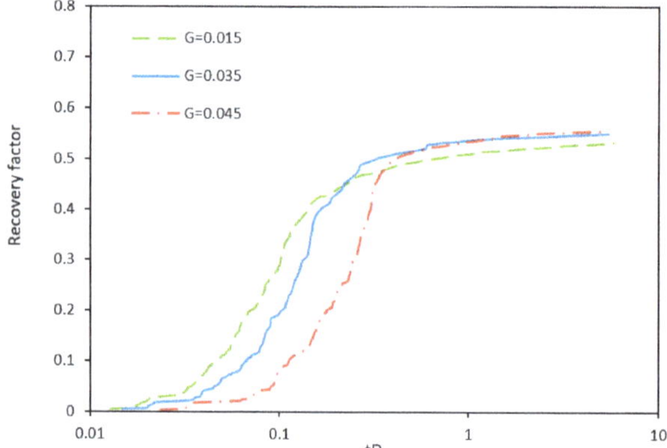

Figure 15. Water imbibition recovery factor with different shape factors.

4. Summary and Conclusions

In this paper, a dynamic pore network model is proposed, including the pore network generation, local capillary pressure function, conductance calculation and boundary conditions for water imbibition. The proposed model is applied to Barnett shale formations, and the corresponding laboratory imbibition experiments are matched using the dynamic established pore network model. Then, systematical sensitivity studies are conducted. According to the simulated results using our dynamic pore network model, wettability dominates the water imbibition characteristics. Besides this, the viscous effects including viscosity, initial capillary pressure and micro-fractures increase the imbibition rate, while the final recovery factor is controlled to a greater extent by the capillarity effect including the cross-area shape factor, contact angle and aspect ratio.

Author Contributions: X.W. proposed the model and did the simulation work under the supervision and help of J.J.S. All authors have read and agreed to the published version of the manuscript.

Funding: This work is supported by the Foundation of China Postdoctoral Science (2019M660933), Foundation of State Key Laboratory of Petroleum Resources and Prospecting (PRP/indep-4-1814) and the Opening Fund of Key Laboratory of Unconventional Oil & Gas Development (19CX05005A-3).

Conflicts of Interest: The authors declare no conflict of interest.

Nomenclature

A	Cross-area of a pore element
$B(G, \theta)$	Local capillary pressure function coefficient
$C(G, \theta)$	Coefficient accounting extra resistance of wetting layers
$E(G, \theta)$	Coefficient of piston-like displacement
$F(G, \theta)$	Coefficient of snap-off
G	Shape factor of cross-area
G_c	Corner shape factor
K	Absolute conductance of a pore element
K^r	Relative conductance of a pore element
S_w	Water saturation
p_c	Capillary pressure
L	The characteristic length of imbibition
P	The perimeter of the cross-area of a pore element
V	Pore body volume
p	Phase pressure
r	Pore or throat radius
β	Half angle of the corners
θ	Contact angle
μ	Fluid viscosity
σ	Interfacial tension
ϕ	Porosity
Δt	Time step size
t_D	Dimensionless time for imbibition upscaling
Superscript	
n	Non-wetting phase
w	Wetting phase
l	Current time step
$l+1$	Next time step
Subscript	
i	ith pore element
j	jth pore element
ij	Connecting throat between ith and jth pore elements

References

1. Shanley, K.W.; Cluff, R.M.; Robinson, J.W. Factors controlling prolific gas production from low-permeability sandstone reservoirs: Implications for resource assessment, prospect development, and risk analysis. *AAPG Bull.* **2004**, *88*, 1083–1121. [CrossRef]
2. Shaoul, J.R.; van Zelm, L.F.; De Pater, C.J. Damage mechanisms in unconventional-gas-well Stimulation–a new look at an old problem. *SPE Prod. Oper.* **2011**, *26*, 388–400.
3. Hirasaki, G.; Zhang, D.L. Surface chemistry of oil recovery from fractured, oil-wet, carbonate formations. *SPE J.* **2004**, *9*, 151–162. [CrossRef]
4. Zhang, P.; Austad, T. Wettability and oil recovery from carbonates: Effects of temperature and potential determining ions. *Colloids Surf. A Physicochem. Eng. Asp.* **2006**, *279*, 179–187. [CrossRef]
5. Sheng, J.J. Review of surfactant enhanced oil recovery in carbonate reservoirs. *Adv. Pet. Explor. Dev.* **2013**, *6*, 1–10.
6. Wang, D.; Butler, R.; Zhang, J.; Seright, R. Wettability survey in Bakken shale with surfactant-formulation imbibition. *SPE Reserv. Eval. Eng.* **2012**, *15*, 695–705. [CrossRef]
7. Alvarez, J.O.; Schechter, D.S. Wettability alteration and spontaneous imbibition in unconventional liquid reservoirs by surfactant additives. *SPE Reserv. Eval. Eng.* **2017**, *20*, 10117. [CrossRef]
8. Sheng, J.J. What type of surfactants should be used to enhance spontaneous imbibition in shale and tight reservoirs? *J. Pet. Sci. Eng.* **2017**, *159*, 635–643. [CrossRef]

9. Zhang, X.; Morrow, N.R.; Ma, S. Experimental verification of a modified scaling group for spontaneous imbibition. *SPE Reserv. Eng.* **1996**, *11*, 280–285. [CrossRef]

10. Akin, S.; Schembre, J.M.; Bhat, S.K.; Kovscek, A.R. Spontaneous imbibition characteristics of diatomite. *J. Pet. Sci. Eng.* **2000**, *25*, 149–165. [CrossRef]

11. Zhou, X.; Morrow, N.R.; Ma, S. Interrelationship of wettability, initial water saturation, aging time, and oil recovery by spontaneous imbibition and waterflooding. *SPE J.* **2000**, *5*, 199–207. [CrossRef]

12. Morrow, N.R.; Mason, G. Recovery of oil by spontaneous imbibition. *Curr. Opin. Colloid Interface Sci.* **2001**, *6*, 321–337. [CrossRef]

13. Fischer, H.; Morrow, N.R. Scaling of oil recovery by spontaneous imbibition for wide variation in aqueous phase viscosity with glycerol as the viscosifying agent. *J. Pet. Sci. Eng.* **2006**, *52*, 35–53. [CrossRef]

14. Takahashi, S.; Kovscek, A.R. Spontaneous countercurrent imbibition and forced displacement characteristics of low-permeability, siliceous shale rocks. *J. Pet. Sci. Eng.* **2010**, *71*, 47–55. [CrossRef]

15. Roychaudhuri, B.; Tsotsis, T.T.; Jessen, K. An experimental investigation of spontaneous imbibition in gas shales. *J. Pet. Sci. Eng.* **2013**, *111*, 87–97. [CrossRef]

16. Morsy, S.; Sheng, J.J. Imbibition characteristics of the barnett shale formation. In Proceedings of the SPE Unconventional Resources Conference, The Woodlands, TX, USA, 1–3 April 2014. SPE-168984-MS.

17. Lucas, R. Rate of capillary ascension of liquids. *Kolloid Z* **1918**, *23*, 15–22. [CrossRef]

18. Li, K.; Horne, R. Characterization of Spontaneous Water Imbibition into Gas-Saturated Rocks. In Proceedings of the SPE/AAPG Western Regional Meeting, Long Beach, CA, USA, 19–22 June 2000.

19. Cai, J.; Perfect, E.; Cheng, C.L.; Hu, X. Generalized modeling of spontaneous imbibition based on Hagen–Poiseuille flow in tortuous capillaries with variably shaped apertures. *Langmuir* **2014**, *30*, 5142–5151. [CrossRef]

20. Mason, G.; Morrow, N.R. Developments in spontaneous imbibition and possibilities for future work. *J. Pet. Sci. Eng.* **2013**, *110*, 268–293. [CrossRef]

21. Øren, P.E.; Bakke, S.; Arntzen, O.J. Extending predictive capabilities to network models. *SPE J.* **1998**, *3*, 324–336. [CrossRef]

22. Valvatne, P.H.; Blunt, M.J. Predictive pore-scale modeling of two-phase flow in mixed wet media. *Water Resour. Res.* **2004**, *40*, W07406. [CrossRef]

23. Behbahani, H.; Blunt, M.J. Analysis of imbibition in mixed-wet rocks using pore-scale modeling. *SPE J.* **2005**, *10*, 466–474. [CrossRef]

24. Behbahani, H.S.; Di Donato, G.; Blunt, M.J. Simulation of counter-current imbibition in water-wet fractured reservoirs. *J. Pet. Sci. Eng.* **2006**, *50*, 21–39. [CrossRef]

25. Wang, X.; Sheng, J.J. Spontaneous imbibition analysis in shale reservoirs based on pore network modeling. *J. Pet. Sci. Eng.* **2018**, *169*, 663–672. [CrossRef]

26. Wang, X.; Sheng, J.J. A self-similar analytical solution of spontaneous and forced imbibition in porous media. *Adv. Geo-Energ. Res.* **2018**, *2*, 260–268. [CrossRef]

27. Nguyen, V.H.; Sheppard, A.P.; Knackstedt, M.A.; Pinczewski, W.V. The effect of displacement rate on imbibition relative permeability and residual saturation. *J. Pet. Sci. Eng.* **2006**, *52*, 54–70. [CrossRef]

28. Blunt, M.J. *Multiphase Flow in Permeable Media: A Pore-Scale Perspective*; Cambridge University Press: Cambridge, UK, 2017.

29. Sheng, Q.; Thompson, K. A unified pore-network algorithm for dynamic two-phase flow. *Adv. Water Resour.* **2016**, *95*, 92–108. [CrossRef]

30. Blunt, M.J.; Scher, H. Pore-level modeling of wetting. *Phys. Rev. E* **1995**, *52*, 6387. [CrossRef]

31. Hughes, R.G.; Blunt, M.J. Pore scale modeling of rate effects in imbibition. *Transp. Porous Media* **2000**, *40*, 295–322. [CrossRef]

32. Idowu, N.A.; Blunt, M.J. Pore-scale modelling of rate effects in waterflooding. *Transp. Porous Media* **2010**, *83*, 151–169. [CrossRef]

33. Aghaei, A.; Piri, M. Direct pore-to-core up-scaling of displacement processes: Dynamic pore network modeling and experimentation. *J. Hydrol.* **2015**, *522*, 488–509. [CrossRef]

34. Aker, E.; Måløy, K.J.; Hansen, A.; Batrouni, G.G. A two-dimensional network simulator for two-phase flow in porous media. *Transp. Porous Media* **1998**, *32*, 163–186. [CrossRef]

35. Al-Gharbi, M.S.; Blunt, M.J. Dynamic network modeling of two-phase drainage in porous media. *Phys. Rev. E* **2005**, *71*, 016308. [CrossRef] [PubMed]

36. Tørå, G.; Øren, P.E.; Hansen, A. A dynamic network model for two-phase flow in porous media. *Transp. Porous Media* **2012**, *92*, 145–164. [CrossRef]

37. Khayrat, K. Modeling Hysteresis for Two-Phase Flow in Porous Media: From Micro to Macro Scale. Ph.D. Thesis, ETH Zurich, Zurich, Switzerland, 2016.

38. Joekar-Niasar, V.; Hassanizadeh, S.M.; Dahle, H.K. Non-equilibrium effects in capillarity and interfacial area in two-phase flow: Dynamic pore-network modelling. *J. Fluid Mech.* **2010**, *655*, 38–71. [CrossRef]

39. Khayrat, K.; Jenny, P. A multi-scale network method for two-phase flow in porous media. *J. Comput. Phys.* **2017**, *342*, 194–210. [CrossRef]

40. Mason, G.; Morrow, N.R. Capillary behavior of a perfectly wetting liquid in irregular triangular tubes. *J. Colloid Interface Sci.* **1991**, *141*, 262–274. [CrossRef]

41. Ransohoff, T.C.; Radke, C.J. Laminar flow of a wetting liquid along the corners of a predominantly gas-occupied noncircular pore. *J. Colloid Interface Sci.* **1988**, *121*, 392–401. [CrossRef]

42. Bakke, S.; Øren, P.E. 3-D pore-scale modeling of sandstones and flow simulations in the pore networks. *SPE J.* **1997**, *2*, 136–149. [CrossRef]

43. Patzek, T.W. Verification of a complete pore network simulator of drainage and imbibition. In Proceedings of the SPE/DOE Improved Oil Recovery Symposium, Tulsa, OK, USA, 3–5 April 2000.

44. Andersen, P.Ø.; Skjæveland, S.M.; Standnes, D.C. A Novel Bounded Capillary Pressure Correlation with Application to Both Mixed and Strongly Wetted Porous Media. In Proceedings of the SPE Abu Dhabi International Petroleum Exhibition & Conference, 13–16 November 2017. SPE-188291-MS.

45. Aziz, K.; Settari, A. *Petroleum Reservoir Simulation*; Chapman & Hall: London, UK, 1979.

46. Moghaddam, R.N.; Jamiolahmady, M. Fluid transport in shale gas reservoirs: Simultaneous effects of stress and slippage on matrix permeability. *Int. J. Coal Geol.* **2016**, *163*, 87–99. [CrossRef]

47. Wang, X.; Sheng, J.J. Pore network modeling of the non-Darcy flows in shale and tight formations. *J. Pet. Sci. Eng.* **2018**, *163*, 511–518. [CrossRef]

48. Morrow, N.R. The effects of surface roughness on contact: Angle with special reference to petroleum recovery. *J. Can. Pet. Technol.* **1975**, *14*, 42–53. [CrossRef]

49. Ma, S.; Zhang, X.; Morrow, N.R. Influence of fluid viscosity on mass transfer between rock matrix and fractures. *J. Can. Pet. Technol.* **1999**, *38*, 25–30. [CrossRef]

 *energies*

Article

Patent Analysis on the Development of the Shale Petroleum Industry Based on a Network of Technological Indices

Jong-Hyun Kim and Yong-Gil Lee *

Department of Energy Resources Engineering, Inha University, 100 Inharo, Nam-gu, Incheon 22212, Korea; kjhpov@gmail.com
* Correspondence: leedomingo@inha.ac.kr; Tel.: +82-32-860-7555

Received: 2 November 2020; Accepted: 16 December 2020; Published: 21 December 2020

Abstract: This study investigated the technological developments in the shale petroleum industry by analyzing patent data using a network of technological indices. The technological developments were promoted by the beginning of the shale industry, and after the first five years, it showed a more complex development pattern with the convergence of critical technologies. This paper described progress in the shale petroleum technologies as changes in relatedness networks of technological components. The relatedness represents degree of convergence between technological components, and betweenness centrality of network represents priority of technological components. In the results, the progress of the critical technologies such as directional drilling, increasing permeability, and smart systems, were actively carried out from 2012 to 2016. Especially, unconverged technology of increasing permeability and the converged technology of directional drilling and smart system has been intensively developed. Some technological components of the critical technologies are more significant in the form of converged technology.

Keywords: shale gas; tight oil; shale petroleum; technological development; patent; network analysis

---

1. Introduction

Unconventional petroleum has gathered great interest in recent times, especially with growing concerns over the depletion of conventional petroleum sources. In a few countries, unconventional petroleum is already being produced economically along with a steady growth in the industry.

In the US, unconventional petroleum, such as shale oil and tight gas, has seen an increase in production amounts as energy resources. In 2008, shale gas and tight oil constituted as much as 16 and 12% of production of natural gas and crude oil in the US, respectively. In 2018, these production amounts reached 70 and 60%, respectively [1]. This economical production of shale petroleum (shale gas and tight oil) can be considered as an achievement of technological advancement in shale petroleum. Despite the economic benefit of shale petroleum, the environmental problem is the most important factor that could pose a threat to shale petroleum production. According to Cooper et al. [2], shale petroleum causes greenhouse gas (GHG) emissions, water overuse, and local issues around the production site.

According to Holditch [3], the mechanisms for producing unconventional petroleum versus conventional petroleum are distinguishable in two ways. First, increasing the permeability of underground formation is a critical mechanism to produce unconventional petroleum such as shale gas and tight oil, which typically are un-permeable formations that contain petroleum.Second, reducing the viscosity of hydrocarbons is a critical mechanism to produce unconventional petroleum such as heavy oil and oil shale. Permeable underground formations contain excessively viscous petroleum, making it difficult for this petroleum to flow into an apparatus or equipment. Therefore, producing unconventional petroleum of a high viscosity requires a method that makes petroleum flowable by using thermal energy and other methods. For example, steam-assisted gravity drainage (SAGD) supplies thermal energy to petroleum, such as bitumen or heavy oil, by injecting it with high temperature steam [4]. Furthermore, producing unconventional petroleum in tight formation requires a method that makes the formation permeable by connecting the production well with the reservoir. For example, shale gas and tight oil have been economically produced by adopting horizontal drilling and hydraulic fracturing [5].

Geny [5] suggested that the factors responsible for the successful and economical production of shale petroleum are the advanced technologies of hydraulic fracturing and horizontal drilling, and the combination of the two technologies. Construction of horizontal wells began in 2003 in the US. The industrial growth of shale petroleum started approximately 4 years since the start of construction. Furthermore, this technological development, which initiated the industrial growth of shale petroleum, has contributed toward associated results such as diversification in the operation strategies of energy companies [6], and regional growth in income and employment [7].

According to previous studies [8–15], production technologies of shale petroleum have grown since production began. Kim and Lee [8] argued that productivity grew by 1.9%, while cumulative production of shale gas doubled from 2008 to 2016. However, the prices of natural gas and crude oil declined in late 2008, and the price of crude oil declined in late 2014. Moreover, the price of natural gas has settled down under 4 dollars per million British thermal unit [16,17]. Due to the decline in the prices of natural gas and crude oil, questions regarding the economic feasibility of producing shale petroleum have cropped up. Even with the decline in the prices of natural gas and crude oil, the productivity gain by developing technologies seemed stagnant until 2013 [9]. In addition, during the period with low commodity prices, some producers tried to gain productivity by changing the proportion of oil (or gas) in the production portfolio [10], or cutting the service costs [11]. Analyzing data from North Dakota's Bakken shale formation, Covert [12] argued that shale petroleum producers have slowly and insufficiently improved their production skills until 2011. Nevertheless, the shale petroleum industry has continued to increase its annual production. Moreover, some factors, such as improvements in the decline curves and increase in the lateral length of the wellbore, indicate that the development of production technologies has affected productivity from 2013 to 2016. The optimization of the production of a well could be conducted after its completion by refracturing the reservoir [14]. West [15] reviewed the improvement in productivity and the issues in production technologies by analyzing the topics of technical papers on shale petroleum from 2018 to 2019. The study found promising technologies such as enhanced recovery, digitalization of instruments, machine learning, advanced modeling, and method or apparatus for solving parent-child issues.

Furthermore, recent research [18–22] showed the various ways of technological development in the shale industry. Davarpanah [18] tested and suggested rheologically effective formate fluids for shale formation by composition. Davarpanah and Mirshekari [19] suggested enhancing gas recovery for shale formations based on a model with improved prediction accuracy on diffusivity in carbon dioxide and methane akinetic absorption. Hu et al. [20] empirically showed improved oil-recovery enhancement for tight reservoirs from an optimized injection method of foam and brine components. Hu et al. [21] also showed a trade-off between carbon dioxide-injection and oil-recovery for enhancing recovery. Jin and Davarpanah [22] suggested water treatment techniques that reduced water use by at least 70% for the enhanced oil-recovery method.

Owing to the arguments calling for productivity gains in the shale petroleum industry, research on the technological development of shale gas has been of interest and has been carried out primarily via patent analysis. International Science and Economic Development Canada (ISED) [23] called for the need for a technological field of shale petroleum production and for a major developer in the technological field. The author summarized that major petroleum companies in the US specialize mainly in technologies such as well casing drilling, fracturing formation, drilling formation, data detection, and determining data. Wei et al. [24] observed that US petroleum companies hold the majority of the technologies, such as equipment and device for drilling, extraction exploration, feed purification, and technology for digital simulation. Kim and Lee [25] argued that the critical technological aspects of the shale petroleum industry, such as obtaining resources, investigation, and data processing, have been actively developed from 2010 to 2016.

This study describes the development of technologies of directional drilling (DD) and increasing permeability (IP), which are core technologies of the shale industry, in terms of convergence with smart systems (SS). This is considering that the shale industry's technological developments were focused on productivity improvement. This study found the following: First, the intensity of technological development, measured by the incidence of patents, increased significantly from 2012 to 2016. This point is concordant with the results of Sandrea [9] who observed that there was progress in the productivity of shale petroleum technology. Second, since 2012, a high proportion of technological developments of DD have been developed in the form of convergence with SS. Third, IP technology had a low tendency to converge with DD and SS, but both the number of patents developed and the complexity of the technology were the highest. Furthermore, "reinforcing fractures by using prop" appeared to be the most critical field of IP technologies since 2012. This result is consistent with the results of Shah et al. [14] who also found that productivity improvement was achieved through reinforcing in the shale industry.

2. Data and Methodology

This study investigated the development of technologies of shale petroleum by analyzing patent data. This study utilized patent data related to production technologies of shale petroleum, such as IP, DD, and SS, from 1997 to 2016. For its analysis, this study calculated the association strength and betweenness centrality by utilizing the most finely distinguished scope (full digit) of technological index (TI) of patent, such as international patent classification (IPC) and cooperative patent classification (CPC).

The analysis of this study has some features. This study focuses only on a portion of technologies of the shale petroleum industry from the data collection stage, focusing on only 26 of the approximately 3900 technology indices. Thus, there are limitations to presenting comparative analysis of various technologies, and to presenting new technologies in an exploratory manner. Still, this study has some advantages. It focuses on the critical technologies of the shale petroleum industry and has the advantage of using association strength, instead of cosine similarity, as the similarity measure. According to Eck and Waltman [26], association strength is an unbiased measure compared to other similarity measures, such as cosine similarity. This is because association strength is not substantially correlated to the occurrence of input data. However, cosine similarity is positively correlated to the occurrence of input data. That is, a frequently occurring TI tends to have higher similarity than a less frequently occurring TI. Lastly, this study distinguishes and presents the results by technological domains, and suggests the results in a numerical and visualized form.

2.1. Data

We retrieved patent data from the Korea intellectual property right information service (KIPRIS), an online patent database system of the Korea intellectual property organization (KIPO) [27]. The retrieved data set includes patents for the technologies of American shale petroleum. The retrieval process was as follows: First, we built a searching query by focusing on three critical technologies of unconventional petroleum, namely "directional drilling," "stimulating production by increasing permeability," and "smart system for control, surveying, or testing." This is because, as described in Section 1, DD and the stimulation technologies of unconventional petroleum have taken the critical role of production and initiated the industrial growth of unconventional petroleum in the US. Furthermore, SS are required to facilitate productive operation, advance the apparatus or method of DD, and increase permeability [28]. Second, we focused on the patents applied to the US patent office. This is because only the US has advanced in the growth of an unconventional petroleum (shale gas and tight oil) industry since 2007.

To collect patent data, we built a searching query that comprised the TIs of three technological domains: directional drilling (DD), stimulating production by increasing permeability (IP), and smart systems for control, surveying, or testing (SS). The TIs pertaining to the three technological domains are shown in Table 1. In Table 1, the abbreviations "DD," "IP," and "SS" indicate TIs "E21B 43/," "E21B 7/," and "E21B 44/." For example, TI IP26 indicates E21B 43/26. The descriptions of TIs are provided in Table 1.

Table 1. Description of technological indices (TIs) by technological domains [29].

Technological Domains	Technological Index	
	DD04	Directional drilling
	DD043	Directional drilling with underwater environment
	DD046	Horizontal drilling
	DD06	Deflecting direction of borehole
	DD061	Tools such as shaft, advancing relative to the guide
Directional drilling (DD: E24B)	DD062	Tools such as shaft, rotating inside a non-rotating guide and traveling with shaft
	DD064	Tools adapted with drill bits
	DD065	Tools using fluid jets
	DD067	Tools locking sections of a pipe or the guide for a shaft
	DD068	Drilling by using down-hole drilling motor
	DD10	Correction of deflected borehole
	IP26	Forming crevices or fractures
	IP2605	Forming crevices or fractures by using gas or liquefied gas
	IP2607	Surface equipment for fracturing operation
Stimulating production by increasing permeability (IP)	IP261	Separated process of completion (1) cementing, plugging, or consolidating, and (2) fracturing or forming formation
	IP263	Forming crevices or fractures by using explosives
	IP2635	Forming crevices or fractures by using nuclear energy
	IP267	Reinforcing fractures by using prop
	IP27	Forming crevices or fractures by using eroding chemicals (acids)
	SS00	Automatic control system for drilling
	SS005	Underground automatic control system
	SS02	Automatic control of the tool feed
Smart system for control, surveying, or testing (SS)	SS04	Tool feed's automatic control responding to the torque of the drive
	SS06	Tool feed's automatic control responding to the flow or pressure of the motive fluid of the drive
	SS08	Tool feed's automatic control responding to the amplitude of the movement of the percussion tool
	SS10	Arrangements of the automatic control stopping process when the tool is lifted from the working face
Convergence of DD and IP (CDI)	Includes technological indices from both DD and IP simultaneously	
Convergence of DD and SS (CDS)	Includes technological indices from both DD and SS simultaneously	
Convergence of IP and SS (CIS)	Includes technological indices from both IP and SS simultaneously	

As shown in Table 1, this study classifies the retrieved patents into six kinds of technologies such as DD, technologies that stimulate production by IP, SS technologies, and three kinds of converged technologies, such as convergence of DD and IP (CDI), convergence of DD and SS (CDS), and convergence of IP and SS (CIS) that involve the TIs within DD, IP, and SS, respectively.

Through the searching query, we found 12,964 applied patents from 1960 to 2019. However, this study focused only on the 6421 granted patents, which were applied from 1997 to 2016, as shown in Figure 1. In Figure 1, the blue line represents the number of applied patents, the orange line represents the number of granted patents, and the gray dash represents the granted ratio. The granted patent count is ordered by the application date.

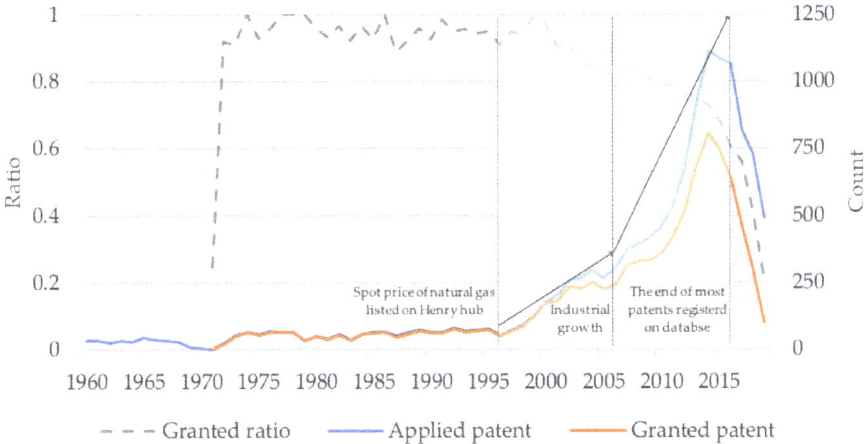

Figure 1. Annual patent count for the whole dataset—1967 to 2019 [27].

In Figure 1, the trend of the blue line increases twice around 1997 and 2007. The supposed reasons that affect the trend are that the spot prices of natural gas were listed in Henry hub in January 1997, and the shale petroleum industry started commercial production in early 2007. In addition, the applied patent counts, the granted patent counts, and the granted ratio have rapidly decreased since 2017. The reason for the sharp decline in observations (blue and orange lines) is considered as an incomplete aggregation of the database. Thus, this study excludes the observations applied since 2017. Moreover, the reason for this study to use observation as an input of analysis is that the occurrence of technology is recognized only when the patent data occurred consecutively over at least 5 years. Thus, the discontinued observations are excluded.

2.2. Methodology

This study utilizes association strength as the technological relatedness measure between TIs of patent [30], and betweenness centrality as a priority measure of TI [31]. The calculation process of betweenness centrality was performed by using the software package "networkX" [32].

2.2.1. Technological Relatedness: Association Strength

In this study, we calculated the association strength similarity. The calculation processes of the two measures were undertaken by following the formulas provided below [30].

Equations (1)–(5) show the calculation process of association strength similarity.

$$cti_{ij} = \sum_{p=1}^{m} ti_{pi}ti_{pj}, \ for \ i \neq j, \tag{1}$$

In Equation (1), cti_{ij} is the number of co-occurrence of TIs i and j. ti_{pi} and ti_{pj} are terms for counting the number of TIs in the patent. $ti_{p\cdot}$ is one ($ti_{p\cdot} = 1$) when the TIs i or j exist in patent p, and $ti_{p\cdot}$ is zero $\left(ti_{p\cdot} = 0\right)$ when either TIs do not exist in patent p. Thus, cti_{ij} becomes one when both ti_{pi} and ti_{pj} are one. m is number of total patents for a five-year research period.

$$sti_i = \sum_{j=1,j\neq i}^{m} cti_{ij}, \tag{2}$$

$$sti_j = \sum_{i=1,j\neq i}^{n} cti_{ij}, \tag{3}$$

$$T = \sum_{i=1}^{n} \sum_{j=1}^{n} cti_{ij}, \tag{4}$$

In Equations (2) and (3), $sti_{i\ or\ j}$ is the total co-occurred number of TIs i or j for a five-year research period. T is sum of total co-occurred number of whole TIs.

$$S_{C_{ij}} = \frac{\frac{cti_{ij}}{T}}{\frac{sti_i}{T}\frac{sti_j}{T}}, \tag{5}$$

In Equation (5), $S_{C_{ij}}$ is the similarity measure between TIs i and j, which occur over a 5-year research period.

2.2.2. Betweenness Centrality

This study uses betweenness centrality to determine the comparative importance of TIs in the graph of shale petroleum technologies [31].

$$C_B(k) = \sum_{\substack{v \neq k \neq w \\ k,v,w \in K}} \frac{\sigma_{vw}(k)}{\sigma_{vw}}, \tag{6}$$

In Equation (6), $C_B(k)$ is the betweenness centrality of TI k. k is an element of K, a set of whole TIs. Each TI is a node in the graph, which comprises TIs and their similarities. σ_{vw} denotes the number of shortest paths from nodes v and w. $\sigma_{vw}(k)$ denotes number of shortest paths through node k.

3. Results

3.1. Development of Unconventional Petroleum

Section 3 focuses on describing the technological development and convergence of the unconventional petroleum technologies. First, we can easily identify differences in the extent of technological development of unconventional petroleum by validating the annual patent counts of each technological domain. Figure 2 presents information concerning the granted patent counts of three technological domains and their converged technologies, and the weight of converged technology of DD, IP, and SS.

In Figure 2A, the orange, gray, and yellow lines represent the annual counts of granted patent, including the TIs of DD, IP, and SS with their converged technologies. Broadly, the granted patent counts of the three technological domains increased from 1997 to 2014. In particular, from 2009 to 2014, patents related with IP (gray line) rapidly expanded from 170 to 464 patents per annum. From 2011 to 2014, patents related to DD (orange line) rapidly expanded from 99 to 186 patents per annum, and patents related with SS (yellow line) rapidly expanded from 62 to 186 patents per annum.

Figure 2B shows the weight of converged technology of the three technological domains (DD, IP, and SS). Interestingly, the weight of converged technology of IP (gray line) shows a very low level of weight in converged technology. While the weight of converged technologies of DD and SS (orange and gray lines, respectively) have fluctuated around 0.2 from 1990s to 2000s, they have increased from 0.2 to 0.4 since 2011.

Figure 2C shows the annual counts of granted patents in converged technologies. The convergence of DD and SS (orange line) always shows higher annual counts than others and have expanded from 2011 to 2014. The convergence of IP and SS (gray line) showed a zero count before 2009, and then showed 11 patents per annum at its peak point in 2014. The convergence of DD and IP (yellow line) also showed a zero count before 2004; it expanded from 2 patents per annum in 2011 to 14 patents per annum in 2016.

Figure 2D shows the trend in the annual count of granted patents for DD, IP, and SS, which is very similar to the orange, gray, and yellow lines of Figure 2A. This is because the weights of the converged technologies are quite stable for the research period, as shown in Figure 2B.

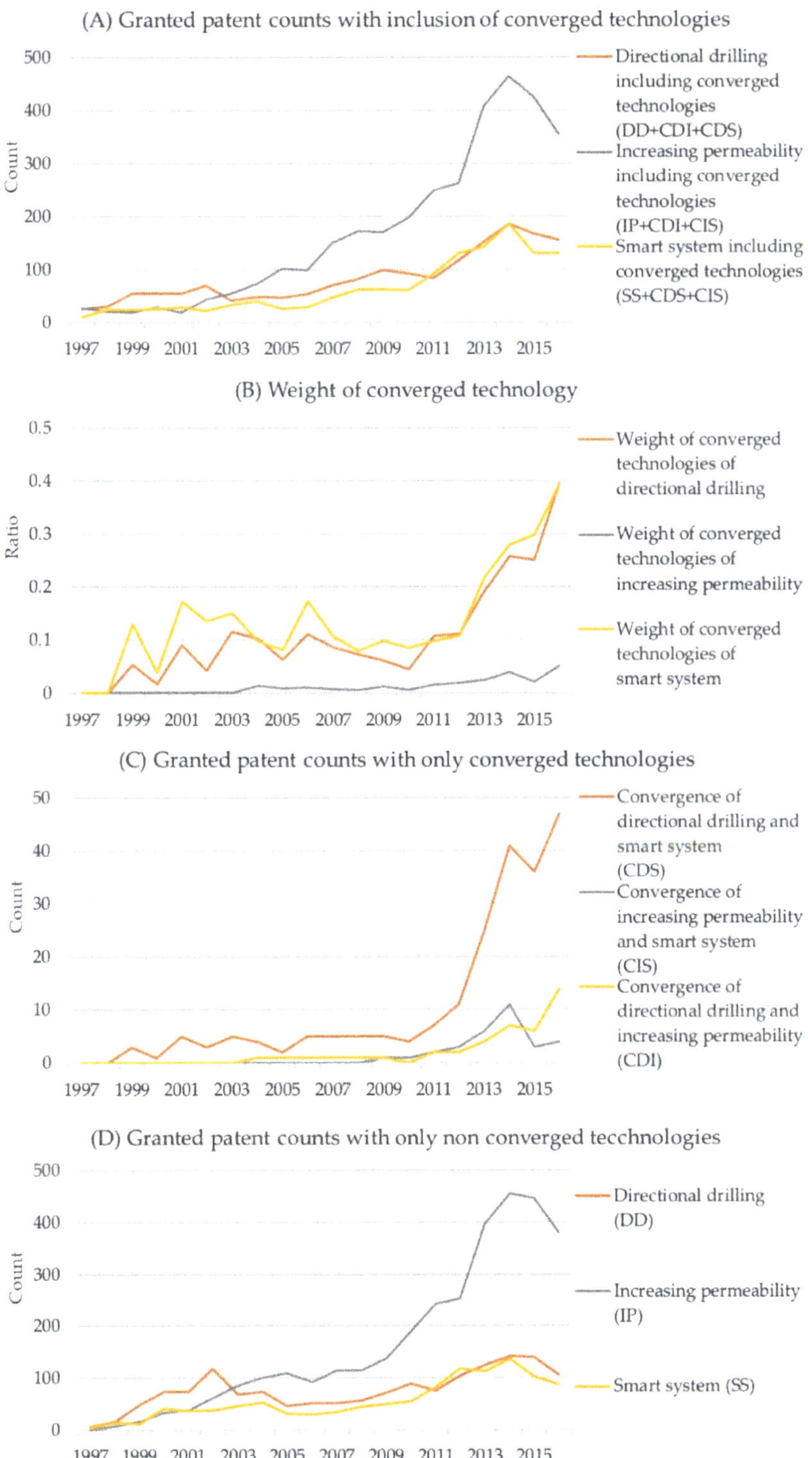

Figure 2. Annual patent counts by technological domains–1997 to 2016 [27].

In summary, the intensity of technological development has increased in the last 20 years. Moreover, in the past 10 years, converged technologies such as CDS, CIS, and CDI have been developed. Technologies related to DD and SS show a lower extent of technological development with a relatively higher weight of converged technology than IP. Technology related to IP shows a higher intensity of technological development with a lower weight of converged technology. Only two patent technological domains, those of CDS and CDI, rebounded in their annual count of granted patents in 2016. In the next subsections, this study presents technological development from the network aspect of technological relatedness.

3.2. Network of Shale Petroleum Technologies

This subsection attempts to describe the technological development of shale petroleum by presenting network properties and visualizing the networks of technological relatedness. The network properties show the development of the patent set from the aspect of a network of technological relatedness. Table 2 presents the network properties of patent technological relatedness in 5-year periods.

Table 2. Network properties of shale petroleum technologies' patents.

Properties	Period			
	1997–2001	2002–2006	2007–2011	2012–2016
Patent count	453	786	1721	3254
Number of nodes	649	995	1930	3877
Number of edges	6197	10,001	24,781	64,598
Ratio between edges and nodes	9.55	10.05	12.84	16.66

Table 2 describes the network development for different research periods. The patent counts grew by 94% on average in each period, while the number of nodes grew by 83%, the number of edges by 123%, and the ratio between edges and nodes by 21%. As described in the previous subsection, the values of Table 2 also show that the intensity of development in shale gas technology has increased. In addition, the number of nodes, which means the number of TIs, has also increased. Meanwhile, the connection between nodes has increased more rapidly. The increased edges and nodes in these networks could mean that these patents contain more combinations of TIs than before. The increased combination of TIs means an increased combination of new technological components, and thus, the emergence of a new technology is expected.

Figure 3 shows networks of technological relatedness to help understand the growth of the networks. The visualized networks show only the network of technological relatedness in the first and last periods. The intermediate process is omitted because the networks show a steadily increasing trend during the research period. Figure 3A,B show the visualized networks of technological relatedness from 1997 to 2001 and from 2012 to 2016, respectively. Figure 3 was drawn using Gephi, a visualization network software [33].

Figure 3 shows networks of technological relatedness of shale petroleum technologies from 1997 to 2001 (Figure 3A) and from 2012 to 2016 (Figure 3B). In Figure 3, the letters in orange, red, and blue represent the TIs of IP, DD, and SS.

When Figure 3A is compared to Figure 3B, the latter has a more compact shape than the former. This difference in the visualized results is due to the quantitative difference in patent count, the number of nodes and edges between the two patent groups as shown in Table 2, and the difference in ratio between the number of edges and nodes. In addition, the distance between the TIs of DD and SS have become closer. The TIs of IP are still separated from other TIs. This point seems to be influenced by the high number of patents of CDS, as shown in Figure 2C. These differences in the distance between TIs can be described by the association strength similarity of TI, which represents technological relatedness. The technological relatedness between technological domains is presented in Table 3.

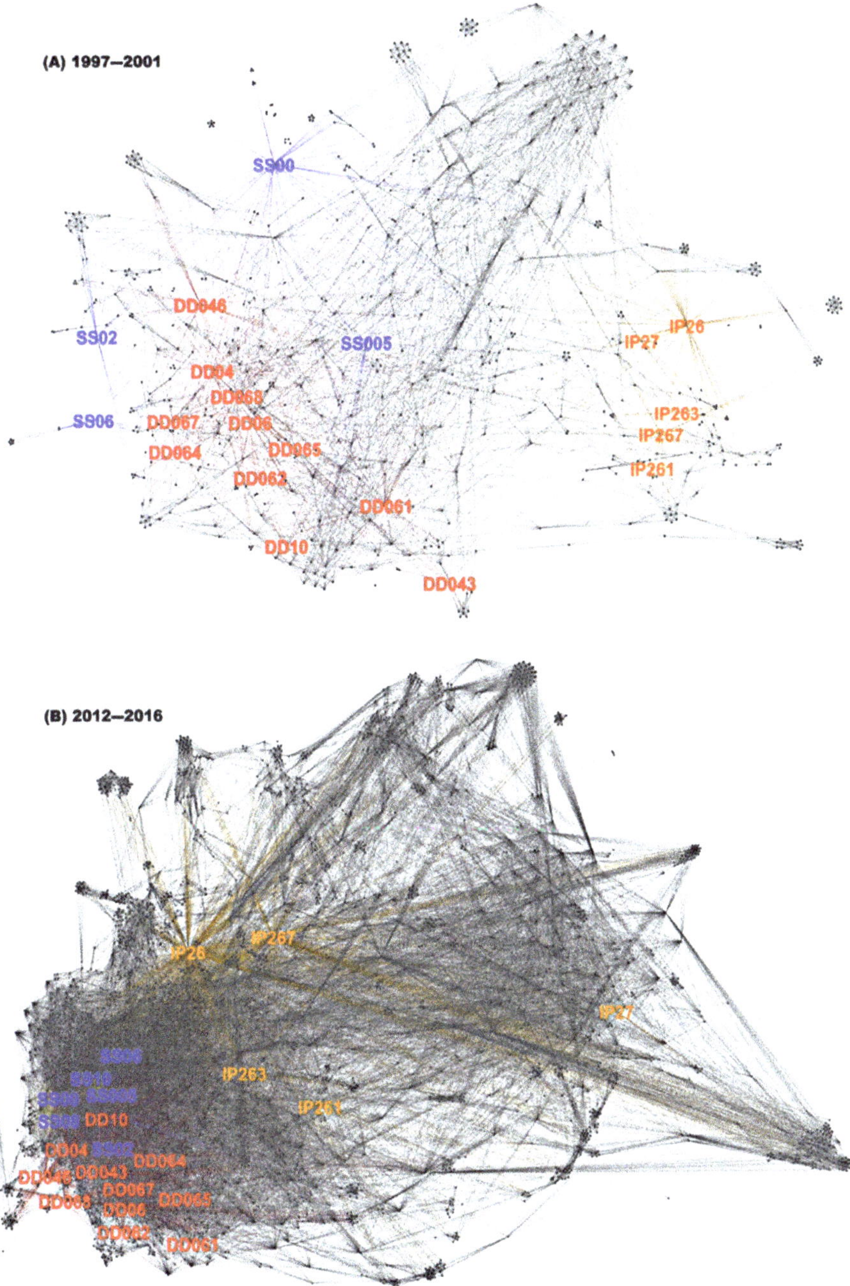

Figure 3. The network of technological relatedness of shale petroleum technologies for 1997 to 2001 (**A**) and 2012 to 2016 (**B**).

Table 3. Technological relatedness between technological domains.

Measure	Technological Domains	Period			
		1997–2001	2002–2006	2007–2011	2012–2016
Association strength between the center nodes of technological domains	DD and SS (DD04 and SS00)	2.82×10^{-2}	3.16×10^{-2}	3.41×10^{-2}	2.90×10^{-2}
	IP and SS (IP26 and SS00)	-	-	1.72×10^{-3}	2.34×10^{-3}
	DD and IP (DD04 and IP26)	-	1.20×10^{-3}	1.05×10^{-3}	2.67×10^{-3}
Number of combinations of technological indices between technological domains	DD and SS	19	18	28	43
	IP and SS	-	-	2	5
	DD and IP	-	3	6	13
Sum of association strength of each combination	DD and SS	2.83×10^{-1}	2.85×10^{-1}	2.93×10^{-1}	3.79×10^{-1}
	IP and SS	-	-	2.31×10^{-3}	6.35×10^{-3}
	DD and IP	-	4.59×10^{-3}	1.29×10^{-2}	1.86×10^{-2}

Table 3 presents the association strength between the center nodes of the technological domains, the number of edges between technological domains, and the sum of association strength similarity of edges between technological domains.

Lines 1–3 in Table 3 show association strength, which represents the weight of the edge between the center nodes of technological domains such as DD, IP, and SS. These association strength similarities show the relatedness between the technological domains. In addition, the higher the association strength, the closer the distance between the converged technologies in the network of TIs. In lines 4–6 of Table 3, the number of combinations of TIs between technological domains represents the number of ways technologies converge. Thus, the convergence of DD and SS represents CDS, the convergence of IP and SS represents CIS, and the convergence of DD and IP represents CDI. In the lines 7–9 of Table 3, the sum of association strength in each combination between the TIs shows the changes in the aggregate quantitative relatedness between technologies.

In the case of CDS, the association strength similarity between DD (DD04) and SS (SS00) has increased by 2.80% for approximately 20 years. The number of combinations also increased by approximately two times over the same period. Moreover, the sum of the association strength increased by 33.82% over the same period. These changes in values are in concordance with the results from Figure 3. These changes show that the convergence in technology is conducted in a more detailed way. Thus, it implies an improvement in the level of technological development.

In the case of CIS, the association strength similarity between IP (IP26) and SS (SS00) occurred for the last two periods. Compared to the initial state, the association strength from 2012 to 2016 increased by 35.61%. The sum of association strength increased by 175.33%. The number of combinations has more than doubled. However, the number of edges was the lowest compared to the other converged technologies. Furthermore, compared to other technologies, CIS occurred recently and has not grown yet. In Table 4, while the granted patent counts of CIS increased, the ratio between the edges and nodes of CIS decreased compared to the initial state in the period from 2012 to 2016. Information pertaining to granted patent counts and other network properties by technological domains are summarized in Table 4.

In the case of CDI, the association strength between DD (DD04) and IP (IP26) occurred for the last three periods. Compared to the initial state, the association strength increased by 121.53%, while the sum of association strength increased by 305.76%. Interestingly, the number of edges increased two times for each of these periods. These results show that CDI has developed in a manner that has increased the various ways in which technology converges. In particular, the newly emerged ways of technological convergence seem to increase the technological relatedness between DD and IP. In Table 4, the granted patent count for CDI is six to ten times bigger in the period 2012 to 2016 compared to the previous periods. Furthermore, the ratio between edges and nodes increased two times in the period 2012 to 2016 compared to the previous periods.

Table 4. Network properties by technological domains.

Technological Domain	Property	Period			
		1997–2001	2002–2006	2007–2011	2012–2016
Directional drilling (DD)	Patent count	212	243	396	585
	Number of nodes	271	345	532	899
	Number of edges	1902	2285	4946	10,811
	Ratio between edges and nodes	7.02	6.62	9.30	12.03
Increasing permeability (IP)	Patent count	111	371	931	1856
	Number of nodes	245	560	1196	2662
	Number of edges	2118	5831	15,658	42,416
	Ratio between edges and nodes	8.64	10.41	13.09	15.93
Smart system (SS)	Patent count	103	131	294	534
	Number of nodes	242	273	547	1017
	Number of edges	1494	1480	3891	11,021
	Ratio between edges and nodes	6.17	5.42	7.11	10.84
Convergence of DD and SS (CDS)	Patent count	27	38	91	219
	Number of nodes	107	106	145	358
	Number of edges	1032	775	1048	4640
	Ratio between edges and nodes	9.64	7.31	7.23	12.96
Convergence of IP and SS (CIS)	Patent count	-	-	4	27
	Number of nodes	-	-	31	94
	Number of edges	-	-	222	609
	Ratio between edges and nodes	-	-	7.16	6.48
Convergence of DD and IP (CDI)	Patent count	-	3	5	33
	Number of nodes	-	15	15	133
	Number of edges	-	57	62	1324
	Ratio between edges and nodes	-	3.80	4.13	9.95

Furthermore, Table 4 shows that IP (15.93) is the highest for the ratio between edges and nodes by technology area, followed by CDS (12.96), DD (12.03), SS (10.84), CDI (9.95), and CIS (6.48). Interestingly, although the IP has converged less with DD and SS, the technological development of IP shows that the combination of technology elements has progressed in a more complex pattern (high connectivity).

As described before, not only have shale petroleum technologies developed, the relationships between technologies have also changed. Next, this study attempts to determine the priorities of technological components, which are assumed to have changed with the development of technology, by presenting the betweenness centrality of TI.

3.3. Priority of Technological Components

In this section, the betweenness centrality is presented to validate the priority of technological components by technological domains. Table 5 presents the betweenness centrality of 10 TIs from the top for the four periods between 1997 and 2016.

Table 5. Betweenness centrality of technological indices (TIs) by period.

Order	Period							
	1997–2001		2002–2006		2007–2011		2012–2016	
	TI	Betweenness Centrality	TI	Betweenness Centrality	TI	Betweenness Centrality	TI	Betweenness Centrality
1	SS00	0.45	DD04	0.38	SS00	0.60	SS00	0.43
2	DD061	0.18	SS00	0.33	IP267	0.32	IP267	0.24
3	DD046	0.11	DD061	0.27	DD04	0.16	DD04	0.17
4	IP267	0.11	IP267	0.23	DD046	0.07	DD06	0.11
5	DD04	0.09	DD046	0.19	DD06	0.05	SS02	0.02
6	SS005	0.06	SS06	0.05	IP27	0.04	IP263	0.02
7	DD068	0.06	DD068	0.04	DD068	0.03	DD068	0.01
8	DD06	0.05	DD06	0.03	IP263	0.02	DD046	0.01
9	IP263	0.03	IP27	0.01	DD061	0.01	DD064	3×10^{-3}
10	SS02	0.02	SS02	0.01	SS02	0.01	SS005	2×10^{-3}

In Table 5, SS00 (automatic control system for drilling) is always ranked at the top. This result implicates that the technological development of the shale petroleum industry has focused on the smart system, which also has a high weight of convergence with DD for Figure 1. IP267 (reinforcing fractures by using prop) has been ranked the second highest from 2007 to 2016. This result is in concordance with the results of Shah et al. [14] Furthermore, DD04 (directional drilling), DD046 (horizontal drilling), DD06 (deflecting direction of borehole), and DD061 (tools such as shaft rotating inside a non-rotating guiding traveling) have been ranked between second and fifth. Interestingly, DD046 was ranked eighth in the last period, while DD06 was still ranked fourth. This result may implicate that the optimization of the production process seems to be more dependent on the technologies related to deflecting the direction of borehole than horizontal drilling, which is a recent technological development. SS005 (underground automatic control system), SS06 (tool feeds' automatic control, which responds to the flow or pressure of the motive fluid of drive), SS02 (automatic control of the tool feed), and DD068 (drilling by using down-hole drilling motor) have been frequently ranked between sixth and tenth. These TIs have a common point, that is, they can be applied to underground drilling systems. Lastly, IP263 (fracturing by using explosive) has seen an increase in its betweenness centrality rank since 2007. These increases in rank may indicate that the diversification of fracturing method has gathered interest since the year the industry came into existence.

To describe the primary focus of technological domains in the recent period, this study presents the betweenness centrality of TI by six technological domains from 2012 to 2016 in Table 6. Results bearing values less than 0.001 are omitted. As shown, there is a big difference in the betweenness centralities of DD06 and DD046. The betweenness centrality of DD06 is approximately seven times bigger than that of DD046. Thus, the unconverged DD technologies have focused recently on the deflecting direction technologies. The unconverged technology of IP mainly focuses on IP267, reinforcing fractures (refracturing), in the recent period. In the case of the unconverged technologies of SS, SS00 is approximately 11 times bigger than SS02, which is ranked the second highest. The detailed component of the technology is not important for the development of the unconverged technology of SS. In the case of CDS, the betweenness centralities of the top two TIs are relatively bigger than that of the others. The top two Tis do not relate to the detailed function of the technologies. This shows that although the technological development of CDS is undertaken with a higher intensity than that of other converged technologies, the form of technological development is less related to the details of the technological component. CIS has only one TI, which shows that CIS has not grown yet. In the case of CDI, the betweenness centrality of Tis is quite similar. Interestingly, a more general level of TI has the lowest ranking value. Thus, it can be presumed that although CDI has a low intensity of technological development, these developments have shown relatively specific functions. In addition, converged technologies, such as CIS and CDI, show lower levels in the variety of priory technological components that pose high betweenness centrality.

Table 6. Betweenness centrality by technological domains—2012 to 2016.

Technological Domain	Order	TI	Betweenness Centrality	Technological Domain	Order	TI	Betweenness Centrality
Directional drilling (DD)	1	DD04	0.7043	Convergence of DD and SS	1	SS00	0.616545
	2	DD06	0.527745		2	DD04	0.27495
	3	DD046	0.074425		3	DD06	0.105687
	4	DD068	0.039533		4	DD067	0.04935
	5	DD061	0.024794		5	SS005	0.017216
	6	DD064	0.017748		6	DD046	0.011913
					7	DD068	0.003399
					8	SS02	0.00288
Increasing permeability (IP)	1	IP267	0.250449	Convergence of IP and SS	1	SS00	0.12202
	2	IP263	0.026008				
Smart system (SS)	1	SS00	0.900833	Convergence of DD and IP	1	DD046	0.057483
	2	SS02	0.083788		2	DD06	0.042737
	3	SS005	0.016659		3	DD04	0.026139
	4	SS04	0.012174				
	5	SS06	0.002649				

Interestingly, some technology components such as DD046, DD067, and SS005 were relatively high within the converged technological domain compared to the order of priority within the unconverged technological domain. DD046 shows a lower priority than DD04 and DD06 within DD in Table 6. Meanwhile, DD046 shows the highest priority within CDI in Table 6. DD067 is not shown within DD, but DD067 shows the fourth priority within CDS in Table 6. SS005 shows lower priority than SS00 and SS02 within SS, but SS005 shows a higher priority than SS02 within CDS in Table 6. Thus, DD046 is an important component in the form of CDI technology. Furthermore, when DD046 is developed in the form of CDI, it is considered more important as a technological component. Moreover, DD067 and SS005 are considered as important technological components in the form of CDS technology.

4. Conclusions

This study attempted to shed light on technological development of the US shale petroleum industry. Here, this study focused on developments and convergences of critical technologies such as directional drilling, increasing permeability, and smart system. To analyze the technological progress, this study measured association strength as relatedness of technological components, described technologies as network of technological components, and measured betweenness centrality as priority of technological components. The results can be stylized as follows: First, the technological developments have been intensively conducted since 2012. Second, the development of DD technologies has been closely related to SS from 2012 to 2016. Except for CDS, the development of converged technologies is lower, considering their intensity and variety, compared to unconverged technologies. Third, the IP technologies are less converged with the other technological domains of direction drilling and smart system. However, IP technologies have intensively developed with higher complexity, more components, and number of inventions than other technological domains. Fourth, some technologies are more significant in the form of converged technology. Horizontal drilling (DD046) is significant in the form of CDI. Tools locking sections of underground apparatus (DD067) and underground automatic control systems (SS005) are significant in the form of CDS.

This study has limitations as its focus is restricted to only some technologies of the shale petroleum industry. However, this study suggests specific information for the respective technologies by using the most specific level of data (full digit of TI). Thus, further investigation is required that analyzes other key technologies of the shale petroleum industry by using the most specific level of data.

Author Contributions: Conceptualization, J.-H.K. and Y.-G.L.; methodology, J.-H.K.; software, J.-H.K.; validation, J.-H.K. and Y.-G.L.; formal analysis, J.-H.K.; investigation, Y.-G.L.; resources, Y.-G.L.; data curation, J.-H.K.; writing—original draft preparation, J.-H.K.; writing—review and editing, Y.-G.L.; visualization, J.-H.K.; supervision, Y.-G.L.; project administration, Y.-G.L.; funding acquisition, Y.-G.L. All authors have read and agreed to the published version of the manuscript.

Funding: This paper was funded by Inha University and National Research Foundation (NRF-2018S1A3A2075175).

Conflicts of Interest: The authors declare no conflict of interest.

Abbreviations

The following abbreviations and symbols used in this manuscript:

CDI	Convergence of directional drilling and stimulating production by increasing permeability
CDS	Convergence of directional drilling and smart system for controlling, surveying, or testing
CIS	Convergence of increasing permeability and smart system for controlling, surveying, or testing
CPC	Cooperative patent classification
DD	Directional drilling, technological index which begins with "E21B 7/".
	Ex) E21B 7/04, E21B 7/04, E21B 7/046, E21B 7/06, E21B 7/061, E21B 7/062, E21B 7/064, E21B 7/065, E21B 7/067, E21B 7/068, E21B 7/10
IP	Stimulating production by increasing permeability, technological index which begins with "E21B 43/".
	Ex) E21B 43/26, E21B 43/2605, E21B 43/2607, E21B 43/261, E21B 43/263, E21B 43/2635, E21B 43/267, E21B 43/27
IPC	International patent classification

ISED	International Science and Economic Development Canada
KIPO	Korea intellectual property organization
KIPRIS	Korea intellectual property right information service
SS	Smart system for controlling, surveying, or testing, a technological index which begins with "E21B 44." Ex) E21B 44/00, E21B 44/005, E21B 44/02, E21B 44/04, E21B 44/06, E21B 44/08, E21B 44/10
TI	Technological index
US	United States

Symbols:

cti_{ij}	Number of co-occurrence of technological index i and j
i, j	Technological index in a patent p
m, n	Number of patents for some five-year research period
ti_{pi}, ti_{pj}	Term for counting the number of technological indices in a patent
p	One out of many patents for some research period
sti_i, sti_j	Total co-occurred number of technological index i or j for some research period
T	Sum of total co-occurred number of whole technological index
$S_{C_{ij}}$	Similarity measure between technical indices i and j
$C_B(k)$	Betweenness centrality of technological index
$\sigma_{vw}(k)$	Number of shortest paths through node k
σ_{vw}	Number of shortest paths from nodes v and w
K	Set of whole technological indices
k, v, m	Technological index consisting of a node of association similarity (technological relatedness) network

References

1. EIA Adds New Play Production Data to Shale Gas and Tight Oil Reports. Available online: https://www.eia.gov/todayinenergy/detail.php?id=38372 (accessed on 17 October 2020).
2. Cooper, J.; Stamford, L.; Azapagic, A. Shale gas: A review of the economic, environmental, and social sustainability. *Energy Technol.* **2016**, *4*, 772–792. [CrossRef]
3. Holditch, S.A. Unconventional oil and gas resource development–Let's do it right. *J. Unconv. Oil Gas Resour.* **2013**, *1*, 2–8. [CrossRef]
4. Gates, I.D. Oil phase viscosity behaviour in expanding-solvent steam-assisted gravity drainage. *J. Pet. Sci. Eng.* **2007**, *59*, 123–134. [CrossRef]
5. Gény, F. *Can Unconventional Gas Be a Game Changer in European Gas Markets?* Oxford Institute for Energy Studies: Oxford, UK, 2010.
6. Kim, J.; Lee, Y. Co-development strategy of gas and electricity in developing countries. *Geosyst. Eng.* **2018**, *22*, 239–250. [CrossRef]
7. Paredes, D.; Komarek, T.; Loveridge, S. Income and employment effects of shale gas extraction windfalls: Evidence from the Marcellus region. *Energy Econ.* **2015**, *47*, 112–120. [CrossRef]
8. Kim, J.-H.; Lee, Y.-G. Analyzing the Learning Path of US Shale Players by Using the Learning Curve Method. *Sustainability* **2017**, *9*, 2232.
9. Sandrea, I. US Shale Gas and Tight Oil Industry Performance: Challenges and Opportunities. Oxford Institute for Energy Studies. 2014. Available online: https://www.oxfordenergy.org/wpcms/wp-content/uploads/2014/03/US-shale-gas-and-tight-oil-industry-performance-challenges-and-opportunities.pdf (accessed on 17 October 2020).
10. Kim, J.; Lee, Y. Learning Curve, Change in Industrial Environment, and Dynamics of Production Activities in Unconventional Energy Resources. *Sustainability* **2018**, *10*, 3322. [CrossRef]
11. Curtis, T. US Shale Oil Dynamics in a Low Price Environment. Oxrford Institute for Energy Studies. 2015. Available online: https://www.oxfordenergy.org/wpcms/wp-content/uploads/2015/11/WPM-62.pdf (accessed on 17 October 2020).
12. Covert, T. Experiential and Social Learning in Firms: The Case of Hydraulic Fracturing in the Bakken Shale. Available online: https://ssrn.com/abstract=2481321.2015 (accessed on 17 October 2020).
13. Curtis, T. Unravelling the US Shale Productivity Gains. Oxford Institute for Energy Studies. 2016. Available online: https://www.oxfordenergy.org/wpcms/wp-content/uploads/2016/11/Unravelling-the-US-Shale-Productivity-Gains-WPM-69.pdf (accessed on 17 October 2020).

14. Shah, M.; Shah, S.; Sircar, A. A comprehensive overview on recent developments in refracturing technique for shale gas reservoirs. *J. Nat. Gas Sci. Eng.* **2017**, *46*, 350–364. [CrossRef]

15. West, R. Prospects for US Shale Productivity Gains. Oxford Institute for Energy Study. 2019. Available online: https://www.oxfordenergy.org/wpcms/wp-content/uploads/2019/10/Prospects-for-USshale-productivity-gains.pdf (accessed on 17 October 2020).

16. Natural Gas Prices (Dollars per Thousand Cubic Feet). Available online: https://www.eia.gov/dnav/ng/ng_pri_sum_dcu_nus_m.htm (accessed on 17 October 2020).

17. Spot Prices (Crude Oil in Dollars per Barrel, Products in Dollars per Gallon). Available online: https://www.eia.gov/dnav/pet/pet_pri_spt_s1_d.htm (accessed on 17 October 2020).

18. Davarpanah, A. The feasible visual laboratory investigation of formate fluids on the rheological properties of a shale formation. *Int. J. Environ. Sci. Technol.* **2019**, *16*, 4783–4792. [CrossRef]

19. Davarpanah, A.; Mirshekari, B. Experimental investigation and mathematical modeling of gas diffusivity by carbon dioxide and methane kinetic adsorption. *Ind. Eng. Chem. Res.* **2019**, *58*, 12392–12400. [CrossRef]

20. Hu, X.; Xie, J.; Cai, W.; Wang, R.; Davarpanah, A. Thermodynamic effects of cycling carbon dioxide injectivity in shale reservoirs. *J. Pet. Sci. Eng.* **2020**, *195*, 107717. [CrossRef]

21. Hu, X.; Li, M.; Peng, C.; Davarpanah, A. Hybrid thermal-chemical enhanced oil recovery methods; an experimental study for tight reservoirs. *Symmetry* **2020**, *12*, 947. [CrossRef]

22. Jin, Y.; Davarpanah, A. Using photo-fenton and floatation techniques for the sustainable management of flow-back produced water reuse in shale reservoirs exploration. *Water Air Soil Pollut.* **2020**, *231*, 1–8. [CrossRef]

23. ISED. Patent Landscape Report: Shale Oil and Gas. 2016. Available online: https://www.ic.gc.ca/eic/site/cipointernet-internetopic.nsf/vwapj/Shale-Oil-Gas-report-May-2017.pdf/$file/Shale-Oil-Gas-report-May-2017.pdf (accessed on 17 October 2020).

24. Wei, Y.; Kang, J.; Yu, B.; Liao, H.; Du, Y. A dynamic forward-citation full path model for technology monitoring: An empirical study from shale gas industry. *Appl. Energy* **2017**, *205*, 769–780. [CrossRef]

25. Kim, J.; Lee, Y. Progress of Technological Innovation of the United States' Shale Petroleum Industry Based on Patent Data Association Rules. *Sustainability* **2020**, *12*, 6628. [CrossRef]

26. Eck, N.J.V.; Waltman, L. How to normalize cooccurrence data? An analysis of some well-known similarity measures. *J. Am. Soc. Inf. Sci. Technol.* **2009**, *60*, 1635–1651. [CrossRef]

27. KIPRIS Web Database Service of Patent Information. Available online: http://www.kipris.or.kr/khome/main.jsp (accessed on 17 October 2020).

28. Xie, J. Rapid shale gas development accelerated by the progress in key technologies: A case study of the Changning–Weiyuan National Shale Gas Demonstration Zone. *Nat. Gas Ind. B* **2018**, *5*, 283–292. [CrossRef]

29. IPC Scheme Viewer. Available online: https://www.wipo.int/classifications/ipc/en/ITsupport/Version20200101/transformations/viewer/index.htm (accessed on 17 October 2020).

30. Balland, P.A. Economic Geography in R: Introduction to the EconGeo Package. Available online: https://ssrn.com/abstract=2962146 (accessed on 17 October 2020).

31. Kwon, Y.; Jeong, D. Technology relevance analysis between wind power energy-fuel cell-green car using network analysis, IPC map. *Collnet J. Scientometr. Inf. Manag.* **2014**, *8*, 109–121. [CrossRef]

32. Hagberg, A.; Swart, P.; Chult, D.S. Exploring network structure, dynamics, and function using NetworkX. In Proceedings of the 7th Python in Science Conference (SciPy2008), Pasadena, CA, USA, 19–24 August 2008; pp. 11–15.

33. Bastian, M.; Heymann, S.; Jacomy, M. Gephi: An open source software for exploring and manipulating networks. In Proceedings of the Third International ICWSM Conference (2009), San Jose, CA, USA, 17–20 May 2009; Volume 8, pp. 361–362.

Publisher's Note: MDPI stays neutral with regard to jurisdictional claims in published maps and institutional affiliations.

Article

An Analytical Model for Production Analysis of Hydraulically Fractured Shale Gas Reservoirs Considering Irregular Stimulated Regions

Kaixuan Qiu [1] and Heng Li [2,3,4,*]

[1] Department of Energy and Resources Engineering, College of Engineering, Peking University, Beijing 100871, China; qiukaixuan@pku.edu.cn
[2] Key Laboratory of Theory and Technology of Petroleum Exploration and Development in Hubei Province, Wuhan 430074, China
[3] Key Laboratory of Tectonics and Petroleum Resources, Ministry of Education, Wuhan 430074, China
[4] School of Earth Resources, China University of Geosciences, Wuhan 430074, China
* Correspondence: liheng@cug.edu.cn

Received: 23 September 2020; Accepted: 6 November 2020; Published: 12 November 2020

Abstract: Shale gas reservoirs are typically developed by multistage, propped hydraulic fractures. The induced fractures have a complex geometry and can be represented by a high permeability region near each fracture, also called stimulated region. In this paper, a new integrative analytical solution coupled with gas adsorption, non-Darcy flow effect is derived for shale gas reservoirs. The modified pseudo-pressure and pseudo-time are defined to linearize the nonlinear partial differential equations (PDEs) and thus the governing PDEs are transformed into ordinary differential equations (ODEs) by integration, instead of the Laplace transform. The rate vs. pseudo-time solution in real-time space can be obtained, instead of using the numerical inversion for Laplace transform. The analytical model is validated by comparison with the numerical model. According to the fitting results, the calculation accuracy of analytic solution is almost 99%. Besides the computational convenience, another advantage of the model is that it has been validated to be feasible to estimate the pore volume of hydraulic region, stimulated region, and matrix region, and even the shape of regions is irregular and asymmetrical for multifractured horizontal wells. The relative error between calculated volume and given volume is less than 10%, which meets the engineering requirements. The model is finally applied to field production data for history matching and forecasting.

Keywords: shale gas; multifractured horizontal wells; production analysis; irregular stimulated region

1. Introduction

Unconventional hydrocarbon resources such as tight and shale oil/gas store in tight formations with ultra-low permeability. With the development of hydraulic fracturing technologies, multifractured horizontal wells (MFHWs) have rapidly emerged as the primary means for exploiting this type of resource. Meanwhile, some technologies are utilized, such as foam injection and carbon dioxide injection [1,2] on recovery enhancement, and photo-Fenton and floatation on sustainable management of flow-back water after hydraulic fracturing [3]. In unconventional reservoirs, due to the propagation of fractures in different directions, branch fractures will be created around the main hydraulic fractures, which have a significant impact on the pressure and rate transient analysis for the fluid flow in the reservoirs.

In order to analyze production data and make long-term forecasts, analytical and numerical tools have been developed. Among them, a large number of numerical approaches [4–8], such as finite element method and boundary element method, are adopted to study the multiple flow mechanisms and fracture characteristics in shale gas reservoirs. Compared with the numerical models, the analytical models are simpler to set up and require fewer original data. Therefore, they are widely used in practice for field applications. Extensive studies have focused on analyzing production data of transient flow period for such complex systems [9–14]. Lee et al. [15] derived a tri-linear model to represent the transient flow behavior within a single fracture in an infinite homogeneous reservoir. Wattenbarger et al. [16] proposed an analytical solution accounting for transient linear flow contributions from each region for a fracture with infinite conductivity in a closed reservoir. El-Banbi [17] and Bello et al. [18] extended previous analytical models to develop a series of linear dual-porosity solutions for fluid flow in tight reservoirs, where constant-rate and constant-pressure inner boundaries were considered. Ozkan et al. [19] and Brown et al. [20] extended the concept of tri-linear flow and presented a solution to simulate the pressure-transient behavior in three flow regions, including the hydraulic fracture, the inner region between hydraulic fractures, and the outer region beyond the tip of hydraulic fractures. Stalgorova et al. [21] and Heidari et al. [22] introduced an enhanced-fracture-region (EFR) model to consider the cases where branch fractures cover the interfracture region. Mahdi et al. [23] firstly provided the transient shape factor among different regions in the triple-porosity model.

The models represent an excellent attempt to capture reservoir parameters and flow characteristics for the complex system associated with the MFHWs. However, most analytical solutions were derived by Laplace transformation [17–23] and Boltzmann's transformation [24]. Numerical inversion algorithm [25] is needed to transform the Laplace space into real-time space. In order to avoid the inconvenience of Laplace transformation, several approximated analytical solutions have been presented. Shahamat et al. [26] introduced the concept of continuous succession of pseudo-steady states based on the capacitance-resistance methodology to analyze the production data in tight and shale reservoirs. In their work, the gas adsorption was ignored, which would underestimate the shale gas production. Ogunyomi [27,28] introduced a model with two compartments in which the first compartment represents the volume of the enhanced fracture region and the second one is the volume of the matrix. The approximate solution was obtained by converting governing partial-differential equations into a system of ordinary differential equations through integration. Then the system of ordinary differential equations was solved by calculating the eigenvalues and eigenvectors. Qiu et al. [29,30] extended this method to a triple-porosity model for production analysis. Nevertheless, they both did not take the non-Darcy flow into account and just derived the approximate solutions based on the oil phase rather than gas phase with the pressure dependent parameters. In this work, we extended the method to consider the gas flow and even account for non-Darcy flow in the hydraulic fractures, and gas adsorption and slippage in the matrix.

In this work, our goal is to derive a new integrative analytical solution for shale gas reservoirs bypassing the Laplace transformation, which can effectively account for non-Darcy flow in the hydraulic fractures, and gas adsorption and slippage in the matrix. The analytical model is created on the basis of a three-region model, which consists of the hydraulic fracture region, the inner region also called stimulated region connected to the hydraulic fracture, and the outer region representing the low permeability matrix. Then we introduce modified pseudo-variables [31,32], i.e., pseudo-pressure and pseudo-time, to eliminate the nonlinearity of governing equations. The whole derivation is based on the integration in real-time space bypassing Laplace transform and numerical inversion. The derived analytical solution is validated with numerical simulations. Further, several numerical cases are designed to verify the applicability in an irregular stimulated region using the new model. The proposed method is finally applied for shale gas production analysis and forecasting in the field.

2. Method

Production of shale gas requires horizontal wells with multistage hydraulic fracturing. The propagation of fractures is uncertain and branch fractures will be created around the main hydraulic fractures, resulting in a stimulated region around each main hydraulic fracture. In order to model such a complex system, the reservoir is simplified as a three-region system, where the first region is the hydraulic fracture region which is regarded as the sole connection to the well. The second one represents the stimulated region with aggregated volume of all the microfractures and the third one is the adjacent ultra-low-permeability matrix directly connected to the stimulated region. Figure 1a shows the schematic 3D model. Three regions are contained in the model: region 1 is the hydraulic fracture, region 2 (darker color) with higher permeability around each hydraulic fracture, and region 3 (lighter color) with lower permeability connected to the region 2. The arrows represent the flow directions. For this model, the flow directions are parallel, and the system is symmetrical with respect to the hydraulic fracture and horizontal-well. Thus, it is feasible to use one quarter of the reservoir shown in Figure 1b to replace the whole reservoir in order to simplify the derivation process. According to Anderson et al. [33], they verified that when the permeability of region 2 is less than or equal to 500 nD, the contribution of the region beyond fractures can be neglected for the 20-years production. For the case that the distance from the fracture face to permeability boundary (x_1) is less than the half distance between fractures (x_2), the contribution would be smaller. Meanwhile, Stalgorova et al. [21] also set numerical models to illustrate the negligible contribution of region beyond fractures, and they obtained that the difference of 20-years production is negligible after comparing the results of numerical simulation with and without region beyond fractures. In the work of Heidari et al. [22], they also did not take the region beyond fractures into account. Therefore, the contribution of the region beyond the tips of the hydraulic fractures is assumed to be negligible in this work.

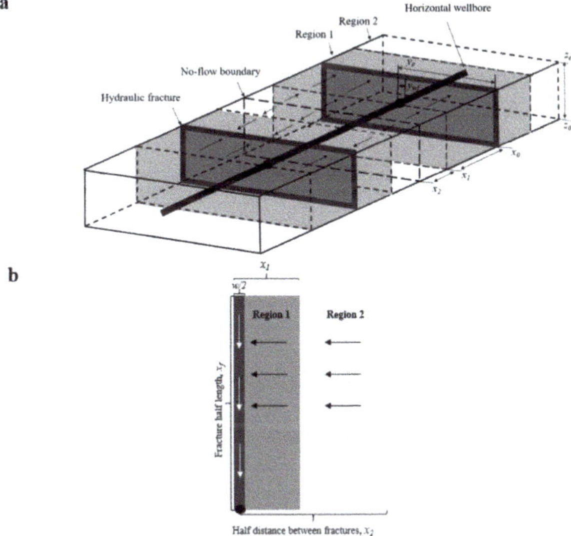

Figure 1. Schematic of a multistage fractured horizontal well with two regions around the hydraulic fracture. (**a**) The 3D model. (**b**) The plan view.

In this work, our analytical model is derived under the following assumptions:

1. The reservoir is homogeneous, isopachous and isothermal in each region.
2. Flow process is 1-D linear in each region.
3. Flow is single gas phase.
4. The high-velocity non-Darcy flow in the hydraulic fracture is considered.
5. The bottom-hole pressure is constant.
6. The impact of gravity is neglected.
7. Gas desorption meets the Langmuir isotherm adsorption equation.

2.1. Gas Adsorption/Desorption Effect

In contrast to conventional gas reservoirs, gas adsorption is an important feature of shale gas reservoir. The Langmuir isotherm adsorption equation [34] is widely used to calculate the shale gas adsorption and its expression is as follows:

$$V = V_L \frac{P}{P_L + P} \tag{1}$$

where V is the volume of the adsorbed gas, and P is pressure. V_L and P_L stand for Langmuir volume and pressure, respectively. When considering the gas adsorption, the effect of adsorption on compressibility of the reservoir is essential. According to Bumb et al. [35], the new compressibility equation can be expressed as,

$$C_t^* = C_f + C_w S_w + C_g (1 - S_w) + C_{gd} \tag{2}$$

$$C_{gd} = \frac{0.031214 \rho_m V_L \overline{B_g} P_L}{\phi (\overline{p} + P_L)^2} \tag{3}$$

where C_f is rock compressibility, C_w is water compressibility, C_g is free gas compressibility and C_{gd} is adsorbed gas compressibility, S_w is water saturation, ρ_m is matrix density, $\overline{B_g}$ is gas reservoir volume factor, and ϕ is the porosity.

Another important change is that the compressibility factor is modified by King [36],

$$z^* = \frac{z}{(1 - S_w) + 0.031214 \rho_m \frac{zT}{\phi} \left(\frac{p_{sc}}{T_{sc}}\right) V_L \frac{1}{p + p_L}} \tag{4}$$

where P_{sc} is the standard condition pressure, T_{sc} is the standard condition temperature, z is compressibility factor, T is the reservoir temperature.

2.2. High-Velocity Non-Darcy Flow Effect

For gas flow in the hydraulic fracture, high-velocity non-Darcy effect is considered in this study. Forchheimer [37] proposed that Darcy's law is inadequate to describe high-velocity gas flow without adding an inertial effect, which is proportional to the square of the flow velocity. To account for the non-Darcy flow effect, an inertial term must be included. The Forchheimer's flow equation is given as,

$$-\nabla p = \frac{\mu}{k} \vec{v} + \beta \rho \left|\vec{v}\right| \vec{v} \tag{5}$$

In order to reduce the nonlinearity, the equivalent permeability is introduced to obtain the extended Darcy's law [37],

$$-\nabla p = \frac{\mu}{k_{eq}} \vec{v} \tag{6}$$

Substituting Equation (6) into Equation (5), the equivalent permeability of hydraulic fracture yields,

$$k_{eq}^F = \frac{\mu k_{iF}}{\mu + \beta \rho \left|\vec{v}\right| k_{iF}} \tag{7}$$

where

$$\beta = \frac{4.1 \times 10^{11}}{(k_{iF})^{1.5}} \tag{8}$$

among which, β is the non-Darcy flow coefficient, k_{iF} is the hydraulic fracture permeability, k_{eq}^F is the equivalent permeability of hydraulic fracture, v is the flow velocity, and ρ is the gas density.

2.3. Non-Darcy Flow in the Matrix

For the nano-porous media in shale reservoir, Darcy's law has difficulty describing the actual gas behavior. The gas flow can be classified as four flow conditions, such as continuum flow, slip flow, transition flow, and free-molecular flow. According to the previous publications [38], the four flow conditions can be qualified by the Knudsen number. However, the Knudsen number varies from 10^{-3} to 1 in most shale reservoirs. In order to represent the non-Darcy gas flow in matrix, the apparent permeability is presented by the following general form:

$$k_a^m = k_\infty^m f(Kn) \tag{9}$$

where k_∞^m is the intrinsic permeability of the porous medium, $f(Kn)$ is the correlation term expressed as a function of the Knudsen number, which is modeled as [38],

$$f(Kn) = (1 + \alpha Kn)\left(1 + \frac{4Kn}{1 + Kn}\right) \tag{10}$$

in which,

$$\alpha = \frac{128}{15\pi^2} \tan^{-1}\left(4Kn^{0.4}\right) \tag{11}$$

Meanwhile, for a capillary tube of radius, r, the intrinsic permeability can be derived [39],

$$k_\infty^m = \frac{r^2}{8} \tag{12}$$

The Knudsen number Kn is defined as the ratio of the molecular mean free path length and the pore radius in shale matrix,

$$Kn = \frac{\lambda}{r} \tag{13}$$

The mean free path can also be calculated,

$$\lambda = \frac{\mu_g}{P}\sqrt{\frac{\pi RT}{2M}} \tag{14}$$

Substituting Equation (14) into Equation (13) on basis of the real gas condition, we can obtain:

$$Kn = \frac{\mu_g z}{rP}\sqrt{\frac{\pi RT}{2M}} \tag{15}$$

where μ_g is the gas viscosity, z is compressibility factor, r is the pore radius, P is the reservoir pressure, R is the universal gas constant, T is the reservoir temperature, and M is the gas molecular weight.

2.4. Derivation of Linearized Gas Diffusivity Equation.

For the flow of shale gas, the gas diffusivity equation will be nonlinear which makes deriving the analytical solution difficult. On one hand, with the reduction in average reservoir pressure, the gas properties like the gas viscosity (μ), gas compressibility (C_t), and gas compressibility factor (z) will change with the pressure. On the other hand, when incorporating the significant mechanisms in shale gas reservoir like gas adsorption and non-Darcy flow effect, the permeability is a variate with pressure rather than a constant.

In order to deal with this problem, the pseudo-pressure and pseudo-time instead of the pressure and time are adopted to linearize the equations. We set a general real gas diffusivity equation in three-dimensional Cartesian coordinate system as example. When the non-Darcy effect is coupled, the

$k(p)$ can be calculated by Equation (7) or Equation (9). Certainly, the $c_t(p)$ can be replaced by $c_t^*(p)$ to represent the gas adsorption effect.

$$\frac{\partial}{\partial x}\left(k(p)\frac{p}{\mu(p)z(p)}\frac{\partial p}{\partial x}\right) + \frac{\partial}{\partial y}\left(k(p)\frac{p}{\mu(p)z(p)}\frac{\partial p}{\partial y}\right) + \frac{\partial}{\partial z}\left(k(p)\frac{p}{\mu(p)z(p)}\frac{\partial p}{\partial z}\right) = \phi c_t(p)\frac{p}{z(p)}\frac{\partial p}{\partial t} \tag{16}$$

where $k(p)$ is the pressure-dependent permeability, $\mu(p)$ is the pressure-dependent viscosity, $c_t(p)$ is the pressure-dependent compressibility.

Considering the pressure dependence of the permeability, we define a general modified pseudo-pressure transformation. Surely, $k(p)$ can be calculated by Equation (7) or Equation (9) to couple with the non-Darcy effect. $z(p)$ can be replaced with $z^*(p)$ calculated by Equation (4) to consider the gas adsorption.

$$m(p) = \frac{1}{k_i}\int_{p_b}^{p} k(p)\frac{p}{\mu(p)z(p)}dp \tag{17}$$

where $m(p)$ is the pseudo-pressure, k_i is the intrinsic permeability, $z(p)$ is the pressure-dependent compressibility factor.

Substituting Equation (17) into Equation (16), the right side of the partial differential equation is still nonlinear.

$$\frac{\partial^2 m(p)}{\partial x^2} + \frac{\partial^2 m(p)}{\partial y^2} + \frac{\partial^2 m(p)}{\partial z^2} = \frac{\phi\mu(p)c_t(p)}{k(p)}\frac{\partial m(p)}{\partial t} \tag{18}$$

To linearize Equation (18), a general modified pseudo-time transformation is defined [24].

$$t_a = \frac{1}{k_i}\int_0^t k(p)\frac{(\mu c_t)_i}{\mu(p)c_t(p)}dt \tag{19}$$

where t_a is the pseudo-time, μ_i is the initial viscosity, c_{ti} is the initial compressibility.

After substituting the Equation (19) into Equation (18), the general linear partial differential equation is derived as,

$$\frac{\partial^2 m(p)}{\partial x^2} + \frac{\partial^2 m(p)}{\partial y^2} + \frac{\partial^2 m(p)}{\partial z^2} = \frac{\phi\mu_i c_{ti}}{k_i}\frac{\partial m(p)}{\partial t_a} \tag{20}$$

Considering the permeability and compressibility in three regions are different, therefore, the pseudo-time in three regions can be shown as follows.

$$t_{aF} = \frac{1}{k_{iF}}\int_0^t k_{eq}^F(p)\frac{(\mu c_t)_{iF}}{\mu_F(p)c_{tF}(p)}dt \tag{21}$$

$$t_{a1} = \frac{1}{k_{i1}}\int_0^t k_1(p)\frac{(\mu c_t)_{i1}}{\mu_1(p)c_{t1}(p)}dt \tag{22}$$

$$t_{a2} = \frac{1}{k_{i2}}\int_0^t k_a^m(p)\frac{(\mu c_t^*)_{i2}}{\mu_2(p)c_{t2}^*(p)}dt \tag{23}$$

where t_{aF}, t_{aF}, t_{a2} are the pseudo-time in the hydraulic fracture region, region 1, and region 2 respectively.

Therefore, we can obtain the linearized gas diffusivity equation in three regions respectively. Firstly, we consider the gas adsorption and gas slippage in matrix and the high-velocity non-Darcy flow in hydraulic fracture region. Thus, we use a new total compressibility coupling the gas adsorption effect in diffusivity equation of region 2. Finally, through the linearization by the modified pseudo-pressure and pseudo-time, the non-Darcy flow effect for matrix and fracture are included in the governing linear equations in different regions.

2.5. Model Description in Different Regions.

2.5.1. Model Description in Matrix Region (Region 2)

The system of equations based on the conceptual model is presented as follows. For the low-permeability matrix region (region 2), the diffusivity equation for gas flow is derived as,

$$\frac{\partial^2 m_2(p)}{\partial x^2} + \frac{\partial^2 m_2(p)}{\partial y^2} + \frac{\partial^2 m_2(p)}{\partial z^2} = \frac{(\phi \mu_i c_{ti}^*)_2}{k_{i2}} \frac{\partial m_2(p)}{\partial t_{a2}} \tag{24}$$

where $m_2(p)$ is the pseudo-pressure in region 2, x, y, z are the Cartesian coordinates, t_{a2} is the pseudo-time for gas flow, c_{ti2}^* and μ_{i2} are the initial modified total compressibility and viscosity in region 2, k_{i2} and ϕ_{i2} are the permeability and porosity in region 2, respectively.

The initial condition for the region is that the pseudo-pressure of the region is equal to initial pseudo-pressure at $t = 0$.

$$m_2(p)(x, y, z, 0) = m(p_i) \tag{25}$$

where $m(p_i)$ is the initial pseudo-pressure.

The boundary conditions are defined as no-flow at top and bottom of the reservoir. Additionally, the both ends of y-direction can also be regarded as no-flow boundaries.

$$\left. \frac{\partial m_2(p)}{\partial y} \right|_{y=0, x_f} = 0 \tag{26}$$

$$\left. \frac{\partial m_2(p)}{\partial z} \right|_{z=z_0, z_e} = 0 \tag{27}$$

Due to the plane of symmetry between adjacent fractures, the location $x = x_2$ is also a no-flow boundary.

$$\left. \frac{\partial m_2(p)}{\partial x} \right|_{x=x_2} = 0 \tag{28}$$

The continuity of flux and pressure across the boundaries between the regions is assumed. In Ogunyomi's [27] and Qiu's [29] work, only one is chosen as a boundary condition. Considering the average pseudo-pressure will be adopted, the continuity of flux is chosen as the last boundary condition which can be given by,

$$\left. k_{i2} \frac{\partial m_2(p)}{\partial x} \right|_{x=x_1} = \left. k_{i1} \frac{\partial m_1(p)}{\partial x} \right|_{x=x_1} \tag{29}$$

2.5.2. Model Description in Stimulated Region Volume (Region 1)

For the stimulated region volume (region 1), the diffusivity equation for gas flow is also expressed as,

$$\frac{\partial^2 m_1(p)}{\partial x^2} + \frac{\partial^2 m_1(p)}{\partial y^2} + \frac{\partial^2 m_1(p)}{\partial z^2} = \frac{(\phi \mu_i c_{ti})_1}{k_{i1}} \frac{\partial m_1(p)}{\partial t_{a1}} \tag{30}$$

where $m_1(p)$ is the pseudo-pressure in region 1, x, y, z are the Cartesian coordinates, t_{a1} is the pseudo-time for gas flow, c_{ti1} and μ_{i1} are the initial total compressibility and viscosity in region 1, k_{i1} and ϕ_1 are the permeability and porosity in region 1, respectively.

For the whole model, the initial condition remains the same.

$$m_1(p)(x, y, z, 0) = m(p_i) \tag{31}$$

The flow in region 1 is also the 1D linear in x-direction, therefore, the outer boundary conditions are identical with those of region 2.

$$\left. \frac{\partial m_1(p)}{\partial y} \right|_{y=0, x_f} = 0 \tag{32}$$

$$\left. \frac{\partial m_1(p)}{\partial z} \right|_{z=z_0, z_e} = 0 \tag{33}$$

According to the continuity hypothesis, the boundary condition is expressed as,

$$k_{i1} \left. \frac{\partial m_1(p)}{\partial x} \right|_{x=x_1} = k_{i2} \left. \frac{\partial m_2(p)}{\partial x} \right|_{x=x_1} \tag{34}$$

$$k_{i1} \left. \frac{\partial m_1(p)}{\partial x} \right|_{x=\frac{w}{2}} = k_{iF} \left. \frac{\partial m_F(p)}{\partial x} \right|_{x=\frac{w}{2}} \tag{35}$$

2.5.3. Model Description in Hydraulic Fracture Region

For the hydraulic fracture region, the diffusivity equation for gas flow is also expressed as,

$$\frac{\partial^2 m_F(p)}{\partial x^2} + \frac{\partial^2 m_F(p)}{\partial y^2} + \frac{\partial^2 m_F(p)}{\partial z^2} = \frac{(\phi \mu_i c_{ti})_F}{k_{iF}} \frac{\partial m_F(p)}{\partial t_{aF}} \tag{36}$$

where $m_F(p)$ is the pseudo-pressure in hydraulic fracture region, x, y, z are the Cartesian coordinates, t_{aF} is the pseudo-time for gas flow, c_{tiF} and μ_{iF} are the initial total compressibility and viscosity in hydraulic fracture region, k_{iF} and ϕ_F are the permeability and porosity in hydraulic fracture region, respectively.

For the whole model, the initial condition remains the same.

$$m_F(p)(x, y, z, 0) = m(p_i) \tag{37}$$

With the assumption of the constant bottom-hole pressure, at the location of $y = 0$, the pressure is the same as the bottom-hole pressure at any time.

$$m_F(p)(x, y = 0, z, t_a) = m(p_{wf}) \tag{38}$$

The flow in hydraulic fracture region is also the 1D linear in the y-direction, therefore, the outer boundary conditions is expressed as,

$$\left. \frac{\partial m_F(p)}{\partial x} \right|_{x=0} = 0 \tag{39}$$

$$\left. \frac{\partial m_F(p)}{\partial y} \right|_{y=x_f} = 0 \tag{40}$$

$$\left. \frac{\partial m_F(p)}{\partial z} \right|_{z=z_0, z_e} = 0 \tag{41}$$

According to the continuity hypothesis, the boundary condition is expressed as,

$$k_{iF}\frac{\partial m_F(p)}{\partial x}\bigg|_{x=\frac{w}{2}} = k_{i1}\frac{\partial m_1(p)}{\partial x}\bigg|_{x=\frac{w}{2}} \tag{42}$$

3. Results

3.1. Derivation of Analytical Solution.

The diffusivity equations in the mathematical model are all PDEs. In order to obtain the analytical solution, the first step is to transform the sets of equations into ODEs. In this work, we adopted the integral method other than the common Laplace Transform. By integrating over the spatial domain, the spatial dependence is eliminated, which is even feasible to irregular regions as will be demonstrated later. The pressure in different regions is assumed as average value in our model. Therefore, the pseudo-time is supposed to be independent with space. Integrating the equations with respect to spatial coordinates,

$$\int_{x_1}^{x_2}\int_0^{x_f}\int_{z_0}^{z_e}\frac{\partial}{\partial x}\Big(\frac{\partial m_2(p)}{\partial x}\Big)dxdydz + \int_{x_1}^{x_2}\int_0^{x_f}\int_{z_0}^{z_e}\frac{\partial}{\partial y}\Big(\frac{\partial m_2(p)}{\partial y}\Big)dxdydz + \int_{x_1}^{x_2}\int_0^{x_f}\int_{z_0}^{z_e}\frac{\partial}{\partial z}\Big(\frac{\partial m_2(p)}{\partial z}\Big)dxdydz$$
$$= \frac{(\phi\mu_i c_{ti}^*)_2}{k_{i2}}\frac{\partial}{\partial t_{a2}}\int_{x_1}^{x_2}\int_0^{x_f}\int_{z_0}^{z_e} m_2(p)dxdydz \tag{43}$$

In Equation (43), we move the pseudo-time outside the spatial integral since the pseudo-time is independent of the spatial coordinates. In order to get a simplified equation, the average pseudo-pressure and the effective pore volume is defined as,

$$\overline{m}(p) = \frac{\iiint m(p)}{\iiint dxdydz} = \frac{\iiint m(p)}{V_b} \tag{44}$$

$$V_p = \phi V_b \tag{45}$$

where V_b is the volume of the region and V_p is the pore volume of the region.

Equation (43) can be rewritten as,

$$\int_0^{x_f}\int_{z_0}^{z_e}\Big(\frac{\partial m_2(p)}{\partial x}\Big|_{x_2} - \frac{\partial m_2(p)}{\partial x}\Big|_{x_1}\Big)dydz + \int_{x_1}^{x_2}\int_{z_0}^{z_e}\Big(\frac{\partial m_2(p)}{\partial y}\Big|_{x_f} - \frac{\partial m_2(p)}{\partial y}\Big|_0\Big)dxdz + \int_{x_1}^{x_2}\int_0^{x_f}\Big(\frac{\partial m_2(p)}{\partial z}\Big|_{z_e} - \frac{\partial m_2(p)}{\partial y}\Big|_{z_0}\Big)dxdy$$
$$= \frac{(\phi\mu_i c_{ti}^*)_2 V_{b_2}}{k_{i2}}\frac{d\overline{m}_2(p)}{dt_{a2}} \tag{46}$$

Using the initial condition and boundary conditions, Equation (46) is simplified to

$$-\int_0^{x_f}\int_{z_0}^{z_e}\Big(k_{i2}\frac{\partial m_2(p)}{\partial y}\Big|_{x_1}\Big)dydz = (V_p\mu_i c_{ti}^*)_2\frac{d\overline{m}_2(p)}{dt_{a2}} \tag{47}$$

According to the equivalent Darcy's law,

$$q_2 = \frac{T_{sc}Z_{sc}}{2TP_{sc}}\int_0^{x_f}\int_{z_0}^{z_e}k_{i2}\frac{\partial m_2(p)}{\partial y}\bigg|_{x_1} dydz \tag{48}$$

Define

$$C_2 = \frac{T_{sc}Z_{sc}}{2TP_{sc}}(\mu_i c_{ti}^*)_2 \tag{49}$$

Therefore, Equation (48) can be rewritten as,

$$C_2 V_{p2} \frac{d\overline{m_2}(p)}{dt_{a2}} = -q_2 \tag{50}$$

where V_{p2} is the pore volume of region 2, q_2 is the flow rate in region 2.

For region 1, we also use the integration method to deal with the equation,

$$\int_0^{x_f} \int_{z_0}^{z_e} \left(\frac{\partial m_1(p)}{\partial x}\bigg|_{x_1} - \frac{\partial m_1(p)}{\partial x}\bigg|_{\frac{w}{2}} \right) dydz + \int_{w/2}^{x_1} \int_{z_0}^{z_e} \left(\frac{\partial m_1(p)}{\partial y}\bigg|_{x_f} - \frac{\partial m_1(p)}{\partial y}\bigg|_0 \right) dxdz + \int_{w/2}^{x_1} \int_0^{x_f} \left(\frac{\partial m_1(p)}{\partial z}\bigg|_{z_e} - \frac{\partial m_1(p)}{\partial y}\bigg|_{z_0} \right) dxdy$$
$$= \frac{(\phi\mu_i c_{ti})_1 V_{b_1}}{k_{i1}} \frac{d\overline{m}_1(p)}{dt_{a1}} \tag{51}$$

After applying the initial condition and boundary conditions, Equation (51) can be rewritten as,

$$-\int_0^{x_f} \int_{z_0}^{z_e} \left(k_{i1} \frac{\partial m_1(p)}{\partial x}\bigg|_{x_1} - k_{i1} \frac{\partial m_1(p)}{\partial x}\bigg|_{x_0} \right) dydz = (V_p \mu_i c_{ti})_1 \frac{d\overline{m}_1(p)}{dt_{a1}} \tag{52}$$

Note that,

$$q_1 = \frac{T_{sc} Z_{sc}}{2TP_{sc}} \int_0^{x_f} \int_{z_0}^{z_e} \left(k_{i1} \frac{\partial m_1(p)}{\partial x}\bigg|_{x_0} \right) dydz \tag{53}$$

Defining

$$C_1 = \frac{T_{sc} Z_{sc}}{2TP_{sc}} (\mu_i c_{ti})_1 \tag{54}$$

Substituting Equations (30), (45), (49), and (50) into Equation (48) results in,

$$C_1 V_{p1} \frac{d\overline{m}_1(p)}{dt_{a1}} = -q_1 + q_2 \tag{55}$$

where V_{p1} is the pore volume of region 1, q_1 is the flow rate in region 1.

For hydraulic fracture region, we also use the integration method to deal with the diffusion equation,

$$\int_0^{x_f} \int_{z_0}^{z_e} \left(\frac{\partial m_F(p)}{\partial x}\bigg|_{\frac{w}{2}} - \frac{\partial m_F(p)}{\partial x}\bigg|_0 \right) dydz + \int_0^{w/2} \int_{z_0}^{z_e} \left(\frac{\partial m_F(p)}{\partial y}\bigg|_{x_f} - \frac{\partial m_F(p)}{\partial y}\bigg|_0 \right) dxdz + \int_0^{w/2} \int_0^{x_f} \left(\frac{\partial m_F(p)}{\partial z}\bigg|_{z_e} - \frac{\partial m_F(p)}{\partial y}\bigg|_{z_0} \right) dxdy$$
$$= \frac{(\phi\mu_i c_{ti})_F V_{b_F}}{k_{iF}} \frac{d\overline{m}_F(p)}{dt_{aF}} \tag{56}$$

After applying the initial condition and boundary conditions, Equation (56) can be rewritten as,

$$\int_0^{x_f} \int_{z_0}^{z_e} \left(k_{iF} \frac{\partial m_F(p)}{\partial x}\bigg|_{\frac{w}{2}} \right) dydz + \int_0^{w/2} \int_{z_0}^{z_e} \left(-k_{iF} \frac{\partial m_F(p)}{\partial y}\bigg|_0 \right) dxdz = (\phi\mu_i c_{ti})_F V_{b_F} \frac{d\overline{m}_F(p)}{dt_{aF}} \tag{57}$$

The flow rate in hydraulic fracture can be expressed as,

$$q_F = \frac{T_{sc} Z_{sc}}{2TP_{sc}} \int_0^{x_f} \int_{z_0}^{z_e} \left(k_{iF} \frac{\partial m_F(p)}{\partial x}\bigg|_0 \right) dydz \tag{58}$$

Defining

$$C_F = \frac{T_{sc} Z_{sc}}{2TP_{sc}} (\mu_i c_{ti})_F \tag{59}$$

Substituting Equations (42), (48), (58), and (59) into Equation (57) results in,

$$C_F V_{pF} \frac{d\overline{m}_F(p)}{dt_{aF}} = -q_F + q_1 \tag{60}$$

The next step is to substitute the average pseudo-pressure with the relationship between pressure and the flow rate. Since it is assumed that gas flows sequentially from region 2 to region 1 then to hydraulic fracture, a general analytical solution for one-dimensional linear gas flow is derived to solve the problem (details are shown in Appendix A), which is given by:

$$\overline{m}(p) = m(p_{wf}) + \frac{4}{\pi^2}[m(p_i) - m(p_{wf})] \sum_{n=1}^{\infty} \frac{q_{Dn}}{(2n-1)^2} \tag{61}$$

where $\overline{m}(p)$ is the average pseudo-pressure, q_{Dn} is the dimensionless production from the nth mode. Combining the assumptions, the average pseudo-pressure in each region can be expressed as,

$$\overline{m}_F(p) = m(p_{wf}) + \frac{4}{\pi^2}[m(p_i) - m(p_{wf})] \sum_{n=1}^{\infty} \frac{q_{DFn}}{(2n-1)^2} \tag{62}$$

$$\overline{m}_1(p) = \overline{m}_F(p) + \frac{4}{\pi^2}[m(p_i) - \overline{m}_F(p)] \sum_{n=1}^{\infty} \frac{q_{D1n}}{(2n-1)^2} \tag{63}$$

$$\overline{m}_2(p) = \overline{m}_1(p) + \frac{4}{\pi^2}[m(p_i) - \overline{m}_1(p)] \sum_{n=1}^{\infty} \frac{q_{D2n}}{(2n-1)^2} \tag{64}$$

We define the productivity index (J) and transmissibility (T_{r1F} and T_{r21}) as,

$$J = \frac{\pi^2}{4} \frac{q_{iF}}{m(p_i) - m(p_{wf})} \tag{65}$$

$$T_{r1F} = \frac{\pi^2}{4} \frac{q_{i1}}{m(p_i) - \overline{m}_1(p)} \tag{66}$$

$$T_{r21} = \frac{\pi^2}{4} \frac{q_{i2}}{m(p_i) - \overline{m}_2(p)} \tag{67}$$

where $\overline{m}_1(p)$, $\overline{m}_2(p)$ is the average pseudo-pressure in hydraulic fracture region, region 1, and region 2, respectively. $q_{iF} = \frac{T_{sc}Z_{sc}}{2TP_{sc}} k_{iF} A_F \frac{m(p_i)}{L_F}$ is the initial production rate from the hydraulic fracture region. $q_{i1} = \frac{T_{sc}Z_{sc}}{2TP_{sc}} k_{i1} A_1 \frac{\overline{m}_1(p)}{L_1}$ is the initial production rate from the region 1. $q_{i2} = \frac{T_{sc}Z_{sc}}{2TP_{sc}} k_{i2} A_2 \frac{\overline{m}_2(p)}{L_2}$ is the initial production rate from the region 2.

Substituting Equations (62)–(64) into Equations (50), (55), and (60), there are six ordinary differential terms on the left side of the system of equations, such as $\frac{dq_{Fn}}{dt_{aF}}, \frac{dq_{Fn}}{dt_{a1}}, \frac{dq_{Fn}}{dt_{a2}}, \frac{dq_{1n}}{dt_{a1}}, \frac{dq_{1n}}{dt_{a2}}, \frac{dq_{2n}}{dt_{a2}}$. In order to solve the system of ordinary differential equations, an approximate pseudo-time $\tilde{t}_a$ is introduced. Therefore, the Equations (50), (55), and (60) can be rewritten as follows.

$$\sum_{n=1}^{\infty} \frac{dq_{2n}}{d\tilde{t}_a} = -\left(\frac{T_{r21}}{C_2 V_{p2}} + \frac{T_{r21}}{C_1 V_{p1}}\right) \sum_{n=1}^{\infty} (2n-1)^2 q_{2n} + \frac{T_{r21}}{C_1 V_{p1}} \sum_{n=1}^{\infty} (2n-1)^2 q_{1n} \tag{68}$$

$$\sum_{n=1}^{\infty} \frac{dq_{1n}}{d\tilde{t}_a} = \frac{T_{r1F}}{C_1 V_{p1}} \sum_{n=1}^{\infty} (2n-1)^2 q_{2n} - \left(\frac{T_{r1F}}{C_1 V_{p1}} + \frac{T_{r1F}}{C_F V_{pF}}\right) \sum_{n=1}^{\infty} (2n-1)^2 q_{1n} + \frac{T_{r1F}}{C_F V_{pF}} \sum_{n=1}^{\infty} (2n-1)^2 q_{Fn} \tag{69}$$

$$\sum_{n=1}^{\infty} \frac{dq_{F_n}}{d\widetilde{t_a}} = -\frac{J}{C_F V_{pF}} \sum_{n=1}^{\infty} (2n-1)^2 q_{F_n} + \frac{J}{C_F V_{pF}} \sum_{n=1}^{\infty} (2n-1)^2 q_{1_n} \tag{70}$$

where q_{DF_n}, q_{D1_n}, and q_{D2_n} are the initial production rate from the nth mode in hydraulic fracture region, region 1, and region 2, respectively. $\widetilde{t_a}$ is an approximate pseudo-time, which is defined as the average of the t_{a1}, t_{a2}, and t_{aF}.

Then we define three time-constant parameters as,

$$\tau_F = \frac{C_F V_{pF}}{J} \tag{71}$$

$$\tau_1 = \frac{C_1 V_{p1}}{T_{r1F}} \tag{72}$$

$$\tau_2 = \frac{C_2 V_{p2}}{T_{r21}} \tag{73}$$

We can rewrite this set of ODEs in matrix form,

$$\begin{pmatrix} \frac{dq_{2_n}}{dt_a} \\ \frac{dq_{1_n}}{dt_a} \\ \frac{dq_{F_n}}{dt_a} \end{pmatrix} = (2n-1)^2 \begin{pmatrix} -\left(\frac{1}{\tau_2} + \frac{T_{r21}}{\tau_1 T_{r1F}}\right) & \frac{T_{r21}}{\tau_1 T_{r1F}} & 0 \\ \frac{1}{\tau_1} & -\left(\frac{1}{\tau_1} + \frac{T_{r1F}}{\tau_F J}\right) & \frac{T_{r1F}}{\tau_F J} \\ 0 & \frac{1}{\tau_F} & -\frac{1}{\tau_F} \end{pmatrix} \begin{pmatrix} q_{2_n} \\ q_{1_n} \\ q_{F_n} \end{pmatrix} \tag{74}$$

where the initial conditions for the system of equations are,

$$q_F(t_a = 0) = q_{iF} \tag{75}$$

$$q_1(t_a = 0) = 0 \tag{76}$$

$$q_2(t_a = 0) = 0 \tag{77}$$

After solving Equation (70), we can get the *n*th flow rate in combination with the initial conditions. The flow rate is the summation of them. By converting the summation to an indefinite integral, the analytical solution in real-time space can be derived as follows (details are shown in Appendix B),

$$q_F = \beta_3 q_{iF} a_3 e^{\lambda_1 \widetilde{t_a}} - \beta_2 q_{iF} a_6 e^{\lambda_2 \widetilde{t_a}} + \beta_1 q_{iF} a_9 e^{\lambda_3 \widetilde{t_a}} + \frac{\beta_3 a_3 q_{iF} \sqrt{\pi}}{4 \sqrt{|\lambda_1 \widetilde{t_a}|}} erfc\left(3 \sqrt{|\lambda_1 \widetilde{t_a}|}\right)$$
$$- \frac{\beta_2 a_6 q_{iF} \sqrt{\pi}}{4 \sqrt{|\lambda_2 \widetilde{t_a}|}} erfc\left(3 \sqrt{|\lambda_2 \widetilde{t_a}|}\right) + \frac{\beta_1 a_9 q_{iF} \sqrt{\pi}}{4 \sqrt{|\lambda_3 \widetilde{t_a}|}} erfc\left(3 \sqrt{|\lambda_3 \widetilde{t_a}|}\right) \tag{78}$$

In Equation (78), the defined parameters are,

$$\beta_1 = \frac{a_1(a_1 a_5 - a_2 a_4)}{(a_1 a_9 - a_3 a_7)(a_1 a_5 - a_2 a_4) - (a_1 a_8 - a_2 a_7)(a_1 a_6 - a_3 a_4)} \tag{79}$$

$$\beta_2 = \frac{a_1 a_8 - a_2 a_7}{a_1 a_5 - a_2 a_4} \beta_1 \tag{80}$$

$$\beta_3 = \frac{\beta_2 a_4 - a_7}{a_1} \beta_1 \tag{81}$$

where λ_1, λ_2, and λ_3 are the eigenvalues of matrix in the Equation (74), $a_1 \sim a_9$ are all the values in the eigenvector of matrix in the Equation (75), β_1, β_2, and β_3 are all the coefficients.

In our model, shale gas in region 2 must firstly flow into region 1 and then into the hydraulic fracture region and finally into the wellbore. Therefore, the flow rate in hydraulic fracture is equal to that in the wellbore. From the final solution, we can find that the flow rate is related to six variables.

Through fitting the production data, these variables can be obtained and then the solution can be used for production analysis and forecasting.

3.2. Model Validation with Numerical Models

In order to verify the derived analytical solution, a numerical model is built with the commercial Eclipse reservoir simulator for comparing with the previous physical model, which is one-quarter of the volume around one hydraulic fracture. The numerical model has 27 grid cells in the x-direction, 50 grid cells in the y-direction, and only one grid cell in the z-direction. In order to capture the transient flow towards the hydraulic fracture, local grid refinement in x-direction is constructed. The top view of the model is shown in Figure 2, where the first column of grids represents half-length of hydraulic fracture, and the horizontal well located in the first row of the grids along the x-direction. The blue region represents the region 2 in which the gas adsorption plays an important role, while the red one is the region 1. Table 1 summarizes the input parameters used in the numerical models, which include the reservoir conditions, hydraulic conductivity, gas adsorption, and non-Darcy flow parameters.

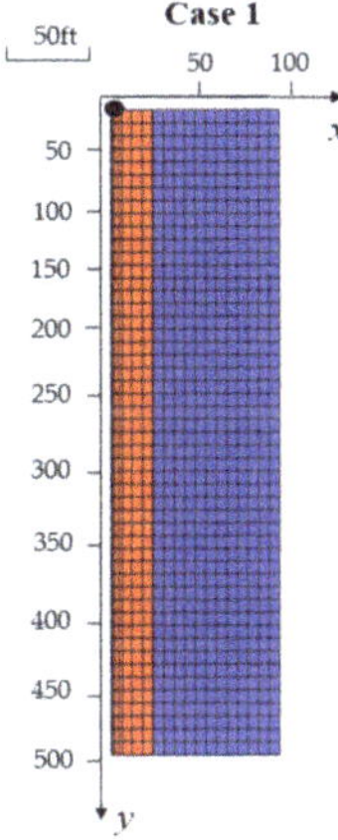

Figure 2. Reservoir grid for the numerical model. The red grids are the high-permeability region and the blue ones are the low-permeability region.

Table 1. Summary of input parameters for the synthetic numerical model.

Parameter	Value
Model dimension ($X \times Y \times Z$) (ft)	$80 \times 500 \times 10$
Initial pressure (Psi)	2500
Bottom-hole pressure (Psi)	500
Temperature (K)	318.15
Langmuir volume (Mscf/ton)	0.096
Langmuir pressure (Psi)	650
Compressibility (10^{-6} Psi^{-1})	7.5
Matrix porosity	0.06
Fracture porosity	0.08
Permeability of hydraulic fracture (mD)	500
Permeability in region 2 (mD)	0.0005
Permeability in region 1 (mD)	0.01
Fracture width (ft)	0.1
D-factor (Day/Mscf)	0.0012

In deriving the new analytical solution in this study, the pressure dependence of gas properties is considered using pseudo-pressure and pseudo-time. However, the results from the numerical models are gas rate versus real time. Therefore, the necessary step is to transform the numerical simulation results of gas rate with time into pseudo-time and then fit with the new model. Figure 3 presents the result analysis. According to the previous definition, the relationship between the pseudo-time and real time is shown in Figure 3a. The plot of 1/q vs. square root pseudo-time for regime diagnosis is shown in Figure 3b. The comparison of production rates obtained from numerical simulation and our analytical model is presented in Figure 3c. The blue dotted line represents the relationship between gas rate and pseudo-time, whereas the red one indicates that from the derived analytical solution. It can be seen that the results from the simulator and the analytical solution agree well with each other. Due to the high velocity gas flow in hydraulic facture region, the time constant in hydraulic fracture (0.001 days) is too short to be observed. Therefore, four flow regimes are identified in Figure 4. Regime 1 exhibits a half-slope straight line on the log-log plot and represents the transient linear flow in region 1. The permeability in this region is higher and hence the time constant in this region is shorter and nearly 22 days. Then an exponential curve of regime 2 indicates that the boundary of region 1 is reached, which is called the boundary-dominated flow in region 1 or inner-boundary-dominated flow. After that, the pressure propagates into the region 2. As for regime 3, it still presents an expected straight line with a half-slope signature. In our model, the permeability of region 2 is low-permeability matrix and thus the time constant is relatively longer (209 days). Regime 4 is the outer-boundary-dominated flow, which is affected by the boundary of region 2. Table 2 summarizes the four model parameters after fitting.

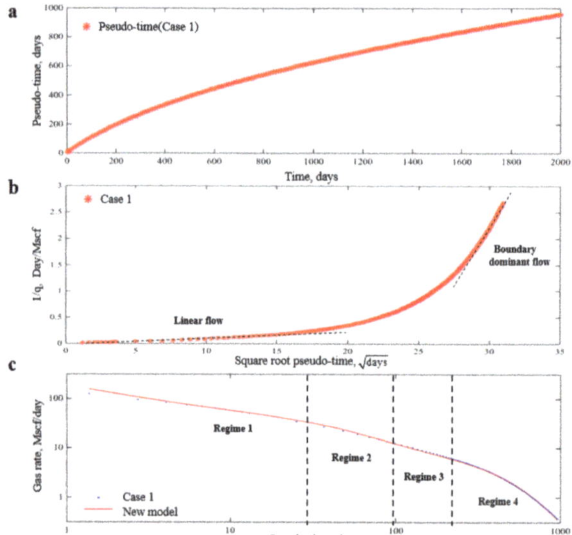

Figure 3. The result analysis for Case 1. (**a**) Comparison of the relationship between pseudo-time and time. (**b**) The plot 1/q vs. square root pseudo-time for regime diagnosis. (**c**) Comparison of the relationship between gas rate and pseudo-time. (Regime 1: transient linear flow in region 1 with slope -1/2. Regime 2: boundary-dominated flow for region 1. Regime 3: transient linear flow in region 2 with slope -1/2. Regime 4: boundary-dominated flow for region 2).

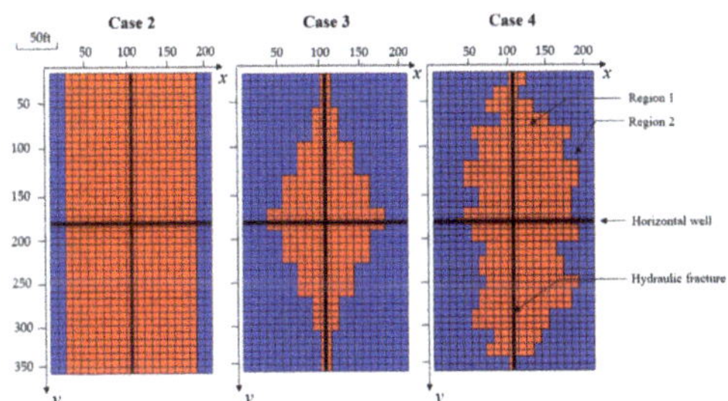

Figure 4. Reservoir grid for the numerical cases showing different shapes of stimulated region. The red grids are the stimulated region and the blue ones are the low-permeability region.

Table 2. Model parameters used in the analytical model for validation against numerical simulation.

Parameter	Value
τ_F (days)	0.001
τ_1 (days)	22
τ_2 (days)	209
T_{r21}/T_{r1F}	0.09
T_{r1F}/J	1.18
q_{iF} (Mscf/D)	109

Based on the output parameters, we can predict the values of physical parameters to further validate our model according to the following step-by-step procedure:

Step 1: Calculate the productivity index.

As defined above, we can calculate the productivity index J combined with the values of initial rate q_{iF} initial pseudo-pressure $m(p_i)$ and pseudo-bottom-hole pressure $m(p_{wf})$.

Step 2: Calculate the transmissibility between fractures and region 1 and region 1 and region 2.

Among the output parameters, we can obtain the ratio of transmissibility $(T_{r21}/T_{r1F}, T_{r1F}/J)$. Considering the calculated value in Step 1, the transmissibility can be calculated (T_{r21}, T_{r1F}).

Step 3: Calculate the pore volume of hydraulic fracture region, region 1, and region 2.

By transforming the formula that we defined in Equations (71)–(73), the pore volumes of different regions (V_{pF}, V_{p1}, V_{p2}) can be obtained by:

$$V_{pF} = \frac{J\tau_F}{C_F} \quad V_{p1} = \frac{\tau_1 T_{r1F}}{C_1} \quad V_{p2} = \frac{\tau_2 T_{r21}}{C_2} \tag{82}$$

Following the above steps, we calculate the physical parameters in the numerical case as shown in Table 3 to compare with the given data. It shows that our model solution is correct within the accepted error bound. Among them, V_g and V_c express the given volume and calculated volume, respectively.

Table 3. Inferred parameter values according to the fitting results.

Parameter	Given Data	Calculated Value	Relative Error $(Vg-Vc)/Vg \times 100\%$
V_{pF} (ft^3)	40	38.37	4.1%
V_{p1} (ft^3)	5970	5954	0.3%
V_{p2} (ft^3)	18,000	17,455	3.0%

3.3. Irregular Stimulated Region with One Hydraulic Fracture

In this section, three numerical cases with only one hydraulic fracture are designed for investigating the applicability of the analytical model for irregular stimulated region in Figure 4. For the three numerical cases, the input parameters are the same with the Case 1 except for the model dimensions. These three cases are identical with 31 grid cells in the x-direction, 51 grid cells in the y-direction, and only one grid cell in the z-direction, which represent the volume of $214 \times 521 \times 10$ ft^3. For Case 2, there is a rectangular stimulated region (region 1). As for Case 3, the stimulated region is irregular but symmetrical. In order to better describe the real condition, the stimulated region in Case 4 is designed to be neither regular nor symmetrical. The D-factor in hydraulic fracture is set as 0.0012 to represent the high-velocity non-Darcy flow.

In general, the results from the simulator and the analytical solution fit very well, as shown in Figure 5. There are also four flow regimes in the three cases. Comparing with Case 1 and Case 2, there are deviations for the cases with irregular stimulated region. The deviations are caused by the irregular inner boundary. Due to the irregular shapes of region 1, the time when the pressure propagates to inner boundaries changes with the distance away from the wellbore in different parts of inner boundary. However, the analytical solution is derived based on the regular inner boundary conditions. For the same reason, the variable starting time in regime 2 results in the deviation in the early time of the third flow regime. The flow time in regime 3 is long enough, thus, the curves in this regime fit well with each other at late time. For the three cases, the outer boundaries are identical and therefore the curves in the fourth regime also agree well with each other. According to the results summarized in Table 4, we can observe that the estimated results shown in Figure 5 are acceptable within engineering accuracy. In conclusion, our analytical solution is feasible for both regular and irregular region conditions.

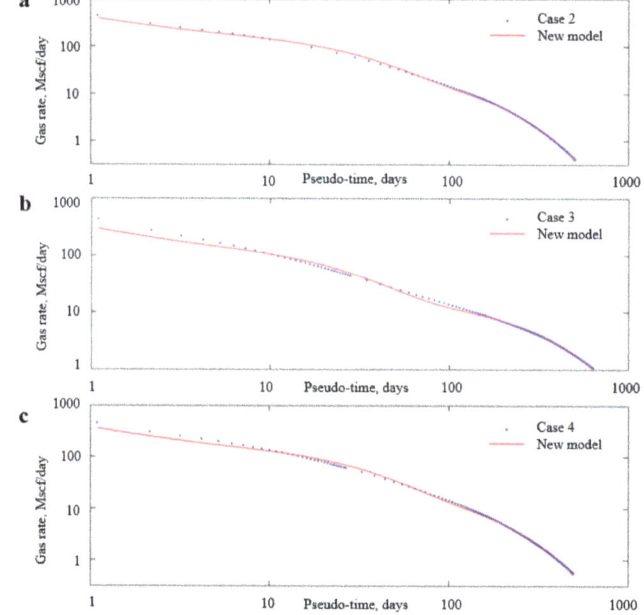

Figure 5. Comparison of the relationship of gas rate vs. pseudo-time from the numerical cases and new analytical solution. (**a**) Comparison of the relationship of gas rate vs. pseudo-time from the Case 2 and new model. (**b**) Comparison of the relationship of gas rate vs. pseudo-time from the Case 3 and new model. (**c**) Comparison of the relationship of gas rate vs. pseudo-time from the Case 4 and new model.

Table 4. Inferred parameter values according to fitting results.

Case	τ_F	τ_1	τ_2	T_{r21}/T_{r1F}	T_{r1F}/J	q_{iF} (Mscf/D)	Given Data (ft^3)	Calculated Data (ft^3)	Relative Error
Case 2	0.02	9	97	0.04	1.6	327	$V_{pF} = 31$	$V_{pF} = 32$	3%
							$V_{p1} = 26{,}258$	$V_{p1} = 25{,}137$	4.3%
							$V_{p2} = 37{,}512$	$V_{p2} = 35{,}343$	5.8%
Case 3	0.01	9	189	0.05	1.4	252	$V_{pF} = 31$	$V_{pF} = 29$	6.5%
							$V_{p1} = 20{,}084$	$V_{p1} = 21{,}123$	5.2%
							$V_{p2} = 43{,}686$	$V_{p2} = 42{,}387$	3%
Case 4	0.02	11	106	0.05	1.3	316	$V_{pF} = 31$	$V_{pF} = 30$	3.2%
							$V_{p1} = 24{,}145$	$V_{p1} = 25{,}155$	4.2%
							$V_{p2} = 39{,}625$	$V_{p2} = 40{,}135$	1.3%

3.4. Irregular Stimulated Regions with Several Hydraulic Fractures

In order to further validate the applicability in irregular region of the new derived model, three more numerical cases of multifractured horizontal wells are designed, as shown in Figure 6. These three cases have the identical dimension with 163 grid cells in the x-direction, 63 grid cells in the y-direction, and only one grid cell of 10 ft in the z-direction. The length of the horizontal well is 762 ft with 10 hydraulic fractures equally spaced along the x-direction. For Case 5, the stimulated regions are all the same regular and symmetrical regions. As a comparison, the stimulated regions in Case 6 are all irregular but symmetrical regions. Meanwhile, there is no interference between fractures in Case 5 and Case 6 and the length of the hydraulic fractures is 352 ft in Case 5 and 464 ft in Case 6. As for Case 7, all the stimulated regions around each hydraulic fracture are different, irregular, and asymmetric. The length of 10 hydraulic fractures ranges from 288 to 432 ft and there is interference between fractures. The other input parameters are the same with Case 1.

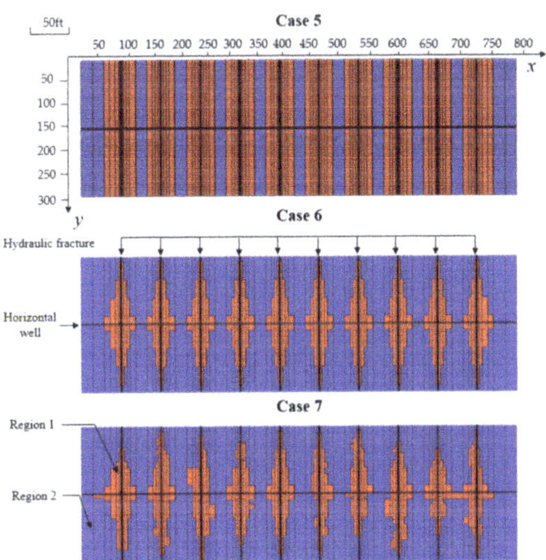

Figure 6. Reservoir grid for multifractured horizontal well (MFHWs).

Figure 7 shows the comparison of gas rate versus pseudo-time obtained from numerical simulations and the new analytical solutions. It can be seen that the results agree very well. Four regimes are also identified. The linear flow time in region 1 based on the fitting results is 9, 7, and 5 days for the three cases, respectively. Then it displays the boundary flow of region 1, which lasts until about the 100th day. The time constant in region 2 is 72, 79, and 91 days, respectively. Finally, the boundary flow of region 2 lasts for hundreds of days. We can obtain that the decline rate of gas production is faster during regimes 1 and 2 and thus the regimes 3 and 4 can last for a longer time. Therefore, for the shale gas reservoir, it is crucial to enhance the volume of stimulated regions to keep a longer high production period. Meanwhile, the contribution of gas adsorption mainly reflects in regimes 3 and 4. It helps to extend the stable production period. According to the output parameters after fitting with the numerical cases, we can calculate the volume of the hydraulic fracture region, region 1, and region 2, as shown in Table 5. Considering that the relative errors meet the requirement of engineering accuracy, our new model is also suitable for the multifractured horizontal well.

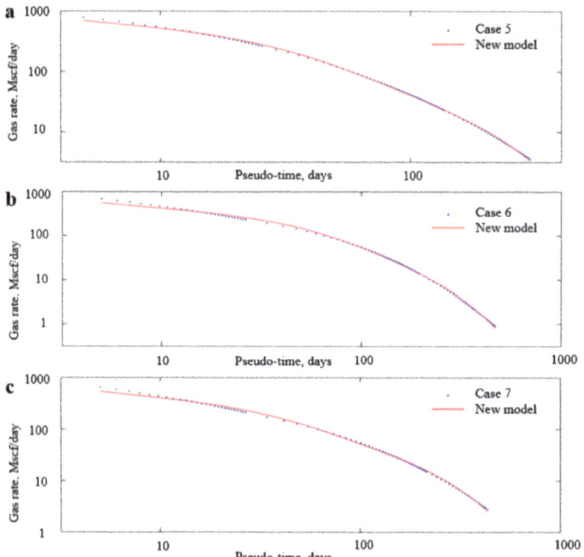

Figure 7. Comparison of the relationship of gas rate vs. pseudo-time from the numerical cases and new analytical solution. (**a**) Comparison of the relationship of gas rate vs. pseudo-time from the Case 5 and new model. (**b**) Comparison of the relationship of gas rate vs. pseudo-time from the Case 6 and new model. (**c**) Comparison of the relationship of gas rate vs. pseudo-time from the Case 7 and new model.

Table 5. Inferred parameters value according to fitting results.

Case	τ_F	τ_1	τ_2	T_{r21}/T_{r1F}	T_{r1F}/J	q_{iF} (Mscf/D)	Given Data (ft³)	Calculated Data (ft³)	Relative Error
Case 5	0.02	9	72	0.03	1.6	990	$V_{pF} = 21$	$V_{pF} = 23$	9.5%
							$V_{p1} = 106{,}022$	$V_{p1} = 103{,}820$	2.1%
							$V_{p2} = 105{,}281$	$V_{p2} = 98{,}054$	6.7%
Case 6	0.03	7	79	0.03	2.4	937	$V_{pF} = 28$	$V_{pF} = 27$	6.5%
							$V_{p1} = 81{,}062$	$V_{p1} = 79{,}302$	2.2%
							$V_{p2} = 130{,}234$	$V_{p2} = 125{,}139$	3.8%
Case 7	0.02	5	91	0.05	1.9	911	$V_{pF} = 24$	$V_{pF} = 23$	4.2%
							$V_{p1} = 67{,}264$	$V_{p1} = 66{,}097$	1.7%
							$V_{p2} = 144{,}036$	$V_{p2} = 151{,}950$	5.5%

3.5. Application to Field Case

The previous section demonstrated the accuracy of the derived analytical solution for production analysis. In this section, we apply the method to history matching and forecasting of field data in a shale gas reservoir. The gas well is selected for its relatively long production history and availability of pressure data. The main work flow is as follows:

- Apply the given parameters and gas material balance equation to transform the time into pseudo-time and pressure into pseudo-pressure.
- Make a diagnostic plot of production rate vs. pseudo-time.
- Analyze the diagnostic plot to identify flow regimes.
- Fit Equation (75) to the production data to obtain the model parameters τ_F, τ_1, τ_2, T_{r21}/T_{r1F}, T_{r1F}/J, and q_{iF}.
- Output the model parameters until a satisfactory matching is obtained.
- Following the step-by-step procedure to calculate the volume of hydraulic fracture region, region 1, and region 2.
- Make forecast of production rate with the model parameters.

We chose a horizontal well called Well B-15 with multiple fracture stages from Barnett shale [40]. The general data of the well for analysis is listed in Table 6. The well was produced under constant pressure and even for short early variable p/q the constant pressure was also applicable [41]. Firstly, we transformed the time into pseudo-time and the relationship between pseudo-time and time is shown in Figure 8a. The plot of 1/q vs. square root pseudo-time for regime diagnosis is shown in Figure 8b. Then the log-log diagnostic plot of gas rate versus pseudo-time was obtained, which exhibits a half-slope straight line in Figure 8c. According to the previous analysis, the flow time in matrix is longer and the decline rate in region 2 is slower. Therefore, the half-slope linear flow was diagnosed as the regime 3. Following the work flow, the next step was to match our model to the production data. The results of history matching and forecasting is shown in Figure 8c. The red marks in Figure 8c represent the production data and the green ones represent the analytical solution for history matching. It indicates a good matching. The six model parameters obtained from history matching are summarized in Table 7. Based on the parameters, we forecast the gas rate represented by the black markers in Figure 8c. According to the step-by-step procedure, we calculated the volume of the hydraulic fracture region as 89 ft^3, the volume of stimulated region as 84.2 MMscf, and the volume of region 2 as 1784.2 MMscf. It is essential for increasing high production period by extending the volume of the stimulated region and decreasing the decline rate in the stimulated region.

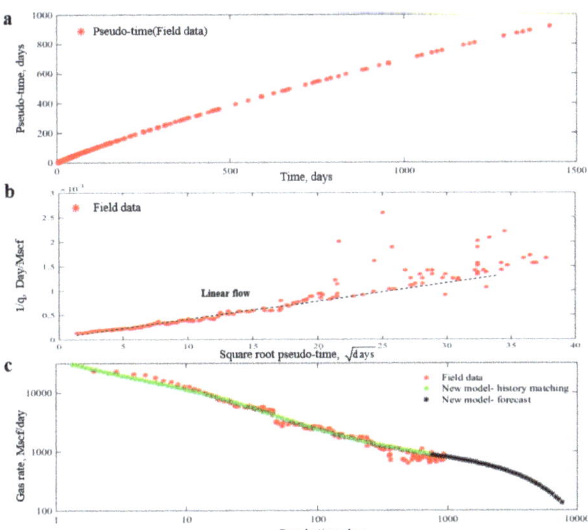

Figure 8. Summary of production profiles for the field example. (**a**) The relationship between time and pseudo-time. (**b**) The plot 1/q vs. square root pseudo-time for regime diagnosis. (**c**) The results of history-matching and forecasting on log-log plot.

Table 6. Summary of the reservoir parameters for analysis.

Parameter	Value
Initial pressure (Psi)	2950
Bottom-hole pressure (Psi)	480
Temperature (K)	344.3
Langmuir volume (Scf/ton)	96
Langmuir pressure (Psi)	650
Compressibility (Psi^{-1})	4×10^{-6}
Porosity	0.06

Table 7. Model parameters used in the analytical model for application in the field example.

Parameter	Value
τ_F (days)	0.01
τ_1 (days)	11
τ_2 (days)	3105
T_{r21}/T_{r1F}	0.075
T_{r1F}/J	0.86
q_{iF} (Mscf/D)	19,041

4. Discussion

The results of this study can be summarized as four types: (1) derivation of the approximate analytical solution; (2) validation of the solution against different numerical models; (3) introducing a step-by-step procedure to predict the values of physical parameters; (4) application of the analytical model in the field case.

Result 1 is one of the novelty of the article. The model contains three regions and effectively accounts for non-Darcy flow in the hydraulic fractures, and gas adsorption and slippage in the matrix. During the derivation process, the governing PDEs are transformed into ordinary differential equations (ODEs) by integration, instead of the Laplace transform. There is no doubt that Laplace transform works extremely well. However, there are still some problems: (1) the first step for Laplace transform is dimensionless transformation. For some dimensionless variables such as dimensionless time, dimensionless pressure, dimensionless production, and so on, more inputs are needed and many are even estimated, which will lead to calculation error. (2) The numerical inversion is an essential step for converting Laplace space into real-time space. Among them, Stefest numerical inversion algorithm is most commonly used. In this algorithm, the number of inversion items N is uncertain. The improper value will result in deviations in real time. Certainly, Result 1 also has some imperfections. For example, a dual-porosity model with Knudsen diffusion [6] would be more representative and the heterogeneity deserves further study.

Result 2 is the key section in the article. Seven numerical cases were set for verification. According to the fitting results, it shows that the analytical solution is feasible for the irregular and asymmetric stimulated regions in a multifractured horizontal well. Considering that the shapes of the enhanced fracture regions are unknown in real cases, we creatively set three types of stimulated regions: regular region, irregular region, and very irregular region. At least, the validation results show that our model is robust.

Result 3 is another novelty of our work. It introduced a step-by-step procedure to calculate inversion parameters. Comparing with the given volumes and calculated volumes from case 1 to case 7, the model is also verified to meet the engineering requirements. Considering the simplifying assumptions, the model may need to be improved and the accurate microseismic data would be required to make further verification.

Result 4 is the most important section. Considering the decline characteristics of shale gas, the constant rate case is not as important as the constant pressure case for long term performance of tight/shale formations. Therefore, our model is derived based on the constant bottom-hole pressure condition, and namely one of limitations for our model is that it cannot be applied in a constant rate case. The second limitation is the data continuity, because the prerequisite for the model to make accurate prediction is great fitting. As for the discontinuous data like data missing, shut-in, or pressure/production jump, our model is not applicable. Combined with the assumption of single-phase flow in this work, the multiphase flow problem cannot be solved by this analytical solution. Gas–water two-phase flow is the most common in shale gas reservoirs, so our future work is to derive the new analytical solution for gas–water two-phase flow.

5. Conclusions

In this paper, we presented a practical analytical model to study the performance of MFHWs from shale gas reservoirs. Numerical models have been used to validate the analytical solutions and an excellent agreement was obtained. The following conclusions are drawn from this work:

- A simple rate versus pseudo-time relationship is presented to account for transient linear and boundary-dominated flow periods in shale gas formation.
- Incorporating the effect of gas adsorption, non-Darcy flow, and slippage flow in the analytical model by defining the modified pseudo-pressure and pseudo-time, accuracy is improved in production forecast in shale gas reservoir.
- Comparing to the Laplace-transform solution, our analytical model is derived in real-time space and it is unnecessary to undertake dimensionless transformation and numerical inversion. It is more applicable in field scale.
- Through the model parameters obtained from history matching the field data, the production rate and cumulative production can be forecasted. In addition, the pore volume of different regions can also be calculated by step-by-step procedure, which was validated to be feasible for the irregular and asymmetric stimulated regions in multifractured horizontal wells. According to the results, the calculation accuracy is less than 10% and meets the engineering requirements.

Author Contributions: Conceptualization, K.Q. and H.L.; methodology and investigation K.Q.; writing—original draft preparation K.Q.; writing—review and editing, H.L.; supervision, H.L.; project administration, H.L.; funding acquisition, H.L. All authors have read and agreed to the published version of the manuscript.

Funding: National Science and Technology Major Project of China: 2017ZX05039-005.

Acknowledgments: This work is funded by the National Science and Technology Major Project of China (Grant no. 2017ZX05039-005).

Conflicts of Interest: The authors declare no conflict of interest.

Nomenclature

V	the volume of the adsorbed gas, ft^3
P	the reservoir pressure, Psi
V_L	Langmuir volume
P_L	Langmuir pressure
C_f	rock compressibility, Psi^{-1}
C_w	water compressibility, Psi^{-1}
C_g	free gas compressibility, Psi^{-1}
C_{gd}	adsorbed gas compressibility, Psi^{-1}
C_t	total compressibility, Psi^{-1}
C_t^*	modified total compressibility, Psi^{-1}
$\overline{B_g}$	gas reservoir volume factor
z^*	modified compressibility factor
z	compressibility factor
Z_{sc}	standard compressibility factor

T	reservoir temperature, K
T_{sc}	standard condition temperature, K
P_{sc}	standard condition pressure, Psi
$\bar{p}$	average reservoir pressure, Psi
k_{iF}	hydraulic fracture permeability, mD
k_{eq}^{F}	equivalent hydraulic fracture permeability, mD
k_{a}^{m}	matrix permeability, mD
k_{∞}^{m}	intrinsic permeability, mD
k_{i1}	region 1 permeability, mD
k_{i2}	region 2 permeability, mD
r	pore radius, ft
R	the universal gas constant
M	the gas molecular weight
Kn	Knudsen number
V_{b}	volume of the region
V_{p}	pore volume of the region
q_{iF}	initial production rate from the hydraulic fracture region
q_{i1}	initial production rate from region 1
q_{i2}	initial production rate from region 2
t	production time, days
t_{a}	pseudo-time, days
t_{aF}	pseudo-time in fracture region, days
t_{aF}	pseudo-time in region 1, days
t_{a2}	pseudo-time in region 2, days
$\tilde{t}_{a}$	approximate pseudo-time, days
$m_{2}(p)$	pseudo-pressure in region 2
$m_{1}(p)$	pseudo-pressure in region 1
$m_{F}(p)$	pseudo-pressure in hydraulic region
$\overline{m}_{2}(p)$	average pseudo-pressure in region 2
$\overline{m}_{1}(p)$	average pseudo-pressure in region 1
$\overline{m}_{F}(p)$	average pseudo-pressure in fracture region
$w/2$	half-width of hydraulic fracture, ft
x_{1}	region 1 impact distance, ft
x_{2}	half distance between fractures, ft
y_{e}	half-length of macro-fracture, ft
z_{e}	depth of top reservoir, ft
z_{0}	depth of bottom reservoir, ft
τ_{1}	region 1 constant time, days
τ_{2}	region 2 constant time, days
τ_{F}	hydraulic region constant time, days
T_{r21}	transmissibility between region 1 and region 2, STB/D/psi
T_{r1F}	transmissibility between microfracture and matrix, STB/D/psi
J	hydraulic fracture region production index, STB/D/psi
V_{pF}	pore volumes of hydraulic region
V_{p1}	pore volumes of region 1
V_{p2}	pore volumes of region 2
α	correlation parameter
β	non-Darcy flow coefficient
ϕ	porosity
ρ_{m}	matrix density, g/cm^{3}
μ	fluid viscosity, cp
v	gas flow velocity

Appendix A Derivation of the Average Pseudo-Pressure

The solution to the 1-D linear oil flow with constant-pressure boundary condition is given by [27]:

$$p_D = 1 + \frac{4}{\pi} \sum_{n=1}^{\infty} \frac{(-1)^n}{2n-1} e^{-\left(\frac{(2n-1)\pi}{2}\right)^2 t_D} \cos\left(\frac{(2n-1)\pi}{2} x_D\right) \tag{A1}$$

For the shale gas case, the pressure and time should be replaced by pseudo-pressure and pseudo-time shown in Equations (17) and (19). Following the same step, we derive the general solution applied for the 1-D linear gas flow with constant-pressure boundary condition.

$$m_D(p) = 1 + \frac{4}{\pi} \sum_{n=1}^{\infty} \frac{(-1)^n}{2n-1} e^{-\left(\frac{(2n-1)\pi}{2}\right)^2 t_{aD}} \cos\left(\frac{(2n-1)\pi}{2} x_D\right) \tag{A2}$$

where

$m_D = \frac{m(p_i)-m(p)}{m(p_i)-m(p_{wf})}$ is the dimensionless reservoir pseudo-pressure.

$t_{aD} = \frac{k t_a}{\phi \mu c_i L}$ is the dimensionless pseudo-time.

$x_D = \frac{x_e - x}{x_e - x_{wf}}$ is the dimensionless distance in the x-direction.

Similarly, the production rate can be obtained as,

$$q_D = 2 e^{-\left(\frac{(2n-1)\pi}{2}\right)^2 t_{aD}} \tag{A3}$$

According to the definition of volume-weighted average pseudo-pressure,

$$\overline{m}(p) = \frac{\int\limits_V m(p)dV}{\int\limits_V dV} \tag{A4}$$

Therefore,

$$\overline{m}_D(p) = \frac{\int_1^{x_D} 1 + \frac{4}{\pi} \sum_{n=1}^{\infty} \frac{(-1)^n}{2n-1} e^{-\left(\frac{(2n-1)\pi}{2}\right)^2 t_{aD}} \cos\left(\frac{(2n-1)\pi}{2} x_D\right) dx_D}{\int_1^{x_D} dx_D} \tag{A5}$$

Performing the integration results in,

$$\overline{m}_D(p) = 1 + \frac{8}{\pi^2} \sum_{n=1}^{\infty} \frac{(-1)^n}{(2n-1)^2} e^{-\frac{(2n-1)^2 \pi^2}{4} t_{aD}} \sin\frac{(2n-1)\pi}{2} \tag{A6}$$

For odd values of n, $\sin\frac{(2n-1)\pi}{2}$ is positive unity and $(-1)^n$ is negative unity. As a result, the product of these two coefficients is always negative. Equation (A6) can be rewritten as,

$$\overline{m}_D(p) = 1 - \frac{8}{\pi^2} \sum_{n=1}^{\infty} \frac{1}{(2n-1)^2} e^{-\frac{(2n-1)^2 \pi^2}{4} t_{aD}} \tag{A7}$$

Substituting the Equation (A3) into the Equation (A7), we can obtain that,

$$\overline{m}_D(p) = 1 - \frac{4}{\pi^2} \sum_{n=1}^{\infty} \frac{q_D}{(2n-1)^2} \tag{A8}$$

Transforming the Equation (A8) into dimensional form, Equation (A8) can be rewritten as,

$$\overline{m}(p) = m(p_{wf}) + \frac{4}{\pi^2} [m(p_i) - m(p_{wf})] \sum_{n=1}^{\infty} \frac{q_{D_n}}{(2n-1)^2} \tag{A9}$$

Appendix B Solution of the System of ODEs

To solve the systems of ODEs in Equation (71), we extract the coefficient matrix as,

$$A_n = \begin{pmatrix} -\left(\frac{1}{\tau_2} + \frac{T_{r21}}{\tau_1 T_{r1F}}\right) & \frac{T_{r21}}{\tau_1 T_{r1F}} & 0 \\ \frac{1}{\tau_1} & -\left(\frac{1}{\tau_1} + \frac{T_{r1F}}{\tau_F J}\right) & \frac{T_{r1F}}{\tau_F J} \\ 0 & \frac{1}{\tau_F} & -\frac{1}{\tau_F} \end{pmatrix}_{3\times 3} \tag{A10}$$

For the 3×3 matrix, we can obtain three eigenvalues λ_1, λ_2, λ_3 and three eigenvectors ξ_1, ξ_2, ξ_3 where

$$\xi_1 = \begin{pmatrix} a_1 \\ a_2 \\ a_3 \end{pmatrix} \xi_2 = \begin{pmatrix} a_4 \\ a_5 \\ a_6 \end{pmatrix} \xi_3 = \begin{pmatrix} a_7 \\ a_8 \\ a_9 \end{pmatrix} \tag{A11}$$

Combining the eigenvalues and eigenvectors, we obtain the solution of the ODEs,

$$\begin{pmatrix} q_{2n} \\ q_{1n} \\ q_{Fn} \end{pmatrix} = (2n-1)^2 \left[C_1 \begin{pmatrix} a_1 \\ a_2 \\ a_3 \end{pmatrix} e^{(2n-1)^2 \lambda_1 t_a} + C_2 \begin{pmatrix} a_4 \\ a_5 \\ a_6 \end{pmatrix} e^{(2n-1)^2 \lambda_2 t_a} + C_3 \begin{pmatrix} a_7 \\ a_8 \\ a_9 \end{pmatrix} e^{(2n-1)^2 \lambda_3 t_a} \right] \tag{A12}$$

where C_1, C_2, and C_3 are unknown constants.

According to the initial assumption in Equations (72)–(74), we can calculate the value of C_1, C_2, and

$$C_3 C_1 = \beta_3 q_{iF} \quad C_2 = -\beta_2 q_{iF} \quad C_3 = \beta_1 q_{iF} \tag{A13}$$

Where

$$\beta_1 = \frac{a_1 (a_1 a_5 - a_2 a_4)}{(a_1 a_9 - a_3 a_7)(a_1 a_5 - a_2 a_4) - (a_1 a_8 - a_2 a_7)(a_1 a_6 - a_3 a_4)}$$

$$\beta_2 = \frac{a_1 a_8 - a_2 a_7}{a_1 a_5 - a_2 a_4} \beta_1$$

$$\beta_3 = \frac{\beta_2 a_4 - a_7}{a_1} \beta_1$$

Therefore, we obtain the function of the final rate,

$$q_{Fn} = (2n-1)^2 \left(\beta_3 a_3 q_{iF} e^{(2n-1)^2 \lambda_1 t_a} - \beta_2 a_6 q_{iF} e^{(2n-1)^2 \lambda_2 t_a} + \beta_1 a_9 q_{iF} e^{(2n-1)^2 \lambda_3 t_a} \right) \tag{A14}$$

To obtain the total production rate, we summate the Equation (A14),

$$q_F = \sum_{n=1}^{\infty} q_{Fn} = \beta_3 a_3 q_{iF} e^{\lambda_1 t_a} - \beta_2 a_6 q_{iF} e^{\lambda_2 t_a} + \beta_1 a_9 q_{iF} e^{\lambda_3 t_a}$$
$$+ \sum_{n=1}^{\infty} (2n-1)^2 \left(\beta_3 a_3 q_{iF} e^{(2n-1)^2 \lambda_1 t_a} - \beta_2 a_6 q_{iF} e^{(2n-1)^2 \lambda_2 t_a} + \beta_1 a_9 q_{iF} e^{(2n-1)^2 \lambda_3 t_a} \right) \tag{A15}$$

Through converting the summation to an integral,

$$q_F = \beta_3 a_3 q_{iF} e^{\lambda_1 t_a} - \beta_2 a_6 q_{iF} e^{\lambda_2 t_a} + \beta_1 a_9 q_{iF} e^{\lambda_3 t_a}$$
$$+ \lim_{z \to \infty} \int_{3\sqrt{|\lambda_1| t_a}}^{z} \left(\beta_3 a_3 q_{iF} e^{-z^2 t_a} - \beta_2 a_6 q_{iF} e^{\frac{\lambda_2}{|\lambda_1|} t_a} + \beta_1 a_9 q_{iF} e^{\frac{\lambda_3}{|\lambda_1|} t_a} \right) \frac{dz}{2\sqrt{|\lambda_1| t_a}} \tag{A16}$$

Adopting the complementary error function, the rate versus pseudo-time function can be

$$q_F = \beta_3 a_3 q_{iF} e^{\lambda_1 t_a} - \beta_2 a_6 q_{iF} e^{\lambda_2 t_a} + \beta_1 a_9 q_{iF} e^{\lambda_3 t_a} + \frac{\beta_3 a_3 q_{iF} \sqrt{\pi}}{4\sqrt{|\lambda_1 t_a|}} erfc\left(3\sqrt{|\lambda_1 t_a|}\right)$$
$$- \frac{\beta_2 a_6 q_{iF} \sqrt{\pi}}{4\sqrt{|\lambda_2 t_a|}} erfc\left(3\sqrt{|\lambda_2 t_a|}\right) + \frac{\beta_1 a_9 q_{iF} \sqrt{\pi}}{4\sqrt{|\lambda_3 t_a|}} erfc\left(3\sqrt{|\lambda_3 t_a|}\right) \tag{A17}$$

References

1. Hu, X.Y.; Li, M.T.; Peng, C.G.; Davarpanah, A. Hybrid Thermal-Chemical Enhanced Oil Recovery Methods—An Experimental Study for Tight Reservoirs. *Symmetry* **2020**, *12*, 947. [CrossRef]
2. Jin, Y.; Davarpanah, A. Using Photo-Fenton and Floatation Techniques for the Sustainable Management of Flow-Back Produced Water Reuse in Shale Reservoirs Exploration. *Water Air Soil Pollut.* **2020**, *231*, 1–8. [CrossRef]
3. Hu, X.; Xie, J.; Cai, W.; Wang, R.; Davarpanah, A. Thermodynamic Effects of Cycling Carbon Dioxide Injectivity in Shale Reservoirs. *J. Pet. Sci. Eng.* **2020**, *195*, 107717. [CrossRef]
4. Wu, M.L.; Ding, M.C.; Yao, J.; Li, C.; Huang, Z.; Xu, S. Production-Performance Analysis of Composite Shale-Gas reservoirs by the Boundary-Element Method. *SPE Reserv. Eval. Eng.* **2019**, *22*, 238–252. [CrossRef]
5. Sun, S.H.; Zhou, M.H.; Lu, W.; Davarpanah, A. Application of Symmetry Law in Numerical Modeling of Hydraulic Fracturing by Finite Element Method. *Symmetry* **2020**, *12*, 1122. [CrossRef]
6. Sun, H.; Chawathe, A.; Hoteit, H.; Shi, X.; Li, L. Understanding Shale Gas Flow Behavior Using Numerical Simulation. *SPE J.* **2015**, *20*, 142–154. [CrossRef]
7. Zhang, M.; Sun, Q.; Ayala, L.F. Semi-Analytical Modeling of Multi-Mechanism Gas Transport in Shale Reservoirs with Complex Hydraulic-Fracture Geometries by the Boundary Element Method. In Proceedings of the SPE Annual Technical Conference and Exhibition, Calgary, AB, Canada, 30 September–2 October 2019. SPE Paper 195902-MS.
8. Davarpanah, A.; Shirmohammadi, R.; Mirshekari, B.; Aslani, A. Analysis of hydraulic fracturing techniques: Hybrid fuzzy approaches. *Arab. J. Geosci.* **2019**, *12*, 402. [CrossRef]
9. Nobakht, M.; Clarkson, C.R. A New Analytical Method for Analyzing Linear Flow in Tight/Shale Gas Reservoirs: Constant-Flowing-Pressure Boundary Condition. *SPE Reserv. Eval. Eng.* **2012**, *15*, 370–384. [CrossRef]
10. Zhao, Y.L.; Zhang, L.H.; Luo, J.X.; Zhang, B.-N. Performance of fractured horizontal well with stimulated reservoir volume in unconventional gas reservoir. *J. Hydrol.* **2014**, *512*, 447–456. [CrossRef]
11. Clarkson, C.R.; Qanbari, F. An Approximate Semianalytical Multiphase Forecasting Method for Multifractured Tight Light-Oil Wells With Complex Fracture Geometry. *J. Can. Pet. Technol.* **2015**, *54*, 489–508. [CrossRef]
12. Wei, S.M.; Xia, Y.; Jin, Y. Quantitative study in shale gas behaviors using a coupled triple-continuum and discrete fracture model. *J. Pet. Sci. Eng.* **2019**, *174*, 49–69. [CrossRef]
13. Wang, Q.; Wan, J.F.; Mu, L.F.; Shen, R.; Jurado, M.J.; Ye, Y. An Analytical Solution for Transient Productivity Prediction of Multi-Fractured Horizontal Wells in Tight Gas Reservoirs Considering Nonlinear Porous Flow Mechanisms. *Energies* **2020**, *13*, 1066. [CrossRef]
14. Zhang, Q.; Jiang, S.; Wu, X.Y. Development and Calibration of a Semianalytic Model for Shale Wells with Nonuniform Distribution of Induced Fractures Based on ES-MDA Method. *Energies* **2020**, *13*, 3718. [CrossRef]
15. Lee, S.T.; Brockenbrough, J. A New Analytic Solution for Finite Conductivity Vertical Fractures With Real Time and Laplace Space Parameter Estimation. In Proceedings of the SPE Annual Technical Conference and Exhibition, San Francisco, CA, USA, 5–8 October 1983. SPE Paper 12013-MS.
16. Wattenbarger, R.A.; El-Banbi, A.H.; Villegas, M.E.; Maggard, J.B. Production Analysis of Linear Flow into Fractured Tight Gas Wells. In Proceedings of the SPE Rocky Mountain Regional/Low-Permeability Reservoirs Symposium, Denver, CO, USA, 5–8 April 1998. SPE Paper 39931-MS.
17. El-Banbi, A.H.; Wattenbarger, R.A. Analysis of Linear Flow in Gas Well Production. In Proceedings of the SPE Gas Technology Symposium, Calgary, AB, Canada, 15–18 March 1998. SPE Paper 39972-MS.
18. Bello, R.O.; Wattenbarger, R.A. Multi-Stage Hydraulically Fractured Horizontal Shale Gas Well Rate Transient Analysis. In Proceedings of the North Africa Technical Conference and Exhibition, Cairo, Egypt, 14–17 February 2010. SPE Paper 126754-MS.
19. Ozkan, E.; Brown, M.L.; Raghavan, R.S.; Kazemi, H. Comparison of Fractured Horizontal-Well Performance in Conventional and Unconventional Reservoirs. In Proceedings of the SPE Western Regional Meeting, San Jose, CA, USA, 24–26 March 2009. SPE Paper 121290-MS.
20. Brown, M.; Ozkan, E.; Raghavan, R.; Kazemi, H. Practical Solutions for Pressure-Transient Responses of Fractured Horizontal Wells in Unconventional Shale Reservoirs. *SPE Reserv. Eval. Eng.* **2011**, *14*, 663–676. [CrossRef]
21. Stalgorova, E.; Mattar, L. Practical Analytical Model to Simulate Production of Horizontal Wells with Branch Fractures. In Proceedings of the SPE Canadian Unconventional Resources Conference, Calgary, AB, Canada, 30 October–1 November 2012. SPE Paper 162515-MS.
22. Heidari, S.M.; Clarkson, C.R. An Analytical Model for Analyzing and Forecasting Production from Multi-fractured Horizontal Wells with Complex Branched-Fracture Geometry. *SPE Reserv. Eval. Eng.* **2015**, *18*, 356–374. [CrossRef]

23. Abassi, M.; Sharifi, M.; Kazemi, A. Development of New Analytical Model for Series and Parallel Triple Porosity Models and Providing Transient Shape Factor between Different Regions. *J. Hydrol.* **2019**, *574*, 683–698. [CrossRef]

24. Yao, F. *Production Evaluation of Fracturing Horizontal Wells for Solution Gas Drive in Tight Oil Reservoirs*; China University of Petroleum (Beijing): Beijing, China, 2018.

25. Stehfest, H. Algorithm 368: Numerical inversion of Laplace transforms [D5]. *Commun. ACM* **1970**, *13*, 47–49. [CrossRef]

26. Shahamat, M.S.; Mattar, L.; Aguilera, R. A Physics-Based Method for Production Data Analysis of Tight and Shale Petroleum Reservoirs Using Succession of Pseudo-Steady States. In Proceedings of the SPE/EAGE European Unconventional Resources Conference and Exhibition, Vienna, Austria, 25–27 February 2014. SPE Paper 167686-MS.

27. Ogunyomi, B.A. *Simple Mechanistic Modeling of Recovery from Unconventional Oil Reservoirs*; The University of Texas at Austin: College Station, TX, USA, 2015.

28. Ogunyomi, B.A.; Patzek, T.W.; Lake, L.W.; Kabir, C.S. History Matching and Rate Forecasting in Unconventional Oil Reservoirs With an Approximate Analytical Solution to the Double-Porosity Model. *SPE Reserv. Eval. Eng.* **2016**, *19*, 70–82. [CrossRef]

29. Qiu, K.X.; Li, H. A New Analytical Solution of the Triple-Porosity Model for History Matching and Performance Forecasting in Unconventional Oil Reservoirs. *SPE J.* **2018**, *23*, 2060–2079. [CrossRef]

30. Qiu, K.X.; Li, H. A New Analytical Model for Production Forecasting in Unconventional Reservior Considering the Simultaneous Matrix-fracture flow. In Proceedings of the Asia Pacific Unconventional Resource Technology Conference, Brisbane, Australia, 18–19 November 2019. SPE Paper URTEC-198266-MS.

31. Tabatabaie, S.H. *Unconventional Reservoirs: Mathematical Modeling of Some Non-linear Problems*; University of Calgary: College Station, AB, Canada, 2014.

32. Behmanesh, H.; Hamdi, H.; Clarkson, C.R.; Thompson, J.M.; Anderson, D.M. Analytical Modeling of Linear Flow in Single-Phase Tight Oil and Tight Gas Reservoirs. *J. Pet. Sci. Eng.* **2018**, *171*, 1084–1098. [CrossRef]

33. Anderson, D.M.; Nobakht, M.; Moghadam, S.; Mattar, L. Analysis of Production Data from Fractured Shale Gas Wells. In Proceedings of the SPE Unconventional Gas Conference, Pittsburgh, PA, USA, 23–25 February 2010. SPE Paper 131787.

34. Langmuir, I. The Constitution and Fundamental Properties of Solids and Liquids. *J. Am. Chem. Soc.* **1916**, *38*, 2221–2295. [CrossRef]

35. Bumb, A.C.; McKee, C.R. Gas-Well Testing in the Presence of Desorption for Coal bed Methane and Devonian Shale. In Proceedings of the SPE Unconventional Gas Technology Symposium, Louisville, KY, USA, 18–21 May 1986. SPE Paper 15227-MS.

36. King, G.R. Material Balance Techniques for Coal Seam and Devonian Shale Gas Reservoirs. In Proceedings of the SPE Annual Technical Conference and Exhibition, New Orleans, Louisiana, 23–26 September 1990. SPE Paper 20730-MS.

37. Forchheimer, P. Wasserbewegung Durch Boden. *Z. Ver. Deutsch Ing.* **1901**, *45*, 1782–1788.

38. Wang, J.; Lou, H.S.; Liu, H.Q. An Intergrative Model to Simulate Gas Transport and Production Coupled With Gas Adsorption, Non-Darcy Flow, Surface Diffusion, and Stress Dependence in Organic-Shale Reservoir. *SPE J.* **2017**, *22*, 244–264. [CrossRef]

39. Wang, H.Y.; Porcu, M.M. Impact of Shale-Gas Apparent Permeability on Production: Combined Effects of Non-Darcy Flow/Gas Slippage, Desorption and Geomechanics. *SPE Reserv. Eval. Eng.* **2015**, *18*, 495–507. [CrossRef]

40. Samandarli, O. *A New Method for History Matching and Forecasting Shale Gas/Oil Reservoir Production Performance with Dual and Triple Porosity Models*; Texas A&M University: College Station, TX, USA, 2011.

41. Hamdi, H.; Behmanesh, H.; Clarkson, C.R. A Semi-Analytical Approach for Analysis of Wells Exhibiting Multi-Phase Transient Linear Flow: Application to Field Data. In Proceedings of the SPE Annual Technical Conference and Exhibition, Calgary, AB, Canada, 30 September–2 October 2019. SPE Paper 196164-MS.

Publisher's Note: MDPI stays neutral with regard to jurisdictional claims in published maps and institutional affiliations.

Article

Development and Calibration of a Semianalytic Model for Shale Wells with Nonuniform Distribution of Induced Fractures Based on ES-MDA Method

Qi Zhang [1,2]**, Shu Jiang** [1,2,*]**, Xinyue Wu** [1,2]**, Yan Wang** [3,4] **and Qingbang Meng** [1,2]

[1]   Key Laboratory of Tectonics and Petroleum Resources, Ministry of Education, China University of Geosciences, Wuhan 430074, China; qizhang@upc.edu.cn (Q.Z.); wxy.xzl@cug.edu.cn (X.W.); mengqb@cug.edu.cn (Q.M.)

[2]   School of Earth Resources, China University of Geosciences, Wuhan 430074, China

[3]   State Key Laboratory of Geomechanics and Geotechnical Engineering, Institute of Rock and Soil Mechanics, Chinese Academy of Sciences, Wuhan 430071, China; ywang@whrsm.ac.cn

[4]   Institute of Geophysics and Geomatics, China University of Geosciences, Wuhan 430074, China

*   Correspondence: jiangsu@cug.edu.cn

Received: 22 June 2020; Accepted: 16 July 2020; Published: 20 July 2020

Abstract: Given reliable parameters, a newly developed semianalytic model could offer an efficient option to predict the performance of the multi-fractured horizontal wells (MFHWs) in unconventional gas reservoirs. However, two major challenges come from the accurate description and significant parameters uncertainty of stimulated reservoir volume (SRV). The objective of this work is to develop and calibrate a semianalytic model using the ensemble smoother with multiple data assimilation (ES-MDA) method for the uncertainty reduction in the description and forecasting of MFHWs with nonuniform distribution of induced fractures. The fractal dimensions of induced-fracture spacing (d_{fs}) and aperture (d_{fa}) and tortuosity index of induced-fracture system (θ) are included based on fractal theory to describe the properties of SRV region. Additionally, for shale gas reservoirs, gas transport mechanisms, e.g., viscous flow with slippage, Knudsen diffusion, and surface diffusion, among multi-media including porous kerogen, inorganic matter, and fracture system are taken into account and the model is verified. Then, the effects of the fractal dimensions and tortuosity index of induced fractures on MFHWs performances are analyzed. What follows is employing the ES-MDA method with the presented model to reduce uncertainty in the forecasting of gas production rate for MFHWs in unconventional gas reservoirs using a synthetic case for the tight gas reservoir and a real field case for the shale gas reservoir. The results show that when the fractal dimensions of induced-fracture spacing and aperture is smaller than 2.0 or the tortuosity index of induced-fracture system is larger than 0, the permeability of induced-fracture system decreases with the increase of the distance from hydraulic fractures (HFs) in SRV region. The large d_{fs} or small θ causes the small average permeability of the induced-fracture system, which results in large dimensionless pseudo-pressure and small dimensionless production rate. The matching results indicate that the proposed method could enrich the application of the semianalytic model in the practical field.

Keywords: history matching; semianalytic model; unconventional gas reservoirs; multistage fractured horizontal wells; fractal theory

---

1. Introduction

Unconventional reservoirs, especially the tight sand and shale reservoirs, have become more and more significant with the development of the horizontal wells and hydraulic fracturing technologies. The complex fracture network surrounding the multistage fractured horizontal wells (MFHWs) called stimulated reservoir volume (SRV), has obvious influences on the production performance [1–3]. Currently, a lot of effort has been done to model the pressure transient and rate transient of MFHWs aiming to forecast the production accurately. Through numerical methods, analytical methods, and empirical methods [4–7], production data analysis in unconventional reservoirs, such as Barnett, Bakken, and Eagle, shows that the linear flow in formation, especially in hydraulic fractures (HFs) and SRV, dominates the production of MFHWs. Meanwhile, the advantage of analytic models is their simplicity and less parameters compared with numerical and empirical models, although they cannot characterize the over-complex heterogeneity properties and fracture system. Thus, considering the main flow regimes in the life of MFHWs in unconventional gas reservoirs, various linear flow models based on linear-flow assumptions have been developed. Ozkan et al. [8] and Brown et al. [9] reported a trilinear model to study the performance of MFHWs in unconventional reservoirs, which divides the formation into three main regions. In order to describe the SRV size between HFs more reasonably, Stalgorova and Matter [10] improved the trilinear model to a five-linear model by simplifying SRV in a region with limited width. On this basis, many scholars studied the production rate or pressure transient performances of MFHWs in unconventional reservoirs [11,12]. Although the models mentioned above can simulate the main linear-flow regimes in both SRV and unstimulated reservoir volume (USRV), the heterogeneous distribution of complex induced fractures are not considered in SRV region, which is the major distinct properties between SRV and USRV as shown in Figure 1.

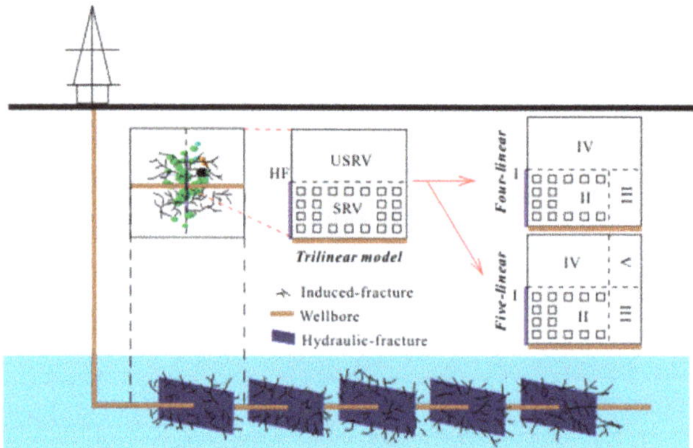

Figure 1. Simplified physical model and three typical presented linear-flow models for a multi-fractured horizontal wells (MFHW) in unconventional reservoirs.

The distribution of induced fractures system in SRV affected by HFs and the pre-existing closed natural fractures (NFs) with self-similar characterization can be described by fractal theory [13–15]. Chang and Yortsos [16] quantified the heterogeneous permeability and porosity of NFs using fractal theory. Then, Cossio et al. [17] introduced the fractal permeability/porosity relation [18] into the fractal diffusivity equation (FDE), which proved to be more fundamental than the classical diffusivity equation. Wang et al. [19] combined FDE with dual-porosity media model in the trilinear model to describe oil flow in SRV region of MFHWs in tight reservoirs, and their model provided a suitable way to study the flow behaviors in heterogeneous induced fractures systems. Sheng et al. [20] extended the fractal trilinear model to shale gas reservoirs coupling multi-porosity media (porous kerogen, inorganic matter, and induced-fracture) with multiple gas transport mechanisms (viscous flow with slippage, Knudsen diffusion, and surface diffusion). However, they did not consider the unstimulated reservoir volume (USRV) between HFs, which cannot accurately describe the microseismic results. In addition, the fracture porosity and permeability distribution affected by the distribution of induced-fracture spacing and aperture, but there are limited reports about the calculation methods for the non-uniform distribution of induced fractures spacing and aperture using fractal theory [21].

History matching to estimate the critical parameters of formation and hydraulic fracturing and rate prediction of MFHWs plays important roles in unconventional reservoirs based on the numerical and analytic models. Samandarli et al. [22] presented a semianalytic method to estimate reservoir parameters by history matching the production data of hydraulically fractured shale gas wells using a least absolute value regression. Correction for gas properties changing with time and desorbed gas at low pressure was considered in the model. According to the algebraic equations, bilinear flow regime and linear flow regime were used to match both synthetic data and field data by doing regression on matrix permeability, fracture permeability, and length. Hategan and Hawkes [23] used a simple analytical model based on the solution of the pseudo-steady state equation to analyze the production of MFHWs in shale gas reservoirs, and thought most shale gas reservoirs fit different type curves by normalizing to reservoir pressure, permeability, and number of fracture stages. Considering the effect of pressure-dependent natural-fracture permeability on production from shale gas wells, Cho et al. [24] modified the trilinear flow model to match the production data of Haynesville and Barnett shale gas wells. The reservoir properties and the correlation coefficients used for the pressure-dependent permeability were matched ignoring the non-uniqueness issues caused by the large number of parameters. Ogunyomi et al. [25] developed an approximate analytical solution to the double-porosity model for MFHWs in unconventional oil reservoirs. Due to the solution obtained in real-time space, the model was used to match the field data with the least-square method mainly by changing the parameters related to time, and forecast the production rates. Clarkson et al. [26] combined analytical, semi-analytical, and empirical methods for history matching and production forecasting for MFHWs in tight and shale gas reservoirs. Linear-to-boundary (LTB) model [27], composite model [9,10], and semi-analytical model (SAM) [26] were used to match the field data by fitting the parameters of SRV and unstimulated region. Chen et al. [28] proposed a comprehensive semi-analytical model for MFHWs considering the shale gas flow mechanisms. The unknowns including the well length, fracture number, conductivity, and length were determined by history matching. Yao et al. [7] developed an analytical multi-linear model based on linear flow and radial flow assumption. Permeability and length of different regions, and fracture length were estimated by history matching with several field data. Therefore, the semianalytic models coupling with history matching methods can be applied for not only obtaining some critical parameters for MFHWs but also predicting the production rate.

The objective of this paper is development and calibration of a semianalytic model for shale wells with nonuniform distribution of induced fractures based on ensemble smoother with multiple data assimilation (ES-MDA). The fractal dimensions of induced-fracture spacing and aperture and tortuosity index of induced-fracture system are included based on fractal theory to describe the properties of SRV region. For shale gas reservoirs, gas transport mechanisms (viscous flow with slippage, Knudsen diffusion, and surface diffusion) among multi-media (porous kerogen, inorganic matter, and fracture system) are taken into account and the model is verified. The effects of the fractal dimensions and tortuosity index of induced fractures on MFHWs performances are analyzed. Then, we employ the ES-MDA method with the presented model to reduce uncertainty in the forecasting of gas production rate for MFHWs in unconventional gas reservoirs using a synthetic case for the tight gas reservoir and a real filed case for the shale gas reservoir. The matching results indicate that the proposed approach could enrich the application of the semianalytic model in the practical field. Finally, some conclusions are drawn and presented based on the results and analysis.

2. Fractal Semianalytic Model for Mfhws in Unconventional Gas Reservoirs

As shown in Figure 2, the basic workflow of a new semianalytic model construction in this study consisted of two main parts: firstly, for SRV, which dominates the productivity of MFHWs in unconventional gas reservoirs, the fractal fracture spacing distribution (FFSD) was taken into account to describe its heterogeneous properties instead of classical idealized dual-media model; secondly, based on the linear flow assumption, combining with the linear models for unstimulated reservoir volume (USRV) (original formation), a coupled fractal multi-linear flow model (FMFM) for MFHWs was developed. The detailed derivations of the FMFM for MFHWs in unconventional gas reservoirs are presented in following sections.

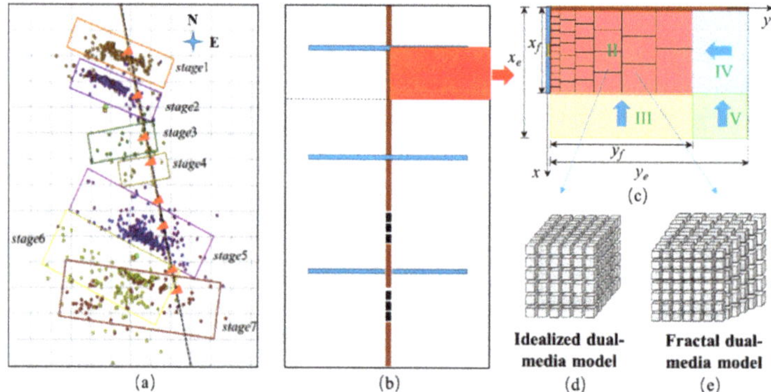

Figure 2. Development of physical and mathematical models. (**a**) Map of micro-seismic monitoring results; (**b**) simplified physical model presented in this study; (**c**) multi-linear mathematical model presented in this study (modified from Sheng et al. [20]); (**d**) idealized homogeneous dual-media model (IDMM) for stimulated reservoir volume (SRV) (modified from Warren and Root [29]); and (**e**) fractal heterogeneous dual-media model (FDMM) for SRV.

2.1. Fractal Fracture Network Permeability and Porosity in SRV

Hydraulic fracturing creates the fracture network surround the horizontal well-bore and hydraulic fracture. Due to differences in situ stress and fracture perforation position, the distribution of induced-fracture properties including induced-fracture spacing and aperture are heterogeneous [21], which affects the distribution of permeability and porosity of fracture network directly. Therefore, the fractal dimensions for both induced-fracture spacing and aperture (d_{fs} and d_{fa}) are first discussed in this section. Then the fracture network permeability and porosity are proposed (details are shown in Appendix A). As shown in Figure 2c, using the fractal theory; thus, the fracture permeability can be obtained by

$$k_f(y) = k_{i-f}(y)n(y)\frac{A_p}{A_r}\frac{\mathrm{d}x}{\mathrm{d}L} = k_w\left(\frac{y}{y_w}\right)^{3(d_{fa}-2)+(d_{fs}-2)-\theta} , \tag{1}$$

where k_f is the fracture permeability, m²; μ is the fluid viscosity, Pa·s; A_r is the sectional-area of a REV, m²; A_p is the sectional-area of induced-fracture, m²; dL is the flow length in the induced-fracture, m; k_w is the fracture permeability at the reference point, m²; y_w is the position of reference point, m; n is the number of fracture sites per unit thickness, m⁻¹; d_{fs} is the fractal dimension of induced-fracture spacing; d_{fa} is the fractal dimension of induced-fracture aperture; and θ is the tortuosity index of induced fracture [30].

Moreover, the fracture porosity is given by [21]

$$\phi_f(y) = \phi_{i-f}\frac{b_f(y)}{b_f(y)+s_f(y)} = \phi_w\left(\frac{y}{y_w}\right)^{d_{fs}+d_{fa}-4} , \tag{2}$$

where ϕ_f is the fracture porosity; ϕ_{i-f} is the single induced-fracture porosity; ϕ_w is the fracture porosity at the reference point; s_f is induced-fracture spacing, m; and b_f is induced-fracture aperture, m.

Then, Equations (1) and (2) are the basic forms of fracture permeability and porosity considering the fractal distribution in SRV of MFHWs in unconventional reservoirs. When $y_w = 0.1$ m and $k_w = 10 \times 10^{-15}$ m², the fracture permeability distribution with different fractal dimensions of induced-fracture spacing and aperture (d_{fs} and d_{fa}) and tortuosity index of induced-fracture (θ) are shown in Figure 3. The fracture permeability k_f decreases if $d_{fs} < 2$ and increases if $d_{fs} > 2$ with the increase of the distance from the reference point y_w as shown in Figure 3a. Meanwhile, the fracture permeability k_f decreases if $d_{fa} < 2$ and increases if $d_{fa} > 2$ with the increase of the distance from the reference point y_w as shown in Figure 3b. Figure 3c shows that the tortuosity index of induced-fracture θ decreases when $\theta < 0$ and increases when $\theta > 0$ with the increase of the distance from the reference point y_w [31]. The reasons can be illustrated as follows: If $d_{fs} < 2$ or $d_{fa} < 2$, as the distance from the reference point increases, the induced-fracture number or aperture decreases and the induced-fracture spacing increases; if $d_{fs} = 2$ or $d_{fa} = 2$, as the distance from the reference point increases, the induced-fracture number or aperture and the induced-fracture spacing are constants; if $d_{fs} > 2$ or $d_{fa} > 2$, as the distance from the reference point increases, the induced-fracture number or aperture increases and the induced-fracture spacing decreases; if $\theta > 0$, the tortuosity of pores is more complex than that when $\theta < 0$. Therefore, for SRV, the fractal dimensions of nature fracture are usually reported from 1.3 to 1.7 [32].

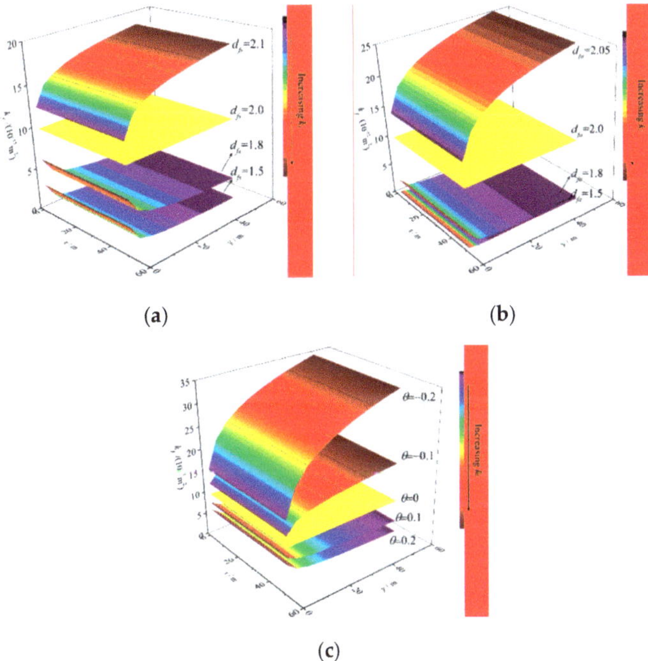

Figure 3. Fracture permeability distribution with different fractal dimensions of induced-fracture spacing and aperture (d_{fs} and d_{fa}) and tortuosity index of induced-fracture (θ). (**a**) Fracture permeability distribution in SRV with different values of d_{fs} when $d_{fa} = 2$ and $\theta = 0$; (**b**) fracture permeability distribution in SRV with different values of d_{fa} when $d_{fs} = 2$ and $\theta = 0$; and (**c**) fracture permeability distribution in SRV with different values of θ when $d_{fs} = 2$ and $d_{fa} = 2$.

2.2. Transient Diffusivity Equations in SRV And USRV

Based on the trilinear model, a newly coupled FMFM was developed for MFHWs in unconventional gas reservoirs as shown in Figure 2c, which followed the basic linear flow assumptions and some simplifying assumptions: (1) the reservoir formation consisted of five regions including the hydraulic fracture, the effective SRV and three USRV regions; (2) the fractal triple-media model (FTMM) was applied to describe the heterogeneous characteristics of induced fractures in SRV, which consisted of three media (fracture network, kerogen, and inorganic matter); (3) idealized dual-media model (IDMM) was used to characterize the homogeneous properties of formation in USRV, which included porous kerogen and inorganic matter; (4) porous kerogen and inorganic matter are uniform and continuous media in both SRV and USRV regions; (5) for shale gas reservoirs, viscous flow, Knudsen diffusion, surface diffusion, and slip factor were taken into account; (6) single-phase fluid flowed into the horizontal wellbore from reservoir only by the hydraulic fractures, and frictional loss within the horizontal well and effects of gravity and capillary forces were ignored in the reservoir; (7) the hydraulic fractures had identical properties, and were evenly distributed along the horizontal well; and (8) there were the no-flow boundaries parallel to the hydraulic fractures at the middle of two fractures in closed reservoir.

2.2.1. Diffusivity Equations of FTMM in SRV (Region II)

Since the induced-fracture system, porous kerogen and inorganic matter coexist in SRV, there are two types of cross-flow flux happening: one is from porous kerogen to inorganic matter, and another one is from inorganic matter to induced-fracture system, both of which can be described by Warren–Root pseudo-steady model. Thus, combining Equations (1) and (2), the diffusivity equation of fluid flow in the induced-fracture system can be expressed as

$$\frac{\partial\left[\phi_w(y/y_w)^{d_{fs}+d_{fa}-4}(p_fM/ZRT)\right]}{\partial t} = \frac{\partial}{\partial y}\left[\frac{p_fM}{ZRT\mu_g}k_w\left(\frac{y}{y_w}\right)^{3(d_{fa}-2)+(d_{fs}-2)-\theta}\frac{\partial p_f}{\partial y}\right] + \frac{1}{x_f}\frac{p_fM}{ZRT\mu_g}k_m\frac{\partial p_{3m}}{\partial x}\bigg|_{x=x_f} + q'_{m-f}, \tag{3}$$

where p_f is the induced-fracture pressure in SRV, Pa; Z is the gas compressibility factor; R is the universal gas constant, 8.314 J/(K·mol); μ_g is the gas viscosity, Pa·s; k_m is the apparent permeability of the inorganic matter, m^2; p_m is the inorganic matter pressure, Pa; x_f is the hydraulic fracture half-length, m; subscript, 3 represents region III; and q'_{m-f} is the mass transfer from inorganic matter to the induced-fracture system, kg/m^3·s.

For porous kerogen, if the Knudsen diffusion, surface diffusion, viscous flow and slippage effect are included, the apparent permeability can be given by (Sheng et al., 2019b)

$$k_k = \frac{\phi_k}{\tau_k}\frac{2r_k}{3}\sqrt{\frac{8ZRT}{\pi M}}C_g\mu_g + \varepsilon_{ks}(1-\phi_k)D_s\frac{c_{\mu s}ZRT}{(p_L+p_k)^2}\mu_g + \frac{\phi_k}{\tau_k}\frac{r_k^2}{8}\left[1+\sqrt{\frac{8ZRT}{M}}\frac{\mu_g}{p_kr_k}\left(\frac{2}{f}-1\right)\right], \tag{4}$$

where k_k is the apparent permeability of the porous kerogen, m^2; ϕ_k is the porous kerogen porosity; τ_k is the porous kerogen tortuosity; r_k is the pore size in porous kerogen, m; C_g is the gas compressibility, Pa^{-1}; ε_{ks} is proportion of solid kerogen volume in total interconnected matrix; D_s is the surface diffusion coefficient, m^2/s; $c_{\mu s}$ is the Langmuir volume on the porous kerogen, mol/m^3; p_L is the Langmuir pressure, Pa; p_k is the porous kerogen pressure, Pa; and f is fraction of molecules striking pore wall which are diffusely reflected.

Here, the Warren–Root pseudo-steady-state model is used to describe the cross-flow between porous kerogen and inorganic matter. Therefore, the diffusivity equation for the porous kerogen can be obtained by

$$\frac{\partial[\phi_k(p_{2k}M/ZRT) + \varepsilon_{ks}(1-\phi_k)c_s]}{\partial t} = \sigma_k\frac{p_{2k}Mk_k}{ZRT\mu_g}(p_{2m}-p_{2k}), \tag{5}$$

where $c_{\mu s}$ is the adsorbed gas volume on the porous kerogen, mol/m^3, which follows Langmuir isotherm equation; σ_k is the pseudo-steady-state shape factor for the porous kerogen, 1/m^2; and subscript, 2 represents region II.

Similarly, if the Warren–Root pseudo-steady-state model is also used to describe the cross-flow between inorganic matter and induced-fracture system, the diffusivity equation for the porous kerogen can be expressed as

$$\frac{\partial(\phi_m p_{2m}M/ZRT)}{\partial t} = \sigma_k\frac{p_{2k}Mk_k}{ZRT\mu_g}(p_{2k}-p_{2m}) - \sigma_m\frac{p_{2m}Mk_m}{ZRT\mu_g}(p_f-p_{2m}), \tag{6}$$

where ϕ_m is the inorganic matter porosity; σ_m is the pseudo-steady-state shape factor for the inorganic matter, 1/m^2; k_m is the apparent permeability of the inorganic matter, m^2, which includes the Knudsen diffusion, viscous flow, and the slippage effect are included, and can be given by

$$k_m = \frac{\phi_m}{\tau_m}\frac{2r_m}{3}\sqrt{\frac{8ZRT}{\pi M}}C_g\mu_g + \frac{\phi_m}{\tau_m}\frac{r_m^2}{8}\left[1+\sqrt{\frac{8ZRT}{M}}\frac{\mu_g}{p_m r_m}\left(\frac{2}{f}-1\right)\right],\tag{7}$$

where τ_m is the inorganic matter tortuosity; r_m is the pore size in the inorganic matter, m.

Outer boundary conditions: flux continuity between SRV (region II) and USRV (region IV)

$$\left.\frac{k_w}{\mu_g}\left(\frac{y}{y_w}\right)^{3(d_{fa}-2)+(d_{fs}-2)-\theta}\frac{p_f M}{ZRT}\frac{\partial p_f}{\partial y}\right|_{y=y_f} = \left.\frac{k_m}{\mu_g}\frac{p_{4m}M}{ZRT}\frac{\partial p_{4m}}{\partial y}\right|_{y=y_f}.\tag{8a}$$

Inner boundary conditions: pressure continuity between SRV (region II) and HF (region I)

$$p_f\big|_{y=y_w} = p_{HF}\big|_{y=y_w}.\tag{8b}$$

In Equation (8a,b), y_w is assumed as hydraulic fracture half-width, m; y_f is the half-width of SRV in y-direction, m; and subscript, 4 represents region IV and HF represents hydraulic fracture.

2.2.2. Diffusivity Equations of FTMM in USRV (Region III, Region IV and Region V)

The fluid flow in both region III and region V can be described as linear flow in x-direction. Thus, according to Equations (5) and (6), the diffusivity equations of porous kerogen and inorganic matter in these two regions can be expressed as, respectively,

$$\begin{cases}\frac{\partial[\phi_k(p_{3k}M/ZRT)+M\varepsilon_{ks}(1-\phi_k)c_s]}{\partial t} = \sigma_k\frac{p_{3k}Mk_k}{ZRT\mu_g}(p_{3m}-p_{3k}) \\ \frac{\partial(\phi_m p_{3m}M/ZRT)}{\partial t} = \frac{\partial}{\partial x}\left(\frac{p_{3m}Mk_m}{ZRT\mu_g}\frac{\partial p_{3m}}{\partial x}\right) + \sigma_k\frac{p_{3k}Mk_k}{ZRT\mu_g}(p_{3k}-p_{3m})\end{cases}, \text{ for region III,}\tag{9}$$

$$\begin{cases}\frac{\partial[\phi_k(p_{5k}M/ZRT)+M\varepsilon_{ks}(1-\phi_k)c_s]}{\partial t} = \sigma_k\frac{p_{5k}Mk_k}{ZRT\mu_g}(p_{5m}-p_{5k}) \\ \frac{\partial(\phi_m p_{5m}M/ZRT)}{\partial t} = \frac{\partial}{\partial x}\left(\frac{p_{5m}Mk_m}{ZRT\mu_g}\frac{\partial p_{5m}}{\partial x}\right) + \sigma_k\frac{p_{5k}Mk_k}{ZRT\mu_g}(p_{5k}-p_{5m})\end{cases}, \text{ for region V,}\tag{10}$$

where subscript 5 represents region V.

Outer boundary conditions: no-flow conditions for region III and region V

$$\left.\frac{\partial p_{3m}}{\partial x}\right|_{x=x_e} = 0,\tag{11a}$$

$$\left.\frac{\partial p_{5m}}{\partial x}\right|_{x=x_e} = 0.\tag{11b}$$

Inner boundary conditions: pressure continuity for region III—SRV (region II), and region V—region IV

$$p_{3m}\big|_{x=x_f} = p_f\big|_{x=x_f},\tag{11c}$$

$$p_{5m}\big|_{x=x_f} = p_{4m}\big|_{x=x_f}.\tag{11d}$$

In Equation (11a–d), x_e is the reservoirs half-size in x-direction, m; subscript 4 represents region IV. Whereas, the fluid flow in y-direction in region IV, and the diffusivity equation can be given by

$$\begin{cases}\frac{\partial[\phi_k(p_{4k}M/ZRT)+M\varepsilon_{ks}(1-\phi_k)c_s]}{\partial t} = \sigma_k\frac{p_{4k}Mk_k}{ZRT\mu_g}(p_{4m}-p_{4k}) \\ \frac{\partial(\phi_m p_{4m}M/ZRT)}{\partial t} = \frac{\partial}{\partial y}\left(\frac{p_{4m}Mk_m}{ZRT\mu_g}\frac{\partial p_{4m}}{\partial y}\right) + \sigma_k\frac{p_{4k}Mk_k}{ZRT\mu_g}(p_{4k}-p_{4m}) + \frac{1}{x_f}\frac{p_{4m}M}{ZRT\mu_g}k_m\left.\frac{\partial p_{5m}}{\partial x}\right|_{x=x_f}\end{cases}\tag{12}$$

Outer boundary conditions: no-flow conditions for region IV

$$\left.\frac{\partial p_{4m}}{\partial y}\right|_{y=y_e} = 0. \tag{13a}$$

Inner boundary conditions: pressure continuity for region IV—SRV (region II)

$$\left.p_{4m}\right|_{y=y_f} = \left.p_f\right|_{y=y_f}. \tag{13b}$$

In Equation (13a,b), y_e is the half-size of HF spacing in y-direction, m.

2.2.3. Diffusivity Equations of FTMM in HF (Region I)

Linear flow occurs in the hydraulic fractures with closed tip and constant production in x-direction, and the diffusivity equation can be expressed as

$$\frac{\partial(\phi_F p_F M/ZRT)}{\partial t} = \frac{\partial}{\partial x}\left(\frac{p_F M/ZRT}{\mu_g}k_F\frac{\partial p_F}{\partial x}\right) + \frac{1}{y_w}\frac{p_F M}{ZRT\mu_g}k_w\left(\frac{y}{y_w}\right)^{3(d_{fa}-2)+(d_{fs}-2)-\theta}\left.\frac{\partial p_f}{\partial y}\right|_{y=y_w}, \tag{14}$$

where ϕ_F is the HF porosity; p_F is the HF pressure, Pa; and k_F is the HF permeability, m^2.

Outer boundary conditions: no-flow conditions for region I

$$\left.\frac{\partial p_F}{\partial x}\right|_{x=x_f} = 0. \tag{15a}$$

Inner boundary conditions: constant production without wellbore storage and skin effect

$$\left.\frac{\partial p_F}{\partial x}\right|_{x=0} = -\frac{q_f ZRT\mu_g}{2Mk_F h y_w \left.p_F\right|_{x=0}}, \tag{15b}$$

where q_f is constant well production from HF, kg/s.

For solving the FMFM as shown in Appendix B, we can obtain the pseudo-pressure in well bottom-hole in Laplace domain as

$$\overline{\varphi_{fD}} = \left.\overline{\varphi_{FD}}\right|_{x_D=0} = \frac{\pi k_w x_f}{Zk_F y_w s\cdot \tan\left(\sqrt{\frac{k_f}{k_F}\left(\omega_F s - \frac{x_f}{y_w}a_1\right)}\right)\sqrt{\frac{k_f}{k_F}\left(\omega_F s - \frac{x_f}{y_w}a_1\right)}}, \tag{16}$$

where, $a_1 = \sqrt{a_2}y_{wD}^{\frac{1+m_1-2n_1}{2}}\left(\dfrac{I_{\frac{-1-m_1}{2+m_1-n_1}}\left(\frac{2\sqrt{a_2}}{2+m_1-n_1}y_{wD}^{\frac{2+m_1-2n_1}{2}}\right) - \frac{h_1}{h_2}K_{\frac{-1-m_1}{2+m_1-n_1}}\left(\frac{2\sqrt{a_2}}{2+m_1-n_1}y_{wD}^{\frac{2+m_1-2n_1}{2}}\right)}{y_{wD}^{\frac{1-n_1}{2}}\left(I_{\frac{1-n_1}{2+m_1-n_1}}\left(\frac{2\sqrt{a_2}}{2+m_1-n_1}y_{wD}^{\frac{2+m_1-2n_1}{2}}\right) + \frac{h_1}{h_2}K_{\frac{1-n_1}{2+m_1-n_1}}\left(\frac{2\sqrt{a_2}}{2+m_1-n_1}y_{wD}^{\frac{2+m_1-2n_1}{2}}\right)\right)}\right)$ and other variables are shown in Appendix B.

Based on Equation (16), if the wellbore storage and skin factor are not considered, the dimensionless flow rate can be derived as [33,34]

$$\overline{q_D} = \frac{Zk_F y_w\cdot \tan\left(\sqrt{\frac{k_f}{k_F}\left(\omega_F s - \frac{x_f}{y_w}a_1\right)}\right)\sqrt{\frac{k_f}{k_F}\left(\omega_F s - \frac{x_f}{y_w}a_1\right)}}{\pi s k_w x_f} \tag{17}$$

If we hope to obtain the solution in the real-time domain, here, the Stehfest algorithm is applied, and Equation (17) becomes

$$q_D = \frac{\ln 2}{t_D} \sum_{i=1}^{N} (-1)^{\frac{N}{2}+1} \times \left\{ \begin{array}{l} \left[\sum_{k=\frac{i+1}{2}}^{\min(i,\frac{N}{2})} \frac{k^{\frac{N}{2}}(2k)!}{\left(\frac{N}{2}-k\right)! \, k! \, (k-1)! \, (i-k)! \, (2k-i)!} \right] \\ \times \frac{Zk_F y_w \cdot \tan\left(\sqrt{\frac{k_f}{k_F}\left(\omega_F s - \frac{x_f}{y_w}a_1\right)}\right)\sqrt{\frac{k_f}{k_F}\left(\omega_F s - \frac{x_f}{y_w}a_1\right)}}{\pi s k_w x_f} \times \left(\frac{\ln 2}{t_D}i\right) \end{array} \right\}. \tag{18}$$

3. Model Verification and Discussion

3.1. Model Verification

The SRV size was assumed to be the same as with the half-length of HF spacing ($y_f = y_e$), the presented FMFM model can be simplified as the model proposed by Sheng et al. [20]. The basic parameters used for FMFM and the Sheng's model are listed in Table 1. The comparison of two models are shown as Figure 4. The results indicate that the production rate and cumulative production calculated by the FMFM presented in this study can fit well with those of Sheng's model. The classical trilinear flow model can be simplified by FMFM when the SRV size between HFs and the fractal distribution of induced-fracture are not considered. In addition, if the shale gas transport mechanisms are ignored in FMFM, the model can be used to predict the production of MFHWs in tight gas and even in tight oil reservoirs fast.

Table 1. Parameters used in multi-linear flow model (FMFM) and Sheng's model [20].

Parameters	Value	Parameters	Value
Fractal dimension of induced-fracture spacing, d_{fs}	1.95	Pore size in porous kerogen, r_k (m)	10×10^{-9}
Fractal dimension of induced-fracture aperture, d_{fa}	2.0	Portion of kerogen volume, ε_{ks}	0.5
Tortuosity index of induced-fracture, θ	-0.05	Porous kerogen porosity, ϕ_k	0.1
Porosity of induced-fracture, ϕ_f	10^{-4}	Porous kerogen tortuosity, τ_k	5
HF half-spacing, y_e (m)	100	Langmuir pressure, p_L (MPa)	13.78
HF half-length, x_f (m)	50	Gas viscosity, μ_g (mPa·s)	0.0184
HF permeability, k_F (m^2)	$10^3 \times 10^{-15}$	Gas compressibility, C_g (MPa^{-1})	5×10^{-2}
HF half-width, y_w (m)	0.01	Surface diffusion coefficient, D_s (m^2/s)	1×10^{-5}
HF porosity, ϕ_F	10^{-3}	Molecular mass of shale gas, M (kg/mol)	0.016
Total compressibility of the inorganic matter, C_{tm} (MPa^{-1})	7.5×10^{-3}	Langmuir volume on kerogen surface, $c_{\mu s}$ (mol/m^3)	700
Total compressibility of induced-fracture, C_{tf} (MPa^{-1})	4×10^{-3}	Fraction of molecules striking pore wall which are diffusely reflected, f	0.8
Inorganic matter porosity, ϕ_m	0.1	Total compressibility of the porous kerogen, C_{tk} (MPa^{-1})	7.5×10^{-3}
Inorganic matter tortuosity, τ_m	5	Reservoir thickness, h (m)	19
Induced-fracture permeability at y_w, k_w (m^2)	2×10^{-15}	Initial pressure, p_i (MPa)	17
Pore size in inorganic matter, r_m (m)	20×10^{-9}	Well bottom pressure, p_{wf} (MPa)	12
Formation temperature, T (K)	338	Reservoir half-width, x_e (m)	200

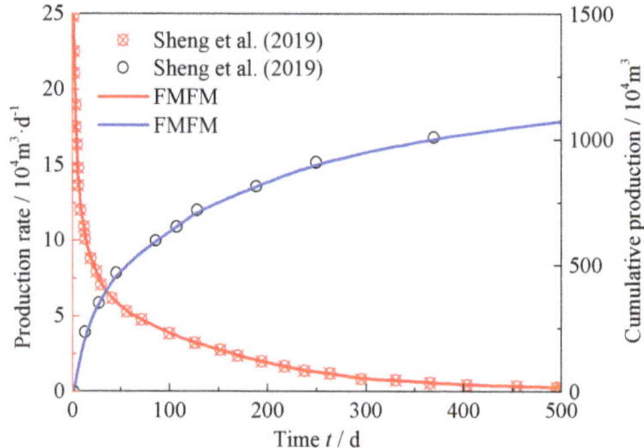

Figure 4. Comparison of production rate and cumulative production between FMFM and Sheng's model [20].

3.2. Sensitivity Analysis of Properties of SRV

Figures 5 and 6 show that the dimensionless pseudo-pressure and dimensionless production rate with different tortuosity index and fractal dimensions of induced-fracture spacing by FMFM, which reflects the various properties of SRV for MFHWs in unconventional gas reservoirs. Figure 5 shows that the dimensionless pseudo-pressure is larger and the dimensionless production rate is smaller when the tortuosity index of induced-fracture θ increases. The larger tortuosity index of induced-fracture θ means the longer flow path of gas transport in SRV. When the tortuosity index of induced-fracture θ is larger than 0, the properties of induced-fracture will decrease far from HF. When the tortuosity index of induced-fracture θ is equal to 0, the properties of induced-fracture is homogeneous in SRV.

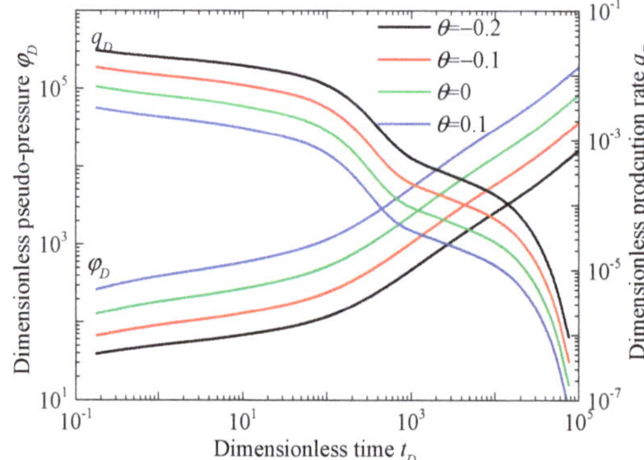

Figure 5. Comparison of production rate and pseudo-pressure with different tortuosity index of induced-fracture θ by FMFM ($d_{fs} = 2$ and $d_{fa} = 2$).

Figure 6 indicates that the dimensionless pseudo-pressure is smaller and the dimensionless production rate is larger when the fractal dimension of induced-fracture d_{fs} increases. The larger fractal dimension of induced-fracture d_{fs} leads to a larger equivalent permeability in SRV. In addition, the linear low in SRV can make the WHFWs produce more gas from the formation, which suggests that the induced-fracture system can improve the development effect for unconventional reservoirs.

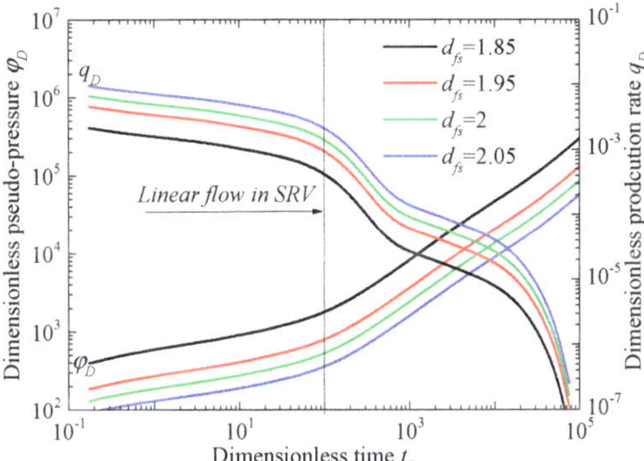

Figure 6. Comparison of production rate and pseudo-pressure with different fractal dimension of induced-fracture spacing d_{fs} by FMFM (d_{fa} = 2 and θ = −0.05).

4. Application of FMFM with ES-MDA Method

4.1. ES-MDA Method

In order to avoid the model update and parameter-state inconsistency, which makes the ensemble Kalman filter (EnKF) complicated, and achieve a better data matching with the ensemble smoother (ES), Emerick and Reynolds [35] reported ES-MDA by assimilating the same observed data multiple times with the covariance matrix of the measurement errors simultaneously. Numerical experiments and examples have shown that the ES-MDA method can provide a better data matching and reduce the uncertainty of model description than the ES method and the EnKF method. Based on the synthetic case and field case data, the application of ES-MDA history matching technology and FMFM for MFHWs is discussed in this section.

The reservoir simulation models are typically stable functions of the rock property fields, which is different with the oceanic and atmospheric models. Based on the common assumption in history-matching problems that the model uncertainty is ignored, we just need to take the parameter-estimation problem into account for ES. Thus, the analyzed vector of model parameters can be given by

$$m_j^a = m_j + C_{MD}(C_{DD} + C_D)^{-1}\left(d_{uc,j} - d_j\right), \; for \; j = 1, \cdots, N_e, \tag{19}$$

where C_{MD} is the cross-covariance matrix between the prior vector of model parameters m_j and the vector of predicted data d; C_{DD} is the N_d—dimension auto-covariance matrix of predicted data; N_d is the total number of measurements assimilated; $d_{uc} \sim N(d_{obs}, C_D)$; d_{obs} is the N_d—dimensional vector of observed data; C_D is the N_d—dimension covariance matrix of observed data measurement errors; and N_e is the number of ensemble member.

Emerick and Reynolds [35] have mentioned that ES-MDA can be used in the nonlinear case because ES is equivalent to a single Gauss–Newton iteration with a full step and an average sensitivity estimated from the prior ensemble, in which case the MDA can be interpreted as an "iterative" ensemble smoother. The ES-MDA algorithm is presented as follows:

(1)   Estimate the number of data assimilations N_a, and the coefficients α_i ($\sum_{i=1}^{N_a} \frac{1}{\alpha_i} = 1$), for $i = 1, \cdots, N_a$.

(2)   For $i = 1 \; to \; N_a$

    a.   Run the ensemble from time zero for obtaining the vector of predicted data

$$d_j = g\left(m_j\right), \; for \; j = 1, \cdots, N_e, \tag{20}$$

where $g(\cdot)$ is the nonlinear forward model, d_j is assignment the model parameters to vector m_j at time zero.

b. For each ensemble member, perturb the observation vector by

$$d_{uc,j} = d_{obs} + \sqrt{\alpha_i}C_D^{1/2}z_{d,j}, \ for \ j = 1,\cdots,N_e,$$ (21)

where $z_{d,j} \sim \mathcal{N}\left(0, I_{N_d}\right)$.

c. Update the ensemble

$$m_j^a = m_j + C_{MD}(C_{DD} + \alpha_i C_D)^{-1}\left(d_{uc,j} - d_j\right), \ for \ j = 1,\cdots,N_e,$$ (22)

where $C = C_{DD} + \alpha_i C_D \equiv \begin{bmatrix} C_{DD} + \alpha_1 C_D & C_{DD} & \cdots & C_{DD} \\ C_{DD} & C_{DD} + \alpha_2 C_D & \cdots & C_{DD} \\ & \vdots & \ddots & \vdots \\ C_{DD} & \cdots & & C_{DD} + \alpha_{N_a} C_D \end{bmatrix}$;

$C_{MD} = \frac{1}{N_e-1}\sum_{j=1}^{N_e}\left(m_j - \overline{m}\right)\left(d_j - \overline{d}\right)^T$; $C_{DD} = \frac{1}{N_e-1}\sum_{j=1}^{N_e}\left(d_j - \overline{d}\right)\left(d_j - \overline{d}\right)^T$; $\overline{m}$ and $\overline{d}$ represent the average values of model parameters and prediction parameter ensembles respectively.

In the procedure of MDA algorithm mentioned above, the data are assimilated N_a times continuously, and the ensemble needs to be rerun before each data assimilation. At the same time, in the ES-MDA algorithm, every time the data is assimilated repeatedly, the disturbance observation vector is resampled. The procedure will reduce the sampling problems due to the matching outliers that may occur when the observation data is perturbed.

4.2. Synthetic Case

Based on the simulator Eclipse, the numerical model of a MFHW in the unconventional gas reservoir is established. The numbers of grids are $X \times Y \times Z = 80 \times 260 \times 3$ and the grid steps are $X \times Y \times Z = 5\,m \times 5\,m \times 5\,m$, respectively. A dual-porosity media model is used to describe the SRV (DUALPORO keyword), and a single-porosity medium model is used to describe the USRV. Hydraulic fractures are described by LGR keyword as shown in Figure 7. If only the gas viscous flow with slippage effect is considered, then the FMFM can be simplified as a semianalytic model for a MFHW with the homogeneous SRV in the tight gas reservoir. We set the same modeling parameters of FMFM with those of numerical model and then discuss the applicability of FMFM with ES-MDA.

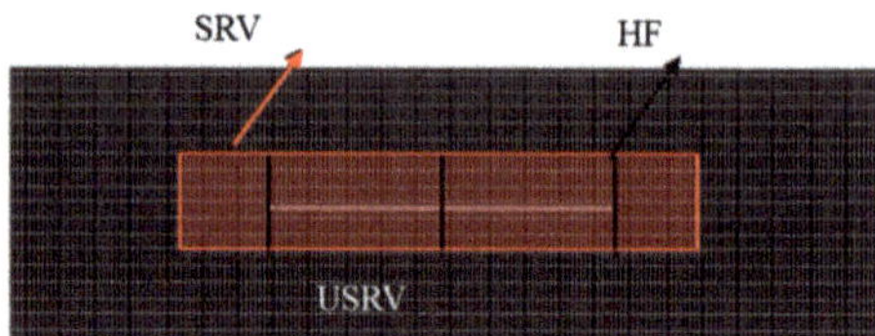

Figure 7. Schematic diagram of a MFHW numerical model ($y_l = y_e$).

Firstly, according to the dimensionless parameters listed in Table A1, the production rate of the numerical model can be non-dimensionalized as $q_D = pq_f/\left(2\pi k_w h c_i\left(\varphi - \varphi_{wf}\right)\right)$.

Then, the N_d—dimensional vector of observed data can be chosen as:

$$d_{obs} = \left[d_s^T, q_o^T\right]_{N_d \times 1}^T,$$ (23)

where d_s donates the N_d-dimensional vector of static parameters of the reservoir and MFHW; $\mathbf{q}_0$ donates the dimensionless production rate of numerical model.

The $N_d \times N_d$ covariance matrix of observed data measurement errors is given by

$$
C_D = \begin{bmatrix} \sigma^2_{d_1} & & \\ & \ddots & \\ & & \sigma^2_{d_{N_d}} \end{bmatrix}_{N_d \times N_d},
\tag{24}
$$

where σ_d is equal to 0.3 for real production rate.

The matching results of dimensionless production rate by ES-MDA after 4 iterations and error analysis are shown in Figure 8. The values of some variables are listed as $N_p = 23$, $N_d = 115$, $N_e = 100$, and $\sqrt{\alpha_i} = 2$. Meanwhile, the root-mean-square error and the average objective function in Figure 7 can be expressed as, respectively,

$$
RMSE = \sqrt{\frac{\sum_{j=1}^{Ne}\left(d_{sobs} - d_{sj}\right)^2}{N_p}},
\tag{25}
$$

$$
O(m) = \frac{1}{N_d}[g(m) - d_{obs}]^T C_D^{-1}[g(m) - d_{obs}].
\tag{26}
$$

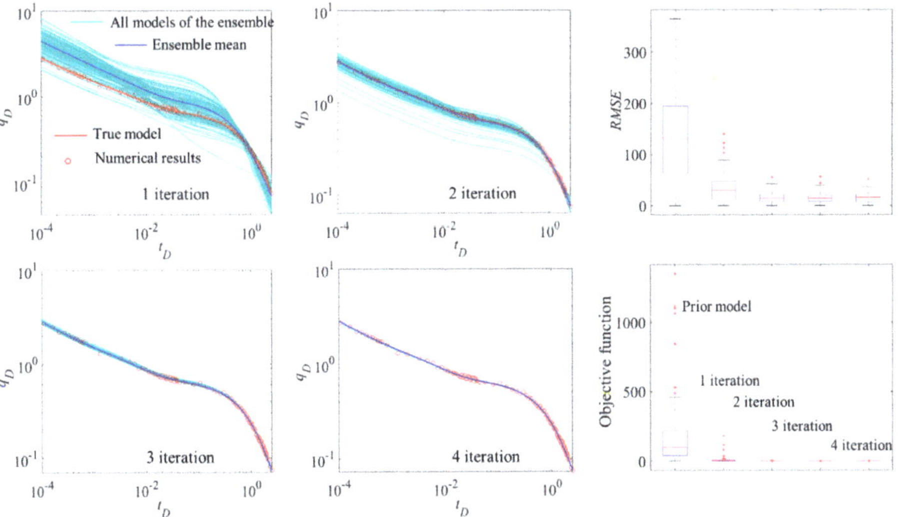

Figure 8. The matching results of dimensionless production rate by ensemble smoother with multiple data assimilation (ES-MDA) after 4 iterations and error analysis.

Figure 8 shows that the error between the ensemble mean model and that of true model is large with one iteration, and the distribution range of ensemble members is large. All models of the ensemble (light blue curves) represent the results calculated by FMFM, the ensemble mean (blue curve) represents the calculated results matching with unperturbed observation data (average value of ensemble members), the true value (red curve) is the true dimensionless production rate obtained by FMFM, and the numerical results (red point) is the observed data in Figure 8. We can see that with two iterations, the distribution range of ensemble members becomes small, and ensemble mean model basically coincides with the true model. With three and four iteration, the distribution range of ensemble members is further reduced around the true model, and the ensemble mean model and the observed data are well-matching, which is almost the same with the true model. In addition, the RMSE and objective function values of the prior model are large, indicating that the model errors and data errors are large. After one iteration, the error decreases. The RMSE and objective function values becomes small and basically the same after three and four iterations, which indicates that the matching results are good as shown in Table 2. Therefore, the discussion mentioned above suggests that parameters of FMFM can be obtained by automatically matching production data of the numerical model by ES-MDA method. The technique can also obtain reservoir physical properties and fracturing parameters by matching the actual production data of the oilfield.

Table 2. Comparison of matching results of FMFM and numerical model parameters by ES-MDA.

Parameters	Eclipse	FMFM	Parameters	Eclipse	FMFM
Total compressibility coefficient, MPa^{-1}	5.5×10^{-3}	5.2714×10^{-3}	USRV matrix permeability, m^2	1×10^{-19}	0.887×10^{-18}
HF permeability, m^2	15×10^{-13}	7.82×10^{-13}	Induced-fracture permeability, m^2	1×10^{-16}	1.33×10^{-16}
SRV matrix permeability, m^2	1×10^{-18}	0.77×10^{-18}	Other parameters	Equal values	

4.3. Field Case

The presented FMFM model with ES-MDA history matching method was applied based the shale gas production data of a MFHW in western China [36]. Compared with the presented models based on different simulating conditions, the matching results are shown in Figure 9. Noting that the actual shale gas production rate of a MFHW in western China is represented by red points; when $d_{fs} = d_{fa} = 2$, $\theta = 0$, $y_{fD} = y_{eD}$, the FMFM can be assumed as a typical trilinear model with homogenous SRV (black line); the results of Sheng's FMPM model [20] with fractal SRV are drawn by the light green line, and they calculated the $d_{fs} = 1.90$ by the box-counting method and the fractal random-fracture-network algorithm and $\theta = -0.05$ by the random walk method; based on the presented FMFM with ES-MDA, the fractal dimensions, tortuosity index, and SRV size can be obtained as shown by the blue line. Figure 9 shows that the fractal dimension and tortuosity index of induced-fracture system matched by FMFM model based on ES-MDA approximate the results of Sheng's FMPM model. The unmatched early-time data was caused by the early-time flow back process. In addition, the results dimensionless production rate calculated by FMFM were smaller but matched better with actual data than Sheng's FMPM model when the SRV size was taken into account. This section mainly provides an application case of our presented approach.

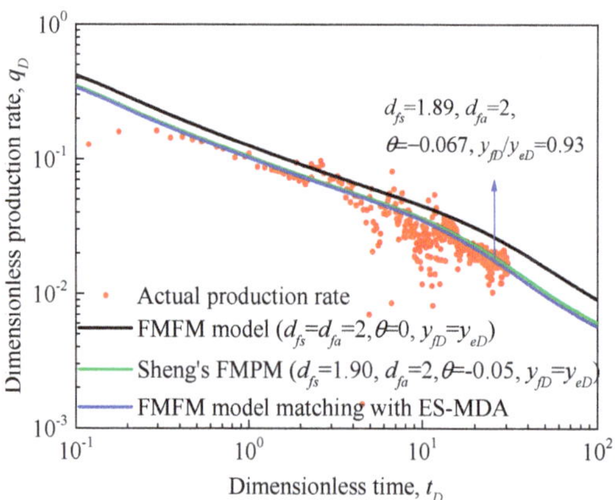

Figure 9. The matching results of actual shale gas production rate and different models.

5. Conclusions

In this paper, a semianalytic fractal multi-linear flow model (FMFM) for MFHWs in unconventional gas reservoirs with consideration of the heterogeneous distribution of the properties of induced-fracture system is proposed. Fractal dimensions of induced-fracture spacing and aperture and tortuosity index of induced-fracture system are included based on fractal theory to describe the properties of SRV region. For shale gas reservoirs, gas transport mechanisms (viscous flow with slippage, Knudsen diffusion, and surface diffusion) among various medium including porous kerogen, inorganic matter, and fracture system are take into account in FMFM. Then, combining with ES-MDA, the FMFM can be applied not only for prediction of gas production rate for MFHWs in unconventional gas reservoirs but also for main uncertainty parameters matching. Some findings can be drawn as follows:

(1) The permeability and porosity of the induced-fracture system are affected by the heterogeneous distribution of induced-fractures spacing and aperture. When the fractal dimensions of induced-fracture spacing (d_{fs}) and aperture (d_{fa}) are smaller than 2.0, the permeability of the induced-fracture system decreases with the increase of the distance from HFs in SRV region, which will become increase if $d_{fs} > 2.0$ or $d_{fa} > 2.0$. When the tortuosity index of the induced-fracture system θ is larger than 0, the permeability of induced-fracture system also decreases with the increase of the distance from HFs in SRV region.

(2) The FMFM divides the formation into three types of porous media in shale reservoirs (porous kerogen, inorganic matter, and fracture system). Triple-porosity media and dual-porosity media are used to describe the fractal SRV region and USRV region, respectively. Multiple gas transport mechanisms such as viscous flow with slippage, Knudsen diffusion, and surface diffusion are considered, and gas flow behaviors in different regions are coupled by pseudo-steady cross-flow among various media. The FMFM is verified by other presented model and the results show that the large d_{fs} or small θ causes the small average permeability of the induced-fracture system, which results in large dimensionless pseudo-pressure and small dimensionless production rate.

(3) Combining the FMFM with ES-MDA history-matching method, the synthetic case for the tight gas reservoir and field case for the shale gas reservoir are discussed. Various main parameters inversions of HFs and NFs are conducted in SRV region. Thus, the presented method can be applied for gas production predicting and history-matching in unconventional gas reservoirs.

Author Contributions: Q.Z. contributed to the conception and drafting the article; S.J. contributed to the study design and manuscript editing. X.W. contributed to programming and analysis; Y.W. contributed to the literature research; Q.M. contributed to the manuscript revision. All authors have read and agreed to the published version of the manuscript.

Funding: This research was funded by the National Natural Science Foundation of China (No.51904279), Fundamental Research Funds for the Central Universities (China University of Geosciences, Wuhan) (No. CUGGC04), and Foundation of Key Laboratory of Tectonics and Petroleum Resources (China University of Geosciences) (No. TPR-2019-03).

Acknowledgments: The authors thank Guanglong Sheng, School of Petroleum Engineering, Yangtze University.

Conflicts of Interest: The authors declare no conflict of interest.

Appendix A. Derivation of Fractal Permeability and Porosity of Induced-Fracture

As shown in Figure 2c, using the fractal theory, the total number of induced-fracture, which is relevant to the fractal dimension of induced-fracture spacing d_{fs}, can be expressed as [37]

$$\int_0^{L_{x,y}} n\left(L_{x,y}\right)dL_{x,y} \propto L_{x,y}^{d_{fs}}, \tag{A1}$$

where $L_{x,y}$ is the position of system, m.

Then, Chang and Yortsos [16] presented the expression of the distribution of fracture site as

$$n\left(L_{x,y}\right) = \alpha L_{x,y}^{d_{fs}-D},$$
(A2)

where D is the Euclidean dimension, which is equal to 2 in Cartesian coordinate system; α is a constant related to fracture porosity.

For the fracture site, we can obtain

$$s_f\left(L_{x,y}\right)n\left(L_{x,y}\right) + b_f\left(L_{x,y}\right)n\left(L_{x,y}\right) = A_{SRV},$$
(A3)

where A_{SRV} is the hydraulic fracture half-length, m.

Substituting Equation (A2) into Equation (A3), Equation (A3) can be rewritten as

$$1 + \frac{b_f\left(L_{x,y}\right)}{s_f\left(L_{x,y}\right)} = \frac{x_f}{\alpha s_f\left(L_{x,y}\right)}L_{x,y}^{2-d_{fs}}.$$
(A4)

In the actual unconventional reservoirs, the value of induced-fracture aperture is commonly far less than that of induced-fracture spacing ($b_f << s_f$), and thus Equation (A4) becomes

$$s_f\left(L_{x,y}\right)\Big|_{\frac{b_f(L_{x,y})}{s_f(L_{x,y})}\to 0} = \frac{x_f}{\alpha}L_{x,y}^{2-d_{fs}}.$$
(A5)

If the induced-fracture spacing at the plane of HF is known, which can be chosen as a reference point, the FFSD in Cartesian coordinate system can be obtained by Equation (A5)

$$s_f(y) = s_{fw}\left(\frac{y}{y_w}\right)^{2-d_{fs}},$$
(A6)

where s_{fw} is the induced-fracture spacing at reference point, m.

For the distribution of induced-fracture aperture (FFAD), the basic form can also be obtained according to Equation (A6) [21]

$$b_f(y) = b_{fw}\left(\frac{y}{y_w}\right)^{d_{fa}-2},$$
(A7)

where b_{fw} is the induced-fracture aperture at reference point, m.

Bai and Elsworth [38] reported the expression of a fracture with a slab shape as

$$k_{i-f} = \frac{b_f(y)^2}{12}.$$
(A8)

In a representative elementary volume (REV) of fracture network, the fluid flux in the fracture system is given by

$$v = A_r\frac{k_f(y)}{\mu}\frac{dp}{dx} = n(y)A_p\frac{k_{i-f}(y)}{\mu}\frac{dp}{dL},$$
(A9)

where v is flux of induced-fracture system in a REV, m^3/s; μ is the fluid viscosity, Pa·s; and dp is the pressure-difference of a REV in y direction, Pa.

Thus, the fracture permeability and porosity can be obtained as Equations (1) and (2).

Appendix B. Solution for the FMFM Model

For solving the FMFM, the basic dimensionless parameters are defined in Table A1.

Table A1. Basic parameters used in FMFM.

Parameters	Expression
Pseudo-pressure for gas phase	$\varphi(p) = 2\int_{p_i}^{p}\left(p/\mu_g Z\right)dp$
Dimensionless pseudo-pressure	$\varphi_D = 2\pi k_w hc_i(\varphi_i - \varphi)/\left(p_i q_f\right)$
Dimensionless time	$t_D = k_w t/\left((\phi C_t)_{f-m-k}\mu_g x_f^2\right)$
Dimensionless length	$x_D = x/x_f,\ y_D = y/y_w,\ x_{eD} = x_e/x_f,$ $y_{eD} = y_e/x_f,\ y_{fD} = y_f/x_f,\ y_{wD} = y_w/x_f$
Total compressibility coefficient of porous kerogen	$C_{tk} = C_g + \left(\varepsilon_{ks}c_{\mu s}ZRT\right)/\left(\phi_k(p_L + p_k)^2\right)$
Transfer coefficient from porous kerogen to inorganic matter	$\lambda_k = \sigma_k k_k x_f^2/k_w$
Transfer coefficient from inorganic matter to induced-fracture	$\lambda_m = \sigma_m k_m x_f^2/k_w$
Storage coefficient of porous kerogen	$\omega_k = (\phi_k C_{tk})/(\phi C_t)_{f-m-k}$
Storage coefficient of inorganic matter	$\omega_m = \left(\phi_m C_g\right)/(\phi C_t)_{f-m-k}$
Storage coefficient of HFs	$\omega_F = \left(\phi_F C_g\right)/(\phi C_t)_{f-m-k}$

Subscript i represents the initial conditions; f, m, k represents fracture, inorganic matter, and porous kerogen respectively.

Therefore, transforming the FMFM into Laplace space, the dimensionless forms and solutions for HF, SRV, and USRV can be obtained as follows:

(1) Dimensionless forms and solutions for USRV.

According to Equations (9) and (10), the dimensionless diffusivity equations in Laplace domain for region III and V can be given by

$$\begin{cases} \overline{\varphi_{3kD}} = \dfrac{\sigma_k k_k x_f^2}{\sigma_k k_k x_f^2 + k_w s\omega_k}\overline{\varphi_{3mD}} \\ \omega_m s\overline{\varphi_{3mD}} = \dfrac{k_m}{k_w}\dfrac{\partial^2\overline{\varphi_{3m}}}{\partial x_D^2} + \sigma_k \dfrac{k_k}{k_w}x_f^2\left(\overline{\varphi_{3kD}} - \overline{\varphi_{3mD}}\right) \end{cases},\ \text{for region III,} \qquad (A10a)$$

$$\begin{cases} \overline{\varphi_{5kD}} = \dfrac{\sigma_k k_k x_f^2}{\sigma_k k_k x_f^2 + k_w s\omega_k}\overline{\varphi_{5mD}} \\ \omega_m s\overline{\varphi_{5mD}} = \dfrac{k_m}{k_w}\dfrac{\partial^2\overline{\varphi_{5m}}}{\partial x_D^2} + \sigma_k \dfrac{k_k}{k_w}x_f^2\left(\overline{\varphi_{5kD}} - \overline{\varphi_{5mD}}\right) \end{cases},\ \text{for region V.} \qquad (A10b)$$

Solving Equation (A10a,b) with boundary conditions, we can obtain

$$\left.\frac{\partial\overline{\varphi_{3mD}}}{\partial x_D}\right|_{x_D=1} = \sqrt{a_3}\left(\frac{\left\{K_{-(1/2)}\left(\sqrt{a_3}x_{eD}\right)I_{-(1/2)}\left(\sqrt{a_3}\right) - I_{-(1/2)}\left(\sqrt{a_3}x_{eD}\right)K_{-(1/2)}\left(\sqrt{a_3}\right)\right\}}{\left\{I_{(1/2)}\left(\sqrt{a_3}\right)K_{-(1/2)}\left(\sqrt{a_3}x_{eD}\right) + I_{-(1/2)}\left(\sqrt{a_3}x_{eD}\right)K_{(1/2)}\left(\sqrt{a_3}\right)\right\}}\right)\left.\overline{\varphi_{fD}}\right|_{x_D=1}, \qquad (A11a)$$

$$\left.\frac{\partial\overline{\varphi_{5mD}}}{\partial x_D}\right|_{x_D=1} = \sqrt{a_5}\left(\frac{\left\{K_{-(1/2)}\left(\sqrt{a_5}x_{eD}\right)I_{-(1/2)}\left(\sqrt{a_5}\right) - I_{-(1/2)}\left(\sqrt{a_5}x_{eD}\right)K_{-(1/2)}\left(\sqrt{a_5}\right)\right\}}{\left\{I_{(1/2)}\left(\sqrt{a_5}\right)K_{-(1/2)}\left(\sqrt{a_5}x_{eD}\right) + I_{-(1/2)}\left(\sqrt{a_5}x_{eD}\right)K_{(1/2)}\left(\sqrt{a_5}\right)\right\}}\right)\left.\overline{\varphi_{4mD}}\right|_{x_D=1}, \qquad (A11b)$$

where $a_3 = a_5 = \dfrac{k_w}{k_m}\left(\omega_m s + \lambda_k - \dfrac{\lambda_k^2}{\omega_k s + \lambda_k}\right)$. According to Equation (3), the dimensionless diffusivity equations in Laplace domain for region IV can be given by

$$\begin{cases} \overline{\varphi_{4kD}} = \dfrac{\sigma_k k_k x_f^2}{\sigma_k k_k x_f^2 + k_w s\omega_k}\overline{\varphi_{4mD}} \\ \omega_m s\overline{\varphi_{4mD}} = \dfrac{k_m}{k_w}\dfrac{1}{y_{wD}^2}\dfrac{\partial^2\overline{\varphi_{4mD}}}{\partial y_D^2} + \sigma_k \dfrac{k_k}{k_w}x_f^2\left(\overline{\varphi_{4kD}} - \overline{\varphi_{4mD}}\right) + \dfrac{k_k}{k_w}\left.\dfrac{\overline{\varphi_{5mD}}}{\partial x_D}\right|_{x_D=1} \end{cases}. \qquad (A12)$$

Solving Equation (A12) with boundary conditions, we can obtain

$$\left.\frac{\partial \overline{\varphi_{4mD}}}{\partial y_D}\right|_{y_D=y_{lD}} = \sqrt{a_4}\left(\frac{\{K_{-(1/2)}(\sqrt{a_4}\,y_{eD})I_{-(1/2)}(\sqrt{a_4}y_{lD})-I_{-(1/2)}(\sqrt{a_4}y_{eD})K_{-(1/2)}(\sqrt{a_4}y_{lD})\}}{\{I_{(1/2)}(\sqrt{a_4}y_{lD})K_{-(1/2)}(\sqrt{a_4}y_{eD})+I_{-(1/2)}(\sqrt{a_4}y_{eD})K_{(1/2)}(\sqrt{a_4}y_{lD})\}}\right)\left.\overline{\varphi_{fD}}\right|_{y_D=y_{lD}} , \tag{A13}$$

where $a_4 = a_5 - \frac{k_w}{k_m}y_{wD}^2 \sqrt{a_5}\tan h\big((y_{lD}-y_{eD})\sqrt{a_5}\big)$.

(2)   Dimensionless forms and solutions for SRV.

According to Equations (3), (5), and (6), the dimensionless diffusivity equations in Laplace domain for region SRV can be given by

$$\left\{\begin{array}{l}\omega_f s y_D^{\theta-2(d_{fa}-2)}\overline{\varphi_{fD}} = \dfrac{1}{y_{wD}^2}\dfrac{\partial^2 \overline{\varphi_{fD}}}{\partial y_D^2} + \dfrac{1}{y_{wD}^2}\dfrac{3(d_{fa}-2)+(d_{fs}-2)-\theta}{y_D}\dfrac{\partial\overline{\varphi_{fD}}}{\partial y_D} + y_D^{\theta-2(d_{fa}-2)}\dfrac{k_m}{k_w}\left.\dfrac{\partial\overline{\varphi_{3m}}}{\partial x_D}\right|_{x_D=1}\\[2mm] +y_D^{\theta-2(d_{fa}-2)}\lambda_m\big(\overline{\varphi_{2mD}}-\overline{\varphi_{fD}}\big)\\[2mm] \overline{\varphi_{2kD}} = \dfrac{\lambda_k}{\lambda_k+s\omega_k}\overline{\varphi_{2mD}}\\[2mm] \omega_m s\overline{\varphi_{2mD}} = \lambda_k\big(\overline{\varphi_{2kD}}-\overline{\varphi_{2mD}}\big) - \lambda_m\big(\overline{\varphi_{2mD}}-\overline{\varphi_{fD}}\big)\end{array}\right. . \tag{A14}$$

Solving Equation (A14) with boundary conditions, we can obtain

$$\left.\frac{\partial \overline{\varphi_{fD}}}{\partial y_D}\right|_{y_D=y_{wD}}$$

$$= \sqrt{a_2}y_{wD}^{\frac{1+m_1-2n_1}{2}}\left(\frac{I_{\frac{-1-m_1}{2+m_1-n_1}}\left(\frac{2\sqrt{a_2}}{2+m_1-n_1}y_{wD}^{\frac{2+m_1-2n_1}{2}}\right)-\frac{h_1}{h_2}K_{\frac{-1-m_1}{2+m_1-n_1}}\left(\frac{2\sqrt{a_2}}{2+m_1-n_1}y_{wD}^{\frac{2+m_1-2n_1}{2}}\right)}{y_{wD}^{\frac{1-n_1}{2}}\left(I_{\frac{1-n_1}{2+m_1-n_1}}\left(\frac{2\sqrt{a_2}}{2+m_1-n_1}y_{wD}^{\frac{2+m_1-2n_1}{2}}\right)+\frac{h_1}{h_2}K_{\frac{1-n_1}{2+m_1-n_1}}\left(\frac{2\sqrt{a_2}}{2+m_1-n_1}y_{wD}^{\frac{2+m_1-2n_1}{2}}\right)\right)}\right)\left.\overline{\varphi_{FD}}\right|_{y_D=y_{wD}} , \tag{A15}$$

where $a_2 = y_{wD}^2\left\{\omega_f s - \dfrac{k_m}{k_w}\sqrt{a_3}\left(\dfrac{\{K_{-(\frac{1}{2})}(\sqrt{a_3}x_{eD})I_{-(\frac{1}{2})}(\sqrt{a_3})-I_{-(\frac{1}{2})}(\sqrt{a_3}x_{eD})K_{-(\frac{1}{2})}(\sqrt{a_3})\}}{\{I_{(\frac{1}{2})}(\sqrt{a_3})K_{-(\frac{1}{2})}(\sqrt{a_3}x_{eD})+I_{-(\frac{1}{2})}(\sqrt{a_3}x_{eD})K_{(\frac{1}{2})}(\sqrt{a_3})\}}\right)+\lambda_m\dfrac{\omega_m s+\frac{\lambda_k\omega_k s}{\lambda_k+\omega_k s}}{\omega_m s+\frac{\lambda_k\omega_k s}{\lambda_k+\omega_k s}+\lambda_m}\right\}$,

$$m_1 = d_{fa}+d_{fs}-4, \qquad n_1 = 3(d_{fa}-2)+(d_{fs}-2)-\theta, \qquad h_1 =$$

$$\frac{\sqrt{a_2}}{b_2}y_{fD}^{\frac{m_1-n_1}{2}}I_{\frac{-1-m_1}{2+m_1-n_1}}\left(\frac{2\sqrt{a_2}}{2+m_1-n_1}y_{fD}^{\frac{2+m_1-2n_1}{2}}\right) - I_{\frac{1-n_1}{2+m_1-n_1}}\left(\frac{2\sqrt{a_2}}{2+m_1-n_1}y_{fD}^{\frac{2+m_1-2n_1}{2}}\right), \qquad h_2 =$$

$$K_{\frac{1-n_1}{2+m_1-n_1}}\left(\frac{2\sqrt{a_2}}{2+m_1-n_1}y_{fD}^{\frac{2+m_1-2n_1}{2}}\right) + \frac{\sqrt{a_2}}{b_2}y_{fD}^{\frac{m_1-n_1}{2}}K_{\frac{-1-m_1}{2+m_1-n_1}}\left(\frac{2\sqrt{a_2}}{2+m_1-n_1}y_{fD}^{\frac{2+m_1-2n_1}{2}}\right), \qquad b_2 =$$

$$\sqrt{a_5}\left(\frac{\{K_{-(1/2)}(\sqrt{a_5}x_{eD})I_{-(1/2)}(\sqrt{a_5})-I_{-(1/2)}(\sqrt{a_5}x_{eD})K_{-(1/2)}(\sqrt{a_5})\}}{\{I_{(1/2)}(\sqrt{a_5})K_{-(1/2)}(\sqrt{a_5}x_{eD})+I_{-(1/2)}(\sqrt{a_5}x_{eD})K_{(1/2)}(\sqrt{a_5})\}}\right).$$

(3)   Dimensionless forms and solutions for HFs.

According to Equation (14), the dimensionless diffusivity equations in Laplace domain for HFs can be given by

$$\omega_F s\overline{\varphi_{FD}} = \frac{k_F}{k_w}\frac{\partial^2\overline{\varphi_{FD}}}{\partial x_D^2} + \frac{x_f^2}{y_w^2}\left.\frac{\partial\overline{\varphi_{FD}}}{\partial y_D}\right|_{y=y_{wD}} , \tag{A16}$$

Solving Equation (A16) with boundary conditions, we can obtain the pseudo-pressure in well bottom-hole in Laplace domain as Equation (16).

References

1.   Yuan, B.; Su, Y.; Moghanloo, R.G.; Rui, Z.; Wang, W.; Shang, Y. A new analytical multi-linear solution for gas flow toward fractured horizontal wells with different fracture intensity. *J. Nat. Gas Sci. Eng.* **2015**, *23*, 227–238. [CrossRef]

2.   Sheng, G.; Zhao, H.; Su, Y.; Javadpour, F.; Wang, C.; Zhou, Y.; Liu, J.; Wang, H. An analytical model to couple gas storage and transport capacity in organic matter with noncircular pores. *Fuel* **2020**, *268*, 117288. [CrossRef]

3. Sheng, G.; Su, Y.; Zhao, H.; Liu, J. A unified apparent porosity/permeability model of organic porous media: Coupling complex pore structure and multi-migration mechanism. *Adv. Geo-Energy Res.* **2020**, *4*, 115–125. [CrossRef]
4. Arevalo-Villagran, J.A.; Wattenbarger, R.A.; Samaniego-Verduzco, F.; Pham, T.T. Production analysis of long-term linear flow in tight gas reservoirs: Case histories. In Proceedings of the SPE Annual Technical Conference and Exhibition, Society of Petroleum Engineers, New Orleans, LA, USA, 30 September–3 October 2001.
5. Al Ahmadi, H.A.; Almarzooq, A.M.; Wattenbarger, R.A. Application of linear flow analysis to shale gas wells-field cases. In Proceedings of the SPE Unconventional Gas Conference, Society of Petroleum Engineers, Pittsburgh, PA, USA, 23–25 February 2010.
6. Yu, W. Developments in Modeling and Optimization of Production in Unconventional Oil and Gas Reservoirs. Ph.D. Thesis, The University of Texas at Austin, Austin, TX, USA, 2015.
7. Yao, S.; Wang, X.; Zeng, F.; Li, M.; Ju, N. A composite model for multi-stage fractured horizontal wells in heterogeneous reservoirs. In Proceedings of the SPE Russian Petroleum Technology Conference and Exhibition, Society of Petroleum Engineers, Moscow, Russia, 24–26 October 2016.
8. Ozkan, E.; Brown, M.L.; Raghavan, R.S.; Kazemi, H. Comparison of fractured horizontal-well performance in conventional and unconventional reservoirs. In Proceedings of the SPE Western Regional Meeting, Society of Petroleum Engineers, San Jose, CA, USA, 24–26 March 2009.
9. Brown, M.; Ozkan, E.; Raghavan, R.; Kazemi, H. Practical solutions for pressure-transient responses of fractured horizontal wells in unconventional shale reservoirs. *SPE Reserv. Eval. Eng.* **2011**, *14*, 663–676. [CrossRef]
10. Stalgorova, K.; Mattar, L. Analytical model for unconventional multifractured composite systems. *SPE Reserv. Eval. Eng.* **2013**, *16*, 246–256. [CrossRef]
11. Wang, W.; Zheng, D.; Sheng, G.; Zhang, Q.; Su, Y. A review of stimulated reservoir volume characterization for multiple fractured horizontal well in unconventional reservoirs. *Adv. Geo-Energy Res.* **2017**, *1*, 54–63. [CrossRef]
12. Luo, W.; Tang, C.; Zhou, Y.; Ning, B.; Cai, J. A new semi-analytical method for calculating well productivity near discrete fractures. *J. Nat. Gas Sci. Eng.* **2018**, *57*, 216–223. [CrossRef]
13. Cai, J.; Yu, B. Prediction of maximum pore size of porous media based on fractal geometry. *Fractals* **2010**, *18*, 417–423. [CrossRef]
14. Yang, F.; Ning, Z.; Liu, H. Fractal characteristics of shales from a shale gas reservoir in the Sichuan Basin, China. *Fuel* **2014**, *115*, 378–384. [CrossRef]
15. Sheng, G.; Su, Y.; Wang, W.; Javadpour, F.; Tang, M. Application of fractal geometry in evaluation of effective stimulated reservoir volume in shale gas reservoirs. *Fractals* **2017**, *25*, 1740007. [CrossRef]
16. Chang, J.; Yortsos, Y.C. Pressure Transient Analysis of Fractal Reservoirs. *SPE Form. Eval.* **1990**, *5*, 31–38. [CrossRef]
17. Cossio, M.; Moridis, G.; Blasingame, T.A. A semianalytic aolution for flow in finite-conductivity vertical fractures by use of fractal theory. *SPE J.* **2013**, *18*, 83–96. [CrossRef]
18. Acuna, J.A.; Ershaghi, I.; Yortsos, Y.C. Practical application of fractal pressure transient analysis of naturally fractured reservoirs. *SPE Form. Eval.* **1995**, *10*, 173–179. [CrossRef]
19. Wang, W.; Shahvali, M.; Su, Y. A Semi-Analytical Fractal Model for Production from Tight Oil Reservoirs with Hydraulically Fractured Horizontal Wells. *Fuel* **2015**, *158*, 612–618. [CrossRef]
20. Sheng, G.; Xu, T.; Gou, F.; Su, Y.; Wang, W.; Lu, M.; Zhan, S. Performance analysis of multi-fractured horizontal wells with complex fracture networks in shale gas reservoirs. *J. Porous Media* **2019**, *22*, 299–320. [CrossRef]
21. Sheng, G.; Su, Y.; Wang, W. A new fractal approach for describing induced-fracture porosity/permeability/compressibility in stimulated unconventional reservoirs. *J. Pet. Sci. Eng.* **2019**, *179*, 855–866. [CrossRef]
22. Samandarli, O.; Al Ahmadi, H.A.; Wattenbarger, R.A. A semi-analytic method for history matching fractured shale gas reservoirs. In Proceedings of the SPE Western North American Region Meeting, Society of Petroleum Engineers, Anchorage, AK, USA, 7–11 May 2011.

23. Hategan, F.; Hawkes, V.R. Well production performance analysis for unconventional shale gas reservoirs: A conventional approach. In Proceedings of the SPE Canadian Unconventional Resources Conference, Society of Petroleum Engineers, Calgary, AB, Canada, 30 October–1 November 2012.

24. Cho, Y.; Ozkan, E.; Apaydin, O.G. Pressure-dependent natural-fracture permeability in shale and its effect on shale-gas well production. *SPE Reserv. Eval. Eng.* **2013**, *16*, 216–228. [CrossRef]

25. Ogunyomi, B.A.; Patzek, T.W.; Lake, L.W.; Kabir, C.S. History matching and rate forecasting in unconventional oil reservoirs using an approximate analytical solution to the double porosity model. In Proceedings of the SPE Eastern Regional Meeting, Society of Petroleum Engineers, Charleston, WV, USA, 21–23 October 2014.

26. Clarkson, C.R.; Williams-Kovacs, J.D.; Qanbari, F.; Behmanesh, H.; Sureshjani, M.H. History-matching and forecasting tight/shale gas condensate wells using combined analytical, semi-analytical, and empirical methods. *J. Nat. Gas Sci. Eng.* **2015**, *26*, 1620–1647. [CrossRef]

27. Nobakht, M.; Mattar, L. Analyzing production data from unconventional gas reservoirs with linear flow and apparent skin. *J. Can. Pet. Technol.* **2012**, *51*, 52–59. [CrossRef]

28. Chen, Z.M.; Liao, X.W.; Zhao, X.L.; Liu, H.M.; Tian, J.; Zhu, L.T.; Wang, L.; Zhao, H.J. A comprehensive model for performance forecast of multiple fractured horizontal well in unconventional reservoirs: A case study. In Proceedings of the SPE Trinidad and Tobago Section Energy Resources Conference, Society of Petroleum Engineers, Port of Spain, Trinidad and Tobago, 13–15 June 2016.

29. Warren, J.R.; Root, P.J. The behavior of naturally fractured reservoirs. *Soc. Pet. Eng.* **1963**, *3*, 245–255. [CrossRef]

30. Bour, O.; Davy, P. On the connectivity of three-dimensional fault networks. *Water Resour. Res.* **1998**, *34*, 2611–2622. [CrossRef]

31. Zhang, Q.; Su, Y.; Zhao, H.; Wang, W.; Zhang, K.; Lu, M. Analytic evaluation method of fractal effective stimulated reservoir volume for fractured wells in unconventional gas reservoirs. *Fractals* **2018**, *26*, 1850097. [CrossRef]

32. Fan, D.; Ettehadtavakkol, A. Semi-Analytical Modeling of Shale Gas Flow through Fractal Induced Fracture Networks with Microseismic Data. *Fuel* **2017**, *193*, 444–459. [CrossRef]

33. Mukherjee, H.; Economides, M.J. A parametric comparison of horizontal and vertical well performance. *SPE Form. Eval.* **1990**, *6*, 209–216. [CrossRef]

34. Van Everdingen, A.F.; Hurst, W. The Application of the Laplace Transformation to Flow Problems in Reservoirs. *J. Pet. Technol.* **1949**, *1*, 305–324. [CrossRef]

35. Emerick, A.A.; Reynolds, A.C. Investigation of the sampling performance of ensemble-based methods with a simple reservoir model. *Comput. Geosci.* **2013**, *17*, 325–350. [CrossRef]

36. Xu, J.; Guo, C.; Wei, M. Production performance analysis for composite shale gas reservoir considering multiple transport mechanisms. *J. Nat. Gas Sci. Eng.* **2015**, *26*, 382–395. [CrossRef]

37. O'Shaughnessy, B.; Procaccia, I. Analytical solutions for diffusion on fractal objects. *Phys. Rev. Lett.* **1985**, *54*, 455–458. [CrossRef]

38. Bai, M.; Elsworth, D. Modeling of subsidence and stress-dependent hydraulic conductivity for intact and fractured porous media. *Rock Mech. Rock Eng.* **1994**, *27*, 209–234. [CrossRef]

Article

Experimental Investigation of the Impacts of Fracturing Fluid on the Evolution of Fluid Composition and Shale Characteristics: A Case Study of the Niutitang Shale in Hunan Province, South China

Jingqiang Tan [1,2], Guolai Li [1,2], Ruining Hu [1,2], Lei Li [1,2,*], Qiao Lyu [1,2] and Jeffrey Dick [1,2]

[1] Key Laboratory of Metallogenic Prediction of Nonferrous Metals and Geological Environment Monitoring, Ministry of Education, School of Geosciences and Info-Physics, Central South University, Changsha 410083, China; tanjingqiang@aliyun.com (J.T.); 185011090@csu.edu.cn (G.L.); huruining611@163.com (R.H.); lvqiao@csu.edu.cn (Q.L.); jeff@chnosz.net (J.D.)

[2] Hunan Key Laboratory of Nonferrous Resources and Geological Hazards Exploration, Changsha 410083, China

* Correspondence: leili08@csu.edu.cn; Tel.: +86-731-8883-6153

Received: 11 June 2020; Accepted: 26 June 2020; Published: 29 June 2020

Abstract: Hydraulic fracturing is a widely used technique for oil and gas extraction from ultra-low porosity and permeability shale reservoirs. During the hydraulic fracturing process, large amounts of water along with specific chemical additives are injected into the shale reservoirs, causing a series of reactions the influence the fluid composition and shale characteristics. This paper is focused on the investigation of the geochemical reactions between shale and fracturing fluid by conducting comparative experiments on different samples at different time scales. By tracking the temporal changes of fluid composition and shale characteristics, we identify the key geochemical reactions during the experiments. The preliminary results show that the dissolution of the relatively unstable minerals in shale, including feldspar, pyrite and carbonate minerals, occurred quickly. During the process of mineral dissolution, a large number of metal elements, such as U, Pb, Ba, Sr, etc., are released, which makes the fluid highly polluted. The fluid–rock reactions also generate many pores, which are mainly caused by dissolution of feldspar and calcite, and potentially can enhance the extraction of shale gas. However, precipitation of secondary minerals like Fe-(oxy) hydroxides and $CaSO_4$ were also observed in our experiments, which on the one hand can restrict the migration of metal elements by adsorption or co-precipitation and on the other hand can occlude the pores, therefore influencing the recovery of hydrocarbon. The different results between the experiments of different samples revealed that mineralogical texture and composition strongly affect the fluid-rock reactions. Therefore, the identification of the shale mineralogical characteristics is essential to formulate fracturing fluid with the lowest chemical reactivity to avoid the contamination released by flowback waters.

Keywords: hydraulic fracturing; fracturing fluids; fluids-rock interaction; environmental implication

---

1. Introduction

Shale gas, as an unconventional resource, has been widely developed in the United States, Canada, China, and Argentina in the past decades to meet the increasing demands for geo-energy. China is the largest holder of shale gas resources worldwide, with estimates ranging from 12.8 to 31.2 trillion m^3 [1–4]. In addition, more than two thirds of the estimated resources are stored in marine shales, in particular in south China. Therefore, the marine shale reserves have raised much attention.

The lower Cambrian Niutitang shale is a prolific gas play distributed in many places across South China, including Sichuan, Chongqing, Guizhou, Hunan, and Hubei provinces. This formation has been widely analyzed and evaluated over the past decade [5–10]: the Niutitang shale is mainly composed by quartz and clay minerals, which ranges from 35% to 77% and 6.2% to 37%, respectively. Carbonate and feldspar are also common in this formation, with ranges from 0% to 27% and 2% to 22%, respectively. The total organic carbon (TOC) ranges from 0.5% to 10% and the equivalent vitrinite reflectance (equal-Ro) ranges from about 1.5% to 4.6%. The largest thickness of this formation in Hunan Province can reach more than 300 m. Studies have showed that the Niutitang shale contains a very large quantity of shale gas resources. The achievements made in the analysis of pores and mineral composition in shale reservoirs [6,11–13] revealed the favorable pore systems for shale gas storage and a high brittleness that favors fracturing. More recently, wells targeting the Niutitang shale in the western Hubei province have already shown industrial gas flow [14].

However, due to the ultra-low porosity and permeability of unconventional reservoirs (e.g., shale and coal-bed methane reservoirs), horizontal drilling and hydraulic fracturing techniques are needed for effective extraction of these unconventional resources [15,16]. During the hydraulic fracturing process, from 7500 to 15,000 cubic metres of water along with specific chemical additives, which are acidic and oxidative, are injected into subsurface shale formations for one typical shale gas well [17,18]. From 5 to 85% of the injected fluid (generally 30–50%) flows back to the near surface and has a salinity three times greater than that of the initial water [19,20]. These flowback waters generally contain lots of toxic and radioactive elements (e.g., U, Pb, Sr), and therefore could lead to contamination of the surface water and shallow aquifers or accumulation of heavy metals in soil when spills and leaks occur [21,22]. There is much research evaluating the influence of shale gas development on groundwater quality in several shale regions in the U.S [23–26] and in the UK [27]. However, when it comes to China, the research is relatively rare. Although some scholars [28,29] have done research about the flowback waters of Longmaxi Shale in the Sichuan Basin, the Niutitang Shale, as a newly developed shale gas reservoir, has not received much attention yet. In addition, most of the research did not explain what caused the high salinity of flowback water. The geochemical interaction between shale and fracturing fluid, which affects not only physical properties of the shale but also the composition of the waters produced during shale gas production, is an important controlling factor [13] but rarely has been discussed [30]. Therefore, more studies about the geochemical processes between fracturing fluids and the shale are required to evaluate and minimize the potential environmental impacts.

By conducting simulation experiments and studying the changes of fracturing fluid, mineral, and physical properties during the experiments, key reactions occurring during the fracturing process in the shale reservoirs can be extrapolated. In this study, the Niutitang shale samples from different depths were exposed to and reacted with fracturing fluids. The purposes of this article are: (1) to evaluate the release of major and minor cations and ions as well as trace metal contaminants during hydraulic fracturing, (2) to assess mineralogical and petrophysical changes of the shale during shale gas production, and (3) to investigate the key reactions between the shale samples and fracturing fluids. If the reactions that release metals can be identified, steps to minimize those reactions could be taken, making the flowback waters less toxic, radioactive, and easier to be treated.

2. Materials and Methods

2.1. Samples and Fracturing Fluid

Samples were cored from an exploration well drilled in Anhua County, Hunan Province (Figure 1). Two groups of samples from different depths were used in the experiment, one from a depth of about 80 m and the other from a depth of about 795 m, to investigate the influence of the initial shale mineral compositions and physical properties. The information about initial shale samples is presented in Table 1. The fracturing fluid used in our experiments was from a chemical company called Rong Sheng Chemical Co., Ltd. (Shenzhen, China) and this kind of fracturing fluid has been used for shale gas exploitation in field. It contains 99.7 wt% water, 0.15 wt% guar gum, 0.03 wt% acid (mainly acetic acid and hydrochloric acid), 0.01 wt% pH adjustor (NaOH, NaHCO$_3$), and some other additives like K$^+$ and Ca^{2+} salts. In each reactor, approximately 8 g of shale sample was exposed to 50 mL fracturing fluid. Most of the shale samples were powder samples, ground to between 100 to 200 mesh, and others were thin sections, for which the main purpose was to provide a surface for field emission scanning electron microscope (FE-SEM) observation.

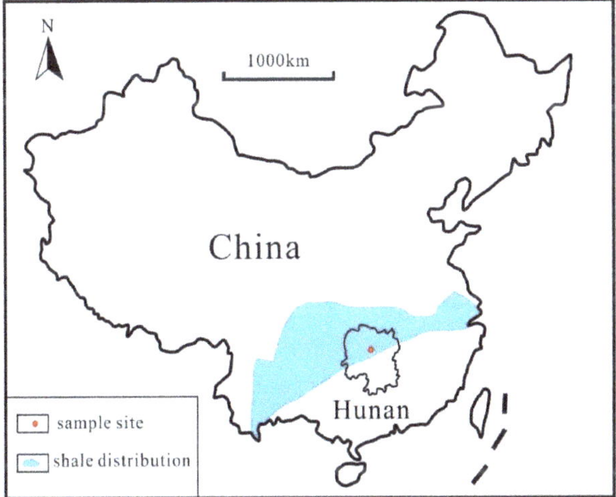

Figure 1. Location of the shale gas well in Anhua (the red spot), where the samples were obtained. The blue area represents the distribution of Lower Cambrian Niutitang Formation in South China (Modified from [31]).

2.2. Experiment Design

During the hydraulic fracturing process, there is a shut-in period (several days to several weeks), which is hypothesized to be the main stage of reactions between shale and the injected fluid. A part of the fracturing fluid flows back to the surface during shale gas production, but it is estimated that more than 50% of the fracturing fluid remains underground and continues to react with the shale [32,33]. Therefore, two sets of experiments with different timescales were designed. One group of experiments was conducted for two weeks, referred to as short-term experiments. The other group of experiments was conducted for one year, referred to as long-term experiments. Each group had two series of experiments, in which samples from different depths (Table 1) were used to investigate the impact of shale characteristics on fracturing process. The purpose of short-term experiments (S1 and S2 series) is to simulate the initial stage of fluid injection into shale formation for hydraulic fracturing, i.e., the main stage of reaction, while that of the long-term experiments (S3 and S4 series) is to investigate the interactions between fracturing fluid that remains underground and the shale reservoirs over a long time.

The short-term experimental temperature was adjusted to 50 °C to simulate the reservoir temperature and minimize the difference between the experimental and field results, while the long-term experimental temperature was difficult to be strictly controlled and we adopted the annual average local temperature (about 20 °C). The temperature of long-term experiments could have an effect on experimental results, but it can still provide insights into the investigation of the geochemical reactions between shale and fluids. The experiments were conducted at atmospheric pressure rather than reservoir pressure, but the influence of pressure on the interactions between fracturing fluid and shale is much smaller than that of temperature [15,34].

The fluid samples were taken at certain intervals, i.e., 12 h, 24 h, 48 h, 3 days, 6 days, 10 days, and 15 days for short-term experiments and 15 days, 30 days, 90 days, 180 days, and 360 days for long-term experiments, to trace the evolution of fracturing fluid compositions. Solids were collected at the end of experiments, dried at 75 °C for 24 h and then analyzed to assess the mineralogical and petrophysical changes after being exposed to fracturing fluid.

Table 1. Mineral composition and physical property of initial shale samples.

Sample	Quartz	Feldspar	Calcite	Pyrite	Clay	TOC	Porosity	Density
S1 [a]	67.9	12.0	nd	10.7	9.3	3.1%	9.4	2.51
S2	65.7	11.5	3.9	7.1	11.9	8.6%	0.7	2.72
S3	65.5	12.1	nd	9.4	13.0	2.7%	9.7	2.46
S4	66.8	11.6	3.2	6.9	11.4	8.3%	0.8	2.70

[a]: Sample 1 (S1) and sample 3 (S3) are from 80 m, while sample 2 (S2) and sample 4 (S4) are from 795 m. Samples S1 and S2 are used in short-term experiments, while S3 and S4 are used in long-term experiments.

2.3. Analytical Technique

Fluid samples were measured using inductively coupled plasma-optical emission spectrometry (ICP-OES: ICAP7400 Radial from Thermo Fisher Scientific, Waltham, MA, USA) and inductively coupled plasma-mass spectrometry (ICP-MS: iCAP RQ from Thermo Fisher Scientific, Waltham, MA, USA). ICP-OES was used to determine the concentrations of major cations, e.g., Mg, Ca, and Fe, while ICP-MS is more accurate in measuring minor cations and trace elements, e.g., Al, Pb, U, Sr, and Ni. The concentrations of dissolved anions like sulfate ion were measured by ion chromatography (IC: ICS-1500 from DIONEX, Sunnyvale, CA, USA). In addition, the pH of the fluid was also analyzed and recorded using AZ8685A type pH meter.

The mineral composition of the powder samples was analyzed using an X-ray powder diffractometer (XRD: Advance D8 from Bruker, Billerica, MA, USA). The porosity was measured with a mercury intrusion meter. Observation of thin sections by FE-SEM coupled with energy-dispersive spectroscopy (EDS) (FE-SEM: MIRA3 LMH from TESCAN, Czech Republic + EDS: X-MAX20 from Oxford Instruments, Abingdon, UK) was performed to provide supplementary information to assess the mineralogical and petrophysical changes of the shale.

3. Results and Discussion

3.1. Aqueous Inorganic Geochemistry

3.1.1. Evolution of pH and Fluid Composition

The initial pH of all groups of experiments is 6. However, the pH values show different variation trends with the increase of reaction time. The pH values of the S1 series drop rapidly in the first 24 h and then become relatively stable. Afterwards, the pH values show small fluctuations around 3.5 (Figure 2a). Concerning the S2 series, the pH values decrease very slightly in the first 24 h, followed by a slow rise to the 72 h, and remain relatively stable at about 6.0 afterwards (Figure 2a). The changes of pH in S3 and S4 series are generally close to those in the short-term experiments. In S3 series the pH values decrease and stabilize at about 3.5, while in S4 series the pH remains close to 6.5, which is consistent with the results of short-term experiments (Figure 3a). However, after nearly three months, the pH of both long-term experiments begins to rise slowly (Figure 3a).

The concentrations of the major ions, i.e., Ca^{2+}, Mg^{2+}, Fe^{3+}, Al^{3+}, and SO_4^{2-}, show significant differences between the two sets of short-term experiments. In the S2 series, Ca^{2+} is released into the fracturing fluid to the greatest extent among all of the measured ions. Ca increases rapidly in the first 72 h, reaching the maximum value of 140 mg/L, and after that the concentrations of Ca^{2+} remain stable (Figure 2b). For the S1 series, the Ca^{2+} concentrations increase to a maximum value of 40 mg/L in the early stage, which is much less than that of the S2 experiments (Figure 2b). The trends of Mg^{2+} release into the fluids are similar to those of Ca^{2+}. In both S1 and S2 experiments, the concentrations of Mg^{2+} increase within the first 72 h and are a little higher in S2 series (Figure 2b).

The situation of the concentrations of Fe^{3+} and SO_4^{2-} is different from that of Mg^{2+} and Ca^{2+}. For the S1 experiments, the concentrations of Fe^{3+} and SO_4^{2-} increase greatly within the first 24 h, after which they are maintained at approximately at 470 mg/L and 2400 mg/L respectively (Figure 2c). The concentrations of aqueous Fe^{3+} and SO_4^{2-} in the S2 experiments increase slightly with a maximum value of 8.5 mg/L and 124 mg/L, respectively (Figure 2c). The trend of aqueous Al^{3+} is similar to that of Fe^{3+}. In the S1 experiments, the concentrations of Al^{3+} increase from less than 1 mg/L to about 85 mg/L within 72 h, while those of the S2 series are always low (Figure 2d).

The changes of concentrations of Ca^{2+}, Mg^{2+}, Fe^{3+}, SO_4^{2-}, and Al^{3+} in long-term S3 and S4 experiments are similar to those of the short-term S1 and S2 experiments. The fluids of the S4 experiments contain much more Ca^{2+} and Mg^{2+}, but the increase of Fe^{3+}, SO_4^{2-} and Al^{3+} is far less compared to S3 (Figure 3).

The variation of major elements in the two sets of experiment indicates that the main reactions that occurred in these sets of experiments are different. In the analyzed shale, the main Fe and S bearing phase is pyrite, which can be easily oxidized. During the process of pyrite oxidation, sulfuric acid is generated [35]. The tremendous increase of Fe^{3+} and SO_4^{2-} within 24 h in the S1 experiments indicates a great generation of sulfuric acid, which results in the rapid drop of pH in the first 24 h. However, in the S2 experiments the increase of Fe^{3+} and SO_4^{2-} is much less and that could account for the slight pH decrease with the first 24 h. The Ca^{2+} and Mg^{2+} are mainly released by dissolution of carbonate minerals including calcite [$CaCO_3$] and dolomite [Ca, Mg $(CO_3)_2$]. They can act as a pH buffer. In the S2 experiments, the finding of much higher concentrations of Ca^{2+} and Mg^{2+} suggests that the main reaction is dissolution of carbonate minerals, which leads to the slow rise of pH in the first 72 h. The slight decrease in the first 24 h caused by pyrite oxidation and the following rise caused by calcite dissolution in the S2 experiments also indicates that the dissolution of calcite is probably triggered by the sulfuric acid generated by pyrite oxidation.

The case of S3 and S4 experiments is similar to the short-term experiments. The main reaction that occurred is pyrite oxidation in S3 and carbonate mineral dissolution in S4. The slight rise of pH after 3 months (Figure 3a) might be caused by clay minerals. The reaction between fracturing fluid and clay minerals would generate some substances that affect pH like silicate ions which are alkaline after hydrolysis [36], causing the rise of pH of the solutions. Furthermore, after a long duration of experiments, the emergence of microorganisms and the degradation of some organic matter [15] will also have an impact on the evolution of the fluids, making the situation more complicated and unpredictable.

In the shale, the Al-bearing minerals include clay minerals and feldspar. According to the results of XRD analysis of shale samples (Table 2), the main source of Al in our experiments is feldspar. The content of feldspar in the S1 and S3 experiments is reduced by about 30% while in the S2 and S4 experiments there is almost no change of feldspar content. The different degrees of feldspar dissolution are mainly caused by pH, as feldspar can be corroded more easily under acidic conditions than under neutral conditions. Clay minerals also affect the release and transport of ions in shale. Many exchangeable ions attach to the surface of clay minerals, and when in contact with fluid, they are separated and diffused into water, which therefore influences the release of ions [18]. Previous studies also have revealed that under acidic conditions clay minerals like illite release more Si and Al than under neutral and alkaline conditions [37], which could also contribute to the different results of the S1(3) and S2(4) experiments. However, to what extent clay minerals will affect the release of ions is not clear yet and further study is needed.

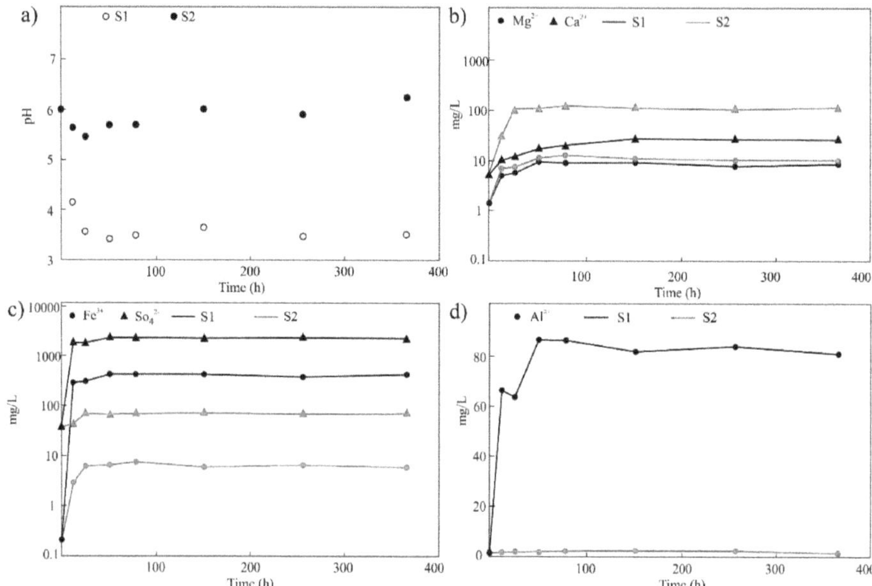

Figure 2. pH (**a**) and concentrations of Ca^{2+} and Mg^{2+} (**b**), Fe^{3+} and SO_4^{2-} (**c**), and Al^{3+} (**d**) in short-term experiments.

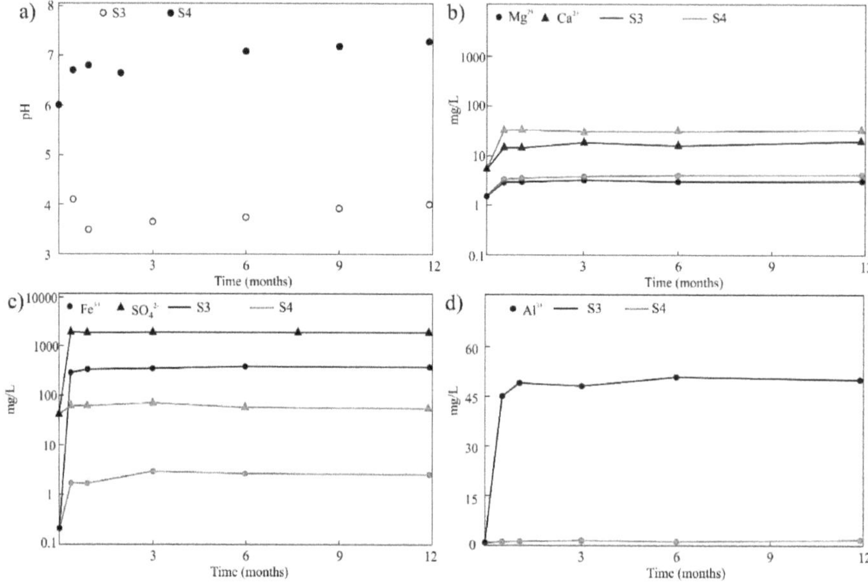

Figure 3. pH (**a**) and concentrations of Ca^{2+} and Mg^{2+} (**b**), Fe^{3+} and SO_4^{2-} (**c**), and Al^{3+} (**d**) in long-term experiments.

Table 2. Mineral compositions of shale samples recovered after experiments.

Sample	Quartz	Feldspar	Pyrite	Clay	Calcite
S1	76.2	9.5	4.7	9.5	nd [a]
S2	69.9	11.2	6.3	12.1	nd
S3	74.3	8.8	4.5	12.2	nd
S4	71.0	12.2	6.1	10.3	0.4

[a]: nd = non detected.

3.1.2. Trace Elements and Contaminant Release

A major environmental concern is that hydraulic fracturing may lead to pollution of underground or surface waters [21]. It has been reported that the fluid recovered during hydraulic fracturing contains many toxic or radioactive metals, e.g., As, Sr, Pb and U, most of which are released from the shale reservoirs through dissolution of shale minerals or degradation of organic matter [38–40]. In this study, the concentrations of trace metals were analyzed to evaluate their source and amount of release.

As shown in Tables 3 and 4, the concentrations of most of the metals are much higher in the S1 experiments than in S2. For example, the concentrations of Ni in the S1 experiments reach a maximum value of 4500 µg/L, while in S2 experiments it only reaches about 170 µg/L, far less than in S1. In addition, in both series of experiments, these metal elements are released into fluids within 24 h rapidly and remain relatively stable. The concentrations of these elements in S3 and S4 series are similar to the short-term experiments (Tables 5 and 6). A very similar phenomenon is also exhibited by Fe^{3+} and SO_4^{2-}, which are also much higher in the S1 experiments and released in the first 24 h. The similarity indicates that pyrite is the main source of metal elements in shales, which is consistent with the observation of previous studies [41,42]. Heavy metals can be incorporated into pyrite by different surface processes, including precipitation, co-precipitation, chemical or physical adsorption, ionic replacement, or redox reactions between dissolved cations and the surface of pyrite [42–45]. For example, Co and Ni can substitute for the Fe(II) in pyrite due to their similar ionic radii [46], and elements (e.g., Cr, Pb, and Zn) that have higher water-exchange reaction kinetics than Fe(II) form metallic sulfides and co-precipitate with pyrite [47]. When pyrite is oxidized, the incorporated metals are released along with Fe^{3+} and SO_4^{2-}, leading to the different results of the experiments.

Table 3. Concentrations (µg/L) of trace metals in short-term fluid sample of S1.

S1	Cr	Co	Ni	Cu	Ba	Pb	U	Sr
0 h	1.6	0.2	2.3	7.2	29.4	1.5	1.4	84.9
12 h	535	445	3320	5000	48.6	17.68	65.5	94.9
24 h	421	445	3240	4230	50.6	15.32	56.8	99.0
48 h	482	596	4110	5220	43.6	14.48	60.4	113
72 h	400	591	4040	5090	39.0	14.68	61.3	111
D6	438	576	4312	5796	29.2	1.44	51.4	116
D10	91	529	4509	6257	127	1.2	48.1	102
D15	573	555	4016	5283	26.9	2.2	42.9	110

Table 4. Concentrations (µg/L) of trace metals in short-term fluid sample of S2.

S2	Cr	Co	Ni	Cu	Ba	Pb	U	Sr
0 h	1.6	0.2	2.3	7.2	29.4	1.5	1.4	84.9
12 h	1.0	23.3	86	33.2	168	11.2	107	242
24 h	6.8	34.7	128	63.7	186	12	117	574
48 h	12.6	39.2	174	99	143	6	42	582
72 h	5.9	47.3	177	80.5	329	21.2	264	877
D6	2.0	25.9	113	34.2	140	2.4	32	620
D10	7.7	42.8	155	61.1	108	3.6	44	525
D15	1.3	28.6	146	38.9	129	2.01	26	868

Table 5. Concentrations (µg/L) of trace metals in long-term fluid sample of S3.

S3	Cr	Co	Ni	Cu	Ba	Pb	U	Sr
0 h	1.6	0.2	2.3	7.2	29.4	1.5	1.4	84.9
D15	256	362	2436	5080	28.6	2.36	31.3	106.8
D30	277	436	2956	5200	25.9	3.01	35.2	100
D90	585	411	2840	4530	43.0	11.4	34.1	145
D180	904	447	3000	4210	24.9	8.4	28.3	105
D360	848	435	2943	2726	26.5	10.9	24.5	106

Table 6. Concentrations (µg/L) of trace metals in long-term fluid sample of S4.

S4	Cr	Co	Ni	Cu	Ba	Pb	U	Sr
0 h	1.6	0.2	2.3	7.2	29.4	1.5	1.4	84.9
D15	4.4	3.56	236	15.76	380	2.08	31.1	484
D30	3.72	2.68	232	14.12	436	1.92	30.5	536
D90	6.09	7.64	220	44.2	507	1.43	132	470
D180	4.12	7.05	189	44.3	361	1.88	246	467
D360	2.01	0.95	144	3.08	448	1.2	340	530

The concentrations of other elements, i.e., Ba, U and Sr, are much higher in the S2 and S4 experiments than in S1 and S3 (Tables 3–6). However, Ba, as an element released from pyrite dissolution, is predicted to have greater concentrations in the S1 and S3 experiments. Considering the much higher concentrations of SO_4^{2-} in S1, this could be caused by the precipitation of $BaSO_4$. The source of U is more complicated. A study on Marcellus shale shows most U is held in silicate minerals and about 20% is hosted in carbonate minerals [38]. Another study states that U is often found in association with organic matter because they can form complexes in low-energy environments [42]. In this study, the concentrations of U are higher in the S2 and S4 experiments. This indicates that the source of U in our experiments can be the organic matter, which is also higher in S2 and S4 (Table 1). In S4 experiments, there is a great increase of U concentration after 3 months. This could be a result of the degradation of organic matter in shale, a phenomenon observed by Marcon et al. (2017) [15]. The concentrations of Sr exhibit an initially rapid increase for both the S1 and S2 experiments and reach their maximum values at about 72 h. This temporal trend is comparable to the observed behavior of Ca and Mg, indicating that the main source of Sr can be the carbonate minerals. Sr isotopes can be used as a tracer of fluid-rock interactions to help identify sources and flow paths of flowback and produced waters, but the Sr released from carbonate minerals dissolution can influence the composition and must also be taken into consideration [35,48,49].

Although only some trace metals were investigated, the fracturing fluid recovered after experiments belongs to Class IV and V water according to GB/T 14848-2017 "Groundwater Quality Standard", which means highly polluted and chemical toxic water. The elevated concentration of these metals is attributed to several acid-base and oxidation-reduction processes. Among these processes, attention should be paid to pyrite oxidation because it is the main source of heavy metals, and the sulfuric acid generated by it can also trigger carbonate minerals and feldspar dissolution. By minimizing the oxidation of pyrite, the concentrations of these toxic and radioactive elements can also be controlled.

3.2. Geochemical Changes in Shale

3.2.1. Dissolution of Minerals in Shale

The investigation of the element concentration in the fluids shows that many elements in the water sample increased to a certain degree, which reflects the dissolution of minerals in shale. The changes of mineral composition in shale samples are determined by XRD and FE-SEM coupled with EDS.

The pyrite dissolution has always been of concern because it can release trace heavy metals. Therefore, it is important to understand the extent of pyrite reaction in order to constrain the release of contaminants. The comparison between the XRD results of initial samples (Table 1) and recovered samples (Table 2) reveals that in the S1 and S3 series of experiments the pyrite is significantly reduced by up to more than 50%, while in S2 and S4 the pyrite content changes very little. The images of short-term chip samples observed by FE-SEM exhibited similar results to that of XRD. In the S1 experiments, the pyrite is obviously corroded, but in the S2 experiments the pyrite remains relatively stable (Figure 4). The results of XRD and FE-SEM analysis in shales are consistent with that of ions analysis in fluids. There are two possible reasons for this different result: (I) The pyrite content is relatively higher in S1, where the mean value is approximately 10%, and a small amount of pyrite particles can be observed on the surface, while in S2 the pyrite content is about 6–7% and relatively lower; (II) The degree of weathering of the two groups of samples differs from each other due to different sampling depths. The sampling position of S1 is about 100 m and closer to the surface, resulting in the vulnerability to groundwater erosion and a greater porosity (Table 1). As shown in FE-SEM images, there are many cracks observed on the surface of pyrite crystals in S1, which may be caused by weathering (Figures 4a and 5). The S2 samples were taken from the shale formation 800 m deep. Therefore, due to the pressure of the overlying strata, the structure of S2 shale formation would be less weathered and have fewer pores compared to S1 (Table 1). The pyrite grains in S2 are also better preserved and cracks do not develop on the surface (Figures 4d and 5). The loose structure and cracks in S1 provide more area exposed to fracturing fluid, leading to a greater degree of reactions.

XRD analysis also shows that the pyrite content is similar between the long-term and short-term experiments. This indicates the inhibition of pyrite dissolution over time. One potential explanation is the surface of pyrite being covered by the precipitation of secondary minerals (Figure 4c), preventing pyrite from contacting the fluid. Another possible reason is that the oxidants in fracturing fluid, namely dissolved oxygen and chemical additives, are exhausted. This study is conducted under open conditions, so exhaustion of oxidant is not a main factor constraining pyrite dissolution. However, at field scale during hydraulic fracturing operation, this may not be the case. Considering the highly reducing conditions underground, limited dissolved oxygen and chemical additives without supplement [35], and a relatively large amount of pyrite in shale, the pyrite reaction would be limited seriously by the oxidant supply. However, due to the high rock to fluid ratio in the field, even only a little dissolution of pyrite would alter the fluid composition greatly [35].

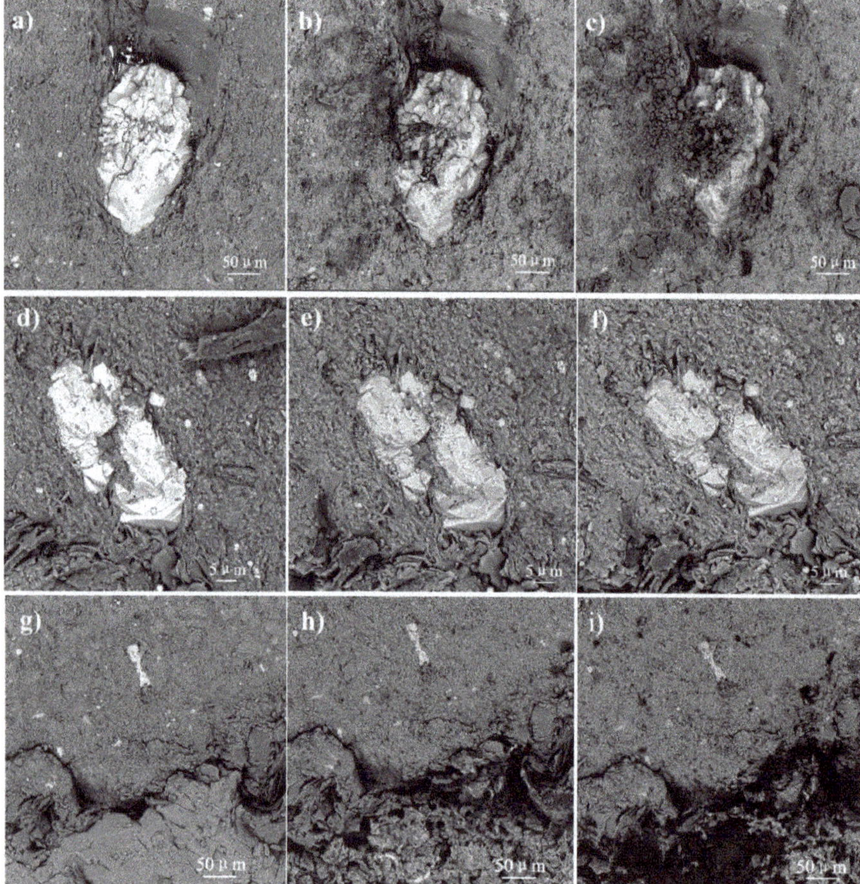

Figure 4. SEM images taken to compare shales before and after experiments. (**a**) is a pyrite grain observed in the S1 initial sample. (**b,c**) are images taken after 5 days and 10 days of reaction respectively. The pyrite was corroded in the middle, where the cracks developed. After 10 days, the surface of the pyrite was covered by precipitates, which may inhibit pyrite dissolution. (**d–f**) are images of S2 sample. No cracks were observed on the surface of the pyrite in the initial sample (**d**). After 5 days (**e**) and 10 days (**f**) of reaction, the pyrite grain did not change significantly. (**g–i**) are images of calcite in S2 sample. After reaction, the calcite was severely dissolved.

The XRD analysis of shales following the reaction confirms a reduction of calcite content in the S2 and S4 experiments. The calcite content in the initial samples is about 4% on average, while there is almost no calcite in samples after the experiments (Tables 1 and 2). The initial content of calcite in S1 and S3 is very low, which may be caused by groundwater erosion, leading to relatively lower Ca^{2+} concentrations in the fluids. The calcite dissolution is also observed on the surface of S2 thin section by FE-SEM, which is not found on the surface of S1 (Figure 4). In addition, the decline of feldspar content in S1 and S3 experiments is also observed (Tables 1 and 2). The feldspar content decreases about 30% after experiments. However, in S2 and S4, it remains unchanged. On the surface of S1 section, dissolution of minerals, which are very likely to be feldspar according to EDS analysis, has been observed (Figure 6). This difference is mainly caused by the different pH values, which depend on the amount of pyrite oxidation.

Figure 5. SEM images of pyrite grains in two samples. In S1, the pyrite grains are fragmented and many cracks develop, while in S2 the pyrite grains are better preserved. This phenomenon may show that S1 has been weathered.

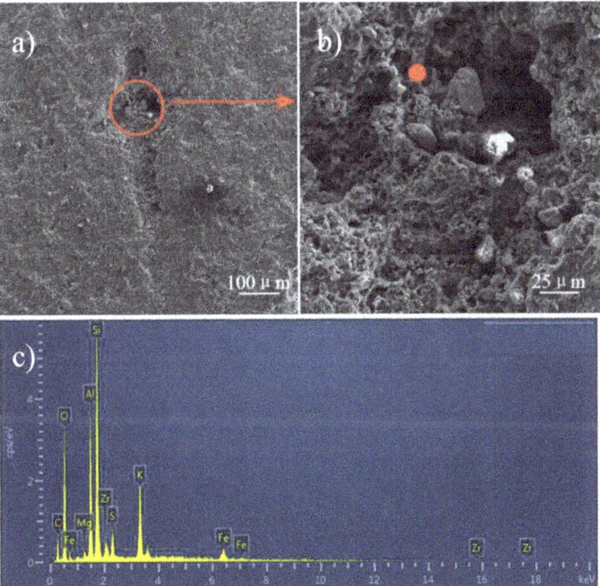

Figure 6. SEM images of dissolved surfaces of minerals at different magnification (**a,b**), showing the formation of a depression. (**c**) The EDS analysis of the mineral residues revealed that the main components were K, Al, Si and O, indicating that the dissolving minerals are very likely to be feldspar.

3.2.2. Precipitation of Secondary Minerals

Pyrite is the main source of aqueous Fe^{3+} and SO_4^{2-}, therefore theoretically the molar ratio of Fe^{3+} to SO_4^{2-} should be close to 0.5. The deviation between the actual ratio and the theoretical value indicates the precipitation of secondary minerals which removes one ion preferentially [30]. Based on the measured SO_4^{2-} concentration, the iron concentrations produced from the dissolution of stoichiometric pyrite are calculated (Figure 7a). Measured Fe concentrations in both the S1 and S2 experiments are less than calculated Fe concentrations, which is indicative of the removal of Fe by precipitation of Fe-bearing phases. In addition, the saturation indices of Fe(III)-bearing phases are calculated (Figure 7b). All the saturation indices are greater than 0, indicating that the fluid is supersaturated for these Fe(III)-bearing phases. What we observed in this study is consistent with the conclusion that the oxidative dissolution of pyrite is accompanied by precipitation of secondary Fe(III)-bearing (hydr)oxide phases [50]. During the precipitation of these secondary (oxy)hydroxide phases, some trace metals like U and Pb are likely to be adsorbed at the surface of or co-precipitate with these minerals [51,52]. This may account for the downward trend of concentrations of U and Pb after about 72 h in the S1 and S2 experiments (Tables 3 and 4), indicating that the precipitation of Fe(III)-bearing (hydr)oxide can restrict the migration of some toxic metals.

Considering the high concentration of SO_4^{2-} in these experiments, there is a high probability of barite ($BaSO_4$) and gypsum ($CaSO_4$) precipitation. The saturation indices (SI) are calculated and presented in Table 7. The SI of barite in both experiments is greater than 0, which means the fluid is supersaturated for BaSO4. Considering that the concentrations of SO_4^{2-} are much higher in S1 than in S2, the Ba concentrations in S2 would be predicted to be lower, but in fact they are relatively high. At the surface of SEM thin sections of both S1 and S2 experiments, highly crystallized secondary minerals are observed, which are confirmed to be $CaSO_4$ by EDS (Figure 8). Most of the gypsum crystals observed are accumulated on pyrite grains, which may be caused by the higher concentration of SO_4^{2-} near the pyrite that is being oxidized. However, the SI of $CaSO_4$ is below 0, indicating that $CaSO_4$ is undersaturated (Table 7). Therefore, the gypsum crystals observed in our experiments might be precipitated during the drying process of SEM samples. Nevertheless, during the process of hydraulic fracturing in field, due to the much higher shale to fluid ratio, the precipitation of $CaSO_4$ is very likely to occur. Gypsum is a common scale mineral and difficult to dissolve. The precipitation of gypsum has the potential to occlude the pores or throats in shale and thus affect the exploitation of shale gas.

Table 7. Calculated saturation indices of barite and gypsum.

Minerals	24 h	48 h	72 h	D6	D15	D30	D90	D220	D360
Barite in S1&3	0.36	0.29	0.25	0.12	0.11	0.11	0.07	0.05	0.05
Gypsum in S1&3	−1.67	−1.44	−1.43	−1.08	−1.18	−1.46	−1.51	−1.57	−1.52
Barite in S2&4	0.52	0.39	0.48	0.38	0.44	0.69	0.84	0.54	0.69
Gypsum in S2&4	−1.21	−1.17	−1.22	−1.22	−0.98	−1.8	−1.85	−2.01	−1.83

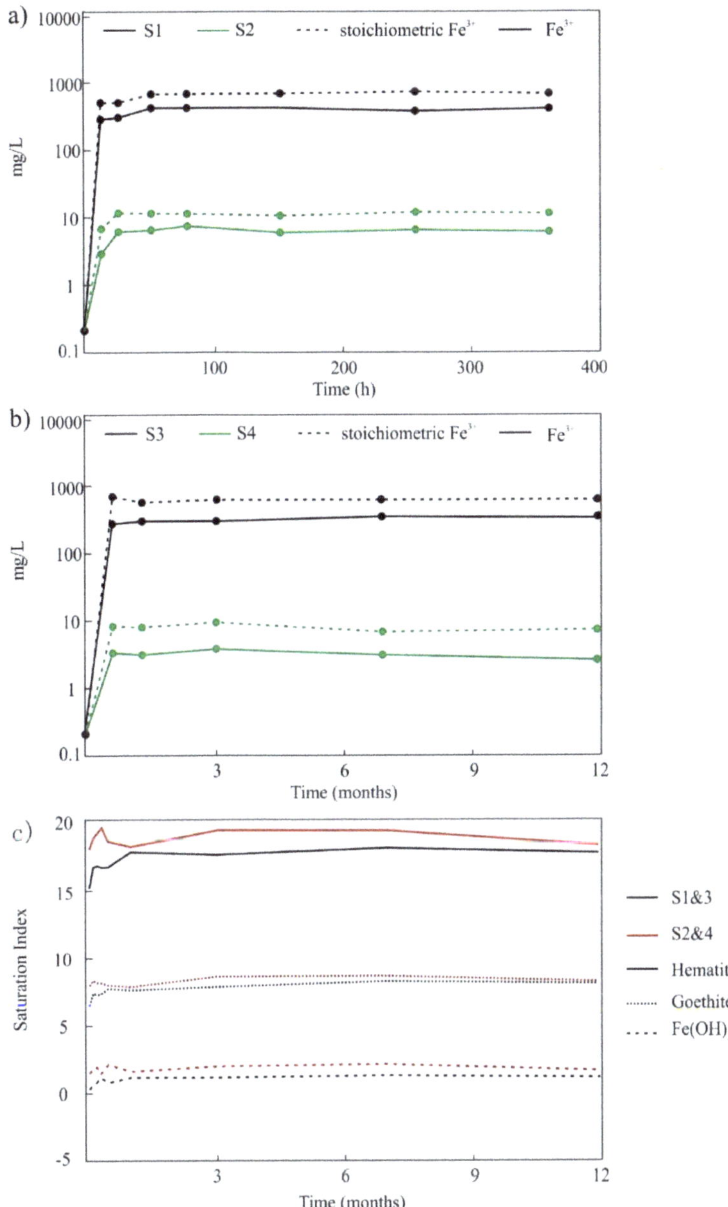

Figure 7. (**a,b**) are images of the comparison between stoichiometric and actual Fe concentrations in short-term and long-term experiments respectively. (**c**) Saturation indices of secondary Fe(III)-bearing phases versus reaction time in the experiments. All the saturation indices were greater than 0, which means the fluid is supersaturated for all these secondary Fe(III)-bearing phases.

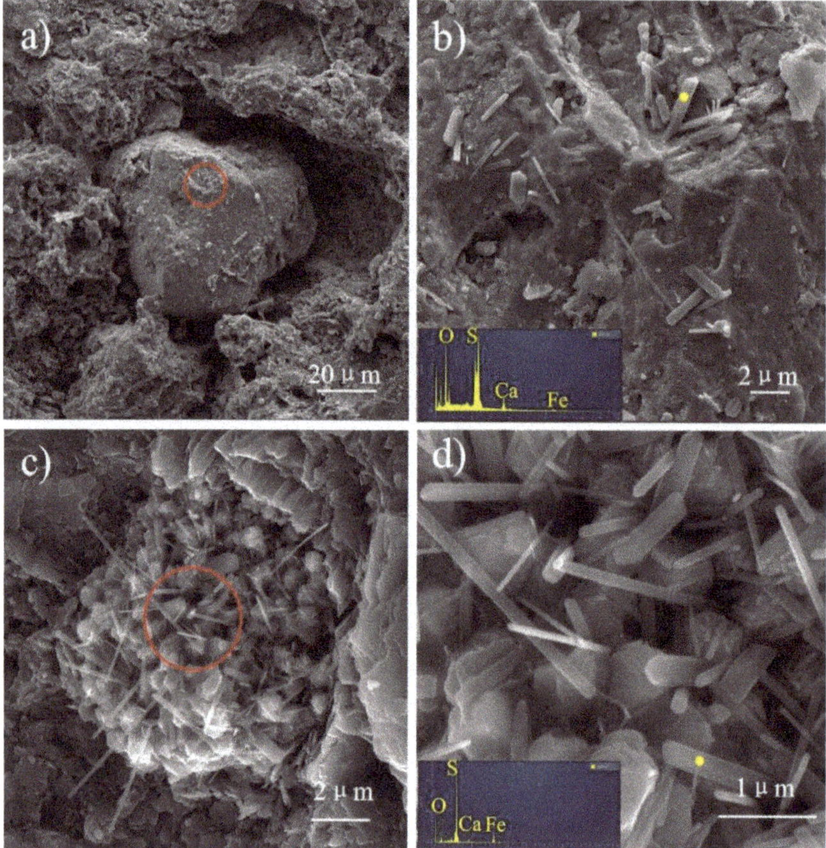

Figure 8. SEM images of the CaSO^4 precipitates observed on the surface of samples in both experiments. (**a,c**) are images of S3 and S4 respectively. (**b,d**) are magnified images of the red zones of (**a,c**) respectively. The main elements in the crystals (the yellow spots) are Fe, S and Ca through EDS analysis, which suggests a high potential of gypsum precipitates.

3.2.3. Alteration of Porosity

During the process of hydraulic fracturing, mineral dissolution generates pores, and the precipitation of secondary minerals may cause pore blockage, which could influence the long-term productivity of shale gas wells. Therefore, it is necessary to investigate the alteration of porosity.

On the surface of the S1 thin sections recovered after experiments, many discrete holes about several microns in diameter are observed through FE-SEM, which are not found in the original sample (Figure 9). Most of these pores are isolated, and the microscopic image shows the dissolution of minerals clearly. The unstable minerals like feldspar and some clay minerals are dissolved, while the more stable minerals like quartz remain undissolved. However, in the S2 experiments, the reacted samples observed through SEM are not much different from the intact sample (Figure 10). Although pores are observed on the surface of S2 thin sections, there are much less of them than in S1. But many irregular pores generated by calcite dissolution are observed in S2, forming a mottled texture over the calcite residue (Figure 11). Unlike other isolated pores, these pores are interconnected and have a high potential to affect the transportation of fluid. The porosity of shale before and after the experiments is also measured (Table 8). The initial porosity of S2 is 0.8%, and after a reaction between shale and fluid, the porosity increased to 1.1%. The initial porosity of S1 is 9.7%, which is much higher than that of S2. After experiments, the porosity increased to 20.0%, which is twice that of the original sample. As shown in Figure 6, the EDS analysis of the mineral residues remaining in the discrete pore reveals that the pores may be caused by feldspar dissolution, which explains the great difference between S3 and S4 experiments.

Table 8. Measured porosity of initial samples and recovered samples.

Samples	Porosity (%)	Density (g/cm^3)
Initial S3	9.7	2.46
Recovered S3	20.0	2.18
Initial S4	0.8	2.70
Recovered S4	1.1	2.68

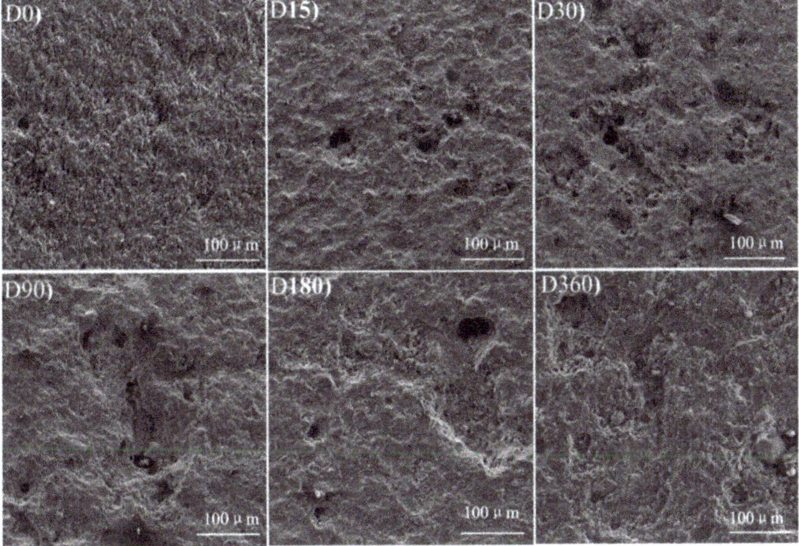

Figure 9. SEM images showing changes of porosity in S3 from time 0 to 360 days. On the surface of the initial sample, no pores were observed, while after exposure to fracturing fluid, many discrete holes were generated.

As shown in Figure 4c, the surface of pyrite grain in S1 is covered with a layer of precipitates, part of which develops on cracks and blocks them. In S2, on the surface of calcite residue after dissolution, some black substances are observed (Figure 4i), which are not found in initial samples. Although the composition of these precipitates was not determined, they all act to occlude pores to some extent. In addition, gypsum crystals developed in pores are also observed (Figure 11). Although this crystal is very small and cannot block the pore in our experiments, it shows that pores occluded by $CaSO_4$ are very likely to occur in the field. It is known that even slight changes in porosity can affect the permeability of the rock [52,53], but the effect of the porosity blockage on fluid transport and long-term productivity of shale gas is not clear yet and needs further study.

According to the results of our experiments, there are several key geochemical reactions caused by the fluid during the fracturing process. When shale is exposed to fracturing fluid, the first reaction is the oxidation of pyrite, during which a large amount of metal elements and sulfuric acid are released, resulting in acidification of the fracturing fluid. The acidic fracturing fluid dissolves carbonate minerals, which tends to neutralize the pH of the solution and generate pore networks. Subsequently, if the shale is not abundant in carbonate minerals, the feldspar is also dissolved due to the vulnerability to acidic solution, which generates many pores and increase the porosity greatly. At the same time, some secondary minerals are also precipitated. For example, $Fe(OH)_3$ precipitates at a pH greater than 3.7, and very high concentrations of SO_4^{2-} cause precipitation of $CaSO_4$ and $BaSO_4$ that may occlude the pores and affect the transportation of hydrocarbons, and ultimately influence of the recovery of shale gas.

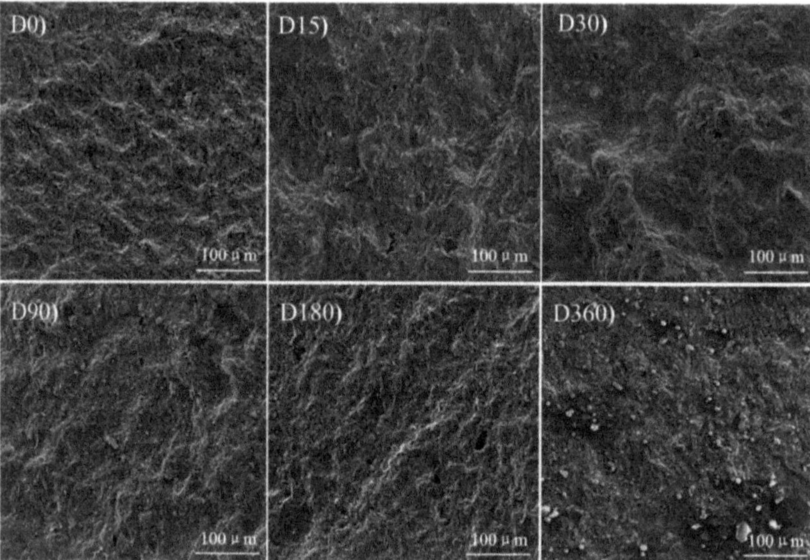

Figure 10. SEM images illustrating changes of porosity in S4. Although several pores were observed, there was no big difference between the initial and recovered samples.

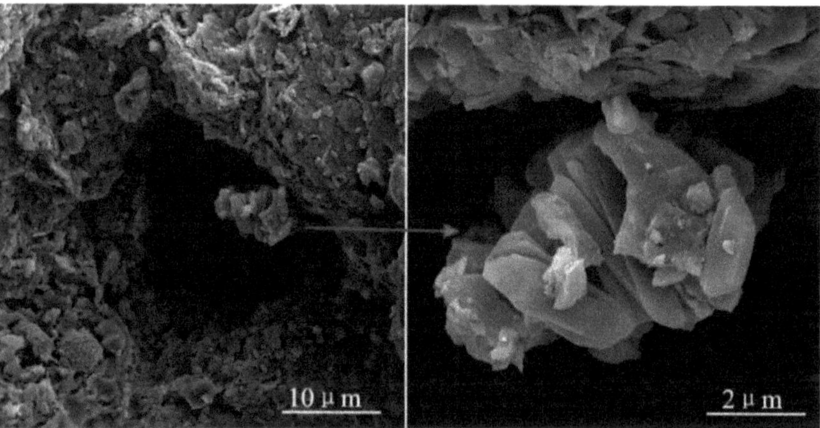

Figure 11. A gypsum crystal developed in a pore in S2, which shows the possibility of gypsum blocking pores.

4. Conclusions and Field Implication

This study reveals that the exposure of shale to hydraulic fracturing fluid causes a series of reactions that are strongly affected by the mineralogy and physical properties of the shale. Fracturing fluid can erode the relatively unstable minerals in shale like calcite and pyrite, thus influencing the evolution of fluid composition. In pyrite-rich shale formations with a loose structure, the pH of fracturing fluid decreases due to the greater extent of pyrite dissolution, while in carbonate mineral-rich shale reservoirs with a tight structure, pyrite dissolves less and calcite can act as a pH buffer, leading to the smaller pH change. The dissolution of these minerals in shale generates porosity, which can increase the shale permeability and then probably enhance hydrocarbon transport. But the precipitation of secondary phases blocks pores or cover cracks, affecting the exploitation of shale gas. Mineral dissolution also releases many trace metal contaminants into fracturing fluid that are highly likely to contaminate surface waters after being brought to the surface by flowback or produced water. Therefore, treatment and monitoring of flowback and produced water are essential for safe disposal or re-use. However, some of the metal contaminants are removed from solution through adsorption by or co-precipitation with secondary phases, which therefore can mitigate migration of contaminants to some extent.

The dissolution of minerals is also related to the additives in fracturing fluid, so the degree of pollution can be minimized by adjusting the composition of solution. The precipitation of secondary Fe-(oxy)hydroxide and the adsorption of metal contaminants would be both enhanced at higher pH. Therefore, the use of low pH fracturing fluid in carbonate mineral-poor shale reservoirs requires more treatment of flowback and produced water and is not suggested. The dissolution of pyrite is mainly due to oxidants in fracturing fluid. The utilization of fracturing fluid with a lower content of oxidants could minimize the release of trace metal contaminants and formation of secondary phases. However, the effect of the high underground pressure and lower temperature in long-term experiments on the reactions between shale and solution, which may impact mineral dissolution and precipitation, has not been characterized in this case and requires further studies. Furthermore, during the process of shale gas production in field, the formation waters in reservoirs typically contain high concentrations of ions. These formation waters may affect the saturation state of secondary minerals and the adsorption of trace metal, which also requires further investigation. Despite the drawbacks, our experiments confirmed that a series of reactions between shale and fluid can be induced, and understanding these processes and their effects on shale gas production can help us to predict the geochemical changes during exploitation and minimize the contaminants.

Author Contributions: Conceptualization, J.T. and Q.L.; Investigation, J.T., G.L., R.H., and L.L.; Resources, J.T. and Q.L.; Data Curation, J.T. and G.L.; Writing—Original Draft Preparation, G.L.; Writing—Review & Editing, J.T., G.L., L.L., and J.D.; Supervision, J.T., L.L., and Q.L.; Funding Acquisition, J.T. and L.L. All authors have read and agreed to the published version of the manuscript.

Funding: This work was funded by the National Natural Science Foundation of China grant number 41872151, the Innovation driven Project of Central South University grant number 502501005, Natural Science Foundation of Hunan Province, grant number 2019JJ50762, and China Postdoctoral Science Foundation, grant number 2019M652803.

Conflicts of Interest: The authors declare no conflict of interest.

References

1. Dong, D.; Wang, Y.; Li, X.; Zou, C.; Guan, Q.; Zhang, C.; Huang, J.; Wang, S.; Wang, H.; Liu, H.; et al. Breakthrough and prospect of shale gas exploration and development in China. *Nat. Gas Ind. B* **2016**, *3*, 12–26. [CrossRef]

2. Dong, D.; Zou, C.; Dai, J.; Huang, S.; Zheng, J.; Gong, J.; Wang, Y.; Li, X.; Guan, Q.; Zhang, C.; et al. Suggestions on the development strategy of shale gas in China. *J. Nat. Gas Geosci.* **2016**, *1*, 413–423. [CrossRef]

3. Zou, C.; Dong, D.; Wang, Y.; Li, X.; Huang, J.; Wang, S.; Guan, Q.; Zhang, C.; Wang, H.; Liu, H.; et al. Shale gas in China: Characteristics, challenges and prospects (I). *Petrol. Explor. Dev.* **2015**, *42*, 753–767. [CrossRef]

4. Zou, C.; Dong, D.; Wang, Y.; Li, X.; Huang, J.; Wang, S.; Guan, Q.; Zhang, C.; Wang, H.; Liu, H.; et al. Shale gas in China: Characteristics, challenges and prospects (II). *Petrol. Explor. Dev.* **2016**, *43*, 182–196. [CrossRef]

5. Zou, C.; Dong, D.; Wang, S.; Li, J.; Li, X.; Wang, Y.; Li, D.; Cheng, K. Geological characteristics and resource potential of shale gas in China. *Petrol. Explor. Dev.* **2010**, *37*, 641–653. [CrossRef]

6. Tan, J.; Horsfield, B.; Fink, R.; Krooss, B.; Schulz, H.-M.; Rybacki, E.; Zhang, J.; Boreham, C.J.; van Graas, G.; Tocher, B.A. Shale gas potential of the major marine shale formations in the upper yangtze platform, South China, Part III: Mineralogical, lithofacial, petrophysical, and rock mechanical properties. *Energy Fuel* **2014**, *28*, 2322–2342. [CrossRef]

7. Tan, J.; Horsfield, B.; Mahlstedt, N.; Zhang, J.; di Primio, R.; Vu, T.A.T.; Boreham, C.J.; van Graas, G.; Tocher, B.A. Physical properties of petroleum formed during maturation of Lower Cambrian shale in the upper Yangtze Platform, South China, as inferred from PhaseKinetics modelling. *Mar. Petrol. Geol.* **2013**, *48*, 47–56. [CrossRef]

8. Tan, J.; Weniger, P.; Krooss, B.; Merkel, A.; Horsfield, B.; Zhang, J.; Boreham, C.J.; van Graas, G.; Tocher, B.A. Shale gas potential of the major marine shale formations in the Upper Yangtze Platform, South China, Part II: Methane sorption capacity. *Fuel* **2014**, *129*, 204–218. [CrossRef]

9. Yang, X.; Fan, T.; Wu, Y. Lithofacies and cyclicity of the Lower Cambrian Niutitang shale in the Mayang Basin of western Hunan, South China. *J. Nat. Gas Sci. Eng.* **2016**, *28*, 74–86. [CrossRef]

10. Xing, L.; Xi, Y.; Jiehui, Z.; Honglin, S. Reservoir forming conditions and favorable exploration zones of shale gas in the Weixin Sag, Dianqianbei Depression. *Petrol. Explor. Dev.* **2011**, *38*, 693–699. [CrossRef]

11. Wang, R.; Ding, W.; Wang, X.; Cui, Z.; Wang, X.; Chen, E.; Sun, Y.; Xiao, Z. Logging identification for the Lower Cambrian Niutitang shale reservoir in the Upper Yangtze region, China: A case study of the Cengong block, Guizhou Province. *J. Nat. Gas Geosci.* **2016**, *1*, 231–241. [CrossRef]

12. Tian, T.; Zhou, S.; Fu, D.; Yang, F.; Li, J. Characterization and controlling factors of pores in the Lower Cambrian Niutitang shale of the Micangshan Tectonic Zone, SW China. *Arab. J. Geosci.* **2019**, *12*, 251. [CrossRef]

13. Tan, J.; Hu, R.; Wang, W.; Dick, J. Palynological analysis of the late Ordovician—Early Silurian black shales in South China provides new insights for the investigation of pore systems in shale gas reservoirs. *Mar. Petrol. Geol.* **2020**, *116*, 104145. [CrossRef]

14. Tan, J.; Wang, Z.; Wang, W.; Hilton, J.; Guo, J.; Wang, X. Depositional environment and hydrothermal controls on organic matter enrichment in the Lower Cambrian Niutitang shale, South China. *AAPG Bull.* **2020**, accepted manuscript.

15. Marcon, V.; Joseph, C.; Carter, K.E.; Hedges, S.W.; Lopano, C.L.; Guthrie, G.D.; Hakala, J.A. Experimental insights into geochemical changes in hydraulically fractured Marcellus Shale. *Appl. Geochem.* **2017**, *76*, 36–50. [CrossRef]

16. Chen, B.; Stuart, F.M.; Xu, S.; Györe, D.; Liu, C. Evolution of coal-bed methane in Southeast Qinshui Basin, China: Insights from stable and noble gas isotopes. *Chem. Geol.* **2019**, *529*, 119298. [CrossRef]

17. Lyu, Q.; Long, X.; Ranjith, P.G.; Tan, J.; Kang, Y.; Wang, Z. Experimental investigation on the mechanical properties of a low-clay shale with different adsorption times in sub-/super-critical CO_2. *Energy* **2018**, *147*, 1288–1298. [CrossRef]

18. Yang, L.; Shi, X.; Ge, H.; Liu, D.; Qu, X.; Gao, J.; Li, L.; Xu, P.; Tan, X. Quantitative investigation on the characteristics of ions transport into water in gas shale: Marine and continental shale as comparative study. *J. Nat. Gas Sci. Eng.* **2017**, *46*, 251–264. [CrossRef]

19. Gregory, K.B.; Vidic, R.D.; Dzombak, D.A. Water management challenges associated with the production of shale gas by Hydraulic Fracturing. *Elements* **2011**, *7*, 181–186. [CrossRef]

20. Stringfellow, W.T.; Domen, J.K.; Camarillo, M.K.; Sandelin, W.L.; Borglin, S. Physical, chemical, and biological characteristics of compounds used in hydraulic fracturing. *J. Hazard. Mater.* **2014**, *275*, 37–54. [CrossRef] [PubMed]

21. Vengosh, A.; Jackson, R.B.; Warner, N.; Darrah, T.H.; Kondash, A. A critical review of the risks to water resources from unconventional shale gas development and hydraulic fracturing in the United States. *Environ. Sci. Technol.* **2014**, *48*, 8334–8348. [CrossRef]

22. Shrestha, N.; Chilkoor, G.; Wilder, J.; Gadhamshetty, V.; Stone, J.J. Potential water resource impacts of hydraulic fracturing from unconventional oil production in the Bakken shale. *Water Res.* **2017**, *108*, 1–24. [CrossRef] [PubMed]

23. Fontenot, B.E.; Hunt, L.R.; Hildenbrand, Z.L.; Carlton, D.D.; Oka, H.; Walton, J.L.; Hopkins, D.; Osorio, A.; Bjorndal, B.; Hu, Q.H.; et al. An Evaluation of Water Quality in Private Drinking Water Wells Near Natural Gas Extraction Sites in the Barnett Shale Formation. *Environ. Sci. Technol.* **2013**, *47*, 10032–10040. [CrossRef] [PubMed]

24. Harkness, J.S.; Darrah, T.H.; Warner, N.R.; Whyte, C.J.; Moore, M.T.; Millot, R.; Kloppmann, W.; Jackson, R.B.; Vengosh, A. The geochemistry of naturally occurring methane and saline groundwater in an area of unconventional shale gas development. *Geochim. Cosmochim. Acta* **2017**, *208*, 302–334. [CrossRef]

25. Darrah, T.H.; Vengosh, A.; Jackson, R.B.; Warner, N.R.; Poreda, R.J. Noble gases identify the mechanisms of fugitive gas contamination in drinking-water wells overlying the Marcellus and Barnett Shales. *Proc. Natl. Acad. Sci. USA* **2014**, *111*, 14076–14081. [CrossRef]

26. Drollette, B.D.; Hoelzer, K.; Warner, N.R.; Darrah, T.H.; Karatum, O.; O'Connor, M.P.; Nelson, R.K.; Fernandez, L.A.; Reddy, C.M.; Vengosh, A.; et al. Elevated levels of diesel range organic compounds in groundwater near Marcellus gas operations are derived from surface activities. *Proc. Natl. Acad. Sci. USA* **2019**, *116*, 9135. [CrossRef] [PubMed]

27. Györe, D.; McKavney, R.; Gilfillan, S.M.V.; Stuart, F.M. Fingerprinting coal-derived gases from the UK. *Chem. Geol.* **2018**, *480*, 75–85. [CrossRef]

28. Alvarez, R.A.; Zavala-Araiza, D.; Lyon, D.R.; Allen, D.T.; Barkley, Z.R.; Brandt, A.R.; Davis, K.J.; Herndon, S.C.; Jacob, D.J.; Karion, A.; et al. Assessment of methane emissions from the U.S. oil and gas supply chain. *Science* **2018**, eaar7204. [CrossRef]

29. Gao, J.; Zou, C.; Li, W.; Ni, Y.; Liao, F.; Yao, L.; Sui, J.; Vengosh, A. Hydrochemistry of flowback water from Changning shale gas field and associated shallow groundwater in Southern Sichuan Basin, China: Implications for the possible impact of shale gas development on groundwater quality. *Sci. Total. Environ.* **2020**, *713*, 136591. [CrossRef]

30. Zeng, L.; Reid, N.; Lu, Y.; Hossain, M.M.; Saeedi, A.; Xie, Q. Effect of the Fluid–Shale Interaction on Salinity: Implications for High-Salinity Flowback Water during Hydraulic Fracturing in Shales. *Energy Fuel* **2020**, *34*, 3031–3040. [CrossRef]

31. Tan, J.; Horsfield, B.; Mahlstedt, N.; Zhang, J.; Boreham, C.J.; Hippler, D.; van Graas, G.; Tocher, B.A. Natural gas potential of Neoproterozoic and lower Palaeozoic marine shales in the Upper Yangtze Platform, South China: Geological and organic geochemical characterization. *Int. Geol. Rev.* **2015**, *57*, 305–326. [CrossRef]

32. Roychaudhuri, B.; Tsotsis, T.T.; Jessen, K. An experimental investigation of spontaneous imbibition in gas shales. *J. Petrol. Sci. Eng.* **2013**, *111*, 87–97. [CrossRef]

33. Balashov, V.N.; Engelder, T.; Gu, X.; Fantle, M.S.; Brantley, S.L. A model describing flowback chemistry changes with time after Marcellus Shale hydraulic fracturing. *AAPG Bull.* **2015**, *99*, 143–154. [CrossRef]

34. Wangersky, J. Principles and applications of aquatic chemistry. *J. Hydrol.* **1994**, *155*, 293–294. [CrossRef]

35. Harrison, A.L.; Jew, A.D.; Dustin, M.K.; Thomas, D.L.; Joe-Wong, C.M.; Bargar, J.R.; Johnson, N.; Brown, G.E.; Maher, K. Element release and reaction-induced porosity alteration during shale-hydraulic fracturing fluid interactions. *Appl. Geochem.* **2017**, *82*, 47–62. [CrossRef]

36. Tian, X. Simulation of Fracturing Fluid-Methane-Minerals Interactions during Hydraulic Fracturing of Shale Gas. Master's Thesis, Chian University of Geoscience, Beijing, China, 2015.

37. Ni, X.; Yu, Y.; Wang, Y.; Gao, S. Dissolution kinetics of Si/Al elements of illites in carbonic acid solutions. *Nat. Gas Ind.* **2014**, *34*, 20–26. [CrossRef]

38. Phan, T.T.; Capo, R.C.; Stewart, B.W.; Graney, J.R.; Johnson, J.D.; Sharma, S.; Toro, J. Trace metal distribution and mobility in drill cuttings and produced waters from Marcellus Shale gas extraction: Uranium, arsenic, barium. *Appl. Geochem.* **2015**, *60*, 89–103. [CrossRef]

39. Chermak, J.A.; Schreiber, M.E. Mineralogy and trace element geochemistry of gas shales in the United States: Environmental implications. *Int. J. Coal Geol.* **2014**, *126*, 32–44. [CrossRef]

40. Jin, L.; Mathur, R.; Rother, G.; Cole, D.; Bazilevskaya, E.; Williams, J.; Carone, A.; Brantley, S. Evolution of porosity and geochemistry in Marcellus Formation black shale during weathering. *Chem. Geol.* **2013**, *356*, 50–63. [CrossRef]

41. Alvarez, R.A.; Pacala, S.W.; Winebrake, J.J.; Chameides, W.L.; Hamburg, S.P. Greater focus needed on methane leakage from natural gas infrastructure. *Proc. Natl. Acad. Sci. USA* **2012**, *109*, 6435–6440. [CrossRef]

42. Lerat, J.G.; Sterpenich, J.; Mosser-Ruck, R.; Lorgeoux, C.; Bihannic, I.; Fialips, C.I.; Schovsbo, N.H.; Pironon, J.; Gaucher, É.C. Metals and radionuclides (MaR) in the Alum Shale of Denmark: Identification of MaR-bearing phases for the better management of hydraulic fracturing waters. *J. Nat. Gas Sci. Eng.* **2018**, *53*, 139–152. [CrossRef]

43. Al, T.A.; Blowes, D.W.; Martin, C.J.; Cabri, L.J.; Jambor, J.L. Aqueous geochemistry and analysis of pyrite surfaces in sulfide-rich mine tailings. *Geochim. Cosmochim. Acta* **1997**, *61*, 2353–2366. [CrossRef]

44. Morse, J.W.; Arakaki, T. Adsorption and coprecipitation of divalent metals with mackinawite (FeS). *Geochim. Cosmochim. Acta* **1993**, *57*, 3635–3640. [CrossRef]
45. Gregory, D.D.; Large, R.R.; Halpin, J.A.; Baturina, E.L.; Lyons, T.W.; Wu, S.; Danyushevsky, L.; Sack, P.J.; Chappaz, A.; Maslennikov, V.V.; et al. Trace Element Content of Sedimentary Pyrite in Black Shales. *Econ. Geol.* **2015**, *110*, 1389–1410. [CrossRef]
46. Sternbeck, J.; Sohlenius, G.; Hallberg, R.O. Sedimentary trace elements as proxies to depositional changes induced by a Holocene fresh-brackish water transition. *Aquat. Geochem.* **2000**, *6*, 325–345. [CrossRef]
47. Morse, J.W.; Luther, G.W. Chemical influences on trace metal-sulfide interactions in anoxic sediments. *Geochim. Cosmochim. Acta* **1999**, *63*, 3373–3378. [CrossRef]
48. Chapman, E.C.; Capo, R.C.; Stewart, B.W.; Kirby, C.S.; Hammack, R.W.; Schroeder, K.T.; Edenborn, H.M. Geochemical and Strontium Isotope Characterization of Produced Waters from Marcellus Shale Natural Gas Extraction. *Environ. Sci. Technol.* **2012**, *46*, 3545–3553. [CrossRef]
49. Capo, R.C.; Stewart, B.W.; Rowan, E.L.; Kolesar Kohl, C.A.; Wall, A.J.; Chapman, E.C.; Hammack, R.W.; Schroeder, K.T. The strontium isotopic evolution of Marcellus Formation produced waters, southwestern Pennsylvania. *Int. J. Coal. Geol.* **2014**, *126*, 57–63. [CrossRef]
50. Blowes, D.W.; Jambor, J.L.; Hanton-Fong, C.J.; Lortie, L.; Gould, W.D. Geochemical, mineralogical and microbiological characterization of a sulphide-bearing carbonate-rich gold-mine tailings impoundment, Joutel, Québec. *Appl. Geochem.* **1998**, *13*, 687–705. [CrossRef]
51. Benjamin, M.M.; Leckie, J.O. Multiple-site adsorption of Cd, Cu, Zn, and Pb on amorphous iron oxyhydroxide. *J. Colloid Interface Sci.* **1981**, *79*, 209–221. [CrossRef]
52. Duff, M.C.; Coughlin, J.U.; Hunter, D.B. Uranium co-precipitation with iron oxide minerals. *Geochim. Cosmochim. Acta* **2002**, *66*, 3533–3547. [CrossRef]
53. Luquot, L.; Gouze, P. Experimental determination of porosity and permeability changes induced by injection of CO_2 into carbonate rocks. *Chem. Geol.* **2009**, *265*, 148–159. [CrossRef]

Article

Reservoir Characteristics of the Lower Jurassic Lacustrine Shale in the Eastern Sichuan Basin and Its Effect on Gas Properties: An Integrated Approach

Jianhua He [1,2,*], Hucheng Deng [2,*], Ruolong Ma [2], Ruyue Wang [3], Yuanyuan Wang [2] and Ang Li [4]

[1]   State Key Laboratory of Shale Oil and Gas Enrichment Mechanisms and Effective Development, Beijing 100083, China

[2]   College of Energy, Chengdu University of Technology, Chengdu 610059, China; lugang19@cdut.edu.cn (R.M.); wangyuanyuan1919@outlook.com (Y.W.)

[3]   Petroleum Exploration and Production Research Institute, China Petroleum and Chemical Corporation, Beijing 100083, China; wangruyue.syky@sinopec.com

[4]   Wuxi Research Institute of Petroleum Geology, Sinopec Petroleum & Production Research Institute, Wuxi 214000, China; syky@sinopec.com

*   Correspondence: hejianhuadizhi@163.com (J.H.); denghucheng@cdut.cn (H.D.)

Received: 14 July 2020; Accepted: 25 August 2020; Published: 31 August 2020

Abstract: The exploration of shale gas in Fuling area achieved great success, but the reservoir characteristics and gas content of the lower Jurassic lacustrine in the northern Fuling areas remain unknown. We conducted organic geochemical analyses, Field Emission Scanning Electron Microscope (FE-SEM), X-ray diffraction (XRD) analysis, high-pressure mercury intrusion (MIP) and CH_4 adsorption experimental methods, as well as NMR logging, to study mineral composition, geochemical, pore structure characteristics of organic-rich shales and their effects on the methane adsorption capacity. The Da'anzhai shale member is generally a set of relatively thick (avg. 75 m) and high carbonate-content (avg. 56.89%) lacustrine sediments with moderate total organic carbon (TOC) (avg. 1.12%) and thermal maturation (Vitrinite reflectance (VR): avg. 1.19%). Five types of lithofacies can be classified: marl (ML), calcareous shale (CS), argillaceous shale (AS), muddy siltstone (MS), and silty shale (SS). CS has good reservoir quality with a high porosity (avg. 4.72%). The small pores with the transverse relaxation time of 0.6–1 ms and 1–3 ms comprised the major part of the porosity in the most lithofacies from Nuclear magnetic resonance (NMR) data, while the large pore (>300 ms) accounts for a small porosity proportion in the CS. The pores mainly constitute of mesopores (avg. 23.2 nm). The clay minerals with a large number of interparticle pores in the SEM contributes most to surface area in the shale lithofacies with a moderate TOC. The adsorption potential of shale samples is huge with an average adsorption capacity of 4.38 mL/g and also has high gas content (avg. 1.04 m^3/t). The adsorption capacity of shale samples increases when TOC increases and temperature decreases. Considered reservoir properties and gas properties, CS with the laminated structures in the medium-upper section of the Da'anzhai member is the most advantage lithofacies for shale gas exploitation.

Keywords: shale gas; reservoir characteristics; gas content; eastern Sichuan Basin; the Da'anzhai member

---

1. Introduction

A great success of shale gas exploration has been made in a Fuling gas field. Recently, many more oil and gas discoveries in lacustrine shale of the Lower Jurassic succession (e.g., Da'anzhai member shale) in the northern Fuling area gives a rise to the possibility of nonmarine shale gas accumulation in neighboring areas [1,2]. Recent studies focus on the pore structure and geochemical characteristics of the Da'anzhai member lacustrine shale [3–5]. There is little research concerning the influence of lacustrine shale reservoir characteristics on the adsorption capacity and total gas content. However, gas content is not only a key factor of favorable shale gas area evaluation, but also controls whether shale reservoir has a commercial exploration value or not [2,6].

In this study, we present a comprehensive approach of mineral composition and geochemical tests, high-resolution imaging investigation, high-pressure mercury intrusion (MIP) and CH_4 adsorption experiments, as well as NMR well logging, to (1) determine the mineral composition and geochemical, pore structure and petrophysical characteristics of the different shale lithofacies in the Da'anzhai members; (2) study the influence of continental reservoir characteristics on shale gas properties; (3) optimize the advantageous shale lithofacies to provide guidance for selecting the target shale gas layers.

2. Geological Setting

The study area is located in the eastern part of the Sichuan Basin, which is one of the major petroliferous basins in China. This area is dominated by the huge thrust-fold belts with an NNE or NE striking. It includes the narrow Datianchi, Huangnitang, Dachigan, Fangdoushan anticlines and gentle Liangping, Bashansi synclines between these structures from east to west [7]. The North Fuling gas field is located in the Bashansi Synclines (Figure 1A,B). The Ziliujing formation in the Lower Jurassic is divided into Da'anzhai, Maanzhan, and Dongyuemiao members. The target layer of theDa'anzhai member consists of three lithological members: the shell limestone in the lower layer (J_1zDa^1); the interbed in the middle layer with black shale, silty shale, and shell limestone (J_1zDa^2); and the shell limestone and marl in the top (J_1zDa^3) (Figure 1C). The middle part of the Da'anzhai member (J_1zDa^2) is the most important source rock with a thickness of 30–80 m. This layer was widely deposited when the lacustrine system was rapidly expanded and the water within the lake was deepened [8], which produced a positive impact on the accumulation and preservation of organic matter.

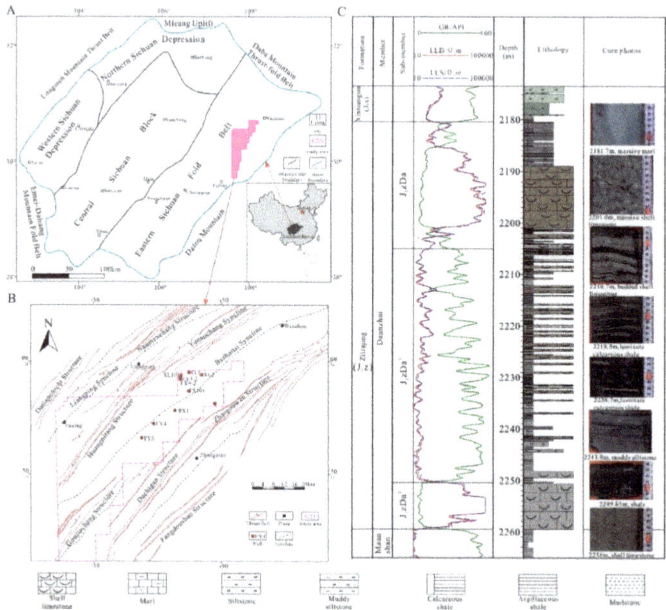

Figure 1. Study location (**A**), geological map (**B**), and simplified stratigraphic column (**C**) of the Jurassic Da'anzhai member in the Eastern Sichuan Basin.

3. Samples and Experiments

In this paper, all the samples in the Da'anzhai member were selected from fresh cores in five wells, including FY1, FY4, FY5, FY3-2, and XL101 wells. Most of the samples were analyzed in the following methods: Total organic carbon determination, Rock-Eval pyrolysis analyses, kerogen vitrinite reflectance analysis, kerogen maceral composition microscopy, X-ray diffraction (XRD) testing, and high-pressure mercury intrusion (MIP), that were performed by the Marine Geology Lab, PetroChina Jianghan Oilfield Company. High-pressure intrusion experiments were conducted by AutoPore IV 9520 instruments, which can obtain a pore size range of 0.003–1000 μm. The pore micromorphology of 15 samples were determined at the State Key Laboratory of Oil and Gas Reservoir Geology and Exploitation of Chengdu University of Technology. These samples were investigated using an "FEI Quanta model 200F from FEI corporation, Netherlands" field-emission scanning electron microscopy (FE-SEM) under a working condition of 20 kv and distance of 8–9 mm. The methane adsorption isotherm experiments of five samples were conducted by the Shale Gas Lab, Sichuan Coal Geology Bureau.

Nuclear magnetic resonance (NMR) well logging is an approach that confirms the nuclear spin states of hydrogen in water and oil within a thin annulus situated deeply into the formation with several inches [9]. Under the pulsed and static magnetic field, the transverse relaxation time (T_2) of protons generally shows a positive relationship with the amount of hydrogen protons in the pore fluid [10]. The NMR well logging curves generally shows 2 to 3 main peaks and the peaks becomes larger with higher relaxation time: the first peak in the shale reservoir normally is the predominant one indicating clay-bound water and capillary-bound water, while free-fluid volume and micro-fractures are only revealed in some small peaks [11].

Field desorption is a relatively accurate method to directly obtain the data of gas content, including desorption gas content, losing gas content and residual gas content. Desorption gas content was measured by the ISO-300 system under a temperature of 80 °C (Figure 2) when the sealed shale samples were placed in the desorption tank. Losing gas content was calculated by the United States Bureau of Mine (USBM) [12]. The samples (100–200 g) were placed in the residual gas tank, and then crushed for 30 min. Finally, residual gas content can be measured by the ISO-300 system (Figure 2).

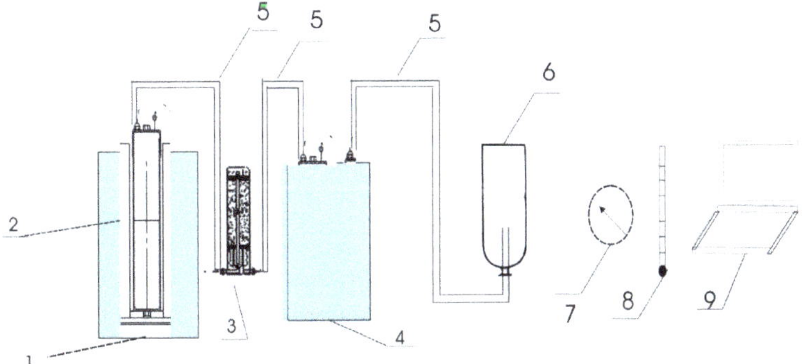

Figure 2. The ISO-300 filed testing system of desorption gas content in the Da'anzhai shale member.1-thermostatic apparatus; 2-desorption tank; 3-dehumidication apparatus; 4-testing instruments of gas flow; 5-flexible pipe; 6-device for collecting gas; 7-pressure gauge; 8-thermometer; 9-data-processing system.

4. Results and Discussions

4.1. Organic Geochemical Characteristics

According to the TOC determination and Rock-Eval data of 143 samples in wells FY3-2 and XL101, The TOC of the Da'anzhai member widely ranged from 0.05 wt.% to 3.08 wt.% (avg. 1.12%)

(Figure 3 and Table 1). Some 65% of samples have a TOC value of more than (>0.5 wt.%). The TOC level over 1.0 wt.% and 2.0 wt.% account for 45.1% and 8.9% of all samples, respectively. The variation of TOC value in vertical is related to lithology (Figure 3). The laminated calcareous shale and massive argillaceous shale in the upper part of the J_1zDa^2 have high TOC value of avg. 1.58 wt.% [13], while the massive or bedded marl and silty mudstone in the J_1zDa^1 and J_1zDa^3 have a low TOC of <0.50 wt.%. The potential yield (P), defined as the sum of Rock-Eval S1 and S2 values, is generally used to evaluate the genetic potential of a source rock [14]. P (S1+S2) and TOC are the most important indicators of organic matter abundance [15], and they show a good positive correlation ($R^2 = 0.82$) for all samples (Figure 3). P has a range of 1.03–6.84 mg/g (avg. 2.34 mg/g). The chloroform bitumen"A" content of 20 samples vary between 0.01% and 0.19% with an average value of 0.11% (Figure 3). Based on the thresholds of nonmarine source rocks from [15], the Da'anzhai member shale is a fair-good source rock and provides a good quality for hydrocarbon generation.

The kerogen composition analysis of 42 samples indicate all samples are dominated by sapropelinite (65–74%), along with moderate vitrinite (25–51.7%) (Figure 4). The sapropel organic matter is mostly correlated with initial production of algae or bacterial phytoplanktonic sources at the surface of lake [16]. The plot of HI vs. Tmax (Figure 4) indicates organic matter (OM) in the Da'anzhai member shale is dominated by type II, mainly typeII$_2$ [17]. The OM type in vertical transitioned from mainly typeII$_2$ in J_1zDa^3 to typeII$_1$ in the J_1zDa^2 and J_1zDa^1. This suggests the change of the OM origins in vertical result in the diversity of kerogen compositions and OM types.

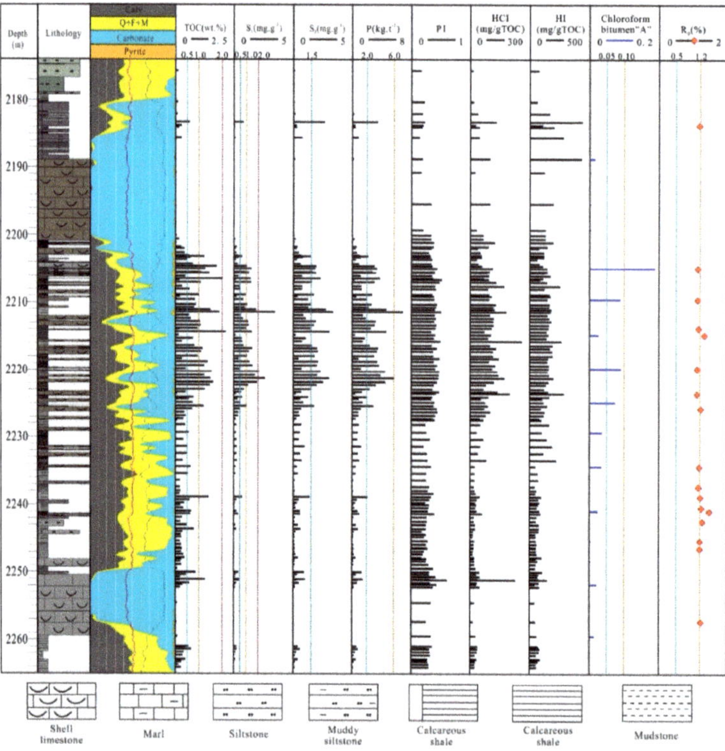

Figure 3. The geochemical and mineralogy characteristics for well FY3-2 in the Jurassic Da'anzhai member. Q: quartz; F: feldspar; M: mica. P = S_1+S_2; PI = $S_1/(S_1+S_2)$; HCl = $S_1/TOC \times 100$; HI = $S_2/TOC \times 100$. The blue, orange, and red dash lines show the classification borders of general, good, and excellent qualities of source rock.

Table 1. The mineral composition and pore structure characteristics of shale samples.

Sample	TOC (wt.%)	Clay (wt.%)	Carbonate (wt.%)	Quartz (wt.%)	Pyrite (wt.%)	I/S (wt.%)	Illite (wt.%)	Kaolinite (wt.%)	Chlorite (wt.%)	Surface Area (m^2/g)	Pore Volume (cm^3/100g)	Porosity (%)	Permeability (mD)	Average Pore Size (nm)	Medium Pore Size (nm)
FY-1	0.56	8.7	75.6	7.4	1.6	2	3.04	1.83	1.83	1.76	1.09	0.8	0.103	62.50	41.7
FY-2	1.31	27.4	47.5	15.6	6.9	8.22	9.86	4.38	4.93	5.69	3.33	2.6	0.965	49.27	4.7
FY-3	1.89	22.7	55	15.7	3.8	6.13	6.81	5.22	4.54	3.80	2.55	2	0.097	49.80	36
FY-4	1.36	50.6	11.4	26.4	4.6	24.79	15.18	5.06	5.57	8.79	0.35	0.6	28.79	41.51	18
FY-5	0.81	9.5	79.5	6.8	2.7	2.57	3.61	1.52	1.81	2.19	2.09	1.6	0.083	80	22.5
FY-6	0.96	37.6	28.3	27	2.1	13.54	11.66	4.89	7.52	2.65	2.58	2	0.638	57.97	15.4
FY-7	1.23	22.8	54.8	15.8	2.2	6.384	9.58	3.88	2.96	3.36	2.72	2.1	0.961	170.21	102.3
FY-8	0.77	35.7	20.1	35.3	2.4	13.21	9.99	7.14	5.36	2.86	0.9	0.7	0.142	38.67	9.9
FY-9	1.21	50	19	24.5	1.4	15.5	15	10.5	9	3.71	4.44	3.5	9.79	35.00	7.2
FY-10	0.41	48.3	0	32.4	0	16.91	11.59	11.59	8.21	1.66	3.05	2.4	1.07	39.71	4.9
FY-11	0.33	12.7	76.1	11.2	0	3.18	2.54	3.05	3.94	1.42	2.16	1.7	0.565	31.06	9.7
FY-12	0.2	51.1	17.7	31.2	0	20.44	15.84	9.71	5.11	13.92	0.67	0.7	0.205	21.85	4.6
FY-13	0.61	46.3	10.3	40.5	0	19.45	15.28	6.95	4.63	12.62	1.12	0.9	0.073	17.78	4.2
FY-14	0.67	44.7	10.5	42.6	0	19.22	13.41	6.71	5.36	4.47	7.01	5.6	0.323	9	6.5
FY-15	0.49	45.3	18.6	32.2	1.8	19.48	16.31	5.44	4.08	9.47	4.7	5.9	9.02	163.73	100.9
FY-16	0.27	21.9	17.8	55.2	0	7.67	5.69	4.59	3.94	5.51	2.56	2	0.126	25	2.7
FY-17	0.26	48.3	18.6	31.4	0	18.84	14.97	9.66	4.83	11.91	2.25	1.7	0.154	27.44	4.5
FY-18	1.29	62.4	8.8	23.7	1.6	21.84	24.96	10.61	4.99	8.73	3.22	2.5	4.65	46.94	6.5

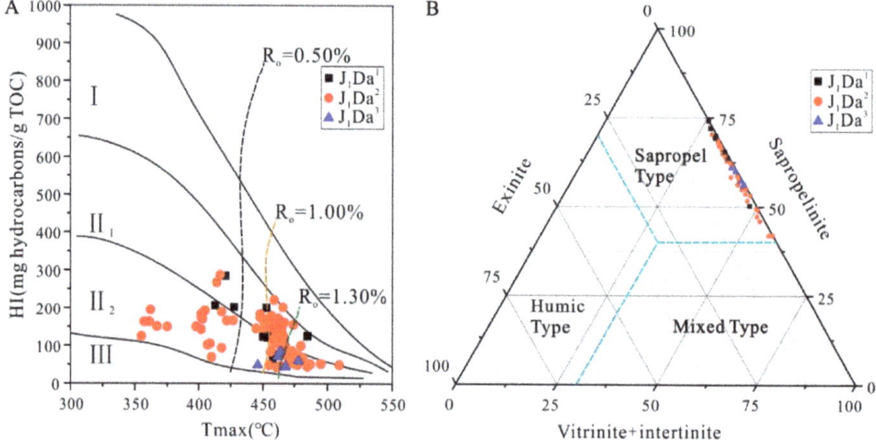

Figure 4. The organic matter type of Da'anzhai member shale. (**A**) Cross-plot between hydrogen index and Tmax showing the kerogen-type and source potential; (**B**) ternary chart of kerogen maceral composition.

The vitrinite reflectance (R_o) values of 17 samples ranged from 1.0% to 1.46% and averaged 1.19% (Figure 3). The temperature of maximum development of the S2 peak (T_{max}) mainly ranges from 439 °C to 510 °C (avg. 452 °C). This indicates most of the samples reach a moderate-high maturity with a high potential for shale gas generation [18].

4.2. Reservoir Characteristics

4.2.1. Mineral Composition and Reservoir Properties

X-ray diffraction analysis of core samples from the two wells showed that the Da'anzhai Member shale is dominated by carbonate minerals (avg. 56.8%), followed by clay minerals (avg. 29.8%), quartz and feldspars (avg. 19.6%), and pyrite (avg. 3.05%) (Figure 3 and Table 1). These kinds of mineral compositions confirmed that the Da'anzhai member shale is not a typical shale of mainly terrigenous origin (generally defined by clay contents greater than 75%). The brittle mineral content (avg. 68.5%) of the Da'anzhai member shale has reached at the minimum standards of fracturing feasibility (40%) [19]. Illite and mixed-layerillite/smectite (avg. 65.8%) are the predominant clay minerals from the corresponding clay mineral data (Table 1) with comparatively small amounts of chlorite (avg. 15.8%) and kaolinite (avg. 18.2%). According to the mineral compositions, sedimentary genesis, and structure, the lithofacies of the Da'anzhai member can be classified as five groups: marl (ML), calcareous shale (CS), argillaceous shale (AS), muddy siltstone (MS), and silty shale (SS).

Porosity and permeability analyses from MIP in well FY3-2 indicates that the differences of reservoir properties between several lithofacies are dramatic. There is a weakly positive relationship between Porosity and permeability (R^2 = 0.31) (Figure 5). The permeability of different lithofacies influenced by microfractures shows a wide range of 0.07–28.76 mD (Table 1). A permeability of 0.1–1 mD account for 66.7% in all samples. The permeability of CS with the laminated structures is relatively high. Moreover, its porosity (avg. 4.72%) is also higher than AS (avg. 2.72%), followed by MS (avg. 1.89%) and SS (avg. 1.75), whereas porosity of ML is relatively poor (avg. 1.02%). The information of pore throat size, the sorting and connectivity within pores also can be obtained from the MIP data and curves (Figure 6). The distribution of pore throat size indicates that AS, CS, and SS contains two types of pores: small pores (8–100 nm) and large pores (1–15 μm) with a good connectivity (Figure 7). But the tight ML and MS composed of a large number of small pores (<20 nm).

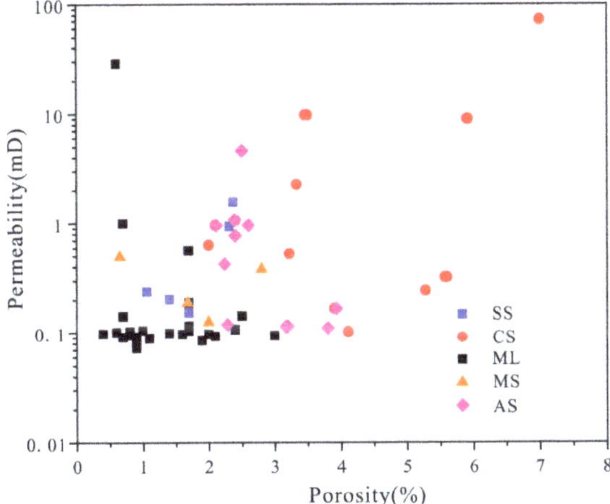

Figure 5. The relationship between porosity and permeability of 62 core samples. These samples were classified based on lithofacies. Note: SS: silty shale; CS: calcareous shale; ML: marl; AS: argillaceous shale; MS: muddy siltstone.

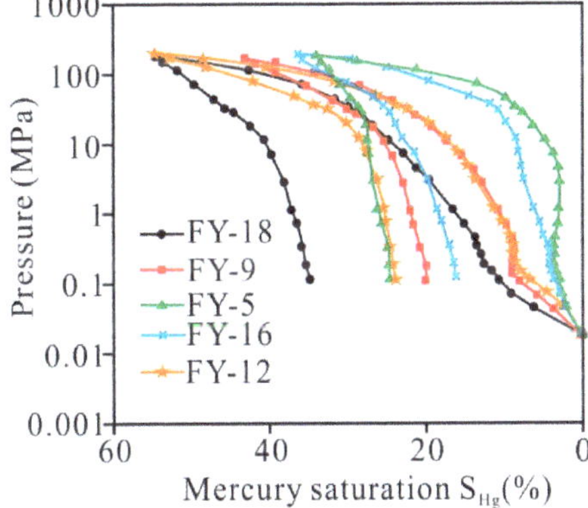

Figure 6. The capillary pressure curves of five typical types of lithofacies. Sample FY-5 of ML; sample FY-12 of SS; sample FY-9 of AS; sample FY-16 of MS; sample FY-9 of CS.

NMR is also an effective tool to determine the probe reservoir characteristics, including bound fluid, free fluid porosity, effective porosity, total porosity, and hydrocarbon existence [20]. The reservoir evaluation results of the Da'anzhai member from NMR well logging were shown in Figure 8 and Table 2. The heterogeneity of shale lithofacies strongly influenced the T_2 spectrum and porosity characteristics from NMR logging data. CS generally constitute of 2–3 peaks in the T_2 spectrum: The first peak (around 3 ms) is the largest peak, followed by a second peak (around 60 ms), while the third (1000 ms) peak is generally missing or it is not significant. This indicates that the pore types mainly consist of micropores and meso/macro pore. Sometimes, the fractures can be developed in these kinds of lithofacies. This is consistent with the distribution of pore throat size from the MIP results. The porosity of bound water is larger than that of free fluid. The T_2 spectrum of AS, SS, and MS shows an amplitude peak and a minor peak. However, the value of T_2 in the AS is generally higher than that of SS and MS. This indicates the AS has better reservoir quality. The T_2 spectrum of ML differs significantly from other lithofacies. It only has one peak around 1 ms, indicating that micropores prevail. Interestingly, free water volume does not exist in the ML because of high carbonate cementation. This is further confirmed by the carbonate content from XRD results.

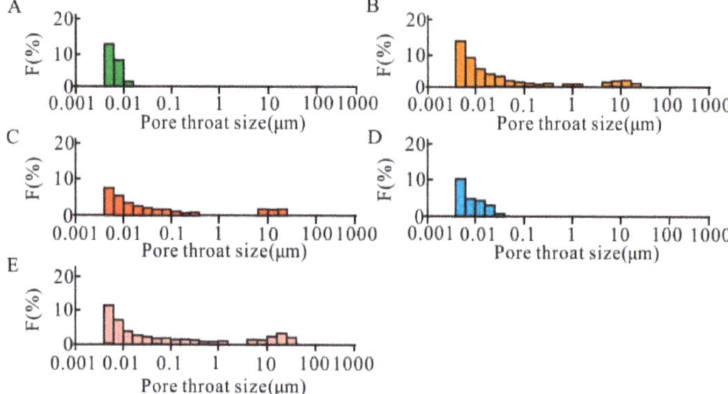

Figure 7. The distribution characteristics of pore throat size in five types of lithofacies. (**A**) for sample FY-5 of marl; (**B**) for sample FY-12 of silty shale; (**C**) for sample FY-9 of argillaceous shale; (**D**) for sample FY-16 of muddy siltstone; (**E**) for sample FY-9 of calcareous shale. F: Frequency.

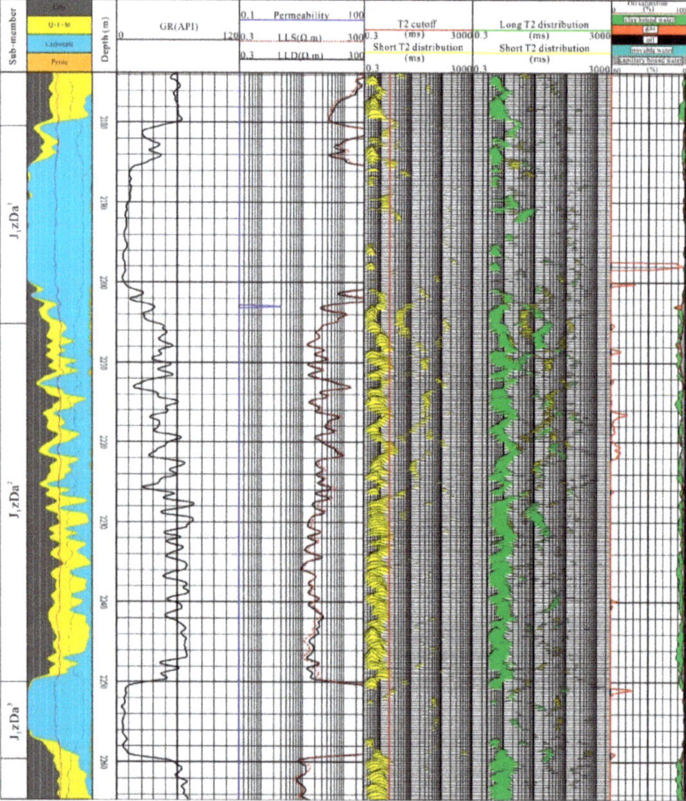

Figure 8. The reservoir evaluation from NMR logging data in the Da'anzhai member, well FY3-2.

Table 2. The reservoir evaluation results of five types of shale lithofacies in the Da'anzhai member from NMR well logging data.

Lithfacies Type	The Peak of T_2 Distribution (ms)	Capillary Bound Water Volume (avg. %)	Movable Water Volume (avg. %)	Effective Porosity (avg. %)	Total Porosity (avg. %)
CS	3–30, 60–300, 1000	2.3	0.6	2.5	4.9
AS	3–30, 30–100	1.5	0.4	1.8	2.1
SS	1–3 3–30	1.2	0.3	1.3	1.8
MS	0.3–3, 3–10	0.6	0.2	0.9	1.2
ML	0.3–3	0.3	0	0.5	0.9

4.2.2. Pore Morphology and Pore Structure Analysis

FE-SEM images are the techniques that are generally used to qualitatively determine the pore morphology characteristics of shales (e.g., the shapes, sizes, and distributions) [21–23]. The FE-SEM images show that a large quantity of micro-nano pores and microfracture exits may occur in the lacustrine shale samples. Based on the relationship between pores and minerals, pores can be classified into the three groups: organic matter-hosted pores (OM pores), the pores in the framework minerals (FM pores), and clay minerals-associated pores (CM pores). OM pores usually form during the thermal evolution and hydrocarbon generation process of kerogen and other OM. They mainly have aspherical, ellipsoidal-shape with a pore size of 2–100 nm (Figure 9A,B). Compared with the marine shale in the Fuling areas [5], OM pores are scarce in the Da'anzhai Lacustrine shale because of a moderate maturity (R_o: avg. 1.19%) and low TOC content. A large number of clay mineral pores was investigated in the SEM. The intrapartical and interlayer pores were developed between the clay mineral. The pore size of the observed wedge-shaped pore is 50 nm to 700 nm, and the length of the slits is up to 6 μm (Figure 9C,E,F). These pores provide a large surface for adsorbed gas. FM pores are mostly a certain amount of nano-scale pyrite framboids or calcite intercrystalline pores (Figure 9G,H). These pores are generally irregular with poor connectivity. The FM pores also contain the nano-scale pores generated by the dissolution of minerals (e.g., quartz and calcite) (Figure 9I). Moreover, a large number of microcracks could be investigated in the CS and AS. These shrinkage cracks were generally caused by dehydration of clay minerals and thermal pressurization of hydrocarbons [24,25], forming a crack-pore network system (Figure 9A,D). The shrinkage crack and the edge of the mineral particle form a bend-ridge-like distribution, with crack spacing of 20–30 nm and length of 8–40 μm. These microcracks provide important channels for gas migration.

Pore structure parameters measured by MIP are shown in Table 1. The average pore size and medium pore size show a main range of 17.8–57.9 nm and 2.7–18 nm, respectively. This indicates the pore size in the Da'anzhai member shale is dominated by mesopores, which is in consistent with the full pore-size distribution results of previous studies in Da'anzhai member shale of the central or northeastern Sichuan Basin [3–5]. The specific surface area varies from 1.42 to 13.92 m^2/g (avg. 5.81 m^2/g). The pore volume ranges from 0.35 to 7.01 cm^3/100g (avg. 2.59 cm^3/100g). Compared with the marine Longmaxi Formation shale in the same area (i.e., Fuling area; [5]), the surface area and pore volume is much lower because a lower TOC of lacustrine shale in Da'anzhai member contributes less to the surface and pore volume than marine shale. In Figure 10, TOC values are positively correlated with surface area and pore volume with a coefficient of 0.46 and 0.23, respectively. However, this correlation is much poorer because of a low-medium TOC value (avg. 1.12 wt.%) and moderate maturity. The clay mineral and surface area exit a strong positive correlation with a coefficient of 0.62. This indicates clay mineral play a more important role on the surface area [26] and pore volume of lacustrine shale with a low TOC (Figure 10). Many interparticle pores among clay minerals are also investigated in the SEM (Figure 9C,E,F). On thecontrary, the carbonate content is negatively correlate with surface area and pore volume, because high carbonate minerals (e.g., calcite) with a strong cementation can greatly influence reservoir quality and usually have adverse impacts. The quartz minerals show a weakly positive relationship with pore volume ($R^2 = 0.49$). The total pore volume increases with quartz and feldspar content in the lacustrine shale due to the generation of pores by dissolution and weakening of compaction [27].

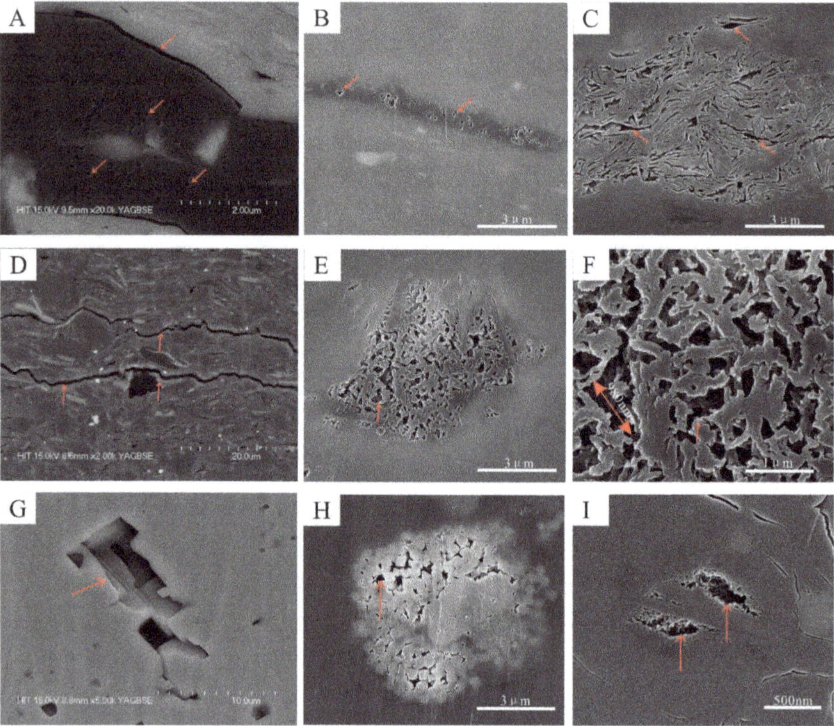

Figure 9. Pore types of the Da'anzhai member shale in the FE-SEM images. (**A**) organic matters-hosted pores and microfractures, CS, 2599.95 m, Well FY1; (**B**) organic matters-hosted pores, AS, 1736.51 m, well FY4; (**C**) interparticle pores between clay minerals, AS, 2543.73 m, well FY5; (**D**) microfractures, 2615 m, CS, well FY1; (**E**) and (**F**) floccule intercrystal pores in flocculated clay microfabric, CS, 2544.36 m, well FY4; (**G**) intercrystalline pores in calcite mineral, ML, 2639.8 m, well FY1; (**H**) intercrystal pores in pyrite framboids, SS, 1736.51 m, well FY4; (**I**) dissolution pores in calcite minerals, MS, 2672.04 m, well 5.

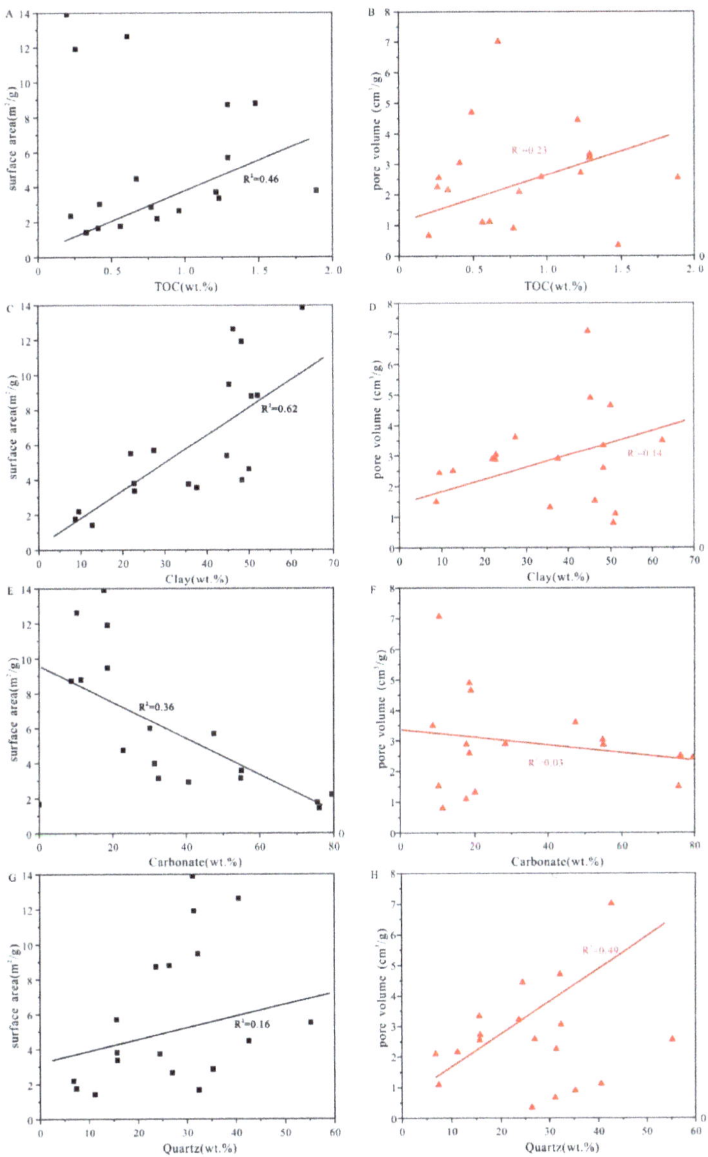

Figure 10. Correlation plots between specific surface area and pore volume and TOC and mineral composition. (**A**) The relationship between TOC and surface area; (**B**) The relationship between TOC and pore volume; (**C**) The relationship between clay and surface area; (**D**) The relationship between clay and pore volume; (**E**) The relationship between carbonate and surface area; (**F**) The relationship between carbonate and pore volume; (**G**) The relationship between Quartz and surface area; (**H**) The relationship between Quartz and pore volume.

4.3. Methane Sorption and Adsorption Potential

The methane adsorption isotherms of five shale samples measured at different pressure and temperatures are shown in Figure 11. We set a temperature group of 30 °C, 80 °C, 100 °C. The pressure values were measured at the 13 balance pressure points. The balancing time at each spot was set by 12.0 h. The amount of methane adsorbed into the shale at early stage of <5 MPa rapidly increased under different temperature and reached 85% of the maximum adsorption capacity. However, the shale with a high TOC content at same temperature (80 °C) has much faster methane adsorption rate with a large gradient (Figure 11A), because a high TOC of organic-rich shale provide a large amount of surface area for methane adsorption. The increased amount of methane adsorption became less from 5 MPa up to 12 MPa, because the methane adsorption reached equilibrium state. This process can be described by the Langmuir isotherm equation [28]:

$$V = \frac{VLbp}{1 + bp} \tag{1}$$

where V is the absorbed methane volume, cm^3/g; V_L is the Langmuir volume, cm^3/g; b is the reciprocal of Langmuir pressure, MPa^{-1}; p is the pressure, MPa. Langmuir volume of Da'anzhai member shale ranges from 1.34 mL/g to 2.28 mL/G (avg. 1.70 mL/g), indicating medium-high adsorbed capacity (Table 3). However, shale with different TOC content has quite different V_L value at the same temperature (80 °C). When TOC content increased from 0.41 wt.% to 1.48 wt.%, the Langmuir volume of shale changed from 1.34 mL/g to 1.77 mL/g (Figure 11). Even with a similar TOC level, the Langmuir volume decreased rapidly at a high temperature. When the experimental temperature jumped from 30 °C to 100 °C, the Langmuir volume of shale decreased from 2.28 mL/g to 1.44 mL/g. This can be interpreted by adsorption potential theory.

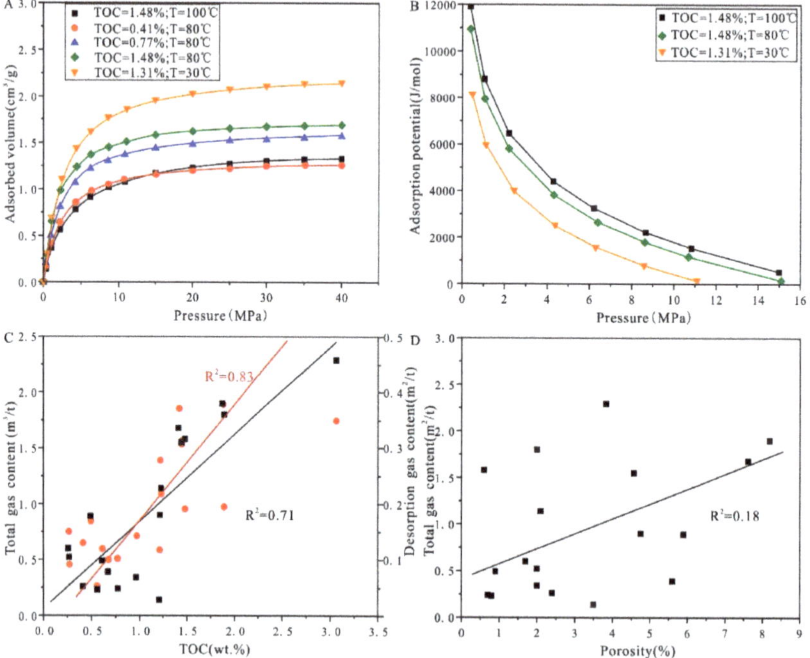

Figure 11. The gas potential of the Da'anzhai member shale. (**A**) and (**B**) isothermal adsorption curves and adsorption potential at different temperatures; (**C**) and (**D**) relationship between TOC, porosity and total gas content.

Table 3. Methane adsorption results and field desorption results of shale samples.

Sample	TOC (wt.%)	Porosity (%)	Langmuir Volume (mL/g)	Langmuir Pressure (MPa)	Desorption Gas Content (m^3/t)	Lost Gas Content (m^3/t)	Residual Gas Content (m^3/t)	Total Gas Content (m^3/t)
FY-1	0.56	0.8	/	/	0.04	0.19	0	0.23
FY-3	1.89	2.0	/	/	0.19	1.51	0.1	1.8
FY-4	1.48	0.6	1.77	2.01	0.19	1.27	0.12	1.58
FY-6	0.96	2.0	/	/	0.14	0.11	0.09	0.34
FY-7	1.31	2.1	2.28	2.46	0.22	0.85	0.07	1.14
FY-8	0.77	0.7	1.66	2.39	0.1	0.06	0.08	0.24
FY-9	1.21	3.5	/	/	0.11	0.02	0.01	0.14
FY-10	0.41	2.4	1.34	2.42	0.13	0.07	0.06	0.26
FY-13	0.61	0.9	/	/	0.12	0.32	0.05	0.49
FY-14	0.67	5.6	/	/	0.1	0.24	0.05	0.39
FY-15	0.49	5.9	/	/	0.17	0.66	0.06	0.89
FY-16	0.27	2.0	/	/	0.09	0.39	0.04	0.52
FY-17	0.26	1.7	/	/	0.15	0.45	0	0.6
XL-1	1.44	4.6	/	/	0.31	0.08	1.16	1.55
XL-2	1.87	8.2	/	/	0.38	0.15	1.37	1.9
XL-3	3.06	3.8	/	/	0.35	0.03	1.91	2.29
XL-4	1.41	7.6	/	/	0.38	0.06	1.24	1.68
XL-5	1.22	4.8	/	/	0.28	0.13	0.49	0.9

Note: /: no data.

Based on Polanyi's theory of adsorption potential [29], the adsorption potential correlated with pressure can be defined as:

$$\varepsilon = \int_{Pi}^{P0} \frac{RT}{P} dP = RT \ln \frac{P0}{Pi} \tag{2}$$

where ε is the adsorption potential, J/mol; P_i is the equilibrium pressure of methane at temperature T, MPa; P_0 is the saturated vapor pressure of methane at T, MPa; R is a constant value of 8.3114, J/(mol·K); T is the thermodynamic temperature, K; P is the pressure, MPa; under a certain temperature, the saturated vapor pressure of methane keeps constant. For example, the saturated vapor pressure of methane is generally 15.8 MPa at 80 °C. So according to Equation (2), the adsorption potential of shale samples can be easily determined from isothermal adsorption experiment (Figure 11B). When the temperature increased under the same pressure, the adsorption potential obviously increased. It indicates the result of an increase of temperature and gas molecular kinetic energy. The increase of adsorption capacity cannot keep more gas molecules absorbed by the adsorbent, because the increasing rate of energy change is much lower than the kinetic energy. Therefore, the adsorption capacity increases with the decrease of temperature.

4.4. Gas Content of Field Desorption

Field desorption analysis of 18 samples indicates total gas content of the Da'anzhai member shale range from 0.14 to 2.29 m^3/t (avg. 1.04 m^3/t). It is much greater than the minimum standard of 0.5 m^3/t for a favorable lacustrine shale gas [2]. Moreover, the CES in the medium-upper section of the J$_1$zDa^2 has high gas content with a range of 1.15–2.29 m^3/t (avg. 1.57 m^3/t). Desorption gas content accounts for 10.56–60.44% (avg. 26.75%) of total gas content. This is much lower than the ratio of desorption gas content to total gas content due to a moderate TOC and maturity. TOC is positively related to the total gas content with a correlation coefficient of 0.71 (Figure 11C). It shows a strongly positive relationship between TOC and desorption gas content (R^2 = 0.83). This suggests OM contributes most to desorption gas content. There is also a weakly positive correlation between porosity and total gas content (Figure 11D), because shale with a high porosity can provide more sites for free gas.

5. Conclusions

(1) The lacustrine Da'anzhai member shale in the north Fuling area has a medium-high OM content (avg. TOC 1.12 wt.%). The laminated calcareous shale in the upper part of the J_1zDa^2 has high TOC value of avg. 1.58 wt.%. The average potential yield P (S1 + S2) and chloroform bitumen content "A" are 2.34 mg hydrocarbons per g TOC and 0.11% with a medium hydrogen generation potential. The kerogen type of mainly typeII$_2$ and a moderate OM maturity (VR = avg. 1.19) indicate a huge potential of gas generation. The mineral content is dominated by calcite, clay and quartz. The clay minerals mainly consist of illite and mixed-layerillite/smectite. The lithofacies of the Da'anzhai member includes marl (ML), calcareous shale (CS), argillaceous shale (AS), muddy siltstone (MS), and silty shale (SS).

(2) Heterogeneity of shale lithofacies strongly influences the reservoir quality. The porosity and permeability of shale is overall low, but the laminated calcareous shale has good reservoir properties with a high porosity (avg. 4.72%). The pore size is dominated by mesopores, and the medium pore diameter is 23.2 nm. Better NMR data show that the small pores with the transverse relaxation time of 0.6–1 ms and 1–3 ms comprised most of the porosity of the Da'anzhai shale member, while the larger pores with a T_2 value of 300–1000 ms and 1000–3000 ms only accounts for a small porosity proportion. However, it is very important for gas migration and generally showed a third peak in the NMR spectrum of CS, and this peak generally missed after centrifuge processing.

(3) The clay minerals in Da'anzhai member shale contributes most to surface area. The average Langmuir volume and total gas content is 1.7 mL/g and 1.04 m^3/t. The adsorption capacity of methane decreases with decreasing TOC and increasing temperature. The total gas content of lacustrine shale is controlled by TOC and porosity. CS have the highest gas content (avg. 1.57 m^3/t) than other shale lithofacies. The reservoir properties, mineral composition and gas content data suggest the laminated calcareous shale in the medium-upper section of the J_1zDa^2 are the most advantage lithofacies for shale gas potential production.

Author Contributions: Conceptualization, J.H. and H.D.; investigation, J.H., H.D., R.M., R.W.; methodology, J.H., H.D.; formal analysis, J.H., R.W., A.L.; resource, H.D., J.H.; data curation, J.H., R.M., Y.W.; Writing—Original draft preparation, J.H.; writing—Review and editing, J.H., H.D., Y.W.; visualization, J.H.; project administration, D.H., R.W.; funding acquisition, J.H., R.W. All authors have read and agreed to the published version of the manuscript.

Funding: This research was supported by supported by State Key Laboratory of Shale Oil and Gas Enrichment Mechanisms and Effective Development (G5800-19-ZS-KFGY006).

Conflicts of Interest: The authors declare no conflict of interest.

References

1. Wei, X.F.; Huang, J.; Li, Y.P.; Wang, Q.B.; Liu, R.B.; Wen, Z.D. The main factors controlling the enrichment and high production of Da'anzhai member continental shale gas in Yuanba area. *Geolog. China* **2014**, *41*, 970–981.

2. Nie, H.K.; He, Z.L.; Liu, G.X.; Zhang, G.R.; Lu, Z.Y.; Li, D.H.; Sun, C.X. Status and direction of shale gas exploration and development in China. *J. China Univ. Mining Technol.* **2020**, *49*, 13–35.

3. Xu, Q.L.; Liu, B.; Ma, Y.S.; Song, X.; Wang, Y.; Chen, Z. Geological and geochemical characterization of lacustrine shale: A case study of the Jurassic Da'anzhai member shale in the central Sichuan Basin, southwest China. *J. Nat. Gas. Sci. Eng.* **2017**, *47*, 124–139. [CrossRef]

4. Long, L.Y.; Zhang, Y.Z.; Wang, Y.J.; Wang, L.G. The pore structure of tight limestone-Jurassic Ziliujing Formation, Central Sichuan Basin, China. *App. Geophys.* **2018**, *15*, 165–174.

5. Chen, L.; Jiang, Z.X.; Liu, Q.X.; Jiang, S.; Liu, K.; Tan, J.; Gao, F. Mechanism od shale gas occurrence: Insights from comparative study on pore structures of marine and lacustrine shales. *Mar. Pet. Geol.* **2019**, *104*, 200–216. [CrossRef]

6. Xu, H.; Zhou, W.; Hu, Q.H.; Xia, X.H.; Zhang, C.; Zhang, H.T. Fluid distribution and gas adsorption behaviors in over-mature shales in southern China. *Mar. Pet. Geol.* **2019**, *109*, 223–232. [CrossRef]

7. Zhai, G. *Petroleum Geology of China*; Petroleum Industry Press: Beijing, China, 1997.

8. Gao, J.; Wang, X.; He, S.; Guo, X.; Zhang, B.; Chen, X. Geochemical characteristics and source correlation of natural gas in Jurassic shales in the North Fuling area, Eastern Sichuan Basin, China. *J. Pet. Sci. Eng.* **2017**, *158*, 284–292. [CrossRef]

9. Coates, G.R.; Xiao, L.; Prammer, M.G. *Nmr Logging: Principles and Applications*; Haliburton Energy Services: Houston, TX, USA, 1999.

10. Neto, A.C.; Guinea, F.; Peres, N.M.; Novoselov, K.S.; Geim, A.K. The electronic properties of graphene. *Rev. Mod. Phys.* **2009**, *81*, 109. [CrossRef]

11. Testamanti, M.N.; Rezaee, R. Determination of Nmr T_2 cut-off for clay bound water in shales: A case study of Carynginia Formation, Perth Basin, Western Australia. *J. Pet. Sci. Eng.* **2017**, *149*, 497–503. [CrossRef]

12. Yao, G.H.; Wang, X.Q.; Du, H.Y.; Yi, W.; Guo, M.; Xiang, R.; Li, Z. Applicability of USBM method in the test on shale gas content. *Acta Pet. Sin.* **2016**, *37*, 802–806.

13. Milad, B.; Slatt, R. Outcrop subsurface reservoir characterization of the Mississippian Sycamore/Meramec play in the SCOOP area, Arbuckle mountains, Oklahoma, USA. In Proceedings of the Unconventional Resource Technology Conference, Denver, CO, USA, 22–24 July 2019.

14. Tissot, B.P.; Welte, D.H. *Petroleum Formation and Occurrence*, 2nd ed.; Springer-Verlag: Heidelberg/Berlin, Germany, 1984; p. 699.

15. Huang, D.F.; Li, J.C.; Zhang, D.J. *Evolution and Hydrocarbon Generation Mechanisms of Terrestrial Organic Matter*; Petroleum Industry Press: Beijing, China, 1984; pp. 1–228.

16. Schnyder, J.; Dejax, J.; Keppens, E.; Tu, T.T.N.; Spagna, P.; Boulila, S.; Galbrun, B.; Riboulleau, A.; Tshibangu, J.P.; Yans, J. An Early Cretaceous lacustrine record: Organic matter and organic carbon isotopes at Bernissart (Mons Basin, Belgium). *Palaeogeogr. Palaeoclimatol. Palaeoecol.* **2009**, *281*, 79–91. [CrossRef]

17. Taylor, G.H.; Teichmüller, M.; Davis, A.; Diessel, C.F.K.; Littke, R.; Robert, P. *Organic Petrology*; Schweitzerbart: Stuttgart, Germany, 1998; pp. 1–704.

18. Xu, H.; Zhou, W.; Zhang, R.; Liu, S.M.; Zhou, Q.M. Characterizations of pore, mineral and petrographic properties of marine shale using multiple techniques and their implications on gas storage capability for Sichuan longmaxi gas shale field in China. *Fuel* **2019**, *241*, 360–371. [CrossRef]

19. Rickman, R.; Mullen, M.J.; Petre, J.E.; Grieser, W.V.; Kundert, D. *A Practical Use of Shale Petrophysics for Stimulation Design Optimization: All Shale Plays Are Not Clones of the Barnett Shale*; Society of Petroleum Engineers: Denver, CO, USA, 2008.

20. Kapur, G.; Findeisen, M.; Berger, S. Analysis of hydrocarbon mixtures by diffusion ordered Nmr spectroscopy. *Fuel* **2000**, *79*, 1347–1351. [CrossRef]

21. Loucks, R.G.; Reed, R.M.; Ruppel, S.C.; Hammes, U. Spectrum of pore types and networks in mudrocks and a descriptive classification for matrix-related mudrock pores. *AAPG Bull.* **2012**, *96*, 1071–1098. [CrossRef]

22. Tian, H.; Pan, L.; Xiao, X.M.; Wilkins, R.W.T.; Meng, Z.P.; Huang, B.J. A preliminary study on the pore characterization of Lower Silurian black shales in the Chuandong Thrust Fold Belt, southwestern China using low pressure N2 adsorption and FE-SEM methods. *Mar. Pet. Geol.* **2013**, *48*, 8–19. [CrossRef]

23. Milad, B.; Slatt, R.; Fuge, Z. Lithology, stratigraphy, chemostratigraphy, and depositional environment of the Mississippian Sycamore rock in the SCOOP and STACK area, Oklahoma, USA: Field, lab, and machine learning studies on outcrops and subsurface wells. *Mar. Pet. Geol.* **2020**, *115*, 1–18. [CrossRef]

24. Jarvie, D.M.; Hill, R.J.; Ruble, T.E.; Pollastro, R.M. Unconventional shale-gas systems: The Mississippian Barnett Shale of north-central Texas as one model for thermogenic shale-gas assessment. *AAPG Bull.* **2007**, *91*, 475–499. [CrossRef]

25. Ross, D.J.K.; Bustin, R.M. The importance of shale composition and pore structure upon gas storage potential of shale gas reservoirs. *Mar. Petrol. Geol.* **2009**, *26*, 916–927. [CrossRef]

26. Ji, L.; Zhang, T.; Milliken, K.L.; Qu, J.; Zhang, X. Experimental investigation of main controls to methane adsorption in clay-rich rocks. *Appl. Geochem.* **2012**, *27*, 2533–2545. [CrossRef]

27. Wang, Y.; Liu, L.F.; Li, S.T.; Ji, H.T.; Xu, Z.J.; Luo, Z.H.; Xu, T.; Li, L.Z. The forming mechanism and process of tight oil sand reservoirs: A case study of Chang 8 oil layers of the Upper Triassic Yanchang Formation in the western Jiyuan area of the Ordos Basin, China. *J. Pet. Sci. Eng.* **2017**, *158*, 29–46. [CrossRef]

28. Langmuir, I. The adsorption of gases on plane surfaces of glass, mica and platinum. *J. Am. Chem. Soc.* **1918**, *143*, 1361–1403. [CrossRef]
29. Polanyi, M. The potential theory of adsorption. *Science* **1963**, *141*, 1010–1013. [CrossRef] [PubMed]

Article

Controls on Pore Structures and Permeability of Tight Gas Reservoirs in the Xujiaweizi Rift, Northern Songliao Basin

Luchuan Zhang [1,2,3,4], **Shu Jiang** [1,2,3,*], **Dianshi Xiao** [4,*], **Shuangfang Lu** [4,*], **Ren Zhang** [3], **Guohui Chen** [1,2,3], **Yinglun Qin** [5] and **Yonghe Sun** [6]

[1] Key Laboratory of Tectonics and Petroleum Resources, Ministry of Education, China University of Geosciences, Wuhan 430074, China; zhangluchuan@cug.edu.cn (L.Z.); chenguohui@cug.edu.cn (G.C.)

[2] Key Laboratory of Theory and Technology of Petroleum Exploration and Development in Hubei Province, China University of Geosciences, Wuhan 430074, China

[3] School of Earth Resources, China University of Geosciences, Wuhan 430074, China; lyp_15826745239@cug.edu.cn

[4] School of Geosciences, China University of Petroleum (East China), Qingdao 266580, China

[5] Oil and Gas Survey, China Geology Survey, Beijing 100083, China; qinyinglun@mail.cgs.gov.cn

[6] School of Earth Sciences, Northeast Petroleum University, Daqing 163318, China; syh79218@nepu.edu.cn

[*] Correspondence: jiangsu@cug.edu.cn (S.J.); xiaods@upc.edu.cn (D.X.); lushuangfang@upc.edu.cn (S.L.)

Received: 28 July 2020; Accepted: 29 September 2020; Published: 5 October 2020

Abstract: As significant components of tight gas reservoirs, clay minerals with ultrafine dimensions play a crucial role in controlling pore structures and permeability. XRD (X-ray diffraction), SEM (scanning electron microscopy), N_2GA (nitrogen gas adsorption), and RMIP (rate-controlled mercury injection porosimetry) experiments were executed to uncover the effects of clay minerals on pore structures and the permeability of tight gas reservoirs, taking tight rock samples collected from the Lower Cretaceous Dengloukou and Shahezi Formations in the Xujiaweizi Rift of the northern Songliao Basin as an example. The results show that the pore space of tight gas reservoirs primarily comprises intragranular-dominant pore networks and intergranular-dominant pore networks according to fractal theory and mercury intrusion features. The former is interpreted as a conventional pore-throat structure where large pores are connected by wide throats, mainly consisting of intergranular pores and dissolution pores, and the latter corresponds to a tree-like pore structure in which the narrower throats are connected to the upper-level wider throats like tree branches, primarily constituting intercrystalline pores within clay minerals. Intragranular-dominant pore networks contribute more to total pore space, with a proportion of 57.79%–90.56%, averaging 72.55%. However, intergranular-dominant pores make more contribution to permeability of tight gas reservoirs, with a percentage of 62.73%–93.40%. The intragranular-dominant pore networks gradually evolve from intergranular-dominant pore networks as rising clay mineral content, especially authigenic chlorite, and this process has limited effect on the total pore space but can evidently lower permeability. The specific surface area (SSA) of tight gas reservoirs is primarily derived from clay minerals, in the order of I/S (mixed-layer illite/smectite) > chlorite > illite > framework minerals. The impact of clay minerals on pore structures of tight gas reservoirs is correlated to their types, owing to different dispersed models and morphologies, and chlorite has more strict control on the reduction of throat radius of tight rocks.

Keywords: clay minerals; pore structures; permeability; tight gas reservoirs; Xujiaweizi Rift; Northern Songliao Basin

1. Introduction

Enormous success in the exploration and exploitation of tight sandstone gas, a kind of unconventional clean natural gas resources, has been achieved worldwide [1–4]. Works on pore systems in tight gas sandstone reservoirs, including pore types, pore-throat sizes and distributions, etc., have drawn great attention because they have significant effects on the storage and flow capacities of tight rocks [5–10]. Many fluid invasion and radiation methods have been applied to disclose pore systems of tight gas sandstones [8,11,12]. The fluid invasion methods primarily include mercury injection porosimetry (MIP), low field nuclear magnetic resonance (NMR), low-pressure gas adsorption (N_2 and CO_2), etc., and the radiation methods principally employ scanning electron microscopy (SEM), small/ultrasmall angle neutron scattering (SANS/USANS), X-ray computer tomography (XCT), etc. Owing to the broad scope of pore-throat dimensions, various techniques need to be integrated to uncover the complete pore systems of tight gas sandstones [6,13].

The pore systems and seepage capacity of tight gas sandstones are simultaneously controlled by multiple factors, such as primary sedimentary environment (particle size, sorting, rounding, mineral compositions, etc.), diagenesis (mechanical compaction, cementation, recrystallization, dissolution, etc.), and regional tectonic movement (fractures) [14–16]. As an important matrix component of tight gas sandstones, clay minerals are generally characterized by ultrafine particle sizes, special morphological structure, and physicochemical properties [17,18]. Currently, research about clay minerals in unconventional reservoirs primarily focuses on their evolution, methane (CH_4) adsorption characteristic, specific surface area, and pore-throat size distributions [18–20]. For instance, Neasham [21] proposed that clay minerals are generally distributed in a pore-throat space in the form of discrete particles, intergrown crystal linings on pore inner surface, or crystals bridging across pores, and they often block throats, resulting in the destruction of pore structure and poor permeability [22]. Ji et al. [23] argued that smectite has the strongest adsorption capacity for CH_4, and physical adsorption is dominant in this process. Chen et al. [17] reported that pores contained in clay minerals are multiscaled from micron to nanosize, which are further classified into interlayer pores, intergranular pores, pores and fractures related to organic matter, pores and fractures related to other types of minerals, dissolution pores, and microfractures. Cao et al. [24] and Yang et al. [25] declared that the total content of clay minerals correlates with pore structure parameters derived from nitrogen gas adsorption (N_2GA) and MIP, such as specific surface area, median pore-throat radius, and maximum pore-throat radius. Ola et al. [18] described in detail the evolution of clay minerals during burial process, especially the conversion of smectite to illite, and they further discussed the relationships between clay mineral diagenesis, shown as I/S ordering, and thermal maturity indicators. However, clay minerals in tight gas sandstones usually contain many types, including kaolinite, illite, chlorite, smectite, etc., and their particle sizes, crystal morphologies, and petrophysical properties show tremendous differences. Therefore, various types of clay minerals possess diverse influence degrees on the pore systems and permeability of tight gas sandstones. More importantly, the quantitative characterization of the impacts of clay minerals on pore structure, porosity, and permeability is still weak at present, and this is the main concern of this work.

In this study, nine tight rock samples with different clay mineral content and permeability are selected from the Lower Cretaceous Shahezi and Denglouku Formations in the Xujiaweizi Rift, Northern Songliao Basin, to carry out XRD, SEM, N_2GA, and rate-controlled mercury injection porosimetry (RMIP) analyses. The objectives of this study are (1) to briefly characterize pore structures and introduce clay mineral morphologies of tight gas reservoirs; (2) to quantitatively evaluate different types of pore networks and their contributions to pore space and permeability; and (3) to discuss the controls of clay minerals on pore structures and permeability of tight gas reservoirs.

2. Geological Setting

The Xujiaweizi Rift is known as one of the largest and most significant gas-bearing rifts in the northern Songliao Basin in Northeastern China (Figure 1A) [26,27], occupying an area of approximately

0.54×10^4 km^2. Structurally, the Xujiaweizi Rift is located in the Southeast Fault Depression, a first-order tectonic unit of the northern Songliao Basin, which is mainly composed of four third-order tectonic units: the Anda–Shengping uplift belt, the Xudong sag, the Xuxi sag, and the Xudong slope belt (Figure 1B) [28]. The study area appears as a dustpan-like shape extending along near the north–south direction on the whole, characterized by faulting in the west and overlapping in the east, under the comprehensive influence of late compression and strike-slip [26,29].

The studied intervals are the Lower Cretaceous Denglouku and Shahezi Formations, separated by the Yingcheng Formation (Figure 1C), and these two strata have proven to be the most favorable reservoirs for tight gas accumulations over the past decade [30,31]. Specifically, the Shahezi Formation was primarily deposited in fan deltas, braided river deltas, and semideep to deep lacustrine facies environments, forming interbedded thick dark mudstone/coal and coarse-grained clastic rocks, such as conglomerate, sandy conglomerate, and gritstone (Figure 1C) [32,33]. The reservoir quality of the Shahezi tight gas reservoirs is relatively poor, with a porosity of 0.4%–10.7%, mainly <6.0%, and a permeability of 0.01×10^{-15} to 11.20×10^{-15} m^2, mainly less than 0.1×10^{-15} m^2 [27], primarily due to the relatively great burial depth of 3000–4500 m. The Denglouku Formation was deposited during a transition period from a rifting stage to depression stage of the Songliao Basin, corresponding to a sedimentary environment of braided rivers, braided river deltas, and lacustrine facies [34], with interbedded sandstones and thin mudstones (Figure 1C). The thickness of the Denglouku Formation primarily ranges from 500 to 1000 m, with a burial depth of primarily shallower than 3500 m [35].

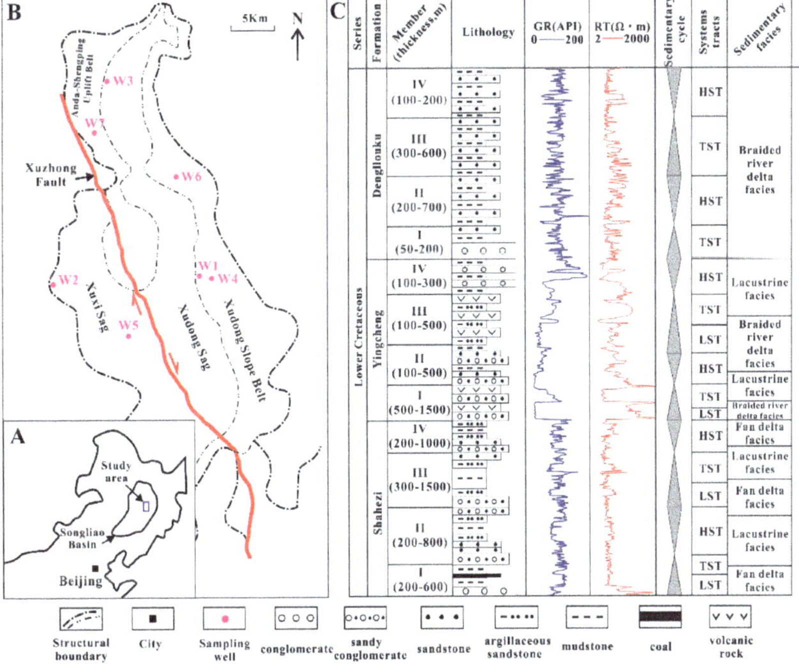

Figure 1. Diagrams showing (**A**) locations of the Songliao Basin and the study area, (**B**) main tectonic units of the Xujiawei Rift and locations of sampling wells, and (**C**) stratigraphic sequence of the Lower Cretaceous Shahezi, Yingcheng, and Denglouku Formations in Xujiaweizi Rift.

3. Experimental Works

3.1. Sample Collection

Nine tight rock samples, with various clay mineral contents of 2–24 wt.% and a permeability of 0.0352×10^{-15} to 2.35×10^{-15} m^2 (Table 1), were assembled from the Lower Cretaceous Dengloukou and Shahezi Formations in the northern Songliao Basin. As tight rocks with relatively lower porosity and relatively higher clay mineral content are usually hard to serve as high-quality natural gas reservoirs [31], the porosities and clay mineral contents of the studied samples are primarily greater than 6.0% and lower than 25 wt.% (Table 1), respectively. All of the nine regular core plugs, with ≈2.5 cm in diameter and 3–6 cm in length, were drilled from the relatively homogeneous section parallel to the sedimentary stratigraphy, and each sample was adequately cleaned in the solution of ethyl alcohol and chloroform in advance to remove residual bitumen. The complete core plug samples were first employed to measure porosity and permeability, and then, each core plug sample was cut into two parts, with a length of ≈1 cm and ≈2–5 cm, respectively. The larger part was first designed for SEM experiments on fresh and polished surfaces, before a 3–5 g powdered sample (60–80 mesh corresponding to particle size of 180–250 μm) was utilized for the N$_2$GA analysis in advance. Then, all of the larger part sample was crushed into finer powder to carry out an XRD analysis. The smaller part was directly applied to the RMIP analysis. The specific experimental procedures are shown in the following section.

Table 1. Petrophysical properties and mineral compositions of nine tight rock samples in the Xujiaweizi Rift.

Sample ID	Formation	Well ID	Depth (m)	Porosity (%)	Permeability ($\times10^{-15}$ m^2)	Mineral Compositions Obtained from XRD (wt.%)								
						Clay Minerals	Chlorite	Illite	I/S	%S	Quartz	Feldspar	Siderite	Calcite
#1	Denglouku	W1	2872.48	6.99	0.93	2	0.26	0.40	1.34	15	63	31	4	0
#2	Denglouku	W2	3012.47	10.0	2.35	4	1.16	0.64	2.20	15	45	45	6	0
#3	Denglouku	W2	3029.14	6.3	0.31	4	1.16	0.80	2.04	15	50	38	8	0
#4	Denglouku	W3	3079.47	8.8	0.26	8	2.88	1.44	3.68	15	54	38	0	0
#5	Denglouku	W3	3085.06	6.5	0.22	16	4.48	2.88	8.64	15	46	38	0	0
#6	Shahezi	W4	3938.31	8.7	0.13	6	4.50	0.78	0.72	10	30	64	0	0
#7	Shahezi	W5	4529.42	7.9	0.09	3	2.76	0.24	0.00	0	41	56	0	0
#8	Shahezi	W6	2772.21	8.8	0.05	24	4.08	3.36	16.56	15	54	19	0	3
#9	Shahezi	W7	3414.61	6.1	0.0352	22	5.50	3.08	13.42	20	60	13	0	5

Clay minerals = chlorite + illite + I/S; I/S = mixed-layer illite/smectite; %S = proportion of smectite in I/S.

3.2. Experimental Methods

3.2.1. RMIP

RMIP analyses were carried out using an ASPE-730 Automatic Mercury Intrusion Porosimeter (Coretest Systems, Inc., Closter, NJ, USA) at the Exploration and Development Research Institute of Daqing Oilfield Company Ltd. Tight rock samples were first cut into cubes, with a volume of <2.0 mL or a weight of <5.0 g, to avoid excessive experimental time, and then were dried to a constant weight at 105 °C before the RMIP experiment. To keep a quasistatic mercury intrusion, the intrusion rate of mercury was set as an extremely low fixed value of $\approx 5 \times 10^{-5}$ mL/min. The mercury intrusion pressure ranged from ≈ 0.025 MPa to ≈ 6.2 MPa, matching a throat radius range of ≈ 0.12–29.4 μm based on the Washburn equation that is shown as follows [36]:

$$r = \frac{2\sigma \cos \theta}{P_c} \tag{1}$$

where r is the throat radius; P_c is the mercury intrusion pressure; σ is the mercury interfacial tension, with a value of 485 mN/m [5]; and θ is the mercury wetting angle, with a value of 140° [5].

As shown in Figure 2A,B, the most important process that occurs repeatedly in the RMIP experiment is the drop and increase of mercury intrusion pressure. The sudden drop of mercury intrusion pressure, marked as rheon by Yuan and Swanson [37], represents the passage of nonwetting mercury from the narrower throats into the wider pores, and the pore-throat radius can be obtained from the initial pressure at the beginning of rheon process using Equation (1). The process of increasing mercury intrusion pressure can be further divided into two subprocesses, namely subison and rison, according to Yuan and Swanson [37]. The subison process mainly refers to the region of rising mercury intrusion pressure immediately after its sudden drop, and the increased pressure needs to be below the pressure at which the rheon process occurred (Figure 2A,B). Thus, a subison process represents a connected void space with lower capillary pressure, i.e., pore systems. Rison primarily refers to a region of continuously increasing mercury intrusion pressure, when the subison process finished (Figure 2A,B), until the next rheon occurs, and the increasing capillary pressure has never been reached previously. Therefore, rison usually corresponds to the filling of a connected void space with greater capillary pressure, i.e., throat systems. Based on the above process, we can obtain mercury intrusion curves within pores, throats, and total space (pores + throats), respectively, according to the fluctuations of mercury intrusion pressure. Several commonly used parameters, such as the average ratio of pore to throat radius (RPT_a) and mercury intrusion saturation in pores or throats (S_{pore} or S_{throat}) can be calculated using the following equation:

$$\begin{cases} RPT_a = \sum\limits_{i=1}^{n} RPT_i f_i \\ S_{pore} = \dfrac{V_{pore}}{\varphi V_{sample}}, \ S_{throat} = \dfrac{V_{throat}}{\varphi V_{sample}} \\ S_{total} = S_{pore} + S_{throat} \end{cases} \tag{2}$$

where f_i is the normalized distribution frequency of RPT_i; V_{pore} and V_{throat} are the mercury volumes intruded into pores and throats, respectively; φ is the porosity of tight rocks; and V_{sample} is the volume of tight rock sample used in the RMIP experiment.

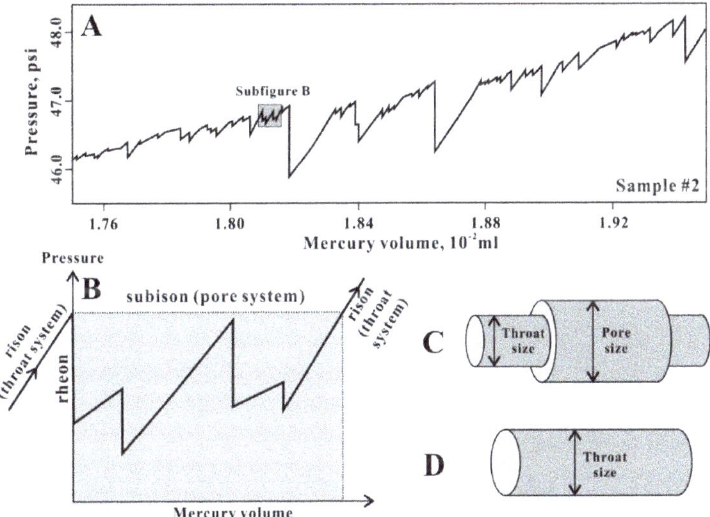

Figure 2. A small piece of raw RMIP data for sample #2 (**A**), mercury intrusion process analysis of RMIP (**B**), and the pore-throat structure models assumed in RMIP (**C**) and PMIP (pressure-controlled mercury intrusion porosimetry) (**D**) methods, respectively.

Compared with the RMIP method, pressure-controlled mercury intrusion porosimetry (PMIP) can only obtain one mercury intrusion curve [5], with a maximum mercury intrusion pressure of up to ≈400 MPa corresponding to a pore size of ≈3.7 nm [36]. The testing time used in the PMIP method is usually 2–3 h, evidently shorter than that spent on RMIP analyses (i.e., 1–2 d). In addition, in the RMIP method, porous media are usually assumed to be composed of pores and throats with different diameters (Figure 2C), which is more consistent with the fine pore structure characteristics of tight reservoirs [38]. However, the porous media primarily consist of capillary bundles with various diameters in the PMIP method (Figure 2D), and thus, this method cannot recognize the difference between pore and throat diameters.

3.2.2. N_2GA

N_2GA analyses were executed using the QUADRASORB-SI Surface Area and Pore Size Analyzer (Quantachrome Instruments Corp., Boynton Beach, FL, USA). The core plug samples were first powdered to 60–80 mesh, with a particle size of 180–250 μm, and then 3–5 g powdered samples were selected to desiccate under a vacuum environment at 110 °C for approximately 12 h to remove the vapor and capillary water ahead of the N_2GA experiments. The static adsorption capacity method was used to measure the amount of adsorbed nitrogen at 77.3 K. The experimental data were interpreted using multipoint BET (Brunauer–Emmett–Teller) and BJH (Barrett–Joyner–Halenda) to obtain specific surface area and pore size distribution, respectively, as comprehensively described by [39].

The BJH model depicts the capillary condensation phenomenon in a cylindrical pore based on the Kelvin equation [40], which shows the relationship between the relative pressure of nitrogen (P/P_0) and Kelvin radius (r_K), as exhibited in the following equation:

$$\ln\left(\frac{P}{P_0}\right) = -\frac{2\sigma V_L}{RTr_K} \tag{3}$$

where P/P_0 is the relative pressure of nitrogen; σ is the surface tension of liquid nitrogen at 8.85 mN/m [40]; V_L is the liquid molar volume of nitrogen at 34.65 mL/mol [40]; r_K is the Kelvin radius; R is the gas constant (8.314 J/mol/K); and T is the absolute temperature (77.3K).

When the specific parameter values used in the N$_2$GA experiment are substituted into Equation (3), we can obtain the following simplified equation:

$$r_K = -\frac{0.9543}{\ln(P/P_0)} \tag{4}$$

The average thickness of the adsorbed layer was determined by the Harkins–Jura equation for nitrogen:

$$t = 0.1 \times \left[\frac{13.99}{0.034 - \lg(P/P_0)}\right]^{1/2} \tag{5}$$

where t represents the average thickness of adsorbed layer.

Finally, we can obtain the pore size according to the following equation:

$$D_p = 2 \times (r_K + t) \tag{6}$$

where D_p represents the pore size.

3.2.3. XRD, SEM, Porosity, and Permeability

A Bruker D8 DISCOVER Advanced X-ray Diffractometer (Bruker Nano Inc., Madison, WI, USA) was used to examine the mineral components of tight rock samples, according to the Petroleum and Gas Industry Standard of China: SY/T 5163-2010. The determination of various mineral contents is based on their different X-ray diffraction peak intensity. SEM tests were measured using a TESCAN-MIRA-3XMU Scanning Electron Microscope (TESCAN, Brno, Czech Republic) to directly view the geometric shape of pores and clay minerals. Helium-derived porosity and nitrogen-derived permeability were measured on the core samples by the CoreLab CMS-300 automatic analyzer (Core Laboratories N.V., Houston, TX, USA) at a confining pressure of around 30 MPa, in order to simulate the actual formation conditions as much as possible, according to the SY/T 6385-2016 standard. The core samples were dried in a vacuum oven at 80 °C for approximately 8 h based on SY/T 5336-2006 in order to generate a minimum damage to tight rock samples and to ensure the accuracy of permeability measurement.

4. Results

4.1. Mineralogical and Petrophysical Properties

Porosity, permeability, and mineral components of nine tight rock samples are listed in Table 1. The results display that tight gas rock samples are primarily composed of quartz, feldspar, and clay minerals, and a small amount of siderite and calcite was also found. Quartz and feldspar are the predominant constituents (Table 1), with contents of 30–63 wt.% and 13–64 wt.%, respectively. The content of clay minerals is between 2 wt.% and 24 wt.% (Table 1), and further analyses show that clay minerals mainly contain I/S, chlorite, and illite, corresponding to contents of 0–16.56 wt.%, 0.26–5.5 wt.%, and 0.24–3.36 wt.%, respectively. Due to the great burial depth of the studied samples, the proportion of smectite in I/S is mainly ≈15% (Table 1), which is in an evolution process from R1 I/S (35% smectite) to R3 I/S (10% smectite), corresponding to a temperature of 170–180 °C [41,42]. Almost no kaolinite was detected in these samples, which may be attributed to the chemical reaction between kaolinite and K-feldspar at a relatively high formation temperature of >100 °C to form illite and aqueous silica [32,43,44], and the latter may precipitate to generate authigenic quartz aggregates or quartz overgrowth, when the formation conditions change. In the study area, the geothermal gradient is ≈4.4 °C/100 m based on Zhou et al. [45], and the average surface temperature is determined as ≈5 °C. Thus, the formation temperature obviously exceeds 100 °C when the burial depth of these selected samples is mainly >3000 m (Table 1). Noticeably, the average clay mineral content of the Denglouku Formation is generally lower than that of the Shahezi Formation, which is probably related to their different sedimentary environments.

The single crystal of illite in studied samples is present as ribbon or featheriness with a relatively regular arrangement (Figure 3A,B), generally between 0.15 and 0.5 μm, whereas the shape of their aggregates is mostly lamellar or fibrous being attached to pore walls, extending far into or completely across pores. In addition, the illite often coexists with feldspar, as a result of chemical reaction between feldspar and organic acid fluids [46]. The single crystal of chlorite is mainly needle-shaped with a dimension from 2 to 3 μm, while their aggregates appear as pompon-, flake-, or rose-like, and wrap the particle surface forming chlorite film (Figure 3C,D). This is favorable for the preservation of storage space in tight reservoirs by inhibiting the overgrowth of quartz [47]. I/S commonly occurs as honeycomb or cotton, primarily distributed in pores with discrete particles (Figure 3E,F).

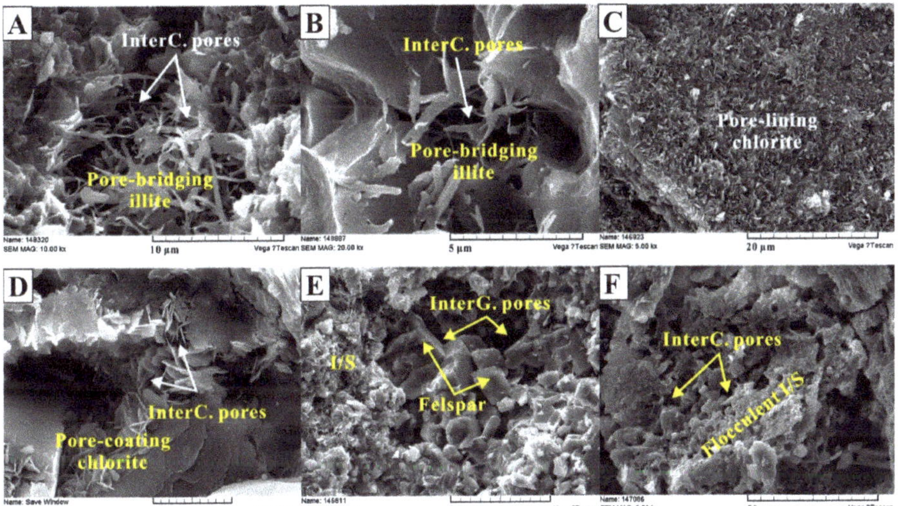

Figure 3. Morphologies of various types of clay minerals and pore types in the tight gas reservoirs of the Xujiaweizi Rift. InterC. = intercrystalline; InterG. = intergranular; I/S = mixed-layer illite/smectite. (**A**) Sample #6; (**B**) sample #6; (**C**) sample #8; (**D**) sample #8; (**E**) sample #1; (**F**) sample #2.

Under a confining pressure of ≈30 MPa, the helium porosities of tight gas rock samples are relatively low, mainly ranging from 6.1% to 10.0%, and the nitrogen permeability primarily varies from 0.0352×10^{-15} to 2.35×10^{-15} m^2. What calls for special attention is that no obvious positive correlation between porosity and permeability exists in studied samples (Table 1), unlike conventional sandstone reservoirs, generally characterized by better positive correlation between porosity and permeability [48]. Furthermore, we also found that the lower content of clay minerals agrees with a higher permeability (Table 1).

4.2. Storage Spaces

The morphology of pores in tight gas reservoirs is extremely complicated due to intense mechanical compaction and cementation. As exhibited in SEM images (Figures 3 and 4), the storage space of tight gas reservoirs in the Xujiaweizi Rift can be divided into three categories: intergranular pores, dissolution pores, and intercrystalline pores. Intergranular pores are primarily distributed between rigid particles (Figures 3E and 4A), such as quartz and feldspar, which have good resistance to compaction. In a polished SEM image, these pores are commonly presented as triangle or polygon with straight and smooth edges (Figure 4A), and they can be easily formed into residual intergranular pores due to the filling of argillaceous and siliceous cement, such as grain-coating chlorite and authigenic quartz (Figure 4B). Dissolution pores are principally related to unstable components (Figure 4C,D), including feldspar and carbonates, which are susceptible to organic acids. Compared with intergranular pores, these pores have the characteristics of a fairly anomalous pore shape and relatively good connectivity due to the presence of dissolution channels. Intercrystalline pores are mostly found within clay minerals (Figure 3), including I/S, chlorite, and illite. This kind of pores are numerous but are generally below 2 μm in size. These pores can contribute to a certain amount of storage space, but they contribute rather less to the permeability of tight gas reservoirs.

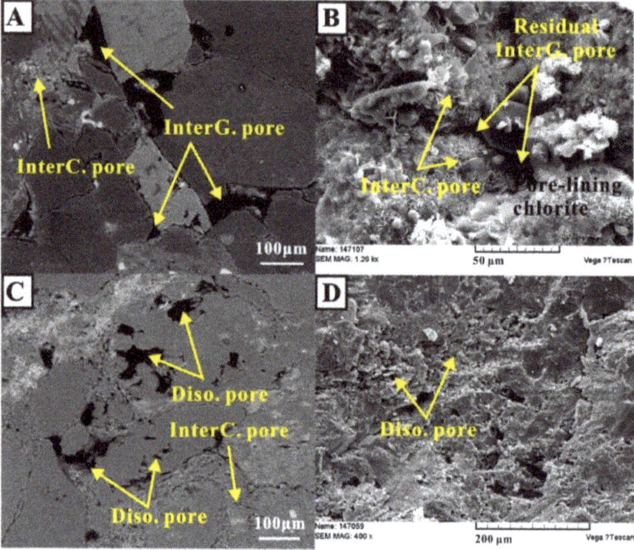

Figure 4. Pore types of tight gas reservoirs in the Xujiaweizi Rift, identified by SEM images. InterC. = intercrystalline; InterG. = intergranular; Diso. = Dissolution. (**A**) Sample #1; (**B**) sample #3; (**C**) sample #6; (**D**) sample #6.

4.3. Pore Structure Derived from N_2GA and RMIP Analyses

As shown in Figure 5, nitrogen adsorption isotherms of nine tight rock samples are attached to Type IV isotherm based on the classification scheme proposed by Brunauer et al. [49], which is a symbol of mesoporous materials (2–50 nm). The hysteresis loops produced by capillary condensation in pores larger than 2 nm are suitable for Type H3 according to IUPAC (International Union of Pure and Applied Chemistry) classification [50], which are usually interpreted to slit-shaped pores given by aggregates of ductile plate-like particles, mainly clay minerals, as exhibited in Figure 6. The above features are consistent with the observations of tight rocks from the Yanchang Formation in the Ordos Basin [19]. The pore volumes and specific surface areas of nine tight rock samples from the N_2GA analyses present a wide range, as listed in Table 2. Sample #8 shows the highest surface area at 3.41 m^2/g, while sample #7 shows the lowest at 0.213 m^2/g, with a mean of 1.23 m^2/g. The values of specific surface area are quite similar with the tight gas siltstone samples studied by Clarkson et al. [11]. Sample #8 corresponds to the largest pore volume at 0.010213 cm^3/g, whereas sample #3 has the smallest value of 0.002324 cm^3/g.

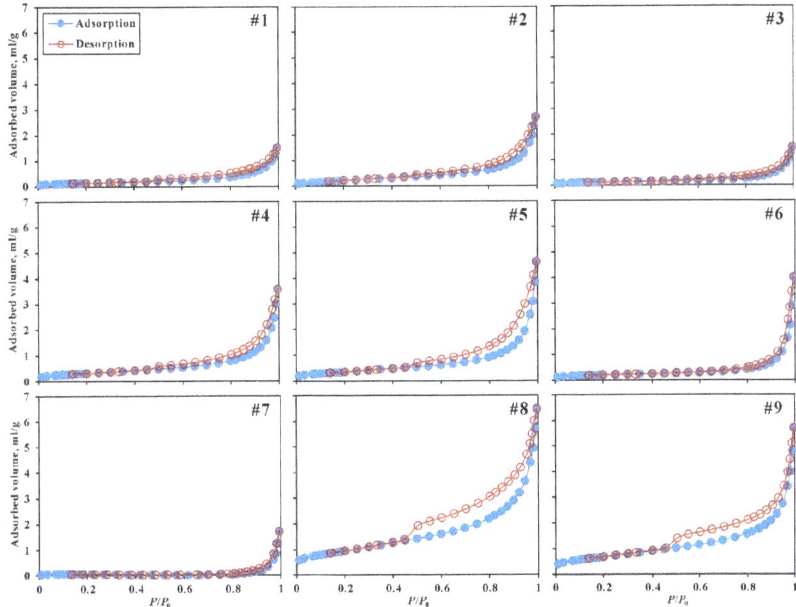

Figure 5. Nitrogen (N_2) adsorption–desorption isotherms of nine tight rock samples in the Xujiaweizi Rift.

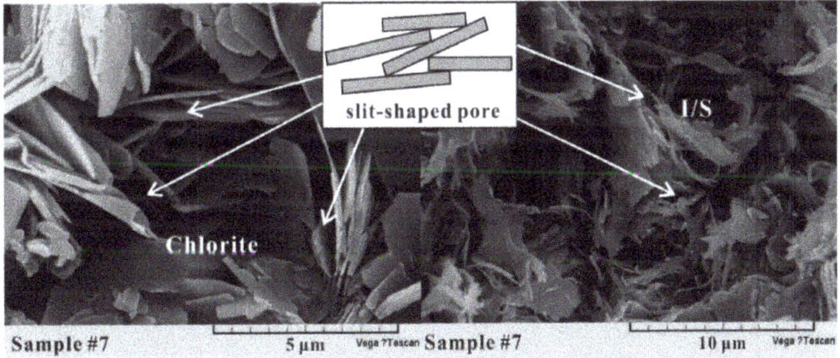

Figure 6. The slit-shaped pores related to the aggregates of plated-like clay minerals.

Table 2. Pore structure properties derived from N_2GA and RMIP experiments for nine tight rock samples in the Xujiaweizi Rift.

Sample ID	N_2GA Experiment		RMIP Experiment							
	SSA, m^2/g	Pore Volume, mL/g	S_{total}, %	S_{pore}, %	S_{throat}, %	RPT_a	r_a, µm	r_d, µm	r_m, µm	P_d, MPa
#1	0.5460	0.002369	65.70	25.44	40.26	86.78	2.179	2.682	1.011	0.274
#2	0.8351	0.004179	59.96	17.36	42.61	94.00	1.931	2.822	1.693	0.260
#3	0.4111	0.002324	63.73	9.46	54.27	81.10	1.918	2.913	0.121	0.252
#4	1.1573	0.005595	65.08	28.13	36.95	199.09	0.854	1.032	0.322	0.712
#5	1.3059	0.007224	49.77	8.02	41.75	235.40	0.671	0.658	0.276	1.118
#6	0.7477	0.006175	45.67	8.94	36.73	203.90	1.077	0.913	0.015	0.805
#7	0.2132	0.002652	54.77	19.11	35.66	296.64	0.560	0.766	0.235	0.959
#8	3.4067	0.010213	43.11	7.26	35.85	332.32	0.555	0.633	0.016	1.161
#9	2.4681	0.008955	26.25	1.65	24.60	178.56	0.611	0.307	0.021	2.396

SSA = specific surface area; S_{total} = total mercury intrusion saturation; S_{pore} = mercury intrusion saturation in pores; S_{throat} = mercury intrusion saturation in throats; RPT_a = ratio of pore to throat radius; r_a = average throat radius; r_d = maximum connected throat radius corresponding to displacement pressure; r_m = mainstream throat radius; P_d = displacement pressure.

The RMIP curves of the nine studied tight sandstones are displayed in Figure 7. At the initial period of mercury intrusion, the shape of total intrusion curves is more consistent with that of throat intrusion curves, and this coincident trend will be more evident with decreasing permeability (Figure 7). The total mercury intrusion saturation derived from the RMIP method ranges from 26.25% to 65.70%, averaging 52.67% (Table 2), and the mercury intrusion saturation in throats mainly ranges from 24.60% to 54.27% (38.74% on average), evidently higher than that in pores corresponding to 1.65%–28.13% (13.93% on average). Owing to the relatively low mercury intrusion pressure ($\approx$6.2 MPa), a large proportion of pores below 0.12 μm in radius cannot be well revealed by this method.

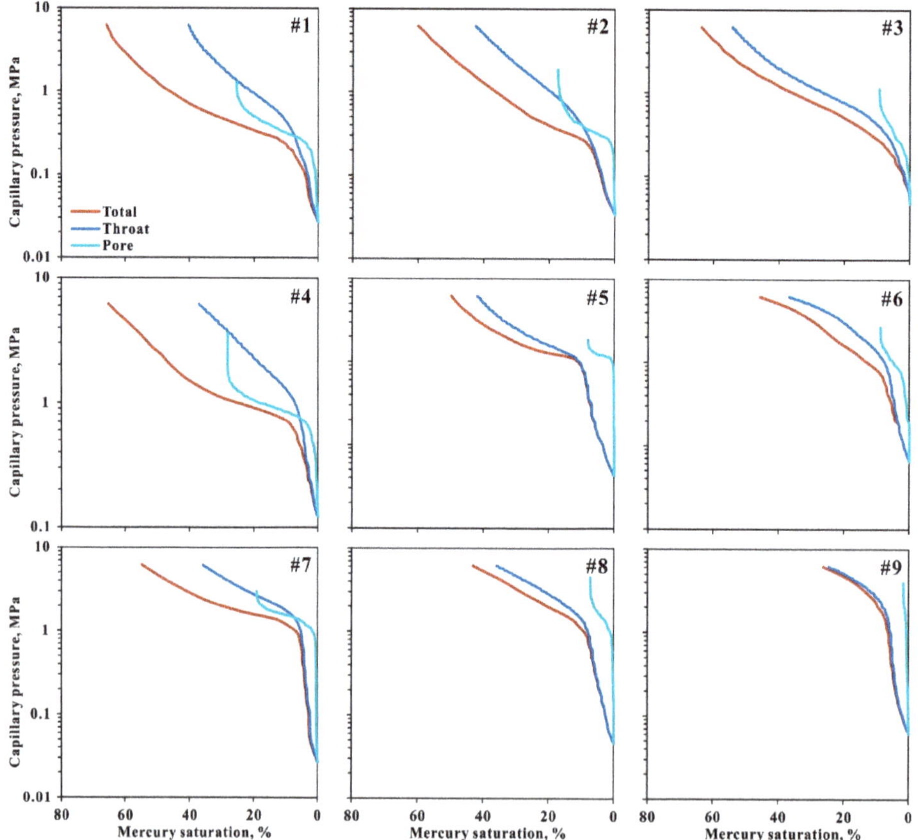

Figure 7. RMIP curves of total (pores + throats), pores and throats for the studied tight rock samples in the Xujiaweizi Rift.

As shown in Figure 8, the pore size of tight rocks obtained from N_2GA method mainly ranges from several nm to $\approx$200 nm (red columns). For the RMIP method, all of the tight rock samples exhibit similar pore size distributions primarily ranging from 200 to 600 μm in dimensions (blue columns), and the pore volumes show a sharp decrease with increasing clay mineral content (Figure 8), whereas their throat size distributions witness considerable differences (green columns), mainly between 0.24 and 6 μm (Figure 8). The average RPT_a is between 81.10 and 332.32 (Table 2), exhibiting a positive correlation with clay mineral content. Based on Figure 8, we also found that with increasing clay mineral contents and decreasing permeability, the throat distribution curves (black solid lines) are narrower, and the values of throat size drop gradually. In addition, several relevant throat radius parameters, such as r_a (average throat radius), r_d (maximum connected throat radius corresponding to displacement pressure), and r_m (mainstream throat radius), are positively correlated with permeability, as Figure 9 exhibits, which further indicates that the throat radius rather than pore radius controls the flow properties of tight rocks.

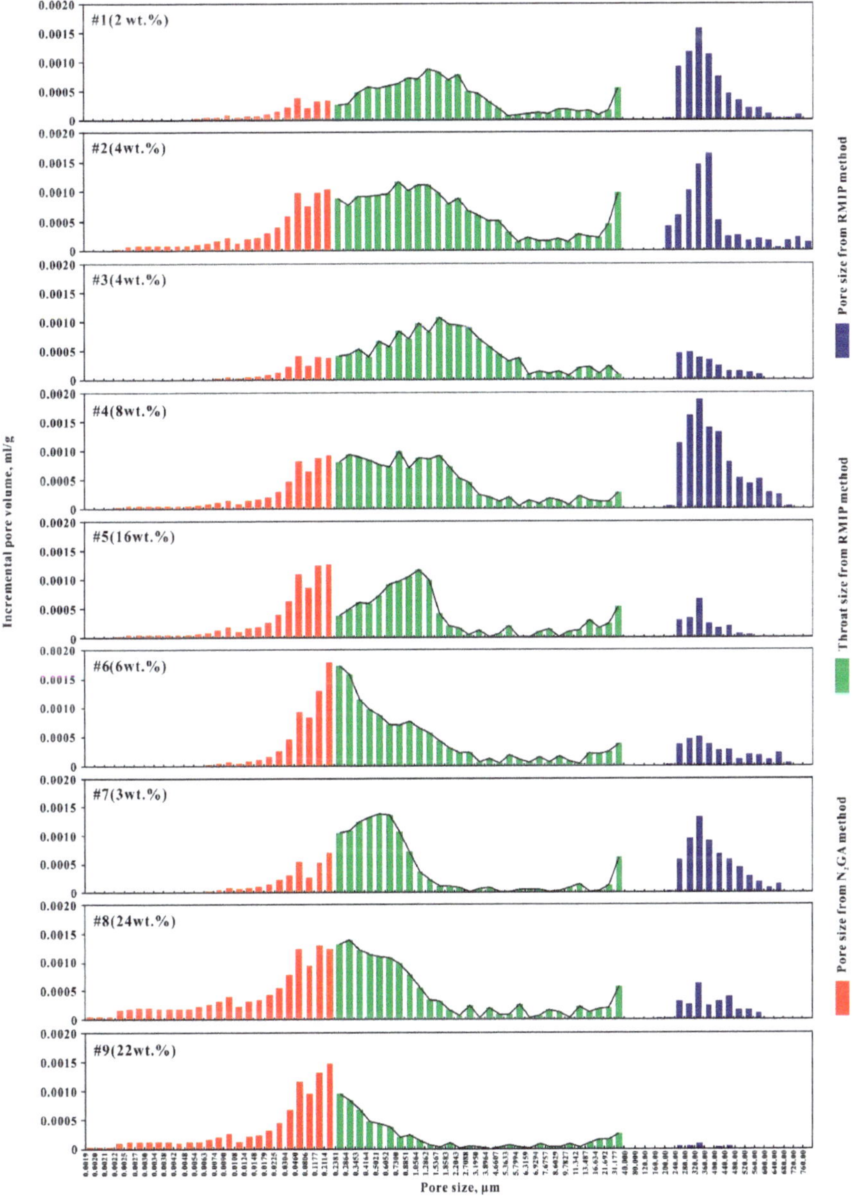

Figure 8. Pore size distributions of nine tight rock samples with various contents of clay minerals and permeability analyzed by nitrogen gas adsorption (N_2GA) and RMIP experiments. The number inside the parentheses is the content of clay minerals.

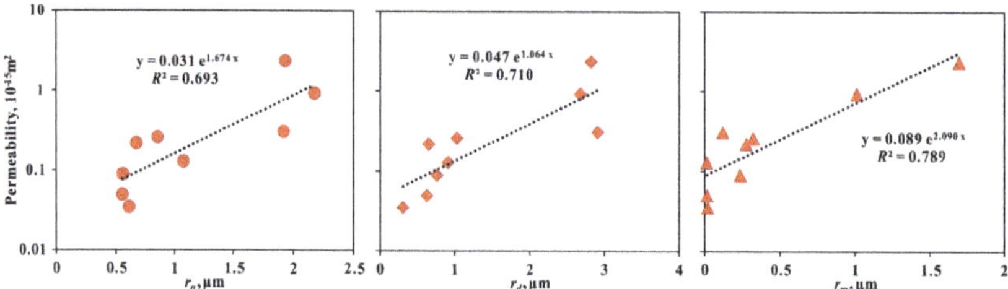

Figure 9. Correlations between permeability and r_a (average throat radius), r_d (maximum connected throat radius corresponding to displacement pressure), r_m (mainstream throat radius) obtained from the RMIP method.

5. Discussions

5.1. Classification of Pore Networks Based on Fractal Theory

Many studies have demonstrated that pore networks of clastic rocks have statistical self-similarity, characterized by their constantness with various scales [51–53]. Pfeifer and Avnir [54] proposed an equation for calculating surface fractal dimension using mercury intrusion data as follows:

$$\log\left(-\frac{dV_r}{dr}\right) \propto (2 - D_s)\log(r) \tag{7}$$

where V_r is the accumulated mercury intrusion volume when the throat radius larger than r, and D_s is defined as the surface fractal dimension. By plotting the dV_r/dr vs. r under double logarithmic coordinates, D_s can be determined through the slope of the fitting line.

The dV_r/dr vs. r obtained from RMIP data shows evident linear relationships (Figure 10A–C), illustrating that the pore-throat networks of studied tight rocks are generally fractal. Noticeably, the dV_r/dr vs. r curves can be divided into two segments at various throat radii for all tight rock samples, and the fractal dimension (D_{s2}) at the low-pressure section is generally greater than that at the high-pressure section (i.e., D_{s1}), which is consistent with studies of Lai and Wang [53]. This phenomenon is possibly attributed to (1) the skin impact related to the rough surface, (2) the presence of microfractures, or (3) the oversimplified assumption of cylindrical pores [53,55]. Evidently, the pore-throat networks corresponding to these two fractal segments cover disparate mercury intrusion features.

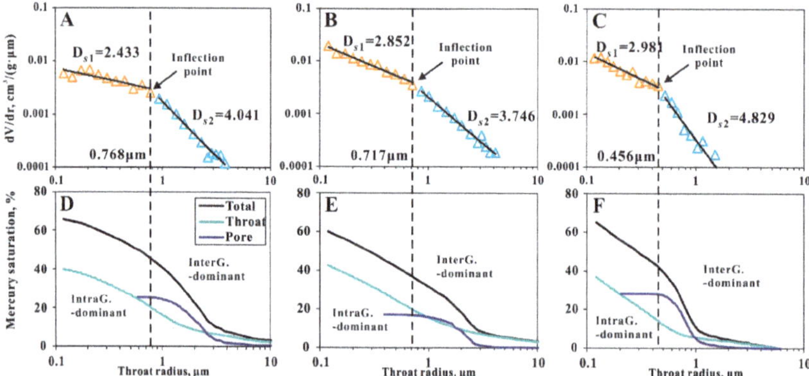

Figure 10. Surface fractal structure of throats (**A–C**) and mercury intrusion curves of total (pores + throats), pores and throats (**D–F**), derived from the RMIP experiment of three typical tight rock samples. (**A,D**): Sample #1, 0.93 mD; (**B,E**): sample #2, 2.35 mD; (**C,F**): sample #4, 0.26 mD.

In the wider throat part, the total mercury intrusion saturation goes up rapidly covering a confined scope of capillary pressure (Figure 10D–F). For example, the mercury intrusion saturation increases by 32.51%, comprising 49.49% of total mercury intrusion saturation, from 13.12% to 45.63%, when the intrusion pressure rises from 0.27 to 0.96 MPa in sample #1 (Figure 10D). Furthermore, the mercury intrusion process in pores is primarily completed at this stage, and the amount of mercury intruded into the pores in this stage accounts for 76% of total amount of mercury intruded into the pores for sample #1. Sakhaee-Pour and Bryant [56] regarded this type of pore-throat networks as conventional intergranular-dominant networks corresponding to large pores with wide throats (Figure 11A). As demonstrated by the polished SEM image in Figure 11C, intergranular-dominant networks mainly contain intergranular pores and dissolution pores with relatively larger dimensions, and slim pores between grains are treated as throats.

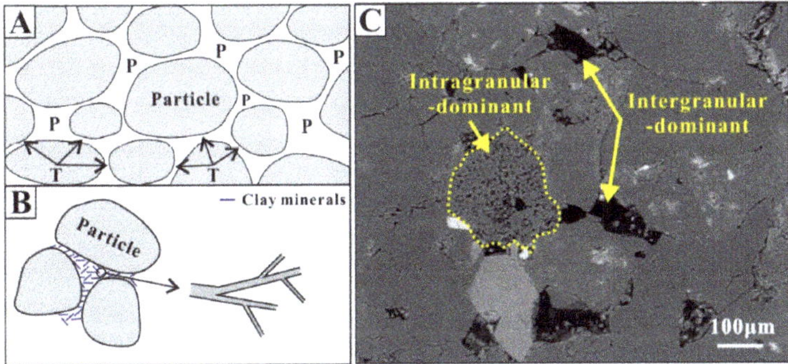

Figure 11. Schematic diagrams showing (**A**) conventional pore-throat structures that larger pores are connected by wider throats, and (**B**) tree-like pore structures that the narrower throats are connected to the wider throats like tree branches. (**C**) SEM image of sample #4 exhibiting an intergranular-dominant and intragranular-dominant pore network. P-pore; T-throat.

In the narrower throat part, the total mercury intrusion saturation grows almost exponentially with the capillary pressure at relatively high pressures, displaying nearly a straight line under semilogarithmic coordinates (Figure 10D–F). The intruded mercury in this stage is almost contributed by the mercury intrusion in throats, with nearly no increase of mercury invaded in the pores. For instance, the mercury intrusion saturation of throats makes up 99.28% of the total mercury intrusion in sample #1 at this stage, i.e., there is no longer any significant discrepancy in pores and throats. Intragranular-dominant pore networks were used to depict this region instead of the abovementioned intergranular-dominant networks, and moreover, Sakhaee-Pour and Bryant [56] adopted a tree-like pore network for the intragranular-dominant region. Intragranular-dominant networks mainly consist of throats with different sizes and lengths, and the narrower throats are connected to the wider throats like tree branches (Figure 11B). As shown in the SEM image in Figure 11C, intragranular-dominant networks predominantly correspond to intercrystalline pores within clay minerals.

The proportions of intergranular-dominant pore networks, determined by RMIP analyses using fractal theory, are listed in Table 3. The proportion of intergranular-dominant pore networks of studied tight samples is between 9.44% and 42.21%, with a mean of 27.45%, implying that intragranular-dominant pores contribute more to pore space. The proportion of intragranular-dominant pore networks is relatively high even for samples with a low clay minerals content. For example, the clay mineral contents of sample #1 and #7 are 2 wt.% and 3 wt.%, respectively, however, their intragranular-dominant pore network percentages still exceed 50%, up to 50.79% and 70.1%, respectively. This may be related to the fact that clay minerals are generally distributed in the intergranular and dissolution pores, resulting in a considerable number of pores associated with clay minerals.

Table 3. Lists of the contributions of intergranular-dominant pore networks to total pore space and permeability and the contributions of various compositions to SSA calculated from Equation (8) for the studied nine tight rocks.

Sample ID	r_{ip}, µm	Contribution of InterG.-dominant Pores to Pore Space, %	Contribution of InterG.-dominant Pores to Permeability, %	Measured SSA, m²/g	Calculated SSA, m²/g	Contributions of Various Compositions to SSA, %			
						Illite	Chlorite	I/S	FM
#1	0.768	42.21	88.14	0.5460	0.2578	9.53	7.48	80.67	2.32
#2	0.718	33.99	89.63	0.8351	0.4726	8.32	18.19	72.25	1.24
#3	0.634	35.07	93.40	0.4111	0.4576	10.74	18.79	69.19	1.28
#4	0.457	33.90	70.21	1.1573	0.8787	10.07	24.29	65.00	0.64
#5	0.467	21.96	62.73	1.3059	1.8550	9.54	17.90	72.29	0.28
#6	0.419	21.47	69.71	0.7477	0.4989	9.60	66.85	22.40	1.15
#7	0.305	29.90	75.19	0.2132	0.2252	6.55	90.83	0.00	2.63
#8	0.326	19.08	67.91	3.4067	3.0836	6.69	9.81	83.35	0.15
#9	0.307	9.44	14.45	2.4681	2.6844	7.05	15.19	77.59	0.18

r_{ip} = throat radius at inflection point; InterG.-dominant = intergranular-dominant; SSA = specific surface area; I/S = mixed-layer illite/smectite; FM = framework minerals.

The total clay minerals and various types of clay minerals are also correlated to the proportion of intergranular-dominant pore networks, as shown in Figure 12 and as the contents of total clay minerals, chlorite, illite, and I/S increase, the proportions of intergranular-dominant pore networks drop. Previous studies revealed that clay minerals are commonly derived from terrigenous detritus, namely primary clay minerals, and the chemical reaction between fluids and unstable minerals or the transformation between clay minerals, namely authigenic clay minerals [46,57]. Primary clay minerals are principally distributed in intergranular pores during the initial deposition stage, while authigenic clay minerals are dispersed in intergranular pores and dissolution pores during middle diagenesis stage. The increase of primary or authigenic clay mineral contents will obviously promote the evolution of intergranular-dominant pore networks (intergranular or dissolution pores) to intragranular-dominant pore networks (intercrystalline pores). We also found that chlorite is more closely correlated to the proportion of intergranular-dominant pores compared with illite and I/S (Figure 12B–D), illustrating that chlorite may be more effective in promoting the above pore evolution process, and thus may make more contribution to intragranular-dominant porosity. Although the increase of clay mineral content substantially promotes the evolution of intergranular-dominant pore networks to intragranular-dominant pore networks, and also it changes the relative proportions of these two types of pore networks, this does not necessarily result in an evident change in total storage space, due to the fact that there is unclear link between clay mineral content and total porosity (Table 1).

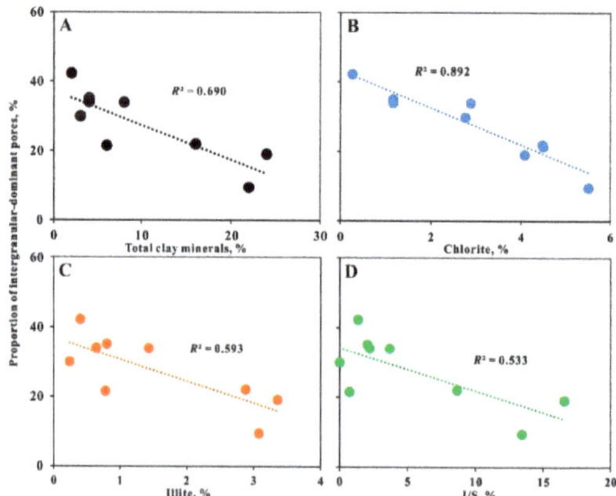

Figure 12. Relationships between contents of total clay mineral (**A**), chlorite (**B**), illite (**C**), I/S (**D**), and proportions of intergranular-dominant pores.

The throat radius at inflection point (r_{ip}), boundary of intergranular-dominant pores and intragranular-dominant pores, ranges from 0.305 to 0.768 μm (Table 3), indicating that intragranular-dominant pore networks of different samples correspond to inconsistent throat size distributions due to multiple degrees of diagenesis and different rock compositions [17,58,59]. The values of r_{ip} are negatively correlated with contents of total clay minerals, chlorite, illite, and I/S (Figure 13), and furthermore, chlorite has the relatively stronger control on r_{ip} (Figure 13B), while the correlations between r_{ip} and illite, I/S are fairly poor (Figure 13C,D).This implies that chlorite is more efficient in lowering values of r_{ip} in relative to illite and I/S, due to the fact that chlorite contributes more to intragranular-dominant pore networks, as evidenced by Figure 12.

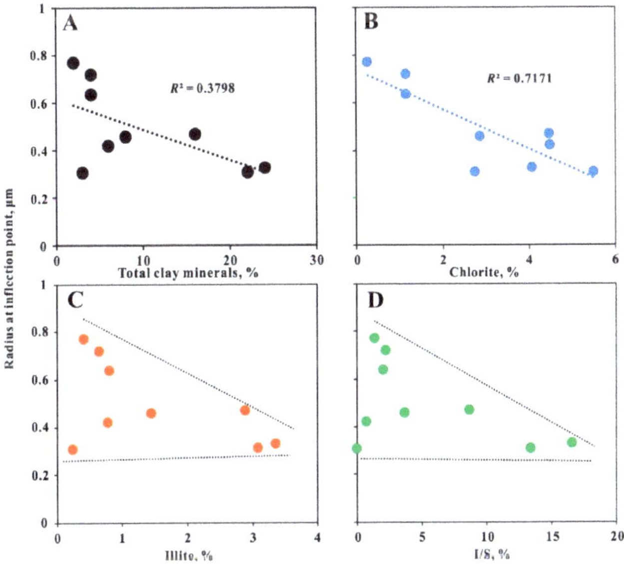

Figure 13. Throat radius at inflection point (r_{ip}) vs. contents of total clay minerals (**A**), chlorite (**B**), illite (**C**), and I/S (**D**).

5.2. Effect of Clay Minerals on Pore Structure Properties

The positive correlation between total clay minerals and SSA is pretty good, as exhibited in Figure 14A, indicating the strong control of clay minerals on SSA of tight rock samples. Specifically, both illite and I/S contents exhibit fine positive correlations with SSA, with a R^2 of 0.81 and 0.94 (Figure 14C,D), respectively, while there is a relatively poor positive relationship between chlorite and SSA (R^2 = 0.37, Figure 14B). In contrast to framework particles, such as quartz, feldspar, and calcite, clay minerals generally have colossal surface area due to their lamellar and plate-like crystal structure and small particles [19,20,60]. Thus, clay minerals are the main contributor to SSA of tight rocks, and the greater the clay mineral content, the greater the SSA. Furthermore, due to the existence of internal surface area, smectite usually corresponds to a total SSA of up to 800 m²/g, whereas the total SSA of illite and chlorite are relatively low at 30 m²/g and 15 m²/g, respectively, as reported by Passey et al. [20]. We can speculate that I/S, which is the intermediate product from smectite to illite [61], should also possess a greater SSA than that of illite and chlorite. Thus, the contribution of various clay minerals to the total SSA is distinct, which well interprets the differences in the correlation coefficients between different types of clay minerals and SSA (Figure 14B–D).

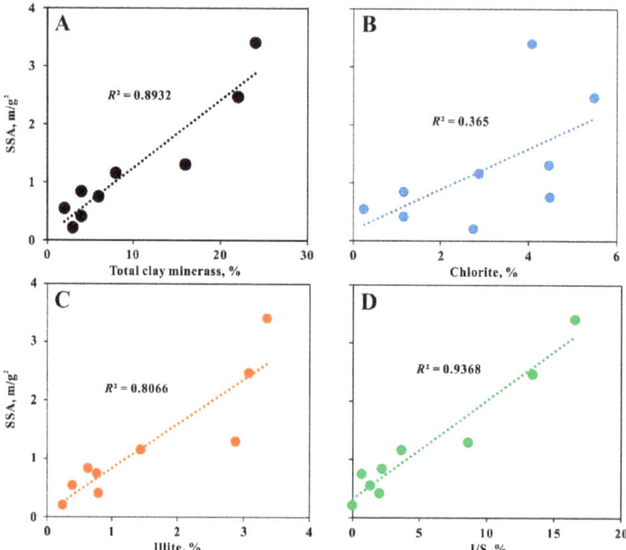

Figure 14. Positive correlations between SSA (specific surface area) derived from N_2GA experiments and contents of total clay minerals (**A**), chlorite (**B**), illite (**C**), I/S (**D**).

The total SSA can be regarded as the sum of SSAs contributed by various tight rock compositions, and therefore, this work proposes a mathematical model to quantitatively reveal the contributions of various clay minerals to total SSA based on the study of Wang et al. [62]. The compositions of tight rocks are first simplified to four categories: chlorite, illite, I/S, and framework minerals, considering that the SSAs of clay minerals are generally far greater than those of framework minerals. Then, the mathematical model can be written as follows:

$$\begin{cases} \sum_{i=1}^{n} \left(SSA_{Ch}W_{(Ch)i} + SSA_{I}W_{(I)i} + SSA_{I/S}W_{(I/S)i} + SSA_{FM}W_{(FM)i} \right) = SSA_i \\ W_{(Ch)i} + W_{(I)i} + W_{(I/S)i} + W_{(FM)i} = 1 \\ SSA_{Ch} > 0,\ SSA_{I} > 0,\ SSA_{I/S} > 0,\ SSA_{FM} > 0 \end{cases} \tag{8}$$

where SSA_{Ch}, SSA_{I}, $SSA_{I/S}$, and SSA_{FM} are the SSAs of chlorite, illite, I/S, and framework minerals per unit weight, respectively; W_{Ch}, W_{I}, $W_{I/S}$, and W_{FM} are the normalized weights of chlorite, illite, I/S, and framework minerals, respectively; and SSA_i is the total SSA of the ith tight rock samples obtained from the N_2GA experiment.

Based on the multiple linear regression method, we can obtain the optimized SSA_{Ch}, SSA_{I}, $SSA_{I/S}$, and SSA_{BM} of 7.412 m²/g, 6.143 m²/g, 15.520 m²/g, and 0.0061 m²/g, respectively, for the nine studied tight rock samples. The estimated SSAs show fine correlation with measured SSAs, with a R^2 of up to 0.92 (Figure 15), indicating that the proposed mathematical model is feasible and reasonable. Specifically, the contribution of I/S to total SSA is the largest, ranging from 0–83.35% with an average of 60.31% (Table 3), consistent with the good correlation between I/S and SSA in Figure 14D. The next is chlorite and illite, corresponding to a proportion of 7.48%–90.83% and 6.55%–10.74%, averaging 29.92% and 8.68% (Table 3), respectively. Framework minerals make the least contribution to total SSA, which is between 0.15% and 2.63%, with a mean of only 1.09% (Table 3).

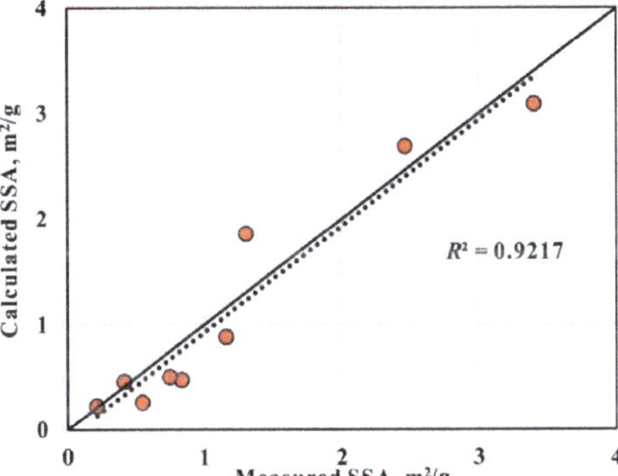

Figure 15. Correlation between measured and calculated SSAs for tight rock samples from the Xujiaweizi Rift. The black solid line represents the best match (1:1 line), and the black dashed line is the fitting line of measured and calculated SSAs. SSA = specific surface area.

Total clay minerals, illite, I/S, and chlorite are all positively associated with pore volumes derived from the N_2GA experiments, as evidenced by Figure 16. This is because the N_2GA technique can effectively reveal pore networks below 200 nm in diameter, which are dominantly linked with clay minerals, as identified by SEM images. It is worth noting that the R^2 of chlorite and pore volume is slightly lower than that of illite and of smectite with their respective pore volumes (Figure 16B–D), indicating that illite and I/S may have a more evident effect on the development of intragranular-dominant pore networks with diameter of smaller than 200 nm [23,59].

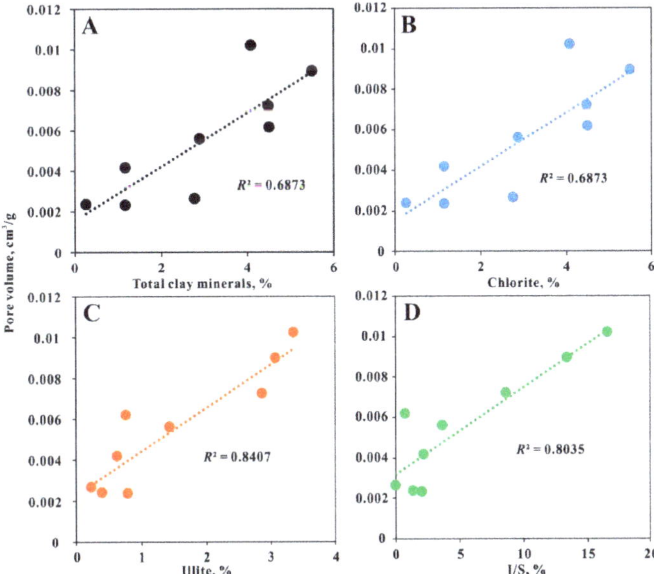

Figure 16. Positive correlations between pore volume derived from N_2GA experiments and contents of total clay minerals (**A**), chlorite (**B**), illite (**C**), I/S (**D**).

Both the average throat radius (r_a) and maximum connected throat radius (r_d, throat radius corresponding to displacement pressure) derived from the RMIP experiments are also negatively correlated with three types of clay minerals (Figure 17A–F). Illite usually occurs as pore-bridging in pore-throat space and can effectively segment pores and throats [21]. Pore-lining chlorite wraps the surface of primary intergranular pores resulting in a significant reduction of the pore-throat radius [21]. I/S distributes in pores/throats with discrete particles and can also block the throats to some extent [21]. Hence, all of these three kinds of clay minerals can lower the value of r_d. Evidently, chlorite is the most effective one to reduce r_a and r_d derived from the RMIP experiments. The RMIP experiment is more suitable for revealing pores larger than 240 nm in diameter in this study, and these pores are more closely related to chlorite comprising relatively larger pores.

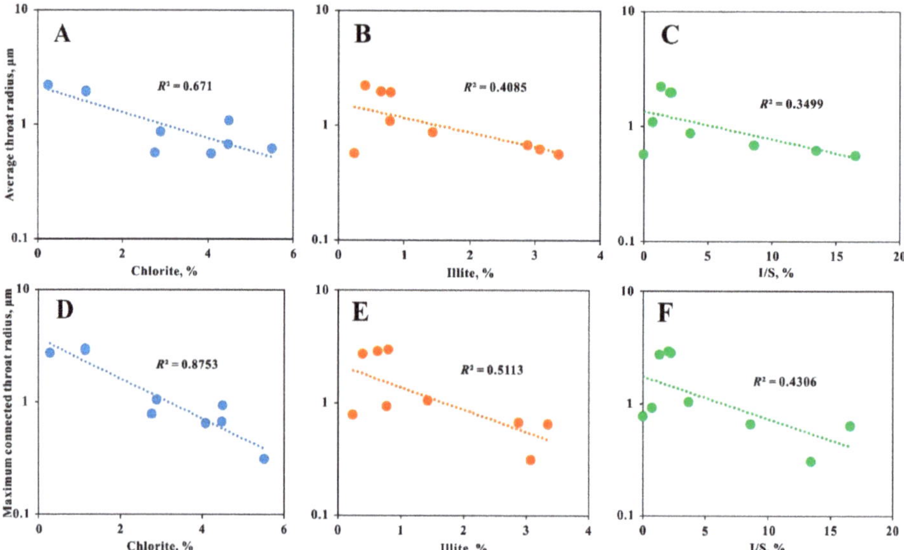

Figure 17. Negative relationships between average throat radius derived from RMIP experiments and contents of chlorite, illite, I/S (**A–C**), and negative associations between maximum connected throat radius derived from RMIP experiments and contents of chlorite, illite, I/S (**D–F**).

The integration of N_2GA and RMIP is more advantageous in uncovering the pore volumes of multiple scales, which can give a more reasonable result for this work (Figure 8). With the rising clay mineral contents from #1 to #8, the pore size distribution curves exhibit an increasing trend in the amplitude of a smaller pores section and a declining trend in that of a larger pores section but no evident change in the coverage areas of pore size distribution curves. This again confirms the evolution from intergranular-dominant pore networks to intragranular-dominant pore networks with increasing clay mineral content discussed in the above section. The diminution in the proportion of larger pores and the elevation in that of smaller pores will obviously damage the connectivity of pore networks, resulting in a worse seepage capacity of tight rocks [53]. Compared with clay mineral cements that give rise to the changes of the proportion of different pore networks, mechanical compaction will obviously reduce the absolute size and volume of all pores. For instance, for samples #7, intensely mechanical compaction results in a less proportion of larger pores (>1 µm in diameter), but the proportion of smaller pores (<0.1 µm in diameter) is comparable with that of tight rock samples with similar clay mineral contents (Figure 8), illustrating that the effect of mechanical compaction on smaller pores (<0.1 µm in diameter) within clay minerals is not significant, due to the protection of rigid particle support [63].

5.3. Effect of Clay Minerals on Permeability

Neasham [21] has demonstrated that the presence of clay minerals can significantly block pores and throats, which is considered to obviously reduce the permeability of the samples. Total clay minerals, chlorite, illite, and I/S are all negatively related to permeability for studied tight rock samples, as shown in Figure 18, and the effect of chlorite on the decrease of permeability is more evident, compared with illite and I/S. Previous studies have demonstrated that throat size rather than pore size controls the seepage capacity of tight reservoirs, and the permeability is primarily contributed by a small part of relatively larger pore-throat networks [12]. In addition, chlorite generally has a stronger control on the throat radius (Figure 17), and it makes a higher contribution to the intragranular-dominant pore networks of tight gas reservoirs (Figure 12), and thus, its effect on permeability is more obvious.

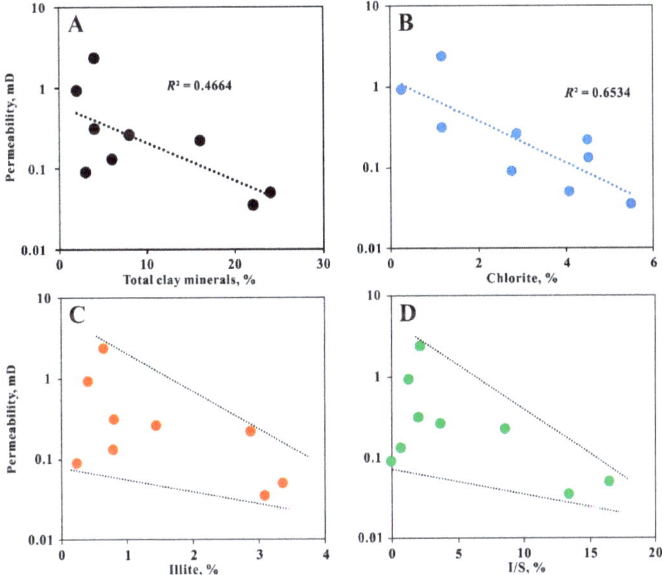

Figure 18. Relationships between contents of total clay minerals (**A**), chlorite (**B**), illite (**C**), I/S (**D**), and permeability.

Purcell et al. [64] proposed a method that was used to calculate the contribution of various scales of throats to the permeability based on mercury intrusion data, and this equation can be written as:

$$
K_j = \frac{\int_{S_j}^{S_{j+1}} r_{(S)}^2 \, dS}{\int_0^{S_{max}} r_{(S)}^2 \, dS}
\tag{9}
$$

where K_j is the contribution of throat radius at r_j to permeability; S_j and S_{j+1} are the mercury intrusion saturations at r_j and r_{j+1}, respectively; $r_{(S)}$ is the throat distribution functions; and dS represents the mercury intrusion saturation from r_j to r_{j+1}. S_{max} is the maximum mercury saturation.

The displacement pressure represents the minimum pressure at which the nonwetting phase fluids begin to significantly displace the wetting phase fluids in porous materials. When the mercury intrusion pressure is lower than displacement pressure, mercury will intrude into pores connected to the external surface of rock samples, which cannot form a continuous flow cluster in these isolated and discontinuous pores [65,66]. When the mercury intrusion pressure exceeds displacement pressure, incremental pores will be filled by mercury, generating increasingly continuous flow cluster. Therefore, the throat radii below the maximum connected throat radius are the primary contributors to the permeability of tight rocks.

The contribution of intergranular-dominant pore networks to the permeability of studied tight rock samples employing Equation (9) is listed in Table 3. For all samples except #9, intergranular-dominant pore networks rather than intragranular-dominant pore networks are the primary contributor to permeability, with a contribution ranging from 62.73% to 93.40%, which agrees with the study of Xi et al. [12]. We also found that with an increasing clay mineral content, the permeability of tight rocks decreases rapidly, while the contribution of intragranular-dominant pore networks to permeability rises speedily. The existence of excess clay minerals will bring about a quick decline of the number of interconnected intergranular-dominant pores, and the fluid has to select the intragranular-dominant pores as their main flow path. This will induce intragranular-dominant pores to play an increasingly significant role in the seepage process, but the absolute permeability value of tight rocks will decrease obviously. For example, the clay mineral content of sample #9 is up to 22 wt.%, with an intergranular-dominant pore proportion of 9.44% (Tables 1 and 3). Although the contribution of intragranular-dominant pore networks to permeability can reach 85.55%, its absolute permeability is only 0.0352 mD (Tables 1 and 3). Thus, compared with intragranular-dominant pore networks, intergranular-dominant pore networks are more crucial in generating a high permeability of tight gas reservoirs.

Based on the above discussion, clay minerals, especially authigenic clay minerals precipitated in intergranular pores and dissolution pores, significantly decreases the permeability of tight gas reservoirs. Nadeau et al. [67] proposed that less than 5% of authigenic I/S can effectively block the pore network and thus obviously reduce the reservoir permeability. Another possible consequence of the precipitation of authigenic clay minerals is that the relatively poor pore networks will result in an increasing risk of formation overpressure, especially for fine-grained sandstone and shale reservoirs. Formation overpressure can help preserve pore space of tight gas reservoirs to some extent, which may contribute to the relatively high porosity of studied tight rock samples with a burial depth of >3500 m (Table 1).

Many permeability prediction equations have been established based on pore structure parameters [68–70], which were obtained from PMIP (pressure-controlled mercury injection porosimetry) and NMR methods, and these prediction equations can be summarized as the following formula:

$$\text{Log}(K) = A + B\text{Log}(\varphi) + C\text{Log}(r_i) \tag{10}$$

where K is permeability; φ is porosity; and r_i is throat radius corresponding to the various total mercury saturation ($i = 10\%$–50%). A, B, and C can be determined according to multiple linear regression.

Evidently, there is no significant correlation between total porosity and permeability for the studied tight rocks (Figure 19A). According to the above discussions, permeability of tight rock samples is dominantly contributed by intergranular-dominant pore networks (Table 3), and the positive relationship between intergranular-dominant porosity ($\varphi_{interG.}$), and permeability is closer (Figure 19A). In addition, for tight rocks with various pore structures, the throat dimension corresponding to the same mercury intrusion saturation is significantly different, and it has different control levels on permeability. Thus, r_{ip} (throat radius at inflection point), instead of r_m (maximum connected throat radius), and r_a (average throat radius) were selected to substitute r_i mentioned above to estimate permeability, due to their well positive correlations with permeability (Figure 19B). Based on the multiple linear regression, the permeability estimation results are shown as follows:

$$\text{Log}(K) = 0.076 + 0.689\text{Log}(\varphi_{interG.}) + 2.841\text{Log}(r_{ip}) \qquad R^2 = 0.91 \tag{11}$$

$$\text{Log}(K) = -0.823 + 0.349\text{Log}(\varphi_{interG.}) + 1.293\text{Log}(r_m) \qquad R^2 = 0.77 \tag{12}$$

$$\text{Log}(K) = -0.96 + 0.991\text{Log}(\varphi_{interG.}) + 1.451\text{Log}(r_a) \qquad R^2 = 0.81 \tag{13}$$

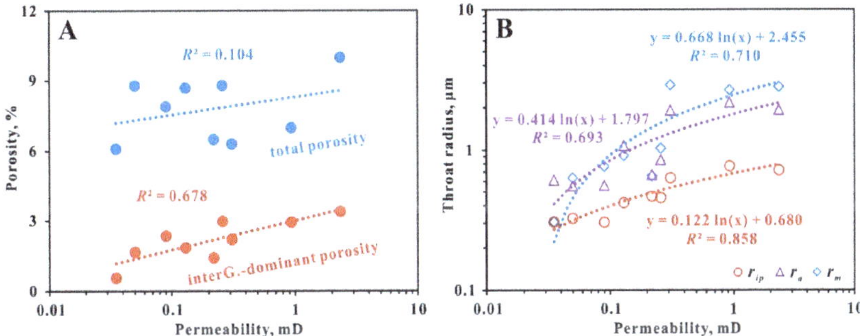

Figure 19. Correlations between permeability and porosity (**A**) and throat radius (**B**) parameters. InterG.-dominant = intergranular-dominant.

These three permeability estimation equations, Equations (11) to (13), indicate that r_{ip} is more appropriate than r_a and r_m in predicting permeability, as shown in Figure 20. Evidently, r_{ip} is the throat size boundary of intergranular-dominant and intragranular-dominant pore networks, as discussed in Section 5.1, representing the change from conventional pore-throat structures to tree-like pore structures of tight rocks, and the seepage capability of tight sandstones worsens rapidly. r_m and r_a also can reflect throat size distribution characteristic of tight rocks to some extent, and can also be applied to estimated permeability, with a R^2 of 0.77 and 0.81, respectively. However, these two parameters fail to reveal the critical throat size that changes pore structure and the permeability of tight rocks, and thus, their estimation accuracy is worse than that of r_{ip} (Figure 20).

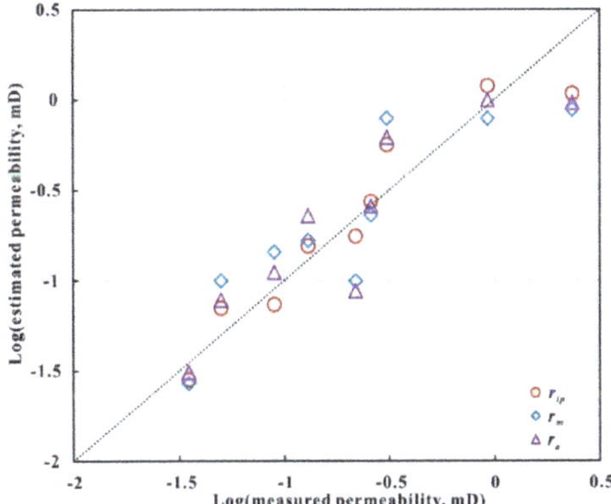

Figure 20. Measured permeability versus estimated permeability using r_{ip}, r_m, and r_a, respectively. The 1:1 line (black dashed line) exhibits the perfect match.

6. Conclusions

(1) Pore networks in tight gas reservoirs can be divided into intergranular-dominant and intragranular-dominant pore networks, based on surface fractal theory and mercury intrusion features. Intergranular-dominant pore networks correspond to the conventional pore-throat structure model that large pores are connected by wide throats, while intragranular-dominant pore networks are characterized by a tree-like pore structure that the narrower throats are connected to the upper-level wider throats like tree branches, and there is no significant difference between pores and throats.

(2)   Clay minerals are the primary contributor to total specific surface area (SSA) of tight gas reservoirs, among which I/S (mixed-layer illite/smectite) contributes to the most, followed by chlorite and illite, and the contribution of framework minerals is the least. Different types of clay minerals exert diverse degrees of influence on pore structures of tight gas reservoirs due to their dispersed model and morphology, and chlorite has the most evident effect on the reduction of the throat radius of tight rocks.

(3)   For tight gas reservoirs, intragranular-dominant pore networks contribute more to the total pore space, while intergranular-dominant pore networks control permeability. Clay minerals, especially authigenic chlorite, can effectively promote the evolution of intergranular-dominant to intragranular-dominant pore networks. Although the above process has little effect on the total pore space, it can significantly decrease the absolute value of permeability.

Author Contributions: Conceptualization, D.X.; funding acquisition, D.X., S.L., and G.C.; investigation, L.Z.; methodology, L.Z. and Y.Q.; project administration, S.L.; resources, S.L.; software, L.Z. and R.Z.; supervision, D.X., S.L. and S.J.; writing—original draft, L.Z.; writing—review and editing, S.J. and Y.S. All authors have read and agreed to the published version of the manuscript.

Funding: This research was funded by the National Natural Science Foundation of China (No.: 41802157, 41972139, 41902127 and 42072174), the National Science and Technology Major Project Foundation of China (No.: 2016ZX05061), the Fundamental Research Funds for the Central Universities (No.: 18CX02069A), the Initiative Postdocs Supporting Program (No.: BX20190303), the Postdoctoral Innovative Research Position of Hubei Province (No.: 247419), the Research Project of Guangxi Investment Group Co., LTD. (No.: NYHT2020-016), the Open Fund of Key Laboratory of Tectonics and Petroleum Resources (No.: TPR-2020-12), and the Project of Excellent Young Talents in Heilongjiang Province (No.: 140119002).

Conflicts of Interest: The authors declare no conflict of interest.

References

1.   Higgs, K.E.; Zwingmann, H.; Reyes, A.G.; Funnell, R.H. Diagenesis, Porosity Evolution, and Petroleum Emplacement in Tight Gas Reservoirs, Taranaki Basin, New Zealand. *J. Sediment. Res.* **2007**, *77*, 1003–1025. [CrossRef]

2.   Dai, J.; Ni, Y.; Wu, X. Tight gas in China and its significance in exploration and exploitation. *Pet. Explor. Dev.* **2012**, *39*, 277–284. [CrossRef]

3.   Zou, C.; Zhu, R.; Liu, K.; Su, L.; Bai, B.; Zhang, X.; Yuan, X.; Wang, J. Tight gas sandstone reservoirs in China: Characteristics and recognition criteria. *J. Pet. Sci. Eng.* **2012**, *88*, 82–91. [CrossRef]

4.   Zhang, L.; Xiao, D.; Lu, S.; Jiang, S.; Chen, L.; Guo, T.; Wu, L. Pore development of the Lower Longmaxi shale in the southeastern Sichuan Basin and its adjacent areas: Insights from lithofacies identification and organic matter. *Mar. Pet. Geol.* **2020**, *104662*, 104662. [CrossRef]

5.   Schmitt, M.; Fernandes, C.; Wolf, F.G.; Neto, J.A.B.D.C.; Rahner, C.P.; Dos Santos, V.S.S.; Rahner, M.S. Characterization of Brazilian tight gas sandstones relating permeability and Angstrom-to micron-scale pore structures. *J. Nat. Gas Sci. Eng.* **2015**, *27*, 785–807. [CrossRef]

6.   Nelson, P.H. Pore-throat sizes in sandstones, tight sandstones, and shales. *AAPG Bull.* **2009**, *93*, 329–340. [CrossRef]

7.   Desbois, G.; Urai, J.L.; Kukla, P.A.; Konstanty, J.; Baerle, C. High-resolution 3D fabric and porosity model in a tight gas sandstone reservoir:A new approach to investigate microstructures from mm- to nm-scale combining argon beam cross-sectioning and SEM imaging. *J. Pet. Sci. Eng.* **2011**, *78*, 243–257. [CrossRef]

8.   Rezaee, M.; Saeedi, A.; Clennell, B. Tight gas sands permeability estimation from mercury injection capillary pressure and nuclear magnetic resonance data. *J. Pet. Sci. Eng.* **2012**, *88*, 92–99. [CrossRef]

9.   Xiao, D.; Jiang, S.; Thul, D.; Lu, S.; Zhang, L.; Li, B. Impacts of clay on pore structure, storage and percolation of tight sandstones from the Songliao Basin, China: Implications for genetic classification of tight sandstone reservoirs. *Fuel* **2018**, *211*, 390–404. [CrossRef]

10.   Lai, J.; Wang, G.; Wang, Z.; Chen, J.; Pang, X.; Wang, S.; Zhou, Z.; He, Z.; Qin, Z.; Fan, X. A review on pore structure characterization in tight sandstones. *Earth-Sci. Rev.* **2018**, *177*, 436–457. [CrossRef]

11. Clarkson, C.; Solano, N.; Bustin, R.; Bustin, A.; Chalmers, G.R.; He, L.; Melnichenko, Y.; Radlinski, A.; Blach, T. Pore structure characterization of North American shale gas reservoirs using USANS/SANS, gas adsorption, and mercury intrusion. *Fuel* **2013**, *103*, 606–616. [CrossRef]

12. Xi, K.; Cao, Y.; Haile, B.G.; Zhu, R.; Jahren, J.; Bjørlykke, K.; Zhang, X.; Hellevang, H. How does the pore-throat size control the reservoir quality and oiliness of tight sandstones? The case of the Lower Cretaceous Quantou Formation in the southern Songliao Basin, China. *Mar. Pet. Geol.* **2016**, *76*, 1–15. [CrossRef]

13. Zhao, H.; Ning, Z.; Wang, Q.; Zhang, R.; Zhao, T.; Niu, T.; Zeng, Y. Petrophysical characterization of tight oil reservoirs using pressure-controlled porosimetry combined with rate-controlled porosimetry. *Fuel* **2015**, *154*, 233–242. [CrossRef]

14. Zhang, L.; Li, B.; Jiang, S.; Xiao, D.; Lu, S.; Zhang, Y.; Gong, C.; Chen, L. Heterogeneity characterization of the lower Silurian Longmaxi marine shale in the Pengshui area, South China. *Int. J. Coal Geol.* **2018**, *195*, 250–266. [CrossRef]

15. Zhang, L.; Lu, S.; Jiang, S.; Xiao, D.; Chen, L.; Liu, Y.; Zhang, Y.; Li, B.; Gong, C. Effect of Shale Lithofacies on Pore Structure of the Wufeng–Longmaxi Shale in Southeast Chongqing, China. *Energy Fuels* **2018**, *32*, 6603–6618. [CrossRef]

16. Chen, L.; Jiang, Z.; Liu, K.; Tan, J.; Gao, F.; Wang, P. Pore structure characterization for organic-rich Lower Silurian shale in the Upper Yangtze Platform, South China: A possible mechanism for pore development. *J. Nat. Gas Sci. Eng.* **2017**, *46*, 1–15. [CrossRef]

17. Chen, S.; Han, Y.; Fu, C.; Zhang, H.; Zhu, Y.; Zuo, Z.; Shangbin, C.; Yufu, H.; Changqin, F.; Han, Z.; et al. Micro and nano-size pores of clay minerals in shale reservoirs: Implication for the accumulation of shale gas. *Sediment. Geol.* **2016**, *342*, 180–190. [CrossRef]

18. Ola, P.S.; Aidi, A.K.; Bankole, O.M. Clay mineral diagenesis and source rock assessment in the Bornu Basin, Nigeria: Implications for thermal maturity and source rock potential. *Mar. Pet. Geol.* **2018**, *89*, 653–664. [CrossRef]

19. Ross, D.J.; Bustin, R.M. The importance of shale composition and pore structure upon gas storage potential of shale gas reservoirs. *Mar. Pet. Geol.* **2009**, *26*, 916–927. [CrossRef]

20. Passey, Q.R.; Bohacs, K.; Esch, W.L.; Klimentidis, R.; Sinha, S. From oil-prone source rock to gas-producing shale reservoir-geologic and petrophysical characterization of unconventional shale gas reservoirs. In Proceedings of the International Oil and Gas Conference and Exhibition, Beijing, China, 8–10 June 2010.

21. Neasham, J.W. The morphology of dispersed clay in sandstone reservoirs and its effect on sandstone shaliness, pore space and fluid flow properties. In Proceedings of the Annual Fall Technical Conference and Exhibition, Denver, CO, USA, 9–12 October 1977.

22. Rahman, M.J.J.; Worden, R.H. Diagenesis and its impact on the reservoir quality of Miocene sandstones (Surma Group) from the Bengal Basin, Bangladesh. *Mar. Pet. Geol.* **2016**, *77*, 898–915. [CrossRef]

23. Ji, L.; Zhang, T.; Milliken, K.L.; Qu, J.; Zhang, X. Experimental investigation of main controls to methane adsorption in clay-rich rocks. *Appl. Geochem.* **2012**, *27*, 2533–2545. [CrossRef]

24. Cao, Z.; Liu, G.; Zhan, H.; Li, C.; You, Y.; Yang, C.; Jiang, H. Pore structure characterization of Chang-7 tight sandstone using MICP combined with N2GA techniques and its geological control factors. *Sci. Rep.* **2016**, *6*, 36919. [CrossRef] [PubMed]

25. Yang, F.; Ning, Z.F.; Zhang, S.D.; Hu, C.P.; Du, L.H.; Liu, H.Q. Characterization of pore structures in shales through nitrogen adsorption experiment. *Nat. Gas Ind.* **2013**, *33*, 135–140.

26. Feng, Z.; Yin, C.; Lu, J.; Zhu, Y. Formation and accumulation of tight sandy conglomerate gas: A case from the Lower Cretaceous Yingcheng Formation of Xujiaweizi fault depression, Songliao Basin. *Pet. Explor. Dev.* **2013**, *40*, 696–703. [CrossRef]

27. Zhao, Z.; Xu, S.; Jiang, X.; Lin, C.; Cheng, H.; Cui, J.; Jia, L. Deep strata geologic structure and tight sandy conglomerate gas exploration in Songliao Basin, East China. *Pet. Explor. Dev.* **2016**, *43*, 13–25. [CrossRef]

28. Zhang, E.H.; Jiang, C.J.; Zhang, Y.G.; Li, Z.A.; Feng, X.Y.; Wu, J. Study on the formation and evolution of deep structure of Xujiaweizi fault depression. *Acta Petrol. Sin.* **2010**, *26*, 149–157, (In Chinese with English abstract).

29. Fu, G.; Wu, W.; Li, N. Controlling effect of three major faults on gas accumulation in the Xujiaweizi faulted depression, Songliao Basin. *Nat. Gas Ind. B* **2014**, *1*, 159–164. [CrossRef]

30. Cai, Q.; Hu, M.; Ngia, N.R.; Hu, Z. Sequence stratigraphy, sedimentary systems and implica-tions for hydrocarbon exploration in the northern Xujiaweizi Fault Depression, Songliao Basin, NE China. *J. Pet. Sci. Eng.* **2017**, *152*, 471–494. [CrossRef]

31. Liu, C.; Yin, C.; Lu, J.; Sun, L.; Wang, Y.; Hu, B.; Li, J. Pore structure and physical properties of sandy conglomerate reservoirs in the Xujiaweizi depression, northern Songliao Basin, China. *J. Pet. Sci. Eng.* **2020**, *192*, 107217. [CrossRef]

32. Lu, S.F.; Gu, M.W.; Zhang, F.F.; Zhang, L.C.; Xiao, D.S. Hydrocarbon accumulation stages and type division of Shahezi Fm. tight glutenite gas reservoirs in the Xujiaweizi Fault Depression, Songliao Basin. *Nat. Gas Ind.* **2017**, *37*, 12–21.

33. Li, R.F.; Yang, Y.Q.; Zhang, G.X.; Ruan, X.F. Systematic restoration of eroded thickness by unconformity in Cretaceous of Xujiaweizi in the north of Songliao Basin. *Earth Sci.* **2012**, *37*, 47–54. (In Chinese with English abstract)

34. Feng, Z.; Cheng-Zao, J.; Xi-Nong, X.; Shun, Z.; Zi-Hui, F.; Cross, T.A. Tectonostratigraphic units and stratigraphic sequences of the nonmarine Songliao basin, northeast China. *Basin Res.* **2010**, *22*, 79–95. [CrossRef]

35. Zhang, P.; Lee, Y.I.; Zhang, J. Diagenesis of tight-gas sandstones in the lower cretaceous denglouku formation, songliao basin, ne china: Implications for reservoir quality. *J. Pet. Geol.* **2014**, *38*, 99–114. [CrossRef]

36. Washburn, E.W. The Dynamics of Capillary Flow. *Phys. Rev.* **1921**, *17*, 273–283. [CrossRef]

37. Yuan, H.H.; Swanson, B.F. Resolving pore-space characteristics by rate-controlled po-rosimetry. *SPE Form. Eval.* **1989**, *4*, 17–24. [CrossRef]

38. He, S.L.; Jiao, C.Y.; Wang, J.G.; Luo, F.P.; Zou, L. Discussion on the differences between constant-speed mercury injection and conventional mercury injection techniques. *Fault-Block Oil Gas Field* **2011**, *18*, 235–237. (In Chinese with English abstract)

39. Gregg, S.J.; Sing, K.S.W.; Salzberg, H.W. *Adsorption, Surface Area, and Porosity*, 2nd ed.; Academic Press: New York, NY, USA, 1982.

40. Barrett, E.P.; Joyner, L.G.; Halenda, P.P. The determination of pore volume and area distributions in porous substances. I. Computations from nitrogen isotherms. *J. Am. Chem. Soc.* **1951**, *73*, 373–380. [CrossRef]

41. Hoffman, J.; Hower, J.; Scholle, P.A.; Schluger, P.R. Clay mineral assemblages as low grade metamorphic geothermometers: application to the thrust faulted disturbed belt of montana, U.S.A. *SEPM Spec. Publ.* **1979**, *26*, 55–79. [CrossRef]

42. Pollastro, R.M. Considerations and Applications of the Illite/Smectite Geothermometer in Hydrocarbon-Bearing Rocks of Miocene to Mississippian Age. *Clays Clay Miner.* **1993**, *41*, 119–133. [CrossRef]

43. Bjorkum, N.G.P.A. An Isochemical Model for Formation of Authigenic Kaolinite, K-Feldspar and Illite in Sediments. *J. Sediment. Res.* **1988**, *58*. [CrossRef]

44. Ehrenberg, S.N.; Nadeau, P.H. Formation of Diagenetic Illite in Sandstones of the Garn Formation, Haltenbanken Area, Mid-Norwegian Continental Shelf. *Clay Miner.* **1989**, *24*, 233–253. [CrossRef]

45. Zhou, Q.; Feng, Z.; Men, G. Present geothermal characteristics and their relationship with natural gas generation in the Xujiaweizi Rift of the northern Songliao Basin. *Sci. China Earth Sci.* **2007**, *37*, 177–188.

46. Yuan, G.; Gluyas, J.; Cao, Y.; Oxtoby, N.H.; Jia, Z.; Wang, Y.; Xi, K.; Li, X. Diagenesis and reservoir quality evolution of the Eocene sandstones in the northern Dongying Sag, Bohai Bay Basin, East China. *Mar. Pet. Geol.* **2015**, *62*, 77–89. [CrossRef]

47. Aagaard, P.; Jahren, J.S.; Harstad, A.O.; Nilsen, O.; Ramm, M. Formation of grain-coating chlorite in sandstones. Laboratory synthesized vs. natural occurrences. *Clay Miner.* **2000**, *35*, 261–269. [CrossRef]

48. Pang, X.J.; Wang, Q.B.; Feng, C.; Zhao, M.; Liu, Z.B. Differences and Genesis of High-quality Reservoirs in Es1+2 at the Northern Margin of the Huanghekou Sag, Bohai Sea. *Acta Sedimentol. Sin.* **2020**. (In Chinese with English abstract) [CrossRef]

49. Brunauer, S.; Deming, L.S.; Deming, W.E.; Teller, E. On a Theory of the van der Waals Adsorption of Gases. *J. Am. Chem. Soc.* **1940**, *62*, 1723–1732. [CrossRef]

50. Sing, K.S.; Everett, D.H.; Haul, R.A.W.; Moscou, L.; Pierotti, R.A.; Rouquerol, J.; Siemieniewsha, T. Reporting physisorption data for gas/solid systems with special reference to the determination of surface area and porosity (Recommendations 1984). *Pure Appl. Chem.* **1985**, *57*, 603–619. [CrossRef]

51. Yao, Y.; Liu, D.; Tang, D.; Tang, S.; Huang, W.; Liu, Z.; Che, Y. Fractal characterization of seepage-pores of coals from China: An investigation on permeability of coals. *Comput. Geosci.* **2009**, *35*, 1159–1166. [CrossRef]

52. Wang, H.; Liu, Y.; Song, Y.; Zhao, Y.; Zhao, J.; Wang, D. Fractal analysis and its impact factors on pore structure of artificial cores based on the images obtained using magnetic resonance imaging. *J. Appl. Geophys.* **2012**, *86*, 70–81. [CrossRef]

53. Lai, J.; Wang, G. Fractal analysis of tight gas sandstones using high-pressure mercury intrusion techniques. *J. Nat. Gas Sci. Eng.* **2015**, *24*, 185–196. [CrossRef]

54. Pfeifer, P.; Avnir, D. Chemistry in non integer dimensions between 2 and 3, I: Fractal theory of heterogenous surface. *J. Chem. Phys.* **1983**, *79*, 3558–3565. [CrossRef]

55. Li, K.; Horne, R.N. Fractal modeling of capillary pressure curves for The Geysers rocks. *Geothermics* **2006**, *35*, 198–207. [CrossRef]

56. Sakhaee-Pour, A.; Bryant, S.L. Effect of pore structure on the producibility of tight-gas sandstones. *AAPG Bull.* **2014**, *98*, 663–694. [CrossRef]

57. Yuan, G.H.; Cao, Y.C.; Xi, K.L.; Wang, Y.Z.; Li, X.Y.; Yang, T. Feldspar dissolution and its impact on physical properties of Paleogene clastic reservoirs in the northern slope zone of the Dongying Sag. *Acta Pet. Sin.* **2013**, *34*, 853–866, (In Chinese with English abstract).

58. Wang, X. Characteristics of clay minerals and their effects on production capacity of the Cretaceous sandstone reservoirs of Songliao Basin, China. *Sci. Géologiques Bull. Mémoires* **1990**, *88*, 105–114.

59. Wang, M.Z.; Liu, S.B.; Ren, Y.J.; Tian, H. Pore characteristics and methane adsorption of clay minerals in shale gas reservoir. *Geol. Rev.* **2015**, *61*, 207–216, (In Chinese with English abstract).

60. Omotoso, O.E.; Mikula, R.J. High surface areas caused by smectitic interstratification of kaolinite and illite in athabasca oil sands. *Appl. Clay Sci.* **2004**, *25*, 37–47. [CrossRef]

61. Boles, S.G.F.J.R. Clay Diagenesis in Wilcox Sandstones of Southwest Texas: Implications of Smectite Diagenesis on Sandstone Cementation. *J. Sediment. Res.* **1979**, *49*. [CrossRef]

62. Wang, Y.; Dong, D.; Yang, H.; He, L.; Wang, S.; Huang, J.; Pu, B.; Wang, S. Quantitative characterization of reservoir space in the Lower Silurian Longmaxi Shale, southern Sichuan, China. *Sci. China Earth Sci.* **2013**, *57*, 313–322. [CrossRef]

63. Mbia, E.N.; Fabricius, I.L.; Krogsbøll, A.; Frykman, P.; Dalhoff, F. Permeability, compressibility and porosity of Jurassic shale from the Norwegian–Danish Basin. *Pet. Geosci.* **2014**, *20*, 257–281. [CrossRef]

64. Purcell, W. Capillary Pressures—Their Measurement Using Mercury and the Calculation of Permeability Therefrom. *J. Pet. Technol.* **1949**, *1*, 39–48. [CrossRef]

65. Katz, A.J.; Thompson, A.H. Quantitative prediction of permeability in porous rock. *Phys. Rev. B* **1986**, *34*, 8179–8181. [CrossRef]

66. Daigle, H.; Johnson, A. Combining Mercury Intrusion and Nuclear Magnetic Resonance Measurements Using Percolation Theory. *Transp. Porous Media* **2015**, *111*, 669–679. [CrossRef]

67. Nadeau, P.H.; Peacor, D.R.; Yan, J.; Hillier, S. IS precipitation in pore space as the cause of geopressuring in Mesozoic mudstones, Egersund Basin, Norwegian continental shelf. *Am. Miner.* **2002**, *87*, 1580–1589. [CrossRef]

68. Pittman, E.D. Relationship of Porosity and Permeability to Various Parameters Derived from Mercury Injection-Capillary Pressure Curves for Sandstone (1). *AAPG Bull.* **1992**, *76*, 191–198. [CrossRef]

69. Coates, G.R.; Xiao, L.; Prammer, M.G. *NMR Logging Principles and Applications, Halliburton Energy Services Publication H02308*; Halliburton: Houston, TX, USA, 1999.

70. Xiao, L.; Mao, Z.-Q.; Wang, Z.-N.; Jin, Y. Application of NMR logs in tight gas reservoirs for formation evaluation: A case study of Sichuan basin in China. *J. Pet. Sci. Eng.* **2012**, *81*, 182–195. [CrossRef]

Article

Investigation of the Pore Structure of Tight Sandstone Based on Multifractal Analysis from NMR Measurement: A Case from the Lower Permian Taiyuan Formation in the Southern North China Basin

Kaixuan Qu [1,2] and Shaobin Guo [1,2,*]

[1] School of Energy Resources, China University of Geosciences (Beijing), Beijing 100083, China; Qukx159357@163.com
[2] Key Laboratory of Strategy Evaluation for Shale Gas, Ministry of Land and Resources, Beijing 100083, China
* Correspondence: guosb58@cugb.edu.cn

Received: 28 June 2020; Accepted: 30 July 2020; Published: 6 August 2020

Abstract: Understanding the pore structure can help us acquire a deep insight into the fluid transport properties and storage capacity of tight sandstone reservoirs. In this work, a series of methods, including X-ray diffraction (XRD) analysis, casting thin sections, scanning electron microscope (SEM), nuclear magnetic resonance (NMR) experiment and multifractal theory were employed to investigate the pore structure and multifractal characteristics of tight sandstones from the Taiyuan Formation in the southern North China Basin. The relationships between petrophysical properties, pore structure, mineral compositions and NMR multifractal parameters were also discussed. Results show that the tight sandstones are characterized by complex and heterogenous pore structure, with apparent multifractal features. The main pore types include clay-dominated micropores and inter- and intragranular dissolution pores. Multifractal parameters of sandstone samples were acquired by NMR and applied to quantitatively describe the pore heterogeneity in higher and lower probability density regions (with respect to small and large pore-scale pore system, respectively). The multifractal parameter (D_{-10}) of lower probability density areas has better correlation with the petrophysical parameters, which is more suitable for evaluating the reservoir properties of tight sandstone. However, the multifractal parameter (D_{10}) of higher probability density areas is more conducive to characterize the pore structure of tight sandstone. Additionally, the mineral compositions of sandstone have a complex effect on multifractal characteristics of pores in different probability density areas. The D_{10} increases with the decrease of quartz content and increase in clay mineral content, whereas D_{-10} decreases with the increase in clay minerals and decrease of authigenic quartz content and feldspar content.

Keywords: tight sandstone; pore structure; heterogeneity; NMR measurements; multifractal analysis

---

1. Introduction

Due to their huge geological reserves and wide distributions, tight sandstone gas greatly alleviates the contradiction between the world's increasing demand on energy and the depletion of conventional resources [1]. In China, tight sandstone gas reservoirs are characterized as one with the porosity and permeability less than 10% and 0.1 mD, respectively [2], and these tight reservoirs require extensive hydraulic fracture or special gas extraction techniques to achieve commercial production [3]. In comparison with conventional sandstones, the pore structures of tight sandstones are usually more complex and heterogeneous, because of their various pore size (nano-scale to micro-scale), poor connectivity and irregular pore geometry [4–7]. The pore structure replaces porosity and permeability as the critical parameter for the evaluation of reservoir properties, as it not only controls gas storage and transport mechanisms, but also determines the displacement efficiency and recovery rate of hydrocarbon in the tight sandstone reservoirs [6–9]. Consequently, detailed characterizations of pore structure are necessary procedures for understanding the producibility of tight sandstone gas reservoirs.

A variety of analytical methods have been geared to qualitatively and quantitatively characterize the pore structure of tight sandstone, such as casting thin sections (CAT), scanning electron microscopy (SEM), mercury injection porosimetry (MIP), gas absorption [10,11], X-ray computed tomography (XCT) [12] and nuclear magnetic resonance (NMR). Among them, NMR is a convenient and non-damaging method for characterizing the pore structure and pore fluids in the rock. In NMR experiments, proton ^{1}H (abundant in water and hydrocarbon) can generate a dipole moment in the presence of an external magnetic field. The signal amplitude of dipole moment is in proportion to the number of protons within pore fluids [13,14]. Thus, NMR is able to characterize petrophysical properties, the saturation of different types of pore fluids (movable fluids, bound fluids and hydrocarbons) and pore size distributions (PSD) of the reservoir rocks [15].

In the past few decades, numerous studies demonstrated that fractal theory is another robust tool for the characterization of pore structure in the porous media, including sedimentary rock, soil, concrete and other materials [16–21]. Fractal geometry successfully builds a bridge between petrophysical properties and microstructures of reservoir rocks, and a single fractal dimension, D, have been introduced to describe the fractal behavior and geometric irregularity of pore structure [22–24]. As a complex geological material, the PSD curves of tight sandstones often show "fluctuations" and "jumps" at different pore size intervals, and fractal characteristics vary greatly among different pore size intervals, which cannot be explained by a single fractal dimension [25]. The reason is that the single fractal dimension can only describe the irregularity and complexity within the specific pore size intervals. Multifractal theory is an extension of single fractal theory, which can offer more precise information about pore structure by resolving multifractal structure into a set of intertwined fractal subsets [26]. Nowadays, multifractal analysis has become a widely used mean to analyze the pore structure of coal, shale, and carbonate reservoirs [25,27–31]. Nevertheless, only a little attention has been given to the pore structure of tight sandstone [32,33]. Hence, multifractal analysis based on NMR experiment provides a brand-new perspective on the nature of pore structure and heterogeneity in tight sandstone reservoirs.

In this work, a case study from the Lower Permian Taiyuan Formation (P_1t) tight sandstones in the southern North China Basin was conducted to investigate pore structure features by various methods. Multifractal analysis was implemented to quantify the heterogeneity of pore structure in different probability density areas, based on NMR T_2 distributions. Additionally, the relationships between multifractal parameters of different probability density areas and mineralogical compositions, pore structure parameters and petrophysical properties of tight sandstone samples were investigated. This study provides new insights into the microscopic pore structure and heterogeneity of tight sandstone with similar geological conditions.

2. Geological Setting

The southern North China basin (SNCB) is a large meso-cenozoic superimposed basin, located in the southern part of the North China plate, with an area of 150,000 km^2 (Figure 1) [34]. The basin mainly consists of five secondary tectonic units from north to south, and the present tectonic pattern shows the NW-WNW (Figure 1).

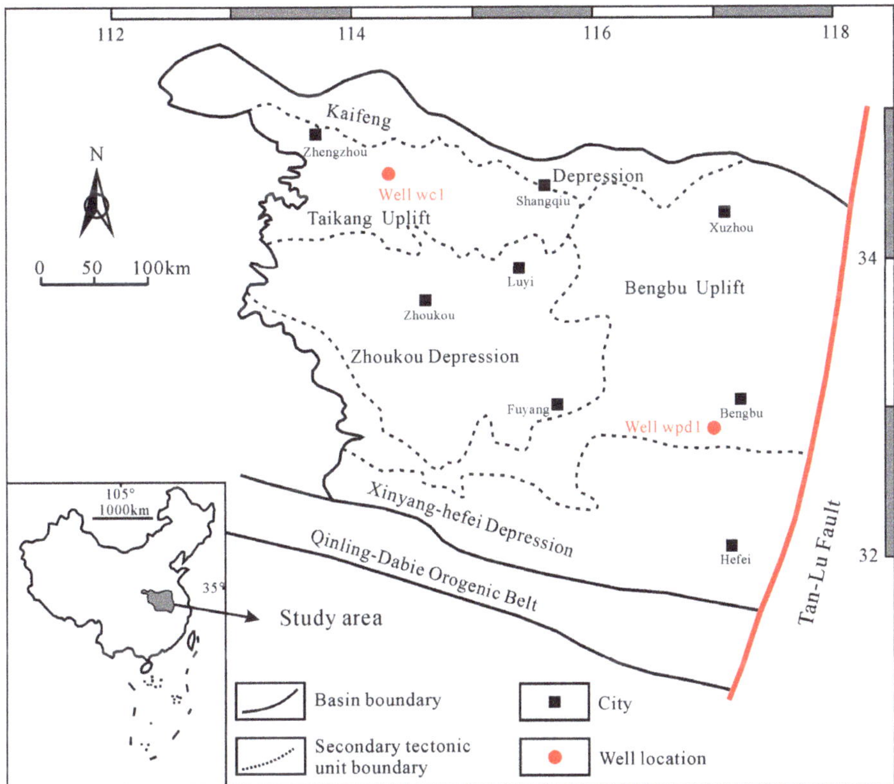

Figure 1. Tectonic geology condition of southern North China basin (SNCB) and investigated well location (adapted from [34]).

The continuous uplift of the entire North China platform during the Caledonian movement led to the absence of the Upper Ordovician, Silurian and Devonian, and Lower Carboniferous strata [35]. From the Late Carboniferous to the Late Permian, affected by the regional tectonic movement, the study area underwent the depositional process from the epicontinental sea to the continental basin (Figure 2) [36]. The Taiyuan Formation (P_1t), which is composed of black shale, coal, limestone and sandstone, was deposited during the early Permian, with a thickness of about 30–175 m (Figure 2) [37]. Sedimentary facies of the Taiyuan Formation are mainly lagoon, tidal flat, barrier island, and carbonate platform (Figure 2). A previous study has shown that shale and coal from Taiyuan Formation possess high organic matter content (TOC), moderate maturity, mainly type III kerogen, and high gas generation potential, which can serve as good source rocks for tight sandstone reservoirs, as well as the potential reservoirs for shale gas and coal bed methane [38].

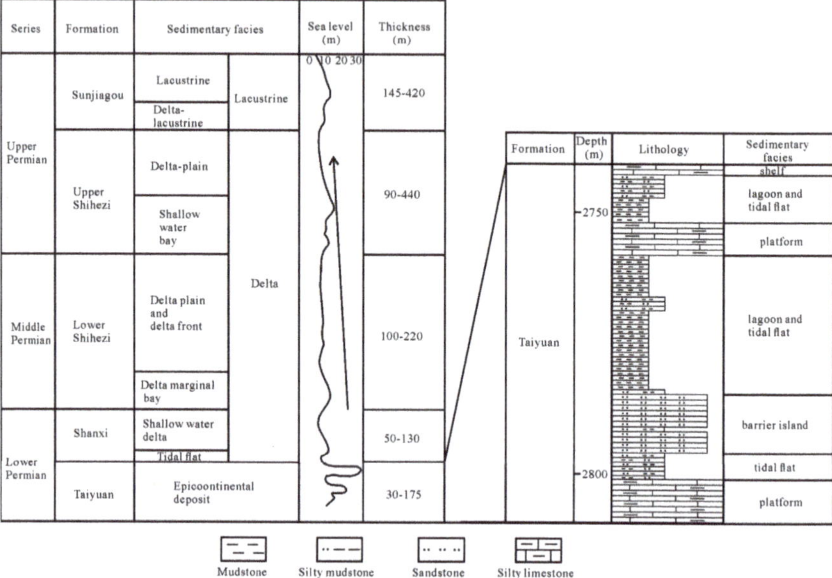

Figure 2. Map showing the stratigraphic column of Permian successions in the SNCB (modified from [38]) and more details of Taiyuan Formation (well wc1), including burial depth, lithology, sedimentary facies.

3. Experiments and Methods

3.1. Samples and Experiments

A total of 5 tight sandstone samples were selected from cores obtained in wells wc1 and wpd1 (Figure 1). Cylindrical plugs, 2.5 cm in diameter and 3–4 cm in length, were cut from the cores. The core plugs were carefully cleaned and dried, to remove contamination from remnant hydrocarbon and drilling fluids. The permeability of samples was first measured using a calibrated instrument DX-07G, with a steady flow of N_2 based on the Chinese Core Measurement and Analysis Method Standard (GB/T29172-2012). Secondly, the samples were placed in the brine with a salinity (~1200 mg/L), similar to formation water under vacuum for at least 24 h, until the brine saturated state. NMR signals were generated using the brine saturated samples. The samples were put in a pulsed magnetic field, and then after a brief pulse, the NMR signal gradually decayed, with a characteristic relaxation time T_2. During this process, the T_2 spectra under saturated conditions were recorded with a low-field NMR instrument Rec core 2500. Magnetic field strength and resonance frequency were 1200 G and 2.38 MHz, respectively. Thereafter, the saturated samples were centrifuged via an HR2500-2 apparatus, to obtain an ideal irreducible water state at a rotational speed of 5000 r/min. The experimental parameters were set as follows: the echo number 2048, the echo space 300 μs, the waiting time 6 s, and the test temperature was 25 °C.

Subsequently, the samples were dried again, and divided into three sections, for analysis of pore structure and mineralogical compositions, respectively. Thin sections, impregnated with blue epoxy resin under vacuum, were prepared to observe the type, geometric shapes, and distribution of the pores. In the SEM experiment, the freshly broken samples, which were polished and coated with carbon in advance, were examined with a ZEISS SIGMA field emission scanning electron microscope equipped with energy-dispersive X-ray spectra (EDX), to identify clay minerals' type, the morphologies of clay occurrence within the pore spaces and micropores associated with clay minerals [39]. The mineral compositions of the samples were determined by XRD analysis, following the Oil and Gas Industry

Standards (SY/T5463-2010). Samples were initially crushed to 100 mesh size, and then they were mixed with ethanol, ground into mortar, and placed on glass slides. The measurement was carried out using the Ultima IV X-ray diffractometer. The mineral content was quantified using Jade software.

3.2. Theory and Methods

3.2.1. NMR Theory

NMR can effectively reveal the important information about pore structure of rock, based on the T_2 transverse relaxation time [15]. The total transverse relaxation T_2 time is associated with three relaxation members: bulk relaxation T_{2B}, surface relaxation T_{2S}, and diffusion of pore fluid T_{2D}, described as [40].

$$\frac{1}{T_2} = \frac{1}{T_{2B}} + \frac{1}{T_{2S}} + \frac{1}{T_{2D}},\tag{1}$$

Generally, the bulk relaxation and diffusion relaxation are usually ignored when magnetic field is uniform. In this case, T_2 can be related directly to pore size: [16].

$$\frac{1}{T_2} \approx \frac{1}{T_{2S}} = \rho\frac{S}{V} = \rho\frac{a}{r},\tag{2}$$

where ρ ($\mu m/s$) is the transversal surface relaxation rate; S (μm^2) and V (μm^3) are the surface area and fluid volume of pore space, respectively; the surface/volume ratio (S/V) is a function of pore radius r (μm), and a is the pore shape factor ($a = 3$ for spherical pore, while $a = 2$ for tubular pore).

Thus, the T_2 distribution under fully brine-saturated conditions can be converted to the curve of pore size distribution by Equation (2), with the help of ρ. Details of this method have been shown in the previous studies [41].

3.2.2. Multifractal Methods Based on NMR T_2 Distributions

Numerous studies have introduced the algorithm of multifractal theory in detail [17,25,42,43]. In this study, the popular box counting method [31] was employed for the implement of multifractal algorithm on the basis of the 100% water-saturated T_2 distributions of the samples. T_2 distributions are split into N square boxes of size r, here $r = 2^m$ ($m = 0, 1, 2 \dots$). The probability mass distribution function $P_i(r)$ of the ith box could be represented as

$$P_i(r) = \frac{M_i(r)}{\sum_{i=1}^{N(r)} M_i(r)},\tag{3}$$

where M_i is the pore volume in the ith box and $\sum_{i=1}^{N(r)} M_i(r)$ is the total porosity. If porosity has a multifractal distribution, and then $P_i(r)$ has a power exponent relationship to r, as follows:

$$P_i(r) \propto r^{\alpha_i},\tag{4}$$

where α_i is the singularity strength for boxes [23]. Furthermore, the number of boxes with a similar α value is defined as $N_\alpha(r)$, by the relationship:

$$N_\alpha(r) \propto r^{-f(\alpha)},\tag{5}$$

where $f(\alpha)$ is a multifractal or singularity spectrum, expressing the fractal dimension of boxes with similar values of α [19]. Furthermore, $f(\alpha)$ could reach its maximum value when the following conditions are met:

$$\frac{df(\alpha)}{d\alpha(q)} = 0,\tag{6}$$

In order to accurately acquire the distribution properties, the partition function is expressed as:

$$X(q,r) = \sum_{i=1}^{N(r)} P_i^q(r) \propto r^{\tau(q)}, \tag{7}$$

where q is a moment expressing the contribution to $X(q, r)$ of boxes with diverse $P_i(r)$, which is commonly defined as $[-10, 10]$; When $q < 0$, $X(q, r)$ represents the density probability of the area with low concentration of porosity; when $q > 0$, $X(q, r)$ denotes the density probability of the area with high concentration of porosity [44]. Moreover, $\tau(q)$, known as mass exponent, could be depicted by:

$$\tau(q) = -\lim_{r \to 0} \frac{log \sum_{i=1}^{N(r)} P_i^q(r)}{logr}, \tag{8}$$

The generalized multifractal dimension D_q, another way to characterize singularity, which can be defined as:

$$D_q = \frac{\tau(q)}{q-1}, \tag{9}$$

On the other hand, the $\alpha(q), f(\alpha)$ can also be determined from $\tau(q)$ with the Legendre transformation, respectively [43]:

$$\alpha(q) = \frac{d\tau(q)}{dq}, \tag{10}$$

$$f(\alpha) = q\alpha(q) - \tau(q), \tag{11}$$

Generally, four sets of parameters, such as D_q, $\alpha(q)$, $f(\alpha)$ and $\Delta\alpha$ $(=\alpha_{max}-\alpha_{min})$, are commonly applied to characterize pore structure heterogeneity. The more complex and nonhomogeneous pore structure corresponds to the larger value of D_q and $\Delta\alpha$.

4. Results

4.1. Mineralogical Compositions of Tight Sandstone

The XRD analysis results of samples are shown in Table 1. Overall, the studied P_1t tight sandstone samples mainly consist of quartz and clay. Therein, quartz content ranges from 58.9% to 74.1% average as 68.42%, and clay content varies from 16.3% to 35.8%, with an average of 24.32%. The clay minerals are dominated by the mixed illite/smectite (I/S) and illite, attended by a few kaolinites, and chlorites (Table 1). The high content of clay minerals may result from the great heterogenous composition and strong alteration of detrital feldspars in the tight sandstones [45]. There is only trace amount of feldspar (0.9–8.3%, averaging 5.22%). The lack of feldspar content is possibly due to the dissolution caused by factors such as basin subsidence, thermal events and acid produced by coal-bearing strata [46]. Other minerals, including ankerite, pyrite and calcite, can be identified only in individual samples, and their contents are extremely low (<2.5%).

Table 1. The compositions of minerals from the P_1t tight sandstone samples.

Sample	Depth (m)	Mineralogical Composition (%)						Clay Composition (%)			
		Quartz	Feldspar	Calcite	Pyrite	Ankerite	Clay	I/S	I	K	C
1	1484.2	69	8.3	0.5	0.4	0	21.8	13.73	5.23	2.62	0.22
2	1485.1	68.2	7.5	0	1.2	0	23.1	14.32	5.08	3.23	0.47
3	1486.5	74.1	7.7	0.9	0	1	16.3	11.9	3.59	0.65	0.16
4	2769.6	58.9	1.7	0	1.1	2.5	35.8	26.49	7.88	1.07	0.36
5	2794.21	71.9	0.9	0	1.7	0.9	24.6	17.47	4.92	1.97	0.24

I/S: Illite/Smectite mixed layer; I: Illite; K: Kaolinite; C: Chlorite.

4.2. Pore Type and Characteristics

Casting thin section and SEM image analysis results show that there are three dominant pore types in the P_1t tight sandstone samples, including micropores associated with clay minerals, secondary intergranular and intragranular dissolution pores (Figure 3). Primary intergranular pores are rarely observed. Quartz overgrowth and authigenic quartz grains are commonly developed in P_1t tight sandstones (Figure 3a,b,d), and pyrite crystals are also observed in some pores (Figure 3e). Secondary dissolution pores mostly occur on detrital feldspars, as a result of partial to complete dissolution (Figure 3a,b). These dissolution pores are typically enveloped by authigenic clay minerals derived from the dissolution of detrital feldspars (Figure 3f). Additionally, a few dissolution pores occur on detrital grain boundaries (Figure 3c).

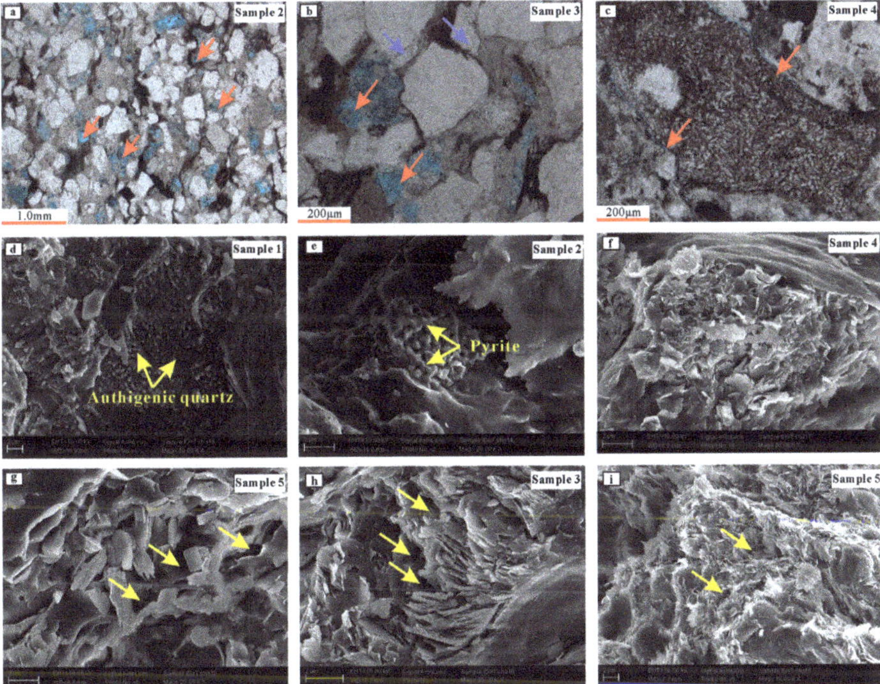

Figure 3. Photomicrographs showing the microscopic pore structure characteristics of P_1t tight sandstone. (**a**) sandstones with intergranular and intragranular dissolution pores (red arrow) and extensive quartz overgrowths; (**b**) intragranular dissolution pores (red arrow) and quartz overgrowths (blue arrow); (**c**) dissolution pores occur on detrital grain edges (red arrow); (**d**) fine authigenic quartz grains; (**e**) pore-filling pyrites; (**f**) authigenic clay minerals fill the dissolution pores; (**g**) slablike kaolinite filling in the pore; (**h**) flaky illite filling in pore; (**i**) hair-like illite/smectite mixed layers filling in pore (yellow arrow, micropores associated with clay minerals).

The pore structures of samples are severely impacted from clay cementation, because the pore throat system is mainly filled by a large amount of authigenic clay minerals such as the slablike or booklet kaolinite (Figure 3g), and the flaky illite and hair-like mixed illite/smectite (I/S) (Figure 3h,i). SEM image analysis showed that abundant micropores are developed within these clay minerals (Figure 3g–i). These micropores are continuously distributed with a multi-scale pore size (mainly < 1 μm), which provide the necessary percolation path, connecting other relatively larger pores for tight sandstone reservoirs to a certain extent.

4.3. Petrophysical Properties and NMR T_2 Distributions

The porosity of the five samples ranges from 1.95% to 3.41%, with an average of 2.7%, and permeability varies from 0.037 mD to 0.494 mD (Table 2). The petrophysical properties of samples are lower than those observed in the Chang 7 reservoir, with an average value of porosity of 7.2% and permeability of 0.18 mD. The Chang 7 reservoir is an important tight oil reservoir from Yanchang Formation in the Ordos basin [47]. This indicates that the samples have relatively poorer pore structures and reservoir quality. Nevertheless, a positive exponential relationship can be observed between porosity and permeability of samples (Figure 4).

Table 2. The petrophysical parameters and pore structure parameters of the tight sandstone samples from NMR measurements.

Sample	φ (%)	K (mD)	φ_m (%)	φ_b (%)	RQI (μm)	FZI (μm)	$T_{2cutoff}$ (ms)	T_{2gm} (ms)	T_{35} (ms)	T_{50} (ms)
1	2.63	0.044	0.38	2.25	0.041	0.151	3.5	1.93	1.42	1.86
2	2.98	0.064	1.5	1.48	0.046	0.149	2.08	2.7	1.46	2.04
3	1.95	0.037	0.67	1.28	0.044	0.219	2.45	2.37	1.18	1.5
4	2.53	0.134	1.45	1.08	0.073	0.280	1.4	1.73	1.28	1.7
5	3.41	0.494	1.85	1.56	0.12	0.341	1.86	2.67	1.48	2.1

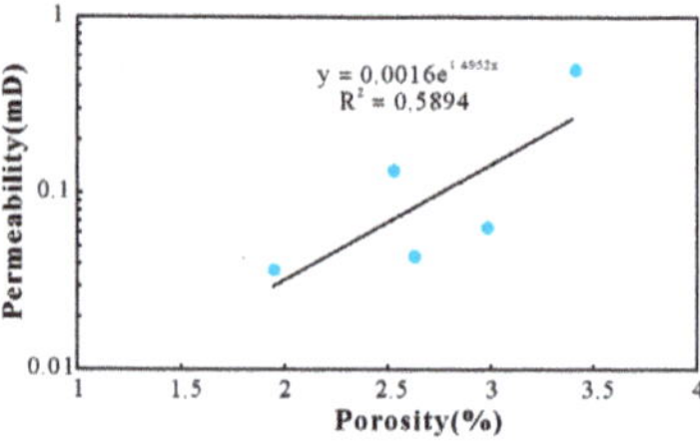

Figure 4. The correlation between porosity and permeability of P_1t sandstone.

Reservoir quality index (RQI) and flow zone indicator (FZI) are two ideal macroscopic petrophysical parameters used to evaluate the micro pore structure and reservoir properties of tight sandstone [48]. RQI and FZI were calculated by the following formulas, respectively [33]:

$$RQI = 0.0316 \times \sqrt{\frac{K}{\varphi}}, \tag{12}$$

$$FZI = RQI \times \frac{100 - \varphi}{\varphi}, \tag{13}$$

where K is the permeability, mD; φ is the porosity, %.

As shown in Table 2, the values of RQI vary from 0.0041 μm to 0.012 μm, while FZI values range from 0.1498 μm to 0.341 μm, averaging as 0.228 μm. These values are close to the tight oil reservoir researched by Zhao et al., 2017, whereas they are lower than the Chang 7 tight reservoir [47].

For the fully water-saturated rock, the T_2 distributions provide information about the pore size distributions. The T_2 relaxation time is in proportion to the pore size [15,49], and the signal amplitude of T_2 distributions reflect the pore fluid content and pore volume. The 100% brine-saturated T_2 spectra of five samples are shown in Figure 5. Except for sample 1 (unimodal T_2 spectrum), all samples show the bimodal characteristics of T_2 spectra, and almost all pore sizes of tight sandstones present in the range from 0.1 to 100 ms. There are no pores with the relaxation time larger than 100 ms, attributed to the absence of residually large intergranular pores. The main peaks are distributed between 0.1 ms and 10 ms, and their signal amplitudes are far larger than the secondary peak. The relative amplitudes of T_2 peaks indicate that sample porosities are dominated by smaller pore sizes, and the larger porosity are relatively few. The pores of sample 1 are all smaller pores.

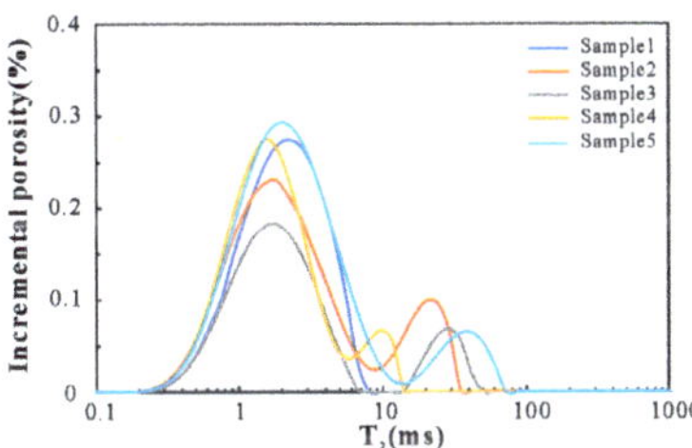

Figure 5. The saturated NMR T_2 spectra of samples.

By analyzing NMR T_2 data, some NMR pore structure parameters, including movable-fluid porosity (φ_m), bound-fluid porosity (φ_b), $T_{2cutoff}$, T_{2gm} (amplitude weighted logarithmic mean), T_{35} and T_{50}, are also summarized in Table 2. Porosity in the rock can be separated into bound-fluid porosity ($T_2 < T_{2cutoff}$) and movable-fluid porosity ($T_2 > T_{2cutoff}$) in the cumulative T_2 spectrum by $T_{2cutoff}$ value (Figure 6). The bound-fluid porosity of tight sandstone usually exists in clay-dominated micropores, which contain capillary and clay-bound water, while movable-fluid porosity tends to reside in large pores which are connected by effective pore throat [18]. Then, the movable-fluid porosity is determined by removing the proportion of the bound fluid from the 100% brine-saturated NMR signal, varying from 0.38% to 1.85%. Compared to movable-fluid porosity, bound-fluid porosity is commonly high, with the range of 1.08%–2.25%, indicating that the sandstone samples have the complex pore structure with poor pore connectivity. T_{35}, T_{50} are corresponding to the T_2 value, where the samples reach 35% and 50% brine saturation in the cumulative T_2 distributions, respectively [3]. Overall, compared to Chang 7 and Xujiahe tight sandstone reservoirs [48,50], $T_{2cutoff}$, T_{2lm}, T_{35}, and T_{50} of samples are characterized by relatively lower values, indicating a narrower pore size distribution in the samples.

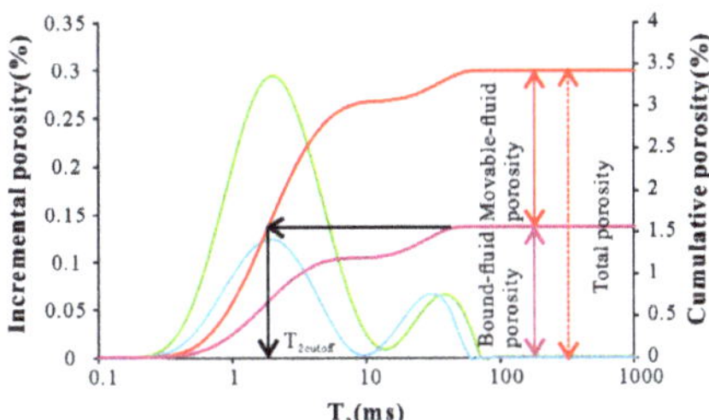

Figure 6. The cumulative and incremental T_2 spectrum of sample 5.

4.4. Multifractal Characteristics

In this study, multifractal characteristics of pore structures were obtained from 100% brine-saturated T_2 spectra. The range of moments q is defined in the interval from −10 to 10. The generalize dimension spectra ($D_q\sim q$) and the relationship of mass exponent $\tau(q)$ versus q are presented in Figures 7 and 8, respectively. The D_q with respect to variable q shows an inverse S-shaped curve. D_q has a larger variation for $q < 0$, while a minor variation for $q > 0$. Moreover, $\tau(q)$ follows a monotone increase as q increase, and the increasing trend gradually becomes smoother with increasing q. Figure 9 represents the multifractal spectra or singularity spectra of samples, where $\alpha(q)$ are also strongly correlated to the variable q. Overall, these spectra of different samples all show two different variation trends, that reveal that the pore size distributions of tight sandstone samples are multifractal. Therefore, the heterogeneity of pore volume distribution can be represented via the generalized dimension and singularity spectra shape and its characteristic parameters, which further reveal the local differences in the whole.

Generally, the heterogeneity of the whole pore size distribution is assessed by the total width of singularity spectra α_{-10}–α_{10} and generalized dimension spectra D_{-10}–D_{10} [19]. Higher values of α_{-10}–α_{10} and D_{-10}–D_{10} usually suggest a more heterogeneous pore size distribution within samples, and vice versa. The right part of the generalized dimension spectra and the singular spectra ($q > 0$) corresponds to the areas with higher probability density of porosity distribution (concentrated areas). However, the left part of the generalized dimension spectra and the singular spectra ($q < 0$) represent the areas with lower probability density (sparse areas) [32,51]. Therefore, multifractality parameters D_{10}, D_0–D_{10}, α_{10}, and α_0–α_{10} can describe the pore characteristics in higher probability areas, and the parameters D_{-10}, D_{-10}–D_0, α_{-10}, and α_{-10}–α_0 play the same role in lower probability areas. For the studied sandstone samples, the porosities in higher probability density areas mainly consist of the smaller pores, such as clay-dominated micropores, whereas the porosities in lower probability density areas mainly refer to the larger pores, such as intergranular dissolution pores, with a relatively larger pore size.

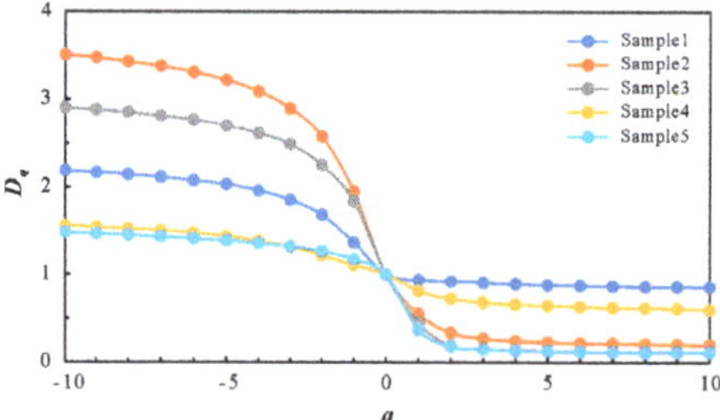

Figure 7. Generalized multifractal dimension spectra.

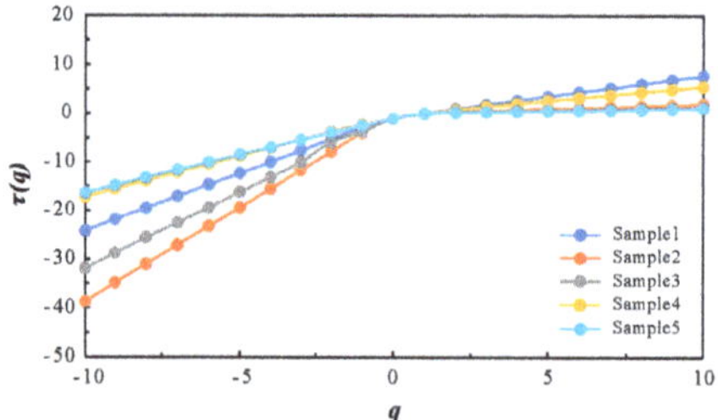

Figure 8. Generalized mass exponent τ(q) versus variable q.

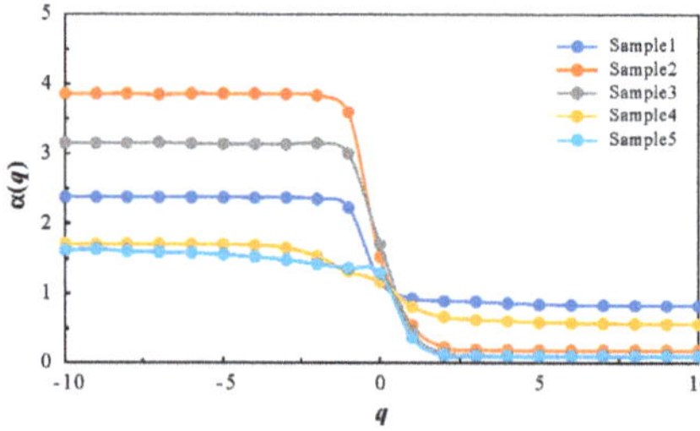

Figure 9. Multifractal spectra of five samples.

D_0 is defined as a capacity dimension or box-counting dimension. D_1, as an information dimension, is a measure of concentration degree of pore size distribution [28]. D_2 is the correlation dimension, which explains the scaling behavior of the second sampling moments [25]. For a monofractal structure, $D_0 = D_1 = D_2$ [44]. However, as shown in Table 3, all samples show a same order of $D_0 > D_1 > D_2$, suggesting that the pore size distribution of every studied tight sandstone sample has a tendency toward a multifractal type of scaling. Additionally, the calculated D_{-10}, D_{10}, $D_{-10}-D_0$, D_0-D_{10} and $D_{-10}-D_{10}$ are also listed in Table 3. For all samples, $D_{-10} > D_{10}$ and $D_{-10}-D_0 > D_0-D_{10}$, indicating that the pore distributions of higher probability areas may be more homogeneous than that of lower probability areas.

Table 3. The generalized dimensions of samples.

Sample	D_{-10}	D_0	D_1	D_2	D_{10}	$D_{-10}-D_0$	D_0-D_{10}	$D_{-10}-D_{10}$
1	2.19	1	0.95	0.94	0.86	1.19	0.142	1.33
2	3.5	1	0.55	0.34	0.2	2.5	0.8	3.3
3	2.9	1	0.45	0.188	0.12	1.9	0.88	2.78
4	1.57	1	0.82	0.72	0.6	0.57	0.4	0.97
5	1.49	1	0.37	0.18	0.11	0.49	0.89	1.38

For singularity parameters, the similar trend, $\alpha_{-10} > \alpha_{10}$ and $\alpha_{-10}-\alpha_0 > \alpha_0-\alpha_{10}$, are also found in Table 4, indicating that the pore system in lower probability density areas owns more obvious multifractal characteristics than that in a higher probability density area. As presented in Table 4, the values of $\alpha_{10}-\alpha_{-10}$ are in the range of 1.15–3.69, indicating that the pore structures of samples are highly heterogeneous. The parameter $A = (\alpha_0-\alpha_{10})/(\alpha_{-10}-\alpha_0)$ is referred to express the asymmetry of singularity spectrum, and $A > 1$ demonstrates a strong fluctuation in pore size distribution. The values of A for sample 1 and sample 2 are lower than 1, which exhibit more stable pore size distributions compared to those of other samples. The multifractal analysis of tight sandstone pores shows that the pore distribution is complicated, multifractal and heterogeneous.

Table 4. The singularity parameters of samples.

Sample	α_{-10}	α_{10}	α_0	$\alpha_{10}-\alpha_0$	$\alpha_0-\alpha_{-10}$	$\alpha_{10}-\alpha_{-10}$	A
1	2.38	0.83	1.17	1.21	0.34	1.55	0.12
2	3.87	0.18	1.52	2.35	1.34	3.69	0.57
3	3.15	0.1	1.7	1.45	1.6	3.05	1.1
4	1.71	0.57	1.17	0.54	0.6	1.14	1.12
5	1.62	0.1	1.3	0.32	1.2	1.52	3.74

5. Discussion

The results shown in the preceding section manifest that several multifractal parameters can be utilized to characterize the pore heterogeneity. In this study, only two multifractal parameters (D_{10} and D_{-10}) are selected to describe multifractal characteristics and evaluate pore heterogeneity in different probability measure areas within tight sandstones. The multifractal parameter D_{10} are used to account for multifractal behaviors of pore network in higher probability density areas, while the parameter D_{-10} represents the multifractal characteristics of pore network in lower probability density areas.

5.1. Relationship between Petrophysical Parameters of Tight Sandstone and Multifractal Parameters

Petrophysical property is the most direct performance of the pore structure of tight sandstone which can significantly affect fractal characteristics of pores. Figure 10 illustrates that multifractal parameters (D_{10} and D_{-10}) show a negative correlation with permeability, RQI and FZI, but they have no obvious relationship with porosity. Theoretically, porosity is mainly affected by the content of pore in the rock, especially large pore, which is independent of the complexity of pore distribution. Nevertheless, multifractal parameters mainly reflect the irregularity and complexity of pore geometry and pore network, with a significant influence on permeability, RQI and FZI. Meanwhile, not all pores are suitable for fractal analysis, and fractal analysis from a NMR T_2 spectrum only focus on pores with the pore size larger than dozens of nanometers or hundreds of nanometers [7], because the relaxation mechanism of fluid is quite complex in the smaller pore. In this case, the influence of bulk relaxation and diffusion relaxation should be considered, and T_2 cannot be directly simplified to surface relaxation T_{2s}. Therefore, the porosity of samples as a quantity cannot effectively constrain the heterogeneity of the pore structure [29].

Besides, D_{-10} shows a more sensitive response to permeability, RQI and FZI than D_{10}, indicating that pore structure in lower probability density areas has a significant impact on petrophysical properties of tight sandstone reservoirs. From the T_2 distributions and petrographic observations of sandstone samples, small-scale clay-dominated micropores associated with short T_2 components ($T_2 < 10$ ms) constitute the majority of the pore system, and only a few proportions are composed of the large-scale dissolution pores associated with long T_2 components ($T_2 > 10$ ms). Hence, pore system in lower probability density areas mainly consists of dissolution pores with low content; micropores dominate pore system in higher probability density areas. The formation of dissolution pores with larger pore scale, however, greatly improve reservoir properties. Therefore, the complexity of pore system composed of dissolution pores play a more important role in the petrophysical properties of sandstones.

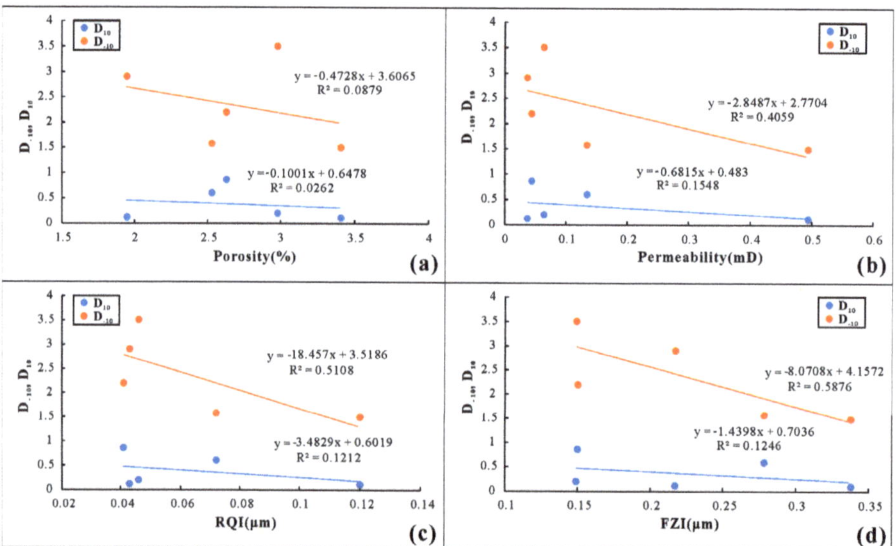

Figure 10. The correlations between petrophysical parameters with multifractal parameters. (**a**) the correlations between D_{-10}, D_{10} with porosity; (**b**) the correlations between D_{-10}, D_{10} with permeability; (**c**) the correlations between D_{-10}, D_{10} with RQI; (**d**) the correlations between D_{-10}, D_{10} with FZI.

Furthermore, the correlation coefficients of multifractal parameters with permeability, RQI and FZI show an increasing trend (Figure 10). This can be explained by the fact that RQI and FZI integrate porosity and permeability, which is the better petrophysical parameters for characterizing the pore structure of tight sandstone. Meanwhile, FZI is the combination of several microscopic pore structure properties, e.g., pore specific surface area, morphology and tortuosity, and therefore the highest correlation coefficient [33]. Hence, FZI is a superior indicator of the pore structure heterogeneity of tight sandstone, especially in the lower probability density area.

5.2. Relationship between Pore Structure of Tight Sandstone and Multifractal Parameters

The relationships between multifractal parameters and NMR pore structure parameters of tight sandstone, including movable-fluid porosity (φ_m), bound-fluid porosity (φ_b), $T_{2cutoff}$, T_{2gm}, T_{35} and T_{50}, are also analyzed, and the correlation coefficients are summarized in Table 5. D_{10} is highly associated with movable-fluid porosity, bound-fluid porosity, $T_{2cutoff}$ and T_{2gm}, but it has no apparent correlation with T_{35} and T_{50} (Figure 11). Additionally, D_{-10} shows few or no relationships with pore structure parameters (Table 5). This may be because small-scale micropores in higher probability density areas dominate the entire pore system of tight sandstone reservoirs, and thus, multifractal parameters of higher probability density regions are more useful for pore structure characterization of tight sandstone.

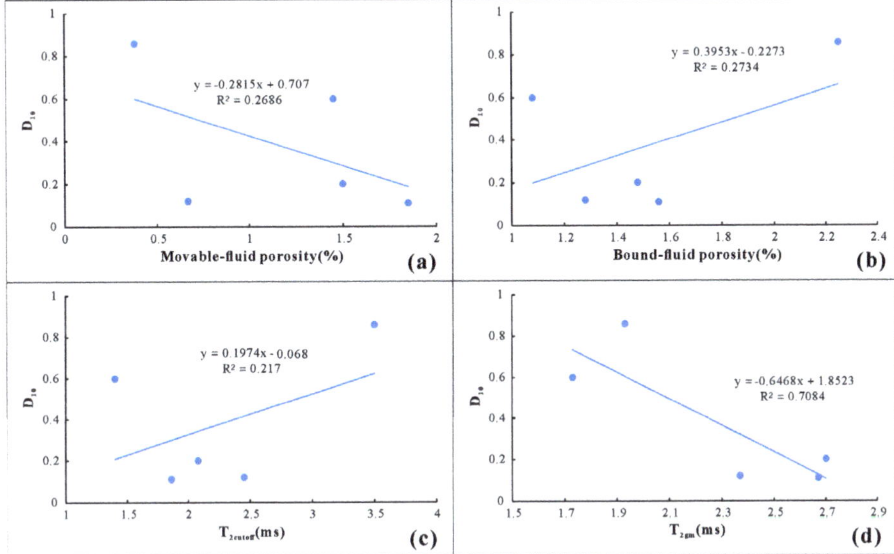

Figure 11. The correlations between multifractal parameter D_{10} and NMR pore structure parameters. (a) the correlation between D_{10} with movable-fluid porosity; (b) the correlation between D_{10} with bound-fluid porosity; (c) the correlation between D_{10} with $T_{2cutoff}$; (d) the correlation between D_{10} with T_{2gm}.

The $T_{2cutoff}$ is the efficient boundary which divides the total pore volume into movable-fluid pores and bound-fluid pores, according to whether the fluid within them can flow or not. For the studied sandstone samples, immovable bound-fluid mainly exists in clay-dominated micropores with poor connectivity, while movable fluid tends to exist in those dissolution pores which are connected by effective pore throats. $T_{2cutoff}$ value severely affects the proportion of different type of pores in tight sandstones. The larger the $T_{2cutoff}$ value, the more bound-fluid pores, and the fewer movable-fluid pores (Figure 12), which increase the heterogeneity of pore network in higher probability density areas.

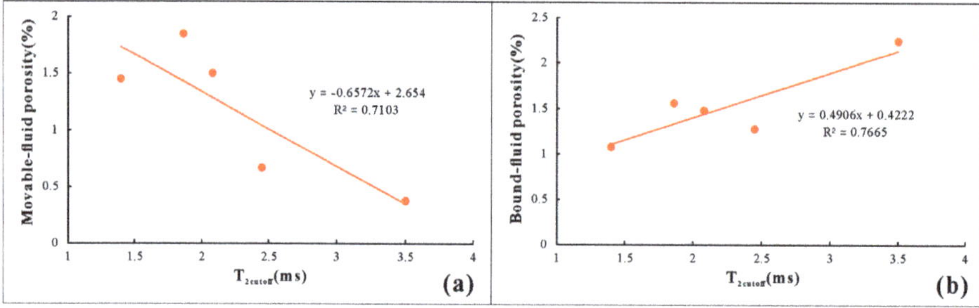

Figure 12. The correlations between movable-fluid porosity, bound-fluid porosity and $T_{2cutoff}$. (**a**) the correlation between $T_{2cutoff}$ with movable-fluid porosity; (**b**) the correlation between $T_{2cutoff}$ with bound-fluid porosity.

Figure 11d illustrates that T_{2gm} has the negative correlation with D_{10}. The T_{2gm} is a comprehensive performance for the whole pore size distribution of rock, and the large T_{2gm} value means a good pore structure in the tight sandstone reservoirs [52]. In this study, a large amount of short T_2 components related to micropores not only dominate the whole T_2 distributions of sandstone samples, but also constitute the main peak. Therefore, T_{2gm} mainly reflects the characteristics of main peak corresponding to pore system in higher probability density areas. Therefore, the heterogeneity of pore network of the high probability density areas tends to decrease as T_{2gm} increases.

Table 5. The correlation coefficients between multifractal parameters and pore structure parameters.

	φ_m (%)	φ_b (%)	$T_{2cutoff}$ (ms)	T_{2gm} (ms)	T_{35} (ms)	T_{50} (ms)
D_{10}	−0.2686	0.2734	0.217	−0.7084	0.0019	−0.0139
D_{-10}	−0.0717	0.0001	0.0678	0.2039	−0.0058	−0.0076

5.3. Effect of miNeral Compositions of Tight Sandstone on Multifractal Characteristics

Several studies have testified that mineral compositions, mineral content and contact between minerals have a significant impact on the pore structure of rock, and the influence of each mineral on pores in distinct lithology may also be different [5,8,30,45,53].

As shown in Figure 13a, quartz content shows the negative correlation with D_{10} and the weak negative relationship with D_{-10}. This result demonstrates that quartz plays different roles in the pore structure of different probability density areas. Sandstone petrographic observations show that the studied tight sandstone samples mainly consist of sedimentary quartz grains (Figure 3a,b), which act as the main skeleton minerals of the tight sandstone to resist compaction for the preservation of pores. In general, the higher content of quartz contributes to the higher textural maturity of sandstone and regular pore structure, which decrease the complexity of the pore systems in higher probability density areas. Nevertheless, the tight sandstone samples contain some authigenic quartz grains filling parts of large dissolution pores, which lead to the irregular pore geometry in lower probability density areas and thus larger values of D_{-10}.

According to Figure 13b, feldspar content is well correlated to D_{-10}, but has no correlation with D_{10}. The above relationships indicate that tight sandstone with high feldspar content tends to have relatively complex and anisotropic pore structure in lower probability density areas. The content of feldspar in tight sandstone is associated with the diagenesis process. After deposition, the Taiyuan Formation experienced a long period of deep burial and diagenetic processes, and the chemical environment of fluid changed since a large amount of organic acids and CO_2 were generated during the early hydrocarbon generation of the organic matter in the source rock, which causes the extensive dissolution of chemically unstable aluminosilicates, such as feldspar. In this study, feldspar contents of all samples are low (average 5.22%), and secondary dissolution porosities as a result of partial/complete dissolution of feldspars are well observed in the sandstone samples (Figure 3a,b), indicating that feldspar experienced the strong alteration. These dissolution pores provide extra pore space for gas storage, while feldspar dissolution is always accompanied by by-products (e.g., authigenic quartz, kaolinite, illite) which precipitate in situ or in adjacent pores, resulting in a rather complex pore network in lower probability density areas.

Different from quartz and feldspar, clay content shows the weak positive correlation with D_{10}, whereas it has a better negative relationship with D_{-10} (Figure 13c). This finding indicates that the effect of clay minerals on the pore structure in different probability density areas is rather complex. A high content of clay can not only reduce complexity of pore in lower probability density areas, but also increases the pore heterogeneity of higher probability density areas. This may be interpreted that tight sandstone samples have the high clay content (average 24.32%), and micropores associated with clay minerals predominate the pore system of the high probability measure areas. The SEM image analysis results show that a large amount of clay minerals fill in between quartz grains with slablike, flaky, and hair-like forms (Figure 3g–i), blocking the pore throat system, causing some large pores to be closed and semi-closed. The micropores gradually replace the large pores, and reduce the amounts of large pores with increasing clay content, which decreases the complexity of the pore network of lower probability density areas. However, clay-dominated micropores have different characteristics within various types of clay minerals, which are also characterized by multi-type pores with multi-scale pore size ranging from nanoscale to microscale, such as intracrystalline micropores (mainly < 2 nm), within aluminosilicate layers, inter-crystalline micropores (2–50 nm) between clay particles, and interparticle micropores (>0 nm) between aggregated clay particles [54]. These factors complicate the pore network in higher probability density areas, resulting in higher D_{10}.

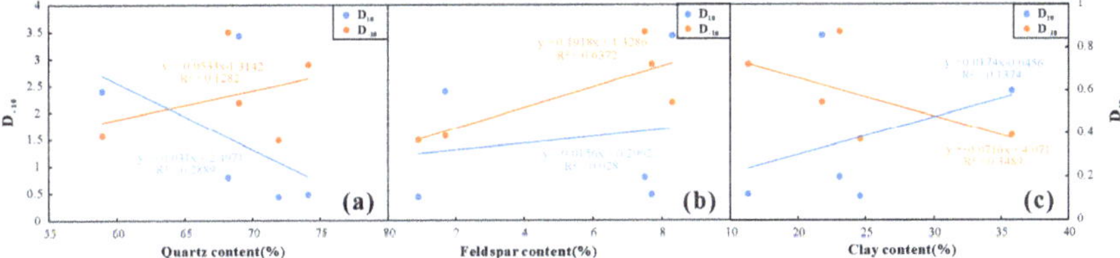

Figure 13. The correlations between mineral compositions and multifractal parameters. (a) the correlations between D_{-10}, D_{10} with quartz content; (b) the correlations between D_{-10}, D_{10} with feldspar content; (c) the correlations between D_{-10}, D_{10} with clay content.

6. Conclusions

In the presented research, the pore structure and its heterogeneity of the Lower Permian Taiyuan Formation tight sandstone in the south China North basin were investigated by a series of experiments and multifractal methods. These conclusions could be obtained:

(1) The tight sandstones from Taiyuan Formation are characterized by a high content of quartz and clay minerals, low porosity (1.95–3.41%) and low permeability (0.037–0.494 mD). Pore types mainly include micropores associated with clay minerals, intergranular and intragranular dissolution pores, whereas primary pores are rarely observed. The NMR T_2 distributions of samples mainly show bimodal characteristics with less long T_2 components, since clay-dominated micropores associated with short T_2 components predominate the whole pore system of the tight sandstones.

(2) The multifractal spectrum, generalized dimensions and mass exponent spectrum demonstrate that pore structures of tight sandstone samples have multifractal characteristics. Meanwhile, the multifractal parameters are useful to quantify the pore heterogeneity in the areas with different probability densities.

(3) Multifractal parameter (D_{-10}) of lower probability density areas is more appropriate to determine petrophysical properties. However, the multifractal parameter (D_{10}) of higher probability density areas is the available indicators for evaluating pore structure.

(4) Mineral compositions have various effects on pore structure in different probability density areas. The decrease of quartz content and increase in clay mineral content contribute to high degree of pore anisotropy of higher probability density areas, whereas the decrease in clay minerals and the increase of quartz content and feldspar content can result in more irregular and nonhomogeneous pore network in lower probability density areas.

(5) Overall, the pore structure in lower probability density areas consists of larger dissolution pores, which determine the reservoir properties. Gas storage and seepage usually occur at those pores of lower probability density areas. However, micropores dominate pore structure in higher probability density areas, and strong complexity and heterogeneity make pores unfavorable for gas storage and seepage there.

Author Contributions: K.Q. conceived of the presented idea, designed the experiments and verified the data along with S.G. The article is originally written by K.Q. and revised by the corresponding author S.G. All authors have read and agreed to the published version of the manuscript.

Funding: This work was funded by Ministry of Science and Technology of the People's Republic of China (Grant No. 2016ZX05034).

Acknowledgments: This study received financial support from the Ministry of Science and Technology of the People's Republic of China (Grant No. 2016ZX05034). We would like to thank the anonymous reviewers for their careful reviews and detailed suggestions.

Conflicts of Interest: The authors declare no conflicts of interest.

References

1. Desbois, G.; Urai, J.L.; Kukla, P.A.; Konstanty, J.; Baerle, C. High-Resolution 3D fabric and porosity model in a tight gas sandstone reservoir: A new approach to investigate microstructures from mm- to nm-scale combining argon beam cross-sectioning and SEM imaging. *J. Pet. Sci. Eng.* **2011**, *78*, 243–257. [CrossRef]
2. Zou, C.; Zhu, R.; Liu, K.; Su, L.; Bai, B.; Zhang, X.; Yuan, X.; Wang, J. Tight gas sandstone reservoirs in China: Characteristics and recognition criteria. *J. Pet. Sci. Eng.* **2012**, *88–89*, 82–91. [CrossRef]
3. Rezaee, R.; Saeedi, A.; Clennell, B. Tight gas sands permeability estimation from mercury injection capillary pressure and nuclear magnetic resonance data. *J. Pet. Sci. Eng.* **2012**, *88–89*, 92–99. [CrossRef]
4. Zhang, L.; Lu, S.; Xiao, D.; Li, B. Pore structure characteristics of tight sandstones in the northern Songliao Basin, China. *Mar. Pet. Geol.* **2017**, *88*, 170–180. [CrossRef]
5. Loucks, R.G.; Reed, R.M.; Ruppel, S.C.; Hammes, U. Spectrum of pore types and networks in mudrocks and a descriptive classification for matrix-related mudrock pores. *AAPG Bull.* **2012**, *96*, 1071–1098. [CrossRef]
6. Xi, K.; Cao, Y.; Haile, B.G.; Zhu, R.; Jahren, J.; Bjørlykke, K.; Zhang, X.; Hellevang, H. How does the pore-throat size control the reservoir quality and oiliness of tight sandstones? The case of the lower cretaceous quantou formation in the southern Songliao Basin, China. *Mar. Pet. Geol.* **2016**, *76*, 1–15. [CrossRef]
7. Zhu, F.; Hu, W.; Cao, J.; Sun, F.; Liu, Y.; Sun, Z. Micro/Nanoscale pore structure and fractal characteristics of tight gas sandstone: A case study from the Yuanba area, northeast Sichuan Basin, China. *Mar. Pet. Geol.* **2018**, *98*, 116–132. [CrossRef]
8. Sun, W.; Zuo, Y.; Wu, Z.; Liu, H.; Xi, S.; Shui, Y.; Wang, J.; Liu, R.; Lin, J. Fractal analysis of pores and the pore structure of the lower cambrian niutitang shale in northern Guizhou province: Investigations using NMR, SEM and image analyses. *Mar. Pet. Geol.* **2019**, *99*, 416–428. [CrossRef]
9. Clarkson, C.R.; Freeman, M.; He, L.; Agamalian, M.; Melnichenko, Y.B.; Mastalerz, M.; Bustin, R.M.; Radliński, A.P.; Blach, T.P. Characterization of tight gas reservoir pore structure using USANS/SANS and gas adsorption analysis. *Fuel* **2012**, *95*, 371–385. [CrossRef]
10. Clarkson, C.R.; Solano, N.; Bustin, R.M.; Bustin, A.M.M.; Chalmers, G.R.L.; He, L.; Melnichenko, Y.B.; Radliński, A.P.; Blach, T.P. Pore structure characterization of North American shale gas reservoirs using USANS/SANS, gas adsorption, and mercury intrusion. *Fuel* **2013**, *103*, 606–616. [CrossRef]

11. Schmitt, M.; Fernandes, C.P.; Wolf, F.G.; Bellini Da Cunha Neto, J.A.; Rahner, C.P.; Santiago Dos Santos, V.S. Characterization of Brazilian tight gas sandstones relating permeability and angstrom-to micron-scale pore structures. *J. Nat. Gas Sci. Eng.* **2015**, *27*, 785–807. [CrossRef]

12. Lai, J.; Wang, G.; Fan, Z.; Chen, J.; Qin, Z.; Xiao, C.; Wang, S.; Fan, X. Three-Dimensional quantitative fracture analysis of tight gas sandstones using industrial computed tomography. *Sci. Rep. UK* **2017**, *7*, 1–12. [CrossRef] [PubMed]

13. Yao, Y.; Liu, D. Comparison of low-field NMR and mercury intrusion porosimetry in characterizing pore size distributions of coals. *Fuel* **2012**, *95*, 152–158. [CrossRef]

14. Dillinger, A.; Esteban, L. Experimental evaluation of reservoir quality in Mesozoic formations of the Perth Basin (Western Australia) by using a laboratory low field nuclear magnetic resonance. *Mar. Pet. Geol.* **2014**, *57*, 455–469. [CrossRef]

15. Daigle, H.; Johnson, A.; Thomas, B. Determining fractal dimension from nuclear magnetic resonance data in rocks with internal magnetic field gradients. *Geophysics* **2014**, *79*, D425–D431. [CrossRef]

16. Lai, J.; Wang, G.; Wang, Z.; Chen, J.; Pang, X.; Wang, S.; Zhou, Z.; He, Z.; Qin, Z.; Fan, X. A review on pore structure characterization in tight sandstones. *Earth Sci. Rev.* **2018**, *177*, 436–457. [CrossRef]

17. Paz Ferreiro, J.; Vidal Vázquez, E. Multifractal analysis of Hg pore size distributions in soils with contrasting structural stability. *Geoderma* **2010**, *160*, 64–73. [CrossRef]

18. Lai, J.; Wang, G.; Cao, J.; Xiao, C.; Wang, S.; Pang, X.; Dai, Q.; He, Z.; Fan, X.; Yang, L.; et al. Investigation of pore structure and petrophysical property in tight sandstones. *Mar. Pet. Geol.* **2018**, *91*, 179–189. [CrossRef]

19. Martínez, F.S.J.; Caniego, F.J.; García-Gutiérrez, C.; Espejo, R. Representative elementary area for multifractal analysis of soil porosity using entropy dimension. *Nonlinear Proc. Geophys.* **2007**, *14*, 503–511. [CrossRef]

20. Mandelbrot, B.B. Multifractal measures, especially for the geophysicist. *Pure Appl. Geophys. PAGEOPH* **1989**, *131*, 5–42. [CrossRef]

21. Muller, J.; McCauley, J.L. Implication of fractal geometry for fluid flow properties of sedimentary rocks. *Transp. Porous Media* **1992**, *8*, 133–147. [CrossRef]

22. Yao, Y.; Liu, D.; Tang, D.; Tang, S.; Huang, W.; Liu, Z.; Che, Y. Fractal characterization of seepage-pores of coals from China: An investigation on permeability of coals. *Comput. Geosci. UK* **2009**, *35*, 1159–1166. [CrossRef]

23. Yu, S.; Bo, J.; Pei, S.; Jiahao, W. Matrix compression and multifractal characterization for tectonically deformed coals by Hg porosimetry. *Fuel* **2018**, *211*, 661–675. [CrossRef]

24. Li, W.; Liu, H.; Song, X. Multifractal analysis of Hg pore size distributions of tectonically deformed coals. *Int. J. Coal Geol.* **2015**, *144–145*, 138–152. [CrossRef]

25. Liu, K.; Ostadhassan, M.; Zou, J.; Gentzis, T.; Rezaee, R.; Bubach, B.; Carvajal-Ortiz, H. Multifractal analysis of gas adsorption isotherms for pore structure characterization of the Bakken Shale. *Fuel* **2018**, *219*, 296–311. [CrossRef]

26. Ge, X.; Fan, Y.; Li, J.; Aleem Zahid, M. Pore structure characterization and classification using multifractal theory—An application in Santanghu basin of western China. *J. Pet. Sci. Eng.* **2015**, *127*, 297–304. [CrossRef]

27. Zheng, S.; Yao, Y.; Liu, D.; Cai, Y.; Liu, Y.; Li, X. Nuclear magnetic resonance T2 cutoffs of coals: A novel method by multifractal analysis theory. *Fuel* **2019**, *241*, 715–724. [CrossRef]

28. Zhao, P.; Wang, X.; Cai, J.; Luo, M.; Zhang, J.; Liu, Y.; Rabiei, M.; Li, C. Multifractal analysis of pore structure of middle Bakken formation using low temperature N2 adsorption and NMR measurements. *J. Pet. Sci. Eng.* **2019**, *176*, 312–320. [CrossRef]

29. Liu, K.; Ostadhassan, M.; Kong, L. Multifractal characteristics of Longmaxi Shale pore structures by N2 adsorption: A model comparison. *J. Pet. Sci. Eng.* **2018**, *168*, 330–341. [CrossRef]

30. Guan, M.; Liu, X.; Jin, Z.; Lai, J. The heterogeneity of pore structure in lacustrine shales: Insights from multifractal analysis using N2 adsorption and mercury intrusion. *Mar. Pet. Geol.* **2020**, *114*, 104150. [CrossRef]

31. Liu, K.; Ostadhassan, M.; Kong, L. Fractal and multifractal characteristics of pore throats in the Bakken Shale. *Transp. Porous Media* **2019**, *126*, 579–598. [CrossRef]

32. Hou, X.; Zhu, Y.; Chen, S.; Wang, Y.; Liu, Y. Investigation on pore structure and multifractal of tight sandstone reservoirs in coal bearing strata using LF-NMR measurements. *J. Pet. Sci. Eng.* **2020**, *187*, 106757. [CrossRef]

33. Zhao, P.; Wang, Z.; Sun, Z.; Cai, J.; Wang, L. Investigation on the pore structure and multifractal characteristics of tight oil reservoirs using NMR measurements: Permian lucaogou formation in jimusaer sag, junggar basin. *Mar. Pet. Geol.* **2017**, *86*, 1067–1081. [CrossRef]

34. Liu, Y.; Zhang, J.; Tang, X. Predicting the proportion of free and adsorbed gas by isotopic geochemical data: A case study from lower Permian shale in the southern North China basin (SNCB). *Int. J. Coal Geol.* **2016**, *156*, 25–35. [CrossRef]

35. Tang, S.; Zhang, J.; Elsworth, D.; Tang, X.; Li, Z.; Du, X.; Yang, X. Lithofacies and pore characterization of the Lower Permian Shanxi and Taiyuan shales in the southern North China Basin. *J. Nat. Gas Sci. Eng.* **2016**, *36*, 644–661. [CrossRef]

36. Diao, Y.; Wei, J.; Li, Z.; Cao, H.; Li, X. Late carboniferous-early permian sequence stratigraphy and paleogeography in the southern North China Basin. *J. Stratigr.* **2011**, *35*, 88–94.

37. Liu, Y.; Tang, X.; Zhang, J.; Mo, X.; Huang, H.; Liu, Z. Geochemical characteristics of the extremely high thermal maturity transitional shale gas in the Southern North China Basin (SNCB) and its differences with marine shale gas. *Int. J. Coal Geol.* **2018**, *194*, 33–44. [CrossRef]

38. Dang, W.; Zhang, J.; Tang, X.; Chen, Q.; Han, S.; Li, Z.; Du, X.; Wei, X.; Zhang, M.; Liu, J.; et al. Shale gas potential of lower Permian marine-continental transitional black Shales in the Southern North China Basin, central China: Characterization of organic geochemistry. *J. Nat. Gas Sci. Eng.* **2016**, *28*, 639–650. [CrossRef]

39. Lai, J.; Wang, G.; Chai, Y.; Xin, Y.; Wu, Q.; Zhang, X.; Sun, Y. Deep burial diagenesis and reservoir quality evolution of high-temperature, high-pressure sandstones: Examples from Lower Cretaceous Bashijiqike formation in Keshen area, Kuqa depression, Tarim basin of China. *AAPG Bull.* **2017**, *101*, 829–862. [CrossRef]

40. Daigle, H.; Johnson, A. Combining mercury intrusion and nuclear magnetic resonance measurements using percolation theory. *Transp. Porous Media* **2016**, *111*, 669–679. [CrossRef]

41. Sigal, R.F. Pore-Size distributions for organic-Shale—Reservoir rocks from nuclear-magnetic- resonance spectra combined with adsorption measurements. *SPE J.* **2015**, *20*, 824–830. [CrossRef]

42. Caniego, F.J.; Martí, M.A.; San José, F. Rényi dimensions of soil pore size distribution. *Geoderma* **2003**, *112*, 205–216. [CrossRef]

43. Ge, X.; Fan, Y.; Zhu, X.; Chen, Y.; Li, R. Determination of nuclear magnetic resonance T2 cutoff value based on multifractal theory—An application in sandstone with complex pore structure. *Geophysics* **2015**, *80*, D11–D21. [CrossRef]

44. San José Martínez, F.; Martín, M.A.; Caniego, F.J.; Tuller, M.; Guber, A.; Pachepsky, Y.; García-Gutiérrez, C. Multifractal analysis of discretized X-ray CT images for the characterization of soil macropore structures. *Geoderma* **2010**, *156*, 32–42. [CrossRef]

45. Shao, X.; Pang, X.; Li, H.; Zhang, X. fractal analysis of pore network in tight gas sandstones using NMR method: A case study from the Ordos Basin, China. *Energy Fuel* **2017**, *31*, 10358–10368. [CrossRef]

46. Wang, R.; Shi, W.; Xie, X.; Zhang, W.; Qin, S.; Liu, K.; Busbey, A.B. Clay mineral content, type, and their effects on pore throat structure and reservoir properties: Insight from the Permian tight sandstones in the Hangjinqi area, north Ordos Basin, China. *Mar. Pet. Geol.* **2020**, *115*, 104281. [CrossRef]

47. Li, P.; Jia, C.; Jin, Z.; Liu, Q.; Zheng, M.; Huang, Z. The characteristics of movable fluid in the Triassic lacustrine tight oil reservoir: A case study of the Chang 7 member of Xin'anbian Block, Ordos Basin, China. *Mar. Pet. Geol.* **2019**, *102*, 126–137. [CrossRef]

48. Lai, J.; Wang, G.; Fan, Z.; Chen, J.; Wang, S.; Zhou, Z.; Fan, X. Insight into the pore structure of tight sandstones using NMR and HPMI measurements. *Energy Fuel* **2016**, *30*, 10200–10214. [CrossRef]

49. Zhang, P.; Lu, S.; Li, J.; Chen, C.; Xue, H.; Zhang, J. Petrophysical characterization of oil-bearing Shales by low-field nuclear magnetic resonance (NMR). *Mar. Pet. Geol.* **2018**, *89*, 775–785. [CrossRef]

50. Gao, H.; Li, H. Determination of movable fluid percentage and movable fluid porosity in ultra-low permeability sandstone using nuclear magnetic resonance (NMR) technique. *J. Pet. Sci. Eng.* **2015**, *133*, 258–267. [CrossRef]

51. Liu, M.; Xie, R.; Guo, J.; Jin, G. Characterization of pore structures of tight sandstone reservoirs by multifractal analysis of the NMRT2 distribution. *Energy Fuel* **2018**, *32*, 12218–12230. [CrossRef]

52. Lai, J.; Wang, G.; Fan, Z.; Zhou, Z.; Chen, J.; Wang, S. Fractal analysis of tight shaly sandstones using nuclear magnetic resonance measurements. *AAPG Bull.* **2018**, *102*, 175–193. [CrossRef]

53. Loucks, R.G.; Reed, R.M.; Ruppel, S.C.; Jarvie, D.M. Morphology, genesis, and distribution of nanometer-scale pores in siliceous mudstones of the Mississippian Barnett Shale. *J. Sediment. Res.* **2009**, *79*, 848–861. [CrossRef]
54. Xiao, D.; Jiang, S.; Thul, D.; Lu, S.; Zhang, L.; Li, B. Impacts of clay on pore structure, storage and percolation of tight sandstones from the Songliao Basin, China: Implications for genetic classification of tight sandstone reservoirs. *Fuel* **2018**, *211*, 390–404. [CrossRef]

Article

Finite Element Simulation of Multi-Scale Bedding Fractures in Tight Sandstone Oil Reservoir

Qianyou Wang [1,2,3], Yaohua Li [4], Wei Yang [1,2,3,*], Zhenxue Jiang [1,2,3,*], Yan Song [1,2,3,*], Shu Jiang [5,6], Qun Luo [1,2,3] and Dan Liu [7]

[1] State Key Laboratory of Petroleum Resources and Prospecting, China University of Petroleum, Beijing 102249, China; tsianyou@126.com (Q.W.); luoqun2002@263.net (Q.L.)
[2] Unconventional Natural Gas Institute, China University of Petroleum, Beijing 102249, China
[3] Unconventional Gas Collaborative Innovation Center, China University of Petroleum, Beijing 102249, China
[4] Oil and Gas Survey, China Geological Survey, Beijing 100083, China; majing06@126.com
[5] Key Laboratory of Tectonics and Petroleum Resources of Ministry of Education, Faculty of Earth Resources, China University of Geosciences, Wuhan 430074, China; sjiang@egi.utah.edu
[6] Energy & Geoscience Institute, University of Utah, Salt Lake City, UT 84108, USA
[7] The Institute of Exploration Techniques, Chinese Academy of Geological Sciences, Langfang 065000, China; liudanbelief@163.com
*   Correspondence: yangw@pku.edu.cn (W.Y.); jiangzx@cup.edu.cn (Z.J.); sya@petrochina.com.cn (Y.S.)

Received: 28 November 2019; Accepted: 23 December 2019; Published: 26 December 2019

Abstract: Multi-scale bedding fractures, i.e., km-scale regional bedding fractures and cm-scale lamina-induced fractures, have been the focus of unconventional oil and gas exploration and play an important role in resource exploration and drilling practice for tight oil and gas. It is challenging to conduct numerical simulations of bedding fractures due to the strong heterogeneity without a proper mechanical criterion to predict failure behaviors. This research modified the Tien–Kuo (T–K) criterion by using four critical parameters (i.e., the maximum principal stress (σ_1), minimum principal stress (σ_3), lamina angle (θ), and lamina friction coefficient (μ_{lamina})). The modified criterion was compared to other bedding failure criteria to make a rational finite element simulation constrained by the four variables. This work conducted triaxial compression tests of 18 column samples with different lamina angles to verify the modified rock failure criterion, which contributes to the simulation work on the multi-scale bedding fractures in the statics module of the ANSYS workbench. The cm-scale laminated rock samples and the km-scale Yanchang Formation in the Ordos Basin were included in the multi-scale geo-models. The simulated results indicate that stress is prone to concentrate on lamina when the lamina angle is in an effective range. The low-angle lamina always induces fractures in an open state with bigger failure apertures, while the medium-angle lamina tends to induce fractures in a shear sliding trend. In addition, the regional bedding fractures of the Yanchang Formation in the Himalayan tectonic period tend to propagate under the conditions of lower maximum principal stress, higher minimum principal stress, and larger stratigraphic dip.

Keywords: bedding fractures; failure criterion; lamina; tight oil; tight sandstone; finite element simulation; numerical simulation; unconventional reservoir

---

1. Introduction

Multi-scale bedding fractures, including km-scale regional bedding fractures and cm-scale lamina-induced fractures, are caused by lamina dissolution or induced by regional tectonic stress [1–10]. The formation of sedimentary lamina and bedding underground exerts a strong control on the following fracture propagation [11–13], so the term lamina-induced fractures is used to indicate the cm-scale fractures forming along the core lamina under the influence of external forces and internal rock mechanical properties [14]. Moreover, the lamina-induced fractures also correspond to the km-scale regional bedding fractures, which are induced by the formation beddings. These fractures have been the focus of conventional and unconventional oil and gas reservoir characterization including shale, tight oil [14–23], and Carboniferous rocks [24]. The fractures have also played an important role in methane gas emissions from coal seams [25–27]. It has been found that the open state of bedding fractures in complicated tectonic zones could play an essential role in oil and gas diffusion, emission, migration, and accumulation [14,17,28–33]. In addition, the lamina or bedding in the tight reservoirs coupled with the hydraulic fractures could induce a more complicated in-situ fracture network [34]. However, the quantitative simulation work of bedding fractures, despite the qualitative description, as mentioned above, is still lacking due to the complicated stress distribution caused by the lamina heterogeneity. It is challenging to distinguish the lamina or bedding from surrounding rocks and predict the failure behaviors under the effects of heterogeneity.

Multiple methods and softwares have been used to conduct numerical simulations and make predictions on the energy resources, e.g., artificial neural networks, ant tracking algorithms, petrophysical logging, and microseisms [35–38]. The ant tracking algorithm can be applied to detect small faults, but it is difficult to extract detailed formation of the fractures due to the extremely low resolution and low coherence of data [39]. Even though the microseism is often used for artificial fracture detection and modeling in the hydraulic fracturing process of the horizontal wells, its roles are confined to the oil-gas exploration and development period [40]. In addition, the accuracy of logging interpretation strongly depends on the data amount, and the resolution of seismic data is too low to conduct a precise fracture simulation. Therefore, it is necessary to find out an efficient method to make an accurate prediction for the multi-scale fractures without the need for a great amount of data.

In this study, a finite element simulation of bedding fractures was conducted based on the dynamic propagation condition for the tectonic bedding fractures. The finite element simulation is widely used to predict the present geological stress and fracture index, because it can provide a platform for researchers to focus on the geological or mechanical model, with the adjustability of the rock failure criterion and boundary stress conditions [41–43]. In addition, this study focused on the modified failure criterion of tectonic bedding fractures in the dynamic simulation environment.

The most challenging work for bedding fracture simulation is to build rational failure criteria in the finite element simulation environment considering four factors, i.e., maximum stress, minimum stress, lamina angle (or the km-scale regional stratigraphic dip), and differences between the lamina and surrounding rock in different lamina lithofacies. Although some failure criteria have been proposed for the bedding fractures based on the mechanical tests of laminated rock, as shown in Table 1 [6,44–47], a coefficient should be additionally proposed to indicate the different lamina lithofacies and the difference between the lamina and the surrounding rock in the laminated cores or the rock formations with bedding fractures.

Table 1. Failure criteria for the bedding fractures proposed by different researchers.

Criterion	Literature Source
$\sigma_1 = \sigma_3 + \dfrac{2\left(c_j + \sigma_3 \tan \phi_j\right)}{\left(1 - \tan \phi_j \tan \beta\right) \sin 2\beta}$	[44]
$\dfrac{\sigma_{1(\beta)} - \sigma_3}{\sigma_{1(\beta=90)} - \sigma_3} = \dfrac{k}{\cos^4 \beta + k \sin^4 \beta + 2nk \cos^2 \beta \sin^2 \beta}$	[6]
$\sigma_1 = \sigma_3 + \left(\dfrac{2 \sin \varphi}{1 - \sin \varphi} - 2S_0 \sqrt{\dfrac{1 + \sin \varphi}{1 - \sin \varphi}}\right) \dfrac{k}{\sin^4 \theta + k \cos^4 \theta + 2nk \cos^2 \theta \sin^2 \theta}$	[47]

In summary, the innovations of the multi-scale fracture (km-scale regional bedding fractures and cm-scale lamina-induced fractures in cores) simulations conducted in this work are reflected in the following aspects: (1) A modified Tien–Kuo (T–K) bedding failure criterion was proposed based on the stress equilibrium equation along the lamina, where two critical parameters (the lamina angle (θ) and lamina friction coefficient (μ_{lamina})) were proposed to precisely characterize the laminated rock.

The μ_{lamina} values were tested based on the triaxial compression tests of different lamina lithofacies. (2) A finite element simulation was conducted based on the modified T–K criterion to study the stress distribution in the laminated rock model composed of the independent lamina and the surrounding rock bodies. (3) The regional bedding fractures distribution of the Upper Triassic Yanchang Formation in the Ordos Basin was clarified based on the regional stratigraphic dip distribution, as well as the regional laminated rock lithofacies distribution proposed in this work, and the regional stress field, which was simulated in our prior simulation work [14].

2. Materials and Methods

2.1. Materials

The six laminated samples studied in this work were collected from the Chang 8 to Chang 6 member, the principal hydrocarbon-accumulating intervals in the Upper Triassic Yanchang Formation of the southern Ordos Basin. These samples could represent the different petrophysical facies in the delta-lacustrine environment. Core-1 and Core-2 are the tight siltstone- and sandstone-deposited distributary channel facies in tractive current hydrodynamic conditions. Core-1 represents the tight channel siltstone with the inconspicuous lamina, i.e., micro-cross-bedding lamina. Core-2 represents the tight channel sandstone with the macro-cross-bedding lamina. Core-3 represents the tight argillaceous siltstone with the horizontal lamina. Core-4 and Core-5 represent the tight sheet siltstone with the obvious lamina, i.e., wavy or lenticular lamina. Core-6 represents the tight siltstone deposited in the lacustrine turbidite fan facies with slumping-induced deformation lamina. Generally, the six laminated cores are typical representatives for all tight sandstone oil reservoirs, especially for the tight sandstone oil reservoir of the Ordos Basin in China.

2.2. Triaxial Compression Tests

To indicate the heterogeneity of laminated rocks and lamina lithofacies in the geo-models, we tested 18 column samples with different lamina angles from the six different laminated cores in triaxial compression experiments. The triaxial compression tests were performed with an RTR apparatus under a 20 MPa confining pressure. The 20 MPa approximates the underground pressure at a 2000 m depth, so it can be used to simulate real geological conditions. Sample treatment was conducted in strict accordance with the International Society of Rock Mechanics (ISRM) rock triaxial test requirements and the China National Standard (GB/T 50266-99).

2.3. Finite Element Simulation

Finite element simulation was performed in the statics module of the ANSYS workbench. In this method, the geological model is divided into different specific elements linked to nodes. In addition, the approximate value of the node displacement could be calculated using the element functions in the equilibrium state. Based on the computed result, we can obtain the stress and strain of these elements [48–52]. For the geological model building work, it is generally assumed that the stress and strain of the shallow crust rocks are linearly dependent during the period of rock elastic deformation [53,54]. In this study, the principle of the finite element approach is composed of the following three relationships: (1) The strain–displacement relationship; (2) strain–stress relationship; and (3) stress–external force relationship. The principle and its detailed explanations can be found in our previous study [17].

In this work, to calculate the boundary stress conditions for the finite element simulation, we modified the Newberry formula to make it more suitable for the tensional and compressive stress fields. Although the rock spatial stress computed by the Newberry formula is appropriate for the tight reservoir simulation [55], the equation is only defined in compressive stress fields and leaves the tensional stress field out of consideration [56–58]. Hence, the positive sign of the compressive stress field in the Newberry formula is changed to a negative sign for the tensional stress field. In addition, this work constrained the calculative process of the boundary stress state, based on the assumption that the 3D geo-model can be regarded as a point mass with the corresponding well burial history area. In this way, we can calculate the boundary stress state by combining the regional tectonic stress, tectonic direction, and rock density. The detailed modification of the Newberry formula and the calculative process of the boundary stress state can also be found in our previous study [17].

3. Results

3.1. Stress Balance along the Lamina

A finite element simulation of lamina-induced fractures is conducted in the statics module of the ANSYS workbench based on two geo-bodies, i.e., the lamina and surrounding rock. The contact relationship was settled to the frictional model characterized by a friction coefficient representing the difference between the two separated geo-bodies. Here, this study defined the coefficient as the lamina friction coefficient (μ_{lamina}) in the laminated rock, where μ_{lamina} could be used to indicate different lamina lithofacies and modify the T–K failure criterion in the finite element simulation. The laminated rock model is shown in Figure 1, which was constructed with the laminated body and the surrounding rock body. Two parameters were introduced into the geo-model to characterize the laminated rock, i.e., the lamina angle (θ) and μ_{lamina}. It should be emphasized that μ_{lamina} of the laminated rock differs from the internal friction coefficient of intact rocks. The μ_{lamina} is not only a physical factor representing the physical sliding behavior in simulation but also a geological factor indicating different lamina lithofacies. Therefore, the range of μ_{lamina} may vary from the traditional internal friction coefficient. Its detailed values for different laminated rock types were tested using the triaxial compression technologies in Section 2.2.

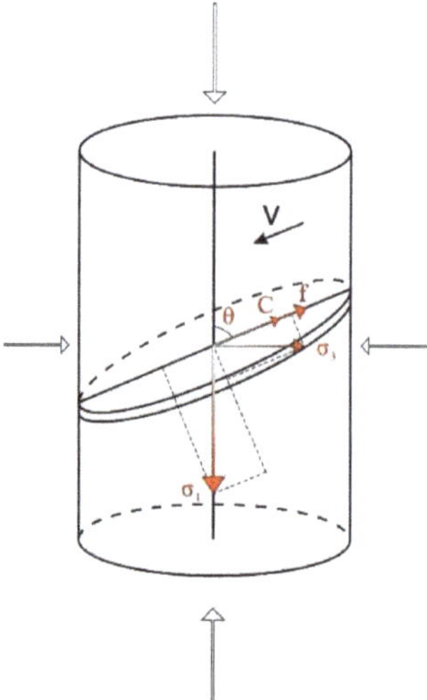

Figure 1. Laminated rock model composed of the lamina and the surrounding rock. Red arrows indicate the decomposition of stress along the lamina surface, where θ indicates the lamina angle, σ_1 indicates the axial stress, σ_3 indicates the confining pressure, C_{lamina} indicates the lamina cohesion, and f indicates the friction force between the lamina surface and surrounding rock.

Another critical factor for the failure criterion is the lamina angle (θ), indicating the intersection angle between the axial stress direction and the lamina dip direction, which is the complementary angle to the β angle (the acute angle between the direction of maximum principal stress and the discontinuity) in previous works [6,47]. A stress equilibrium equation was built based on the decomposition of stress along the lamina surface (Equation (1), Figure 1). In this equation, θ indicates the lamina angle, σ_1 indicates the axial stress, σ_3 indicates the confining pressure, and C_{lamina} indicates the lamina cohesion, which is different from the cohesion of homogeneous rock.

$$\sigma_1 \cos \theta - \sigma_3 \sin \theta = \mu_{lamina}(\sigma_3 \cos \theta + \sigma_1 \sin \theta) + C_{lamina} \tag{1}$$

The Mohr–Coulomb rock failure criterion (Equation (2)) is usually used to explain the propagation condition for shear fractures. Here, an assumption was made that the propagation condition for bedding fractures meets the Mohr–Coulomb rock failure criterion. In other words, the lamina-induced fractures could be regarded as specially compressed fractures along the lamina surface in the intact rock. Then, Equation (3) was achieved when we solved the simultaneous Equations (1) and (2). It describes the relationship among the internal friction angle (φ) of intact rock, the lamina friction coefficient (μ_{lamina}), and the lamina angle (θ) of laminated rock. Equation (4) describes the relationship among the cohesion (S_0), the internal friction angle (φ) of intact rock, and the lamina angle (θ) of laminated rock. The final relationship (Equation (5)) was built among the cohesion (S_0), the internal friction angle (φ) of intact rock, the lamina angle (θ), and the lamina friction coefficient (μ_{lamina}) of laminated rock, when Equation (3) was put into Equation (4). Thus, an important function (Equation (5)) was used to modify the T–K criterion in the following section. Furthermore, the detailed derivation and physical interpretations are shown in Appendix A.

$$\sigma_3 = 2S_0 \frac{\cos \varphi}{1 - \sin \varphi} + \sigma_3 \frac{1 + \sin \varphi}{1 - \sin \varphi} \tag{2}$$

$$\varphi = \arcsin \frac{\mu_{lamina} \cos \theta + \sin \theta}{(\mu_{lamina} + 1) \cos \theta + (1 - \mu_{lamina}) \sin \theta} \tag{3}$$

$$S_0 = \frac{1}{2} \cdot \frac{1 - \sin \varphi}{\cos \varphi} \cdot \frac{C_{lamina}}{\cos \theta - \mu_{lamina} \sin \theta} \tag{4}$$

$$S_0 = \frac{1}{2} \cdot \frac{\dfrac{\cos \theta - \mu_{lamina} \sin \theta}{(\mu_{lamina} + 1) \cos \theta + (1 - \mu_{lamina}) \sin \theta}}{\cos \arcsin \dfrac{\mu_{lamina} \cos \theta + \sin \theta}{(\mu_{lamina} + 1) \cos \theta + (1 - \mu_{lamina}) \sin \theta}} \cdot \frac{C_{lamina}}{\cos \theta - \mu_{lamina} \sin \theta} \tag{5}$$

3.2. The Failure Criterion of Bedding Fractures

Tien and Kuo (2001) proposed a common failure criterion (Equation (6)) for the intact bedded rocks, which demonstrates the characteristics of strength anisotropy revealed by the laboratory experiments, where k and n indicate the elastic constants of laminated rocks [6]. In addition, the T–K criterion is based on the nonlinear Hoek–Brown criterion for the homogeneous rock with no lamina.

$$\frac{\sigma_{1(\theta)} - \sigma_3}{\sigma_{1(\theta=90)} - \sigma_3} = \frac{k}{\cos^4 \theta + k \sin^4 \theta + 2nk \sin^2 \theta \cos^2 \theta} \tag{6}$$

Zhou et al. (2017) proposed a more efficient modified T–K criterion (Equation (7)) to replace the nonlinear Hoek–Brown criterion with the linear Mohr–Coulomb criterion, where φ indicates the cohesion of surrounding rock beyond the lamina and S_0 indicates the cohesion of laminated rocks [47]. The main problem for Zhou et al.'s modified criterion is that the cohesion of laminated rocks (S_0) was a constant in Equation (7), but it actually varied with the lamina angle (θ) and the lamina friction coefficient (μ_{lamina}) when the compressed fractures primarily propagated along the lamina surface in the laminated rocks. Thus, θ and μ_{lamina} could be used together, as two modified factors of S_0 in the

T–K criterion, to predict the simulated failure behaviors in cm-scale laminated cores to the km-scale bedding formation.

$$\sigma_1 = \sigma_3 + \left(\frac{2\sin\varphi}{1-\sin\varphi} - 2S_0 \sqrt{\frac{1+\sin\varphi}{1-\sin\varphi}} \right) \frac{k}{\sin^4\theta + k\cos^4\theta + 2nk\cos^2\theta\sin^2\theta} \tag{7}$$

Here, Equation (5) was put into Equation (7) to modify the T–K criterion. Thus, a modified T–K criterion is proposed by Equation (8), where φ indicates the surrounding rock properties and the other parameters indicate the lamina properties. The modified criterion is composed of four critical variables, i.e., the maximum principal stress (σ_1), minimum principal stress (σ_3), lamina angle (θ), and lamina friction coefficient (μ_{lamina}), which are advantageous and could also be used in the ANSYS finite element simulation.

$$\sigma_1 - \sigma_3 = \left(\frac{2\sin\varphi}{1-\sin\varphi} - \frac{\frac{\cos\theta - \mu_{lamina}\sin\theta}{(\mu_{lamina}+1)\cos\theta + (1-\mu_{lamina})\sin\theta}}{\cos\arcsin\frac{\mu_{lamina}\cos\theta + \sin\theta}{(\mu_{lamina}+1)\cos\theta + (1-\mu_{lamina})\sin\theta}} \times \frac{C_{lamina}}{\cos\theta - \mu_{lamina}\sin\theta}\sqrt{\frac{1+\sin\varphi}{1-\sin\varphi}} \right) \times \frac{k}{\sin^4\theta + k\cos^4\theta + 2nk\cos^2\theta\sin^2\theta} \tag{8}$$

3.3. Index of Bedding Fractures

Li et al. (2018) simulated the distribution of the structural fracture of the Upper Triassic Yanchang Formation in the Ordos Basin by the ANSYS software with a simplified fracture index set. The simulation was proposed based on the relationship between the maximum principal stress and the failure strength, and represents the total fracture possibility based on the distance between the envelope line and the stress state point (Equation (9)) [14]. In the bedding fracture simulation, σ_1' (the maximum normal stress satisfied the critical rupture condition) is replaced by the failure criterion for the bedding fractures (i.e., Equation (8)) to indicate the index of bedding fractures, as shown in Equation (10).

$$\begin{cases} f = \frac{\sigma_1 - \sigma_1'}{\sigma_1}, \sigma_3 > -T_0 \\ where \quad \sigma_1' = \sigma_3 + \frac{4}{\sqrt{1+\mu^2}-\mu}T_0 + \mu\sigma_3' \\ f_{LF} = 1, \sigma_3 < -T_0 \end{cases} \tag{9}$$

$$\begin{cases} f_{LF} = \frac{\sigma_1 - \sigma_1'}{\sigma_1}, \sigma_3 > -T_{0-lamina} \\ where \quad \sigma_1' = \sigma_3 + \left(\frac{2\sin\varphi}{1-\sin\varphi} - \frac{\frac{\cos\theta - \mu_{lamina}\sin\theta}{(\mu_{lamina}+1)\cos\theta + (1-\mu_{lamina})\sin\theta}}{\cos\arcsin\frac{(\mu_{lamina}-1)\cos\theta + (1+\mu_{lamina})\sin\theta}{(\mu_{lamina}+1)\cos\theta + (1-\mu_{lamina})\sin\theta}} \times \frac{C_{lamina}}{\cos\theta - \mu_{lamina}\sin\theta}\sqrt{\frac{1+\sin\varphi}{1-\sin\varphi}} \right) \times \frac{k}{\sin^4\theta + k\cos^4\theta + 2nk\cos^2\theta\sin^2\theta} \\ f_{LF} = 1, \sigma_3 < -T_{0-lamina} \end{cases} \tag{10}$$

where T_0 is the tensile strength, $T_{0-lamina}$ is the tensile strength of laminated sandstone; μ is the coefficient of internal friction; f_{LF} is the fracture index of lamina induced fractures.

4. Discussion

4.1. Comparisons of Different Failure Criteria For Bedding Fractures

To verify the modified T–K criterion, the testing data were compared to the calculated data using three criteria, i.e., the modified T–K criterion in this study, Zhou et al.'s modified T–K criterion, and the Jaeger criterion (Table 1). Figure 2 shows the triaxial compression test curves in the confining pressures of 10 and 20 MPa, where the samples are the artificial laminated rock in Zhang et al. (2017) (Figure 2a,b) and the naturally laminated rock in Zhou et al. (2017) (Figure 2c,d) [47,59]. The detailed criterion functions are listed in Table 1, and the detailed values of each criterion's parameters are listed in Table 2. Detailed peak strength values are listed in Table 3, where samples a-0 to b-90 are achieved from Zhang et al. (2017) (Figure 2a,b), samples of c-0 to d-90 are referenced from Zhou et al. (2017) (Figure 2c,d), and samples of '1-1' to '6-3' are tested in this work, as shown in Figure 3. The lamina angle in this work was the complementary angle of the β angle in Zhou et al. (2017) [47,59].

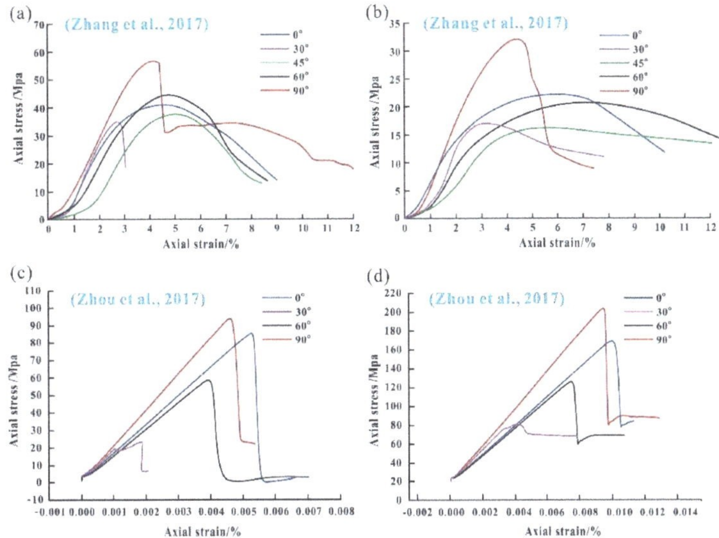

Figure 2. Stress–strain curves of laminated rock tested in previous works. (**a**) Curves tested in Zhang et al. (2017) [59]; samples are artificial laminated cores tested at 10 MPa confining pressure. (**b**) Curves tested in Zhang et al. (2017); samples are artificial laminated cores tested at 20 MPa confining pressure. (**c**) Curves tested in Zhou et al. (2017) [47]. The peak strength values of these curves are also used to verify the modified Tien–Kuo (T–K) criterion, and the samples are the naturally laminated cores tested at 10 MPa confining pressure. (**d**) Curves tested in Zhou et al. (2017). The peak strength values of these curves are also used to verify the modified T–K criterion, and samples are the naturally laminated cores tested at 20 MPa confining pressure.

Table 2. Parameters used to fit the function curves of the failure criteria.

Criteria	n	k	$\Phi/°$	μ_{lamina}	σ_3/MPa	C_{lamina}/MPa	S_0/MPa	C_j */MPa	$\varphi_j/°$ **
The modified T–K criterion	8	0.5	63.5	0.1	10	20	-	-	-
in this study	3.3	1.1	63.5	0.9	20	20	-	-	-
Zhou et al. (2017)'s	11	3.3	63.5	-	10	-	11	-	-
modified T–K criterion [47]	2.2	1.3	63.5	-	20	-	8	-	-
Jaeger criterion [44]	-	-	-	-	10	-	-	5	12
	-	-	-	-	20	-	-	2	39.5

*: The lamina cohesion in the Jaeger criterion; **: The internal friction angle in the Jaeger criterion.

Table 3. Peak strength values of the stress–strain curves in Figures 2 and 4.

Samples	Lamina Angle/°	Peak Strength/MPa	Confining Pressure/MPa
a-0	0	40.5	0
a-30	30	35	0
a-45	45	37.2	0
a-60	60	44.1	0
a-90	90	56.8	0
b-0	0	22.2	0
b-30	30	17.6	0
b-45	45	17.1	0

Table 3. *Cont.*

Samples	Lamina Angle/°	Peak Strength/MPa	Confining Pressure/MPa
b-60	60	20.5	0
b-90	90	32.3	0
c-0	0	85.4	10
c-30	30	23.2	10
c-60	60	59.4	10
c-90	90	94.1	10
d-0	0	168	20
d-30	30	80.4	20
d-60	60	128	20
d-90	90	205	20
1-1	80	221	20
1-2	30	220	20
1-3	60	239	20
2-1	90	222	20
2-2	5	195	20
2-3	40	124	20
3-1	75	208	20
3-2	10	189	20
3-3	45	122	20
4-1	85	300	20
4-2	3	234	20
4-3	40	175	20
5-1	70	241	20
5-2	10	218	20
5-3	30	173	20
6-1	70	146	20
6-2	10	162	20
6-3	30	121	20

As shown in Figure 2, there was a much lower peak strength for those samples with the lamina angle of 30°. Notably, black square points in Figure 3 indicate the compression peak strength values tested with the 20 MPa confining pressure in Zhou et al. (2017) (Figure 2d) [47]. Black round points in Figure 3 indicate the compressive peak strength values tested with the 10 MPa confining pressure in Zhou et al. (2017) (Figure 2c) [47]. In Figure 3, function curve 1 indicates the peak strength values fitted in Zhou et al. (2017)'s modified T–K criterion with a 10 MPa confining pressure [47]. Function curve 2 indicates the peak strength values fitted in Zhou et al. (2017)'s modified T–K criterion with a 20 MPa confining pressure [47]. Function curves 3 and 4 indicate the peak strength values fitted in the Jaeger failure criterion with 10 and 20 MPa confining pressures, respectively [44]. In addition, function curves 5 and 6 indicate the peak strength values fitted in our modified T–K criterion with 10 and 20 MPa confining pressures, respectively.

Generally, the modified T–K failure criterion has almost the same trend as the T–K criterion and even agrees better with the testing data in Zhang et al. (2017), as shown in Figure 3, especially for the medium-low-angle range (nearly 20°–50°) [59]. Moreover, the four variables in this criterion are in accordance with the ANSYS parameter environment and could provide us with a better failure criterion in the finite element simulation.

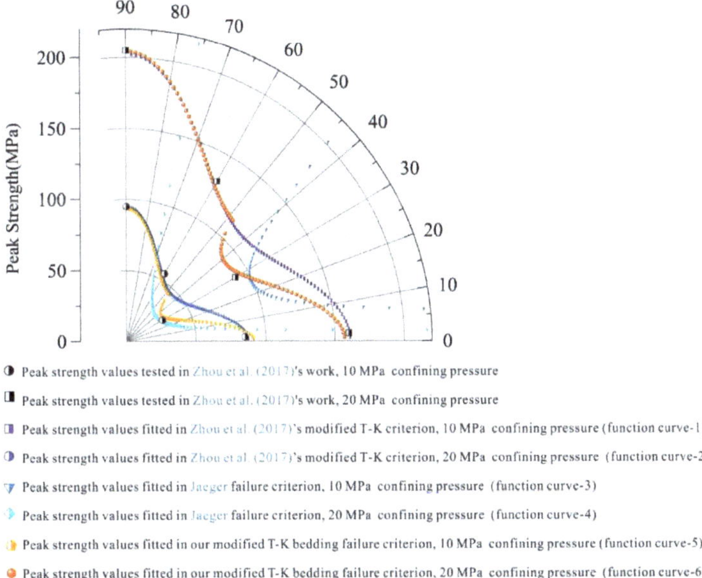

Peak strength values tested in Zhou et al. (2017)'s work, 10 MPa confining pressure

Peak strength values tested in Zhou et al. (2017)'s work, 20 MPa confining pressure

Peak strength values fitted in Zhou et al. (2017)'s modified T-K criterion, 10 MPa confining pressure (function curve-1)

Peak strength values fitted in Zhou et al. (2017)'s modified T-K criterion, 20 MPa confining pressure (function curve-2)

Peak strength values fitted in Jaeger failure criterion, 10 MPa confining pressure (function curve-3)

Peak strength values fitted in Jaeger failure criterion, 20 MPa confining pressure (function curve-4)

Peak strength values fitted in our modified T-K bedding failure criterion, 10 MPa confining pressure (function curve-5)

Peak strength values fitted in our modified T-K bedding failure criterion, 20 MPa confining pressure (function curve-6)

Figure 3. Function curves of the different failure criteria for the bedding fractures, i.e., the modified T–K criterion in this work, Zhou et al. (2017)'s modified T–K criterion, and the Jaeger criterion [44,47]. Black points indicate the peak strength values of the laminated rock in different confining pressures tested in Zhou et al. (2017).

4.2. Lamina Friction Coefficients of Different Lamina Lithofacies

It is necessary to use the lamina friction coefficient (μ_{lamina}) to indicate the regional heterogeneity of lamina lithofacies in the geo-simulation work. In order to test the lamina friction coefficient (μ_{lamina}) of the different lamina rock types, we tested 18 column samples with different lamina angles from six different laminated cores in triaxial compression experiments. Triaxial compression stress–strain curves are shown in Figure 4. Peak strength values obviously varied between the samples with the different lamina angles, except Core-1, which was similar to the intact core without lamina. The peak strength values were compared to the calculated values using the modified T–K criterion (Figure 5). Black points with the different shapes indicate the peak strength values of the six cores shown in Figures 4 and 5. Colored dotted lines indicate the function curve of the modified T–K criterion, where the different colors indicate the calculated data in different lamina friction coefficients (μ_{lamina}). The values of the lamina friction coefficient (μ_{lamina}) are 0.01, 0.1, 0.2, 0.3, 0.4, 0.5, 0.6, 0.7, 0.8, 0.9, and 1. In addition, the value of φ is 58°, σ_3 is 20 MPa, n is 2.20, k is 0.9, and C_{lamina} is 20 MPa.

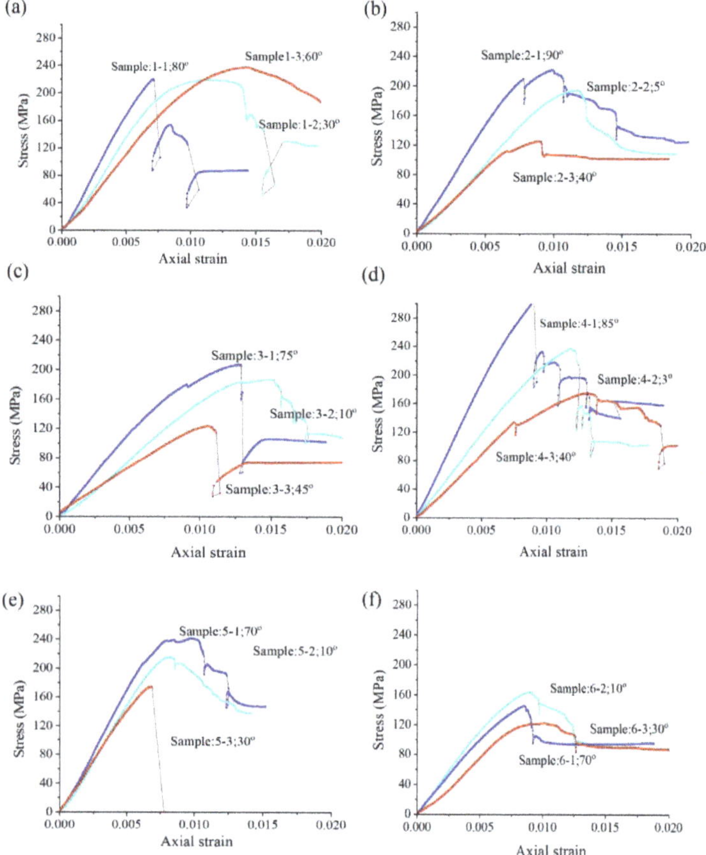

Figure 4. Stress–strain curves of the six laminated core samples with different lamina angles at a 20 MPa confining pressure. (**a**) Core-1; (**b**) Core-2; (**c**) Core-3; (**d**) Core-4; (**e**) Core-5; (**f**) Core-6. Blue lines indicate those samples in a high lamina angle range, cyan lines indicate those samples in a low lamina angle range, and red lines indicate those samples in a medium lamina angle range.

As indicated in Figure 5, the following results could be drawn. (1) Different lamina friction coefficients (μ_{lamina}) indicate different laminated core types. The cores more strongly characterized by the obvious lamina, such as Core-4, Core-5, and Core-6, are typically correlated with a lower μ_{lamina}. (2) The fitting data agreed well with the testing data, except Core-1, which is characterized by the inconspicuous and micro-cross-bedding lamina. Here, the strength values of Core-1 are nearly located in the same range due to the weak influence of these inconspicuous and micro-cross-bedding lamina within Core-1. In other words, the smaller the difference between the lamina and surrounding rock, the weaker the influence of the lamina in different angle ranges. Furthermore, the values of μ_{lamina} of Core-4 and Core-5 are around 0.5, which are the laminated siltstone deposited by traction currents in the delta front sedimentary environment. The value of μ_{lamina} of Core-6 is around 0.3, which is the laminated siltstone sedimented in the lacustrine turbidite environment. The value of μ_{lamina} of Core-3 is around 0.9, which is the laminated mudstone sedimented in a still water environment. From the perspective of the regional finite element simulation, the μ_{lamina}, as a critical variable, should be assigned differently according to the laminated rock types deposited in the different sedimentary environments during Yanchang Formation in the Ordos Basin.

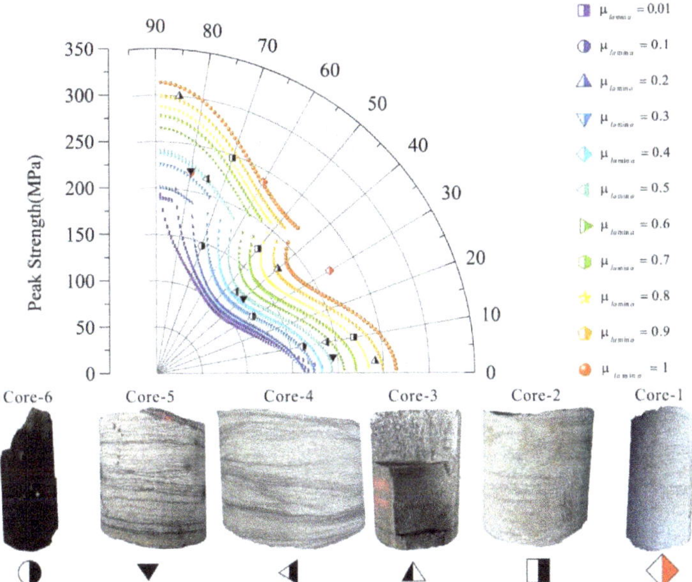

Figure 5. Function curves calculated with the modified T–K criterion using different lamina coefficients (μ_{lamina}). Black points in different shapes indicate the different laminated rocks. Notably, peak strength values of Core-1 (with inconspicuous lamina) in the red points do not agree with the calculated function curves.

4.3. Stress Distribution in the Laminated Rock

The laminated core model (cm-scale) in the finite element simulation is composed of two bodies, as shown in Figure 6, where the contact part between two bodies is settled as frictional with a sliding friction coefficient in accordance with the lamina friction coefficient in the failure criteria of this study. The failure behavior of the laminated rock in the special lamina angle range is determined by the stress distribution constrained by the constant value of μ_{lamina}. Therefore, it is necessary to study the stress distribution of the laminated rock using the finite element simulation technology, especially for the different lamina angle ranges, i.e., the low, medium, and high lamina angle ranges.

Researchers have tried to simulate the compression behavior of tight intact rock and the laminated rock using finite element simulation technology [60,61]. To better understand the stress distribution during the compression process, the failure distribution around the compression samples was simulated by using the modified criterion in the ANSYS environment. The simulated samples were divided into two parts: The surrounding rock and the lamina body, with the lamina surface acting as the sliding surface. Additionally, the mechanical parameters of the surrounding rock used in the simulation were different from those of the laminated rock. Here, the surrounding rock was set as siltstone, while the laminated rock was claystone in the ANSYS software. In addition, the contact relationship, i.e., the lamina sliding surface, was set as frictional in the ANSYS software with the fractional coefficient value of 0.2. Another critical parameter, the lamina angle, was designed in the computer model, as shown in Figure 6a,c,e.

Accordingly, the failure criterion was defined according to the modified T–K criterion in the ANSYS software. Figure 6a,c,e show the simulated results, in contrast to the CT (computerized tomography) results of fractured samples (Figure 6b,d,f) with the low, medium, and high lamina angle ranges, respectively.

By contrast, for the high lamina angle (Figure 6a,b) outside the effective lamina angle range, stress was prone to concentrate on the surrounding rock body compared to the samples simulated in the medium and low lamina angle ranges (Figure 6c,e). Therefore, the typical X-conjugate fractures were prone to propagate without associated bedding fractures. Notably, the effective lamina angle indicates that those compressed fractures were generally induced by the lamina body on which the stress was primarily concentrated.

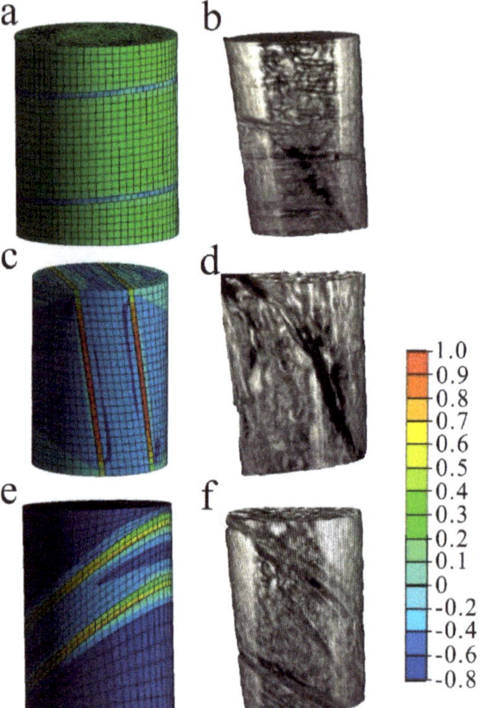

Figure 6. Finite element simulation results of the bedding fractures (i.e., lamina-induced fractures) and the computerized tomography (CT) photos of the fractured laminated samples. The color code for the finite element simulation results of (**a,c,e**) indicates the fractured index based on the modified T–K criterion. (**b,d,f**) show the CT photos of fractured samples, and the darker color indicates a bigger fracture aperture.

For the medium lamina angle in the effective lamina angle range, a shear fracture propagated between the two bedding fractures, as shown in the CT photo (Figure 6c,d), and was accompanied by some shear fractures between the laminated bodies. The fracture aperture was smaller than that in the core with a small lamina angle and was characterized by the shear sliding distance, as shown in Figure 6d. In addition, stress was prone to concentrate around the rock body rather than the sample simulated at a low lamina angle, which could be indicated by the CT photos (Figure 6a,c,d).

For the low lamina angle (Figure 6e,f) in the effective lamina angle range, the CT photo showed a bigger fracture aperture in the darker color. In addition, the simulated fracture index of the lamina was much higher than that of the surrounding rock, indicating a stronger propagation tendency for the bedding fractures caused by the stress concentration on the lamina.

Overall, stress in these compression samples was prone to concentrate on the lamina when the lamina angle was in the effective lamina angle (low- and medium-angle range). The low-angle lamina always induces fractures in an extensively open state with a bigger failure aperture, and the medium-angle lamina always induces fractures in a shear sliding trend. On the contrary, stress was prone to concentrate on the surrounding rock body when the lamina angle was outside of the effective range, and contributed less to the compressed bedding fractures in the laminated rock.

4.4. Distribution of the Regional Bedding Fractures

In order to simulate the regional bedding fractures generated in a particular tectonic period using the modified criteria, previous simulation studies of the conventional structural fractures at the regional scale are used as the tectonic stress field [14,19,20,62]. Thus, the bedding fractures of the Yanchang Formation in the Ordos Basin generated during the latest orogenic episode (i.e., Himalayan episode) have been simulated by ameliorating Li et al. (2018) [14]. Specifically, as a result, the bedding fractures were calculated (Figure 7d) based on the improved rock failure criteria for bedding fractures (Equation (10)), the stress field simulated in Li et al. (2018) (Figure 7a,b) [14], the stratigraphic dip distribution map (Figure 7c), and the lamina fraction coefficient in different lithofacies (Figure 7d). Here, the stratigraphic dip at the regional scale (i.e., the intersection angle between the horizontal direction and the dip direction at the regional scale) is equal to the lamina angle at the cm-scale, because the maximum principal stress is in the horizontal direction during the orogenic periods, where the stratigraphic dips of the Yanchang Formation are valued in the relatively effective range for the bedding fracture failure criteria, as mentioned before (Figure 7c). In addition, μ_{lamina} is assigned different values according to the different fitting results in Figure 5. The laminated siltstone sedimented in a delta front environment (Figures 5 and 7d) is assigned 0.5, the laminated turbidite sandstone sedimented in a lacustrine turbidite environment (Figures 5 and 7d) is assigned 0.3, and the mud sedimented in a still water environment (Figures 5 and 7d) is assigned 0.9.

The finite element simulation environment, including the boundary condition, Himalayan tensile stress field, geological model, and petrophysical model, remained consistent with those in Li et al. (2018) [14]. The bedding fracture index in the Yanchang Formation was calculated based on four variables, i.e., the maximum principal stress (σ_1), minimum principal stress (σ_3), lamina angle (θ), and lamina friction coefficient (μ_{lamina}). The calculated bedding fracture distribution, as shown in Figure 8, is broadly consistent with the structural fractured regions simulated by Li et al. (2018) [14], such as the typical oilfield regions of "Dingbian," "Xin'anbian," and "Zhiluo" (Figure 7c). The difference is that the simulated bedding fractures spread over an even larger area, which reveals that bedding fractures are more easily induced than the conventional structural fractures (i.e., the tensile, shear, and hybrid fractures) under the same stress condition. For the Himalayan tensile stress field of the Yanchang Formation in the Ordos Basin, it is obvious that the bedding fractures tend to propagate under the conditions of lower maximum principal stress, higher minimum principal stress, and higher stratigraphic dip values (Figure 7a–d). Generally, the bedding fractures distribution simulated in this work (Figure 7d) would be a supplement for the structural fractures simulated by Li et al. (2018) [14], and both of the two research works contribute to the further study of the natural fracture network underground (including the bedding fractures and the conventional structural fractures) for the Yanchang Formation of the Ordos Basin.

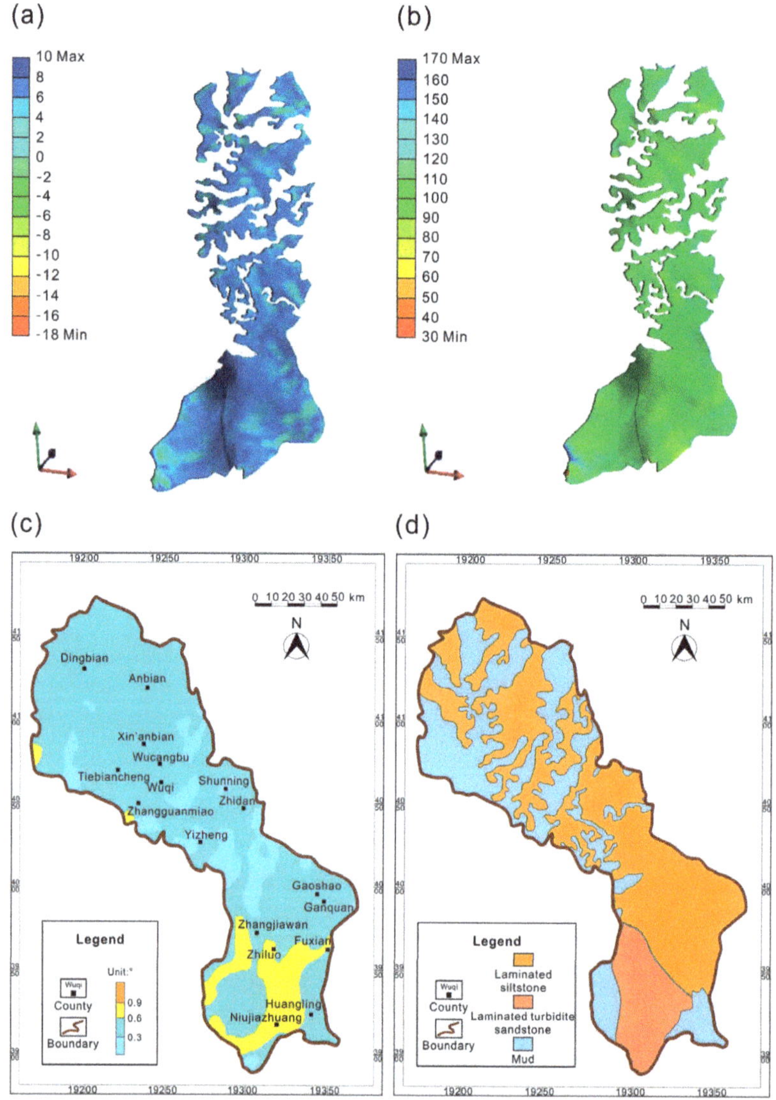

Figure 7. Calculation parameters used in the bedding fractures simulation. (**a,b**) Himalayan stress field during the Yanchang Formation simulated in Li et al. (2018) [14]; (**c**) stratigraphic dip map of tight reservoirs of Yanchang Formation; (**d**) lamina lithofacies distribution of tight reservoirs of Yanchang Formation.

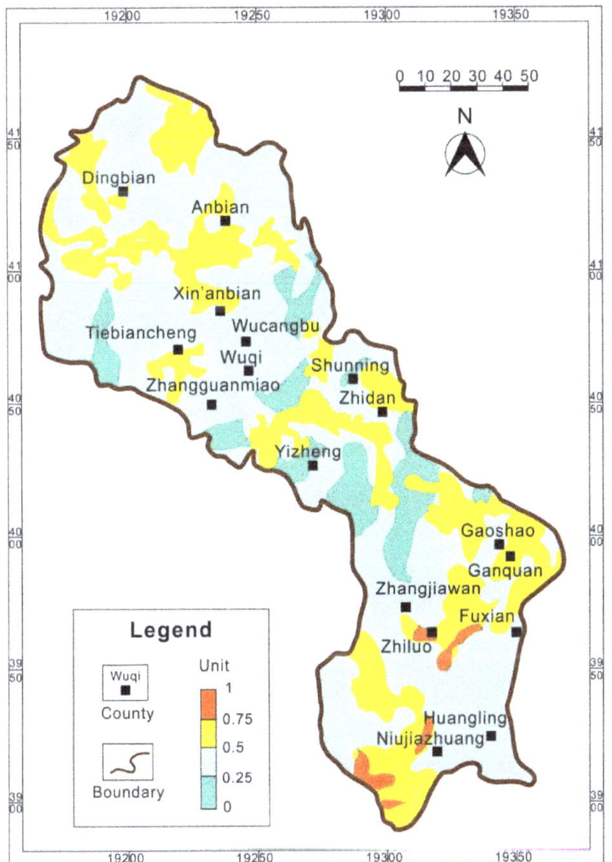

Figure 8. Bedding fractures distribution of tight sandstone reservoirs of Yanchang Formation in Ordos Basin during the Himalayan episode.

5. Conclusions

1. Two critical parameters were introduced to modify the T–K criterion, i.e., the lamina angle (θ) based on the stress equilibrium equation along the lamina surface, and the lamina friction coefficient (μ_{lamina}) related to the different laminated rock type.

2. Stress is prone to concentrate on the lamina when the lamina angle is in the effective low and medium range. The low-angle lamina always induces fractures in the expansively open state with the bigger failure aperture, and the medium-angle lamina always induces fractures in the shear sliding trend. On the contrary, stress is prone to concentrate on the surrounding rock body, when the lamina angle is outside of the effective range, and contributes less to the compressed fractures in the laminated rock.

3. The bedding fractures of the Yanchang Formation in the Himalayan stress field were simulated based on the new rock failure criteria proposed in this work. It is concluded that these bedding fractures tend to propagate under the conditions of lower maximum principal stress, higher minimum principal stress, and larger stratigraphic dip.

Author Contributions: Conceptualization, Y.L.; Data curation, Q.W., Z.J. and D.L.; Formal analysis, D.L.; Funding acquisition, W.Y., Z.J. and Y.S.; Investigation, Q.W. and Y.L.; Methodology, Q.W. and Y.L.; Project administration, W.Y., Y.S. and Q.L.; Resources, D.L.; Software, Y.L. and D.L.; Supervision, W.Y., Z.J., Y.S., S.J. and Q.L.; Validation,

Y.L.; Visualization, Y.L.; Writing—original draft, Q.W. and Y.L.; Writing—review & editing, Q.W. and Y.L. All authors have read and agreed to the published version of the manuscript.

Funding: This study was financially supported by the Science Foundation for top-notch innovative talents of China University of Petroleum, Beijing (No. 2462017BJB07), the National Science and Technology Major Project (No. 2016ZX05034-001 and 2017ZX05035-002), and the Key Scientific and Technological Project of Yanchang Oilfield (No. ycsy2015ky-B-01-09).

Acknowledgments: We highly appreciate two anonymous reviewers for their valuable and constructive comments, which are significantly helpful for us to improve the quality of our research achievements. Special acknowledgment is also given to the Scientific Research Institute of Petroleum Exploration and Development from the Shaanxi Yanchang Petroleum (Group) Co., Ltd. for providing the samples used in this study.

Conflicts of Interest: The authors declare no conflict of interest.

Nomenclature

σ_1	the maximum principal stress
σ_3	the minimum principal stress
θ	the lamina angle
μ_{lamina}	the lamina friction coefficient
C_{lamina}	the lamina cohesion
f	the friction force between the lamina surface and surrounding rock
φ	the internal friction angle
S_0	the cohesion of laminated rocks
k and n	the elastic constants of laminated rocks
β	the acute angle between the direction of maximum principal stress and the discontinuity
C_j	the lamina cohesion in the Jaeger criterion
φ_j	the internal friction angle in the Jaeger criterion
T_0	the tensile strength
f_{LF}	the fracture index of lamina induced fractures
$T_{0\text{-}lamina}$	the tensile strength of laminated sandstone
μ	the coefficient of internal friction
σ_1'	the maximum normal stress satisfied the critical rupture condition

Appendix A

The stress balance in Figure 1 is based on shear strength and is expressed as follows:

$$\sigma_1 \cos\theta - \sigma_3 \sin\theta = \mu_{lamina}(\sigma_3 \cos\theta + \sigma_1 \sin\theta) + C_{lamina} \tag{A1}$$

where $f = \mu_{lamina} \times (\sigma_3\cos\theta + \sigma_1\sin\theta)$ and represents the physical and frictional interactions between the lamina and the surrounding rock, as shown in Figure 1. C_{lamina} represents the chemical cementation between the lamina and the surrounding rock, which is treated as a constant parameter in this simulation.

To establish an equation in contrast to the Mohr–Coulomb failure criterion (Equation (A2)), Equation (A3) is achieved when Equation (A1) is transformed into (A2).

$$\sigma_1 = 2S_0 \frac{\cos\varphi}{1 - \sin\varphi} + \frac{1 + \sin\varphi}{1 - \sin\varphi}\sigma_3 \tag{A2}$$

$$\sigma_1 = \frac{C_{lamina}}{\cos\theta - \mu_{lamina}\sin\theta} + \frac{\sin\theta + \mu_{lamina}\cos\theta}{\cos\theta - \mu_{lamina}\sin\theta}\sigma_3 \tag{A3}$$

It is possible to compare Equation (A2) with (A3) when the lamina-induced fractures are considered a special product emerging when shear fractures slide along the lamina surface. Thus, an equivalent relationship between the constant term and the coefficient term of the variables σ_1 and σ_3 can be expressed as Equations (A4) and (A5):

$$\frac{C_{lamina}}{\cos\theta - \mu_{lamina}\sin\theta} = 2S_0\frac{\cos\varphi}{1 - \sin\varphi} \tag{A4}$$

$$\frac{\sin\theta + \mu_{lamina}\cos\theta}{\cos\theta - \mu_{lamina}\sin\theta} = \frac{1 + \sin\varphi}{1 - \sin\varphi} \tag{A5}$$

Then, φ and S_0 can be expressed as Equations (A6) and (A7):

$$\varphi = \arcsin \frac{\mu_{lamina}\cos\theta + \sin\theta}{(\mu_{lamina}+1)\cos\theta + (1-\mu_{lamina})\sin\theta} \tag{A6}$$

$$S_0 = \frac{(1-\sin\varphi)C_{lamina}}{2\cos\varphi(\cos\theta - \mu_{lamina}\sin\theta)} \tag{A7}$$

Thus, S_0 can be further expressed as Equation (A8) when Equation (A6) is put into Equation (A7).

$$S_0 = \frac{1}{2}\cdot\frac{\frac{\cos\theta-\mu_{lamina}\sin\theta}{(\mu_{lamina}+1)\cos\theta+(1-\mu_{lamina})\sin\theta}}{\cos\arcsin\frac{\mu_{lamina}\cos\theta+\sin\theta}{(\mu_{lamina}+1)\cos\theta+(1-\mu_{lamina})\sin\theta}}\cdot\frac{C_{lamina}}{\cos\theta-\mu_{lamina}\sin\theta} \tag{A8}$$

References

1. Cheng, Y.; Lu, Y.; Ge, Z.; Cheng, L.; Zheng, J.; Zhang, W. Experimental study on crack propagation control and mechanism analysis of directional hydraulic fracturing. *Fuel* **2018**, *218*, 316–324. [CrossRef]
2. Desai, C.S.; Zaman, M.M.; Lightner, J.G.; Siriwardane, H.J. Thin-layer element for interfaces and joints. *Int. J. Numer. Anal. Methods Geomech.* **1984**, *8*, 19–43. [CrossRef]
3. Hu, S.-C.; Tan, Y.-L.; Zhou, H.; Guo, W.-Y.; Hu, D.-W.; Meng, F.-Z.; Liu, Z.-G. Impact of Bedding Planes on Mechanical Properties of Sandstone. *Rock Mech. Rock Eng.* **2017**, *50*, 2243–2251. [CrossRef]
4. Ma, L.-J.; Liu, X.-Y.; Wang, M.-Y.; Xu, H.-F.; Hua, R.-P.; Fan, P.-X.; Jiang, S.-R.; Wang, G.-A.; Yi, Q.-K. Experimental investigation of the mechanical properties of rock salt under triaxial cyclic loading. *Int. J. Rock Mech. Min. Sci.* **2013**, *62*, 34–41. [CrossRef]
5. Sampath, K.H.S.M.; Perera, M.S.A.; Elsworth, D.; Ranjith, P.G.; Matthai, S.K.; Rathnaweera, T. Experimental Investigation on the Mechanical Behavior of Victorian Brown Coal under Brine Saturation. *Energy Fuels* **2018**, *32*, 5799–5811. [CrossRef]
6. Tien, Y.M.; Kuo, M.C. A failure criterion for transversely isotropic rocks. *Int. J. Rock Mech. Min. Sci.* **2001**, *38*, 399–412. [CrossRef]
7. Vervoort, A.; Min, K.-B.; Konietzky, H.; Cho, J.-W.; Debecker, B.; Dinh, Q.-D.; Frühwirt, T.; Tavallali, A. Failure of transversely isotropic rock under Brazilian test conditions. *Int. J. Rock Mech. Min. Sci.* **2014**, *70*, 343–352. [CrossRef]
8. Wu, J.-H.; Ohnishi, Y.; Nishiyama, S. Simulation of the mechanical behavior of inclined jointed rock masses during tunnel construction using Discontinuous Deformation Analysis (DDA). *Int. J. Rock Mech. Min. Sci.* **2004**, *41*, 731–743. [CrossRef]
9. Zhang, Z.-H.; Deng, J.-H.; Zhu, J.-B.; Li, L.-R. An Experimental Investigation of the Failure Mechanisms of Jointed and Intact Marble under Compression Based on Quantitative Analysis of Acoustic Emission Waveforms. *Rock Mech. Rock Eng.* **2018**, *51*, 2299–2307. [CrossRef]
10. Zhou, Y.; Wu, S.; Li, Z.; Zhu, R.; Xie, S.; Jing, C.; Lei, L. Multifractal Study of Three-Dimensional Pore Structure of Sand-Conglomerate Reservoir Based on CT Images. *Energy Fuels* **2018**, *32*, 4797–4807. [CrossRef]
11. Cilona, A.; Aydin, A.; Likerman, J.; Parker, B.; Cherry, J. Structural and statistical characterization of joints and multi-scale faults in an alternating sandstone and shale turbidite sequence at the Santa Susana Field Laboratory: Implications for their effects on groundwater flow and contaminant transport. *J. Struct. Geol.* **2016**, *85*, 95–114. [CrossRef]
12. Ishii, E. The role of bedding in the evolution of meso- and microstructural fabrics in fault zones. *J. Struct. Geol.* **2016**, *89*, 130–143. [CrossRef]
13. Zhang, Y.; Zhang, Z.; Sarmadivaleh, M.; Lebedev, M.; Barifcani, A.; Yu, H.; Iglauer, S. Micro-scale fracturing mechanisms in coal induced by adsorption of supercritical CO_2. *Int. J. Coal Geol.* **2017**, *175*, 40–50. [CrossRef]
14. Li, Y.; Song, Y.; Jiang, Z.; Yin, L.; Chen, M.; Liu, D. Major factors controlling lamina induced fractures in the Upper Triassic Yanchang formation tight oil reservoir, Ordos basin, China. *J. Asian Earth Sci.* **2018**, *166*, 107–119. [CrossRef]
15. Espinoza, D.N.; Shovkun, I.; Makni, O.; Lenoir, N. Natural and induced fractures in coal cores imaged through X-ray computed microtomography—Impact on desorption time. *Int. J. Coal Geol.* **2016**, *154–155*, 165–175. [CrossRef]

16. Li, L.; Sheng, J.J.; Su, Y.; Zhan, S. Further Investigation of Effects of Injection Pressure and Imbibition Water on CO2 Huff-n-Puff Performance in Liquid-Rich Shale Reservoirs. *Energy Fuels* **2018**, *32*, 5789–5798. [CrossRef]

17. Li, Y.; Song, Y.; Jiang, Z.; Yin, L.; Luo, Q.; Ge, Y.; Liu, D. Two episodes of structural fractures: Numerical simulation of Yanchang Oilfield in the Ordos basin, northern China. *Mar. Pet. Geol.* **2018**, *97*, 223–240. [CrossRef]

18. Liang, S.; Elsworth, D.; Li, X.; Fu, X.; Sun, B.; Yao, Q. Key strata characteristics controlling the integrity of deep wells in longwall mining areas. *Int. J. Coal Geol.* **2017**, *172*, 31–42. [CrossRef]

19. Liu, J.; Ding, W.; Wang, R.; Yin, S.; Yang, H.; Gu, Y. Simulation of paleotectonic stress fields and quantitative prediction of multi-period fractures in shale reservoirs: A case study of the Niutitang Formation in the Lower Cambrian in the Cen'gong block, South China. *Mar. Pet. Geol.* **2017**, *84*, 289–310. [CrossRef]

20. Liu, J.; Ding, W.; Yang, H.; Wang, R.; Yin, S.; Li, A.; Fu, F. 3D geomechanical modeling and numerical simulation of in-situ stress fields in shale reservoirs: A case study of the lower Cambrian Niutitang formation in the Cen'gong block, South China. *Tectonophysics* **2017**, *712–713*, 663–683. [CrossRef]

21. Weniger, S.; Weniger, P.; Littke, R. Characterizing coal cleats from optical measurements for CBM evaluation. *Int. J. Coal Geol.* **2016**, *154–155*, 176–192. [CrossRef]

22. Widera, M. Lignite cleat studies from the first Middle-Polish (first Lusatian) lignite seam in central Poland. *Int. J. Coal Geol.* **2014**, *131*, 227–238. [CrossRef]

23. Zeng, L.; Li, X. Fractures in sandstone reservoirs with ultra-low permeability: A case study of the Upper Triassic Yanchang Formation in the Ordos Basin, China. *AAPG Bull.* **2009**, *93*, 461–477.

24. Małkowski, P.; Ostrowski, Ł.; Brodny, J. Analysis of Young's modulus for Carboniferous sedimentary rocks and its relationship with uniaxial compressive strength using different methods of modulus determination. *J. Sustain. Min.* **2018**, *17*, 145–157. [CrossRef]

25. Cai, Y.; Liu, D.; Mathews, J.P.; Pan, Z.; Elsworth, D.; Yao, Y.; Li, J.; Guo, X. Permeability evolution in fractured coal—Combining triaxial confinement with X-ray computed tomography, acoustic emission and ultrasonic techniques. *Int. J. Coal Geol.* **2014**, *122*, 91–104. [CrossRef]

26. Kędzior, S.; Dreger, M. Methane occurrence, emissions and hazards in the Upper Silesian Coal Basin, Poland. *Int. J. Coal Geol.* **2019**, *211*, 103226. [CrossRef]

27. Tutak, M.; Brodny, J. Forecasting Methane Emissions from Hard Coal Mines Including the Methane Drainage Process. *Energies* **2019**, *12*, 3840. [CrossRef]

28. Ardakani, O.H.; Sanei, H.; Ghanizadeh, A.; McMechan, M.; Ferri, F.; Clarkson, C.R. Hydrocarbon potential and reservoir characteristics of Lower Cretaceous Garbutt Formation, Liard Basin Canada. *Fuel* **2017**, *209*, 274–289. [CrossRef]

29. Fan, D.; Ettehadtavakkol, A. Semi-analytical modeling of shale gas flow through fractal induced fracture networks with microseismic data. *Fuel* **2017**, *193*, 444–459. [CrossRef]

30. Gong, J.; Rossen, W.R. Characteristic fracture spacing in primary and secondary recovery for naturally fractured reservoirs. *Fuel* **2018**, *223*, 470–485. [CrossRef]

31. Lyu, W.; Zeng, L.; Zhang, B.; Miao, F.; Lyu, P.; Dong, S. Influence of natural fractures on gas accumulation in the Upper Triassic tight gas sandstones in the northwestern Sichuan Basin, China. *Mar. Pet. Geol.* **2017**, *83*, 60–72. [CrossRef]

32. Mehana, M.; Al Salman, M.; Fahes, M. Impact of Salinity and Mineralogy on Slick Water Spontaneous Imbibition and Formation Strength in Shale. *Energy Fuels* **2018**, *32*, 5725–5735. [CrossRef]

33. Zeng, L.; Tang, X.; Wang, T.; Gong, L. The influence of fracture cements in tight Paleogene saline lacustrine carbonate reservoirs, western Qaidam Basin, northwest ChinaGeologic Notes. *AAPG Bull.* **2012**, *96*, 2003–2017. [CrossRef]

34. Parvizi, H.; Rezaei-Gomari, S.; Nabhani, F. Robust and Flexible Hydrocarbon Production Forecasting Considering the Heterogeneity Impact for Hydraulically Fractured Wells. *Energy Fuels* **2017**, *31*, 8481–8488. [CrossRef]

35. Brodny, J.; Tutak, M. Exposure to Harmful Dusts on Fully Powered Longwall Coal Mines in Poland. *Int. J. Environ. Res. Public Health* **2018**, *15*, 3. [CrossRef]

36. Ostojic, J.; Rezaee, R.; Bahrami, H. Production performance of hydraulic fractures in tight gas sands, a numerical simulation approach. *J. Pet. Sci. Eng.* **2012**, *88–89*, 75–81. [CrossRef]

37. Tokhmchi, B.; Memarian, H.; Rezaee, M.R. Estimation of the fracture density in fractured zones using petrophysical logs. *J. Pet. Sci. Eng.* **2010**, *72*, 206–213. [CrossRef]

38. Tutak, M.; Brodny, J. Predicting Methane Concentration in Longwall Regions Using Artificial Neural Networks. *Int. J. Environ. Res. Public Health* **2019**, *16*, 2. [CrossRef]

39. Miller, P.; Dasgupta, S.; Shelander, D. Seismic imaging of migration pathways by advanced attribute analysis, Alaminos Canyon 21, Gulf of Mexico. *Mar. Pet. Geol.* **2012**, *34*, 111–118. [CrossRef]

40. Maxwell, S.C.; Waltman, C.; Warpinski, N.R.; Mayerhofer, M.J.; Boroumand, N. Imaging Seismic Deformation Induced by Hydraulic Fracture Complexity. *SPE Reserv. Eval. Eng.* **2009**, *12*, 48–52. [CrossRef]

41. Ding, W.; Fan, T.; Yu, B.; Huang, X.; Liu, C. Ordovician carbonate reservoir fracture characteristics and fracture distribution forecasting in the Tazhong area of Tarim Basin, Northwest China. *J. Pet. Sci. Eng.* **2012**, *86–87*, 62–70. [CrossRef]

42. Ju, W.; Sun, W. Tectonic fractures in the Lower Cretaceous Xiagou Formation of Qingxi Oilfield, Jiuxi Basin, NW China Part one: Characteristics and controlling factors. *J. Pet. Sci. Eng.* **2016**, *146*, 617–625. [CrossRef]

43. Zeng, L.; Wang, H.; Gong, L.; Liu, B. Impacts of the tectonic stress field on natural gas migration and accumulation: A case study of the Kuqa Depression in the Tarim Basin, China. *Mar. Pet. Geol.* **2010**, *27*, 1616–1627. [CrossRef]

44. Jaeger, J.C. Shear failure of anisotropic rocks. *Geol. Mag.* **1960**, *97*, 65–72. [CrossRef]

45. Zhou, Y.; Feng, X.; Xu, D.; Chen, D.; Li, S. Experimental study of mechanical response of thin-bedded limestone under bending conditions. *Rock Soil Mech.* **2016**, *37*, 1895–1902.

46. Zhou, Y.-Y.; Feng, X.-T.; Xu, D.-P.; Fan, Q.-X. Experimental Investigation of the Mechanical Behavior of Bedded Rocks and Its Implication for High Sidewall Caverns. *Rock Mech. Rock Eng.* **2016**, *49*, 3643–3669. [CrossRef]

47. Zhou, Y.-Y.; Feng, X.-T.; Xu, D.-P.; Fan, Q.-X. An enhanced equivalent continuum model for layered rock mass incorporating bedding structure and stress dependence. *Int. J. Rock Mech. Min. Sci.* **2017**, *97*, 75–98. [CrossRef]

48. Fries, T.-P.; Belytschko, T. The extended/generalized finite element method: An overview of the method and its applications. *Int. J. Numer. Methods Eng.* **2010**. [CrossRef]

49. Geuzaine, C.; Remacle, J.-F. Gmsh: A 3-D finite element mesh generator with built-in pre- and post-processing facilities. *Int. J. Numer. Methods Eng.* **2009**, *79*, 1309–1331. [CrossRef]

50. Liu, G.R.; Dai, K.Y.; Nguyen, T.T. A Smoothed Finite Element Method for Mechanics Problems. *Comput. Mech.* **2006**, *39*, 859–877. [CrossRef]

51. Roters, F.; Eisenlohr, P.; Hantcherli, L.; Tjahjanto, D.D.; Bieler, T.R.; Raabe, D. Overview of constitutive laws, kinematics, homogenization and multiscale methods in crystal plasticity finite-element modeling: Theory, experiments, applications. *Acta Mater.* **2010**, *58*, 1152–1211. [CrossRef]

52. Stefanou, G. The stochastic finite element method: Past, present and future. *Comput. Methods Appl. Mech. Eng.* **2009**, *198*, 1031–1051. [CrossRef]

53. Wang, H. Numerical modeling of non-planar hydraulic fracture propagation in brittle and ductile rocks using XFEM with cohesive zone method. *J. Pet. Sci. Eng.* **2015**, *135*, 127–140. [CrossRef]

54. Wang, X.; Shi, F.; Liu, H.; Wu, H. Numerical simulation of hydraulic fracturing in orthotropic formation based on the extended finite element method. *J. Nat. Gas Sci. Eng.* **2016**, *33*, 56–69. [CrossRef]

55. Newberry, N.R.; Nicoll, R.A. Comparison of the action of baclofen with gamma-aminobutyric acid on rat hippocampal pyramidal cells in vitro. *J. Physiol.* **1985**, *360*, 161–185. [CrossRef] [PubMed]

56. Fan, X.; Gong, M.; Zhang, Q.; Wang, J.; Bai, L.; Chen, Y. Prediction of the horizontal stress of the tight sandstone formation in eastern Sulige of China. *J. Pet. Sci. Eng.* **2014**, *113*, 72–80. [CrossRef]

57. Fang, Y.; den Hartog, S.A.M.; Elsworth, D.; Marone, C.; Cladouhos, T. Anomalous distribution of microearthquakes in the Newberry Geothermal Reservoir: Mechanisms and implications. *Geothermics* **2016**, *63*, 62–73. [CrossRef]

58. Rust, A.C.; Cashman, K.V. Multiple origins of obsidian pyroclasts and implications for changes in the dynamics of the 1300 B.P. eruption of Newberry Volcano, USA. *Bull. Volcanol.* **2007**, *69*, 825–845. [CrossRef]

59. Zhang, G.; Wang, L.; Wu, Y.; Li, Y.; Yu, S. Failure mechanism of bedded salt formations surrounding salt caverns for underground gas storage. *Bull. Eng. Geol. Environ.* **2017**, *76*, 1609–1625. [CrossRef]

60. Hajiabdolmajid, V.; Kaiser, P.K.; Martin, C.D. Modelling brittle failure of rock. *Int. J. Rock Mech. Min. Sci.* **2002**, *39*, 731–741. [CrossRef]

61. Jing, L.; Nordlund, E.; Stephansson, O. A 3-D constitutive model for rock joints with anisotropic friction and stress dependency in shear stiffness. *Int. J. Rock Mech. Min. Sci. Geomech. Abstr.* **1994**, *31*, 173–178. [CrossRef]
62. Liu, J.; Ding, W.; Yang, H.; Jiu, K.; Wang, Z.; Li, A. Quantitative prediction of fractures using the finite element method: A case study of the lower Silurian Longmaxi Formation in northern Guizhou, South China. *J. Asian Earth Sci.* **2018**, *154*, 397–418. [CrossRef]

Article

Occurrence, Classification and Formation Mechanisms of the Organic-Rich Clasts in the Upper Paleozoic Coal-Bearing Tight Sandstone, Northeastern Margin of the Ordos Basin, China

Guanqun Yang [1,2,3], Wenhui Huang [1,2,3,*], Jianhua Zhong [4] and Ningliang Sun [4]

[1] School of Energy Resources, China University of Geosciences (Beijing), Beijing 100083, China; yq19890923@126.com

[2] Key Laboratory of Marine Reservoir Evolution and Hydrocarbon Enrichment Mechanism, Ministry of Education, Beijing 100083, China

[3] Beijing Key Laboratory of Unconventional Natural Gas Geological Evaluation and Development Engineering, Beijing 100083, China

[4] School of Geosciences, China University of Petroleum, Qingdao 266580, China; zhongjianhua57@163.com (J.Z.); sunningliangll@163.com (N.S.)

* Correspondence: huangwh@cugb.edu.cn

Received: 24 April 2020; Accepted: 23 May 2020; Published: 27 May 2020

Abstract: The detailed characteristics and formation mechanisms of organic-rich clasts (ORCs) in the Upper Paleozoic tight sandstone in the northeastern margin of the Ordos Basin were analyzed through 818-m-long drilling cores and logging data from 28 wells. In general, compared with soft-sediment clasts documented in other sedimentary environments, organic-rich clasts in coal-bearing tight sandstone have not been adequately investigated in the literature. ORCs are widely developed in various sedimentary environments of coal-bearing sandstone, including fluvial channels, crevasse splays, tidal channels, sand flats, and subaqueous debris flow deposits. In addition to being controlled by the water flow energy and transportation processes, the fragmentation degree and morphology of ORCs are also related to their content of higher plants organic matter. The change in water flow energy during transportation makes the ORCs show obvious mechanical depositional differentiation. Four main types of ORC can be recognized in the deposits: diamictic organic-rich clasts, floating organic-rich clasts, loaded lamellar organic-rich clasts, and thin interlayer organic-rich clasts. The relationship between energy variation and ORCs deposition continuity is rarely studied so far. Based on the different handling processes under the control of water flow energy changes, we propose two ORCs formation mechanisms: the long-term altering of continuous water flow and the short-term water flow acting triggered by sudden events.

Keywords: coal-bearing tight sandstone; organic-rich clasts; occurrence; classifications; formation mechanisms; Ordos Basin

---

1. Introduction

The soft-sediment clasts (SSCs), as aggregations of the fine sediments formed at the syngenetic sedimentary stage, are often dispersedly preserved in the water-transported sandy hosting sediments [1,2]. SSCs are widely developed in a variety of modern and ancient sedimentary settings, including glacial, alluvial, fluvial, estuarine, coastal, shoreline and deep-water environments [3–12]. The terms used by the researchers to characterize sedimentary features are varied, for instance clay balls [3], armored or unarmored mud balls [4,7,13–18], mud pebbles [4], mud clasts [6,8,12,19,20], rip-up clasts [21–25], blocks [26–28], intraformational clasts [29]. SSCs are also commonly regarded as the palaeoenvironmental indicators to obtain geological information on clasts transportation, deposition and deformation processes, and regional-scale palaeoenvironmental setting [2,30]. Therefore, numerous targeted researches have been carried out on the morphology, classification, formation and transportation mechanism of different SSCs under particular settings [1,12]. In addition to the above types, a special kind of soft sediment clasts is developed in coal-bearing strata, consisting of coaly fragments, carbonaceous mud clasts and coalified plant remains, etc., typically marked by enriching in organic matter. Here, the term "organic-rich clasts (ORCs)" is used for describing this outstanding feature.

As early as the beginning of the last century, the organic-rich clasts in coal-bearing strata had been observed [31]. Since then, the organic-rich clasts were widely found in the coal-bearing strata of the major coal-bearing basins in the world, mostly distributed in the sandstone either close to or far above the top of the coal seams [31–37]. Researchers often interpreted the significance of the organic-rich clasts from the perspective of coal petrology, including morphology, composition, degree of thermal evolution, time of coalification [37–39]. In recent years, coal-derived gas (especially tight sandstone gas) has been developed on a large commercial scale [40,41]. For example, in the Upper Paleozoic tight sandstone gas-bearing strata of the Ordos Basin in China, several tight sandstone gas fields with a reserve of over 1×10^8 m^3 have been discovered, like the Sulige gas field [42,43]. Meanwhile, the influence of organic-rich clasts in the coal-bearing strata on tight sandstone gas reservoirs has also received attention, involving diagenesis, property and gas accumulation [20,44,45]. Moreover, it must also be mentioned that organic-rich clasts in sandstone (mainly terrestrial higher plant) can supply abundant hydrocarbon [46,47]. Despite these geological implications, however, few studies have ever focused on the formation and transportation mechanisms of organic-rich clast in the coal-bearing sandstone strata.

Taking the Upper Paleozoic coal-bearing tight sandstone formation in the northeast margin of the Ordos Basin as an example, the present paper has three main objectives: (1) to identify the morphology and distribution of organic-rich clasts in different sedimentary environments; (2) to investigate the classification and sedimentary sequence of organic-rich clasts in the coal-bearing strata; (3) to systematically analyze the formation mechanisms of organic-rich clasts in the coal-bearing tight sandstone. The results of this paper will be conducive to deepen the understanding of organic-rich clasts.

2. Geological Setting

The Ordos Basin, located in the western part of the Sino-Korean plate (Figure 1A), covering an area of about 25×10^4 km^2, is an intracratonic basin developed from the Archaean to Early Proterozoic [48,49]. As a large polycyclic superimposed basin, the Ordos Basin underwent a multi-stage tectonic movement during a long geological history. According to the current tectonic features of the Ordos Basin, six tectonic units were identified, which are composed of the Yimeng uplift in the north, the Weibei uplift in the south, the Western margin thrust belt and the Tianhuan depression in the west, the Jinxi fault-fold belt in the east, and the Shanbei slope in the central region [50]. The study area is located in the northeastern margin of the Ordos Basin, presenting a gentle monoclinal structure towards the west on the whole (Figure 1B). In addition, many small folds and faults have developed on it [51].

The Upper Paleozoic strata in the Ordos Basin consist of the Upper Carboniferous Benxi Formation, the Lower Permian Taiyuan Formation and the Shanxi Formation, the Middle Permian Lower and Upper Shihezi Formation, and the Upper Permian Shiqianfeng Formation (Figure 1C, D and Figure 2). Benxi Formation, Taiyuan Formation, and Shanxi Formation are coal-bearing strata, which are the interest intervals of this study. In the Late Carboniferous-Early Permian, a set of coal-bearing sedimentary formation deposited based on the Ordovician paleo crust of weathering with the Yinshan paleo-land supplying sources [52]. From the Benxi Formation to the Taiyuan Formation, a barrier coastal sedimentary system developed in the background of the epeiric sea, mainly including barrier islands, lagoons, tidal flats, and carbonate platforms [53]. Nevertheless, the fluvial-delta sedimentary system developed in the Shanxi Formation [54–56]. A warm and humid climate and lush vegetation promoted the development of organic-rich sediments composed of coal, carbonaceous mudstone and dark mudstone, with a cumulative thickness of > 200 m [57]. Moreover, the coal-bearing sequences are the main source rock characterized by a high content of thermally mature total organic carbon, providing a sufficient source of gas for the tight sandstone gas reservoirs in the Upper Paleozoic [58,59]. Besides, barrier island sandstone and channel sandstones constitute the favorable reservoirs adjacent to the source rocks [60].

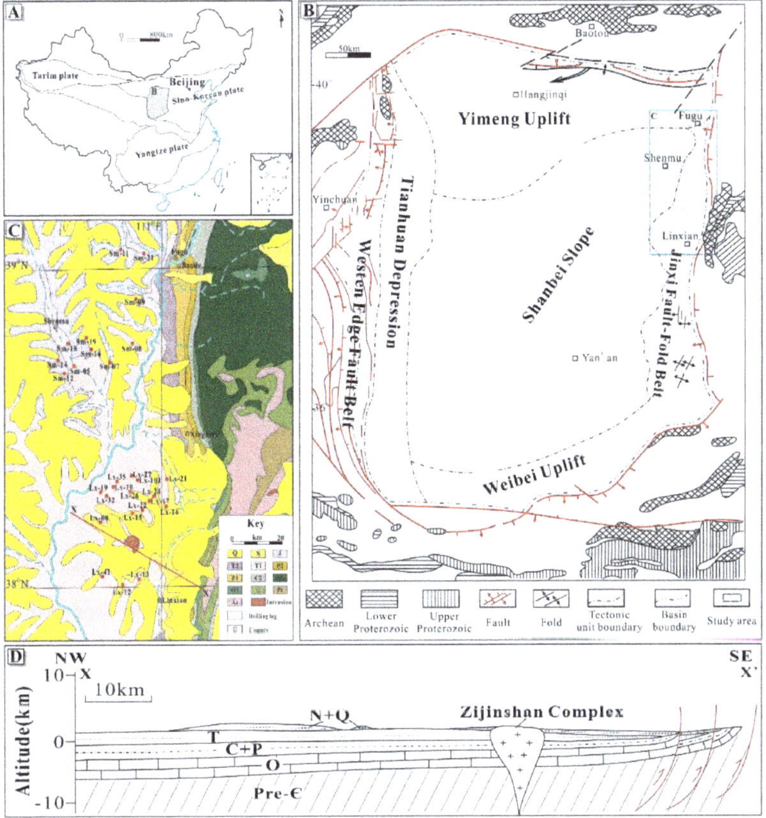

Figure 1. (**A**) Location of the Ordos Basin in simplified tectonic map of China. (**B**) Tectonic division of the Ordos Basin (modified after Luo et al.) [61], showing the location of the study area. (**C**) Schematic geological map of the northeastern margin of the Ordos Basin showing the drilling sites. (**D**) Sketch cross section through the south of study area (modified from Liao et al.) [62], position located on (**B**).

Period	Epoch	Formation	Thickness (m)	Lithology	Sedimentary facies
Permian	Upper	Shiqianfeng Formation	190~260		Flood plain
					Fluvial channel
					Flood plain
					Fluvial channel
	Middle	Shangshihezi Formation	170~210		Flood plain
					Fluvial channel
					Flood plain
					Fluvial channel
					Flood plain
					Fluvial channel
					Flood plain
					Fluvial channel
		Xiashihezi Formation	180~220		Flood plain
					Braided channel
					Flood plain
					Braided channel
	Lower	Shanxi Formation	60~120		Inter-distributary bay
					Distributary channel
					Inter-distributary bay
					Distributary channel
					Inter-distributary bay
					Distributary channel
Carboniferous		Taiyuan Formation	50~100		Tidal flat
					Tidal channel
					Tidal flat
					Barrier island
	Upper	Benxi Formation	40~60		Lagoon
					Tidal channel
					Carbonate platform
					Tidal flat
	Ordovician				

Legend: Pebbled coarse sandstone; Coarse sandstone; Middle sandstone; Fine sandstone; Siltstone; Silty mudstone; Mudstone; Carbonaceous mudstone; Shale; Coal; Limestone; Marlstone; Dolomite; Weathering crust.

Figure 2. Lithochronostratigraphical column of the Upper Paleozoic strata in the study area. Here, we described the development characteristics of organic-rich clasts in coal-bearing sandstones in detail through core observation, and the descriptions contain lithology, morphology, distribution, and deformation. Thirty-six typical hand specimens were selected from the cores to prepare thin sections; meanwhile, the microscopic feature information of ORC was captured by the ZEISS optical microscope. Twelve whole-rock polished blocks were prepared to carry out maceral analysis. Maceral analysis was performed under a Zeiss Axio imager microscope equipped with an oil immersion objective and a white incident and a blue light source where >800 points were considered for each sample. Maceral was classified by the ICCP (International Committee for Coal Petrology) System 1994 [63–65]. The lithofacies codes are named according to [66]; some codes are added and modified following [67]. Based on the facies data derived from core descriptions and logging interpretations, a series of detailed comparative analysis was made on the organic-rich clasts, in order to find out the origins and control factors of ORCs developed in different sedimentary environments.

3. Methodology

A total of 28 exploration wells were selected as research objects in this study, drilled from 2013 to 2017 by the China United Coalbed Methane Corporation. In total, more than 818-m-long conventional cores were received from 28 wells, diffusely covering the developed intervals of the tight sandstone gas reservoirs in coal-bearing formation; additionally, all the wells were logged with a set of comprehensive wireline surveys and cuttings. The cores have been macroscopically examined and described at 1-cm scale to obtain information on lithology, sedimentary structures, and geometric features of the organic-rich clasts. Samples have been collected from the shallow marine shelf, barrier coast and fluvial-delta deposits in all the three coal-bearing formations, thus providing sufficient geological information to undertake a comprehensive study of ORCs.

4. Results

Eleven types of lithofacies related to organic-rich clasts are recognized within the Carboniferous-Permian coal-bearing strata (Table 1). These have been divided into 5 types of facies associations standing for different sedimentary processes and settings, which have been identified in cores in conjunction with the interpretation of logging curves. These facies associations are mainly interpreted as channels (fluvial channels and delta distributary channels), crevasse splays, sand flats, tidal channels, and subaqueous debris flow deposits. The occurrence of organic-rich clasts will be described in detail below, representing the variation of water-flow energy during sediment transport.

Table 1. Main lithofacies of the Upper Paleozoic Benxi, Taiyuan and Shanxi Formations in the northeastern margin of the Ordos Basin.

Facies Code	Lithofacies	Description	Depositional Environment	Example
Gm	Matrix-supported conglomerate	Mixed of gravels, sands and detrital clays, mainly gravels, poorly sorted and angular, weak grading, massive bedding, erosional base, thin thickness	Fluvial channel, crevasse splay	
Gc	Clast-supported conglomerate	Mainly gravels, with little matrix, sub-angular to sub-rounded, moderate sorted, graded bedding, erosional base, ranging in thickness from a few centimeters to 20 cm	Fluvial channel, distributary channel, tidal channel	
Sm	Medium- to coarse-grained sandstone	Moderate to well sorted and rounded, massive bedding, structureless, may be pebbly,	Fluvial channel, distributary channel, tidal channel, debris flow deposits, crevasse splay	
St	Fine- to coarse-grained sandstone	Trough cross-bedding, moderate sorted	Fluvial channel, distributary channel, tidal channel	
Sp	Fine- to coarse-grained sandstone	Planar cross-bedding, well sorted and rounded	Fluvial channel, distributary channel	

Table 1. *Cont.*

Facies Code	Lithofacies	Description	Depositional Environment	Example
Sr	Very fine- to coarse-grained sandstone	Ripple cross-lamination, moderate sorted	Fluvial channel, distributary channel, tidal channel, crevasse splay, sand flat	
Sh	Fine- to coarse-grained sandstone	Parallel bedding, well sorted and rounded	Fluvial channel, distributary channel, tidal channel, debris flow deposits	
Si	Siltstone and muddy sandstone	Fine lamination, small ripples, well sorted and rounded, with more detrital clays	Abandon channel, flood plain, mud and mixed flat, lagoon, shallow marine deposits	
Fl	Sandstone or siltstone interbedded with mudstone	Horizontal lamination, small ripples, convolute bedding, lenticular bedding, bioturbation	Overbank, abandoned channel, waning crevasse splay, mixed flat	
Fr	Mudstone and shale	Horizontal lamination or massive, plant fragments and roots, bioturbation, pyrite nodules	Flood plain, mud flat, lagoon, shallow marine deposits	
C	Coal and carbonaceous mudstone	Massive, well-developed cleats, plant fragments, containing pyrite nodules	Vegetated swamp deposits, lagoon	

The organic-rich clasts are widely distributed in the coal-bearing tight sandstones of Benxi Formation, Taiyuan Formation and Shanxi Formation in the study area. The lithology of ORCs identified in the core includes carbonaceous clasts, carbonaceous mudstone clasts, dark mudstone clasts, shale clasts, and plant fragments (Figure 3A–D). In terms of color, the organic-rich clasts are mostly black or grayish black, and easy to dye the hosting sandstone. Moreover, the associated pyrite, commonly identified in the core hand specimens, is a characteristic mineral formed from organic-rich clasts in a local reducing microenvironment during diagenesis (Figure 3E). The shapes of ORCs are varied, including blades, discs, plate strips, tearing chips and irregular forms. The grain size is generally gravel size, with a maximum of more than 10 cm. Although strongly deformed, the long axes of most ORCs are parallel or sub-parallel to the stratum (Figure 3A,B).

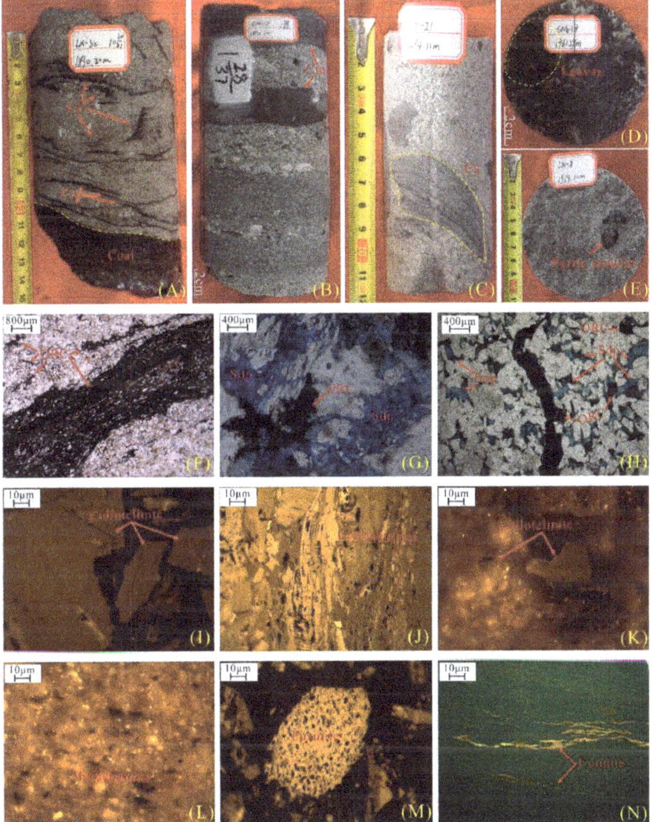

Figure 3. Typical photographs of core (**A–E**), thin sections (**F–H**) and polished blocks (**I–N**) of ORC in the studied coal-bearing sandstones. Abbreviations: Cc = Carbonaceous clasts, Cm = Mudstone clasts, Cs = Shale clasts, Sdp = Secondary dissolution pores, PPL = plane-polarized light, RL = reflected light. (**A**) Scattered coaly clasts with plastic deformation (LX-32, 1890.20 m). (**B**) Cobble dark mudstone clasts in the conglomerate (SM-18, 1871.10 m). (**C**) Shale clasts showing clear bedding (SM-21, 1694.11 m). (**D**) Plant leaves with clear outlines (SM-18, 1961.29 m). (**E**) Associated pyrite nodules of ORC (LX-8, 1969 m). (**F**) Thin-bedded ORC containing fine-grained sediments (PPL, LX-44, 2063.12 m). (**G**) Secondary dissolution pores of feldspars around the ORC (PPL, SM-4, 2134.1 m). (**H**) A strip of ORC with secondary dissolution pores (PPL, LX-103, 1721.5 m). (**I**) Homogenous collotelinite (RL, SM-19, 2116.52 m). (**J**) Cementing collodetrinite (RL, LX-35, 1922.54 m). (**K**) Blocky collotelinite and vitrodetrinite (RL, SM-20, 2072.98 m). (**L**) Scattered vitrodetrinite in the matrix (RL, SM-19, 2083.1 m). (**M**) Clumpy pyro-fusinite and scattered vitrinite (RL, LX-33, 1732.1 m). (**N**) Cutinite with yellow fluorescence (fluorescence-mode, SM-7, 1872.80 m).

Under the microscope, ORCs with small particle size can also be observed widely, which usually show a characteristic of compressional deformation (Figure 3F–H). The results of microscopic observation show that these ORCs are mainly composed of organic matter (OM), fine-grained sediments (Figure 3F–N), and some authigenic inorganic minerals (such as pyrite). As for the organic macerals, there are mainly vitrinite, inertinite and liptinite (Figure 3I–N). Vitrinite (type III kerogen, formed by the gelification of plant remains) is the most abundant organic component, with light gray to gray under the reflected light (Figure 3I–M). Most of these Vitrinite components are collotelinite (Figure 3I,K), collodetrinite (Figure 3J) and vitrodetrinite (Figure 3I,K–M). Vitrodetrinite, especially for the muddy ORCs, is present in the form of small discrete particles in the argillaceous matrix (Figure 3K,L). Inertinite (type-IV kerogen, derived from terrestrial plant tissues) consists of semifusinite, fusinite (Figure 3M), micrinite and inertodetrinite, and has bright white color under the reflected light. Liptinite (type-II kerogen, transformed from plant organs) is relatively rare in the analyzed samples. Only some cutinite (Figure 3N) and bituminite with obvious fluorescence characteristics are seen. Thus, it can be seen that the organic matter found in ORCs is mainly derived from terrestrial higher plants.

4.1. Facies Association I

Description

The facies association I is composed of Gc, Sm, St, Sp, Sr, Fl, Si, Fr, and C (Figure 4A). The sandstones are characterized by large thickness, intercalated with thin layers of siltstone and mudstone. Sedimentary structures such as basal erosional surface, massive bedding, cross-bedding, parallel bedding, and wavy bedding are well developed. Overall, the sandstones of facies association I have a fining-upward trend and a bell-shaped or jugged box-shaped logging curve (Figure 4A). On the erosion surfaces, the organic-rich clasts are most concentrated, which are poorly sorted, angular to sub-angular, and chaotic (Figure 4B,E–G,K). In the middle parts of the sand body, the floating ORC is scattered with long-axis tending towards bed-parallel orientations, mostly the shape of slender and deformed (Figure 4C,J,L). The occurrence of small granule ORC is in the form of aggregates as discontinuous or continuous laminations in the upper parts (Figure 4D,H,I). These clasts are flaky, good sorted and rounded, coexisting with high fine-grained sediment content (Figure 4C,J,M).

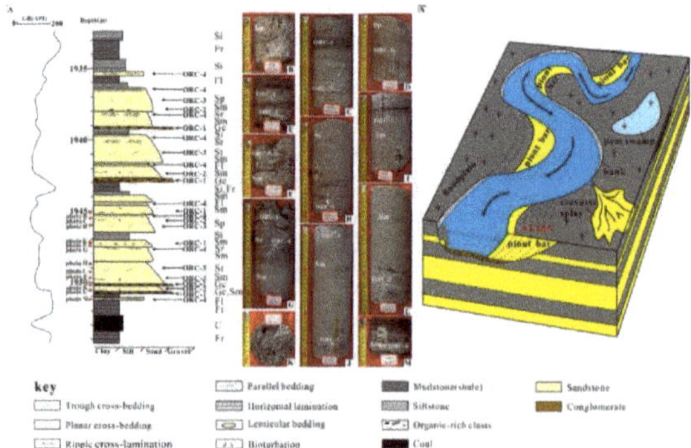

Figure 4. Sedimentary characteristics of the organic-rich clasts in the fluvial channel or distributary channel (log LX-21). (**A**) Schematic sedimentary succession with natural gamma-ray curve in the channel environment, showing the occurrence of ORCs in different sedimentary locations. (**B–M**) Typical core photographs of various ORCs occurring in different lithofacies. (**C**) The position of log A in the schematic diagram, showing the cause of ORC formation in channel: bank collapse and basal erosion.

Interpretation

Thick sand sediments in the facies association I are interpreted as fluvial or distributary channel deposits (Figure 4N), while fine-grained plant remnants in flood plain and mire provide the source for the formation of ORC, all developing in the Shanxi Formation. The ORCs in sandstone are chiefly the products of either: (1) scouring of the bottom underlying semi-consolidated peat by the high energy turbulent flow; (2) lateral erosion or slumping of bank sediments triggered by the water flow (high-density turbidity or quasi-stable flow). The ORC from different parts in sandstone underwent different sedimentation. Only a small amount of produced ORCs is similarly in-situ deposited in the vicinity of the erosion surface or after short-distance rolling transport. However, most of the produced ORCs are carried away and reworked by the water flow and water-transported debris [68]. The scattered floating ORC is the result of hindered setting in a high-density turbulent flow. After intense modification and long-distance transport, the ORC breaks up into small pieces of high maturity and deposits as the energy of the water flow diminishes. The loaded ORC, in the form of weakly continuous cross-lamination, is formed under the condition of transitional to high flow regime. In the low flow regime, the fragmented ORC is shaped into the continuous lamellar aggregate.

4.2. Facies Association II

Description

This facies association is built up of Sm, Sr, Si, Fl, and C (Figure 5A). The thickness of Sm and Sr ranges between 1 and 3 m. Sandstones have an upward-coarsening trend with sharp contacts at the bottom, as well as a finger-shaped logging curve (Figure 5A). Massive bedding, wavy bedding, and lenticular bedding can be observed. Also, roots can be recognized in the mudstone (Figure 5B,D,F,H,I). The ORC in facies association II is generally subrounded, which range in size between granule and pebble (2~64 mm; Figure 5I). The scattered clasts distribute in the massive host sandstone (Sm; Figure 5H,I). In the upper parts, wavy lamina or thin intercalated aggregates of granule ORC are often observed (Figure 5B,C).

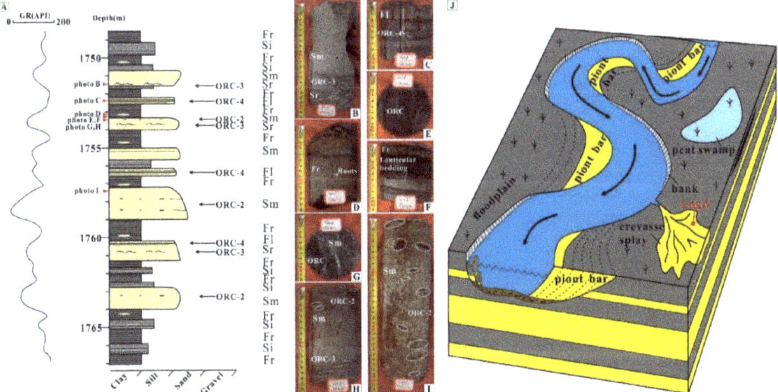

Figure 5. Sedimentary characteristics of the organic-rich clasts in the crevasse splay (log SM-8). (**A**) Schematic sedimentary succession and natural gamma-ray curve response of crevasse splay, showing the positions of different types of ORC. (**B–I**) Typical photographs of lithofacies and ORC types in the crevasse splay, including some typical sedimentary structures. (**J**) The location of logA in the fluvial sedimentary model.

Interpretation

Multilayered thin sandstone within this facies association developed by the channel burst in the effect of flood, identified as the crevasse splay (Figure 5J). The ORCs are derived from the organic sediments at the channel bank or the initial area in which flood flow through (Figure 5D–F). The scouring of a high-density flood, which has a fast flow rate and high energy, is the main reason for the organic-rich sediments turn into ORC in the special sedimentary environment. Due to the strong agitation of high-energy flood, ORC has a high degree of fragmentation and low structural maturity, rarely resulting in large blocks. The energy of flood decreases rapidly as a result of uncontained condition so that the ORC is deposited along with the sandy sediments after short-distance transportation. After the waters receded, the small ORC under the buoyant handling deposit as the thin layers.

4.3. Facies Association III

Description

The facies association III comprises Gm, Sm, Sr, Sh, Fl, Si, Fr, and C (Figure 6A). The sandstone body is developed in a multi-period, with a cumulative thickness of more than 15 m. Massive bedding, wavy bedding, and parallel bedding are main sedimentary structures. The GR logging curve is finger-shaped with some fluctuation (Figure 6A). The ORC in this facies association is characterized by variable size (granules to boulders; 2 to >256 mm) and shape, angular nature and plastic deformation. At the bottom parts of the single sand body, the clasts are disorderly accumulated with gravel or coarse sand matrix (Figure 6H). The floating clasts are isolated occurring in the structureless host sandstone, with strong deformation and long-axis running parallel to bed (Figure 6B,C,E–G,I,K). The small granule clasts are clustered in the form of wavy or parallel laminations (Figure 6J).

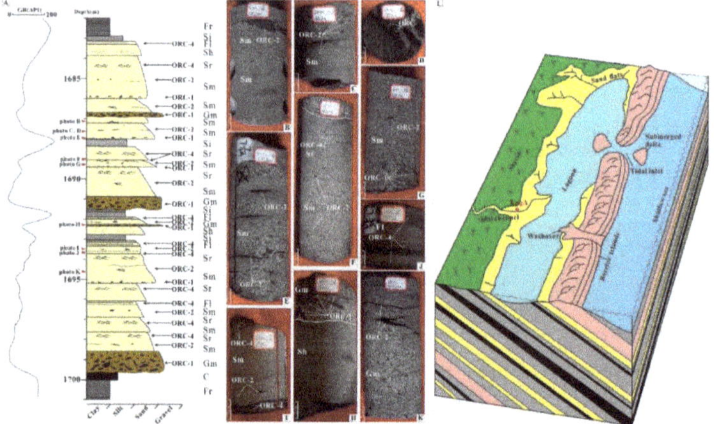

Figure 6. Sedimentary characteristics of the organic-rich clasts in the tidal channel (log LX-10). (**A**) Schematic sedimentary succession and the gamma-ray response of the tidal channel, including the positions of ORC. (**B–K**) Photos of typical lithofacies and ORC types in the tidal channel. (**L**) Schematic sedimentary model of barrier coast, showing the location of logA.

Interpretation

Thick sandstone with ORC in the sequence is interpreted as the deposits of the tidal channel (Figure 6L), mostly distributed in the upper Benxi and Taiyuan Formations. The formation of ORC, whose precursors are mainly organic-rich deposits in marsh or lagoon, is controlled by tidal processes, including basal erosion, bank erosion, and slump. With the changes in the positional relationship between the sun, the moon and the earth, the energy of tide has the characteristic of periodic variations. The channel-bottom deposits are flood dominated, with irregular, poorly sorted ORC mixed with gravel or coarse sand. Some large particles transported by tidal turbulence probably trapped and deposited in singles within the massive sandstone, usually resulting in heavy bending and deforming (Figure 6B,K). When the channel is ebb dominated, tidal energy gradually weakens, and the small granule ORC transported by buoyancy tends to deposit as laminations in wavy bedding sandstone (Figure 6J).

4.4. Facies Association IV

Description

The facies association IV is characterized by more internal structures containing Sr, Fl, Si, Fr, and C (Figure 7A). Fine-grained sediments and coal are developed extensively, while sandy sediments range in thickness between 0.5 and 2 m. It is quite common to see tidal bedding such as flaser bedding, wavy bedding and lenticular bedding (Figure 7G,I–K), as well as various bioturbation structures (Figure 7F,H). The coaly fragments and plant pieces, predominantly sub-rounded to rounded, are common in the sandstone. The size of these clasts are in range from small pebbles to large cobbles (4~256 mm), occasionally boulders (Figure 7C). The most common occurrence of ORC is loaded as the wavy discontinuous lamination (Figure 7B–D).

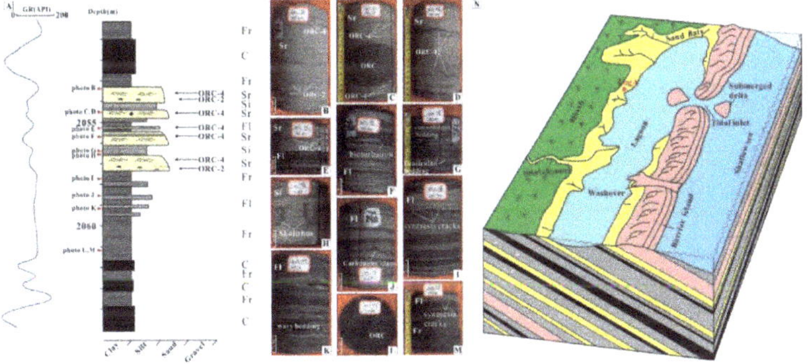

Figure 7. Sedimentary characteristics of the organic-rich clasts in the sand flat (log SM-5). (**A**) Schematic sedimentary succession with gamma-ray response of the sand flat, showing the occurrence of ORC. (**B–J**) Photos of typical lithofacies and ORC types in the sand flat, noting the sedimentary structures and bioturbation. (**K**) Location of logA in the schematic barrier coast sedimentary model.

Interpretation

Thin wavy bedding sandstone of this facies association typifies the sand flat deposit in a tidal flat environment (Figure 7N). The thick coal seams and organic-rich mud shales (Figure 7L,M), which develop in marsh and lagoon, supply sufficient source for ORC. Two main mechanisms exist on the formation of ORC: basal erosion under the effect of tide and detachment due to synaeresis cracking (Figure 7I,M), while the transport of the broken ORC from the origin site is operated by the processes of tide currents. At the level of high tide, some large particles of ORC are trapped by sandy deposits; whereas most ORC would break up repeatedly under the action of tide currents. With the ebb of tide, the residual granule ORC would load and form into wavy lamellar aggregates. The circular coaly ORC, having a grain size of >10 cm in Figure 7C, should be a cross-section of the plant stem, which probably formed directly after depositing in the sand.

4.5. Facies Association V

Description

The fine-to-medium sandy sediments are the major component of facies association V, while the gravels barely exist (Figure 8A). They display bed with a succession of Sm, Sh, Si, and Fr. Massive bedding and parallel bedding can be recognized generally. Moreover, the overall sandstone body develops multi-cycle erosion surfaces with fining-upward. The ORCs are mainly mudstone (shale) breccias (Figure 8C,D,F,G,I), but rarely coaly fragments and plant pieces. The intact boulder ORC retains the horizontal bedding of shale (Figure 8C,G). These clasts are angular to sub-rounded, with the characteristic of imbricate arrangement. The long axis of the particles is generally parallel or sub-parallel to the bedding. On the surface of basal erosion, ORCs are accumulated in variable shape and size (small pebbles to boulders; Figure 8C,F,H). In the middle parts of the single sandstone body, long strip or flaky clasts are sporadically floating in structureless sandstone (Figure 8D,E,I,J). The finely rounded pebble ORCs are deposited in weak continuous parallel laminations at the upper parts (Figure 8B).

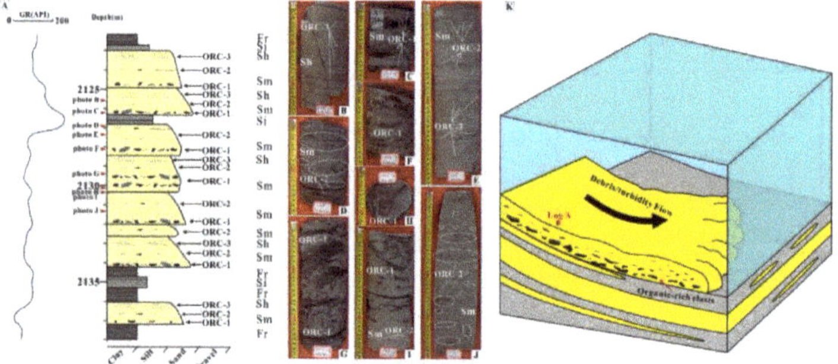

Figure 8. Sedimentary characteristics of the organic-rich clasts in the debris/turbidity flow (log LX-13). (**A**) Schematic sedimentary succession of the subaqueous debris/turbidity flow, occurring with high ORC content. (**B–J**) Photos of lithofacies and ORC types, noting the differences in the morphology and amounts of ORCs in different lithofacies. (**K**) Location of logA in the debris/turbidity flow schematic sedimentary model.

Interpretation

Facies association V is limited in the lower Benxi Formation, representing the deposition of the sandy debris or turbidity flow controlled by the topography of the underlying Ordovician weathering crust locally (Figure 8K). In this case, the ORC is probably owing to flow basal erosion of muddy bottom or flow-triggered failure of muddy slope. The thick clast-supported Sm together with the presence of intact boulder ORC indicates the processes of high-energy water flow. The ORCs carried by rolling movement at the bottom of dense flow deposits first, with a typical feature of very low textural maturity for short-distance transportation (Figure 8C,F–I). Subsequently, some of the ORCs lifted by turbulence will overcome the limitation of buoyancy and deposit individually or in groups, which orient in the long-axis direction (Figure 8D,E,J). The ORCs, carried by high-dense flow far and modified strongly, deposit with the weakening of flow energy and transporting capacity (Figure 8B).

5. Discussion

5.1. Occurrence and Classification

The development of ORCs in coal-bearing sandstone is an important geological phenomenon which cannot be ignored. The occurrence of ORCs is not only influenced by the transportation of water flow, but also related to their own composition [2,12,30]. The hydrodynamic conditions of the water flow control the transportation process of ORCs (including the transportation ways, the transportation distance and the reworking intensity), which determines the final sedimentary style of ORCs, specific performances: (1) Some rolling transported ORCs have a short transportation distance and low degree of re-working within the dense flow, chaotically accumulating above the erosion surface with low texture maturity (Figure 9J–L). (2) The jumping transported ORCs are subjected to the strong hydrodynamic forces. Some large and intact ORCs are caught in the rapidly deposited sandy sediments leading to the severe plastic deformation (Figure 9G–I); some ORCs that "survived" from the actions of water flow and the collisions of grains (mostly sandy sediments) are small-sized but numerous with good sorting and roundness (Figure 9D–F); but the others are completely disintegrated during the transportation. (3) Some suspended and lightweight ORCs are transported over long distance with relatively weak reworking degree, and deposited as thin interlayers together with fine-grained sediments (mainly silts and clays) under weak hydrodynamic conditions (Figure 9A–C). The composition of ORCs also has an important influence on their susceptibility to mechanical disintegration [30]. The ORCs are mainly composed of higher plant debris Under the action of water flow, some muddy ORCs can be completely disintegrated into clay particles after undergoing mechanical attrition and disintegration; while higher plant fragments only change from large to small particles.

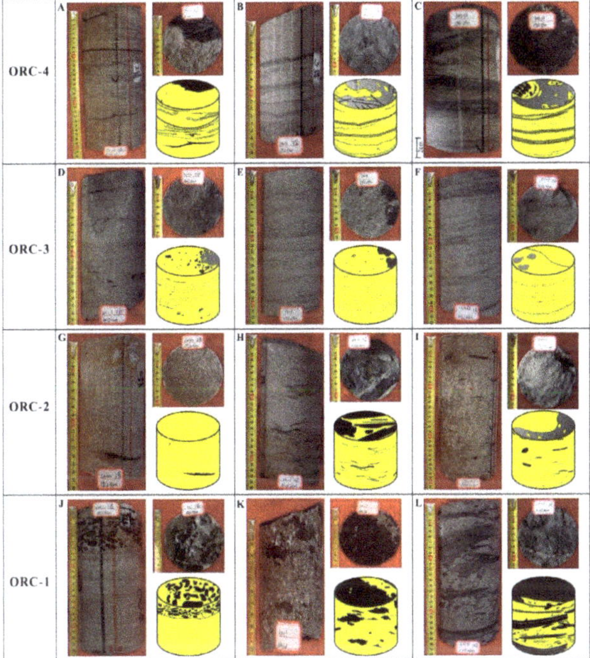

Figure 9. Typical core photos and 3D sketches of 4 types of the organic-rich clasts in the coal-bearing sandstone strata. Abbreviations: ORC-1 = Diamictic organic-rich clasts (**J-K**), ORC-2 = Floating organic-rich clasts (**G-I**), ORC-3 = Loaded lamellar organic-rich clasts (**D-F**), ORC-4 = Thin interlayer organic-rich clasts (**A-C**).

The classification of ORCs proposed here is based on considerations of their occurrence and the characteristics of hosting sediments. The characteristics given in Figure 8 and Table 2 are designed to aid in identification and interpretation with different types of ORCs. The energy of water flow changes dynamically, resulting in its different transportation mechanism at different stages [5,12,69].

Under a range of its processes, the vertical distribution characteristic of ORCs in coal-bearing sandstone confirms to the law of mechanical sedimentary differentiation. As the energy and transporting capacity of single water flow changes from strong to weak, the sedimentary sequences of ORCs can be concluded from the bottom up as follows: diamictic organic-rich clasts (ORC-1), floating organic-rich clasts (ORC-2), loaded lamellar organic-rich clasts (ORC-3) and thin interlayer organic-rich clasts (ORC-4, listed in Table 2, Figure 9). From ORC-1 to ORC-4, the grain size of their hosting sandy sediments shows a change from coarse to fine (Figure 9), which also confirms a vertical differentiation of ORCs. However, these four types of ORCs can occur differently in different facies associations or not developed originally in part, depending on the actual conditions in a special geological environment.

Table 2. Classification of the organic-rich clasts in the studied coal-bearing tight sandstone.

Style Code	Style Name	Features of ORCs in Sediment	Matrix	Transportation Distance	Part of Single Sand Body
ORC-1	Diamictic organic-rich clasts	**Shape**: irregular, lath-shaped, rip-up, angular to sub-rounded **Distribution**: chaotic distribution **Size**: particle size varies from granule to cobble	A complete range from gravel to clay, poorly sorted, massive	In-situ or a close distance	Bottom part
ORC-2	Floating organic-rich clasts	**Shape**: angular to sub-rounded, commonly irregular deformation as wrapped or squeezed **Distribution**: isolated or scattered, imbrication, long axis of clasts parallel to sub-parallel to bedding **Size**: pebble to cobble, particle size greater than 1 cm	Medium to coarse-grained, well sorted, Clay-poor, unstratified or structureless	A short distance	Middle and lower part
ORC-3	Loaded lamellar organic-rich clasts	**Shape**: sub-rounded to rounded, high sphericity **Distribution**: Distributed at the bottom of the lamina, poor continuity **Size**: granule to pebble	Fine to coarse-grained, well sorted, Clay-poor, parallel bedding, cross bedding, wavy bedding	A moderate distance	Middle and upper part
ORC-4	Thin interlayer organic-rich clasts	**Shape**: sub-angular to rounded, flakelet **Distribution**: in the form of multiple lamellar aggregates, wavy or horizontal, good continuity **Size**: mainly granule, occasional centimeter-sized clasts are plant fragments (such as stems or leaves)	Very Fine to fine grained, well sorted, clay-rich, lamination	A long distance	Top part

5.2. Formation Mechanism

The Upper Paleozoic sedimentary period in the Ordos Basin is in the background of transgression, and ORC is generated under the water action from rivers, tides, waves, and episodically floods and debris flows (Figure 10). The formation process of ORC underwater flow can be distinguished into four stages in sequence: (1) Original sedimentation of organic matter and fine-grained sediments to form source; (2) detachment of organic-rich sediments under hydrodynamism, which includes two main mechanisms: one is the basal erosion resulting in high-energy water flow, and the other is the instability slope failure triggered by water flow vibration; (3) transportation and reworking by water flow and water-transported debris; among them, the reworking is a dynamic process, including multiple crushing, deformation (such as squeezing squashing folding pressing), sorting, rounding; (4) allogenic re-deposition [1,2]. The initial failure products of organic-rich sediments are irregular ORC blocks, while transportation by water flow is the determinant of the difference in ORC deposition characteristics [1,2,30]. Based on this, we propose that the formation mechanisms of ORC could be classified into: long-term altering of continuous water flow (such as fluvial water, tide, etc.) and short-term water flow acting triggered by sudden events (such as flood, debris flow, etc.).

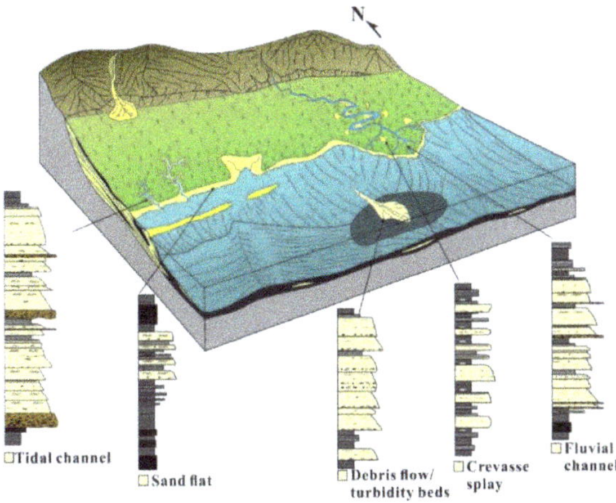

Figure 10. Schematic sedimentary model of the organic-rich clasts in the coal-bearing sandstone strata in the northeastern margin of the Ordos Basin.

Regarding the first instance, the sedimentary characteristics of the residual ORC in the coal-bearing sandstone are the results of modification by the continuous or periodic water flow within a long period. In the process of transportation, ORCs undergo a series of dynamic processes including liquefying, crushing, squeezing, squashing, winnowing, folding and pressing in the long-term and repeatedly, resulting in broken-up of most pre-formed ORC into smaller fragments. The vertical differentiation between sandy sediments and organic-rich debris is obvious, which suggests that the evolution of water flow from high-density, high-energy to low-density; low-energy turbulence is a slow but gradual process. In the lower flow regime, the suspended ORC often deposits as thin interlayers with fine-grained muddy sediments, usually neglecting the role of ORC previously.

Interestingly, the modification degree of transport on ORC in the short-term water flow is less than the first type, typically low textual maturity. The temporary water flow is often triggered by a sudden event, which may be flood, earthquake, storm or tsunami. This water flow tends to be high-energy, high-density and dissipates energy rapidly (often lasting only a few hours) for breaking through the confined state, resulting in the weak re-working on ORC during transportation. In this study, the thick-bedded diamictic ORC deposited in the early stage of the high-energy water flow energy weakening. Where after, some ORCs were captured during the deposition of sandy sediments, transported in skipping. Finally, the ORCs contained in the suspension transport formed thin interlayers in a lower flow regime, but it is seldom recognized. There are two major reasons: the basal erosion by the new sudden water flow or continuous deposition with fine-grained deposits in the absence of new sudden water flow.

6. Conclusions

(1) It is proposed to use the descriptive vocabulary organic-rich clasts (ORC) to collectively represent carbonaceous fragments, carbonaceous mudstone clasts, shale clasts, dark mudstone clasts and plant fragments developed in coal-bearing sandstones.

(2) A total of five sedimentary environments of ORC were identified, including fluvial channels, crevasse fans, tidal channels, sand flats, and subaqueous debris flow deposits.

(3) The occurrence of ORCs in coal-bearing sandstones is not only controlled by the changes of the water flow during transportation, but also related to the decomposition resistance of their components. After a series of processes during water flow transport, ORCs shows the characteristic of obvious mechanical differentiation in the vertical direction. Based on this, we propose that ORCs can be classified into four types: diamictic organic-rich clasts, floating organic-rich clasts, loaded lamellar organic-rich clasts, and thin interlayer organic-rich clasts.

(4) The changes in water flow energy during transportation play a controlling role in the formation of ORCs. We have summarized two formation mechanisms of ORCs in coal-bearing sandstones, including the long-term altering of continuous water flow and the short-term water flow acting triggered by sudden events.

Author Contributions: Conceptualization, G.Y.; methodology, G.Y.; software, G.Y.; validation, G.Y., W.H., J.Z. and N.S.; formal analysis, G.Y.; investigation, N.S.; resources, G.Y.; data curation, G.Y.; writing—original draft preparation, G.Y.; writing—review and editing, W.H. and J.Z.; visualization, G.Y.; supervision, W.H.; project administration, W.H.; funding acquisition, W.H. All authors have read and agreed to the published version of the manuscript.

Funding: This research was financially supported by the National Major Science and Technology Projects of China (No.2016ZX05066001-003) and National Natural Science Foundation of China (No. U1910205).

Acknowledgments: Special thanks are given to China United Coalbed Methane Corporation for providing the studied samples and data.

Conflicts of Interest: The authors declare no conflicts of interest.

References

1. Knight, J. Morphology and palaeoenvironmental interpretation of deformed soft-sediment clasts: Examples from with Late Pleistocene glacial outwash, Tempo Valley, Northern Ireland. *Sediment. Geol.* **1999**, *128*, 293–306. [CrossRef]

2. Knight, J. Significance of soft-sediment clasts in glacial outwash, Puget Sound, USA. *Sediment. Geol.* **2009**, *220*, 126–133. [CrossRef]

3. Haas, W. Formation of Clay Balls. *J. Geol.* **1927**, *35*, 150–157. [CrossRef]

4. Karcz, I. Mud pebbles in a flash floods environment. *J. Sediment. Res.* **1969**, *39*, 333–337. [CrossRef]

5. Kneller, B.; Branney, M. Sustained high-density turbidity currents and the deposition of thick massive sands. *Sedimentology* **1995**, *42*, 607–616. [CrossRef]

6. Allen, J. Reworking of muddy intertidal sediments in the severn estuary, Southwestern U.K.-A preliminary survey. *Sediment. Geol.* **1987**, *50*, 1–23. [CrossRef]

7. Goldschmidt, P. Armoured and unarmoured till balls from the Greenland Sea floor. *Mar. Geol.* **1994**, *121*, 121–128. [CrossRef]

8. Selby, I.; Evans, N. Origins of mud clasts and suspensions on the seabed in Hong Kong. *Cont. Shelf Res.* **1997**, *17*, 57–78. [CrossRef]

9. Knight, J. Processes of soft-sediment clast formation in the intertidal zone. *Sediment. Geol.* **2005**, *181*, 207–214. [CrossRef]

10. Southern, S.; Patacci, M.; Felletti, F.; McCaffrey, W. Influence of flow containment and substrate entrainment upon sandy hybrid event beds containing a co-genetic mud-clast-rich division. *Sediment. Geol.* **2015**, *321*, 105–122. [CrossRef]

11. Gao, Z.; Zhou, C.; Feng, J.; Wu, H.; Li, W. Mechanism and sedimentary environment of the muddy gravel concomitant with thick layer sandstone of Cretaceous in Kuqa depression. *Acta Petrol. Sin.* **2016**, *37*, 996–1010. (In Chinese) [CrossRef]

12. Li, S.; Li, S.; Shan, X.; Gong, C.; Yu, X. Classification, formation, and transport mechanisms of mud clasts. *Int. Geol. Rev.* **2017**, *59*, 1–12. [CrossRef]
13. Bell, H. Armored Mud Balls: Their Origin, Properties, and Role in Sedimentation. *J. Geol.* **1940**, *48*, 1–31. [CrossRef]
14. Dickas, A.; Lunking, W. The origin and destruction of armored mud balls in a fresh-water lacustrine environment, Lake Superior. *J. Sediment. Res.* **1968**, *38*, 1366–1370. [CrossRef]
15. Little, R. Lithified Armored Mud Balls of the Lower Jurassic Turners Falls Sandstone, North-Central Massachusetts. *J. Geol.* **1982**, *90*, 203–207. [CrossRef]
16. Diffendal, R. Armored Mud Balls and Friable Sand Megaclasts from a Complex Early Pleistocene Alluvial Fill, Southwestern Morrill County, Nebraska. *J. Geol.* **1984**, *92*, 325–330. [CrossRef]
17. Mather, A.; Stokes, M.; Pirrie, D.; Hartley, R. Generation, transport and preservation of armoured mudballs in an ephemeral gully system. *Geomorphology* **2008**, *100*, 104–119. [CrossRef]
18. Wani, H.; Dar, S.; Mondal, M. Occurrence of Unusual Unarmored, Unlithified Fossil Mud Balls in Plio-Pleistocene Lacustrine Sediments, Kashmir, India. *J. Geol.* **2017**, *125*, 479–486. [CrossRef]
19. Ghandour, I.; Al-Washmi, H.; Haredy, R. Gravel-Sized Mud Clasts on an Arid Microtidal Sandy Beach: Example from the Northeastern Red Sea, South Al-Wajh, Saudi Arabia. *J. Coastal Res.* **2013**, *291*, 110–117. [CrossRef]
20. Henares, S.; Arribas, J.; Cultrone, G.; Viseras, C. Muddy and dolomitic rip-up clasts in Triassic fluvial sandstones: Origin and impact on potential reservoir properties (Argana Basin, Morocco). *Sediment. Geol.* **2016**, *339*, 218–233. [CrossRef]
21. Allen, J. *Sedimentary Structures, their Character and Physical Basis*; Elsevier: Amsterdam, The Netherlands, 1982.
22. Kortekaas, S.; Dawson, A. Distinguishing tsunami and storm deposits: An example from Martinhal, SW Portugal. *Sediment. Geol.* **2007**, *200*, 208–221. [CrossRef]
23. Goff, J.; Chagué-Goff, C.; Nichol, S.; Jaffe, B.; Dominey-Howes, D. Progress in palaeotsunami research. *Sediment. Geol.* **2012**, *243–244*, 70–88. [CrossRef]
24. Gibbard, P.; Boreham, S.; Roe, H.; Burger, A. Middle Pleistocene lacustrine deposits in eastern Essex, England and their paleogeographical implications. *J. Quat. Sci.* **2015**, *11*, 281–298. [CrossRef]
25. Ishizawa, T.; Goto, K.; Yokoyama, Y.; Miyairi, Y.; Sawada, C.; Takada, K. Reducing the age range of tsunami deposits by, 14 C dating of rip-up clasts. *Sediment. Geol.* **2018**, *364*, 334–341. [CrossRef]
26. Plint, A. Slump blocks, intraformational conglomerates and associated erosional structures in Pennsylvanian fluvial strata of eastern Canada. *Sedimentology* **1986**, *33*, 387–399. [CrossRef]
27. Poll, H.; Patel, I. Slump blocks, intraformational conglomerates and associated erosional structures in Pennsylvanian fluvial strata of eastern Canada. *Sedimentology* **1989**, *36*, 137–145. [CrossRef]
28. Gibling, M.; Tandon, S. Erosional marks on consolidated banks and slump blocks in the Rupen River, north-west India. *Sedimentology* **1997**, *44*, 339–348. [CrossRef]
29. Smith, N. Flume experiments on the durability of mud clasts. *J. Sediment. Res.* **1972**, *42*, 378–383. [CrossRef]
30. Woźniak, P.; Pisarska-JamroźY, M. Debris flows with soft-sediment clasts in a Pleistocene glaciolacustrine fan (Gdańsk Bay, Poland). *Catena* **2018**, *165*, 178–191. [CrossRef]
31. Petrascheck, W. Das Vorkommen von SteinkohlengerIllen in einen Karbonsandstein. *Verh. Kais. Königlich Geol. Reichsanst.* **1909**, *16*, 380–386. (In German)
32. Brune, A. Einlagerungen fremder Gesteine in Steinkohlenflo"tzen unter besonderer Berfcksichtigung Ausffllung von Erosionhohlra"umen. *Glfckauf* **1930**, *66*, 1157–1165. (In German)
33. Cross, A. The geology of Pittsburgh coal: Stratigraphy, petrology, origin and composition, and geologic interpretation of mining problems. In Proceedings of the Second Conference on Origin and Constitution of Coal, Crystall Cliffs, NS, Canada, 18–20 June 1952; pp. 32–111.
34. Dembowski, Z.; Jachowitz, A. Otoczaki i okruchy we gla na wto'rnym zloyu w piaskowcach warstw orzeskich i yaziskich wiercenia Mie dzyrzecze IG 2. *Biuletin Inst. Geol.* **1964**, *184*, 125–176, (In Polish with English summary).
35. Gayer, R.; Pešek, J. Cannibalisation of coal measures in the south Wales coalfield-significance for foreland basin evolution. *Proc. Ussher Soc.* **1992**, *8*, 44–49.

36. Daněk, V.; Pešek, J.; Valterová, P. Coal clasts in the Bolsovian (Westphalian C) sequence of the Kladno-Rakovník continental basin (Czech Republic): Implications for the timing of maturation. *Pol. Geol. Inst. Spec. Pap.* **2002**, *7*, 63–78.

37. Pešek, J.; Sýkorová, I. A review of the timing of coalification in the light of coal seam erosion, clastic dykes and coal clasts. *Int. J. Coal Geol.* **2006**, *66*, 13–34. [CrossRef]

38. Paszkowski, M.; Lachowicz, H.; Michalik, M.; Teller, L.; Uchman, A.; Urbanek, Z. Composition, age and provenance gravel-sized clast from the Upper Carboniferous of the Upper Silesian Coal basin (Poland). *Stud. Geol. Pol.* **1995**, *108*, 45–126.

39. Gayer, R.; Pešek, J.; Sýkorová, I.; Valterová, P. Coal clasts in the upper Westphalian sequence of the South Wales coal basin: Implication for the timing of maturation and fracture permeability. *Geol. Soc. Lond. Spec. Publ.* **1996**, *109*, 103–120. [CrossRef]

40. Dai, J. Coal-derived gas theory and its discrimination. *Chin. Sci. Bull.* **2018**, *63*, 1290–1305. [CrossRef]

41. Dai, J.; Ni, Y.; Liao, F.; Hong, F.; Yao, L. The significance of coal-derived gas in major gas producing countries. *Petrol. Explor. Dev.* **2019**, *46*, 435–450. [CrossRef]

42. Zou, C. *Unconventional Petrolume Geology*; Elsevier: Amsterdam, The Netherlands, 2013.

43. Fu, J.; Fan, L.; Liu, X.; Huang, D. Gas accumulation conditions and key technologies for exploration and development of Sulige gas field. *Petrol. Res.* **2018**, *3*, 91–109. [CrossRef]

44. Yang, Y.; Zhong, J.; Sun, Y.; Wang, J.; Fan, L.; Ni, L.; Zhao, Y. Discussion on characteristics and accumulation mechanisms of "proximal-generation and proximal-storage" type tight sandstone gas accumulations in Shuixigou Group, Turpan-Hami Basin. *J. China Univ. Petrol.* **2014**, *38*, 34–41. (In Chinese) [CrossRef]

45. Qin, Y. Research progress of symbiotic accumulation of coal measure gas in China. *Nat. Gas Ind. B* **2018**, *5*, 466–474. [CrossRef]

46. Saller, A.; Lin, R.; Dunham, J. Leaves in turbidite sands: The main source of oil and gas in the deep-water Kutei Basin, Indonesia. *AAPG Bull.* **2006**, *90*, 1585–1608. [CrossRef]

47. Baudin, F.; Disnar, J.R.; Martinez, P.; Dennielou, B. Distribution of the organic matter in the channel-levees systems of the Congo mud-rich deep-sea fan (West Africa). Implication for deep offshore petroleum source rocks and global carbon cycle. *Mar. Petrol. Geol.* **2010**, *27*, 995–1010. [CrossRef]

48. Wang, H.; Mo, X. An outline of the tectonic evolution of China. *Episodes* **1995**, *18*, 6–16. [CrossRef] [PubMed]

49. Zhao, J.; Zhang, W.; Li, J.; Cao, Q.; Fan, Y. Genesis of tight sand gas in the Ordos Basin, China. *Org. Geochem.* **2014**, *74*, 76–84. [CrossRef]

50. Yang, Y.; Li, W.; Ma, L. Tectonic and stratigraphic controls of hydrocarbon systems in the Ordos basin: A multicycle cratonic basin in Central China. *AAPG Bull.* **2005**, *89*, 255–269. [CrossRef]

51. Gao, X.; Wang, Y.; Ni, X.; Li, Y.; Wu, X.; Zhao, S.; Yu, Y. Recovery of tectonic traces and its influence on coalbed methane reservoirs: A case study in the Linxing area, eastern Ordos Basin, China. *J. Nat. Gas Sci. Eng.* **2018**, *56*, 414–427. [CrossRef]

52. Yang, M.; Liu, C.; Lan, C.; Liu, L.; Wang, J. Sequence Stratigraphy of Late Paleozoic Coal-Bearing Measures in Northeastern Ordos Basin. *Acta Sediment. Sin.* **2008**, *12*, 1005–1013. (In Chinese)

53. Xue, C.; Wu, J.; Zhong, J.; Zhang, S.; Zhang, B.; Hao, B.; Wang, D. Characteristics of the marine-terrigenous interdepositional shale: A case study of Taiyuan formation in Linxing area of Ordos basin. *J. China Univ. Min. Technol.* **2019**, *48*, 870–881. (In Chinese) [CrossRef]

54. Guo, Y.; Liu, H.; Quan, B.; Wang, Z.; Qian, H. Late Paleozoic sedimentary system and paleogeographic evolution of Ordos area. *Acta Sediment. Sin.* **1998**, *16*, 44–51. (In Chinese)

55. Guo, D. The Study of Sedimentary System of Upper Paleozoic in East Ordos Basin. Ph.D. Thesis, Northwest University, Xi'an, China, 2009. (In Chinese).

56. Wang, Y.; Chen, S.; Liang, H.; Wang, J.; Li, T. Sedimentary Facies and Their Evolution in the Upper Paleozoic of Baode Area,Ordos Basin. *Acta Sediment. Sin.* **2016**, *34*, 775–784. (In Chinese) [CrossRef]

57. Shi, J.; Huang, W.; Lv, C.; Cui, X. Geochemical characteristics and geological significance of the Upper Paleozoic mudstones from Linxing area in Ordos Basin. *Acta Petrol. Sin.* **2018**, *39*, 876–889. (In Chinese) [CrossRef]

58. Fu, N.; Yang, S.; He, Q.; Xu, W.; Lin, Q. High-efficiency reservoir formation conditions of tight sandstone gas in Linxing-Shenfu blocks on the east margin of Ordos Basin. *Acta Petrol. Sin.* **2016**, *37* (Suppl. 1), 111–120. (In Chinese) [CrossRef]

59. Li, Y.; Wang, Z.; Pan, Z.; Niu, X.; Yu, Y.; Meng, S. Pore structure and its fractal dimensions of transitional shale: A cross-section from east margin of the Ordos Basin, China. *Fuel* **2019**, *241*, 417–431. [CrossRef]

60. Xie, Y.; Meng, S.; Gao, L.; Sun, X.; Duan, C.; Wang, H. Assessments on potential resources of deep coalbed methane and compact sandstone gas in Linxing Area. *Coal Sci. Technol.* **2015**, *43*, 21–24, 28. (In Chinese) [CrossRef]

61. Luo, J.; Shi, C.; Li, B.; Li, Z.; Li, J.; Han, Y.; Zhao, J.; Du, J.; Dai, Y.; Yang, B. Sediment provenance of Chang 6 and Chang 8 oil-bearing of Yanchang formation in the Xi Feng area and peripheral of Ordos Basin: Evidence from rock geochemistry. *Sci. China Ser. D* **2007**, *37*, 62–72. (In Chinese) [CrossRef]

62. Liao, C.; Zhang, Y.; Wen, C. Structural Styles of the Eastern Boundary Zone of the Ordos Basin and Its Regional Tectonic Significance. *Acta Geol. Sin.* **2007**, *81*, 466–474. (In Chinese)

63. International Committee for Coal and Organic Petrology (ICCP). The new inertinite classification (ICCP System 1994). *Fuel* **2001**, *80*, 459–471. [CrossRef]

64. Sykorova, I.; Pickel, W.; Christanis, K.; Wolf, M.; Taylor, G.H.; Flores, D. Classification of huminite—ICCP system 1994. *Int. J. Coal Geol.* **2005**, *62*, 85–106. [CrossRef]

65. Pickel, W.; Kus, J.; Flores, D.; Kalaitzidis, S.; Christanis, K.; Cardott, B.J.; Misz-Kennan, M.; Rodrigues, S.; Hentschel, A.; Hamor-Vido, M.; et al. Classification of liptinite—ICCP system 1994. *Int. J. Coal Geol.* **2017**, *169*, 40–61. [CrossRef]

66. Miall, A. A review of the braided-river depositional environment. *Earth Sci. Rev.* **1977**, *13*, 1–62. [CrossRef]

67. Wang, S.; Shao, L.; Wang, D.; Sun, Q.; Sun, B.; Lu, J. Sequence stratigraphy and coal accumulation of Lower Cretaceous coal-bearing series in Erlian Basin, northeastern China. *AAPG Bull.* **2019**, *103*, 1653–1690. [CrossRef]

68. Zhong, J.; Wen, Z.; Wang, G.; Wang, X.; Rao, M.; Li, Y.; Ni, J.; Shen, X. Influences of the Current Breaking of the Yellow River on the Anomalous Vertical Development and Evolution of the River Course. *Geol. Rev.* **2003**, *49*, 616–621. (In Chinese)

69. Pisarska-Jamroży, M. Transitional deposits between the end moraine and outwash plain in the Pomeranian glaciomarginal zone of NW Poland: A missing component of ice-contact sedimentary models. *Boreas* **2006**, *35*, 126–141. [CrossRef]

Article

An Analytical Solution for Transient Productivity Prediction of Multi-Fractured Horizontal Wells in Tight Gas Reservoirs Considering Nonlinear Porous Flow Mechanisms

Qiang Wang [1,2], Jifang Wan [1,3,*], Langfeng Mu [4,5], Ruichen Shen [3], Maria Jose Jurado [6,*] and Yufeng Ye [7]

1   Research Institute of Petroleum Exploration and Development, Beijing 100083, China; wangqiang.xxsy@sinopec.com
2   Sinopec Star (Beijing) New Energy Research Institute Co., Ltd., Beijing 100083, China
3   CNPC Engineering Technology R&D Company Limited, Beijing 102206, China; srcdri@cnpc.com.cn
4   Missouri University of Science and Technology, Rolla, MO 65409, USA; lmqk7@mst.edu
5   Chinese Academy of Geological Sciences, Beijing 100037, China
6   Institute of Earth Sciences Jaume Almera, CSIC, 08028 Barcelona, Spain
7   National Oil and Gas Exploration and Development Company Limited, Beijing 100034, China; yeyufeng@cnpcint.com
*   Correspondence: wanjifang@126.com (J.W.); mjjurado@ictja.csic.es (M.J.J.); Tel.: +86-01082312171 (J.W.)

Received: 16 January 2020; Accepted: 25 February 2020; Published: 1 March 2020

Abstract: Multi-fractured horizontal wells (MFHW) is one of the most effective technologies to develop tight gas reservoirs. The gas seepage from tight formations in MFHW can be divided into three stages: early stage with high productivity, transitional stage with declined productivity, and final stage with stable productivity. Considering the characteristics and mechanisms of porous flows in different regions and at different stages, we derive three coupled equations, namely the equations of porous flow from matrix to fracture, from fracture to near wellbore region, and from new wellbore region to wellbore then an unstable productivity prediction model for a MFHW in a tight gas reservoir is well established. Then, the reliability of this new model, which considers the multi-fracture interference, is verified using a commercial simulator (CMG). Finally, using this transient productivity prediction model, the sensitivity of horizontal well's productivity to several relevant factors is analyzed. The results illustrate that threshold pressure gradient has the most significant influence on well productivity, followed by stress sensitivity, turbulence flow, and slippage flow. To summarize, the proposed model has demonstrated a potential practical usage to predict the productivity of multi-stage fractured horizontal wells and to analyze the effects of certain factors on gas production in tight gas reservoirs.

Keywords: tight gas reservoir; multi-fractured horizontal well (MFHW); unstable productivity model; productivity forecast; influencing factor analysis

---

1. Introduction

Natural gas is an indispensable energy resource which plays an important role on the world's energy map. As global demand continues to increase, conventional sources of natural gas will not be able to meet the world's requirements. Consequently, the oil industry has shifted its research focus from conventional gas to unconventional natural gas, especially tight gas. Recently, alongside continuous developments in petroleum geology and the boost to oil and gas industry technologies, the exploitation and development of tight gas reservoirs has made considerable breakthrough and become the primary growth point for oil and gas production [1–6]. Because of its strong heterogeneity, complicated microscopic pore structure, and poor connectivity of the effective sand bodies, the porous flow mechanism of tight gas reservoirs is obviously distinct from those of conventional reservoirs. Moreover, because of poor reservoir physical properties, small discharging radius, lack of or low natural productivity, high development difficulty, and some other potential disadvantages, a MFHW with hydraulic stimulations is extensively used for developing tight gas reservoirs [7–16].

To date, many scholars are dedicated to the transient productivity of fractured wells in unconventional hydrocarbon reservoirs, considering the different characteristics of nonlinear flows compared to multi-scale flows. The methods to calculate the transient productivity of fractured horizontal wells primarily include analytical and numerical solutions, among which the analytical solution includes the complex potential theory, conformal transformation, and the equivalent flow resistance method [14,17,18]. Research on the transient productivity of tight gas reservoirs is increasingly wide. Through an iterative algorithm, Zheng et al. [19] quantified the correlation between saturation and pressure of the infinitesimal coal. They used the Corey relative permeability model to describe relative gas/water permeability as a function of the pressure. By applying the inter-porosity flow equation based on a pseudo-pressure, Zhenzihao et al. [20] developed a density-based rate-transient analysis technique, using which they predicted the gas production rate and were able to assess initial gas content in the system. This methodology is also capable of accurately forecasting gas production rates by converting the response of the corresponding liquid (separating pressure-dependent effects), which was performed using dimensionless depletion-driven parameters. Feng et al. [21] also developed a model thoroughly analyzing MFHWs in heterogeneous gas reservoirs. For this purpose, they used source functions, Green's solution, and the element method to constrain a model in Laplace coordinates. However, by combining the pseudo-function method and material balance to describe homogeneous gas reservoirs, one cannot eliminate the nonlinearity of non-homogeneous gas reservoirs. To solve this issue, they calculated material balance by taking into account heterogeneous gas reservoirs from different regions. Jie et al. [22] created a new analytical model, which considers fracture damage and complex gas transport mechanisms as well as the shale gas stimulated reservoir volume (SRV). Berawala et al. [23] presented a mathematical 1D + 1D model to investigate the main controlling factors during a continuum-flow regime in shale-gas production in the context of well-induced fractures. Wenxi et al. [24] simulated shale gas production using a multi-stage fractured horizontal well using another model established in their study, which was based on a trilinear flow and did not assume that the secondary fracture length was equal to half the distance between hydraulic fractures. Jinzhou et al. [9] calculated gas productivity for the multi-staged horizontal wells in fractured tight sandstone gas reservoirs using a semi-analytical mathematical approach. Their model considered how flow in the fracture network (natural fracture system) affected the outcome. This helped to create a production forecast model of complex hydraulics with simulated impacts arising from natural fractures. As a result, after using the Gaussian elimination method, they obtained a semi-analytical solution.

Most previous studies simplified the nonlinear flow mechanisms of tight reservoirs to some extent. It is difficult to accurately and comprehensively reflect the actual flow state of fluid flowing into horizontal wells. To fill this gap, considering the seepage properties of the tight gas reservoirs and the steady-state seepage of fractured horizontal wells, we introduce the transient model of a matrix gas discharge radius. Considering the influence of various factors of tight reservoirs on the productivity, we also introduce the equivalent well diameter model. Under the same conditions, the pressure distribution equation of a single fracture is equivalent to the vertical well pressure distribution equation. Moreover, the transient productivity model describing a multi-stage fractured horizontal well is obtained. The actual data from a gas field in China were applied to verify the productivity model. It was demonstrated that two results obtained by the method proposed in this paper and dominant commercial software have an error of less than 1.62%, and the single parameter sensitivity analysis was also carried out. This paper provides a useful and feasible tool for reservoir engineers to predict, evaluate, and optimize the productivity of multi-fractured horizontal wells in tight gas reservoirs.

2. Reservoir Characteristics Analysis and Physical Model Assumptions

2.1. Productivity Forecast Model for Individual Horizontal Wells

As shown in Figure 1, for fractured horizontal wells, different production stages occur in different seepage areas within different flow media. Within the control area of individual wells, gas located in the fracture first flows to the wellbore. Next, the matrix near the wellbore area re-supplies the fracture. Therefore, the gas in the matrix flows toward the wellbore through the fracture. Then, the matrix far from the wellbore area supplies the matrix near the wellbore, and gas in the far wellbore area flows to the fracture through the matrix throat near the wellbore.

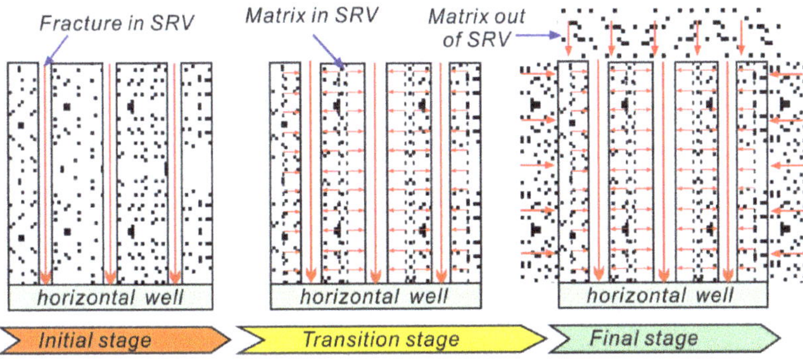

Figure 1. Schematics for each production period of the fractured horizontal well.

To achieve an accurate productivity prediction, based on the characteristics of different production stages of fractured horizontal wells in tight gas reservoirs, different flow media in each seepage area, and different seepage mechanisms, it is therefore necessary to establish corresponding productivity equations by considering the influence of factors such as: (a) threshold pressure gradient; (b) stress sensitivity effect; (c) slippage effect; and (d) high-speed non-Darcy effect. This comprehensive model is able to predict the productivity of the fractured horizontal wells accurately.

In the initial high-production stage, the artificial fractures are the primary flow medium, and hence the gas seepage occurs primarily in the artificial fractures. During this stage, the high-speed non-Darcy seepage cannot be ignored. Moreover, both stress-sensitive effects in artificial fractures and slippage effects play important roles. In the transition period, the gas seepage area is primarily in the proximity of the wellbore (SRV) and the flow medium is the matrix near the wellbore. In this stage, low-speed non-Darcy seepage plays a significant role, which is primarily affected by the threshold pressure gradient and the stress sensitive effect in the matrix. In the final stage of production, the seepage area is far from the wellbore (outside the SRV), and the flow medium is primarily the matrix far from the wellbore. In this stage, low-speed non-Darcy seepage plays a predominant role. However, the influence of the threshold pressure gradient, slippage effect, and stress sensitive effect in the matrix should be considered as well.

2.2. Physical Model Assumption

As shown in Figure 2, a fractured horizontal well with perforation completion is bored in a horizontal, infinite, homogeneous, and equal thickness tight gas reservoir. Some basic assumptions made for multi-stage fractured horizontal wells are as follows:

(1) This model is applicable for isothermal single-phase unstable flows and the influence of gravity is neglected.

(2) Fractures completely penetrate the target zone. Fractures and wellbores are arranged both symmetrically and equidistantly, and the fractures are perpendicular to the horizontal wellbore.

(3) Gas flows evenly into fracture along the fracture wall and then into the horizontal wellbore through the fracture.

(4) Mutual interference exists between the fractures and the pressure loss in the wellbore is neglected.

(5) Contamination of the fracture wall is neglected.

(6) Pressure loss in the horizontal wellbore is neglected.

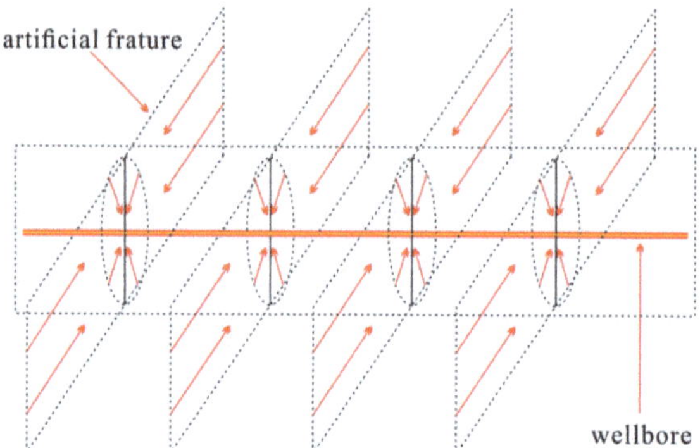

Figure 2. The schematic of seepage field in fissures of fractured horizontal well.

3. Mathematical Model

3.1. Mathematical Model for Nonlinear Flow Mechanisms

Considering the threshold pressure gradient, stress sensitive effect, and slippage effect, the nonlinear flow mathematical model of tight gas reservoirs can be derived through the generalized form of Darcy's Law [25,26].

3.1.1. Threshold Pressure Gradient Effect

Because of the differences of porosity structure characteristics, when gas receive the effect of surface molecules force, this will in turn contribute to the threshold pressure phenomena. Consequently, gas flow in low tight gas reservoirs obviously differs from that in conventional gas reservoirs with medium or high permeability [27]. The threshold pressure gradient, which is associated with non-Darcy flow in low-permeability reservoirs, is defined as the level of pressure gradient that must be attained to enable the fluid to flow.

$$v_g = \frac{K_g}{\mu_g}\left(\frac{dp}{dy} - \lambda\right) \tag{1}$$

where v_g is velocity of gas, m/s; K_g is permeability measured by gas, mD; μ_g is gas viscosity, mPa·s; p is the pressure in the matrix, MPa; and λ is the threshold pressure gradient factor, MPa.

3.1.2. Stress Sensitivity Effect

Pressure depletion easily causes the deformation of the pores across different scales. When affected by the failure of the propellant, the artificial fracture is deformed or even completely closed [28,29]. The relation of permeability stress sensitivity because of matrix pore throat deformation is as follows [30]:

$$K_m = K_{m0}e^{-\alpha_F(p_i-p)} \tag{2}$$

where K_{m} is permeability of matrix, mD; K_{m0} is the initial permeability of matrix, mD; a_F is the coefficient of stress sensitivity, MPa^{-1}; p_i is the initial formation pressure, MPa; and p is current pressure, MPa.

3.1.3. Gas Slippage Effect

Slippage is a phenomenon in which natural gas from a reservoir bypasses crude oil and water that is released from the capillary opening of porous oil formations and approaches to the mean free path of the natural gas. Gas slippage can be defined as the gas movement through liquid phase of the reservoir front. The gas slippage effect is an important factor affecting gas flow in compact porous media [31–33]. We used the Klinkenberg equation to describe its dynamic characteristics [34,35]:

$$K_{\mathrm{g}} = K(1 + b/\bar{p})$$ (3)

where K_{g} is the permeability measured by gas, mD; b is the slippage factor, MPa; $\bar{p}$ is the average gas reservoir pressure, MPa; and K is the permeability of a different medium, mD.

3.2. Productivity Model for Porous Flow between Matrix and Fractures

3.2.1. Steady-state Productivity Model

There are two different porous flow zones: the matrix porous flow zone and the fracture porous flow zone. In the process of porous flow from matrix to fracture, the porous flow medium is the matrix. Each fracture can be simplified as a linear source. The two-dimensional non-Darcy elliptic flow model is assumed to occur in the formation; i.e., the conjugate isopiestic elliptic cylinder and hyperboloid streamline with the well as the center and fracture endpoint as focus is achieved [36]. When a well starts producing, an equal pressure elliptic cylinder is created in the formation (Figure 3), and the relationship between the rectangular coordinate and elliptical coordinate is defined as follows:

$$x = a\cos\eta,\, y = b\sin\eta$$ (4)

$$a = x_F\cosh\xi,\, b = \sinh\xi$$ (5)

where x and y are Cartesian coordinates; a and b are major axis and minor axis, respectively; x_F is the half length of fracture, m; and η and ξ are elliptical coordinates, m.

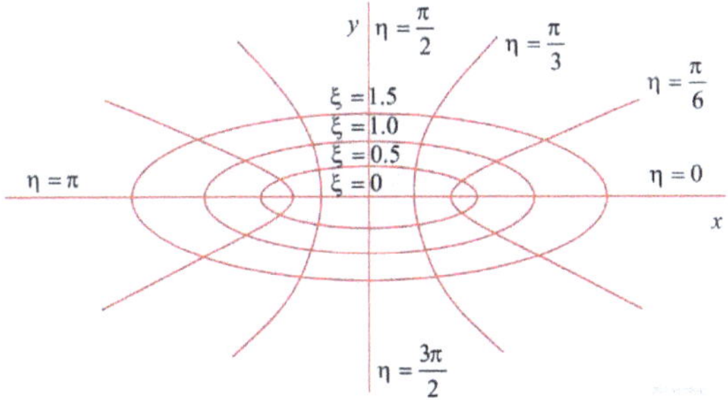

Figure 3. Flow field of the elliptic cylinder with a constant pressure.

By combining Equations (4) and (5), an isobaric elliptical equation and hyperbolic streamline equations can be expressed as Equation (6) and (7), and its flow field is illustrated in Figure 3:

$$\frac{x^2}{a^2} + \frac{y^2}{b^2} = 1$$ (6)

$$\frac{x^2}{(x_F \cos \eta)^2} - \frac{y^2}{(x_F \sin \eta)^2} = 1 \tag{7}$$

Combining Equations (4) and (5), the relationship between p and y under Cartesian coordinates can be transformed to the relationship between p and ξ under elliptical coordinates.

$$\frac{dp}{dy} = \frac{dp}{d\xi} \cdot \frac{d\xi}{dy} = \frac{\pi}{2x_F \cosh \xi} \cdot \frac{dp}{d\xi} \tag{8}$$

The average mass flow rate of the cross section of the elliptical column in y direction is as follows:

$$q = \frac{45\pi a}{32} h \cdot v_{\mathrm{m}} \tag{9}$$

where q is the gas flow rate, m^3/s; h is the reservoir thickness, m; and v_{m} is the gas velocity, m/s.

By combining Equations (8)–(10), a porous flow equation considering stress sensitivity, threshold pressure gradient, and slippage effect can be expressed as follows:

$$\frac{K_{\mathrm{m0}} \exp[-\alpha(p_i - p)]}{\mu} \cdot (1 + b/\overline{p}) \cdot \left(\frac{\pi}{2x_F \cosh \xi} \cdot \frac{dp}{d\xi} - \lambda \right) = \frac{ZT}{p} \frac{p_{sc}}{Z_{sc}T_{sc}} \frac{32q_{sc}}{45\pi x_F h \cosh \xi} \tag{10}$$

where Z is the gas compressibility factor, dimensionless; T is the reservoir temperature, K; Z_{sc} is the gas compressibility factor under standard conditions, dimensionless; T_{sc} is the standard temperature, K; and p_{sc} is the standard pressure, MPa.

Assuming $f(p) = \frac{p \exp^{\alpha(p - p_i)}}{\mu Z}$, the pseudo-pressure function equation is obtained.

$$\psi(p) = \int_{p_0}^{p} f(\delta) d\delta \tag{11}$$

where δ is variable, MPa; p_0 is the initial pressure, MPa; and p is the current pressure, MPa.

By substituting Equation (11) into Equation (10), a simplified porous flow equation is derived as follows:

$$\psi(p_i) - \psi(p) = \frac{64p_{sc}T \cdot \int_{\xi(0)}^{\xi(r_e)} d\xi}{45\pi^2 K_{\mathrm{m0}} Z_{sc} T_{sc} (1 + b/\overline{p}) h} \cdot q_{sc} + \frac{2x_F}{\pi} \int_{\xi(0)}^{\xi(r_e)} \lambda f(\overline{p}) \cosh \xi d\xi \tag{12}$$

where $\psi(p_i)$ is the initial pseudo-pressure, $\mathrm{MPa}^2/(\mathrm{mPa \cdot s})$; $\xi(r_e)$ is variable upper limit, dimensionless; and $\xi(0)$ is variable lower limit, dimensionless.

3.2.2. Transient Model of Discharge Radius in the Matrix

The discharge radius of the tight gas reservoir matrix has a non-transient effect and will expand as the pressure wave propagates. The entire area of gas flow at each instant actually involves the whole formation. Thus, the gas reservoir can be divided into two zones: the pressure-related zone and the un-flushed zone. The radius of the excitation zone (i.e., the radius of the matrix's deflation) increases with time. The motion law of the dynamic zone and the non-excited zone can be obtained using the material balance equation with boundary conditions.

From Equations (4) and (5), the relationship between ξ and a can be expressed as follows:

$$\xi = \ln\left[(a + \sqrt{a^2 - x_F^2}) / x_F \right] \tag{13}$$

Figure 4 is the schematic of elliptical flow area which illustrates the relationship between the major axis of ellipse a and the discharge radius of matrix $R(t)$. Note that a and $R(t)$ can be expressed as follows:

$$a = x_F + R(t) \tag{14}$$

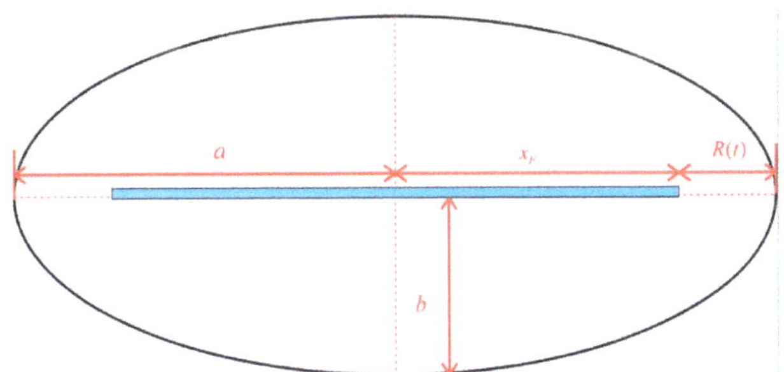

Figure 4. Schematic of the elliptical seepage area.

Note that the discharge radius of matrix $R(t)$ is a transient value, which increases over time t, and the material balance equation is then used to determine $R(t)$.

The original gas reserves in the region with radius $R(t)$ can be calculated as follows:

$$M_0 = \pi[R^2(t) - r_w^2]h\phi\frac{\rho_{aT}}{p_{aT}}p_a \tag{15}$$

where M_0 is the initial reserve in place, kg; r_w is the well radius, m; ϕ is the porosity, dimensionless; ρ_{aT} is the gas density at standard condition, kg/m^3; p_{aT} is the standard pressure, MPa; and p_a is the real formation pressure, MPa.

The current reserves can be expressed using the average formation pressure $\bar{p}$:

$$M_t = \pi[R^2(t) - r_w^2]h\phi\frac{\rho_{aT}}{p_{aT}}\bar{p} \tag{16}$$

Moreover, the average formation pressure $\bar{p}$ can be calculated using the following equation:

$$\bar{p} = p_i - \frac{p_i^2 - p_{wf}^2}{4p_i \ln[R(t)/r_w]} \tag{17}$$

where p_{wf} is the bottom hole flow pressure, MPa.

Because the gas production rate is constant at Q_{aT}, the total mass of production gas is equal to $\rho_{aT}Q_{aT}t$ at time t. Therefore, the conservation law of matter can be expressed as follows:

$$M_0 - M_t = \rho_{aT}Q_{aT}t \tag{18}$$

Furthermore, the productivity equation of steady state seepage of tight gas reservoirs considering the threshold pressure gradient can be obtained as follows:

$$Q_{aT} = \frac{\pi k h\left[p_i^2 - p_{wf}^2 - \lambda\bar{p}(R(t) - r_w)\right]}{\mu p_{aT} \ln[R(t)/r_w]} \tag{19}$$

Combining Equations (16)–(19), the discharge radius equation, which considers the threshold pressure gradient, can be obtained as follows:

$$R^2(t) - r_w^2 = \frac{4kp_i}{u\phi} \cdot \frac{p_i^2 - p_{wf}^2}{\lambda \bar{p}[R(t) - r_w]} t \tag{20}$$

Moreover, taking the stress sensitivity effect into consideration, the relationship between $R(t)$ and t is expressed as:

$$R^2(t) - r_w^2 = \frac{4k_{m0}e^{-\alpha(p_i - \bar{p})}p_i}{u\phi} \cdot \frac{(p_i^2 - p_{wf}^2) - \lambda\bar{p}[R(t) - r_w]}{p_i^2 - p_{wf}^2} t \tag{21}$$

Placing $R(t) = e^{\xi(t)}$ and $r_w = 1$ into Equation (21), the relationship between $\xi(t)$ and t can thus be written as:

$$e^{2\xi_e(t)} - 1 = \frac{4k_{m0}e^{-\alpha(p_i - \bar{p})}p_i}{u\phi} \cdot \frac{(p_i^2 - p_{wf}^2) - \lambda\bar{p}[e^{\xi_e(t)} - 1]}{p_i^2 - p_{wf}^2} t \tag{22}$$

$$\bar{p} = p_i - \frac{p_i^2 - p_{wf}^2}{4p_i\xi_e(t)} \tag{23}$$

Combining Equations (12), (22), and (23), an unstable-state productivity model from matrix to fracture can be obtained as follows:

$$\psi(p_i) - \psi(p) = \frac{64p_{sc}T \cdot \int_0^{\xi_e(t)} d\xi}{45\pi^2 K_{m0}Z_{sc}T_{sc}(1 + b/\bar{p})h} \cdot q_{sc} + \frac{2x_F}{\pi} \int_0^{\xi_e(t)} \lambda f(\bar{p}) \cosh \xi d\xi \tag{24}$$

3.2.3. Transient Productivity Model for Fractured Horizontal Wells

The transient productivity model for fractured horizontal wells can be obtained by introducing the transient model of the discharge radius in the matrix. Assuming that the flow rate of gas in the gas reservoir-fracture is q_{sc1} (converted to ground standard condition), the pressure at the edge of the fracture is p_1 and the corresponding pseudo-pressure is $\psi(p_1)$. Moreover, the pseudo-pressure distribution equation of transient seepage flow in gas reservoir-fracture is expressed as [37–39]:

$$\psi(p_i) - \psi(p_1) = \frac{64p_{sc}T \cdot \xi_e(t)}{45\pi^2 K_{m0}Z_{sc}T_{sc}(1 + b/\bar{p})h} \cdot q_{sc1} + \frac{2x_F}{\pi} \int_0^{\xi_e(t)} \lambda f(\bar{p}) \cosh \xi d\xi \tag{25}$$

3.3. Productivity Model for Porous Flows between the Fracture and Near the Wellbore Area

In the fracture that is near wellbore seepage, the flowing medium is the fracture itself. Because of a high permeability of the fracture and the high velocity of gas flow in the fracture, Darcy's equation is not applicable, and rather the Forchheimer equation is required to describe the gas flow. Considering the effects of high velocity turbulence, stress sensitivity, and slippage, the mathematical model of gas seepage in rocks can be calculated using the following equations.

Because of high conductivity in artificial fractures, the gas flows quickly inside the fracture with characteristics of a turbulent flow, which conforms to the high-speed non-Darcy seepage principle [8]. The kinematic equation is then obtained as follows:

$$\frac{dp}{dx} = \frac{\mu}{K_F} v_F + \beta_g \rho_g v_F^2 \tag{26}$$

where K_F is the permeability of fracture, mD; v_F is the seepage velocity of gas in fractures, m/s; β_g is the turbulence coefficient, m^{-1}; and ρ_g is the gas density, g/m^3.

The stress-sensitive equation of fractures can be expressed as follows:

$$K_F = K_{F0}e^{-\alpha_F(p_i-p)} \tag{27}$$

where K_F is the permeability of the fracture, mD; K_{F0} is the initial permeability of the fracture, mD; and α_F is the coefficient of stress sensitivity, MPa^{-1}.

The velocity of gas flows in the fracture can be obtained as follows:

$$v_F = \frac{q_2}{2w_Fh} = \frac{Z}{P}\cdot\frac{p_{sc}T}{Z_{sc}T_{sc}}\cdot\frac{q_{sc2}}{2w_Fh} \tag{28}$$

where q_{sc2} is the flow velocity of the linear flow zone in the fracture, m^3/d; and w_F is the width of fracture, m.

Assuming that the flow rate in linear flow zone is q_{sc2} (converted to standard surface conditions), the pressure at the interface between the linear flow zone and the radial flow zone (radius is h/2) is p_2, and the pseudo-pressure is $\psi(p_2)$, the pseudo-pressure distribution equation of the gas reservoir near the wellbore unstable seepage flow is obtained as follows:

$$\psi(p_1) - \psi(p_2) = \frac{p_{sc}T(x_F - h/2)}{2K_{F0}w_FhZ_{sc}T_{sc}}\cdot q_{sc2} + \beta_g\frac{MTZp_{sc}^2(x_F - h/2)f(\bar{p})}{4R\bar{p}w_F^2h^2Z_{sc}^2T_{sc}^2}\cdot q_{sc2}^2 \tag{29}$$

where q_{sc2} is the flow rate in linear flow zone under standard surface conditions, m^3/d.

$\beta_g\frac{MTZp_{sc}^2(x_F-h/2)f(\bar{p})}{4R\bar{p}w_F^2h^2Z_{sc}^2T_{sc}^2}\cdot q_{sc2}^2$ is the additional pseudo-pressure drop caused by the high-speed turbulent velocity.

3.4. Radial Porous Flow Model from Fracture to Wellbore

When gas accumulates around the wellbore at the edge of fracture, the radial flow leads to an additional pressure drop near the wellbore. As shown in Figure 5, this phenomenon is called the radial concentration effect.

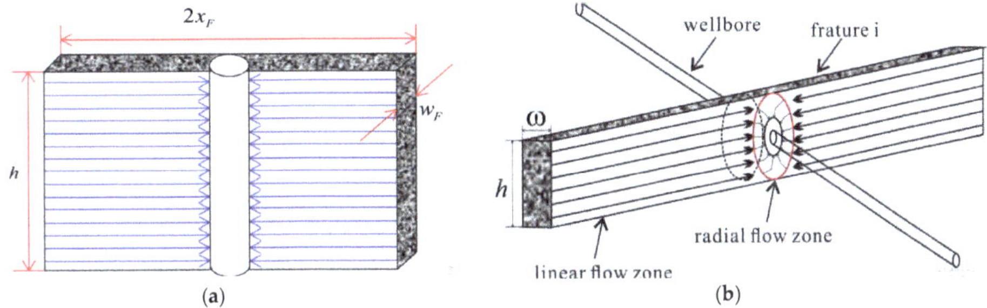

Figure 5. Schematic of porous flow from fracture to wellbore: (**a**) porous flow in one fracture of a vertical well; and (**b**) porous flow in one fracture of a horizontal well.

Assuming that the flow rate in radial flow zone is q_{sc3} (converted to standard surface conditions), the velocity of the radial flow can be calculated as follows:

$$v_3 = \frac{q_3}{2\pi rw_F} = \frac{Z}{p}\cdot\frac{p_{sc}T}{Z_{sc}T_{sc}}\cdot\frac{q_{sc3}}{2\pi rw_F} \tag{30}$$

where v_3 is the velocity of radial flow in the fracture, m/s; q_3 is the flow rate in the fracture under reservoir conditions, m^3/d; and q_{sc3} is the flow rate in the fracture under surface condition, m^3/d.

Combining Equations (25)–(30), the pseudo-pressure equation of fracture-wellbore transient flow is obtained.

$$\psi(p_2) - \psi(p_{wf}) = \frac{p_{sc}T\ln(h/2r_w)}{2\pi K_{F0}w_F Z_{sc}T_{sc}}\cdot q_{sc3} + \beta_g \frac{MTZ(1/r_w - 2/h)p_{sc}^2}{4\pi^2 R\bar{p}Z_{sc}^2 T_{sc}^2 w_F^2}\cdot q_{sc3}^2 \tag{31}$$

3.5. Productivity Model of a Single Fracture in Horizontal Wells

The flow field of the fractured horizontal wells can be divided into two parts: (a) external seepage field (reservoir-fracture); and (b) internal seepage field (fracture-horizontal wellbore). According to the principle of hydropower similarity, the external seepage field will continuously supply gas to the internal seepage field. Therefore, the interfacial pressure and rate will be equal.

Considering the slippage effect, pressure sensitive effect, threshold pressure gradient, and high-speed non-Darcy turbulence effect, the productivity equation can be expressed by Combining Equations (25), (29), and (31).

$$\psi(p_i) - \psi(p_{wf}) = \frac{2x_F}{\pi}\int_0^{\xi_e(t)} \lambda f(\bar{p})\cosh\xi d\xi + \frac{p_{sc}T}{Z_{sc}T_{sc}}\left[\frac{64\xi_e(t)}{45\pi^2 K_{m0}(1+b/\bar{p})h} + \frac{x_F-h/2}{2K_{F0}w_F h} + \frac{\ln(h/2r_w)}{2\pi K_{F0}w_F}\right]\cdot q_{sc}$$
$$+\beta_g \frac{MTZp_{sc}^2}{4R\bar{p}w_F^2 Z_{sc}^2 T_{sc}^2}\left[\frac{(x_F-h/2)f(\bar{p})}{h^2} + \frac{(1/r_w - 2/h)}{\pi^2}\right]\cdot q_{sc}^2 \tag{32}$$

3.6. Equivalent Wellbore Radius Model

The productivity equation of a complex well pattern under complex conditions, e.g., non-Darcy, can be obtained analogous to ordinary vertical wells under Darcy seepage conditions. Therefore, the equivalent diameter of a single fracture in a horizontal well with transverse fracturing can be obtained by combining Equation (32) with the ordinary vertical well productivity equation after considering the threshold pressure gradient, stress sensitivity effect, and slippage effect under generalized Darcy percolation conditions [40].

The pressure distribution equation of normal vertical wells is obtained from the generalized Darcy equation, stress sensitivity, and the slippage effect equation:

$$\psi(p_i) - \psi(p) = \frac{p_{sc}T\ln(R_e/r)\cdot q_{sc}}{2\pi K_{F0}Z_{sc}T_{sc}(1+b/\bar{p})h} + \lambda f(\bar{p})(R_e - r) \tag{33}$$

The pressure distribution equation of a normal vertical well can be expressed as follows:

$$\psi(p_i) - \psi(p_{wf}) = \frac{p_{sc}T\ln(R_e/r_{equ})\cdot q_{sc}}{2\pi K_{F0}Z_{sc}T_{sc}(1+b/\bar{p})h} + \lambda f(\bar{p})(R_e - r_{equ}) \tag{34}$$

Combining Equations (33) and (34), the equivalent diameter, r_{equ}, of a single fracture in the horizontal well is obtained.

3.7. Productivity Model of Multi-fractured Horizontal Wells

Multiplied vertical fractures in a horizontal well will interfere with each other, and the degree of mutual influence depends on the location of the fracture as well.

According to the superposition theory of the pseudo-pressure drop, MFHWs are equivalent to multiple vertical wells using the equivalent radius model [41,42]. Thus, the seepage flow of MFHW is transformed into the superposition of multi-fractured vertical wells. According to the pseudo-pressure superposition principle, the pseudo-pressure of each fracture at fracture j is as follows:

$$
\left\{
\begin{array}{l}
\psi(p_i) - \psi(p_{wf1}) = \Delta\psi(p)_{11}(q_{sc1}) + \Delta\psi(p)_{21}(q_{sc2}) + \Delta\psi(p)_{31}(q_{sc3}) + \cdots + \Delta\psi(p)_{n1}(q_{scn}) \\
\psi(p_i) - \psi(p_{wf2}) = \Delta\psi(p)_{12}(q_{sc1}) + \Delta\psi(p)_{22}(q_{sc2}) + \Delta\psi(p)_{32}(q_{sc3}) + \cdots + \Delta\psi(p)_{n2}(q_{scn}) \\
\psi(p_i) - \psi(p_{wf3}) = \Delta\psi(p)_{13}(q_{sc1}) + \Delta\psi(p)_{23}(q_{sc2}) + \Delta\psi(p)_{33}(q_{sc3}) + \cdots + \Delta\psi(p)_{n3}(q_{scn}) \\
\qquad\qquad\qquad\qquad\vdots \\
\psi(p_i) - \psi(p_{wfn}) = \Delta\psi(p)_{1n}(q_{sc1}) + \Delta\psi(p)_{2n}(q_{sc2}) + \Delta\psi(p)_{3n}(q_{sc3}) + \cdots + \Delta\psi(p)_{nn}(q_{scn})
\end{array}
\right.
\tag{35}
$$

Assuming that gas seepage in the horizontal wellbore is infinite and that pressure in the horizontal wellbore is balanced,

$$
\psi(p_{wf1}) = \psi(p_{wf2}) = \ldots = \psi(p_{wfn})
\tag{36}
$$

Thus, the productivity of a fractured horizontal well is as follows:

$$
q_{sc} = q_{sc1} + q_{sc2} + \ldots + q_{sci} = \sum_{i=1}^{n} q_{sci}
\tag{37}
$$

Rewriting Equation (35) for each fracture allows us to form a closed system equation, which can be solved to obtain the gas productivity of a multi-stage fractured horizontal well in a tight gas reservoir.

4. Results and Analysis

4.1. Model Validation

To verify the accuracy of the transient productivity forecast model, a numerical model of MFHW was established using CMG. By assuming the same parameters for a certain tight gas reservoir (Table 1), the daily production rate, cumulative production of the model established in this study, and CMG were obtained.

Table 1. Fundamental parameters of a gas field at Jilin Oilfield in China.

Parameters (Unit)	Value
Initial formation pressure (MPa)	52
Bottom hole flow pressure (MPa)	44
Porosity (-)	0.06
Initial permeability of matrix ($10^{-3}\,\mu m^2$)	0.4
Initial permeability of fracture ($10^{-3}\,\mu m^2$)	5000
Viscosity (mPa·s)	0.8
z-factor (-)	1.2
Thickness of formation (m)	15
Horizontal length (m)	850
Width of fracture (m)	0.003
Spacing of fracture (m)	80
Molecular mass (g/mol)	17.28
Comprehensive compression coefficient of formation (MPa^{-1})	0.0023
Stress sensitivity coefficient of matrix (MPa^{-1})	0.3
Stress sensitivity coefficient of fracture (MPa^{-1})	0.3
Gas slippage factor (MPa)	2
Reservoir temperature (K)	293
Borehole radius (m)	0.1

Figure 6 shows the daily production rate and cumulative production simulated by the model proposed in this paper and CMG, as well as the field production data. The data simulated by the proposed model and CMG are in good agreement with the field data. Meanwhile, the curves of the daily production rate simulated by the proposed model and CMG have a high degree of conformity and a consistent declining trend. At the end of simulation (1500 days), cumulative production simulated by the proposed model and CMG is $11{,}222 \times 10^4$ m^3 $11{,}404 \times 10^4$ m^3, respectively. The results obtained by the two methods have an error of less than 1.62%. In this way, we demonstrated that the proposed model is accurate enough to simulate production of a multi-stage fractured horizontal well in tight gas reservoirs. During the field production process, the gas well is sometimes shut down. Therefore, when production begins again, the actual production rate is slightly higher than the simulated results because of the recovery of formation pressure facilitated by the shutdown.

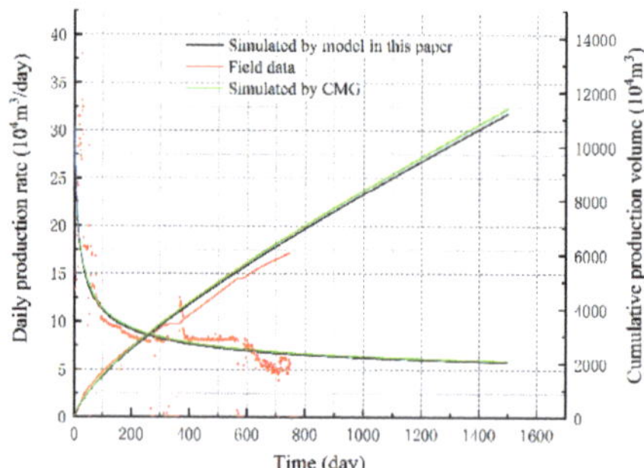

Figure 6. Comparison of the gas production rate of the model in this study and the CMG numerical model.

Figure 7 shows the reservoir pressure distribution of the transient productivity model after 1800 days of production. The gas seepage shows an elliptical flow pattern around the fractures. With continuous production, the pressure wave continues to spread outwards and the discharge radius continues to expand, gradually forming an elliptical stimulated zone that considers the horizontal wellbore as the horizontal major axis and the fractures as the vertical minor axis.

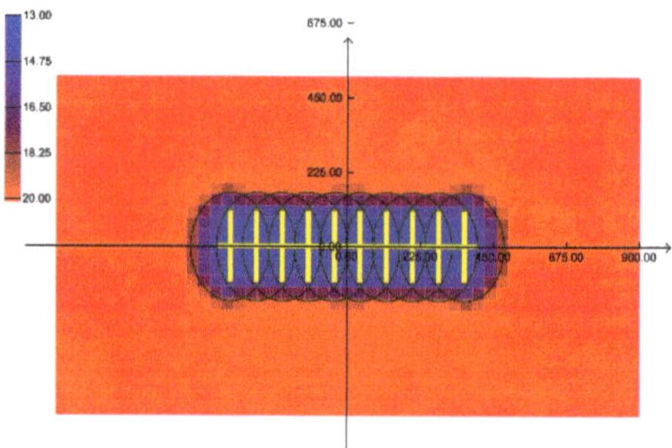

Figure 7. Reservoir pressure distribution of the multi-stage fracture well production.

In this case, we introduce a new concept called the *contribution degree* to evaluate the influence of different factors to productivity growth. Contribution degree (a dimensionless quantity) is defined as the increase in production rate caused by certain factors divided by the production rate when the factor is neglected. The contribution degree of different factors to productivity growth at different times was analyzed, as illustrated in Figure 8. Except for the seepage effect, all of these factors have a negative effect on the output growth. In the early stage of production, the turbulence effect and the stress sensitive effect have considerable influence over gas production, while the effect of a threshold pressure gradient is insignificant. With continuous gas production, the effect of the threshold pressure gradient becomes significant, while the turbulence and stress sensitive effects weaken. It is practical to consider the turbulence effect due to high-speed seepage in the artificial fractures. In general, the increasing influence of contributing factors of production is as follows: (a) threshold pressure gradient; (b) stress sensitivity; (c) turbulence effect; and (d) slippage effect. The slippage effect of gas only works in the low-pressure stage, and its influence is small.

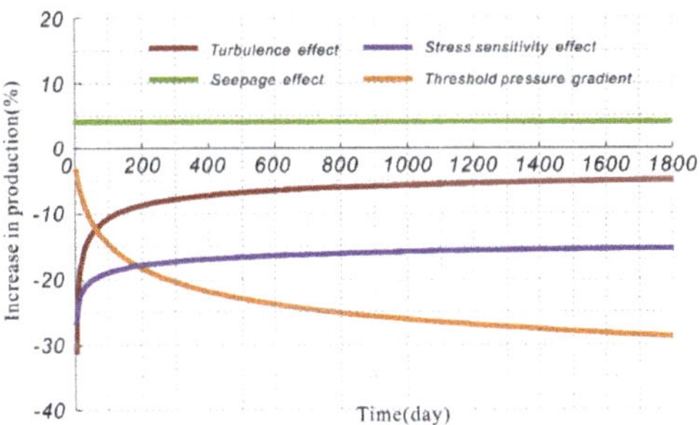

Figure 8. A schematic of the contribution of different factors to productivity at different periods.

4.2. The Influence of Seepage Mechanisms on Gas Production

Different threshold pressure gradients have different effects on gas well productivity. Figure 9 shows that the larger is the threshold pressure gradient, the more significant is the impact on gas well productivity. When the threshold pressure gradient ranges from 0 to 0.6 MPa/m, it has relatively little effect on gas production, compared with a range around 0.6–0.8 MPa/m. While the gas production is primarily derived from fractures, there is no threshold pressure gradient in the initial stage; therefore, the daily production rate at the beginning is almost the same as that at 46×10^4 m^3/d. With a gradual decrease of production until the stable period, or the final stage, the influence of the threshold pressure gradient becomes increasingly significant. When the production time is 500 days, the daily gas production (without considering the threshold pressure gradient) is 34×10^4 m^3/d; however, a daily gas production with a threshold pressure gradient of 0.8 MPa/mis 10×10^4 m^3/d. At the end of the production, the daily gas production (without considering the threshold gradient) is stable at about 33×10^4 m^3/d, and the cumulative gas production is 5.2×10^8 m^3. When the threshold pressure gradient is 0.8 MPa/m, the daily gas production is about 8.2×10^4 m^3/d, and the cumulative gas production is 1.59×10^8 m^3. Thus, the threshold pressure gradient has a considerable influence on productivity in the middle and final stages and hence cannot be ignored for predicting the production of tight gas reservoirs.

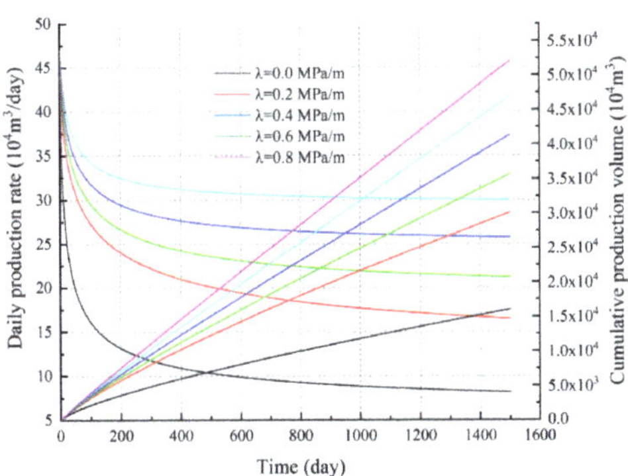

Figure 9. The effect of different slippage factors on productivity.

Because the pore size of the tight reservoir is extremely small, the seepage capacity of the medium is obviously sensitive to the pressure, and the deformation of the medium has considerable influences on the reservoir's properties. Figure 10 shows the effect of matrix stress sensitivity on gas productivity.

The figure indicates that the matrix stress sensitivity coefficient has negative effects on daily gas production; the larger is the pressure sensitivity coefficient, the smaller is the impact on productivity. When the stress sensitivity of the matrix is not considered, the gas production rate at the start is 43.7×10^4 m^3/d, and the high-yield period is considerably long, reaching up to 100 days before entering the transition stage. When production time reaches 1000 days, production gradually transforms to the steady flow stage. At this time, the daily gas production remains stable at 8.5×10^4 m^3/d, and the cumulative gas production at the end of production can reach 1.03×10^8 m^3. When the coefficient of matrix stress sensitivity is 0.8 MPa^{-1}, the initial production rate is only 6.5×10^4 m^3/d, indicating that the stress sensitivity coefficient has particularly considerable impacts in the initial stage. When the production time reaches 100 days, the well productivity stabilizes at 1.8×10^4 m^3/d.

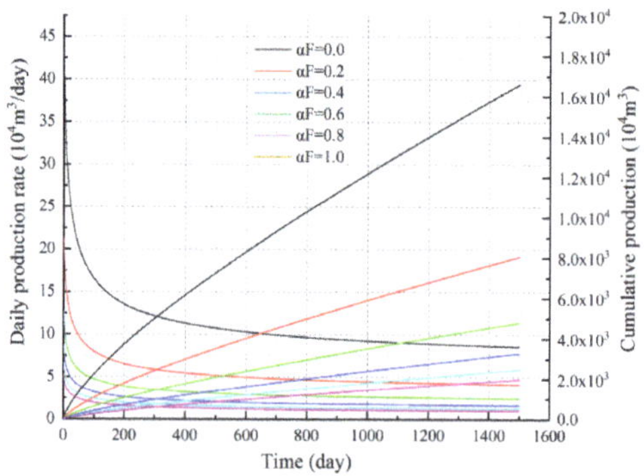

Figure 10. The effect of different matrix stress sensitivity coefficients on productivity.

4.3. The Influence of Formation Properties on Gas Production

The matrix permeability of different tight reservoirs varies. Therefore, it is important to evaluate the impact of matrix permeability on productivity, as illustrated in Figure 11. Below, we discuss the influence of matrix permeability of 0.05, 0.1, 0.15, 0.2, and 0.25×10^{-3} μm^2 on daily and cumulative gas production. Because gas is supplied by large fractures in the initial stage of production (without the participation of the matrix in the seepage process), the daily gas production at the initial stage of production remains virtually constant, i.e., 46×10^4 m^3/d. At the middle and end of the production stage, gas is primarily supplied by pores in the matrix; therefore, the influence of matrix permeability on productivity gradually appears. As matrix permeability increases, the influence on gas wells becomes considerable. When the matrix permeability is 0.25×10^{-3} μm^2, the daily production rate is 20.3×10^4 m^3/d at the final stage of production. When the matrix permeability is 0.05×10^4 m^3/d, the daily production rate stabilizes at 9.5×10^{-3} μm^2, which is reduced by 53.2% compared to permeability of 0.25×10^{-3} μm^2. Moreover, the final cumulative gas production is only 1.83×10^4 m^3/d/d, which is reduced by 88.5% compared to permeability of 0.2×10^4 m^3/d. These results show that matrix permeability has considerable effects on productivity, and the influence of the matrix permeability should be considered in the production allocation and productivity prediction of new wells.

Figure 12 demonstrates that formation thickness has a considerable influence on gas productivity in the initial and stable production stages. A larger formation thickness induces larger gas productions at the initial and stable production stages. With the gradual increase in thickness, the previously increased range of stable productivity remains the same. Figure 12 shows the changing curves for daily production rate and cumulative production as a function of time—corresponding to different formation thicknesses, e.g., 5, 14, 23, 32, 25, 41, and 50 m. When formation thickness is 5 m, the initial production rate is 40×10^4 m^3/d, and then is quickly reduced to the stable gas production rate. This indicates that the gas supply capacity of artificial fractures is insufficient, and the matrix rapidly begins to supply gas. Note that the daily gas production rate is 4.5×10^4 m^3/d, and the cumulative gas production is 0.87×10^8 m^3 at the end of production. When the formation thickness is 50 m, it is large enough to be a sufficient gas source. Specifically, the initial stages of production can reach 41.2×10^4 m^3/d, and the stable gas production in the final stage of production can reach 13.9×10^4 m^3/d; the total gas production reaches 2.66×10^8 m^3.

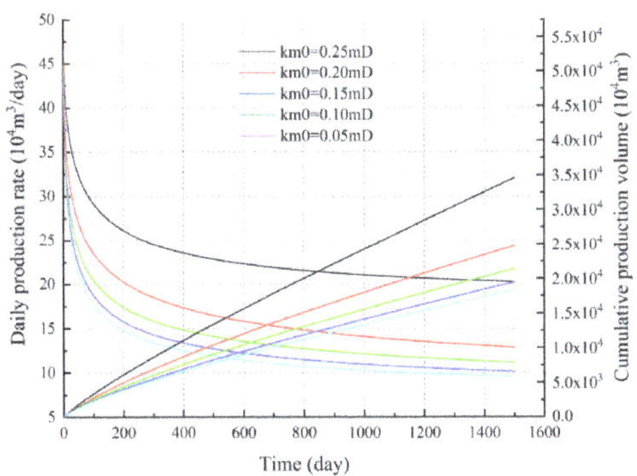

Figure 11. The effect of different matrix permeability on productivity.

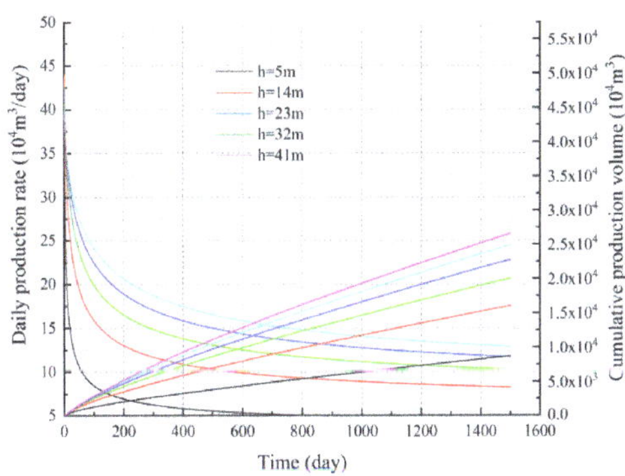

Figure 12. The effect of different formation thicknesses on productivity.

4.4. Influence of Fracture Length on Gas Production

The primary function of artificial fractures is to increase formation conductivity, and hence effectively improve the production capacity of reservoirs. Therefore, the length of the artificial fracture has a direct impact on both daily gas production and stable production capacity. Figure 13 shows its impact on the daily production rate and cumulative production when the half-lengths of artificial fractures are 50, 100, 150, 200, and 250 m. It can be seen that, at the initial stage of production, the production rate changes little, i.e. it is almost 46×10^4 m^3/d for all fracture lengths. The high-productivity period varies proportionally to the length of the artificial fractures, which might be caused by the fact that the length of these fractures can promote high-speed turbulence within them. When the half-length of the artificial fracture is 50 and 250 m, the production rate at the end of the stage is 8.7×10^4 m^3/d and 10.9×10^4 m^3/d, respectively. As the artificial fracture length increases, the drainage radius and area increase; therefore, the production capacity of reservoirs and the controlled reserves of fractured horizontal wells improve. When the gas supply is sufficient, the daily gas production rate must be increased during the stage of stable production.

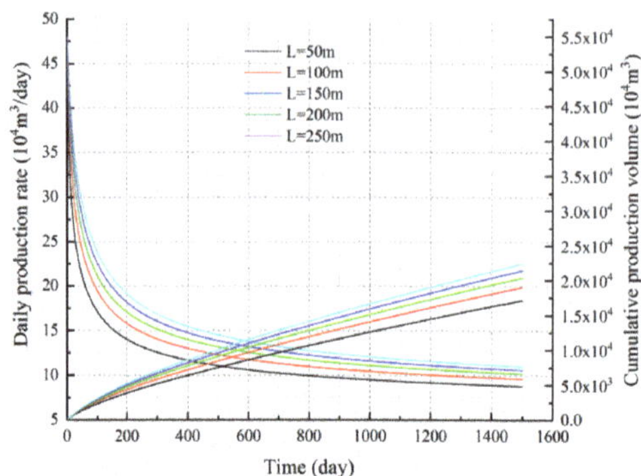

Figure 13. The effect of different artificial fracture lengths on productivity.

In contrast to the fracture length, the artificial fracture permeability primarily affects daily production in the early stage, but has little influence on the seepage of the matrix in the final stage. Figure 14 shows the impact on productivity due to fractured horizontal wells, which correspond to artificial fracture permeability of 200, 600, 1000, 1400, and 1800×10^{-3} μm^2. Here, we see that the artificial fracture permeability has a significant influence on initial gas production. When the artificial fracture permeability is 200×10^{-3} μm^2, the initial gas production rate is only 6.45×10^4 m^3/d; however the daily production rate can reach 51.38×10^4 m^3/d for an artificial fracture permeability of $1800 \times 10^{-3}\mu m^2$, i.e., an increase of 696%. This demonstrates that the higher artificial fracture permeability induces a higher gas seepage rate and hence a higher gas production rate. Moreover, because of the high permeability, the gas source cannot be fully supplied, which leads to a sharp decrease in gas production. When the production time is less than 100 days, the daily production rate drops to 18×10^4 m^3/d. During the stable production period, the artificial fracture permeability has little effect on productivity because gas is supplied by the matrix; and fractures act as channels but cannot supply gas themselves. When the artificial permeability is 200 and 1800×10^{-3} μm^2, the final gas production rate is 2.98 and 9.28×10^4 m^3/d, respectively. The cumulative production is correspondingly 0.56×10^8 m^3 and 1.79×10^8 m^3.

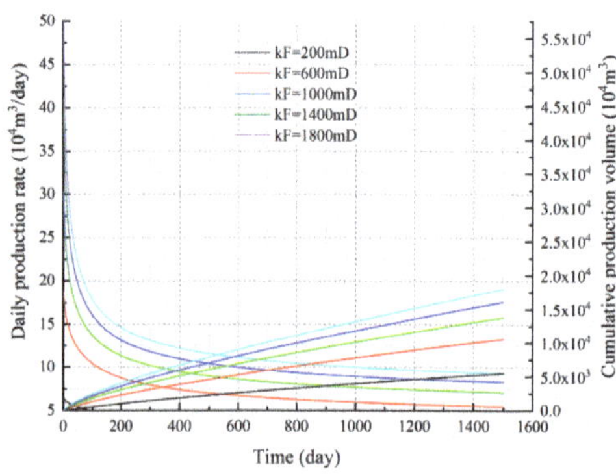

Figure 14. The effect of different artificial fracture permeability on productivity.

We attribute the influence of pressure on productivity as primarily due to the effect of a pressure differential. When the bottom flow pressure is fixed, formation pressure is greater, production pressure difference is greater, the driving force of gas seepage is greater, and the gas production rate is likewise higher. When the formation pressure is fixed, the bottom flow pressure is greater, the production pressure differential is smaller, and the driving force of gas is smaller, making the gas production rate smaller Figure 15 shows the daily and cumulative production rates when the formation pressure is 50, 55, 60, 65, and 70 MPa. The figure shows that the higher is the formation pressure, the greater is the initial production rate, in addition to the stable production rate at the final stage. When formation pressure is 50 MPa, the initial production rate is 31.2×10^4 m^3/d. The gas production rate and cumulative gas production volume are 6.18×10^4 m^3/d and 1.19×10^8 m^3, respectively. When the formation pressure is 70 MPa, the initial production considerably increases to 137.1×10^4 m^3/d. The production rate and cumulative production at the end are 26.3×10^4 m^3/d and 5.15×10^8 m^3, respectively.

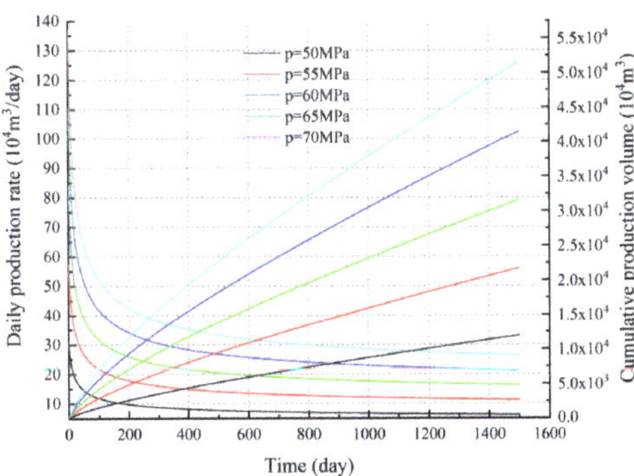

Figure 15. The effect of different formation pressures on productivity.

Figure 16 shows the daily production rate and the cumulative production volume corresponding to different bottom flow pressures, i.e., 0, 10, 30, 40, and 50 MPa. The larger is the bottom hole flow pressure, the smaller are the initial and stable production rates. When the bottom hole flow pressure is zero (an ideal condition that is actually impossible), the initial production rate is 203×10^4 m^3/d, and the production rate in the middle and final stage is 37×10^4 m^3/d, while cumulative production volume at the end of production is 7.2×10^8 m^3. When the bottom hole flow pressure is 40 MPa, the initial production rate is relatively small (64×10^4m^3/d) at the initial stage, but production decreases quickly and enters into the low production period almost instantly. Note that the production rate in the middle and later stages is 12.4×10^4 m^3/d, and the cumulative production volume at the end of production is 2.4×10^8 m^3.

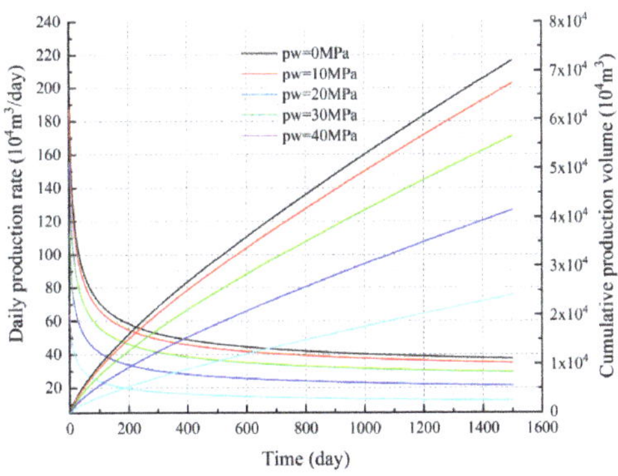

Figure 16. The effect of different WBHPs on productivity.

5. Conclusion

This paper presents a comprehensive mathematical model to predict the gas productivity of MFHWs in tight gas reservoirs. A commercial simulator (CMG) was used to verify this new model. Our major conclusions can be summarized as follows:

(1) The typical production process of fractured horizontal wells in tight gas reservoirs can be divided into three stages based on different seepage areas, flow media, and different seepage characteristics. In the initial stage, the linear and radial flow of gas in fractures shows the tell-tale characteristics of high-speed non-Darcy seepage; in the transitional stage, the gas in the matrix flows in the elliptical seepage area corresponding to each fracture in the near well area; and in the final stage, gas in the matrix flows in the radial seepage area far from the well, both of which show the characteristics of low-speed non-Darcy seepage.

(2) We establish the full cycle productivity prediction model of a multi-stage fractured horizontal well in tight gas reservoirs based on: (a) the different seepage mechanisms of different production stages; and (b) the seepage areas of the horizontal wells in tight gas reservoirs. This is accomplished by considering nonlinear seepage mechanisms, such as the gas slippage effect, threshold pressure gradient, stress sensitive effect, and the confluence of multiple interferences within these fractures.

(3) Based on the actual gas field data, we compared and analyzed the productivity prediction model established in this study using CMG. The results obtained by the two methods have an error of less than 1.62%. We demonstrated that the proposed model is accurate enough to simulate production of a multi-stage fractured horizontal well in a tight gas reservoir.

(4) The significance of four influencing parameters to contribution degree of productivity was analyzed. Except for the seepage effect, the three other factors, namely turbulence effect, stress sensitivity, and threshold pressure gradient effect, have a negative effect on productivity. The increasing influence of contribution factors of production is as follows: threshold pressure gradient, stress sensitivity, turbulence effect, and slippage effect. At the end of production, each contribution degree of these parameters is −29.3%, −15.2%, −5.4%, and 4.8%.

(5) According to the model proposed in this study, the sensitivity analysis of the productivity of fractured horizontal wells was carried out by employing the characteristics of seepage mechanisms, reservoir physical properties, and techniques. Different parameters have different effects on the initial production, stable production, stable production span, and final production of gas wells. These factors need to be comprehensively considered while optimizing any future gas field plan.

Author Contributions: All authors have contributed to this work. Conceptualization, Q.W. and J.W.; methodology, J.W. and L.M.; software, J.W. and R.S.; validation, Y.Y. and L.M.; data curation, Q.W. and R.S.; writing—original draft preparation, Q.W. and M.J.J.; and writing—review and editing, M.J.J. All authors have read and agreed to the published version of the manuscript.

Funding: This research was funded by the National Program on Key Basic Research Project of China (Grant No. 2015CB250900) and the National Oil and Gas Major Project of China (Grant No. 2017ZX05013-006-004).

Acknowledgments: The authors acknowledge the Research Institute of Petroleum Exploration and Development, Petrochina for permission to publish this paper. We are also grateful to all the anonymous readers for their constructive comments.

References

1. Ma, Y.; Cai, X.; Zhao, P. China's shale gas exploration and development: Understanding and practice. *Pet. Explor. Dev.* **2018**, *45*, 589–603. [CrossRef]

2. Zhiming, C.; Xinwei, L.; Chenghui, H.; Xiaoliang, Z.; Langtao, Z.; Yizhou, C.; Heng, Y.; Zhenhua, C. Productivity estimations for vertically fractured wells with asymmetrical multiple fractures. *J. Nat. Gas Sci. Eng.* **2014**, *21*, 1048–1060. [CrossRef]

3. Wang, W.; Fan, D.; Sheng, G.; Chen, Z.; Su, Y. A review of analytical and semi-analytical fluid flow models for ultra-tight hydrocarbon reservoirs. *Fuel* **2019**, *256*, 115737. [CrossRef]

4. Zhao, Y.; Zhang, L.; Luo, J.; Zhang, B. Performance of fractured horizontal well with stimulated reservoir volume in unconventional gas reservoir. *J. Hydrol.* **2014**, *512*, 447–456. [CrossRef]

5. Yang, Y.; Liu, Z.; Yao, J.; Zhang, L.; Ma, J.; Hejazi, S.; Luquot, L.; Ngarta, T. Flow simulation of artificially induced microfractures using digital rock and lattice Boltzmann methods. *Energies* **2018**, *11*, 2145. [CrossRef]

6. Yang, Y.; Yao, J.; Wang, C.; Gao, Y.; Zhang, Q.; An, S.; Song, W. New pore space characterization method of shale matrix formation by considering organic and inorganic pores. *J. Nat. Gas Sci. Eng.* **2015**, *27*, 496–503. [CrossRef]

7. Qiang, W.; Min, T.; Zhanguo, W.; Yi, W. An Unsteady Productivity Prediction Method of Multi-fractured Horizontal Well in Tight Volcanic Rock Reservoir. *J. Southwest Pet. Univ. (Sci. Technol. Ed.)* **2014**, *36*, 107–115.

8. Zou, C.; Zhu, R.; Wu, S.; Yang, Z.; Tao, S.; Yuan, X. Types, characteristics, genesis and prospects of conventional and unconventional hydrocarbon accumulations:taking tight oil and tight gas in China as an instance. *Acta Pet. Sin.* **2012**, *33*, 173–187.

9. Zhao, J.; Pu, X.; Li, Y.; He, X. A semi-analytical mathematical model for predicting well performance of a multistage hydraulically fractured horizontal well in naturally fractured tight sandstone gas reservoir. *J. Nat. Gas Sci. Eng.* **2016**, *32*, 273–291. [CrossRef]

10. Wang, Z.; Ran, B.; Tong, M.; Wang, C.; Yue, Q. Forecast of fractured horizontal well productivity in dual permeability layers in volcanic gas reservoirs. *Pet. Explor. Dev.* **2014**, *41*, 642–647. [CrossRef]

11. Sun, H.; Ouyang, W.; Zhang, M.; Tang, H.; Chen, C.; Ma, X.; Fu, Z. Advanced production decline analysis of tight gas wells with variable fracture conductivity. *Pet. Explor. Dev.* **2018**, *45*, 472–480. [CrossRef]

12. Yang, Y.; Zhang, W.; Gao, Y.; Wan, Y.; Su, Y.; An, S.; Sun, H.; Zhang, L.; Zhao, J.; Liu, L.; et al. Influence of stress sensitivity on microscopic pore structure and fluid flow in porous media. *J. Nat. Gas Sci. Eng.* **2016**, *36*, 20–31. [CrossRef]

13. Wan, J.; Peng, T.; Shen, R.; Jurado, M.J. Numerical model and program development of TWH salt cavern construction for UGS. *J. Pet. Sci. Eng.* **2019**, *179*, 930–940. [CrossRef]

14. Bagher Asadi, M.; Dejam, M.; Zendehboudi, S. Semi-Analytical Solution for Productivity Evaluation of a Multi-Fractured Horizontal Well in a Bounded Dual-Porosity Reservoir. *J. Hydrol.* **2019**, 124288.

15. Dong, J.; Chen, M.; Jin, Y.; Hong, G.; Zaman, M.; Li, Y. Study on micro-scale properties of cohesive zone in shale. *Int. J. Solids Struct.* **2019**, *163*, 178–193. [CrossRef]

16. Qiang, W.; Mengni, Y.; Ning, L.; Yufeng, Y.; Jiaxin, D. Research progress of numerical simulation models for shale gas reservoirs. *Geol. China* **2019**, *46*, 1284–1299.

17. Xiao, C.; Meng, Z.; Tian, L. Semi-analytical modeling of productivity analysis for five-spot well pattern scheme in methane hydrocarbon reservoirs. *Int. J. Hydrog. Energy* **2019**, *44*, 26955–26969. [CrossRef]

18. Luo, W.; Tang, C.; Zhou, Y.; Ning, B.; Cai, J. A new semi-analytical method for calculating well productivity near discrete fractures. *J. Nat. Gas Sci. Eng.* **2018**, *57*, 216–223. [CrossRef]

19. Sun, Z.; Shi, J.; Zhang, T.; Wu, K.; Feng, D.; Sun, F.; Huang, L.; Hou, C.; Li, X. A fully-coupled semi-analytical model for effective gas/water phase permeability during coal-bed methane production. *Fuel* **2018**, *223*, 44–52. [CrossRef]

20. Zhang, Z.; Ayala H, L.F. Analytical dual-porosity gas model for reserve evaluation of naturally fractured gas reservoirs using a density-based approach. *J. Nat. Gas Sci. Eng.* **2018**, *59*, 224–236. [CrossRef]

21. Tian, F.; Wang, X.; Xu, W. A semi-analytical model for multiple-fractured horizontal wells in heterogeneous gas reservoirs. *J. Pet. Sci. Eng.* **2019**, *183*, 106369. [CrossRef]

22. Zeng, J.; Li, W.; Liu, J.; Leong, Y.-K.; Elsworth, D.; Tian, J.; Guo, J.; Zeng, F. Analytical solutions for multi-stage fractured shale gas reservoirs with damaged fractures and stimulated reservoir volumes. *J. Pet. Sci. Eng.* **2019**, 106686. [CrossRef]

23. Berawala, D.S.; Andersen, P.Ø.; Ursin, J.R. Controlling Parameters During Continuum Flow in Shale-Gas Production: A Fracture/Matrix-Modeling Approach. *SPE-190843-PA* **2019**, *24*, 1378–1394. [CrossRef]

24. Ren, W.; Lau, H.C. Analytical modeling and probabilistic evaluation of gas production from a hydraulically fractured shale reservoir using a quad-linear flow model. *J. Pet. Sci. Eng.* **2020**, *184*, 106516. [CrossRef]

25. Garra, R.; Salusti, E. Application of the nonlocal Darcy law to the propagation of nonlinear thermoelastic waves in fluid saturated porous media. *Phys. D Nonlinear Phenom.* **2013**, *250*, 52–57. [CrossRef]

26. Balankin, A.S.; Valdivia, J.-C.; Marquez, J.; Susarrey, O.; Solorio-Avila, M.A. Anomalous diffusion of fluid momentum and Darcy-like law for laminar flow in media with fractal porosity. *Phys. Lett. A* **2016**, *380*, 2767–2773. [CrossRef]

27. Hao, F.; Cheng, L.S.; Hassan, O.; Hou, J.; Liu, C.Z.; Feng, J.D. Threshold Pressure Gradient in Ultra-low Permeability Reservoirs. *Pet. Sci. Technol.* **2008**, *26*, 1024–1035. [CrossRef]

28. Zeng, F.; Peng, F.; Guo, J.; Xiang, J.; Wang, Q.; Zhen, J. A Transient Productivity Model of Fractured Wells in Shale Reservoirs Based on the Succession Pseudo-Steady State Method. *Energies* **2018**, *11*, 2335. [CrossRef]

29. Wang, W.; Su, Y.; Yuan, B.; Wang, K.; Cao, X. Numerical Simulation of Fluid Flow through Fractal-Based Discrete Fractured Network. *Energies* **2018**, *11*, 286. [CrossRef]

30. Soliman, M.Y.; Hunt, J.L.; Azari, M. Fracturing Horizontal Wells in Gas Reservoirs. *Spe Prod. Facil.* **1999**, *14*, 277–283. [CrossRef]

31. Rubin, C.; Zamirian, M.; Takbiri-Borujeni, A.; Gu, M. Investigation of gas slippage effect and matrix compaction effect on shale gas production evaluation and hydraulic fracturing design based on experiment and reservoir simulation. *Fuel* **2019**, *241*, 12–24. [CrossRef]

32. Zhao, T.; Zhao, H.; Ning, Z.; Li, X.; Wang, Q. Permeability prediction of numerical reconstructed multiscale tight porous media using the representative elementary volume scale lattice Boltzmann method. *Int. J. Heat Mass Transf.* **2018**, *118*, 368–377. [CrossRef]

33. Cao, P.; Liu, J.; Leong, Y.-K. Combined impact of flow regimes and effective stress on the evolution of shale apparent permeability. *J. Unconv. Oil Gas Resour.* **2016**, *14*, 32–43. [CrossRef]

34. Li, J.; Sultan, A.S. Klinkenberg slippage effect in the permeability computations of shale gas by the pore-scale simulations. *J. Nat. Gas Sci. Eng.* **2017**, *48*, 197–202. [CrossRef]

35. Firouzi, M.; Alnoaimi, K.; Kovscek, A.; Wilcox, J. Klinkenberg effect on predicting and measuring helium permeability in gas shales. *Int. J. Coal Geol.* **2014**, *123*, 62–68. [CrossRef]

36. Liu, Y.-W.; Liu, C.-Q. New Analysis Method for the Vertical Fracture Well. In *International Meeting on Petroleum Engineering*; Society of Petroleum Engineers: Beijing, China, 1995; p. 10.

37. Ren, Z.; Yan, R.; Huang, X.; Liu, W.; Yuan, S.; Xu, J.; Jiang, H.; Zhang, J.; Yan, R.; Qu, Z. The transient pressure behavior model of multiple horizontal wells with complex fracture networks in tight oil reservoir. *J. Pet. Sci. Eng.* **2019**, *173*, 650–665. [CrossRef]

38. Zhao, Y.; Zhang, L.; Xiong, Y.; Zhou, Y.; Liu, Q.; Chen, D. Pressure response and production performance for multi-fractured horizontal wells with complex seepage mechanism in box-shaped shale gas reservoir. *J. Nat. Gas Sci. Eng.* **2016**, *32*, 66–80. [CrossRef]

39. Zhu, W.; Qi, Q.; Ma, Q.; Deng, J.; Yue, M.; Liu, Y. Unstable seepage modeling and pressure propagation of shale gas reservoirs. *Pet. Explor. Dev.* **2016**, *43*, 285–292. [CrossRef]

40. Feng, Z.; ShiMin, D.; JianMin, Q.; SiQing, D.; WeiGang, S.; JiGui, Y. Productivity of the Horizontal Well with Finite-conductivity Fractures. *Nat. Gas Geosci.* **2009**, *20*, 817–821.

41. Lamarche, L. Short-time analysis of vertical boreholes, new analytic solutions and choice of equivalent radius. *Int. J. Heat Mass Transf.* **2015**, *91*, 800–807. [CrossRef]

42. Mbia, E.N.; Fabricius, I.L.; Oji, C.O. Equivalent pore radius and velocity of elastic waves in shale. Skjold Flank-1 Well, Danish North Sea. *J. Pet. Sci. Eng.* **2013**, *109*, 280–290. [CrossRef]

Article

Pore-Structural Characteristics of Tight Fractured-Vuggy Carbonates and Its Effects on the P- and S-Wave Velocity: A Micro-CT Study on Full-Diameter Cores

Wei Li *, Xiangjun Liu, Lixi Liang, Yinan Zhang, Xiansheng Li and Jian Xiong

State Key Laboratory of Oil and Gas Reservoir Geology and Exploitation, Southwest Petroleum University, Chengdu 610500, China; liuxj@swpu.edu.cn (X.L.); lianglixi@swpu.edu.cn (L.L.); 201721000587@stu.swpu.edu.cn (Y.Z.); 201821000182@stu.swpu.edu.cn (X.L.); 201599010137@swpu.edu.cn (J.X.)
* Correspondence: liwei2014@cugb.edu.cn

Received: 8 October 2020; Accepted: 17 November 2020; Published: 23 November 2020

Abstract: Pore structure has been widely observed to affect the seismic wave velocity of rocks. Although taking lab measurements on 1.0-inch core plugs is popular, it is not representative of the fractured-vuggy carbonates because many fractures and vugs are on a scale up to several hundred microns (and greater) and are spatially heterogeneous. To overcome this shortage, we carried out the lab measurements on full-diameter cores (about 6.5–7.5 cm in diameter). The micro-CT (micro computed tomography) scanning technique is used to characterize the pore space of the carbonates and image processing methods are applied to filter the noise and enhance the responses of the fractures so that the constructed pore spaces are reliable. The wave velocities of P- and S-waves are determined then and the effects of the pore structure on the velocity are analyzed. The results show that the proposed image processing method is effective in constructing and quantitatively characterizing the pore space of the full-diameter fractured-vuggy carbonates. The porosity of all the collected tight carbonate samples is less than 4%. Fractures and vugs are well-developed and the spatial distributions of them are heterogeneous causing, even the samples having similar porosity, the pore structure characteristics of the samples being significantly different. The pores and vugs mainly contribute to the porosity of the samples and the fractures contribute to the change in the wave velocities more than pores and vugs.

Keywords: fracture; vug; micro CT; carbonate; pore structure; wave velocity

---

1. Introduction

During the past 20 years, deep buried carbonate formations have become one of the major sources of natural gas resources in China. Unlike the carbonate formations discovered in the middle-east having a porosity of about 8–25% buried in depths of 2000–4500 m underground, lots of the carbonate gas formations discovered in China are buried over 5000 m in depth and consequently, these formations are usually 'tight' having a porosity of less than 6% [1]. The pore structural characteristics of these deep-buried tight carbonates are different from those buried at shallower depths and referred studies are rare. In the seismic exploration, as well as the acoustic-logging evaluation, of the tight fractured-vuggy carbonate formations, a widely-aware challenge is that the seismic and acoustic properties of the carbonate formations depend, not only on the minerals and the porosity, but also the pore structural characteristics, especially for the tight deeply buried carbonates [2–7]. This issue becomes further complicated considering the fractures and the vugs developed in the carbonate rocks [8,9]. These challenges aroused interest in characterizing both the pore-structures and elastic properties of the tight fractured-vuggy carbonates in the lab before field formation evaluations.

Thin section observation, mercury injection, gas absorption, nuclear magnetic resonance (NMR), and micro-CT scanning are the popular methods for characterizing the pore structure in lab [10–18]. Among these methods, micro-CT scanning has a unique advantage in offering a visible spatial distribution of the pore space, including both the fractures and vugs. Characterizing the pore space of a rock using micro-CT scanning usually contains three major steps: segmenting the 2D pore space from each CT-scanned 2D image, constructing the 3D pore space with all the 2D images, and extracting pore structural parameters to quantitatively characterize the pore structure [19–22]. Histogram shape, clustering, entropy, object attribute, spatial and local methods are the commonly used threshold segmenting method [23,24]. Parameters widely used to characterize the pore structure are the porosity, tortuosity, connectivity, pore radius distribution, ratio of pore radius over throat radius, fracture orientation, etc. [25–32]. These pore structural parameters along with porosity are then usually used to interpret the behavior of the elastic waves propagating in carbonates [33–35]. It has been shown that the porosity–velocity relationship is related to the pore structure, especially those possessing fractures and vugs [36–38]. Previous studies are usually carried out on 1.0-inch cores that might not be representative for tight fractured-vuggy carbonates. Moving the lab study from 1.0-inch core plugs to full diameter cores could provide more representative results but adds to the difficulties in acquiring and interpreting the lab data.

One difficulty is the relatively low quality of the micro-CT scanning images. For the full-diameter cores having a diameter of, for example, 6.5 cm, it is relatively hard for the x-ray to penetrate the rock. This fact causes noised micro-CT images and low contrasts between the fractures and the rock matrix. Consequently, image processing, for example, filtering and enhancement, are necessary before segmenting and constructing the pore space of fractured-vuggy carbonates. Averaging and median filtering are commonly used filtering methods [39–41]. For the filtering of the micro-CT images of the fractured-vuggy carbonates, the challenge involved in denoising the images are that the method should maintain the details of the pore structure or at least avoid blurring the image. In this consideration, filtering methods, for example, nonlocal means and anisotropic diffusion filtering, that tried to maintain the structural boundaries during filtering might be potentially better [42,43]. For the issue of the low contrasts between the fractures and the matrix, image enhancement techniques are possible solutions [44,45]. One issue involved here is that the 'background energy' of the figure might not be uniform and, in this consideration, the methods that can calibrate the background energy difference might be potentially better, for example, the top-bottom hat method [46,47].

Another aspect of difficulties is in linking the wave speeds to the porosity and pore-structural parameters. The pore space of 1.0-inch core plugs contains limited amounts of fractures and vugs while that of the full-diameter cores is usually composed of pores, fractures, and vugs that affect the wave speeds of carbonates comprehensively. Thus, the cross plot of wave speeds over porosity might be scattered [38,48]. It is not clear which pore-structural parameter is the key in linking the wave velocity to the porosity, especially for the fractured-vuggy carbonates.

In this paper, we first develop an image-processing method to improve the quality of each the micro-CT scanned image of the full diameter cores, construct the pore space, and extract the pore structural characteristics. Then, these pore structural parameters are related to the wave speeds to analyze the effect of pore structure on the wave velocity of fractured-vuggy carbonates.

2. Scanned Micro-CT Images of the Full-Diameter Carbonate Cores

We collected 18 carbonate samples, having a diameter of 6.5–7.5 cm and a length of 5.0–10.0 cm, from a burial depth of over 5000 m in the DY Group, Sichuan Basin, Southwest China. Fractures and vugs can be observed with naked eyes on the surface of the sample. According to the standard of the study area, the pore spaces having a ratio of length over width higher than 10 are defined as the fractures, the pore spaces having a diameter longer than 2.0 mm are defined as the vugs, and shorter than 2.0 mm are defined as the pores [49]. From the naked-eye observation of the fractures and vugs appearance on the surface of the collected carbonate cores, the samples can be roughly summarized into three groups: fractured-vuggy carbonates containing relatively developed fractures and vugs, fractural carbonates possessing relatively developed fractures, and vuggy carbonates having relatively developed vugs. The exact classification of the samples was done according to the quantitative analysis of the pore spaces constructed from the micro-CT scanned images.

The samples are scanned using an industrial micro-nano CT instrument—phoenix v|tome|x M manufactured by General Motors Corporation. The High power 300 kV micro-scale X-ray source was used. The micro-CT scanning was taken with an X-ray tube voltage of 200 kV and a tube current of 180 mA. Each sample was scanned for 8 h to achieve the best physical resolution of the instrument. A resolution of 40–50 µm/pixel was achieved and about 1100–1700 slices, depending on the lengths of the scanned samples, having about 1680 × 1695 pixels were obtained for each sample. The resolution and number of slices of the micro-CT scanning of all the samples are listed in Table 1. Some of the obtained micro-CT images are shown in Figure 8. According to the observation of micro-CT scanned images, the length of the fractures varies obviously from partly through the sample (e.g., sample Num. 10) to fully penetrate the sample, e.g., sample Num. 12. The orientations of the fractures are different. Some of the fractures, in the sample, e.g., Num. 16, are parallel while others intersect each other, e.g., Num. 12. The spatial distributions and the sizes of the vugs are significantly heterogeneous. The pores are widely distributed throughout the sample.

Table 1. The resolution, pixels per scanned image, and the amount of the scanned images, and the size information of all the samples.

Num.	Resolution (um/pix)	Pixels Per Scanned Image	Number of Scanned Images	Sample Radius (um)	Sample Length (um)
1	50	1450 × 1454	1600	36,300	80,000
2	50	1442 × 1448	1298	36,125	64,900
3	45	1607 × 1644	1427	36,157	64,215
4	50	1456 × 1437	1470	36,162	73,500
5	50	1402 × 1409	1400	35,137	70,000
6	50	1457 × 1485	1203	36,775	60,150
7	45	1423 × 1427	1400	32,062	63,000
8	43	1489 × 1514	1124	32,282	48,332
9	45	1384 × 1363	1400	31,140	63,000
10	40	1613 × 1641	1445	32,540	57,800
11	43	1405 × 1404	1600	30,196	68,800
12	42	1463 × 1451	1445	30,723	60,690
13	40	1682 × 1682	1465	33,640	58,600
14	41	1586 × 1583	1600	32,482	65,600
15	50	1418 × 1435	1400	35,662	70,000
16	45	1394 × 1396	1400	31,387	63,000
17	40	1548 × 1565	1742	31,130	69,680
18	46	1314 × 1340	1400	30,521	64,400

3. Difficulties in Constructing the Pore Space

If we zoom in each 2D micro-CT image, it is obvious that the speckled random noise and the Gaussian noise are involved in the image (Figure 1a). To give a relatively straightforward view of the noises, we used the false coloring technique to turn a grayscale image into an RGB color image (Figure 1b,c). After that, the speckled random noise is as shown as the red dots; the Gaussian noise makes the matrix (colored in yellow) into yellow and green messy patterns. It is important to filter the noise before constructing the pore space, especially when image enhancement is necessary, to avoid mistakenly considering the noise as parts of the pore space. The pepper and speckled random noise can be removed by median filtering. However, the most commonly used averaging filtering is not applicable for removing the Gaussian noise because it blurs the image and lowers the contrasts between the pore space and the matrix. To avoid this issue, the Gaussian noise was filtered using nonlocal mean filtering. After filtering, the noise involved in the image is decreased and the image is only slightly blurred that will not affect the acquisition of the pore space (Figure 1d).

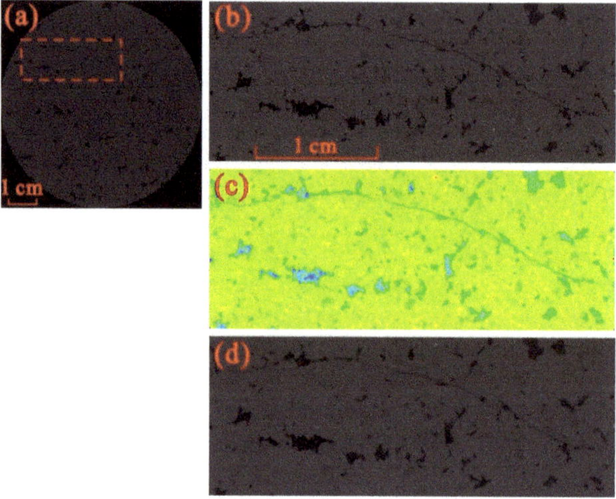

Figure 1. The application of image filtering method to the micro-CT scanned image: (**a**) the micro-CT scanning image, (**b**) the enlarged image, (**c**) the false-colored image, and (**d**) the filtered image. After the false coloring, the pore space is in sky blue and the matrix is in yellow. The speckled random noises are the red dots and Gaussian noises make the matrix a mix of green and yellow.

After filtering the image, another difficulty is the lower contrast between the fractures and the matrix compared to the contrast between the vugs or pores with the matrix (Figure 2). A consequence of this lower contrast is that, if one uses the commonly used binary thresholding segmentation, a low threshold can segment the pores and the vugs well but it cannot segment the fractures (Figure 2b), and a high threshold can continuously segment the fractures, but the volumes of the acquired pores and vugs are larger than they should be and the shapes of the pores and vugs are partly distorted (Figure 2c).

We applied three methods, including the watershed segmentation [50], histogram equalization enhancement [44], and top-bottom hat algorithm, to enhance and acquire the fractures (Figure 3). The watershed algorithm is self-adaptive thresholding that segments the pore space basing on the local minimum instead of a single threshold. The acquired pore space using watershed is as shown in Figure 3b. Although the acquired pores and vugs are clear, the watershed method cannot acquire the fractures. After enhancing the image with the histogram equalization algorithm (Figure 3c) and filtering the remained noise enhanced by the algorithm (Figure 3d), the acquired pore space is similar to that segmented using a high threshold in that the fractures are acquired but the volume of the pores and vugs are larger compared to the original figure and the shapes of the pores and vugs are distorted to some extent. Compared to the former two methods, the top-bottom hat algorithm is better in that the fracture is acquired and it avoids the 'overflow' of the pores and vugs and maintains the majority of the shape details (Figure 3e,f). Basing on the above studies about the methods of filtering and enhancement applied to the micro-CT images of tight fractured-vuggy carbonates, a workflow is established to acquire and link the pore structural characteristics to the wave velocities.

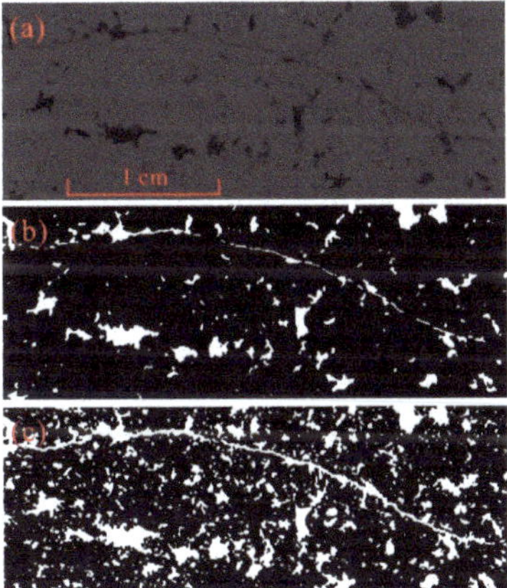

Figure 2. Binary threshold image segmentation: (**a**) the original figure, (**b**) the segmented image using a low threshold, and (**c**) the segmented image using a high threshold. The fractures cannot be segmented as shown in the subgraph (**b**) and the volumes of the pores and the vugs spilled out in the subgraph (**c**).

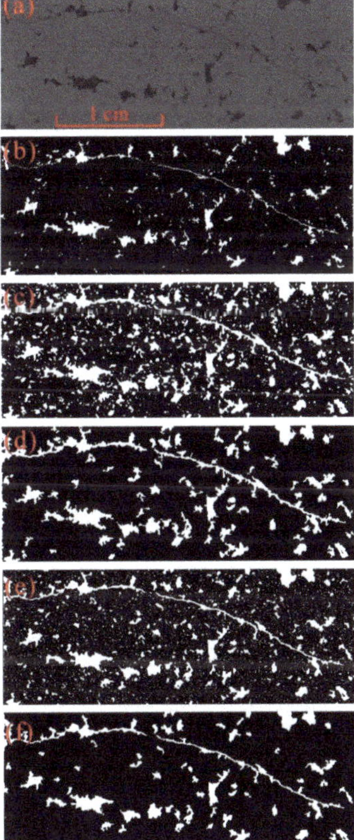

Figure 3. Different methods applied to fracture acquirement: (**a**) the filtered image, (**b**) the segmented pore space by watershed method, (**c**) the segmented pore space after histogram equalization enhancement, (**d**) the filtered image of the subgraph (**c**), (**e**) the segmented pore space after top-bottom hat algorithm, (**f**) the filtered image of the subgraph (**e**).

The image processing methods mentioned above to contour the issue of the low contrast between the pore space and the matrix is convenient to use. However, if the contrast between the pore space and the matrix is too low, physical method might be necessary to enhance the contrast, for example, the difference map method [51]. The difference map method is to scan the sample twice under the conditions of dry and saturated with fluids containing X-ray dense agent, respectively. The pore space then can be highlighted by subtracting the dry image from the saturated image. This method is effective, but the disadvantage is that one has to make sure that the pore space can be fully saturated with the fluid. For the rocks having relatively high porosity, is relatively easier to saturate the pore space. However, for tight rocks, for example, the rocks having a porosity of less than 6%, it is hard to fully saturate the sample. For our samples, the image processing method is enough for extracting the pore space and thus, we did not apply physical method to enhance the contrast between the pores and the matrix.

4. Methods

The method is mainly composed of the acquisition and analysis of the three experimental measurements: the micro-CT, helium porosity, and the pulse transmission measurements. The overall method is shown in Figure 4. First, the collected carbonate samples were sawed into cylindrical plugs and scanned using micro-CT instruments. The obtained 2D images of micro-CT were filtered and enhanced to improve quality. Then, the pore space in each 2D image was acquired and the 3D pore spaces of the samples were constructed using the software AVIZO (manufactured by Visualization Sciences Group, FEI Co., Hillsboro, OR, USA) basing on the 2D pore space. After that, the pore space was divided into pores, vugs, and fractures according to the geometry and the diameter of each individual pore space. Finally, the pore structure parameters were calculated to quantify the characteristics of the pore space and analyze its effect on wave speeds. The density and porosity were measured with helium and the wave speeds were determined using the pulse transmission technique. The following two subsections provide details of pore space acquisition, helium porosity measurements, and the pulse transmission measurements.

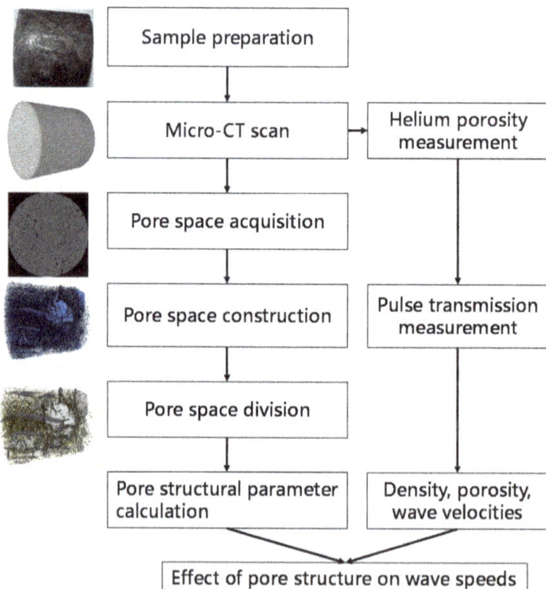

Figure 4. Workflow chat for characterizing the pore space and analyzing its effect on wave speeds. After sawing the samples into cylindrical plugs, the samples are scanned using micro-CT. Then, the quality

of each acquired 2D micro-CT image is improved using image filtering and enhancement techniques. The pore spaces of the samples are constructed and divided into pores, fractures, and vugs basing on the geometric characteristics of each individual pore space. Finally, the pore-structural parameters are acquired and related to the density, porosity, and wave velocities to analyze its effect on wave speeds.

4.1. Pore Space Construction and Division Method

Median filtering and nonlocal means filtering were applied to the micro-CT images first to remove the speckled random noises and the Gaussian noises. Then, the micro-CT image was enhanced with the top-bottom hat transformation. Finally, the processed image was filtered again to remove the remaining noises that were also enhanced after top-bottom hat transformation.

For the division of the pore space, the total pore space was segmented into individual volumes using the watershed algorithm and then the individual volumes were classified into pores, fractures, and vugs according to its geometry and diameter. Usually, the fracture is identified from the pores and vugs using the ratio of the length and width over thickness and the vugs are separated from the pores by the diameter of the isovolumetric sphere. It should be noted that using only the ratio of length over the thickness may mistakenly classify a throat into a fracture (Figure 5). Thus, we used both the ratio of length over thickness and the ratio of width over thickness to avoid this issue. According to the standard of the studies area, the volumes having both the ratios of length over thickness and width over thickness higher than 10 were defined as fractures. The remaining volumes having an equivalent diameter longer than 2.0 mm were defined as vugs and those having an equivalent diameter shorter than 2.0 mm were defined as pores.

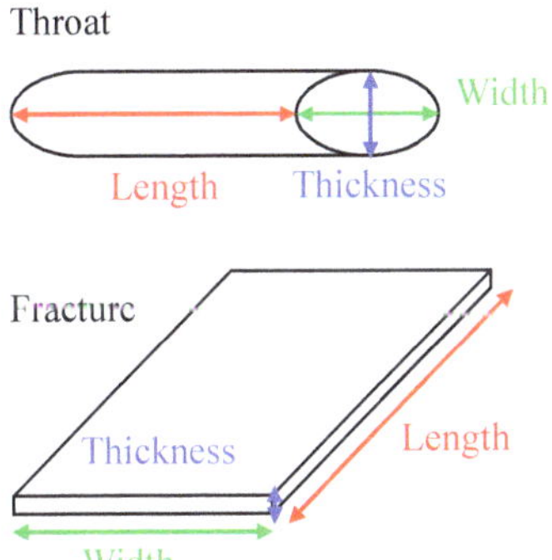

Figure 5. Illustration of the difference between a throat and a fracture in geometry. The lengths of a throat and a fracture are both usually significantly longer than their widths and thicknesses. The difference is that the width and the thickness of a throat are close while the width of a fracture is usually significantly longer than the thickness of it. To avoid mistakenly acquire a throat as a fracture, both the ratios of length over thickness and width over thickness should be large.

After dividing the total pore space into pores, vugs, and fractures the pore structural parameters are quantified. The porosity of the whole pore space, the pores, the vugs, and the fractures were calculated by dividing the total amount of the pixels in the whole pore space, the pores, the vugs, and the fractures over that of the whole sample, respectively. The equivalent pore diameter was calculated by averaging the diameters of the isovolumetric spheres of the pores and the vugs. The orientation of a fracture was defined by the normal vector of the fracture surface.

4.2. Porosity and Wave Velocity Measurements

The porosity of the samples was determined with the single-cell He-gas filling method to offer a reference for the micro-CT pore-space construction. The pore volume was determined using Boyle's

Law under room temperature and a confining pressure of 3 MPa applied to the external surface of the sample jacket. The diameter and the length of the samples are both measured three times for each sample using a caliper and the averaged diameter and length were used to calculate the volume of the cylindrical samples. The porosimeter apparatus was built based on API RP40 (1998) [52] as shown in Figure 6. Before measuring the porosity of the samples, the measurement system was calibrated using a standard sample to calibrate the system dead volume. The total pore volume of the samples was determined following API RP40 (1998) and the porosity was obtained by dividing the volume of the pore space over that of the sample.

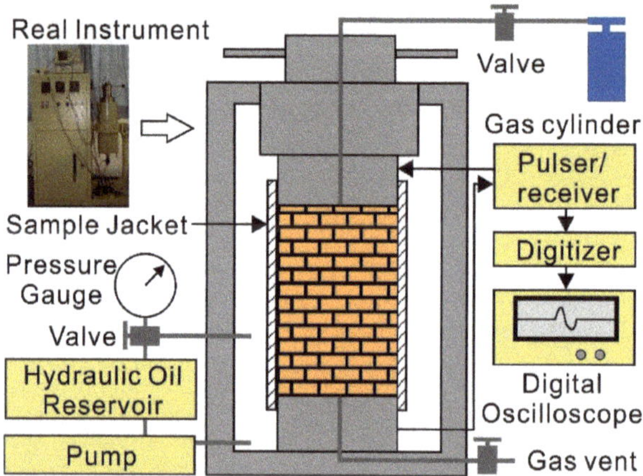

Figure 6. Scheme of the wave velocity measurement system. The system is mainly composed of three parts that provide confining pressure, porosity measurement, and pulse transmission measurement, respectively.

The wave velocities of the P- and S-waves of the collected carbonate samples were determined with the ultrasonic pulse transmission technique. The voltage step was periodically applied to the piezoelectric ceramic to generate a pulse. The generated pulse transmits throughout the sample and encountered the piezoelectric ceramic used to receive the pulse by converting the vibration back to an electrical voltage. The voltage was recorded into the computer by an 8-bit digitizer and a digital oscilloscope programmed using LabVIEW software. The sampling rate was 10.0 ns. The averaged receiving signal was stacked over 300 times and then collected. The transit time was picked at the first amplitude peak of the received waveform. The calibration of the transducer delay was determined from the measurements taken on a set of cylindrical aluminum plugs (6061-T6) with different lengths following Melendez-Martinez (2014) [53]. By plotting the transit time against cylinder length, the excitation delay, equaling 16.16 μs for the longitudinal-mode piezoelectric discs or 8.82 μs for the transverse mode piezoelectric plates, was obtained from the non-zero intercept of the fitting line. The measurement was taken under a confining pressure of 70 MPa (the in-situ confining pressure).

5. Results

5.1. Pore Structure Characteristics of the Tight Carbonates

After processing each 2D micro-CT image, the processed 2D images were imported into the software AVIZO to construct and divide the pore spaces of the samples. Taking sample Num. 10 as an example, the total pore space (Figure 7a) was separated into individual volumes using a watershed algorithm (Figure 7b). The colors used in Figure 7b help identify the individual volumes. These individual volumes were classified into pores (colored in yellow), vugs (in blue), and fractures (in gray) according to the geometry of the individual volumes as shown in Figure 7c. It can be seen in Figure 7 that the proposed method can effectively divide the pore spaces into pores, vugs, and fractures. The pore structure parameters were acquired basing on the characteristics of the constructed spaces of pores, vugs, and fractures.

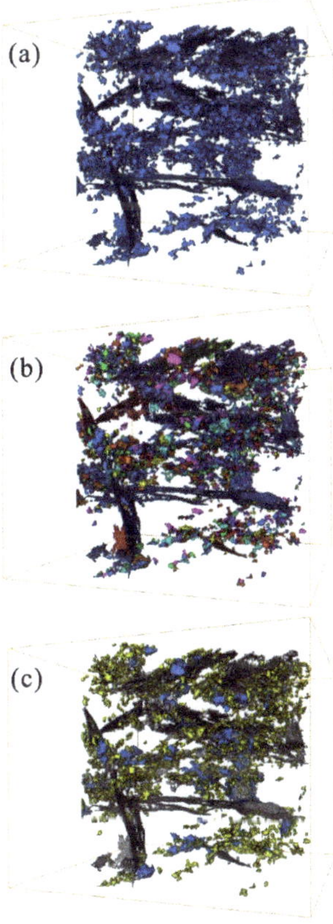

Figure 7. Illustration of (**a**) the acquired pore space, (**b**) segmented individual volumes with watershed method, and (**c**) divided pores, vugs, and fractures according to the geometry of individual volumes.

The structural characteristics of the tight carbonate samples vary significantly, although the values of the porosity of the samples are close being less than 4%. The results of six samples are shown in the 6th and 7th columns of Figure 8 and the rest are shown in Appendix A. Inside the pore space, bubble-like pores and vugs and sheet-like fractures are observed. The pores are separately distributed, and the fractures are intersected and widely distributed inside the core. Fractured-vuggy, fractured, and vuggy carbonates are observed. For the fractured-vuggy carbonates, both the fractures and the vugs are well developed inside the sample. Although, the fractional volumes of the fractures are relatively low compared to the sum of those of the vugs and pores, the fractures penetrating the samples and each other connecting the pore spaces. For the fractured carbonates of the collected samples, the pore space is mainly composed of the fractures and the total porosity of them are all less than 2%.

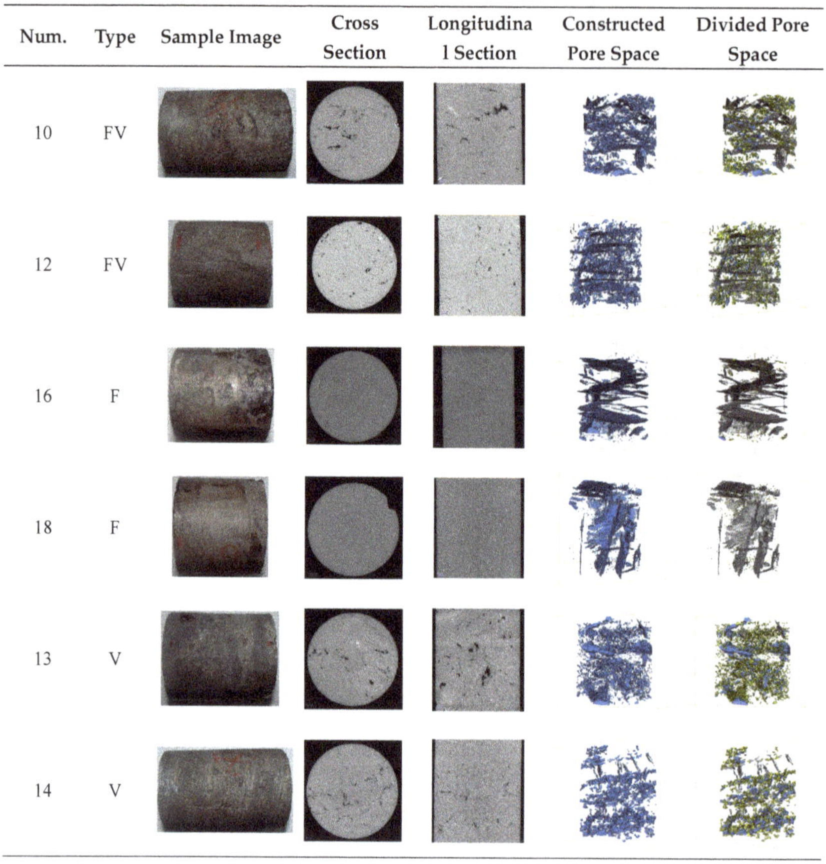

Num.	Type	Sample Image	Cross Section	Longitudinal Section	Constructed Pore Space	Divided Pore Space
10	FV					
12	FV					
16	F					
18	F					
13	V					
14	V					

Figure 8. Examples of the sample, cross section, and longitudinal section of the CT scanning image, constructed pore space, and divided pore space of the carbonate samples. 'FV', 'F', and 'V' represent fractured-vuggy carbonate, fractured carbonate, and vuggy carbonate, respectively. The total pore spaces of the carbonate samples are in blue as shown in the sixth column. The pore space after division is shown in the seventh column with the fractures colored in gray, vugs in blue, and pores in yellow. The rest of the constructed and divided pore spaces of the samples are provided in the supplementary file. The subgraphs are too small that a scale would be hard to see if it is directly marked on the subgraph. Thus, instead of a scale, we provide the physical size including the radius and the length of all the samples in Table 1.

The acquired pore structural parameters are listed in Table 2. The micro-CT porosity of the collected samples ranges from about 1% to 4%. The averaged equivalent pore diameter ranges from about 100 to 400 µm. The vugs are the main contribution of the large pores. The volumetric fractions of pores, vugs, and fractures vary significantly with the samples be 5.16–57.63%, 0–75.22%, and 1.77–94.01%, respectively. The majority of the samples contain both the vugs and fractures. The orientations of the fractures change a lot from being parallel to intersect with each other. The intersection of the fractures with pores and vugs form the main flow channel of the tight carbonates.

Table 2. Pore structural parameters obtained from the constructed pore space. 'VF' represents volume fraction. The VF of pores, vugs, or fractures equals the ratio of the volume of them over that of the total pore space, respectively. The orientation of the fractures parallel or perpendicular to the ends of the core plug is defined as $0°$ or $90°$, respectively. The VR of oriented fractures equals the ratio of the volume of the fractures oriented in a certain angle range, for example, $0–30°$, over that of all fractures.

Num.	CT Porosity %	VF of Pores %	VF of Vugs %	VF of Fractures %	Averaged Equivalent Pore Diameter μm	VF of Oriented Fractures 0–30° %	30–60° %	60–90° %
1	2.41	24.61	17.79	57.60	413.25	58.80	13.85	27.34
2	1.18	53.49	27.40	19.11	332.85	14.59	40.52	44.89
3	2.31	31.20	25.07	43.73	398.99	51.75	20.18	28.07
4	3.29	14.82	63.99	21.19	699.49	52.99	27.49	19.52
5	2.14	28.68	30.93	40.39	586.22	42.63	23.77	33.59
6	3.27	30.39	44.93	24.68	341.34	28.11	25.60	46.30
7	2.27	29.98	7.51	62.50	392.12	27.85	21.88	50.27
8	3.17	15.09	32.18	52.73	248.23	84.32	15.68	0.00
9	2.78	29.70	27.47	42.83	217.91	30.44	33.29	36.28
10	4.06	20.92	50.33	28.75	317.20	11.41	33.09	55.50
11	1.37	45.82	31.92	22.27	293.2	28.79	48.56	22.65
12	2.24	43.97	19.66	36.37	232.04	48.53	20.16	31.31
13	3.56	24.51	73.72	1.77	278.98	21.52	45.11	33.37
14	1.67	57.63	35.46	6.91	380.38	31.93	17.61	50.46
15	3.01	15.93	75.22	8.84	316.76	48.20	29.90	21.90
16	1.62	5.16	0.82	94.01	244.98	89.05	10.95	0.00
17	1.39	14.86	0.70	84.44	417.90	59.04	1.56	39.40
18	1.27	46.92	0.00	53.08	254.17	13.78	15.70	70.52

5.2. Porosity and Wave Velocities of the Tight Carbonates

The He gas porosity and wave speeds of both the P- and S-waves of the tight carbonate samples are shown in Figures 9 and 10. The samples are collected at a burial depth of over 5000 m and, thus, the porosity of the tight carbonate samples is less than 5%. The porosity acquired from the micro-CT scanning is close to that obtained from He gas filling demonstrating that the acquired pore space from micro-CT is reasonable (Figure 9). The difference in the acquired porosity between the micro-CT and He gas filling might be caused by the isolated pores or the resolution of the micro-CT that pores less than 40 μm are too small to be detected. The wave speeds are measured under an in-situ confining pressure of about 70 MPa. The P- and S-wave velocities of the collected samples range from about 5.6 to 6.7 km/s and 2.6 to 3.2 km/s, respectively. The ratio of V_p over V_s ranges from 1.95 to 2.41. The cross plots of the wave velocities of both P- and S-waves against the He gas porosity are scattered (Figure 10).

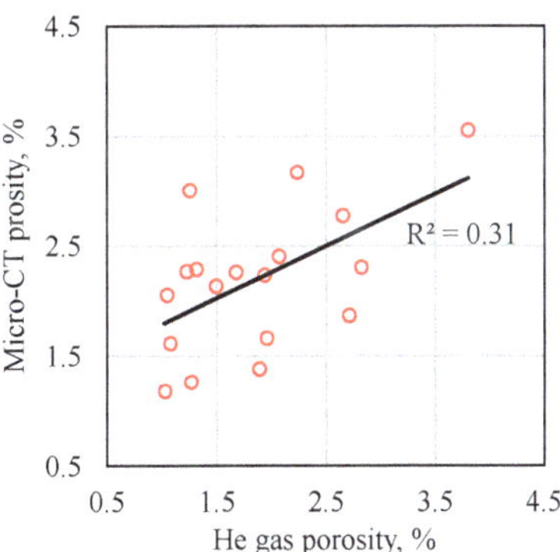

Figure 9. Cross plot of the micro-CT porosity against He gas filling porosity. It can be seen that the porosity acquired from micro-CT is close to that from He gas.

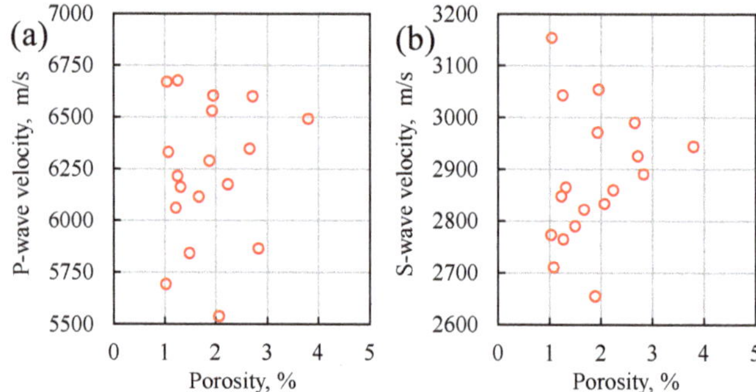

Figure 10. Cross plots of (**a**) P-wave and (**b**) S-wave speeds against He gas porosity. The data referring to both P- and S-waves are scattered.

6. Discussion

The pore structure of the tight carbonates is involved to help analyze and explain the relationship between the porosity and the wave speeds of both P- and S-waves. Seen from the constructed pore space of the tight carbonates, the pore spaces of the tight carbonate samples vary significantly and fractures and vugs are well developed causing the relationship between the porosity and the wave speeds to be scattered.

The pores and vugs possess the majority of the pore space for most of the samples, except those fractured tight carbonates. Thus, the porosity of the samples is positively related to the sum of the volume fraction of the vugs and pores (Figure 11). The existence of vugs and pores improves the porosity of the tight carbonate samples. However, considering the wave speeds, the fractures are key compared to the pores and the vugs because the vugs are distributed heterogeneously in the samples, and some of them are on or near the sidewall of the core plugs that might not affect the path of the waves.

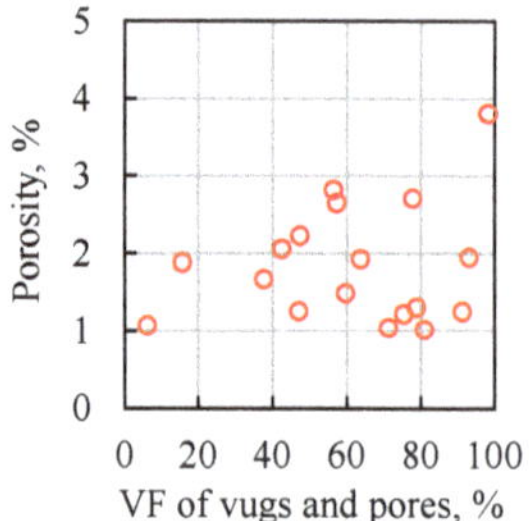

Figure 11. The cross plot of the sum of the volume fractions (VF) of vugs and pores against the He gas porosity.

If a fracture is perpendicular to the wave path, the wave has to transit through the fracture and consequently, its speed is slower. However, a fracture can merely affect the wave speed if it is parallel to the wave path. This kind of fracture adds to the total porosity but will not reduce the wave speed. Thus, the heterogeneity in the spatial distribution of the vugs and the fractures with high orientation angles are the main causes of the scatterings in the porosity and wave velocity relationship. On this consideration, the wave speeds of both the P- and S-waves are related to the porosity of the fractures as shown in Figure 12. Removing the samples with a limited amount of low-oriented fractures (the fractures having their surface perpendicular or nearly perpendicular to the direction of wave propagation), the wave velocity of the remaining samples is negatively related to the fracture porosity indicating that the wave velocity of the tight carbonate samples is more sensitive to the fractures, especially those intersecting its wave path. The Pearson coefficient of S-wave velocity with fracture porosity is about two times that of the P-wave velocity with fracture porosity indicating that the S-wave is more sensitive to the fracture porosity compared to the P-wave.

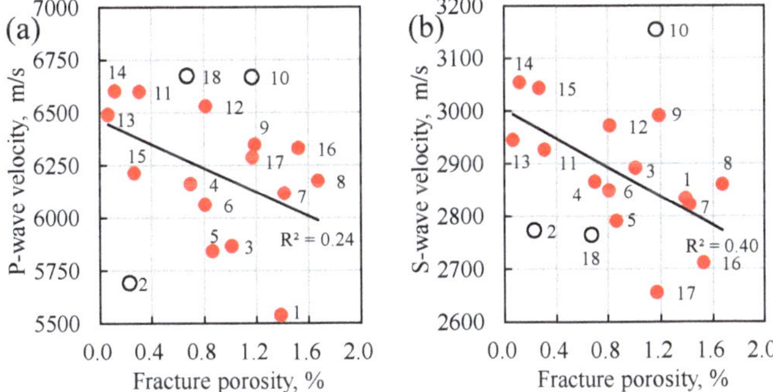

Figure 12. The cross plots of (**a**) P- and (**b**) S-wave porosity against fracture porosity. For the convenience in referring the dots to the micro-CT images, the number of the samples are marked on the figure. The red solid and black empty dots represent the data collected from the samples possessing fractures with low and high orientation angles, respectively. Decreasing trends of the wave speeds with porosity are observed for both P- and S-waves in the samples mainly containing low orientation fractures.

7. Conclusions

Carbonate rocks, buried at depths deeper than 5 km, are usually tight. Characterizing the pore structure characteristics of the tight carbonates are key in evaluating its physical properties. To be representative, micro-CT scanning and pulse transmission measurements were carried out on full-diameter drilling cores instead of 1.0-inch core plugs. The two main challenges in constructing the pore space of a full-diameter carbonate core sample basing on micro-CT are the noisy image and the low contrast between the fractures and the matrix. The proposed micro-CT workflow can effectively construct the pore space and divide the pore space into pores, vugs, and fractures so that the pore structural parameters can be quantitatively acquired.

Both the micro-CT acquired and He gas-filling measured porosity show that the porosity of the collected tight carbonate samples is less than 5%. Although the values of the porosity of the tight carbonates are similar, the pore structure varies significantly from sample to sample. The majority of the samples possess well-developed fractures and vugs. The spatial distribution of fractures, vugs, and pores is strongly heterogeneous. Both parallel and intersected fractures are observed in the constructed pore space. The porosity of the samples is positively related to the volume fraction of the vugs and pores. Due to the complex pore structure of the tight carbonates, the relationships between the porosity and the wave speeds of both P- and S-waves are scattered. The wave velocity is related to the fracture porosity for the samples having the majority of the fractures aligned perpendicular to the wave path.

The micro-CT scanning taken on full-diameter core plugs do not contain any information about the pores having a diameter of less than 50 microns. A combination of the study on full-diameter core plugs and those on smaller samples or other pore structure measurements, for example, SEM, NMR, and mercury injection, is an interesting further work.

Author Contributions: Conceptualization, W.L., X.L. (Xiangjun Liu), and L.L.; methodology, W.L., X.L. (Xiangjun Liu), and L.L.; software, W.L. and Y.Z.; lab measurement, X.L. (Xiansheng Li), and J.X. All authors have read and agreed to the published version of the manuscript.

Funding: This research received no external funding.

Conflicts of Interest: The authors declare no conflict of interest.

Appendix A. Constructed and Divided Pore Spaces of the Rest Tight Carbonates

The constructed pore spaces and the pores, vugs, and fractures acquired by dividing the pore spaces of six samples are listed in Figure 8. The results referring to the remaining samples are shown in Table A1.

Table A1. Photos, cross and longitudinal sections of the CT scanning image, constructed pore spaces, and divided pore spaces of the remaining carbonate samples. This figure together with Figure 8 provides the constructed and divided pores spaces of all the collected tight carbonate samples. 'FV', 'F', and 'V' represent fractured-vuggy carbonate, fractured carbonate, and vuggy carbonate, respectively. In the last column, the fractures, vugs, and pores are colored in gray, blue, and yellow, respectively. The subgraphs are too small that a scale would be hard to see if it is directly marked on the subgraph. Thus, instead of a scale, we provide the physical size including the radius and the length of all the samples in Table 1.

Num.	Type	Sample Image	Cross Section	Longitudinal Section	Constructed Pore Space	Divided Pore Space
1	FV					
2	FV					
3	FV					
4	FV					
5	FV					
6	FV					
7	FV					
8	FV					

Table A1. *Cont.*

Num.	Type	Sample Image	Cross Section	Longitudinal Section	Constructed Pore Space	Divided Pore Space
9	FV					
11	FV					
15	V					
17	F					

References

1. Du, J.; Zou, C.; Xu, C.; He, H.; Shen, P.; Yang, Y.; Li, Y.; Wei, G.; Wang, Z.; Yang, Y. Theoretical and technical innovations in strategic discovery of a giant gas field in Cambrian Longwangmiao Formation of central Sichuan paleo-uplift, Sichuan Basin. *Pet. Explor. Dev.* **2014**, *41*, 294–305. [CrossRef]
2. Kazemzadeh, E.; Nabi-Bidhendi, M.; Moezabad, M.K.; Rezaee, M.R.; Saadat, K. A new approach for the determination of cementation exponent in different petrofacies with velocity deviation logs and petrographical studies in the carbonate Asmari formation. *J. Geophys. Eng.* **2007**, *4*, 160–170. [CrossRef]
3. Weger, R.J.; Eberli, G.P.; Baechle, G.T.; Massaferro, J.L.; Sun, Y.-F. Quantification of pore structure and its effect on sonic velocity and permeability in carbonates. *Aapg Bull.* **2009**, *93*, 1297–1317. [CrossRef]
4. Claes, S.; Soete, J.; Cnudde, V.; Swennen, R. A three-dimensional classification for mathematical pore shape description in complex carbonate reservoir rocks. *Math. Geol.* **2016**, *48*, 619–639. [CrossRef]
5. Zhu, L.; Zhang, C.; Wei, Y.; Zhou, X.; Huang, Y.; Zhang, C. Inversion of the permeability of a tight gas reservoir with the combination of a deep Boltzmann kernel extreme learning machine and nuclear magnetic resonance logging transverse relaxation time spectrum data. *Interpretation* **2017**, *5*, T341–T350. [CrossRef]
6. Zhu, L.-Q.; Zhang, C.; Wei, Y.; Zhang, C.-M. Permeability Prediction of the Tight Sandstone Reservoirs Using Hybrid Intelligent Algorithm and Nuclear Magnetic Resonance Logging Data. *Arab. J. Sci. Eng.* **2016**, *42*, 1643–1654. [CrossRef]
7. Li, H.; Zhang, J. Well log and seismic data analysis for complex pore-structure carbonate reservoir using 3D rock physics templates. *J. Appl. Geophys.* **2018**, *151*, 175–183. [CrossRef]
8. Bing, X.; Li, B.; Zhao, A.L.; Zhang, Y.H.; Wang, Y. Application of Sonic Scanner logging to fracture effectiveness evaluation of carbonate reservoir: A case from Sinian in Sichuan Basin. *Lithol. Reserv.* **2017**, *29*, 117–123. (In Chinese)
9. He, S.; Qin, Q.R.; Wang, J.S.; Li, F.; Duan, W. Fracture Properties and Development Mechanisms of Sinian Dengying—4 Member in Central Sichuan. *Lithol. Reserv.* **2020**, *27*, 60–66. (In Chinese)
10. Okabe, H.; Blunt, M.J. Pore space reconstruction of vuggy carbonates using microtomography and multiple-point statistics. *Water Resour. Res.* **2007**, *43*, 1–5. [CrossRef]
11. Rezaee, M.; Motiei, H.; Kazemzadeh, E. A new method to acquire m exponent and tortuosity factor for microscopically heterogeneous carbonates. *J. Pet. Sci. Eng.* **2007**, *56*, 241–251. [CrossRef]

12. Tiwari, P.; Deo, M.; Lin, C.; Miller, J. Characterization of oil shale pore structure before and after pyrolysis by using X-ray micro CT. *Fuel* **2013**, *107*, 547–554. [CrossRef]
13. Zhang, J. Experimental Study and Modeling for CO2Diffusion in Coals with Different Particle Sizes: Based on Gas Absorption (Imbibition) and Pore Structure. *Energy Fuels* **2016**, *30*, 531–543. [CrossRef]
14. Li, W.; Zou, C.; Wang, H.; Peng, C. A model for calculating the formation resistivity factor in low and middle porosity sandstone formations considering the effect of pore geometry. *J. Pet. Sci. Eng.* **2017**, *152*, 193–203. [CrossRef]
15. Wang, L.; Zhao, N.; Sima, L.; Meng, F.; Guo, Y. Pore Structure Characterization of the Tight Reservoir: Systematic Integration of Mercury Injection and Nuclear Magnetic Resonance. *Energy Fuels* **2018**, *32*, 7471–7484. [CrossRef]
16. Yuan, Y.; Rezaee, R.; Verrall, M.; Hu, S.-Y.; Zou, J.; Testmanti, N. Pore characterization and clay bound water assessment in shale with a combination of NMR and low-pressure nitrogen gas adsorption. *Int. J. Coal Geol.* **2018**, *194*, 11–21. [CrossRef]
17. Dong, S.; Zeng, L.; Xu, C.; Dowd, P.; Gao, Z.; Mao, Z.; Wang, A. A novel method for extracting information on pores from cast thin-section images. *Comput. Geosci.* **2019**, *130*, 69–83. [CrossRef]
18. Zhang, F.; Jiang, Z.; Sun, W.; Li, Y.; Zhang, X.; Zhu, L.; Wen, M. A multiscale comprehensive study on pore structure of tight sandstone reservoir realized by nuclear magnetic resonance, high pressure mercury injection and constant-rate mercury injection penetration test. *Mar. Pet. Geol.* **2019**, *109*, 208–222. [CrossRef]
19. Pal, N.R.; Pal, S.K. A review on image segmentation techniques. *Pattern Recognit.* **1993**, *26*, 1277–1294. [CrossRef]
20. Martin, J.B.; Branko, B.; Hu, D.; Oussama, G.; Stefan, I.; Peyman, M.; Adriana, P.; Christopher, P. Pore-scale imaging and modeling. *Adv. Water Resour.* **2013**, *51*, 197–216.
21. Liang, L.; Yongkoo, S.; Karl, J. Pore-Scale Visualization of Methane Hydrate-Bearing Sediments with Micro-CT. *Geophys. Res. Lett.* **2018**, *11*, 5417–5426.
22. Njiekak, G.; Schmitt, D.R.; Kofman, R.S. Pore systems in carbonate formations, Weyburn field, Saskatchewan, Canada: Micro-tomography, helium porosimetry and mercury intrusion porosimetry characterization. *J. Pet. Sci. Eng.* **2018**, *171*, 1496–1513. [CrossRef]
23. Sankur, B. Survey over image thresholding techniques and quantitative performance evaluation. *J. Electron. Imaging* **2004**, *13*, 146–168. [CrossRef]
24. Noiriel, C. Resolving Time-dependent Evolution of Pore-Scale Structure, Permeability and Reactivity using X-ray Microtomography. *Rev. Miner. Geochem.* **2015**, *80*, 247–285. [CrossRef]
25. Sok, R.M.; Varslot, T.; Ghous, A.; Latham, S.; Knackstedt, M.A. Pore scale characterization of carbonates at multiple scales: Integration of micro-ct, bsem, and fibsem. *Petrophysics* **2010**, *51*, 379–387.
26. Wildenschild, D.; Sheppard, A.P. X-ray imaging and analysis techniques for quantifying pore-scale structure and processes in subsurface porous medium systems. *Adv. Water Resour.* **2013**, *51*, 217–246. [CrossRef]
27. Neto, I.A.L.; Misságia, R.M.; Ceia, M.A.; Archilha, N.L.; Oliveira, L.C. Carbonate pore system evaluation using the velocity–porosity–pressure relationship, digital image analysis, and differential effective medium theory. *J. Appl. Geophys.* **2014**, *110*, 23–33. [CrossRef]
28. Tonietto, S.N.; Smoot, M.Z.; Pope, M. Pore type characterization and classification in carbonate reservoirs. In Proceedings of the AAPG Annual Convention and Exhibition, Houston, TX, USA, 6–9 April 2014.
29. Promentilla, M.A.B.; Takafumi, S.; Takashi, H.; Nobufumi, T. Characterizing the 3d pore structure of hardened cement paste with synchrotron microtomography. *J. Adv. Concr. Technol.* **2008**, *6*, 273–286. [CrossRef]
30. Promentilla, M.A.B.; Cortez, S.M.; Papel, R.A.D.; Tablada, B.M.; Sugiyama, T. Evaluation of Microstructure and Transport Properties of Deteriorated Cementitious Materials from Their X-ray Computed Tomography (CT) Images. *Materials* **2016**, *9*, 388. [CrossRef]
31. Li, B.; Tan, X.; Wang, F.; Lian, P.; Gao, W.; Li, Y. Fracture and vug characterization and carbonate rock type automatic classification using X-ray CT images. *J. Pet. Sci. Eng.* **2017**, *153*, 88–96. [CrossRef]
32. Xu, Z.; Lin, M.; Jiang, W.; Cao, G.; Yi, Z. Identifying the comprehensive pore structure characteristics of a rock from 3D images. *J. Pet. Sci. Eng.* **2020**, *187*, 106764. [CrossRef]
33. Eberli, G.P.; Baechle, G.T.; Anselmetti, F.S.; Incze, M.L. Factors controlling elastic properties in carbonate sediments and rocks. *Geophysics* **2003**, *22*, 654–660. [CrossRef]
34. Ralf, J.W. Quantitative Pore/Rock Type Parameters in Carbonates and Their Relationship to Velocity Deviations. Ph.D. Thesis, University of Miami, Coral Gables, FL, USA, 2006.

35. Li, T.; Wang, R.; Wang, Z. A method of rough pore surface model and application in elastic wave propagation. *Appl. Acoust.* **2019**, *143*, 100–111. [CrossRef]

36. Lubis, L.; Harith, Z.Z.T. Pore Type Classification on Carbonate Reservoir in Offshore Sarawak using Rock Physics Model and Rock Digital Images. *IOP Conf. Ser. Earth Environ. Sci.* **2014**, *19*, 12003. [CrossRef]

37. Zambrano, M.; Tondi, E.; Mancini, L.; Arzilli, F.; Lanzafame, G.; Materazzi, M.; Torrieri, S. 3D Pore-network quantitative analysis in deformed carbonate grainstones. *Mar. Pet. Geol.* **2017**, *82*, 251–264. [CrossRef]

38. Regnet, J.-B.; Fortin, J.; Nicolas, A.; Pellerin, M.; Guéguen, Y. Elastic properties of continental carbonates: From controlling factors to an applicable model for acoustic-velocity predictions. *Geophysics* **2019**, *84*, MR45–MR59. [CrossRef]

39. Huang, T.S.; Yang, G.J.; Tang, G.Y. A fast two-dimensional median filtering algorithm. *IEEE Trans. Acoust. Speech Signal Process.* **1979**, *27*, 13–18. [CrossRef]

40. Yin, G. Adaptive Filtering with Averaging. In *Directions in Robust Statistics and Diagnostics*; Springer Science and Business Media LLC: Berlin/Heidelberg, Germany, 1995; Volume 74, pp. 375–396.

41. Lu, C.-T.; Chou, T.-C. Denoising of salt-and-pepper noise corrupted image using modified directional-weighted-median filter. *Pattern Recognit. Lett.* **2012**, *33*, 1287–1295. [CrossRef]

42. Vargas, J.I.D.L.R.; Villa, J.J.; Gonzalez, E.; Cortez, J. A tour of nonlocal means techniques for image filtering. In Proceedings of the 2016 International Conference on Electronics, Communications and Computers (CONIELECOMP), Cholula, Mexico, 24–26 February 2016; pp. 32–39.

43. Wang, X.; Shen, S.; Shi, G.; Xu, Y.; Zhang, P. Iterative non-local means filter for salt and pepper noise removal. *J. Vis. Commun. Image Represent.* **2016**, *38*, 440–450. [CrossRef]

44. Wang, Y.; Chen, Q.; Zhang, B. Image enhancement based on equal area dualistic sub-image histogram equalization method. *IEEE Trans. Consum. Electron.* **1999**, *45*, 68–75. [CrossRef]

45. Xu, H.; Zhai, G.; Wu, X.; Yang, X. Generalized Equalization Model for Image Enhancement. *IEEE Trans. Multimed.* **2013**, *16*, 68–82. [CrossRef]

46. Kushol, R.; Kabir, H.; Salekin, S.; Rahman, A.B.M.A. Contrast Enhancement by Top-Hat and Bottom-Hat Transform with Optimal Structuring Element: Application to Retinal Vessel Segmentation. In Proceedings of the International Conference Image Analysis and Recognition, Montreal, QC, Canada, 5–7 July 2017; Springer: Cham, Switzerland, 2017.

47. Wang, W.; Wang, W.; Hu, Z. Segmenting retinal vessels with revised top-bottom-hat transformation and flattening of minimum circumscribed ellipse. *Med Biol. Eng. Comput.* **2019**, *57*, 1481–1496. [CrossRef] [PubMed]

48. Wang, Z.; Wang, R.; Wang, F.; Qiu, H.; Li, T. Experiment study of pore structure effects on velocities in synthetic carbonate rocks. *Geophysics* **2015**, *80*, D207–D219. [CrossRef]

49. Li, L.X.; Chang, C.; Xu, W.; Yan, Y.J.; Yang, L.; Deng, H. Reconstruction of porosity and permeability characteristics of fracture-vug reservoirs by using digital core together with imaging logging. *Nat. Gas Explor. Dev.* **2017**, *40*, 16–23. (In Chinese)

50. Roerdink, J.B.; Meijster, A. The Watershed Transform: Definitions, Algorithms and Parallelization Strategies. *Fundam. Inform.* **2000**, *41*, 187–228. [CrossRef]

51. Shabaninejad, M.; Middleton, J.; Fogden, A. Systematic pore-scale study of low salinity recovery from Berea sandstone analyzed by micro-CT. *J. Pet. Sci. Eng.* **2018**, *163*, 283–294. [CrossRef]

52. API RP40. *Recommended Practices for Core Analysis*; American Petroleum Institute (API): Washington, WA, USA, 1998.

53. Melendez-Martinez, J. Elastic Properties of Sedimentary Rocks. Ph.D. Thesis, University of Alberta, Edmonton, AB, Canada, 2014.

Publisher's Note: MDPI stays neutral with regard to jurisdictional claims in published maps and institutional affiliations.

Article

Reservoir Properties of Low-Permeable Carbonate Rocks: Experimental Features

Aliya Mukhametdinova [1,*], Andrey Kazak [1], Tagir Karamov [1], Natalia Bogdanovich [1], Maksim Serkin [2], Sergey Melekhin [2] and Alexey Cheremisin [1]

[1] Center for Hydrocarbon Recovery, Skolkovo Institute of Science and Technology, Bolshoy Boulevard 30, 121205 Moscow, Russia; A.Kazak@skoltech.ru (A.K.); Tagir.Karamov@skoltech.ru (T.K.); N.Bogdanovich@skoltech.ru (N.B.); A.Cheremisin@skoltech.ru (A.C.);

[2] PermNIPIneft Branch, LUKOIL Engineering LLC, Soviet Army Street, 614066 Perm, Russia; Maksim.Serkin@pnn.lukoil.com (M.S.); Sergej.Melehin@pnn.lukoil.com (S.M.)

* Correspondence: A.Mukhametdinova@skoltech.ru

Received: 12 March 2020; Accepted: 28 April 2020; Published: 3 May 2020

Abstract: This paper presents an integrated petrophysical characterization of a representative set of complex carbonate reservoir rock samples with a porosity of less than 3% and permeability of less than 1 mD. Laboratory methods used in this study included both bulk measurements and multiscale void space characterization. Bulk techniques included gas volumetric nuclear magnetic resonance (NMR), liquid saturation (LS), porosity, pressure-pulse decay (PDP), and pseudo-steady-state permeability (PSS). Imaging consisted of thin-section petrography, computed X-ray macro- and microtomography, and scanning electron microscopy (SEM). Mercury injection capillary pressure (MICP) porosimetry was a proxy technique between bulk measurements and imaging. The target set of rock samples included whole cores, core plugs, mini cores, rock chips, and crushed rock. The research yielded several findings for the target rock samples. NMR was the most appropriate technique for total porosity determination. MICP porosity matched both NMR and imaging results and highlighted the different effects of solvent extraction on throat size distribution. PDP core-plug gas permeability measurements were consistent but overestimated in comparison to PSS results, with the difference reaching two orders of magnitude. SEM proved to be the only feasible method for void-scale imaging with a spatial resolution up to 5 nm. The results confirmed the presence of natural voids of two major types. The first type was organic matter (OM)-hosted pores, with dimensions of less than 500 nm. The second type was sporadic voids in the mineral matrix (biogenic clasts), rarely larger than 250 nm. Comparisons between whole-core and core-plug reservoir properties showed substantial differences in both porosity (by a factor of 2) and permeability (up to 4 orders of magnitude) caused by spatial heterogeneity and scaling.

Keywords: reservoir properties; void space structure; porosity; permeability; complex rocks; NMR; MICP; CT; SEM

---

1. Introduction

Worldwide, oil reserves estimated in traditional reservoirs are in a consistent decline, while triggering complications and enhancements of technologies of hydrocarbon exploration and characterization. One of the promising solutions to maintain production at current levels is the development of hard-to-recover reserves, including low-permeable (tight) carbonate reservoirs [1,2].

The correct determination of reservoir properties for the case remains challenging for oil and gas companies despite an increase in their interest in developing such formations and deposits [3–5]. Both porosity and permeability contribute significantly to resource assessment and reserves' estimation of a target asset.

Tight carbonate rocks feature low porosity and permeability, the absence of structural, and stratigraphic control of the distribution of oil-containing intervals, which do not allow distinguishing them from well-logging data. All mentioned factors lead to a low oil recovery factor (<10–20%). Therefore, one critical task is the development of methods for estimating the reservoir properties of tight rocks.

Conventional methods for petrophysical core analysis were established during the 1970–1980s, when the development of conventional reserves had reached its peak [3]. Notably, the measured porosity varied in the range of 10–30% and permeability 10–200 mD.

The main obstacles to the proper determination of the reservoir properties (porosity and permeability) of tight reservoir rocks are their low porosity (<3–5%), and low permeability (<1 mD), as well as a high degree of heterogeneity of void space structure (VSS). However, conventional laboratory methods were developed for the characterization of conventional highly porous and permeable reservoir rocks, such as well-sorted sandstones. Complex reservoirs reside at the edge of their operating envelope in terms of porosity and permeability [6]. For example, conventional Dean–Stark extraction tends to overestimate water saturation in tight reservoirs, and reliable results require new approaches [7]. Indeed, conventional porosity techniques may overestimate and implicitly add uncertainty to the results of both geological and hydrodynamic modeling of field-scale processes [8].

In other words, the target formations fall out of the operational domain for conventional methods. Therefore, companies (operators and oilfield service divisions) tend to develop or adopt new techniques and equipment for such measurements.

At the current stage, laboratory petrophysical methods targeting tight reservoirs fall into two groups. The first group includes bulk methods, delivering a single integral property value (the result of determination) for a target rock sample. The second group improves the understanding of bulk values by direct (explicit) imaging of the void space structure, as well as mineral and organic matrix.

The petrophysical properties of the target rocks suggest a complicated inhomogeneous microstructure. The dimensions of voids fall below the micrometer scale. They require novel methods for evaluating the reservoir properties and high-resolution imaging, which obtains three-dimensional (3D) images of the void space structure with resolution up to 50 nm [9,10].

Multiple publications characterize the porosity [11,12], the porous structure, and permeability of tight sands, shales, and carbonates [13–15]. Based on published papers, we summarized conventional methods (basic in core analysis) and unconventional methods (widely applied for tight rock characterization) (Table 1).

Table 1. Methods applied for reservoir properties' laboratory assessment.

Measured Parameter	Conventional Method	Unconventional Method
Porosity	Gravimetric or Liquid Saturation (LS) Pulse-Decay (PDP)	Nuclear Magnetic Resonance (NMR) Mercury Injection Capillary Pressure (MICP) Gas adsorption
Permeability	Steady-State Pulse-Decay (PDP)	Pulse-Decay Permeametry (PDP) Pseudo-Steady-State (PSS) Oscillating Pulse Technique (OPT) Gas Research Institute (GRI)
Void Space Structure	Thin-sections	Mercury Injection Capillary Pressure (MICP) X-ray Computed Tomography (CT) Scanning Electron Microscopy (SEM)

Multiscale research studies on tight reservoirs suggest utilizing novel laboratory methods for characterizing the reservoir properties and the void space structure [16–20]. Nuclear magnetic resonance (NMR) and mercury injection capillary pressure (MICP) methods have been introduced as primary tools for defining the porosity and pore size, whereas the X-ray computed tomography (CT) and SEM (or FIB-SEM) often visualize the void space of tight sands and carbonates.

First of all, the use of NMR relaxometry in the study of low-porous sand and clay rocks falls into a separate group [21,22]. In the presence of clays or at low porosity, the most commonly used T2 cutoff values ultimately give inaccurate estimates of permeability due to reservoir heterogeneity. X-ray tomography (CT) in a modern petrophysical laboratory is becoming an increasingly popular method in the study of rocks, complementing routine, and special core analysis [23]. X-ray microtomography often joins mercury porosimetry to study shale and tight rocks [24,25]. Different industries have implemented scanning electron microscopy (SEM) for more than 55 years. One of the key aims has been to characterize various industrial materials at micro- and nanoscales [26]. Today's increased interest from petroleum engineers to detailed characterization of tight reservoirs has raised interest in the application of SEM for shale investigations at very high magnifications [27]. Various microstructural techniques, including X-ray CT, petrographical microscopy, and SEM, identify and characterize macro- to nanoscale voids in tight carbonate rocks.

However, the application of the listed unconventional methods is complicated. Firstly, published case studies mainly cover only a few methods for characterizing the reservoir properties. The majority of publications present success stories with excellent and reliable results, even in cases of low- and ultra-low permeable reservoir rocks [8,16,17,19]. Therefore, recommended workflows often do not include failure scenarios and do not highlight the limitations of conventional methods. Secondly, all existing workflows present case studies for a particular type of reservoir (tight sands and tight shales) and rarely provide the application of the same procedure to different rock types. Therefore, it remains a challenge for us to construct an experimental workflow, which would combine the most mentioned methods in the publications and obtain real-time data for the ultra-low-permeable rock.

In this work, we applied a number of advanced laboratory methods for the characterization of an ultra-low-permeable carbonate Tlaynchy-Tamakian Formation. We employed an integrated analysis to investigate the advantages and limitations of the methods for estimating porosity and permeability. The study eventually discusses an optimal suite of methods for gathering quality datasets.

2. Materials and Methods

2.1. Materials

The study covered a representative collection of rock samples of an organic-rich carbonate Tlyanchy-Tamakian Formation of an Upper Devonian Frasnian age. The formation resides in the Volga-Ural and Timan-Pechora basins in the European part of Russia (Figure 1). Previous extensive studies characterized the formation in terms of geochemical properties, including oil generation potential [28–31]. The formation is one of the promising sources of oil in the basin [32,33]. Assessments of oil resources in the formation vary from hundreds of millions to billions of tons [30,31,34]. However, the highly heterogeneous rock texture and complex structure of voids essentially constrain the recovery of oil and gas.

The target collection of rock samples related to mostly carbonate, carbonate-siliceous thin-layered rocks with a moderate total organic carbon (TOC) content up to 10 wt.%. Carbonates were predominantly represented by organic-rich wackestones, and, in some cases, by packstones and crystalline limestones. Target rock samples came from a well located in the south of the Perm region of the Volga-Ural Province, Russian Federation (Figure 1). The entire collection of rock samples provided by the operator included 32 whole cores and more than 300 drilled core plugs corresponding to three main horizons—overlay Trudolubian (D3tr), main Tlyanchy-Tamakian (D3tch), and underlay Sargaevian (D3srg) unit (Figure 2).

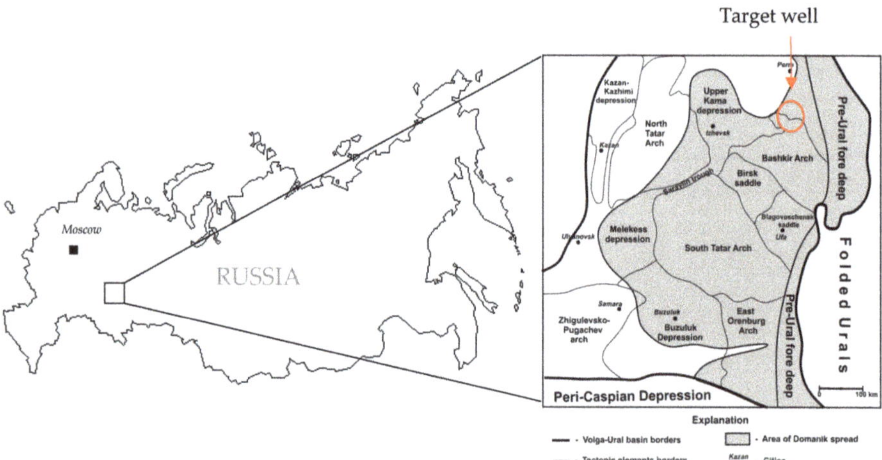

Figure 1. The geographic location of the target formation (modified after [35]).

Period	Epoch	Stage	Formation	Lithology
Devonian	Late	Famennian	Baituganian (D₃bt)	
		Frasnian	Yanchikonian (D₃yn)	
			Alparian (D₃alp)	
			Trudolubian (D₃tr)	
			Tlyanchy-tamakian (D₃tch)	
			Sargaevian (D₃srg)	
			Prekazanian (D₃pkz)	

- clastic limestone
- siliceous limestone
- organic-rich limestone breccia
- sandstone
- organic-rich limestone
- organic-rich siliceous limestone
- argillaceous limestone

Figure 2. Tlyanchy-Tamakian Formation in the regional stratigraphic chart.

Together with the core samples, the operator supplied the basic petrophysical information on the core material, so-called an a priori dataset. It included the porosity and permeability data for the entire collection of whole cores and core plugs using an automated gas permeameter Ekogeos DarcyMeter (Russia). The pressure pulse-decay or pulse-decay permeameter (PDP) technique with nitrogen provided all the measurements.

Received data showed porosity <1.6% (Figure 3) and permeability <1 mD (Figure 4). The reservoir properties ranged as follows: carbonate content 60–99%; clay content 0–10%; TOC content 0.5–10 wt.%.

Figure 3 shows a clear difference in the porosity levels of the whole cores and core plugs. We see the main reason in the different volumes of investigation boosted by a high degree of heterogeneity of the target formation in a wellbore scale. Based on the porosity data for the core plugs, we selected the core plugs based on the porosity threshold of 1.2% (to the left from the histogram peak in Figure 3b).

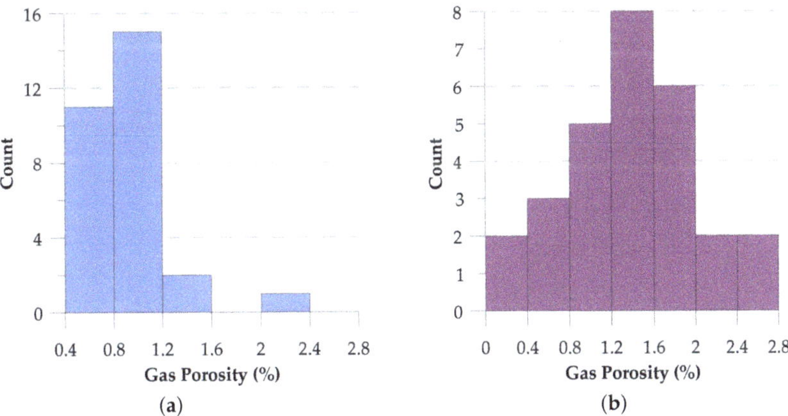

Figure 3. The porosity of received collection: (**a**) Ø100 × 200 mm whole cores; (**b**) Ø30 × 30 mm core plugs (courtesy of LUKOIL LLC, a priori data).

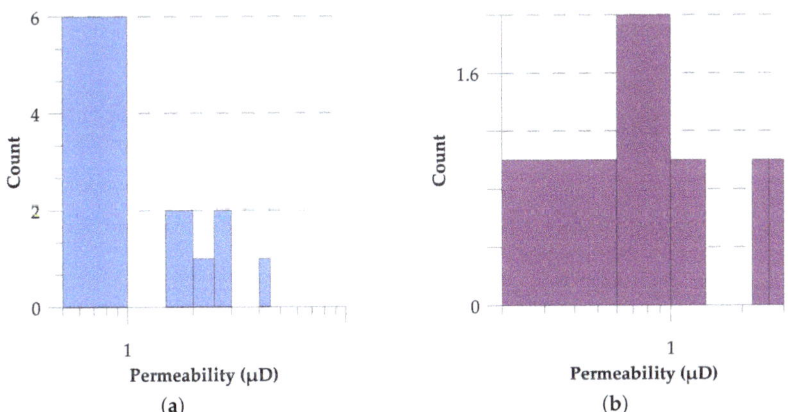

Figure 4. Permeability of received collection: (**a**) Ø100 × 200 mm whole cores; (**b**) Ø30 × 30 mm core plugs (courtesy of LUKOIL LLC, a priori data).

We studied a subset of samples corresponding to the Tlyanchy-Tamakian horizon, 27 m thickness. The horizon was split into upper and lower units. The target collection of core samples included 2 whole cores and 27 core plugs with accompanying rock chips. We selected the samples based on the porosity (Figure 3) and permeability (Figure 4), and the availability of the rock chips.

The samples represented distinct lithological types and were organic-rich mixed rocks predominantly composed of carbonate and siliceous components. According to the rock texture and mineral composition, we distinguished four groups of rock types using the modified Dunham classification [36]: organic-rich wackestone, crystalline limestone with detritus, carbonaceous silicites with biogenic detritus, and grainstone.

2.1.1. Whole Core Samples

The whole cores #545 and #551 represented the Upper-D3tch and Lower-D3tch rocks correspondingly. Each core had a length of 200 and a diameter of 100 mm (isolated from environmental exposure using layers of foil and paraffin [37]).

Samples showed high heterogeneity and anisotropy due to intense thin-bed layering (Appendix A, Figure A1). Various imprints of shells (presumably brachiopods) contributed to both rock heterogeneity and anisotropy.

CT on whole cores was used to locate homogeneous areas with a certain amount of organic matter and separate the parts with artificial fractures and other imperfections. Multiscale characterization of the void space structure persuaded us to separate the whole core into a set of subsamples (Figure 5).

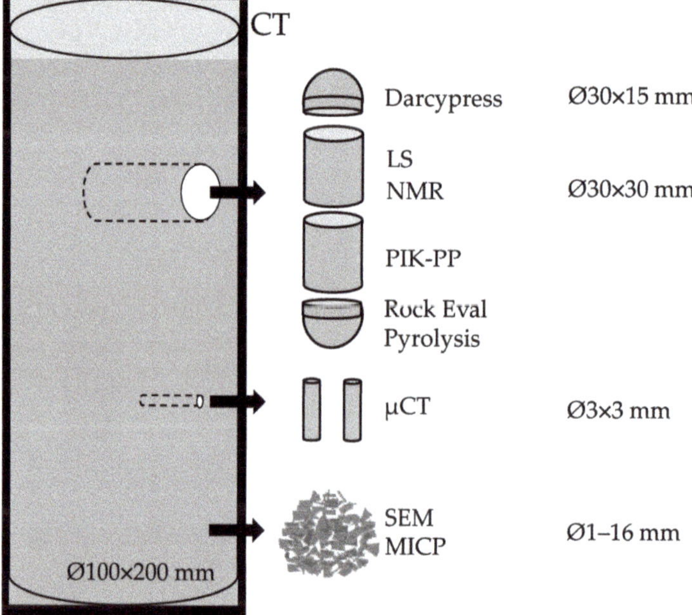

Figure 5. Schematic diagram of the whole-core breakdown into subsamples.

2.1.2. Core Plugs and Mini Cores

The rock sample set included 33 samples in total: 27 core plugs (Ø30 × 30 mm) initially provided by the operator, 4 additionally drilled core plugs (Ø30 × 30 mm), and 2 mini cores (Ø3 × 3 mm) drilled out the whole cores. The collection of plugs had a high level of heterogeneity; shell inclusions (traces) made the core fragile and sensitive to destructive methods. Each core plug went through measurements by gas porosimetry, liquid saturation, NMR, and µCT (on mini-cores).

2.1.3. Rock Chips and Caps

Each core plug came with corresponding rock chips of an irregular shape with average dimensions of 20–50 mm and rock caps of Ø30 × 15 mm (received while drilling the whole core). The rock chips participated in the experiment, including Darcypress pseudo-steady-state decay, Rock-Eval pyrolysis, MICP, and SEM. Darcypress required square or round shaped fragments of the core with an approximate size of 1 cm^3. SEM and MICP used rock chips of a smaller size. While SEM did not require a specific size/shape, MICP tests consisted of measurements on 4 different fractions: mesh sizes of 1–2; 2–4; 4–8; and 8–16 mm.

2.1.4. Petrographic Thin Sections

We prepared a set of 27 thin sections (20 × 20 mm) for rock typing. Initially, the thin sections had a standard thickness of 30 μm; then we thinned them to 10–20 μm to maximize of textural information output.

2.2. Methods & Techniques

We used a suite of modern methods and equipment to study the reservoir properties and microstructure of the target rock samples. Bulk methods include gas porosimetry, nuclear magnetic resonance (NMR), and Darcypress. Microstructural methods stand for visualizing techniques such as thin-section optical microscopy, CT, μCT, and SEM. A separate method is MICP, which belongs to the transition group, i.e., provides porosity measurement and describes the void space structure as well. Appendix B described common shared techniques.

2.2.1. Conventional Gas Porosity and Permeability of Plugs

Porosity and permeability measurements were conducted using an automated gas permeameter-porosimeter Geologika PIK-PP (Russia) [38]. The tests used nitrogen as a probe gas, while permeability measurements were obtained using the PDP technique. The confining pressure was 3.4 MPa. The stabilization criteria for pressure decay were 0.1%/min. Data reliability required performing the tests three times and calculating an average value. We calibrated the unit with supplied artificial samples with known parameters.

2.2.2. Nuclear Magnetic Resonance

We employed nuclear magnetic resonance relaxometry (NMR) to determine the porosity of the target rock samples. A low-field NMR unit, Oxford Instruments Geospec 2/53 (UK), estimated saturation and porosity at each step of the experimental workflow. NMR analysis involves measuring the polarization and relaxation of hydrogen atoms in a magnetic field of 0.05 T and a radiofrequency of 2.28 MHz [39–41]. Analysis and interpretation of the results employed principles characteristic of NMR application in petrophysics [42]. The instrument's calibration used reference samples supplied by green imaging technologies (GIT) with known parameters, including NMR, liquid volume, length of 90 and 180 ° pulse, etc. T2-relaxation curve measurements resulted from the Carr-Purcell-Meybum-Gill (CPMG) pulse sequence with the time-echo (TE = 2τ) set to 0.1 ms. Preliminary tests justified the number of trains (accumulation) in the pulse sequence (90°–τ–180°–2τ–180°– . . . –180°) and estimated the highest signal-to-noise ratio (SNR), as well as the minimum number of scans—128. Data reduction, analysis, and interpretation involved using a GIT Systems Advanced 7.5.1 software.

2.2.3. Darcypress Permeability

The Cydarex DarcyPress analyzer's design allowed us to measure the permeability of shale or other rock samples in an extensive range of permeabilities from 10^{-6} mD to several Darcy units [43]. We ran the setup in a pseudo-steady-state (PSS) mode. Sample preparation included molding a small rock sample with a diameter of less than 1 cm into epoxy resin, and cutting a disk with a thickness of 2–5 mm. The permeability measurement procedure of the DarcyPress analyzer is similar to the standard measurement procedure for transient state measurements [44]. To characterize the target samples, we followed both single- and multi-step pseudo-steady-state procedures. The CYDAR 2017 software enabled data reduction, analysis, and interpretation.

2.2.4. Mercury Injection Porosimetry

The mercury injection capillary pressure (MICP) porosimetry characterized the void space structure and determined pore throat size distributions [45,46]. For this purpose, we measured Hg intrusion–extrusion versus pressure at 25 °C in the laboratory. Experimental activities involved

Micromeritics AutoPore V 9605 and AutoPore IV 9520 instruments driven by Micromeritics MicroActive AutoPore V 9600 2.03.00 software.

Sample preparation included rock crushing in fractions of 8–16, 4–8, 2–4, and 1–2 mm fractions, followed by drying in a vacuum cabinet in a vacuum level of 1 mm Hg at a temperature of 110 °C. Before and after experiments, the rock samples resided in a desiccator to avoid capturing moisture from the environment.

Before and after each series of measurements, we ran blank (empty-penetrometer) tests. The Hg pressure table included more than 100-gauge pressure points evenly distributed in a range of 0–60 kpsi (0–414 MPa), filling voids with throats down to 3 nm. The Hg pressure equilibration time was equal to the recommended value of 10 s. We applied both blank and material compressibility corrections [47] to the raw data to achieve the highest quality of interpretation.

Experimental data reduction included Akima spline interpolation [48] of the corresponding intrusion curves, followed by the calculation of pore size distributions (PSDs) [49]. Additional outputs included total intrusion volume (TIV), median pore diameter (MPD), and volume (MPV) [45].

We validated the quality of the data using test measurements on the standard reference silica-alumina sample (part number 004-16822-00, lot A-501-52).

2.2.5. Computed Tomography Scan Imaging

We imaged the microstructure of the target rock samples in 3D using both computed tomography (CT) and micro-CT [50,51]. For this purpose, we used the GE phoenix v|tome|x L240, versatile high-resolution microfocus system for 2D and 3D computed tomography, and 2D non-destructive X-ray inspection.

The device includes a combination of unipolar 240 microfocus and 180 kV high-power nanofocus X-ray tubes, and handles large samples up to 500 × 1300 mm. We used an accelerating voltage of 70–200 kV and a current of 140–580 µA depending on the size of the sample, yielding a voxel size of 3–117 µm. Scanning of the whole cores required the application of a 0.5-mm-thick tin filter. A detector based on a 2024 × 2024 px photodiode array with an element size of 0.2 × 0.2 mm in combination with a CsI scintillator captured the target X-ray projections. We acquired the raw data using GE datos|x acquisition 2.6.1-RTM software. Reconstruction of a set of 1000–2400 2D radiographic projections into 3D X-ray density images involved the application of GE datos|x reconstruction 2.6.1-RTM software and processed the models using FEI PerGeos 1.5 software [52].

2.2.6. Scanning Electron Microscopy

For high-resolution 2D imaging and micromorphological characterization, we used scanning electron microscopy (SEM). The Thermo Fisher Scientific I Quattro S instrument analyzed small (2–5 mm) rock specimens chipped from each sample. The instrument allowed us to perform investigations with an electron beam current range from 1 pA to 200 nA with accelerating voltage 200 V to 30 kV. The minimum spatial resolution was 1 nm (at 30 kV).

Sample preparation included multi-step polishing—starting with grinding paper to polishing cloth with 1-µm diamond suspension. We then attached the polished samples to a holder with carbon tape, coated them with gold (coating thickness of less than 20 nm), and loaded these into the instrument.

Scanning involved secondary electrons (SE) and backscattered electrons (BSE), magnification ×500–250k, acceleration voltage 10–15 kV, and working distance 9–11 mm with approximately 1–2 nm maximum pixel size [53]. The resulting images had dimensions of 1536 × 1094 px.

3. Results

3.1. Porosity and Permeability

Table 2 shows the porosity data from gas porosimetry, liquid saturation, and NMR, as well as gas-injection and Darcypress permeability. Because of the limited amount of the core material, as well as the specific probe requirements for each method, we observed the gaps in the received dataset at multiple depths. It complicated the integrated analysis of reservoir properties and disabled a direct comparison of the utilized methods. Nevertheless, Table 2 describes each technique in terms of measuring range and average numbers for the target rock samples.

Table 2. Porosity and permeability of the core plugs (including a priori Darcymeter data) and core chips.

Sample ID	DarcyMeter Porosity (%)	PIK-PP Porosity (%)	NMR Porosity (%)	Liquid Porosity (%)	DarcyMeter Permeability (mD)	PIK-PP Permeability (mD)	Darcypress Permeability (mD)
545	0.74	0.62	1.20	1.08	9.02×10^{-1}	6.20×10^{-2}	9.07×10^{-4}
551	1.31	0.84	2.30	1.22	2.30	1.13×10^{-2}	4.00×10^{-4}
696	1.23	0.22	2.10		3.85×10^{-3}	3.00×10^{3}	
697	1.17	0.17	1.50		2.51×10^{-3}	2.00×10^{-3}	
700	2.24	0.83	1.50			7.18	
701	1.77		1.70		5.35×10^{-2}		
703	0.92		1.80	0.83	8.48×10^{-3}		2.36×10^{-3}
705	1.30	0.35	1.10		3.61×10^{-3}	1.00×10^{-3}	
707	2.63		1.10		1.97×10^{-2}		
708	1.72		1.50		2.24×10^{-2}		2.49×10^{-2}
709	0.74				9.30×10^{-1}		2.74×10^{-2}
710	0.24		1.10		1.05		
711	1.35		1.40	1.04	5.19×10^{-2}		4.32×10^{-3}
712	1.04	0.57	0.91		2.19×10^{-2}	1.20×10^{-2}	
714	1.92	0.73	1.70		8.48×10^{-1}	2.73×10^{-1}	
717	1.38		2.30		7.56×10^{-3}		
719	2.12	0.61	0.88		2.55	1.43	
721	0.75		2.40	1.19	3.17×10^{-3}		4.82×10^{-3}
728	0.77	0.22	1.60		3.58×10^{-1}	1.68×10^{-1}	3.19×10^{-3}
740	0.30		1.60	0.81	2.01×10^{-3}		1.65×10^{-2}
747	1.73	0.33	1.20		2.40×10^{-2}	9.00×10^{-3}	
748	1.60		2.90	1.29	2.38×10^{-2}		2.52×10^{-3}
752	1.31				2.87		3.20×10^{-2}
753	1.75	0.40	2.00		3.49×10^{-2}	1.60×10^{-2}	
755	1.03	0.33	1.30		8.86×10^{-3}	1.00×10^{-3}	
762	1.49		2.50				9.15×10^{-2}
763	2.61				3.21×10^{-3}		
766	1.10		2.20		8.46×10^{-2}		
768	1.43				3.85×10^{-3}		1.56×10^{-2}
Min.	0.24	0.17	0.88	0.81	2.01×10^{-3}	1.00×10^{-3}	4.00×10^{-4}
Avg.	1.37	0.49	1.67	1.06	4.69×10^{-1}	7.06×10^{-1}	1.74×10^{-2}
Max.	2.63	0.84	2.90	1.29	2.87	7.18	9.15×10^{-2}

During the study, we faced another essential issue—the necessity for core cleaning or solvent extraction. To identify the role of extraction in the core analysis workflow, we repeated some of the tests (particularly NMR for identifying the porosity and Darcypress for permeability) before and after the extraction (Figure 6). The obtained results show the core cleaning did not adversely affect either NMR porosity or permeability. The permeability of the cleaned core radically increased for only three samples, which can be explained by the presence of artificial fractures that occurred during sample preparation (Figure 6b). TOC was around 5 wt.% and less than 10 wt.% for the target rock samples [54].

Thus, we continue the core analysis and interpretation of the results for the core in a non-extracted (non-cleaned) state.

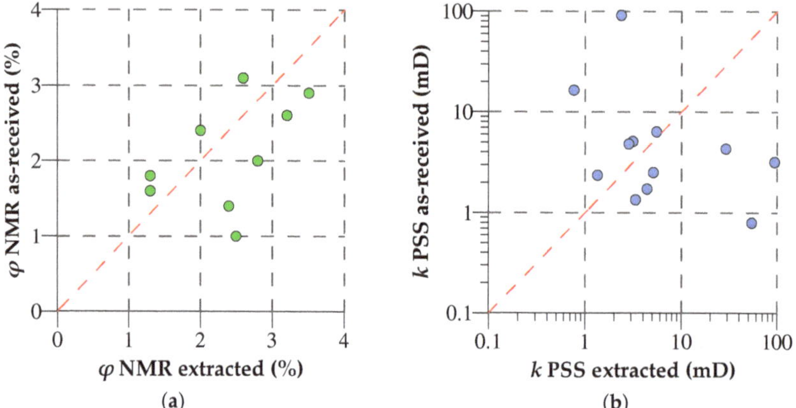

Figure 6. NMR porosity (**a**) and pseudo-steady-state permeability (PSS) (Darcypress) permeability (**b**) for the as-received versus extracted samples.

In addition to the core plugs and chips, two whole core plugs passed through to the proposed laboratory workflow (Figure 5).

3.2. Microstructural Characterization

According to the Dunham classification of carbonate rocks [36], sample #545 was a crystalline limestone with detritus (Figure 7) because it was composed of predominantly crystalline calcite (sometimes with biogenic structure) and calcite remains of tentaculite shells. Sample #551 was an organic-rich wackestone (Figure 7) because the matrix held the immersed calcite detritus.

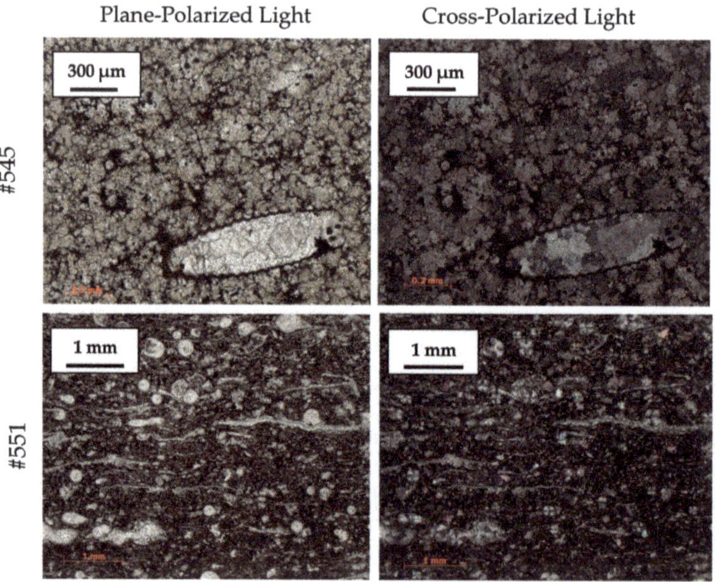

Figure 7. Petrographical analysis of thin-sections located at the closest depth to samples #545 and #551.

Our experience and observations showed that the petrographical description of the formation distinguished lithotypes, but did not provide reliable information on the void space structure [55,56]. Following down along the scale bar, we attempted to characterize the void space using whole-core CT and mini-plug micro-CT. We found that, similarly to the petrographic analysis of thin-sections, CT did not reveal natural voids and, thus, turned out to be a meaningless tool for the target rock (Figure 8).

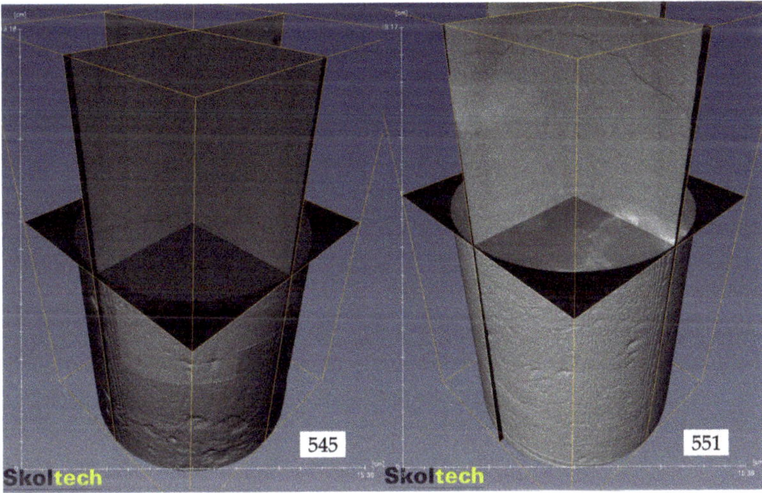

Figure 8. Visualization of 3D X-ray density models resulting from whole-core CT of the target rock samples.

The whole-core CT enabled the assessment of the degree of sample preservation, including the characteristics of artificial (technogenic) fractures, single large inclusions of relatively less dense organic matter, and the spatial inhomogeneity of the X-ray density. The studied rock samples featured a low contrast of the X-ray density of the rock-forming minerals. The mineral composition of the target rock samples made the three-dimensional tomographic models much less contrasting in X-ray density than they looked as real samples in daylight or petrographic thin sections.

We obtained similar results and drew the same conclusion from an analysis of mini-core micro-CT data. Micro-CT at a voxel size down to 3 µm did not fully resolve the natural-genesis void space within the density models. We separated distinct pores without revealing the connection between them (Figure 9).

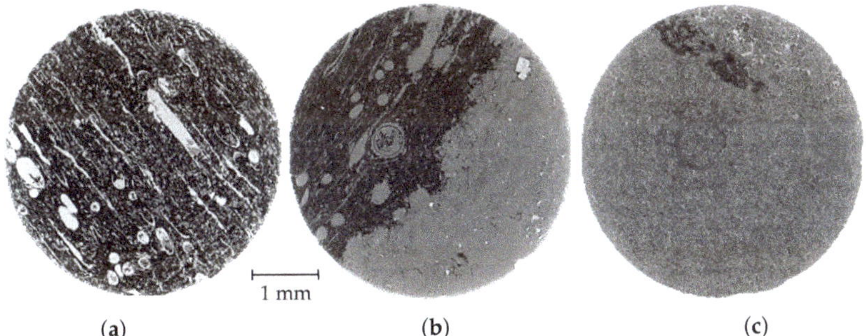

Figure 9. Representative micro-CT slices of mini-cores taken from the whole core #551: (**a**) detritus (tentaculites) in calcite matrix with organic matter—organic-rich wackestone; (**b**) contact of organic-rich wackestone (dark) and crystalline limestone (light); (**c**) crystalline limestone.

The micro-CT results only complemented the description of rock heterogeneity and a spatial distribution of the potential organic matter (OM) inclusions. The analysis of mini-core micro-CT data revealed several structural and textural features at the microscale. In general, the studied rock samples represented various combinations of a relatively dense finely-crystalline mass of limestone and a relatively less dense carbonate-siliceous mass (Figure 9). However, the identified features did not pertain to the void space structure; therefore, their detailed description fell out of the paper's scope.

The void space structure characterization at the highest resolution involved SEM imaging. Magnification of ×10k allowed us to identify pores due to its tiny size. Pores with specific dimensions of less than 1 μm, in the majority of cases, resided in the OM. In rare (individual) cases, the mineral matrix also contained voids (Figures 10 and 11). SEM images for samples #545 (Figure 10) and #551 (Figure 11) illustrated the multiscale variation with the maximum magnification on the last picture in the sequence.

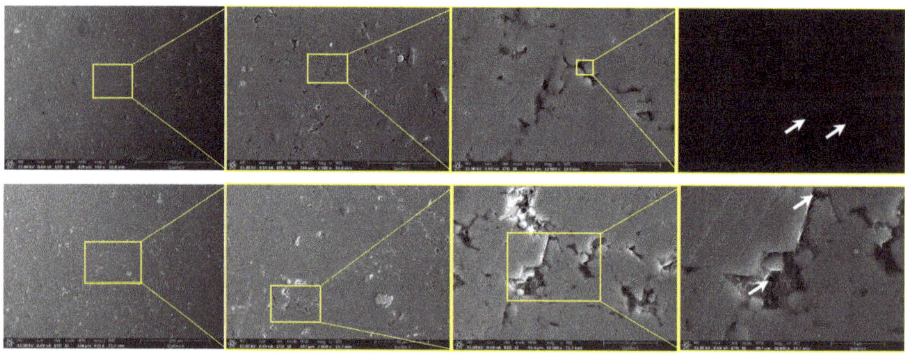

Figure 10. SEM images for samples #545-1 and #545-2 (white arrows indicate voids).

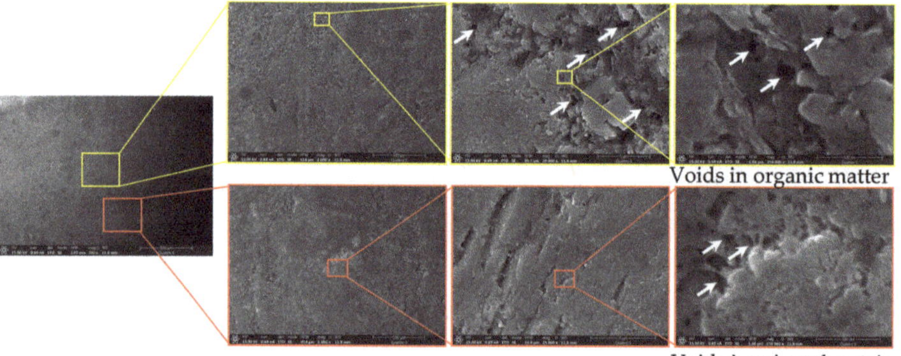

Figure 11. SEM images for sample #551-1 (white arrows indicate voids).

Sample #545 was composed of predominantly crystalline limestone with biogenic texture (Figure 7). The OM filled the space between calcite crystals with tiny rare voids less than 1 μm (Figure 10). Sample #551 encompassed a significant number of crystals immersed in a calcite matrix with OM (Figure 7). SEM images showed that there were two major types of voids (Figure 11). The first type had specific dimensions of less than 500 nm and resided in the OM. The second type was sporadic voids in the mineral matrix (biogenic clasts), with the size rarely larger than 250 nm.

3.3. Pore Size Distributions

PSDs derived from MICP data reduction for both whole cores showed the actual independence of the differential properties of the void space structure on the rock fraction size (Figures 12 and 13). Sample #545 (taken from the upper formation interval) showed almost identical PSDs before and after extraction, while sample #551 (taken from the lower BF interval) demonstrated the substantial difference in void space structure while maintaining the size range.

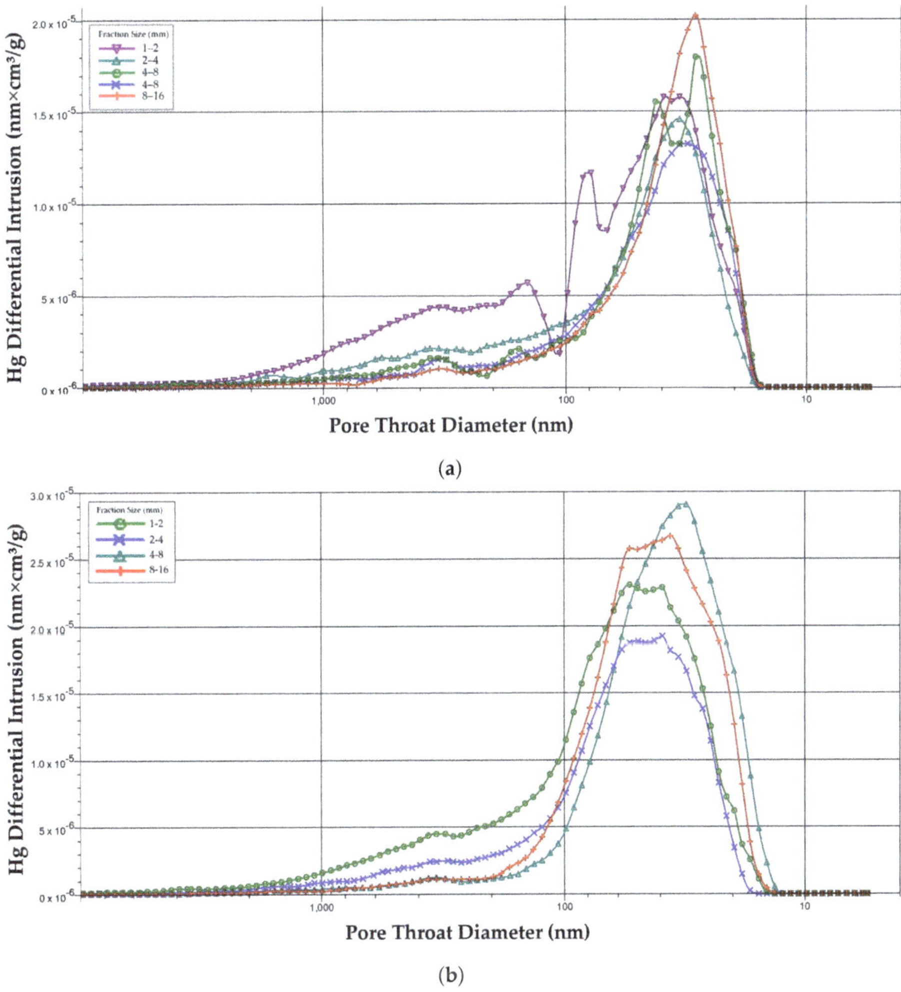

Figure 12. Mercury injection capillary pressure (MICP) pore size distributions (PSDs) for the whole core #545 depend on the rock fraction size (**a**) before and (**b**) after extraction.

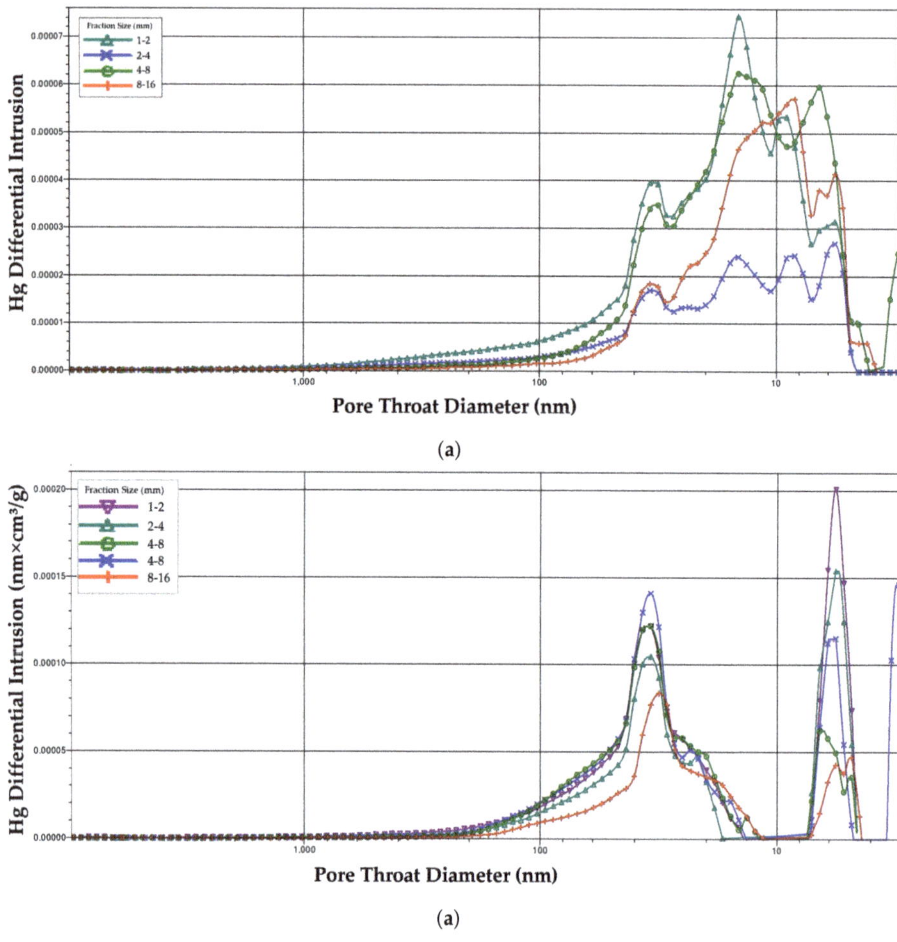

Figure 13. MICP PSDs for the whole core #551 depend on the rock fraction size before and after extraction.

For both samples, the pore throat size spanned in the range of 10–100 nm. The whole core #545 (from the upper formation interval) showed a unimodal distribution with a median of around 50 nm. In comparison, the sample #551 from the lower formation unit showed a bimodal distribution with modes at approximately 50 and 8 nm correspondingly.

4. Discussion

The study included a full-featured characterization of the tight carbonate reservoir rock with a complex void space structure. The laboratory workflow included state-of-the-art techniques for unconventional core analysis. However, our experience showed that given a limited amount of core material, the solution was non-trivial, and the results required careful analysis and understanding of data quality.

Here we discuss the obtained results in the form of a connected story ending with definite scientific conclusions. We start from the quality of porosity and permeability data. Notably, we go through the relationships between porosity and/or permeability obtained using different methods. Eventually, we come up with a suite of laboratory methods required to understand the reservoir properties of the target rock.

4.1. Limitations of Pressure Pulse-Decay Technique to Measure Core-Plug Gas Porosity and Permeability

At the initial stage, we compared the gas porosity (Figure 3) and permeability (Figure 4) between the obtained data and the (a priori) dataset provided by the operator. The comparison (Figure 14) revealed two features. Firstly, PIK-PP gas porosity weakly correlated with a priori data (r^2 = 0.54). Generally, a priori values were 2–3 times higher than those in the obtained laboratory results. Secondly, the comparison of permeability demonstrated a particular correlation. Nevertheless, the PIK-PP setup underestimated both gas porosity and permeability in comparison to the a priori dataset.

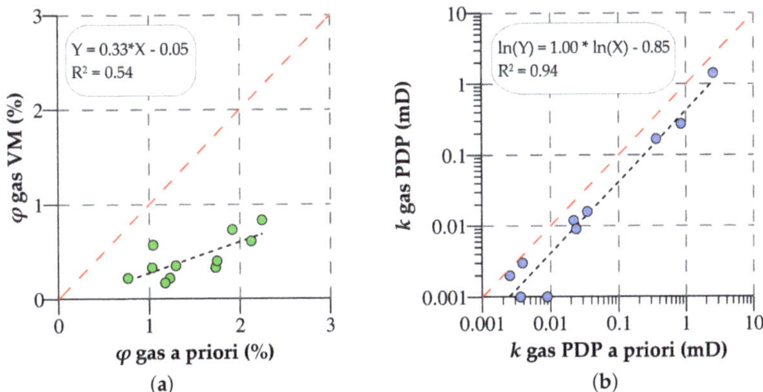

Figure 14. Gas volumetric (VM) porosity (**a**) and gas permeability (**b**) cross plots for core plugs show a comparison between the obtained data and the legacy dataset.

Both comparisons showed that the PIK-PP device (and potentially, its analogs) was not applicable for investigating the target rock samples. The most probable reason for this was the technical limitations of the method in terms of measuring porosity and permeability. The minimum limit of permeability for the PIK-PP device was 0.1 mD [38]. Therefore, the range of the observed values suggested that the porosity of the samples fell close to a lower detection limit of the device—0.6% for porosity. The mismatch in permeability (r^2 = 0.94 in log-log scale) was much lower than that in porosity (Figure 14b). The behavior was related to the physical process of permeability measurement, that is, pressure drop versus time observation. Unlike porosity, permeability resulted from the reduction of a pressure decline curve. Thus, determining permeability was much more robust in terms of stabilization time since it did not require pressure monitoring until the end of gas propagation.

The observed results did not confirm the applicability of the PDP technique, implemented in both PIK-PP and DarcyMeter instruments for the target rock samples. Nevertheless, the quality determination of permeability required a set of certified low-permeable samples for calibration. To date, a technology for the manufacturing of calibration samples is in a development stage with no commercial offerings [57].

The lack of applicable routine core analysis methods required the application of advanced petrophysical techniques for characterizing porosity and permeability. Techniques describing the void space structure complemented the bulk methods and included µCT and SEM, NMR, and MICP.

4.2. NMR Delivers the Highest Porosity

The advanced part of the presented petrophysical research included liquid-saturation NMR for getting core-plug porosity, as well as Darcypress for defining the permeability of rock samples embedded in epoxy resin disks. Cross plots between NMR, LS, and a priori gas porosity demonstrated several features (Figure 15).

Firstly, we considered LS porosity and a priori gas porosity to be the same within the uncertainty of the method. The LS, also referred to as the gravimetric method, provided reliable results for rocks with a distinct number of open pores. In our case, the pore space mainly consisted of hard-to-saturate tiny pores. Thus, the amount of liquid remaining on the rock surface had a significant impact on the weighing results and led to an overestimation of the total porosity.

Secondly, a comparison of NMR vs. a priori showed no clear correlation between NMR and a priori (gas porosity). We explained the behavior by low reliability of the a priori gas porosity.

In summary, we considered that PIK-PP (and its analogs based on the standard PDP techniques) provided uncertain gas porosity. The NMR delivered the highest porosity values.

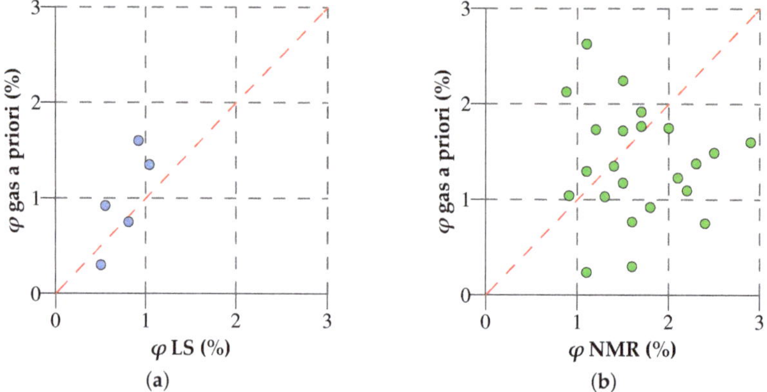

Figure 15. Liquid saturation (LS) (**a**) and NMR porosity (**b**) cross plots showing comparison with a priori data for core plugs.

Fourthly, LS results were similar to the ones by gas porosity. A potential explanation of such behavior was the initially low open porosity, which complicated the porosity characterization for both of these methods. Due to the limited data on LS and the invasiveness of the technique, we considered its results as complementary information.

Thirdly, NMR delivered the highest porosity values (Figure 16). We explained this observation by the technical ability of NMR to detect a wide range of pores—from nano-sized pores filled with high-viscous components to large voids occupied by mobile (free) fluids. Voids captured by low-field NMR typically had dimensions in the range of 10 nm–10 µm [58,59]. Moreover, unlike gas and liquid-station techniques, NMR delivered total porosity, including isolated hydrocarbon-filled voids and organic-hosted pores.

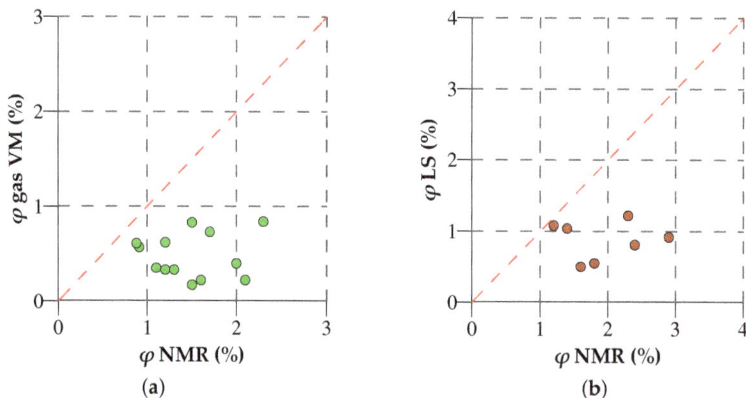

Figure 16. Comparison of NMR vs. pulse-pressure decay (PDP) gas volumetric porosity (**a**) and LS results (**b**) on core plugs.

4.3. Pseudo-Steady-State Technique Delivers the Lowest Permeability

To understand which laboratory method provided reliable permeability data, we compared the Darcypress results for non-extracted samples and a priori gas permeability (Figure 17). The cross plot showed the difference in permeability ranges in which the maximum value reached by Darcypress did not exceed 0.1 mD. Previously, the Darcypress instrument proved to provide reliable permeability data in the range from 1 nD to several Darcies [43]. The observed difference in permeability by two orders of magnitude suggested that at least a priori gas permeability data was unreliable for target rock samples. The same fact also implied the low quality of data provided by the PIK-PP unit.

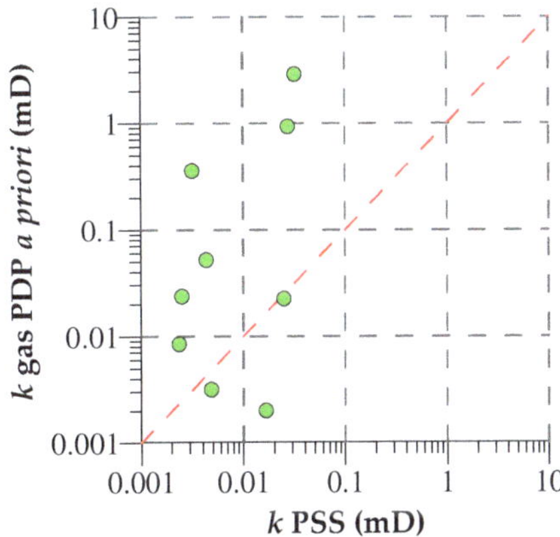

Figure 17. A cross plot of gas permeability by Darcypress versus a priori data.

In this case, core-plug dimensions were nearly 3 times larger than those for rock chips for Darcypress analysis. This fact may explain the difference in results. The difference in specimen preparation could also impact the quality of permeability measurements. A high number of artificial fractures (cracks) induced by core plugging may have led to an overestimation of rock permeability.

General recommendations for the characterization of tight-rock permeability included the method developed by the Gas Research Institute (GRI) [60]. Previously, we attempted to measure the matrix permeability of the target rock samples using a commercially available Core Lab SMP-200 instrument. However, our experience showed that matrix permeability varied in the range from nano- to pico-Darcy [54] and did not explain the reservoir properties observed by the bulk methods.

In summary, the PSS technique implemented in the Darcypress instrument seemed to deliver the most reliable and precise permeability data. Further research on low permeability determination requires the application of reference samples possessing nano- and micro-Darcy range permeability [57].

4.4. Standard Solvent-Cleaning Protocols Adversely Affect Reservoir Properties

A principal question for petrophysical laboratory core analysis is the requirement of core extraction or solvent cleaning. Generally, the core of conventional reservoirs rock should be extracted before reservoir properties' determination [6]. In the case of complex tight reservoirs, that requirement is often questionable for several reasons. One of the reasons is the excessively long time required to extract rock samples fully. Here we try to understand the effect of core solvent cleaning on the target rock samples.

We observed non-essential changes in reservoir properties after the core extraction (Figure 6). Firstly, NMR porosity before extraction corresponded to that after extraction within the range of 1–2%, with 50% of data pointing to evidence of a porosity increase, and the remaining 50% telling the opposite. The growth of porosity was mainly driven by the removal of OM from the void space with micro- and nanometer-size pores.

Secondly, the Darcypress permeability remained the same (before and after extraction) for 50% of the tested samples. Core extraction also led to core disintegration and the development of artificial fractures due to core fragility and thin-layered structure. The artificial fractures (Appendix A, Figure A2) boosted permeability.

Thirdly, we observed that the extraction tended to decrease the porosity and permeability for selected rock samples (highlighted points in Figure 6), which was an unexpected behavior. Our literature research delivered few results on this topic, but we may consider several options. The first is the precipitation of organic solvents and components in voids during the extraction [61]. The second is an alteration of the void space structure due to its exposure to the solvent and corresponding physical–chemical interactions between the mineral–organic matrix and the solvent or damage to the minerals [62]. Removal of solid-phase OM led to creating empty pores and overestimating the total porosity [63]. A complete understanding of this behavior requires further research efforts involving Rock-Eval pyrolysis on core samples before and after extraction. For this reason, in this study, we tended to skip core cleaning and remove this step from the applied core analysis workflow.

4.5. Void Space Structure Characterization Explains Low Reservoir Properties

To understand the quality of the obtained reservoir properties, we characterized a void space structure using three levels of visualization—2D petrography for thin-sections, whole-core 3D CT, mini-core 3D µCT, and 2D SEM of core chips with maximal spatial resolution.

Optical microscopy was appropriate only for typing but did not provide essential information on the void space structure for two reasons. The first reason was the nanoscale size of voids limited their visibility (half of the light wavelength). The second reason was the thin-section manufacturing procedure. Moreover, thin-section preparation became complicated when injecting epoxy into nano- and micropores [64]. SEM allowed us to overcome these restrictions during studying micro- and nanoscale voids in tight rocks.

The low X-ray density contrast of both CT and µCT data suggested that one can achieve the complete resolution of individual minerals and their associations using a dual-energy CT [65,66]. A voxel size of 116–122 µm did not provide a proper spatial resolution to visualize VSS elements for the target whole cores. Moreover, reliable detection of a void object in a digital rock model required it to have a Feret diameter of at least 2–3 vx corresponding (at the obtained voxel size) to 232–366 µm, or about a third of a millimeter. However, with an open porosity of the samples of 1–2% (Table 2), we could not justify the presence of such large voids distributed uniformly throughout the volume of the sample.

Generally, the voxel size around 2–3 µm did not allow to resolve spatially and, therefore, visualize and quantify the elements of the void space structure for the studied rock samples. The reason was that a correct segmentation of a void object in a digital model requires at least 2–3 vx in each Feret dimension that, at the obtained voxel size, corresponds to a characteristic void size of 8–12 µm. Moreover, the MICP results for samples #545 and #551 showed that the pore throats had typical dimensions of less than the first hundreds of nanometers. Thus, micro-CT, even with a voxel size of 0.5 µm corresponding to the spatial resolution of 1.5–2 µm, would not be able to image the void space structure of the target rock samples (Figure 9).

Images obtained with SEM demonstrated that mainly sporadic small (<3 µm) pores in organic matter and mineral matrix were present in the investigated samples (Figures 10 and 11). Recent studies on the application of nanoscale-resolution 3D imaging show that a relatively small number of connected pores with pore diameter greater than 150 nm sustain most of the hydrocarbon production [8]. Thus, SEM enabled quality 2D imaging of micro- and nanoscale voids in the target rock samples. In summary, our results showed that SEM was the only reliable method among all tested that resolved micro- and nanovoids in the studied rocks samples.

MICP data highlighted the different effects of solvent extraction on pore size distributions (PSDs). The observed difference in the PSD reaction to solvent extraction between the lower and upper formation depth intervals implied heterogeneity in the void space structure. For sample #545 (upper interval), we saw an extension of PSDs in a pore-throat range of less than 100 nm (Figure 12). For sample #551, the extraction led to a radical change of the void space structure (Figure 13). Extraction vanished pore throats with sizes in the range 9–50 nm, and the liberated void volume redistributed into two ranges: 3–8 and 50–200 nm. We explained this by a relatively high content of solid OM in the lower interval. We assumed that OM blocked the void space and thereby led to an underestimation of the effective pore-throat size in the mineral matrix. The solvent extraction at least partially removed OM from the voids in the mineral matrix and thereby boosted their effective size and volume. In summary, we established that the sample cleaning alters the void space structure differently for different depth intervals. This effect was essential for understanding the measured porosity and permeability properties of tight carbonates similar to the target formation and should be accounted for during planning petrophysical core analysis.

In summary, the results of the VSS study showed that SEM was one of the few laboratory methods capable of imaging voids in the target rock samples. Sporadic micrometer-to-nanometer pores imaged in 2D by SEM explained the low porosity range. The narrow pore-throat distributions by MICP with peaks at around 70 nm justified the low permeability range. MICP also revealed a complex rock response to solvent cleaning, in which distinct rock units may exhibit a different change in pore space structure.

4.6. Reservoir Properties of Whole Cores

We summarized both the a priori and obtained porosity and permeability data collected for the target whole cores (Table 3). The a priori dataset included both gas porosity and permeability measured on whole cores (Section 2.1), while our results related to core plugs and rock chips. Nevertheless, we attempted to integrate both datasets. Initially, we managed to drill 2 core plugs from each whole core. One core plug was characterized in terms of gas porosity and permeability, and another core plug was saturated with kerosene and characterized in terms of NMR and liquid saturation techniques. The remaining parts of the whole cores were distributed in accordance with the sample preparation scheme (Figure 5).

Table 3. Porosity and permeability of the whole core samples.

Fluid, Technique, Sample Form-Factor & Property	Whole Core	
	#545	#551
Gas Whole-Core Porosity (a priori) (%)	0.74	1.31
Kerosene NMR Core-Plug Porosity (%)	1.20	2.30
Hg MICP Crushed-Rock Porosity (%)	1.60	1.80
Gas Volumetric Core-Plug Porosity (%)	0.62	0.84
Kerosene Liquid-Staturation Core-Plug Porosity (%)	1.08	1.22
Gas PDP Whole-Core Permeability (a priori) (mD)	9.02×10^{-1}	2.30
Gas PDP Core-Plug Permeability (mD)	6.20×10^{-2}	1.13×10^{-2}
Gas PSS Rock-Chip Permeability (mD)	9.07×10^{-4}	4.00×10^{-4}

Both NMR and MICP delivered the highest porosity values among all methods. In the case of sample #545, MICP porosity was larger (1.6%) than that by NMR (1.2%). For sample #551, the trend was opposite—1.8% by MICP versus 2.3% by NMR. We explained the behavior by a high percentage of nanopores (Figure 12). At the same time, NMR was not able to resolve voids with specific dimensions of less than 10 nm [59]. We explained the increased NMR porosity for sample #551 by the presence of viscous hydrocarbon components and OM-hosted pores captured by low-field NMR. In contrast to

NMR, MICP relied on Hg intrusion into pores via throats. In our case, Hg could not penetrate a part of the void space due to potential pore-throat blockage by the viscous components.

Pulse-decay based instruments tended to underestimate the gas porosity for both whole cores. Particularly, the comparison showed a substantial difference between the obtained gas porosity (0.84%) and the corresponding a priori data (1.31%) for sample #551 (Table 3). We explained the difference due to several reasons. Firstly, a priori data resulted from whole-core measurements (Section 2.1), while the obtained results related to the core plug. However, [67] suggested a large variability in properties in the length-scale between ones and tens of centimeters. Secondly, a priori whole-core measurements may have been affected by the presence of larger pores and vugs, which could be physically destroyed during plugging. Therefore, the porosity difference of 0.47% presumably illustrated the effects of heterogeneity and scaling.

The gas permeability of both whole cores ranged as follows: whole cores yielded the highest values, while rock chips—the lowest values. The highest permeability difference reached 4 orders of magnitude. We assumed that the discrepancy of 1–2 orders of magnitude between PDP results was probably caused by the effects of scaling and rock heterogeneity. Section 4.3 explained the difference between the PDP and PSS results.

5. Conclusions

In the presented study, we implemented an integrated approach to characterize the ultralow-permeable carbonate reservoir rocks of the Tlyanchy-Tamakian Formation, Volga-Ural oil-gas province in Russia.

The suite of laboratory methods included both bulk measurements and multiscale void space characterization. Bulk techniques included gas volumetric, NMR, LS porosity and PDP, and PSS permeability, while imaging consisted of thin-section petrography, CT and µCT, and SEM. MICP was a proxy technique between bulk measurements and imaging. The target set of rock samples included whole cores, core plugs, mini cores, rock chips, and crushed rock.

The study resulted in the following significant findings for the target rock samples:

1. The gas volumetric core-plug gas-porosity results did not confirm the applicability of the PDP technique for the target rock samples. NMR delivered the highest porosity values due to its physical principle of non-invasive sensing. NMR–LS, NMR–gas porosity comparisons showed that NMR was the most appropriate technique for total porosity determination;
2. MICP porosity matched both NMR and imaging results and highlighted the different effects of solvent extraction on the throat size distribution of the target rock samples. The first was vanishing pore throats with sizes in the range 9–50 nm, followed by a redistribution of the liberated void volume into two ranges: 3–8 and 50–200 nm. The second was a partial removal of OM from voids in the mineral matrix, thereby boosting their effective size and volume;
3. PDP core-plug gas permeability measurements were consistent but overestimated in comparison to PSS results. We observed the difference reaching two orders of magnitude;
4. petrographic thin-section analysis, as well as CT and µCT, did not resolve the void space structure of the target rock samples. SEM proved to be the only feasible method for void-scale imaging with a spatial resolution up to 5 nm. The results confirmed the presence of natural voids of two major types. The first type was OM-hosted pores, with dimensions of less than 500 nm. The second type was sporadic voids in the mineral matrix (biogenic clasts), rarely larger than 250 nm;
5. comparisons between whole-core and core-plug reservoir properties showed substantial differences in both porosity (by a factor of 2) and permeability (up to 4 orders of magnitude) caused by spatial heterogeneity and scaling.

The observed experimental features are essential for understanding the measured porosity and permeability properties of tight carbonates similar to the target formation and are critical during the planning of petrophysical core analysis.

Author Contributions: Conceptualization, A.M., A.K., and A.C.; methodology, A.M., A.K., N.B., A.C., and M.S.; investigation, A.M., A.K., and T.K.; supervision, N.B., S.M., M.S., and A.C.; formal analysis, A.M. and A.K.; resources, N.B., A.C., M.S., and S.M.; data curation, A.M., A.K., T.K., and A.C.; writing and editing, A.M., A.K., T.K, and A.C. All authors have read and agreed to the published version of the manuscript.

Funding: This research received funding within the framework of an industrial project with LUKOIL Engineering LLC.

Acknowledgments: The authors would like to express gratitude to the PermNIPIneft branch of LUKOIL-Engineering LLC for providing the core material for testing. The authors thank their colleagues from Skoltech's Center for Hydrocarbon Recovery: Mikhail Yu. Spasennykh for the opportunity to conduct the research; Victor Nachev, Evgeny Shilov, Philipp Denisenko, and Denis Orlov for help in conducting experiments and discussions. We appreciate the major contribution of Maxim S. Mel'gunov (Boreskov Institute of Catalysis SB RAS), as well as Ivan V. Myakshin (LUKOIL-Engineering PermNIPIneft) with planning and conducting mercury injection porosimetry experiments. Finally, the authors owe to anonymous peer reviewers for their time and effort to make this paper even better.

Conflicts of Interest: The authors declare no conflict of interest.

Appendix A. Supplementary Figures

Figure A1. 3D view photo of the whole core #545.

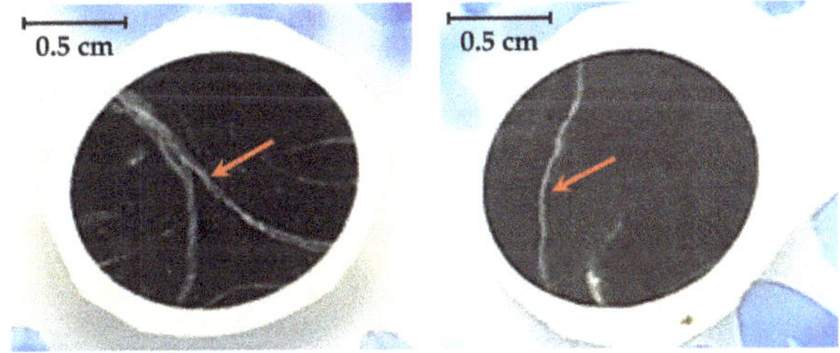

Figure A2. Fractures occurred after extraction on an example of the Darcypress probes.

Appendix B. Common Shared Techniques

The core analysis workflows employed the main recommendations for rock analysis [6].

We determined the open porosity of the target rock samples using the standard liquid saturation or gravimetric method. The technique consisted of saturating a rock sample with a liquid (usually kerosene or water) and determining its volume by immersion in the saturating fluid utilizing precise laboratory scales A&D GH-202GH with an AD1653 gravimetric console.

The core saturation procedure consisted of vacuuming the samples, capillary imbibition, and injection of a saturating fluid under a pressure of 15 MPa using an automatic saturating unit Geologika PIK-SK.

Core crushing and probe preparation employed the crushing machine ASCS Scientific Jaw Crusher JC-300-ST-Q. We separated the core fractions by mesh size using the vibratory sieve shaker RETSCH AS 450 control.

Core cleaning (extraction) included cleaning the rock samples (core plugs and core chips) with chloroform in a Soxhlet apparatus. We controlled the extraction quality by both visual inspection with a UV lamp and by measuring the TOC content on rock specimens every 24 h. The time of extraction averaged 150 h.

After extraction, we dried the rock samples at a temperature of 70 °C in the laboratory heating oven Memmert VO400 until attaining a constant weight.

The Wildcat Technology HAWK RW instrument provided Rock-Eval pyrolysis measurements for source rock geochemical analysis and data interpretation [68].

Optical microscopy is the most common and universal method for studying mineral composition and textural features of sedimentary rocks. We imaged a set of thin sections (10–20 μm) using the Carl Zeiss Imager A2m polarizing microscope with transmitted light (12 V halogen lamp, 100 W). The resulting spatial resolution depended on many factors related to sample preparation. However, we managed to achieve the image pixel size of about 10 μm.

References

1. Zhang, K.; Sebakhy, K.; Wu, K.; Jing, G.; Chen, N.; Chen, Z.; Hong, A.; Torsæter, O. Future Trends for Tight Oil Exploitation. In Proceedings of the SPE North Africa Technical Conference and Exhibition, Cairo, Egypt, 14 September 2015; p. 14.
2. Sheng, J.J. Critical Review of Field EOR Projects in Shale and Tight Reservoirs. *J. Pet. Sci. Eng.* **2017**, *159*, 654–665. [CrossRef]
3. *Carbonate Petroleum Reservoirs*; Roehl, P.O.; Choquette, P.W. (Eds.) Springer: New York, NY, USA, 1985; p. 622.
4. Ahr, W.M. *Geology of Carbonate Reservoirs: The Identification, Description, and Characterization of Hydrocarbon Reservoirs in Carbonate Rocks*; Wiley: Hoboken, NJ, USA, 2008; p. xvi, 277.
5. Burchette, T. Carbonate Rocks and Petroleum Reservoirs: A Geological Perspective From the Industry. *Geol. Soc. London Spec. Publ.* **2012**, *370*, 17–37. [CrossRef]
6. *Recommended Practices for Core Analysis. Recommended Practice 40*, 2nd ed.; American Petroleum Institute (API): Washington, DC, USA, 1998; p. 220.
7. Kazak, E.S.; Kazak, A.V. A Novel Laboratory Method for Reliable Water Content Determination of Shale Reservoir Rocks. *J. Pet. Sci. Eng.* **2019**, *183*, 106301. [CrossRef]
8. Goral, J.; Walton, I.; Andrew, M.; Deo, M. Pore System Characterization of Organic-Rich Shales using Nanoscale-Resolution 3D Imaging. *Fuel* **2019**, *258*, 116049. [CrossRef]
9. Nelson, P.H. Pore-Throat Sizes in Sandstones, Tight Sandstones, and Shales. Geologic Note. *AAPG Bull.* **2009**, *93*, 329–340. [CrossRef]
10. Kazak, A.; Chugunov, S.; Chashkov, A. Integration of Large-Area Scanning-Electron-Microscopy Imaging and Automated Mineralogy/Petrography Data for Selection of Nanoscale Pore-Space Characterization Sites. *SPE-191369-PA* **2018**, *21*, 821–836. [CrossRef]
11. Saidian, M.; Kuila, U.; Godinez, L.; Rivera, S.; Prasad, M. A comparative study of porosity measurement in mudrocks. *SEG Tech. Program Expand. Abstr.* **2014**, 2433–2438. [CrossRef]
12. Labani, M.; Rezae, R. Petrophysical Evaluation of Gas Shale Reservoirs. In *Fundamentals of Gas Shale Reservoirs*; Wiley & Sons: Hoboken, NJ, USA, 2015; pp. 117–137. [CrossRef]
13. Moghadam, A.A.; Chalaturnyk, R. Laboratory Investigation of Shale Permeability. In Proceedings of the SPE/CSUR Unconventional Resources Conference, Calgary, AB, Canada, 20 October 2015; p. 27.
14. Profice, S.; Hamon, G.; Nicot, B. Low-Permeability Measurements: Insights. *Petrophysics* **2016**, *57*, 30–40.
15. Sander, R.; Pan, Z.; Connell, L.D. Laboratory Measurement of Low Permeability Unconventional Gas Reservoir Rocks: A Review of Experimental Methods. *J. Nat. Gas Sci. Eng.* **2017**, *37*, 248–279. [CrossRef]

16. Jiang, Z.; Mao, Z.; Shi, Y.; Wang, D. Multifractal Characteristics and Classification of Tight Sandstone Reservoirs: A Case Study from the Triassic Yanchang Formation, Ordos Basin, China. *Energies* **2018**, *11*, 2242. [CrossRef]

17. Krakowska, P.; Puskarczyk, E.; Jędrychowski, M.; Habrat, M.; Madejski, P.; Dohnalik, M. Innovative Characterization of Tight Sandstones From Paleozoic Basins in Poland Using X-ray Computed Tomography Supported by Nuclear Magnetic Resonance and Mercury Porosimetry. *J. Pet. Sci. Eng.* **2018**, *166*, 389–405. [CrossRef]

18. Adebayo, A.R.; Babalola, L.; Hussaini, S.R.; Alqubalee, A.; Babu, R.S. Insight into the Pore Characteristics of a Saudi Arabian Tight Gas Sand Reservoir. *Energies* **2019**, *12*, 4302. [CrossRef]

19. Smodej, J.; Lemmens, L.; Reuning, L.; Hiller, T.; Klitzsch, N.; Claes, S.; Kukla, P.A. Nano- to Millimeter Scale Morphology of Connected and Isolated Porosity in the Permo-Triassic Khuff Formation of Oman. *Geosciences* **2019**, *10*, 7. [CrossRef]

20. Yuan, Y.; Rezaee, R. Comparative Porosity and Pore Structure Assessment in Shales: Measurement Techniques, Influencing Factors and Implications for Reservoir Characterization. *Energies* **2019**, *12*, 2094. [CrossRef]

21. Jiang, T.; Rylander, E.; Singer, P.M.; Lewis, R.E.; Sinclair, S.M. Integrated Petrophysical Interpretation of Eagle Ford Shale with 1-D and 2-D Nuclear Magnetic Resonance (NMR). In Proceedings of the SPWLA 54th Annual Logging Symposium, New Orleans, LA, USA, 22–26 June 2013.

22. Gao, H.; Li, H. Determination of Movable Fluid Percentage and Movable Fluid Porosity in Ultra-Low Permeability Sandstone Using Nuclear Magnetic Resonance (NMR) Technique. *J. Pet. Sci. Eng.* **2015**, *133*, 258–267. [CrossRef]

23. Schmitt, M.; Halisch, M.; Fernandes, C.P.; Santos, V.S.S.d.; Weller, A. Fractal Dimension: An Indicator to Characterize the Microstructure of Shale and Tight Gas Sands Considering Distinct Techniques and Phenomena. In Proceedings of the International Symposium of the Society of Core Analysts, Snowmass, CO, USA, 21–26 August 2016; p. 6.

24. Comisky, J.T.; Santiago, M.; McCollom, B.; Buddhala, A.; Newsham, K.E. Sample Size Effects on the Application of Mercury Injection Capillary Pressure for Determining the Storage Capacity of Tight Gas and Oil Shales. In Proceedings of the Canadian Unconventional Resources Conference, Calgary, AB, Canada, 15–17 November 2011.

25. Hu, Q.; Gao, X.; Gao, Z.; Ewing, R.; Dultz, S.; Kaufmann, J. Pore Accessibility and Connectivity of Mineral and Kerogen Phases in Shales. In Proceedings of the SPE/AAPG/SEG Unconventional Resources Technology Conference, Denver, CO, USA, 25–27 August 2014.

26. Goldstein, J.; Newbury, D.; Joy, D. *Scanning Electron Microscopy and X-ray Microanalysis*, 3rd ed.; Kluwer Academic, Plenum Publishers: New York, NY, USA, 2003.

27. Camp, W.K. Electron Microscopy of Shale Hydrocarbon Reservoirs. In *AAPG Memoir 102*; American Association of Petroleum Geologists: Tulsa, OK, USA, 2013.

28. Gabnasyrov, A.V.; Lyadova, N.A.; Putilov, I.S.; Solovyev, S.I. Domanik Shale Oil: Unlocking Potential. In Proceedings of the SPE Russian Petroleum Technology Conference and Exhibition, Moscow, Russia, 24–26 October 2016.

29. Bushnev, D.A.; Burdel'naya, N.S. Modeling of Oil Generation by Domanik Carbonaceous Shale. *Pet. Chem.* **2013**, *53*, 145–151. [CrossRef]

30. Fadeeva, N.P.; Kozlova, E.V.; Poludetkina, E.N.; Shardanova, T.A.; Pronina, N.V.; Stupakova, A.V.; Kalmykov, G.A.; Khomyak, A.N. The Hydrocarbon-Generation Potential of the Domanik Rocks in the Volga–Ural Petroliferous Basin. *Mosc. Univ. Geol. Bull.* **2016**, *71*, 41–49. [CrossRef]

31. Stupakova, A.V.; Kalmikov, G.A.; Korobova, N.I.; Fadeeva, N.P.; Gotovskiy, Y.A. The Domanic Formation of the Volga-Ural Basin—Types of the Section, Formation Conditions and Hydrocarbon Potential. *Georesursy* **2017**, *1*, 112–124. [CrossRef]

32. Requejo, A.G.; Sassen, R.; Kennicutt, M.C.; Kvedchuk, I.; McDonald, T.; Denoux, G.; Comet, P.; Brooks, J.M. Geochemistry of Oils from The Northern Timan-Pechora Basin, Russia. *Org. Geochem.* **1995**, *23*, 205–222. [CrossRef]

33. Abrams, M.A.; Apanel, A.M.; Timoshenko, O.M.; Kosenkova, N.N. Oil Families and Their Potential Sources in the Northeastern Timan Pechora Basin, Russia. *AAPG Bull.* **1999**, *83*, 553–577.

34. Prishchepa, O.M.; Averianova, O.Y.; Ilyinskiy, A.A.; Morariu, D. *Tight Oil and Gas Formations—Russia's Hydrocarbons Future Resources*; VNIGRI: Saint-Petersburg, Russia, 2014.

35. Kadyrov, R.; Galiullin, B.; Statsenko, E. The Porous Space Structure of Domanik Shales in the East of Russian Plate. In Proceedings of the International Multidisciplinary Scientific GeoConference-SGEM, Albena, Bulgaria, 2–8 July 2018. 2018; pp. 907–914.

36. Dunham, R.J.; Ham, W.E. Classification of Carbonate Rocks According to Depositional Texture. In *Classification of Carbonate Rocks—A Symposium*; American Association of Petroleum Geologists: Tulsa, OK, USA, 1962; Volume 1, pp. 108–121.

37. McPhee, C.; Reed, J.; Zubizarreta, I. *Core Analysis: A Best Practice Guide*, 1st ed.; Elsevier: Amsterdam, The Netherlands, 2015.

38. Geologika Manual. *PIK-PP Automated Unit for Measurung Porosity and Permeability of Rock Samples:User Manual*; Geologika Internal Publishing: Novosibirsk, Russia, 2016.

39. Abragam, A. *The Principles of Nuclear Magnetism*; Clarendon Press: Oxford, UK, 1961; p. 599.

40. Callaghan, P. *Principles of Nuclear Magnetic Resonance Microscopy*; Clarendon Press: Oxford, UK, 1991.

41. Bloembergen, N.; Purcell, E.M.; Pound, R.V. Relaxation Effects in Nuclear Magnetic Resonance Absorption. *Phys. Rev.* **1948**, *73*, 679–712. [CrossRef]

42. Straley, C.; Rossini, D.; Vinegar, H.J.; Tutunjan, P.; Morriss, C.E. Core Analysis by Low-Field NMR. *Log Anal.* **1997**, *38*, 84–94.

43. Lenormand, R.; Bauget, F.; Ringot, G. Permeability Measurement on Small Rock Samples. In Proceedings of the International Symposium of the Society of Core Analysts, Halifax, Canada, 4–7 October 2010.

44. Cydarex. *DarcyPress User Manual*; Cydarex Internal Publishing: Paris, France, 2017.

45. Webb, P.A. *An Introduction to the Physical Characterization of Materials by Mercury Intrusion Porosimetry with Emphasis on Reduction and Presentation of Experimental Data*; Micromeritics Instrument Corp.: Norcross, GA, USA, 2001; p. 23.

46. Wong, P.-Z. *Methods in the Physics of Porous Media*; Wong, P.-Z., Ed.; Academic Press: Cambridge, MA, USA, 1999; Volume 35, p. 485.

47. Micromeritics. *AutoPore V Series Operator Manual Nov 2017 (Rev D)*; Micromeritics Instrument Corporation: Norcross, GA, USA, 2017.

48. Peters, E.J. *Advanced Petrophysics—Volume 2: Dispersion, Interfacial Phenomena/Wettability, Capillarity/Capillary Pressure, Relative Permeability*, 1st ed.; Live Oak Book Company: Austin, TX, USA, 2012; Volume 2, p. 276.

49. Akima, H. A New Method of Interpolation and Smooth Curve Fitting Based on Local Procedures. *J. Assoc. Comput. Mach.* **1970**, *17*, 589–602. [CrossRef]

50. Cnudde, V.; Boone, M.N. High-Resolution X-ray Computed Tomography in Geosciences: A Review of the Current Technology and Applications. *Earth-Sci. Rev.* **2013**, *123*, 1–17. [CrossRef]

51. Remeysen, K.; Swennen, R. Application of Microfocus Computed Tomography in Carbonate Reservoir Characterization: Possibilities and Limitations. *Mar. Pet. Geol.* **2008**, *25*, 486–499. [CrossRef]

52. Thermo Fisher Scientific. PerGeos Software Instructions. 2018. Available online: https://www.thermofisher.com/ru/ru/home/industrial/electron-microscopy/electron-microscopy-instruments-workflow-solutions/3d-visualization-analysis-software/pergeos-digital-rock-analysis.html (accessed on 15 June 2018).

53. Erdman, N.; Drenzek, N. Integrated Preparation and Imaging Techniques for the Microstructural and Geochemical Characterization of Shale by Scanning Electron Microscopy. In *AAPG Memoir. Electron Microscopy of Shale Hydrocarbon Reservoirs*; Camp, W., Diaz, E., Wawak, B., Eds.; AAPG: Tulsa, OK, USA, 2013; Volume 102, pp. 7–14.

54. Mukhametdinova, A.; Shilov, E.; Nachev, V.; Bogdanovich, N.; Cheremisin, A. A Complex of Laboratory Studies of Reservoir Properties of Domanik Formation Rocks. In Proceedings of the Carbonate Reservoirs, Moscow, Russia, 16–18 October 2018.

55. Karamov, T.; Mukhametdinova, A.; Bogdanovich, N.; Plotnikov, V.; Khakimova, Z. Pore Structure Investigation of Upper Devonian Organic-Rich Shales within the Verkhnekamsk Depression. In Proceedings of the International Multidisciplinary Scientific GeoConference-SGEM, Albena, Bulgaria, 28 June–7 July 2019; pp. 1045–1052.

56. Mukhametdinova, A.; Karamov, T.; Bogdanovich, N.; Cheremisin, A.; Plotnikov, V. Complex Characterization of Organic-Rich Carbonate Shales Saturation. *Int. Multidiscip. Sci. GeoConference-SGEM* **2019**, *19*, 719–726. [CrossRef]

57. Chugunov, S.; Kazak, A.; Amro, M.; Freese, C.; Akhatov, I. Towards Creation of Ceramic-Based Low Permeability Reference Standards. *Materials* **2019**, *12*, 3886. [CrossRef]
58. Yao, Y.; Liu, D. Comparison of Low-Field NMR and Mercury Intrusion Porosimetry in Characterizing Pore Size Distributions of Coals. *Fuel* **2012**, *95*, 152–158. [CrossRef]
59. Lyu, C.; Ning, Z.; Wang, Q.; Chen, M. Application of NMR T2 to Pore Size Distribution and Movable Fluid Distribution in Tight Sandstones. *Energy Fuels* **2018**, *32*, 1395–1405. [CrossRef]
60. Guidry, K.; Luffel, D.; Curtis, J. *Development of Laboratory and Petrophysical Techniques for Evaluating Shale Reservoirs*; Gas Technology Institite: Des Plaines, IL, USA, 1996; p. 304.
61. Simpson, G.A.; Fishman, N.S. Unconventional Tight Oil Reservoirs: A Call For New Standardized Core Analysis Workflows And Research. In Proceedings of the International Symposium of the Society of Core Analysts, St. John's, NL, Canada, 16–21 August 2015.
62. Byrne, M.; Patey, I. Core Sample Preparation—An Insight to New Procedures. In Proceedings of the International Symposium of the Society of Core Analysts, Abu Dhabi, UAE, 5–9 October 2004.
63. Burger, J.; McCarty, D.; Peacher, R.; Fischer, T. Sample Preparation for Unconventional Analysis: A Case Against Solvent Extraction. In Proceedings of the International Symposium of the Society of Core Analysts, Avignon, France, 8–11 September 2014.
64. Lazar, O.R.; Bohacs, K.M.; Schieber, J.; Macquaker, J.H.; Demko, T.M. *Mudstone Primer: Lithofacies Variations, Diagnostic Criteria, and Sedimentologic-Stratigraphic Implications at Lamina to Bedset Scales*; SEPM (Society for Sedimentary Geology): Tulsa, OK, USA, 2015.
65. Teles, A.P.; Lima, I.; Topes, R.T. Rock Porosity Quantification by Dual-Energy X-ray Computed Microtomography. *Micron* **2016**, *83*, 72–78. [CrossRef]
66. Tsuchiyama, A.; Nakano, T.; Uesugi, K.; Uesugi, M.; Takeuchi, A.; Suzuki, Y.; Noguchi, R.; Matsumoto, T.; Matsuno, J.; Nagano, T.; et al. Analytical Dual-Energy Microtomography: A New Method for Obtaining Three-Dimensional Mineral Phase Images and Its Application to Hayabusa Samples. *Geochim. Cosmochim. Acta* **2013**, *116*, 5–16. [CrossRef]
67. Ringrose, P.S.; Martinius, A.W.; Alvestad, J. Multiscale Geological Reservoir Modelling in Practice. *Geol. Soc. London Spec. Publ.* **2008**, *309*, 123–134. [CrossRef]
68. Peters, K.E. Guidelines for Evaluating Petroleum Source Rock Using Programmed Pyrolysis. *Am. Assoc. Pet. Geol. Bull.* **1986**, *70*, 318–329.

Article

Natural Fractures in Carbonate Basement Reservoirs of the Jizhong Sub-Basin, Bohai Bay Basin, China: Key Aspects Favoring Oil Production

Guoping Liu [1,2], Lianbo Zeng [1,2,*], Chunyuan Han [3], Mehdi Ostadhassan [4], Wenya Lyu [1,2], Qiqi Wang [5], Jiangwei Zhu [1,2] and Fengxiang Hou [3]

[1] State Key Laboratory of Petroleum Resources and Prospecting, China University of Petroleum (Beijing), Beijing 102249, China; liugpzy228@gmail.com (G.L.); wylvwenwen@163.com (W.L.); zhujianghan@gmail.com (J.Z.)
[2] College of Geosciences, China University of Petroleum (Beijing), Beijing 102249, China
[3] PetroChina Huabei Oilfield Company, Cangzhou 062552, China; yjy_hcy@petrochina.com.cn (C.H.); yjy_hfx@petrochina.com.cn (F.H.)
[4] Department of Petroleum Engineering, University of North Dakota, Grand Forks, ND 58202, USA; mehdi.ostadhassan@gmail.com
[5] Jackson School of Geosciences, University of Texas at Austin, Austin, TX 78713, USA; wangqiqi@utexas.edu
*   Correspondence: lbzeng@sina.com; Tel.: +86-010-8973-1288

Received: 16 August 2020; Accepted: 3 September 2020; Published: 7 September 2020

Abstract: Analysis of natural fractures is essential for understanding the heterogeneity of basement reservoirs with carbonate rocks since natural fractures significantly control key attributes such as porosity and permeability. Based on the observations and analyses of outcrops, cores, borehole image logs, and thin sections from the Mesoproterozoic to Lower Paleozoic in the Jizhong Sub-Basin, natural fractures are found to be abundant in genetic types (tectonic, pressure-solution, and dissolution) in these reservoirs. Tectonic fractures are dominant in such reservoirs, and lithology, mechanical stratigraphy, and faults are major influencing factors for the development of fractures. Dolostones with higher dolomite content are more likely to have tectonic fractures than limestones with higher calcite content. Most tectonic fractures are developed inside mechanical units and terminate at the unit interface at nearly perpendicular or high angles. Also, where a thinner mechanical unit is observed, tectonic fractures are more frequent with a small height. Furthermore, the dominant direction of tectonic fractures is sub-parallel to the fault direction or oblique at a small angle. In addition, integrating diverse characteristics of opening-mode fractures and well-testing data with oil production shows that, in perforated intervals where dolostone and limestone are interstratified or dolostone is the main lithologic composition, fractures are developed well, and the oil production is higher. Moreover, fractures with a larger dip angle have bigger apertures and contribute more to oil production. Collectively, this investigation provides a future reference for understanding the importance of natural fractures and their impact on oil production in the carbonate basement reservoirs.

Keywords: natural fracture; influencing factor; oil production; carbonate rock; basement reservoir; Jizhong Sub-basin

---

1. Introduction

Basement reservoirs usually refer to traps that accumulate oil and gas in topographic uplifts of basement rocks under unconformities, which are covered by younger sediments [1–4]. Based on their location in topographic uplifts, these reservoirs are divided into weathering crusts and inner reservoirs [5–7]. Basement rocks include metamorphic rocks and some volcanic and carbonate rocks [4,8–10].

As an important type of unconventional petroleum system for further exploration and development, basement reservoirs are receiving increasing attention worldwide, and hydrocarbons are being exploited from them in areas such as China, North Africa, the USA, and Southeast Asia [1,2,11,12]. Considering recent advancements in production technology and demand for more resources, research to extract oil and gas from the deep inner reservoirs of basements is growing as well [6,13,14]. In the Bohai Bay Basin, basement reservoirs are the primary type of petroleum traps, particularly in the Jizhong Sub-Basin, which is considered as one of the structural units in the Bohai Bay Basin, where 29 out of 43 oil and gas fields are basement reservoirs or related to basement rocks [15]. Volumetric calculations showed proven reserves of 526.83 million tons (3861.66 million bbl) of original oil in place (OOIP) in basement reservoirs, accounting for 51.66% of the total OOIP in the Jizhong Sub-Basin [9]. Carbonate rocks are the primary basement rocks in the Jizhong Sub-basin [16].

Previous studies have shown that primary pores are less in these basement reservoirs, and secondary pores dominate the storage space, mainly composed of dissolution pores and natural fractures [17–19]. Since basement rocks have experienced multiple periods of tectonism and diagenesis, various natural fractures are generally developed, causing significant heterogeneity in such reservoirs [13,20,21]. Like other types of fractured reservoirs, natural fractures are the main flow pathways for hydrocarbons in basement reservoirs [8,12,22,23]. Moreover, natural fractures are a significant component of storage space for petroleum accumulation in these reservoirs as well [24–26], where fractures at different scales can connect pores to control the quality of the reservoir [3,13,27,28].

Despite the vital role that natural fractures play in the reservoir quality as explained above, limited studies are carried out to delineate their significance entirely in basement reservoirs with carbonate rocks. The development of fractures, the main factors affecting their existence, and their contribution to the performance are unclear in such reservoirs and demand further investigation.

This study's primary purpose is to characterize natural fractures and understand their role in oil production in the deep inner reservoirs of carbonate basements in the Jizhong Sub-Basin of the Bohai Bay Basin. Therefore, outcrops, cores, borehole image logs, and thin sections were used in a detailed analysis for fracture characterizations, including orientation, dip angle, height, length, density, aperture, and filling. Based on these, lithology, mechanical stratigraphy, and faults were also analyzed to reveal how they would affect the growth mechanism of fractures. Moreover, the significance of opening-mode fractures on oil production was presented as well. These efforts will enable us to understand the heterogeneity of petrophysical properties and reservoir performances in these basements.

2. Geological Setting

2.1. Location and Structure

The Bohai Bay Basin, which is in the eastern part of China, is a Cenozoic rift basin developed in the basement of the North China Platform [9,29,30] (Figure 1a). The Jizhong Sub-Basin in the northwest is one of the sub-basins in the Bohai Bay Basin. It is distributed in an NNE–SSW direction and covers an area of 32,000 km^2 (12,355 mile^2) [31,32] (Figure 1b). A number of NNE–SSW-oriented depressions, uplifts, and slopes have been developed within the sub-basin. Additionally, two transfer zones—Xushui-Anxin-Wenan and Wuji-Shenxian-Hengshui—are developed in the Jizhong Sub-Basin, near E–W and NW–SE strikes, respectively (Figure 1b). The two transfer zones have divided the sub-basin into three regions: northeast, center, and southwest. The extensional faults formed by multistage tectonic movements are developed in this sub-basin, while grabens and half-grabens, from these faults, have created various topographic uplift structures in this sub-basin [33,34] (Figure 2).

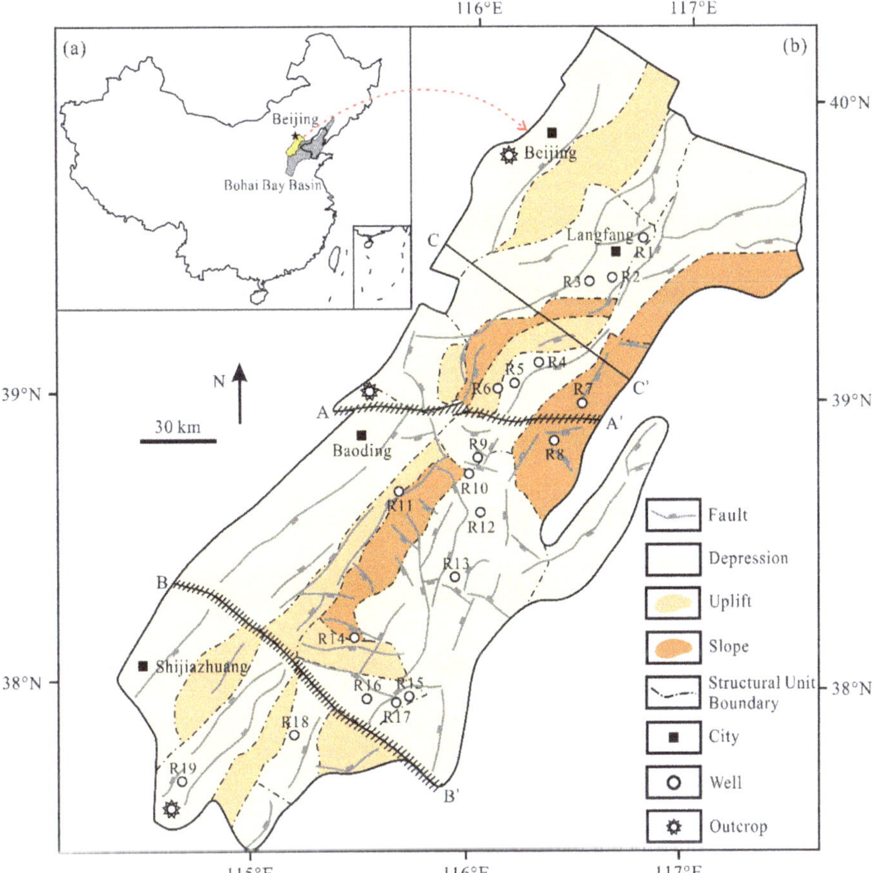

Figure 1. (**a**) Location of the Jizhong Sub-Basin in the Bohai Bay Basin in China (modified from Zhao et al., 2015) [9]. (**b**) Map showing the faults, structural units, transfer zones, wells, and outcrops in the Jizhong Sub-Basin (modified from Chang et al., 2016) [35]. The fault data was modified from the Huabei Oilfield database. A-A' is the Xushui-Anxin-Wenan transfer zone, and B-B' is the Wuji-Shenxian-Hengshui transfer zone. C-C' represents the position of the cross-section in Figure 2.

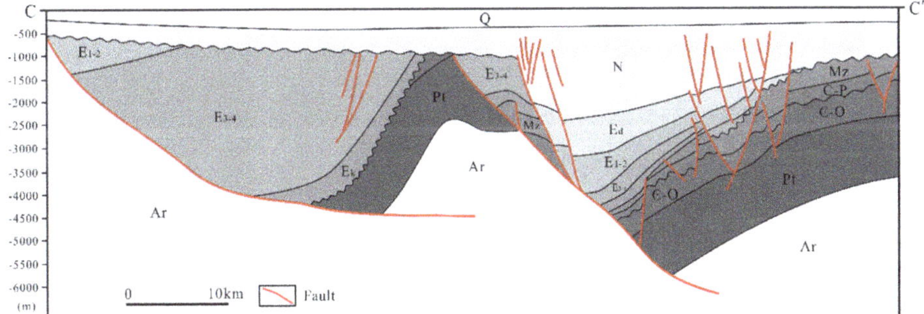

Figure 2. A geological cross-section in the Jizhong Sub-Basin, showing the structural pattern of the strata (modified according to He et al., 2017) [36]. The position of the cross-section is shown by C-C' in Figure 1.

2.2. Stratigraphy

Basement rocks in the Jizhong Sub-Basin are developed in the Archean, Proterozoic, Paleozoic, and Mesozoic from bottom to top (Figures 2 and 3) [19,37]. The Archean and Paleoproterozoic are composed of metamorphic rocks, while the Changcheng, Jixian, and Qingbaikou systems, which are developed in the Mesoproterozoic and Neoproterozoic, are a set of sediments made up of marine carbonate rocks. The Lower Paleozoic develops the Cambrian and Ordovician, which are typical sediments dominated by carbonate rocks of neritic platform facies. The Carboniferous and Permian in the Upper Paleozoic are clastic rocks of continental and transitional facies, where coal-bearing strata are relatively abundant. The Mesozoic is mainly Jurassic and Cretaceous, which are continental clastic rocks containing volcanic rocks. Due to the erosion and intermittent deposition caused by multistage structural uplifts, the Jizhong Sub-Basin strata are absent from the Silurian, Devonian, Early Carboniferous, and Triassic periods [29]. The Cenozoic strata consist of a set of interbedded depositions that are mainly sandstone and mudstone of lacustrine and deltaic facies [38].

The target layer in this study is the basements of carbonate rocks in the Mesoproterozoic, Neoproterozoic, and Lower Paleozoic periods. These basement rocks have undergone complex sedimentation, diagenesis, and tectonism, forming various reservoirs with proven economic hydrocarbon accumulations [39]. Source rocks are mainly dark mudstones of lacustrine facies in the Paleogene, along with coal and dark mudstones of transitional facies in the Carboniferous and Permian periods [40]. These source rocks are often adjacent to the basements, where hydrocarbons have migrated into the reservoir through faults and unconformities [34,41]. It should be noted that these coal and dark mudstones are also good regional caprocks for the carbonate basement reservoirs in this sub-basin. Additionally, many interlayers with high argillaceous content exhibit high-quality caprocks in the inner reservoirs of these basements [36,42]. As a result, combinations of source rocks, reservoirs, and caprocks have created a suitable petroleum system in the carbonate basements of the Jizhong Sub-Basin and have provided great potential for oil and gas exploration and development in this area [9].

2.3. Reservoir

The carbonate rocks that constitute the basement reservoirs in the Mesoproterozoic and Neoproterozoic are mainly dolostone, while in the Lower Paleozoic are dominated by limestone [38]. In these reservoirs, a network of pores and natural fractures have provided the storage space for the economic accumulation of hydrocarbons [17–19]. The pores are mainly secondary and are formed by dissolution, and natural fractures are more frequent and have multiple genetic types. These storage spaces provide favorable conditions for oil and gas accumulation and migration in these reservoirs. The carbonate rocks have an average matrix porosity lower than 12%, while the porosity analysis from cores has shown that samples with a porosity higher than 3% account for more than 70% of all samples [29,43]. The permeability of these basement reservoirs is relatively high, and samples with an air permeability greater than 1 mD account for more than 60% of all measured samples [43].

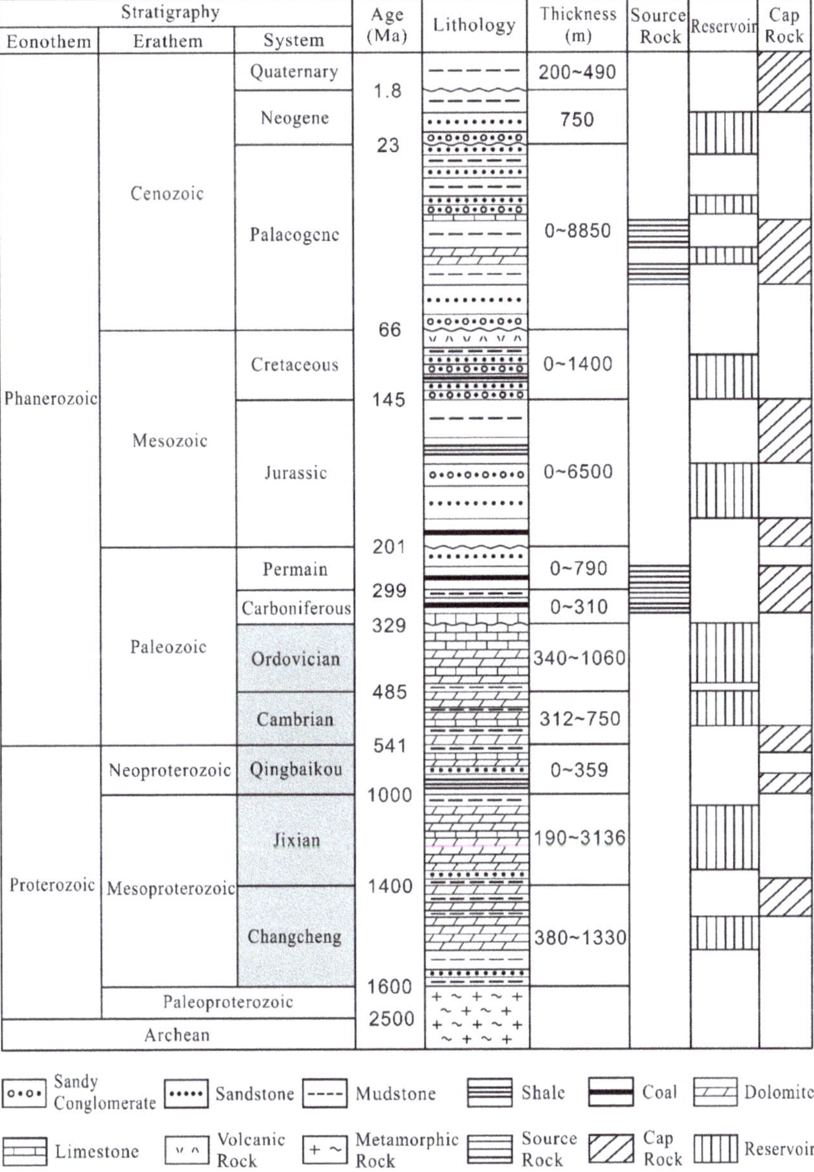

Figure 3. The stratigraphic column of the Jizhong Sub-Basin, showing the major petroleum systems (source rocks, reservoirs, and cap rocks) (modified according to Wu et al., 2011, and Liu et al., 2017) [19,37]. The gray mark in the system is the target layer of this study.

3. Dataset and Methodology

In this investigation, we collected data and samples from outcrops, cores, thin sections, and borehole image logs in the basement reservoirs of carbonate rocks in the Mesoproterozoic to Lower Paleozoic in the Jizhong Sub-Basin of the Bohai Bay Basin in eastern China. The outcrops and subsurface targets are from the same formation and have experienced similar tectonic movements and diageneses [39,44]. Cores and borehole image logs from 19 wells from these basement reservoirs were analyzed as well.

The length of the cores is 123.8 m (406.2 ft). Thin sections are 119 pieces coming from cores. In addition, other useful information such as lithology, fault, perforated interval, and oil production, were collected from the Renqiu Oilfield database and relevant literature. We combined these different sources of data to comprehensively analyze and study the natural fractures in deep inner reservoirs of these basements from macro to micro-scale.

Natural fracture characteristics, including height, length, orientation, dip angle, density, fracture zone, spacing, aperture, and filling, were examined closely from the sources mentioned above. The fracture density in this study refers to the linear density that was measured based on the number of fractures per unit length. The fracture zone is defined as multiple sets of tectonic fractures that are developed in rocks and usually are interwoven into a network, which makes it difficult to measure and count individually. These parameters in outcrops and cores were identified and measured on-site, while in borehole image logs, they were manually picked and interpreted [45,46]. It should be pointed out that in the cores, there are some unnatural fractures caused by the drilling activity and pressure release. Usually, the surface of fractures produced during the drilling activity is uneven, and these fractures do not have obvious directionality. However, the natural fracture surface is relatively flat or even smooth and has a strong directionality. Moreover, because there was no underground fluid flowing through, the surface of fractures produced during the drilling activity and formed by pressure release is very new. Based on these characteristics, we distinguished such unnatural fractures from natural fractures in the observation and statistics of fractures in cores, in order to minimize their influence on the real data.

Natural fractures in borehole image logs usually appear as sinusoidal curves, making it possible to quantify their orientation, dip angle, density, and aperture [47,48]. Moreover, in borehole image logs with water-based mud, opening-mode fractures are usually filled with mud filtrate or low-resistance minerals and appear as dark sinusoidal curves, while filled fractures with high resistance minerals (such as dolomite and calcite) often present as light or white sinusoidal curves [49]. Thin sections with a thickness of 30 μm were made with blue-dye resin to highlight natural fractures and pores, and some of them were impregnated with alizarin red to distinguish calcite and dolomite [50]. These thin sections are divided into two types, vertical and parallel to the wellbore. The specific directions of each thin sections are marked in the captions of Figures. Fractures in these thin sections were observed and measured by the Olycia g3 software from Olympus, Japan [51].

By studying the variation law of the characteristics and intensity of natural fractures, the factors controlling fracture development in these reservoirs were determined. Furthermore, by comparing the lithology, fracture characteristics, and oil production in six perforated intervals, we evaluated the role of natural fractures on oil production and proposed ideas to optimize development plans in the carbonate basement reservoirs to enhance production. It should be noted that, in this analysis, the fracture density refers to the linear density of opening-mode fractures, and oil production refers to the daily production of oil during the well-testing stage.

4. Results

4.1. Fracture Characteristics

This study distinguished three genetic types of natural fractures, including tectonic, pressure-solution, and dissolution fractures, in the basement reservoirs of carbonate rocks in the Jizhong Sub-Basin [39,52,53]. Among these, tectonic fractures have a higher development degree than others and are the dominant type in these reservoirs.

4.1.1. Tectonic Fractures

Tectonic fractures in the outcrops appear in sets, and their occurrences are stable (Figure 4). Statistical analysis of outcrops confirms that tectonic fractures are developed in three major sets of NNE–SSW, NW–SE, and near E–W strikes, while less developed in other directions (Figure 5). On the cross-section, tectonic fractures may pass through the rock formation interface and have a height of several meters (Figure 4a). Other tectonic fractures within the rock formation with a height of less than a few tens of centimeters were also observed (Figure 4b). On the horizontal plane, tectonic fractures show mutual crosscutting relations, and their lengths vary considerably, from several centimeters to meters (Figure 4c). The dip angles of these fractures are mainly high (>60°) and near-horizontal (<15°). Ultimately, the number of tectonic fractures with oblique dip angles (15°–60°) was found less than 20% of the total (1326).

Based on core observations, tectonic fractures usually exhibit a fracture plane with steps (Figure 6a). In mudstones and argillaceous carbonate rocks, these fractures demonstrate clear striaes along the direction of fractures propagation, with a very smooth surface (Figure 6b). Borehole image log interpretation indicates that tectonic fractures appear as sinusoidal curves and are randomly distributed (Figure 7a). Some tectonic fractures are intertwined to form a fracture network (Figure 7b). The dip angels of these fractures are concentrated in the range of 60° to 85°, followed by those with angles less than 30°. In particular, the dip angels of fractures in mudstones are mainly less than 20°. The linear density of tectonic fractures in a single well varies notably, ranging from $1.2\,\mathrm{m}^{-1}$ to $8.6\,\mathrm{m}^{-1}$. Furthermore, borehole image logs confirm the existence of fracture sets similar to those in the outcrops, mainly in the NNE–SSW, NW–SE, and near E–W strikes. More than 65% of tectonic fractures are opening-mode ones in which minerals are not filled, while others are entirely or partially filled by calcite, hydrocarbons, and clay minerals (Figures 6c and 7c).

Figure 4. Tectonic fractures in outcrops. (**a**) Fractures are developed with a height of several meters on the cross-section. (**b**) Fractures are developed within the rock formation with a height of a few tens of centimeters. (**c**) Fractures show mutual crosscutting relations on the horizontal plane where Set C arrests Set D.

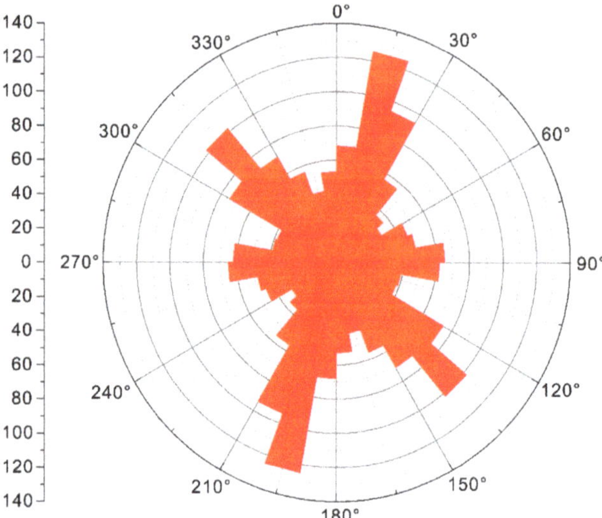

Figure 5. Orientations of tectonic fractures in outcrops of carbonate rocks in the Jizhong Sub-Basin (*n* = 1105).

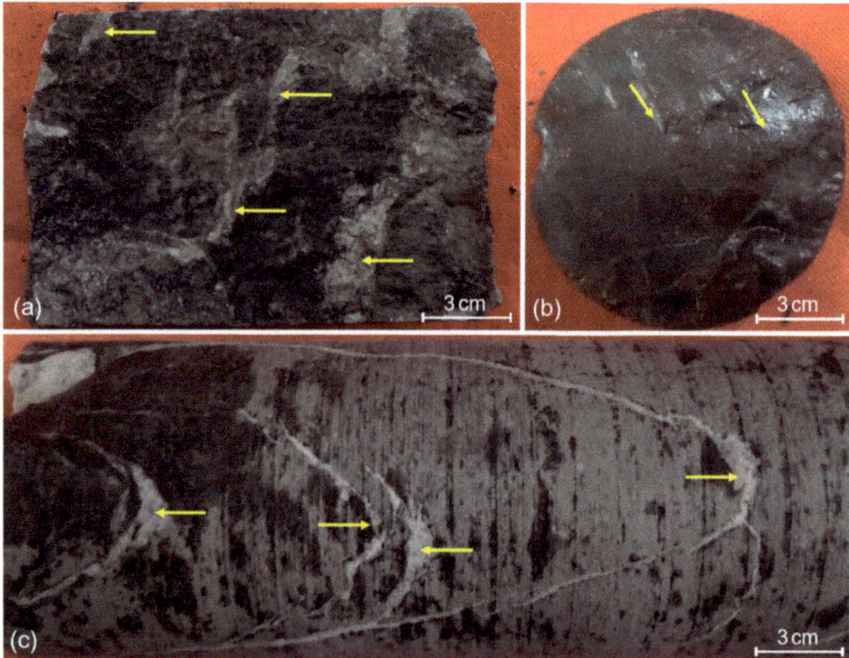

Figure 6. Tectonic fractures in cores. (**a**) Fractures in Well R6, depth 5583.18 m (18,317.52 ft). (**b**) Fractures in Well R14, depth 4010.37 m (13,157.38 ft). The dip angle of this fracture is 6°. (**c**) Fractures in Well R7, depth 4301.19 m (14,111.52 ft). Calcite entirely filled fractures.

The inspection of thin sections reveals that tectonic fractures are widely distributed in these carbonate rocks (Figure 8). Two sets of tectonic fractures can cut through or arrest each other (Figure 8b,c), and multiple sets are interwoven to form a network (Figure 8d). The development of these fractures does not exhibit any relationship with the bedding plane of carbonate rocks. If early developed fractures are filled with minerals such as calcite, pyrite, clay, or hydrocarbons, they will not provide effective storage space and seepage channels for the reservoir and become ineffective (Figure 8d–f) [54,55]. In this regard, the same fracture could be filled multiple times with different minerals (Figure 8e,f). More than 60% of tectonic fractures in thin sections are open and do not show any mineral fillings. The apertures of these fractures vary, while most of them are less than 100 μm and are concentrated below 60 μm.

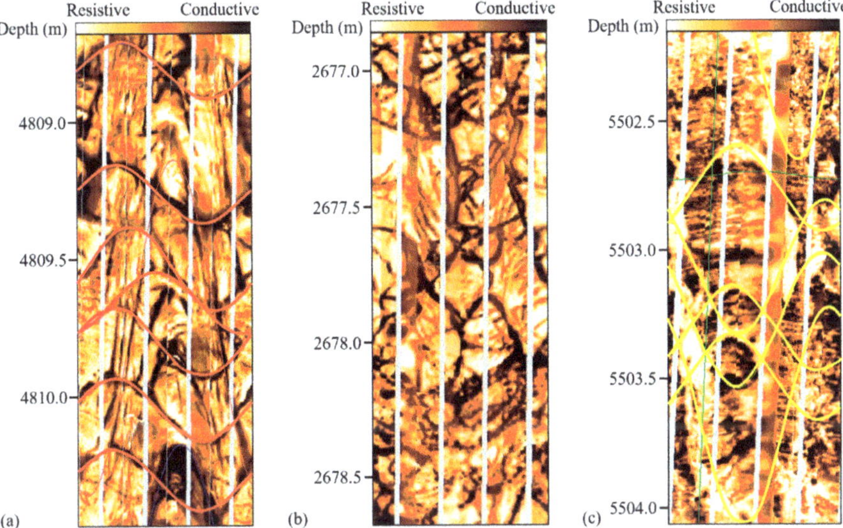

Figure 7. Tectonic fractures in borehole image logs. (**a**) Tectonic fractures are conductive in Well R8. The red sinusoidal curves represent recognized fractures. (**b**) Tectonic fractures that are not filled link together like a network in Well R15. (**c**) Tectonic fractures in Well R3. The yellow sinusoidal curves represent the fractures filled by minerals.

4.1.2. Pressure-Solution Fractures

Pressure-solution fractures are formed during structural and diagenetic processes due to pressure solution [56]. The pressure-solution fractures in these reservoirs are composed of bed-parallel fractures and stylolites. The bed-parallel fractures are formed along the depositional interface under various geological conditions, with distinguishable characteristics such as bending, discontinuity, and branching (Figures 9a–c and 10a) [57]. These fractures exhibit small dip angles and are nearly parallel to the depositional interface. There are normally insoluble mineral residues recognized in them, such as clay minerals, while they can also be filled with hydrocarbons or other minerals. Stylolites are generally irregularly wavy or serrated, parallel or sub-parallel to the horizontal plane, with a small number intersecting the horizontal plane at a small angle (Figures 9c,d and 10b). Iron argillaceous minerals and hydrocarbons can fill some of these fractures. These pressure-solution fractures are poor in lateral continuity, and their apertures in the thin sections are commonly less than 35 μm.

4.1.3. Dissolution Fractures

Dissolution fractures are formed through long-term underground fluid, including the new fracture formed after the dissolution transformed the earlier fracture and the fracture formed when the dissolution connected a lot of pores [58,59]. When dissolution fractures are formed from earlier ones, their fracture walls are rough and uneven, and their apertures are larger than the previous stage fractures (Figure 11a). Although newly dissolved pores are preserved inside or at the edges of the original fractures and the shape of the initial sets are changed after the dissolution process, the original distribution of these fractures can still be discerned. Dissolution fractures that are formed when multiple pores are connected like a string of beads will become an effective storage space for hydrocarbons in the reservoir (Figure 11b). Besides, fractures filled with minerals can also become dissolution fractures when unstable filling minerals like calcite entirely or partially are dissolved via acidic water leaching or groundwater scouring (Figure 11c,d). Overall, dissolution fractures are irregular in shape, often in the pattern of snake-like and anastomosing (Figures 10c and 11) [60]. The apertures of dissolution fractures measured in thin sections vary significantly and are between 40 μm and 80 μm and sometimes become relatively large (up to 200 μm).

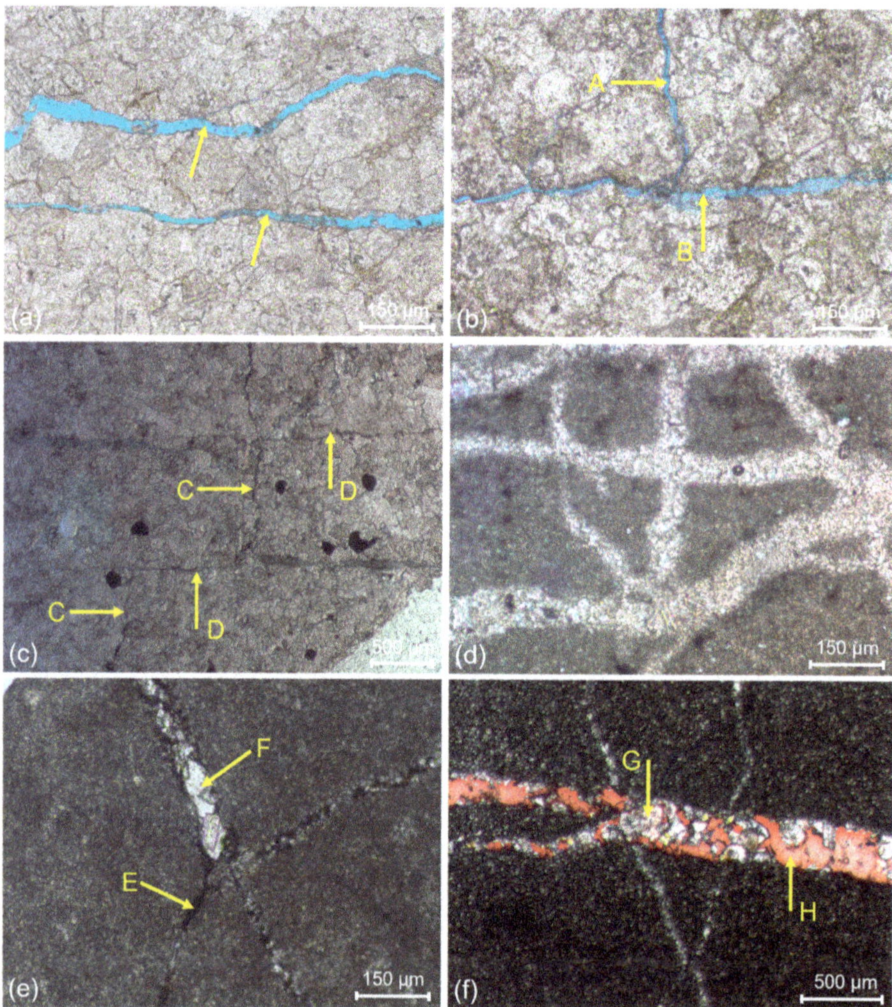

Figure 8. Tectonic fractures in thin sections. (**a**) Fractures in Well R2, depth 5039.60 m (16,534.12 ft). (**b**) Fractures in Well R2, depth 5039.35 m (16,533.30 ft). Group B terminated Group A. (**c**) Fractures in Well R5, depth 5916.02 m (19,409.51 ft). Group D terminated Group C. (**d**) Fractures are filled with calcite in Well R9, depth 4548.10 m (14,921.59 ft). (**e**) Fractures in Well R5, depth 5728.31 m (18,793.67 ft). E shows hydrocarbons, and F is calcite. (**f**) Fractures in Well R8, depth 4703.26 m (15,430.64 ft). G is dolomite, and H is calcite. The directions of these thin sections are vertical to the wellbore.

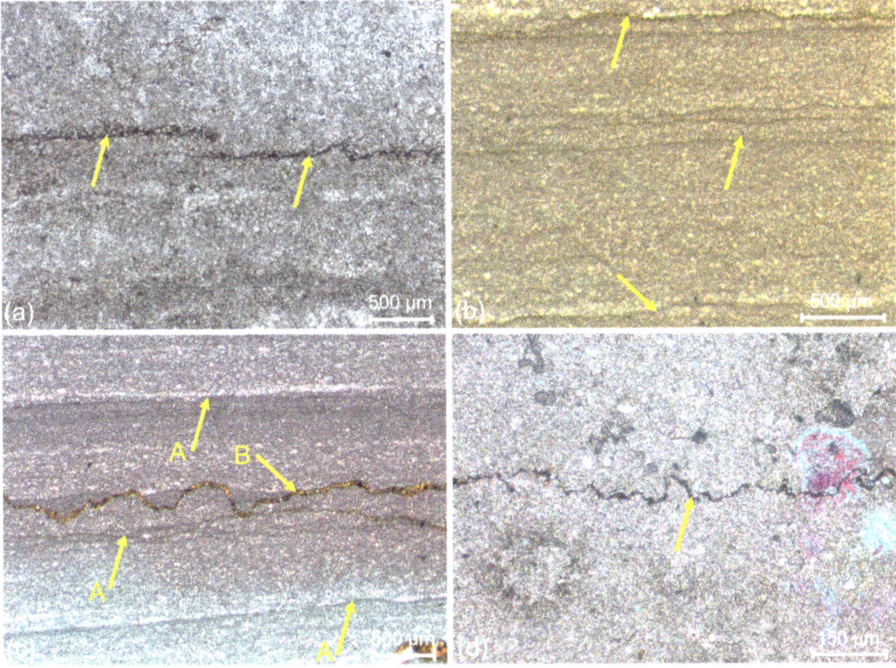

Figure 9. Pressure-solution fractures in thin sections. (**a**) Bed-parallel fractures in Well R2, depth 5041.27 m (16,539.60 ft). (**b**) Bed-parallel fractures in Well R8, depth 4861.80 m (15,950.79 ft). (**c**) Fractures in Well R8, depth 4862.12 m (15,951.84 ft). A is the bed-parallel fracture. B is the stylolite. (**d**) Stylolite in Well R9, depth 4548.53 m (14,923.00 ft). The directions of these thin sections are parallel to the wellbore.

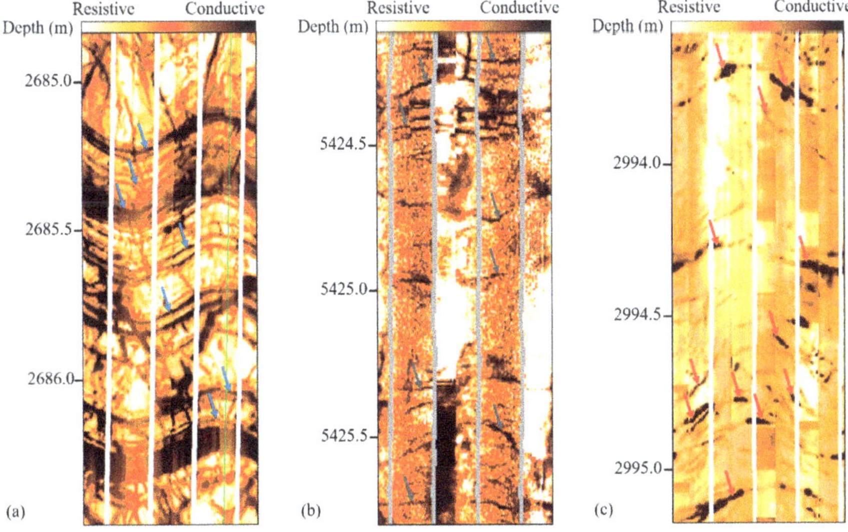

Figure 10. Fractures in borehole image logs. (**a**) Bed-parallel fractures in Well R15. (**b**) Stylolites in Well R1. (**c**) Dissolution fractures in Well R16. The arrows mark the identified fractures.

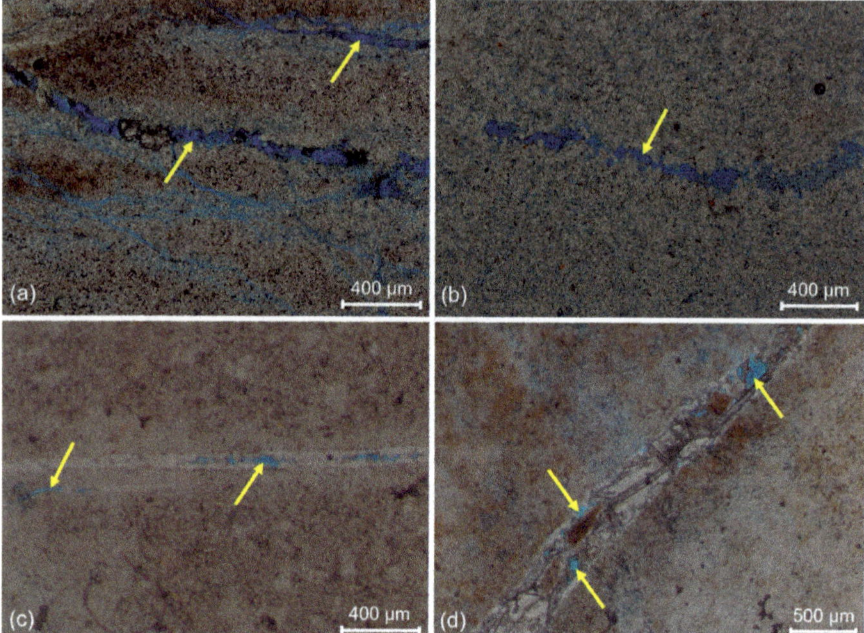

Figure 11. Dissolution fractures in thin sections. (**a**) Fractures in Well R12, depth 3105.18 m (10,187.60 ft). (**b**) Fractures in Well R12, depth 3112.35 m (10,211.12 ft). (**c**) Fractures in Well R17, depth 2764.60 m (9070.21 ft). (**d**) Fractures in Well R17, depth 2764.00 m (9068.24 ft). The directions of these thin sections are vertical to the wellbore. The minerals dissolved in these filled fractures are calcite.

4.2. Factors Influencing Fracture Development

4.2.1. Lithology

The lithology in the basement reservoirs of carbonate rocks includes dolostone, limestone, and a small amount of mudstone [39]. The development of tectonic fractures is divided into two categories: (1) only one or several sets of tectonic fractures that are developed in the rocks, and they are often regular and can be accurately measured and counted (Figure 7a); and (2) multiple sets of tectonic fractures that are interwoven into a network, which makes it difficult to measure and count each fracture, creating a fracture zone (Figure 7b). Observations and statistical analysis of different lithologies depict that the fracture zone is developed in both dolostone and limestone, but barely in mudstone (Figure 12). Based on borehole image logs, the ratio of the fracture zone thickness to the rock thickness in dolostone is 18.1%, while in limestone, it is 12.5%. Moreover, the linear density of tectonic fractures in the borehole image logs, excluding the fracture zone, was also analyzed (Figure 12). Hereof, the average linear density of tectonic fractures in dolostone is found the largest, which can reach 6.8 m^{-1}, followed by limestone, and finally mudstone, 4.1 m^{-1} and 1.3 m^{-1}, respectively. Also, in the outcrops, tectonic fractures are more developed in dolostone than limestone, while tectonic fractures in mudstone have the least development degree (Figure 13). All of these indicate that lithology controls the abundance of tectonic fractures in the target layers.

Lithology controlling fracture development is essentially due to the rock composition, particle size, and particle arrangement [61,62]. Both dolomite and calcite are the major minerals in carbonate rocks, but the Young's modulus of dolomite (8.71×10^4–14.18×10^4 MPa) is higher than calcite (5.69×10^4–8.82×10^4 MPa) [63–65]. Therefore, under the same stress conditions, dolostone with higher dolomite content is more likely to develop tectonic fractures than limestone with higher calcite content. However, the abundance of these minerals (calcite and dolomite) in the mudstone is very low, hence tectonic fractures are developed poorly. Based on thin sections, limestone is more likely to host dissolution fractures than dolostone and mudstone due to its higher calcite content [66].

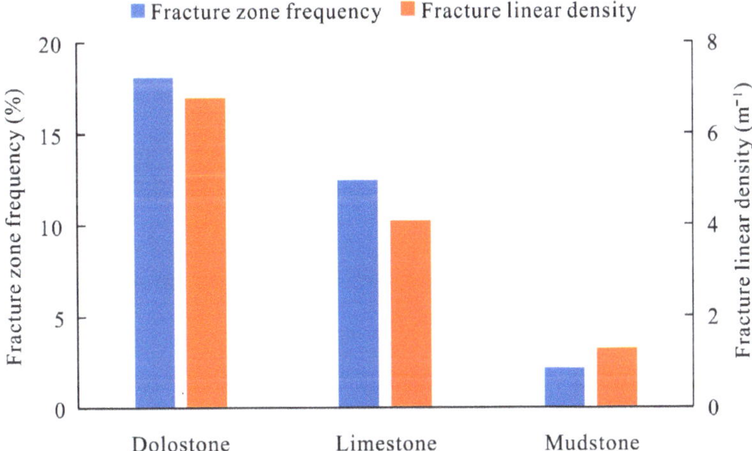

Figure 12. Schematic diagram comparing the fracture zone frequency and fracture linear densities of tectonic fractures in different lithologies. The fracture zone frequency refers to the ratio of the fracture zone thickness to the rock thickness based on borehole image logs.

Figure 13. Tectonic fractures in different lithologies of carbonate rocks in the outcrops.

4.2.2. Mechanical Stratigraphy

Mechanical stratigraphy subdivides stratified rock into discrete mechanical units consisting of one or more stratified rock units with consistent or similar rock mechanical properties such as brittleness, tensile strength, elastic stiffness, and fracture mechanics properties [49,67,68]. These mechanical units are generally, but not always, one layer with uniform lithology, which is not exactly the same as the lithologic layer. Further analysis in these basements reveals that mechanical stratigraphy which controls tectonic fractures can work in two ways: (1) the interface of the mechanical unit controls the growth and termination of tectonic fractures, and (2) the thickness of the mechanical unit controls the development degree and height of tectonic fractures.

Tectonic fractures in the outcrops mainly are developed inside the mechanical unit, and most of them cut through the entire mechanical unit and end at the interface of two separate mechanical units (Figure 14). These fractures are nearly perpendicular to or are inclined at a high angle to the interface of mechanical units. As the mechanical unit thickness increases, the height of tectonic fractures increases as well. Based on core observations, tectonic fractures only are developed in the same mechanical unit and terminated at the interface when lithological variations dictate a different mechanical unit (Figure 15). Moreover, tectonic fractures of the same set are developed at approximately equal intervals in the same mechanical unit. The outcrop observations show that in the mechanical units that are limited in thickness, the mean spacing of tectonic fractures displays a strong linear relationship with the thickness of mechanical units, which means the mean spacing of tectonic fractures increases as the thickness of mechanical units increases (Figure 16) [45,69,70]. Consequently, the stress regime has caused tectonic fractures in the thinner mechanical unit to be more frequent and with smaller heights, unlike thicker mechanical units.

Figure 14. The outcrop section shows tectonic fractures in different mechanical units. I to V refer to the number of mechanical units. The linear density of tectonic fractures is 4.3 m^{-1} in I, 10.6 m^{-1} in II, 9.3 m^{-1} in III, 3.6 m^{-1} in IV, and 6.7 m^{-1} in V.

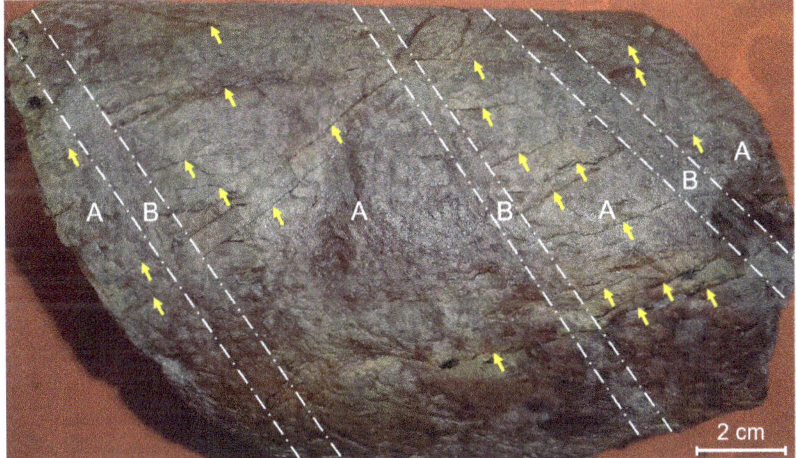

Figure 15. Tectonic fractures in the cores of Well R15, depth 2598.08 m (8523.88 ft). The fractures are mainly developed in layer A and end at the interface between separate layers A and B.

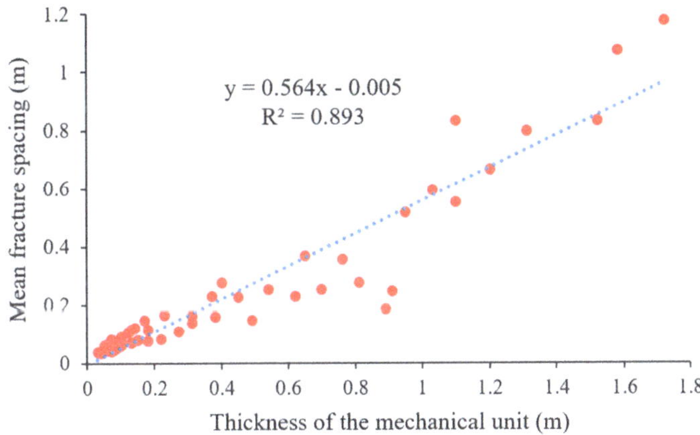

Figure 16. The relationship between the mean fracture spacing and the thickness of mechanical units from 52 data points on outcrops.

4.2.3. Fault

Since the Paleozoic, the Jizhong Sub-basin has experienced three large-scale tectonic movements: Indosinian, Yanshan, and Himalayan movements, to form a typical multi-period structural superimposed sub-basin [39]. The sub-basin mainly has caused two-set extensional fault systems of NNE-SSW and NW-SE strikes, in addition to some near E–W strikes [34,42] (Figure 17). The characteristics of these faults indicate that the faults with NNE–SSW strikes are major ones that control the orientations of deep tectonic belts, and the faults with NW–SE and near E–W strikes are the lateral faults which segment the tectonic belts and control the formation and scale of deep reservoirs [44]. Interpretations of borehole image logs explain that tectonic fractures in these reservoirs are mainly NNE–SSW, NW–SE, and near E–W strikes overall. Nevertheless, in different locations of the sub-basin, fracture directions vary notably (Figure 17). In the northeastern and southwestern regions of this sub-basin, the dominant direction of tectonic fractures is the NNE–SSW strikes, while in the central part and near the transfer zones, the dominant direction of these tectonic fractures becomes more complicated where all the three sets (NNE–SSW, NW–SE and near E–W strikes) can appear.

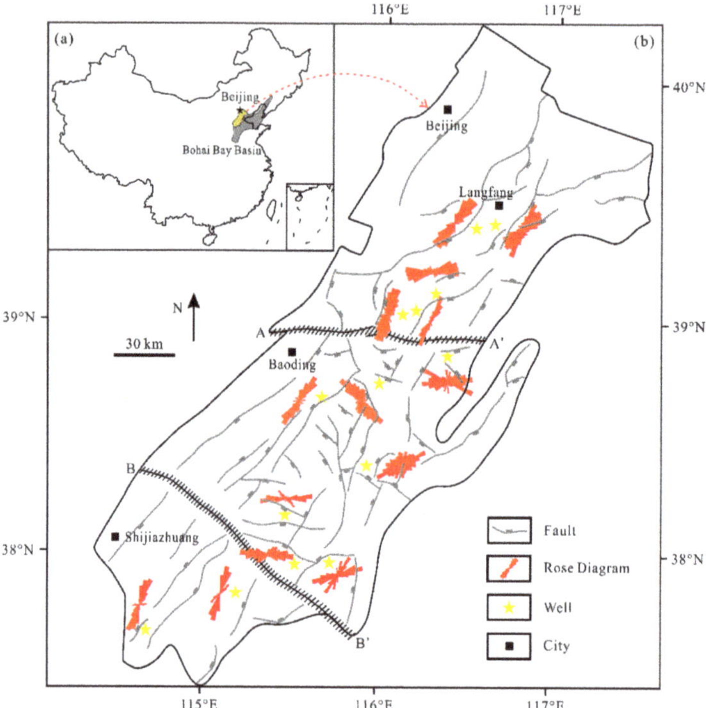

Figure 17. The relationship between fault and fracture strikes. (**a**) Location of the Jizhong Sub-Basin in the Bohai Bay Basin in China (modified from Zhao et al., 2015) [9]. (**b**) Faults and tectonic fracture orientations in the carbonate rocks of the Jizhong Sub-basin. The fault data was modified from the Huabei Oilfield database. The data regarding fracture orientations is derived from the borehole image logs of 14 wells.

The main reason is that the NNE–SSW strike faults are the only one set that is developed in the northeastern and southwestern regions of this sub-basin and all the NNE–SSW, NW–SE, and near E–W strike faults are developed in the central area of the sub-basin. The orientation of the fractures and faults is consistent, and the dominant direction of these fractures is sub-parallel to the direction of faults or oblique at a smaller angle. Therefore, one can conclude that tectonic fractures in this sub-basin are associated with faults, and these faults have a significant influence on the fracture direction.

5. Discussion

The basement reservoirs of carbonate rocks that are discovered in the Jizhong Sub-Basin in the Bohai Bay Basin are mainly dolostones and limestones from the Mesoproterozoic, Neoproterozoic, and Lower Paleozoic periods [38]. Reservoir properties determine the oil and gas accumulation in these reservoirs [17,71,72]. Initial structures and pores of carbonate rocks in these basements are mostly transformed by recrystallization, dolomitization, and tectonism, and has formed the interconnected composite reservoir system consisting of pores and fractures [18,19]. Qiao et al. (2002) studied the relationship between the porosity and burial depth of 32 carbonate basement reservoirs in the Bohai Bay Basin [73]. They found that the porosity does not decrease significantly with the increase of burial depth, while the average porosity generally varies between 5% and 6%, with a negligible change. This infers that because the height variation is not significant in these reservoirs, pores are less affected by the buried depth, and a relatively larger amount of porosity can stay intact to provide a suitable reservoir quality [71]. Moreover, a single basement reservoir has a relatively uniform pressure system and small changes in fluid properties, and the lithology distribution in the reservoir is relatively stable [9,18,19,38]. However, the above research results show that the development of fractures in these reservoirs has strong heterogeneity. Observations from outcrops, cores, borehole image logs, and thin sections show that these fractures are usually connected (Figures 4 and 6–8). We speculate that the development of opening-mode fractures improves the effectiveness of the storage space, thereby influencing the reservoir's effective permeability and oil production in these carbonate basement reservoirs.

Through a comprehensive analysis of lithology, opening-mode fractures, and well-testing data, we found that geological characteristics and oil production in the carbonate basement reservoirs vary regionally within the reservoir unit. Considering Well R10, which is drilled in a typical basement inner reservoir with a total of 6 perforated intervals at a depth of 4095–4142 m (13,435.0–13,589.2 ft), significant differences in oil production from each perforated interval is reported (Figure 18). Among them, the V and VI perforated intervals displayed higher oil production, 16.62 and 45.58 tons per day (124.30 barrels and 340.95 barrels per day), respectively, making them the main oil production section of Well R10. These intervals are followed by the II and III perforated intervals, with oil production of 4.94 tons and 6.53 tons per day (36.93 barrels and 48.88 barrels per day), respectively. Finally, perforated intervals I and IV are dry layers without any oil production. The lithologic comparison of different perforated intervals indicates that the I and IV perforated intervals without any oil production are mainly limestones, especially the I perforated interval, which is only limestones. In this regard, other perforated intervals with oil production have different degrees of dolostone content where in the V and VI perforated intervals with the largest quantities of oil production, dolostone and limestone are interstratified (Figure 18).

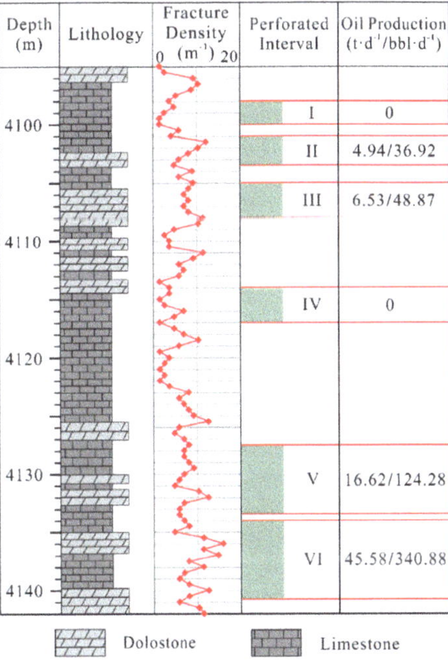

Figure 18. Schematic diagram showing the lithology combination, fracture density, perforated interval, and oil production in Well R10. Oil production refers to the daily production of oil in the well-testing stage. Lithology and fracture data were derived from the borehole image logs, and the data pertaining to perforated intervals and oil production were collected from the Huabei Oilfield database. I to VI refer to the name of perforated intervals, which correspond to the names also shown separately in Figure 19. Unit t·d⁻¹ refers to the average number of tons of oil produced per day.

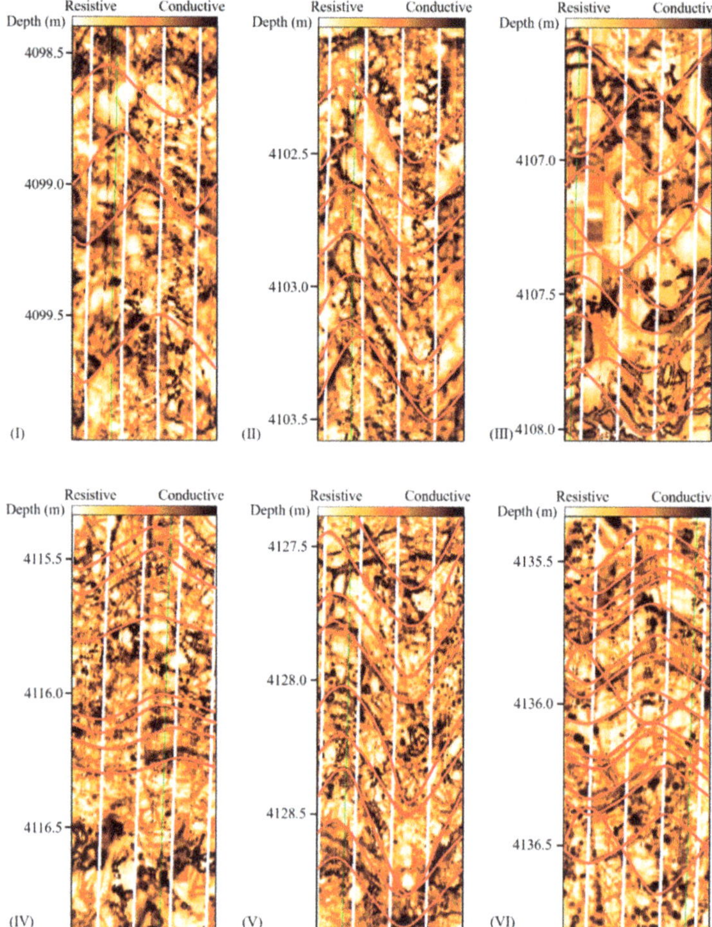

Figure 19. Tectonic fractures in the borehole image logs from Well R10. I to VI represent a part of each perforated interval, respectively. Perforated interval numbers are similar to Figure 18. Red lines represent tectonic fractures that appear as sinusoidal curves.

Fracture interpretations from borehole image logs reveal that opening-mode fractures are well developed at a depth of 4095 m to 4142 m (13,435.0 ft to 13,589.2 ft) in the Well R10, and the average fracture linear density can reach 6.78 m^{-1}. Besides, the development degree and dip angle of these fractures in different perforated intervals demonstrate a large discrepancy (Figure 19). In this aspect, comparing the oil production with the fracture linear density of each perforated intervals, a positive correlation can be found, which means higher fracture linear density can lead to better oil production (Figures 18 and 19). Considering II, III, V, and VI perforated intervals with higher oil production, opening-mode fractures are developed better, and their dip angles are usually greater than 45° (Figure 19II–VI). However, opening-mode fractures are developed less in the I and IV perforated intervals where layers are dry (Figure 19I,IV). In particular, the dip angles of these fractures in IV perforated interval are lower than 45°, and some are even near horizontal.

Geological characteristics of these perforated intervals reveal that the primary reason for different oil production is the varying degree of regional development of opening-mode fractures. Specifically, oil production grows with the increasing of fractures density (Figure 20). In the intervals where the dolostone and limestone are interstratified or dolostone makes up the primary lithology, fractures are generally more developed, thus the oil production is higher. However, in the intervals with higher quantities of limestone, the development of opening-mode fractures is relatively poor, and the oil production is lower as well. In addition, tectonic fractures are the dominant type of natural fractures in these basement reservoirs, and their effectiveness varies with the dip angles where tectonic fractures with smaller dip angles exhibit smaller aperture because of the overburden stress. This has caused their connectivity to become relatively poor, thus, their contribution becomes less to oil production [50,74,75]. Also, when the dip angle of tectonic fractures becomes larger, aperture increases and their contribution to oil production enhances. Therefore, it is deduced that tectonic fractures play a major role in the quality of basement reservoirs with carbonate rocks in the Jizhong Sub-basin, and their development has a significant impact on the oil production from these reservoirs.

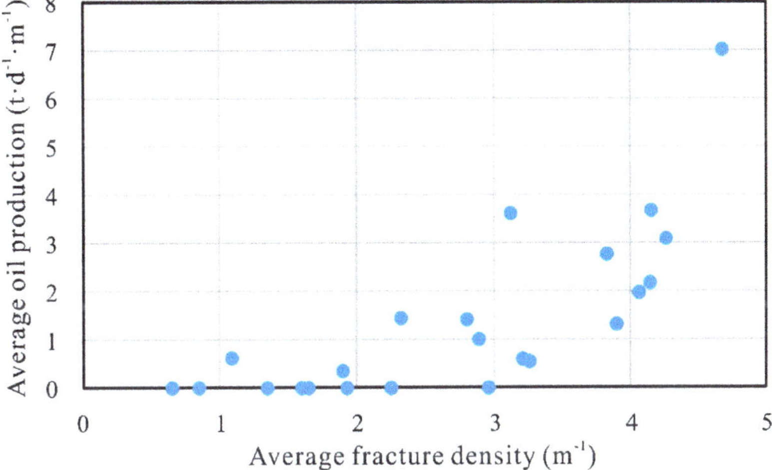

Figure 20. The relationship between the average oil production and the average fractures density in the perforated intervals. The average oil production refers to the daily production of oil in the well-testing stage in every meter perforated interval. The average fracture density is the linear density of opening-mode fractures in the perforated interval from the borehole image logs. Unit $t \cdot d^{-1} \cdot m^{-1}$ refers to the average number of tons of oil produced per day in one-meter perforated interval.

6. Conclusions

This study indicates that natural fractures are abundant in genetic types, including tectonic, pressure-solution, and dissolution ones, in the carbonate basement reservoirs in the Jizhong Sub-Basin, where tectonic fractures are the dominant type. Macroscopic fractures vary in height and length from several centimeters to meters, with a wide range of linear densities and dip angles. These fractures are found in three sets of NNE–SSW, NW–SE, and near E–W strikes. The apertures of microscopic fractures are concentrated below 60 μm, and more than 60% of them are opening-mode ones without mineral fillings.

Dolostones with higher dolomite content are more likely to develop tectonic fractures than limestones with higher calcite content, and tectonic fractures have the least development intensity in the mudstone. Most tectonic fractures are developed inside the mechanical units and end almost perpendicularly or with a higher angle to the mechanical unit interface. In a thinner mechanical unit, these fractures are more frequent and have a smaller height. The dominant orientation of fractures changes regularly and is sub-parallel to faults or oblique at a small angle.

The analysis of opening-mode fractures and well-testing data in perforated intervals reveals that oil production is related to fracture characteristics. In intervals where dolostone and limestone are interstratified or dolostone is the main lithologic composition, fractures are generally more developed, and oil production is higher, unlike intervals where limestone is the main constituent. Moreover, tectonic fractures with larger dip angles present a bigger aperture and contribute more to oil production. Finally, it is concluded that natural fractures are the main controlling factor of oil production to create favorable reservoir conditions in these basements. This study provides reference and future guidance for a better understanding of natural fractures and reservoir heterogeneity in the basement reservoirs of carbonate rocks.

Author Contributions: Conceptualization, G.L. and L.Z.; Data curation, C.H. and F.H.; Formal analysis, G.L.; Investigation, W.L. and J.Z.; Resources, C.H. and F.H.; Software, J.Z.; Supervision, L.Z.; Validation, W.L. and J.Z.; Writing—original draft, G.L.; Writing—review and editing, L.Z., M.O. and Q.W. All authors have read and agreed to the published version of the manuscript.

Funding: This research was funded by the National Major Science and Technology Projects of China, grant number 2017ZX05008-006-002-005.

Acknowledgments: The authors greatly acknowledge the Petro China Huabei Oilfield Company for providing information and other support. The authors would also like to thank the work made by Li Jian and Lyu Peng in the China University of Petroleum-Beijing, which enabled our study to proceed successfully. We are also particularly grateful to Veljko Pajovic and three anonymous reviewers for providing constructive comments and suggestions, which improve our manuscript significantly.

Conflicts of Interest: The authors declare no conflict of interest.

References

1. P'an, C.H. Petroleum in basement rocks. *AAPG Bull.* **1982**, *66*, 1597–1643.
2. Petford, N.; McCaffery, K.J.W. Hydrocarbons in Crystalline Rocks: An introduction. *Geol. Soc. Lond. Spec. Publ.* **2003**, *214*, 1–5. [CrossRef]
3. Sircar, A. Hydrocarbon production from fractured basement formations. *Curr. Sci.* **2004**, *87*, 147–151.
4. Luo, J.; Morad, S.; Liang, Z.; Zhu, Y. Controls on the quality of Archean metamorphic and Jurassic volcanic reservoir rocks from the Xinglongtai buried hill, western depression of Liaohe basin, China. *AAPG Bull.* **2005**, *89*, 1319–1346. [CrossRef]
5. Price, L.C. Thermal stability of hydrocarbons in nature: Limits, evidence, characteristics, and possible controls. *Geochem. Cosmochem. Acta* **1993**, *57*, 3261–3280. [CrossRef]
6. Gao, X.; Pang, X.; Li, X.; Chen, Z.; Shan, J.; Liu, F.; Zou, Z.; Li, W. Complex petroleum accumulating process and accumulation series in the buried-hill trend in the rift basin: An example of Xinglongtai structure trend, Liaohe subbasin. *Sci. China Ser. D Earth Sci.* **2008**, *51*, 108–116. [CrossRef]
7. Jin, Q.; Zhao, X.; Jin, F.; Ma, P.; Wang, Q.; Wang, J. Generation and accumulation of hydrocarbons in a deep "buried hill" structure in the baxian depression, bohai bay basin, eastern China. *J. Pet. Geol.* **2014**, *37*, 391–404.
8. Schubert, F.; Diamond, L.W.; Tóth, T.M. Fluid-inclusion evidence of petroleum migration through a buried metamorphic dome in the Pannonian Basin, Hungary. *Chem. Geol.* **2007**, *244*, 357–381. [CrossRef]
9. Zhao, X.; Jin, F.; Wang, Q.; Bai, G. Buried-hill play, Jizhong subbasin, Bohai Bay basin: A review and future prospectivity. *AAPG Bull.* **2015**, *99*, 1–26. [CrossRef]
10. Wang, Q.; Laubach, S.E.; Gale, J.; Ramos, M.J. Quantified fracture (joint) clustering in Archean basement, Wyoming: Application of Normalized Correlation Count method. *Pet. Geosci.* **2019**, *25*, 415–428. [CrossRef]
11. Trice, R. Basement exploration, West of Shetlands: Progress in opening a new play on the UKCS. *Geol. Soc. Lond. Spec. Publ.* **2014**, *397*, 81–105. [CrossRef]
12. Parnell, J.; Baba, M.; Bowden, S.; Muirhead, D. Subsurface biodegradation of crude oil in a fractured basement reservoir, Shropshire, UK. *J. Geol. Soc.* **2017**, *174*, 655–666. [CrossRef]
13. Cuong, T.X.; Warren, J.K. Bach Ho field, a fractured granitic basement reservoir, Cuu Long basin, offshore Se Vietnam: A "Buried-hill" play. *J. Pet. Geol.* **2009**, *32*, 129–156. [CrossRef]
14. Jin, Q.; Mao, J.; Du, Y.; Huang, X. Fracture filling mechanisms in the carbonate buried-hill of Futai Oilfield in Bohai Bay Basin, East China. *Pet. Explor. Dev.* **2015**, *42*, 497–506. [CrossRef]
15. Zang, M.; Wu, K.; Cui, Y.; Du, W. Types of buried hill and its hydrocarbon accumulation in Jizhong Depression. *J. Oil Gas Technol.* **2009**, *31*, 166–169.

16. Deng, J.; Shi, H.; Wang, B.; Wang, G.; Chen, G. Characteristics and main controlling factors of Paleozoic carbonate buried-hill gas reservoirs in Bozhong Sag. *Pet. Geol. Oilfield Dev. Daqing* **2015**, *34*, 15–20.

17. Kosa, E.; Hunt, D.W. Heterogeneity in Fill and Properties of Karst-Modified Syndepositional Faults and Fractures: Upper Permian Capitan Platform, New Mexico, U.S.A. *J. Sediment. Res.* **2006**, *76*, 131–151. [CrossRef]

18. Wang, J.; Zhao, L.; Zhang, X.; Yang, Z.; Cao, H.; Chen, L.; Shan, F.; Liu, M. Buried hill karst reservoirs and their controls on productivity. *Pet. Explor. Dev.* **2015**, *42*, 852–860. [CrossRef]

19. Liu, N.; Qiu, N.; Chang, J.; Shen, F.; Wu, H. Hydrocarbon migration and accumulation of the Suqiao buried-hill zone in Wen'an Slope, Jizhong Subbasin, Bohai Bay Basin, China. *Mar. Pet. Geol.* **2017**, *86*, 512–525. [CrossRef]

20. Eig, K.; Bergh, S.G. Late Cretaceous-Cenozoic fracturing in Lofoten, North Norway: Tectonic significance, fracture mechanisms and controlling factors. *Tectonophysics* **2011**, *499*, 190–205. [CrossRef]

21. Joudaki, M.; Farzipour-Saein, A.; Nilfouroushan, F. Kinematics and surface fracture pattern of the Anaran basement fault zone in NW of the Zagros fold-thrust belt. *Int. J. Earth. Sci.* **2016**, *105*, 869–883. [CrossRef]

22. Gong, L.; Fu, X.; Gao, S.; Zhao, P.; Luo, Q.; Zeng, L.; Yue, W.; Zhang, B.; Liu, B. Characterization and prediction of complex natural fractures in the tight conglomerate reservoirs: A fractal method. *Energies* **2018**, *11*, 2311. [CrossRef]

23. Loza Espejel, R.; Alves, T.M.; Blenkinsop, T.G. Distribution and growth styles of isolated carbonate platforms as a function of fault propagation. *Mar. Pet. Geol.* **2019**, *107*, 484–507. [CrossRef]

24. Ding, W.; Zhu, D.; Cai, J.; Gong, M.; Chen, F. Analysis of the developmental characteristics and major regulating factors of fractures in marine-continental transitional shale-gas reservoirs: A case study of the Carboniferous-Permian strata in the southeastern Ordos Basin, central China. *Mar. Pet. Geol.* **2013**, *45*, 121–133. [CrossRef]

25. Nabawy, B.S. Estimating porosity and permeability using Digital Image Analysis (DIA) technique for highly porous sandstones. *Arab. J. Geosci.* **2014**, *7*, 889–898. [CrossRef]

26. Guo, P.; Lei, L.; Ren, D. Simulation of three-dimensional tectonic stress fields and quantitative prediction of tectonic fracture within the Damintun Depression, Liaohe Basin, northeast China. *J. Struct. Geol.* **2016**, *86*, 211–223. [CrossRef]

27. Milad, B.; Slatt, R. Impact of lithofacies variations and structural changes on natural fracture distributions. *Interpretation* **2018**, *6*, T873–T887. [CrossRef]

28. Milad, B.; Ghosh, S.; Slatt, R.; Marfurt, K.; Fahes, M. Practical aspects of upscaling geocellular geological models for reservoir fluid flow simulations: A case study in integrating geology, geophysics, and petroleum engineering multiscale data from the Hunton Group. *Energies* **2020**, *13*, 1604. [CrossRef]

29. Gao, C.; Zhang, X.; Zha, M.; Wang, X. Characteristics of buried hill reservoir in Jizhong Depression. *Lithol. Reserv.* **2011**, *23*, 6–12.

30. Ju, W.; Wu, C.F.; Wang, K.; Sun, W.F.; Li, C.; Chang, X.X. Prediction of tectonic fractures in low permeability sandstone reservoirs: A case study of the Es-3(m) reservoir in the Block Shishen 100 and adjacent regions, Dongying Depression. *J. Pet. Sci. Eng.* **2017**, *156*, 884–895. [CrossRef]

31. Zhao, L.; Jiang, Y.; Liu, H.; Fan, B. Thermal evolution of Paleogene source rocks and relationship with reservoir distribution in Raoyang Sag, Bohai Bay Basin. *Pet. Geol. Recover. Effic.* **2012**, *19*, 1–4.

32. Wu, X.; Niu, J.; Wu, F.; Liu, X. Major control factors of hydrocarbon accumulation in Ordovician interior buried hills, Bohaiwan Basin. *Mar. Orig. Pet. Geol.* **2013**, *18*, 1–12.

33. Yang, M.; Liu, C.; Yang, B.; Zhao, H. Extensional structures of the Paleogene in the central Hebei Basin, China. *Geol. Rev.* **2002**, *48*, 58–67.

34. Su, L.; Luo, P.; Zou, H.; Shi, B.; Zheng, X. Hydrocarbon accumulation conditions of the Ordovician buried hills in the slope zone of the Jizhong Depression. *Geotecton. Metallog.* **2003**, *27*, 191–196.

35. Chang, J.; Qiu, N.; Zhao, X.; Xu, W.; Xu, Q.; Jin, F.; Han, C.; Ma, X.; Dong, X.; Liang, X. Present-day geothermal regime of the Jizhong Depression in Bohai Bay Basin, East China. *Chin. J. Geophys.* **2016**, *59*, 1003–1016.

36. He, D.F.; Cui, Y.Q.; Zhang, Y.Y.; Shan, S.Q.; Xiao, Y.; Zhang, C.B.; Zhou, C.A.; Gao, Y. Structural genetic types of paleoburied hill in Jizhong depression, Bohai Bay Basin. *Acta Petrol. Sin.* **2017**, *33*, 1338–1356.

37. Wu, X.; Lv, Y.; Tian, J.; Guo, Y. Characteristics and evaluation of carbonate buried hill inner cap rock in Jizhong Depression. *Lithol. Reserv.* **2011**, *23*, 49–54.

38. Zhang, W.; Yang, D.; Chen, Y.; Qian, Z.; Zhang, C.; Liu, H. Sedimentary Structural Characteristics and Hydrocarbon Distributed Rules of Jizhong Depression. *Acta Geol. Sin.* **2008**, *82*, 1103–1112.

39. Xiao, Y.; Liu, G.; Han, C.; Zhu, J.; Zhou, C.; Lv, W.; Gao, Y.; Zeng, L.; Ma, X. Development characteristics and main controlling factors of natural fractures in deep carbonate reservoirs in the Jizhong Depression. *Nat. Gas Ind.* **2018**, *38*, 33–42.

40. Du, J.; Zou, W.; Fei, B.; Lei, H.; Zhang, F.; Zhang, Y. *Buried Hill Oil and Gas Complex Zone in Jizhong Sub-Basin*; Petroleum Industry Press: Beijing, China, 2002.

41. Yang, K.; Dang, C.; Dai, F. Paleo-source/paleo-reservoir typed reservoirs: Cases of anticlinal paleo-reservoir and buried hill reservoirs in Bohaiwan Basin region. *Mar. Orig. Pet. Geol.* **2007**, *12*, 27–32.

42. Gao, X.; Wu, W.; Lu, X.; Cui, Z.; Kong, L.; Jia, L.; Wang, H. Multiplicity of hydrocarbon reservoir and accumulation controlling factors within buried hills in Jizhong depression. *J. China Univ. Pet.* **2011**, *35*, 31–35.

43. Yu, H.; Wang, D.; Niu, C.; Li, L. Characteristics and formation mechanisms of buried hill carbonate reservoirs in Bonan Low Uplift, Bohai Bay. *Pet. Geol. Exp.* **2015**, *37*, 150–156.

44. Chen, Q.; Lao, H.; Wu, K.; Wu, Z.; Cui, Y. Favorable hydrocarbon accumulation conditions for carbonate reservoirs in deep-buried hills in the Jizhong Depression, Bohai Bay Basin. *Nat. Gas Ind.* **2013**, *33*, 32–39.

45. Cilona, A.; Aydin, A.; Likerman, J.; Parker, B.; Cherry, J. Structural and statistical characterization of joints and multi-scale faults in an alternating sandstone and shale turbidite sequence at the Santa Susana Field Laboratory: Implications for their effects on groundwater flow and contaminant transport. *J. Struct. Geol.* **2016**, *85*, 95–114. [CrossRef]

46. Lai, J.; Li, D.; Wang, G.; Xiao, C.; Hao, X.; Luo, Q.; Lai, L.; Qin, Z. Earth stress and reservoir quality evaluation in high and steep structure: The Lower Cretaceous in the Kuqa Depression, Tarim Basin, China. *Mar. Pet. Geol.* **2019**, *101*, 43–54. [CrossRef]

47. Zeng, L.; Su, H.; Tang, X.; Peng, Y.; Gong, L. Fractured tight sandstone oil and gas reservoirs: A new play type in the Dongpu depression, Bohai Bay Basin, China. *AAPG Bull.* **2013**, *97*, 363–377. [CrossRef]

48. Milad, B.; Ghosh, S.; Suliman, M.; Slatt, R. Upscaled DFN Models to Understand the Effects of Natural Fracture Properties on Fluid Flow in the Hunton Group Tight Limestone. In Proceedings of the Unconventional Resources Technology Conference (URTeC), Houston, TX, USA, 23–25 July 2018.

49. Dashti, R.; Rahimpour-Bonaba, H.; Zeinali, M. Fracture and mechanical stratigraphy in naturally fractured carbonate reservoirs-A case study from Zagros region. *Mar. Pet. Geol.* **2018**, *97*, 466–479. [CrossRef]

50. Lyu, W.; Zeng, L.; Zhang, B.; Miao, F.; Lyu, P.; Dong, S. Influence of natural fractures on gas accumulation in the Upper Triassic tight gas sandstones in the northwestern Sichuan Basin, China. *Mar. Pet. Geol.* **2017**, *83*, 60–72. [CrossRef]

51. Liu, G.P.; Zeng, L.B.; Li, H.N.; Mehdi, O.; Minou, R. Natural fractures in metamorphic basement reservoirs in the Liaohe Basin, China. *Mar. Pet. Geol.* **2020**, *119*, 1–15. [CrossRef]

52. Ju, W.; Hou, G.; Feng, S.; Zhao, W.; Zhang, J.; You, Y.; Zhan, Y.; Yu, X. Quantitative prediction of the Yanchang Formation Chang 63 reservoir tectonic fracture in the Qingcheng-Heshui Area, Ordos Basin. *Earth Sci. Front.* **2014**, *21*, 310–320.

53. Liu, G.P.; Zeng, L.B.; Sun, G.; Zu, K.; Qin, L.; Mao, Z.; Mehdi, O. Natural fractures in tight gas volcanic reservoirs and their influences on production in the Xujiaweizi depression, Songliao Basin, China. *AAPG Bull.* **2020**. Available online: http://archives.datapages.com/data/bulletns/aop/2020-06-22/aapgbltn17169aop.html (accessed on 10 August 2020).

54. Gale, J.F.W.; Lander, R.H.; Reed, R.M.; Laubach, S.E. Modeling fracture porosity evolution in dolostone. *J. Struct. Geol.* **2010**, *32*, 1201–1211. [CrossRef]

55. Li, J.; Zeng, L.; Li, W.; Zhang, Y.; Cai, Z. Controls of the Himalayan deformation on hydrocarbon accumulation in the western Qaidam Basin, Northwest China. China. *J. Asian Earth Sci.* **2019**, *174*, 294–310. [CrossRef]

56. Dunham, J.B.; Larter, S. Association of Stylolitic Carbonates and Organic Matter: Implications for Temperature Control on Stylolite Formation. *AAPG Bull.* **1981**, *65*, 922.

57. Zeng, L.; Tang, X.; Wang, T.; Gong, L. The influence of fracture cements in tight Paleogene saline lacustrine carbonate reservoirs, western Qaidam Basin, northwest China. *AAPG Bull.* **2012**, *96*, 2003–2017. [CrossRef]

58. Dijk, P.E.; Berkowitz, B.; Yechieli, Y. Measurement and analysis of dissolution patterns in rock fractures. *Water Resour. Res.* **2002**, *38*, 1–12. [CrossRef]

59. Chaudhuri, A.; Rajaram, H.; Viswanathan, H. Alteration of fractures by precipitation and dissolution in gradient reaction environments: Computational results and stochastic analysis. *Water Resour. Res.* **2008**, *44*, 1–19. [CrossRef]

60. Rustichelli, A.; Tondi, E.; Agosta, F.; Cilona, A.; Giorgioni, M. Development and distribution of bed-parallel compaction bands and pressure solution seams in carbonates (Bolognano Formation, Majella Mountain, Italy). *J. Struct. Geol.* **2012**, *37*, 181–199. [CrossRef]

61. Göktan, R.M. Brittleness and micro-scale rock cutting efficiency. *Min. Sci. Technol.* **1991**, *13*, 237–241. [CrossRef]

62. Lorenz, J.C.; Sterling, J.L.; Schechter, D.S.; Whigham, C.L.; Jensen, J.L. Natural fractures in the Spraberry Formation, Midland Basin, Texas: The effects of mechanical stratigraphy on fracture variability and reservoir behavior. *AAPG Bull.* **2002**, *86*, 505–524.

63. Mavko, G.; Mukerji, T.; Dvorkin, J. *Elasticity and Hooke's Law in the Rock Physics Handbook*; Cambridge University Press: Cambridge, UK, 2009; pp. 169–228.

64. Zhang, C.; Dong, D.; Wang, Y.; Guan, Q. Brittleness evaluation of the Upper Ordovician Wufeng-Lower Silurian Longmaxi shale in southern Sichuan Basin, China. *Energy Explor. Exploit.* **2017**, *35*, 430–443. [CrossRef]

65. Bakri, Z.; Zaoui, A. Structural and mechanical properties of dolomite rock under high pressure conditions: A first-principles study. *Phys. Status Solid.* **2011**, *248*, 1894–1900. [CrossRef]

66. Zhao, X.; Hu, X.; Xiao, K.; Jia, Y. Characteristics and major control factors of natural fractures in carbonate reservoirs of Leikoupo Formation in Pengzhou area, western Sichuan Basin. *Oil Gas Geol.* **2018**, *39*, 30–39.

67. Laubach, E.S.; Olson, E.J.; Gross, R.M. Mechanical and fracture stratigraphy. *AAPG Bull.* **2009**, *93*, 1413–1426. [CrossRef]

68. McGinnis, N.R.; Ferrill, A.D.; Morris, P.A.; Smart, J.K. Mechanical stratigraphic controls on natural fracture spacing and penetration. *J. Struct. Geol.* **2017**, *95*, 160–170. [CrossRef]

69. Gillespie, P.A.; Howard, C.B.; Walsh, J.J.; Watterson, J. Measurement and characterization of spatial distributions of fractures. *Tectonophysics* **1993**, *226*, 113–141. [CrossRef]

70. Gross, M.R.; Fischer, M.P.; Engelder, T.; Greenfield, R.J. Factors controlling joint spacing in interbedded sedimentary rocks: Integrating numerical models with field observations from the Monterey formation, USA. *Geol. Soc. Lond. Spec. Publ.* **1995**, *92*, 215–233. [CrossRef]

71. Zhao, X.; Wang, Q.; Jin, F.; Wang, H.; Luo, J.; Zeng, J.; Fan, B. Main controlling factors and exploration practice of subtle buried hill hydrocarbon reservoir in Jizhong depression. *Acta Pet. Sin.* **2012**, *33*, 71–79.

72. Gong, L.; Gao, S.; Fu, X.; Chen, S.; Lvu, B.; Yao, J. Fracture characteristics and their effects on hydrocarbon migration and accumulation in tight volcanic reservoirs: A case study of the Xujiaweizi fault depression, Songliao Basin, China. *Interpretation* **2017**, *5*, 57–70. [CrossRef]

73. Qiao, H.; Fang, C.; Niu, J.; Guan, D. *Petroleum Geology of Deep Horizon in Bohaiwan Basin*; Petroleum Industry Press: Beijing, China, 2002.

74. Mattila, J.; Tammisto, E. Stress-controlled fluid flow in fractures at the site of a potential nuclear waste repository, Finland. *Geology* **2012**, *40*, 299–302. [CrossRef]

75. Zeng, L.; Li, X. Fractures in sandstone reservoirs with ultra-low permeability: A case study of the Upper Triassic Yanchang Formation in the Ordos Basin, China. *AAPG Bull.* **2009**, *93*, 461–477.

Article

Multi-Phase Tectonic Movements and Their Controls on Coalbed Methane: A Case Study of No. 9 Coal Seam from Eastern Yunnan, SW China

Ming Li [1,2], Bo Jiang [1,2,*], Qi Miao [2,3], Geoff Wang [4], Zhenjiang You [4] and Fengjuan Lan [1,2]

[1] Key Laboratory of Coalbed Methane Resources and Reservoir Formation Process, Ministry of Education (China University of Mining and Technology), Xuzhou 221116, China; cumtmingli@cumt.edu.cn (M.L.); lanfj1986@cumt.edu.cn (F.L.)

[2] School of Resources and Earth Science, China University of Mining and Technology, Xuzhou 221116, China; miaoqi198@cumt.edu.cn

[3] No. 198 Coalfield Geology Prospecting Team of Yunnan Province, Kunming 650208, China

[4] School of Chemical Engineering, The University of Queensland, Brisbane, QLD 4072, Australia; gxwang@uq.edu.au (G.W.); z.you@uq.edu.au (Z.Y.)

* Correspondence: jiangbo@cumt.edu.cn; Tel.: +86-139-5135-9055

Received: 22 October 2020; Accepted: 16 November 2020; Published: 17 November 2020

Abstract: Multi-phase tectonic movements and complex geological structures limit the exploration and hotspot prediction of coalbed methane (CBM) in structurally complex areas. This scientific problem is still not fully understood, particularly in the Bumu region, Southwest China. The present paper analyses the occurrence characteristics and distribution of CBM based on the comprehensive analysis of CBM data. In combination with the analysis of the regional tectonics setting, geological structure features and tectonic evolution. The control action of multi-phase tectonic movements on CBM occurrence are further discussed. Results show that the Indosinian local deformation, Yanshanian intense deformation, and Himalayan secondary derived deformation formed the current tectonic framework of Enhong synclinorium. The intense tectonic compression and dextral shear action in the Yanshanian and Himalayan movements caused the complex geological structures in Bumu region, composed of the Enhong syncline, associated reverse faults and late derived normal fault. The CBM distribution is complex, which has the central and western NNE-trending high gas content zones along the syncline hinge zone and the reverse faults. The geological structure controls on CBM enrichment are definite and important. Based on geological structure features and responses of gas content, methane concentration, and gas content gradient, the gas controlling patterns of geological structure are determined and can be classified into five types: the reverse fault sealing, syncline sealing, monoclinal enrichment, normal fault dispersion, and buried floor fault dispersion types. The structural compression above the neutral surface plays an important role in the syncline sealing process, which is indicated by an increase in gas content gradient. The EW-trending tectonic intense compression and dextral shear action in the Himalayan movement avoided the negative inversion of NNE-trending Yanshanian compressive structure and its destruction of CBM reservoir. However, the chronic uplift and derived normal fault during Himalayan period caused the constant dissipation of CBM.

Keywords: coalbed methane; geological structure; gas controlling pattern; neutral surface; tectonic movement; Bumu region

---

1. Introduction

Coalbed methane (CBM) resources are abundant in China, where early-stage large-scale CBM development is undertaken actively [1]. However, the complex geological settings of CBM need further investigation in order to achieve large-scale commercial production efficiently [1,2]. The coal-bearing basins in China experienced multi-phase tectonic movements in geologic history and developed complex geological structures [3]. Geological structure plays an important role in the process of CBM generation, occurrence, migration, and enrichment in coal seams [4–11]. Especially, the influence on CBM occurrence from the geological structure modification of the coal reservoir physical properties, such as porosity, adsorption, and permeability [12–20]. Research has been focused on the variation of gas content with geological structure; generally, fold and fault. CBM resources within depth less than 2000 m in China are estimated to be 36.81 trillion m^3, of which more than 84% occur in nine large-scale basins, such as the Junggar, Tianshan, Tuha, Santanghu, Ordos, Qinshui, Erenhot, Hailar, and western Guizhou–eastern Yunnan basins [1]. Most of these basins exhibit the syncline or synclinorium structure. Syncline structure is a favorable area in terms of CBM reservoir formation, mainly due to the increase in coal seam buried depth and gas content [7,21]. On the other hand, the compression above the neutral surface and the extension below the neutral surface of the syncline are also important for the CBM occurrence [22–24]. However, there is still a lack of effective methods for determining the neutral surface and its effects on CBM. The influence of fault structure on CBM is generally summarized as gas contents are usually higher near the reverse faults and lower adjacent to the normal faults [1,25]. The normal faults can also increase the gas content in closed and occluded types [9]. The early faults are mainly modified and reversed during the late tectonic movement, especially the early compression and late extension. The evolution of fault and the effects of geological structure on other gas-bearing characteristics, such as the gas composition, gas pressure, and gas content gradient, are still not well understood.

The Enhong synclinorium is located in the eastern Yunnan region, Southwest China, with abundant coal resources (Figure 1) [26]. The coal-bearing stratum is the Upper Permian Changxing Formation (P_3c) and Longtan Formation (P_3l). The estimated coal-bearing area is about 485 km^2. The geological coal reserves with burial depth less than 2000 m is 5.25×10^9 tons, which makes this synclinorium one of the largest coking coal producing areas in China [27]. The amount of CBM resources with burial depth less than 2000 m is 6.13×10^{10} m^3, and 82% of the volume is located at a depth less than 1000 m, showing a good prospect for CBM exploitation [28]. 12 CBM wells have been drilled in the research area. However, the Enhong synclinorium is characterized by the complex geological structure, which is composed of various scales and orientations of dense folds and faults. The complex structure restricts the exploration and commercial exploitation of CBM. Considering the complexity of the geological structure, as well as the insufficient and uneven CBM data for the whole region, this work takes the Bumu region as the research area because of the rich geological and CBM data available, which is located in the central part of the Enhong synclinorium.

This article presents the occurrence characteristics and distribution of CBM based on the comprehensive analysis of those CBM data. Multi-phase tectonic movements in structurally complex CBM fields are determined according to the analysis of the regional tectonics setting, geological structure features and structural evolution. The gas controlling patterns of geological structure are jointly proposed by the multiple CBM parameters analysis. The multi-phase tectonic movements controls on CBM occurrence are further discussed. The results of this study provide new insights and improved understanding of the influence of fold neutral surface and fault sealing on CBM, which may have significant implications for CBM exploitation in structurally complex areas.

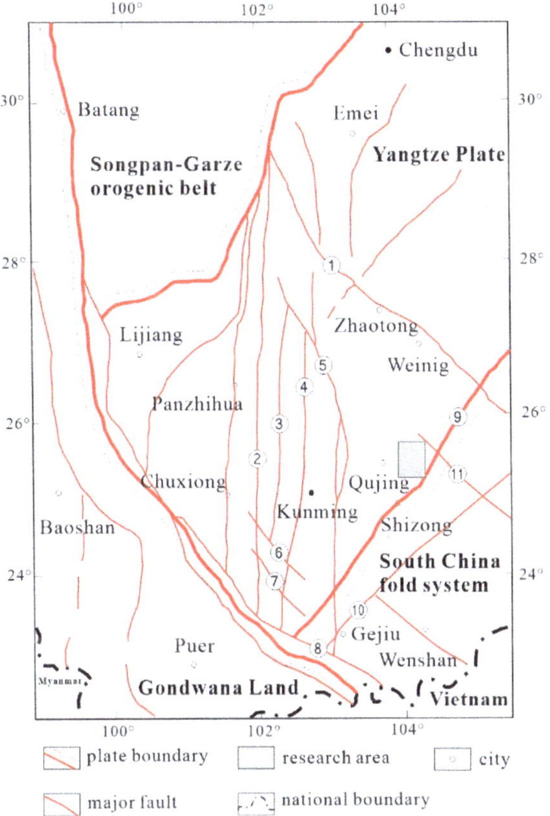

Figure 1. Tectonic location of the study area.1-Weining-Ziyun-Nandan fault, 2-Anninghe-Lvzhijiang fault, 3-Luoci-Yimen fault, 4-Puduhe-Dianchi fault, 5-Xiaojiang fault, 6-E'shan-Tonghai fault, 7-Huanian fault, 8-Red river fault, 9-Mi'le-Shizong fault, 10-Nanpanjiang fault, 11-Youjiang fault.

2. Geological Setting

2.1. Regional Structure

The research area is tectonically located in the southwestern margin of the Yangtze plate (Figure 1). It occurs at the boundary between the Tethys-Himalaya tectonic domain and the circum Pacific tectonic domain [29,30]. This tectonic unit is bordered by the NE-trending Mi'le-Shizong fault in the south, the NS-trending Xiaojiang fault in the west and the NW-trending Youjiang fault in the northeast [31,32].

The Bumu region is located in the central part of the Enhong synclinorium, which is the main geological structure in this area (Figure 2). The Enhong synclinorium, with a length of 53 km and width of 6–12 km, is located between the NE-trending Fuyuan-Mi'le fault and the NS-trending Pingguan-Agang fault [33]. The Enhong synclinorium is mainly composed of Enhong syncline and many secondary anticlines and synclines, which extend in the direction of NE, NNE, NS, and NNW from the southwest to the northeast of the Enhong synclinorium. The deformation strength of the fold decreases gradually from west to east and from north to south, with the stratigraphic dip angle of about 10–25°. Meanwhile, the NE-plunging Enhong synclinorium is strongly destroyed by numerous multilevel multidirectional faults (Figure 2). The secondary derived drag folds, derived faults, and associated faults developed well in the northwestern limb of Enhong synclinorium. The derived brachy drag folds in the south are mainly arranged in echelons to the right row, and the axial traces form an acute angle with the Fuyuan-Mi'le fault, indicating that the deformation mechanism of the folds is derived from the right-lateral slipping of the Fuyuan-Mi'le fault (Figure 2). The NE-trending secondary associated folds are strongly deformed and extend near the fault zone of Fuyuan-Mi'le fault. The secondary faults in the northwestern limb are mostly the NNE-trending derived faults and the NE-trending associated branch faults of the Fuyuan-Mi'le fault.

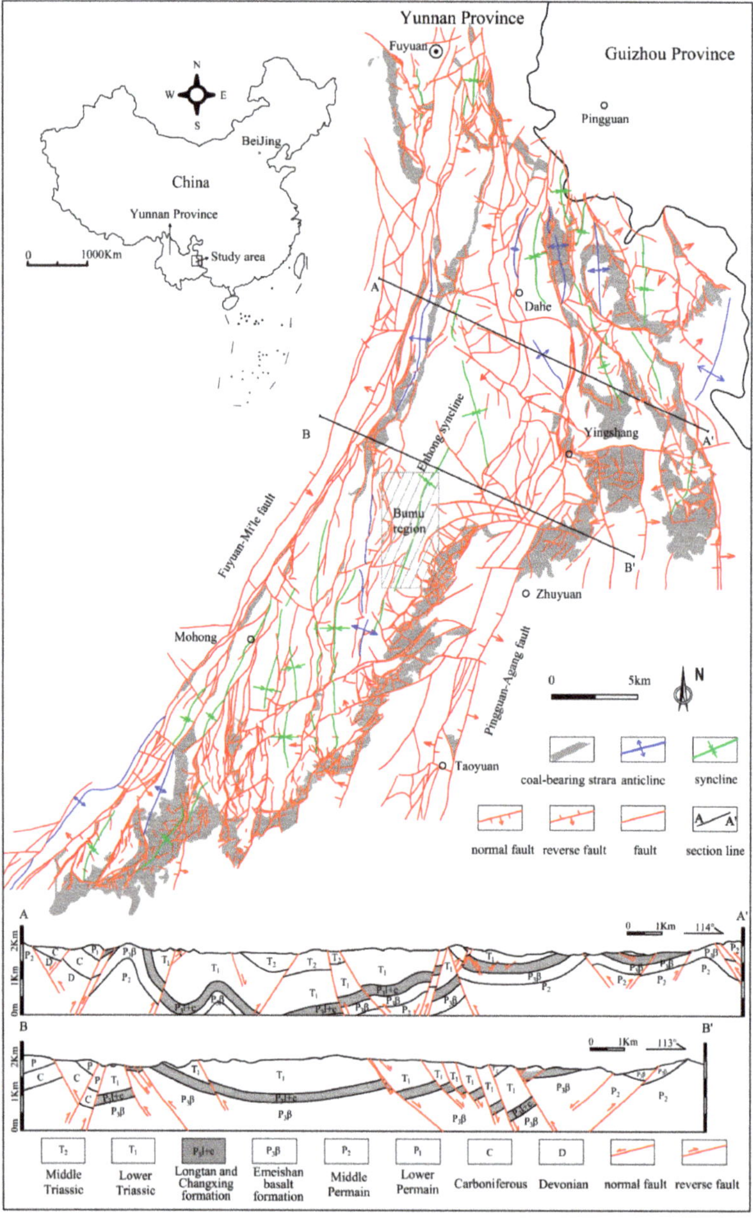

Figure 2. Geological sketch map and structural cross section of Enhong synclinorium.

The NS- and NNW-trending folds developed well in the north part of the southeastern limb of the Enhong synclinorium (Figure 2). The secondary faults in the southeastern limb are commonly the NS-trending associated faults and the NE-, EW-, and NW-trending strike-slip normal derived faults, which are mostly arranged as an echelon pattern and are restricted by the Pingguan-Agang fault at a large angle. Especially the arcuate fault zone, composed by the right-lateral strike-slip normal faults, is formed near the eastern part of the Bumu region. Additionally, the latter small oblique strike-slip faults, mainly the NNW-trending left-lateral strike-slip faults and NEE-trending right-lateral strike-slip faults, cut across the major structure and complicate geological structure.

2.2. Coal-Bearing Strata and Coal Seam

The basement of the Yangtze plate in this area is the Mesoproterozoic Kunyang Group. The sedimentary stratum are Siluric, Devonian, Carboniferous, Permian, Triassic, Paleogene, Neogene, and Quaternary [29]. The exposure strata in the research area are mainly the limestone of Yongningzhen Formation (T_1y) and the siltstone and sandstone of Feixianguan Formation (T_1f). The buried strata, exposed by drilling, are Kayitou Formation (T_1k), Changxing Formation (P_3c), Longtan Formation (P_3l) and Emeishan basalt Formation ($P_3\beta$) (Figure 3a). The Changxing Formation (P_3c) and Longtan Formation (P_3l) are the major coal-bearing strata. They were measured between 214.33 m to 278.55 m in thickness and consist of 28 to 46 coal seams (Figure 3a). The total thickness of the coal seams ranges from 20.23 m to 35.54 m with an average thickness of 30.22 m. The minable coal seams are the coal seams Nos. 4, 7–9, 12, 13, 15, 16, and 18–24, with an average thickness of 19.57 m, and the $R_{o,max}$ ranges from 1.07% to 1.15% (Table 1). The thickness, coal quality, and lateral stability of coal seams in the upper segment of Longtan Formation (P_3l^2) are favorable for the exploitation of coal and CBM. The No. 9 coal seam, with a thickness of 1.22–5.80 m and an average thickness of 3.28 m, is stable and the unique regional minable coal seam in the research area.

Table 1. Characteristic and gas content of main coal seam in the Bumu region.

Stratum	Coal Seam	Thickness (m)	$R_{o,max}$ (%)	Coal Maceral (%)			V_{daf} (%)	Gas Composition (%)			Gas Content (m^3/t)
				V	I	L		CH$_4$	C$_{2+}$	N$_2$	
P_3c	4	1.06	1.08	67.9	9.4	22.7	30.92	63.33	5.31	30.07	9.25
P_3l^2	7	1.18	–	71.0	6.7	22.3	28.85	66.70	9.60	22.07	11.06
P_3l^2	8	0.90	1.09	69.0	5.0	26.0	28.40	–	–	–	–
P_3l^2	9	3.28	1.10	64.9	10.9	24.2	29.74	66.93	9.54	22.72	11.21
P_3l^2	12	1.75	1.08	66.4	9.8	23.8	27.71	59.91	13.48	24.65	9.15
P_3l^2	13	0.73	–	66.3	9.6	24.1	25.78	71.33	7.46	20.19	13.17
P_3l^2	15	1.64	1.07	66.2	8.7	25.1	27.37	65.86	11.73	20.95	14.00
P_3l^1	16	1.43	1.08	66.6	9.5	23.9	26.89	67.92	10.98	19.17	10.64
P_3l^1	18	0.87	1.09	66.6	11.1	22.3	26.96	65.74	10.83	21.48	13.52
P_3l^1	19	0.98	1.08	64.8	14.2	21.0	25.35	–	–	–	–
P_3l^1	20	0.83	1.07	68.2	10.1	21.7	25.53	71.32	12.37	15.03	12.34
P_3l^1	21	1.40	1.11	63.1	11.2	25.7	26.41	73.93	11.12	13.50	12.31
P_3l^1	22	1.32	–	68.1	7.9	24.0	26.19	71.85	14.75	12.45	12.58
P_3l^1	23	1.26	1.15	62.6	12.7	24.7	26.41	46.57	3.63	48.52	7.94
P_3l^1	24	1.11	1.09	67.2	12.5	20.3	26.50	–	–	–	–

Note: $R_{o,max}$—maximum vitrinite reflectance, V_{daf}—volatile yield (dry ash-free base), *V*—vitrinite, *I*—inertinite, *L*—liptinite.

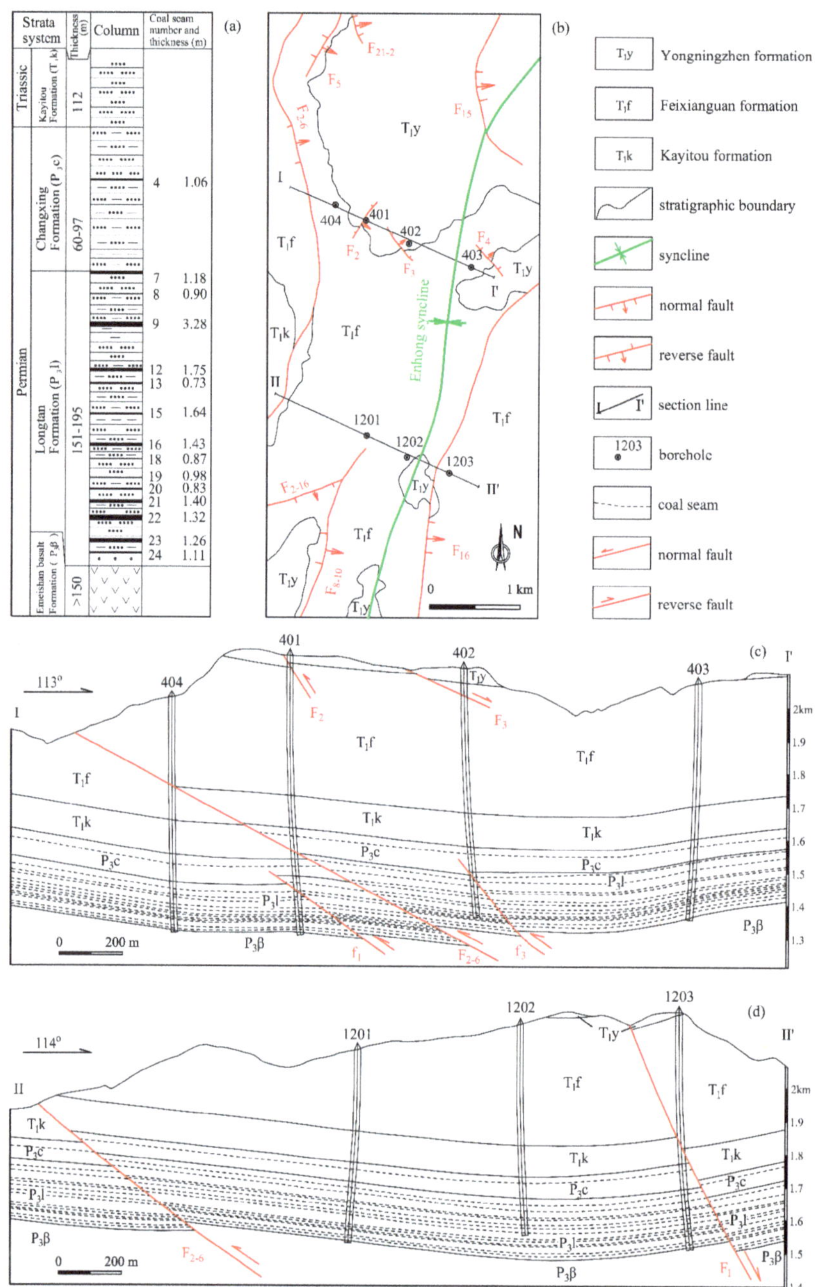

Figure 3. Stratigraphic column of coal-bearing strata (**a**), structural outline map (**b**) and structural cross sections (**c,d**) of Bumu region.

3. Material and Experimental Methodology

The geological structures and structural features in Bumu region were analyzed and determined according to the field and drilling geological surveys. The structural features, distribution law, structural pattern, and mechanical causes of numerous multilevel folds and faults of Enhong synclinorium in the region were systematically studied based on the theory of modern structural geology and geomechanical analysis. In combination with the analysis of regional tectonic setting, the multi-phase tectonic evolution and deformation mechanism in the research area were discussed and determined.

The CBM gas content and gas components of 108 coal samples from different coal seams in Bumu region were measured according to the desorption method and gas chromatography method that following Chinese National Standards GB/T 19559-2008 and GB/T 13610-2003, respectively (Table 1). The No. 9 coal seam has the most CBM data, obtained from 27 boreholes (Table 2). The maximum vitrinite reflectance measurement, coal maceral composition analysis, and proximate analysis were performed following the Chinese National Standards GB/T6948-2008, GB/T8899-1998, and GB/T212-2008, respectively (Table 1). These CBM data form the basis for the discussion of the occurrence characteristics and distribution of CBM in Bumu region. Based on the response and variation of CBM gas content, gas component and gas content gradient to the different types and periods geological structures, the control action of different geological structures, fold neutral surface action and fault sealing in multi-phase tectonic movements on CBM occurrence were further discussed.

Table 2. List of gas-bearing property parameters and structure position of No. 9 coal seam.

Borehole	Depth (m)	Gas Content (m^3/t)	CH_4 Concentration (%)	Gas Content Gradient (m^3/t/100 m)	Structure Position	Gas Controlling Pattern *
0103	472.92	15.30	75.37	3.24	Hanging wall of F_{8-10}	I
103	503.85	13.41	72.54	2.66	Hanging wall of F_{8-10}	I
301	531.76	19.27	83.40	3.62	Footwall of F_{2-6}	I
301-1	402.67	15.30	73.90	3.80	Hanging wall of F_{2-6}	I
404	587.74	13.61	91.09	2.32	Footwall of F_{2-6}	I
0101	489.24	9.72	71.66	1.99	Hinge zone of syncline	II
101	578.86	10.50	67.03	1.81	Hinge zone of syncline	II
1202	564.99	15.03	78.98	2.66	Hinge zone of syncline	II
403	564.77	10.97	67.95	1.94	Hinge zone of syncline	II
504	688.21	14.65	92.02	2.13	Hinge zone of syncline	II
505	683.71	9.11	90.86	1.33	Hinge zone of syncline	II
602	838.12	10.93	92.02	1.30	Hinge zone of syncline	II
501	687.98	7.31	61.62	1.06	NWW limb of syncline	III
502	767.14	9.10	53.76	1.19	NWW limb of syncline	III
503	731.10	11.71	51.58	1.60	NWW limb of syncline	III
1201	435.25	11.20	71.97	2.57	NWW limb of syncline	III
2301	490.30	10.25	66.81	2.09	NWW limb of syncline	III
EH01	265.07	7.79	89.85	2.94	SEE limb of syncline	III
0102	468.32	8.39	70.19	1.79	Footwall of F_{16}	IV
102	509.22	9.59	75.82	1.88	Footwall of F_{16}	IV
1203	574.00	10.31	49.67	1.80	Footwall of F_{16}	IV
2302	566.00	9.81	57.13	1.73	Footwall of F_{16}	IV
303	554.54	9.38	57.55	1.69	Footwall of F_{16}	IV
401	688.09	11.50	63.74	1.67	Hanging wall of f_1	V
402	646.40	9.70	42.16	1.50	Footwall of f_3	V
4501	715.42	9.91	31.72	1.39	Hanging wall of F_2	V
4503	623.72	11.1	45.65	1.78	Footwall of f_4	V

* Gas controlling pattern: I—reverse fault sealing type, II—syncline sealing type, III—monoclinal enrichment type, IV—normal fault dispersion type, V—buried floor fault dispersion type.

4. Geological Structure and Tectonic Evolution

The geological structures in Bumu region are mainly the Enhong syncline, reverse, and normal faults (Figure 3b). The Enhong syncline is the main geological structure, which controls the overall distribution, orientation, and burial depth of coal seams (Figure 3c,d). The Enhong syncline is an open NNE-plunging symmetrical syncline, with the hinge line plunges in the direction of NE17° and the plunge angle of 15–30°. The coal seams are generally distributed in NE and NW direction, with a dip angle of about 5–20° in the south and about 15–30° in the north.

Geological survey has discovered 20 faults, developed well in this area with the directions of NNE, NE, and NW (Figure 3b). The faults are mainly reverse faults, with a few normal faults. The reverse faults are mostly in NNE direction and the normal faults are mostly in NE and NW directions. The imbricate large reverse faults of F_{2-5}, F_{2-6}, F_{8-10}, and F_5, with the fault throw of 40–55 m, are located in the NWW limb of Enhong syncline (Figure 3c). The normal faults are limited by the NNE-trending reverse faults and that indicates the normal faults are formed later. Normal faults generally extend unsteadily, especially the F_{16} fault, as the southern section of the arcuate strike-slip normal derived faults, extends in NE–NEE direction. There are seven faults, F_{2-6}, F_{8-10}, F_{16}, f_1, f_2, f_3, and f_4 in No. 9 coal seam.

Since the formation of coal measures in the Late Permian, the study area has mainly experienced Indosinian movement, Yanshanian movement, and Himalayan movement. Affected by the NW-trending basement fault on the northeast, the NW-trending folds and reverse faults were formed beside the basement faults under the NS-trending tectonic compression in the Indosinian [34]. In other areas of the Enhong synclinorium, including the Bumu region, the influence of the Indosinian movement was very weak and the Indosinian geological structure was rare. During the late Jurassic and early Cretaceous, the subduction of the Pacific plate and the compression of the South China block led to the intense Yanshanian tectonic movement [35]. The intense tectonic compression in the direction of NW–SE caused formation of the Enhong syncline, Fuyuan-Mi'le fault, Pingguan-Agang fault, and the associated branch faults. In the Bumu region, the NNE-trending large reverse fault of F_{2-6} and F_{8-10}, as well as small buried reverse fault of f_1, f_2, f_3, and f_4 were formed (Figure 3). The former Indosinian NW-trending structures ware modified intensely and the Yanshanian NE-trending structures were formed rarely in the northeast of the Enhong synclinorium. During the Eocene, the northward subduction of the Indian plate and the westward subduction of the Pacific plate for the Eurasian plate caused the Himalayan movement, EW-trending tectonic compression and dextral shear action [36–38]. It caused a right-lateral shear slip and property transformation of the Fuyuan-Mi'le fault and the Pingguan-Agang fault. Meanwhile, the formation of the NNE-trending secondary derived drag folds and derived compression-shear faults in the NW limb, combined with the NE- and EW-trending strike-slip normal derived faults in the SE limb complicated and formed the current tectonic framework of the Enhong synclinorium. The derived normal fault of F_{16} were formed in the southeast of the Bumu region. The Indosinian NW-trending structures in the northeast, the Yanshanian NNE-trending Enhong syncline, large boundary, and associated branch reverse faults, combined with the Himalayan secondary derived faults and drag folds, formed the current tectonic framework of the Enhong synclinorium.

5. Gas Occurrence Features

5.1. Gas Composition

In the Bumu region, the composition of CBM is complex, mainly composed of CH_4, N_2, CO_2, and heavy hydrocarbon gases. The heavy hydrocarbon gases (C_{2+}) include ethane and propane in the research area. The concentration of CH_4 varies greatly and ranges from 9.17% to 95.32%, with an average value of 65.60%. The N_2 and CO_2 account for 21.76% and 1.61%, respectively. The concentration of heavy hydrocarbon gas is abnormally high, 0.30–25.51%, and the average value is up to 10.47%. The CBM $\delta^{13}C_1$ ranges from −51.69‰ to −43.43‰ with an average of −46.57‰, indicating that CBM is mainly the pyrolysis product of humic organic matter and possibly the secondary biogenic gas in the shallow coal seam [39]. The local high concentration of heavy hydrocarbon gas is mainly the complete result of the coal maceral, microbial degradation, and coal pore structure [29,33].

There is no apparent linear relationship between the CBM composition and coal burial depth, indicating that the geological influence factors of CBM are complex. The concentrations of methane, heavy hydrocarbon and alkane gas generally tend to increase with coal burial depth (Figure 4a–c), while the N_2 concentration decreases with the increase in coal burial depth (Figure 4d).

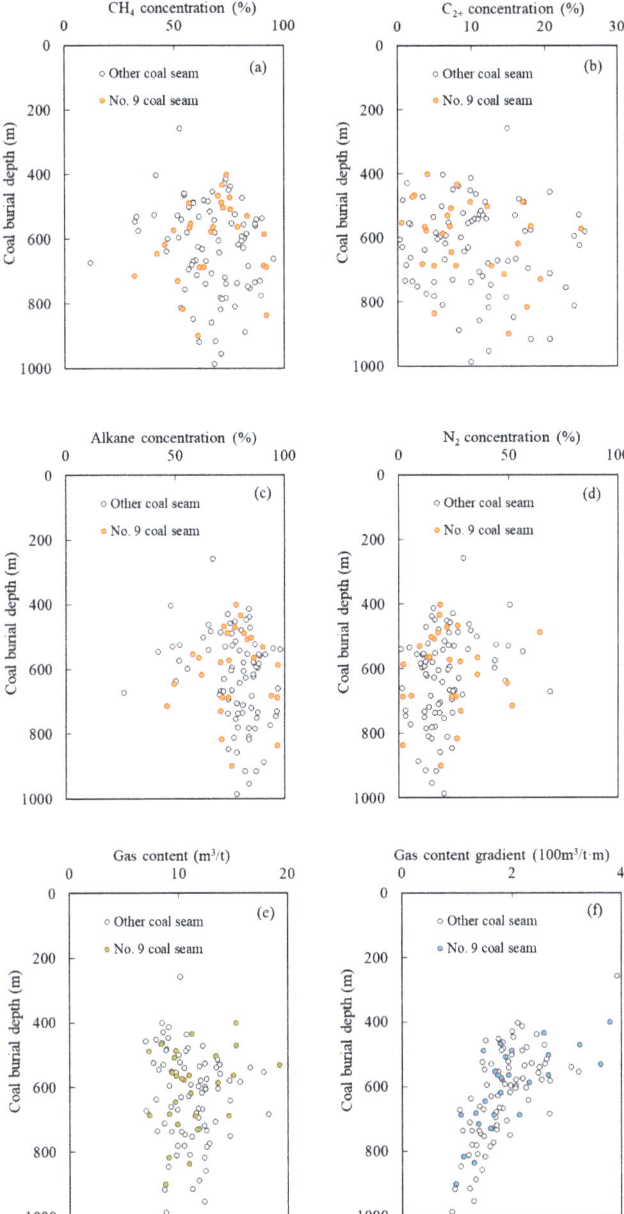

Figure 4. The gas-bearing property parameters vary with the coal burial depth. (**a**) methane concentration, (**b**) heavy hydrocarbon gas concentration, (**c**) alkane concentration, (**d**) N_2 concentration, (**e**) gas content, and (**f**) gas content gradient.

In the Bumu region, the methane concentration of CBM decrease while as the N_2 and heavy hydrocarbon gas concentration increase. The dissipation of methane, as well as its mutual exchange and mixture with surface atmosphere are mostly the main geological cause. It indicates that the CBM occurrence space in the study area has relatively low sealing capacity, which may have resulted from destruction of fault in the CBM reservoir. This is consistent with the complicated geological structure in the Bumu region.

5.2. Gas Content and Its Distribution

In the Bumu region, the gas content of coal seam ranges from 6.94 to 19.27 m^3/t and is mainly between 8 and 13 m^3/t. The average value of gas content is 11.19 m^3/t. Generally, the CBM content tends to increase linearly with coal burial depth and the gas content gradient tends to decrease with coal burial depth following the power function [40]. With the increase in the coal burial depth, the gas content gradient exhibits an overall decreasing trend, but there is no significant positive correlation with the gas content in the Bumu region (Figure 4e,f). The overall relatively low methane concentrations and gas contents of coal seam indicate the widely distributed open and semi-open reservoir space of CBM in the study area. The influence of geological structure on the sealing of CBM reservoir is important and will be discussed in greater detail later in this paper. The gas content of different coal seams fluctuate significantly with stratigraphic sequence and these observations were previously interpreted as a result of multiple unattached CBM-bearing systems in the vertical section [41,42]. Therefore, this study selects the No. 9 main coal seam with stable development and largely available gas data as the research object to discuss the gas distribution features (Table 2, Figure 5).

Gas content of No. 9 coal seam in the Bumu region is relatively high, ranging from 7.31 to 19.27 m^3/t and with an average value of 11.29 m^3/t. The distribution pattern of gas content is complex and has an overall banding distribution in the direction of NNE. The CBM enrichment region, with the gas content higher than 10 m^3/t and about half of the study area, is distributed along the hinge zone of the Enhong syncline. The gas contents of the two limbs of syncline decrease and form two low gas content zones in the northwest and southeast of the study area. At the same time, the development of fault structures complicates the geological structure and the occurrence of CBM in the study area, resulting in the further differentiation of CBM content distribution, particularly in the west (Figure 5). With the gas content of 10 m^3/t as the boundary, the central NNE-trending high gas content zone, western NNE-trending high gas content zone, northwest, and southeast low gas content zones can be further divided as follows.

(1) Central NNE-trending high gas content zone, with a length of 6500 m in the NNE direction and a width about 1500 m. The gas content is 10.25–15.30 m^3/t, with an average of 12.11 m^3/t. Meanwhile, two secondary high gas content zones have been formed in the southwest and northeast, with the gas content generally higher than 13 m^3/t and 11 m^3/t, respectively. The distribution of gas content gradient is similar to that of gas content, and there is an overall trend that the gas content gradient decreases from the southwest to the northeast, in the direction of the Enhong syncline plunging. However, the methane concentration of CBM is high in the southwest and northeast while is low in the central region.

(2) In the western NNE-trending high gas content zone, the gas content, gas content gradient, and methane concentration vary greatly and distribute complexly. The gas content is 9.70–19.27 m^3/t and the highest gas content zone forms in the study area. The gas content gradient is 1.50–3.62 m^3/t/100 m and the methane concentration is 42.16–83.40%.

(3) In the northwest low gas content zone, the gas content, gas content gradient, and methane concentration are all low, with the mean value of 8.77 m^3/t, 1.21 m^3/t/100 m and 49.03%, respectively. Such a situation is very unfavourable for the CBM accumulation.

(4) In the southeast low gas content zone, the gas content is 7.79–9.81 m^3/t, with its contour lines extending in NNE direction and decreasing in SEE direction. In general, the contour lines of gas content gradient and methane concentration also extend in NNE direction, while decreasing in NWW direction.

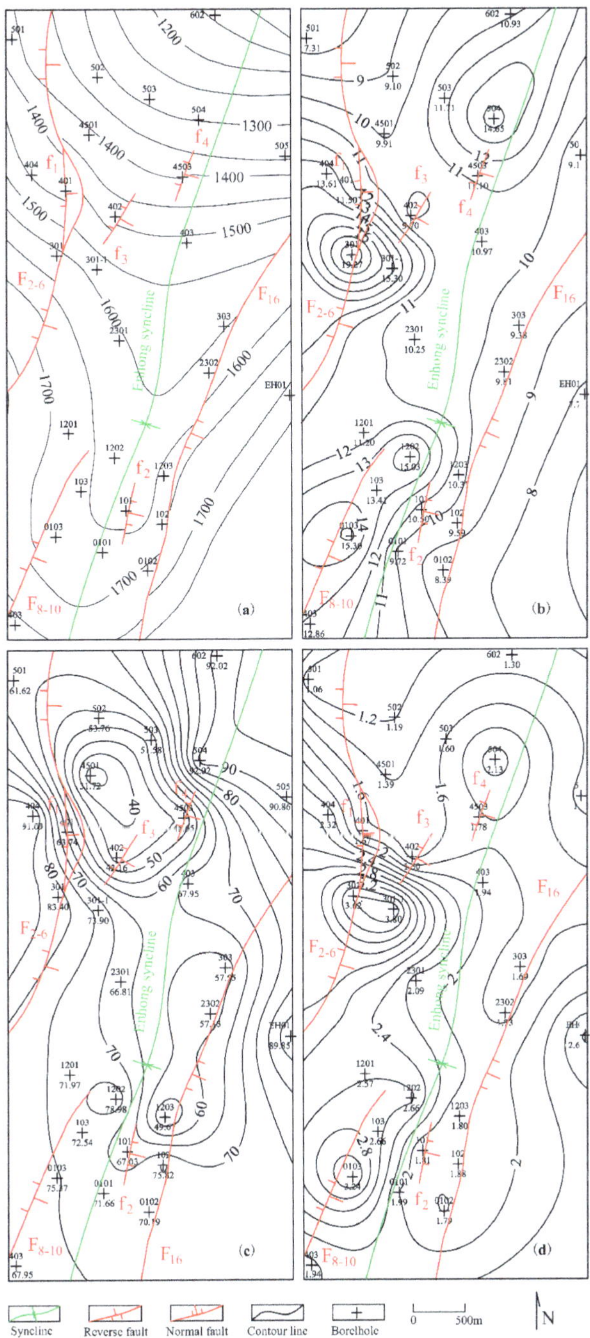

Figure 5. Contour maps of No. 9 coal seam floor, structural sketch and gas-bearing property parameters. (**a**) coal seam floor and structural sketch; (**b**) gas content; (**c**) methane concentration; (**d**) gas content gradient.

6. Gas Controlling Pattern of Geological Structure

In the Bumu region, as a complex geologic structural area with well-developed syncline, normal fault and reverse fault structures, geologic structures play an important role in the occurrence and distribution of CBM, especially the effects of multi-phase tectonic movements after the formation of coal seam. A systematic study on the structural characteristics, tectonic evolution, and CBM occurrence shows that the gas distribution pattern is mainly controlled by the development of geological structures. The development of the Enhong syncline leads to the deep burial and structural compression of the coal seam in the syncline hinge zone, which could allow gas enrichment and shows that the central NNE-trending high gas content zone spreads along the Enhong syncline axis and an overall decreasing trend from the hinge zone of the Enhong syncline, in the central NNE-trending high gas content zone, to its limbs in the northwest and southeast low gas content zones. The development of different types of fault structures further complicates the geological structure and CBM content distribution in the study area, especially in the western NNE-trending high gas content zone.

According to the variation of gas content, gas content gradient, and methane concentration in different structure parts, the effects of different geological structures on CBM occurrence are determined as the following five types: reverse fault sealing type, syncline sealing type, monoclinal enrichment type, normal fault dispersion type, and buried floor fault dispersion type (Figure 6).

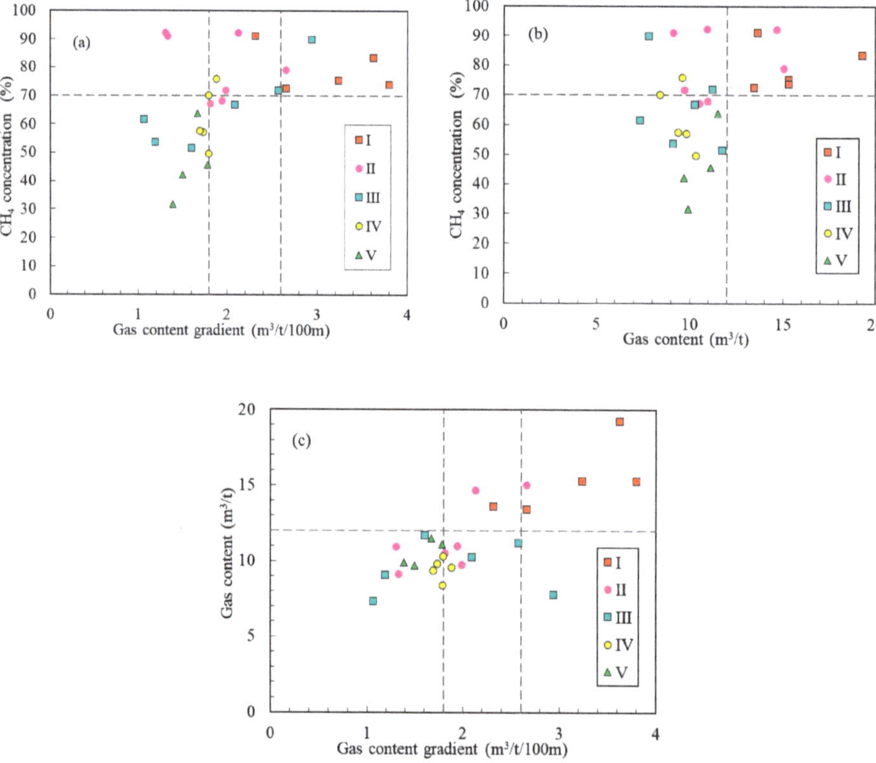

Figure 6. Scatter diagram of gas-bearing property parameters of 5 gas controlling pattern. (**a**) plot of methane concentration to gas content gradient; (**b**) plot of methane concentration to gas content; (**c**) plot of gas content to gas content gradient. Gas controlling pattern: I—reverse fault sealing type, II—syncline sealing type, III—monoclinal enrichment type, IV—normal fault dispersion type, V—buried floor fault dispersion type.

(1) Reverse fault sealing type: No. 9 coal seam from boreholes Nos. 0103, 103, 301, 301-1, and 404 is characterized by the high value of gas content, methane concentration, and gas content gradient, which is generally higher than 12 m^3/t, 70% and 2.6 m^3/t/100 m, respectively. The mean values of those properties in the test boreholes are 15.38 m^3/t, 79.26% and 3.13 m^3/t/100 m, respectively. Those boreholes are located near the large reverse fault of F$_{2-6}$ and F$_{8-10}$, which are formed during the Yanshanian period. Compared with the Yanshanian reverse faults in the North China plate that are mostly reversed to the normal faults in the later period [7], the Yanshanian large reverse fault, even underwent the strike-slip structural modification during the Himalayan period, still have good seal capacity and are beneficial for the preservation of CBM.

(2) Synclinal sealing type: No. 9 coal seam is characterized by the high gas content, methane concentration and the medium gas content gradient, which is generally higher than 10 m^3/t, 70%, and about 2.0 m^3/t/100 m, respectively (Table 2). The CBM of No. 9 coal seam from boreholes Nos. 0101, 101, 1202, 403, 504, 505, and 602, located in the hinge zone of the Enhong syncline, has a mean value of 11.56 m^3/t, 80.07% and 1.88 m^3/t/100 m, respectively. The development of syncline leads to the increase in coal seam burial depth, which is conducive to the accumulation of CBM and the increase in gas content from limb to hinge zone. In contrast, the gas content gradient tends to decrease with the increase in the burial depth [40]. Therefore, the increase in the gas content gradient of coal seam in the Enhong syncline hinge zone, compared to the limbs, is strong evidence that the structural compression above the neutral surface is conducive to the accumulation of CBM.

(3) Monoclinal enrichment type: No. 9 coal seam located in syncline limbs, such as Nos. 501, 502, 503, 1201, 2301, and EH01 boreholes, is characterized by the medium gas content, methane concentration and the low gas content gradient, which is generally lower than 12 m^3/t, 70% and about 1.8 m^3/t/100 m, respectively (Table 2). The gas content gradient varies greatly, ranges from 1.06 to 2.94 m^3/t/100 m, which is caused by its decrease with the increase in burial depth.

(4) Normal fault dispersion type: No. 9 coal seam located in the footwall of F$_1$ normal fault, such as Nos. 0102, 102, 1203, 2302, and 303 boreholes, is characterized by the low gas content, methane concentration, and gas content gradient, which is generally lower than 10 m^3/t, 70%, and about 1.8 m^3/t/100 m, respectively (Table 2). The right-lateral slipping of the Pingguan-Agang fault derives the F$_{16}$ normal fault during the Himalayan period. The extensional fault destroys the CBM reservoir and provides seepage channels for CBM dissipation as well as its mutual exchange with the surface atmosphere.

(5) Buried floor fault dispersion type: No. 9 coal seam located near the buried f$_1$, f$_3$, and f$_4$ reverse faults is characterized by the low gas content, methane concentration, and gas content gradient, which is generally lower than 10 m^3/t, 60%, and about 1.8 m^3/t/100 m, respectively (Table 2). The scale and displacement of these buried floor faults are small, which fail to form a good sealing property for CBM. In addition, the faults penetrate the Emeishan basalt formation underlying the coal-bearing strata, and it has been reported that natural gas can be migrated and be stored in the Emeishan basalt [43,44]. The well-developed fractures in the basalt also serve as an escaping path and as a reservoir of CBM, respectively [40]. These fractures destroy the sealing of the fault and cause the dispersion of CBM.

For No. 9 coal seam in the Bumu region, methane concentration is sensitive to the sealing of CBM reservoir and is generally larger than 70% in reverse fault and syncline sealing types (Figure 6). Gas content and gas content gradient vary greatly with burial depth resulted from the different geological structure control action (Figure 7).

Gas content gradient can indicate the influence of geological structure on CBM reservoir from the buried depth, which is lower than 1.8 m^3/t/100 m in normal and buried floor fault dispersions while is higher than 2.4 m^3/t/100 m in reverse fault sealing type (Figures 6 and 7). Gas content is an important parameter for the enrichment of CBM, which is higher than 12 m^3/t in reverse fault type and around 12 m^3/t in syncline sealing type.

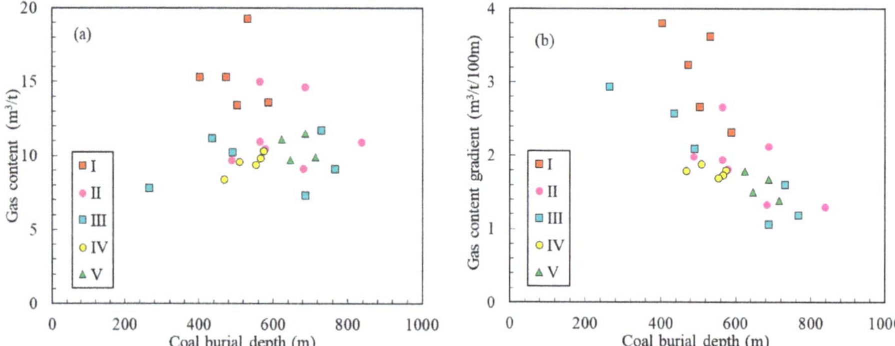

Figure 7. Gas content (**a**) and gas content gradient (**b**) vary with coal burial depth. Gas controlling pattern: I—reverse fault sealing type, II—syncline sealing type, III—monoclinal enrichment type, IV—normal fault dispersion type, V—buried floor fault dispersion type.

The tectonic movement was dominated by the continuous subsidence movements after the coal-bearing strata deposited in the Late Permian until the Late Jurassic. According to the previous numerical simulation research about the burial and thermal histories of coal seams based on the analysis of vitrinite reflectance and homogenization temperature of fluid inclusion, the largest ancient burial depth of the coal seams can reach 3800 m in the research areas [45–47]. Under the effect of paleotemperature, the coal seams experienced deep metamorphism and reached the fat coal stage, accompanied with the massive CBM generation [45,46]. The influence of the Indosinian movement was very weak and the Indosinian geological structure was rare in the Bumu region. The intense tectonic compression and the formation of compressive structures during the Yanshanian movement, as well as the peak-period of gas generation, are conducive to CBM enrichment. The gas-controlling pattern is manifested as the reverse fault sealing, syncline sealing, and monoclinal enrichment of CBM, which are the main controlling factors to the central and western NNE-trending high gas content zones. The tectonic intense compression and dextral shear action in the Himalayan movement avoided the inversion of compressive structure to extensional structure and the destruction of CBM reservoir, which are common in the North China plate [48]. However, the chronic uplift and derived normal fault during the Himalayan period caused the constant dissipation of CBM. The normal fault dispersion and buried floor fault dispersion are the main reasons for the northwest and southeast low gas content zones in the Bumu region.

7. Conclusions

Based on the largely available geological and CBM data, a comprehensive analysis is performed on the Bumu region, one of the hotspots of CBM development in Southwest China. This study has drawn several conclusions, as follows:

(1) The Indosinian local deformation, Yanshanian intense deformation, and Himalayan secondary derived deformation formed the current tectonic framework of Enhong synclinorium. The intense tectonic compression and dextral shear action in the Yanshanian and Himalayan movements caused the complex geological structures in Bumu region, which are composed of the Enhong syncline, associated reverse faults and late derived normal fault.

(2) The methane concentration of CBM decreases while the nitrogen and heavy hydrocarbon gas concentrations increase in Bumu region. The relatively low methane concentration and gas content imply the open to semi-open reservoir space of CBM. The CBM distribution is complex, which has the central and western NNE-trending high gas content zones along the syncline hinge

zone and the reverse faults. The geological structure controls on CBM enrichment are definite and important.

(3) Based on geological structure features and responses of gas content, methane concentration and gas content gradient, the gas controlling pattern of geological structure are determined and can be classified into five types: the reverse fault sealing, syncline sealing, monoclinal enrichment, normal fault dispersion, and buried floor fault dispersion types.

(4) The structural compression above the neutral surface plays an important role in the syncline sealing process, which is indicated by the increase in gas content gradient. The EW-trending tectonic intense compression and dextral shear action in the Himalayan movement avoided the negative inversion of NNE-trending Yanshanian compressive structure and its destruction of CBM reservoir. However, the chronic uplift and derived normal fault during Himalayan period caused the constant dissipation of CBM.

Author Contributions: B.J. and Q.M. conceived the ideas; M.L. analyzed the data and wrote the paper; G.W., Z.Y. and F.L. provided language and technical support. All authors have read and agreed to the published version of the manuscript.

Funding: This research was funded by National Science and Technology Major Project (2016ZX05044), National Natural Science Foundation of China (Nos. 41402136, 41702172) and Natural Science Foundation of Jiangsu Province (No. BK20140183).

Acknowledgments: The Author Ming Li thanks the China Scholarship Council (CSC) providing financial support for the collaborative research visiting in School of Chemical Engineering at the University of Queensland Australia. We would like to thank the reviewers and editors for their constructive comments and suggestions.

Conflicts of Interest: The authors declare no conflict of interest.

References

1. Qin, Y.; Moore, T.; Shen, J.; Yang, Z.B.; Shen, Y.L.; Wang, G. Resources and geology of coalbed methane in China: A review. *Int. Geol. Rev.* **2018**, *60*, 777–812. [CrossRef]

2. Tao, S.; Chen, S.D.; Pan, Z.J. Current status, challenges, and policy suggestions for coalbed methane industry development in China: A review. *Energy Sci. Eng.* **2019**, *7*, 1–16. [CrossRef]

3. Cao, D.Y.; Ning, S.Z.; Guo, A.J.; Li, H.T.; Chen, L.M.; Liu, K.; Tan, J.Q. *Tectonic Framework of Coalfields and Tectonic Control of Coal Seams in China*; Science Press: Beijing, China, 2017.

4. Pashin, J.C. Stratigraphy and structure of coalbed methane reservoirs in the United States: An overview. *Int. J. Coal Geol.* **1998**, *35*, 209–240. [CrossRef]

5. Pashin, J.C. Geology of North American coalbed methane reservoirs. In *Coal Bed Methane: Theories and Applications*; Elsevier: Amsterdam, The Netherlands, 2020.

6. Groshong, R.H.; Pashin, J.C.; Mcintyre, M.R. Structural controls on fractured coal reservoirs in the southern Appalachian Black Warrior foreland basin. *J. Struct. Geol.* **2009**, *31*, 874–886. [CrossRef]

7. Jiang, B.; Qu, Z.H.; Wang, G.; Li, M. Effects of structural deformation on formation of coalbed methane reservoirs in Huaibei coalfield, China. *Int. J. Coal Geol.* **2010**, *82*, 175–183. [CrossRef]

8. Jing, B.; Li, M.; Song, Y.; Cheng, G.X.; Zhu, G.Y. *Tectonically Deformed Coal and Its Gas Geological Significance*; Science Press: Beijing, China, 2020.

9. Li, M.; Jiang, B.; Lin, S.F.; Lan, F.J.; Wang, J.L. Structural controls on coalbed methane reservoirs in Faer coal mine, Southwest China. *J. Earth Sci.* **2013**, *24*, 437–448. [CrossRef]

10. Chen, Y.; Tang, D.Z.; Xu, H.; Li, Y.; Meng, Y.J. Structural controls on coalbed methane accumulation and high production models in the eastern margin of Ordos Basin, China. *J. Nat. Gas Sci. Eng.* **2015**, *23*, 524–537. [CrossRef]

11. Zhu, H.Z.; Liu, P.; Chen, P.; Kang, J.N. Analysis of coalbed methane occurrence in Shuicheng Coalfield, southwestern China. *J. Nat. Gas Sci. Eng.* **2017**, *47*, 140–153. [CrossRef]

12. Guo, X.J.; Huan, X.; Huan, H.H. Structural characteristics of deformed coals with different deformation degrees and their effects on gas adsorption. *Energy Fuels* **2017**, *31*, 13374–13381. [CrossRef]

13. Liu, H.W.; Jiang, B. Geochemical alteration and mineralogy of coals under the influence of fault motion: A case study of Qi'nan colliery, China. *Minerals* **2019**, *9*, 389. [CrossRef]

14. Lu, Y.J.; Liu, D.M.; Cai, Y.D.; Li, Q.; Jia, Q.F. Pore-fractures of coalbed methane reservoir restricted by coal facies in Sangjiang-Muling coal-bearing Basins, Northeast China. *Minerals* **2020**, *13*, 1196. [CrossRef]

15. Hower, J.C.; Davis, A. Vitrinite reflectance anisotropy as a tectonic fabric element. *Geology* **1981**, *9*, 165–168. [CrossRef]

16. Li, Z.; Liang, Z.K.; Jiang, Z.X.; Yu, H.L.; Yang, Y.D.; Xiao, L. Pore connectivity characterization of lacustrine shales in Changling fault depression, Songliao Basin, China: Insights into the effects of mineral compositions on connected pores. *Minerals* **2019**, *9*, 198. [CrossRef]

17. Liu, X.F.; He, X.Q. Effect of pore characteristics on coalbed methane adsorption in middle-high rank coals. *Adsorption* **2017**, *23*, 3–12. [CrossRef]

18. Mastalerz, M.; He, L.; Melnichenko, Y.B.; Rupp, J.A. Porosity of coal and shale: Insights from gas adsorption and SANS/USANS techniques. *Energy Fuels* **2012**, *26*, 5109–5120. [CrossRef]

19. Yao, H.F.; Kang, Z.Q.; Li, W. Deformation and reservoir properties of tectonically deformed coals. *Petrol. Explor. Dev.* **2014**, *41*, 460–467. [CrossRef]

20. Karayiğit, A.İ.; Mastalerz, M.; Oskay, R.G.; Buzkan, İ. Bituminous coal seams from underground mines in the Zonguldak Basin (NW Turkey): Insights from mineralogy, coal petrography, Rock-Eval pyrolysis, and meso-and microporosity. *Int. J. Coal Geol.* **2018**, *199*, 91–112. [CrossRef]

21. Wei, C.T.; Qin, Y.; Wang, G.X.; Fu, X.H.; Zhang, Z.Q. Numerical simulation of coalbed methane generation, dissipation and retention in SE edge of Ordos Basin, China. *Int. J. Coal Geol.* **2010**, *82*, 147–159. [CrossRef]

22. Casey, M.; Butler, R.W.H. Modelling approaches to understanding fold development: Implications for hydrocarbon reservoirs. *Mar. Petrol. Geol.* **2004**, *21*, 933–946. [CrossRef]

23. Shen, J.; Fu, X.H.; Qin, Y.; Liu, Z. Control actions of structural curvature of coal-seam floor on coalbed gas in the No. 8 coal mine of Pingdingshan. *J. China Coal Soc.* **2010**, *35*, 586–589.

24. Fu, H.J.; Tang, D.Z.; Xu, T.; Xu, H.; Tao, S.; Zhao, J.L.; Chen, B.L.; Yin, Z.Y. Preliminary research on CBM enrichment models of low-rank coal and its geological controls: A case study in the middle of the southern Junggar Basin, NW China. *Mar. Petrol. Geol.* **2017**, *83*, 97–110. [CrossRef]

25. Pashin, J. Variable gas saturation in coalbed methane reservoirs of the Black Warrior Basin: Implications for exploration and production. *Int. J. Coal Geol.* **2010**, *82*, 135–146. [CrossRef]

26. Zhang, Z.G.; Qin, Y.; Yi, T.S.; You, Z.J.; Yang, Z.B. Pore structure characteristics of coal and their geological controlling factors in the eastern Yunnan and western Guizhou, China. *ACS Omega* **2020**, *5*, 19565–19578. [CrossRef]

27. Zhang, Z.M.; Wu, Y. *China Coalmine Gas-Geologic Laws and Mapping*; China University of Mining and Technology Press: Xuzhou, China, 2014.

28. Li, S.; Tang, D.Z.; Pan, Z.J.; Xu, H.; Guo, L.L. Evaluation of coalbed methane potential of different reservoirs in western Guizhou and eastern Yunnan, China. *Fuel* **2015**, *139*, 257–267. [CrossRef]

29. Lan, F.J.; Qin, Y.; Wang, A.K.; Li, M.; Wang, G.X. The origin of high and variable concentrations of heavy hydrocarbon gases in coal from the Enhong syncline of Yunnan, China. *J. Nat. Gas Sci. Eng.* **2020**, *76*, 1–11. [CrossRef]

30. Cheng, G.X.; Jiang, B.; Li, M.; Li, F.L.; Xu, S.C. Quantitative characterization of fracture structure in coal based on image processing and multifractal theory. *Int. J. Coal Geol.* **2020**, *228*, 1–20.

31. Zhang, G.W.; Guo, A.L.; Wang, Y.J.; Li, S.Z.; Dong, Y.P.; Liu, S.F.; He, D.F.; Cheng, S.Y.; Lu, R.K.; Yao, A.P. Tectonics of South China continent and its implications. *Sci. China-Earth Sci.* **2013**, *56*, 1804–1828. [CrossRef]

32. Qiu, L.; Yan, D.P.; Yang, W.X.; Wang, J.B.; Tang, X.L.; Ariser, S. Early to Middle Triassic sedimentary records in the Youjiang Basin, South China: Implications for Indosinian orogenesis. *J. Asian Earth Sci.* **2017**, *141*, 125–139. [CrossRef]

33. Lan, F.J.; Qin, Y.; Li, M.; Lin, Y.C.; Wang, A.K.; Shen, J. Abnormal concentration and origin of heavy hydrocarbon in upper Permian coal seams from Enhong syncline, Yunnan, China. *J. Earth Sci.* **2012**, *23*, 842–853. [CrossRef]

34. Li, M.; Jiang, B.; Liu, J.G.; Zhu, P.; Cheng, G.X. Geological models and structural controls of tectonically deformed coal in Tucheng syncline, western Guizhou Province. *J. China Coal Soc.* **2018**, *42*, 726–731.

35. Wan, T.F. *The Tectonics of China: Data, Maps and Evolution*; Higher Education Press: Beijing, China, 2011.

36. Hou, Y.G.; He, S.; Tang, D.Q. Tectonic reverse of Cenozoic basins and its relationship with the biogas accumulation in north-east of Yunnan Province. *J. Cent. South Univ.* **2012**, *43*, 2238–2246.

37. Spiro, B.F.; Liu, J.J.; Dai, S.F.; Zeng, R.S.; Large, D.; French, D. Marine derived $^{87}Sr/^{86}Sr$ in coal, a new key to geochronology and palaeoenvironment: Elucidation of the India-Eurasia and China-Indochina collisions in Yunnan, China. *Int. J. Coal Geol.* **2019**, *215*, 103304. [CrossRef]

38. Liu, H.P.; Wang, Z.C.; Xiong, B.X.; Li, Y.L.; Liu, L.Q.; Zhang, J.Z. Coupling analysis of Mesozoic-Cenozoic foreland basin and mountain system in central and western China. *Geosci. Front.* **2000**, *7*, 55–72.

39. Lan, F.J.; Qin, Y.; Li, M.; Tang, Y.H.; Guo, C.; Zhang, F. Microbial degradation and its influence on components of coalbed gases in Enhong syncline, China. *Int. J. Min. Sci. Technol.* **2013**, *23*, 293–299. [CrossRef]

40. Li, M.; Jiang, B.; Lin, S.F.; Lan, F.J.; Zhang, G.S. Characteristics of coalbed methane reservoirs in Faer Coalfield, Western Guizhou. *Energy Explor. Exploit.* **2013**, *31*, 411–428. [CrossRef]

41. Qin, Y.; Xiong, M.H.; Yi, T.S.; Yang, Z.B.; Wu, C.F. On unattached multiple superposed coalbed-methane system: In a case of the Shuigonghe syncline, Zhijin-Nayong coalfield, Guizhou. *Geol. Rev.* **2008**, *54*, 65–70.

42. Shen, Y.L.; Qin, Y.; Guo, Y.H.; Yi, T.S.; Yuan, X.X.; Shao, Y.B. Characteristics and sedimentary control of a coalbed methane-bearing system in Lopingian (late Permian) coal bearing strata of western Guizhou Province. *J. Nat. Gas Sci. Eng.* **2016**, *33*, 8–17. [CrossRef]

43. Wang, Y.Z.; Cao, Y.C.; Wang, S.P.; Song, Y.B. Advances in research of spatial structures of unconformity. *Geotecton. Metallog.* **2006**, *30*, 326–330.

44. Zhang, R.X.; Wang, X.Z.; Lan, D.Q.; Kang, B.P. Reservoir evaluation of Emeishan basalts in Southwest Sichuan. *Nat. Gas Explor. Dev.* **2006**, *29*, 17–20.

45. Tang, S.L.; Tang, D.Z.; Xu, H.; Tao, S.; Li, S.; Geng, Y.G. Geological mechanisms of the accumulation of coalbed methane induced by hydrothermal fluids in the western Guizhou and eastern Yunnan regions. *J. Nat. Gas Sci. Eng.* **2016**, *33*, 644–656. [CrossRef]

46. Tao, S.; Tang, D.Z.; Qin, Y.; Xu, H.; Li, S.; Cai, J.L. Analysis on thermal history of coal strata of typical mining areas in western Guizhou and eastern Yunnan. *Coal Geol. Explor.* **2010**, *38*, 17–21.

47. Dou, X.Z.; Jiang, B.; Qin, Y.; Qu, Z.H.; Yang, Z.B.; Wu, Y.Y. Pattern and mechanism of metamorphism of late Permian coal in western Guizhou. *J. China Coal Soc.* **2012**, *37*, 424–429.

48. Song, Y.; Zhao, M.J.; Liu, S.B.; Wang, H.Y.; Chen, Z.H. The influence of tectonic evolution on the accumulation and enrichment of coalbed methane (CBM). *Chin. Sci. Bull.* **2005**, *50*, 1–6. [CrossRef]

Article

A Novel Data-Driven Method to Estimate Methane Adsorption Isotherm on Coals Using the Gradient Boosting Decision Tree: A Case Study in the Qinshui Basin, China

Jiyuan Zhang [1,2], Qihong Feng [1,2,*], Xianmin Zhang [1,2], Qiujia Hu [3], Jiaosheng Yang [4] and Ning Wang [3]

[1]   Key Laboratory of Unconventional Oil & Gas Development, China University of Petroleum (East China), Qingdao 266580, China; 20180075@upc.edu.cn (J.Z.); spemin@126.com (X.Z.)
[2]   School of Petroleum Engineering, China University of Petroleum (East China), Qingdao 266580, China
[3]   PetroChina Huabei Oilfield Co., Renqiu 062552, China; mcq_hqj@petrochina.com.cn (Q.H.); yjy_wangning@petrochina.com.cn (N.W.)
[4]   PetroChina Research Institute of Petroleum Exploration and Development, Langfang Branch, Langfang 065007, China; yangjs69@petrochina.com.cn
*   Correspondence: fqihong@upc.edu.cn; Tel.: +86-532-86981229

Received: 28 September 2020; Accepted: 12 October 2020; Published: 15 October 2020

Abstract: The accurate determination of methane adsorption isotherms in coals is crucial for both the evaluation of underground coalbed methane (CBM) reserves and design of development strategies for enhancing CBM recovery. However, the experimental measurement of high-pressure methane adsorption isotherms is extremely tedious and time-consuming. This paper proposed the use of an ensemble machine learning (ML) method, namely the gradient boosting decision tree (GBDT), in order to accurately estimate methane adsorption isotherms based on coal properties in the Qinshui basin, China. The GBDT method was trained to correlate the adsorption amount with coal properties (ash, fixed carbon, moisture, vitrinite, and vitrinite reflectance) and experimental conditions (pressure, equilibrium moisture, and temperature). The results show that the estimated adsorption amounts agree well with the experimental ones, which prove the accuracy and robustness of the GBDT method. A comparison of the GBDT with two commonly used ML methods, namely the artificial neural network (ANN) and support vector machine (SVM), confirms the superiority of GBDT in terms of generalization capability and robustness. Furthermore, relative importance scanning and univariate analysis based on the constructed GBDT model were conducted, which showed that the fixed carbon and ash contents are primary factors that significantly affect the adsorption isotherms for the coal samples in this study.

Keywords: methane adsorption isotherm; coal properties; machine learning; gradient boosting decision tree; estimation model

---

1. Introduction

As an unconventional hydrocarbon resource, coalbed methane (CBM) has been unlocked for commercial development in the USA, China, Australia, Canada, and India [1]. The recovery of CBM from coal seams has multiple favorable effects, such as the reduction of greenhouse gas release into the atmosphere, enhancement of underground coal mining safety, and addition to natural gas supply [2,3]. It is commonly believed that the majority of methane exists within coal seams via physical adsorption [4,5]. The accurate characterization of methane adsorption isotherms in coals is crucial for the successful development of CBM resources, because the isotherm determines the in situ level of gas saturation, which significantly affects CBM production rates [6].

To date, experimental methods that were commonly used for measuring high-pressure methane adsorption isotherms have included the manometric, the gravimetric, and the volumetric methods [7]. Although these methods differ in the means by which the adsorption amount is determined, they all require indispensable procedures that typically include sample preparation, adsorption equilibrium, and data deduction. Such tedious experimental procedures are not only time-consuming, but they also may result in varying sources of uncertainties. Previous studies [8,9] showed that adsorption isotherms on a same sample measured in different laboratories may exhibit noticeable discrepancies, which may be attributed to uncertainties that stem from e.g., the determination of reference/pump and void volume [10,11], the choice of equation of state (EoS) [8] and impurities in the measurement gas [9]. As such, it is pointed out by [12] that extremely tedious procedures, including through the calibration of the instrument, careful operations, and check of the repeatability, are needed in order to ensure the accuracy and consistency in adsorption isotherm measurements.

When compared with the measurement of adsorption isotherms, determinations of coal properties (e.g., proximate analysis, maceral group identification, vitrinite reflectance measurement) are much easier and faster. Numerous studies have shown that the methane adsorption capacity on coals is potentially affected by the coal properties (e.g., ash, fixed carbon and inherent moisture contents, maceral group, vitrinite reflectance) and experimental conditions (e.g., sample particle size, equilibrium moisture, and temperature) [13–15]. As such, it is reasonable and should be viable to estimate/predict the adsorption isotherm while using mathematical regression techniques that are based on these influencing factors. Feng et al. [16] quantitively correlated the Langmuir volume (V_L) with vitrinite reflectance, proximate parameters, vitrinite content, and temperature while using the alternating conditional expectation (ACE) algorithm. More recently, Chattaraj et al. [1] applied the multiple regression analysis method to develop a predictive model for VL based on proximate and ultimate parameters. It should be noted that only V_L was estimated in Feng et al. and Chattarj et al.; neither study considered the estimation of Langmuir pressure (P_L), which determines the curvature of an adsorption isotherm. In other words, the models that were proposed by [1,16] can only predict the maximum adsorption capacity instead of the adsorption isotherms. The difficulty in the accurate estimation of P_L may be due to the uncertainties in its correlations with coal properties. For example, Laxminarayana and Crosdale [17] and Dutta et al. [18] found that Langmuir pressure decreases with the increase in vitrinite reflectance for Australian and Indian coals. However, Busch et al.'s [13] statistics on $\approx$1000 coal samples show a very scattered pattern between P_L and vitrinite reflectance. Zhang et al. [19] proposed the use of deep neural network (DNN) in order to predict CO_2 adsorption on porous carbon based on surface area, micropore, and mesopore volumes. However, gas adsorption behavior on coals is more complicated than on porous carbons due to the higher degree of chemical and physical complexity of coals.

Having addressed these issues, it should be of practical significance to accurately estimate the adsorption isotherm from parameters that are easy and fast to determine in order to reduce the time-consuming and expensive work of adsorption isotherm measurement. This paper proposed the use of the gradient boosting decision trees (GBDT) [20,21] in order to accurately estimate adsorption isotherms that are based on coal properties (ash, fixed carbon, moisture, vitrinite, and vitrinite reflectance) and experimental condition (equilibrium moisture and temperature) for coal samples acquired from the Qinshui basin. The GBDT is an ensemble method that combines a number of base estimators (decision trees) with the gradient boosting algorithm in order to improve the robustness over a single estimator. The GBDT has empirically proven to be highly efficient and promising for solving various regression and classification problems in the field of energy and petroleum engineering [22,23]. However, to the best knowledge of the authors, the application of GBDT in estimating the adsorption isotherm has not yet been reported. The superiority of the GBDT in terms of accuracy and robustness was then confirmed by comparison with the back-propagation artificial neural network (BP-ANN) and support vector machine (SVM). Sensitivity analysis was then conducted while using the constructed GBDT model to analyze the effect of each input variable on the adsorption isotherm.

2. Materials and Methods

2.1. Geological Background of the Study Area

The study area is the Anze block in the southern Qinshui basin, North China (Figure 1), where commercial developments of CBM resources have been ongoing since more than two decades ago. The Qinshui Basin is a large compound synclinal basin surrounded by the uplifts of Wutai Mountain, Taihang Mountain, Zhongtiao Mountain, and Huo Mountain [24]. The study area consists of the Pennsylvanian Benxi and Taiyuan formations, Permian Shanxi, Xiashihezi, Shangshihezi and Shiqianfeng formations, and Triassic to Quaternary deposits. The primary CBM-bearing formations are No. 3 coal seam in the Shanxi formation and No. 15 coal seam of the Taiyuan formation (Figure 2). The No. 3 and No. 5 coals are characterized with high metamorphism. The coal ranks are in the range of low volatile bituminous to anthracite with $R_{o,m}$ high up to 4.5% [25]. Maceral compositions are dominated by vitrinite and subordinate inertinite, while liptinite is microscopically unrecognizable. The Lithotypes are primarily semi-bright and bright coals.

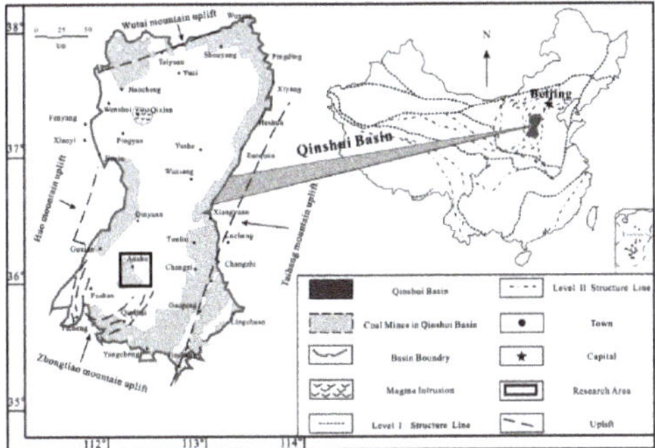

Figure 1. Illustration of the study area where coal samples were retrieved. Reprint with permission [24]; 2011, Elsevier Ltd.

2.2. Samples and Experiments

A number of 165 coal samples were acquired while using the downhole coring technique from 72 CBM wellbores in No. 3 and No. 5 coal seams. After being transported to the laboratory in sealed tanks, the coal samples were subjected to proximate analysis, vitrinite reflectance, and adsorption isotherm measurements in order to develop a database that is used for machine learning. Proximate analysis was conducted following the Chinese standard GB/T 212-2008 [26]. The maceral group was identified at 50× magnification under plane polarized reflected light with a fluorescence illuminator, following the Chinese standard GB/T 8899-2013 [27]. Vitrinite reflectance (Romax) was measured according to Chinese standard GB/T 6948-2008 [28] at a magnification of 500 × oil immersions. More details on the analysis procedure are given in [29]. Methane adsorption isotherms were measured on 60–80 mesh moisture-equilibrium coal powders while using the manometric method [7]. For each sample, the experimental temperature was set to be identical with the in situ temperature where the sample was retrieved. Each adsorption isotherm is comprised of eight equilibrium pressures (ranging from ≈0.5 to ≈8.5 MPa) with corresponding adsorption amounts, which results in a total number of 8 × 165 = 1320 data points in the database. Table 1 summarizes experimental data for the samples.

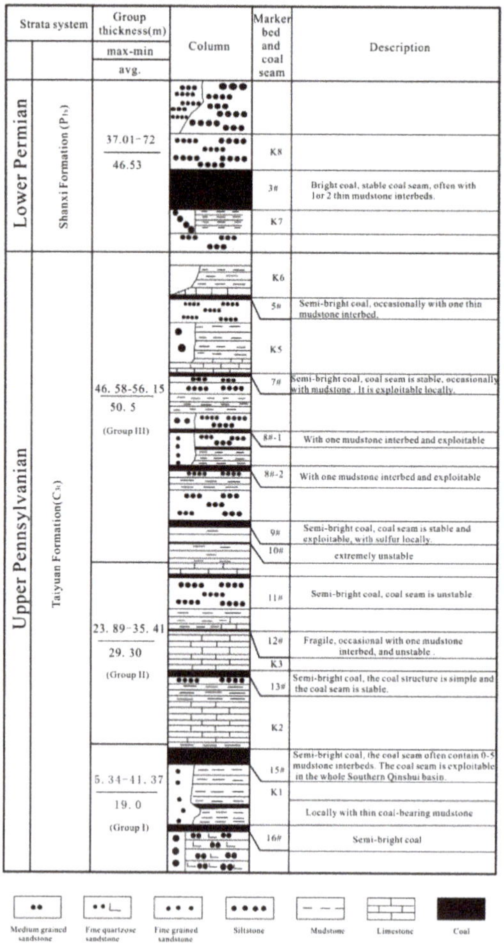

Figure 2. Stratigraphy of the coal-bearing strata in the study area. Reprint with permission [24]; 2011, Elsevier Ltd.

Table 1. Summary of the experimental data.

Property	Maximum	Minimum	Average
Ash (a.d.), %	49.59	4.85	18.70
Moisture (a.r.), %	2.20	0.34	1.10
Fixed carbon (d.a.f.), %	93.08	78.15	87.74
Vitrinite (m.m.f), %	97.80	47.50	80.77
Vitrinite reflectance, %	3.18	1.67	2.39
Equilibrium moisture, %	33.90	6.00	14.22
Temperature, °C	45.0	24.0	33.71
Langmuir volume, m^3/t	37.26	12.53	24.25
Langmuir pressure, MPa	2.90	1.52	2.03

Note: a.r.—as received; a.d.—air dry; d.a.f—dry ash free; m.m.f—mineral matter free.

2.3. Basics of GBDT

The basic philosophy behind the GBDT is to use an ensemble of classification and regression trees (CARTs) to fit the training data samples through minimizing a regularized objective function. Each CART is comprised of a number of leaf nodes and each leaf node is associated with a binary decision rule structure and a continuous score. In GBDT, a number of CARTs are developed in a sequential manner in order to form an accurate ensemble model. For the completeness of this paper, the GBDT algorithm is briefly addressed, as follows. Readers are referred to [20,21] for more details on the GBDT algorithm.

For a given data set with d dimensions and n examples $\mathcal{D} = \{(\mathbf{x}_i, y_i)\}, (\mathbf{X}_i \in \mathbb{R}^d, y_i \in \mathbb{R}, i = 1, 2, \ldots, n)$, the output F is predicted as the sum of K additive functions, which is written as

$$F(\mathbf{x}) = \sum_{m=0}^{M} \beta_m h(\mathbf{x}; \{R_{lm}\}_l^L) \tag{1}$$

where h represents a tree with a number of L nodes; R_{lm} represents partitioned region that is defined by the terminal node l of the mth tree; $\{\beta_m\}_0^M$ are expansion coefficients that are jointly fit with $\{R_{lm}\}_l^L$ to the training data set by minimizing a regularized objective function:

$$\mathcal{L} = \sum_i^n \psi(y_i, F(\mathbf{x}_i)) \tag{2}$$

where ψ is a differentiable loss function, which was assigned as the squared error in this study.

The minimization of the loss function is achieved by iteratively adding leaf nodes that result in the steepest decent [21], which is mathematically expressed as:

$$\gamma_{lm} = \underset{\gamma}{\arg\min} \sum_{\mathbf{x}_i \in R_{lm}} \psi(y_i, F_{m-1}(\mathbf{x}_i) + \gamma) \tag{3}$$

$$F_m(\mathbf{x}) = F_{m-1}(\mathbf{x}) + \upsilon \gamma_{lm} 1(\mathbf{x}_i \in R_{lm}) \tag{4}$$

where υ is the shrinkage factor in the range of (0, 1] that controls the learning rate of the training process. Empirically, small values of υ are beneficial in conserving the model and, thus, help in increasing the generalization capability [22].

2.4. Construction of the GBDT Estimation Model

2.4.1. Input Features

For constructing a reliable regression model, a first key step is to properly identify input features (or independent variables) [30]. A most popular method for identifying the input features is to conduct univariate correlation regression, and features with high degree of correlation (e.g., in terms of high correlation coefficient) with the output are fed into the estimation model [31,32]. The primary drawback of this method is that feature(s) with weak, but certain, underlying correlations with the output may be excluded from the model, which may tend to decrease the modeling accuracy. Beker et al. [33] argued that all features with either explicit (strong) or implicit (weak) correlations with the output variables should be included in a machine learning model in order to attain high modeling accuracy. In this regard, we assigned features as the input of the model that have been demonstrated empirically to exert potential effects on the adsorption amount, and that are less expensive and more rapid to be experimentally measured than the adsorption isotherm. Section 3 will discuss the effect of the inclusion of these "less significant" features on the model accuracy.

In this study, the adsorption isotherm is represented with a series of discrete (adsorption amount versus equilibrium pressure) data points (Figure 3). Thus, the estimation of adsorption isotherm is, in fact, transformed to the estimation of adsorption amounts at given equilibrium pressures. In this way, the equilibrium pressure is an essential input variable for the construction of the estimation model. An alternative option to estimate the adsorption isotherm would be to use an adsorption model (e.g., the Langmuir type model) to represent the isotherm and then correlate the adsorption model parameters (e.g., Langmuir volume and Langmuir pressure) with certain input features. However, our preliminary evaluation of this alternative option turned out to fail in accurately reproducing the adsorption isotherm, which is probably due to the weak correlation of Langmuir pressure with input features, such as coal properties and experimental conditions, as mentioned earlier.

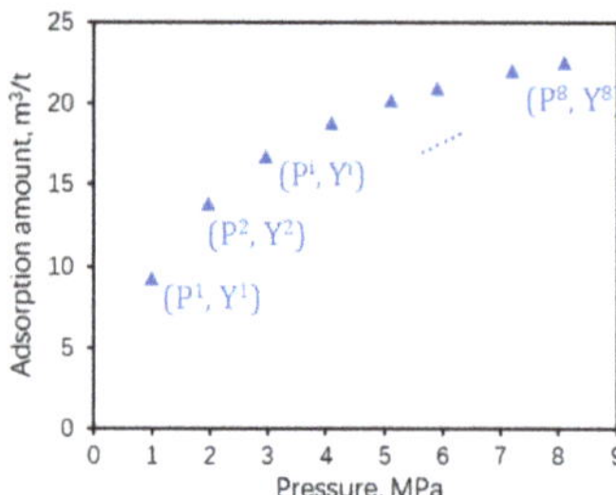

Figure 3. An example of the experimentally measured adsorption isotherm represented with discrete equilibrium points. P^i and Y^i represent the equilibrium pressure and the corresponding adsorption amount for the ith equilibrium points.

For the coal samples that were used in this study, coal properties that exhibit strong control on methane adsorption capacity are ash (Figure 4a) and fixed carbon (Figure 4b) (with $R^2 \geq 0.6$), which, therefore, are assigned as the input features. The vitrinite reflectance (Figure 4c) exhibits a generally linear positive effect on the adsorption although the correlation is relatively loose ($R^2 = 0.36$), which is also included in the input feature bank. Other factors, including inherent and equilibrium moisture, vitrinite content, and experimental temperature, which show weak correlations (with $R^2 \leq 0.1$) with the adsorption capacity (Figure 4d through 3g), are also included in the model construction given numerous documentations of their potential effect on adsorption isotherm (e.g., [1,4,15–18]).

As mentioned earlier, our goal is to develop an estimation model that is based on data that are less expensive and less time-consuming to obtain, so that the adsorption isotherms can be fast estimated with reasonable accuracies. Therefore, other coal properties that may exert potential influence on methane adsorption isotherms, such as micro-pore surface area/volume [34–36] and surface functional groups [13,37] of coals, are not considered because such information requires experimental endeavors that inevitably bring in additional expenses. Besides, the experimental determination of the pore characteristics is rather complicated while using techniques, such as gas (N_2/CO_2) adsorption [38], focused ion beam scanning electron microscopy (FIB-SEM, [39]) and small-angle neutron scattering (SANS, [40]), which require special experimental apparatus and they may be even more time-consuming than the measurement of adsorption isotherms.

As a short summary, the input features for constructing the estimation model for the adsorption isotherm are: coal properties (including ash, fixed carbon, inherent moisture, vitrinite, and vitrinite reflectance) and experimental conditions (equilibrium pressure, equilibrium moisture, and temperature).

2.4.2. Determination of Optimal GBDT Hyperparameters

Prior to conducting GBDT regressions, the whole database comprising of 165 samples and adsorption isotherms is randomly divided into three sub-sets, namely the training (99 samples, 60%), validation (33 samples, 20%), and testing (33 samples, 20%) sets (Figure 5). The training set was used for training the GBDT network, while the validation set was for monitoring the performance and for determining the optimal model parameters (which is to be addressed in the following paragraph).

The testing set was assumed to be "unseen" during the model construction process and used for testing the generalization capability of the constructed regression model. It should be noted again that each adsorption isotherm is represented with eight (adsorption amount versus equilibrium pressure) discrete data points, and, thus, the training, validation, and testing sets are, in effect, constituted with a number of $99 \times 8 = 792$, $33 \times 8 = 264$, and $33 \times 8 = 264$ data points, respectively (Figure 5).

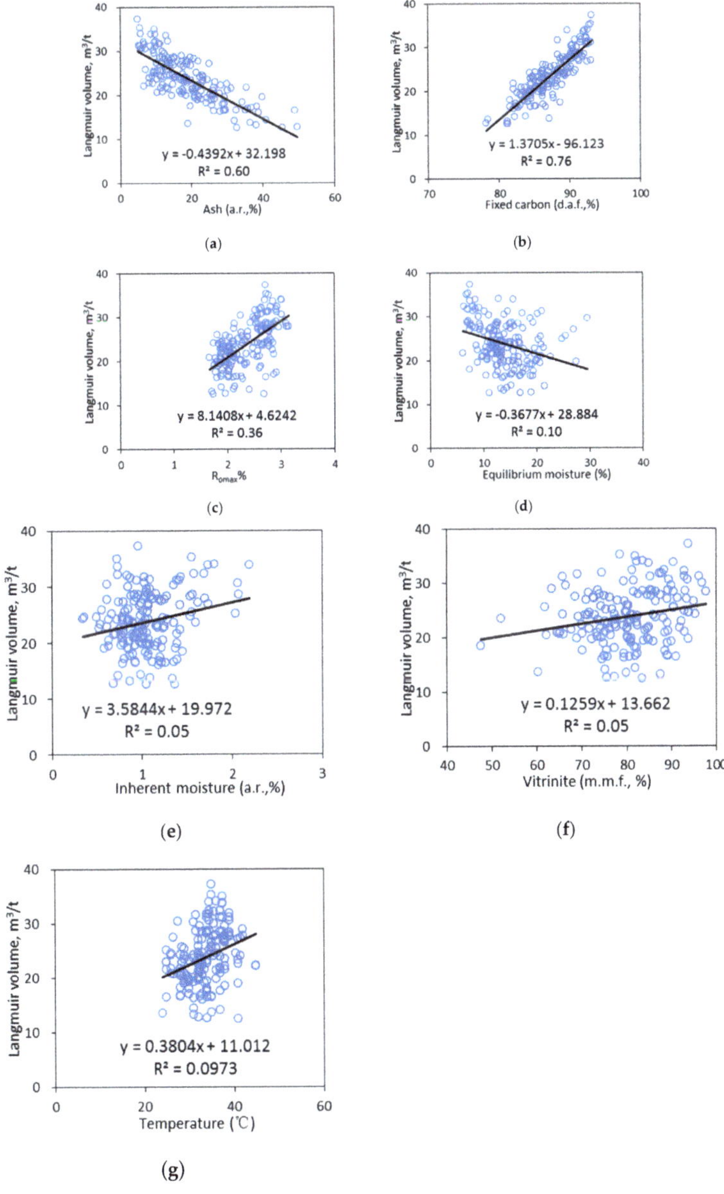

Figure 4. Correlation analysis of input features with the adsorption capacity represented with Langmuir volume. (**a**) ash (a.r.); (**b**) fixed carbon (d.a.f.); (**c**) vitrinite reflectance; (**d**) inherent moisture (a.r.); (**e**) equilibrium moisture; (**f**) vitrinite (m.m.f) and (**g**) temperature.

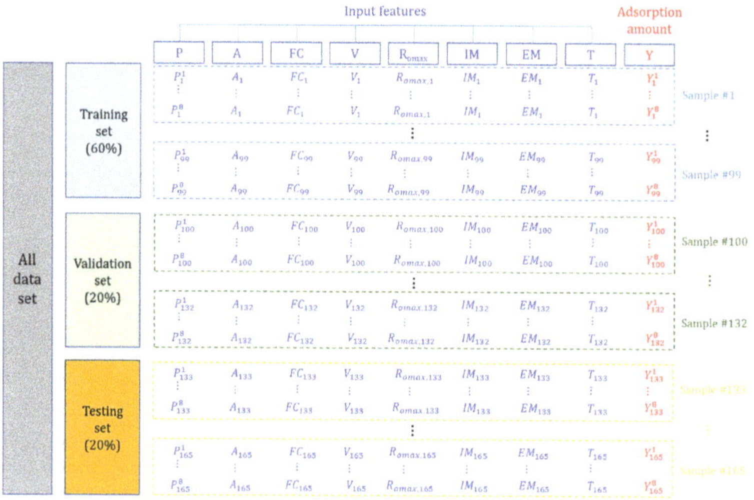

Input features — P, A, FC, V, R_{omax}, IM, EM, T · Adsorption amount — Y

	P	A	FC	V	R_{omax}	IM	EM	T	Y	
Training set (60%)	P_1^1	A_1	FC_1	V_1	$R_{omax,1}$	IM_1	EM_1	T_1	Y_1^1	Sample #1
	⋮	⋮	⋮	⋮	⋮	⋮	⋮	⋮	⋮	
	P_1^8	A_1	FC_1	V_1	$R_{omax,1}$	IM_1	EM_1	T_1	Y_1^8	
					⋮					⋮
	P_{99}^1	A_{99}	FC_{99}	V_{99}	$R_{omax,99}$	IM_{99}	EM_{99}	T_{99}	Y_{99}^1	Sample #99
	⋮	⋮	⋮	⋮	⋮	⋮	⋮	⋮	⋮	
	P_{99}^8	A_{99}	FC_{99}	V_{99}	$R_{omax,99}$	IM_{99}	EM_{99}	T_{99}	Y_{99}^8	
Validation set (20%)	P_{100}^1	A_{100}	FC_{100}	V_{100}	$R_{omax,100}$	IM_{100}	EM_{100}	T_{100}	Y_{100}^1	Sample #100
	⋮	⋮	⋮	⋮	⋮	⋮	⋮	⋮	⋮	
	P_{100}^8	A_{100}	FC_{100}	V_{100}	$R_{omax,100}$	IM_{100}	EM_{100}	T_{100}	Y_{100}^8	
					⋮					⋮
	P_{132}^1	A_{132}	FC_{132}	V_{132}	$R_{omax,132}$	IM_{132}	EM_{132}	T_{132}	Y_{132}^1	Sample #132
	⋮	⋮	⋮	⋮	⋮	⋮	⋮	⋮	⋮	
	P_{132}^8	A_{132}	FC_{132}	V_{132}	$R_{omax,132}$	IM_{132}	EM_{132}	T_{132}	Y_{132}^8	
Testing set (20%)	P_{133}^1	A_{133}	FC_{133}	V_{133}	$R_{omax,133}$	IM_{133}	EM_{133}	T_{133}	Y_{133}^1	Sample #133
	⋮	⋮	⋮	⋮	⋮	⋮	⋮	⋮	⋮	
	P_{133}^8	A_{133}	FC_{133}	V_{133}	$R_{omax,133}$	IM_{133}	EM_{133}	T_{133}	Y_{133}^8	
					⋮					
	P_{165}^1	A_{165}	FC_{165}	V_{165}	$R_{omax,165}$	IM_{165}	EM_{165}	T_{165}	Y_{165}^1	Sample #165
	⋮	⋮	⋮	⋮	⋮	⋮	⋮	⋮	⋮	
	P_{165}^8	A_{165}	FC_{165}	V_{165}	$R_{omax,165}$	IM_{165}	EM_{165}	T_{165}	Y_{165}^8	

All data set

Figure 5. Illustration of the database structure and division of the database into the training, validation and testing sets. P—equilibrium pressure; A—ash; FC—fixed carbon; V—vitrinite; R_{omax}—vitrinite reflectance; IM—inherent moisture; EM—equilibrium moisture; T—temperature; Y—adsorption amount. Subscript j denotes the jth sample; Superscript I on "P" and "Y" denote the ith equilibrium point on the adsorption isotherm.

The empirical results from [41,42] demonstrate that the accuracy and generalization capability of the GBDT can be significantly influenced by three parameters, namely the number of estimators, the shrinkage factor, and the maximal tree depth. As such, these parameters should be optimized in order to ensure the accuracy and robustness of the GBDT. In this study, the optimal values for the three parameters were determined through the exhaustive grid search method [43]. That is, all possible combinations of the parameter values were run sequentially, and the optimal parameterization was determined to be the one that results in the lowest root mean squared error (RMSE) for the validation set. Previous studies [41,42] suggested that a satisfactory performance of GBDT can be obtained with relatively small shrinkage factors (<0.1) and low-level tree complexity (with tree depth<6). As such, the shrinkage factor was varied from 0.005 to 0.105 with a step of 0.01, and the maximum tree depth was varied from two to seven with a step of 1 in this study. The optimal number of trees is a problem-dependent hyperparameter, which was set to vary from 500 to 5000 with a step of 500.

2.4.3. Evaluation Matrices

The performance of the GBDT estimation was quantitatively evaluated through four metrics, namely average absolute error (AAE), average relative error (ARE), root mean square error (RMSE), and determination coefficient (R^2). The definitions for these metrics are:

$$AAE = \frac{1}{N} \sum_{i=1}^{N} |y_i - f_i| \tag{5}$$

$$ARE = \frac{1}{N} \sum_{i=1}^{N} \left| \frac{y_i - f_i}{y_i} \right| \tag{6}$$

$$RMSE = \frac{1}{N} \sum_{i=1}^{N} (y_i - f_i)^2 \tag{7}$$

$$R^2 = \frac{\sum_{i=1}^{N}(y_i - f_i)^2}{\sum_{i=1}^{N}(y_i - \overline{y})^2} \tag{8}$$

where y and f are the measured and estimated adsorption amounts, respectively; $\overline{y}$ is the mean value of the measured adsorption amount; and, N is the number of data points.

2.5. Comparison with BP-ANN and SVM

The BP-ANN and SVM are powerful supervised machine learning algorithms that have been successfully applied in solving nonlinear regression problems in a variety of fields [32,44–46]. A most popular version of the BP-ANN is the multilayer perception network (MLPN), which is comprised of one input layer, one or more hidden layer, and one output layer. The training of a MLPN is, in essence, an iterative process of updating the weights and biases of the nodes by using the back propagation algorithm in order to minimize an error function. The basic philosophy behind the SVM is to convert the nonlinear regression problem in the true space into linear approximations in a higher dimensional feature space by minimizing a regularized loss function. Mathematical details on the BP-ANN and SVM have been extensively addressed previously (see e.g., [24,39]), which, therefore, are not repeated in this paper.

The LIBSVM pakage [47] and the neural network module that were implemented in the Matlab (V2019) were used for conducting SVM and BP-ANN regressions, respectively. The data points and input variables are identical with that in the GBDT regression. A BP-ANN with three layers (one input, one hidden, and on output layer) has proven to be capable of approximating any continuous function with any accuracy [32], which, therefore, was adopted in this study. It should be noted that (i) the performance of a BP-ANN can be significantly influenced by the number of neurons in the hidden layer [44] and (ii) for an SVM with a kernel of radial basis function (RBF, which is most frequently used for regression), the regression accuracy is associated with regulation and error goal parameters [48,49]. In order to attain a fair comparison, parameters that may affect the BP-ANN and SVM performance were tuned and optimized while using the exhaustive grid search, in a similar manner with the GBDT. Table 2 shows the optimal key model parameters for the BP-ANN and SVM.

Table 2. Modeling parameters for the back-propagation artificial neural network (BP-ANN) and support vector machine (SVM).

Method	Property	Value
BP-ANN	No. of hidden layers	1
	No. of nodes in each hidden layer	20
	Activation function for hidden layer(s)	Tangent
	Activation function for output layer	Linear
SVM	Activation function	RBF
	Regulation parameter	86
	Error goal parameter	0.005

3. Results

3.1. Performance of the GBDT Estimation Model

The optimal hyperparameter values for the GBDT, as determined with the exhaustive grid search method, were 0.01 for the shrinkage factor, 3 for the tree depth, and 1500 for the number of trees, respectively. Figure 6 depicts the GBDT estimation results for the training, validation, and testing sets. It is shown in Figure 6a that all of the training data points are grouped closely around the 45-degree line. Table 3 demonstrates extremely low error matrices (an AAE of 0.33 m^3/t, an ARE of 2.31%, and a RMSE of 0.42 m^3/t) and a remarkably high R^2 of 0.993 for the training set. These evaluation matrices prove that the GBDT is capable of accurately reproducing the adsorption amount that is based on the input variables. For the validation and testing set, although the cross plots of the measured versus estimated values show a more scattered pattern than the training set, the majority of the data points are distributed around the 45-degree line and the deviations are within small ranges (Figure 6b,c). The AAE, ARE, RMSE, and determination coefficient (R^2) are calculated to be 0.83 m^3/t, 5.97%, 1.00 m^3/t, and 0.950 for the validation, and 0.85 m^3/t, 6.35%, 1.06 m^3/t, and 0.946 for the testing sets, respectively. The error matrices for the validation and testing are quite comparable, suggesting strong robustness of the constructed model (Table 3). In this regard, the GBDT model can be considered to have a strong generalization capability, as indicated by the relatively low error matrices and high R^2.

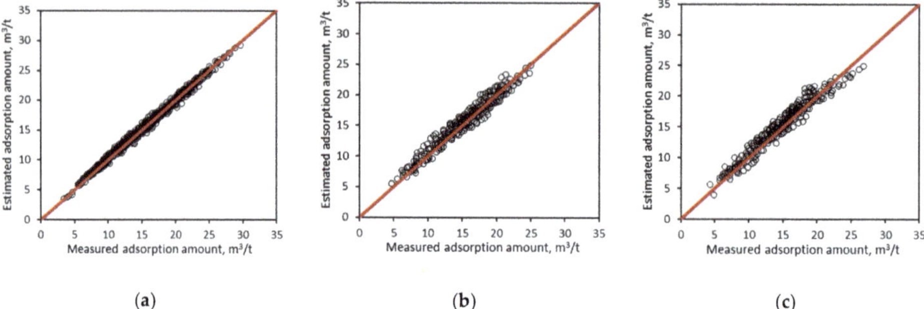

(a)   (b)   (c)

Figure 6. Cross plot of the gradient boosting decision trees (GBDT) estimated versus measured adsorption amount for the (**a**) training (**b**) validation; and, (**c**) testing sets. Open circles are data points; red lines are 45-degree lines.

Table 3. Error statistics of the GBDT, BP-ANN, and SVM models.

Data Set	Error Matrices	GBDT	ANN	SVM
Training set	AAE, m^3/t	0.33	0.21	0.71
	ARE, %	2.31	1.62	5.58
	RMSE, m^3/t	0.42	0.28	1.01
	R^2, fraction	0.993	0.997	0.959
Validation set	AAE, m^3/t	0.83	1.14	1.11
	ARE, %	5.97	8.10	9.12
	RMSE, m^3/t	1.00	1.45	1.57
	R^2, fraction	0.950	0.895	0.877
Testing set	AAE, m^3/t	0.85	1.26	0.96
	ARE, %	6.35	9.25	7.81
	RMSE, m^3/t	1.06	1.81	1.23
	R^2, fraction	0.946	0.842	0.927
Whole set	AAE, m^3/t	0.53	0.61	0.84
	ARE, %	3.85	4.44	6.74
	RMSE, m^3/t	0.73	1.06	1.19
	R^2, fraction	0.977	0.952	0.940

The comparison between the estimated and measured adsorption isotherms for typical samples in the testing set was conducted in order to further demonstrate the accuracy of the GBDT model in reproducing the adsorption isotherm for an individual coal sample. The methane adsorption capacity on the coal samples is predominantly controlled by the ash and fixed carbon contents, as mentioned in Section 2.4.1. Therefore, four typical samples—one with the highest ash content, one with the lowest ash content, one with the highest fixed carbon content and one with the lowest fixed carbon content—among all samples in the testing set were selected for illustrating the model accuracy.

For the two samples with respective ash contents of 9.6% and 39.96% and one sample with low fixed carbon content (83.88%), the estimated adsorption isotherms are in excellent agreement with the measured ones, as can be seen from Figure 7. For the sample with high fixed carbon content (91.54%), the adsorption equilibrium points at lower pressures (≤≈4.0 MPa) agrees well with the measured ones, whereas certain deviations exist for the equilibrium points at higher pressures (>≈4.0MPa). The maximum error occurs at an equilibrium pressure of ≈8.0 MPa, with the estimated and measured adsorption amounts being 23.71 and 25.23 m^3/t, respectively. Such discrepancy, as we note, can be considered to be acceptable given the uncertainties that are associated with sample preparation, data acquisition, and measurement operations [12]. Previous reproducibility tests [50,51] showed that discrepancies in the adsorption isotherm measurement may reach high, up to 10–15% on a same coal sample, which are even higher than the GBDT estimation results. It should also be pointed out that the estimated adsorption amount follows a monotonically increasing trend with increasing pressure (which is basic characteristics for methane adsorption isotherms on coals), although no specific constraint was applied in the training process to compel such monotonicity. These results confirm the reliability of the constructed GBDT model in estimating the methane adsorption isotherms on coals with reasonable accuracies.

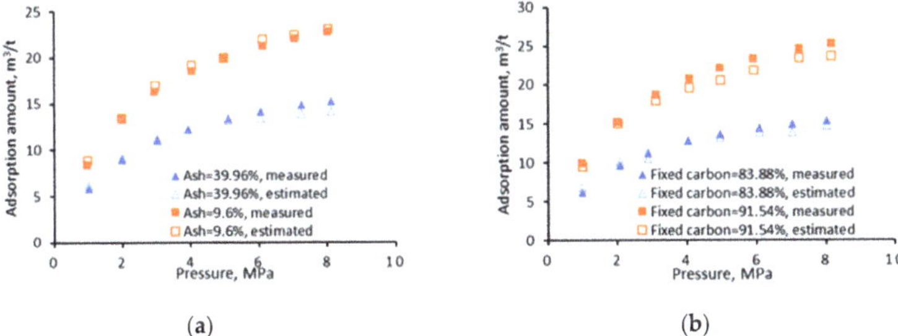

Figure 7. Comparison of the estimated with measured adsorption isotherms for samples with (**a**) ash contents of 9.6% and 39.96%, respectively, and (**b**) fixed carbon contents of 83.88% and 91.54%, respectively.

3.2. Comparison with BP-ANN and SVM

Figure 8 shows the cross plots of BP-ANN estimated with measured adsorption amounts for the training, validation, and testing sets. All of the data points are generally located on the 45-degree line, which suggests that BP-ANN has an extraordinary capability to accurately correlate the output with input variables for the training set, as can be seen from Figure 8a. Table 3 demonstrates that the BP-ANN outperforms the GBDT in terms of error matrices for the training set. However, Figure 8b,c demonstrate that a noticeable number of data points deviate severely from the 45-degree line for both the validation and testing sets, resulting in higher errors (AAE, ARE, and RMSE) and lower R^2 than the GBDT (Table 3). These observations suggest that the generalization capability of BP-ANN is highly questionable and severe over-fitting issue occurs. As such, the BP-ANN should not be considered to be suitable for accurately estimating the adsorption isotherms.

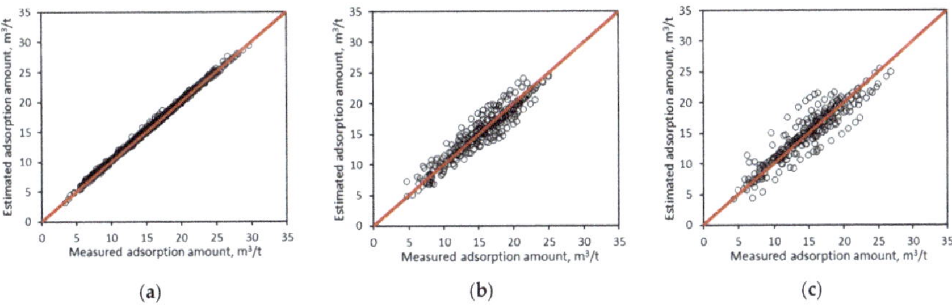

Figure 8. Cross plot of the BP-ANN estimated versus measured adsorption amounts for the (**a**) training, (**b**) validation, and (**c**) testing sets. Open circles are data points; red lines are 45-degree lines.

Figure 9 depicts the estimation results of SVM regression. As shown, there is a noticeable number of data points that severely deviate from the 45-degree line for the training, validation, and testing sets. Thus, it is concluded that the SVM is neither capable of accurately learning the underlying correlations between the output and input variables nor capable of giving reasonable predictions. Comparisons of the evaluation matrices for the SVM with those for the GBDT and BP-ANN (Table 3) suggest that the SVM has better generalization capability than the BP-ANN, but performs worse than the GBDT.

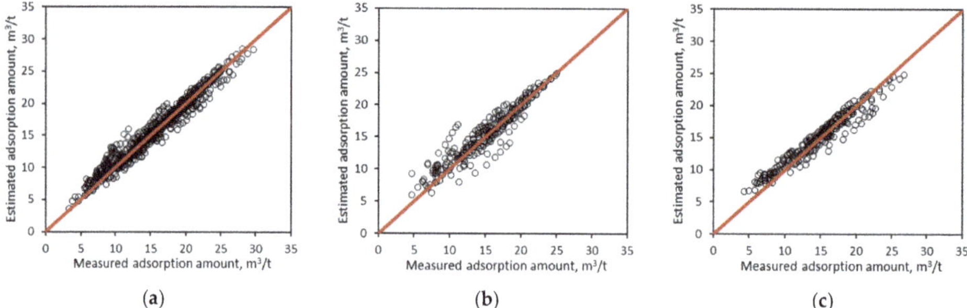

Figure 9. Cross plot of the SVM estimated versus measured adsorption amounts for the (**a**) training, (**b**) validation, and (**c**) testing sets. Open circles are data points; red lines are 45-degree lines.

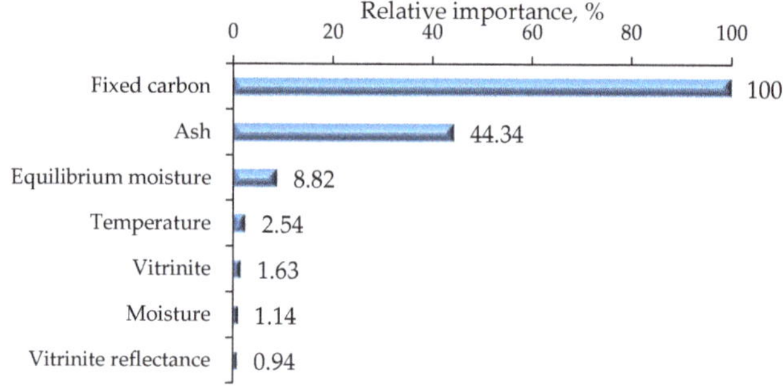

Figure 10. Relative importance of the input variables to the adsorption isotherm.

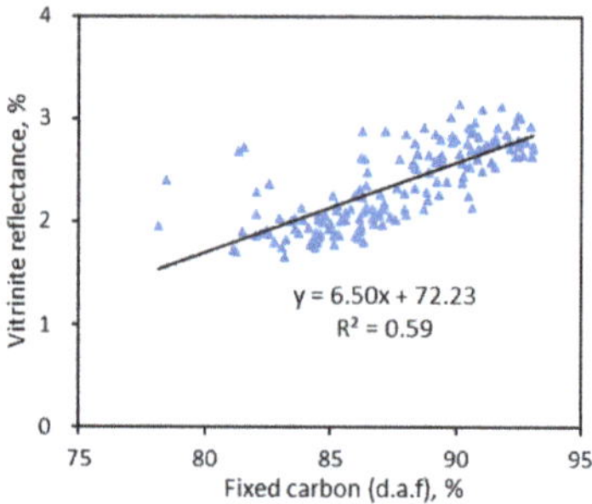

Figure 11. Dependence of vitrinite reflectance on fixed carbon of the coal samples.

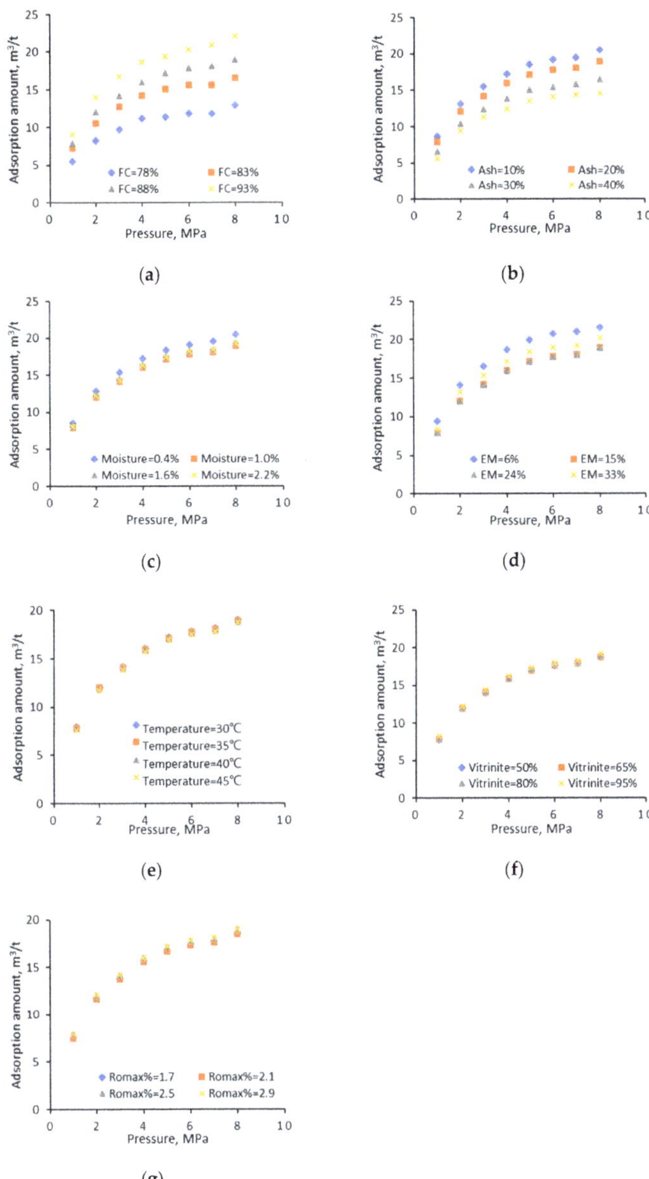

Figure 12. Calculated adsorption isotherms using the constructed GBDT model with reference to varying (**a**) fixed carbon (d.a.f), (**b**) ash (a.d.), (**c**) inherent moisture (a.r.), (**d**) equilibrium moisture, (**e**) temperature, (**f**) vitrinite (m.m.f), and (**g**) vitrinite reflectance.

4. Discussion

4.1. Analyses of Effects of Input Features on Adsorption Isotherms

4.1.1. Relative Importance of Input Features

Once the estimation model has been constructed, it should be of practical meaning to quantify the effect of each input feature on the adsorption isotherm. In this section, the relative importance of each input variable is quantified while using the mean decrease impurity importance (MDI) [52,53]. A most significant advantage of the MDI over conventional Pearson or Spearman coefficients is that the MDI does not require a priory assumption of linear or monotonic dependence of the output on the input features, which, therefore, should be more accurate in quantifying the effects of each input feature [54]. Figure 10 shows that fixed carbon and ash are three key factors that control the adsorption amount. The equilibrium moisture has a relative importance of ≈8.8%, while the remaining factors (temperature, vitrinite, vitrinite reflectance, and inherent moisture) have relative importance of less than 3.0%, which suggests the very minor or even negligible influences of these factors on the adsorption amount. Here, it is noted the effect of vitrinite reflectance is significantly diluted when compared with the correlation analysis in Section 2.3, which is possibly due to the collinearity between the vitrinite reflectance and fixed carbon for the coal samples (Figure 11). The existence of collinearity may result in the abnormal response of the output to one or several of the collinear inputs [55]. Fixed carbon demonstrates an obviously stronger correlation on the adsorption capacity than vitrinite reflectance does and, thus, the effect of the vitrinite reflectance has a high risk of being overridden by the fixed carbon considering their collinearity, as can be seen from Figure 4b,c.

4.1.2. Univariate Analyses

The univariate analysis was conducted using the constructed GBDT model to further demonstrate how the adsorption isotherms are affected by the input features. The base value was set to be 20%, 1.0%, 88%, 80%, 2.5%, 15%, and 35°C for ash, inherent moisture, fixed carbon, vitrinite, vitrinite reflectance, equilibrium moisture, and temperature, respectively. These values were set as approximately the averaged ones that are shown in Table 1. Each input variable was tuned at four values (that are within the range of all the coal samples in this study) and the corresponding adsorption isotherms (at pressures of 1 to 8 MPa with a step of 1 Mpa) were sequentially calculated, which are shown in Figure 12.

- Fixed carbon

Figure 12a depicts the adsorption isotherm with reference to varying fixed carbon. It is well demonstrated that the isotherm tends to move upwards as fixed carbon increases. Previous studies [16,56] observed that the methane adsorption capacity follows a first decreasing and then increasing trend with increasing fixed carbon, with the minimum occurring at ≈60–80%. This parabolic trend may be attributed to the variations in the micro-pore surface areas that are associated with the coalification jump that occurs approximately in the range of 75–85% fixed carbon [17]. More recently, Chattaraj et al. [1] showed that, for Indian coals with fixed carbon content of >75%, methane adsorption capacity is in positive linear correlation with fixed carbon. It is interesting to note that the coal samples used in this study have a generally high fix carbon contents of >77%, which suggests that our findings are in line with these previous studies.

- Ash

Figure 12b illustrates that the adsorption isotherm exhibits an obvious negative correlation with ash at all pressures. It is well understood that an increase in ash content tends to decrease the adsorption isotherm on coals, because (i) ash has no affinity to methane adsorption [7,18] and (ii) ash-rich samples are generally associated with lower micro porosities [57] and, therefore, provide less adsorption space to accommodate gas molecules.

- Moisture

Variations in adsorption isotherms caused by inherent and equilibrium moistures are obviously less significant than that by fixed carbon or ash, which is consistent with the ranking of relative importance, as shown in Figure 12c,d. Additionally, it is noticeable that the adsorption capacity does not follow a monotonous decreasing trend with the increase in either inherent or equilibrium moistures. It has been extensively addressed in previous studies [14,58,59] that a coal sample in the moisture-equilibrium state has a significantly lower adsorption capability than in the dry state. This is due to the occupation of some adsorption sites on the coal surface by water molecules because coals have a preferential affinity to water over methane [7]. However, for a coal sample that is already in a moisture-equilibrated state, a further increment in moisture content does not affect the adsorption capacity to gas [14,60]. Besides, as stated in [7,13], the moisture content may be predominated by the coal rank. Thus, the effect of moisture content on the adsorption isotherm may possibly be overridden by the coal rank indicators such as fixed carbon for the coal samples in this study.

- Temperature

Figure 12e shows that there is no significant change in the adsorption isotherm with elevating temperature. Most previous studies [61,62] conclude that the elevation in temperature may result in a noticeable reduction in methane adsorption capacity, because the sorptive surface coverage at a specific gas pressure decreases with increasing temperature, as derived from thermodynamics [7]. However, Crosdale et al.'s [60] experiments on moist coals showed no significant dependence of adsorption capacity on temperatures. More recently, Guan et al. [63] showed that the adsorption capacities for both methane and CO_2 remained constant as the temperatures were elevated from 323 to 343 K. Our observations are consistent with [60], which may be attributed to the compensation by water molecule release for the reduction in the sorptive surface coverage caused by temperature elevation [6].

- Vitrinite

To date, there are still controversies regarding the effect of vitrinite content on the methane adsorption capacity. Some studies [1,4,17] showed that vitrinite-rich (bright) coals have a higher methane adsorption capacity than the inertinite-rich (dull) ones with equivalent ranks, which may be attributed to the existence of more micro-pores in vitrinite that is favorable to accommodation of gas molecules [64]. Dutta et al. [18] and Feng et al. [16] stated that methane adsorption capacity follows a "U-shaped" trend with vitrinite content. Other authors [13,65,66] found no obvious correlation between the adsorption capacity and vitrinite content, which holds valid for the coal samples in this study (Figure 12f).

- Vitrinite reflectance

Vitrinite reflectance is a commonly used indicator of the coal rank (maturity), which numerous previous studies [15,16,18] have demonstrated to be closely correlated with the methane adsorption capacities in coals. For the coal samples that were investigated in this study, the vitrinite reflectance exerts a negligible effect on the adsorption isotherm (Figure 12g). This is in line with [67], who argued that the vitrinite reflectance alone cannot control the maximum sorption capacities and simple lithotype analysis is insufficient for evaluating the effects of coal type. One explanation for this observation is that the influence of vitrinite reflectance on methane adsorption capacity is caused by the variations in macromolecular [68] and pore [56] structures during the coalification process as coal maturity increases. Besides, it is again noted that there exists a dependence of vitirnite reflectance on the fixed carbon for the coals in this study (Figure 11). Thus, the effect of vitrinite reflectance may be overridden by that of the fixed carbon from the standpoint of statistical regressions.

The univariate analyses based on the GBDT model are in well accordance with numerous previous studies, which further confirms the validity of the constructed model, as can be seen from the above discussion. It can be also concluded that the GBDT has a remarkably capability of "automatically" identifying the true important features and properly finding the underlying correlations between the output and each input feature, even though both of the features with collinearity and features exerting minor/negligible effects on the output were included in the model.

4.2. Influence of Input Features on the Model Accuracy

The constructed model includes not only features with convincing control on the adsorption capacity (equilibrium pressure, ash content, fixed carbon content, and vitrinite reflectance), but also features showing minor or negligible relevance with the output (vitrinite content, inherent moisture, equilibrium moisture, and temperature), as mentioned earlier in Section 2.4.1. To demonstrate the influence of input feature selection on the model accuracy, several estimation models with different scenarios of input features (Table 4) were separately constructed, following the same procedure described in Section 2.4.2.

Table 4. Input feature scenarios for analyzing the estimation accuracy.

Scenario No.	Input Features *
1	P, A, R_o, FC
2	P, A, R_o, FC, EM
3	P, A, R_o, FC, IM
4	P, A, R_o, FC, V
5	P, A, R_o, FC, T
6	P, A, R_o, FC, EM, IM
7	P, A, R_o, FC, EM, V
8	P, A, R_o, FC, EM, T
9	P, A, R_o, FC, EM, IM, V
10	P, A, R_o, FC, EM, IM, T
11	P, A, R_o, FC, EM, V, T
12	P, A, R_o, FC, EM, IM, V, T

* Abbreviations: P—equilibrium pressure; A—ash; FC—fixed carbon; IM—inherent moisture; R_o—vitrinite reflectance; EM—equilibrium moisture; V—vitrinite; T—temperature.

We began the analysis by including only equilibrium pressure and three coal property parameters (fixed carbon, ash and vitrinite reflectance) that show relatively strong correlations with adsorption capacity (Figure 4) in order to estimate the adsorption isotherm (Scenario#1 in Table 4). It can be seen from Figure 13 that this scenario produces an estimation result with the lowest accuracy in terms of all the evaluation matrices, suggesting that using only these four key features are not sufficient for accurate estimation of the isotherm. With these four parameters held in the model, we then added one of the remaining less significant features (inherent moisture, equilibrium moisture, vitrinite, temperature) at a time into the model. It is shown (Figure 13) that the inclusion of equilibrium moisture in the model (Scenarios#2) results in a most noticeable reduction in the estimation error than that of any of the other features (Scenarios#3, #4, and #5). In order to honor the contribution of equilibrium moisture to estimation accuracy improvement, we fixed equilibrium moisture together with the aforementioned four key parameters in the input feature bank; the feature bank was then expanded by adding one (Scenarios#6, #7, and #8) or two (Scenarios#9, #10, and #11) out of the remaining features sequentially in order to further examine the effect of input feature scenarios on the estimation results. It is depicted in Figure 13 that the estimation accuracy exhibits a general decreasing trend with more input features being included in the model. The model that incorporates all available input features (the one addressed in Section 3.1, which is assigned as Scenario#12 in this section) demonstrates the highest estimation accuracy among all of the scenarios investigated.

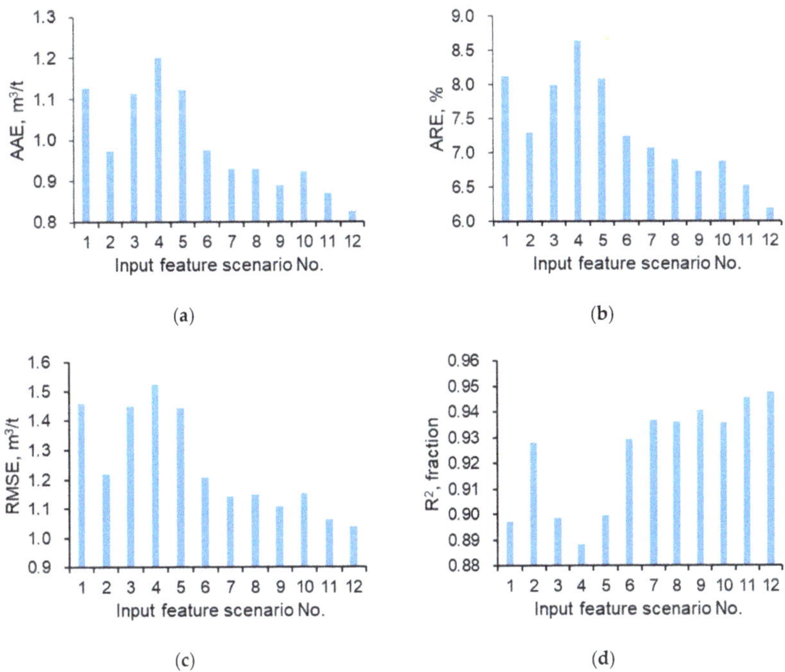

Figure 13. Error matrices of (**a**) average absolute error (AAE), (**b**) average relative error (ARE), (**c**) root mean squared error (RMSE) and (**d**) R^2 for different input feature scenarios.

All available features that may potentially affect the isotherm should be incorporated in the construction of the estimation model for the adsorption isotherm, as indicated from the above results. The exclusion of insignificant features identified with correlation coefficient is highly questionable and tends to decrease the estimation accuracy. This finding is well supported by Beker et al. [33]. It is reiterated that the GBDT is highly robust to interferences from insignificant features and it has a strong capability to properly find the underlying correlations between the input features and the adsorption amount.

It should be noted that feeding more input features into the estimation model requires more efforts to obtain the associated feature information. Generally, the proximate analysis parameters (ash, fixed carbon, and inherent moisture contents) are less expensive and easier to be experimentally measured than the maceral analysis parameters (e.g., vitrinite content). Therefore, it should be of practical significance to use as less maceral features as possible while ensuring relatively high modeling accuracies. Scenarios#7, #8, # 9, and #11 result in high modeling accuracies when compared with Scenario #12, as can be seen from Figure 13. Among these four scenarios, only Scenario#8 does not include the vitrinite, which is a required input feature for all of the remaining scenarios (Table 4). As such, Scenario#8 may be the most "cost-effective" ones when considering the less input features and reasonably high modeling accuracy.

5. Conclusions

This paper proposed the use of a machine learning algorithm, namely GBDT, in order to estimate methane adsorption isotherm on coals that are based on coal properties (ash, fixed carbon, inherent moisture, and vitrinite contents and vitrinite reflectance), equilibrium moisture content and temperature. Laboratory tests, including proximate analysis, maceral group identification, vitrinite reflectance determination, and adsorption isotherm measurements, were conducted on 165 coal samples retrieved from the Qinshui basin in China in order to develop a database for regression. It has been demonstrated that the GBDT is capable of not only reproducing the adsorption isotherms with reasonable accuracies, but also properly recovering the underlying relation between the input and output variables. As a comparison, the BP-ANN is associated with the over-fitting problem, whereas the SVM has difficulties in accurately estimating the adsorption isotherms in both the training and testing stages. Such observations confirmed the superiority of the GBDT over other ML tools in solving the specific regression problem in this study. Furthermore, the relative importance scanning and univariate analysis based on the constructed GBDT model showed that the adsorption isotherms are primarily controlled by the fixed carbon and ash contents for the coals that were investigated in this study. Other factors, including vitrinite, inherent and equilibrium moistures, vitrinite reflectance, and temperature, exert minor or even negligible effects on the adsorption isotherm.

Author Contributions: Conceptualization, Q.F.; methodology, J.Z.; software, X.Z.; validation, N.W.; formal analysis, X.Z.; investigation, J.Z.; data curation, J.Y.; writing—original draft preparation, J.Z.; writing—review and editing, X.Z.; visualization, J.Y.; supervision, Q.F.; project administration, Q.H.; funding acquisition, J.Z. and X.Z. All authors have read and agreed to the published version of the manuscript.

Funding: This research was funded by China National Natural Science Foundation, grant numbers 51904319 and U1810105, China Postdoctoral Science Foundation, grant number 2018M642727.

Conflicts of Interest: The authors declare no conflict of interest.

References

1. Chattaraj, S.; Mohanty, D.; Kumar, T.; Halder, G.; Mishra, K. Comparative study on sorption characteristics of coal seams from Barakar and Raniganj formations of Damodar Valley Basin, India. *Int. J. Coal Geol.* **2019**, *212*, 103202. [CrossRef]
2. Zhang, J.; Feng, Q.; Zhang, X.; Hu, Q.; Wen, S.; Chen, D.; Zhai, Y.; Yan, X. Multi-fractured horizontal well for improved coalbed methane production in eastern Ordos basin, China: Field observations and numerical simulations. *J. Pet. Sci. Eng.* **2020**, *194*, 107488. [CrossRef]
3. Zhang, J.; Feng, Q.; Zhang, X.; Bai, J.; Karacan, C.Ö.; Wang, Y.; Elsworth, D. A two-stage step-wise framework for fast optimization of well placement in coalbed methane reservoirs. *Int. J. Coal Geol.* **2020**, *225*. [CrossRef]
4. Crosdale, P.J.; Beamish, B.B.; Valix, M. Coalbed methane sorption related to coal composition. *Int. J. Coal Geol.* **1998**, *35*, 147–158. [CrossRef]
5. Kim, D.; Seo, Y.; Kim, J.; Han, J.; Lee, Y. Experimental and simulation studies on adsorption and diffusion characteristics of coalbed methane. *Energies* **2019**, *12*, 3445. [CrossRef]
6. Peng, Z.; Liu, S.; Li, Y.-J.; Deng, Z.; Feng, H. Pore-scale lattice Boltzmann simulation of gas diffusion–adsorption kinetics considering adsorption-induced diffusivity change. *Energies* **2020**, *13*, 4927. [CrossRef]
7. Busch, A.; Gensterblum, Y. CBM and CO_2-ECBM related sorption processes in coal: A review. *Int. J. Coal Geol.* **2011**, *87*, 49–71. [CrossRef]
8. Gasparik, M.; Rexer, T.F.; Aplin, A.C.; Billemont, P.; De Weireld, G.; Gensterblum, Y.; Henry, M.; Krooss, B.M.; Liu, S.; Ma, X.; et al. First international inter-laboratory comparison of high-pressure CH_4, CO_2 and C_2H_6 sorption isotherms on carbonaceous shales. *Int. J. Coal Geol.* **2014**, *132*, 131–146. [CrossRef]
9. Gensterblum, Y.; van Hemert, P.; Billemont, P.; Battistutta, E.; Busch, A.; Krooss, B.M.; De Weireld, G.; Wolf, K.-H. European inter-laboratory comparison of high pressure CO_2 sorption isotherms ii: Natural coals. *Int. J. Coal Geol.* **2010**, *84*, 115–124. [CrossRef]
10. Mavor, M.J.; Hartman, C.; Pratt, T.J. Uncertainty in sorption isotherm measurements. In Proceedings of the International Coalbed Methane Symposium, Tuscaloosa, AL, USA, 12–14 May 2004.
11. van Hemert, P.; Rudolph-Floter, S.; Wolf, K.-H.A.; Bruining, J. Estimate of equation of state uncertainty for manometric sorption experiments: Case study with Helium and carbon dioxide. *SPE J.* **2010**, *15*, 146–151. [CrossRef]

12. Zlotea, C.; Moretto, P.; Steriotis, T. A round robin characterisation of the hydrogen sorption properties of a carbon based material. *Int. J. Hydrogen Energy* **2009**, *34*, 3044–3057. [CrossRef]

13. Busch, A.; Han, F.; Magill, C.R. Paleofloral dependence of coal methane sorption capacity. *Int. J. Coal Geol.* **2019**, *211*. [CrossRef]

14. Day, S.; Sakurovs, R.; Weir, S. Supercritical gas sorption on moist coals. *Int. J. Coal Geol.* **2008**, *74*, 203–214. [CrossRef]

15. Weniger, P.; Kalkreuth, W.; Busch, A.; Krooss, B.M. High-pressure methane and carbon dioxide sorption on coal and shale samples from the Paraná Basin, Brazil. *Int. J. Coal Geol.* **2010**, *84*, 190–205. [CrossRef]

16. Feng, Q.; Zhang, J.; Zhang, X.; Shu, C.; Wen, S.; Wang, S.; Li, J. The use of alternating conditional expectation to predict methane sorption capacity on coal. *Int. J. Coal Geol.* **2014**, *121*, 137–147. [CrossRef]

17. Laxminarayana, C.; Crosdale, P.J. Role of coal type and rank on methane sorption characteristics of Bowen Basin, Australia coals. *Int. J. Coal Geol.* **1999**, *40*, 309–325. [CrossRef]

18. Dutta, P.; Bhowmik, S.; Das, S. Methane and carbon dioxide sorption on a set of coals from India. *Int. J. Coal Geol.* **2011**, *85*, 289–299. [CrossRef]

19. Zhang, Z.; Schott, J.A.; Liu, M.; Chen, H.; Lu, X.; Sumpter, B.G.; Fu, J.; Dai, S. Prediction of carbon dioxide adsorption via deep learning. *Angew. Chem. Int. Ed.* **2019**, *131*, 265–269. [CrossRef]

20. Friedman, J.H. Greedy function approximation: A gradient boosting machine. *Ann. Stat.* **2001**, *29*, 1189–1232. [CrossRef]

21. Friedman, J.H. Stochastic gradient boosting. *Comput. Stat. Data Anal.* **2002**, *38*, 367–378. [CrossRef]

22. Luo, Z.; Sun, Z.; Ma, F.; Qin, Y.; Ma, S. Power Optimization for Wind Turbines Based on Stacking Model and Pitch Angle Adjustment. *Energies* **2020**, *13*, 4158. [CrossRef]

23. Amar, M.; Shateri, M.; Hemmati-Sarapardeh, A.; Alamatsaz, A. Modeling oil-brine interfacial tension at high pressure and high salinity conditions. *J. Petrol. Sci. Eng.* **2019**, *183*, 106413. [CrossRef]

24. Cai, Y.; Liu, D.; Yao, Y.; Li, J.; Qiu, Y. Geological controls on prediction of coalbed methane of No. 3 coal seam in Southern Qinshui Basin, North China. *Int. J. Coal Geol.* **2011**, *88*, 101–112. [CrossRef]

25. Song, Y.; Ma, X.; Liu, S.; Jiang, L.; Hong, F.; Qin, Y. Accumulation conditions and key technologies for exploration and development of Qinshui coalbed methane field. *Pet. Res.* **2018**, *3*, 320–335. [CrossRef]

26. Chinese National Standard GB/T 212-2008. *Proximate Analysis of Coal*; Standardization Administration of China: Beijing, China, 2008.

27. Chinese National Standard GB/T 8899-2013. *Determination of Maceral Composition and Minerals in Coal*; Standardization Administration of China: Beijing, China, 2013.

28. Chinese National Standard GB/T 6948-2008. *Method of Determining Microscopically the Reflectance of Vitrinite in Coal*; Standardization Administration of China: Beijing, China, 2008.

29. Sanders, M.; Rimmer, S. Revisiting the thermally metamorphosed coals of the Transantarctic Mountains, Antarctica. *Int. J. Coal Geol.* **2020**, *228*, 103550. [CrossRef]

30. Zhang, J.; Feng, Q.; Zhang, X.; Shu, C.; Wang, S.; Wu, K. A supervised learning approach for accurate modeling of CO_2-Brine interfacial tension with application in identifying the optimum sequestration depth in saline aquifers. *Energy Fuel.* **2020**, *34*, 7353–7362. [CrossRef]

31. Cai, J.; Luo, J.; Wang, S.; Yang, S. Feature selection in machine learning: A new perspective. *Neurocomputing* **2018**, *300*, 70–79. [CrossRef]

32. Feng, Q.; Zhang, J.; Zhang, X.; Wen, S. Proximate analysis based prediction of gross calorific value of coals: A comparison of support vector machine, alternating conditional expectation and artificial neural network. *Fuel Process. Technol.* **2015**, *129*, 120–129. [CrossRef]

33. Beker, W.; Gajewska, E.P.; Badowski, T.; Grzybowski, B.A. Prediction of major regio-, site-, and diastereoisomers in diels-alder reactions by using machine-learning: The importance of physically meaningful descriptors. *Angew. Chem. Int. Ed.* **2019**, *58*. [CrossRef]

34. An, F.H.; Cheng, Y.P.; Wu, D.M.; Wang, L. The effect of small micropores on methane adsorption of coals from Northern China. *Adsorption* **2013**, *19*, 83–90. [CrossRef]

35. Clarkson, C.R.; Bustin, R.M. The effect of pore structure and gas pressure upon the transport properties of coal: A laboratory and modeling study. 1. Isotherms and pore volume distributions. *Fuel* **1999**, *78*, 1333–1344. [CrossRef]

36. Liu, X.; He, X. Effect of pore characteristics on coalbed methane adsorption in middle-high rank coals. *Adsorption* **2017**, *23*, 3–12. [CrossRef]

37. Hao, S.; Wen, J.; Yu, X.; Wei, C. Effect of the surface oxygen groups on methane adsorption on coals. *Appl. Surf. Sci.* **2013**, *264*, 433–442. [CrossRef]

38. Jiang, J.; Yang, W.; Cheng, Y.; Zhao, K.; Zheng, S. Pore structure characterization of coal particles via MIP, N_2 and CO_2 adsorption: Effect of coalification on nanopores evolution. *Powder Technol.* **2019**, *354*, 136–148. [CrossRef]

39. Li, Z.; Liu, D.; Cai, Y.; Ranjith, P.; Yao, Y. Multi-scale quantitative characterization of 3-D pore-fracture networks in bituminous and anthracite coals using FIB-SEM tomography and X-ray μ-CT. *Fuel* **2017**, *209*, 43–53. [CrossRef]

40. Zhang, R.; Liu, S.; Bahadur, J.; Elsworth, D.; Melnichenko, Y.; He, L.; Wang, Y. Estimation and modeling of coal pore accessibility using small angle neutron scattering. *Fuel* **2015**, *161*, 323–332. [CrossRef]

41. Elith, J.; Leathwick, J.R.; Hastie, T. A working guide to boosted regression trees. *J. Anim. Ecol.* **2008**, *77*, 802–813. [CrossRef]

42. Ridgeway, G. Generalized Boosted Models: A Guide to the GBM Package. Available online: http://citeseerx.ist.psu.edu/viewdoc/summary?doi=10.1.1.151.4024 (accessed on 27 September 2020).

43. Pedregosa, F.; Varoquaux, G.; Gramfort, A.; Michel, V.; Thirion, B.; Grisel, O.; Blondel, M.; Prettenhofer, P.; Weiss, R.; Dubourg, V.; et al. Scikit-learn: Machine learning in python. *J. Mach. Learn. Res.* **2012**, *12*, 2825–2830.

44. Zhang, J.; Feng, Q.; Wang, S.; Zhang, X.; Wang, S. Estimation of CO_2–brine interfacial tension using an artificial neural network. *J. Supercrit. Fluid* **2016**, *107*, 31–37. [CrossRef]

45. Dixit, N.; McColgan, P.; Kusler, K. Machine learning-based probabilistic lithofacies prediction from conventional well Logs: A case from the Umiat Oil Field of Alaska. *Energies* **2020**, *13*, 4862. [CrossRef]

46. Jadidi, M.; Kostic, S.; Zimmer, L.; Dworkin, S.B. An artificial neural network for the low-cost prediction of soot emissions. *Energies* **2020**, *13*, 4787. [CrossRef]

47. Chang, C.; Lin, C. LIBSVM: A library for support vector machines. *ACM Trans. Intell. Syst. Technol.* **2019**, *2*, 1–27. [CrossRef]

48. Khan, P.W.; Byun, Y.-C.; Lee, S.-J.; Kang, D.-H.; Kang, J.-Y.; Park, H.-S. Machine learning-based approach to predict energy consumption of renewable and nonrenewable power sources. *Energies* **2020**, *13*, 4870. [CrossRef]

49. Memon, Z.A.; Trinchero, R.; Manfredi, P.; Canavero, F.G.; Stievano, I.S. Compressed machine learning models for the uncertainty quantification of power distribution Networks. *Energies* **2020**, *13*, 4881. [CrossRef]

50. Busch, A.; Gensterblum, Y.; Krooss, B.M. Methane and CO_2 sorption and desorption measurements on dry Argonne premium coals: Pure components and mixtures. *Int. J. Coal Geol.* **2003**, *55*, 205–224. [CrossRef]

51. Li, D.; Liu, Q.; Weniger, P.; Gensterblum, Y.; Busch, A.; Krooss, B.M. High-pressure sorption isotherms and sorption kinetics of CH_4 and CO_2 on coals. *Fuel* **2010**, *89*, 569–580. [CrossRef]

52. Breiman, L.; Friedman, J.; Stone, C.J.; Olshen, R.A. *Classification and Regression Trees*; Chapman & Hall (Wadsworth, Inc.): New York, NY, USA, 1984.

53. Louppe, G. Understanding Random Forests: From Theory to Practice. Ph.D. Thesis, University of Liège, Liège, Belgium, 2014.

54. Zhang, J.; Sun, Y.; Shang, L.; Feng, Q.; Gong, L.; Wu, K. A unified intelligent model for estimating (gas + n-alkane) interfacial tension. *Fuel* **2020**, *282*, 118783. [CrossRef]

55. Tomaschek, F.; Hendrix, P.; Baayen, R.H. Strategies for addressing collinearity in multivariate linguistic data. *J. Phon.* **2018**, *71*, 249–267. [CrossRef]

56. Levy, J.H.; Day, S.J.; Killingley, J.S. Methane capacities of Bowen Basin coals related to coal properties. *Fuel* **1997**, *76*, 813–819. [CrossRef]

57. Chalmers, G.R.; Bustin, R.M. On the effects of petrographic composition on coalbed methane sorption. *Int. J. Coal Geol.* **2007**, *69*, 288–304. [CrossRef]

58. Clarkson, C.R.; Bustin, R.M. Binary gas adsorption/desorption isotherms: Effect of moisture and coal composition upon carbon dioxide selectivity over methane. *Int. J. Coal Geol.* **2000**, *42*, 241–271. [CrossRef]

59. Weniger, P.; Franců, J.; Hemza, P.; Krooss, B.M. Investigations on the methane and carbon dioxide sorption capacity of coals from the SW Upper Silesian Coal Basin, Czech Republic. *Int. J. Coal Geol.* **2012**, *93*, 23–39. [CrossRef]

60. Crosdale, P.J.; Moore, T.A.; Mares, T.E. Influence of moisture content and temperature on methane adsorption isotherm analysis for coals from a low-rank, biogenically-sourced gas reservoir. *Int. J. Coal Geol.* **2008**, *76*, 166–174. [CrossRef]

61. Krooss, B.V.; Van Bergen, F.; Gensterblum, Y.; Siemons, N.; Pagnier, H.J.M.; David, P. High-pressure methane and carbon dioxide adsorption on dry and moisture-equilibrated Pennsylvanian coals. *Int. J. Coal Geol.* **2002**, *51*, 69–92. [CrossRef]

62. Pan, J.; Hou, Q.; Ju, Y.; Bai, H.; Zhao, Y. Coalbed methane sorption related to coal deformation structures at different temperatures and pressures. *Fuel* **2012**, *102*, 760–765. [CrossRef]

63. Guan, C.; Liu, S.; Li, C.; Wang, Y.; Zhao, Y. The temperature effect on the methane and CO_2 adsorption capacities of Illinois coal. *Fuel* **2018**, *211*, 241–250. [CrossRef]

64. Clarkson, C.R.; Bustin, R.M. Variation in micropore capacity and size distribution with composition in bituminous coal of the Western Canadian Sedimentary Basin: Implications for coalbed methane potential. *Fuel* **1996**, *75*, 1483–1498. [CrossRef]

65. Carroll, R.E.; Pashin, J.C. Relationship of sorption capacity to coal quality: CO_2 sequestration potential of coalbed methane reservoirs in the Black Warrior basin. In Proceedings of the International Coalbed Methane Symposium, University of Alabama, Tuscaloosa, AL, USA, 5–7 May 2003.

66. Faiz, M.; Saghafi, A.; Sherwood, N.; Wang, I. The influence of petrological properties and burial history on coal seam methane reservoir characterisation, Sydney Basin, Australia. *Int. J. Coal Geol.* **2007**, *70*, 193–208. [CrossRef]

67. Laxminarayana, C.; Crosdale, P.J. Controls on methane sorption capacity of Indian coals. *AAPG Bull.* **2002**, *86*, 201–212. [CrossRef]

68. Guo, X.; Huan, X.; Huan, H. Structural characteristics of deformed coals with different deformation degrees and their effects on gas adsorption. *Energy Fuel.* **2017**, *31*, 13374–13381. [CrossRef]

Publisher's Note: MDPI stays neutral with regard to jurisdictional claims in published maps and institutional affiliations.

Article

Development and Performance Evaluation of Solid-Free Drilling Fluid for CBM Reservoir Drilling in Central Hunan

Pinghe Sun [1,2], Meng Han [1,2], Han Cao [1,2,*], Weisheng Liu [1,2], Shaohe Zhang [1,2] and Junyi Zhu [1,2]

[1] Key Laboratory of Metallogenic Prediction of Nonferrous Metals and Geological Environment Monitoring, Ministry of Education, Changsha 410083, China; pinghesun@csu.edu.cn (P.S.); hanmeng@csu.edu.cn (M.H.); weishengliu@csu.edu.cn (W.L.); zhangshaohe@163.com (S.Z.); zhujunyi@csu.edu.cn (J.Z.)

[2] School of Geosciences and Info-Physics, Central South University, Changsha 410083, China

* Correspondence: hancao@csu.edu.cn

Received: 9 August 2020; Accepted: 15 September 2020; Published: 16 September 2020

Abstract: Solid-free drilling fluid is a matter of cardinal significance in the course of Coal bed Methane (CBM) reservoir drilling. This study evaluated the performance of solid-free CBM drilling fluid in central Hunan. Three types of surfactants, namely TX-10 (nonionic), HSB1618 (zwitterionic) and penetrant T (anionic), were added in basic fluid at various concentrations of 0.05, 0.10 and 0.15% (m/m). This study comprised of drilling fluid rheology, sample mineral analysis, sample nuclear magnetic resonance (NMR) scanning, sample wettability, and sample surface micro characteristics tests. The results show that TX-10 and HSB1618 enhance the rheological properties of drilling fluid, such as yield point and gel strength. Penetrant T has opposite effect on it. It was found that the minimum American Petroleum Institute (API) filtration is only 0.3 mL. This study adopted a new method using laser particle size analyzer to evaluate suspension performance. Based on the surface micro characteristics of the sample and the NMR scanning tests, it is found that the residual amount of basic fluid + HSB1618 in the sample is the smallest. The wettability modification curve indicates that three surfactants decrease the sample's hydrophobicity. With the increase of surfactant concentration, all above parameters change regularly. The basic fluid + 0.10% HSB1618 has the strongest hydrophobicity for sample at pH = 10. This study obtained a set of solid-free drilling fluid system, which provides better suspension capacity and large contact angle and reduces residue of drilling fluid in CBM reservoir. Ultimately, it can accelerate the desorption of coal gas and reduce damage to the reservoir.

Keywords: CBM; surfactant; solid-free drilling fluid; CBM reservoir wettability

---

1. Introduction

Compared with conventional oil and natural gas, Coal bed Methane (CBM) is a kind of highly efficient and clean energy which has obvious advantages in safety, economy and environmental protection [1,2]. China is rich in CBM reserves, although having less than Russia and Canada. However, due to the complex geological conditions of domestic CBM reservoirs and development technology is somewhat behind, China's CBM exploration and development is still in the initial stage [3–6]. The deep layer in the middle Hunan area is dominated by conventional natural gas exploration, but the development of CBM plays an important role in the shallow gas development [7–9].

In CBM development, it is very important to use proper drilling fluid system for safe, efficient and environmental protection drilling. First of all, solid phase particles in drilling fluid tend to block the fractures and pores of the CBM reservoir, and then block the gas production channel. To reduce the intrusion of solid phase particles into CBM reservoir, it is advisable to drill with solid-free drilling fluid with good suspension performance [10–12]. It has the characteristics of low density, good suspension performance and small filtration and maximize the control of solid intrusion which can reduce reservoir leakage and damage and is a perfect for low density drilling fluid system [13–15]. Secondly, the reservoir is easy to absorb or adsorb liquids and gases. When drilling fluid invasions into CBM reservoir, it is easy to change its surface wettability, thus affecting the desorption and percolation of CBM, and thereby affecting CBM productivity [16–20].

Foam drilling fluid has a special network structure, which can carry cuttings well. Cai et al. [21] improved the foaming volume of foamed drilling fluid up to 50% by using chemically treated nano-SiO_2 dispersions. By studying the foam's properties, Su et al. [22] found that the appropriate temperature was 40–100 °C and that the foaming performance of hard foam could maintain within 120 °C. Fractures and minerals affect the permeability of CBM reservoirs [23–28]. Cui et al. [29] combined FESEM with X-ray mu-CT together with EDS to quantify the mineral and fracture characteristics. Coal reservoir characteristics may be related to coal rank and maceral as well as mineral content [30]. Yang et al. [31] used NMR tests and got the scattergram of coal pore size at a variety of strain rates. It was found that the crest value in the fractures rose and the crest value for the meso-macropores declined with the fortify of strain rates. The T2 spectrum obtained by NMR can be converted to the pore throat distribution. Adebayo et al. [32] mentioned T2 is connected with the surface relaxivity and the pores surface to volume ratio.

Huang et al. [33] mentioned that SWIA could adjust the core wettability. Li et al. [34] and Shen et al. [35] mentioned that the surfactant can increase the wettability of coal, and further enhance the desorption of methane; it can advance the recovery factor in theory and critical desorption pressure of CBM. Drilling fluid and coal surface contact, its pH value directly affects the wettability, further influencing the permeability of CBM. Through a lot of experiments to study the influence of drilling fluid pH value on the coal wettability, Zheng et al. [36] found that wettability and drilling fluid pH value is related, first reduced, then increased, and finally reduced. The above research results have played an important role in accelerating the development of CBM drilling fluid. However, there is a lack of special evaluation on the rheological, wettability of drilling fluid and the amount of additives for the practical characteristics of CBM drilling, and the relevant research lacks a systematic approach and depth.

Therefore, this study takes this as the starting point to analyze the rock-carrying capacity of the solid-free drilling fluid and study the influence of surfactant concentration and pH in the solid-free drilling fluid on the wettability of the CBM reservoir in central Hunan. This study optimizes the indoor evaluation method of the suspension performance of the solid-free drilling fluid, optimizes the formulation of the solid-free drilling fluid for the central Hunan CBM reservoir, and reveals the wettability mechanism of the central Hunan CBM reservoir when the solid-free drilling fluid penetrates. It is of great significance to optimize the formulation of the solid-free drilling fluid during the development of CBM in central Hunan, which further contributes to the rational development of CBM resources and the improvement of CBM productivity in central Hunan.

2. Experimental Method

2.1. Materials

2.1.1. CBM Reservoir Sample

The reserves of CBM resources are 2.8 billion m^3 and the abundance of resource reserves is 0.43 m^3/km^2 in the Lengshuijiang mining area in central Hunan. The development prospect is broad. CBM reservoir sample of Lengshuijiang mining area was selected and made into φ20 mm × 10 mm standard sample which has the same bedding direction and similar structure. Rock powder was obtained by crushing some samples and passing a 200-mesh sieve.

2.1.2. Surfactant

The mass fraction of nonionic surfactant TX-10 is 99%, which formed by condensation of nonylphenol and ethylene oxide in the presence of a catalyst. Its molecular formula is $C_{15}H_{24}O.(C_2H_4O)_n$, where when n changes, the product has different properties and different applications. Its melting point is 44–46 °C, boiling point is 250 °C, density is 1.06 g/mL, and flash point is 535 °F. It does not exist in an ionic state in water and has good stability, which is not easy to be affected by strong electrolyte, acid and alkali. At the same time, good compatibility with other surfactants. The mass fraction of zwitterionic surfactant HSB1618 is 40% which has both ionic properties and excellent resistance to calcium and magnesium ions. Its free amine content is ≤2%, sodium chloride content is ≤7%, and organochlorine content is 0. The mass fraction of anionic surfactant penetrant T is 40% which is cheap and not resistant to strong acids and bases, reductants and metal salts. The pH value of 1% aqueous solution is 7.0–9.5. It works best when the temperature is less than 40 °C and the pH value is 5–10. Its active part tends to dissociate into negative ions in water, and there is a large organic anion that can interact with alkali to form salt.

2.2. Solid-Free Drilling Fluid

On the basis of previous research, the basic fluid of the solid-free drilling fluid was obtained as Table 1.

Table 1. The basic fluid composition.

Components	KCl	Na$_2$CO$_3$	XC	PAC
The mass fraction	3%	0.1%	0.2%	0.15%
Function	Inhibitor	Alkalinity regulator	Viscosities	Filtration loss reducer, Inhibitor
				Foam stabilizer

2.3. XRD

X-ray diffraction tests were carried out on reservoir samples in accordance with the SY/T5163–2010, and quantitative analysis of mineral composition was conducted using D8 Advance x-ray diffractometer of Bruker Company in Bremen, Germany. The contents of hard and brittle minerals and clay minerals were determined to study the microscopic mechanisms affecting the adsorption properties and wettability of samples.

2.4. Rheological Properties Test

The rheological properties test was conducted according to API standards. The time required to flow up to 500 mL of drilling fluid was measured by the funnel viscometer produced by Qingdao Tongchun Petroleum Instrument Co., Ltd. (QTPI) (Qingdao, China), which is the viscosity of drilling fluid (FV).

The ZNN-SD6 rotary viscometer produced by QTPI was used to obtain the drilling fluid viscosity and shear stress. The drilling fluid was placed in a circular space between two concentric cylinders. The outer cylinder rotates at a constant speed through variable transmission, and the outer cylinder produces a torque through the action of the measured drilling fluid on the inner cylinder, which makes the inner cylinder connected with the torsion spring rotate at a corresponding angle. According to Newton's law, the size of the angle is proportional to the viscosity of the drilling fluid, so the measurement of the viscosity turns to the measurement of the inner cylinder angle. By the sensor display value, readings at the different rpm values, i.e., 600, 300, 200, 100, 6 and 3 were measured, including gel strengths. The apparent viscosity (*AV*), yield point (*YP*) and plastic viscosity (*PV*) can be obtained by Equations (1)–(3).

$$AV = 0.5\varphi_{600} \tag{1}$$

$$PV = \varphi_{600} - \varphi_{300} \tag{2}$$

$$YP = 0.511(2\varphi_{300} - \varphi_{600}) \tag{3}$$

where φ_{600} is a 600-rpm dial reading of the viscometer; φ_{300} is a 300-rpm dial reading of the viscometer.

2.5. Filtration Test

According to the API standard, the filtration test was conducted using the ZNS-5A medium pressure filter press assembly produced by QTPI. First, 500 mL drilling fluid was injected into the cylindrical drilling fluid cup and the lid was tightened; then, the air source was connected to adjust the pressure to 0.69 MPa; and finally, the air valve was opened to let the air source enter the drilling fluid cup. The filtration area of the instrument was 45.60 ± 0.60 cm^2 (1 ± 0.1 in^2). The time of filtration and the amount of filtration (F_{API}) after filtration were measured. Meanwhile, the quality of filter cake was observed to measure the effect of borehole wall protection.

2.6. Laser Particle Size Analysis Test

Added 2 g of rock powder over 200 mesh sieve into 200 g solid-free drilling fluid, then stirred for 5 min and stood for 30 min, and extracted 20 mL of upper liquid. Distilled water was selected as the dispersion medium, and the particle size distribution of suspended rock powder in the liquid was analyzed by Rise-2002 laser particle size analyzer produced by Jinan Rise Science & Technology Co., Ltd. (Jinan, China), so as to further evaluate the suspension performance of the solid-free drilling fluid.

2.7. Wettability Test

Different solid-free drilling fluids were configured by varying the surfactant concentration and drilling fluid pH. The samples were soaked in different drilling fluids for 48 h and dried naturally for 24 h. JCY series contact angle instrument produced by Shanghai Fangrui Instrument Co., Ltd. (Shanghai, China) was used to measure the dynamic contact angle of distilled water on the samples within 0~12 s and obtain the contact angle photos. Then, the wetting modification effect of the solid-free drilling fluid on the samples was analyzed by comparing with the samples that were not soaked or soaked in water.

2.8. Microscope Test

The soaked sample was naturally dried for 24 h, and the adsorption state of drilling fluid on the sample surface was observed with SQ500MF high-power integrated video microscope produced by Shanghai victory & Shuangquan tech. (Shanghai, China). The magnification factor is 100, and the distribution state of different drilling fluids on the sample surface was observed by comparing with the sample soaked in water to further analyze the adsorption principle.

2.9. NMR Scanning Test

The soaked sample was wrapped in PTFE tape to reduce water evaporation. NMR scanning with MacroMR12-150H-I NMR scanner produced by Suzhou Niumag Analytical Instrument Corporation was used to obtain T_2 spectrum and signal imaging. The T_2 spectrum was converted into the pore size distribution to compare the pore size distribution of each sample under the saturated state.

According to the basic principle of NMR scanning imaging, the relaxation characteristics of fluid in porous media can be expressed by Equation (4). According to Equation (4), the relaxation time of NMR T_2 is in direct proportion to the pore size ($D = 2\,r$). If $C = 2F_s\rho_2$, the Equation (5) can be obtained. The T_2 spectrum of rock can be converted into pore size distribution curve by Equation (5).

$$\frac{1}{T_2} \approx \rho_2 \frac{s}{v} = F_s \frac{\rho_2}{r} \tag{4}$$

$$D = CT_2 \tag{5}$$

where T_2 is NMR transverse relaxation time, ms; ρ_2 is rock transverse surface relaxation strength coefficient, nm/ms; S is total pore surface area of rock, nm^2; V is pore volume of rock, nm^3; r is pore radius, nm; F_s is geometric shape factor (spherical pores, $F_s = 3$; Cylindrical pores, $F_s = 2$); D is pore diameter of the rock, nm; C is rock conversion coefficient, nm/ms.

3. Results and Discussions

3.1. Mineral Analysis

XRD test results of samples are shown in Figure 1. The results show that the mineral content in the sample is 25% montmorillonite, 20% pyrophyllite, 16% illite, 13% quartz, 12% kaolin, 11% pyrite and 3% anatase. Clay minerals are mainly montmorillonite and illite with content up to 53.

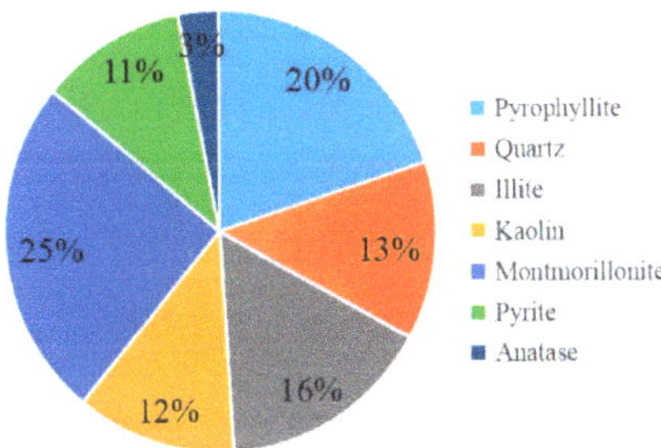

Figure 1. The mineral analysis of coal bed methane (CBM) reservoir sample.

3.2. Statistical Analysis of Rheological Properties

The rheological parameters of the solid-free drilling fluid with different concentrations of TX-10, HSB1618 and penetrant T are shown in Table 2. The apparent viscosity of basic fluid + HSB1618 is the largest, indicating the maximum consistency. When the surfactant concentration is 0.10%, the API filtration of basic fluid + HSB1618 is also the minimum. On the contrary, the apparent viscosity of basic fluid + penetrant T is the smallest, the filtration is the largest, and the performance parameters of basic fluid + TX-10 are in the middle. The funnel viscosity and apparent viscosity of several drilling fluids have the same rule. No matter what kind of surfactant is added, the API filtration is small and the filter cake is thin and tough, which is an ideal drilling fluid parameter and has good wall protection effect.

Table 2. The rheological properties of solid-free drilling fluid.

Drilling Fluid	Basic Fluid + TX-10			Basic Fluid + HSB1618			Basic Fluid + Penetrant T		
$C_{\text{Surfactant}}$/%	0.05	0.10	0.15	0.05	0.10	0.15	0.05	0.10	0.15
FV/s	22	24	36	25	35	42	21	18	17
F_{API}/mL	2.2	1.9	1.8	2.6	0.8	0.3	0.8	3.0	5.2
AV/MPa·s	9.8	14.8	14.9	11.2	14.7	16.2	8.9	5.9	5.7
PV/MPa·s	6.4	10.2	10.1	7.9	9.1	9.3	5.8	4.2	3.6
YP/Pa	3.44	4.61	4.95	3.41	5.76	7.04	3.21	1.81	2.07
pH	9.0	9.0	9.0	9.5	9.5	9.5	8.5	8.5	8.5
ρ/g·cm^{-3}	0.87	0.80	0.77	0.81	0.78	0.75	1.0	1.0	1.0

The horizontal displacement of CBM reservoir is long, and cuttings are not easy to be carried during drilling, so the rheological property of drilling fluid should be adjusted to make it have good suspension and rock carrying capacity. The yield point of the solid-free drilling fluid with different concentrations of surfactants is shown in Figure 2. As can be seen from the figure, the yield point of basic fluid + penetrant T is very small, while basic fluid + HSB1618 and basic fluid + TX-10 have higher yield point. The difference increases with the concentration of surfactant and is very small when the concentration is 0.05%. As the concentration of surfactant increases, the yield point of basic fluid + HSB1618 and basic fluid + TX-10 increases, while basic fluid + penetrant T has an obvious downward trend. In conclusion, the results show that the basal fluid + HSB1618 has higher yield point, better rock-carrying and hole-cleaning capacity.

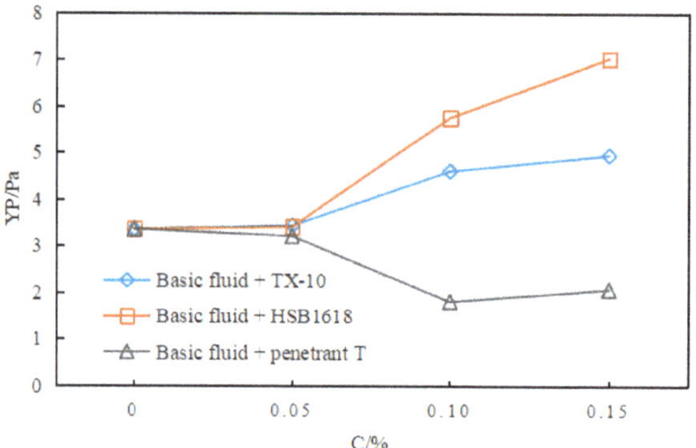

Figure 2. The yield point of solid-free drilling fluid.

The gel strengths of the solid-free drilling fluid with different concentrations of surfactants is shown in Figure 3. It can be seen from the figure that when the concentration of surfactant is 0.10% and 0.15%, the gel 10 min and the gel 10 sec of basic fluid + TX-10 and basic fluid + HSB1618 have sufficient range, while the gel strengths of basic fluid + penetrant T is very small, and the difference between the gel 10 min and the gel 10 sec is very small. With the increase of surfactant concentration, the gel strengths of basic fluid + TX-10 and basic fluid + HSB1618 increased, while that of basic fluid + penetrant T decreased. At the concentration of 0.05%, the gel strengths of basic fluid + HSB1618 is less than that of basic fluid + TX-10. With the increase of the concentration, the gel strengths of basic fluid + HSB1618 is gradually greater than that of basic fluid + TX-10. The results show that the basic fluid + TX-10 and the basic fluid + HSB1618 have good suspension performance at the surfactant concentration of 0.10% and 0.15%, which can effectively reduce the formation of cuttings bed in the horizontal section of CBM reservoir drilling.

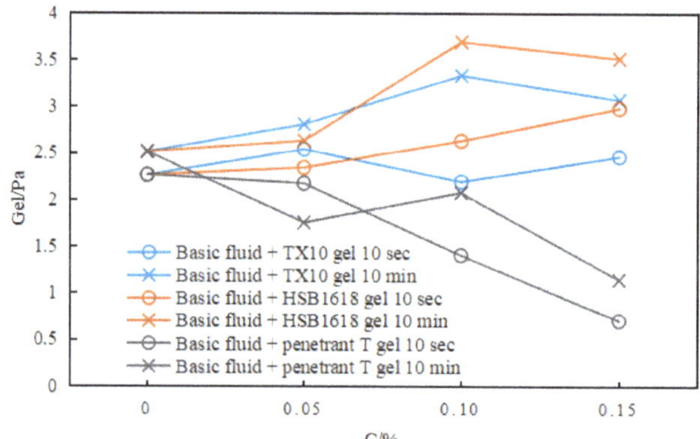

Figure 3. The gel strength of solid-free drilling fluid.

The plastic viscosity of the solid-free drilling fluid with different concentrations of surfactants is shown in Figure 4. The plastic viscosity reflects the internal friction between suspended particles and liquid phase as well as continuous liquid phase in the dynamic equilibrium of the destruction and recovery of network structure in drilling fluid under laminar flow. The main factor affecting the plastic viscosity is the content of solid phase, the higher the content of solid phase, the greater the plastic viscosity. In addition, clay dispersion and polymer viscosifier also have an impact on the plastic viscosity, because they can affect the volume fraction or liquid viscosity. In these drilling fluids, basic fluid + TX-10 and basic fluid + HSB1618 have higher plastic viscosities than basic fluid + penetrant T. Surfactants TX-10 and HSB1618 act as viscosifier in drilling fluids, and penetrant T acts as viscosity reducers. With the increase of surfactant concentration, the plastic viscosity of the basic fluid + TX-10 and the basic fluid + HSB1618 increases, and the basic fluid + penetrant T decreases. The change of surfactant and its concentration change the plastic viscosity of drilling fluid by changing the liquid viscosity.

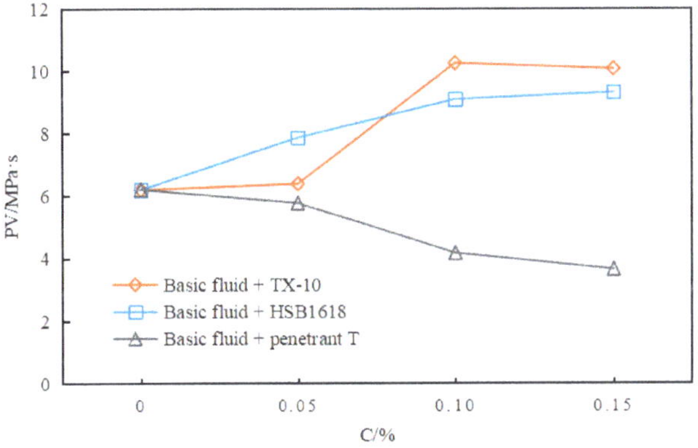

Figure 4. The plastic viscosity of solid-free drilling fluid.

3.3. Micro Analysis of Drilling Fluid Suspension Performance

According to the experimental data of laser particle size analysis, the particle size ranges of basic fluid + TX-10, basic fluid + HSB1618 and basic fluid + penetrant T are 22.664–637.059 µm, 29.703–1118.3 µm and 0.049–2.603 µm, respectively. Basic fluid + HSB1618 suspended rock powder has the largest particle size, followed by basic fluid + TX-10, and basic fluid + penetrant T is the smallest. The particle size of the first 20 differentials is select to draw a graph, and the particle size analysis of the suspended rock powder in the solid-free drilling fluid is shown in Figure 5. With the increase of surfactant concentration, the differentials of rock powder suspended by basic fluid + TX-10 and basic fluid + HSB1618 increases, while the basic fluid + penetrant T does not change much. The maximum differential particle size of basic fluid + 0.15% TX-10, basic fluid + 0.15%HSB1618 and basic fluid + 0.15% penetrant T suspension is 164.744, 309.681 and 0.327 µm, respectively. The maximum differential particle size of basic fluid + 0.05%HSB1618, basic fluid + 0.10%HSB1618 and basic fluid + 0.15%HSB1618 are 150.539, 258.582, 309.681 µm, respectively. Basic fluid + HSB1618 has better suspension capacity, firstly because of its large consistency and gel strengths, and secondly because of the ionization of carboxyl, phenolic hydroxyl group and other functional groups on the rock powder surface, which makes the surface negatively charged. The compression of the double electric layer increases the electrostatic repulsion after HSB1618 is added, which is beneficial to the suspension of rock powder.

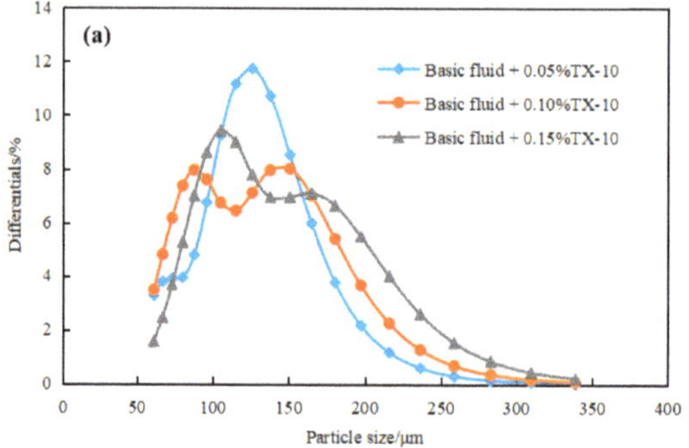

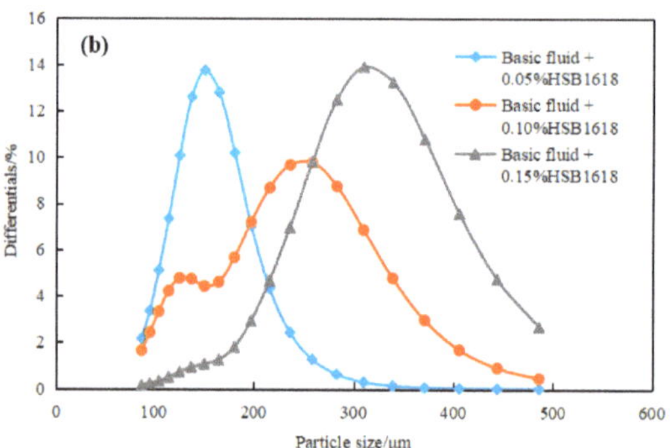

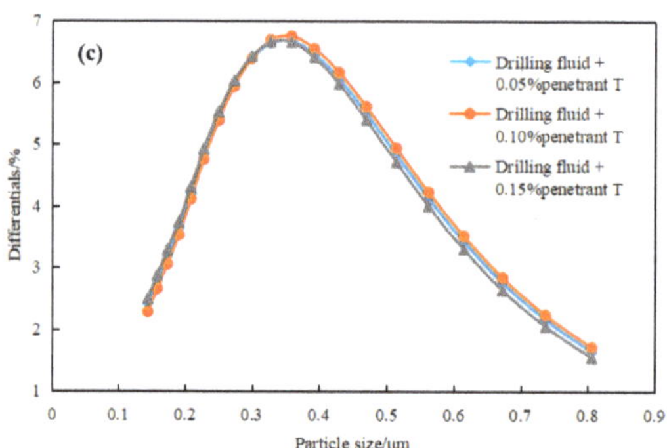

Figure 5. The particle size of the suspended rock powder in (**a**) basic fluid + TX-10, (**b**) basic fluid + HSB1618 and (**c**) basic fluid + penetrant T.

3.4. Adsorption Mechanism of Drilling Fluid

Drilling fluid with 0.10% (a) TX-10, (b) HSB1618, (c) penetrant T added to the basic fluid is shown in Figure 6. It can be seen from the figure that, when the concentration of surfactant is 0.10%, basic fluid + TX-10 and basic fluid + HSB1618 are rich in foam, while basic fluid + penetrant T has poor foamability. The foaming volume of the 200 mL of drilling fluid is 250, 240 and 200 mL, respectively. The basic fluid + TX-10 foam is loose while the basic fluid + HSB1618 foam is fine and uniform.

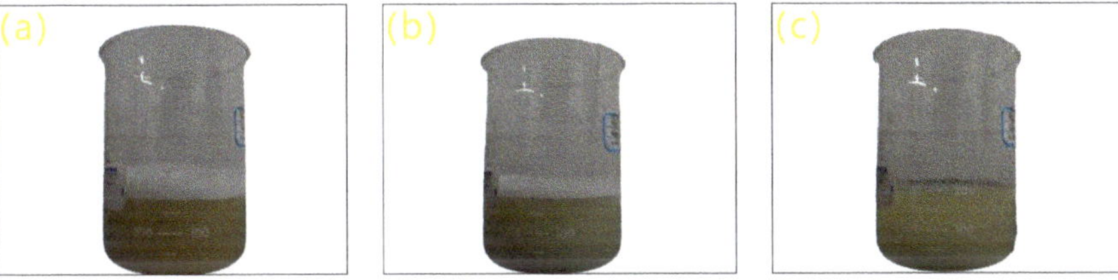

Figure 6. The basic fluid is added with 0.10% (**a**) TX-10, (**b**) HSB1618 and (**c**) penetrant T.

CBM reservoirs are highly absorbent which can absorb various liquids and gases. After different drilling fluids (a) water, (b) basic fluid + 0.10%TX-10, (c) basic fluid + 0.10%HSB1618, (d) basic fluid + 0.10%penetrant T, the microscopic diagram of sample surface is shown in Figure 7. According to the figure, there is no significant change on the sample surface after soaking in water, and fracture can be clearly seen on the surface. The basic fluid + 0.10% TX-10 is block or strip on the sample surface with uneven distribution. The basic fluid + 0.10%HSB1618 is evenly spread on the surface of sample, showing a thin layer, while the basic fluid + 0.10%penetrant T has a small adsorption capacity on the sample surface, and is distributed in a granular or strip form with uneven distribution. In the dense adsorption layer of basic fluid + 0.10%HSB1618, the hydrophilic group of HSB1618 points to the aqueous phase, and the sample surface is changed from hydrophobic to hydrophilic.

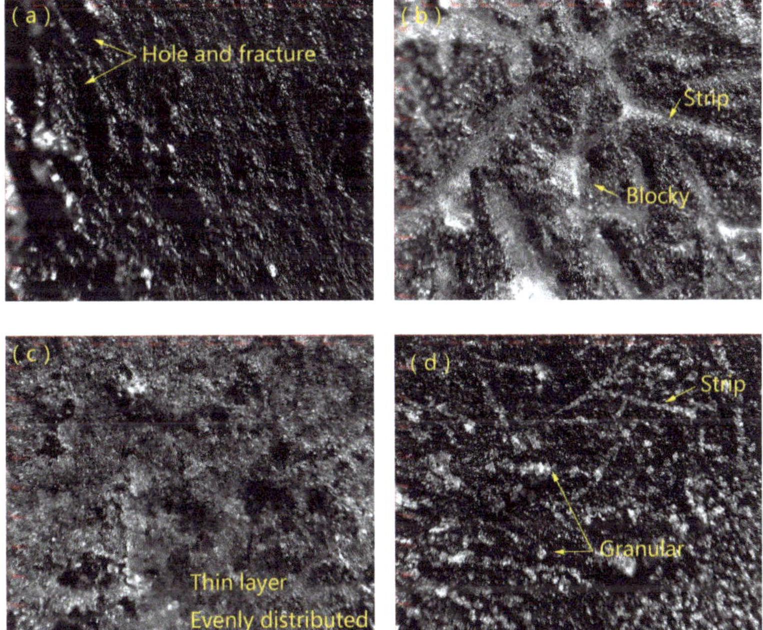

Figure 7. The microscopic photos of sample surface soaked by different drilling fluids: (**a**) water, (**b**) basic fluid + 0.10% TX-10, (**c**) basic fluid + 0.10%HSB1618 and (**d**) basic fluid + 0.10 penetrant T.

Penetrant T is an anionic surfactant with negative charge, while the carboxyl and phenolic hydroxyl group on the CBM reservoir surface ionizes, making the sample surface also negatively charged. The adsorption capacity of anionic surfactant on CBM reservoir surface is minimal due to electrostatic interaction. HSB1618 is a zwitterionic surfactant consisting of anionic and cationic surfactants, in which the cationic part is adsorbed by electrostatic interaction. TX-10 is a nonionic surfactant that has no charge and cannot be adsorbed on the sample surface by electrostatic interaction. However, the hydrophobic non-polar groups such as aliphatic hydrocarbon and aromatic hydrocarbon on the sample surface make it hydrophobic. The main forces between nonionic surfactant and sample surface are hydrophobic and dispersive forces. In the case of minimal adsorption of penetrant T, basic fluid + penetrant T has better wetting effect, because penetrant T has high permeability, fast and uniform permeability, which can quickly fill pores on the sample surface to play a wetting role.

3.5. Application of NMR Pore Throat Distribution Curve

The NMR T2 spectrum of each sample obtained by MRI scanning was converted into pore throat distribution, as shown in Figure 8. The pore throats of the samples are concentrated in the interval 0.01~0.1 µm, and the curves presented a bimodal shape, with the right peak being lower, that is, the proportion of large pore throats is small.

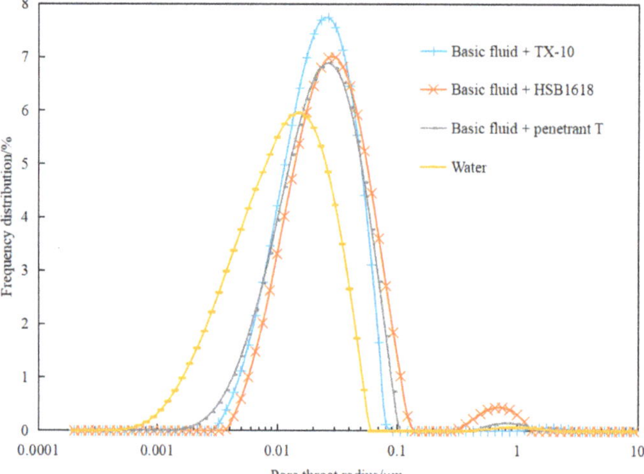

Figure 8. Pore throat distribution of samples soaked by different drilling fluids.

By comparing the distribution of NMR pore throat of samples soaked by different drilling fluids, it is found that the average pore throat of samples soaked by drilling fluids is greater than that of samples soaked by water. As the surfactant in the drilling fluid is hydrophilic, the pore fluid of samples soaked in the drilling fluid is larger than that soaked in water, resulting in a slightly larger pore throat radius. The pore throat frequency distribution of the sample soaked by basic fluid + TX-10 is the highest, and the 0.026-µm pore throat frequency distribution reaches the highest, 7.7%. The hydrophilicity of different surfactants is different, and the pore throat distribution of the samples soaked by three kinds of drilling fluid is different. This difference is also related to the difference in drilling fluid NMR results and samples.

3.6. Transverse and Longitudinal Section Imaging

NMR scan imaging of samples soaked by different drilling fluids is shown in Figure 9. In terms of adsorption degree and residual amount, the cross-sectional imaging showed that the more red parts, the more liquid. The adsorption degree and residual amount in the sample ranged from large to small as basic fluid + TX-10, water, basic fluid + penetrant T and basic fluid + HSB1618. During immersion, the samples were placed longitudinally. In terms of saturation changes at different longitudinal positions, the cross-sectional imaging showed that the longitudinal saturation change from large to small was basic fluid + TX-10, basic fluid + penetrant T, basic fluid + HSB1618 and the water.

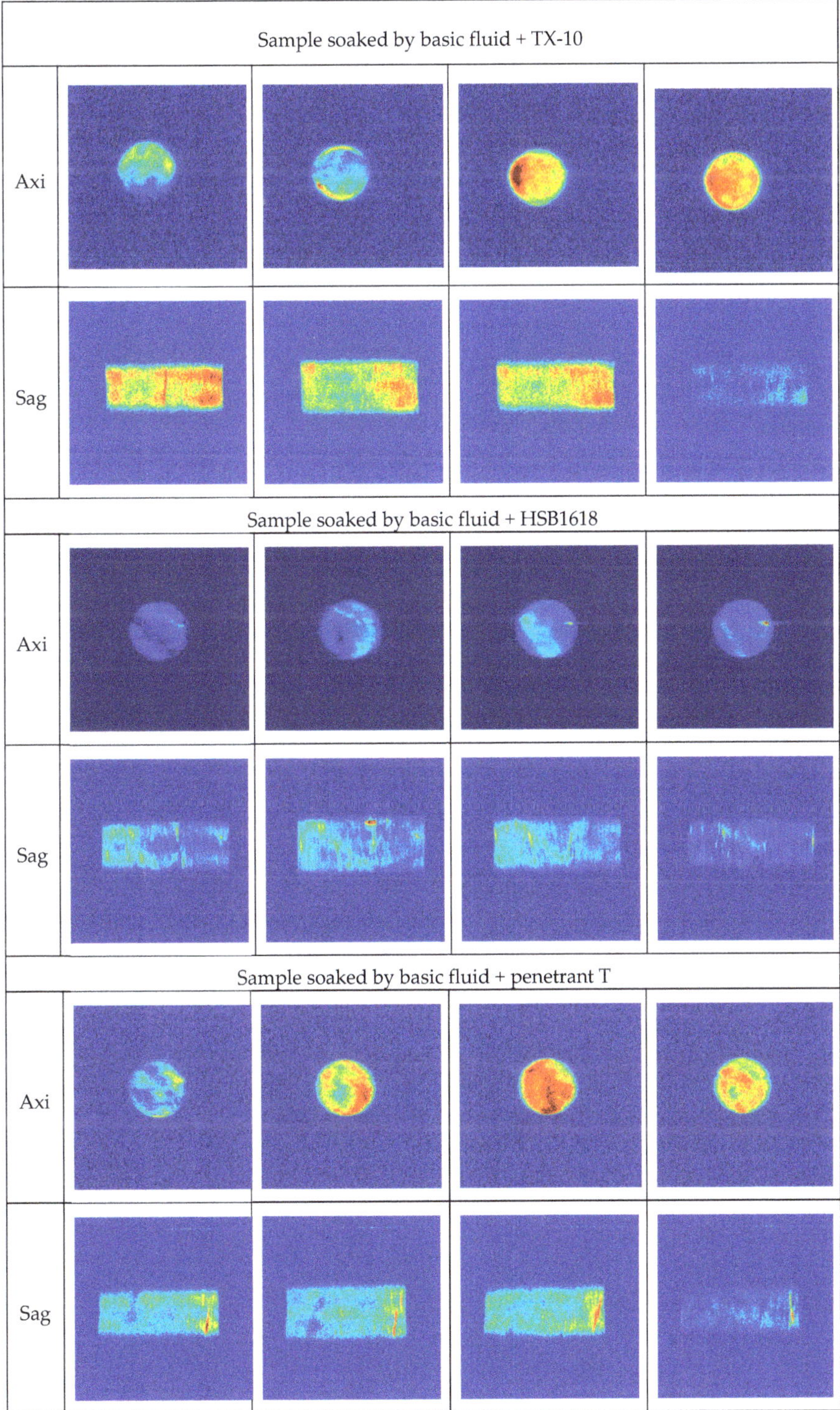

Figure 9. *Cont.*

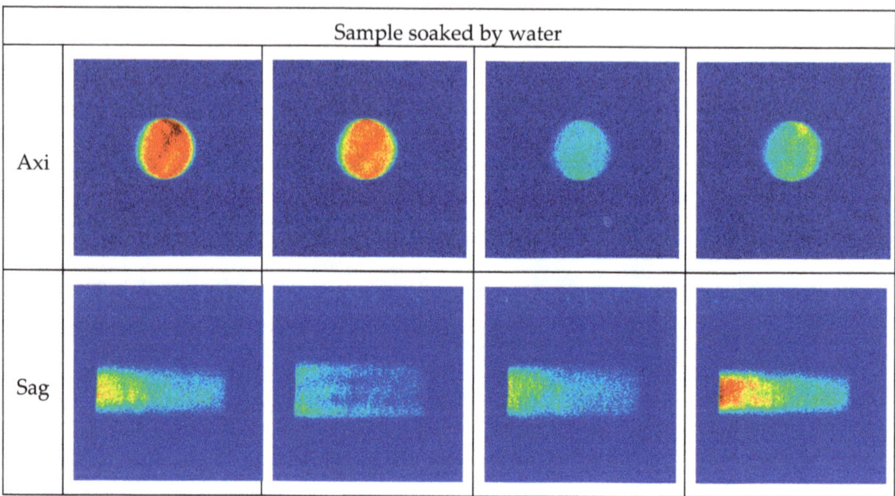

Figure 9. NMR imaging of samples soaked by different drilling fluids.

The basic fluid + HSB1618 should be selected based on the comprehensive comparison of adsorption degree, residual amount and saturation changes at different positions in the longitudinal direction. At the same time, because of the strong heterogeneity of samples, the difference of each sample has an impact on the results. In the drilling process of CBM reservoir in Central Hunan, 3%KCl + 0.1%Na$_2$CO$_3$ + 0.1%HSB1618 + 0.2%XC + 0.1%PAC can reduce the residual drilling fluid in CBM reservoir and further reduce the damage to the reservoir.

3.7. Effect of Solid-Free Drilling Fluid Additive on Wettability of CBM Reservoir

The dynamic and static contact angles of the solid-free drilling fluid to the sample are shown in Figures 10 and 11. It can be seen from the figure that the contact angle of the unsoaked sample is the largest and does not change with time. The contact angle changes very little after water soaking, which indicates that the wettability of water to sample is very weak. The contact angle of sample decreases after soaked in three kinds of solid-free drilling fluids, and the trend of dynamic contact angle decrease significantly, indicating that the three kinds of drilling fluids can greatly increase the moisture of sample. The wetting ability of drilling fluid added with 0.10% surfactant is different to some extent. The wetting ability of drilling fluid added with 0.10% surfactant is as follows: basic fluid + penetrant T, basic fluid + HSB1618 and basic fluid + TX-10.

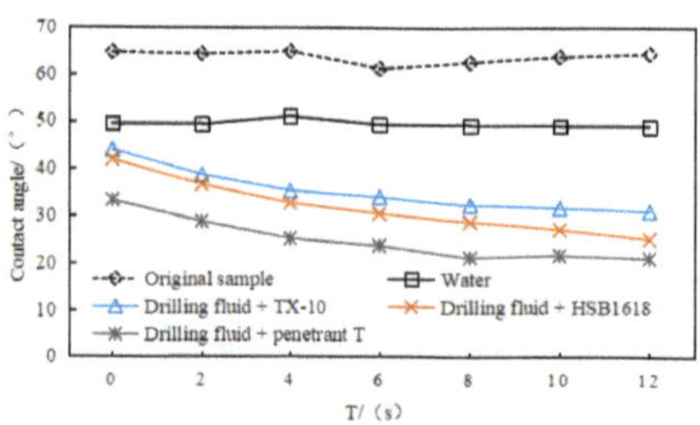

Figure 10. The contact angle variation of solid-free drilling fluid to sample.

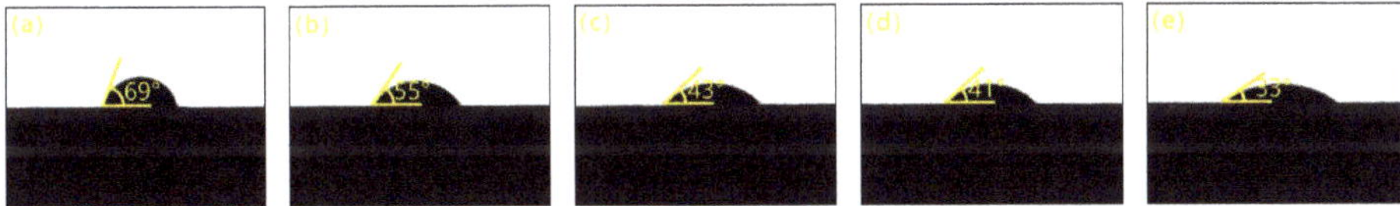

Figure 11. The contact angle of (**a**) original sample and sample soaked by different drilling fluids (**b**) water, (**c**) basic fluid + 0.10% TX-10, (**d**) basic fluid + 0.10%HSB1618 and (**e**) basic fluid + 0.10 penetrant T.

The curve of contact angle of sample with different concentrations (a) TX-10, (b) HSB1618 and (c) penetrant T added to the basic fluid is shown in Figure 12. According to the figure, with the increase of surfactant concentration, the wettability of the three drilling fluids to sample increases to different degrees, in which the wettability of basic fluid + TX-10 and basic fluid + HSB1618 increases significantly. The wettability of basic fluid + 0.05% TX-10 and basic fluid + 0.05%HSB1618 is similar to that of water, while basic fluid + 0.05% penetrant T can significantly increase the wettability of sample. When the surfactant concentration is 0.15%, the dynamic contact angle of the three drilling fluids fluctuates greatly, which greatly improves the moisture of the sample.

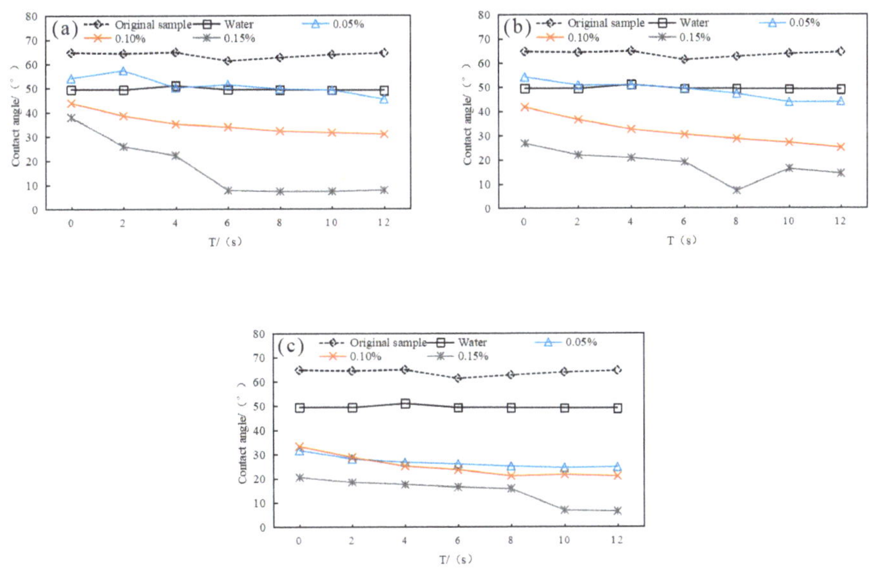

Figure 12. The contact angle variation of different concentrations (**a**) TX-10, (**b**) HSB1618 and (**c**) penetrant T added in basic fluid to sample.

3.8. Effect of Solid-Free Drilling Fluid pH on Wettability of CBM Reservoir

According to the rheological properties and wetting effect of drilling fluid, HSB1618, the surfactant with the best effect, is selected and the pH of drilling fluid is changed at the optimal concentration of 0.10%. The change of contact angle of drilling fluid pH to sample is shown in Figure 13. It can be seen from the figure that the contact angle increases between pH 8–10, that is, the hydrophilicity of sample decreases. pH 10–12 shows a decreasing trend, that is, the hydrophilicity of sample increase. The contact angle is the smallest at pH 12, and the wetting effect of the drilling fluid on the sample is maximized. The contact angle is the largest at pH = 10, and the hydrophobicity of the drilling fluid to the sample is minimal. The change of hydrophilicity of sample is because the change of drilling fluid pH affects the ionization degree of carboxyl and phenol hydroxyl group on sample surface, further changes the sample surface potential, and finally affects the wettability.

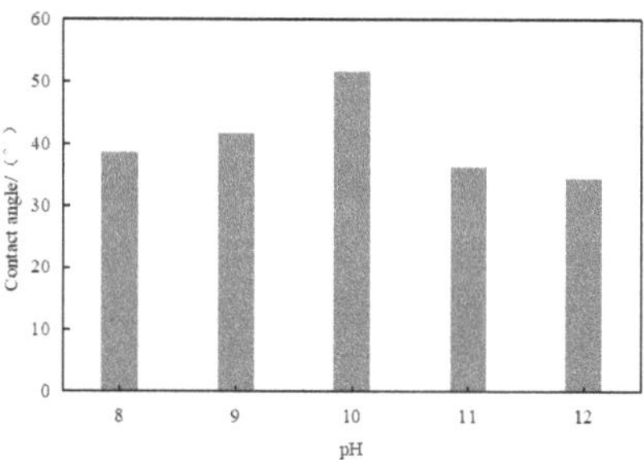

Figure 13. The contact angle variation of solid-free drilling fluid pH to sample.

4. Conclusions

The performance of solid-free CBM drilling fluid in central Hunan have been examined. The results show that TX-10 and HSB1618 enhance the rheological properties of drilling fluid, such as yield point and gel strength. It was found that the minimum API filtration is only 0.3 mL. Experimental results show that the suspended cuttings capacity is successively from large to small: basic fluid + HSB1618, basic fluid + TX-10 and basic fluid + penetrant T in the drilling process of CBM reservoir in central Hunan. With the increase of surfactant concentration, the suspended cuttings capacity of basic fluid + HSB1618 and basic fluid + TX-10 increase. The residual amount of basic fluid + HSB1618 in the sample is the smallest. The longitudinal saturation changes from large to small was basic fluid + TX-10, basic fluid + penetrant T, basic fluid + HSB1618 and the water.

With the increase of surfactant concentration, the wettability of the three drilling fluids to samples increases in different degrees. When the surfactant concentration is 0.15%, the dynamic contact angle of the three drilling fluids fluctuates greatly, which greatly improves the water wettability of the samples. The basic fluid + permeant T has the best ability to improve the wettability of the reservoir surface in the drilling process of CBM reservoir in central Hunan. The hydrophilicity of samples decrease when the basic fluid + HSB1618 is at pH 8–10, while that of samples increase at pH 10–12.

In the process of CBM development in central Hunan, the solid-free drilling fluid formula 3%KCl + 0.1%Na$_2$CO$_3$ + 0.1%HSB1618 + 0.2%XC + 0.1%PAC has strong suspension performance and large moistening contact angle, which is helpful for the flowback of drilling fluid, and can reduce the residual amount of drilling fluid in CBM reservoir, and finally reduce the damage to reservoir. Only three kinds of surfactants are discussed in this paper, and the application of cationic surfactants in solid-free drilling fluid can be studied in the future. In the follow-up studies, it is planned to carry out comparative studies of CBM reservoirs in multiple regions and study the acoustic and mechanical properties of reservoir samples before and after the action of solid-free drilling fluid.

Author Contributions: Conceptualization, P.S., M.H. and H.C.; Methodology, P.S., H.C. and S.Z.; Software, M.H. and J.Z.; Validation, P.S., W.L., M.H. and J.Z.; Writing—Original Draft Preparation, P.S., M.H., S.Z. and H.C.; Writing—Review and Editing, P.S., M.H., S.Z., W.L., and H.C. All authors have read and agreed to the published version of the manuscript.

Funding: This study was supported by Hunan Provincial Innovation Foundation for Postgraduate (No. 2019zzts162), the National Natural Science Foundation of China (No. 41602372), the open fund of the State Key Laboratory of Oil and Gas Reservoir Geology and Exploitation (Southwest Petroleum University, No. PLN201609 & No. PLN201607) and the fund of Ministry of Land and Resources Key Laboratory of Drilling Technology in Complex Conditions (No. DET201612).

Acknowledgments: All of the authors would like to direct their thanks to the professors and students in the Key Laboratory of Metallogenic Prediction of Nonferrous Metals and Geological Environment Monitoring (Central South University), Ministry of Education.

Conflicts of Interest: The authors declare no conflict of interest.

References

1. Golsanami, N.; Sun, J.; Liu, Y.; Yan, W.; Lianjun, C.; Jiang, L.; Dong, H.; Zong, C.; Wang, H. Distinguishing fractures from matrix pores based on the practical application of rock physics inversion and NMR data: A case study from an unconventional coal reservoir in China. *J. Nat. Gas Sci. Eng.* **2019**, *65*, 145–167. [CrossRef]
2. Le, T.D.; Moyne, C.; Murad, M.A.; Panfilova, I. A three-scale poromechanical model for swelling porous media incorporating solvation forces: Application to enhanced coalbed methane recovery. *Mech. Mater.* **2019**, *131*, 47–60. [CrossRef]
3. Zheng, M.; Li, J.; Wu, X.; Wang, S.; Guo, Q.; Yu, J.; Zheng, M.; Chen, N.; Yi, Q. China's conventional and unconventional natural gas resources: Potential and exploration targets. *J. Nat. Gas Geosci.* **2018**, *3*, 295–309. [CrossRef]
4. Song, Y.; Ma, X.; Liu, S.; Jiang, L.; Hong, F.; Qin, Y. Accumulation conditions and key technologies for exploration and development of Qinshui coalbed methane field. *Pet. Res.* **2018**, *3*, 320–335. [CrossRef]
5. Long, Q.; Hu, Q.; Zhang, Z.; Ren, T. On factors affecting coalbed gas content measurement. *Measurement* **2018**, *121*, 47–56. [CrossRef]
6. Hao, C.; Cheng, Y.; Dong, J.; Liu, H.; Jiang, Z.; Tu, Q. Effect of silica sol on the sealing mechanism of a coalbed methane reservoir: New insights into enhancing the methane concentration and utilization rate. *J. Nat. Gas Sci. Eng.* **2018**, *56*, 51–61. [CrossRef]
7. Wang, X.; Li, J.; Xu, S.; Lin, L.; Liu, Z.; Tan, H. Preliminary evaluation of hydraulic fracturing effects of coalbed methane wells in central Hunan. *China Coalbed Methane* **2018**, *15*, 3–8.
8. Zhu, W.; Gu, S. Feasibility and risk analysis of coalbed methane development in central Hunan. *China Saf. Sci. J.* **2010**, *20*, 97–101.
9. Yi, H. Study on characteristics of coalbed methane reservoirs in the lower carboniferous water measurement group in central Hunan. *Coal Geol. China* **2010**, *22*, 22–25.
10. Xu, J.; Zhai, C.; Ranjith, P.G.; Sun, Y.; Qin, L. Petrological and ultrasonic velocity changes of coals caused by thermal cycling of liquid carbon dioxide in coalbed methane recovery. *Fuel* **2019**, *249*, 15–26. [CrossRef]
11. Zheng, C.; Jiang, B.; Xue, S.; Chen, Z.; Li, H. Coalbed methane emissions and drainage methods in underground mining for mining safety and environmental benefits: A review. *Process. Saf. Environ. Prot.* **2019**, *127*, 103–124. [CrossRef]
12. Liu, D.; Wang, Q.; Wang, Y.; Wang, H.; Yu, H.; Yuan, M. Laboratory research on degradable drilling-in fluid for complex structure wells in coalbed methane reservoirs. *Pet. Explor. Dev.* **2013**, *40*, 249–253. [CrossRef]
13. Shi, Z.; Zhao, Y.; Qi, H.; Liu, J.; Hu, Z. Research and Application of Drilling Technology of Extended-reach Horizontally-intersected Well Used to Extract Coalbed Methane. *Procedia Earth Planet. Sci.* **2011**, *3*, 446–454. [CrossRef]
14. Gentzis, T.; Deisman, N.; Chalaturnyk, R.J. Effect of drilling fluids on coal permeability: Impact on horizontal wellbore stability. *Int. J. Coal Geol.* **2009**, *78*, 177–191. [CrossRef]
15. Rafati, R.; Smith, S.R.; Haddad, A.S.; Novara, R.; Hamidi, H. Effect of nanoparticles on the modifications of drilling fluids properties: A review of recent advances. *J. Pet. Sci. Eng.* **2018**, *161*, 61–76. [CrossRef]
16. Chattaraj, S.; Mohanty, D.; Kumar, T.; Halder, G. Thermodynamics, kinetics and modeling of sorption behaviour of coalbed methane—A review. *J. Unconv. Oil Gas Resour.* **2016**, *16*, 14–33. [CrossRef]
17. Xi, X.; Jiang, S.G.; Zhang, W.; Wang, K.; Shao, H.; Wu, Z. An experimental study on the effect of ionic liquids on the structure and wetting characteristics of coal. *Fuel* **2019**, *244*, 176–183. [CrossRef]
18. Roussi, L.; Stihle, J.; Geantet, C.; Uzio, D.; Tayakout-Fayolle, M. Coal-derived liquid asphaltenes diffusion and adsorption in supported hydrotreating catalysts. *Fuel* **2013**, *109*, 167–177. [CrossRef]
19. Shaida, M.A.; Dutta, R.K.; Sen, A.K. Removal of diethyl phthalate via adsorption on mineral rich waste coal modified with chitosan. *J. Mol. Liq.* **2018**, *261*, 271–282. [CrossRef]
20. Jawad, A.H.; Ismail, K.; Ishak, M.A.M.; Wilson, L.D. Conversion of Malaysian low-rank coal to mesoporous activated carbon: Structure characterization and adsorption properties. *Chin. J. Chem. Eng.* **2019**, *27*, 1716–1727. [CrossRef]

21. Cai, J.; Gu, S.; Wang, F.; Yang, X.; Yue, Y.; Wu, X.; Chixotkin, V.F. Decreasing Coalbed Methane Formation Damage Using Microfoamed Drilling Fluid Stabilized by Silica Nanoparticles. *J. Nanomater.* **2016**, *2016*, 1–11. [CrossRef]

22. Su, J.; Dong, W.; Zhou, S.; Deng, M. Synthesis and Assessment of a CO2 -Switchable Foaming Agent Used in Drilling Fluids for Underbalanced Drilling. *J. Surfactants Deterg.* **2018**, *21*, 375–387. [CrossRef]

23. Zhu, H.; Tang, X.; Jiang, S.; Liu, S.; Zhang, B.; Jiang, S.; McLennan, J.D. Permeability stress-sensitivity in 4D flow-geomechanical coupling of Shouyang CBM reservoir, Qinshui Basin, China. *Fuel* **2018**, *232*, 817–832. [CrossRef]

24. Clarkson, C.R.; Salmachi, A. Rate-transient analysis of an undersaturated CBM reservoir in Australia: Accounting for effective permeability changes above and below desorption pressure. *J. Nat. Gas Sci. Eng.* **2017**, *40*, 51–60. [CrossRef]

25. Wang, Z.; Qin, Y. Physical experiments of CBM coproduction: A case study in Laochang district, Yunnan province, China. *Fuel* **2019**, *239*, 964–981. [CrossRef]

26. Akhondzadeh, H.; Keshavarz, A.; Sayyafzadeh, M.; Kalantariasl, A. Investigating the relative impact of key reservoir parameters on performance of coalbed methane reservoirs by an efficient statistical approach. *J. Nat. Gas Sci. Eng.* **2018**, *53*, 416–428. [CrossRef]

27. Ivakhnenko, O.P.; Makhatova, M.N.; Nadirov, K.; Bondarenko, V. Unconventional Coalbed Methane Reservoirs Characterization Using Magnetic Susceptibility. *Energy Procedia* **2016**, *97*, 318–325. [CrossRef]

28. Morad, K. Selected topics in coalbed methane reservoirs. *J. Nat. Gas Sci. Eng.* **2012**, *8*, 99–105. [CrossRef]

29. Cui, J.; Liu, D.; Cai, Y.; Pan, Z.; Zhou, Y. Insights into fractures and minerals in subbituminous and bituminous coals by FESEM-EDS and X-ray μ-CT. *Fuel* **2019**, *237*, 977–988. [CrossRef]

30. Yuan, Y.; Tang, Y.; Shan, Y.; Zhang, J.; Cao, D.; Wang, A. Coalbed methane reservoir evaluation in the Manas mining area, southern Junggar Basin. *Energy Explor. Exploit.* **2018**, *36*, 114–131. [CrossRef]

31. Yang, Z.; Fan, C.; Lan, T.; Li, S.; Wang, G.; Luo, M.; Zhang, H. Dynamic Mechanical and Microstructural Properties of Outburst-Prone Coal Based on Compressive SHPB Tests. *Energies* **2019**, *12*, 4236. [CrossRef]

32. Adebayo, A.R.; Babalola, L.O.; Hussaini, S.R.; Alqubalee, A.; Babu, R.S. Insight into the Pore Characteristics of a Saudi Arabian Tight Gas Sand Reservoir. *Energies* **2019**, *12*, 4302. [CrossRef]

33. Huang, W.; Lei, M.; Qiu, Z.; Leong, Y.-K.; Zhong, H.; Zhang, S. Damage mechanism and protection measures of a coalbed methane reservoir in the Zhengzhuang block. *J. Nat. Gas Sci. Eng.* **2015**, *26*, 683–694. [CrossRef]

34. Li, P.; Ma, D.; Zhang, J.; Tang, X.; Huo, Z.; Li, Z.; Liu, J. Effect of Wettability on Adsorption and Desorption of Coalbed Methane: A Case Study from Low-Rank Coals in the Southwestern Ordos Basin, China. *Ind. Eng. Chem. Res.* **2018**, *57*, 12003–12015. [CrossRef]

35. Shen, J.; Zhao, J.; Qin, Y.; Shen, Y.; Wang, G.G. Water imbibition and drainage of high rank coals in Qinshui Basin, China. *Fuel* **2018**, *211*, 48–59. [CrossRef]

36. Zheng, S.; Huang, Z.; Yao, A. Research on the effect of drilling fluid's pH value on the coal's wettability. *J. Pet. Explor. Prod. Technol.* **2017**, *8*, 849–853. [CrossRef]

Article

Research on the Estimate of Gas Hydrate Saturation Based on LSTM Recurrent Neural Network

Chuanhui Li * and Xuewei Liu

School of Geophysics and Information Technology, China University of Geosciences (Beijing),
Beijing 100083, China; liuxw@cugb.edu.cn
* Correspondence: lichuanhui@cugb.edu.cn

Received: 31 October 2020; Accepted: 3 December 2020; Published: 11 December 2020

Abstract: Gas hydrate saturation is an important index for evaluating gas hydrate reservoirs, and well logs are an effective method for estimating gas hydrate saturation. To use well logs better to estimate gas hydrate saturation, and to establish the deep internal connections and laws of the data, we propose a method of using deep learning technology to estimate gas hydrate saturation from well logs. Considering that well logs have sequential characteristics, we used the long short-term memory (LSTM) recurrent neural network to predict the gas hydrate saturation from the well logs of two sites in the Shenhu area, South China Sea. By constructing an LSTM recurrent layer and two fully connected layers at one site, we used resistivity and acoustic velocity logs that were sensitive to gas hydrate as input. We used the gas hydrate saturation calculated by the chloride concentration of the pore water as output to train the LSTM network. We achieved a good training result. Applying the trained LSTM recurrent neural network to another site in the same area achieved good prediction of gas hydrate saturation, showing the unique advantages of deep learning technology in gas hydrate saturation estimation.

Keywords: gas hydrate; saturation; deep learning; recurrent neural network

---

1. Introduction

Gas hydrate is an ice-like crystalline solid, formed by water molecules and methane molecules under low temperature and high pressure. It is mainly distributed in seabed sediments on continental margins and permafrost regions. Gas hydrate can cause seabed geo-hazards and atmospheric environmental problems [1], but is also a clean energy with huge reserves [2]. Gas hydrate saturation is an important index for evaluating gas hydrate reservoirs. Well logs are widely used to estimate gas hydrate saturation due to their fast speed and low cost. The common methods for estimating the saturation of gas hydrate by using well logs mainly include resistivity methods and velocity methods [3]. Resistivity-based methods use resistivity logs to estimate gas hydrate saturation according to Archie's law [4,5], while velocity-based methods use the theoretical or empirical relationship between gas hydrate saturation and velocity to estimate gas hydrate saturation by using velocity logs. The frequently used relationships between gas hydrate saturation and velocity include time-average equations [6], the effective medium theory [7,8], and three-phase Biot-type equations [9,10].

The close relationship between gas hydrate saturation and well log machine learning technology provides a new idea for using well logs to estimate gas hydrate saturation. Singh et al. [11,12] used different combinations of well logs to predict gas hydrate saturation through unsupervised and supervised machine learning algorithms. They obtained a higher accuracy of gas hydrate saturation than in classic resistivity and velocity methods, showing the advantages of machine learning technology in gas hydrate saturation predictions. As the most vigorous branch of machine learning, deep learning technology can achieve more accurate prediction and classification than traditional technology. This is because it builds a deep neural network model with multiple hidden layers and uses a lot of data to train the model to learn complex and effective information. Therefore, to use well logs better to estimate gas hydrate saturation and to establish the deep internal connections and laws of the data, we propose a method of estimating gas hydrate saturation from well logs by using deep learning technology.

The concept of deep learning first proposed by Hilton et al. [13] has been successfully applied in image, audio, and natural language processing, and its unique advantages have attracted increasing attention from geoscientists. Deep learning technology is being gradually applied to well log interpretation and reservoir prediction, such as in rock facies classification [14–19] and the prediction of shale content [20] and porosity [21]. Well logs are sequence samples, so to estimate the gas hydrate saturation, we adopted the long short-term memory (LSTM) recurrent neural network, which is suitable for processing sequential data to apply to the well logs that are sensitive to gas hydrate. This method brought good application results in the Shenhu area, South China Sea. It demonstrated the unique advantages of deep learning technology in gas hydrate saturation estimates, and laid the foundation for its further application in gas hydrate research.

2. Long Short-Term Memory (LSTM) Recurrent Neural Network

2.1. Recurrent Neural Network (RNN)

A recurrent neural network (RNN) is a neural network model with memory function that can discover the interrelationships between samples. It is especially used to process data with sequential characteristics. Unlike other network structures, an RNN introduces the idea of self-loop, which can input the output of the previous and next samples into the model for operation (Figure 1). The feature information processed by the model contains not only the information of the sequence data before the current sample, but also the information of the current sample itself. However, an RNN cannot effectively deal with long-term dependency problems (neurons that are far away in the hidden layer) because in the process of using the stochastic gradient descent method to train the RNN, the partial derivative of the loss function to the weight matrix will tend toward zero or infinity as the number of input sequence samples increases. This will bring problems of gradient vanishing or gradient exploding, limiting its wide application.

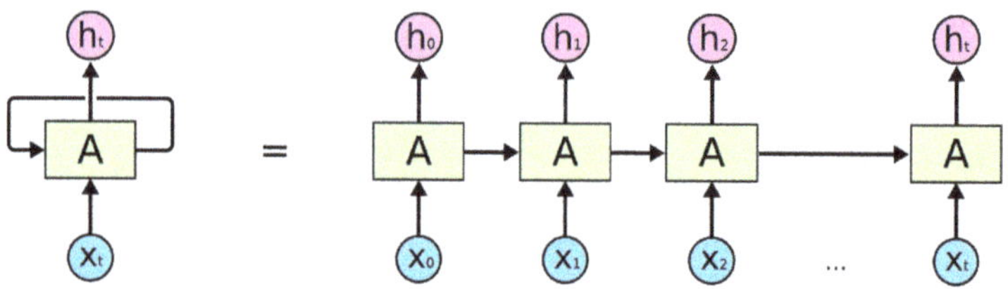

Figure 1. Unfolded form of recurrent neural network [22].

2.2. LSTM Recurrent Neural Network

The LSTM network is a special recurrent neural network proposed by Hochreiter and Schmidhuber in 1997 [23]. It improves and perfects the loop body repeated in a chain in the conventional RNN. By adding a forget gate layer, an input gate layer, and an output gate layer in the network cell, continuous write, read, and reset operations on memory cells can be performed [24]. This enables LSTM to have long-term learning capabilities, and effectively solves the problems of gradient vanishing and gradient exploding, making it one of the most successful RNN networks. Figure 2 shows the basic network structure of LSTM, while Figure 3 shows the structure of an LSTM neuron.

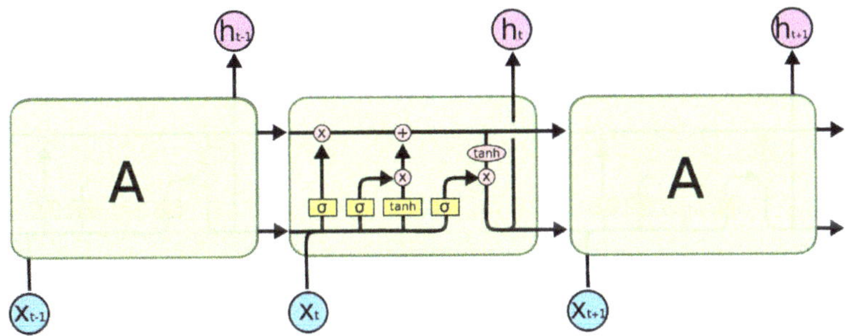

Figure 2. The basic network structure of the long short-term memory (LSTM) network [22].

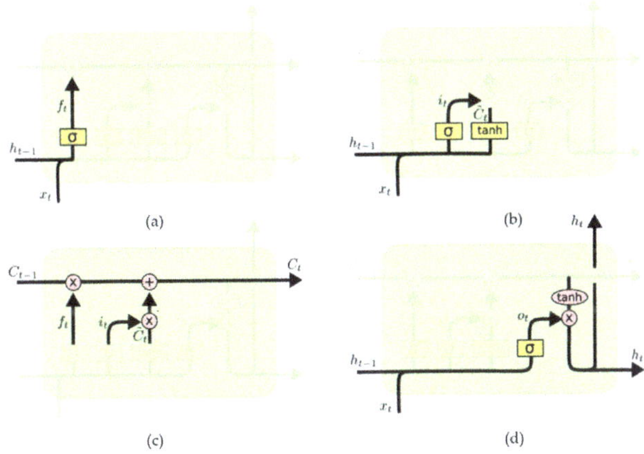

Figure 3. The structure of an LSTM neuron [22]: (**a**) the forget gate layer, (**b**) the input gate layer, (**c**) the cell status, and (**d**) the output gate layer.

The forget gate layer of the LSTM network determines which information needs to be discarded (Figure 3). The expression is:

$$f_t = \sigma\left(W_f \cdot [h_{t-1}, x_t] + b_f\right) \tag{1}$$

The input gate layer determines which new information is stored in the cell state (Figure 3b). The expression is:

$$i_t = \sigma(W_i \cdot [h_{t-1}, x_t] + b_i) \tag{2}$$

$$\widetilde{C}_t = \tanh(W_C \cdot [h_{t-1}, x_t] + b_C) \tag{3}$$

Then, the current cell status (Figure 3c) is updated to:

$$C_t = f_t \cdot C_{t-1} + i_t \cdot \widetilde{C}_t \tag{4}$$

The cell state of LSTM runs through the whole process, so that information is transmitted in a fixed and unchanging way. The output gate layer determines the information that needs to be output at that moment (Figure 3d). The expression is:

$$h_t = \sigma(W_o \cdot [h_{t-1}, x_t] + b_o) \cdot \tanh(C_t) \tag{5}$$

where x_t is the input vector of the LSTM neuron; f_t is the activation vector of the forget gate layer; i_t is the activation vector of the input gate layer; h_t is the output vector of the LSTM neuron; C_t is the neuron cell state vector; W is weight matrix; b is the bias term; σ is the sigmoid function; (tanh) is the hyperbolic tangent function; the subscript t indicates different moments.

3. Gas Hydrate Saturation Estimate

3.1. Geological Background

The Shenhu area is in the Pearl River Mouth Basin, in the middle of the northern slope of the South China Sea (Figure 4), and it is a key area for gas hydrate exploration. The water depth is 500–1500 m, the seabed topography is complicated, and the topographic slope varies greatly [25]. Since the late Miocene, with its gravity flow having developed and its high deposition rate, several kilometers of Mesozoic and Cenozoic sediments have accumulated to form enough organic matter to provide a source for gas hydrates [26]. In previous geological surveys of the area, many geophysical and geochemical markers indicating the existence of gas hydrates were discovered. In 2007, the Guangzhou Marine Geological Survey conducted the first gas hydrate drilling expedition in this area, and successfully drilled gas hydrate samples.

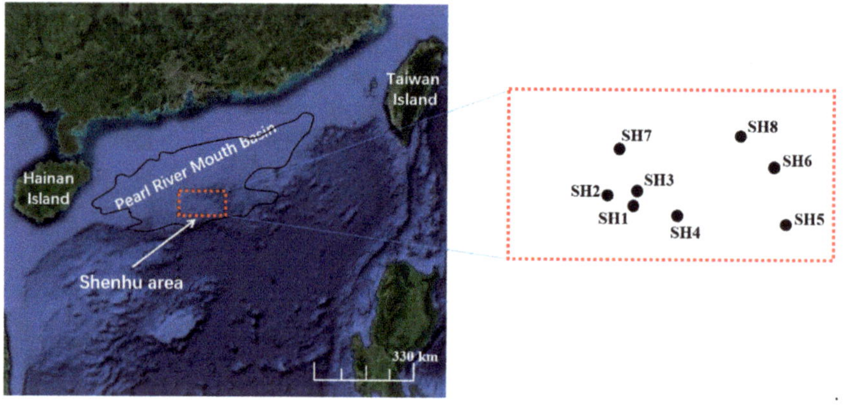

Figure 4. The Pearl River Mouth Basin in the northern slope of the South China Sea; the Shenhu area is shown by the red rectangle.

3.2. Well Logs

Eight sites were drilled in the expedition area in 2007 (Figure 4). Gas hydrates were found in the cores of sites SH2, SH3, and SH7, but no hydrates were found at sites SH1 and SH5. The other three sites, namely, SH4, SH6, and SH9, were drilled for logging without cores.

Figure 5 shows the well logs of site SH2. The cores at this site confirmed that the gas hydrate-bearing sediments were in the range of 190–220 m, and the hydrate saturation could reach 47.3% [27].

In the well logs of site SH2, the resistivity and acoustic velocity in the gas hydrate-bearing formations showed apparent high value anomalies, while the density and gamma showed no obvious changes. The well logs of site SH7 (Figure 6) showed that the depth of the gas hydrate-bearing formation was approximately 152–177 m, and the hydrate saturation could reach 43% [27]. The well log characteristics of the gas hydrate-bearing formation at site SH7 were completely consistent with those at site SH2.

Gas hydrate causes the chloride concentration of the formation pore water to decrease, so the saturation of gas hydrate can be calculated by measuring the chloride concentration of pore water from cores [28] using:

$$S_h = \frac{1}{\rho_h}\left(1 - \frac{Cl_{pw}}{Cl_{sw}}\right) \tag{6}$$

where $\rho_h = 0.924$ is the value of the density of pure gas hydrate in g/cm^3. Here, Cl_{sw} is the in situ baseline pore water chloride concentration and Cl_{pw} is the measured chloride concentration in core water after gas hydrate dissociation. The baseline chloride concentration can be determined by smoothly fitting the chloride data above and below the gas hydrate zone [3].

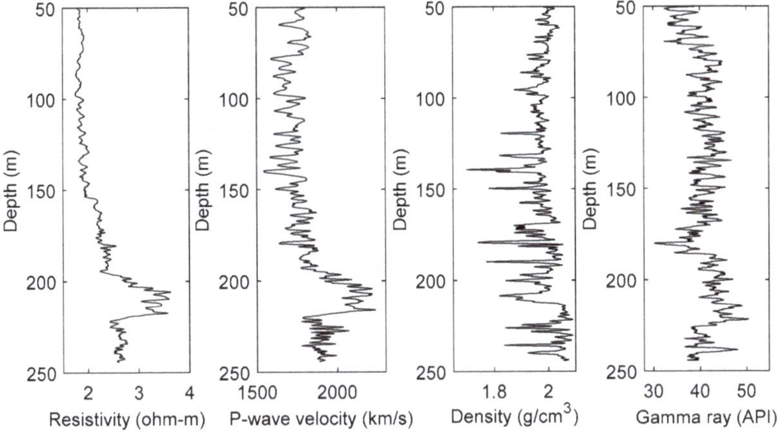

Figure 5. Well logs at site SH2.

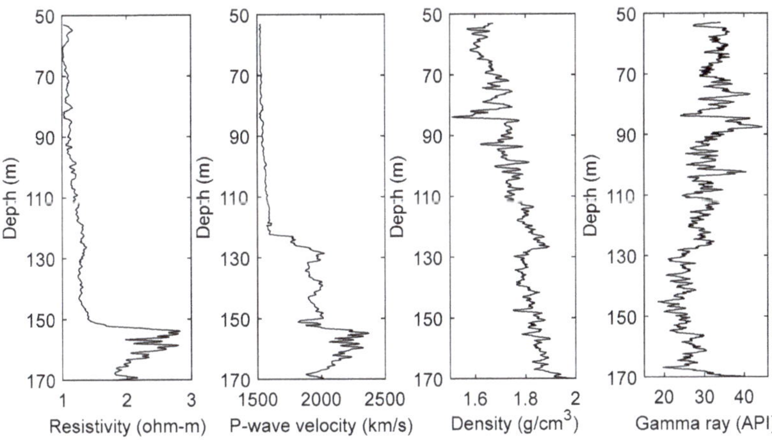

Figure 6. Well logs at site SH7.

Because the chloride concentration of the formation pore water was relatively less disturbed, and the chloride concentration measured by the cores was more accurate, the gas hydrate saturation calculated by the pore water chloride concentration had a higher accuracy [28]. Figure 7 shows the gas hydrate saturations calculated by using the chloride concentration measured by cores in the gas hydrate-bearing formation at sites SH2 and SH7. There were 41 gas hydrate-bearing cores at site SH2, and 21 cores containing gas hydrate at site SH7 [3,29].

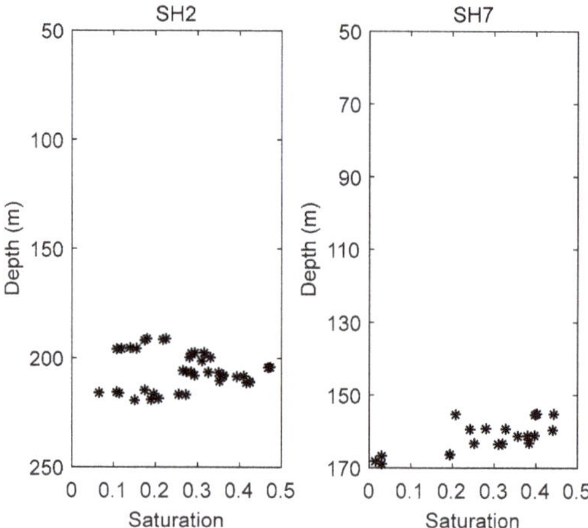

Figure 7. Gas hydrate saturations calculated by the chloride concentration of the pore water from the cores at sites SH2 and SH7.

3.3. Data Preparation

To use the LSTM recurrent neural network to estimate the gas hydrate saturation, site SH2 was used as a training well to train the LSTM recurrent neural network, while site SH7 was used as a verification well to verify the accuracy of the network model. In site SH2, the resistivity and acoustic velocity, which are more sensitive to gas hydrate, were used as the input of the network model. The gas hydrate saturations calculated by the chloride concentration of the pore water in the cores were used as the output to train the LSTM recurrent neural network.

Because there were only 41 gas hydrate saturation values calculated from the chloride concentration at site SH2, too little training data would seriously affect the training effect of the LSTM recurrent neural network model. Therefore, the interpolation of the gas hydrate saturation was performed at the sampling interval of the well logs to obtain 1400 sample datasets in the range of 191–219 m (Figure 8) where the resistivity and the acoustic velocity were the input of the network model, and the interpolated gas hydrate saturation were output. Before the dataset was input to the LSTM recurrent neural network for training, 1000 consecutive samples were selected as the training dataset, with the remaining samples used as the test dataset. To eliminate the dimensional influence between the parameters, and to ensure that each parameter was within a reasonable distribution range, data standardization processing was required. The expression is:

$$z_i = \frac{x_i - \mu_i}{\delta_i} \tag{7}$$

where z_i refers to the log parameters after standardization, x_i refers to the input log parameters, μ_i and δ_i are the mean and standard deviation of the parameters, respectively.

3.4. The Prediction Framework of the LSTM Recurrent Neural Network

We constructed an LSTM network prediction model that included an LSTM recurrent layer and two dense layers (Figure 9), where $\overline{x}_i$ is the standardized input sequence sample of the resistivity and p-wave velocity; $\overline{y}_i$ is the output saturation sample; $LSTM_i$ is the LSTM neuron that makes up the LSTM recurrent layer, which has the exact structure in Figure 3; o_i is the output of the LSTM neuron; C_i and h_i have the same meanings as in Equations (1)–(5). Because the actual data were not particularly complicated, to improve the calculation efficiency, the number of nodes of the two fully connected layers was set to 20 and 10, respectively. The optimization algorithm adopted the Adam algorithm, and the dropout regularization method was used to prevent over-fitting.

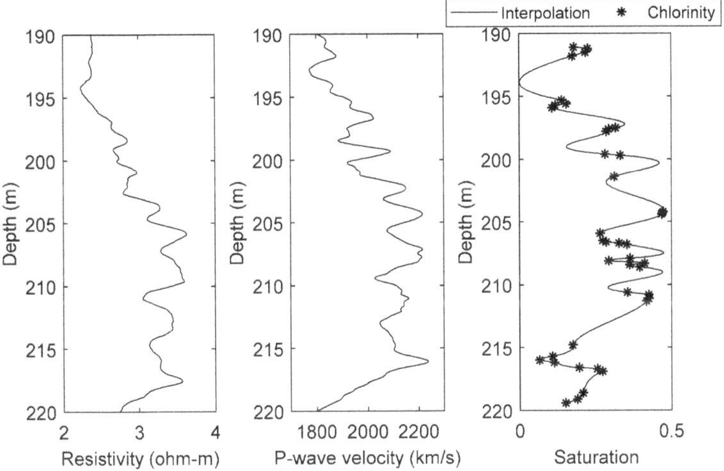

Figure 8. Training dataset of the network model.

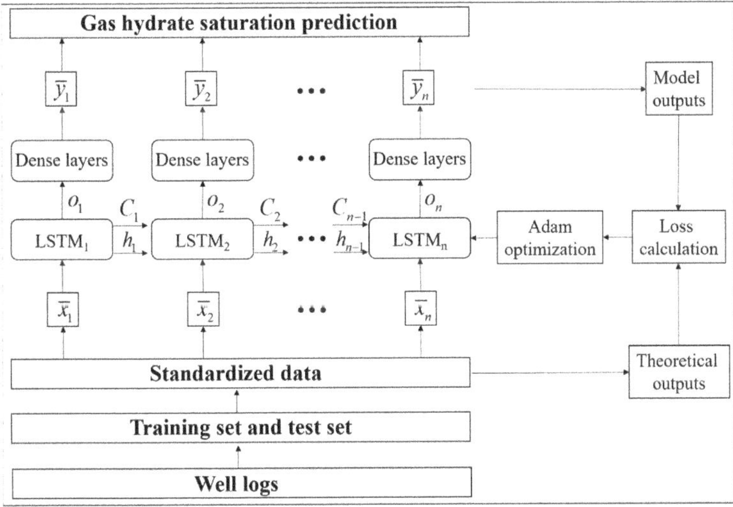

Figure 9. The prediction framework of LSTM recurrent neural network.

The training process of the LSTM recurrent neural network was similar to that of a conventional fully connected neural network, namely: (1) Use feedforward propagation to input training data into the network, calculate the output of the LSTM unit, and then extract features through the two fully connected layers. This trains it layer by layer to the output layer to obtain the predicted estimate of this sample. (2) Back-calculate the error term of each neuron. The backward propagation of the error term of the LSTM recurrent neural network includes two directions: the first is the back propagation along time, that is, starting from the current t time, calculating the error term at each time; the second is propagating the error term to the upper layer. (3) Use the Adam optimization algorithm based on gradient descent to adjust the model parameters by calculating the gradient of each weight according to the corresponding error item, so that the prediction is close to the optimization target. (4) Through the above iterations, train until it meets the required optimization target, then the LSTM recurrent neural network prediction model that meets the error requirements is established.

3.5. Results

Figure 10 shows the training results of the LSTM recurrent neural network using site SH2. The red dotted line shows the predicted saturation of the gas hydrate of the network model, and the blue curve shows the true value input into the model. The calculation shows that the correlation coefficient between the predicted value and the true value was 0.9605, and the root mean square error was 0.0208. The LSTM recurrent neural network achieved a good training effect, so it could be used for the prediction of gas hydrate saturation at site SH7.

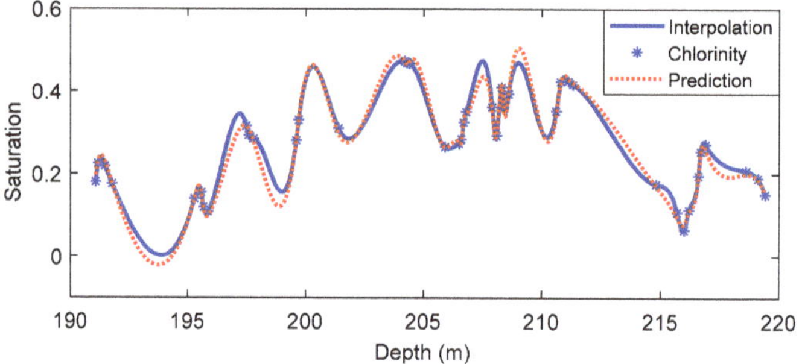

Figure 10. Training results of the LSTM recurrent neural network at site SH2.

We selected the resistivity and acoustic velocity logs of 155–167 m at site SH7, standardized the data, and input the data into the previously trained LSTM recurrent neural network to obtain the prediction of the gas hydrate saturation (Figure 11). The black curve in Figure 11 shows the predicted value, and the black asterisks show the gas hydrate saturations calculated by the chloride concentration of the pore water at site SH7. The overall change trend of the predicted value of gas hydrate saturation obtained by the LSTM recurrent neural network was reasonable, and the prediction was basically consistent with the 21 measured values of site SH7. We picked out the corresponding 21 predicted values of gas hydrate saturation, and calculated the correlation coefficient and root mean square error between the predicted value and the true value. We obtained 0.7085 and 0.1208. We therefore achieved a relatively accurate prediction of gas hydrate saturation using the LSTM recurrent neural network.

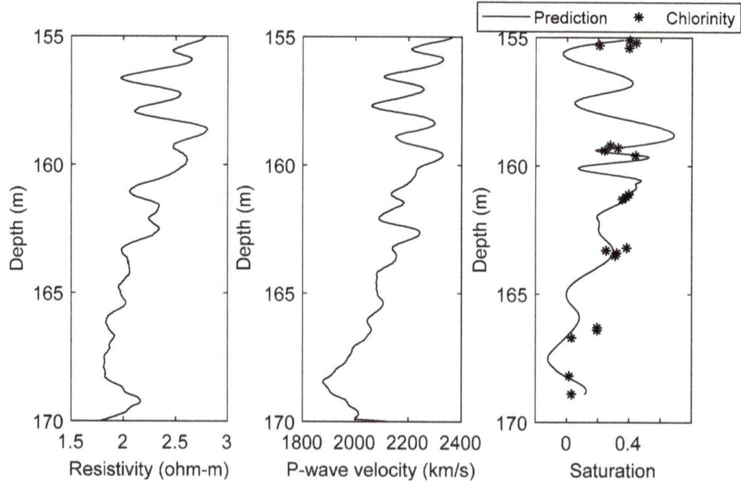

Figure 11. Prediction of the gas hydrate saturation at site SH7.

4. Discussion

The design of the network structure is key to improving the accuracy of a network model. We used an LSTM network prediction model that included one LSTM recurrent layer and two fully connected layers. The number of nodes in the two fully connected layers was 20 and 10, respectively. We did this because the complexity of the actual data was relatively low and because we wanted to improve calculation efficiency. In addition to selecting parameters based on experience, the optimal network structure could also be selected by using the training dataset for repeated experiments. There are many ways to use dropout regularization in LSTM network training [30], either in the loop of LSTM or in the final fully connected layer. We chose to put dropout regularization in the fully connected layer.

The analysis of the cores in the Shenhu area showed that the gas hydrate-bearing sediments consisted of silt (70%), sand (<10%), and clays (15%–30%) [31]. Because the well logs of gas hydrate-bearing sediments were the comprehensive responses of lithology and gas hydrates, the log characteristics of gas hydrate-bearing sediments, with varying lithologies, were different. Therefore, the LSTM network trained by well logs is only suitable for gas hydrate saturation predictions of gas hydrate-bearing sediments with small lithological differences, such as adjacent sites in the same exploration area. For sites that are further apart, or located in other exploration areas, the predictions may have large errors.

5. Conclusions

Based on the successful application of machine learning technology in gas hydrate saturation using well logs, we proposed a method for estimating gas hydrate saturation from well logs using deep learning technology to establish the deep internal connections and laws of the data. Considering that well logs are sequence samples, this method designed the LSTM recurrent neural network to be suitable for processing sequential data, took the resistivity and acoustic velocity logs that are more sensitive to gas hydrates as input, took the gas hydrate saturation calculated by the chloride concentration as the output, and trained the LSTM recurrent neural network to accurately predict the saturation of gas hydrate. This method had higher accuracy prediction of gas hydrate saturation than traditional machine learning methods and achieved good application results in the two studied sites in the Shenhu area, South China Sea. It demonstrated the unique advantages of deep learning technology in gas hydrate saturation estimates, and laid the foundation for its further application in gas hydrate research.

Author Contributions: C.L. designed the experiments and wrote the paper; X.L. proposed the theory. All authors have read and agreed to the published version of the manuscript.

Funding: This research was funded by the National Natural Science Foundation of China (No. 41974153) and the Fundamental Research Funds for the Central Universities (No. 2652019038).

Acknowledgments: The design of this research was done by C.L. while he was a visiting scholar at the College of Earth, Ocean, and Atmospheric Sciences at Oregon State University. C.L. would like to thank Anne Tréhu for her providing the opportunity to work at Oregon State University.

Conflicts of Interest: The authors declare no conflict of interest.

References

1. Ruppel, C.D. Tapping methane hydrates for unconventional natural gas. *Elements* **2007**, *3*, 193–199. [CrossRef]
2. Archer, D. Methane hydrate stability and anthropogenic climate change. *Biogeosci. Discuss.* **2007**, *4*, 993–1057. [CrossRef]
3. Wang, X.J.; Hutchinson, D.R.; Wu, S.G.; Yang, S.X.; Guo, Y.Q. Elevated gas hydrate saturation within silt and silty clay sediments in the Shenhu area, South China Sea. *J. Geophys. Res.* **2011**, *116*, B05102.
4. Collett, T.S.; Ladd, J. Detection of gas hydrate with downhole logs and assessment of gas hydrate concentrations and gas volumes on the Blake Ridge with electrical resistivity log data. *Proc. Ocean Drill. Program Sci. Results* **2000**, *164*, 179–191.

5. Lee, M.W.; Collett, T.S. Gas hydrate saturations estimated from fractured reservoir at Site NGHP-01-10, Krishna-Godavari Basin, India. *J. Geophys. Res.* **2009**, *114*, B07102. [CrossRef]

6. Wood, W.T.; Stoffa, P.L.; Shipley, T.H. Quantitative detection of methane hydrate through high-resolution seismic velocity analysis. *J. Geophys. Res.* **1994**, *99*, 9681–9969. [CrossRef]

7. Helgerud, M.B.; Dvorkin, J.; Nur, A.; Sakai, A.; Collett, T. Elastic-wave velocity in marine sediments with gas hydrates: Effective medium modeling. *Geophys. Res. Lett.* **1999**, *26*, 2021–2024. [CrossRef]

8. Jakobsen, M.; Hudson, J.A.; Minshull, T.A.; Singh, S.C. Elastic properties of hydrate-bearing sediment using effective medium theory. *J. Geophys. Res.* **2000**, *105*, 561–577. [CrossRef]

9. Carcione, J.M.; Gei, D. Gas-hydrate concentration estimated from P- and S-wave velocities at the Mallik 2L-38 research well, Mackenzie Delta, Canada. *J. Appl. Geophys.* **2004**, *56*, 73–78. [CrossRef]

10. Carcione, J.M.; Tinivella, U. Bottom-simulating reflectors: Seismic velocities and AVO effects. *Geophysics* **2000**, *65*, 54–67. [CrossRef]

11. Singh, H.; Seol, Y.; Myshakin, E.M. Prediction of gas hydrate saturation using machine learning and optimail set of well-logs. *Comput. Geosci.* **2020**, 1–17. [CrossRef]

12. Singh, H.; Seol, Y.; Myshakin, E.M. Automated Well-Log Processing and Lithology Classification by Identifying Optimal Features Through Unsupervised and Supervised Machine-Learning Algorithms. *SPE J.* **2020**, *25*. [CrossRef]

13. Hinton, G.E.; Osindero, S.; The, Y.W. A fast learning algorithm for deep belief nets. *Neural Comput.* **2006**, *18*, 1527–1554. [CrossRef] [PubMed]

14. Hall, B. Facies classification using machine learning. *Lead. Edge* **2016**, *35*, 906–909. [CrossRef]

15. Bestagini, P.; Lipari, V.; Tubaro, S. A machine learning approach to facies classification using well logs. *SEG Tech. Program Expand. Abstr.* **2017**, 2137–2142. [CrossRef]

16. Zhang, L.; Zhan, C. Machine learning in rock facies classification—An application of XGBoost. *Int. Geophys. Conf. Qingdao China.* **2017**, 1371–1374. [CrossRef]

17. Hall, M.; Hall, B. Distribution collaborative prediction: Results of the machine learning contest. *Leading Edge* **2017**, *36*, 267–269. [CrossRef]

18. Sidahmed, M.; Roy, A.; Sayed, A. Streamline rock facies classification with deep learning cognitive process. *SPE Annu. Technical Conf. Exhib.* **2017**. [CrossRef]

19. An, P.; Cao, D. Research and application of logging lithology identification based on deep learning. *Prog. Geophys.* **2018**, *33*, 1029–1034. (In Chinese)

20. An, P.; Cao, D. Shale content prediction based on LSTM recurrent neural network. In Proceedings of the SEG 2018 Workshop: SEG Maximizing Asset Value Through Artificial Intelligence and Machine Learning, Beijing, China, 17–19 September 2018; pp. 49–52.

21. An, P.; Cao, D.; Zhao, B.; Yang, X.; Zhang, M. Reservoir physical parameters prediction based on LSTM recurrent neural network. *Prog. Geophys.* **2019**, *34*, 1849–1858. (In Chinese)

22. Sofiyanti, N.; Fitmawati, D.I.; Roza, A.A. Understand LSTM Networks. *GITHUB Colah Blog* **2015**, *22*, 137–141.

23. Hochreiter, S.; Schmidhuber, J. Long short-term memory. *Neural Comput.* **1997**, *9*, 1735–1780. [CrossRef]

24. Graves, A. *Supervised Sequence Labelling with Recurrent Neural Networks*; Springer: Berlin/Heidelberg, Germany, 2012.

25. Briais, A.; Patriat, P.; Tapponnier, P. Updated interpretation of magnetic anomalies and seafloor spreading stages in the South China Sea: Implications for the tertiary tectonics of Southeast Asia. *J. Geophys. Res.* **1993**, *98*, 6299–6328. [CrossRef]

26. Clift, P.; Lin, J.; Barckhausen, U. Evidence of low flexural rigidity and low viscosity lower continental crust during continental break-up in the South China Sea. *Mar. Pet. Geol.* **2002**, *19*, 951–970. [CrossRef]

27. Wang, X.; Collett, T.S.; Lee, M.W.; Yang, S.; Guo, Y.; Wu, S. Geological controls on the occurrence of gas hydrate from core, downhole log, and seismic data in the Shenhu are, South China Sea. *Mar. Geol.* **2014**, *357*, 272–292. [CrossRef]

28. Yuan, T.; Hyndman, R.D.; Spence, G.D.; Desmons, B. Seismic velocity increase and deep-sea gas hydrate concentration above a bottom-simulating reflector on the northern Cascadia continental slope. *J. Geophys. Res.* **1996**, *101*, 655–671. [CrossRef]

29. Chen, Y.; Dunn, K.; Liu, X.; Du, M.; Lei, X. New method for estimating gas hydrate saturation in the Shenhu area. *Geophysics* **2014**, *79*, IM11–IM22. [CrossRef]

30. Gal, Y.; Ghahramani, Z. A theoretically grounded application of dropout in recurrent neural networks. *arXiv* **2015**, arXiv:1512.05287.
31. Chen, F.; Zhou, Y.; Su, X.; Liu, G.; Lu, H.; Wang, J. Gas hydrate saturation and its relation with grain size of the hydrate-bearing sediments in the Shenhu Area of northern South China Sea. *Mar. Geol. Quat. Geol.* **2011**, *31*, 95–100. (In Chinese) [CrossRef]

Article

Fractional Time Derivative Seismic Wave Equation Modeling for Natural Gas Hydrate

Yanfei Wang [1,2,3,*], Yaxin Ning [1,2,3] and Yibo Wang [1,3]

[1] Key Laboratory of Petroleum Resources Research, Institute of Geology and Geophysics, Chinese Academy of Sciences, Beijing 100029, China; nyx@mail.iggcas.ac.cn (Y.N.); wangyibo@mail.igcas.ac.cn (Y.W.)

[2] College of Earth and Planetary Sciences, University of Chinese Academy of Sciences, Beijing 100049, China

[3] Innovation Academy of Earth Science, Chinese Academy of Sciences, Beijing 100029, China

* Correspondence: yfwang@mail.iggcas.ac.cn

Received: 1 October 2020; Accepted: 3 November 2020; Published: 12 November 2020

Abstract: Simulation of the seismic wave propagation in natural gas hydrate (NGH) is of great importance. To finely portray the propagation of seismic wave in NGH, attenuation properties of the earth's medium which causes reduced amplitude and dispersion need to be considered. The traditional viscoacoustic wave equations described by integer-order derivatives can only nearly describe the seismic attenuation. Differently, the fractional time derivative seismic wave-equation, which was rigorously derived from the Kjartansson's constant-Q model, could be used to accurately describe the attenuation behavior in realistic media. We propose a new fractional finite-difference method, which is more accurate and faster with the short memory length. Numerical experiments are performed to show the feasibility of the proposed simulation scheme for NGH, which will be useful for next stage of seismic imaging of NGH.

Keywords: natural gas hydrate; seismic modeling; fractional derivatives; numerical simulation

---

1. Introduction

Seismic exploration is a main technique in natural gas hydrate (NGH) survey. To gain a better estimation of the NGH content, we need to understand the characteristics of the seismic wave propagation in NGH. Traditional seismic modeling and inversion techniques are usually built on perfectly elastic medium model. However, the real underground medium usually has attenuation properties, which will cause amplitude loss and phase distortion of seismic waves. In particular, gas hydrate layer often shows abnormal velocity and attenuation characteristics due to hydrate filling. Ignoring the attenuation of the medium will make the numerical simulation and inversion results different from the real situation [1–3], which will result in the inability to obtain the true and accurate structural features of the underground. Studying the numerical simulation method of viscoacoustic seismic wave will help to obtain the actual propagation of underground seismic wave. Moreover, using the viscoacoustic wave equation for migration and inversion can effectively compensate the loss of amplitude, accelerate the convergence rate, and make the inversion results closer to the actual underground geological characteristics [4].

At present, many methods have been proposed for viscoacoustic seismic wave simulation [5–7]. Aki et al. [8] achieves viscoacoustic wave simulation by introducing complex velocity in the frequency domain, but the calculation is huge. Another method to construct viscoacoustic models is by combining mechanical elements [5,9–11], such as standard linear solid model (SLS) and Maxwell model, which are approximately constant-Q model and require a large amount of memory and computational time [5,7]. In addition to the above integer-order method, the fractional-order wave equation can better describe the amplitude loss and phase distortion of seismic waves. Carcione et al. [12] proposed a fractional time derivative wave equation using the stress-strain relationship proposed by Kjartansson [13]. The equation can accurately describe the constant-Q behavior, but the fractional time derivative will introduce huge memory consumption [14,15]. In order to reduce memory consumption, Chen and Holm [16] proposed using fractional Laplacian to simulate irregular attenuation behavior. Carcione [17] uses fractional Laplacian to give a new wave equation, which effectively reduces memory and describes amplitude attenuation and velocity dispersion in a single term. Song et al. [18] proposed an asymptotic local finite difference based on the truncated difference stencil and applied the method to the fractional wave equation pointed out by Carcione [17]. Zhu and Harris [3] proposed a nearly constant-Q (NCQ) wave equation by approximating the fractional time derivative wave equation. The amplitude attenuation and velocity dispersion are described by two decoupled fractional Laplacians in the NCQ equation, which are helpful to compensate the attenuation loss in the inverse problem. Sun et al. [4,19] further numerically simulated the equation proposed by Zhu and Harris using low-rank one-step wave approximation [20,21]. Yao et al. [22] developed a local-spectral approach to implement the fractional Laplacian in the viscoacoustic wave equation. Wang et al. [23] improved the numerical discretization of the temporal derivatives in the NCQ equation. Xu et al. [24] adopted radial basis collocation method to solve the NCQ equation. The NCQ equation avoids solving the fractional time derivative and therefore saves the amount of calculation and memory, but it is actually an approximate modification of the fractional time derivative wave equation and has a certain loss of accuracy. In this paper we use the fractional time derivative wave equation to simulate the wave field. In recent years, many scholars have proposed different methods for solving fractional time derivatives [25,26], which have promoted its development.

In the previous researches, the realization of the fractional time derivative was approximated by the original Grunwald–Letnikov finite difference method. In this paper, a more unified definition of fractional derivatives is proposed to calculate the fractional time derivative. The rigorous form is the limit of the original finite difference when time step approaches zero. This method is more stable at different memory lengths and has higher accuracy when the memory length is short. We first give a review of the derivation of the fractional time derivative wave equation. Then, we introduce the more unified definition of fractional derivatives with finite difference method and apply this method to calculate the fractional time derivative. Finally, some numerical examples on NGH model illustrate the accuracy and stability of this new method.

2. Materials and Methods

2.1. Definitions of Fractional Derivatives

2.1.1. Grunwald–Letnikov Fractional-Order Derivative

The definition of Grunwald–Letnikov (G-L) fractional-order derivative is an extension of the definition of the limit of integer-order derivative to real-order. Let $\alpha > 0$, the Grunwald–Letnikov α-th order fractional derivative of the function $f(t)$ with respect to t and the terminal value a is given by [12,27,28]

$$_{a}^{GL}D_{t}^{\alpha}f(t) = \lim_{\substack{h \to 0 \\ mh = t - a}} \frac{1}{h^{\alpha}} \sum_{k=0}^{m} \omega_{k}^{(\alpha)} f(t - kh),\tag{1}$$

where $\omega_{k}^{(\alpha)} = (-1)^{k} \binom{\alpha}{k} = \frac{(-1)^{k}\Gamma(\alpha+1)}{\Gamma(k+1)\Gamma(\alpha-k+1)}$, $\Gamma(\cdot)$ denotes gamma function given below and α represents the order of the fractional operator. Binomial coefficients can be generalized to real arguments by means of the gamma function

$$\Gamma(x) = \int_{0}^{\infty} t^{x-1}e^{-t}dt, \ x > 0.\tag{2}$$

For which it is well-known that $\Gamma(1) = 1$ and $\Gamma(x+1) = x\Gamma(x)$, for any $x > 0$.

It is clear that if $\alpha = 1$, $\,_{a}^{GL}D_t^{\alpha}f(t) = \frac{f(t)-f(t-h)}{h}$; and if $\alpha = 2$, $\,_{a}^{GL}D_t^{\alpha}f(t) = \frac{f(t)-2f(t-h)+f(t-2h)}{h^2}$. They correspond to the first-order difference quotient and second-order difference quotient, respectively.

2.1.2. Riemann–Liouville Fractional-Order Derivative

The Riemann–Liouville (R-L) integral of non-integer order $\alpha > 0$ for $f(t)$ in suitable space (e.g., $L^p(a,b)$) is defined by

$$_{a+}I_t^{\alpha}f(t) = \frac{1}{\Gamma(\alpha)} \int_a^t \frac{f(\tau)}{(t-\tau)^{1-\alpha}}d\tau. \tag{3}$$

The Riemann–Liouville derivative with the lower integration limit a would be

$$_{a+}^{RL}D_t^{\alpha}f(t) = \frac{1}{\Gamma(1-\alpha)}\frac{d}{dt}\int_a^t \frac{f(\tau)}{(t-\tau)^{\alpha}}d\tau, \tag{4}$$

which is called the Riemann–Liouville fractional derivative of order α.

2.1.3. Caputo's Fractional-Order Derivative

The Caputo's derivative of fractional order $\alpha > 0$ is defined by

$$_{a+}^{C}D_t^{\alpha}f(t) = \frac{1}{\Gamma(m-\alpha)}\int_a^t \frac{f^{(m)}(\tau)}{(t-\tau)^{\alpha+1-m}}d\tau, \tag{5}$$

where $m = \lceil\alpha\rceil$ is the smallest integer such that $m > \alpha$, $f^{(m)}$ denotes the classical derivative of integer order m.

2.1.4. The Riesz Fractional-Order Derivative

The kernel $k(t)$ of the Riesz fractional-order derivative is given by

$$k(t) = \frac{|t|^{-(\alpha+1)}}{2\Gamma(-\alpha)\cos(\beta)}, \quad \alpha > -1, \tag{6}$$

where $\beta = (\pi/2)\alpha$. Clearly, if $\alpha = 1$, the kernel would be $k(t) = -\frac{|t|^{-2}}{\pi}$. With the above kernel, the α-th order Riesz fractional derivative of function $f(t)$ is defined by

$$D^{\alpha}f(t) = f(t) * k(t) = \frac{1}{2\Gamma(\alpha)\cos\left(\frac{\pi\alpha}{2}\right)}\int_{-\infty}^{\infty}|t-\tau|^{-\alpha-1}f(\tau)d\tau, \tag{7}$$

where the notation "$*$" denotes the convolution operator. Equation (7) can be regarded as the two-sided fractional derivatives. With above definitions, suppose $t \in [a,b]$, for the order of $\alpha > 0$, the Riesz fractional derivative can be written as

$$\frac{\partial^{\alpha}f(t)}{\partial|t|^{\alpha}} = -C_{\alpha}\left(_aD_t^{\alpha}f(t) + _tD_b^{\alpha}f(t)\right), \tag{8}$$

where $C_{\alpha} = \frac{1}{2\cos(\pi\alpha/2)}$, $\alpha \neq 1$, $_aD_t^{\alpha}$ and $_tD_tD_b^{\alpha}$ denotes the left- and right-side Riemann–Liouville derivatives given by (9) and (10), respectively:

$$_aD_t^{\alpha}f(t) = \frac{1}{\Gamma(1-\alpha)}\frac{d}{dt}\int_a^t \frac{f(\tau)}{(t-\tau)^{\alpha}}d\tau, \tag{9}$$

$$_tD_b^{\alpha}f(t) = -\frac{1}{\Gamma(1-\alpha)}\frac{d}{dt}\int_t^b \frac{f(\tau)}{(\tau-t)^{\alpha}}d\tau. \tag{10}$$

In which, we assume that $\alpha \in (0,1)$. If α is greater than 1, e.g., $\alpha \in (1,2]$, the definitions can be made correspondingly as

$$_aD_t^\alpha f(t) = \frac{1}{\Gamma(2-\alpha)} \frac{d^2}{dt^2} \int_a^t \frac{f(\tau)}{(t-\tau)^{\alpha-1}} d\tau,\tag{11}$$

$$_tD_b^\alpha f(t) = \frac{1}{\Gamma(2-\alpha)} \frac{d^2}{dt^2} \int_t^b \frac{f(\tau)}{(\tau-t)^{\alpha-1}} d\tau.\tag{12}$$

2.1.5. Relations between the above Fractional-Order Derivatives

For $\alpha \in (m-1,m)$, $f(t) \in C^{m-1}([a,t])$, and $f^{(m)}$ is continuous on $[a,t]$, there holds

$$_a^{RL}D_t^\alpha f(t) = {}_{a+}^{GL}D_t^\alpha f(t) = \sum_{k=0}^{m-1} \frac{f^{(k)}(a)(x-a)^{k-\alpha}}{\Gamma(k+1-\alpha)} + \frac{1}{\Gamma(m-\alpha)} \int_a^t \frac{f^{(m)}(\xi)(x-a)^{k-\alpha}}{(x-\xi)^{\alpha-m+1}} d\xi,\tag{13}$$

i.e., the definition of R-L is equal to that of G-L for smooth functions.

Clearly, we can see from (13) that the three definitions can be identical if $f(t)$ is smooth enough and satisfies zero initial conditions of $f^{(k)}(a) = 0$ $(k = 0,1,\cdots,m-1)$.

It also indicates from the three definitions that R-L is suitable for general functions, G-L is easy to be utilized for discretization, and Caputo's is often related with continuous processes depending on time.

With above preparations, we now turn to the gas hydrate model establishment and the related seismic wave modeling in fractional-order form.

2.2. Establishing Velocity and Quality Factor Model of Hydrate Layer

Gas hydrate layer often shows abnormal velocity and attenuation characteristics due to hydrate filling. Accurately establishing the velocity and quality factor model is very important for the simulation of the seismic wave field of gas hydrate. White [29] first proposed the concept of a mesoscopic scale theoretical model. The so-called mesoscopic scale is an intermediate scale that is much larger than rock particles but much smaller than the wavelength. The object of White's theoretical research is the non-uniform infiltration phenomenon of immiscible multiphase fluids infiltrating into the porous medium. Many scholars have found through analysis that the mesoscopic scale porous model can better explain the attenuation of the seismic frequency band [30]. Therefore, the White theory is chosen to model the velocity and quality factor of the hydrate layer.

The calculation steps for the velocity and quality factor of the hydrate formation are as follows. First, through the effective medium model [31], the elastic modulus of the solid phase and dry rock skeleton can be calculated from the elastic modulus of each mineral component of the rock, and then the velocity and quality factor of the saturated fluid rock can be calculated by the White theory [29,32]. Through the above method, the velocity and quality factor of stratums with different mineral components and hydrate saturation can be calculated. Therefore, seismic wave field simulations can be performed on different hydrate stratums, which is very helpful for studying the law of seismic wave propagation in hydrate stratums.

2.3. Viscoacoustic Wave Equation

To simulate seismic wave propagation of the gas hydrate stratums, in anelastic media, the stress $\sigma(t)$ can be expressed as a convolution of the variation of the strain $\varepsilon(t)$ and the relaxation function $\varphi(t)$ in the following form:

$$\sigma(t) = \varphi(t) * \dot{\varepsilon}(t),\tag{14}$$

where the symbol "$*$" refers to the convolution operator.

Kjartansson [13] gives a relaxation function, which exactly describes the constant Q characteristic and is widely used in many seismic applications. The relaxation function is written as [3,12]

$$\varphi(t) = \frac{M_0}{\Gamma(1-2\gamma)}\left(\frac{t}{t_0}\right)^{-2\gamma} H(t), \tag{15}$$

where $\gamma = \arctan(1/Q)/\pi$ is a dimensionless parameter and we can know that $0 < \gamma < 1/2$ by the range of $Q(Q > 0)$; $M_0 = \rho c_0^2 \cos^2(\pi\gamma/2)$ is a bulk modulus, Γ is the Euler's Gamma function, $H(t)$ is the Heaviside step function and t_0 is a reference time, e.g., $t_0 = 1/\omega_0$, and c_0 is the phase velocity at reference frequency ω_0.

Combining the first-order conservation equation $\frac{\partial}{\partial t}v = \frac{1}{\rho_0}\nabla\sigma$ (v is the phase velocity, ∇ denotes the gradient operator) with the strain-velocity equation $\frac{\partial}{\partial t}\varepsilon = \nabla\cdot v$ (here $\nabla\cdot$ denotes the divergence operator), along with Equations (14) and (15), the fractional time derivative wave equation with uniform density ρ can be derived as

$$\frac{\partial^{2-2\gamma}\sigma}{\partial t^{2-2\gamma}} = c_0^2\omega_0^{-2\gamma}\cos^2(\pi\gamma/2)\nabla^2\sigma. \tag{16}$$

The above Equation (16) was rigorously derived from Kjartansson's constant Q model [13], hence it could accurately describe the attenuation behavior in realistic media. The Equation (16) reduces to the classical acoustic equation when $\gamma \to 0$ (corresponding to $Q \to \infty$). When $\gamma \to \frac{1}{2}$ (corresponding to $Q \to 0$), this equation describes the propagation of the seismic wave in infinite attenuation medium. When $0 < \gamma < \frac{1}{2}$, this equation describes the propagation of the seismic wave in attenuating medium. Moreover, in discrete calculation, a short-term memory principle shown in [33] is applied. The fractional time wave equation describes the loss mechanisms only by two parameters, i.e., the phase velocity c_0 and the quality Q. Therefore, the method based on the above equation is more accurate and simpler than the other nearly constant Q methods, such as the methods based on SLS model or complex velocity [3,17].

In Zhu and Harris [3], the authors approximated the time fractional wave equation, and expressed the amplitude attenuation and velocity dispersion with two independent fractional Laplacians to obtain an approximately constant Q wave equation. The separated amplitude attenuation and frequency dispersion term are beneficial to compensate for attenuation loss in the inverse problem. The fractional Laplacian form of the wave equation reads as

$$\begin{aligned}
\frac{\partial v(r,t)}{\partial t} &= \frac{1}{\rho_0(r)}\nabla\sigma(r,t) + S, \\
\frac{\partial}{\partial t}\varepsilon(r,t) &= \nabla\cdot v(r,t), \\
\sigma(r,t) &= M_0(r)\left[\eta(r)(-\nabla^2)^{\gamma(r)} + \tau(r)\frac{\partial}{\partial t}(-\nabla^2)^{\gamma(r)-1/2}\right]\varepsilon(r,t),
\end{aligned} \tag{17}$$

where S denotes the source, $\sigma(r,t)$ and $\varepsilon(r,t)$ are the stress and strain fields at position r at time t, $\eta(r)$ and $\tau(r)$ are two coefficients vary in space and are in the form of $\eta(r) = -c_0^{2\gamma(r)}(r)\omega_0^{-2\gamma(r)}\cos(\pi\gamma(r))$ and $\tau(r) = -c_0^{2\gamma(r)-1}(r)\omega_0^{-2\gamma(r)}\cos(\pi\gamma(r))$, respectively, and all other parameters are defined the same as above. In calculation, the spatial fractional Laplace operator adopts fast Fourier transform method. Separating the amplitude attenuation term and dispersion term are beneficial to compensate attenuation loss in the inverse problem. However, $\gamma(r)$ changes with space and is taken as the average of the entire space in the Fourier transform, which does not conform to the actual situation.

2.4. Methods

2.4.1. Fractional-Order Derivatives Approximation

The critical matter of modeling wave propagation by using the Equation (16) is the method for computing the fractional time derivative. There are different techniques for approximating different fractional derivatives. The spatial derivatives $\nabla^2\sigma$ can be calculated by the fast Fourier transform or the finite difference. Carcione et al. [12] computed the fractional time derivative by the original Grunwald–Letnikov finite difference method with second-order accuracy, which is in the following form:

$$\frac{\partial^\beta f(t)}{\partial t^\beta} = f_h^{(\beta)}(t) \approx \frac{1}{h^\beta}\sum_{j=0}^{J}(-1)^j\binom{\beta}{j}f\left(t+\left(\frac{\beta}{2}-j\right)h\right),\tag{18}$$

where β is the order of fractional derivative with $\beta = 2 - 2\gamma$ and hence $1 < \beta < 2$, h is the time sampling step and $J = t/h - 1$, $\binom{\beta}{j}$ is the fractional binomial coefficients, which can be expressed in terms of Euler's Gamma function as

$$\binom{\beta}{j} = \frac{\beta(\beta-1)(\beta-2)\cdots(\beta-j+1)}{j!} = \frac{\Gamma(\beta+1)}{j!\Gamma(\beta-j+1)} = \frac{\Gamma(\beta+1)}{\Gamma(j+1)\Gamma(\beta-j+1)}.\tag{19}$$

Calculation of the fractional derivative using Equation (18) requires storing all the previous values of f. Therefore, the original method will consume a lot of memory and computation, even though we can truncate the memory length in some situations.

If the function $f(t)$ has $m + 1$-order continuous derivative on interval $[a, t]$ and the integer m satisfies $m < \beta < m+1$, the limit (as $h \to 0$) of the Equation (18) can be obtained using the R-L fractional derivative:

$$\begin{aligned}
{}_aD_t^\beta f(t) &= \lim_{h\to 0}f_h^{(\beta)}(t)\\
&= \sum_{k=0}^{m}\frac{f^{(k)}(a)(t-a)^{-\beta+k}}{\Gamma(-\beta+k+1)} + \frac{1}{\Gamma(-\beta+m+1)}\int_a^t(t-\tau)^{m-\beta}f^{(m+1)}(\tau)d\tau
\end{aligned}\tag{20}$$

In this study, we focus on the finite difference discretization of the fractional time derivative $\frac{\partial^\beta\sigma}{\partial t^\beta}$ for $1 < \beta < 2$. Then the fractional derivative can be defined as following a more unified form:

$$\begin{aligned}
\frac{\partial^\beta f(t)}{\partial t^\beta} &\triangleq \lim_{h\to 0}f_h^{(\beta)}(t)\\
&= c1\frac{(t-a)^{-\beta}f(t_0)}{\Gamma(1-\beta)} + c2\frac{(t-a)^{1-\beta}f'(t_0)}{\Gamma(2-\beta)} + \frac{1}{\Gamma(2-\beta)}\int_a^t(t-\tau)^{1-\beta}f^{(2)}(\tau)d\tau,
\end{aligned}\tag{21}$$

where the symbol "$\triangleq$"denotes definition, $c1$ and $c2$ are constants of 0 or 1. When $c1 = c2 = 1$, (21) is equivalent to R-L fractional derivative and G-L fractional derivative. When $c1 = c2 = 0$, (21) is equivalent to the Caputo fractional derivative.

2.4.2. Finite Difference Method for Integer-Order Derivatives

Define the transfer operator T^h, and difference operators ∇_h, Δ_h and δ_h (respectively refers to the backward difference, forward difference and central difference), where $h \in \mathbb{R}$ is the time sampling stepsize. The operator T^h acts on the function $f(t)$, $t \in \mathbb{R}$, gives

$$\begin{aligned}
T^h f(t) &= f(t+h),\\
\nabla_h f(t) &= f(t) - f(t-h),\\
\Delta_h f(t) &= f(t+h) - f(t),\\
\delta_h f(t) &= f\left(t+\tfrac{1}{2}h\right) - f\left(t-\tfrac{1}{2}h\right).
\end{aligned}\tag{22}$$

Assume that $f(t)$ is sufficiently smooth, we have that the m-order derivative of $f(t)$ can be approximated by $f^{(m)}(t) = D^m f(t)$, $m \in \mathbb{N}$, satisfying

$$D^m f(t) = h^{-m} \nabla_h^m f(t) + O(h) = h^{-m}(I - T^{-h})^m f(t) + O(h)$$

or

$$D^m f(t) = h^{-m} \delta_h^m f(t) + O(h^2) = h^{-m}(T^{h/2} - T^{-h/2})^m f(t) + O(h^2)$$

for backward and central difference, respectively; in which, I is the identity. The power m of the two difference operators ∇_h and δ_h can be expanded as

$$\nabla_h^m = \sum_{j=0}^{m} (-1)^j \binom{m}{j} T^{-jh}$$

and

$$\delta_h^m = \sum_{j=0}^{m} (-1)^j \binom{m}{j} T^{(m/2-j)h}$$

respectively. The above integer-order derivative forms can be extended to fractional cases, i.e., for any $\beta \in \mathbb{R}$, the fractional-order operator can be related by

$$
\begin{aligned}
\nabla_h^\beta &= \sum_{j=0}^{\infty} (-1)^j \binom{\beta}{j} T^{-jh}, \\
\delta_h^\beta &= \sum_{j=0}^{\infty} (-1)^j \binom{\beta}{j} T^{(\beta/2-j)h}.
\end{aligned}
\tag{23}
$$

2.4.3. Fractional Differencing Scheme

With the above preparation, we present some details about how to calculate the fractional-order time derivatives. Equation (20) is in the limit case where h approaches 0, which is more rigorous and accurate than Equation (18). Jia and Li [34] using the rigorous form to compute the spatial fractional derivative in the fractional advection-dispersion equation. Here, we consider the time fractional derivative discretization. We discretize time domain by ($i = 0, 1, 2, \cdots, L$) in interval $[a, t]$ ($t_0 = a$). The positive integer L ($L = (t_i - t_0)/h$) is the memory length. Since $1 < \beta < 2$, so $m = 1$. Using the discretization scheme (23) to calculate the fractional time derivative, the Equation (21) can be discretized as follows:

$$
\begin{aligned}
\frac{\partial^\beta f(t)}{\partial t^\beta}\Big|_{t=t_L} &= c1 \frac{(t_L-t_0)^{-\beta} f(t_0)}{\Gamma(1-\beta)} + c2 \frac{(t_L-t_0)^{1-\beta} f'(t_0)}{\Gamma(2-\beta)} + \frac{1}{\Gamma(2-\beta)} \int_{t_0}^{t_L} (t_L - \tau)^{1-\beta} f^{(2)}(\tau) d\tau \\
&= c1 \frac{(t_L-t_0)^{-\beta} f(t_0)}{\Gamma(1-\beta)} + c2 \frac{(t_L-t_0)^{1-\beta} (f(t_1)-f(t_0))}{\Gamma(2-\beta)h} + \\
&\quad \frac{h^{-\beta}}{\Gamma(3-\beta)} \sum_{j=0}^{L-1} [(j+1)^{2-\beta} - (j)^{2-\beta}]\big(f(t_{L-j+1}) - 2f(t_{L-j}) + f(t_{L-j-1})\big) + R_h^L,
\end{aligned}
\tag{24}
$$

where R_h^L is the truncation error of magnitude $O(h)$. It is obvious that the fractional derivative at time t depends on the past value between $[a, t]$. Recalling that we need to numerically solve the Equation (16). Using above fractional difference scheme, an explicit difference scheme for the stress $\sigma(t)$ can be obtained as follows:

$$
\begin{aligned}
\sigma(t_{L+1}) &= \frac{\Gamma(3-\beta)}{h^{-\beta}} c_0^2 \omega_0^{-2\gamma} \cos^2(\pi\gamma/2) \nabla^2 \sigma(t_L) + 2\sigma(t_L) - \sigma(t_{L-1}) \\
&\quad - \sum_{j=1}^{L-1} [(j+1)^{2-\beta} - (j)^{2-\beta}](\sigma(t_{L-j+1}) - 2\sigma(t_{L-j}) + \sigma(t_{L-j-1})) \\
&\quad - c1 \frac{(2-\beta)(1-\beta)(t_L-t_0)^{-\beta} \sigma(t_0)}{h^{-\beta}} - c2 \frac{(2-\beta)(t_L-t_0)^{1-\beta}[\sigma(t_1)-\sigma(t_0)]}{h^{1-\beta}} + S(t_L).
\end{aligned}
\tag{25}
$$

In which, $S(t_L)$ denotes the seismic source at time t_L, which can be chosen by users, e.g., the Ricker wavelet. Equation (25) does not need to compute the term $\begin{pmatrix} \beta \\ j \end{pmatrix}$, and hence the speed of calculation is accelerated. Spatial derivatives are calculated by the fast Fourier transform with staggered grid.

In the following, we will call the simulation method based on the Equation (18), i.e., Grunwald–Letnikov finite difference method, as the "original method", the simulation method based on the Equation (24) as the "new method", where we choose $c1 = c2 = 0$.

2.5. Proposition

In the above fractional differencing scheme, the memory length L is variable and is a fixed number in each differencing calculation. This is particularly important for efficient simulation. Otherwise, the memory length $t - a$ will be a function of the variable t. According to Equations (18) and (24), when L is large, the calculation time with difference scheme (24) will be less than that of difference scheme of (18). However, as the value of L increases, its influence on the derivative will decrease. Therefore, proper choice of the memory length L will balance the accuracy of calculation and the time cost. In addition, the fractional differencing scheme (24) is stable, since the residual R_h^i related to discrete step h is bounded by the infinitesimal equivalent value of h^2, and the fact that for any $1 < \alpha < 2$,

$$
\begin{aligned}
(x+1)^\alpha - x^\alpha &= \sum_{k=1}^{\infty} \frac{\alpha(\alpha-1)\cdots(\alpha-k+1)}{k!} \frac{1}{x^{k-\alpha}} \\
&= \alpha \frac{1}{x^{1-\alpha}} + \frac{\alpha(\alpha-1)}{2} \frac{1}{x^{2-\alpha}} + \sum_{k=3}^{\infty} \frac{\alpha(\alpha-1)\cdots(\alpha-k+1)}{k!} \frac{1}{x^{k-\alpha}}.
\end{aligned}
\tag{26}
$$

The above expression is bounded. In next section, we will show how these values of memory length influence the results.

3. Results

3.1. Different Q Media

First, we investigate the accuracy of the solution of fractional time wave equation using different discrete algorithms in uniform media. The media is discretized on a grid of 151×151 points with the same spatial sampling interval 2 m in x- and z-directions and the time step is 0.05 ms. The phase velocity c_0 is 2200 m/s at a reference frequency $f_0 = \omega_0/(2\pi) = 1500$ Hz. The source is a Ricker wavelet with the central frequency of 100 Hz located at the center of the model. Figures 1–4 show the seismograms calculated by the original method and the new method with different values of $Q = 5, 10, 30, 100$ and different memory lengths of $L = 9, 14, 24, 34, 44$. When Q and L are both small, the deviation of the seismograms calculated by the original method is large. In order not to affect the display effect of other curves, these seismograms are not displayed. The results are also partially magnified for clear presentation of details. The solution calculated by the original method with a long memory length is excellently consistent with the analytical solution [12]. However, when the memory length is smaller, a degraded numerical solution will be obtained. We choose the solution calculated by the original method with all previous values as the reference solution. From Figures 1–4, when L is relatively small, it can be seen that the seismograms calculated by the original method has a large deviation from the reference curve, and the amplitude cannot return to 0 after the propagation of wavelet, which will seriously disturb the wavefield. In particular, when Q is relatively small, this error will be more remarkable. However, the new method is relatively stable when L is small and is closer to the reference curve. Figure 5 shows the comparison of the snapshots and seismograms calculated by different methods at $Q = 10$ and $L = 34$. Receivers are at the same depth as the source. The snapshot and seismogram calculated by the original method have obvious false disturbances, and pseudo hyperbolic fluctuations occur in the seismogram. The original method is inaccurate at a small memory length L. We access the accuracy of numerical solutions by the root-mean-square (rms) error, which is defined by

$$err_{\text{rms}} = \sum_{i=1}^{nt} (d_i^p - d_i^h)^2 / \sum_{i=1}^{nt} (d_i^p)^2,$$

where d_i^p denotes the reference solution calculated by the original method with all previous values and d_i^h denotes the calculated value using the difference method with limited memory length L. The relationship between the rms error and the number of memory length for different values of Q is shown in Figure 6a. The error decreases with the increasing value of L, and the smaller value of Q results in a larger error. The new method has higher accuracy and stability at a small value of L. Table 1 shows the simulation time that is used to solve the fractional time derivative wave equation by different methods. The new method is faster at the same memory length, whose calculation time is about 20% of the original method (round numbers of CPU time is used). Under the same error requirements, we can choose a smaller memory length using the new method, which can further save a lot of calculation time.

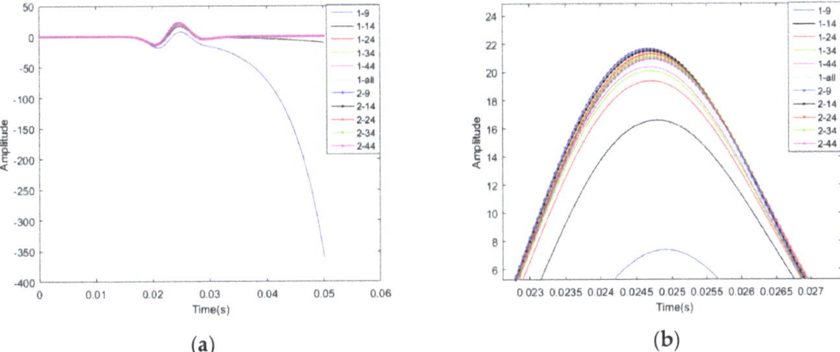

(a)   (b)

Figure 1. (**a**) Comparisons of seismograms at $Q = 100$ and 30 m away from the source. The first number '1' represents the original method, and the first number '2' represents the new method. The second number represents the memory length L. (**b**) Zoomed part of (**a**).

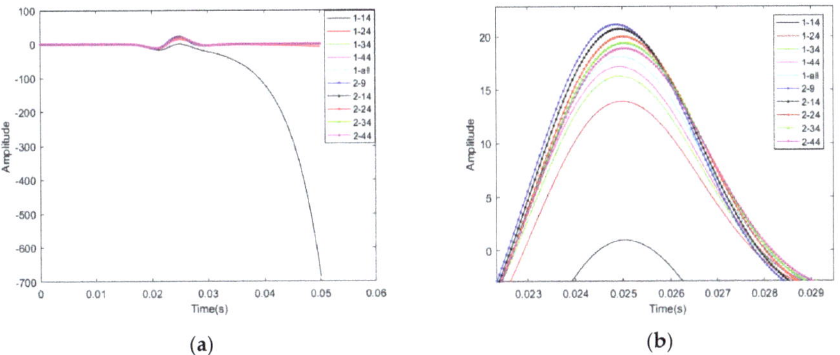

(a)   (b)

Figure 2. (**a**) Comparisons of seismograms at $Q = 30$ and 30 m away from the source. The first number '1' represents the original method, and the first number '2' represents the new method. The second number represents the memory length L. (**b**) Zoomed part of (**a**).

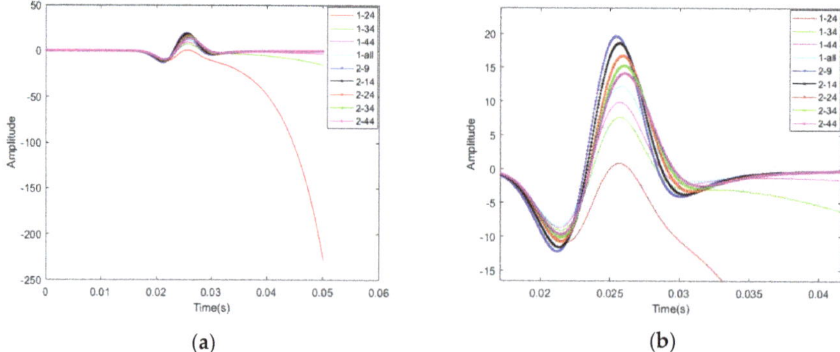

(a)　　　　　　　　　　　　(b)

Figure 3. (**a**) Comparisons of seismograms at $Q = 10$ and 30 m away from the source. The first number '1' represents the original method, and the first number '2' represents the new method. The second number represents the memory length L. (**b**) Zoomed part of (**a**).

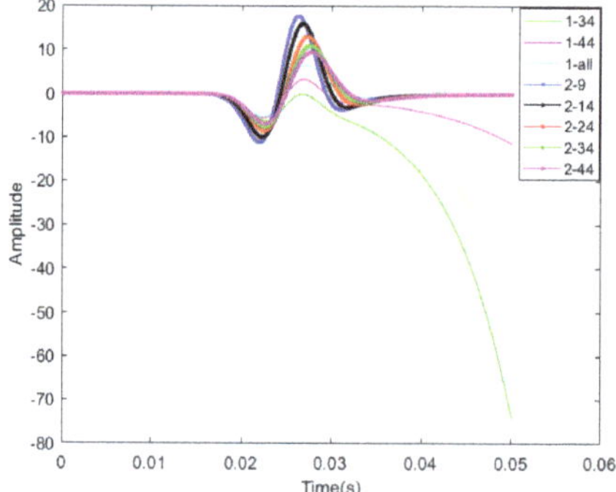

Figure 4. Comparisons of seismograms at $Q = 5$ and 30 m away from the source. The first number '1' represents the original method, and the first number '2' represents the new method. The second number represents the memory length L.

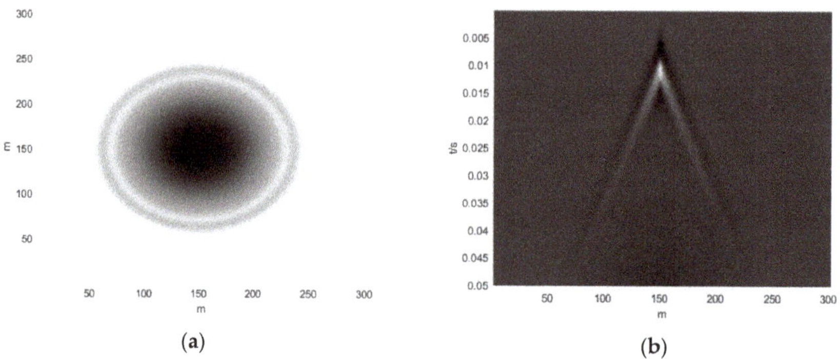

(a)　　　　　　　　　　　　(b)

Figure 5. *Cont.*

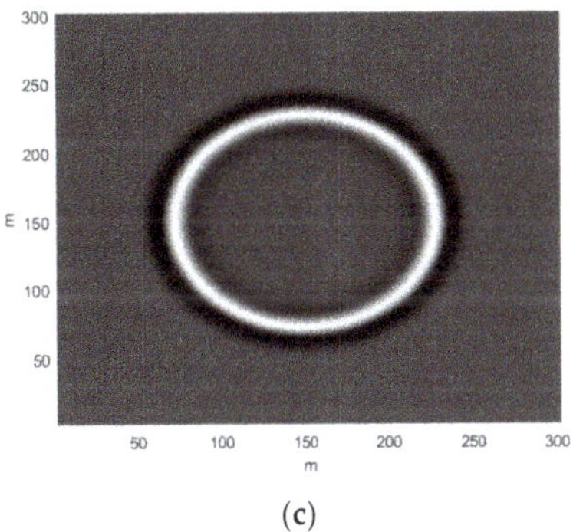
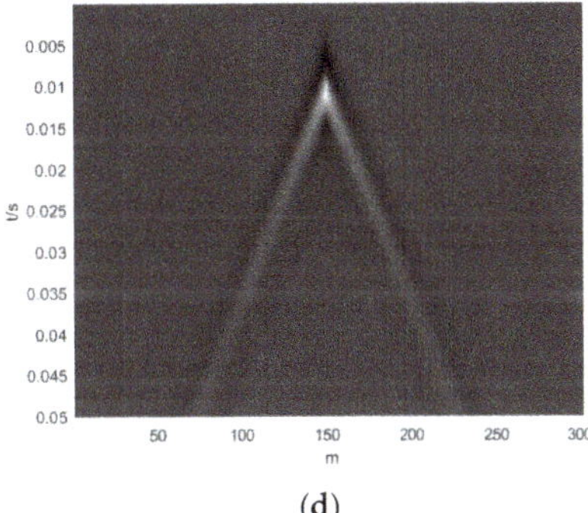

(c)

(d)

Figure 5. Calculation results of $Q = 10$, $L = 34$: (**a**) and (**b**) are the snapshots (0.05 s) and seismogram calculated by the original method, respectively; (**c**) and (**d**) are the snapshots (0.05 s) and seismogram calculated by the new method, respectively.

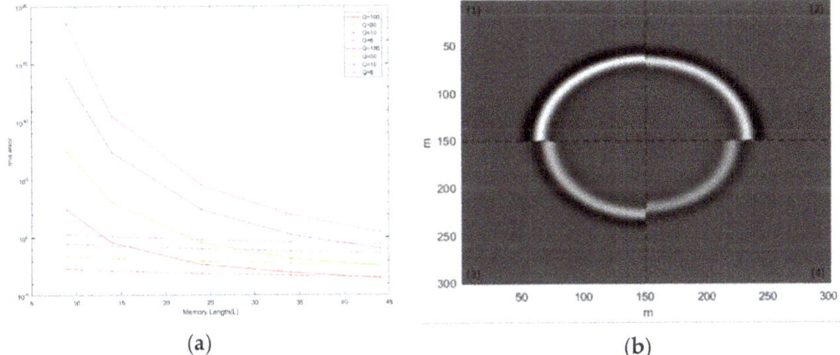

(a)   (b)

Figure 6. (**a**) The relationship between the error and the number of memory length for different values of Q: solid lines represent the original method; dotted lines represent the new method. (**b**) Four snapshot parts calculated by the new methods: (1) $Q = 100$; (2) $Q = 30$; (3) $Q = 10$; (4) $Q = 5$; Snapshots are recorded at 0.05 s.

Table 1. Simulation time with different memory lengths of L.

Memory Lengths		$L = 9$	$L = 14$	$L = 24$	$L = 34$	$L = 44$
CPU Time (s)	Original method	286	440	760	1097	1413
	New method	57	88	152	219	282

Figure 6b (1)–(4) shows the snapshots calculated by the new method at $Q = 100$, 30, 10, 5, respectively. As the value of Q decreases, we can see that the amplitude gradually decreases, and the phase gradually delays.

3.2. Layered Model

To verify the accuracy and stability of the new method in the media with large contrast in velocity and Q, we construct a two-layers model. The velocity and the Q of the top layer are 1200 m/s and 30, respectively. In the bottom layer, the velocity and the Q are 2200 m/s and 300, respectively. The interface is at a depth of 120 m. The media is discretized on a grid of 151 × 151 points with the same spatial sampling interval 2 m in x- and z-directions. The time step is 0.05 ms. A Ricker wavelet with the central frequency of 100 Hz is located at the center of the model (150, 150 m). Receivers are at the same depth as the source. Figure 7 shows the traces at x = 150 m extracted from seismograms calculated by the original method and the new method with different memory length $L = 9, 14, 24$. Similarly, we chose the solution calculated by the original method with all previous values as the reference solution. Using the original method with the memory length $L = 14$ leads to a significant deviation between the numerical solution and the reference curve. However, the solution calculated by the new method with a smaller memory length $L = 9$ is excellently consistent with the reference curve. Figure 8a,b show the snapshot and seismogram calculated by the original method with memory length $L = 14$. Figure 9a,b show the snapshot and seismogram calculated by the new method with memory length $L = 9$. Figure 10a,b show snapshots and seismograms calculated by the original method with all previous values of the length L. Obvious false disturbances can be observed in the Figure 8a,b, which differ greatly from the reference solution Figure 10a,b. Figure 9a,b are almost identical to Figure 10a,b. These numerical examples demonstrate that the solution calculated by the new method is more accurate and stable in such a large contrast velocity and Q model. For the two-layers model, the computation time of the new method is also about 20% of the original method at the same memory length. By the new method, we can choose a smaller memory length and get a better solution, thereby we can further save memory resources and cost of computation.

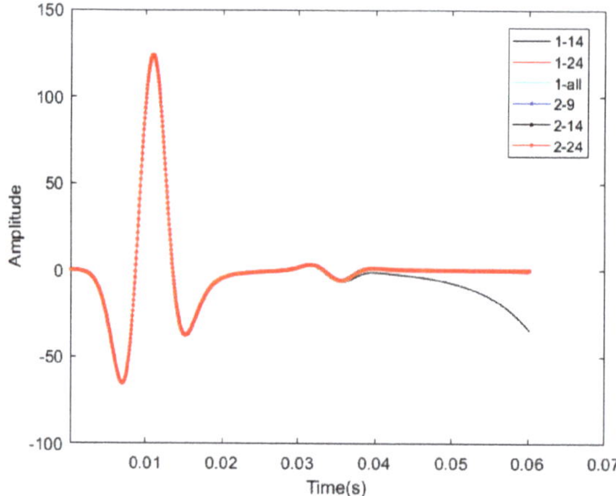

Figure 7. Comparison of seismograms in a two-layers model. The first number '1' represents the original method, and the first number '2' represents the new method. The second number represents the memory length L.

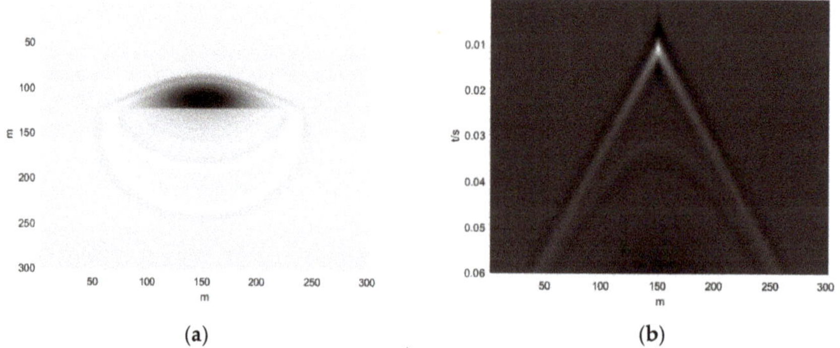

Figure 8. (**a**) Snapshot (0.05 s) and (**b**) seismogram calculated by the original method with memory length $L = 14$.

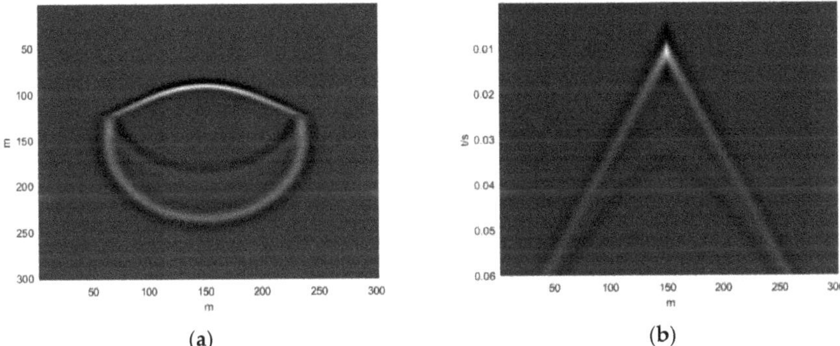

(a) (b)

Figure 9. (a) Snapshot (0.05 s) and (b) seismogram calculated by the new method with memory length $L = 9$.

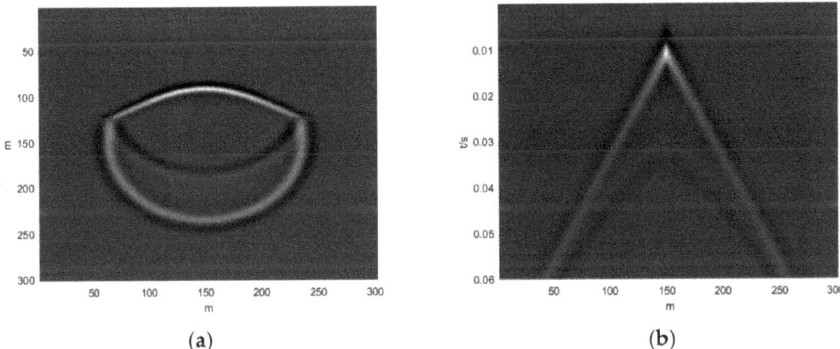

(a) (b)

Figure 10. (a) Snapshot (0.05 s) and (b) seismogram calculated by the original method with L equaling all previous values.

3.3. Simulations on Velocity and Quality Factor Model of Hydrate Layer

The elastic modulus of each mineral component and fluid we used is shown in Table 2. We assume that the rock solid phase is composed of 60% clay and 40% quartz, and the pores can contain different proportions of water, natural gas hydrate, and methane gas. The curve of velocity and inverse quality factor with hydrate saturation is calculated by white theory as shown in Figure 11. The seismic frequency used in the calculation is 35 Hz.

Table 2. Parameters of each rock component.

Rock Parameters	Values
Clay bulk modulus (Pa)	20.9×10^9
Clay shear modulus (Pa)	6.85×10^9
Clay density (kg/m^3)	2580
Quartz bulk modulus (Pa)	36.6×10^9
Quartz shear modulus (Pa)	45×10^9
Quartz density (kg/m^3)	2650
Sea water bulk modulus (Pa)	2.55×10^9
Density of sea water (kg/m^3)	1050
Seawater viscosity coefficient (Pa·s)	0.0018
Methane bulk modulus (Pa)	0.01×10^9
Methane gas density (kg/m^3)	100
Methane viscosity coefficient (Pa·s)	0.00002
Hydrate bulk modulus (Pa)	5.6×10^9
Hydrate shear modulus (Pa)	2.4×10^9
Hydrate density (kg/m^3)	920
Permeability (m^2)	100×10^{-14}
Interparticle connection coefficient	9

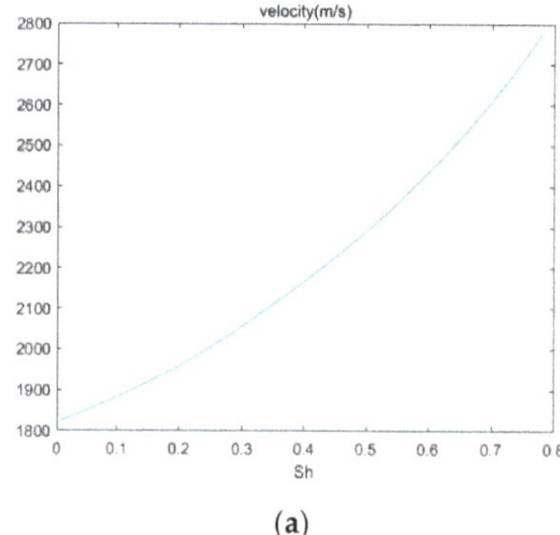
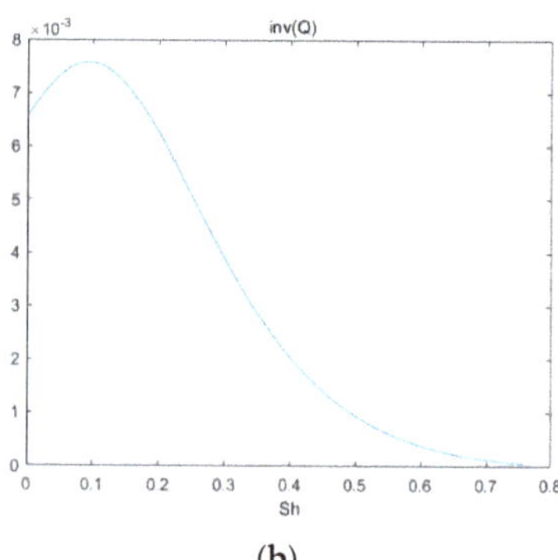

(a) (b)

Figure 11. The curve of velocity (**a**) and inverse quality factor (**b**) with hydrate saturation.

The calculation result of White theory shows that longitudinal wave velocity increases with the increase of hydrate saturation. When the hydrate saturation is small, the longitudinal wave attenuation increases rapidly to the maximum with the increase of saturation, and then the longitudinal wave attenuation decreases rapidly. Priest et al. [35] used hydrate resonance column (Gas hydrate resonant column) device to measure the attenuation value of 13 sandstone samples containing hydrate. The measurement results show that in the range of hydrate saturation less than 3–5%, the longitudinal wave attenuation increases rapidly to the maximum value with the increase of saturation, and then the longitudinal wave attenuation decreases rapidly, and then shows a slow increase trend. Therefore, it shows that there is a certain consistency between the White theoretical model and the experimental measurement results. In all subsequent seismic wave simulations of hydrate stratum, White theory is used to establish the velocity and quality factor models of hydrate stratum.

3.4. Layered NGH Model

Now we consider seismic simulations of NGH model using fractional-order differencing of the constant-Q viscoacoustic wave equation. The amplitude of seismic wave propagation in hydrate layer always is very weak, which is called the amplitude blank zone. In order to simulate this phenomenon, we assume that the saturation of hydrate varies with porosity, and the ratio of hydrate saturation to porosity is constant. We also assume that when the depth increases, the porosity of the rock generally decreases as the pressure increases, and the hydrate saturation also decreases as the porosity decreases.

The first layered NGH model is a high hydrate saturation model and the pores of the underlying layer are filled with free gas. The model parameters are given in Table 3. The velocity model and the quality factor Q are shown in Figure 12a,b, respectively. From which we can see that there are high-speed and high-Q anomalies in the hydrate layer. The seismic simulations with fractional-order differencing is shown in Figures 13 and 14. It can be seen that bottom-simulating reflector (BSR) features are obvious, and its polarity is reversed, and the amplitude is significantly increased. Above the BSR, an amplitude blank zone can be seen in the hydrate layer. The reflection of the upper layer of gas hydrate is relatively weak.

Table 3. High hydrate saturation model with gas bearing layer.

Stratum Serial Number	Depth (m)	Porosity	Hydrate Saturation	Gas Saturation
1 (Seawater)	1300	-	-	-
2 (General sediment)	200	0.5	0	0
3 (Hydrate)	50	0.5	0.4	0
4 (Hydrate)	50	0.45	0.36	0
5 (Hydrate)	50	0.4	0.32	0
6 (Hydrate)	50	0.35	0.28	0
7 (Hydrate)	50	0.35	0.28	0
8 (Free gas)	100	0.42	0	0.01
9 (General sediment)	200	0.42	0	0

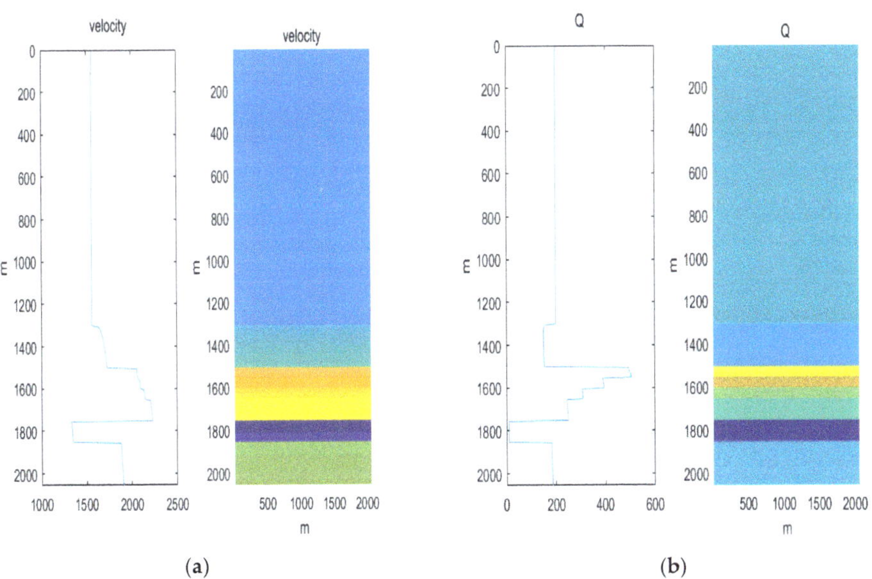

Figure 12. (**a**) Velocity model; (**b**) Quality factor Q.

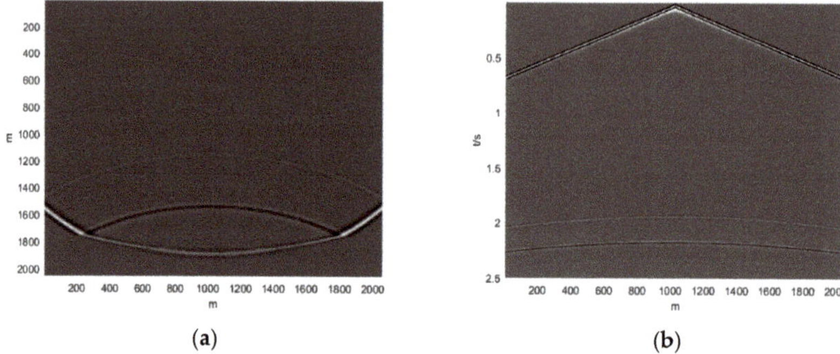

(a) (b)

Figure 13. (**a**) Seismic snapshot; (**b**) Seismic record.

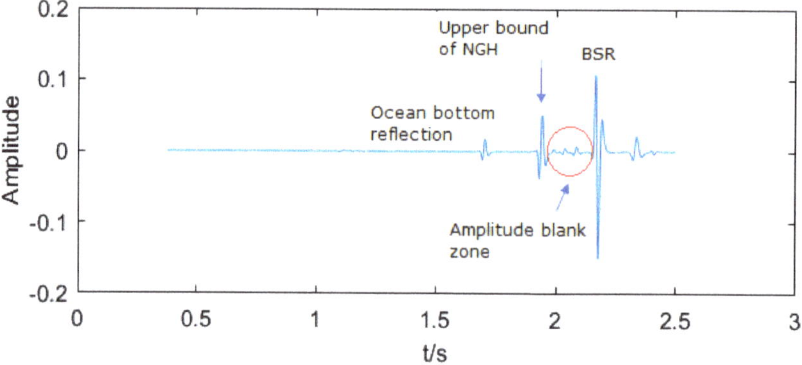

Figure 14. One section of the seismic record.

The second layered NGH model is a low hydrate saturation model and the pores of the underlying layer are also filled with free gas. The model parameters are given in Table 4. The velocity model and the quality factor Q are shown in Figure 15a,b, respectively. From which we can see that there are high-speed and low-Q anomalies in the hydrate layer. The seismic simulations with fractional-order differencing is shown in Figure 16, Figure 17. The amplitude of BSR is weaker than that of the first NGH model, but it is still obvious, and its polarity is reversed. There also exists an amplitude blank zone above the BSR and the relatively weak reflection of the upper layer of gas hydrate.

Table 4. Low hydrate saturation model with gas bearing layer.

Stratum Serial Number	Depth (m)	Porosity	Hydrate Saturation	Gas Saturation
1 (Seawater)	1300	-	-	-
2 (General sediment)	200	0.5	0	0
3 (Hydrate)	50	0.5	0.1429	0
4 (Hydrate)	50	0.45	0.1286	0
5 (Hydrate)	50	0.4	0.114	0
6 (Hydrate)	50	0.35	0.1	0
7 (Hydrate)	50	0.35	0.1	0
8 (Free gas)	100	0.42	0	0.01
9 (General sediment)	200	0.42	0	0

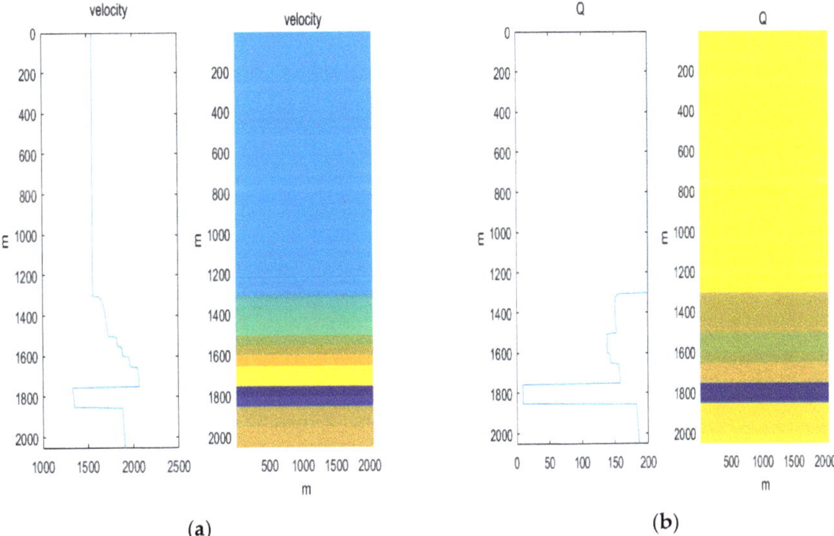

Figure 15. (**a**) Velocity model; (**b**) Quality factor Q.

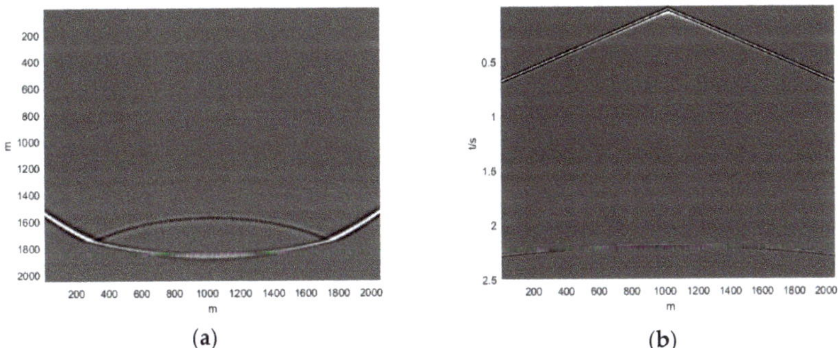

Figure 16. (**a**) Seismic snapshot; (**b**) Seismic record.

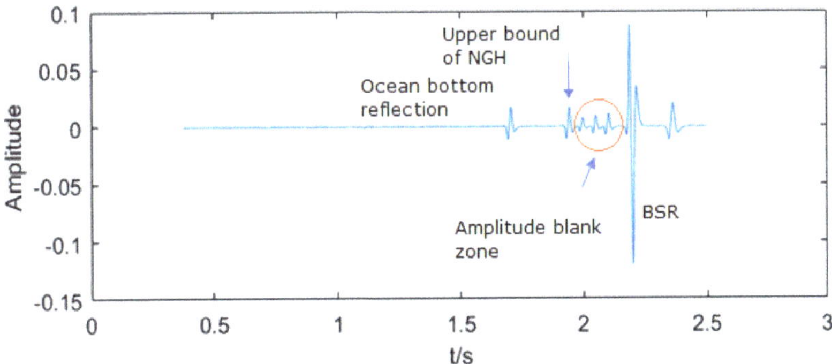

Figure 17. One section of the seismic record.

We also perform numerical experiments on the layered NGH model without gas bearing layer for high hydrate saturation model and low hydrate saturation model, respectively. For the former one, the models, seismic wave propagation and the section of the seismic records are shown in Figures 18–20, respectively; for the later one, the models, seismic wave propagation and the section of the seismic records are shown in Figures 21–23, respectively. Weak BSR can be seen in high hydrate saturation model, but it is almost invisible in low hydrate saturation model. Through the above layered NGH model test, when the underlying layer contains free gas, the BSR phenomenon is more obvious. When the underlying layer does not contain free gas, the BSR phenomenon is relatively weak. However, as the hydrate saturation increases, the BSR phenomenon becomes more obvious.

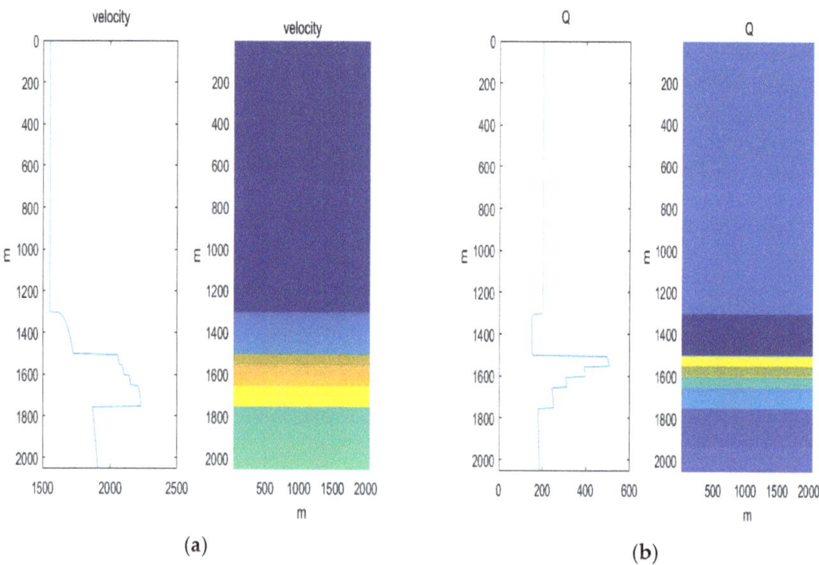

(a) (b)

Figure 18. (a) Velocity model; (b) Quality factor Q.

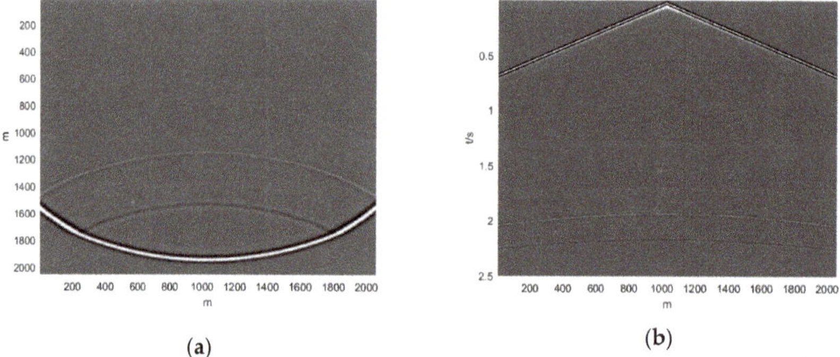

(a) (b)

Figure 19. (a) Seismic snapshot; (b) Seismic record.

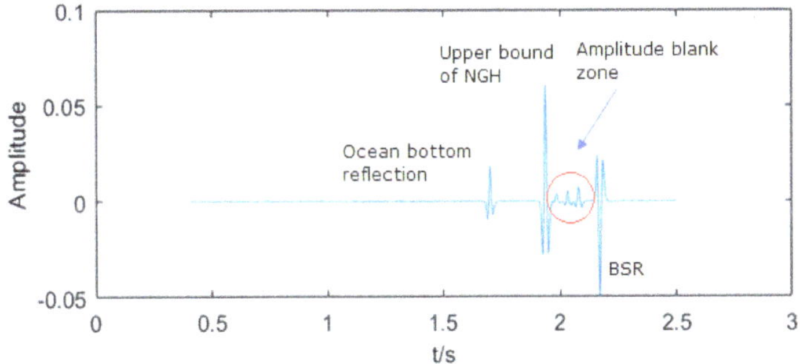

Figure 20. One section of the seismic record.

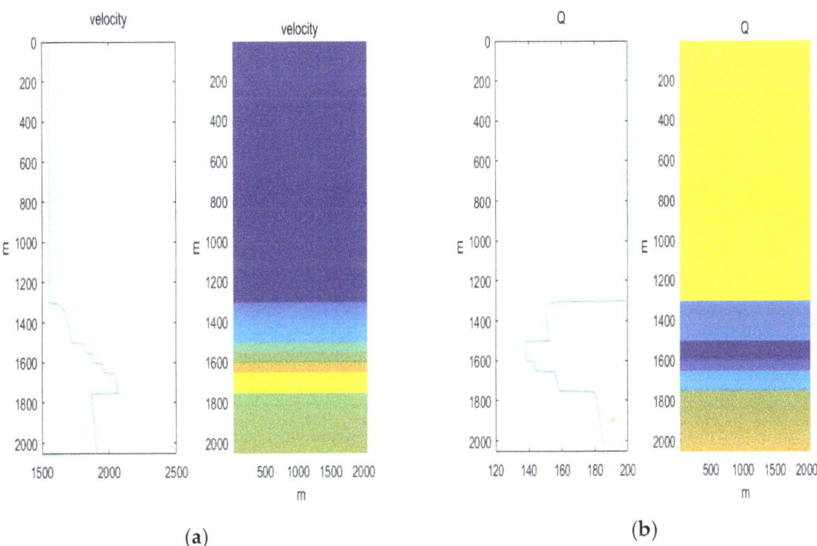

(**a**)

(**b**)

Figure 21. (**a**) Velocity model; (**b**) Quality factor Q.

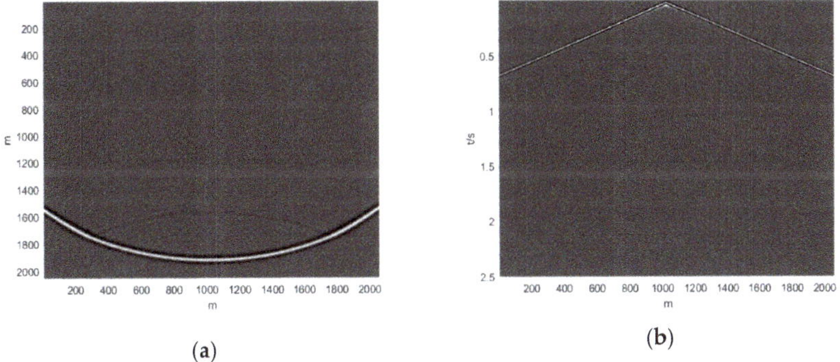

(**a**)

(**b**)

Figure 22. (**a**) Seismic snapshot; (**b**) Seismic record.

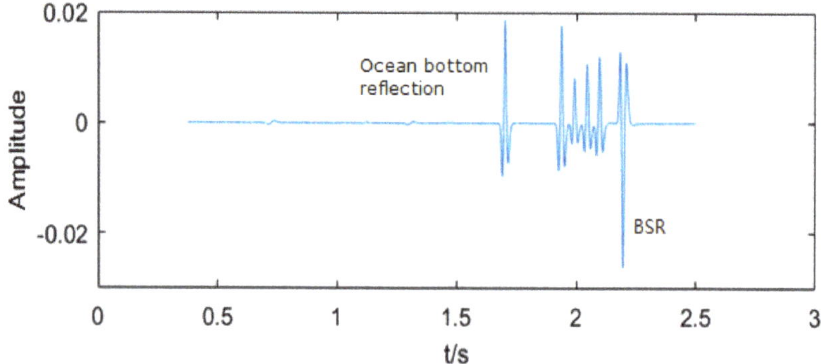

Figure 23. One section of the seismic record.

3.5. Complex Seabed NGH Model

We consider using the White theory to model the hydrate formation, and then the fractional wave equation method is used to simulate the seismic wave of the established hydrate mode. The relationship between the content and elastic parameters of each rock component and the seismic wave velocity, attenuation characteristics, and seismic wave propagation law of the rock as a whole, lays a theoretical foundation for the use of seismic exploration to identify natural gas hydrates and estimate the hydrate content. The geological structure of the seabed is usually complex, and a large number of gas hydrates are distributed on the sea floor. The main identification mark of natural gas hydrate is bottom-simulating reflector (BSR). The position of BSR is usually consistent with the bottom interface of gas hydrate, which reflects that the fluctuation of bottom interface of gas hydrate is similar to that of seabed. We have established a complex seabed model containing natural gas hydrate, and there are a lot of "gas chimney" under the hydrate. The velocity and quality factor models are shown in the Figure 24a,b, respectively.

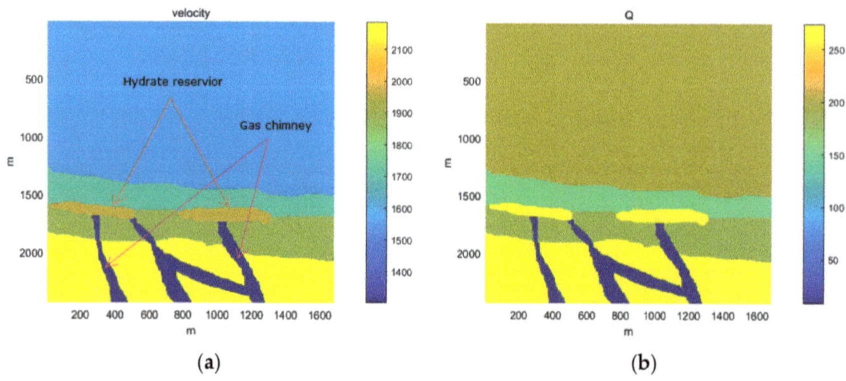

Figure 24. (**a**) The seismic velocity model; (**b**) The quality factor model.

Using the fractional Laplace operator wave Equation (17) to simulate wave field in hydrate formation, meanwhile the time derivative adopts finite difference, and the space derivative adopts the staggered grid fast Fourier transform method. In calculation, the first two first-order conservation equations in Equation (17) apply PML absorbing boundary conditions, and do not perform absorbing boundary processing on the equations containing Laplace operator. The snapshots of wave propagation and the seismic record are shown in Figures 25 and 26, respectively. As shown in Figure 25,

BSR phenomenon can be clearly observed. Its main characteristics are the polarity reversal and the significantly increase in amplitude, especially at the junction of the hydrate bottom interface and the gas chimney. The same phenomenon can be seen in the seismic record Figure 26.

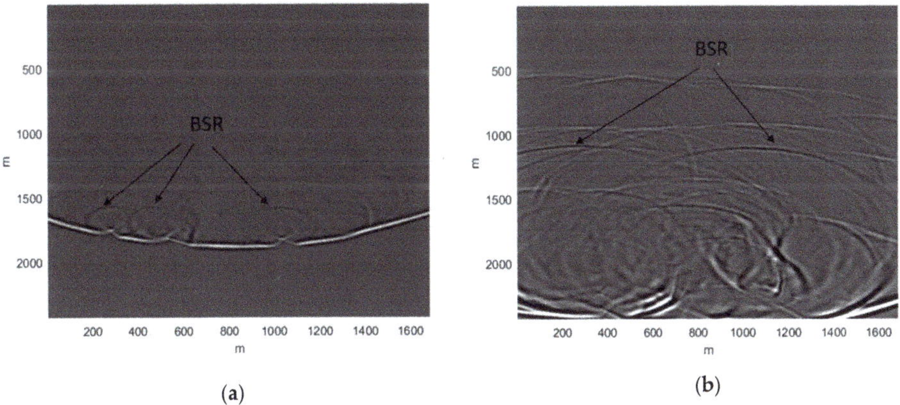

(a)

(b)

Figure 25. (**a**) Seismic snapshots at 1.2 s; (**b**) Seismic snapshots at 1.5 s.

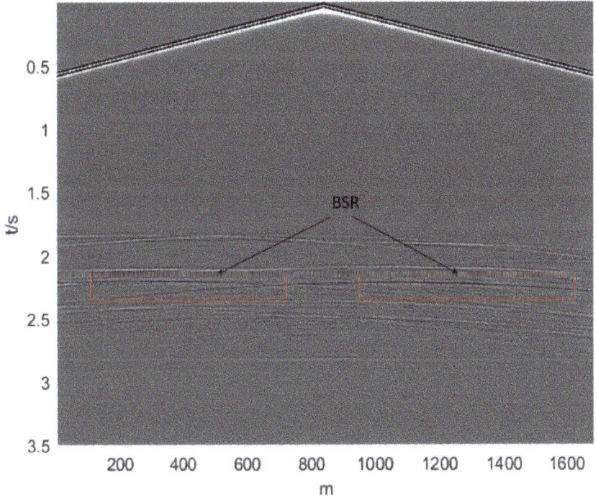

Figure 26. The seismic record.

4. Discussion and Future Research

In order to better find gas hydrate accumulation areas through seismic exploration and estimate the scale of gas hydrate resources and hydrate content, it is necessary to in-depth study of the law and characteristics of seismic wave propagation in gas hydrate-bearing formations and establish the relationship of hydrate formation between seismic attributes and seismic response. For complex oceanary geological conditions, the mathematical analytical solution of seismic wave propagation is difficult to find. Numerical simulation based on the assumption of known underground physical parameters and medium models, uses computers and various numerical simulation methods to obtain seismic waves. The approximate solution of the field is a relatively simple and efficient method.

When the underlying stratum contains free gas, the BSR phenomenon is more obvious, with a significant increase in amplitude and polarity reversal. At the same time, a reflective blank zone in the hydrate layer can be seen above the BSR. When the underlying formation does not contain free gas, the BSR phenomenon is relatively weak. As the hydrate saturation increases, the BSR phenomenon becomes more obvious. On the basis of phenomena, such as BSR and amplitude blank zone, hydrates can be further identified, and the hydrate content can be estimated based on the characteristics of speed and quality factor. The fractional time derivative wave equation can accurately describe the constant-Q behavior, but the fractional time derivative will introduce huge memory consumption. We proposed a new finite-difference method, which can save memory resources and cost of computation. But for large-scale models, the amount of calculation is still relatively large. It is necessary to further study the fast algorithm of fractional time derivative. In addition, the current gas hydrate model is still relatively rough. Establish a more sophisticated model hydrate is still of great importance.

In this study, we only consider the forward simulation of the seismic propagation behavior in NGH. Next stage, we will consider the seismic imaging. The basic idea is performing a least-squares error minimization problem, i.e., if we denote d the simulated data by aforementioned seismic simulation steps with fractional differencing, d_{obs} the observed data, m the velocity model, and L the forward operator that acts on m to generate d, then we will solve the following regularized minimization problem

$$J(m) = \frac{1}{2}\|d_{obs} - d\|_2^2 + \alpha \cdot \Omega(m) = \frac{1}{2}\|d - Lm\|_2^2 + \alpha \cdot \Omega(m) \rightarrow min \tag{27}$$

where $\Omega(m)$ denotes the *a priori* knowledge of the model, $\alpha > 0$ is the regularization parameter. The above optimization problem requires forward simulation and gradient or Hessian information calculation, e.g., least-squares migration (LSM) or full waveform inversion (FWI) can be developed. Of course, some advanced regularization techniques either physics-based or learned with artificial neural network (ANN) should be fully employed [36–38]. It is well-understanding that the easy way of seismic imaging is using reverse time migration (RTM) [39], however performing RTM requires a big deal of computation. We will continue to report our research in this field in future works.

5. Conclusions

We have introduced a more rigorous form of definition of fractional derivative to calculate the fractional time derivative of the viscoasoustic wave equation, which is more accurate and simpler than the other nearly constant-Q methods for describing the propagation of viscoasoustic wave. The rigorous form is the limit of the original definition when time step approaching zero. We discretize the rigorous form and bring it into the fractional time derivative wave equation to obtain the explicit expression of viscoelastic wave extrapolation. The computation time of the new method is merely about 20% of the original method at the same memory length. Some numerical examples demonstrate that the solution calculated by the new method is more accurate and stable in the uniform model or the large contrast velocity and Q model. By the new method, we can choose a smaller memory length at an acceptable accuracy; thereby, we can save memory resources and computation time. For the wavefield simulation of gas hydrate layer, White theory is selected to establish velocity and quality factor models, and then the fractional wave equation is used to simulate the propagation of seismic waves in the hydrate model. Finally, the connection between the elastic parameters and content of each component of the rock in the hydrate layer, the seismic wave speed, attenuation characteristics, and the law of seismic wave propagation are established. The above contents provide theoretical foundation for identifying gas hydrate and estimating hydrate content by seismic exploration.

Author Contributions: Y.W. (Yanfei Wang) designed the study. Y.N. and Y.W. (Yanfei Wang) conducted experiments. Y.W. (Yanfei Wang), Y.N. and Y.W. (Yibo Wang) wrote the paper. All authors contributed to synthetic data interpretation and provided significant input to the final manuscript. All authors have read and agreed to the published version of the manuscript.

Funding: This research was supported by the Original Innovation Program of CAS (grant no. ZDBS-LY-DQC003), the Key Research Program of the Institute of Geology & Geophysics, CAS (grant no. IGGCAS-201903), and the National Key R & D Program of the Ministry of Science and Technology of China (grant nos. 2018YFC0603500 and 2018YFC1504203).

Acknowledgments: We are grateful to four reviewers' important questions and suggestions which make an improvement of our paper.

Conflicts of Interest: The authors declare no conflict of interest.

References

1. Guo, P.; Mcmechan, G.A. Evaluation of three first-order isotropic viscoelastic formulations based on the generalized standard linear solid. *J. Seism. Explor.* **2017**, *26*, 199–226.
2. Kang, I.B.; McMechan, G.A. Separation of intrinsic and scattering Q based on frequency-dependent amplitude ratios of transmitted waves. *J. Geophys. Res.* **1994**, *99*, 23875–23885. [CrossRef]
3. Zhu, T.; Harris, J.M. Modeling acoustic wave propagation in heterogeneous attenuating media using decoupled fractional Laplacians. *Geophysics* **2014**, *79*, T105–T116. [CrossRef]
4. Sun, J.; Fomel, S.; Zhu, T.; Hu, J. Q-compensated least-squares reverse time migration using low-rank one-step wave extrapolation. *Geophysics* **2016**, *81*, S271–S279. [CrossRef]
5. Carcione, J.M. *Wave Fields in Real Media: Wave Propagation in Anisotropic, Anelastic, Porous and Electromagnetic Media*; Elsevier: Amsterdam, The Netherlands, 2007.
6. Štekl, I.; Pratt, R.G. Accurate viscoelastic modeling by frequency-domain finite differences using rotated operators. *Geophysics* **1998**, *63*, 1779–1794. [CrossRef]
7. Zhu, T.; Carcione, J.M.; Harris, J.M. Approximating constant-Q seismic propagation in the time domain. *Geophys. Prospect.* **2013**, *61*, 931–940. [CrossRef]
8. Aki, K.; Richards, P.G. *Quantitative Seismology: Theory and Methods*; W. H. Freeman: San Franccisco, CA, USA, 1980; Volume 1.
9. Blanch, J.O.; Robertsson, J.O.A.; Symes, W.W. Modeling of a constant Q: Methodology and algorithm for an efficient and optimally inexpensive viscoelastic technique. *Geophysics* **1995**, *60*, 176–184. [CrossRef]
10. Blanch, J.O.; Symes, W.W. Efficient iterative viscoacoustic linearized inversion. In *SEG Technical Program Expanded Abstracts 1995*; Society of Exploration Geophysicists: Houston, TX, USA, 1995; pp. 627–630. [CrossRef]
11. Robertsson, J.O.; Blanch, J.O.; Symes, W.W. Viscoacoustic finite-difference modeling. *Geophysics* **1994**, *59*, 1444–1456. [CrossRef]
12. Carcione, J.M.; Cavallini, F.; Mainardi, F.; Hanyga, A. Time-domain seismic modeling of constant-Q wave propagation using fractional derivatives. *Pure Appl. Geophys.* **2002**, *159*, 1719–1736. [CrossRef]
13. Kjartansson, E. Constant Q-wave propagation and attenuation. *J. Geophys. Res. Space Phys.* **1979**, *84*, 4737–4748. [CrossRef]
14. Caputo, M.; Mainardi, F. A new dissipation model based on memory mechanism. *Pure Appl. Geophys.* **1971**, *91*, 134–147. [CrossRef]
15. Carcione, J.M. Theory and modeling of constant-Q P- and S-waves using fractional time derivatives. *Geophysics* **2009**, *74*, T1–T11. [CrossRef]
16. Chen, W.; Holm, S. Fractional Laplacian time-space models for linear and nonlinear lossy media exhibiting arbitrary frequency power-law dependency. *J. Acoust. Soc. Am.* **2004**, *115*, 1424–1430. [CrossRef] [PubMed]
17. Carcione, J.M. A generalization of the Fourier pseudospectral method. *Geophysics* **2010**, *75*, A53–A56. [CrossRef]
18. Song, G.; Zhang, X.; Wang, Z.; Chen, Y.; Chen, P. The asymptotic local finite-difference method of the fractional wave equation and its viscous seismic wavefield simulation. *Geophysics* **2020**, *85*, T179–T189. [CrossRef]
19. Sun, J.; Zhu, T.; Fomel, S. Viscoacoustic modeling and imaging using low-rank approximation. *Geophysics* **2015**, *80*, A103–A108. [CrossRef]
20. Fomel, S.; Ying, L.; Song, X. Seismic wave extrapolation using low rank symbol approximation. *Geophys. Prospect.* **2013**, *61*, 526–536. [CrossRef]
21. Sun, J.; Fomel, S. Low-rank one-step wave extrapolation. *SEG Tech. Program Expand. Abstr.* **2013**, *2013*, 3905–3910. [CrossRef]
22. Yao, J.; Zhu, T.; Hussain, F.; Kouri, D.J. Locally solving fractional Laplacian viscoacoustic wave equation using Hermite distributed approximating functional method. *Geophysics* **2017**, *82*, T59–T67. [CrossRef]

23. Wang, N.; Zhu, T.; Zhou, H.; Chen, H.; Zhao, X.; Tian, Y. Fractional Laplacians viscoacoustic wavefield modeling with k-space-based time-stepping error compensating scheme. *Geophysics* **2020**, *85*, T1–T13. [CrossRef]

24. Xu, Y.; Li, J.-Y.; Chen, X.; Pang, G. Solving fractional Laplacian visco-acoustic wave equations on complex-geometry domains using Grünwald-formula based radial basis collocation method. *Comput. Math. Appl.* **2020**, *79*, 2153–2167. [CrossRef]

25. Abd-Elhameed, W.M.; Youssri, Y.H. A Novel Operational Matrix of Caputo Fractional Derivatives of Fibonacci Polynomials: Spectral Solutions of Fractional Differential Equations. *Entropy* **2016**, *18*, 345. [CrossRef]

26. Youssri, Y.H.; Abd-Elhameed, W.M. Numerical spectral Legendre-Galerkin algorithm for solving time fractional telegraph equation. *Rom. J. Phys.* **2018**, *63*, 3–4.

27. Podlubny, I. *Fractional Differential Equations: Mathematics in Science and Engineering*; Academic Press: Cambridge, MA, USA, 1999; Volume 198.

28. Samko, S.G.; Kilbas, A.A.; Marichev, D.I. *Fractional Integrals and Derivatives: Theory and Applications*; Gordon and Breach Science Publishers: Philadelphia, PA, USA, 1993.

29. White, J.E. Computed seismic speeds and attenuation in rocks with partial gas saturation. *Geophysics* **1975**, *40*, 224–232. [CrossRef]

30. Shapiro, S.A.; Müller, T.M. Seismic signatures of permeability in heterogeneous porous media. *Geophysics* **1999**, *64*, 99–103. [CrossRef]

31. Helgerud, M.B.; Dvorkin, J.; Nur, A.; Sakai, A.; Collett, T. Elastic-wave velocity in marine sediments with gas hydrates: Effective medium modeling. *Geophys. Res. Lett.* **1999**, *26*, 2021–2024. [CrossRef]

32. Carcione, J.M.; Helle, H.B.; Pham, N.H. White's model for wave propagation in partially saturated rocks: Comparison with poroelastic numerical experiments. *Geophysics* **2003**, *68*, 1389–1398. [CrossRef]

33. Sumelka, W.; Voyiadjis, G.Z. A hyperelastic fractional damage material model with memory. *Int. J. Solids Struct.* **2017**, *124*, 151–160. [CrossRef]

34. Jia, X.; Li, G. Numerical simulation and parameters inversion for non-symmetric two-sided fractional advection-dispersion equations. *J. Inverse Ill Posed Probl.* **2016**, *24*, 29–39. [CrossRef]

35. Priest, J.A.; Best, A.I.; Clayton, C.R.I. Attenuation of seismic waves in methane gas hydrate-bearing sand. *Geophys. J. Int.* **2006**, *164*, 149–159. [CrossRef]

36. Geng, Z.; Wang, Y. Automated design of a convolutional neural network with multi-scale filters for cost-efficient seismic data classification. *Nat. Commun.* **2020**, *11*, 3311. [CrossRef] [PubMed]

37. He, Q.; Wang, Y. Reparameterized full waveform inversion using deep neural networks. *Geophysics* **2020**. [CrossRef]

38. Wang, Y.F.; Volkov, V.T.; Yagola, A.G. *Basic Theory of Inverse Problems: Variational Analysis and Geoscientific Applications*, 1st ed.; Science Press: Beijing, China, 2020; in press.

39. Zhu, T.; Harris, J.M.; Biondi, B.L. Q-compensated reverse-time migration. *Geophysics* **2014**, *79*, S77–S87. [CrossRef]

Publisher's Note: MDPI stays neutral with regard to jurisdictional claims in published maps and institutional affiliations.

Article

Reservoir Formation Model and Main Controlling Factors of the Carboniferous Volcanic Reservoir in the Hong-Che Fault Zone, Junggar Basin

Danping Zhu [1,*], **Xuewei Liu** [1] **and Shaobin Guo** [2]

[1] School of Geophysics and Information Technology, China University of Geosciences, Beijing 100083, China; liuxw@cugb.edu.cn

[2] School of Energy Resources, China University of Geosciences, Beijing 100083, China; guosb58@cugb.edu.cn

* Correspondence: zdanping13@126.com

Received: 12 October 2020; Accepted: 16 November 2020; Published: 21 November 2020

Abstract: The Hong-Che Fault Zone is one of the important oil and gas enrichment zones in the Junggar Basin, especially in the Carboniferous. In recent five years, it has been proven that the Carboniferous volcanic rock has 140 million tons of oil reserves, and has built the Carboniferous volcanic reservoir with a capacity of million tons. Practice has proven that the volcanic rocks in this area have great potential for oil and gas exploration and development. To date, Carboniferous volcanic reservoirs have been discovered in well areas such as Che 32, Che 47, Che 91, Chefeng 3, Che 210, and Che 471. The study of drilling, logging, and seismic data shows that the Carboniferous volcanic reservoirs in the Hong-Che Fault Zone are mainly distributed in the hanging wall of the fault zone, and oil and gas have mainly accumulated in the high part of the structure. The reservoirs are controlled by faults and lithofacies in the plane and are vertically distributed within 400 m from the top of the Carboniferous. The Carboniferous of the Hong-Che Fault Zone has experienced weathering leaching and has developed a weathering crust. The vertical zonation characteristics of the weathering crust at the top of the Carboniferous in the area of the Che 210 well are obvious. The soil layer, leached zone, disintegration zone, and parent rock developed from top to bottom. Among these reservoirs, the reservoirs with the best physical properties are mainly developed in the leached zone. Based on a comprehensive analysis of the Carboniferous oil and gas reservoirs in areas of the Chefeng 3 and Che 210 wells, it is believed that the formation of volcanic reservoirs in the Hong-Che Fault Zone was mainly controlled by structures and was also controlled by lithofacies, unconformity surfaces, and physical properties.

Keywords: junggar basin; hong-che fault zone; carboniferous; volcanic reservoir; main controlling factors of hydrocarbon accumulation

---

1. Introduction

Volcanic reservoirs are widely distributed in more than 300 basins or blocks in over 20 countries and five continents and are becoming an important new area for global oil and gas resource exploration and development [1–5] (Table 1).

According to the characteristics of volcanic oil and gas reservoirs, which have already been discovered around the world, these strata have strong epochal and regional characteristics and mainly include Archean, Carboniferous, Permian, Cretaceous, and Paleogene strata. In addition, they are mainly distributed in the circum-Pacific, Mediterranean and Central Asian regions [6]. The circum-Pacific region is the main area of distribution of volcanic oil and gas reservoirs, which includes the United States, Mexico, and Cuba in North America; Venezuela, Brazil, and Argentina in South America; and China, Japan, and Indonesia in Asia. These areas are followed in importance by the Mediterranean region and Central Asia. Some volcanic oil and gas reservoirs have also been found on the African continent, such as in Egypt, Libya, Morocco in North Africa and Angola in South Africa [7].

Table 1. Production statistics of global volcanic oil and gas fields.

Country	Field Name	Basin	Type	Output		Reservoir Rock
				Oil (t/d)	Gas (10^4m^3/d)	
Cuba	Cristales	South Cuba	oil	3425		basaltic tuff
Brazil	Igarape Cuia	Amazonas	oil	68–3425		dolerite sill
Vietnam	15-2-RD 1X	Cuu Long	oil	1370		altered granite
Argentina	YPF Palmar Largo	Noroeste	oil, gas	550	3.4	vuggy basalt
Georgia	Samgori		oil	411		laumontite tuff
United States	West Rozel	North Basin	oil	296		basalt, agglomerate
Venezuela	Totumo	Maracaibo	oil	288		igneous rocks
Argentina	Vega Grande	Neuquen	oil, gas	224	1.1	fractured andesite
New Zealand	Kora	Taranaki	oil	160		andesite tuffs, volcaniclastics
Japan	Yoshii-Kashiwazaki	Niigata	gas		49.5	rhyolite
Brazil	Barra Bonita	Parana	gas		19.98	flood basalt, dolerite sill
Australia	Scotia	Bowen-Surat	gas		17.8	fractured andesite
Indonesia	Jatibarang	NW Java	oil, gas	85		fractured basalt, andesitic tuff, tuff breccia
Mexico	Furbero	Vera Cruz	oil	9		gabbro
Azerbaijan	Muradkhanly	western	oil	12–64		andesite and basalt

Data Source: This table is modified from [1,7].

At present, volcanic reservoirs have been found in 14 sedimentary basins in China (Table 2) [8]. There are three main sets of volcanic strata that developed in the sedimentary basins of China: Carboniferous-Permian, Jurassic-Cretaceous, and Paleogene-Neogene [7]. In terms of spatial locations, the volcanic reservoirs in China can be divided into two parts: Eastern part and western part. The volcanic rocks in eastern China (such as the Bohai Bay Basin) mainly developed in rifted basins, which were controlled by an intracontinental rift environment that was caused by subduction of the Pacific plate under the Chinese mainland since the Mesozoic and Cenozoic [9–11]. The development of volcanic rocks in the western basins, as represented by the Junggar Basin, was closely related to the formation and closure of the Paleo-Asian and Paleo-Tethys Oceans and the orogeny that was induced by them [12–14].

Table 2. Distribution of volcanic rock oil and gas reservoirs in China.

Basin	Region	Reservoir Name	Strata	Reservoir Rock
Bohai Bay	Jiyang Depression	Binnan oilfield	Paleogene	Basalt, andesitic basalt
		Linpan lin 9 fault block	Paleogene, Neogene	Tuff
		Shanghe 3 District	Paleogene, Neogene	Basalt, diabase
	Jizhong Depression	Caojiawu Gas Reservoir	Paleogene	Diabase
	Huanghua Depression	Fenghuadian	Upper Jurassic	Andesite
	Liaohe Depression	Rehetai-Oulituozi	Paleogene	Trachyte
		Niuxintuo	Mesozoic	Rhyolite, andesite, breccia, and tuff
Sichuan	West Sichuan	Zhougongshan	Upper Permian	Basalt
Junggar	Northwestern Margin	Karamay Oilfield District 5, 8	Carboniferous	Basalt
		Hong-Che area	Carboniferous, Permian	Basalt, andesite, and volcanic breccia

Table 2. *Cont.*

Basin	Region	Reservoir Name	Strata	Reservoir Rock
	Central Part	Shixi area	Carboniferous, Permian	Basalt, andesite, diabase, breccia, and tuff
		Kalameili area	Carboniferous	Basalt, andesite, breccia, and tuff
	Eastern part	Wucaiwan area	Carboniferous	Basalt, andesite, rhyolite, volcanic breccia, tuff
Subei	Dongtai Depression	Biandong structure	Paleogene	Basalt
Songliao	Xujiaweizi Fault Depression	Xingcheng	Cretaceous	Rhyolite
	Changling Fault Depression	Haerjin structure	Cretaceous	Rhyolite
Erlian	Manite Depression	Abei	Jurassic	Andesite
Santanghu	Malang Depression	Haerjiawu Formation	Carboniferous	Andesite
Hailar	Beier depression	Budate Group buried hill	Triassic	Altered basalt and andesite

Data Source: This table is modified from [8].

Since the first discovery of volcanic reservoirs in the San Joaquin Basin, California, USA in 1887, more than 300 volcanic-related reservoirs or oil-gas occurrences have been discovered worldwide [2,7]. The research and understanding of volcanic oil and gas exploration can be categorized into three stages: Accidental discovery stage, local exploration stage, and comprehensive exploration stage. Up to now, extensive volcanic reservoir exploration has been conducted worldwide, and many volcanic reservoirs have been discovered. Volcanic rocks have changed from being an initial "forbidden zone" for oil and gas exploration to a "target zone". Volcanic reservoirs in the sedimentary basins of China were first discovered in the northwestern margin of the Junggar Basin in 1957. At present, the exploration and development of volcanic reservoirs in the sedimentary basins of China is being comprehensively carried out, and great progress and breakthroughs have been made.

From a global perspective, nearly all types of igneous rocks, from basic rocks to acidic rocks and from lava to pyroclastic rocks, may have the potential to form effective reservoirs. Compared with conventional reservoirs, the formation of volcanic rock storage conditions is more random, which requires specific analysis based on the actual geology of specific regions [5]. At present, the proven oil and gas reserves in volcanic rocks account for only approximately 1% of the total proven global oil and gas reserves but have great exploration potential [4].

The Junggar Basin is a large-scale composite superimposed basin, which has undergone a complex tectonic evolution process. In the Late Carboniferous to Early Permian, the Paleo-Asian Ocean was completely closed, and the northern Xinjiang was in a post-collisional extensional environment. Large-scale mantle-derived magmatism took place in northern Xinjiang, which included the Junggar region. Volcanic rocks were widely distributed on the uplift structures of the basin [15]. Volcanic rocks in the high parts of structures have undergone long-term weathering, leaching, and dissolution, as well as transformations caused by later tectonic activities, have developed rich secondary pores and fractures, and have become high-quality reservoirs. During the process of oil and gas exploration that has extended to the deep layers of the basin, Carboniferous volcanic rocks have become an important new exploration target. The volcanic eruptions in the Junggar basin are mainly transitional facies between sea and land, with multiple eruptions, mostly small volcanoes, and the complexity of volcanic lithology increases. Most of the volcanic eruptions in foreign countries are marine facies, erupting on the seabed, and the scale of volcanoes is large.

The Hong-Che Fault Zone is located at the southern end of a fault system at the northwestern margin of the Junggar Basin. It is one of the long-term direction areas for oil and gas migration and is rich in oil and gas resources. The Carboniferous in the Hong-Che Fault Zone is an important hydrocarbon enrichment zone in the northwestern margin where various types of oil and gas reservoirs have developed, which have reserves of considerable scale and excellent exploration potential. The distribution system of faults in the Hong-Che Fault Zone is complex, and the lithology and lithofacies of the Carboniferous volcanic rocks are heterogeneous. The distribution of volcanic reservoirs is affected by lithology, and the whole region is oil-bearing but is locally enriched, and the differences are relatively large. At present, there is a lack of systematic research on the formation of volcanic reservoirs in this area, and the distribution laws of oil and gas and the main controlling factors for hydrocarbon accumulations are not clear. Therefore, it is necessary to summarize the characteristics, reservoir forming mechanisms, and main controlling factors of volcanic reservoirs, which is of great practical significance for predicting favorable volcanic reservoir zones and effectively developing volcanic reservoirs in this area. This knowledge can also provide a theoretical basis for the next exploration and development of Carboniferous volcanic rock-based oil and gas in the Junggar Basin.

2. Geological Setting

The northwestern marginal fault system of the Junggar Basin is located in the middle of the Central Asian Orogenic Belt (CAOB) (Figure 1), which is in the coupling region between the Junggar Basin and West Junggar Block. Its formation and evolution were mainly affected by the activity of the West Junggar Orogenic Belt to the west [16]. From north to south, it is divided into the Wu-Xia, Ke-Bai, and Hong-Che Fault Zones [17–20]. The Hong-Che Fault Zone is located at the southern end of the fault system, which is located at the northwestern margin of the Junggar Basin (Figure 1). It strikes nearly north-south and has a length of approximately 80 km and width of approximately 10–20 km. It is adjacent to the Ke-Bai Fault Zone to the north, Chepaizi Uplift to the west, Sikeshu Depression to the south, and the Zhongguai Uplift and Shawan Depression to the east and extends over an area of approximately 1500 km^2 [21,22].

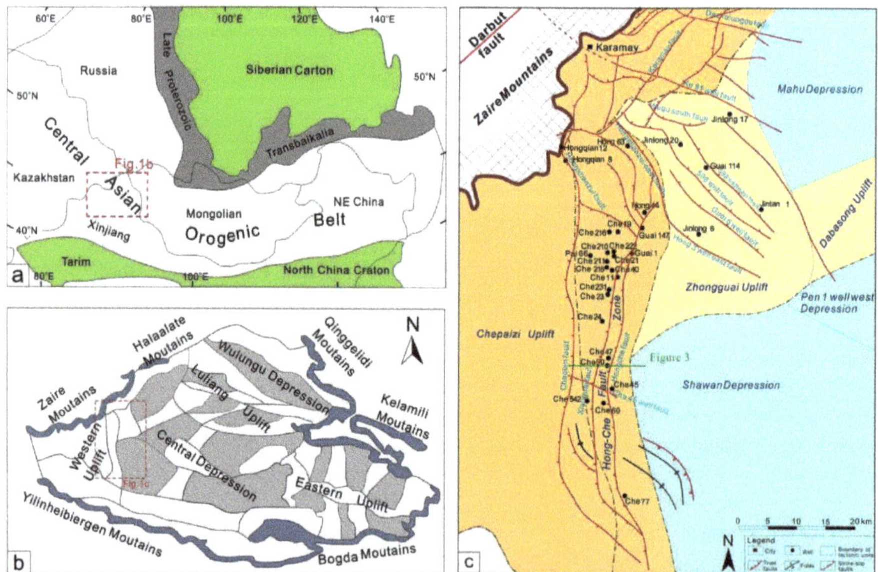

Figure 1. Sketch maps. (**a**) The tectonic location of the Junggar Basin in the CAOB. (**b**) Division of tectonic units in the Junggar Basin. (**c**) The major faults in and around the Hong-Che Fault Zone (modified after [23]).

Studies have shown that the West Junggar region was in a post orogenic extensional environment during the Late Carboniferous to Early Permian [24–27]. The Late Carboniferous-Early Permian volcanic rocks in the West Junggar area and surrounding areas are mainly basalts and rhyolites with less neutral components, which form a typical "bimodal" series. A-type granites are well developed. This tectonic rock association and geochemical characteristics indicate that they formed in an extensional environment of crustal thinning. It is comprehensively judged that the study area has a post orogenic extensional background. During this period, the basin was mainly composed of grabens and half grabens, which were controlled by normal faults with a dual structure of "lower faults and upper depressions" [16,27]. Since the Permian, it has experienced five tectonic stages: Early Permian post orogenic extension, strong Middle-Late Permian compression and thrusting, inherited Triassic thrust superimposition, overall Jurassic-Cretaceous oscillation, and Cenozoic intracontinental foreland [16,28].

According to the drilling and seismic data, the Carboniferous, Permian, Triassic, Jurassic, Cretaceous, Paleogene, Neogene, and Quaternary all developed from bottom to top in the study area (Figure 2). The Lower Permian Jiamuhe Formation (P_1j), Fengcheng Formation (P_1f), Middle Permian Xiazijie Formation (P_2x), Lower Wuerhe Formation (P_2w), and Upper Permian Upper Wuerhe Formation (P_3w) developed from bottom to top. The Baikouquan Formation (T_1b), Karamay Formation (T_2k), and Baijiantan Formation (T_3b) developed from bottom to top in the Triassic. The Jurassic mainly includes the Badaowan Formation (J_1b), Sangonghe Formation (J_1s), Xishanyao Formation (J_2x), Toutunhe Formation (J_2t), and Qigu Formation (J_3q). The Tugulu Group (K_1tg) developed in the Cretaceous. Since the study area has been located at a high position of structures from the Permian to the end of the Jurassic and has been in a state of continuous uplift for a long period, the Permian and Triassic are missing in most areas of the fault zone. From east to west, the Jurassic and Cretaceous overlapped and were deposited on top of the Carboniferous bedrock. Due to the influence of the late Hercynian, Indosinian, Yanshanian, and Himalayan movements, the unconformity surfaces of the lower boundaries of the Upper Wuerhe Formation (P_3w), Jurassic, Cretaceous, and Neogene are developed in the study area.

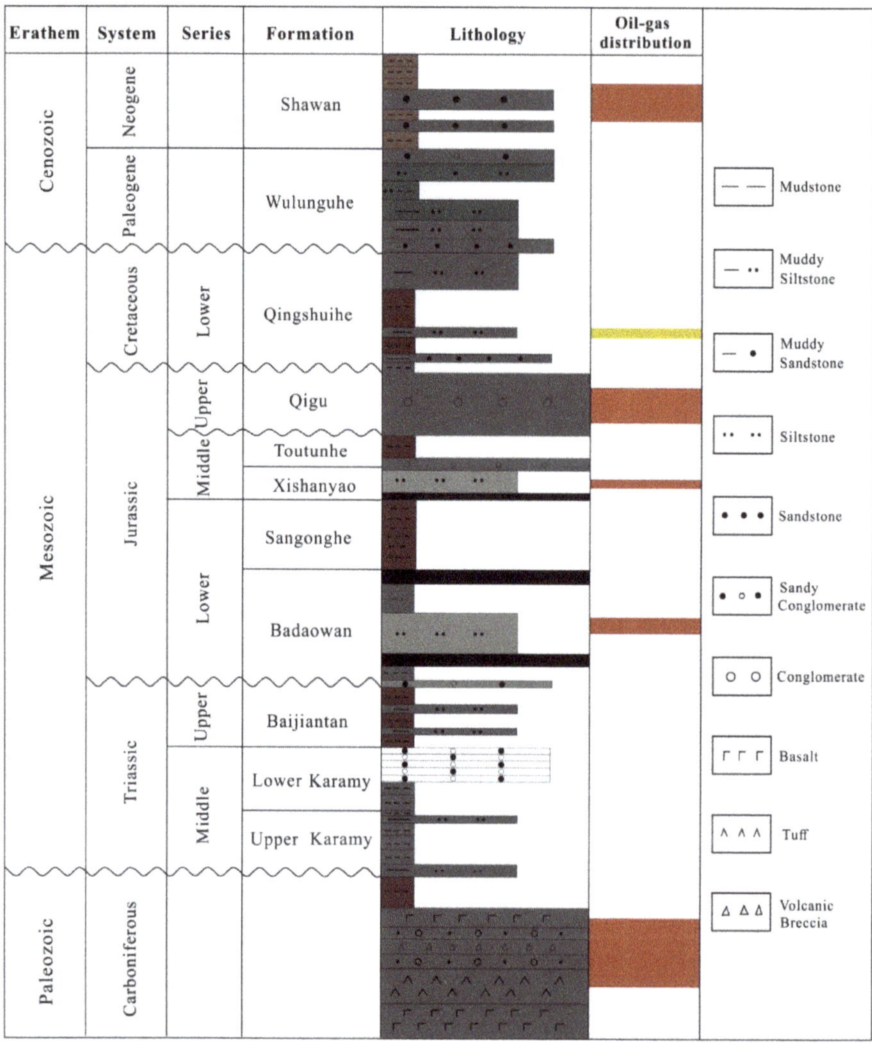

Figure 2. Stratigraphic characteristics of the Hong-Che Fault Zone (modified from [29]).

3. Data and Methods

The research data includes three-dimensional seismic data of the Hong-Che Fault Zone, well logging, core and casting thin section data of exploration wells. The target layer of study is the Carboniferous volcanic rocks. Through the interpretation of seismic data, we analyze the structural characteristics of the reservoir. Based on the analysis of typical oil and gas reservoirs in the Chefeng 3 and Che 210 well blocks, the main controlling factors and hydrocarbon accumulation model of Carboniferous volcanic rocks in the Hong-Che Fault Zone are obtained.

3.1. Structural Characteristics

During the Middle-Late Permian, a strong compressional orogenic activity occurred in the western mountains of the Junggar Basin and the Hong-Che Fault Zone was characterized by a large-scale thrust nappe due to compression. The structural type of the Hong-Che Fault Zone mainly consists of a thrust nappe structure. This fault zone is generally characterized by the development of multiple N-S thrust faults and fault terrace belts uplifted from east to west [30,31]. The fault combination is mainly an imbricate thrust fault combination (Figure 3).

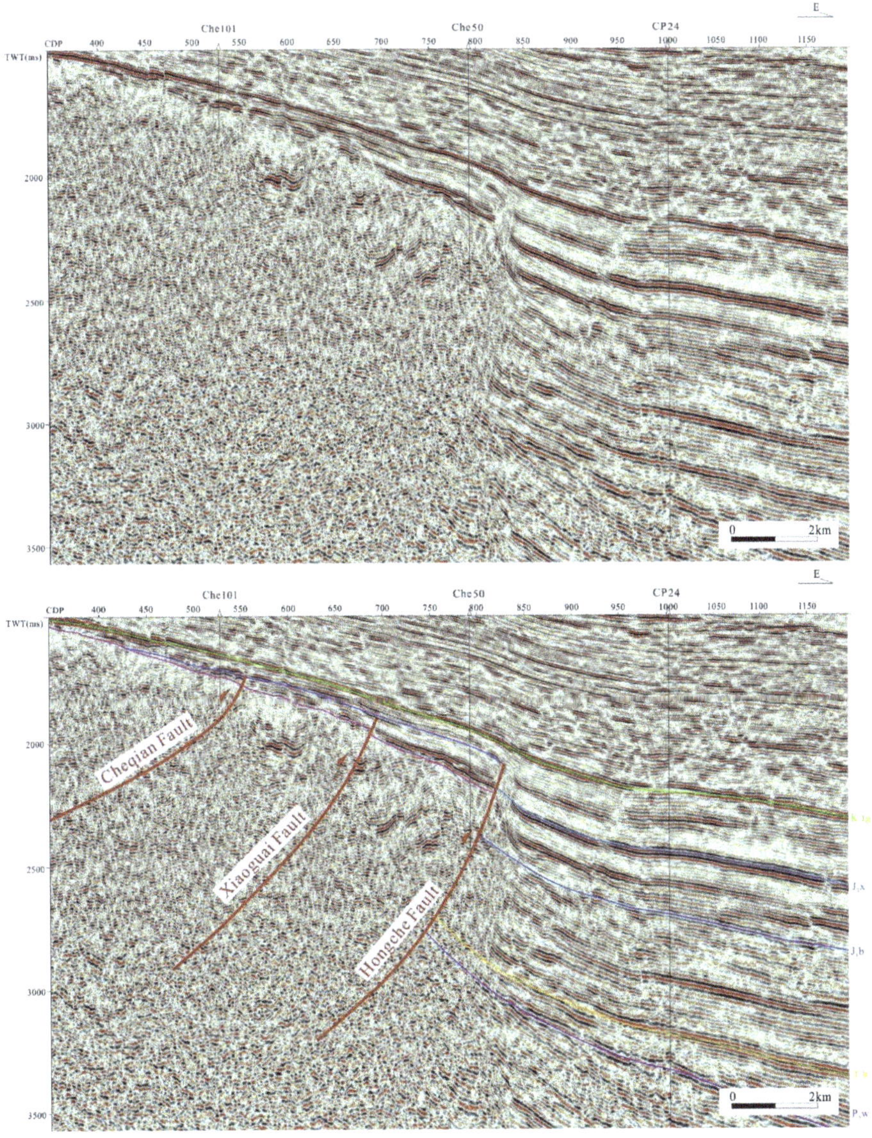

Figure 3. Uninterpreted (**top**) and interpreted (**bottom**) seismic profiles across the Hong-Che Fault Zone. The location of the profile is marked in Figure 1c (modified from [23]).

The strike direction of the thrust fault is nearly north-south and its dip is westward. The dip angle of the upper part is 50–70° and the dip angle of the lower part is 20–30° [17]. The fault strata date is from the Carboniferous to the upper Jurassic boundary. The Hongche Fault has a variable strike and extends approximately 120 km in an arc shape along the plane and can be divided into three sections: South, middle, and north. Among them, the north section tends to trend NNE, the middle section trends nearly NS, and the south section trends NW [23].

3.2. Characteristics of Volcanic Reservoirs

Drilling has shown that the Carboniferous lithology of Hong-Che Fault Zone is dominated by igneous rocks with small amounts of sedimentary rocks [32]. Igneous rocks can be divided into igneous extrusive rocks and igneous intrusive rocks depending on the type of volcanism. Based on the volcanism products, the Carboniferous igneous rocks in the study area can be divided into five types: Volcanic lava, pyroclastic lava, pyroclastic rock, sedimentary pyroclastic rock, and pyroclastic sedimentary rock (Table 3). The volcanic lavas include basalt and andesite and pyroclastic rocks include tuff and volcanic breccia.

Table 3. Lithology statistics of Carboniferous in the Hong-Che Fault Zone.

Major Category	Volcanism Manner	Rock Type	Number of Samples	Thin Section Lithology
Sedimentary rock		Sedimentary rocks	37	Sandstone Sandy conglomerate Conglomerate Mudstone
Igneous rock	Extrusive rock (Igneous rock)	Pyroclastic sedimentary rocks	47	Tuffaceous glutenite Tuffaceous sandstone
		Sedimentary pyroclastic rocks	33	Sedimentary tuff
		Pyroclastic rocks (168)	49	Tuff Basaltic tuff Andesitic tuff Acidic tuff
			30	Basaltic breccia tuff Andesitic breccia tuff
			89	Basaltic tuffaceous volcanic breccia Basaltic volcanic breccia Andesitic volcanic breccia
		Pyroclastic lava	8	Basaltic breccia lava Basaltic tuff lava Basaltic andesitic breccia lava
		Volcanic lava	104	Basalt Amygdaloidal basalt Andesite
	Intrusive rock (Plutonic rocks, hypabyssal rock)	Intermediate intrusive rocks	7	Diorite Fine diorite Amphibolite

Data Source: This table is modified from [32].

Liu (2013) collected thin section data for 404 sample points from 36 wells of the Carboniferous in the Hong-Che Fault Zone and constructed a Carboniferous lithology statistical map (Figure 4) [32]. It can be seen from Figure 4 that igneous rocks mainly developed in this area, the percentage of normal sedimentary rocks is below 10%, and intrusive rocks rarely developed. Among the igneous rocks, pyroclastic rocks and volcanic lavas are most developed, which account for 42% and 26%, respectively, and are followed by pyroclastic sedimentary rocks, which account for 12%.

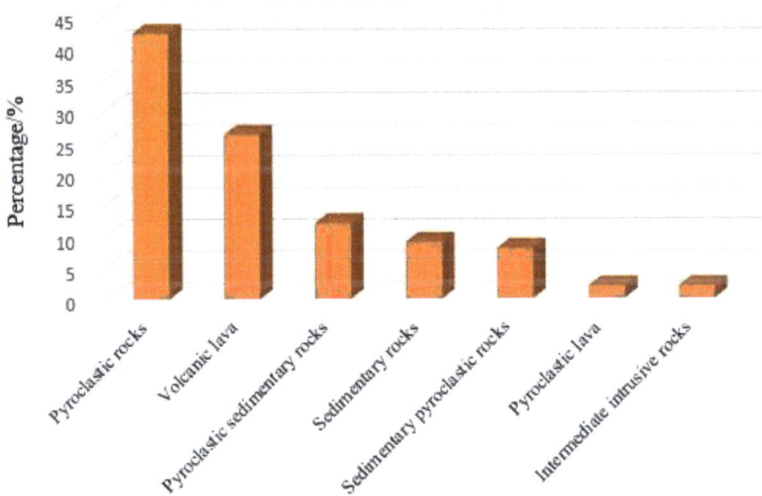

Figure 4. Lithology statistics map of the Carboniferous in the Hong-Che Fault Zone.

The Carboniferous volcanic rocks in the Hong-Che Fault Zone are characterized by multiple stages and intermittent eruptions. There were at least three eruption periods in the Carboniferous and there were multiple eruption cycles in each eruption period. The lithofacies distributions of each eruption period exhibit both similarities and differences [29,32]. The lithofacies of the Carboniferous volcanic rocks gradually changed from volcanic eruptive facies to overflow facies to volcanic sedimentary facies and later to pyroclastic facies (Figure 5). Carboniferous volcanic eruptions mainly occurred along the Hongche Fault, Guaiqian Fault, and other large-scale boundary faults, such as the Che 47, Che 43, Che 46, and Che 72 well volcanic eruption centers. The volcanic eruptions occurred along the faults and were linear fissure eruptions [33].

Oil testing and well logging interpretation data show that the lithofacies of the Carboniferous reservoirs in the Hong-Che Fault Zone are mainly eruptive facies, overflow facies, clastic sedimentary facies, and volcanic sedimentary facies. Among the 46 reservoir samples collected, eruptive facies account for 39.1% and are followed by clastic sedimentary facies, which account for 28.3%. The reservoir samples that developed in the overflow facies account for 23.9% and a small amount (8.7%) developed in volcanic sedimentary facies (Figure 6).

Through core analyses of the volcanic rocks, the relationship between volcanic lithofacies and porosity and permeability parameters was obtained (Table 4) [29,32]. The porosity and permeability of each lithofacies are quite different. The physical properties of the eruptive and clastic sedimentary facies are most favorable, while those of the volcanic sedimentary facies are least favorable.

The reservoir spaces of the Carboniferous reservoirs in the Hong-Che Fault Zone have dual pore media, which include pores and fractures. The primary pores are generally not developed in the Carboniferous and are mainly secondary pores, which include intragranular dissolved pores, intragranular intercrystalline pores, zeolite dissolution pores, and residual intergranular pores, as well as other dissolved pores [34]. There are various types of fractures with complex characteristics in the Carboniferous volcanic reservoirs in the Hong-Che Fault Zone, which include structural fractures, diagenetic fractures, dissolution fractures, and induced fractures. Among these fracture types, structural fractures are the main fracture types. The structural fractures of the Carboniferous volcanic rocks mainly consist of oblique and reticular fractures (Figure 7). The fracture tendency is disorderly and the strikes are NNW and NNE. After the formation of Carboniferous volcanic rocks, the Hong-Che Fault Zone experienced four major tectonic activities, which correspond to the four stages of fracture formation. Vertically, the Carboniferous volcanic rock fractures are widely developed from the top of the Carboniferous downward to a depth of approximately 250 m but the fractures rarely developed below depths of 250 m and are only found in individual wells [35,36].

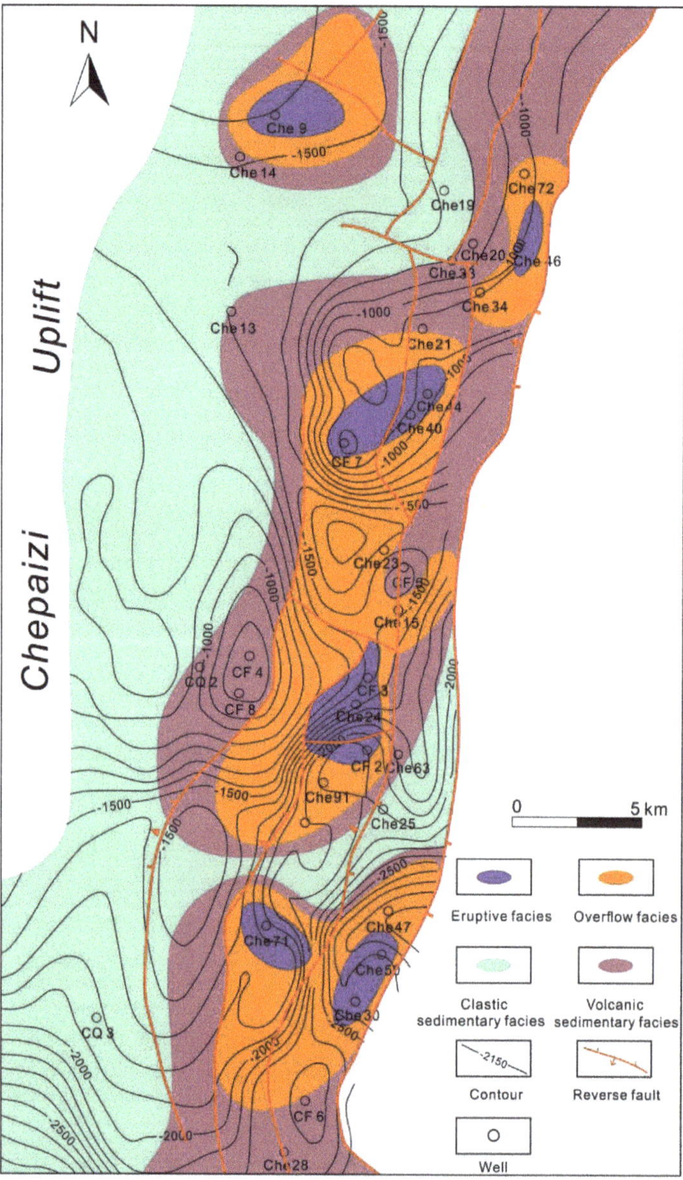

Figure 5. Lithofacies plane distribution map of the Carboniferous volcanic rocks in the Hong-Che Fault Zone (modified from [32]).

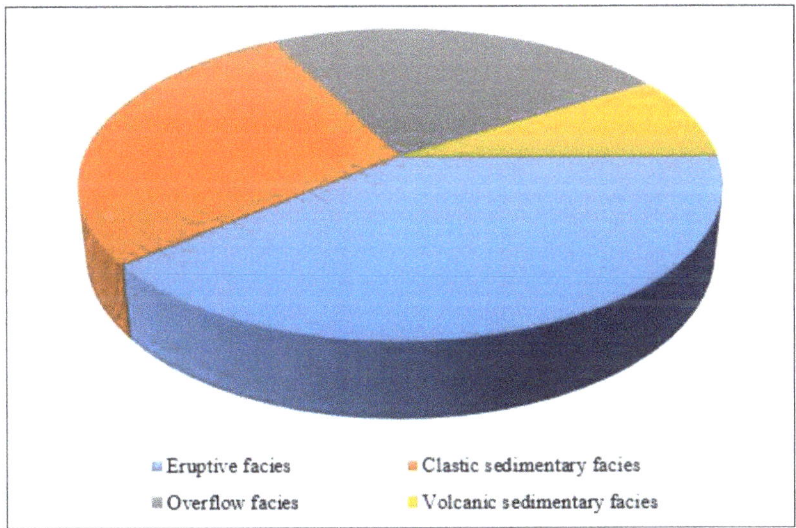

Figure 6. Lithofacies of Carboniferous reservoir in the Hong-Che Fault Zone (modified from [32]).

Table 4. Relationship between lithofacies and physical properties of the Carboniferous volcanic rocks in the Hong-Che Fault Zone.

Lithofacies	Effective Porosity/%	Horizontal Permeability/mD
Eruptive facies	10.52	11.23
Clastic sedimentary facies	14.51	7.44
Overflow facies	8.93	2.83
Volcanic sedimentary facies	4.72	0.98

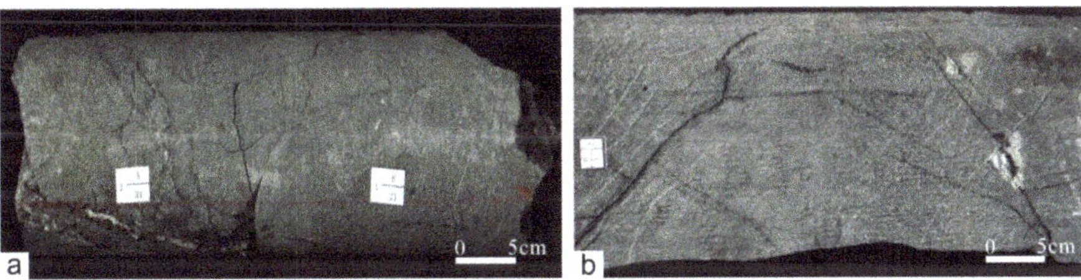

Figure 7. Photos showing the fracture characteristics of the core. (**a**) The tuffaceous sandstone in the Che 211 well at 1176.67–1177.09 m. (**b**) The andesite in the Chefeng 7 well at 1350.22–1350.35 m.

Taking the Che 210 well block as an example, based on core observations and casting thin section data analysis, the pore types of the Carboniferous reservoirs in this area are mainly dissolution pores and micro-fracture pores. By microscopic analysis, the core at 1180.55 m in the Che 222 well consists of tuffaceous fine sandstone with intragranular dissolved pores, which account for 50% of the total pores and micro-fractures, which also account for 50%. There is fine-grained sandstone at 1334.08 m in the Che 222 well and intragranular dissolved pores account for 100% of the total pores (Figure 8).

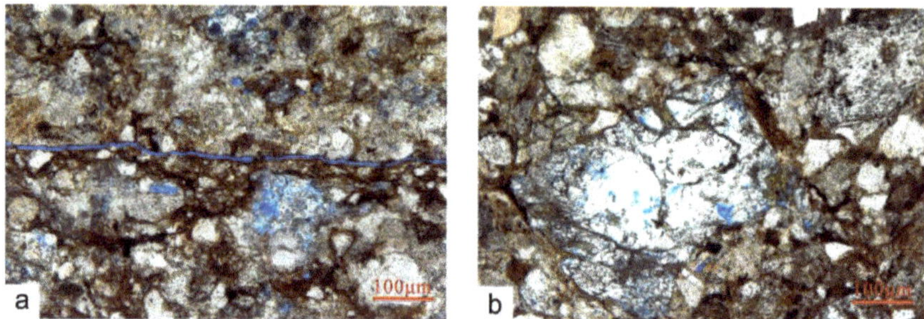

Figure 8. Casting thin section photos of Che 222 well. (**a**) The tuffaceous fine sandstone at 1180.55 m. (**b**) The fine-grained sandstone at 1334.08 m.

According to the analysis of casting thin section data, the tuffaceous sandstone in the area of the Che 210 well mainly developed dissolution pores, which are dominated by intragranular dissolved pores, matrix dissolved pores, and micro-fractures. Volcanic breccias mainly developed dissolution pores and microfractures, tuff mainly developed fractures, and basaltic andesite mainly developed dissolution pores dominated by phenocryst-dissolved pores (Figure 9).

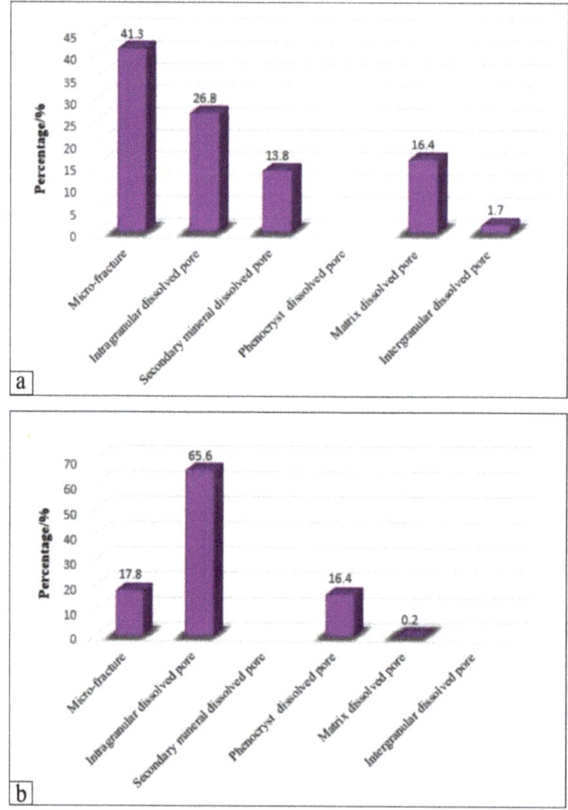

Figure 9. Statistical figure of different lithology pore types in the Che 210 well block. (**a**) The pore types of tuffaceous sandstone. (**b**) The pore types of volcanic breccia.

3.3. Reservoir Characteristics and Types

The Carboniferous reverse faults in the Hong-Che Fault Zone are developed and can be divided into two groups: One group is a nearly north-south trending fault system, which extends farther in the plane direction and is the main fault of the Hong-Che Fault Zone. The other group is nearly an EW trending fault system with short plane extensions and small fault distances. The two groups of faults cut each other to form a fault block group, which formed a series of fault block traps such as the Che 23 well, Che 210 well, Che 91 well, and Che Feng 6 well. The Hong-Che Fault Zone has a variety of favorable fault block traps of different sizes and there are many oil and gas producing locations, which easily formed fault block oil and gas reservoirs. Along the plane, they are mainly distributed along the fault zone in strips and along the profile, they are mainly distributed in the ascending wall of the Hong-Che Fault Zone but there are also some favorable traps in the footwall. The main types of traps are fault block and fault-lithology. The Carboniferous in the Hong-Che Fault Zone mainly developed fault block reservoirs, lithologic reservoirs, and fault-lithologic reservoirs [37].

Taking the reservoir of the Chefeng 3 well block as an example, the Carboniferous fault block in this area can be divided into three secondary structures: The Che 91 well fault block, Che 63 well west fault block, and Che 24 well fault block. It is believed that this area is controlled by fault blocks.

The Carboniferous reservoir in the area of the Che 91 well consists mainly of volcanic rock and its lithology is mostly volcanic breccia, basalt, andesite, and tuff. The physical properties of the explosive facies breccia are most favorable and are followed by broken basalt, andesite, and tuff, which are least favorable. This reservoir is a pore-fracture dual-medium reservoir. Due to the strong tectonic activity, the fractures in the study area are relatively well-developed and are mainly structural fractures, which more effectively change the physical properties of igneous reservoirs in this section.

The reservoir type of the Carboniferous in the area of the Chefeng 3 well is a fault block reservoir, which is controlled by volcanic lithofacies (Figure 10). The reservoir is controlled by faults and volcanic lithofacies along the plane and is controlled by ancient volcanic eruption sequences in the vertical direction. The reservoir lithology is mainly composed of volcanic breccia from eruptive facies, broken basalts of overflow facies, and the upper sedimentary tuff is a good caprock.

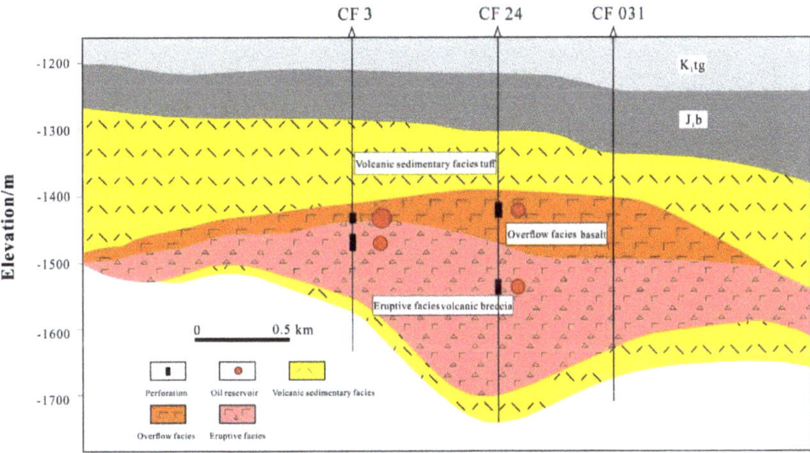

Figure 10. Lithology profile of the Carboniferous reservoir in the Chefeng 3 well area ("CF" represents Chefeng).

Taking the Che 210 well area as an example, two groups of reverse faults developed mainly in the Carboniferous. One group is a near east-west fault with a short plane extension distance, such as Che 212 well north fault, Che 210 well south fault, Che 213 well south fault, and Guai 2 well south fault. There is a group of nearly north-south trending faults with long plane extension distances, such as the Che 36 well east fault, Che 211 well west fault, and Che 11 well west fault, which control the stratigraphic distribution of the Hong-Che Fault Zone and gradually rise from east to west and have resulted in serious denudation of the strata. The two groups of faults cut each other to form multiple fault block traps. The Carboniferous reservoir in the area of the Che 210 well developed in four fault block traps, which are Che 210 well, Che 211 well, Chefeng 7 well, and Che 228 well block trap (Figure 11).

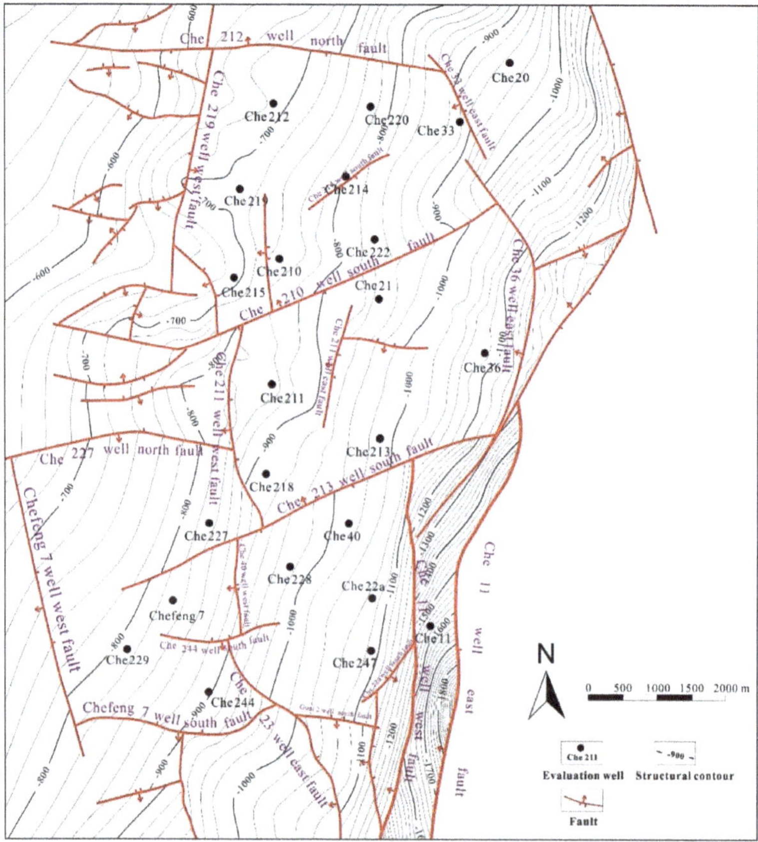

Figure 11. Contour line of the top structure of Carboniferous Formation of the Che 210 well block.

According to a statistical analysis of thin section identifications and core observation data, three types of reservoirs developed in the Carboniferous strata in the area of the Che 210 well: Tuffaceous sandstone, tuff and volcanic breccia, and basaltic andesite in overflow facies (Figure 12). Among them, tuffaceous sandstone is widely distributed in this area and forms the main reservoir, which is followed by tuff and volcanic breccia, while basaltic andesite is less commonly distributed. These properties can be seen in the oil-bearing property statistical histogram of cores from different lithologies of the Carboniferous in the Che 210 well area. Relatively high oil-bearing grades are present in the tuffaceous sandstone and volcanic breccia, and basaltic andesite has poor grades (Figure 13). According to the oil test results and lithology analysis of the Carboniferous in the Che 210 well area, the main lithology of the oil-producing section is tuffaceous sandstone. The oil test results confirm that commercial oil flow can also be obtained from tuff, volcanic breccia, and basaltic andesite but there is only local development in this area.

Figure 12. Photos showing the lithology of Carboniferous Formation in the Che 210 well block. (**a**) The volcanic breccia in the Che 210 well at 1447.5–1447.6 m. (**b**) The volcanic breccia in the Che 22a well at 1866.3–1866.5 m. (**c**) The tuff in the Chefeng 7 well at 1267.5–1267.7 m. (**d**) The tuff in the Che 40 well at 1377.7–1378.0 m. (**e**) The tuffaceous sandstone in the Che 211 well at 1176.6–1177.0 m. (**f**) The tuffaceous sandstone in the Che 44 well at 1600.9–1601.1 m. (**g**) The andesite in the Chefeng 7 well at 1350.2–1350.3 m. (**h**) The basalt in the Chefeng 7 well at 1350.4–1350.6 m.

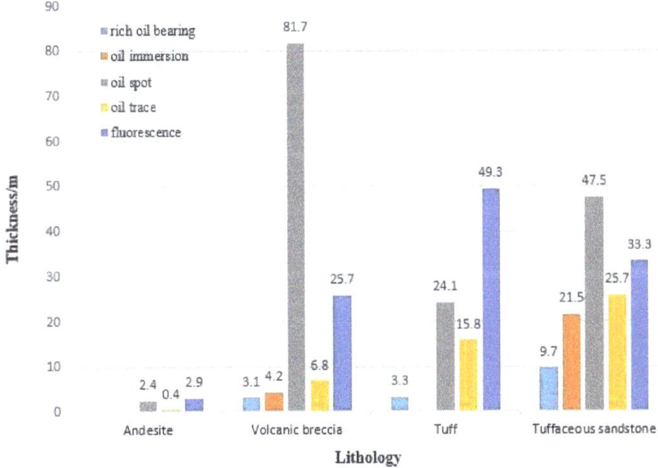

Figure 13. The statistical figure of different lithologies and core oil contents in the Che 210 well block. ("Rich oil bearing" means crude oil can be seen in more than 75% of the observed core section, "oil immersion" means crude oil can be seen in more than 40% of the observed core section, "oil spot"

"means crude oil can be seen in 40%–5% of the observed core section, "oil trace" means crude oil can be seen in less than 5% of the observed core section, and "fluorescence" means the crude oil is not visible to the naked eyes, but the fluorescence detection shows it.).

The Carboniferous in the Che 210 well area was exposed at the surface for a long period, experienced long-term weathering and leaching, and was then directly covered by the Jurassic Badaowan Formation, which lacked Permian and Triassic sediments. Weathering and leaching have further transformed the top of the Carboniferous bedrock into a reservoir, which macroscopically, is the bedrock reservoir controlled by unconformity. The Carboniferous in the area of the Che 210 well mainly contains oil. Reservoir oil layers mainly developed in the leaching zone at the top of the Carboniferous. The widely distributed tuffaceous sandstone has large numbers of matrix pores. In addition, with later leaching, transformation, and fracture communication, various lithologies developed secondary pores and micro-fractures such as intragranular dissolved pores and matrix dissolved pores. Tuffaceous sandstone, tuff, volcanic breccia, and basaltic andesite can all form good volcanic reservoirs and among these, volcanic breccia reservoirs have the best physical properties. The area of the Che 210 well is located in the middle of the Hong-Che Fault Zone, which is associated with violent tectonic movements and well-developed faults. There are two groups of faults in the entire area, which have cut the Carboniferous oil reservoirs into four fault blocks.

The Carboniferous reservoirs in area of the Che 210 well are massive reservoirs, which are controlled by faults and physical properties. The oil reservoir is controlled by fault blocks and lithology along the plane. The reservoir is divided into four fault blocks. Vertically, the oil layers are distributed within 350 m from the top of the Carboniferous and their oil-bearing properties are affected by weathering and leaching.

Due to the complex volcanic lithology and the scattered rock mass, the oil recovery effect after large-scale fracturing is good. The method of supplementing energy does not work well, and neither water injection nor steam injection works.

4. Analysis of Main Controlling Factors of Carboniferous Reservoir

4.1. Structure

In the late Cretaceous, the Hong-Che Fault Zone tilted and the overall structure tilted to the south. The structural pattern of the Carboniferous changed from high in the south and low in the north to high in the north and low in the south. At this time, the oil and gas, which were in the traps in the middle and south migrated along the fault to the north and formed mixed-source oil and gas reservoirs in the north. The Carboniferous oil and gas in the Hong-Che Fault Zone are mainly concentrated in the north, and oil and gas mainly accumulated at the high part of the structure [38,39].

At the same time, the Carboniferous volcanic reservoirs in the Hong-Che Fault Zone were greatly affected by faults. Most of the reservoirs are distributed in strips and blocks along the Hong-Che Fault Zone. The controlling effect of faults on hydrocarbon accumulations is reflected in three ways. The first is the openness of the faults. During periods of fault activity, the faults acted as migration channels for oil and gas and the main migration directions along the faults were north-south and vertical. The second consideration is sealing. When fault zones were in relatively static stages, the faults acted as sealing zones for oil and gas accumulations [40–45]. The third factor is the effect of fractures, which were derived from faults. During active fault periods, large numbers of cracks were produced due to stress release. Fractures connected the storage spaces of various pores and fractures, greatly improved reservoir property and permeability, and even increased the permeability of rocks by several orders of magnitude. There is a positive correlation between fracture density and oil well productivity. The greater the fracture density, the higher the oil well productivity [46].

4.2. Volcanic Lithofacies

Another important factor that controls oil and gas reservoirs is volcanic lithofacies. Reservoirs with different lithofacies have different pore fracture structures and reservoir space combinations. In volcanic breccias, intragranular dissolved pores, intergranular pores, and matrix pores are well developed and the storage performance is best. Basalt pores are well developed but most of them are filled pores, therefore, their storage performance is not ideal. Tuff has many micropores and the reservoir properties are worst, therefore, tuff can act as a local cap rock.

The Carboniferous volcanic rocks in the Hong-Che Fault Zone developed various types of reservoir-caprock assemblages, which mainly include the following four types: (1) Eruptive facies form the reservoirs and the volcanic sedimentary facies tuffs form the caprocks. (2) Eruptive facies function as reservoirs and the dense basalts in the overflow facies function as cap rocks, such as the oil and gas reservoir in the Che 47 well. The Che 47 well is located in the middle section of the Hong-Che Fault Zone. The Carboniferous volcanic reservoir lithology in this well mainly consists of eruptive facies basaltic volcanic breccia, gray tuffaceous volcanic breccia, and variegated andesitic basaltic volcanic breccia. The cap rock is a dense basalt in the overflow facies. (3) The transitional facies between the volcanic and sedimentary rock acts as the reservoir and the volcanic sedimentary facies tuff is the caprock. (4) The overflow facies function as reservoirs and the volcanic sedimentary facies tuffs are cap rocks [21].

On the whole, the physical reservoir properties of the eruptive facies in the Hong-Che Fault Zone are most favorable and are followed by overflow facies, while the volcanic sedimentary facies are least favorable. In addition, there are also tuffaceous sandstone, sandy tuff, and mudstone in the transitional facies zone. Under favorable conditions of oil and gas sources and plugging conditions, tuffaceous sandstone can also migrate and accumulate into reservoirs.

4.3. Unconformity

In the hanging wall of the Hong-Che Fault Zone, the volcanic rocks at the top of Carboniferous are in direct contact with Permian, Jurassic, and Cretaceous strata and form a large-scale unconformity. From east to west, the overlying strata of the Carboniferous change from older to younger. Unconformity surfaces act as migration channels for oil and gas. The Permian oil and gas from the Shawan Depression first migrated along faults and then migrated upward (westward) along an unconformity to effective volcanic traps and then formed hydrocarbon reservoirs [47]. The unconformity surface was weathered and denuded into a weathering crust, which can seal oil and gas to form caprock.

The Carboniferous rocks in the fault terrace zone of the northwestern margin of the Junggar Basin suffered from long-term denudation and weathering leaching and a weathering crust developed. The Carboniferous volcanic rocks and clastic rocks have a weathering crust, which was modeled as five layers, namely, a soil layer, hydrolysis zone, corrosion zone, disintegration zone, and parent rock [48,49]. According to the statistics, the distances between the soil layer, hydrolysis zone, corrosion zone, and disintegration zone to the top boundary of the unconformity surface of the Carboniferous are 15–50, 50–150, 150–250, and 250–450 m, respectively. The thickness of the weathering crust in the fault terrace zone of the northwestern margin is generally approximately 450 m, and the thickness of the weathering crust in the fault development area can be greater than 600 m.

The physical properties of different structures in the weathering crust are quite different. The average porosities of the soil layer, hydrolysis zone, corrosion zone, and disintegration zone are 2.6%, 6.4%, 15.8%, and 12.7%, respectively. Reservoirs in the corrosion zone have the best physical properties and belong to type I reservoirs. The second most favorable physical properties are in the disintegration zone, which belong to type II–III reservoirs. The soil layer and hydrolytic zone are mainly distributed in the lower part and slope area of the paleogeomorphology and the high part is mostly missing. The soil layer is a nonreservoir layer and the hydrolysis zone is a type IV reservoir with visible oil and gas displays but no productivity [48]. The physical properties of the weathering crust reservoir are controlled by the lithology of the parent rock and degree of weathering leaching.

The hydrocarbon accumulations in the Carboniferous in the fault terrace zones of the northwestern margin are characterized as oil-bearing throughout the entire zone and exhibit local enrichment [50,51]. The controlling factors for local enrichment of oil and gas are mainly effective sealing in an upward direction of the oil and gas migration, physical properties of weathering crust reservoirs, and preservation conditions. Among them, the physical properties of weathering crust reservoirs determine the output of the reservoir and the preservation conditions determine the viscosity of the reservoir. The quality of weathering crust reservoirs and their spatial distribution scales are the primary controlling factors for hydrocarbon accumulation.

4.4. Physical Properties

The Chepaizi Uplift, where the Hong-Che Fault Zone is located, is one of the oldest uplifts in the Junggar Basin. The Carboniferous has been exposed to the surface for a long period and has been subjected to weathering and leaching, which gradually transformed the Carboniferous volcanic rocks into favorable reservoirs. Judging from the current cast thin section analysis data from the Che 210 well block in the Hong-Che Fault Zone, the reservoir space mainly consists of secondary dissolved pores and micro fractures, while no primary pores are found, which indicate that weathering and leaching determined the reservoir capacity of the area. According to the relationship between the porosity and permeability data of the actual core analyses in this area and distance from the weathering crust, the vertical zonation characteristics of the weathering crust at the top of the Carboniferous are obvious: The thickness of the soil layer is approximately 0–30 m, the lithology is mainly weathered residual soil, which was formed by weathering and rock erosion, and the regional distribution is relatively stable, therefore, it can function as a regional cap rock. The leached zone is 30–350 m. The reservoir physical properties are good, the average porosity is 7.42%, and the average permeability is 0.071 mD. This is the main reservoir in this area. Based on the current evaluation results, the oil layer mainly developed in the leached zone. Depths of 350–700 m represent the disintegration zone. Compared with the corrosion zone, the reservoir permeability of the disintegration zone is similar to that of the corrosion zone but the porosities are quite different, with an average porosity of 5.07% and average permeability of 0.143 mD. The reservoirs in the disintegration zone are mainly dry layers, which are partially water bearing but not active. The parent rock zone is below 700 m, its reservoir physical properties are poor, its average porosity is 1.60%, and its average permeability is 0.032 mD, which is a lower threshold in this area (Figure 14).

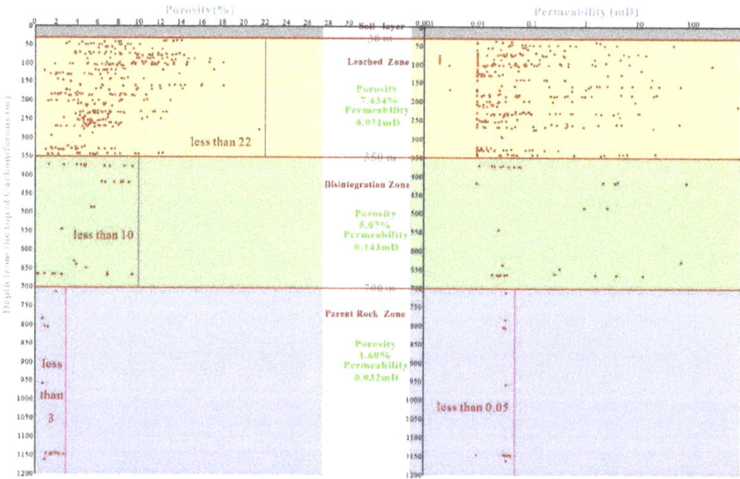

Figure 14. The stratigraphic section of the Che 210 well block.

Vertically, the degree of development of dissolution pores and microfractures in the Carboniferous reservoirs in the Che 210 well block was clearly affected by differences in weathering and leaching. The reservoirs with the best physical properties mainly developed in the leached zone, which is 30–350 m from the top of the Carboniferous. Meanwhile, oil production tests and production tests in this area further indicate that the oil reservoirs are distributed within a range of 350 m below the upper boundary of the Carboniferous. This indicates that the physical reservoir properties control the vertical distribution ranges of the reservoirs.

5. Reservoir Formation Model

The crude oil in the Carboniferous reservoir in the Hong-Che Fault Zone mainly comes from the source rocks of Permian Fengcheng Formation and Lower Wuerhe Formation in the Shawan Depression. Since the Permian, this area has been in a structural pattern of high in the west and low in the east for a long period and has become a favorable area directionally for the migration of oil and gas, which was generated in the Shawan Depression. The oil and gas first migrated along the faults and then moved upward (westward) with the unconformity as the migration channel. The oil and gas mainly migrated along a spatial channel, which consisted of faults-unconformities and entered the trap in the hanging wall from the oil-generating area of the footwall of the Hong-Che fault [52–59]. Oil and gas migrated vertically through faults, laterally along unconformities, gradually migrated to higher parts, and accumulated in the stratigraphic traps of volcanic weathering bodies. The reservoir formation model is as follows (Figure 15).

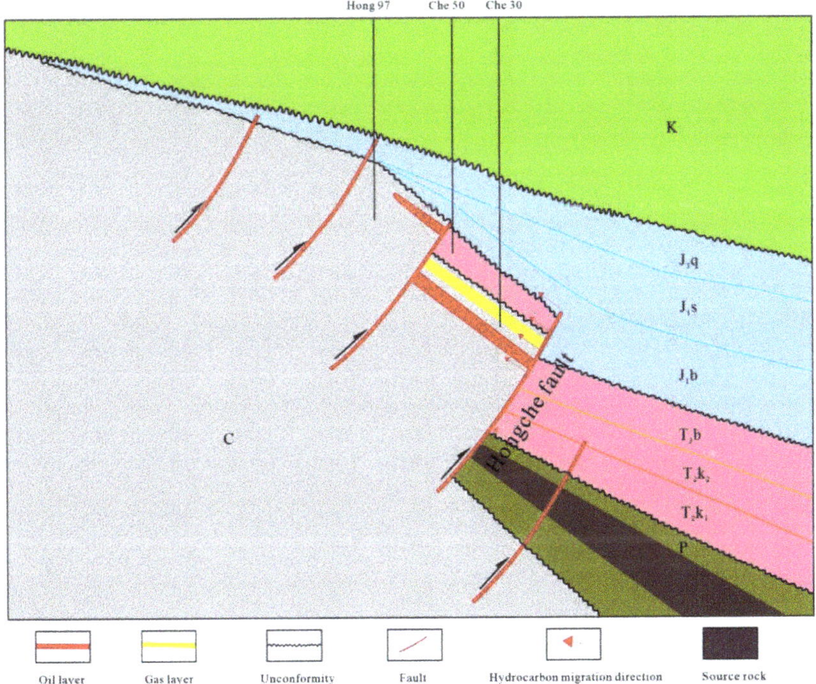

Figure 15. Hydrocarbon accumulation model of Carboniferous in the Hong-Che Fault Zone (adapted from [21]).

6. Conclusions

1. The Carboniferous volcanic reservoir in the Hong-Che Fault Zone is mainly distributed in the hanging wall of the fault zone and oil and gas has mainly accumulated in the high part of the structure. The reservoir is controlled by faults and lithofacies in the plane direction and is vertically distributed within 400 m from the top of the Carboniferous. The formation of the volcanic reservoir was mainly controlled by structures and was also controlled by volcanic lithofacies, unconformity surfaces, and physical properties.

2. The physical reservoir properties of the eruptive facies in the Hong-Che Fault Zone are most favorable and are followed by overflow facies, while the volcanic sedimentary facies are least favorable.

3. The Carboniferous portion of the Hong-Che Fault Zone has been exposed to the surface for a long period, has been subjected to weathering and leaching, and a weathering crust has developed. The vertical zonation characteristics of the weathering crust at the top of the Carboniferous in the area of the Che 210 well are obvious. A soil layer, corrosion zone, disintegration zone, and parent rock developed from top to bottom. Among them, the reservoirs with the best physical properties are developed in the corrosion zone, which are 30–350 m distant from the top of the Carboniferous.

4. The reservoir space of the Carboniferous reservoir in the Hong-Che Fault Zone consists mainly of secondary pores and fractures.

Author Contributions: D.Z. conceived the presented idea and analyzed the research data. The article is originally written by D.Z. and revised by author X.L. and S.G. All authors have read and agreed to the published version of the manuscript.

Funding: This research received no external funding.

Acknowledgments: The author of this article would like to thank the Xinjiang Oilfield Branch of the China National Petroleum Corporation for the information provided and wishes to thank Song Wang for his help in explaining the seismic data.

Conflicts of Interest: The authors declare no conflict of interest.

References

1. Schutter, S.R. Occurrences of Hydrocarbons in and around Igneous Rocks. *Geol. Soc. Lond. Spec. Publ.* **2003**, *214*, 35–68. [CrossRef]

2. Petford, N.; McCaffrey, K. Hydrocarbons in crystalline rocks: An introduction. *Geol. Soc. Lond. Spec. Publ.* **2003**, *214*, 1–5. [CrossRef]

3. Jiang, H.; Shi, Y.; Zhang, Y.; Fan, Z.; Shi, F.; Kou, Y.; Wang, L. Potential of global volcanics-hosted oil-gas resources. *Resour. Ind.* **2009**, *11*, 20–22.

4. Liu, J.; Meng, F.; Cui, Y.; Zhang, Y. Discussion on the formation mechanism of volcanic oil and gas reservoirs. *Acta Pet. Sin.* **2010**, *26*, 1–13.

5. Wang, L.; Li, J.; Shi, Y.; Zhao, Y.; Ma, Y. Review and prospect of global volcanic reservoirs. *Geol Chin.* **2015**, *42*, 1610–1620.

6. Zorin, Y.A. Geodynamics of the western part of the Mongolia-Okhotsk collisional belt, Trans-Baikal region (Russia) and Mongolia. *Tectonophysics* **1999**, *306*, 33–56. [CrossRef]

7. Zou, C.; Zhao, W.; Jia, C.; Zhu, R.; Zhang, G.; Zhao, X.; Yuan, X. Formation and distribution of volcanic hydrocarbon reservoirs in sedimentary basins of China. *Pet. Explor. Dev.* **2008**, *35*, 257–271. [CrossRef]

8. Zhang, Z.; Wu, B. Research status and exploration technology investigation of volcanic oil and gas reservoirs at home and abroad. *Nat. Gas Explor. Dev.* **1994**, *16*, 1–26.

9. Li, S.; Lu, F.; Lin, C. *Meso-Cenozoic Basin Evolution and Geodynamic Environments in Eastern China and Adjacent Areas*; Publishing House of China University of Geosciences: Wuhan, China, 1997.

10. Gu, T.; Dai, J.; Niu, J. Effective reservoir recognition of paleogene trachyte in the middle part of the east sag of Liaohe Depression. *Pet. Explor. Dev.* **2007**, *34*, 310–315.

11. Zhao, W.; Zou, C.; Li, J.; Feng, Z.; Zhang, G.; Hu, S.; Kuang, L.; Zhang, Y. Comparative study on volcanic hydrocarbon accumulations in western and eastern China and its significance. *Pet. Explor. Dev.* **2009**, *36*, 1–11.

12. Long, X.; Sun, M.; Yuan, C.; Xiao, J.; Chen, H.; Zhao, Y.; Cai, K.; Li, J. Genesis of Carboniferous volcanic rocks in the eastern Junggar: Constraints on the closure of the Junggar Ocean. *Acta Pet. Sinica* **2006**, *22*, 31–40.

13. Xu, J.; Mei, H.; Yu, X.; Bai, Z.; Niu, H.; Cheng, F.; Zheng, Z.; Wang, Q. Adakite volcanic related to plate subduction in Late Paleozoic island arc of northern margin Junggar: A product of partial melting of subducting slab. *Chin. Sci. Bull.* **2001**, *46*, 684–688.

14. Tang, Y.; Chen, F.; Peng, P. Characteristics of volcanic rocks in Chinese basins and their relationship with oil-gas reservoir forming process. *Acta Pet. Sinica* **2010**, *26*, 185–194.

15. Zhou, Y. The Gravitational and Magnetic Field Research of Carboniferous Volcanic Rocks in Western-Central Junggar Basin. Ph.D. Thesis, Nanjing University, Nanjing, China, 2018.

16. Sui, F. Tectonic Evolution and Its Relationship with Hydrocarbon Accumulation in the Northwest Margin of Junggar Basin. *Acta Geol. Sin.* **2015**, *89*, 779–793.

17. He, D.; Yin, C.; Du, S.; Shi, X.; Ma, H. Characteristics of structural segmentation of foreland thrust belts—A case study of the fault belts in the northwestern margin of Junggar Basin. *Earth Sci. Front.* **2004**, *11*, 91–101.

18. Yu, Y.; Wang, X.; Rao, G.; Wang, R. Mesozoic reactivated transpressional structures and multi-stage tectonic deformation along the Hong-Che fault zone in the northwestern Junggar Basin, NW China. *Tectonophysics* **2016**, *679*, 156–168. [CrossRef]

19. Qi, L.; Bao, Z.; Xian, B.; Huang, Z. Structural Transform Zone and Its Control of Mesozoic Deposits in Northwestern Margin of Junggar Basin. *Xinjiang Pet. Geol.* **2009**, *30*, 29–32.

20. Tan, K.; Zhang, F.; Zhao, Y.; Tan, J.; Guan, Y.; Yang, Z. Comparative analysis on the segmentation of tectonic characteristic in northwest Junggar basin. *Pet. Geol Eng.* **2008**, *22*, 1–3.

21. Yao, W.; Dang, Y.; Zhang, S.; Zhi, D.; Xing, C.; Shi, J. Formation of Carboniferous Reservoir in Hongche Fault Belt, Northwestern Margin of Junggar Basin. *Nat. Gas Geosci.* **2010**, *21*, 917–923.

22. Zhong, W.; Huang, X.; Zhang, Y.; Jia, C.; Wu, K. Structural characteristics and reservoir forming control of Hongche fault zone in Junggar Basin. *Compet. Hydropower Reserves* **2018**, *11*, 1–5.

23. Liu, Y.; Wu, K.; Wang, X.; Liu, B.; Guo, J.; Du, Y. Architecture of buried reverse fault zone in the sedimentary basin: A case study from the Hong-Che Fault Zone of the Junggar Basin. *J. Struct. Geol.* **2017**, *105*, 1–17. [CrossRef]

24. Han, B.; Ji, J.; Song, B.; Chen, L.; Zhang, L. Late Paleozoic vertical growth of continental crust around the Junggar Basin, Xinjiang, China (Part I): Timing of post-collisional plutonism. *Acta Pet. Sin.* **2006**, *22*, 1077–1086.

25. Su, Y.; Tang, H.; Hou, G.; Liu, C. Geochemistry of aluminous A-type granites along Darabut tectonic belt in West Junggar, Xinjiang. *Geochimica* **2006**, *35*, 55–67.

26. Meng, J.; Guo, Z.; Fang, S. A new insight into the thrust structures at the northwestern margin of Junggar Basin. *Earth Sci. Front.* **2009**, *16*, 171–180.

27. He, D.; Wu, S.; Zhao, L.; Zheng, M.; Li, D.; Lu, Y. Tectono-Depositional Setting and Its Evolution during Permian to Triassic around Mahu Sag, Junggar Basin. *Xinjiang Pet. Geol.* **2018**, *39*, 35–47.

28. Liang, Y. Geological Structure, Formation and Evolution of Chepaizi Uplift in Western Junggar Basin. Ph.D. Dissertation, China University of Geosciences, Beijing, China, 2019.

29. Chen, X.; Kuang, L.; Cha, M.; Shao, Y.; Lei, D.; Yang, D.; Li, L.; Huang, Y.; Chen, Z.; Xu, C.; et al. *Hydrocarbon Accumulation Mechanism and Exploration Technology of Volcanic Rocks: A Case Study of Junggar Basin*; Science Press: Beijing, China, 2014; p. 58.

30. Fan, C.; Qin, Q.; Yuan, Y.; Wang, X.; Zhu, Y. Structure characteristics and fracture development pattern of the Carboniferous in Hongche fracture belt. *Spec. Oil Gas Reserves* **2010**, *17*, 47–49.

31. Dong, D.; Li, L.; Wang, X.; Zhao, L. Structural Evolution and Dislocation Mechanism of Western Margin Chepaizi Uplift of Junggar Basin. *J. Jilin Univ. Earth Sci. Ed.* **2015**, *45*, 1132–1141.

32. Liu, H. Study on Hydrocarbon Accumulation Law for Volcanic Rock Reservoirs in Hongche Fault Belt, Junggar Basin. Master's Thesis, China University of Petroleum (East China), Dongying, China, 2013.

33. Yin, L.; Pan, J.; Tan, K.; Wang, Y.; Wang, B.; Xu, D. Application of volcanic seismic reservoir to oil and gas exploration of Carboniferous in Hongche fault belt in Junggar Basin. *Litho Reserves* **2010**, *22*, 25–30.

34. Gan, X.; Jiang, Y.; Qin, Q.; Song, W. Characteristics of the Carboniferous volcanic reservoir in the Hongche fault zone. *Spec. Oil Gas Reserves* **2011**, *18*, 45–47.

35. Yuan, Y.; Cai, Y.; Fan, Z.; Jiang, Y.; Qin, Q.; Jiang, Q. Fracture characteristics of Carboniferous volcanic reservoirs in Hongche fault belt of Junggar Basin. *Litho Reserves* **2011**, *23*, 47–51.

36. Su, P.; Qin, Q.; Yuan, Y.; Jiang, F. Characteristics of Volcanic Reservoir Fractures in Upper Wall of HongChe Fault Belt. *Xinjiang Pet. Geol.* **2011**, *32*, 457–460.
37. Pan, J.; Hao, F.; Tan, K.; Wei, P.; Ren, P.; Chen, Y.; Yin, L. Characteristics and accumulation of Paleozoic volcanic rock reservoirs in Hongche fault belt, Junggar Basin. *Litho Reserves* **2007**, *19*, 53–56.
38. Wu, K.; Guo, J.; Yao, W.; Liu, Q.; Liu, Y.; Liu, B. Analysis on the structure and accumulation differences of Hongche fault belt in Junggar Basin. *Geol. Resour.* **2019**, *28*, 57–65.
39. Wang, Z.; Ye, J. Modeling of Pool-Forming Dynamics for Jurassic in Che-Guai Area, Northwest Edge of Junggar Basin. *Geol. Sci. Tech. Inform.* **2010**, *29*, 63–67.
40. Shang, E.; Jin, Z.; Ding, W.; Zhang, Y.; Zeng, J.; Wang, H. Study on physical simulation experiment for the controlling of faults to oil-Taking the Hongche faults in the northwest of Junggar Basin as an example. *Pet. Geol. Exp.* **2005**, *27*, 414–418.
41. Chen, S.; Ran, Y.; Lu, J.; Wu, E. The geochemistry research on the sealing feature of fault in Hongche faults. *J. Southwest Pet. Univ. (Sci. Tech. Ed.)* **2008**, *30*, 21–24. [CrossRef]
42. Ji, J.; Wu, K.; Liu, Y.; Pei, Y.; Li, T. Cementing and sealing actions of Hongche Fault Belt in Zhongguai area of Northwest Margin of Junggar Basin. *Pet. Geol. Oilfield Dev. Daqing* **2019**, *38*, 17–25.
43. Jia, C.; Guan, J.; Liang, Z.; Yao, W.; Shi, J. Reservoir-Forming Conditions and the Main Control Factors Analysis of Triassic System in Chepaizi Prominence, Junggar Basin. *Xinjiang Geol.* **2012**, *30*, 434–437.
44. Zhong, W.; Jiang, Y.; Zhang, S.; Wang, Y.; Zhang, Y.; Wu, K. Sealing Evaluation of Hongche Fault Zone in Cheguai Area, Junggar Basin, Northwest China. *Xinjiang Geol.* **2019**, *37*, 368–372.
45. Xu, Y.; Wang, L.; Liu, Z.; Shi, L. Characteristics of fluid inclusions and time frame of hydrocarbon accumulation for volcanic reservoirs in Chepaizi Uplift. *Fau-Block OilGas Fie* **2020**, *27*, 545–550.
46. Hou, L.; Zou, C.; Liu, L.; Wen, B.; Wu, X.; Wei, Y.; Mao, Z. Geologic essential elements for hydrocarbon accumulation within Carboniferous volcanic weathered crusts in northern Xinjiang, China. *Acta Pet. Sin.* **2012**, *33*, 533–540.
47. Zhao, A.; Wang, Z.; Li, W. Hydrocarbon sources and accumulation modes of Hongche fault zone in the northwestern margin of Junggar Basin. *J. Xian Shiyou Univ. (Nat. Sci. Ed.)* **2015**, *30*, 16–22.
48. Liang, S.; Wu, K.; Huang, Y.; Ji, D.; Fu, X. Weathering Crust Characterization and Its Hydrocarbon Geology Significance in the Northwestern Junggar Basin. *Spec. Oil Gas Reserves* **2018**, *25*, 56–59.
49. Zou, C.; Hou, L.; Tao, S.; Yuan, X.; Zhu, R.; Zhang, X.; Li, F.; Pang, Z. Hydrocarbon accumulation mechanism and structure of large-scale volcanic weathering crust of the Carboniferous in northern Xinjiang, China. *Sci. Chin. Earth Sci.* **2011**, *41*, 1613–1626. [CrossRef]
50. Hou, L.; Luo, X.; Wang, J.; Yang, F.; Zhao, X.; Mao, Z. Weathered volcanic crust and its petroleum geologic significance: A case study of the Carboniferous volcanic crust in northern Xinjiang. *Pet. Explor. Dev.* **2013**, *40*, 257–265. [CrossRef]
51. Hou, L.; Zou, C.; Kuang, L.; Wang, J.; Zhang, G.; Kuang, J.; Liu, L. Discussion on controlling factors for Carboniferous hydrocarbon accumulation in the Ke-Bai fractured zone of the northwestern margin in Junggar Basin. *Acta Pet. Sinica* **2009**, *30*, 513–517.
52. Kuang, L.; Xue, X.; Zou, C.; Hou, L. Oil accumulation and concentration regularity of volcanic lithostratigraphic oil reservoir: A case from upper-plate Carboniferous of KA-BAI fracture zone, Junggar Basin. *Pet. Explor. Dev.* **2007**, *34*, 285–290.
53. Zhang, Z.; Liu, H.; Li, W.; Fei, J.; Xiang, K.; Qin, L.; Xi, W.; Zhu, L. Origin and accumulation process of heavy oil in Chepaizi area of Junggar Basin. *J. Earth Sci. Environ.* **2014**, *36*, 18–32.
54. Shi, X.; Zhang, L.; He, D.; Du, S.; Wang, X.; Zhang, C.; Guan, S.; Yang, G. The reservoir formation model in the northwestern margin of Junggar Basin. *Nat. Gas Geosci.* **2005**, *16*, 460–463.
55. Zhuang, X.M. Petroleum geology features and prospecting targets of Chepaizi Uplift, Junggar Basin. *Xinjiang Geol.* **2009**, *27*, 70–74.
56. Liang, Y.; He, D.; Zhen, Y.; Zhang, L.; Tian, A. Tectono-stratigraphic sequence and basin evolution of Shawan Sag in the Junggar Basin. *Oil Gas Geol.* **2018**, *39*, 943–954.
57. Wu, S.; He, D.; Zheng, M.; Liu, D.; Wu, H. Extensional structural feature and of Shawan sag, Junggar in the stage of tectonic evolution Carboniferous—Permian. *Chin. J. Geol.* **2018**, *53*, 185–206.

58. Zheng, S.; Liu, L.; Wang, Q.; Xu, Z.; Chen, H.; Xiao, Y. Sedimentary facies analysis of the Lower Jurassic Badaowan Formation in Chepaizi Area, Junggar Basin. *J. Northeast Pet. Univ.* **2019**, *43*, 21–34.
59. Liang, Y.; Wu, S.; Lu, Y.; Zheng, M.; Liu, D.; Kong, Y.; Wu, H. The structural model in the transitional zone of Chepaizi uplift and Shawan sag. *Chin. J. Geol.* **2018**, *53*, 155–168.

Publisher's Note: MDPI stays neutral with regard to jurisdictional claims in published maps and institutional affiliations.

Article

Productivity-Index Behavior for a Horizontal Well Intercepted by Multiple Finite-Conductivity Fractures Considering Nonlinear Flow Mechanisms under Steady-State Condition

Maojun Cao [1,2,*], Hong Xiao [1] and Caizhi Wang [2]

[1] School of Computer & Information Technology, Northeast Petroleum University, Daqing 163318, China; xh_daqing@126.com
[2] Department of Well Logging & Remote Sensing Technology, Research Institute of Petroleum Exploration and Development, PetroChina, Beijing 100083, China; wangcaizhi@petrochina.com.cn
* Correspondence: caomaojun@nepu.edu.cn; Tel.: +86-137-9698-8520

Received: 23 March 2020; Accepted: 13 April 2020; Published: 17 April 2020

Abstract: In this paper, a mathematical model is proposed to investigate the effect of nonlinear flow mechanisms on productivity-index (PI) behavior in hydraulically fractured reservoirs during steady-state condition. This approach focuses on the fact that PI approaches a constant value at a certain time, indicating the beginning of steady state. In this model, the reservoirs are considered as an elliptical-shaped drainage with constant-pressure boundary, which is depleted by a multiple-fractured horizontal well (MFHW), and various nonlinear flow mechanisms, such as the non-Darcy flow effect and pressure-dependency effect, control flow patterns in the hydraulic fractures. Then, an exact algorithm of solving the resulting nonlinear equations is developed to obtain the PI of MFHW using a semi-analytical approach. Next, type curves are generated to investigate the influences of flow mechanisms and fracture properties on PI. The most interesting points in this study are the following: (1) PI is determined by the properties of MHFW (i.e., dimensions and configuration), the reservoir geometry, and flow mechanism; (2) PI is deteriorated by non-Darcy flow caused by inertial forces; and (3) PI is reduced under the influence of pressure sensitivity caused by the degradation of dynamic conductivity. Generally, this study provides a significant insight into understanding the factors affecting the productivity of a MFHW with nonlinear flow mechanisms.

Keywords: horizontal well with multiple finite-conductivity fractures; elliptical-shaped drainage; productivity index; non-Darcy flow; pressure-dependent conductivity

---

1. Introduction

With the success of Bakken tight oil resource in the United States, tight oil and gas reservoirs have become a significant source of hydrocarbon supply globally [1]. Subsequently, the technology of horizontal drilling combined with hydraulically fracturing has effectively enabled commercial production from the reservoir with extremely low permeability [2,3]. In practice, the characteristics for a stimulated well are identified by flow regimes; for example, the approach of pressure-derivative is used to identify flow-regimes for pressure transient analysis. For a multi-fractured horizontal well, the sequence of flow-regimes may be more complex (Figure 1) [4]. At some point after compound linear flow caused by the fracture interference, a compound elliptical flow regime (regime 5 in Figure 1a) would appear as the transient pressure waves move past the ends of the fractures [5,6]. Put another way, the compound elliptical response is the pseudo-steady state in an infinite drainage area. It is noted that the isopotential line approximates to an elongated elliptical shape for a long horizontal wellbore, as shown in Figure 1b.

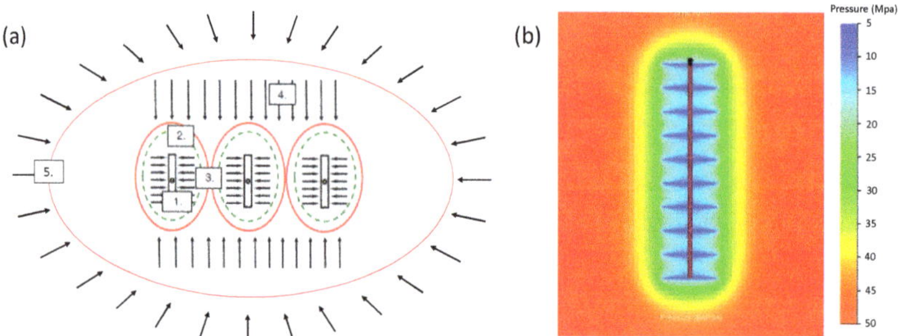

Figure 1. The flow regime characteristics of MHFW, where (**a**) sequence of flow-regimes for a MFHW, completed in a tight sand gas reservoir modified from Clarkson [4]. Lines represent isopotential lines; arrows represent streamlines. Flow-regime 1 corresponds to linear flow; flow-regime 2 corresponds to elliptical flow; flow-regime 3 corresponds to fracture interference; flow-regime 4 corresponds to compound linear, and flow-regime 5 corresponds to compound elliptical flow. (**b**) simulation of flow regimes for a stimulated MFHW in a tight sand gas reservoir. (simulated through numerical simulator).

In physics, the performance of a fractured reservoir is determined by numerous factors such as formation petrophysical properties, fluid properties, and reservoir and wellbore configuration [7]. Numerous methods are provided to capture the physics of fluid flowing in the fractured reservoirs. Originating from the pioneer work suggested by Warren and Root [8], the fractures are represented by a 3D uniformly spaced fracture network, and the matrix blocks are contributing sources. It is a fictitious homogeneous porous medium. As a result, multiple-continuum models are used to compute the transient response of uniform naturally (continuously) fractured reservoirs. In analytical modeling, these models consist of a set of one-dimensional linear flow regions, and the main difference is the way in which these regions are coupled [9–11]. Therefore, it is difficult for the multiple-continuum model to capture the intrinsic behavior of complexity of nonuniform naturally fractured reservoirs, such as hydraulically fractured reservoirs. To capture the strong heterogeneity between fracture and matrix, various improved techniques are presented to subdivise the grids, such as the embedded discrete fracture model (EDFM), Galerkin finite-element method (FEM), and discrete-element method (DEM) [12–14]. However, it is still a huge challenge to obtain accurate transmissibility by computing the irregular connections between fracture segments. Considering the high computational burden and the great difficulties in gridding, the mesh-free semi-analytical method is necessarily developed as an alternative to the numerical simulation method. On the basis of the works suggested by Cinco-Ley et al. [15], various semi-analytical models, which have the flexibility of accounting for the individual fracture properties, are proposed to investigate the influence factors on the performance of the fractured reservoirs [16–19]. It is proven that the semi-analytical methods have huge potential in analyzing transient performance response and evaluating productivity of wells.

From the viewpoint of petroleum engineering, the effects of these parameters on reservoir performance can be represented by the productivity index (PI). It is defined as the ratio of production rate to a certain pressure drop. PI is determined by the flow regime, with different behavior characteristics with time. It starts out with a high value at an early time when the transient response is dominant. PI declines with time to a small value, and approximates to a constant value, known as stabilized PI, at a late time period when the compound elliptical response is reached [20]. Currently, most studies have put the focus on the transient productivity index to identify the flow regimes [21–23]. However, they found that the designed fractured wells do not usually result in the expected performance, which is attributed to the existence of nonlinear flows. From the viewpoint of multi-physics, the flowing in the underground porous media is the result of fluid–thermal–solid coupling [24]. The reservoir production is generally regarded as the isothermal process; therefore, the thermal is not considered in the production simulation. Strictly speaking, the rock deformation may have a significant effect on fluid flow owing to fluid-particle interactions. According to the conclusions presented by Zhang and Tahmasebi [25,26], the effect of solid deformation on fluid flow, especially in hydraulic fractures, could be incorporated by regarding the porosity and permeability as a function of stress changes. In other words, the conductivity of hydraulic fracture changes because of a variety of mechanisms, including fracture closure under high effective stress and permeability loss as a result of proppant embedment and crushing. Besides, non-Darcy flow may occur in the fracture when the well is produced at a high flow rate [27]. Therefore, the nonlinear flow effects have to be taken into account to investigate the effect of nonlinear behaviors on the performance of a vertical well and fractured well in the infinite-acting reservoir. Limited efforts have been made to examine the effects of non-Darcy flow and pressure-sensitive conductivity on the performance of horizontal wells with multiple fractures. Wu et al. [28] used a numerical approach to investigate the non-Darcy effect in an unconventional gas reservoir. Lin and Zhu [29] developed a slab source method to evaluate the performance of horizontal wells with or without fractures using an analytical approach. Besides the non-Darcy effect, pressure sensitivity in the fracture may also exert an important influence on the transient response. A transient pressure analysis for a vertical well intercepting a dynamic-conductivity fracture has been conducted using analytical and numerical methods. Pedroso and Correa [30] developed a new model to investigate the effects of pressure-sensitive permeability on the flow behavior of a fractured well. Zhang et al. [31] presented the results of a study that analyzed the build-up of pressure in a vertically fractured well with a stress-sensitive conductivity in the fracture. Yao et al. [32] studies the effect of dynamic conductivity on the transient performance of multiple fractured horizontal well.

To the best of our knowledge, few papers are published to study the productivity index of a multiple-fractured horizontal well (MFHW) in a finite drainage area with an elliptical boundary under nonlinear flow conditions. The objective of this work is to quantify the effect of non-Darcy flow and pressure sensitivity on the inflow performance of a MFHW. First, we establish a hybrid model that describes Darcy flow in the reservoirs toward hydraulic fractures, as well as non-Darcy flow along the fracture with pressure-dependent conductivity under steady-state conditions. First, a mathematical model is established to describe the fluid flow depleted by a MFHW in a reservoir with elliptical-shaped boundary, with the flexibility of accounting for fracture properties (i.e., fracture dimensions and configuration). Next, nonlinear flow mechanisms in the fracture, including stress-sensitivity effect and non-Darcy flow condition, are taken into account, which results in the nonlinearity in partial differential equation. Then, the technique of dimension transformation is used to render the resulting nonlinear equations amenable to linear treatment, and an effective and efficient algorithm is developed to solve the model. Finally, a study for sensitivity analysis is introduced to investigate the effect of fracture properties and nonlinear flow mechanisms on the productivity-index behavior.

2. Model Assumption

2.1. Physical Model

In this physical model, a horizontal well is located in the center of a homogeneous reservoir with an elongated elliptical boundary. The whole drainage area consists of several different flow patterns in four contiguous flow regions, as sketched in Figure 2, including (1) the hydraulic fractures, (2) the inner stimulated rock volume (SRV) region impacted by hydraulic fracturing, (3) the outer linear region beyond the extent of the SRV, and (4) the outer elliptical region beyond the tips of the horizontal wellbore.

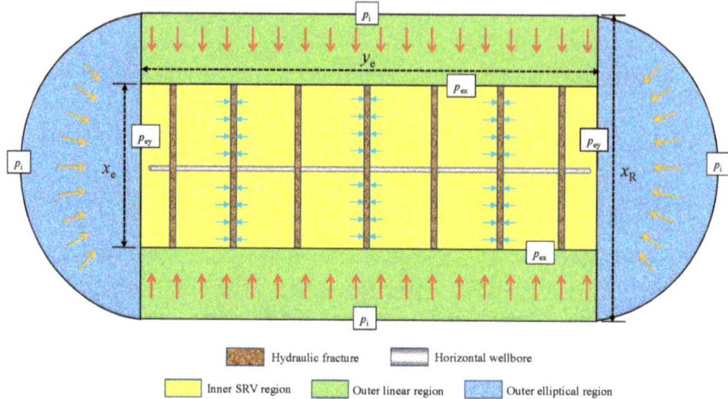

Figure 2. Schematic of the compound elliptical-flow model representing four contiguous regions for a multiple-fractured horizontal well (MFHW). SRV, stimulated rock volume.

Different from the anisotropic characteristics in a naturally fractured reservoir, where a large number of nature fractures are uniformly and globally distributed [33–35], this paper assumes that the reservoir is stimulated by several hydraulic fractures, so the resulting hydraulically fractured reservoir is regarded as isotropic. In addition, there are some necessary assumptions, as follows:

- The reservoir is horizontal and of uniform thickness, with impermeable lower and upper boundaries. The pressure on the outer boundary in the x–y plane keeps constant.
- Along the horizontal wellbore, multiple hydraulic fractures are evenly distributed. Hydraulic fractures are considered to be a finite conductivity.
- The wellbore is produced under constant-rate condition.
- Flow in the reservoir is assumed to be single-phase fluid (i.e., pure gas) with slight compressibility and constant viscosity.
- Fluid flowing in the fracture is assumed to be nonlinear.
- Fluids from the reservoir enter the wellbore only through hydraulic fractures, and the pressure loss in the wellbore is neglected.

2.2. Variables Definitions

To introduce the mathematical model conveniently and concisely, the dimensionless variables are defined, respectively, by

$$p_D = \frac{2\pi k_m h(p_i - p)}{q\mu B}, \ \xi_D = \frac{\xi}{L_{ref}}, \ q_D = \frac{q}{q_{ref}}, \ C_{fD} = \frac{k_f w_f}{k_m L_{ref}} \tag{1}$$

When the flow in the fracture satisfies non-Darcy effect owing to high velocity, the corresponding dimensionless variables are defined as

$$(q_{DND})_f = \frac{k_f \beta}{\mu w_f h} q_{ref}, \ \text{and} \ q_{cD} = \frac{q_c}{q_{ref}}, \tag{2}$$

When the fracture conductivity behaves pressure sensitivity, the corresponding dimensionless variables are defined as

$$\gamma_{fD} = \frac{q\mu B}{2\pi k_m h} \gamma_f \tag{3}$$

where p_m is the pressure in the reservoir, Pa; p_f is the pressure in the fracture, Pa; p_i is the pressure on the elliptical boundary, Pa; p_{ex} is the pressure on the width of inner SRV region, Pa; p_{ey} is the pressure on the length of inner SRV region, Pa; (x_{ofm}, y_{ofm}) is the location of m-th fracture tip, m; x_e is the width of inner SRV region, m; y_e is the length of inner SRV region, m; x_R is the width of the whole drainage region, m³; h is the reservoir thickness, m; w_f is the fracture width, m; L_f is the fracture length, m; r_w is the radius of wellbore, m; L_{ref} is the reference length, m; q_{fm} is the function of inflow distribution along the m-th fracture, m²/s; q_{wm} is the production rate, m³/s; q_c is the influx rate, m³/s; μ is the viscosity, Pa s; B is the volume factor; γ_f is the permeability modulus, Pa⁻¹; and β is the Beta factor, m⁻¹. Here, the subscript D represents dimensionless variables.

3. Mathematical Model

3.1. Fluid Flow in the Inner SRV Region

The SRV region is regarded as a rectangular reservoir with constant-pressure boundary containing a set of hydraulic fractures, as shown in Figure 3. The diffusion equation governing fluid flow is given as the following dimensionless form:

$$\frac{\partial^2 p_{mD}}{\partial x_D^2} + \frac{\partial^2 p_{mD}}{\partial y_D^2} + 2\pi \sum_{m=1}^{N_f} \int_0^{L_{fDm}} q_{fDm}(u_D)\delta(x_D - x_{ofDm} - u_D)\delta(y_D - y_{ofDm})du_D = 0 \tag{4}$$

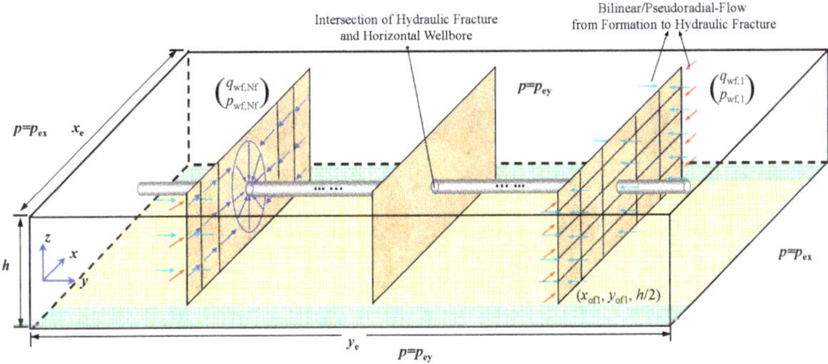

Figure 3. Schematic of MFHW in the inner region with constant-pressure outer boundary.

The outer boundary conditions are satisfied by

$$\begin{cases} p_{mD}(x_D = x_{eD}, y_D) = p_{mD}(x_D = 0, y_D) = p_{eDx} \\ p_{mD}(x_D, y_D = 0) = p_{mD}(x_D, y_D = y_{eD}) = p_{eDy} \end{cases} \tag{5}$$

where there are N_f fractures in the SRV region, the dimensionless length of m-th fracture is L_{fDm}, and the dimensionless coordinate of m-th fracture tip is (x_{ofDm}, y_{ofDm}). The pressure on the outer boundary parallel to the fracture is p_{eDx}, and the pressure on the outer boundary perpendicular to the fracture is p_{eDy}.

Using Fourier transformation and inverse transformation (Appendix A), the dimensionless pressure distribution caused by N_f fracture is given by

$$p_{mD}(x_D, y_D) = p_{eDx}IB(x_D, y_D) + p_{eDy}ID(x_D, y_D) + 2\sum_{m=1}^{N_f} \int_0^{L_{fDm}} q_{fDm}(u_D)\delta p_{uD}(x_D, y_D, u_D)du_D \tag{6}$$

3.2. Fluid Flow in the Outer Region

The fluids stored in the outer region are depleted and then flow into the inner SRV region through the interface (denoted by the yellow block in Figure 4). The streamlines satisfy the hyperbolic shape (orange solid line in Figure 4), which are the counterpart of elliptical-shaped isopotential line (black dashed line). Once the fluids enter the inner region, the streamlines are in a linear shape around the interface. Therefore, the coordinate systems are divided into two subsets: 1D elliptical coordinate in the outer region (blue region), and 1D linear coordinate (yellow region).

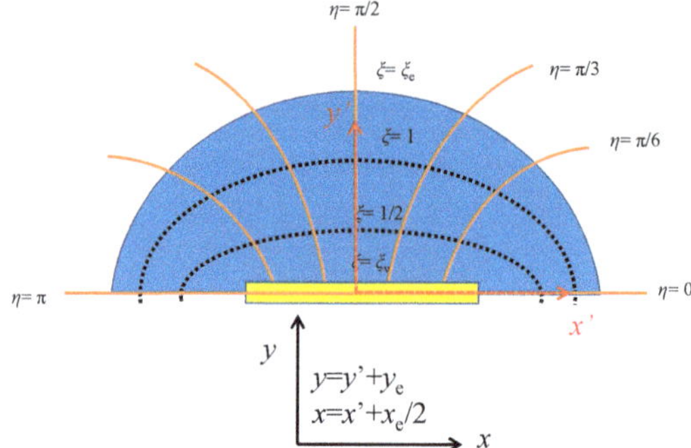

Figure 4. Elliptical coordinate used in the outer elliptical region.

The outer boundary satisfies the constant pressure of p_{eD}.

In the outer elliptic region, as shown in Figure 4, there are two sets of coordinates: the Cartesian coordinates (x',y') and the elliptic coordinates (ξ,η). The relationship is given by

$$x' = (x_e/2)\cosh\xi\cos\eta, \quad y' = (x_e/2)\sinh\xi\sin\eta \tag{7}$$

We used the elliptic coordinates to simulate the flowing process. The outer and inner boundaries of the elliptic region, represented by ξ in the elliptic coordinates, could be expressed using spatial variables in the Cartesian coordinates:

$$\begin{cases} \xi_e = \xi(x' = x_R/2, y' = 0) = \ln[(x_R + \sqrt{x_R^2 - x_e^2})/x_e] \\ \xi_w = \xi(x' = 0, y' = 0) = \ln 1 = 0 \end{cases} \tag{8}$$

The corresponding pressure gradient is written as the following dimensionless expression,

$$\left(\frac{\partial p_{mD}}{\partial \xi_D}\right)_{\xi_D=0} = \frac{p_i - p_{ey}}{\xi_e - \xi_w} = -\frac{p_{eyD}}{\ln[(x_{RD} + \sqrt{x_{RD}^2 - x_{eD}^2})/x_{eD}]} \tag{9}$$

Note that the flux inflow in Equation (9) is continuous on the interface between the outer elliptical region and inner SRV region. Meanwhile, the flowing is modeled using linear coordinates. As a result, the continuous condition is written in the form of the average integral, which is given by

$$\underbrace{\frac{1}{x_{eD}}\int_0^\pi \left.\frac{\partial p_{mD}}{\partial \xi_D}\right|_{\xi_D=0} d\eta_D}_{elliptic\ flow} = \underbrace{\frac{1}{x_{eD}}\int_0^{x_{eD}} \left.\frac{\partial p_{mD}}{\partial y_D}\right|_{y_D=y_{eD}} dx_D}_{linear\ flow} \tag{10}$$

It could be expressed as the function of p_{eDx} and p_{eDy} (Appendix B), which is written as follows:

$$A_2 p_{eDx} + B_2 p_{eyD} = \sum_{m=1}^{N_f} q_{wDm} C_{2,m} \tag{11}$$

Likewise, the continuity condition of flux inflow between outer linear region and inner SRV region contributes to

$$A_1 p_{eDx} + B_1 p_{eyD} = \sum_{m=1}^{N_f} q_{wDm} C_{1,m} \tag{12}$$

We can obtain the dimensionless pressure on the interface by combing Equations (11) and (12),

$$\begin{cases} p_{eDx} = \sum\limits_{m=1}^{N_f} q_{wDm}\left(\dfrac{B_2 C_{1,m} - B_1 C_{2,m}}{A_1 B_2 - A_2 B_1}\right) \\ p_{eDy} = \sum\limits_{m=1}^{N_f} q_{wDm}\left(\dfrac{A_1 C_{2,m} - A_2 C_{1,m}}{A_1 B_2 - A_2 B_1}\right) \end{cases} \tag{13}$$

where the relationship between influx distribution and production rate of m-th fracture is given as.

Substituting Equation (13) into Equation (6) could yield the pressure distribution caused by N_f fractures, which is the function with regard to influx distribution along the fracture.

3.3. Fluid Flow in the Fracture

For hydraulic fracture, the flow pattern is widely assumed to be 1D linear and incompressible [36]. As shown in Figure 5, the governing equation of fluid flow on the m-th fracture is described as the following dimensionless form:

$$\frac{\partial^2 p_{fDn}}{\partial x_{Dm}^2} - \frac{2\pi}{C_{fDm}} q_{fDm}(x_{Dm}) + \frac{2\pi}{C_{fDm}}[q_{wfDm}\delta(x_{Dm} - x_{wDm})] = 0 \tag{14}$$

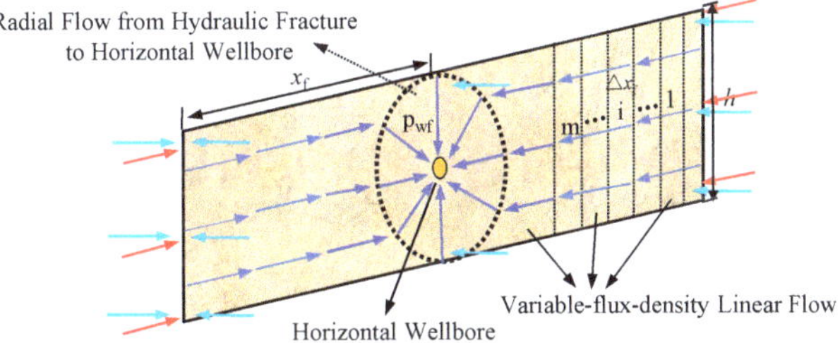

Figure 5. Schematic of the flow pattern in the transverse fracture.

The corresponding boundary conditions are described as

$$\left.\frac{\partial p_{fDm}}{\partial x_{Dm}}\right|_{x_{Dm}=0} = \left.\frac{\partial p_{fD}}{\partial x_{Dm}}\right|_{x_{Dm}=L_{fDm}} = 0 \tag{15}$$

Integrating Equation (14) from 0 to x_{Dm} with respect to x_{Dm} based on Equation (15) twice, the result is given by

$$p_{fDm}(x_{Dm}) = p_{fDm}(0) + \frac{2\pi}{C_{fDm}}\int_0^{x_{Dm}} d\zeta \int_0^{\zeta} q_{fDm}(\xi)d\xi - \frac{2\pi}{C_{fDm}} q_{wDm} H(x_{Dm}, x_{wDm}) \tag{16}$$

where $\delta(x, x_0) = \begin{cases} \infty, & x = x_0 \\ 0, & x \neq x_0 \end{cases}$ is the Dirac function, $H(x, x_0) = \begin{cases} 1, & x \geq x_0 \\ 0, & x < x_0 \end{cases}$ is the Heaviside function,

and $G(x, x_0) = \begin{cases} x - x_0, & x \geq x_0 \\ 0, & x < x_0 \end{cases}$ is the integral of Heaviside function.

As shown in Figure 5, the fluid flow is regarded as a 1D linear pattern in the region far from wellbore, and becomes a 1D radial pattern around the wellbore. It is different from the vertical fracture, where the vertical fracture is in lateral contact with the wellbore and only linear flow is considered. The additional pressure loss owing to the convergence of fluids into the horizontal wellbore should be considered. The effect of flow convergence is regarded as a skin factor, namely convergence flow skin:

$$S_{c,m} = \frac{h_D}{L_{fDm}} \frac{1}{C_{fDm}} \left[\ln\left(\frac{h_D}{2r_{wD}}\right) - \frac{\pi}{2} \right] \tag{17}$$

Equation (16) is rewritten as the final expression by

$$p_{fDm}(x_{Dm}) = p_{fDm}(0) + q_{wDm}S_{c,m} + \frac{2\pi}{C_{fDm}} \int_0^{x_{Dm}} d\zeta \int_0^\zeta q_{fDm}(\xi)d\xi - \frac{2\pi}{C_{fDm}} q_{wDm}G(x_{Dm}, x_{wDm}) \tag{18}$$

3.3.1. Model of a Conductivity Fracture with Non-Darcy Flow

Non-Darcy flow may occur in the fracture when the inertial forces may no longer be neglected compared with viscous forces. It is very common near production wells where local velocities can be very high. Extra pressure drop is required to overcome the inertial forces.

To account for the effect of extra pressure loss caused by high-velocity flow, which is proportional to the square of the velocity, the Forchheimer flow equation is presented [27]:

$$\nabla p = \frac{\mu}{k} v + \beta v |v| \tag{19}$$

The apparent fracture permeability in Equation (19) is defined as

$$k_{f,app} = k_f \frac{1}{1 + k_f \beta(|v|/\mu)} \tag{20}$$

The corresponding apparent permeability is rewritten by

$$k_{f,app} = k_f \frac{1}{1 + (q_{DND})_f |q_{cD}(x_D)|} \tag{21}$$

Substituting Equation (21) into Equation (18) yields the following form:

$$C_{fD,app}(q_{cDm}) \frac{\partial p_{fDm}}{\partial x_{Dm}} + 2\pi q_{cDm} + 2\pi q_{wDm} H(x_{Dm}, x_{wDm}) = 0 \tag{22}$$

where the apparent conductivity is written as $C_{fD,app}(q_{cDm}) = \frac{C_{fDm}(x_D)}{1 + (q_{DND})_f |q_{cDm}|}$, and the flux on the cross section is given as $q_{cDm} = -\frac{C_{fDm}}{2\pi} \frac{\partial p_{fDm}}{\partial x_{Dm}}$. The corresponding integral form is given by

$$q_{cDm} = q_{wDm} H(x_{Dm}, x_{wDm}) - \int_0^{x_{Dm}} q_{fDm}(x')dx' \tag{23}$$

Note that Equation (22) presents a general model for a uniform-conductivity under the non-Darcy-flow condition.

3.3.2. Model of a Conductivity Fracture with Pressure Sensitivity Effect

The variation of permeability caused by a stress change can be expressed as a function of pore pressure. For hydraulic fracture, a general model presented by Zhang et al. [31] is introduced to describe the relationship between fracture permeability and pore pressure next:

$$k_f(p_f) = k_{min} + [k_f(p_i) - k_{fmin}] \cdot \exp[-\gamma_f \cdot (p_i - p_f)] \tag{24}$$

According to the definitions of the dimensionless parameters, the pressure-sensitive conductivity of a fracture is

$$\frac{C_{fD}(p_{fD})}{C_{fDi}} = \left(1 - \frac{C_{fDmin}}{C_{fDi}}\right) \cdot \exp[-\gamma_{fD} \cdot p_{fD}] + \frac{C_{fDmin}}{C_{fDi}} \tag{25}$$

Substituting Equation (25) into Equation (14) yields the following equation governing flow in the m-th fracture flowing:

$$\frac{\partial}{\partial x_{Dm}}\left(C_{fDm}(p_{fDm})\frac{\partial p_{fDm}}{\partial x_{Dm}}\right) - 2\pi q_{fDm}(x_{Dm}) + 2\pi q_{wDm}\delta(x_{Dm}, x_{wDm}) = 0 \tag{26}$$

Note that Equation (26) presents a general model for a uniform-conductivity with the pressure sensitivity effect.

4. Semi-Analytical Solution for Coupled Model

4.1. Dimension Transformation

Compared with the equation of uniform conductivity fracture described by Equations (14), (22) and (26) are subject to nonlinear equations. In the case of considering non-Darcy flow, the fracture conductivity is a function of flowing rate; in the case of considering pressure-sensitivity effect, the fracture conductivity is a function of pressure. In essence, both flow rate and pressure chang with the spatial variable. Here, we attempt to solve the problem in the same way as the linear treatment in Equation (14) by a dimension transformation, which is defined as

$$\begin{cases} \xi_{Dm}(x_{Dm}) = \hat{C}_{fDm} \cdot \int_0^{x_{Dm}} \frac{1}{C_{fDm}(x'_D)} dx'_D \\ \hat{C}_{fDm} = L_{fDm} / \int_0^{L_{fDm}} \frac{1}{C_{fDn}(x'_D)} dx'_D \end{cases} \tag{27}$$

Then, the fracture-flow equations (Equations (22) and (26)) can be written as

$$p_{fDm}(\xi_{Dm}) = p_{fDm}(\xi_{Dm} = 0) + q_{wDm}S_{c,m} + \frac{2\pi}{\hat{C}_{fDm}}\left(\int_0^{\xi_{Dm}} d\zeta \int_0^\zeta \tilde{q}_{fDm}(\varsigma)d\varsigma - q_{wDm}G(\xi_{Dm}, \xi_{wDm})\right) \tag{28}$$

Equation (28) has the same form as the fracture-flow equations for a uniform-conductivity fracture with constant conductivity given by Equation (18). The difference is that the variable ξ_D is a function of the variable x_D and depends on the distribution of conductivity $C_{fD}(x_D)$.

4.2. Discretization

In this study, the productivity index is achieved by discretizing the fracture panel into M segments with equal length Δx_{Di}, as shown in Figure 6a. Note that the equal-length fracture segment would be transformed into an unequal-length segment after using dimension transformation, as shown in Figure 6b.

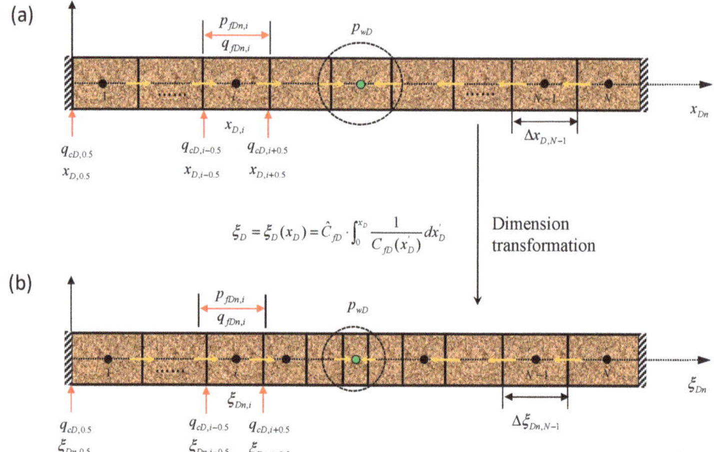

Figure 6. Illustration of dimension transformation and fracture discretization.

In the reservoirs, the dimensionless pressure drop on the j-th segment of the n-th fracture caused by the i-th segment of the m-th fracture is given by

$$p_{mDn,j} = \sum_{m=1}^{N_f} q_{wDm} BD_m^{n,j} + \sum_{m=1}^{N_f} \sum_{i=1}^{M_m} q_{fDm,i} \Delta p_{uDm,i}^{n,j} \frac{\Delta \xi_{Dm,i}}{\Delta x_{Dm,i}} \tag{29}$$

where $BD_m^{n,j} = IB_{n,j}\left(\frac{B_2 C_{1,m} - B_1 C_{2,m}}{A_1 B_2 - A_2 B_1}\right) + ID_{n,j}\left(\frac{A_1 C_{2,m} - A_2 C_{1,m}}{A_1 B_2 - A_2 B_1}\right)$, and $\Delta p_{uDm,i}^{n,j} = \int_{x_{ofDm,i-1}}^{x_{ofDm,i}} \delta p_{uD}(u) du$.

In the fracture, the dimensionless pressure on the j-th segment of the n-th fracture is expressed as

$$p_{wDn} - p_{fDn,j} = \frac{2\pi}{\hat{C}_{fDn}} q_{wDn} G(\xi_{Dn,j}, \xi_{wDn}) - [I(\xi_{Dn,j}) - I(\xi_{wD})] - q_{wDn} S_{c,n} \tag{30}$$

where $I(\xi_D) = \int_0^{\xi_D} d\zeta \int_0^{\zeta} \widetilde{q}_{fDn}(\varsigma) d\varsigma$.

According to the continuity condition that the pressure must be continuous along the fracture surface, the following conditions must hold along the fracture plane:

$$p_{mD}(x_{ofDn} + x_{Dn,j}, y_{ofDn}) = p_{fD}(x_{Dn,j}) \tag{31}$$

Moreover, the sum of all the segments' flow rate equals to the total flow rate at the m-th fracture:

$$q_{wfDm} = \sum_{i=1}^{M_m} \left(\widetilde{q}_{fDm,i} \frac{\Delta \xi_{Dm,i}}{\Delta x_{Dm,i}}\right) \tag{32}$$

Combining Equations (29)–(32), we can obtain the fundamental coupled solutions by computing the apparent linear equations [36,37].

4.3. Computation Consideration

According to the previous statement, the dimensionless conductivity of C_{fDn} is the function of pressure p_{fDn} in the fracture with pressure-dependent conductivity, and the dimensionless conductivity of C_{fDn} is the function of flowing rate q_{cDn} in the fracture under non-Darcy flow. In a mathematical context, C_{fDn} is a function with regard to the spatial variable. At a given k-th step, if the distribution of pressure p_{fDn} or flowing rate q_{cDn} was known, $\hat{C}_{fDn}$ along the fracture would be obtained. Thus, the nonlinear governing equation for the $(k + 1)$-th step could be linearized on the assumption of known conductivity distribution on the k-th step.

In the case of considering the non-Darcy flow effect (Case 1), the formation for the k-th step yields

$$\frac{C_{fDn}}{1 + (q_{DND})_{fn} q_{cDn}^{\langle k \rangle}} \frac{\partial p_{fDn}^{\langle k+1 \rangle}}{\partial x_{Dn}} + 2\pi q_{cDn}^{\langle k+1 \rangle}(x_{Dn}) + 2\pi q_{wDn}^{\langle k+1 \rangle} H(x_{Dn}, x_{wDn}) = 0 \tag{33}$$

In the case of considering the pressure-dependent conductivity effect (Case 2), the formation for the k-th step yields

$$C_{fDn}(p_{fDn}^{\langle k \rangle}, x_D) \frac{\partial p_{fDn}^{\langle k+1 \rangle}}{\partial x_{Dn}} + 2\pi \left[q_{wDn}^{\langle k+1 \rangle} H(x_{Dm}, x_{wDm}) - \int_0^{x_{Dm}} q_{fDm}^{\langle k+1 \rangle}(x')dx' \right] + 2\pi q_{wDn}^{\langle k+1 \rangle} H(x_{Dn}, x_{wDn}) = 0 \tag{34}$$

The flow distribution q_{fDn} and pressure distribution p_{fDn} along the fracture at the k-th time step would be achieved based on coupled solution. Next, the calculated p_{fDn} was used to update the fracture conductivity (Case 1); the calculated q_{cDn} was used to update the fracture conductivity (Case 2). The iterative procedure is repeated until the wellbore-pressure was converged. The calculation procedure is given as follows:

- Step 1: Initial calculation, with $k = 0$, the fracture is assumed to be uniform (Case 1 and Case 2). By combing through Equation (29) to Equation (32), we can obtain $(q_{fDn})^k$. The fracture pressure $(p_{fDn})^k$ (Case 1) and flowing rate $(q_{fDn})^k$ (Case 2) would be then achieved from Equation (28).
- Step 2: Calculating the pressure-sensitive conductivity $C_{fDn}[(p_{fDn})^k]$ (Case 1) and $C_{fDn}[(q_{cDn})^k]$ (Case 2) along fracture, and then transforming x_{Dn} into ξ_{Dn} based on Equation (26).
- Step 3: Solving Equation (33) with the updated $C_{fDn}[(p_{fDn})^k]$ to achieve $(p_{fDn})^{k+1}$ and $(p_{wD})^{k+1}$ in Case 1; solving Equation (34) with the updated $C_{fDn}[(q_{cDn})^k]$ to achieve $(q_{cDn})^{k+1}$ and $(p_{wD})^{k+1}$ in Case 2.
- Step 4: Terminate the iterative procedure if $|(p_{wD})^{k+1} - (p_{wD})^k|/(p_{wD})^k < \varepsilon$; otherwise, update pressure distribution along fracture by setting $(p_{fDn})k = (p_{fDn})^{k+1}$ (Case 1) and flow distribution along fracture by setting $(q_{cDn})^k = (q_{cDn})^{k+1}$ (Case 2), and $k = k + 1$ back to step 2 until the convergence is achieved.

5. Results and Sensitivity Analysis

The productivity index (PI) is defined as the ratio of the production rate to pressure drawdown, as follows:

$$J_D = \frac{\mu B}{2\pi k_m h} \frac{q}{p_i - p_w} = \frac{1}{p_{wD}} = \left(\frac{1}{2} \ln \frac{4A}{e^\gamma C_A r_{we}^2} \right)^{-1} \tag{35}$$

Put another way, the dimensionless PI is the reciprocal of flow resistivity. Here, A is the drainage area, γ is the Eular constant, and C_A is the shape factor. Specially, r_{we} is the effective wellbore radius determined by the geometry of drainage area, well configuration, and nonlinear flow mechanism.

5.1. Influence of Fracture Properties on PI

In this subsection, the nonlinear flow mechanisms are firstly not taken into account, and the focus is put on the impacts of the dimensions and configuration of MFHW on PI behavior. For the purpose of discussion, homogeneous fracture properties and MFHW configuration are assumed; the fractures are evenly spaced along horizontal wellbore, and all fractures are of identical properties. As a result, the PI is mainly determined by five factors, including dimensionless conductivity (C_{fD}), fracture number (N_f), penetration ratio of fracture length to inner SRV width ($I_x = L_f/x_e$), penetration ratio of fractured horizontal wellbore to inner SRV length ($I_y = D_f/y_e$) (D_f is defined as the distance between two outmost fractures), and the drainage ratio of inner drainage to the whole drainage ($I_e = x_e/x_R$).

Figure 7 shows the effect of dimensionless conductivity on PI in the case of different penetration ratios. We further assume that the fracture is symmetrical (i.e., the fracture length on the right hand side equals to the left hand side with regard to wellbore), and all fractures are located in the center of the drainage. The fracture conductivity changes from 0.1 to 1000 (i.e., C_{fD} = 0.1, 0.5, 1, 5, 10, 50, 100, 500, and 1000). As shown in Figure 7a, at a given penetration ratio (I_x), PI is increased with the increase in dimensionless conductivity. However, after a certain dimensionless conductivity, the increase could be miniscule. In other words, beyond a certain dimensionless conductivity, the PI increase with conductivity essentially equals the PI decrease caused by the inner interference within the fracture. Generally, the threshold value is regarded as $C_{fD,threshold}$ = 300. When the C_{fD} is larger than 300, there is no significant pressure drop within the fracture; the PI reaches the maximum, and does not increase with the conductivity. Figure 7b shows the PI derivative with regard to $\ln(C_{fD})$ in the semi-log plot. For each given I_x, the conductivity, corresponding to the maximum PI derivative, is defined as certain conductivity. Taking I_x = 1, for example (J_{Dmax} = 7.25), when the derivative is at a maximum, and that the certain value is $C_{fD,certain}$ = 7.5, the corresponding value of PI is J_D = 4.54. This indicates that the ratio of J_D/J_{Dmax} equals 62.6. This means that PI of MFHW at C_{fD} = 7.5 could reach 62.6% of the maximum, but PI only increases by 37.4% when the C_{fD} is further increased from 7.5 to 300. Therefore, the optimum dimensionless conductivity is defined to as the certain value. The optimum conductivity indicates that the inflow from the reservoir could match the outflow of the fracture. As analyzed from Figure 7b, the larger the I_x, the larger the optimum C_{fD}.

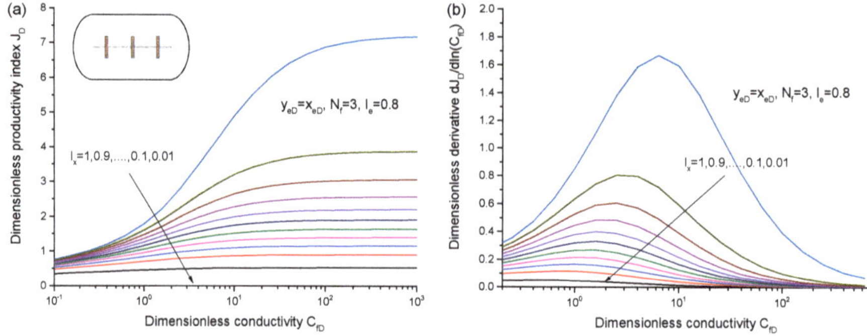

Figure 7. Productivity-index behavior of multiple fractures, including (**a**) the productivity index vs. dimensionless conductivity, and (**b**) the derivative of productivity index vs. dimensionless conductivity.

Figure 8 shows the effect of the number of fractures on PI behavior. As expected, the cases with a larger number exhibit a higher value of PI. That is, increasing the number contributes to an increased connected area of MHFW with the reservoirs, and an increase in the fracture–fracture interference. However, as the number of fractures increases beyond a certain number, the increase of PI could be miniscule. For example, PI increases by 8.97% when the number is increased from N_f = 2 to N_f = 3. In comparison, incremental PI would decline to 0.39% when the fracture number is further increased from N_f = 9 to N_f = 10. Overall, the effect of the increased connected area could offset the effect of the increase of fracture interference, which results from the increased fracture number. Besides, the optimum conductivity is decreased with the increase in fracture number, as shown in Figure 8b.

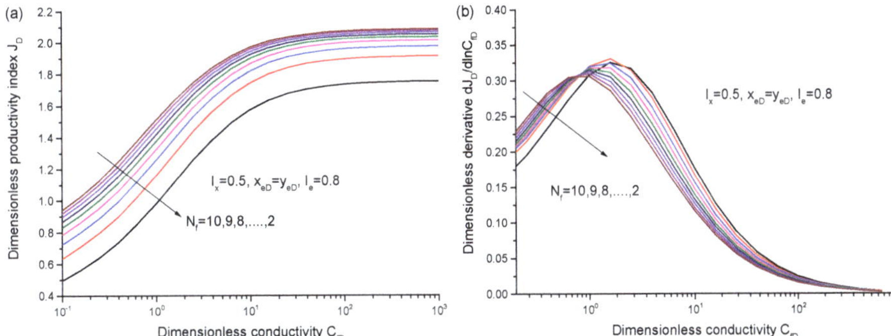

Figure 8. Influence of number of fractures on PI behavior, including (**a**) the productivity index vs. dimensionless conductivity, and (**b**) the derivative of productivity index vs. dimensionless conductivity.

Figure 9 presents the effect of the ratio of inner SRV area to the whole area, i.e., I_e. Here the value that $I_e = 1$ means that the area of inner SRV equals to the whole drainage area. Compared with the PI that $I_e = 1$, the more approaching unity the value is, the larger the PI is, which is caused by the decrease of the proportion of inner SRV. Moreover, the tendency is more noticeable in the condition of high conductivity. When $C_{fD} = 0.1$, the PI at $I_e = 0.5$ is that $J_{D@Ie=0.5} = 0.45$, while the PI at $I_e = 1$ is that $J_{D@Ie=1} = 0.65$. Hence, the ratio of $I_e = 1$ to $I_e = 0.5$ in PI value is 1.44. As a comparison, when $C_{fD} = 1000$, the PI at $I_e = 0.5$ is that $J_{D@Ie=0.5} = 0.72$, while the PI at $I_e = 1$ is that $J_{D@Ie=1} = 1.95$. The ratio in PI is 2.7.

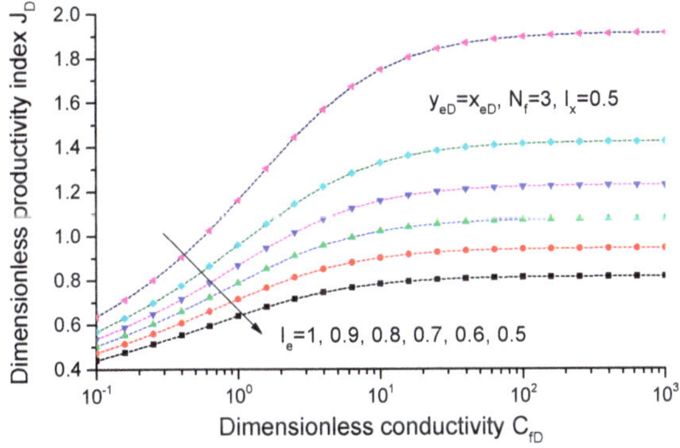

Figure 9. Influence of penetration ratio of inner SRV region with regard to the whole region on productivity-index (PI) behavior.

5.2. Influence of Non-Darcy Flow on PI

In this subsection, the non-Darcy flow mechanism is taken into account, and we investigate the effect of non-Darcy flow on PI behavior. Figure 10 shows the flow distribution along the fracture, where $C_{fD} = 1$. In this figure, the conductivity is set to be relatively low. It is shown that the influx density, which is defined as the influx rate per length along fracture face, is concentrated around the wellbore and fracture tips. It is noted that the concentration region of the influx density is controlled by a larger pressure drop/depletion. The variable of $(q_{DND})_f$, defined in Equation (2), is the Forchheimer number for a given inertial factor β. When the $(q_{DND})_f$ is increased, the influx density is increasingly more concentrated towards wellbore. Put another way, an extra pressure drop is required to offset the effect of non-Darcy flow caused by the inertial force. Figure 11 presents the effect of fracture conductivity on influx-density distribution along the fracture. Without the non-Darcy effect (Forchheimer number equals zero) shown in Figure 11a, the increase in conductivity makes the distribution of fluid influx more gentle. When $C_{fD} > 300$ (i.e., infinite conductivity), which indicates that the pressure on any point in fracture panel is same as the pressure on the wellbore, the inflow flux is concentrated on the fracture tips. With the non-Darcy effect (Figure 11b), when $C_{fD} > 1$, more volume of influx fluid is concentrated on the nearby region of wellbore. The phenomenon is consistent with the characteristics given in Figure 11a. As the conductivity increases, the effect of non-Darcy flow on the distribution of inflow flux becomes weak, until the distribution of inflow flux is independent of the non-Darcy flow effect.

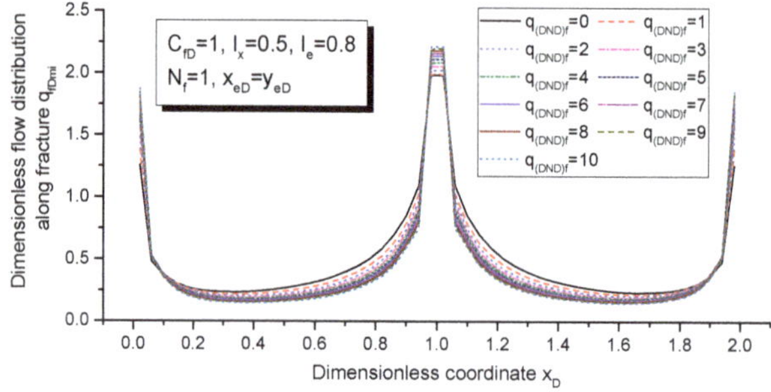

Figure 10. Influx-flow distribution along the fracture under the influence of non-Darcy flow.

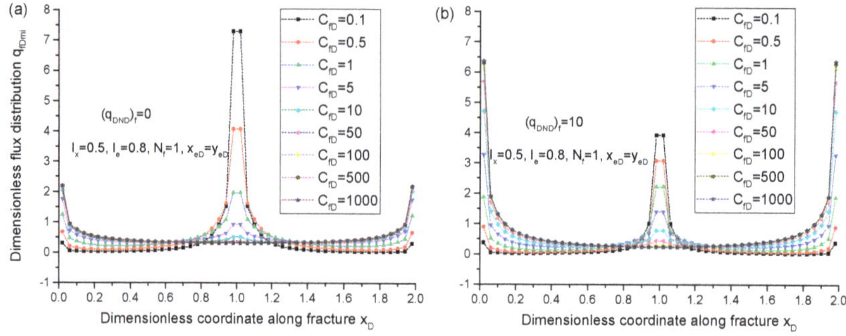

Figure 11. Influx-flow distribution along the fracture under the influence of conductivity (**a**) without and (**b**) with non-Darcy flow effect.

Figure 12 shows the effect of non-Darcy flow on PI behavior. As shown in Figure 12a, at any given conductivity, PI is the highest in the condition of $(q_{DND})_f = 0$. PI increases with the increase of C_{fD}, until it reaches the maximum value (J_{Dmax}) at $C_{fD} = 300$. However, the increasing rate of PI is gradually slowed down with C_{fD}; likewise, the maximum PI is independent of $(q_{DND})_f$. In other words, non-Darcy flow plays a negative role in determining PI for the range of low- and intermediate-conductivity $(C_{fD} < 300)$, but has no effect on the maximum PI, which corresponds to the infinite conductivity. The relationship between PI and the Forchheimer number is further shown in Figure 12b. It is shown that, although PI declined with the Forchheimer number, the decreasing rate slows down.

Figure 13 shows the effect of configuration and dimension of fractures on PI behavior with the consideration of the non-Darcy flow effect. As shown in Figure 13a, PI loss in the case of a lower fracture number is relatively more noticeable than that in the case of a larger number. For example, when $N_f = 2$ and $(q_{DND})_f = 0$, PI is that $J_D = 0.96$; when $N_f = 2$ and $(q_{DND})_f = 10$, PI is that $J_D = 0.75$. The relative loss in PI is up to 22%, which is defined as $1 - J_{D@(qDND)f=10}/J_{D@(qDND)f=0}$. In comparison, the relative loss in PI is only 9%. Figure 13b shows the penetration ratio of fracture length with respect to the inner SRV width (I_x). The effect of Forchheimer number is weaker when the I_x is relatively small, but its effect would be amplified when the I_x is relatively small. Figure 13c shows the penetration ratio of the inner SRV region with regard to whole drainage (I_e). In the small range ($I_e < 0.9$), the relationship between I_e and PI exhibits an approximate linear behavior. When $I_e > 0.9$, PI is increased rapidly with the I_e.

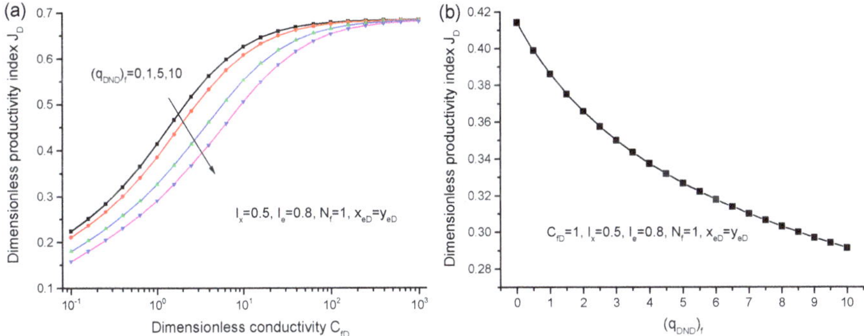

Figure 12. (**a**) Influence of fracture conductivity on PI behavior under the effect of non-Darcy flow, (**b**) influence of Forchheimer number on PI behavior with the consideration of finite conductivity.

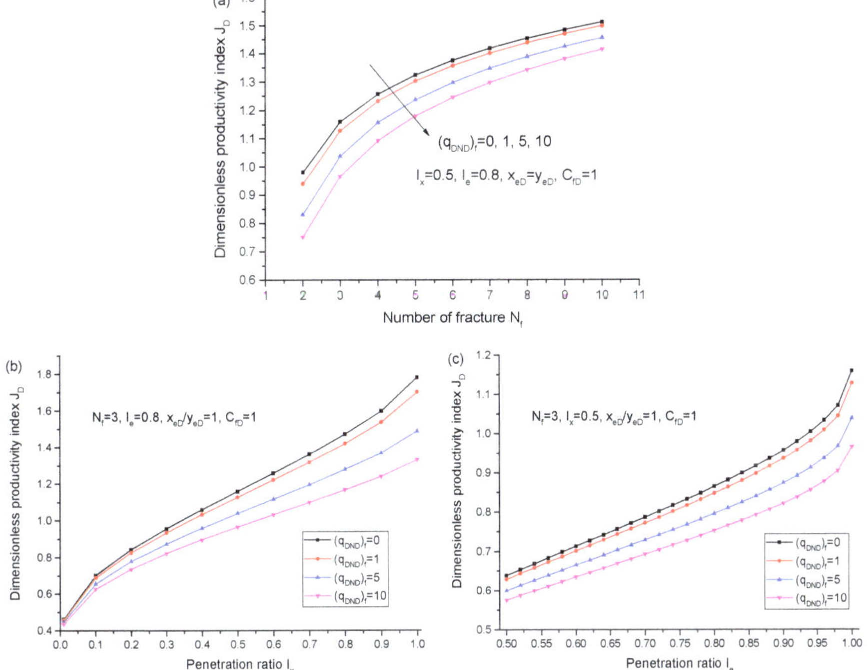

Figure 13. Influence of dimensions of fracture on PI behavior in the consideration of non-Darcy flow, including (**a**) the influence of the number of fractures, (**b**) the influence of penetration ratio of fracture, and (**c**) the influence of penetration ratio of horizontal well.

5.3. Influence of the Pressure-Sensitivity Effect on PI

The characteristic of the PI behavior is similar to the behavior of non-Darcy flow. First, Figure 14 shows the influx-distribution along the fracture with the consideration of the pressure-sensitivity effect. Compared with the nonsensive case (Figure 14a), more influx density is distributed around the wellbore in the sensitive case (Figure 14b), and the tendency would be more remarkable with the decrease of the conductivity. The reason is explained in that the apparent conductivity is degraded and the flow resistance is consequently increased, which is caused by the closing of the fracture under the influence of rock compaction.

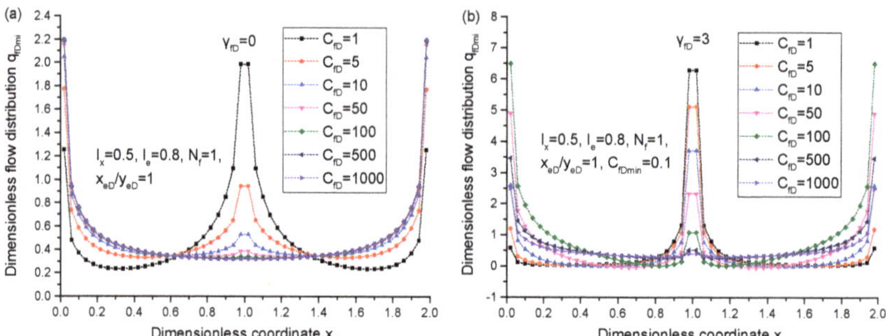

Figure 14. Influence of dimensionless conductivity on influx-flow distribution (**a**) with and (**b**) without the pressure-sensitivity effect.

Figure 15 shows the effect of pressure sensitivity on PI behavior. The intensity is measured by the dimensionless variable γ_{fD}. As shown in Figure 15a, the effect of pressure sensitivity on PI becomes weak with the increase of the conductivity. When the magnitude of the conductivity reaches the level of infinite conductivity, the effect of pressure sensitivity is almost neglected. As expected, as presented in Figure 15b, PI declines until the minimum with the increasing γ_{fD}, and corresponding γ_{fD} is defined as the threshold. Meanwhile, the decreasing rate slows down. For example, if $C_{fD} = 1$, PI reaches the minimum J_{Dmin} when $\gamma_{fDthreshold} = 1.2$; if $C_{fD} = 10$, PI reaches the minimum J_{Dmin} when $\gamma_{fDthreshold} = 2.7$. Thus, a larger conductivity contributes to a larger $\gamma_{fDthreshold}$.

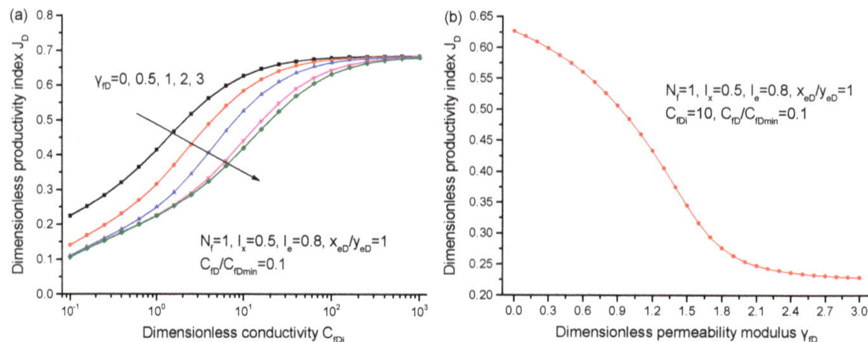

Figure 15. (**a**) Influence of fracture conductivity on PI behavior with the pressure-sensitivity effect, and (**b**) influence of permeability modulus on PI behavior with consideration of finite conductivity.

Figure 16 shows the effect of configuration and dimension of fractures on PI in the consideration of pressure sensitivity in the fracture. The overall features are similar to the case of considering non-Darcy flow, so we will not rewrite again.

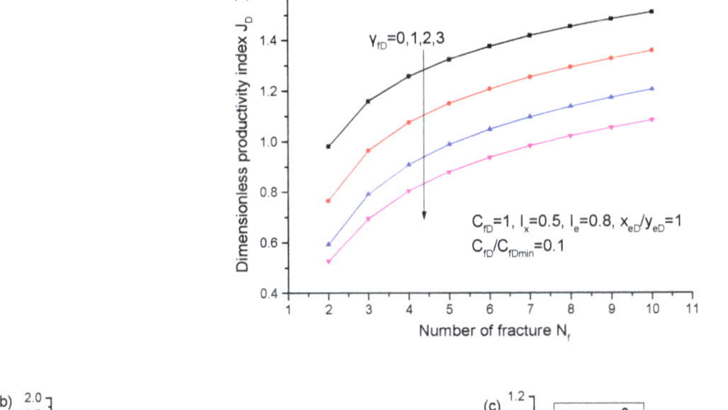

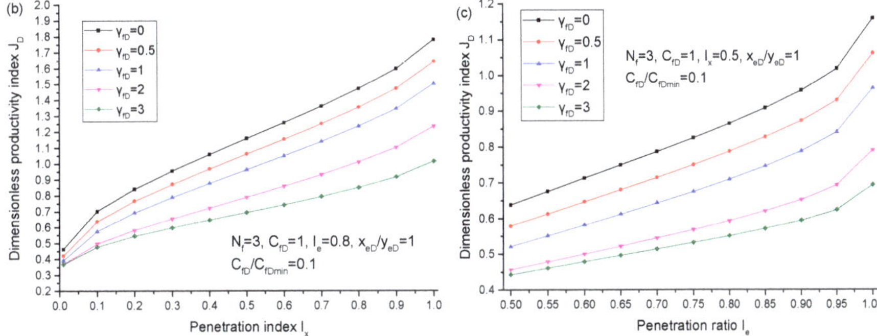

Figure 16. Influence of dimensions of the fracture on PI behavior in the consideration of pressure sensitivity, including (**a**) number of fracture, (**b**) penetration index of fracture, and (**c**) penetration of horizontal wellbore.

6. Conclusions

We derived a new steady-state productivity model for a multi-fractured horizontal well in a reservoir with pseudo-elliptical boundary. We summarize some important findings as follows:

1. PI increases with the increase of C_{fD}, until it reaches the maximum (J_{Dmax}) at $C_{fD} = 300$.
2. PI is deteriorated under the influence of nonlinear flow mechanisms. With the consideration of non-Darcy flow, for the small range of the penetration ratio of the inner SRV region with regard to whole drainage ($I_e < 0.9$), the relationship between I_e and PI exhibits an approximately linear behavior. When $I_e > 0.9$, PI is increased rapidly with the I_e.
3. With the consideration of pressure sensitivity, the apparent fracture is degraded from the initial conductivity to the minimal conductivity, which is caused by fracture closure. As a result, an extra pressure drop is acquired to offset the conductivity degradation, and PI would be declined to the minimal PI.
4. The disadvantage dimensions of fracture (such as small conductivity, penetration ratio, and less fractures) contribute to severe pressure depletion, while, in turn, the severe pressure depletion will strengthen the effect of the nonlinear flow mechanism on PI behavior.
5. If the conductivity in the fracture reaches the level of infinite conductivity, the influence of nonlinear flow mechanism on PI could be neglected.

Author Contributions: M.C. established the model and wrote the manuscript, and H.X. supervised this research and revised the manuscript. C.W. validated the model and conducted the sensitivity analysis. All authors have read and agreed to the published version of the manuscript.

Funding: This work was supported by the National Science and Technology Major Project (No. 2017ZX05019005-006) and Heilongjiang Natural Science Foundation (LH2019F004). Special thanks to the anonymous reviewers and editors for their valuable comments.

Conflicts of Interest: The authors declare no conflict of interest.

Appendix A. Analytical Solution for the Inner SRV Region

The appendix is an optional section that can contain details and data supplemental to the main text. For example, explanations of experimental details that would disrupt the flow of the main text, but nonetheless remain crucial to understanding and reproducing the research shown; figures of replicates for experiments of which representative data is shown in the main text can be added here if brief, or as Supplementary data. Mathematical proofs of results not central to the paper can be added as an appendix.

According to the governing equation and boundary conditions, Fourier sine transformation are used to deal with Equation (4), which is defined as

$$F_x[f(x_D)] = \int_0^{x_{eD}} f(x_D) \sin(\beta_v x_D)dx_D = \overline{f}(\beta_v) \tag{A1}$$

$$F_y[f(y_D)] = \int_0^{y_{eD}} f(y_D) \sin(\gamma_v y_D)dy_D = \overline{f}(\gamma_v) \tag{A2}$$

$$\beta_v = \frac{v\pi}{x_{eD}}, \ v = 0,1,2,3\cdots, \ \gamma_v = \frac{v\pi}{y_{eD}}, \ v = 0,1,2,3\cdots, \tag{A3}$$

Taking the Fourier transformation of Equations (4) and (5), the first term on the left hand side (LHS) of the original equation is given as

$$F_y\left[F_x\left(\frac{\partial^2 p_{mD}}{\partial x_D^2}\right)\right] = \frac{\beta_v p_{eDx}[1-(-1)^v]}{\gamma_v}[1-\cos(\gamma_v y_{eD})] - \beta_v^2 \hat{\overline{p}}_{mD} \tag{A4}$$

Likewise, the second term on the LHS is given as

$$F_x\left[F_y\left(\frac{\partial^2 p_{mD}}{\partial y_D^2}\right)\right] = \frac{\gamma_v p_{eDy}[1-(-1)^v]}{\beta_v}[1-\cos(\beta_v x_{eD})] - \gamma_v^2 \hat{\overline{p}}_{mD} \tag{A5}$$

The third term on the LHS is given as

$$\int_0^{L_{fDm}} q_{fDm}(u_D)\delta(x_D - x_{ofDm} - u_D)\delta(y_D - y_{ofDm})du_D$$
$$= \int_0^{L_{fD}} q_{fD}(u_D)[\sin[\beta_n(x_{ofD} + u_D)]\sin(\gamma_m y_{ofD})]du_D \tag{A6}$$

Finally, the Fourier solution for Equation (4) is given as

$$\frac{\beta_v p_{eDx}[1-(-1)^v][1-\cos(\gamma_v y_{eD})]}{\gamma_v} - \beta_v^2 \hat{\overline{p}}_{mD} + \frac{\gamma_v p_{eDy}[1-(-1)^v][1-\cos(\beta_v x_{eD})]}{\beta_v} - \gamma_v^2 \hat{\overline{p}}_{mD}$$
$$+2\pi \sum_{m=1}^{N_f} \int_0^{L_{fDm}} q_{fDm}(u_D)\{\sin[\beta_v(x_{ofDm} + u_D)]\sin(\gamma_v y_{ofDm})\}du_D = 0 \tag{A7}$$

The inversion Fourier sine transformation is written as

$$\hat{\overline{p}}_{mD} = p_{eDx}\frac{\beta_v[1-(-1)^v][1-\cos(\gamma_v y_{eD})]}{\gamma_v(\beta_v^2+\gamma_v^2)} + p_{eDy}\frac{\gamma_v[1-(-1)^v][1-\cos(\beta_v x_{eD})]}{\beta_v(\beta_v^2+\gamma_v^2)}$$
$$+2\pi \sum_{m=1}^{N_f} \int_0^{L_{fDm}} q_{fDm}(u_D)\left[\frac{\sin[\beta_v(x_{ofDm}+u_D)]\sin(\gamma_v y_{ofDm})}{\beta_v^2+\gamma_v^2}\right]du_D \tag{A8}$$

where

$$N(\beta_v) = \int_0^{x_{eD}} \sin^2(\beta_v x_D)dx_D = \frac{x_{eD}}{2}\left(1 - \frac{\sin(\beta_v x_{eD})\cos(\beta_v x_{eD})}{\beta_v x_{eD}}\right) \tag{A9}$$

$$N(\gamma_v) = \int_0^{y_{eD}} \sin^2(\gamma_v y_D)dy_D = \frac{y_{eD}}{2}\left(1 - \frac{\sin(\gamma_v y_{eD})\cos(\gamma_v y_{eD})}{\gamma_v y_{eD}}\right) \tag{A10}$$

Accordingly, the original solution is expressed in forms of Fourier solution,

$$p_{mD} = \sum_{v=0}^{\infty} \frac{\sin(\beta_v x_D)}{N(\beta_v)}\left[\sum_{v=0}^{\infty} \frac{\sin(\gamma_v y_D)}{N(\gamma_v)}\hat{\bar{p}}_{mD}(\beta_v,\gamma_v)\right] \tag{A11}$$

Equation (A11) could be further expressed as

$$p_{mD} = \frac{4m_{eDx}}{x_{eD}y_{eD}}\sum_{v=1}^{\infty}\frac{\sin(\gamma_v y_D)[1-\cos(\gamma_v y_{eD})]}{\gamma_v}\underbrace{\sum_{v=1}^{\infty}\sin(\beta_n x_D)\frac{\beta_v[1-(-1)^v]}{(\beta_v^2+\gamma_v^2)}}_{V_1}$$

$$+\frac{4m_{eDy}}{x_{eD}y_{eD}}\sum_{v=1}^{\infty}\frac{\sin(\beta_v x_D)[1-\cos(\beta_v x_{eD})]}{\beta_v}\underbrace{\sum_{v=1}^{\infty}\sin(\gamma_v y_D)\frac{\gamma_v[1-(-1)^v]}{(\gamma_v^2+\beta_v^2)}}_{V_2}$$

$$+\frac{8\pi}{x_{eD}y_{eD}}\sum_{m=1}^{N_f}\int_0^{L_{fDm}}q_{fDm}(u_D)\left(\sum_{v=1}^{\infty}\sin(\beta_v x_D)\sin[\beta_v(x_{ofDm}+u_D)]\underbrace{\sum_{v=1}^{\infty}\frac{\sin(\gamma_v y_D)\sin(\gamma_v y_{ofDm})}{\beta_v^2+\gamma_v^2}}_{V_3}\right)du_D \tag{A12}$$

where

$$V_1 = \sum_{v=1}^{\infty}\frac{\beta_v \sin(\beta_v x_D)}{\beta_v^2+\gamma_v^2} + \sum_{v=1}^{\infty}(-1)^{v-1}\frac{\beta_v \sin(\beta_v x_D)}{\beta_v^2+\gamma_v^2} = \frac{x_{eD}}{2}\frac{\sinh[v\pi(x_{eD}-x_D)/y_{eD}]+\sinh(v\pi x_D/y_{eD})}{\sinh(v\pi x_{eD}/y_{eD})} \tag{A13}$$

$$V_2 = \frac{y_{eD}}{2}\frac{\sinh[v\pi(y_{eD}-y_D)/x_{eD}]+\sinh(v\pi y_D/x_{eD})}{\sinh(v\pi y_{eD}/x_{eD})} \tag{A14}$$

$$V_3 = \frac{x_{eD}y_{eD}}{4v\pi}\frac{\cosh[v\pi(y_{eD}-|y_D-y_{ofD}|)/x_{eD}]-\cosh\{v\pi[y_{eD}-(y_D+y_{ofD})]/x_{eD}\}}{\sinh(v\pi y_{eD}/x_{eD})} \tag{A15}$$

Correspondingly, Equation (A12) is finalized by

$$p_{mD}(x_D,y_D) = p_{eDx}IB(x_D,y_D) + p_{eDy}ID(x_D,y_D) + 2\sum_{m=1}^{N_f}\int_0^{L_{fDm}}q_{fDm}(u_D)\delta p_{uD}(x_D,y_D,u_D)du_D \tag{A16}$$

where

$$\begin{cases}
IB = \frac{2}{\pi}\sum_{v=1}^{\infty}\frac{1}{v}\sin\left(\frac{v\pi y_D}{y_{eD}}\right)\frac{[1+(-1)^{v-1}]}{\sinh(v\pi x_{eD}/y_{eD})}\left(\sinh\frac{v\pi(x_{eD}-x_D)}{y_{eD}}+\sinh\frac{v\pi x_D}{y_{eD}}\right)\\[2mm]
ID = \frac{2}{\pi}\sum_{v=1}^{\infty}\frac{1}{v}\sin\left(\frac{v\pi x_D}{x_{eD}}\right)\frac{[1+(-1)^{v-1}]}{\sinh(v\pi y_{eD}/x_{eD})}\left(\sinh\frac{v\pi(y_{eD}-y_D)}{x_{eD}}+\sinh\frac{v\pi y_D}{x_{eD}}\right)\\[2mm]
\delta p_{uD} = \sum_{v=1}^{\infty}\frac{1}{v}\frac{\sin(\beta_v x_D)\sin[\beta_v(x_{ofDm}+u_D)]}{\sinh(v\pi y_{eD}/x_{eD})}\cosh n\pi\frac{y_{eD}-|y_D\pm y_{ofDm}|}{x_{eD}}
\end{cases} \tag{A17}$$

Appendix B. Analytical Solution for the Outer Region

On the interface between the inner SRV region and outer elliptical region, the continuity condition is given as

$$\underbrace{\int_0^{\pi} \left.\frac{\partial p_{mD}}{\partial \xi_D}\right|_{\xi_D=0} d\eta_D}_{elliptic\ flow} = \underbrace{\int_0^{x_{eD}} \left.\frac{\partial p_{mD}}{\partial y_D}\right|_{y_D=y_{eD}} dx_D}_{linear\ flow} \tag{A18}$$

In detail, the right hand side (RHS) of Equation (A18) is expressed as

$$RHS = -p_{eDx}A_2 - p_{eDy}\left(\frac{1}{x_{eD}\ln\left[x_{RD}+\sqrt{x_{RD}^2-x_{eD}^2}/x_{eD}\right]} + B_2\right) + \sum_{m=1}^{N_f} q_{wDm}C_{2,m} \tag{A19}$$

where

$$\begin{cases} A_2 = \frac{4}{\pi x_{eD}}\sum_{v=1}^{\infty}\frac{1}{v}\frac{[1+(-1)^{v-1}]}{\sinh(v\pi x_{eD}/y_{eD})}\left(\cosh\frac{v\pi x_{eD}}{y_{eD}}-1\right) \\[2mm] B_2 = \frac{-1}{x_{eD}\ln\left[(x_{RD}+\sqrt{x_{RD}^2-x_{eD}^2})/x_{eD}\right]} - \frac{2}{\pi x_{eD}}\sum_{v=1}^{\infty}\frac{1}{v}\frac{[1+(-1)^{v-1}]^2}{\sinh(v\pi y_{eD}/x_{eD})}\left(\cosh\frac{v\pi y_{eD}}{x_{eD}}-1\right) \\[2mm] C_{2,m} = -\frac{8}{\pi L_{fDm}}\sum_{v=1}^{\infty}\frac{1}{v^2}\frac{[1-(-1)^v]}{\sinh(v\pi y_{eD}/x_{eD})}\sin\left(\frac{v\pi L_{fDm}}{2x_{eD}}\right)\sin\left(\frac{v\pi(x_{ofDm}+0.5L_{fDm})}{x_{eD}}\right)\sinh v\pi\frac{y_{ofDm}}{x_{eD}} \end{cases} \tag{A20}$$

Thus, Equation (A18) is written as

$$A_2 p_{eDx} + B_2 p_{eyD} = \sum_{m=1}^{N_f} q_{wDm}C_{2,m} \tag{A21}$$

Likewise, on the interface between the inner SRV region and outer linear region, the continuity condition is given as

$$\frac{1}{y_{eD}}\int_0^{y_{eD}} \left.\frac{\partial p_{mD}}{\partial x_D}\right|_{x_D=x_{eD}} dy_D = \frac{1}{y_{eD}}\int_0^{y_{eD}} \left.\frac{\partial p_{mD}}{\partial x_D}\right|_{x_D=x_{eD}} dy_D \tag{A22}$$

In detail, the right hand side (RHS) of Equation (A22) is expressed as

$$RHS = -p_{eDx}\left(\frac{1}{x_{RD}-x_{eD}}-A_1\right) - p_{eDy}B_1 - \sum_{m=1}^{N_f} q_{wDm}C_{1,m} \tag{A23}$$

where

$$\begin{cases} A_1 = \frac{-1}{(x_{RD}-x_{eD})} - \frac{1}{\pi y_{eD}}\sum_{v=1}^{\infty}\frac{1}{v}\frac{[1+(-1)^{v-1}]^2}{\sinh(v\pi x_{eD}/y_{eD})}\left(\cosh\frac{v\pi x_{eD}}{y_{eD}}-1\right) \\[2mm] B_1 = \frac{2}{\pi y_{eD}}\sum_{v=1}^{\infty}\frac{1}{v}\frac{[1+(-1)^{v-1}]}{\sinh(v\pi y_{eD}/x_{eD})}\left(\cosh\frac{v\pi y_{eD}}{x_{eD}}-1\right) \\[2mm] C_{1,m} = -\frac{4x_{eD}}{\pi y_{eD}L_{fDm}}\sum_{v=1}^{\infty}\frac{1}{v^2}\frac{(-1)^v}{\sinh(v\pi y_{eD}/x_{eD})}\sin\left(\frac{v\pi L_{fDm}}{2x_{eD}}\right)\sin\left(\frac{v\pi(x_{ofDm}+0.5L_{fDm})}{x_{eD}}\right)\sinh\frac{v\pi(y_{eD}-y_{ofDm})}{x_{eD}} \end{cases} \tag{A24}$$

Thus, Equation (A22) is written as

$$A_2 p_{eDx} + B_2 p_{eyD} = \sum_{m=1}^{N_f} q_{wDm}C_{2,m} \tag{A25}$$

References

1. Clarkson, C.R.; Bustin, R.M.; Seidle, J.P. Production-data analysis of single-phase (gas) coalbed-methane wells. *SPE Reserv. Eval. Eng.* **2007**, *3*, 312–341. [CrossRef]
2. Wang, J.H.; Wang, X.D.; Dong, W.X. Rate decline curves analysis of multiple-fractured horizontal wells in heterogeneous reservoirs. *J. Hydrol.* **2017**, *553*, 527–539. [CrossRef]
3. Deng, J.; Zhu, W.Y.; Ma, Q. A new seepage model for shale gas reservoir and productivity analysis of fractured well. *Fuel* **2014**, *124*, 232–240. [CrossRef]
4. Clarkson, C.R. Production data analysis of unconventional gas wells: Review of theory and best practices. *Int. J. Coal Geol.* **2013**, *109–110*, 101–146. [CrossRef]
5. Zhang, Q.; Su, Y.L.; Wang, W.D.; Lu, M.; Ren, L. Performance analysis of fractured wells with elliptical SRV in shale reservoirs. *J. Nat. Gas Sci. Eng.* **2017**, *45*, 380–390. [CrossRef]
6. Wang, J.L.; Wei, Y.S.; Luo, W.J. A unified approach to optimize fracture design of a horizontal well intercepted by primary- and secondary-fracture networks. *SPE J.* **2019**, *24*, 1270–1287. [CrossRef]
7. Al-Rbeawi, S. Productivity-index behavior for hydraulically fractured reservoirs depleted by constant production rate considering transient-state and semisteady-state conditions. *SPE Prod. Oper.* **2018**, *33*. [CrossRef]
8. Warren, J.E.; Root, P.J. The behavior of naturally fractured reservoirs. *SPE J.* **1963**, *3*, 245–255. [CrossRef]
9. Obinna, E.D.; Hassan, D. A model for simultaneous matrix depletion into natural and hydraulic fracture networks. *J. Nat. Gas Sci. Eng.* **2014**, *16*, 57–69.
10. Stalgorova, E.; Mattar, L. Analytical model for unconventional multifractured composite systems. *SPE Reserv. Eval. Eng.* **2013**, *16*, 246–256. [CrossRef]
11. Al-Ghamdi, A.; Chen, B.; Behmanesh, H.; Qanbari, F.; Aguilera, R. An improved triple-porosity model for evaluation of naturally fractured reservoirs. *SPE Reserv. Eval. Eng.* **2011**, *8*, 397–404. [CrossRef]
12. Karimi-Fard, M.; Durlofsky, L.J. A general gridding, discretization, and coarsening methodology for modeling flow in porous formations with discrete geological features. *Adv. Water Resour.* **2016**, *96*, 354–372. [CrossRef]
13. Xu, Y.; Cavalcante Filho, J.S.A.; Yu, W. Discrete-fracture modeling of complex hydraulic-fracture geometries in reservoir simulators. *SPE Reserv. Eval. Eng.* **2017**, *20*, 403–422. [CrossRef]
14. Li, L.Y.; Lee, S.H. Efficient field-scale simulation of black oil in a naturally fractured reservoir through discrete fracture networks and homogenized media. *SPE Reserv. Eval. Eng.* **2008**, *8*, 750–758. [CrossRef]
15. Cinco-Ley, H.; Samaniego, F.; Dominguez, N. Transient pressure behavior for a well with a finite-conductivity vertical fracture. *Soc. Pet. Eng. J.* **1978**, *8*, 254–265.
16. Wang, J.L.; Jia, A.L.; Wei, Y.S.; Qi, Y.D. Approximate semi-analytical modeling of transient behavior of horizontal well intercepted by multiple pressure-dependent conductivity fractures in pressure-sensitive reservoir. *J. Pet. Sci. Eng.* **2017**, *153*, 157–177. [CrossRef]
17. Wang, J.L.; Jia, A.L.; Wei, Y.S. A general framework model for simulating transient response of a well with complex fracture network by use of source and Green's function. *J. Nat. Gas Sci. Eng.* **2018**, *55*, 254–275. [CrossRef]
18. Zhou, W.T.; Banerjee, R.; Poe, B. Semianalytical production simulation of complex hydraulic-fracture networks. *SPE J.* **2013**, *19*, 6–18. [CrossRef]
19. Kuchuk, F.; Biryukov, D. Pressure-transient behavior of continuously and discretely fractured reservoirs. *SPE Reserv. Eval. Eng.* **2014**, *17*, 82–97. [CrossRef]
20. Al-Rbeawi, S. The optimal reservoir configuration for maximum productivity index of gas reservoirs depleted by horizontal wells under Darcy and non-Darcy flow conditions. *J. Nat. Gas Sci. Eng.* **2018**, *49*, 179–193. [CrossRef]
21. Chen, Z.M.; Liao, X.W.; Zhao, X.L.; Lyu, S.; Zhu, L. A comprehensive productivity equation for multiple fractured vertical well with non-linear effects under steady-state flow. *J. Pet. Sci. Eng.* **2017**, *149*, 9–24. [CrossRef]
22. Li, J.; Guo, B.; Gao, D.; Ai, C. The effect of fracture-face matrix damage on productivity of fractures with infinite and finite conductivities in shale-gas reservoirs. *SPE Drill. Complet.* **2012**, *9*, 347–353. [CrossRef]
23. Prats, M.; Raghavan, R. Finite horizontal well in a uniform-thickness reservoir crossing a natural fracture normally. *SPE J.* **2013**, *10*, 982–992. [CrossRef]

24. Tahmasebi, P.; Kamrava, S. A pore-scale mathematical modeling of fluid-particle interactions: Thermo-hydro-mechanical coupling. *Int. J. Greenh. Gas Control* **2019**, *83*, 245–255. [CrossRef]

25. Zhang, X.M.; Tahmasebi, P. Effects of grain size on deformation in porous media. *Transp. Porous Media* **2019**, *129*, 321–341. [CrossRef]

26. Zhang, X.M.; Tahmasebi, P. Micromechanical evaluation of rock and fluid interactions. *Int. J. Greenh. Gas Control* **2018**, *76*, 266–277. [CrossRef]

27. Guppy, K.H.; Cinco-Ley, H.; Ramey, H.J. Effect of non-Darcy flow on the constant-pressure production of fractured wells. *Soc. Pet. Eng. J.* **1981**, *12*, 390–400. [CrossRef]

28. Wu, Y.S.; Li, J.F.; Ding, D.; Wang, C.; Di, Y. A generalized framework model for simulation of gas production in unconventional gas reservoir. In Proceedings of the SPE Reservoir Simulation Symposium, The Woodlands, TX, USA, 18–20 February 2013.

29. Lin, J.; Zhu, D. Predicting well performance in complex fracture systems by slab source method. In Proceedings of the SPE Hydraulic Fracturing Technology Conference, The Woodlands, TX, USA, 6–8 February 2012.

30. Pedroso, C.A.; Correa, A.C. A new model of a pressure-dependent-conductivity hydraulic fracture in a finite reservoir: Constant rate production case. In Proceedings of the Latin American and Caribbean Petroleum Engineering Conference, Rio de Janeiro, Brazil, 30 August–3 September 1997.

31. Zhang, Z.; He, S.L.; Liu, G.; Guo, X.; Mo, S. Pressure buildup behavior of vertically fractured wells with stress-sensitive conductivity. *J. Pet. Sci. Eng.* **2014**, *122*, 48–55. [CrossRef]

32. Yao, S.S.; Zeng, F.H.; Liu, H. A semi-analytical model for hydraulically fractured wells with stress-sensitive conductivities. In Proceedings of the SPE Unconventional Resources Conference, Calgary, AB, Canada, 5–7 November 2012.

33. Biryukov, D.; Kuchuk, F.J. Transient pressure behavior of reservoirs with discrete conductive faults and fractures. *Transp. Porous Media* **2012**, *95*, 239–268. [CrossRef]

34. Rasoulzadeh, M.; Kuchuk, F.J. Pressure transient behavior of high-fracture-density reservoirs (dual-porosity models). *Transp. Porous Media* **2019**, *127*, 222–243. [CrossRef]

35. Kuchuk, F.J.; Habashy, T. Pressure behavior of laterally composite reservoirs. *SPE Form. Eval.* **1997**, *3*, 47–56. [CrossRef]

36. Luo, W.J.; Tang, C.F. A semianalytical solution of a vertical fractured well with varying conductivity under non-Darcy-flow condition. *SPE J.* **2015**, *20*, 1028–1040. [CrossRef]

37. Luo, W.J.; Tang, C.F.; Feng, Y. Mechanism of fluid flow along a dynamic conductivity fracture with pressure-dependent permeability under constant wellbore pressure. *J. Pet. Sci. Eng.* **2018**, *166*, 465–475. [CrossRef]

Article

Numerical Investigation of Injection-Induced Fracture Propagation in Brittle Rocks with Two Injection Wells by a Modified Fluid-Mechanical Coupling Model

Song Wang [1,2,3], Jian Zhou [4], Luqing Zhang [1,3,*] and Zhenhua Han [1,3]

[1]   Key Laboratory of Shale Gas and Geoengineering, Institute of Geology and Geophysics, Chinese Academy of Sciences, Beijing 100029, China; wangsong@mail.iggcas.ac.cn (S.W.); hanzhenhua@mail.iggcas.ac.cn (Z.H.)
[2]   College of Earth and Planetary Sciences, University of Chinese Academy of Sciences, Beijing 100049, China
[3]   Innovation Academy for Earth Science, Chinese Academy of Sciences, Beijing 100029, China
[4]   Key Laboratory of Urban Security and Disaster Engineering, Ministry of Education, Beijing University of Technology, Beijing 100124, China; zhoujian@mail.iggcas.ac.cn
*   Correspondence: zhangluqing@mail.iggcas.ac.cn; Tel.: +86-10-8299-8642; Fax: +86-10-6201-0846

Received: 10 August 2020; Accepted: 7 September 2020; Published: 10 September 2020

Abstract: Hydraulic fracturing is a key technical means for stimulating tight and low permeability reservoirs to improve the production, which is widely employed in the development of unconventional energy resources, including shale gas, shale oil, gas hydrate, and dry hot rock. Although significant progress has been made in the simulation of fracturing a single well using two-dimensional Particle Flow Code (PFC2D), the understanding of the multi-well hydraulic fracturing characteristics is still limited. Exploring the mechanisms of fluid-driven fracture initiation, propagation and interaction under multi-well fracturing conditions is of great theoretical significance for creating complex fracture networks in the reservoir. In this study, a series of two-well fracturing simulations by a modified fluid-mechanical coupling algorithm were conducted to systematically investigate the effects of injection sequence and well spacing on breakdown pressure, fracture propagation and stress shadow. The results show that both injection sequence and well spacing make little difference on breakdown pressure but have huge impacts on fracture propagation pressure. Especially under hydrostatic pressure conditions, simultaneous injection and small well spacing increase the pore pressure between two injection wells and reduce the effective stress of rock to achieve lower fracture propagation pressure. The injection sequence can change the propagation direction of hydraulic fractures. When the in-situ stress is hydrostatic pressure, simultaneous injection compels the fractures to deflect and tend to propagate horizontally, which promotes the formation of complex fracture networks between two injection wells. When the maximum in-situ stress is in the horizontal direction, asynchronous injection is more conducive to the parallel propagation of multiple hydraulic fractures. Nevertheless, excessively small or large well spacing reduces the number of fracture branches in fracture networks. In addition, the stress shadow effect is found to be sensitive to both injection sequence and well spacing.

Keywords: hydraulic fracturing; discrete element method; modified fluid-mechanical coupling algorithm; injection sequence; well spacing; stress shadow effect

---

1. Introduction

With the rapid growth of the global economy, the conventional oil and gas resources with decreasing production cannot meet the need for energy. Unconventional energy resources such as shale gas, shale oil, gas hydrate, and dry hot rock geothermal energy are quite abundant, and their economically and technically feasible exploitation is an effective solution to the problem of future energy shortages [1–4]. However, extremely tight pore structure and ultra-low permeability for the unconventional reservoir rocks restrict the flow of energy and efficient production [5,6]. Hydraulic fracturing is an important technical method for reservoir reconstruction in the process of energy resources exploitation. By injecting high-pressure fluid into the reservoir, the rock is fractured, thereby increasing the conductivity [7–9]. Affected by the heterogeneity and anisotropy of the reservoir rock, the initiation and propagation of fractures, the leak-off of the fracturing fluid, the interaction between hydraulic fractures and natural fractures, and many other physical and mechanical conditions, hydraulic fracturing is generally regarded as the multi-field coupling problem under complex in-situ stress [10–13]. Mastering the mechanism of hydraulic fracturing technology is crucial to building a complex fracture network system and forming a high-yield unconventional energy reservoir.

Laboratory testing is a direct and effective method to discover the mechanism of hydraulic fracturing. Various observation and analysis techniques have been utilized by domestic and foreign geologists to study the fracturing behaviors of reservoir rocks under the action of pressure fluids [14–17]. Guo et al. [18] used the true triaxial test system and the high-energy computed tomography (CT) scanning based on linear accelerator to capture the effects of flow rate, in-situ stress and fluid viscosity on fracturing horizontal well in shale. They indicated that the abovementioned multi-aspect factors dominate the initiating and propagating rules of hydraulic fractures, and the interaction between hydraulic fractures and sedimentary bedding. Zhang et al. [19] observed the propagation patterns of supercritical carbon dioxide (SC-CO$_2$) induced hydraulic fractures in shale through CT scanning and digital radiography (DR) scanning. Comparing with hydraulic fracturing, they believed that the percolation effect of SC-CO$_2$ reduces the initiation pressure needed to fractures by more than 50%, and induces more secondary fractures to form complex fracture networks. He et al. [20] investigated the surface characteristics of hydraulic fractures using the scanning electron microscope (SEM), and found that numerous micro cracks propagating along the boundaries of hard minerals and soft organic matters connected the pores and macro hydraulic fractures, thus increasing the permeability of shale. Zhuang et al. [21] performed a set of hydraulic fracturing tests on granites under true triaxial loading with different injection schemes, and analyzed the mineral fracturing behaviors and induced seismicity according to thin section microscopic observations and acoustic emission technology. The results show that hydraulic fractures under six fluid injection schemes are all composed of the intragranular cracks splitting microcline, orthoclase and quartz, but the cyclic pulse pressurization scheme improves the injectivity and decreases induced seismicity.

In addition to these laboratory tests, low-cost and high-efficiency numerical simulation becomes one of the most common ways to solve the hydraulic fracturing problems as the computer science develops [22]. Various numerical simulation methods including the finite-element method (FEM), the finite-difference method (FDM), the discrete element method (DEM), and the discontinuous displacement method (DDM) are successful in reproducing the hydraulic fracturing process of reservoir rocks and revealing the mechanism of hydraulic fracture propagation [23,24]. Li et al. [25] simulated the multi-cluster hydraulic fracturing based on the FEM software ABAQUS, and demonstrated that stress interference from multiple-clusters causes the suppression and transfer of the fracture network. Liao et al. [26] stated that horizontal stress anisotropy plays a key role in the interaction between hydraulic fractures and natural fractures in term of the three-dimensional fracturing simulation by the FDM software FLAC 3D. Zangeneh et al. [27] established a conceptual reservoir model with faults in the DEM software UDEC to confirm the influence of fluid diffusion on triggering fault slip and to evaluate the maximum magnitude of seismic events. Janiszewski et al. [28] discussed the interaction of fractures in granite under different approach angle conditions with the help of the DDM software FRACOD, and concluded that the existence of low dip-angle fractures allows more complex fracture networks.

Each kind of numerical simulation software developed on the basis of the continuous or discontinuous methods has its own unique advantages. As a typical representative of DEM software, the numerical model created in PFC is composed of thousands of basic circular particles bonded together by the contact bonds [29]. Furthermore, the failure of a contact bond corresponds to the generation of a microcrack in the model. Because of the simple topology and the fast updating mechanism of particle information, PFC shows obvious advantages in emulating the initiation, propagation and aggregation of microcracks and the ultimate macro failure for rock and soil mass under complex stress conditions [30–33].

The fluid-mechanical coupling algorithm in PFC was first proposed by Cundall [34], which not only considers the interaction process of pore pressure and fracture aperture, but also realizes the fluid flow in both fractures and rock matrix. Al-Busaidi et al. [35] recreated the laboratory test results of hydraulic fracturing for Lac du Bonnet granite by simulation, which proved the effectiveness of this algorithm. Yoon et al. [36–38] probed into the seismic activity induced by fluid injection during the development of deep granite reservoir. The detailed seismic parameters involving temporal and spatial distribution, moment magnitudes and radiated seismic energy of the induced seismic events were obtained through the reservoir fracturing simulations, which provided constructive guidance for the optimization of field fracturing schemes. Tomac et al. [39,40] extended this algorithm and explored the influences of mineral grains, fluid pressure and temperature on permeable hot granite during injecting pressurized cold fluid. They presented that the thermal damage area of wellbore caused by rock cooling and cold fluid infiltration increases as the fluid dynamic viscosity decreases. In our previous research, the calculation method of fluid pressure in the domain after contact bond failure was modified to make the update of the simulation results more accurate when the numerical model contains a large-volume injection well [41,42]. Following the improved algorithm, we comprehensively studied the hydraulic fracturing mechanism in isotropic and laminated reservoirs under different in-situ stress state, injection rate and fluid viscosity, and validated the ability in modeling the variety of the interaction scenarios between hydraulic fractures and natural fractures such as direct crossing and crossing with an offset [43–47].

Although the PFC has been employed in many hydraulic fracturing studies, most of them only consider the simple case of single injection well and ignore the interference of the multiple injection wells. However, in order to obtain a better fracturing effect in the field of production, it is necessary to arrange multiple injection wells in a suitable space and select a reasonable fracturing sequence. In this study, taking the Alxa porphyritic granite as an example, a two-well fracturing model based on our modified algorithm was established to discuss the effects of injection sequence and well spacing on hydraulic fracturing behaviors. The rest of this paper is organized as follows. First, the characteristics of contact bond and the modified fluid-mechanical coupling algorithm in PFC were introduced. Second, the micromechanical parameters were carefully calibrated and validated according to the experimental results of uniaxial loading and the analytical solutions of the breakdown pressure, respectively. Third, the boundary conditions of the numerical model were transformed with reference to diverse schemes so as to analyze the change in breakdown pressure, fracture propagation pressure, fluid pressure distribution and fracture propagation pattern. Finally, the stress variation along the height direction of hydraulic fracture was extracted to assess the stress shadow effect of hydraulic fractures.

2. Modeling Methodology

2.1. Introduction of Particle Flow Code

PFC is a commercial DEM software that applies the rigid, unbreakable circular particles as the basic elements of the model. The circular particles in contact are available for connection as a whole by giving a contact bond model. During the process of solution, the displacements, contact forces and moments of particles are continuously updated with the calculation time by Newton's second law and force-displacement law until the unbalanced force and unbalanced moment in the model reach the program termination conditions [29].

Plentiful models of contact bonds built into PFC including the Parallel-Bond Model (PBM), the Smooth-Joint Model (SJM), the Flat-Joint Model (FJM) and the Hertz Contact Model (HCM) determine the macro mechanical properties of various solid materials [48]. In particular, envisioned as a collection of springs with constant strength and stiffness (Figure 1a), the PBM is mostly exerted into the simulations of rock materials by assigning the particle-interaction laws. Moreover, the normal component $F_i^n(t + \Delta t)$ and the tangential component $F_i^s(t + \Delta t)$ for the contact force and the contact moment $M(t + \Delta t)$ at time $t + \Delta t$ depend on the corresponding values at the start of the timestep with the increments. The detailed equations used are given by [43]:

$$F_i^n(t + \Delta t) = F_i^n(t) + \Delta F_i^n = F^n(t)n_i + \left(-k^n A \Delta U_i^n\right)n_i \tag{1}$$

$$F_i^s(t + \Delta t) = F_i^s(t) + \Delta F_i^s = F_i^s(t) + \left(-k^s A \Delta U_i^s\right) \tag{2}$$

$$M(t + \Delta t) = M(t) + \Delta M = M(t) + (-k^n I)\left(\omega^{[B]} - \omega^{[A]}\right)\Delta t \tag{3}$$

where k^n and k^s are the normal and shear stiffness of the parallel bond, respectively; ΔU^n and ΔU^s are the relative normal-displacement increment and the relative shear-displacement increment in one timestep Δt, respectively; $\omega^{[A]}$ and $\omega^{[B]}$ are the relative rotation velocities of the two contacting particles, respectively; n_i is the normal vector of each contact ($i = 1$ or 2); A is the cross sectional area of the bond, and I is the moment of inertia of the bond.

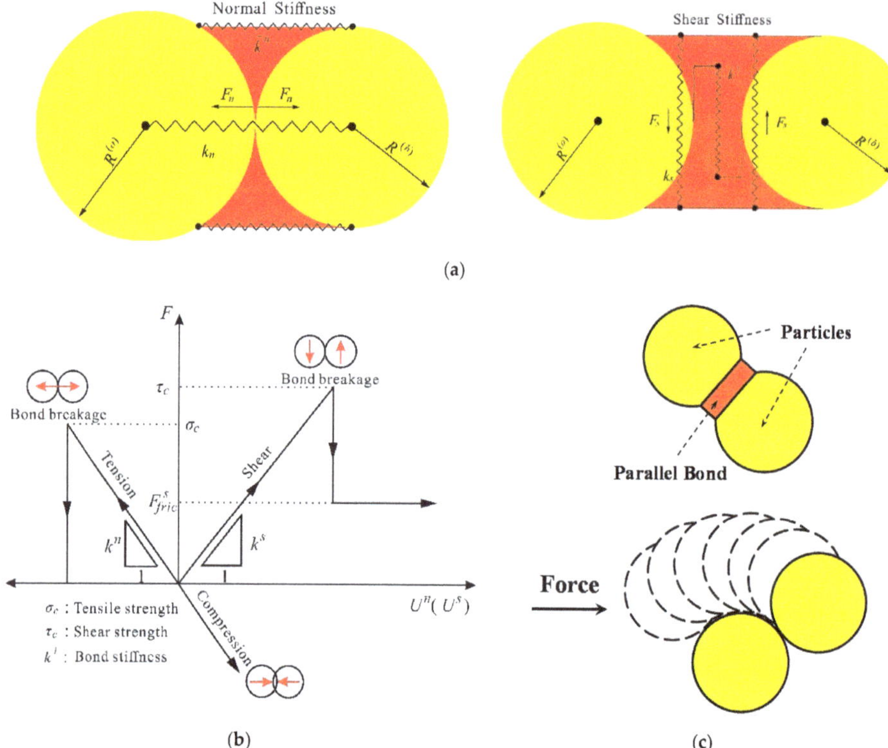

Figure 1. Schematic diagram of the Parallel-Bond contact in PFC2D (modified from Itasca Consulting Group, Inc.: Minneapolis, MN, USA [48]): (**a**) conceptual models; (**b**) the deformation and failure mechanisms; and (**c**) the failure mode.

Then the normal stress σ and shear stress τ distributed on the cross-section of the bond are updated by the following equations [48]:

$$\sigma = \frac{-F^n}{A} + \beta\frac{|M|}{I}R \tag{4}$$

$$\tau = \frac{|F_i^s|}{A} \tag{5}$$

where β is the moment-contribution factor.

At any time of solution, if the absolute value of normal stress $|\sigma|$ inside the bond exceeds the tensile strength of the bond σ_c, or the absolute value of shear stress $|\tau|$ inside the bond exceeds the shear strength of the bond τ_c, the tensile failure or shear failure occurs respectively in the PBM (Figure 1b). Correspondingly, a tensile microcrack or a shear microcrack is generated in the numerical model. After the failure of the parallel bond, the force, moment, and stiffnesses related to the contact are removed, resulting in the rotation of the two particles around each other under external force (Figure 1c). Recent studies have pointed out that the rotation after the bond failure reduces the self-locking effect between the particles, causing an extremely low ratio of uniaxial compressive strength (UCS) to uniaxial tensile strength (UTS) [49,50]. Nonetheless, the PBM revised by potyondy [51] and Ding and Zhang [52] is still employed in this study, in which the moment-contribution factor β is added to the bond failure criterion to overcome the above defects.

2.2. Modified Fluid-Mechanical Coupling Algorithm

Cundall's fluid-mechanical coupling algorithm achieves the flow of viscous fluid in the parallel bond. As illustrated in Figure 2a, the fluid network topology is constructed after the circular particles are assembled by the PBM. More specifically, connecting the centers of the contacting particles by blue lines forms a series of enclosed domains (blue polygons) which are regarded as "reservoirs" to store the fluid pressure. The centers of the adjacent reservoir domains (blue circles) are linked by the virtual pipes (magenta lines) to ensure fluid flow. Any pipe is approximated as two parallel plates existing in the parallel bond. A differential pressure between the two connected reservoir domains drives the flow of fluid from the high-pressure region to the low-pressure region. The volumetric flow rate q is calculated by the cubic law [34]:

$$q = \frac{e^3}{12\mu} \frac{P_2 - P_1}{L} \tag{6}$$

where e is the hydraulic aperture; $P_2 - P_1$ is the fluid pressure difference between the reservoir domains; L represents the length of the flow channel; μ is the fluid dynamic viscosity.

Under the action of confining pressure and fluid pressure, the hydraulic aperture e changes with the normal stress σ of the bond, which is described as [53,54]:

$$e = e_{\text{inf}} + (e_0 - e_{\text{inf}}) \exp(-0.15\sigma) \tag{7}$$

when σ tends to infinity, the aperture decreases gradually to e_{inf}; when $\sigma = 0$, the aperture e_0 represents the residual aperture due to no force interaction between the particles on both sides of the flow channel. Therefore, this mechanism enables the fluid to diffuse into the surrounding reservoirs whether fractures are in the model. According to the Equation (8), the values of e_0 and e_{inf} are determined by the permeability k, total volume V of the reservoirs and flow channel length L [35].

$$k = \frac{1}{12V} \sum_{\text{pipes}} Le^3 \tag{8}$$

In the fluid calculation process of one timestep, the increment of fluid pressure ΔP for a reservoir domain is derived from the apparent volume change of this domain ΔV_d and the fluid flow of surrounding reservoir domains Σq, as shown in Equation (9) [43]:

$$\Delta P = \frac{K_f}{V_d}\left(\sum q\Delta t - \Delta V_d\right) \tag{9}$$

where K_f is bulk modulus of fluid; V_d is the apparent volume of the reservoir domain.

After the fluid calculation step, the fluid pressure acts on the particles in the form of the body force (Figure 2b), and the magnitude of the force is the product of the fluid pressure P_f and the length of the connecting line l_i at the contact point. As the parallel bond between adjacent domains is broken owing to the effect of confining pressure and fluid pressure, Al-Busaidi et al. [35] and Zhao and Young [55] used the average value of the fluid pressures P_{f1} and P_{f2} in the two domains before bond failure to express the fluid pressure P'_f in the new domain:

$$P'_f = \frac{P_{f1} + P_{f2}}{2} \tag{10}$$

Obviously, the Equation (10) is not suitable for the cases of large-volume injection well in the model because of the inaccurate average fluid pressure in the domains around the well. An optimized computing method is as follows [41]:

$$P'_f = \left[\frac{(V_{f1} + V_{f2})}{(V_{r1} + V_{r2})\varphi} - 1 \right] K_f \tag{11}$$

where V_{f1} and V_{f2} are the fluid volume in the two domains before bond failure, respectively; V_r1 and V_r2 are the volume of the two domains under the fluid pressure of 0 MPa; φ is the porosity of the model.

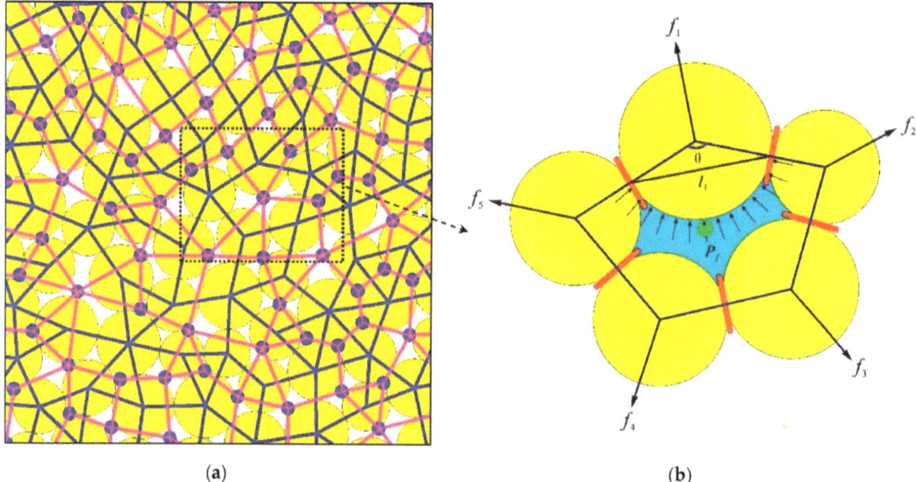

Figure 2. The sketch of the fluid-mechanical coupling model in PFC2D: (**a**) the generation of fluid network; and (**b**) fluid-particle interaction mechanism (yellow circles: solid particles, magenta lines: flow channels, blue polygons: reservoir domains, blue circles: the centers of reservoirs, red lines: Parallel-Bond contacts).

3. Modeling Calibration and Validation

The input micro-parameters involved in the fluid-mechanical coupling model in PFC2D are divided into three categories to characterize the physical and mechanical properties of basic particles, parallel bonds and fluid networks, most of which cannot be directly obtained from the laboratory tests. Debugging the parameters to match the calibration results with experimental results or analytical solutions is absolutely essential for the subsequent reliable numerical simulations. Fortunately, the relationships between the input parameters and macro mechanical characteristics of the models have been presented in several studies [56–59], which is helpful for the quick selection and calibration of the parameters. Combined with the uniaxial compression and tensile test results of Alxa porphyritic granite, a group of input parameters which reproduced the mechanical behaviors of the real granite under uniaxial loading were determined in this section. To verify the accuracy of the parameters in simulating hydraulic fracturing, the granite model with a single injection well were created and contrasted with the analytical solutions of the breakdown pressure.

3.1. Laboratory Test

As one of the preferred candidate locations for geological disposal of high-level radioactive waste, Alxa is located in the west of Inner Mongolia Autonomous Region, with the geographic coordinates of 97°10′ ~ 106°53′ E and 37°24′ ~ 42°47′ N. The NRG-1 deep borehole in Alxa revealed that a large number of porphyritic granites are distributed from the surface to 600 m underground. The granitic cores from different depths of the borehole were processed into the standard specimens with a diameter of 50 mm and a height of 100 mm. Before the uniaxial compression tests, the average density of the specimens was 2650.0 kg/m^3. During the testing, the MTS 815 servo-controlled hydraulic testing machine at the Institute of Crustal Dynamics, China Earthquake Administration was employed to compress the granites with an axial loading speed of 0.06 m/min.

The failure characteristics of Alxa porphyritic granites after the tests and the stress-strain curves are shown in Figures 3 and 4, respectively. It can be seen that all the specimens were cut by numerous splitting cracks, which conforms to the typical brittle fracturing pattern. The macro mechanical parameters of the specimens were extracted from the stress-strain curves, and the average values of the UCS, the elastic modulus (E) and the Poisson's ratio (ν) were 159.4 MPa, 48.0 GPa and 0.23, respectively. The tensile test is not conducted in this research; however, the average UTS of this kind of granites is about 9.5 MPa according to our previous studies [60,61].

Figure 3. The failure characteristics of Alxa porphyritic granites after the uniaxial compressive tests.

3.2. Material Parameters Calibration

A rectangular numerical model with the same size as the test specimen was generated in PFC2D to emulate the uniaxial loading responses. The model was comprised of 10,412 circular particles with a radius distribution of 0.3–0.45 mm and was assigned the contact bonds of PBM. The effective modulus and stiffness of particles can inherit from the corresponding parameters of PBM. After the initial parameters were given, the simulations of uniaxial compression and direct tension were carried out to calculate the UCS, E, ν and UTS. Then the relevant parameters were continuously adjusted by the comparison between the experimental results and the simulated results. The final calibrated parameters are listed in Table 1, by which the simulated macro mechanical properties are in good agreement with the actual values, as shown in Table 2. The stress-strain curves and the spatial distribution of cracks are presented in Figure 4, which also proves the accuracy of parametric calibration.

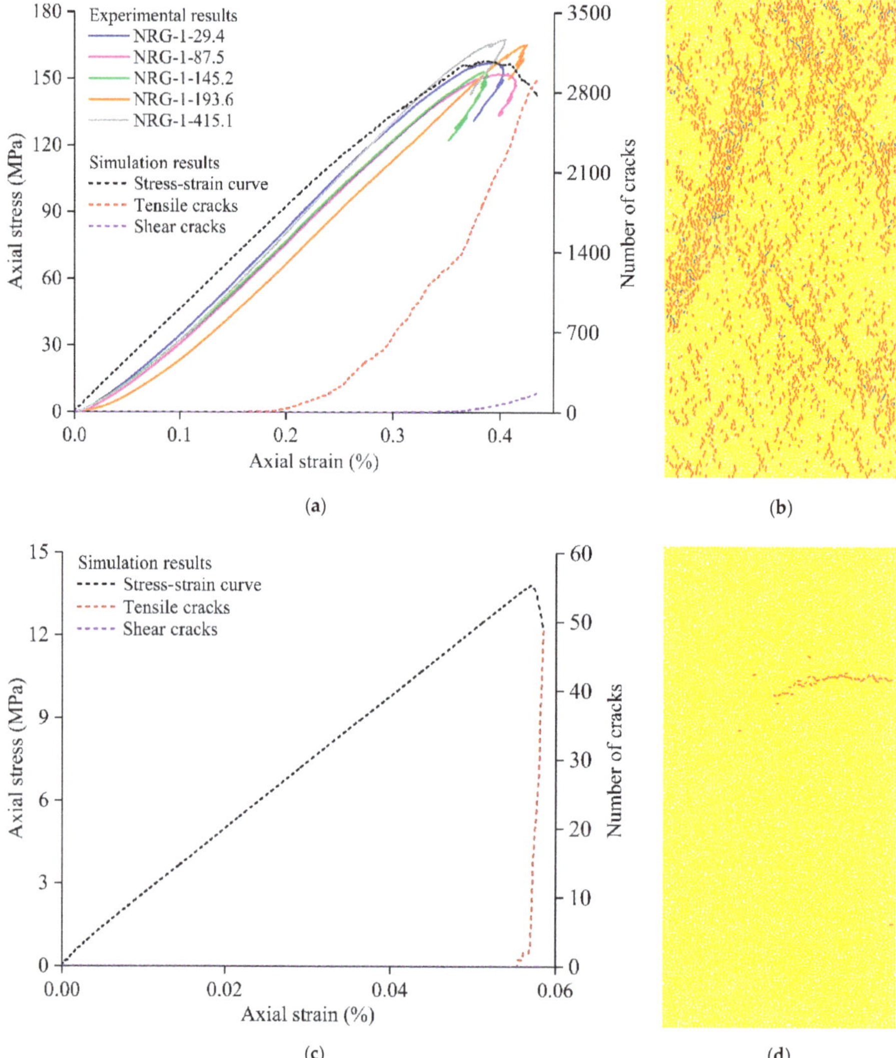

Figure 4. The simulated results under the calibrated parametric condition: (**a**) stress-strain curve and microcrack evolution under uniaxial compression; (**b**) microcrack distribution after uniaxial compression; (**c**) stress-strain curve and microcrack evolution under uniaxial tension; and (**d**) microcrack distribution after uniaxial tension.

Most of the microcracks generated in the numerical granite under the compressive loading are the tension-type modes. Fewer microcracks formed in the middle of model under the tensile loading, and aggregated into a macro fracture.

Table 1. Calibrated Micro-Parameters in the Numerical Models of Alax Porphyritic Granite.

Micro-Parameters of the Basic Particles	Notations	Values
Particle density (kg/m^3)	ϱ	2650.0
Porosity (%)	φ	10
Effective modulus of particle (GPa)	E'	30.0
Contact normal to shear stiffness ratio	k_n/k_s	1.5
Friction coefficient of particle	f	0.8
Micro-Parameters of the Parallel Bonds	**Notations**	**Values**
Effective modulus of the parallel bonds (GPa)	E_c	30.0
Normal to shear stiffness ratio of the parallel bonds	k^n/k^s	1.5
Tensile strength of the parallel bonds (MPa)	σ_c	28.0
Shear strength of the parallel bonds (MPa)	τ_c	120.0
Radius multiplier	λ	1.0
Moment contribution factor	β	0.1
Micro-Parameters of the Hydraulic Properties	**Notations**	**Values**
Initial hydraulic aperture (m)	e_0	2.2×10^{-6}
Infinite hydraulic aperture (m)	e_{inf}	2.2×10^{-7}
Model Permeability (m^2)	k	1.0×10^{-17}
Bulk modulus of the fracturing fluid (GPa)	K_f	2.0
Viscosity of the fracturing fluid (Pa·s)	μ	1.0×10^{-3}

Table 2. Comparison of the Calibrated and Experimental Results of the Macro Mechanical Properties for Alax Porphyritic Granites.

Macro Properties	Average of Experimental Results	Simulation Results
UCS (MPa)	159.4	159.7
UTS (MPa)	9.5 [1]	13.8
E (GPa)	48.0	46.4
ν	0.23	0.21

[1] The UTS of Alax porphyritic granite was acquired from Zhou et al. [60,61].

3.3. Validation of Hydraulic Fracturing Model

The fluid parameters for hydraulic fracturing were decided by the actual working conditions. After constructing the fluid networks and adding the fluid parameters in the PBM, it was still vital to verify whether the fracturing simulation could describe the realistic fracturing behaviors, including the breakdown pressure and the propagation of hydraulic fractures. A classical equation for the breakdown pressure P_{wf} when the fluid is injected into a circular elastic borehole was proposed by Haimson and Fairhurst [62] as following expression:

$$P_{wf} = 3\sigma_v - \sigma_h - P_0 + \sigma_t \tag{12}$$

where σ_v is the confining pressure in vertical direction; σ_h is the confining pressure in horizontal direction; P_0 is initial pore pressure; σ_t is the UTS of the model.

In order to facilitate the comparison of simulated and theoretical breakdown pressures and avoid the stress interference of multiple injection wells, the granite model with a single injection well was established, as shown in Figure 5a. The square fracturing model with a width and height of 1.0 m consisted of 11,778 circular particles with a radius distribution of 4–6 mm. Furthermore, the vertical and horizontal confining pressures were servo-controlled by moving the walls around the model. An injection well with a radius of 50 mm was excavated in the center of the model to inject viscous fluid. It is noted that the particles surrounding the injection well have been replaced by smaller particles to smooth the well surface and prevent stress concentration (Figure 5b). No fluid flow domains were covered on the border of the fracturing model to give the impermeable boundary conditions. The hydraulic apertures e_0 and e_{inf} calculated by Equations (7) and (8) were 2.2×10^{-6} and 2.2×10^{-7}, respectively, which were input into the model as basic hydraulic property parameters (Table 1). The fracturing fluid was the liquid water with a bulk modulus of 2.0 GPa and a viscosity of 1.0×10^{-3}. The numerical model kept the vertical confining pressure σ_v and the horizontal confining pressure σ_h equal. Under the conditions of the initial pore pressure of 0 MPa and the fluid injection rate of 1.0×10^{-5} m³/s, the hydraulic fracturing simulations with confining pressures from 5 MPa to 50 MPa (5-MPa intervals) were implemented.

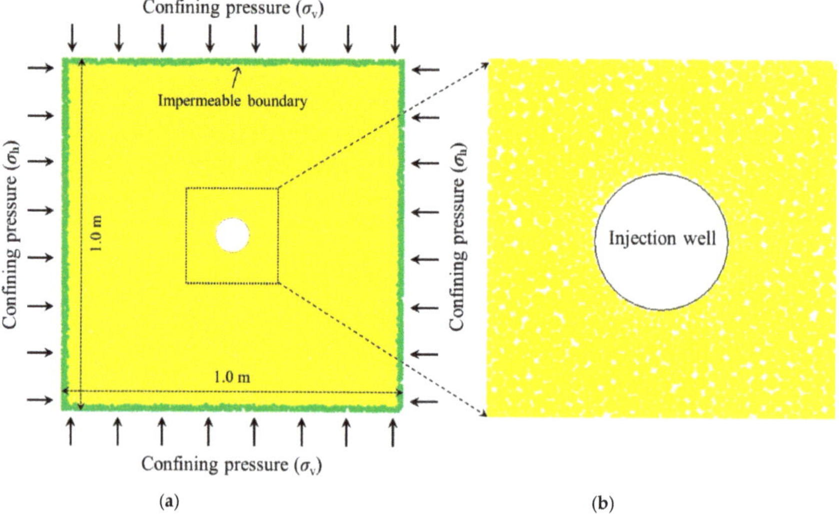

Figure 5. Granite model with a single injection well for hydraulic fracturing: (**a**) the model size and applied boundary conditions; and (**b**) particle densification near the injection well.

The simulated values and theoretical values of breakdown fractures are summarized in Figure 6, and the error between them is less than 20%. The main reason for the deviation is that the analytical model assumes that the entire reservoir is impermeable, while the modified hydraulic fracturing model in PFC considers the leakage of fracturing fluid to the surrounding domains.

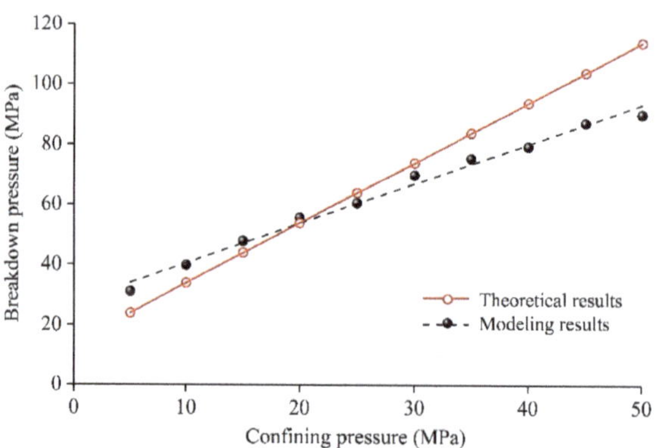

Figure 6. Comparison of breakdown pressures between the simulated values and the analytical solutions.

Figure 7 shows the representative borehole pressure histories, crack spatial distribution and corresponding fluid pressure field in hydraulic fracturing under different confining pressures. It is found that the borehole pressure increases rapidly at the beginning of injection, and as the borehole pressure approaches the breakdown pressure, the borehole pressure increases slowly and non-linearly, which reflects that the leakage of fracturing faster is greater when the borehole pressure is high. The leakage of fluid from the injection well and fractures to the surrounding fluid networks can also be seen from the figures of the fluid pressure distribution.

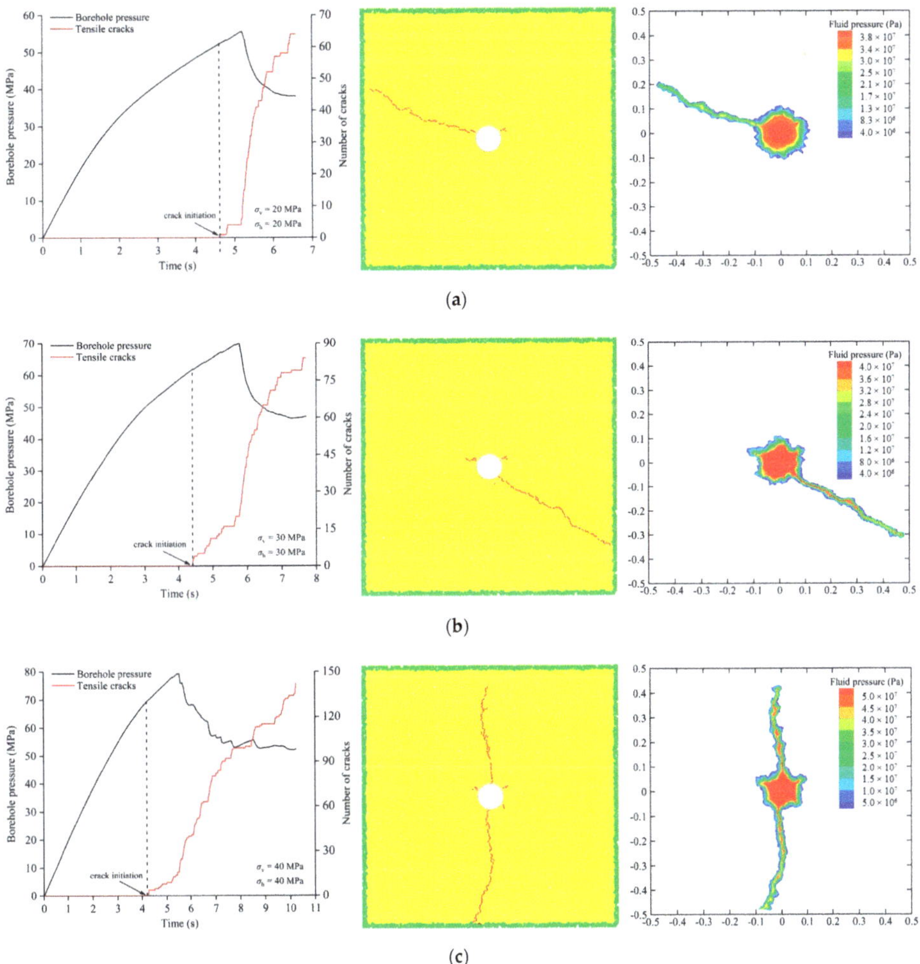

Figure 7. The representative borehole pressure histories, crack spatial distribution and corresponding fluid pressure field in the single-well fracturing under different in-situ stress conditions: (**a**) $\sigma_v = \sigma_h$ = 20 MPa; (**b**) $\sigma_v = \sigma_h$ = 30 MPa; and (**c**) $\sigma_v = \sigma_h$ = 40 MPa.

Before the borehole pressure reaches the breakdown pressure, a few cracks initiated around the wellbore. After increasing to the breakdown pressure, the borehole pressure dropped sharply to the fracture propagation pressure and remained unchanged. As a result, the number of microcracks grew dramatically along certain directions. Hubbert and Willis [63] proposed that the initiation and propagation of deep hydraulic fracturing fractures follow the direction of maximum principal stress.

From Figure 7, we can see that the macro hydraulic fractures propagated along the different directions under various confining pressures. The difficulty in predicting the propagation direction stems from the fact that the micro defects on the surface of the injection well have more significant impact on the initiation and propagation of hydraulic fractures under hydrostatic pressure conditions.

Another noteworthy issue is that almost all the hydraulic fractures acquired from the fluid-mechanical coupling algorithm are made up of tensile cracks, which seems to be distinct from lots of shear-type seismic events observed in hydraulic fracturing tests for granite [64,65]. Al-Busaidi et al. [35] have demonstrated that the shear failures recorded in the laboratory tests are caused by the slippage of the preexisting cracks in the specimens, and the mechanism of injection-induced fracture is predominantly tensile failure. In brief, the simulation results of granite fracturing using the modified fluid-mechanical coupling algorithm are accurate.

4. Hydraulic Fracturing Process in the Specimens with Two Injection Wells

4.1. Modeling Scenarios

The multi-well hydraulic fracturing behaviors of reservoir rocks not only affected by the influencing factors of single-well fracturing (e.g., the mechanical properties of rocks, fluid injection rate, fluid viscosity and in-situ stress), but also by the influencing factors unique to multi-well fracturing (e.g., the number of injection wells, injection sequence and well spacing). Taking the two-well hydraulic fracturing process as the research object, a larger numerical model with a width of 1.5 m and a height of 1.0 m, which was an aggregation of 17,700 circular particles, was constructed based on the fluid-solid coupling method introduced in Section 2 to get enough space to propagate for the fractures (Figure 8a). Two injection wells (Well No.1 and Well No.2) with the radius of 50 mm and the well spacing of L are symmetrical about the central axis of the model. After the calibration parameters in Table 1 were given in the two-well fracturing model, the influences of injection sequence and well spacing under different in-situ stress states on the breakdown pressure, fluid pressure distribution and hydraulic fracture propagation were discussed.

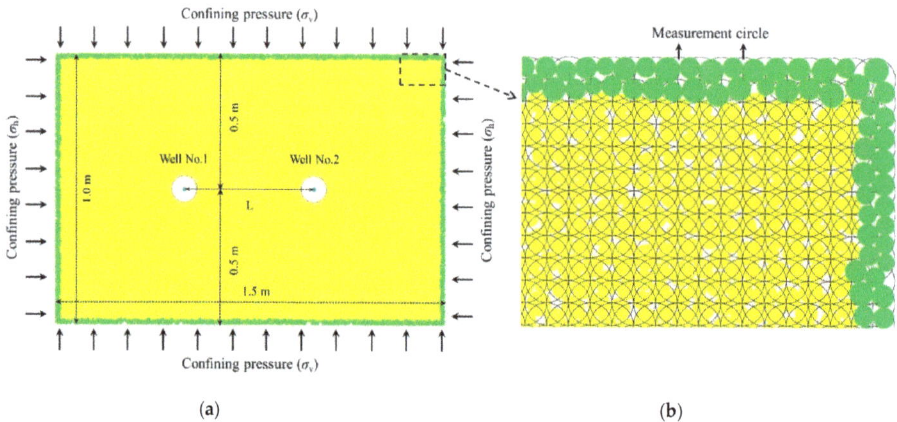

(a)   (b)

Figure 8. Granite model with two injection wells for hydraulic fracturing: (**a**) the model size and applied boundary conditions; and (**b**) arrangement of measurement circles.

4.2. The Influence of Injection Sequence under Different In-Situ Stress Conditions

The well spacing L of the two-well fracturing model was set to 0.5 m. Three injection sequence schemes were designed for two water injection wells, which are: Scheme 1—simultaneous injection in Well No.1 and Well No.2; Scheme 2—first injection in Well No.1 and second injection in Well No.2;

and Scheme 3—first injection in Well No.2 and second injection in Well No.1. The fluid injection rate for each well was always 1.0×10^{-5} m^3/s. If any hydraulic fracture propagates to the boundary of the model, the fluid injection and numerical simulation stop immediately, regardless of the shut-in stage in the field fracturing process. In the asynchronous fracturing of two injection wells, the second injection was conducted after the pore fluid pressure produced by the first injection was removed, so as to eliminate the effect of residual pore pressure.

The influence of different in-situ stresses was also considered in these injection schemes. When the model was under hydrostatic pressure, the vertical and horizontal confining pressures were fixed at 30 MPa; when the maximum in-situ stress was in the horizontal direction, the vertical and horizontal confining pressures were fixed at 15 MPa and 30 MPa, respectively. The three injection schemes have been repeatedly executed under each in-situ stress condition.

Figures 9–11 summarize the borehole pressure histories, crack spatial distribution and corresponding fluid pressure field for the fracturing of two-well granite model with the well spacing of 0.5 m under the hydrostatic pressure condition ($\sigma_v = \sigma_h = 30$ MPa).

For the Scheme 1 of injection sequence, the borehole pressure curves recorded in Well No.1 and Well No.2 were nearly coincident in the process of simultaneous injection, and the breakdown pressures of the two wells were 62.0 MPa and 57.7 MPa (Figure 9). Similar to the fracturing of the single well, microcracks initially originated at the micro defects around the boreholes, and increased rapidly as the borehole pressures reduced to the fracture propagation pressure. However, the hydraulic fractures between the two wells were deflected at a large angle and close to each other during the propagations along the initial microcrack directions. Especially from the corresponding distribution of pore pressure field, it can be seen that the pore pressure field in the surrounding rock resulted from the fluid seepage in the two fractures will be superimposed on the tips. The other two hydraulic fractures were also considerably distorted until they propagated to the calculation boundary. In the end, four fracture branches formed in the model.

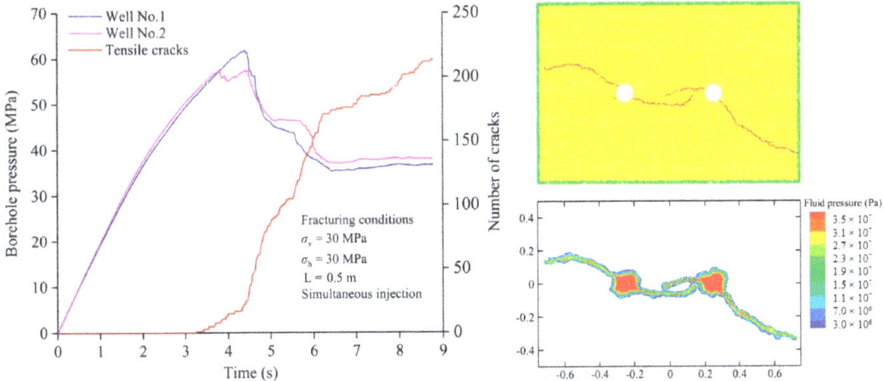

Figure 9. The borehole pressure histories, crack spatial distribution and corresponding fluid pressure field in the simultaneous fracturing under the hydrostatic pressure condition ($\sigma_v = \sigma_h = 30$ MPa).

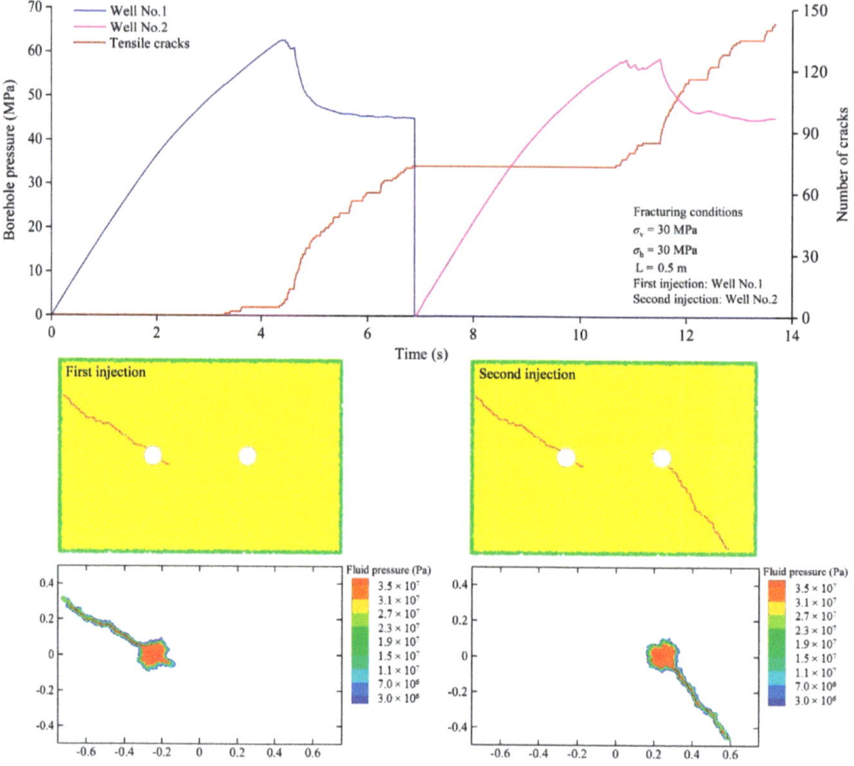

Figure 10. The borehole pressure histories, crack spatial distribution and corresponding fluid pressure field in the asynchronous fracturing (first injection: Well No.1 and second injection: Well No.2) under the hydrostatic pressure condition ($\sigma_v = \sigma_h = 30$ MPa).

For the Scheme 2 of injection sequence, the breakdown pressures of Well No.1 and Well No.2 were 62.7 MPa and 58.8 MPa (Figure 10). After the first injection and the second injection, both hydraulic fractures in the left and right part of the model extended straight to the boundary at a high angle without distortion. Although the microcracks between the two injection wells were also initiated in the model, they did not propagate further.

For the Scheme 3 of injection sequence, the breakdown pressures of the two wells were 64.3 MPa and 58.3 MPa (Figure 11). The final fracture propagation pattern was the same as that in the Scheme 2. The above simulation results may indicate that under the hydrostatic pressure condition, obvious differences between simultaneous injection and asynchronous injection exist in the fracturing behaviors, but the injection sequences in asynchronous fracturing has little effect on the generation of cracks.

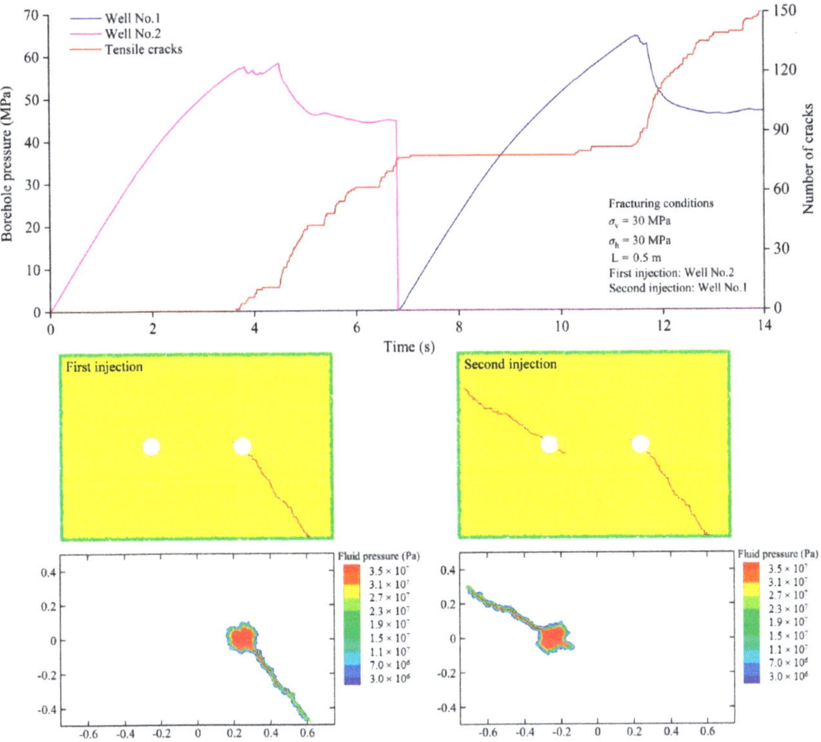

Figure 11. The borehole pressure histories, crack spatial distribution and corresponding fluid pressure field in the asynchronous fracturing (first injection: Well No.2 and second injection: Well No.1) under the hydrostatic pressure condition ($\sigma_v = \sigma_h = 30$ MPa).

Figures 12–14 present the borehole pressure histories, crack spatial distribution and corresponding fluid pressure field under the conditions of the three injection schemes when $\sigma_v = 15$ MPa and $\sigma_h = 30$ MPa. The breakdown pressures of the two wells were slightly reduced compared with those under hydrostatic pressure.

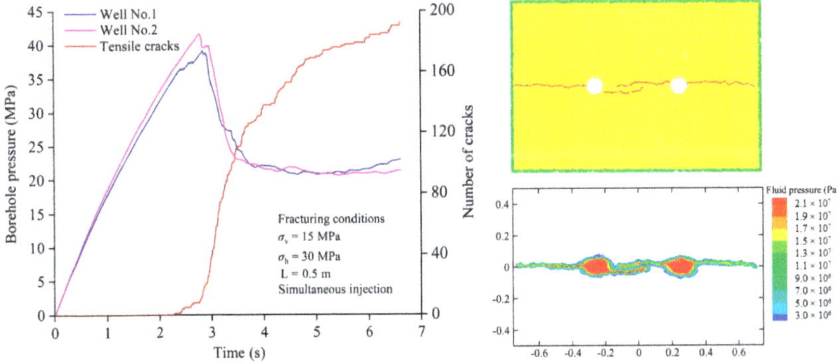

Figure 12. The borehole pressure histories, crack spatial distribution and corresponding fluid pressure field in the simultaneous fracturing under the in-situ differential stress condition ($\sigma_v = 15$ MPa and $\sigma_h = 30$ MPa).

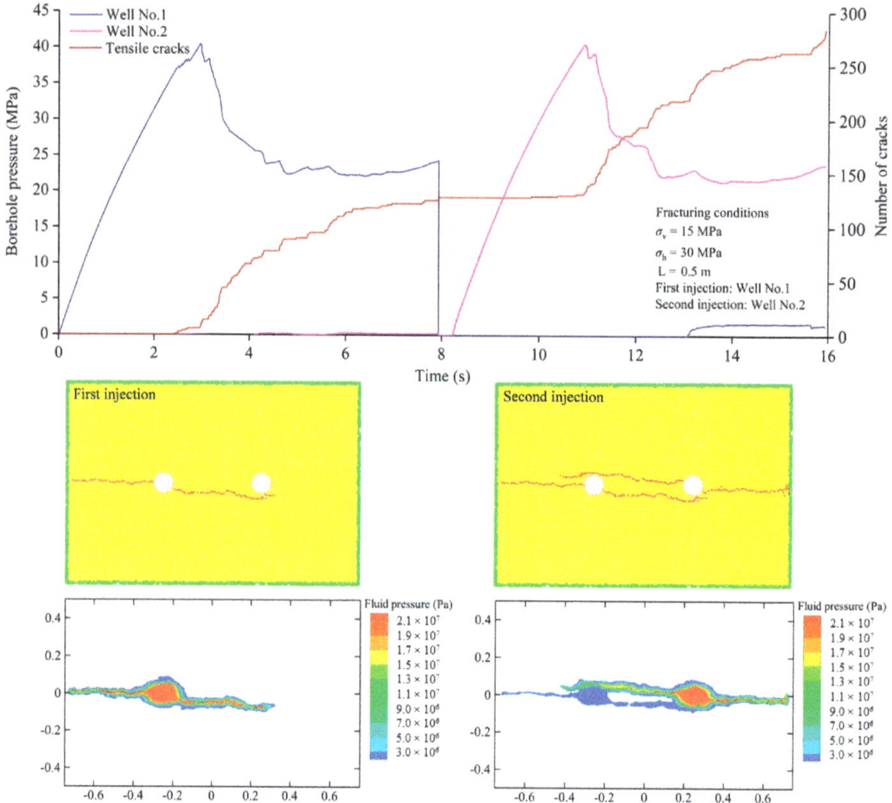

Figure 13. The borehole pressure histories, crack spatial distribution and corresponding fluid pressure field in the asynchronous fracturing (first injection: Well No.1 and second injection: Well No.2) under the in-situ differential stress condition (σ_v = 15 MPa and σ_h = 30 MPa).

In-situ differential stress gave rise to the propagations for the four hydraulic fractures along the horizontal direction in the simultaneous fracturing (Figure 12). The fractures between the two wells were not at the identical height, yet had a tendency to deflect and merge.

After the fracturing of Well No.1 in the Scheme 2, two hydraulic fractures extended along the horizontal direction to both sides of the borehole (Figure 13). The propagation pattern of the left hydraulic fracture was entirely consistent with that in simultaneous fracturing. Interestingly, the right hydraulic fracture was not bent and extended straight through the bottom of Well No.2. From the borehole pressure history and pore pressure distribution, it seems that Well No.2 was isolated from the hydraulic fracture as before. Stopping the injection in Well No.1 and continuing to fracture the Well No.2, the other two hydraulic fractures were produced along the horizontal direction. Due to the interference of the existing hydraulic fractures, the initiation position of the new fracture in the right part of the model has been changed in contrast to the fracture simulated by the Scheme 1. At about 13 s of the simulation time, a fluid pressure of 1.8 MPa was observed in Well No.1, which implies that the new hydraulic fracture between the injection wells extended from Well No.2 has been connected with Well No.1. A large quantity of fracturing fluid flowed into the Well No.1 through this fracture, and further migrated to fill the hydraulic fractures generated by the first fracturing. The four fracture branches in the model aggravated the leakage of fluid to the surrounding formation and prevented the borehole pressure of Well No.1 from increasing.

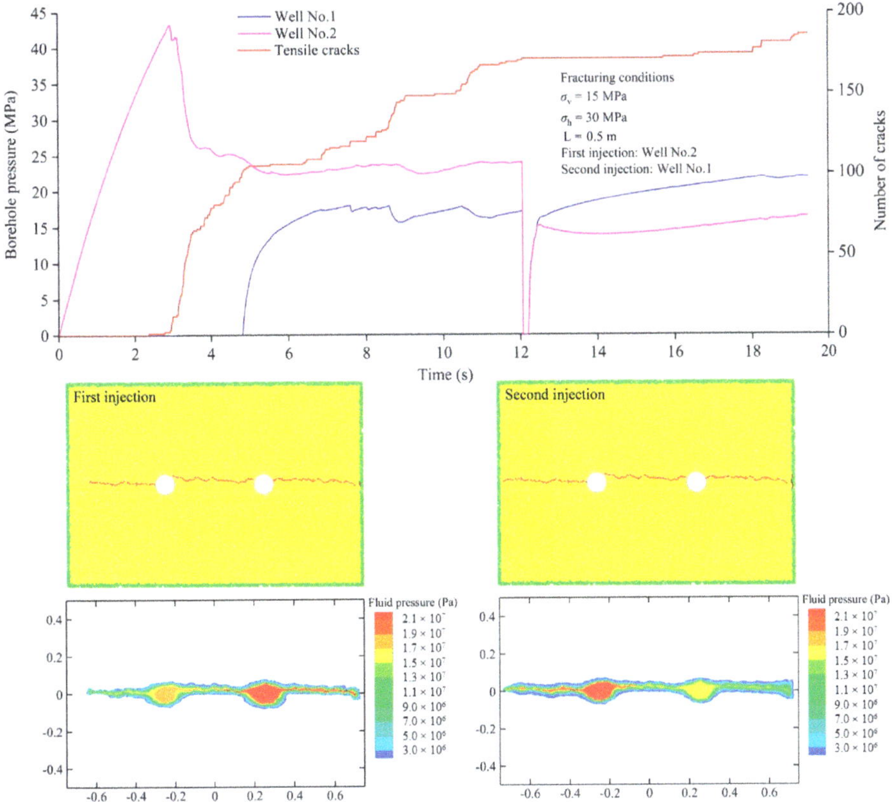

Figure 14. The borehole pressure histories, crack spatial distribution and corresponding fluid pressure field in the asynchronous fracturing (first injection: Well No.2 and second injection: Well No.1) under the in-situ differential stress condition (σ_v = 15 MPa and σ_h = 30 MPa).

The breakdown pressure of the Well No.2 was 43.3 MPa in the Scheme 3 (Figure 14). Unlike the fracturing in the Scheme 1 and Scheme 2, the hydraulic fracture between the two wells directly penetrated into the Well No.1 during the first injection, so that the borehole pressure of Well No.1 gradually rose to approximately 17 MPa. It is noteworthy that a new hydraulic fracture emerged from the outside of Well No.1 and propagated to the left edge of the model without reaching the previous breakdown pressures. This may be attributed to two aspects: on the one hand, the seepage of fracturing fluid has increased the pore fluid pressure and decreased the effective stress of the granite model; on the other hand, the rate of the fluid injection from Well No.2 into Well No.1 through the fracture was relatively low, which provided enough time for the stress of the solid framework around the Well No.1 to adjust. Additionally, the fluid pressure was found to decay steeply when flowing in the fractures on the left side of Well No.2. In the absence of sufficient fluid pressure, the propagation speed of the corresponding fractures was slow. During the second injection, the existing hydraulic fractures induced the fracturing fluid leak into the surrounding rock and limited the growth of borehole pressures for two wells. For example, the borehole pressure of Well No.1 was maintained near the fracture propagation pressure, which promoted the hydraulic fracture in the left part of the model to extend to the edge, but inhibited the generation of new hydraulic fracture.

The breakdown pressure, fracturing propagation pressure, crack number and hydraulic fracture branches after the fracturing under different in-situ stress states are listed in Table 3. In any injection sequence scheme, before the borehole pressure increased to the breakdown pressure, the fluid injected into the wells had no time to migrate in large area. Consequently, no evidence was found for the clear correlation between the breakdown pressure and injection sequence. As more fluid seeped into the rock matrix in the simultaneous injection, the effective stress between particles was markedly weakened, making the fluid pressure required for hydraulic fracture propagation lower than that in asynchronous fracturing. In the process of asynchronous fracturing, if the hydraulic fractures induced in the first injection have penetrated into the well for the second injection, the fluid pressure in the second injection well was hold at the fracture propagation pressure and the fractures extended tardily.

Table 3. Summary of the Fracturing Characteristics for the Two-Well Granite Model with the Well Spacing of 0.5 m under Different Injection Sequences.

In-Situ Stress Conditions	Hydraulic Fracturing Characteristics	Scheme 1		Scheme 2		Scheme 3	
		Well No.1	Well No.2	Well No.1	Well No.2	Well No.1	Well No.2
σ_v = 30 MPa σ_h = 30 MPa	Breakdown pressure (MPa)	62.0	57.7	62.7	58.8	64.3	58.3
	Fracture extension pressure (MPa)	36.7	38.1	45.3	44.8	46.2	44.6
	Crack number	214		145		149	
	Hydraulic fracture branches	4		2		2	
σ_v = 15 MPa σ_h = 30 MPa	Breakdown pressure (MPa)	39.2	41.6	40.4	40.5	-	43.3
	Fracture extension pressure (MPa)	20.8	20.7	22.3	21.7	16.5	23.6
	Crack number	192		285		186	
	Hydraulic fracture branches	4		4		3	

Overall, the propagations of the hydraulic fractures between the two wells can be divided into four modes: (1) extending linearly and directly penetrating the two wells; (2) extending linearly at different heights without connection; (3) extending at certain deflection angles and merging at the tips; (4) suspending the extension after initiation. Mode (2) and (3) are suggested to improve the complexity of fracture networks. By comparing the hydraulic fracture propagation modes for the three schemes, both simultaneous fracturing and asynchronous fracturing under different in-situ stresses were observed to have the ability to form multiple fracture branches, which depended on the superposition of fluid pressures. Under the hydrostatic pressure condition, the initial expansion direction of hydraulic fractures in the asynchronous fracturing is affected by the micro-defects around the borehole, largely leading to no fractures between the two wells. In contrast, simultaneous fracturing reduces the effective stress of the rock matrix and forces the hydraulic fractures to extend to this region. When the maximum in-situ stress is in the horizontal direction, the initial expansion direction of hydraulic fractures is also along the horizontal direction. In this case, simultaneous fracturing intensifies the coalescence of the fractures between the two wells and simplifies the fracture networks. Therefore, selecting the reasonable injection sequence according to the in-situ stress condition is necessary for stimulating reservoir to produce more hydraulic fractures.

4.3. The Influence of Well Spacing under Different In-Situ Stress Conditions

The two-well fracturing models with the well spacing L of 0.3 m, 0.4 m and 0.6 m were established respectively, and the micro mechanical parameters, permeability, injection rate and in-situ stress conditions consistent with those in Section 4.2 were allocated to the models. Simultaneous fracturing was conducted on each model without considering the influence of injection sequence. Figures 15 and 16 show the fracturing simulation results of the granite models with diverse well spacing under different in-situ stress conditions. Refer to Figures 9 and 12 for the simultaneous fracturing results when the well spacing is 0.5 m.

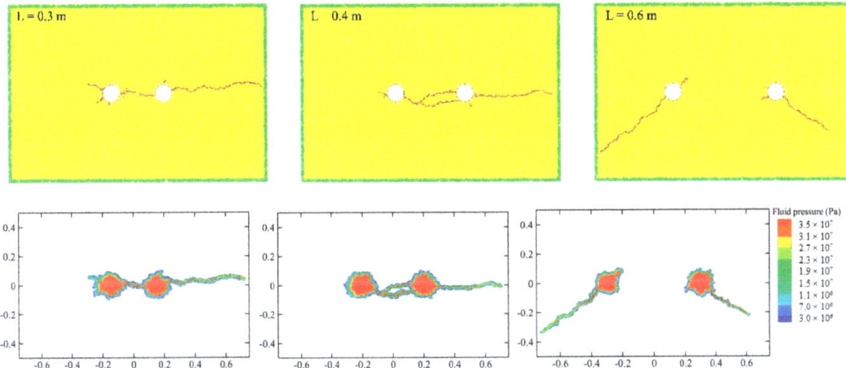

Figure 15. The crack spatial distributions and corresponding fluid pressure fields for the two-well granite models with the well spacing of 0.3 m, 0.4 m and 0.6 m in the simultaneous fracturing under the hydrostatic pressure condition ($\sigma_v = \sigma_h = 30$ MPa).

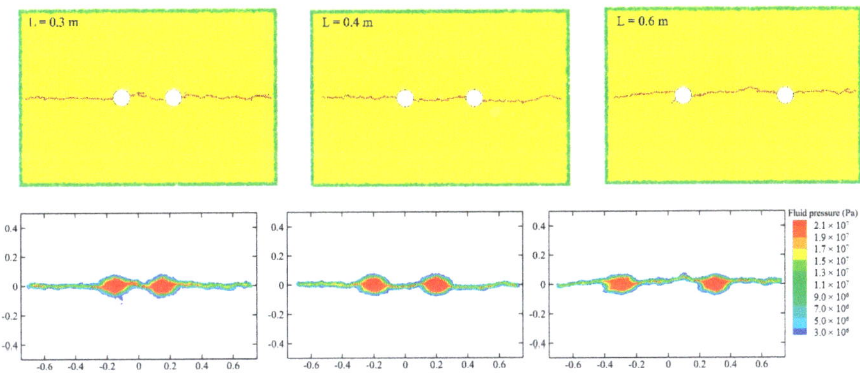

Figure 16. The crack spatial distributions and corresponding fluid pressure fields for the two-well granite models with the well spacing of 0.3 m, 0.4 m and 0.6 m in the simultaneous fracturing under the in-situ differential stress condition ($\sigma_v = 15$ MPa and $\sigma_h = 30$ MPa).

Under the hydrostatic pressure condition, the hydraulic fractures in these models exhibit the interesting propagation patterns. When L = 0.3 m, plenty of microcracks initiated around the Well No.1 and Well No.2. Two hydraulic fractures extended along the horizontal direction and merged into one fracture in the middle of the model. The fracturing fluid injected into Well No.1 flowed towards Well No.2 through the hydraulic fracture and assisted the hydraulic fracture on the right side of Well No.2 to extend to the model boundary, while other microcracks near Well No.1 were not allowed to further extend owing to insufficient fluid pressure. When L = 0.4 m, the hydraulic fractures initiated from the micro defects on the surface of the two wells extended along the initial directions of different heights and deflected to each other. Until the fracture from Well No.1 reached the bottom of Well No.2 and the fracture from Well No.2 was arrested by the fracture from Well No.1, the two wells were more closely connected. Counting the hydraulic fracture extending to the right edge, three fracture branches were developed in the model. When L = 0.5 m, the hydraulic fractures between the two wells were also distorted, but did not meet before the simulation stopped as a consequence of the larger well spacing. When the well spacing grew to 0.6 m, only two hydraulic fractures extended to the left and right edges of the model, and the microcracks between the two wells stagnated.

Under the in-situ differential stress condition, each model after fracturing had the similar propagation pattern for fractures. Initially, four fractures extended horizontally from both sides of the two injection wells. Then the two hydraulic fractures between the wells merged into one fracture. Finally, three major fracture branches crossed the whole model. With the increase of well spacing, the micro defects around the wells control the crack initiation, thereby creating the hydraulic fractures at the different heights to merge after deflection.

The representative fluid pressures and crack information for the models with different well spacing after the fracturing are listed in Table 4. Before the continuous propagation of hydraulic fractures, there were too few microcracks to build the hydraulic connection between the two wells. Not surprisingly, the breakdown pressures under the same in-situ stress condition for these models remained steady as the well spacing changes. By contrast, a positive correlation was found between well spacing and fracture propagation pressure under the hydrostatic pressure condition. In the fracturing of the models with small well spacing, due to the superposition of the leaking fluid in the middle of the model, the effective stress of the rock matrix in this region and the fluid pressure required for hydraulic fracture propagation were correspondingly reduced. Obviously, this superimposed effect of fluid was suppressed when the well spacing was set larger. In addition, this phenomenon of fracture propagation pressure had not been monitored when the hydraulic fractures were controlled by the maximum horizontal in-situ stress. Taken together, these results suggest that there is an association between well spacing and fracture branches. Too small or too large well spacing makes it difficult to construct complex fracture networks in the reservoir. Relatively speaking, to ensure the hydraulic fractures to extend in the area between the two wells, the well spacing under the hydrostatic pressure may be smaller than that under the in-situ differential stress.

Table 4. Summary of the Fracturing Characteristics for the Two-Well Granite Model with Different Well Spacing under the Simultaneous Fracturing.

In-Situ Stress Conditions	Hydraulic Fracturing Characteristics	L = 0.3 m		L = 0.4 m		L = 0.6 m	
		Well No.1	Well No.2	Well No.1	Well No.2	Well No.1	Well No.2
σ_v = 30 MPa σ_h = 30 MPa	Breakdown pressure (MPa)	62.3	61.6	60.1	59.2	61.8	62.4
	Fracture extension pressure (MPa)	36.9	37.2	38.2	40.1	45.6	46.5
	Crack number	138		150		118	
	Hydraulic fracture branches	2		3		2	
σ_v = 15 MPa σ_h = 30 MPa	Breakdown pressure (MPa)	39.8	37.4	40.2	38.7	40.6	39.9
	Fracture extension pressure (MPa)	21.2	20.9	21.0	21.1	20.8	22.8
	Crack number	179		160		163	
	Hydraulic fracture branches	3		3		3	

5. Analysis of Stress Shadow Effect

The fluid net pressure during the propagation of hydraulic fractures has contributed to the increase in the stress of the rock matrix around the fractures along the height direction. This important behavior of fractures is defined as "stress shadow effect", which has non-negligible impacts on the extension directions, opening degrees and shapes [66–68]. Especially in the situation of multiple injection wells in the reservoir, the interaction of the stress shadows for the adjacent hydraulic fractures becomes more intense and complicated.

However, discrete particle elements in the DEM suppress the direct expression of continuum physical quantities including stress and strain rate. The measurement circle is a monitoring mechanism built in PFC software to describe these quantities in a specified circular area by tracking the forces and displacements of particles and their related contacts. Therefore, the measurement circles densely distributed in the model realize the transition from the discrete physical quantities in the micro scale to the continuous ones in the macro scale. In order to obtain the rich data of measurement points, a total of 14,751 measurement circles with a radius of 0.01 m were regularly arranged in 99 rows and 149 columns to cover each two-well fracturing model (Figure 8b). The stress distributions of the models after fracturing were compared with their initial states without fluid injection to obtain the final stress increments.

Typical vertical stress increment distributions ($\Delta\sigma_y$) under different injection sequence and well spacing conditions are presented in Figures 17 and 18, respectively. All the models demonstrated a noticeable rise in the vertical stress on the upper and lower outer sides of the hydraulic fractures, which was in line with the basic law of stress shadow effect and proved the reliability of the modified fluid-mechanical coupling algorithm again. These vertical stress increments gradually decreased with the increasing distances to the fractures. In the area near the model boundaries, the vertical stress was roughly unchanged. Nevertheless, in the small area closest to the hydraulic fractures, the vertical stress was reduced rather than increased because of the leakage of fluid into the rock matrix. The injection of fluid led to an evident increase of vertical stress in the area around the wells without hydraulic fractures as well, which was higher than that caused by the hydraulic fractures.

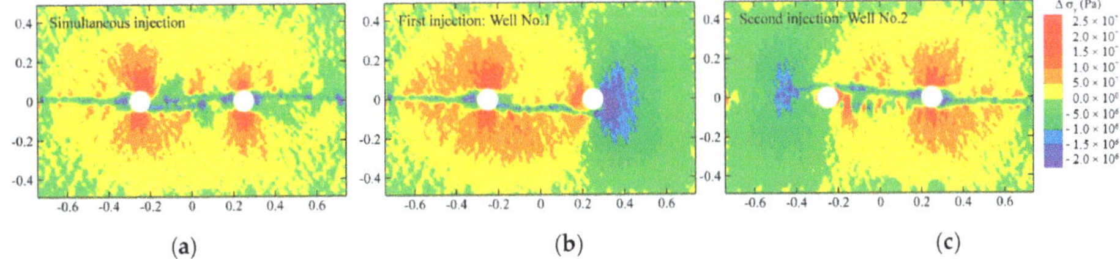

Figure 17. Vertical stress increment distributions for the two-well granite model with the well spacing of 0.5 m under different injection sequences when σ_v = 15 MPa and σ_h = 30 MPa: (**a**) simultaneous fracturing; (**b**) first injection in Well No.1; and (**c**) second injection in Well No.2.

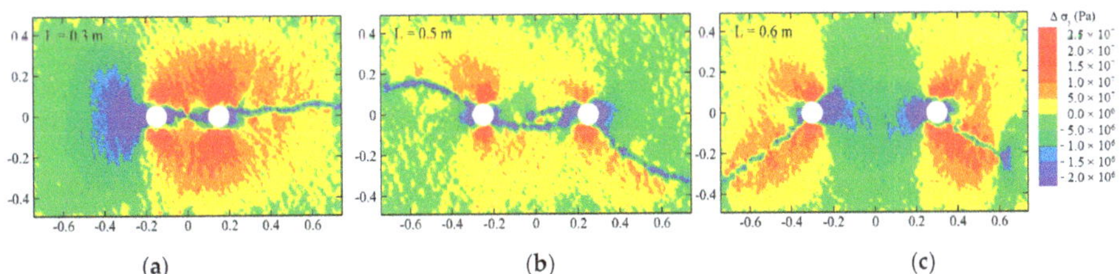

Figure 18. Vertical stress increment distributions for the two-well granite models with the different well spacing in the simultaneous fracturing when σ_v = 30 MPa and σ_h = 30 MPa: (**a**) L = 0.3 m; (**b**) L = 0.5 m; and (**c**) L = 0.6 m.

More concretely, the injection sequence and well spacing are the dominant factors in the stress shadow effect. For the simultaneous fracturing when σ_v = 15 MPa and σ_h = 30 MPa (Figure 17a), no expected superposition of the stress shadow effects existed between the two adjacent hydraulic fractures. Under such superposition effect, the two parallel hydraulic fractures would repel each other and progressively separate [69]. A possible explanation for this is that the tensile stress fields at the fracture tips and the fluid leakage effect both acted on the area between the two fractures, which diminished the vertical stress and promoted the mutual attraction of fractures. Excluding this area, the superposition effect of stress shadows was widely scattered in the middle of the model. For the first injection in Well No.1 (Figure 17b), a large range of vertical compressive stress declined at the tip of fracture on the right side of Well No.2, resulting from the concentration of tensile stress and the climb of pore fluid pressure. The initial compressive stress state in the model have been altered into the tensile stress state near the fracture tip, whereas the decrease of vertical stress in the region far from the tip may not satisfy the alteration of stress state. For the second injection in Well No.2 (Figure 17c), the stress shadow effect of the hydraulic fractures around Well No.1 was weakened stemming from the removal of preceding fluid and the primary vertical stress increment encompassed Well No.2. As a consequence, the new hydraulic fracture did not deflect to the old one, but moved away. The comparison of the fracturing under different injection sequences reveals that the influence range of stress shadow effect in simultaneous fracturing exceeds that in asynchronous fracturing.

When σ_v = 30 MPa and σ_h = 30 MPa, under the condition of small well spacing (Figure 18a), the superposition of stress shadow effect between the two injection wells enhanced greatly and restricted the initiation of hydraulic fractures in other directions. As the well spacing was slightly increased (Figure 18b), the superposition range of stress shadow effect reduced and the high vertical stress on the upper and lower sides of the wells prohibited the fractures in the middle of the model from twisting outwards. When L = 0.6 m (Figure 18c), no interaction between the stress shadow effects was produced by different fractures. Therefore, the superposition degree of stress shadow effects between the two wells is in inverse proportion to their well spacing.

6. Discussion and Conclusions

The two-dimensional fluid-mechanical coupling algorithm in PFC has been extensively used to study the flow of fracturing fluid in rock, hydraulic fracture propagation and induced seismicity [33,35–47,53–55]. In theory, it is practicable to conduct the two-dimensional simplified treatment for the relatively homogeneous and isotropic rocks in the macro scale, for the reason that the two-dimensional models can be regarded as a section of the real three-dimensional domains. Furthermore, two-dimensional simulation has the advantages of simple pre-processing, efficient calculation and intuitive results presentation in solving certain specific problems compared with three-dimensional simulation.

In this paper, the two-dimensional numerical models of Alxa porphyritic granites were strictly generated in PFC based on the PBM and the modified fluid-mechanical coupling algorithm. The micro input parameters in these models were effectively calibrated by the laboratory test results. Moreover, the accuracy of injection-induced fracturing was validated against the analytical solutions. A series of simulations for hydraulic fracturing in two-well granite models were performed to investigate the influences of injection sequence and well spacing on the breakdown pressure, fracture propagation pressure, fracture propagation pattern and stress shadow. The main results of this study are summarized as follows:

(1) The injection sequence and well spacing have significant effects on the fracture propagation pressure instead of the breakdown pressure. Under the condition of hydrostatic pressure, lower fracture propagation pressure is required by the simultaneous fracturing models or small well spacing models. But these effects generally decay as the vertical in-situ stress declines.

(2) Adjusting the injection sequence is a feasible way to control the propagation direction of hydraulic fractures. In the state of hydrostatic pressure, simultaneous fracturing allows more fracturing fluid to penetrate into the surrounding rock and compels the fractures to deflect to the horizontal direction, which is beneficial for the formation of complex fracture networks between the injection wells. The initial expansion of the fractures in asynchronous fracturing is easily disturbed by the micro defects around the boreholes, largely engendering no fractures between these wells. When the maximum in-situ stress is in the horizontal direction, in contrast to the asynchronous fracturing, simultaneous fracturing intensifies the coalescence of horizontal fractures.

(3) The reasonable well spacing in the reservoir affects the growth of fracture branches. Excessively small or large well spacing limits the generation of multiple fracture branches in fracture networks. Comparatively speaking, for getting more fractures between the injection wells, the

expected well spacing under the hydrostatic pressure may be smaller than that under the in-situ differential stress.

(4) The stress shadow effects of hydraulic fractures are also closely related to the injection sequence and well spacing. In the case of simultaneous fracturing or small well spacing, the superposition of the stress shadow effects in the middle of the model strengthens their impact. When the fluid injection decreases or the well spacing increases, this superposition will be suppressed.

In summary, the findings of this study not only reveal the propagation and interaction mechanisms of hydraulic fractures in multi-well fracturing, but also provide valuable reference for the optimization of fracturing technology in field production. However, the two-dimensional models still have an inevitable defect, that is, the hydraulic fractures are forced to expand in the selected section, which fails to represent the three-dimensional expansion patterns. Further works associated with the two-dimensional and three-dimensional multi-well fracturing simulations, such as the interaction between hydraulic and natural fractures, the temperature effect of fracturing fluid, and the proppant migration, should be involved in future research.

Author Contributions: Conceptualization, L.Z. and J.Z.; methodology, S.W.; software, J.Z.; validation, S.W. and Z.H.; investigation, S.W.; resources, L.Z.; data curation, Z.H.; writing—original draft preparation, S.W.; writing—review and editing, J.Z.; visualization, S.W.; supervision, L.Z. All authors have read and agreed to the published version of the manuscript.

Funding: This research was funded by the National Natural Science Foundation of China (Grant Nos. 41672321, 41972287, 41572312), China Postdoctoral Science Foundation (Grant Nos. 2018M630204, 2019T120133) and National Key Research and Development Program (Grant No.2018YFB1501801).

Conflicts of Interest: The authors declare no conflict of interest.

References

1.   Yuan, B.; Wood, D.A. A holistic review of geosystem damage during unconventional oil, gas and geothermal energy recovery. *Fuel* **2018**, *227*, 99–110. [CrossRef]

2.   Soeder, D.J. The successful development of gas and oil resources from shales in North America. *J. Pet. Sci. Eng.* **2018**, *163*, 399–420. [CrossRef]

3.   Kumari, W.G.P.; Ranjith, P.G.; Perera, M.S.A.; Li, X.; Li, L.H.; Chen, B.K.; Isaka, B.L.A.; De Silva, V.R.S. Hydraulic fracturing under high temperature and pressure conditions with micro CT applications: Geothermal energy from hot dry rocks. *Fuel* **2018**, *230*, 138–154. [CrossRef]

4.   Li, F.; Yuan, Q.; Li, T.; Li, Z.; Sun, C.; Chen, G. A review: Enhanced recovery of natural gas hydrate reservoirs. *Chin. J. Chem. Eng.* **2019**, *27*, 2062–2073. [CrossRef]

5.   Yang, Y.; Wang, K.; Zhang, L.; Sun, H.; Zhang, K.; Ma, J. Pore-scale simulation of shale oil flow based on pore network model. *Fuel* **2019**, *251*, 683–692. [CrossRef]

6.   Li, H.; Zhou, L.; Lu, Y.; Yan, F.; Zhou, J.; Tang, J. Changes in Pore Structure of Dry-hot Rock with Supercritical CO_2 Treatment. *Energy Fuels* **2020**, *34*, 6059–6068. [CrossRef]

7.   Warpinski, N.R. Hydraulic fracturing in tight, fissured media. *J. Pet. Technol.* **1991**, *43*. [CrossRef]

8.   Osiptsov, A.A. Fluid Mechanics of Hydraulic Fracturing: A Review. *J. Pet. Sci. Eng.* **2017**, *156*, 513–535. [CrossRef]

9.   Patel, S.M.; Sondergeld, C.H.; Rai, C.S. Laboratory studies of hydraulic fracturing by cyclic injection. *Int. J. Rock Mech. Min. Sci.* **2017**, *95*, 8–15. [CrossRef]

10. Yang, T.H.; Tham, L.G.; Tang, C.A.; Liang, Z.Z.; Tsui, Y. Influence of heterogeneity of mechanical properties on hydraulic fracturing in permeable rocks. *Rock Mech. Rock Eng.* **2004**, *37*, 251–275. [CrossRef]

11. Yarushina, V.M.; Bercovici, D.; Oristaglio, M.L. Rock deformation models and fluid leak-off in hydraulic fracturing. *Geophys. J. Int.* **2013**, *194*, 1514–1526. [CrossRef]

12. Dahi Taleghani, A.; Gonzalez, M.; Shojaei, A. Overview of numerical models for interactions between hydraulic fractures and natural fractures: Challenges and limitations. *Comput. Geotech.* **2016**, *71*, 361–368. [CrossRef]

13. Huang, L.; Liu, J.; Zhang, F.; Dontsov, E.; Damjanac, B. Exploring the influence of rock inherent heterogeneity and grain size on hydraulic fracturing using discrete element modeling. *Int. J. Solids Struct.* **2019**, *176–177*, 207–220. [CrossRef]

14. Ishida, T.; Aoyagi, K.; Niwa, T.; Chen, Y.; Murata, S.; Chen, Q.; Nakayama, Y. Acoustic emission monitoring of hydraulic fracturing laboratory experiment with supercritical and liquid CO_2. *Geophys. Res. Lett.* **2012**, *39*. [CrossRef]

15. Bennour, Z.; Watanabe, S.; Chen, Y.; Ishida, T.; Akai, T. Evaluation of stimulated reservoir volume in laboratory hydraulic fracturing with oil, water and liquid carbon dioxide under microscopy using the fluorescence method. *Geomech. Geophys. Geo-Energy Geo-Resour.* **2018**, *4*, 39–50. [CrossRef]

16. Gonçalves da Silva, B.; Einstein, H. Physical processes involved in the laboratory hydraulic fracturing of granite: Visual observations and interpretation. *Eng. Fract. Mech.* **2018**, *191*, 125–142. [CrossRef]

17. Li, B.Q.; Gonçalves da Silva, B.; Einstein, H. Laboratory hydraulic fracturing of granite: Acoustic emission observations and interpretation. *Eng. Fract. Mech.* **2019**, *209*, 200–220. [CrossRef]

18. Guo, T.; Zhang, S.; Qu, Z.; Zhou, T.; Xiao, Y.; Gao, J. Experimental study of hydraulic fracturing for shale by stimulated reservoir volume. *Fuel* **2014**, *128*, 373–380. [CrossRef]

19. Zhang, X.; Lu, Y.; Tang, J.; Zhou, Z.; Liao, Y. Experimental study on fracture initiation and propagation in shale using supercritical carbon dioxide fracturing. *Fuel* **2017**, *190*, 370–378. [CrossRef]

20. He, J.; Li, X.; Yin, C.; Zhang, Y.; Lin, C. Propagation and characterization of the micro cracks induced by hydraulic fracturing in shale. *Energy* **2020**, *191*, 116449. [CrossRef]

21. Zhuang, L.; Jung, S.G.; Diaz, M.; Kim, K.Y.; Hofmann, H.; Min, K.B.; Zang, A.; Stephansson, O.; Zimmermann, G.; Yoon, J.S. Laboratory True Triaxial Hydraulic Fracturing of Granite Under Six Fluid Injection Schemes and Grain-Scale Fracture Observations. *Rock Mech. Rock Eng.* **2020**, 1–16. [CrossRef]

22. Jing, L. A review of techniques, advances and outstanding issues in numerical modelling for rock mechanics and rock engineering. *Int. J. Rock Mech. Min. Sci.* **2003**, *40*, 283–353. [CrossRef]

23. Jing, L.; Hudson, J.A. Numerical methods in rock mechanics. *Int. J. Rock Mech. Min. Sci.* **2002**, *39*, 409–427. [CrossRef]

24. Zhang, Y.; Wong, L.N.Y. A review of numerical techniques approaching microstructures of crystalline rocks. *Comput. Geosci.* **2018**, *115*, 167–187. [CrossRef]

25. Li, J.; Dong, S.; Hua, W.; Li, X.; Pan, X. Numerical investigation of hydraulic fracture propagation based on cohesive zone model in naturally fractured formations. *Processes* **2019**, *7*, 28. [CrossRef]

26. Liao, J.; Hou, M.Z.; Mehmood, F.; Feng, W. A 3D approach to study the interaction between hydraulic and natural fracture. *Environ. Earth Sci.* **2019**, *78*, 1–18. [CrossRef]

27. Zangeneh, N.; Eberhardt, E.; Bustin, R.M.; Bustin, A. A numerical investigation of fault slip triggered by hydraulic fracturing. In Proceedings of the ISRM International Conference for Effective and Sustainable Hydraulic Fracturing, Brisbane, Australia, 20–22 May 2013; pp. 477–488. [CrossRef]

28. Janiszewski, M.; Shen, B.; Rinne, M. Simulation of the interactions between hydraulic and natural fractures using a fracture mechanics approach. *J. Rock Mech. Geotech. Eng.* **2019**, *11*, 1138–1150. [CrossRef]

29. Potyondy, D.O.; Cundall, P.A. A bonded-particle model for rock. *Int. J. Rock Mech. Min. Sci.* **2004**, *41*, 1329–1364. [CrossRef]

30. Hazzard, J.F.; Young, R.P.; Maxwell, S.C. Micromechanical modeling of cracking and failure in brittle rocks. *J. Geophys. Res. Solid Earth* **2000**, *105*, 16683–16697. [CrossRef]

31. Fakhimi, A.; Carvalho, F.; Ishida, T.; Labuz, J.F. Simulation of failure around a circular opening in rock. *Int. J. Rock Mech. Min. Sci.* **2002**, *39*, 507–515. [CrossRef]

32. Shan, P.; Lai, X. Mesoscopic structure PFC~2D model of soil rock mixture based on digital image. *J. Vis. Commun. Image Represent.* **2019**, *58*, 407–415. [CrossRef]

33. Liu, G.; Sun, W.C.; Lowinger, S.M.; Zhang, Z.H.; Huang, M.; Peng, J. Coupled flow network and discrete element modeling of injection-induced crack propagation and coalescence in brittle rock. *Acta Geotech.* **2019**, *14*, 843–868. [CrossRef]

34. Cundall, P.A. *Fluid Formulation for PFC2D*; Itasca Consulting Group: Minneapolis, MN, USA, 2000.

35. Al-Busaidi, A.; Hazzard, J.F.; Young, R.P. Distinct element modeling of hydraulically fractured Lac du Bonnet granite. *J. Geophys. Res. Solid Earth* **2005**, *110*, 1–14. [CrossRef]

36. Yoon, J.; Backers, T.; Dresen, G. Prototype PFC2D model for simulation of hydraulic fracturing and induced seismicity. In Proceedings of the ISRM International Symposium—EUROCK, Stockholm, Sweden, 28–30 May 2012.

37. Yoon, J.S.; Zimmermann, G.; Zang, A. Discrete element modeling of cyclic rate fluid injection at multiple locations in naturally fractured reservoirs. *Int. J. Rock Mech. Min. Sci.* **2015**, *74*, 15–23. [CrossRef]

38. Yoon, J.S.; Zimmermann, G.; Zang, A.; Stephansson, O. Discrete element modeling of fluid injection–induced seismicity and activation of nearby fault. *Can. Geotech. J.* **2015**, *52*, 1457–1465. [CrossRef]

39. Tomac, I.; Gutierrez, M. Coupled hydro-thermo-mechanical modeling of hydraulic fracturing in quasi-brittle rocks using BPM-DEM. *J. Rock Mech. Geotech. Eng.* **2017**, *9*, 92–104. [CrossRef]

40. Tomac, I.; Gutierrez, M. Micromechanics of Hydro-Thermo-Mechanical Processes in Rock Accounting for Thermal Convection. *J. Eng. Mech.* **2019**, *145*. [CrossRef]

41. Zhou, J.; Zhang, L.; Han, Z. Hydraulic fracturing process by using a modified two-dimensional particle flow code-method and validation. *Prog. Comput. Fluid Dyn.* **2017**, *17*, 52–62. [CrossRef]

42. Zhang, L.; Zhou, J.; Han, Z. Hydraulic fracturing process by using a modified two-dimensional particle flow code-case study. *Prog. Comput. Fluid Dyn.* **2017**, *17*, 13–26. [CrossRef]

43. Zhou, J.; Zhang, L.; Braun, A.; Han, Z. Numerical Modeling and Investigation of Fluid-Driven Fracture Propagation in Reservoirs Based on a Modified Fluid-Mechanically Coupled Model in Two-Dimensional Particle Flow Code. *Energies* **2016**, *9*, 699. [CrossRef]

44. Zhou, J.; Zhang, L.; Pan, Z.; Han, Z. Numerical investigation of fluid-driven near-borehole fracture propagation in laminated reservoir rock using PFC2D. *J. Nat. Gas Sci. Eng.* **2016**, *36*, 719–733. [CrossRef]

45. Zhou, J.; Zhang, L.; Pan, Z.; Han, Z. Numerical studies of interactions between hydraulic and natural fractures by Smooth Joint Model. *J. Nat. Gas Sci. Eng.* **2017**, *46*, 592–602. [CrossRef]

46. Zhou, J.; Zhang, L.; Braun, A.; Han, Z. Investigation of processes of interaction between hydraulic and natural fractures by PFC modeling comparing against laboratory experiments and analytical models. *Energies* **2017**, 1001. [CrossRef]

47. Zhang, L.; Zhou, J.; Braun, A.; Han, Z. Sensitivity analysis on the interaction between hydraulic and natural fractures based on an explicitly coupled hydro-geomechanical model in PFC2D. *J. Pet. Sci. Eng.* **2018**, *167*, 638–653. [CrossRef]

48. Itasca Consulting Group Inc. *Manual PFC2D (Particle Flow Code)*; Version 4.0 Users' Guide; Itasca Consulting Group Inc.: Minneapolis, MN, USA, 2008.

49. Cho, N.; Martin, C.D.; Sego, D.C. A clumped particle model for rock. *Int. J. Rock Mech. Min. Sci.* **2007**, *44*, 997–1010. [CrossRef]

50. Wu, S.; Xu, X. A Study of Three Intrinsic Problems of the Classic Discrete Element Method Using Flat-Joint Model. *Rock Mech. Rock Eng.* **2016**, *49*, 1813–1830. [CrossRef]

51. Potyondy, D.O. Parallel-Bond Refinements to Match Macroproperties of Hard Rock. In Proceedings of the 2nd FLAC/DEM Symposium, Melbourne, Australia, 14–16 February 2011.

52. Ding, X.; Zhang, L. A new contact model to improve the simulated ratio of unconfined compressive strength to tensile strength in bonded particle models. *Int. J. Rock Mech. Min. Sci.* **2014**, *69*, 111–119. [CrossRef]

53. Yoon, J.S.; Zang, A.; Stephansson, O. Numerical investigation on optimized stimulation of intact and naturally fractured deep geothermal reservoirs using hydro-mechanical coupled discrete particles joints model. *Geothermics* **2014**, *52*, 165–184. [CrossRef]

54. Han, Z.; Zhou, J.; Zhang, L. Influence of grain size heterogeneity and in-situ stress on the hydraulic fracturing process by PFC2D Modeling. *Energies* **2018**, *11*, 1413. [CrossRef]

55. Zhao, X.; Paul Young, R. Numerical modeling of seismicity induced by fluid injection in naturally fractured reservoirs. *Geophysics* **2011**, *76*. [CrossRef]

56. Yoon, J. Application of experimental design and optimization to PFC model calibration in uniaxial compression simulation. *Int. J. Rock Mech. Min. Sci.* **2007**, *44*, 871–889. [CrossRef]

57. Schöpfer, M.P.J.; Childs, C.; Walsh, J.J. Two-dimensional distinct element modeling of the structure and growth of normal faults in multilayer sequences: 1. Model calibration, boundary conditions, and selected results. *J. Geophys. Res. Solid Earth* **2007**, *112*. [CrossRef]

58. Ajamzadeh, M.R.; Sarfarazi, V.; Haeri, H.; Dehghani, H. The effect of micro parameters of PFC software on the model calibration. *Smart Struct. Syst.* **2018**, *22*, 643–662. [CrossRef]

59. Wang, Y.; Tonon, F. Modeling Lac du Bonnet granite using a discrete element model. *Int. J. Rock Mech. Min. Sci.* **2009**, *46*, 1124–1135. [CrossRef]

60. Zhou, J.; Zhang, L.; Yang, D.; Braun, A.; Han, Z. Investigation of the quasi-brittle failure of alashan granite viewed from laboratory experiments and grain-based discrete element modeling. *Materials* **2017**, *10*, 835. [CrossRef]

61. Zhou, J.; Lan, H.; Zhang, L.; Yang, D.; Song, J.; Wang, S. Novel grain-based model for simulation of brittle failure of Alxa porphyritic granite. *Eng. Geol.* **2019**, *251*, 100–114. [CrossRef]

62. Haimson, B.; Fairhurst, C. Initiation and Extension of Hydraulic Fractures in Rocks. *Soc. Pet. Eng. J.* **1967**, *7*, 310–318. [CrossRef]

63. Hubbert, M.K.; Willis, D.G. Mechanics of Hydraulic Fracturing. *Pet. Trans. Aime.* **1957**, *210*, 153–168. [CrossRef]

64. Mao, R.; Feng, Z.; Liu, Z.; Zhao, Y. Laboratory hydraulic fracturing test on large-scale pre-cracked granite specimens. *J. Nat. Gas Sci. Eng.* **2017**, *44*, 278–286. [CrossRef]

65. Xing, Y.; Zhang, G.; Luo, T.; Jiang, Y.; Ning, S. Hydraulic fracturing in high-temperature granite characterized by acoustic emission. *J. Pet. Sci. Eng.* **2019**, *178*, 475–484. [CrossRef]

66. Germanovich, L.N.; Astakhov, D.K. Fracture closure in extension and mechanical interaction of parallel joints. *J. Geophys. Res. Solid Earth* **2004**, *109*. [CrossRef]

67. Olson, J.E. Multi-fracture propagation modeling: Applications to hydraulic fracturing in shales and tight gas sands. In Proceedings of the 42nd US Rock Mechanics Symposium and 2nd US-Canada Rock Mechanics Symposium, San Francisco, CA, USA, 29 June–2 July 2008.

68. Zhou, L.; Chen, J.; Gou, Y.; Feng, W. Numerical investigation of the time-dependent and the proppant dominated stress shadow effects in a transverse multiple fracture system and optimization. *Energies* **2017**, *10*, 83. [CrossRef]

69. Zhang, X.; Jeffrey, R.G.; Thiercelin, M. Deflection and propagation of fluid-driven fractures at frictional bedding interfaces: A numerical investigation. *J. Struct. Geol.* **2007**, *29*, 396–410. [CrossRef]

Article

Proppant Transportation in Cross Fractures: Some Findings and Suggestions for Field Engineering

Yan Zhang [1,2], **Xiaobing Lu** [1,2], **Xuhui Zhang** [1,2,*] **and Peng Li** [1,2]

[1]   Institute of Mechanics, Chinese Academy of Sciences, Beijing 100190, China;
     zhangyan162@imech.ac.cn (Y.Z.); xblu@imech.ac.cn (X.L.); lipeng@imech.ac.cn (P.L.)
[2]   School of Engineering Science, University of Chinese Academy of Sciences, Beijing 100049, China
*   Correspondence: zhangxuhui@imech.ac.cn; Tel.: +86-10-82544192

Received: 10 August 2020; Accepted: 18 September 2020; Published: 19 September 2020

Abstract: The proppant transportation is a typical two-phase flow process in a complex cross fracture network during hydraulic fracturing. In this paper, the proppant transportation in cross fractures is investigated by the computational fluid dynamics (CFD) method. The Euler–Euler two-phase flow model and the kinetic theory of granular flow (KTGF) are adopted. The dimensionless controlling parameters are derived by dimensional analysis. The equilibrium proppant height (EPH) and the ratio of the proppant mass (RPM) in the secondary fracture to that in the whole cross fracture network are used to describe the movement and settlement of proppants in the cross fractures. The main features of the proppant transportation in the cross fractures are given, and several relative suggestions are presented for engineering application in the field. The main controlling dimensionless parameters for relative EPH are the proppant Reynolds number and the inlet proppant volume fraction. The dominating dimensionless parameters for RPM are the relative width of the primary and the secondary fracture. Transportation of the proppants with a certain particle size grading into the cross fractures may be a good way for supporting the hydraulic fractures.

Keywords: proppant transportation; cross fractures; CFD simulation; dimensional analysis; equilibrium proppant height

---

1. Introduction

Unconventional energy resources such as low permeability, shale, and tight oil and gas reservoirs account for a larger and larger proportion in the present oil and gas exploration [1–4]. The hydraulic fractures are the main flow channel for these fluid resources due to the natural poor flow capability of the porous media, and it is of great importance to know the effective support range and the distribution of proppants in cross fractures.

Many researchers have studied the proppant transportation in the cross fractures by experiments and numerical simulations. Alotaibi and Miskimins [5] designed a cross fracture system with one primary fracture, three secondary fractures, and two tertiary fractures. They found that the proppants were able to flow into the subsidiary fractures and form a proppant bed. However, they did not realize that the proppants moved not so far in the subsidiary fractures. The transportation distance of the proppant in the subsidiary fractures is important for the production. Sahai et al. [6] investigated the effects of the fracture geometrical complexity, the pumping rates, the proppant concentration, and the proppant size on the proppant transportation. The mechanism of the proppant from the primary fracture into the secondary fracture was also analyzed. McClure [7] analyzed in detail the formation process of the equilibrium proppant height (EPH). As the proppants settle at the bottom of the fractures, the height of the proppant bed gradually grows to EPH during transportation [7]. The velocity of the mixture of water and proppants above the proppant bed in the fractures is called the equilibrium mixture velocity (EMV) when the EPH is reached. It is worth noting that the proppant bed will be eroded if the proppant bed height is greater than the EPH. The proppant bed height decreases over time during erosion until the EPH is reached. The Euler–Lagrange method was used by Hu et al. [8] to study the proppant transportation in a single vertical fracture. They suggested that the coarse proppants may be transported first, followed by the fine proppants. The coarse proppants will form a proppant bed quickly and then the fine proppants are transported far in the fractures. Roostaei et al. [9] combined a proppant transport model with the numerical hydraulic fracture model to study the fracturing response and the effect of proppant injection on the fracture propagation and dimensions. They found that the fluid viscosity is the most important parameter on the proppant transportation.

Little attention is paid to the amount of proppants entering the secondary fracture, which is a very important quantity for the field engineering. In addition, the mechanism of the proppant moving from the primary fracture into the secondary fracture is also not well understood.

In this paper, the proppant transportation behaviors in the cross fractures are investigated in detail based on the previous work [10] of our group. The Euler–Euler two-phase flow model combined with the kinetic theory of granular flow (KTGF) approach is used. Two dimensionless parameters to describe the proppant distribution in the cross fractures are presented. The first one is the relative EPH, representing the ratio of the EPH to the height of the cross fractures. The other one is the ratio of the proppant mass (RPM) in the secondary fracture to that in the whole cross fracture network. In Section 2, the numerical model and the dimensionless parameters relative to the proppant transportation in the cross fractures are introduced. Compared to the previous work [10], the boundary conditions and the fracture width in the numerical model have been changed to capture more information of the flow behaviors and to be practical for the field engineering. In Section 3, the effects of the dimensionless parameters are analyzed. Some suggestions are given for the field engineering application based on the simulated results.

2. Methods

2.1. CFD Simulation

2.1.1. Model Descriptions

The Euler–Euler two-phase flow model is used to simulate the proppant transportation in the cross fractures. This model takes the multitude of particles as an artificial solid phase that can interpenetrate the continuous water phase [11,12]. The KTGF approach is used to simulate the particle collisions. Based on the KTGF approach, an additional equation, i.e., the particle temperature equation, is solved to represent the fluctuations of the particles. This equation leads to additional terms as the particle pressure force and so on. More details about this model were detailed in the previous paper [10,13].

2.1.2. Geometry and Mesh

The geometry of the cross fractures in the numerical simulation is shown in Figure 1. The cross fractures contain one primary fracture and one secondary fracture. The secondary fracture intersects the primary fracture at a certain angle (bypass angle θ). It is assumed that the height of the cross fractures remains constant along the moving direction of the mixture, and the height of the primary fracture equals that of the secondary fracture. The scales of the primary fracture and secondary fracture are length × height = 1000 mm × 150 mm and 600 mm × 150 mm, respectively. The horizontal distance between the entrance of the secondary fracture and the primary fracture is 400 mm. The width and bypass angle (θ) of the primary and secondary fracture are varied in different cases. The geometry references the experiment apparatus of Alotaibi and Miskimins [5], Tong and Mohanty [11], and Patankar et al. [14]. The mixture of the proppants and water enters the primary fracture from an inlet which is simplified as a rectangular opening. In actual hydraulic fractures, the proppants are generally blocked by an unruptured stratum. That means that the front end of the cross fractures is closed. Therefore, a wall is set at the end of the geometry in the numerical model. The mixture is only permitted to flow out of the cross fractures through the outlet on the top of the wall (Figure 1). The flow field is divided into hexahedral structured cells due to the consideration of computation accuracy and convergence. A standard case is first calculated with the proppant density ρ_s = 3300 kg/m^3, the proppant diameter d_s = 0.5 mm, the water density ρ_l = 1000 kg/m^3, the water viscosity μ_l = 0.001 Pa·s, the secondary fracture width w_b = 1.5 mm, the primary fracture width w_a = 5 mm, the injection velocity (U_0) 0.2 m/s, and inlet proppant volume fraction (α_{s0}) 3%. To evaluate the mesh independence, three kinds of mesh with different sizes are performed. The mesh size in the height, length, and width is 4 × 4 × 1 mm (coarse), 2 × 2 × 1 mm (medium), and 2 × 2 × 0.5 mm (fine), respectively. The EPH obtained from the medium grid is similar to that from the fine grid. So, the grid size 2 × 2 × 0.5 mm is used.

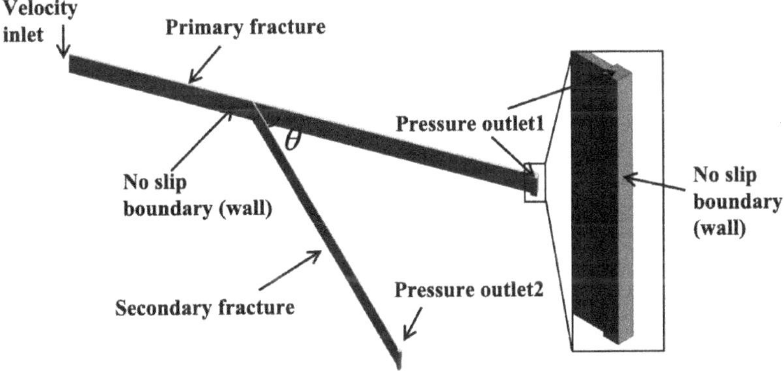

Figure 1. The geometry representing the cross fractures. The cross fractures consist of a primary fracture and a secondary fracture in length × height: 1000 mm × 150 mm and 600 mm × 150 mm, respectively. The width and the bypass angle will be changed in different cases.

2.1.3. Initial and Boundary Conditions

Initially, the cross fractures are filled with water, and then the proppants begin to enter the cross fractures. The velocity inlet is set. Different pumping flow rates and sand ratios are imposed by varying the injection rates and the inlet volume fraction of the proppants. The inner walls of the cross fractures are set as the no-slip wall boundary conditions for each phase. The pressure outlet is set to be zero gauge pressure. The values of the parameters used in the numerical simulation are listed in Table 1. The ANSYS FLUENT software is used for the numerical simulation.

Table 1. Parameter values in the numerical simulation.

Parameters	Units	Value
Proppant diameter	mm	0.5
Water density	kg/m^3	1000
Water viscosity	Pa·s	0.001
Primary fracture: length × height	mm	1000 × 150
Secondary fracture: length × height	mm	600 × 150
Size of grid: length × height × width	mm	2 × 2 × 0.5

2.1.4. Solution Algorithm

The phase coupled Semi-Implicit Method for Pressure Linked Equations (SIMPLE) method is used for the pressure–velocity coupling, and the Green–Gauss cell-based method is used to discretize the gradient. The equation of momentum, the particle temperature equation, the turbulent kinetic energy, and the dissipation rate equation are treated with the second-order upwind scheme. The Quadratic Upstream Interpolation for Convective Kinetics (QUICK) scheme is used to discretize the volume fraction equation.

2.2. Dimensional Analysis

All the variables of the problem relating the proppant transportation in the cross fractures based on the equations and the boundary conditions are shown and explained as follows.

The variables of geometry of the cross fractures: the width of the primary fracture and the secondary fracture (w_a and w_b), the length of the primary and the secondary fracture (L_a and L_b), the height of the cross fractures (H), the distance from the secondary fracture to the primary fracture entrance (l), and the angle between the primary and secondary fracture (bypass angle θ).

The variables of physical properties of the proppant and water: the density of the proppant and the water (ρ_s and ρ_l), the average diameter of the proppant (d_s), and the viscosity of the water (μ_l).

The variables relating the boundary conditions: the injection velocity of the mixture (U_0), the inlet volume fraction of the proppant (α_{s0}), and the gravity acceleration g.

Two parameters to evaluate the quality of hydraulic fracturing are introduced. The first one is the EPH in the primary fracture, and the other one is the RPM. The EPH in the cross fractures is discussed by many researchers [5,13,15,16]. The EPH and the RPM can be written as a causal function of the above variables:

$$(h, R) = f(\alpha_{s0}, U_0, g; \theta, w_a, w_b, L_a, L_b, l, H; \rho_s, d_s, \rho_l, \mu_l) \tag{1}$$

where h and R are the EPH and the RPM, respectively. The dimensionless causal relationship can be written as [17]:

$$\left(\frac{h}{H}, R\right) = f\left(\alpha_{s0}, Re, Ar, \theta, \frac{w_a}{d_s}, \frac{w_b}{d_s}, \frac{L_a}{d_s}, \frac{L_b}{d_s}, \frac{l}{d_s}, \frac{H}{d_s}, \frac{\rho_s}{\rho_l}\right) \tag{2}$$

in which h/H is the relative EPH, $Re = \rho_l d_s U_0 / \mu_l$ is the proppant Reynolds number, $Ar = (\rho_s - \rho_l)\rho_l d_s^3 g / \mu_l^2$ is the Archimedes number, w_a/d_s and w_b/d_s are the relative width of the primary and secondary fracture, L_a/d_s and L_b/d_s are the relative length of the primary and secondary fracture, l/d_s is the relative distance of the secondary fracture to the primary fracture entrance, H/d_s is the relative height of the cross fractures, and ρ_s/ρ_l is the ratio of the proppant density to the water density. Here, the relative length and height of the cross fractures as well as the density ratio of the proppant to water are constant. Equation (2) can be rewritten as:

$$\left(\frac{h}{H}, R\right) = f\left(\alpha_{s0}, Re, Ar, \theta, \frac{w_a}{d_s}, \frac{w_b}{d_s}, \frac{l}{d_s}\right). \tag{3}$$

The effect of the controlling dimensionless independent variables at the left side of Equation (3) on the dependent variables is investigated in this paper. Table A1 in Appendix A shows all the cases adopted in the numerical simulation. The bold part in Table A1 is the standard case. Cases 1–8 are set to study the effects of the inlet proppant volume fraction. Cases 3 and 9–13 are set to study the effects of the proppant Reynolds number. Cases 3 and 14–18 concern the effects of the Archimedes number. Cases 3 and 19–24 concern the effects of the bypass angle. Cases 3 and 25–33 are designed to study the effects of the relative width of the primary fracture and secondary fracture. Cases 3 and 34–39 are designed to study the relative distance of the secondary fracture to the primary fracture entrance.

The slick-water hydraulic fracturing is widely used in the unconventional resources [3,18]. The slick-water consists of water and chemical ingredients such as the friction reducer, the clay stabilizer, and the bactericide. These chemical additives account for no more than 1% content in slick-water. However, it plays an important role in reducing the friction of the side wall. The viscosity of the slick-water is about 0.8–1.2 mPa·s. Generally, the tap water is used instead of the slick-water in the experiment. In this paper, the tap water is also used with a constant viscosity of 0.001 Pa·s.

3. Results and Discussion

Figure 2 shows the deposition form of the proppant in the primary fracture at different time points (Case 3). The proppant distribution is similar to the experiment results of previous researchers [1,5,6,19,20]. The proppants first stack at a certain distance from the entrance after entering the primary fracture. The height of proppant bed increases with little change in length until the EPH is reached. Then, the transportation of proppant tends to be stable, as the height of the proppant bed remains unchanged, and the proppant bed only changes in the fracture length direction. The proppant bed at the bottom of the fracture plays an important role in supporting the fractures after the pressure is released. It resembles a stationary porous medium through which the gas and oil will be extracted.

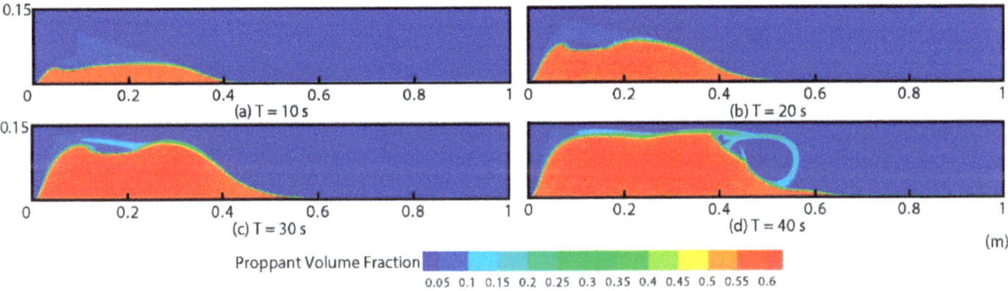

Figure 2. The distribution of the proppant in the primary fracture at time (**a**) 10 s, (**b**) 20 s, (**c**) 30 s, and (**d**) 40 s. (Case 3 in Table A1).

In Figure 3a,b, the velocity vector charts of the proppant when the proppant bed height in the primary fracture is about 10% and 75% of the EPH are given (Case 3). The red part of Figure 3 represents the proppant bed. The velocity vector chart of the proppant when the proppant bed moves ahead in primary fracture with the constant bed height is shown in Figure 3c. The proppants settle quickly to the fracture bottom due to the low viscosity of the water initially. As the proppant bed height increases, the following proppants are resisted by the proppant bed and have to move from the upper part of the proppant bed to the depth of the fractures. From the proppant velocity vector chart, it can be found that the proppant bed at the bottom of the fractures does not move. The front part of the bed consists of the following injected proppants, which move forward from the top of the proppant bed surface. Based on the simulation results, the transportation of the proppants in the fractures can be divided into two distinct zones at the steady state: the proppant bed zone and the mixture zone. The mixture zone is the mixture of the proppant and water above the proppant bed.

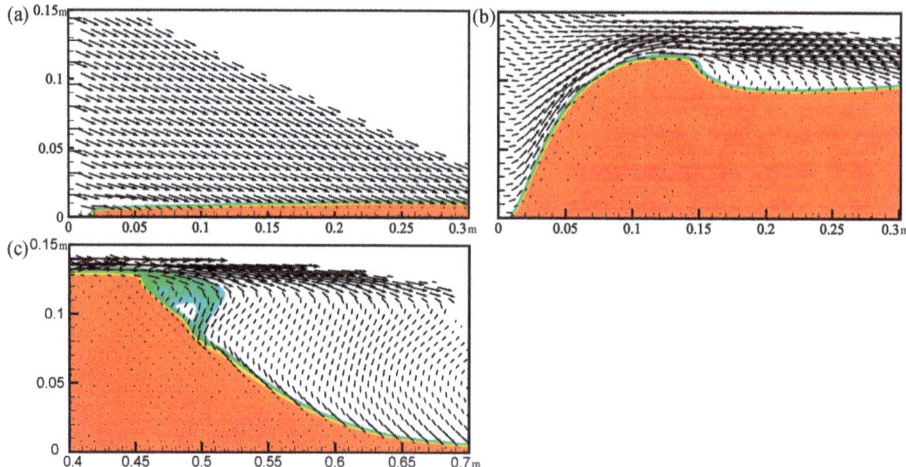

Figure 3. The velocity vector charts of the proppant in the primary fracture, (**a**) the proppant bed height is about 0.1 EPH, (**b**) the proppant bed height is about 0.75 EPH, and (**c**) the proppant bed height has reached the EPH. The red part represents the proppant bed. The length and height of the primary fracture is about 1 m and 0.15 m, respectively. The horizontal coordinates indicate the distance from the entrance of the cross fractures. (Case 3 in Table A1). EPH: equilibrium proppant height.

The inlet proppant volume fraction changes from 1% to 11%. Figure 4 shows the change of the relative EPH (h/H) with the dimensionless parameters given in Equation (3). The relative EPH increases with the increase of α_{s0} at the value of $\alpha_{s0} < 5\%$ (Figure 4a). When the value of α_{s0} is greater than 5%, the relative EPH becomes stable. Although the relative EPH is constant, the time for the proppant bed height to reach the EPH is shorter due to more proppant injection per unit time. For the field engineering application, a larger proppant inlet volume fraction can be used to achieve a faster stability of EPH. However, a large proppant inlet volume fraction may also lead to the blockage of the fractures.

In the field engineering, the sand ratio (ε) is the ratio of the proppant bulk volume to the water volume. The inlet proppant volume fraction (α_{s0}) is the ratio of the proppant volume to the total volume of the proppant and water. The α_{s0} and the ε values satisfy the relationship $\alpha_{s0} = \xi\varepsilon/(\xi\varepsilon + 1)$, where ξ is the ratio of the bulk density (ρ_{sb}) to the real density (ρ_s) of the proppant. The sand ratio of the hydraulic fracturing in the field engineering is about 3–8% [21]. Then, the inlet volume fraction of the proppant is about 1–5%. As a result, it can be also concluded that the relative EPH increases with the increase of the sand ratio. According to the simulated results, the most economical sand ratio is about 8%, because the relative EPH becomes stable when the sand ratio is larger than 8%.

The change of the relative EPH with the Reynolds number is shown in Figure 4b. The relative EPH decreases with the increase of the Reynolds number. The Reynolds number characterizes the ratio of the inertia effect to the viscosity effect. A larger Reynolds number causes a higher inertia effect of the proppant and a larger average mixture velocity above the proppant bed. More proppants will be carried far in the fractures. The relative EPH decreases because more proppants on the bed surface move far away in the fractures.

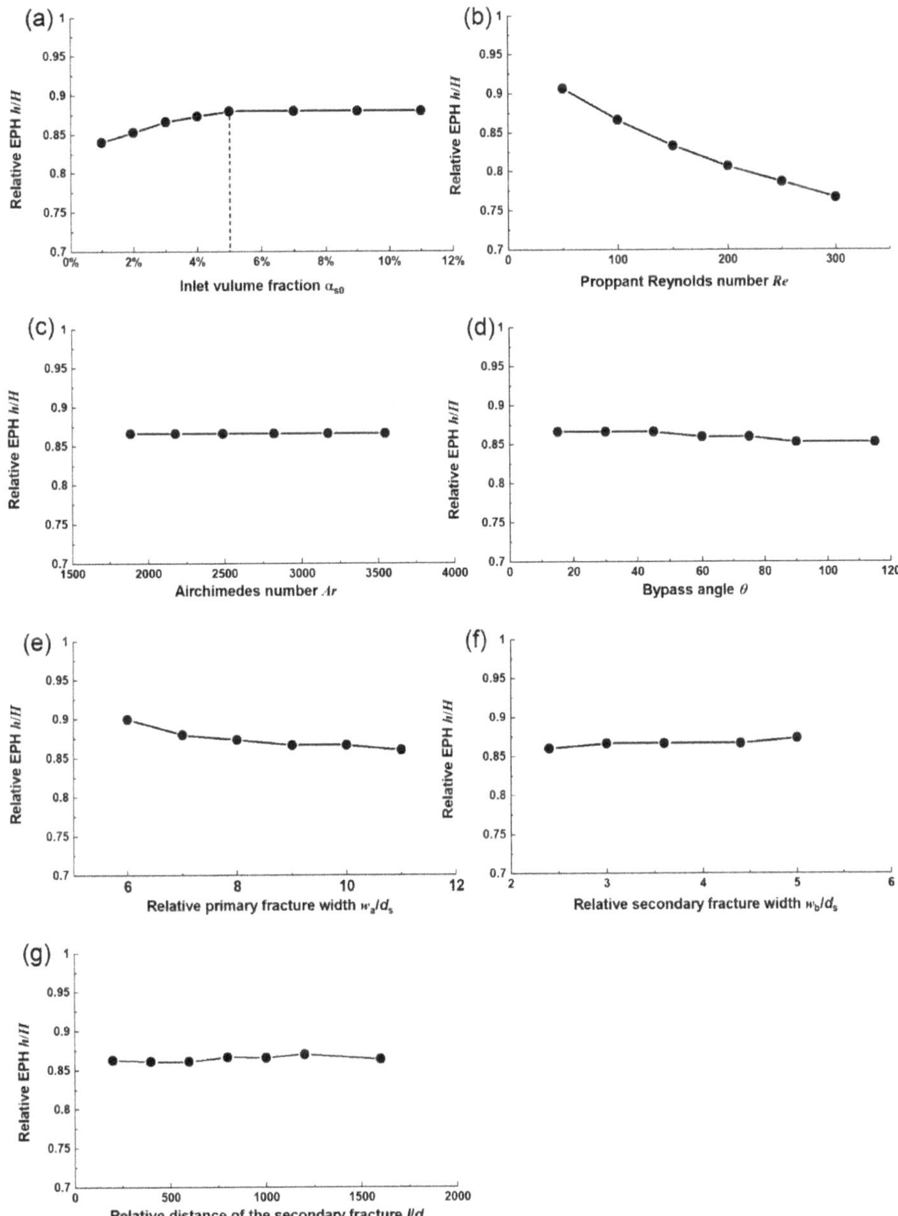

Figure 4. The change of the relative EPH with the dimensionless parameters, (**a**) inlet proppant volume fraction, (**b**) proppant Reynolds number, (**c**) Archimedes number, (**d**) angle between primary and secondary fracture (bypass angle), (**e**) relative width of primary fracture, (**f**) relative width of secondary fracture, and (**g**) relative distance of secondary fracture to primary fracture entrance.

Figure 4c,d shows the EPH development with the Archimedes number *Ar* and the bypass angle θ. The results indicate that the relative EPH changes little with the increase of *Ar* and θ. The form of the Archimedes number can be written as:

$$Ar = \frac{(\rho_s - \rho_l)\rho_l d_s^3 g}{\mu_l^2} = \frac{\rho_l d_s U_s}{\mu_l} \tag{4}$$

where $U_s = (\rho_s - \rho_l)g d_s^2 / \mu_l$ is related to the settling velocity of a single particle in water [22]. As a result, the Archimedes number can also be called as the proppant settling Reynolds number. It represents the settling effect of the proppant, which mainly affects the sedimentation speed of the proppant in the fractures. The proppants quickly settle to the bottom of the fractures due to the low viscosity of the water after injection (Figure 3a). If the settling effect is enhanced, the time for the proppant bed to reach the EPH will be reduced. Table 2 gives the time for reaching the EPH at different values of the Archimedes number. However, the EPH does not change. The change of the relative EPH with the relative width of the primary and secondary fracture as well as the relative distance of the secondary fracture to the primary fracture entrance are shown in Figure 4e–g. With the increase in the values of w_a/d_s, w_b/d_s, and l/d_s, the relative EPH changes slightly. It can be concluded that the inlet proppant volume fraction and the proppant Reynolds number are the main controlling dimensionless parameters for the relative EPH.

Table 2. The time for reaching the EPH at different Archimedes numbers.

Ar	1888	2178	2488	2820	3171	3545
Time/s	50	46	43	40	38	36

The bypass angle, the relative width of the secondary fracture, and the relative distance of the secondary fracture to the primary fracture entrance are the dimensionless parameters related with the secondary fracture. Comparing Figure 4d–g, it is found that the relative EPH is almost constant with the bypass angle, the relative width of the secondary fracture, and the relative distance of the secondary fracture to the primary fracture entrance. In cross fractures, the secondary fracture has little effect on the proppant transportation in the primary fracture. The reason is that the width of the secondary fracture is always small, and the primary fracture is the main channel for the proppant transportation. That means that previous experiments or numerical simulation results in a single fracture can be extend to the cross fractures.

The proppant Reynolds number, which is also called proppant transport Reynolds number, is divided by the proppant settling Reynolds number:

$$\Pi = \frac{Re}{Ar} = \frac{U_0}{U_s} = \frac{U_0 \mu_l}{d_s^2 (\rho_s - \rho_l)g} \tag{5}$$

where U_0 is the injection velocity of the mixture. The dimensionless number Π is the ratio of the transport effect to the settling effect. Taking the secondary fracture as a single fracture, the average proppant velocity entering the secondary fracture from the primary fracture is set as the transport velocity. Substituting the parameters into Equation (5), the maximum value of Π in the secondary fracture is about 0.008 in all the cases, which is much smaller than that in the primary fracture listed in Table A1. This means that the settling effect dominates the movement of proppants in the secondary fracture.

An important issue existing in field engineering is how to transport the proppant from the primary fracture into the subsidiary fracture efficiently. The oil and gas flow through the subsidiary fracture into the primary fracture; then, they are collected in the wellbore. Less proppant transported into the subsidiary fractures will cause the blockage of the seepage flow channel of the oil and gas. It is found that there are two mechanisms for the proppant transporting from the primary fracture into the secondary fracture [6]. The first one is the gravity effect, and the other one is the water-carrying effect. The gravity is along the vertical direction, and it may not drive the proppant movement to other directions directly. The mechanism of the gravity effect may be that the proppants form a high proppant bed in the primary fracture firstly and then enter the secondary fracture under the gravity effect due to the deposition instability. The water-carrying effect is that the drag force on the proppant forms due to the pressure difference between the fracture entrance and the outlet with water entering the secondary fracture at a certain velocity. Figure 5 gives the formation process of the proppant bed in the secondary fracture (Case 3). When the proppants move to the entrance of the secondary fracture, they directly enter the secondary fracture and slowly build up a proppant bed. Hence, it may be inappropriate to use the gravity effects to explain the proppants entering the secondary fracture, and the fluid-carrying effect may be the main controlling factor.

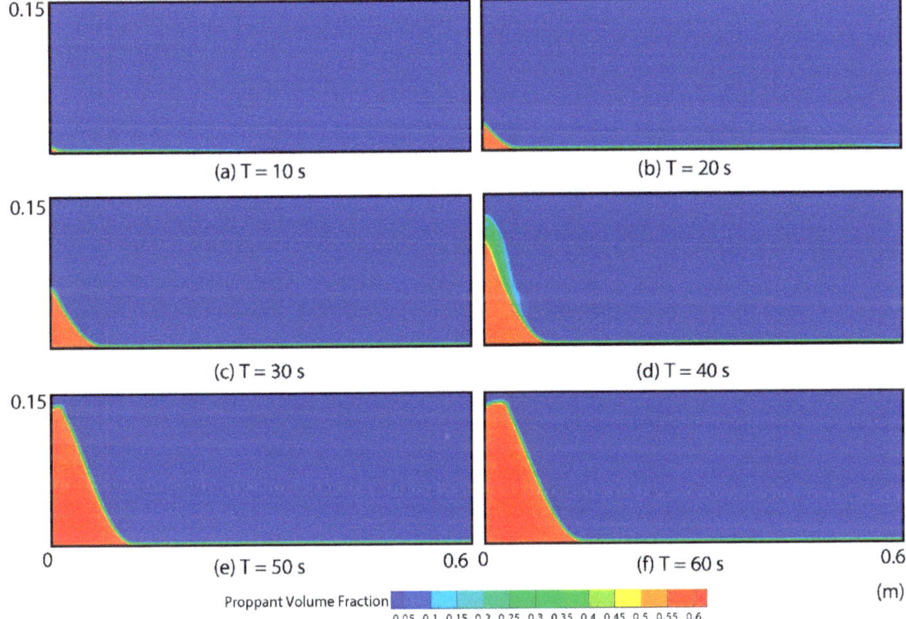

Figure 5. The formation process of the proppant bed in the secondary fracture at time (**a**) 10 s, (**b**) 20 s, (**c**) 30 s, (**d**) 40 s, (**e**) 50 s, and (**f**) 60 s. (Case 3 in Table A1).

Figure 6 is a top view of the cross fractures. At the cross section C, the proppants enter from the primary fracture into the secondary fracture. The mass of the proppant in the whole cross fractures and secondary fracture (shaded part in Figure 6) is calculated by the following equation:

$$m_s = \iiint_V \alpha_s \rho_s dV. \tag{6}$$

The RPM can be written as:

$$RPM = \frac{m_{s,\text{sed}}}{m_{s,\text{who}}} = \frac{\iiint_{V_{\text{sed}}} \alpha_s \rho_s dV}{\iiint_{V_{\text{who}}} \alpha_s \rho_s dV} \tag{7}$$

where $m_{s,\text{sed}}$ and $m_{s,\text{who}}$ are the proppants' mass in the secondary and whole cross fractures, respectively. V_{sed} and V_{who} are the volume of the secondary fracture and whole cross fractures, respectively.

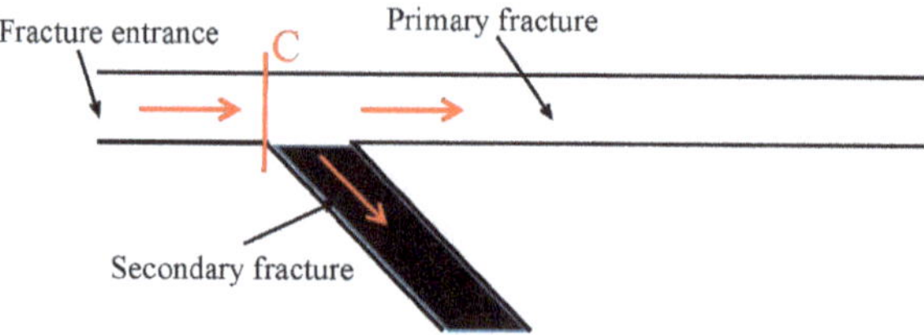

Figure 6. A top view of the cross fractures. The shaded part is the secondary fracture.

Figure 7 gives the change of RPM with the dimensionless parameters mentioned in Equation (7). In order to reflect the distribution of the proppants in the primary fracture and the secondary fracture, the RPM is calculated with the same injected mass of the proppant. The RPM curve with the different inlet proppant volume fraction (α_{s0}) is shown in Figure 7a. When the inlet volume fraction of the proppant increases, the RPM decreases slightly from 6% to 3%. That means that the increase of the inlet volume fraction does not lead to the increase of the amount of the proppants entering the secondary fracture. More proppants stay in the primary fracture at the large inlet proppant volume fraction. The reason may be that the total resistance of the proppants on the water is larger at the larger inlet proppant volume fraction. The water velocity at the inlet of the secondary fracture is reduced accordingly. As a result, the amount of proppant entering the secondary fracture reduced.

Figure 7b shows the RPM curve with Reynolds number. Similar to the effect of the inlet proppant volume fraction, the RPM decreases from 4.6% to 2.5% with the Reynolds number changing from 50 to 300. The proppants hardly enter the secondary fracture when the Reynolds number is large. The RPM changes little with the Archimedes number and the bypass angle (Figure 7c,d). However, the RPM is greatly influenced by the relative width of the primary and the secondary fracture compared to the α_{s0}, Re, Ar, and θ (as shown in Figure 7e,f). When w_a/d_s changes from 6 to 11, the RPM decreases from 16% to 3%. This is because when the relative width of the primary fracture is larger, the primary fracture becomes a more favorable channel for the proppant transportation, and more proppants move into the primary fracture, leading to the decrease of the RPM. The RPM increases from 3% to 9% as the value of w_b/d_s changes from 2.4 to 5. When the relative width of the secondary fracture is larger, the possession of the primary fracture is weakened, and more proppants move into the secondary fracture, resulting in the increase of the RPM. If the proppants consist of particles of different diameters, the coarse particles will tend to stay in the primary fracture, and the fine particles are accessible to the secondary fracture. Sahai et al. [6] also found that the proppants in the secondary fractures are thinner than those in the primary fracture by using the proppants with certain particle size grading to investigate the sorting effect of particles at the intersection of the primary and secondary fracture. In conventional hydraulic fracturing, the naturally ceramsite sand proppants with a certain size grading are used. This may be a good way to improve the hydraulic fracturing and increase the oil and gas recovery. The RPM decreases from 6% to 2.3% with the relative distance of the secondary fracture to the primary fracture entrance increasing from 200 to 1400. This is because the transportation time increases for the proppant entering the secondary fracture when the relative distance of the secondary fracture to the primary fracture entrance increases. When injecting the same mass of proppant, the farther the secondary fracture is from the primary fracture, the less proppant will be transported. It can be concluded that the width of the cross fractures has the greatest impact on the amount of proppant entering the secondary fracture.

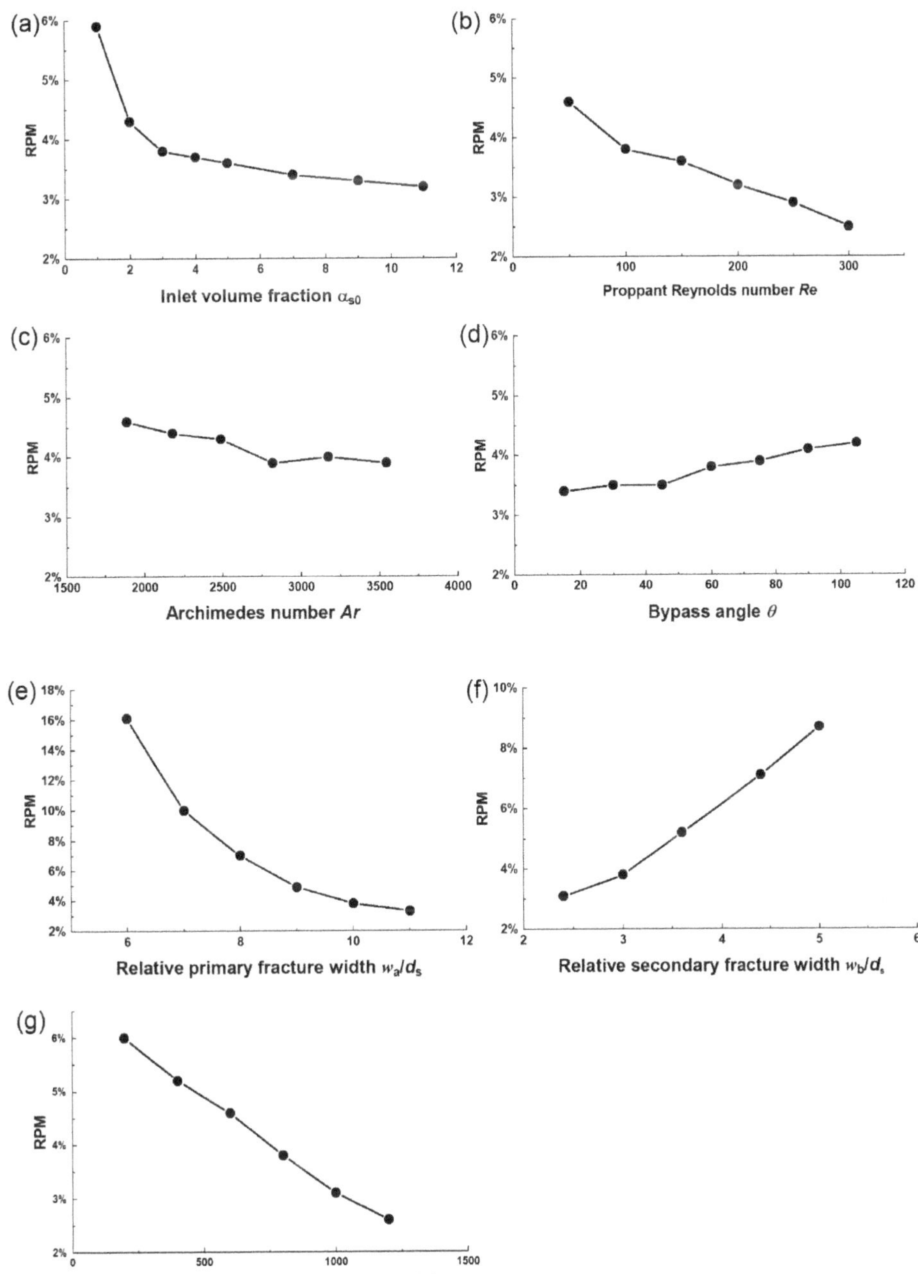

Figure 7. The curve of ratio of the proppant mass (RPM) with the dimensionless parameters, (**a**) inlet proppant volume fraction, (**b**) proppant Reynolds number, (**c**) Archimedes number, (**d**) angle between primary and secondary fracture (bypass angle), (**e**) relative width of primary fracture, (**f**) relative width of secondary fracture, and (**g**) relative distance of secondary fracture to primary fracture entrance.

In order to make the numerical simulation results practical for field engineering, more cases are simulated, which are shown in Table A1 (Cases 40–65), and the contour map of RPM relating to the dimensionless parameters is given in Figure 8. Each black dot in Figure 8 represents a case in Table A1, and the coordinates indicate the value of the dimensionless parameters. The percentage of the proppant entering the secondary fracture can be estimated in field engineering based on Figure 8.

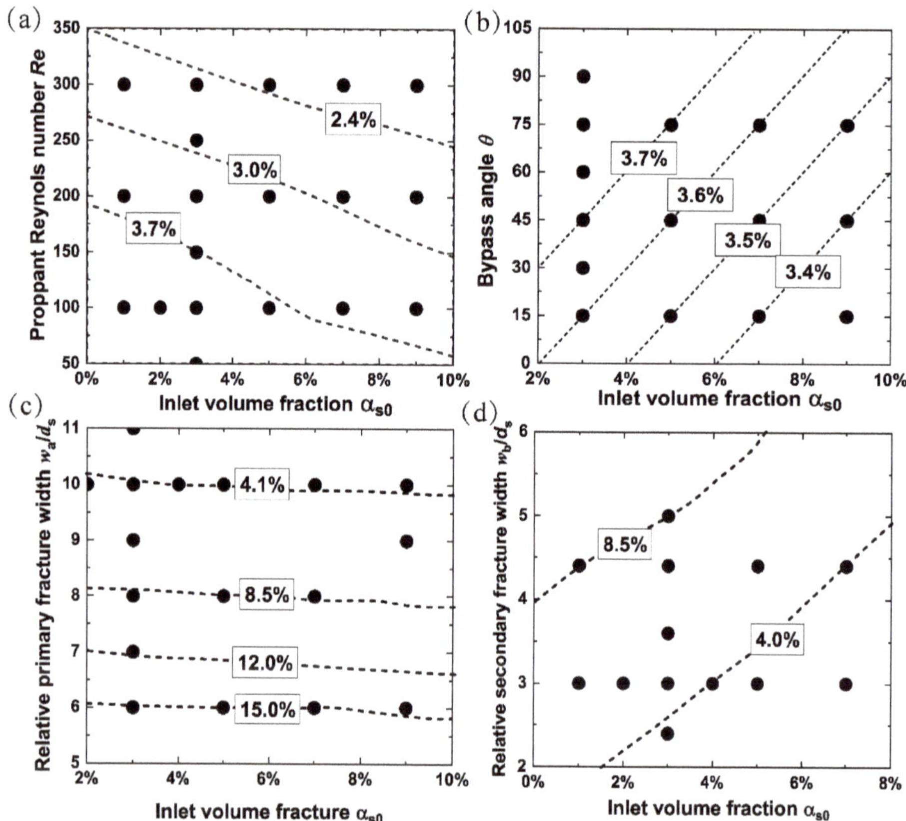

Figure 8. The contour map of RPM relating to the dimensionless parameters, (**a**) inlet proppant volume fraction and the proppant Reynolds number, (**b**) inlet proppant volume fraction and the bypass angle, (**c**) inlet proppant volume fraction and the relative width of the primary fracture, (**d**) inlet proppant volume fraction and the relative width of the secondary fracture.

4. Conclusions

In this paper, the proppant transportation in the cross fractures is investigated by using the computational fluid dynamics (CFD) method. The Euler–Euler two-phase flow model and the KTFG approach are adopted to describe the flow behaviors. The dimensionless parameters relating to the proppant transportation in the cross fractures, such as the inlet proppant volume fraction, the proppant Reynolds number, the Archimedes number, the bypass angle, the relative width of the primary and secondary fracture, and the relative distance of the secondary fracture to the primary fracture entrance, are derived based on dimensional analysis.

Two dimensionless parameters are proposed to evaluate the distribution of proppants in the cross fractures, i.e., the relative EPH and the RPM. The simulation results show that the main controlling dimensionless parameters for the relative EPH are the inlet proppant volume fraction and proppant Reynolds number. The dominating dimensionless parameters for the RPM are the relative width of the primary and the secondary fracture. The relative EPH decreases with the increase of Re, while it increases with the increase of the sand ratio. The admirable sand ratio is about 8% for the field engineering based on the simulation results. When w_a/d_s changes from 6 to 11, the RPM decreases from 16% to 3%. The RPM increases from 3% to 9% when w_b/d_s changes from 2.4 to 5. It is suggested that the proppants with a certain particle size grading may be a good way for improving the hydraulic fracturing and increasing the oil and gas recovery. The settling effect is dominating in the secondary fracture. The proppants enter the secondary fracture mainly under the water-carrying effect. A graph (Figure 8) is given for the engineers to predict the percentage of the proppant entering the secondary fracture.

Author Contributions: Conceptualization, Y.Z., X.L. and X.Z.; methodology, Y.Z. P.L. and X.Z.; software, Y.Z. and P.L.; validation, Y.Z. and P.L.; investigation, Y.Z. and X.L.; data curation, Y.Z.; writing—original draft preparation, Y.Z.; writing—review and editing, X.Z. and X.L.; visualization, Y.Z. and P.L.; supervision, X.Z. and X.L.; funding acquisition, X.Z. All authors have read and agreed to the published version of the manuscript.

Funding: This research was funded by National Major Oil and Gas Projects of China, grant number 2017X05049003-002 and The APC was funded by National Major Oil and Gas Projects of China.

Acknowledgments: This research is supported by the National Major Oil and Gas Projects of China (No.2017X05049003-002). The support is gratefully acknowledged.

Appendix A

Table A1. All the cases in the numerical simulation.

Cases	Proppant		Water		Fractures			U_0/m·s^{-1}	Dimensionless Parameters							
	ρ_s/kg·m^{-3}	d_s/mm	ρ_l/kg·m^{-3}	μ_l/Pa·s	w_b/mm	w_a/mm	l/m		α_{s0}/%	Re	Ar	θ/°	w_a/d_s	w_b/d_s	l/d_s	Re/Ar
1	3300	0.5	1000	1.0×10^{-3}	1.5	5.0	0.4	0.2	1	100	2820	45	10	3	800	0.035
2	3300	0.5	1000	1.0×10^{-3}	1.5	5.0	0.4	0.2	2	100	2820	45	10	3	800	0.035
3	**3300**	**0.5**	**1000**	$\mathbf{1.0 \times 10^{-3}}$	**1.5**	**5.0**	**0.4**	**0.2**	**3**	**100**	**2820**	**45**	**10**	**3**	**800**	**0.035**
4	3300	0.5	1000	1.0×10^{-3}	1.5	5.0	0.4	0.2	4	100	2820	45	10	3	800	0.035
5	3300	0.5	1000	1.0×10^{-3}	1.5	5.0	0.4	0.2	5	100	2820	45	10	3	800	0.035
6	3300	0.5	1000	1.0×10^{-3}	1.5	5.0	0.4	0.2	7	100	2820	45	10	3	800	0.035
7	3300	0.5	1000	1.0×10^{-3}	1.5	5.0	0.4	0.2	9	100	2820	45	10	3	800	0.035
8	3300	0.5	1000	1.0×10^{-3}	1.5	5.0	0.4	0.2	11	100	2820	45	10	3	800	0.035
9	3300	0.5	1000	1.0×10^{-3}	1.5	5.0	0.4	0.1	3	50	2820	45	10	3	800	0.018
10	3300	0.5	1000	1.0×10^{-3}	1.5	5.0	0.4	0.3	3	150	2820	45	10	3	800	0.053
11	3300	0.5	1000	1.0×10^{-3}	1.5	5.0	0.4	0.4	3	200	2820	45	10	3	800	0.071
12	3300	0.5	1000	1.0×10^{-3}	1.5	5.0	0.4	0.5	3	250	2820	45	10	3	800	0.088
13	3300	0.5	1000	1.0×10^{-3}	1.5	5.0	0.4	0.6	3	300	2820	45	10	3	800	0.106
14	2700	0.5	818	1.0×10^{-3}	1.5	5.0	0.4	0.2	3	100	1888	45	10	3	800	0.053
15	2900	0.5	879	1.0×10^{-3}	1.5	5.0	0.4	0.2	3	100	2178	45	10	3	800	0.046
16	3100	0.5	939	1.0×10^{-3}	1.5	5.0	0.4	0.2	3	100	2488	45	10	3	800	0.040
17	3500	0.5	1060	1.0×10^{-3}	1.5	5.0	0.4	0.2	3	100	3171	45	10	3	800	0.031
18	3700	0.5	1121	1.0×10^{-3}	1.5	5.0	0.4	0.2	3	100	3545	45	10	3	800	0.028
19	3300	0.5	1000	1.0×10^{-3}	1.5	5.0	0.4	0.2	3	100	2820	15	10	3	800	0.035
20	3300	0.5	1000	1.0×10^{-3}	1.5	5.0	0.4	0.2	3	100	2820	30	10	3	800	0.035
21	3300	0.5	1000	1.0×10^{-3}	1.5	5.0	0.4	0.2	3	100	2820	60	10	3	800	0.035
22	3300	0.5	1000	1.0×10^{-3}	1.5	5.0	0.4	0.2	3	100	2820	75	10	3	800	0.035
23	3300	0.5	1000	1.0×10^{-3}	1.5	5.0	0.4	0.2	3	100	2820	90	10	3	800	0.035
24	3300	0.5	1000	1.0×10^{-3}	1.5	5.0	0.4	0.2	3	100	2820	105	10	3	800	0.035
25	3300	0.5	1000	1.0×10^{-3}	1.5	3.0	0.4	0.2	3	100	2820	45	6	3	800	0.035
26	3300	0.5	1000	1.0×10^{-3}	1.5	3.5	0.4	0.2	3	100	2820	45	7	3	800	0.035
27	3300	0.5	1000	1.0×10^{-3}	1.5	4.0	0.4	0.2	3	100	2820	45	8	3	800	0.035
28	3300	0.5	1000	1.0×10^{-3}	1.5	4.5	0.4	0.2	3	100	2820	45	9	3	800	0.035
29	3300	0.5	1000	1.0×10^{-3}	1.5	5.5	0.4	0.2	3	100	2820	45	11	3	800	0.035
30	3300	0.5	1000	1.0×10^{-3}	1.2	5.0	0.4	0.2	3	100	2820	45	10	2.4	800	0.035

Table A1. *Cont.*

Cases	Proppant		Water		Fractures			U_0/m·s^{-1}	Dimensionless Parameters							
	ρ_s/kg·m^{-3}	d_s/mm	ρ_l/kg·m^{-3}	μ_l/Pa·s	w_b/mm	w_a/mm	l/m		α_{s0}/%	Re	Ar	θ/°	w_a/d_s	w_b/d_s	l/d_s	Re/Ar
31	3300	0.5	1000	1.0×10^{-3}	1.8	5.0	0.4	0.2	3	100	2820	45	10	3.6	800	0.035
32	3300	0.5	1000	1.0×10^{-3}	2.2	5.0	0.4	0.2	3	100	2820	45	10	4.4	800	0.035
33	3300	0.5	1000	1.0×10^{-3}	2.5	5.0	0.4	0.2	3	100	2820	45	10	5	800	0.035
34	3300	0.5	1000	1.0×10^{-3}	1.5	5.0	0.1	0.2	3	100	2820	45	10	3	200	0.035
35	3300	0.5	1000	1.0×10^{-3}	1.5	5.0	0.2	0.2	3	100	2820	45	10	3	400	0.035
36	3300	0.5	1000	1.0×10^{-3}	1.5	5.0	0.3	0.2	3	100	2820	45	10	3	600	0.035
37	3300	0.5	1000	1.0×10^{-3}	1.5	5.0	0.5	0.2	3	100	2820	45	10	3	1000	0.035
38	3300	0.5	1000	1.0×10^{-3}	1.5	5.0	0.6	0.2	3	100	2820	45	10	3	1200	0.035
39	3300	0.5	1000	1.0×10^{-3}	1.5	5.0	0.8	0.2	3	100	2820	45	10	3	1600	0.035
40	3300	0.5	1000	1.0×10^{-3}	1.5	5.0	0.4	0.4	1	200	2820	45	10	3	800	0.071
41	3300	0.5	1000	1.0×10^{-3}	1.5	5.0	0.4	0.6	1	300	2820	45	10	3	800	0.106
42	3300	0.5	1000	1.0×10^{-3}	1.5	5.0	0.4	0.4	5	200	2820	45	10	3	800	0.071
43	3300	0.5	1000	1.0×10^{-3}	1.5	5.0	0.4	0.6	5	300	2820	45	10	3	800	0.106
44	3300	0.5	1000	1.0×10^{-3}	1.5	5.0	0.4	0.4	7	200	2820	45	10	3	800	0.071
45	3300	0.5	1000	1.0×10^{-3}	1.5	5.0	0.4	0.6	7	300	2820	45	10	3	800	0.106
46	3300	0.5	1000	1.0×10^{-3}	1.5	5.0	0.4	0.4	9	200	2820	45	10	3	800	0.071
47	3300	0.5	1000	1.0×10^{-3}	1.5	5.0	0.4	0.6	9	300	2820	45	10	3	800	0.106
48	3300	0.5	1000	1.0×10^{-3}	1.5	5.0	0.4	0.2	5	100	2820	15	10	3	800	0.035
49	3300	0.5	1000	1.0×10^{-3}	1.5	5.0	0.4	0.2	5	100	2820	45	10	3	800	0.035
50	3300	0.5	1000	1.0×10^{-3}	1.5	5.0	0.4	0.2	5	100	2820	75	10	3	800	0.035
51	3300	0.5	1000	1.0×10^{-3}	1.5	5.0	0.4	0.2	7	100	2820	15	10	3	800	0.035
52	3300	0.5	1000	1.0×10^{-3}	1.5	5.0	0.4	0.2	7	100	2820	45	10	3	800	0.035
53	3300	0.5	1000	1.0×10^{-3}	1.5	5.0	0.4	0.2	7	100	2820	75	10	3	800	0.035
54	3300	0.5	1000	1.0×10^{-3}	1.5	5.0	0.4	0.2	9	100	2820	15	10	3	800	0.035
55	3300	0.5	1000	1.0×10^{-3}	1.5	5.0	0.4	0.2	9	100	2820	45	10	3	800	0.035
56	3300	0.5	1000	1.0×10^{-3}	1.5	5.0	0.4	0.2	9	100	2820	75	10	3	800	0.035
57	3300	0.5	1000	1.0×10^{-3}	1.5	3.0	0.4	0.2	5	100	2820	45	6	3	800	0.035
58	3300	0.5	1000	1.0×10^{-3}	1.5	4.0	0.4	0.2	5	100	2820	45	8	3	800	0.035
59	3300	0.5	1000	1.0×10^{-3}	1.5	3.0	0.4	0.2	7	100	2820	45	6	3	800	0.035
60	3300	0.5	1000	1.0×10^{-3}	1.5	4.0	0.4	0.2	7	100	2820	45	8	3	800	0.035
61	3300	0.5	1000	1.0×10^{-3}	1.5	3.0	0.4	0.2	9	100	2820	45	6	3	800	0.035
62	3300	0.5	1000	1.0×10^{-3}	1.5	4.5	0.4	0.2	9	100	2820	45	9	3	800	0.035
63	3300	0.5	1000	1.0×10^{-3}	2.2	5.0	0.4	0.2	1	100	2820	45	10	4.4	800	0.035
64	3300	0.5	1000	1.0×10^{-3}	2.2	5.0	0.4	0.2	5	100	2820	45	10	4.4	800	0.035
65	3300	0.5	1000	1.0×10^{-3}	2.2	5.0	0.4	0.2	7	100	2820	45	10	4.4	800	0.035

References

1. Tong, S.; Singh, R.; Mohanty, K.K. A Visualization Study of Proppant Transport in Foam Fracturing Fluids. *J. Nat. Gas Sci. Eng.* **2018**, *52*, 235–247. [CrossRef]
2. Osiptsov, A.A. Fluid Mechanics of Hydraulic Fracturing: A Review. *J. Pet. Sci. Eng.* **2017**, *156*, 513–535. [CrossRef]
3. Barati, R.; Liang, J.T. A Review of Fracturing Fluid Systems Used for Hydraulic Fracturing of Oil and Gas Wells. *J. Appl. Polym. Sci.* **2014**, *131*, 40735. [CrossRef]
4. Wang, J.; Elsworth, D. Role of Proppant Distribution on the Evolution of Hydraulic Fracture Conductivity. *J. Pet. Sci. Eng.* **2018**, *166*, 249–262. [CrossRef]
5. Alotaibi, M.; Miskimins, J. Slickwater Proppant Transport in Complex Fractures: New Experimental Findings & Scalable Correlation. In Proceedings of the SPE Annual Technical Conference and Exhibition, Houston, TX, USA, 28–30 September 2015. [CrossRef]
6. Sahai, R.; Miskimins, J.; Olson, K.E. Laboratory Results of Proppant Transport in Complex Fracture Systems. In Proceedings of the SPE Hydraulic Fracturing Technology Conference, The Woodlands, TX, USA, 4–6 February 2014. [CrossRef]
7. McClure, M. Bed Load Proppant Transport During Slickwater Hydraulic Fracturing: Insights from Comparisons between Published Laboratory Data and Correlations for Sediment and Pipeline Slurry Transport. *J. Pet. Sci. Eng.* **2018**, *161*, 599–610. [CrossRef]
8. Hu, X.D.; Wu, K.; Li, G.S.; Tang, J.Z.; Shen, Z.H. Effect of Proppant Addition Schedule on the Proppant Distribution in a Straight Fracture for Slickwater Treatment. *J. Pet. Sci. Eng.* **2018**, *167*, 110–119. [CrossRef]
9. Roostaei, M.; Nouri, A.; Fattahpour, V.; Chan, D. Coupled Hydraulic Fracture and Proppant Transport Simulation. *Energies* **2020**, *13*, 2822. [CrossRef]
10. Li, P.; Zhang, X.H.; Lu, X.B. Numerical Simulation on Solid-Liquid Two-Phase Flow in Cross Fractures. *Chem. Eng. Sci.* **2018**, *181*, 1–18. [CrossRef]
11. Tong, S.; Mohanty, K.K. Proppant Transport Study in Fractures with Intersections. *Fuel* **2016**, *181*, 463–477. [CrossRef]
12. Zhong, W.Q.; Yu, A.B.; Zhou, G.W.; Xie, J.; Zhang, H. CFD Simulation of Dense Particulate Reaction System: Approaches, Recent Advances and Applications. *Chem. Eng. Sci.* **2016**, *140*, 16–43. [CrossRef]
13. Li, P.; Su, J.Z.; Zhang, Y.; Zhang, X.H.; Lu, X.B. The Two Phase Flow of Proppant-Laden Fluid in a Single Fracture. *Mech. Eng.* **2017**, *39*, 135–144. (In Chinese)
14. Patankar, N.A.; Joseph, D.D.; Wang, J.; Barree, R.D.; Conway, M.; Asadi, M. Power Law Correlations for Sediment Transport in Pressure Driven Channel Flows. *Int. J. Multiph. Flow* **2002**, *28*, 1269–1292. [CrossRef]
15. Dogon, D.; Golombok, M. Self-Regulating Solutions for Proppant Transport. *Chem. Eng. Sci.* **2016**, *148*, 219–228. [CrossRef]
16. Liu, Y.; Sharma, M.M. Effect of Fracture Width and Fluid Rheology on Proppant Settling and Retardation: An Experimental Study. In Proceedings of the SPE Annual Technical Conference and Exhibition, Dallas, TX, USA, 9–12 October 2005. [CrossRef]
17. Tan, Q.M. *Dimensional Analysis: With Case Studies in Mechanics*, 1st ed.; Springer: Berlin Heidelberg, Germany, 2011; pp. 7–16.
18. King, G.E. Thirty Years of Gas Shale Fracturing: What Have We Learned? In Proceedings of the SPE Annual Technical Conference and Exhibition, Society of Petroleum Engineers, Florence, Italy, 19–22 September 2010. [CrossRef]
19. Mack, M.; Sun, J.; Khadilkar, C. Quantifying Proppant Transport in Thin Fluids: Theory and Experiments. In Proceedings of the SPE Hydraulic Fracturing Technology Conference, Society of Petroleum Engineers, The Woodlands, TX, USA, 4–6 February 2014. [CrossRef]
20. Huang, X.; Yuan, P.; Zhang, H.; Han, J.H.; Alberto, M.; Bao, J. Numerical Study of Wall Roughness Effect on Proppant Transport in Complex Fracture Geometry. In Proceedings of the SPE Middle East Oil & Gas Show and Conference, Manama, Kingdom of Bahrain, 6–9 March 2017. [CrossRef]

21. Li, N.; Li, J.; Zhao, L.Q.; Luo, Z.F.; Liu, P.L.; Guo, Y.J. Laboratory Testing on Proppant Transport in Complex-Fracture Systems. *SPE Prod. Oper* **2016**, *32*, 4. [CrossRef]
22. Zhang, H.R.; Liang, H.R.; Yan, X.H.; Wang, B.H.; Wang, N. Simulation on Water and Sand Separation from Crude Oil in Settling Tanks Based on the Particle Model. *J. Pet. Sci. Eng.* **2017**, *156*, 366–372. [CrossRef]

Article

A Novel Equivalent Continuum Approach for Modelling Hydraulic Fractures

Eziz Atdayev [1,*]**, Ron C. K. Wong** [1] **and David W. Eaton** [2]

[1]   Department of Civil Engineering, University of Calgary, Calgary, AB T2N 1N4, Canada; rckwong@ucalgary.ca
[2]   Department of Geoscience, University of Calgary, Calgary, AB T2N 1N4, Canada; eatond@ucalgary.ca
*   Correspondence: e.atdayev@gmail.com

Received: 30 October 2020; Accepted: 23 November 2020; Published: 25 November 2020

Abstract: Hydraulic fracturing has transitioned into widespread use over the last few decades. There are a variety of numerical methods available to simulate hydraulic fracturing. However, most current methods require a large number of input parameters, of which the values of some parameters are poorly constrained. This paper proposes a new method of modelling the hydraulically fractured region using void-ratio dependent relation to define the permeability of the fractured region. This approach is computationally efficient and reduces the number of input parameters. By implementing this method with an equivalent continuum representation, uncertainties are reduced arising from heterogeneity and anisotropy of earth materials. The computational efficiency improves modelling performance in stress sensitive zones such as in the vicinity of the injection well or near faults.

Keywords: hydraulic fracturing; void ratio; permeability; FEM; ABAQUS

---

1. Introduction

Hydraulic fracturing is a fluid injection process intended to create fractures in order to increase formation permeability [1]. This process is applicable in both vertical and horizontal wells, but is now performed mainly in extended horizontal wells [2] to allows multiple hydraulically fractured zones to be completed in a single well [3]. Each hydraulically fractured zone is then referred to as a stage, and a single extended horizontal well may have more than 20 stages. Geomechanical analysis is crucial in the pre-assessment of the geological formation to ensure the safety and effectiveness of the practice [4]. Moreover, considering the high cost of drilling and completion programs, modelling is a key assessment tool for hydraulic fracturing. There are a multitude of techniques and methods available for the modelling of hydraulic fracturing [5–11].

Analytical approaches are common for modelling hydraulic fractures. These methods are primarily based on the well established Perkins–Kern–Nordgren (PKN) and Khristianovic– Geertsma–de Klerk (KGD) models [12–17]. These models remain highly useful for a quick hydraulic fracturing assessment at a site. The PKN model is better suited for cases of fractures with length equal to or greater than twice the fracture height. These conditions are generally applied to hydraulic fracturing in shale reservoirs [18]. The KGD method tends to bea better representative of cases where the fracture length is equal to or less than its height. Such fracture conditions are mainly applicable to brownfields applications where fracturing is used to revive production [19,20].

Based on these analytical solutions, there are numerous numerical methods available for hydraulic fracture modelling. Among the most common numerical modelling techniques are the cohesive zone method (CZM), extended finite element method (XFEM), and discrete element method (DEM) [9,21–26].

Hybrid techniques such as the finite element method-discrete element method (FEM-DEM) have also started to emerge. Such techniques attempt to combine the advantages of two methods to improve their performance [27,28].

Another type of modelling approach that has received less attention is a class of methods called equivalent continuum techniques. These methods attempt to represent fractured media through a generalised approach that provides descriptive results for a sufficiently large scale. Previous applications of equivalent continuum approaches have tended to largely focus on relatively shallow depths of analysis [29–32].

2. Literature Review on Hydraulic Fracturing Modelling Methods

The practice of hydraulic fracturing has resulted in an extensive body of literature focused on modelling techniques. Some of these techniques are briefly described below.

2.1. Analytical Models

Analytical solutions tend to be highly generalised and simplified through a set of assumptions. For example, assumptions that underlie the PKN and KGD models are: Homogeneous, isotropic, and linear elastic rock, laminar flow, no effects of additives such as proppant, and simple fracture geometry. The main distinction between PKN and KGD models is the assumed cross-sectional shape of the hydraulic fracture. In the PKN model, the hydraulic fracture is assumed to have an elliptical cross-section, while the KGD model assumes a rectangular shape. The KGD model can easily be converted to radial coordinates, which is useful to represent penny-shaped cracks. General solutions for pressure, width, and length for PKN, KGD, and penny-shaped hydraulic fracture models are widely known [17,33].

The use of analytical solutions can be extended to the formations with natural fractures using the leak-off coefficient. The introduction of leak-off coefficient to the analytical solutions improves the agreement of planar hydraulic fracture geometry with field data. However, analytical solutions are constrained by the simplified definition of approximate hydraulic fracture geometries. On the other hand, numerical models are capable of simulating complex hydraulic fracture geometries and their effects on host medium.

2.2. Cohesive Zone Method (CZM)

The cohesive zone method utilises cohesive elements within a continuum model to represent fractures in the rock. Cohesive elements use traction-separation relation for mechanical strength weakening [34]. In the cohesive zone, this relationship defines the energy release equation, wherein if a specific threshold strain/stress is reached, then the cohesive element is relaxed and allowed to deform more freely. In addition, the permeability relation for the cohesive zone uses a parallel plate theory, which is an effective representation for fracture permeability. As outlined below, a recent enhancement of the CZM is known as XFEM.

2.3. Extended Finite Element Modelling (XFEM)

Extended finite element modelling (XFEM) uses the same fundamental approach as the cohesive zone method, except that the fractured zone is defined differently. In particular, for the XFEM, heaviside, junction, and branch enrichment functions are introduced into the domain. These functions are used to calculate where the fracture tip will propagate and how the deformation of the fracture will be imposed onto surrounding rock [35].

Both CZM and XFEM attempt to model fractures individually, which complicates the computation process due to the increased number of equations in the model. Additionally, the large dimensional difference between the fractures and general modelling domain imposes challenges for model convergence. Another challenge for XFEM approaches is mesh dependency due to the use of enrichment functions, the dimensions of the element specifying the orientation and displacement of fractures become highly mesh dependent. Although the CZM approach does not suffer from the same degree of mesh dependency, it requires a pre-definition of the fracture geometry, which imposes limitations on how a fracture will propagate.

2.4. Discrete Element Method (DEM)

The discrete element method (DEM) is a modelling approach wherein each mesh element is considered as a separate continuum particle and the intervening space between is treated as fractures [36,37]. These fractures use the same type of physics as in the case of CZM, with respect to strength degradation and fracture initiation. A superiority of the DEM over the CZM is an ability for fractures to propagate in any direction in 3-dimensional space. The convergence rate is improved, compared to XFEM, due to a similar dimensionality of the fractures and continuum elements.

The improvements provided by the DEM approach are based on the treatment of fractures as contact points between the elements. However, this method also has high computational requirements, as it generally requires a large number of model elements in order for the model to be realistic and representative.

2.5. FEM-DEM

FEM-DEM is a hybrid approach that combines the advantages of DEM with finite element modelling (FEM) to reduce the computational load. The DEM component is applied within model micro-scale sections, and the stresses resultant from the DEM are then imposed within the FEM, which models the macro-scale domain [38]. This method is most applicable if the domain can be discretised into high- and low-sensitivity subdomains. It should also be mentioned that FEM-DEM codes are available that combine these two methods for different purposes [27,28].

2.6. Equivalent Continuum Methods (ECM)

Equivalent continuum methods assume that within a representative element volume (REV) the medium can be treated as homogeneous. Instead of defining fractures individually and modelling the strength degradation, this means that the properties of the domain are generalised and modelled as a continuum. The region of interest can then be discretised based on the regions of high and low fracture density, allowing for the modelling of the fractured and virgin zones of the reservoir [18,39]. The use of an ECM representation makes it simpler to apply well established numerical approaches such as the finite element method (FEM) to solve governing equations that describe the physical processes and behaviour of the medium.

Most of previous applications of ECM are focused on soils and near-surface materials [29–32], rather than for modelling hydraulic fracturing in rocks at a greater depth. This may be due to the general availability of a wide range of empirical methods to define the properties of fractured materials at shallow depths [40,41].

A novel ECM method introduced in this paper attempts to address the necessity of optimised hydraulic fracturing modelling. The introduced method in this paper uses the dependency of the permeability on void ratio to define the hydraulic fracture propagation. This method allows for an easier convergence due to uniform mesh dimensions and the exploitation of an already present coupled fluid flow and continuum relations.

3. Methodology

3.1. Average Volumetric Strain in Fractured Zone

During hydraulic fracturing, it is a common practice to record microseismicity, injection pressure, and rates. These data are then used to analyse the effectiveness of the hydraulic fracturing program. One type of analysis used in this paper is a comparison of estimates of volumetric strains based on microseismicity and injection pressure. In estimating the volumteric strain using the injection pressure, rock is assumed to be cohesion-less and fractures to open when change in injection pressure equals minimum in-situ stress. Thus, the volumetric strain using injection pressure is given by σ'_{min}/K, where σ'_{min} is the minimum in-situ effective stress and K denotes bulk modulus of the fractured domain. The volumetric strain using microseismicity is estimated using V_{inj}/SRV, where V_{inj} is the injected volume and SRV represents stimulated rock volume (SRV). SRV is defined as a three dimensional rock volume where new fractures and opened natural fractures during hydraulic fracturing are contained [42]. The uniqueness of the strain when fracturing occurs suggests that both of the volumteric strains should exhibit same solution. Thus, this uniqueness allows one to relate the hydraulic fracturing permeability to volumteric strain which then can be related to the void ratio.

3.2. Void Ratio Dependent Permeability

Permeability and porosity of a medium are often closely related. Relationships between these two quantities are widely used in reservoir engineering through semi-log porosity-permeability plots [43]. In our ECM method, a similar approach is used to estimate formation permeability. Unlike conventional porosity-permeability relations, our proposed relation uses a step-like function. This provides a good representation for hydraulic fracturing, due to steep spatial gradients of permeability in the vicinity of a hydraulic fracture [17]. Practicality and robustness of this approach arise from the availability of built-in relationships in ABAQUS FEM between permeability and void ratio [34]. This functionality is described by Equation (1).

$$\eta(e) = a * \left(tanh\left(\left(\frac{e}{e_{thresh}}\right)^{\alpha}\right)\right)^{\alpha} \tag{1}$$

where η is permeability, a is a maximum permeability, e is the void ratio, e_{thresh} is the minimum void ratio after which fracture permeability starts to dominate, and α is a factor related to rock brittleness and fracture density.

Void ratio is a scalar quantity while permeability is a tensor. Therefore, the anisotropy of permeability, i.e., in which it varies with direction, must be handled through transformation. In our implementation, permeability anisotropy is introduced using a normalisation and denormalisation method proposed by [44]. In this method, first the microseismic data are normalised using the dimensions of the SRV using Equation (2),

$$X_{trans} = X * \frac{L_{scal}}{L_{max}}$$
$$Y_{trans} = Y * \frac{L_{scal}}{L_{min}} \tag{2}$$
$$Z_{trans} = Z * \frac{L_{scal}}{L_{int}}$$

where X, Y, and Z are coordinates of detected events in the microseismic cloud along the maximum, minimum, and intermediate stress directions, respectively; L_{max}, L_{min}, and L_{int} specify the SRV length parameters along the maximum, minimum, and intermediate stress directions, respectively;

$L_{scal} = \sqrt[3]{L_{max}L_{min}L_{int}}$, and X_{trans}; Y_{trans}, and Z_{trans} are transformed coordinates of detected events in microsesimic cloud.

To illustrate this transformation, assume a notional SRV characterised by an elliptical microseismic cloud in a 2D plane. The elliptical shape implies that the resultant fracture propagation rate and formation permeability are both anisotropic. After normalising the microseismic data in this canonical example, the resultant microseismic cloud will be circumscribed by a circular boundary. These steps are graphically depicted in Figure 1. This approach allows for a simplification of the analysis into a more straightforward one-dimensional analysis.

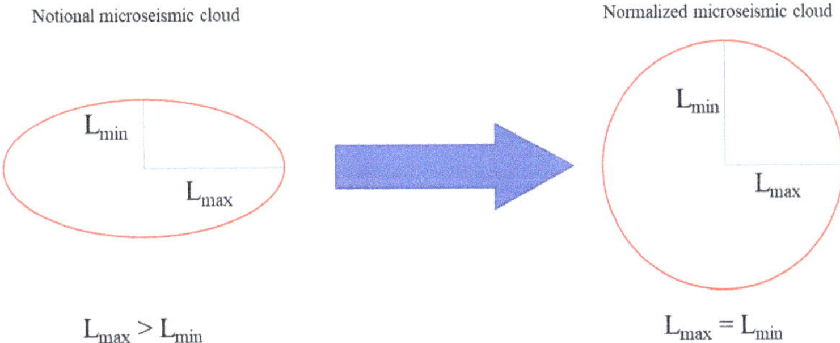

Figure 1. Applying the normalisation method by [44] on a notional stimulated rock volume (SRV) microseismic cloud. The notional microseismic cloud is assumed to be within anisotropic 2-dimensional ellipse. After normalisation, the microseismic cloud can be bounded with a circular shape reducing the analysis to a 1-dimensional scale.

The permeability function that is obtained using the resultant normalised data are denormalised using Equation (3),

$$k_x = (\frac{L_{max}}{L_{scal}})^2 * k_0$$
$$k_y = (\frac{L_{min}}{L_{scal}})^2 * k_0$$
$$k_z = (\frac{L_{int}}{L_{scal}})^2 * k_0$$

$$(3)$$

where, k_0 is the permeability estimated using one-dimensional analysis and k_x, k_y, and k_z are permeabilities in X, Y, and Z directions, respectively.

To provide a case study of our ECM approach, the proposed relation was applied to the Hoadley field data to demonstrate the applicability of this method using real data.

3.3. Case Study—Hoadley Tight Sand Reservoir

The Hoadley gas field is located northwest of Red Deer and southwest of Edmonton, Alberta, Canada. The pay formation is the Glauconite tight sand member of the Cretaceous Mannville Group, which is overlain by the coal-bearing Medicine River Formation. Open-hole, multi-stage hydraulic fracturing was carried out in two horizontal injection wells and monitored using a 12-level geophone array from a nearby vertical well [45]. The configuration of the wells and directions of maximum and minimum in-situ stresses are shown in Figure 2a. In Figure 2a blue and red line and, the big blue dot indicate two injection and vertical observation wells, respectively. The scattered dots are the observed microseismicity during hydraulic fracturing. The different colour is assigned for microseismicity observed at each stage. Simplified stratigraphy of the Hoadley field, as well as a depth histogram showing the number of microseismic events at given depths, is provided in Figure 2b. This paper is focused on a single stage of the hydraulic fracturing program in well 1. The injection pressure and rate profiles for stage 12 hydraulic fracturing in well 1 are plotted in Figure 3.

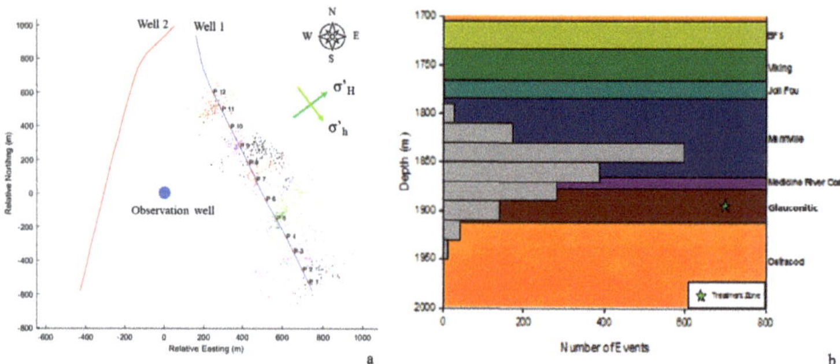

Figure 2. Hoadley field well orientation (**a**) and simplified stratigraphy (**b**). Depth histogram of microseismic events. Gray bars indicate total number of microseismic events recorded at a given depth interval during the hydraulic fracturing [18].

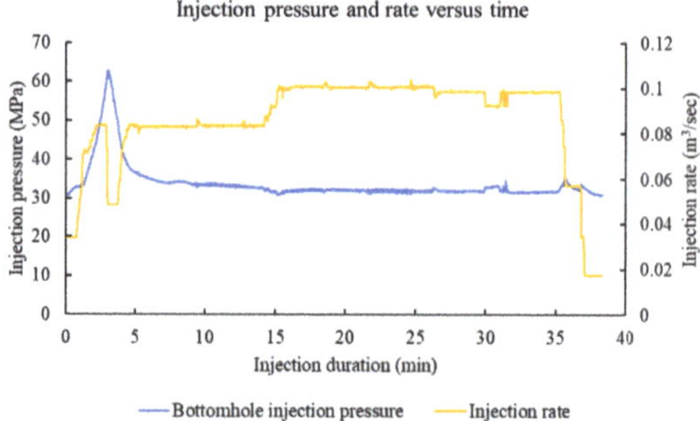

Figure 3. Field recorded injection pressure and rate for stage 12 hydraulic fracturing.

Furthermore, the volumteric strain using the injected volume and injection pressure methods are plotted in Figure 4. The other field data are also included for comparison [45–48]. The trend shown in Figure 4 suggests unique volumetric strain when fractures open, which allows relating the fractured rock permeability to volumetric strain.

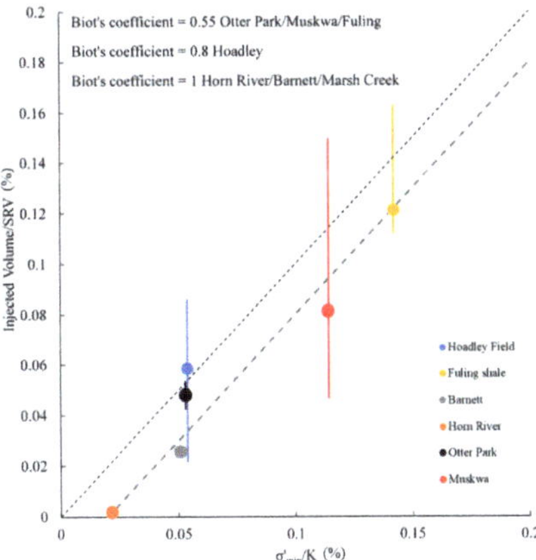

Figure 4. Volumetric strain comparison using the injected volume and minimum in-situ stress approaches. The plotted relation suggests that fracture propagates when strain due to injection equals the strain due to minimum in-situ stress.

Despite most of the microseismicity apparently concentrating above the target unit, this paper models only the Glauconite member. We focus on the in-zone microseismic response due to the abundance of geomechanical property data for Glauconite member, whereas other units are depicted through correlations only. The input parameters of the Glauconite member for ABAQUS FEM are provided in Table 1.

Table 1. The summary of Glauconite data.

Property	Value
Porosity	8–20% [49] 15% [18]
Permeability	0.001–0.01 mD [49] 0.07 mD [18]
Density	2220 kg/m^3 [50]
Depth	2075 m [49] 1900 m [18]
Pore pressure *gradient*	8.5–16.1 kPa/m [49] 4.86 kPa/m [18]
σ_v **gradient**	24.09 kPa/m [18]
σ_H **gradient**	25.79 kPa/m [18]
σ_h **gradient**	11.66 kPa/m [18]
Young's modulus	45.04 GPa [18]
Poisson's ratio	0.23 [18]

The first step in ECM simulation of the hydraulic fracturing is to normalise the microseismic data to reduce the analysis to one-dimensional form. Prior to normalisation, the microseismic data was rotated to align with maximum horizontal stress. Then the coordinates were translated to be relative to the injection port of stage 12. The planar view of rotated and translated microseismic event coordinates for stage 12 are illustrated in Figure 5a. The values for L_{max}, L_{min}, and L_{int} are 150 m, 50 m, and 75 m, respectively. The normalised microseismic cloud for stage 12 is presented in Figure 5b.

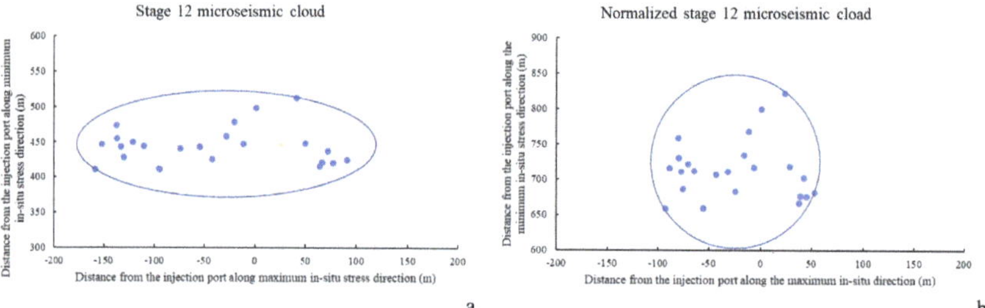

Figure 5. The planar microseismic event coordinates for stage 12 rotated to align with maximum in-situ stress direction and translated to be relative to injection port coordinates (**a**). The normalised stage 12 microseismic event coordinates using $L_{max} = 150$ m, $L_{min} = 50$ m, and $L_{int} = 75$ m (**b**).

The void ratio dependent permeability relation is constructed to match the normalised microseismic cloud. The resulting relation is shown in Figure 6a. As outlined above, this relationship is then denormalised to obtain anisotropic permeability. The resultant permeabilities along the maximum, minimum, and intermediate stress directions, respectively are plotted in Figure 6b.

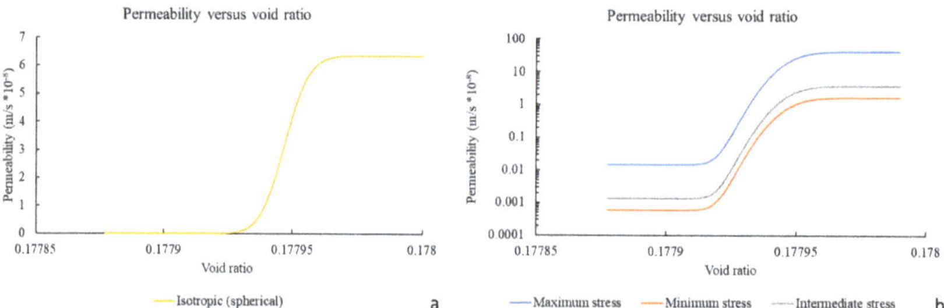

Figure 6. The best matching void ratio dependent permeability relation for normalised and isotropic microseismic cloud (**a**). The denormalised void ratio dependent permeabilities along maximum, minimum, and intermediate in-situ stress directions for stage 12 hydraulic fracturing (**b**).

The anisotropic permeability relation is then used in ABAQUS to combine with the continuity equation, including the stress change effect. The constructed model uses C3D8P three-dimensional pore pressure solid element with an approximate mesh size of 20 m. The modelled region dimensions are 1000 m by 250 m by 225 m at the maximum, minimum and vertical stress directions, respectively. The modelled region dimensions are much larger than the SRV to avoid boundary effects. At the outer and top surfaces of the model, a constant pressure boundary is applied to ensure a constant far-field effective stress effect. This injection rate is only 12.5% of the average injection rate since the constructed model uses symmetry boundary along three surfaces. The inner and bottom surfaces use a symmetry boundary. The injection is defined using concentrated fluid flow load on a single node at a constant rate of 0.015 m³/s. In-situ stresses and pore pressure are introduced through pre-defined fields in ABAQUS. A constant value at

1900 m is applied for the in-situ stresses and pore pressure. The boundary conditions and injection port locations are indicated in Figure 7 using a simplified sketch.

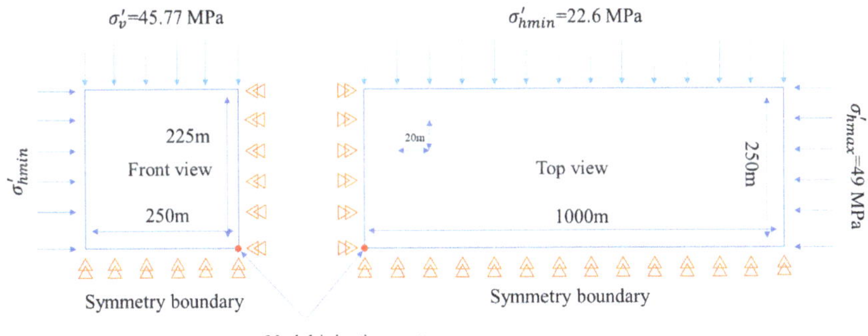

Figure 7. The boundary conditions and injection port location. The single headed arrows triangles indicate loading directions and symmetry surfaces, respectively. Red dot represents the location of injection port.

The results from the ECM are compared to field data, as well as the PKN and KGD models. The injection pressure profile obtained from three models and field observation are provided in Figure 8. The PKN and KGD models account for the intact rock cohesion of 11.55 MPa. Furthermore, to account for the natural fractures in the formation, the leak-off coefficient is introduced into analytical solutions. The leak-off coefficient is estimated based on fracture half-length of 150 m at the end of 38 min injection period. As indicated in Figure 8, the analytical models slightly overestimate injection pressure while ECM shows a good match with reported data.

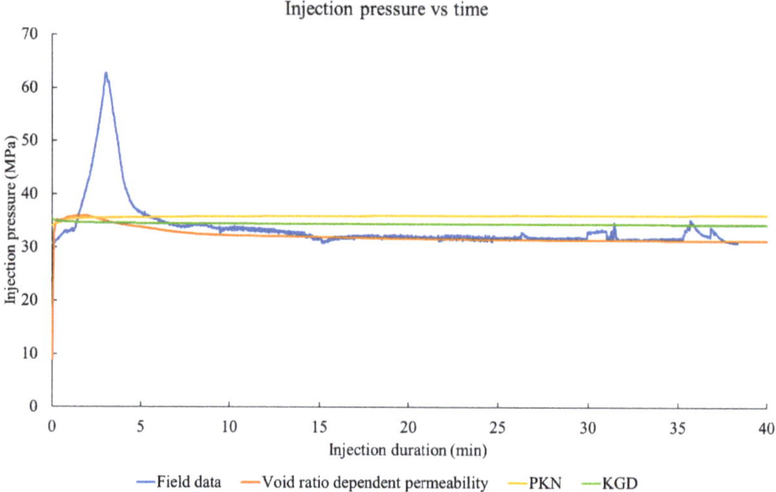

Figure 8. The injection pressure comparison between Equivalent Continuum Methods (ECM), Perkins–Kern–Nordgren (PKN), and Khristianovic–Geertsma–de Klerk (KGD) models against the field observation. As it is observed, the analytical solutions slightly overestimate the injection pressure while ECM shows a good match.

The analytical and ECM estimates for fracture length are compared against observed microseismicity. Fracture length along maximum stress comparison is provided in Figure 9. Besides the previously mentioned methods a diffusivity model estimate is included in Figure 9. The diffusivity model estimates the fracture length growth according to $r(t) = \sqrt{4\pi Dt}$ where $r(t)$ is the fracture length as a function of time in m, D is diffusivity in m^2/s and time in s. The estimated diffusivity using the microseismicity along maximum stress direction is 4 m^2/s. This indicates the case for fracture length increment with assumed constant permeability. As observed from Figure 9, both analytical solutions show the same fracture length elongation profile as the estimation of fluid leak-off coefficient based on the assumed final length of the fracture. For the ECM it is assumed that microseismicity is observed at a given node when pore pressure reaches 12.12 MPa. This satisfies the Mohr Coulomb failure criterion for Glauconite formation [18]. The ECM estimate for fracture length is lower compared to other methods. This is possible due to some of the microseismicity being not due to pore pressure change but due to total stress change around the increased pore pressure region.

Furthermore, the fracture length estimates along minimum in-situ stress direction are compared in Figure 10. As the PKN and KGD models are one-dimensional solutions, they do not show any fracture extend along minimum in-situ stress direction. In addition, the diffusivity model assumes a radial flow. Therefore, it overestimates the fracture extend along the minimum in-situ stress. The ECM approach underestimates the fracture length for the minimum in-situ stress direction, too. This further confirms the possibility of some of the microseismicity due to total stress change effects. The model can be calibrated to match better the microseismic cloud however, the injection pressure profile would differ from the field observation. Considering the higher reliability of injection data compared to microseismicity, it is more preferential to match the injection data.

Another advantage of the proposed ECM has optimised modelling efficiency. The ECM and XFEM models were run using a personal computer with Core i7 2.4 GHz quad-core processor. The ECM required

8768.1 s of CPU time to model 3600 s (or 60 min) of injection period with no divergence problem. However, XFEM required about 28,800 s CPU time before running into convergence problem after 300 s (or 5 min) of the injection period. Therefore, for given generalised data set ECM approach poses as a more efficient modelling method. The XFEM might become a more preferred option if more detailed geomechanical data will allow for a better optimisation of this method.

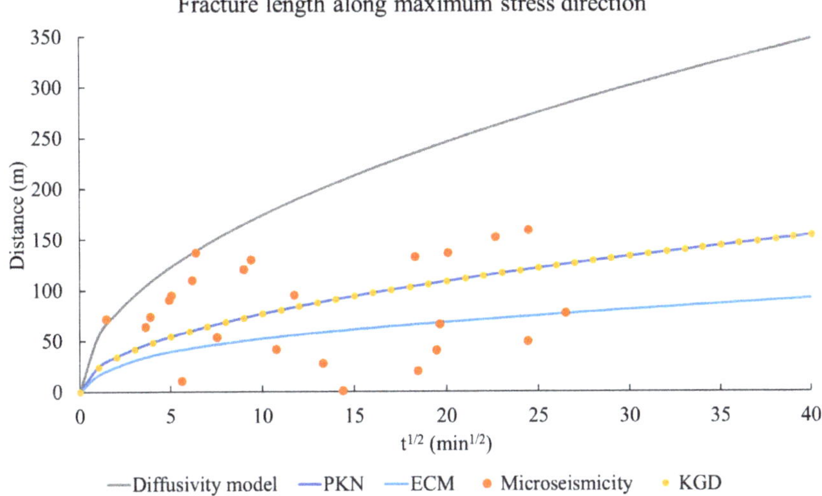

Figure 9. Comparison of fracture length increment along maximum in-situ stress direction using diffusivity, PKN, KGD, and ECM models. The orange dots indicate recorded microseismicity along the maximum in-situ stress direction.

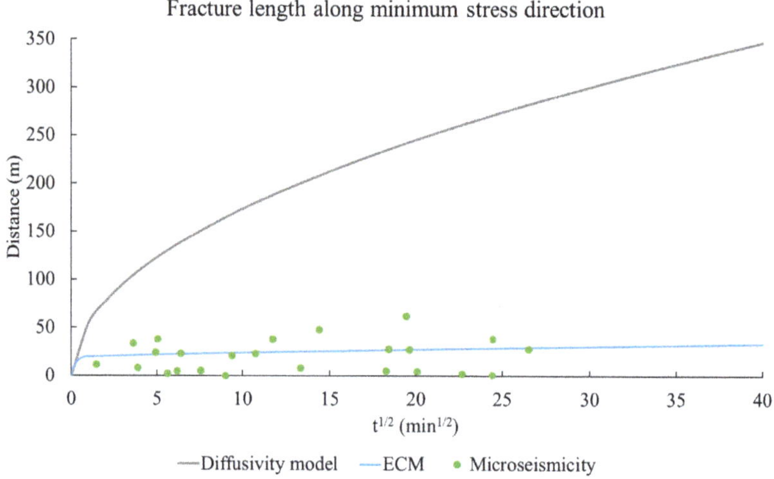

Figure 10. Comparison of fracture length increment along minimum in-situ stress direction using diffusivity and ECM models. The green dots indicate recorded microseismicity along minimum in-situ stress direction.

4. Conclusions

Based on increasing interest in the use of optimised modelling techniques to simulate hydraulic fracturing, we developed a robust and computationally efficient equivalent continuum method. Our ECM method implemented using ABAQUS applies a step-like relationship between void ratio and permeability and incorporates microseismic data to define the stimulated volume of a reservoir. Anisotropy is accounted for by applying a transformation that enables the use of a standard one-dimensional modelling approach with a uniform mesh. This method is particularly useful for cases where there is insufficient data available to define parameters for other numerical approaches, including stress/strain sensitive zones in the target formation. In addition, the use of representative element volume allows the homogenisation of rock elastic and fracture properties for a modelled region. The proposed ECM is more representative compared to PKN, KGD, and diffusivity models and more efficient compared to the XFEM approach. A case study of hydraulic fracturing of tight sand in the Hoadley gas field, Alberta, illustrates the utility of our method for characterising hydraulic fracturing.

Author Contributions: Conceptualisation, E.A.; methodology, E.A.; writing—original draft preparation, E.A.; writing—review and editing, R.C.K.W. and D.W.E.; supervision, R.C.K.W.; funding acquisition, D.W.E. All authors have read and agreed to the published version of the manuscript.

Funding: This research was funded by Natural Sciences and Engineering Research Council of Canada grant number CRDPJ/474748-2014.

Acknowledgments: This research was funded by Natural Sciences and Engineering Research Council of Canada grant number CRDPJ/474748-2014. Sponsors of the Microseismic Industry Consortium are also thanked for their ongoing support of this initiative.

Conflicts of Interest: The authors declare no conflict of interest.

Abbreviations

The following abbreviations are used in this manuscript:

FEM	Finite Element Modelling
CZM	Cohesive Zone Modelling
XFEM	Extended Finite Element Modelling
ECM	Equivalent Continuum Method
SRV	Stimulated Reservoir Volume

References

1. Li, Q.; Xing, H.; Liu, J.; Liu, X. A review on hydraulic fracturing of unconventional reservoir. *Petroleum* **2015**, *1*, 8–15. [CrossRef]
2. Montgomery, C.T.; Smith, M.B. Hydraulic fracturing: History of an enduring technology. *JPT J. Pet. Technol.* **2010**, *62*, 26–32. [CrossRef]
3. Wu, K.; Olson, J.E. A simplified three-dimensional displacement discontinuity method for multiple fracture simulations. *Int. J. Fract.* **2015**, *193*, 191–204. [CrossRef]
4. Warpinski, N. A review of hydraulic-fracture induced microseismicity. In Proceedings of the 48th US Rock Mechanics/Geomechanics Symposium, Minneapolis, MN, USA, 1-4 June 2014; American Rock Mechanics Association. Available online: https://www.onepetro.org/conference-paper/ARMA-2014-7774 (accessed on 24 November 2020).
5. Yao, Y.; Gosavi, S.V.; Searles, K.H.; Ellison, T.K. Cohesive fracture mechanics based analysis to model ductile rock fracture. In Proceedings of the 44th US Rock Mechanics Symposium and 5th US-Canada Rock Mechanics Symposium, Salt Lake City, UT, USA, 27–30 June 2010.
6. Haddad, M.; Sepehrnoori, K. Simulation of multiple-stage fracturing in quasibrittle shale formations using pore pressure cohesive zone model. In Proceedings of the Unconventional Resources Technology Conference, Denver, CO, USA, 25–27 August 2014; pp. 1777–1792. [CrossRef]
7. Haddad, M.; Sepehrnoori, K. Integration of XFEM and CZM to model 3D multiple-stage hydraulic fracturing in quasi-brittle shale formations: Solution-dependent propagation direction. In Proceedings of the AADE National Technical Conference and Exhibition, San Antonio, TX, USA, 8–9 April 2015; pp. 8–9.
8. Haddad, M.; Sepehrnoori, K. Simulation of hydraulic fracturing in quasi-brittle shale formations using characterized cohesive layer: Stimulation controlling factors. *J. Unconv. Oil Gas Resour.* **2015**, *9*, 65–83. [CrossRef]
9. Wang, H. Numerical modeling of non-planar hydraulic fracture propagation in brittle and ductile rocks using XFEM with cohesive zone method. *J. Pet. Sci. Eng.* **2015**, *135*, 127–140. [CrossRef]
10. Zielonka, M.G.; Searles, K.H.; Ning, J.; Buechler, S.R. Development and validation of fully-coupled hydraulic fracturing simulation capabilities. In Proceedings of the SIMULIA Community Conference, Providence, RI, USA, 20–22 May 2014; pp. 19–21.
11. Arndt, S.; Zee, W.V.D.; Hoeink, T.; Hughes, B.; Nie, J. Hydraulic fracturing simulation for fracture networks. In Proceedings of the SIMULIA Community Conference, Berlin, Germany, 18–21 May 2015; Volume 4.
12. Nasirisavadkouhi, A. A comparison study of KGD. *PKN Modif. P3D Model* **2015**. [CrossRef]
13. Yu, G.; Aguilera, R. 3D analytical modeling of hydraulic fracturing stimulated reservoir volume. In *SPE Latin America and Caribbean Petroleum Engineering Conference, Mexico City, Mexico, 16–18 April, 2012*; Society of Petroleum Engineers, SPE-153486-MS, SPE. Available online: https://www.onepetro.org/conference-paper/SPE-153486-MS (accessed on 24 November 2020).
14. Bouteca, M.J. 3D analytical model for hydraulic fracturing: Theory and field test. In Proceedings of the SPE Annual Technical Conference and Exhibition, New Orleans, Louisiana, USA, 27–30 September 1984; doi:10.2523/13276-ms. [CrossRef]
15. Stalgorova, E.; Mattar, L. Analytical model for unconventional multifractured composite systems. *SPE Reserv. Eval. Eng.* **2013**, *16*, 246–256. [CrossRef]
16. Geertsma, J.; De Klerk, F. A rapid method of predicting width and extent of hydraulically induced fractures. *J. Pet. Technol.* **2007**, *21*, 1571–1581. [CrossRef]
17. Nordgren, R. Propagation of a vertical hydraulic fracture. *Soc. Pet. Eng. J.* **1972**, *12*, 306–314. [CrossRef]

18. Maulianda, B. On Hydraulic Fracturing of Tight Gas Reservoir Rock. Ph.D. Thesis, University of Calgary, Calgary, AB, Canada, 2016. [CrossRef]
19. Gonzalez, D.; Holman, R.; Richard, R.; Xue, H.; Morales, A.; Kwok, C.K.; Judd, T. Accounting for production shadow in infill/DUC well hydraulic fracturing modeling and calibration. In *SPE Liquids-Rich Basins Conference-North America, Midland, Texas, USA, 13–14 September, 2017*; Society of Petroleum Engineers. Available online: https://www.onepetro.org/conference-paper/SPE-187479-MS (accessed on 24 November 2020).
20. Valiullin, A.; Makienko, V.; Yudin, A.; Overin, A.; Gromovenko, A. Channel fracturing technique helps to revitalize brown fields in langepas area. In *Society of Petroleum Engineers—SPE Oil and Gas India Conference and Exhibition, Mumbai, India, 24–26 November 2015*; Society of Petroleum Engineers. Available online: https://www.onepetro.org/conference-paper/SPE-178131-MS (accessed on 24 November 2020). [CrossRef]
21. Chen, Z.; Bunger, A.P.; Zhang, X.; Jeffrey, R.G. Cohesive zone finite element-based modeling of hydraulic fractures. *Acta Mech. Solida Sin.* **2009**, *22*, 443–452. [CrossRef]
22. Wang, X.L.; Liu, C.; Wang, H.; Liu, H.; Wu, H.A. Comparison of consecutive and alternate hydraulic fracturing in horizontal wells using XFEM-based cohesive zone method. *J. Pet. Sci. Eng.* **2016**, *143*, 14–25. [CrossRef]
23. Guo, J.; Luo, B.; Lu, C.; Lai, J.; Ren, J. Numerical investigation of hydraulic fracture propagation in a layered reservoir using the cohesive zone method. *Eng. Fract. Mech.* **2017**, *186*, 195–207. [CrossRef]
24. Yao, Y.; Liu, L.; Keer, L.M. Pore pressure cohesive zone modeling of hydraulic fracture in quasi-brittle rocks. *Mech. Mater.* **2015**, *83*, 17–29. [CrossRef]
25. Carrier, B.; Granet, S. Numerical modeling of hydraulic fracture problem in permeable medium using cohesive zone model. *Eng. Fract. Mech.* **2012**, *79*, 312–328. [CrossRef]
26. Yao, Y. Linear elastic and cohesive fracture analysis to model hydraulic fracture in brittle and ductile rocks. *Rock Mech. Rock Eng.* **2012**, *45*, 375–387. [CrossRef]
27. Lavrov, A.; Larsen, I.; Holt, R.M.; Bauer, A.; Pradhan, S. Hybrid FEM/DEM simulation of hydraulic fracturing in naturally-fractured reservoirs. In Proceedingd of the 48th US Rock Mechanics/Geomechanics Symposium, Minneapolis, MN, USA, 1–4 June 2014; pp. 626–633.
28. Munjiza, A.; John, N.W. Mesh size sensitivity of the combined FEM/DEM fracture and fragmentation algorithms. *Eng. Fract. Mech.* **2001**, *69*, 281–295. [CrossRef]
29. Min, K.B. Fractured Rock Masses as Equivalent Continua—A Numerical Study. Ph.D. Thesis, Mark Och Vatten, Stockholm, Sweden, 2004.
30. Oda, M. An equivalent continuum model for coupled stress and fluid flow analysis in jointed rock masses. *Water Resour. Res.* **1986**, *22*, 1845–1856. [CrossRef]
31. Villeneuve, M.C.; Heap, M.J.; Kushnir, A.R.; Qin, T.; Baud, P.; Zhou, G.; Xu, T. Estimating in situ rock mass strength and elastic modulus of granite from the Soultz-sous-Forêts geothermal reservoir (France). *Geotherm. Energy* **2018**, *6*, 1–29. [CrossRef]
32. JianPing, Y.; WeiZhong, C.; DianSen, Y.; JingQiang, Y. Numerical determination of strength and deformability of fractured rock mass by FEM modeling. *Comput. Geotech.* **2015**, *64*, 20–31. [CrossRef]
33. Mack, M.G.; Warpinski, N.R. Mechanics of hydraulic fracturing. In *Reservoir Stimulation*; Wiley: New York, NY, USA, 2000; pp. 1–6.
34. SIMULIA. ABAQUS 6.16-1. 2016. Available online: http://130.149.89.49:2080/v2016/index.html (accessed on 24 November 2020).
35. Youn, D.J.; Griffiths, D.V. Hydro-Mechanical Coupled Simulation of Hydraulic Fracturing Using the Extended Finite Element Method (XFEM). Ph.D Thesis, Colorado School of Mines, Golden, GO, USA, 2016.
36. Kang, Y.; Yu, M.; Miska, S.; Takach, N.E. Wellbore stability: A critical review and introduction to DEM. In Proceedings of the SPE Annual Technical Conference and Exhibition, New Orleans, LA, USA, 4–7 October 2009; Volume 4, pp. 2689–2712. [CrossRef]
37. Zhou, J.; Huang, H.; Mattson, E.; Wang, H.F.; Haimson, B.C.; Doe, T.W.; Oldenburg, C.M.; Dobson, P.F. *Modeling of Hydraulic Fracture Propagation at the kISMET Site Using a Fully Coupled 3D Network-Flow and Quasi-Static Discrete Element Model*; Technical Report; Idaho National Lab (INL): Idaho Falls, ID, USA, 2017.

38. Nguyen, T.K.; Combe, G.; Caillerie, D.; Desrues, J. FEM x DEM modelling of cohesive granular materials: Numerical homogenisation and multi-scale simulations. *Acta Geophys.* **2014**, *62*, 1109–1126. [CrossRef]

39. Maulianda, B.; Savitri, C.D.; Prakasan, A.; Atdayev, E.; Yan, T.W.; Yong, Y.K.; Elrais, K.A.; Barati, R. Recent comprehensive review for extended finite element method (XFEM) based on hydraulic fracturing models for unconventional hydrocarbon reservoirs. *J. Pet. Explor. Prod. Technol.* **2020**. [CrossRef]

40. Hoek, E.; Diederichs, M.S. Empirical estimation of rock mass modulus. *Int. J. Rock Mech. Min. Sci.* **2006**, *43*, 203–215. [CrossRef]

41. Oda, M. Similarity rule of crack geometry in statistically homogeneous rock masses. *Mech. Mater.* **1984**, *3*, 119–129. [CrossRef]

42. Mayerhofer, M.; Lolon, E.; Warpinski, N.; Cipolla, C.; Walser, D.; Rightmire, C. What Is Stimulated Rock Volume? In Proceedings of the SPE Shale Gas Production Conference, Fort Worth, TX, USA, 16–18 November 2009.

43. Ahmed, T. *Reservoir Engineering Handbook*; Gulf Professional Publishing: Oxford, UK, 2018.

44. Shapiro, S.; Hummel, N. Nonlinear diffusion-based interpretation of hydraulic fracturing induced seismicity. In Proceedings of the 75th EAGE Conference & Exhibition-Workshops, London, UK, 10–13 June 2013.

45. Eaton, D.; Caffagni, E.; van der Baan, M.; Matthews, L. Passive seismic monitoring and integrated geomechanical analysis of a tight-sand reservoir during hydraulic-fracture treatment, flowback and production. In Proceedings of the Unconventional Resources Technology Conference, Denver, CO, USA, 25–27 August 2014; pp. 1537–1545.

46. Ren, L.; Lin, R.; Zhao, J. Stimulated reservoir volume estimation and analysis of hydraulic fracturing in shale gas reservoir. *Arab. J. Sci. Eng.* **2018**, *43*, 6429–6444. [CrossRef]

47. Li, Q.; Aguilera, R. Parametric probabilistic models for fluid diffusivity inversion and forward microseismic generation using seismicity rates. *SPE West. Reg. Meet.* **2018**. [CrossRef]

48. Xu, Y.; Adefidipe, O.; Dehghanpour, H. Estimating fracture volume using flowback data from the Horn River Basin: A material balance approach. *J. Nat. Gas Sci. Eng.* **2015**, *25*, 253–270. [CrossRef]

49. Reynolds, M.; Thomson, S.; Peyman, F.; Hung, A.; Quirk, D.; Chen, S. A Direct Comparison of Hydraulic Fracture Geometry and Well Performance between Cemented Liner and Openhole Packer Completed Horizontal Wells in a Tight Gas Reservoir. In Proceedings of the PSPE Hydraulic Fracturing Technology Conference, Woodlands, TX, USA, 6–8 February 2012; pp. 1–19. [CrossRef]

50. Broger, E.K.; Syhlonyk, G.E.; Zaitlin, B.A. Glauconite Sandstone Exploration: A Case Study from the Lake Newell Project, Southern Alberta. In *Petroleum Geology of the Cretaceous Mannville Group Western Canada—Memoir 18*; American Association of Petroleum Geologists, AAPG: Tulsa, OK, USA, 1997; pp. 140–168. Available online: http://archives.datapages.com/data/cspg_sp/data/018/018001/140_cspgsp0180140.htm (accessed on 24 November 2020).

Article

Numerical Investigation on Proppant–Water Mixture Transport in Slot under High Reynolds Number Conditions

Tao Zhang [1],*, Ruoyu Yang [1], Jianchun Guo [1],* and Jie Zeng [2]

[1]   State Key Laboratory of Oil and Gas Reservoir Geology and Exploitation, Southwest Petroleum University, Chengdu 610500, China; 201711000132@stu.swpu.edu.cn

[2]   School of Engineering, The University of Western Australia, Perth, Western Australia 6009, Australia; jie.zeng@research.uwa.edu.au

*   Correspondence: zhangt@swpu.edu.cn (T.Z.); guojianchun@vip.163.com (J.G.)

Received: 29 September 2020; Accepted: 27 October 2020; Published: 29 October 2020

Abstract: Water hydraulic fracturing involves pumping low viscosity fluid and proppant mixture into the artificial fracture under a high pumping rate. In that high Reynolds number conditions (HRNCs, $Re > 2000$), the turbulence effect is one of the key factors affecting proppant transportation and placement. In this paper, a Eulerian multiphase model was used to simulate the proppant particle transport in a parallel slot under HRNCs. Turbulence effects in high pumping rates and frictional stress among the proppant particles were taken into consideration, and the Johnson-Jackson wall boundary conditions were used to describe the particle-wall interaction. The numerical simulation result was validated with laboratory-scale slot experiment results. The simulation results demonstrate that the pattern of the proppant bank is significantly affected by the vortex near the wellbore, and the whole proppant transport process can be divided into four stages under HRNCs. Furthermore, the proppant placement structure and the equilibrium height of proppant dune under HRNCs are comprehensively discussed by a parametrical study, including injection position, velocity, proppant density, concentration, and diameter. As the injection position changes from the lower one to the top one, the unpropped area near the entrance decrease by 7.1 times, and the equilibrium height for the primary dune increase by 5.3%. As the velocity of the slurry jet increases from 2 m/s to 5 m/s (Re = 2000–5000), the vortex becomes stronger, so the non-propped area near the inlet increase by 5.3 times, and the equilibrium height decrease by 5.2%. The change of proppant properties does not significantly change the vortex; however, the equilibrium height is affected by the high-speed flush. Thus, the conventional equilibrium height prediction correlation is not suitable for the HRNCs. Therefore, a modified bi-power law prediction correlation was proposed based on the simulation data, which can be used to accurately predict the equilibrium height of the proppant bank under HRNCs (mean deviation = 3.8%).

Keywords: water fracturing; turbulence effect; Eulerian multiphase modeling; proppant transport mechanism; equilibrium height prediction model

---

1. Introduction

Hydraulic fracturing technology has been widely used in recent years to economically exploit unconventional resources, especially for shale oil and gas [1]. In fracturing process, high pressurized fluid is injected to initial the fracture and propagate it. Once the fracture is created, the high strength particle (proppant) was carried by fracturing fluid and injected in the fracture, to keep the fracture open and ensure an effective flow path for hydrocarbon flow [2]. The artificial fracture geometry is influenced by geological factors and fracturing treatment. Figure 1 shows the comparison of the fracture geometry between water fracturing for shale and conventional fracturing.

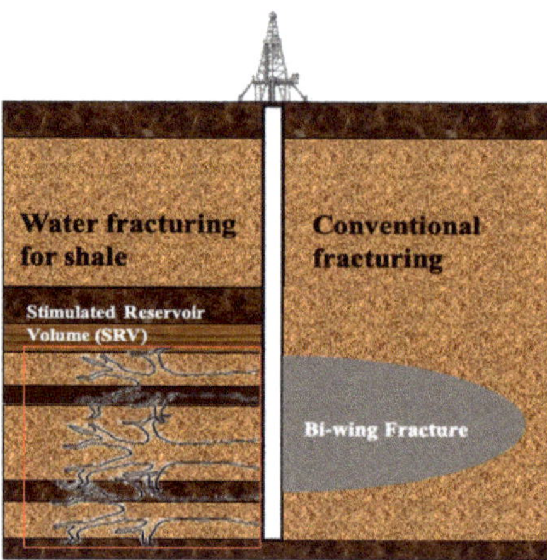

Figure 1. Schematic diagram of different artificial fracture geometry for conventional fracturing and water fracturing.

Different from conventional fracturing, the goal for shale fracturing is breaking the formation and generating larger stimulated reservoir volume (SRV) with more complex fracture, due to the extremely low permeability and porosity of shale rock [3]. Therefore, the low viscosity water is used as the primary fracturing fluid system used for the stimulation of shale formation [4,5]. However, the low viscosity of the fluid affects the proppant transport capability. To address this problem, engineers often pump water–proppant mixture into the fracture at a very high pumping rate. Due to the low-viscosity fluid and high pumping rate, the turbulence effects become an important factor affecting the transport behaviors of the proppant particles in fracture [6,7]. In general, Reynolds number (Re) is a dimensionless number that can be used to characterize fluid flow type. For the general channel flow field, $Re = 2000$ is a critical value for laminar flow and turbulent flow, and we defined high Reynolds number condition (HRNCs) when Re is larger than 2000. The behavior and mechanism of the proppant transport process and placement under that HRNCs in the shale fracturing process are different from conventional fracturing. Because of its complexity and significance, continuous research on this problem is conducted by many scholars.

Numerous laboratory experimental research contributes to the understanding of the transport and settling behaviors in low viscosity fracturing fluids. The first experiment about the sand-water mixture transport in the slot is carried by Kern et al. [8], in which they found the sand quickly settled to the slot bottom and formed a proppant dune near the inlet side because of the poor sand-carrying ability of water. Besides, once the equilibrium height reached, sand injected later moved and settled to the rear of the proppant dune. STIM-LAB has been studied on the proppant transport process for more than 20 years, and the effects of the proppant density, diameter, and volume concentration on the equilibrium height of the proppant bed are comprehensively investigated [9,10]. Liu et al. [11] conducted similar slot experiments to STIM-LAB, and the results showed that the initial position and the equilibrium height of the proppant bed changed with the perforation position and the slurry flow rates. Palisch et al. [12] concluded that as the equilibrium height reached the critical value, the mechanisms of the proppant particles transporting on the top of the proppant dune were dominated by fluidization and sedimentation. The turbulent flow suspended the proppant particles off the proppant dune, and then proppant settled again after being transported to some distances during fluidization.

Based on the experimental results above, some correlations for the proppant particle transport and settling have been built up. Liu et al. [11] developed a fitting equation of the height of the equilibrium gap and the injection flow rate. Patankar et al. [9] and Wang et al. [13] established the empirical correlations in the form of power and bi-power law, which were widely used in the industry. Since there is no accurate method for the proppant particles settling in slick-water fracturing treatments, some revised Stokes' Laws which considered the effects of the proppant volume concentration and the fracture wall are often used to predict the proppant particles settling [14,15].

Due to the ability to solve the flow patterns of liquid-solid two phases and their interaction simultaneously, Computational Fluid Dynamics (CFD) technology provides an alternative method to study the proppant transport and to settle in hydraulic fracture accurately. Patankar et al. [16,17] used a DNS model to study the lift-off of particles in plane Poiseuille flows. The results showed the interactions between the particles in the sedimentation process, ignoring the fracture propagation and the effect of fracturing fluid loss. Tsai et al. [18] employed a large-eddy model to simulate the flow field and tracked the proppant particles in Lagrangian coordinates, where the Wen-Yu drag model was used to couple the interaction of the fracturing fluid and proppant. The results showed that the pumping rate and the proppant parameters have an essential influence on the proppant settling and placement. A Computational Fluid Dynamics coupled with Discrete Element Method (CFD-DEM) was employed by Zeng et al. [6] to study the proppant transport process in a small-scale fracture. Although a representative particle model was used to reduce computational efforts, the time consumed is still considerable [19].

Compared with the CFD method mentioned above, the Eulerian multiphase flow model has the advantages of high efficiency and low computational cost. This method is the priority choice for the fluid–solid multiphase flow simulations of the engineering problems [20]. In the last three decades, the Eulerian multiphase flow model has gained considerable progress [21]. As described by Agrawal et al. [22], the solid phase governing equations was extended from the mono-disperse system to the poly-disperse system. Srivastava et al. [23] developed a solid phase frictional stress model in the dense solid-phase flow by considering the interactions between the particles as the solid volume concentration is greater than the critical concentration. Benyahia et al. [24] evaluated Jenkins-Louge and Johnson-Jackson's solid wall boundary conditions and pointed out their application scope. The simulation results of the fluidized bed and the slurry flow in the pipeline using the Eulerian multiphase flow model were highly consistent with experimental results [25,26]. Recently, this method is also used to investigate the transport and settling behavior of proppant in single fractures [27] and the cross fractures [28].

Although Eulerian multiphase flow has been widely used in proppant transportation research, the systematic study of proppant transport mechanisms and behavior under HRNCs has not been reported before. In this paper, a Eulerian two-fluid model considering turbulence effect and particle friction stress were used to study the proppant transport and settling behaviors under HRNCs. The mechanism during the transport process under HRNCs and the impact of such factors as the inlet velocity, the inlet position, the proppant parameters, and the volume concentration on proppant transport and settlement are comprehensively discussed. The equilibrium height of the proppant dune was also studied, and a modified equilibrium height prediction model was proposed.

2. Eulerian Multiple Flow Model

2.1. Governing Equations

Based on kinetic theory, continuity and momentum equations were established for fluid and solid phase, respectively [21]. Continuity equation of the solid phase (m = s) and the fluid phase (m = l) is given by:

$$\frac{\partial}{\partial t}(\alpha_s \rho_s) + \nabla \cdot (\alpha_s \rho_s \overrightarrow{v}_s) = 0 \tag{1}$$

$$\frac{\partial}{\partial t}(\alpha_l \rho_l) + \nabla \cdot (\alpha_l \rho_l \overrightarrow{v}_l) = 0 \tag{2}$$

where α_s and α_l is the solid and liquid phase volume concentration, dimensionless; ρ_s and ρ_l is the density of solid and liquid phase, kg/m^3; $\overrightarrow{v}_s$ an $\overrightarrow{v}_l$ is the velocity vector of solid and liquid phase, m/s.

Momentum equation for the fluid phase is given by:

$$\frac{\partial}{\partial t}\left(\alpha_l \rho_l \vec{v}_l\right) + \nabla \cdot \left(\alpha_l \rho_l \vec{v}_l \vec{v}_l\right) = -\alpha_l \nabla p_l + \nabla \cdot \bar{\bar{\tau}}_l + \alpha_l \rho_l \vec{g} + \beta\left(\vec{v}_s - \vec{v}_l\right) \tag{3}$$

Momentum equation for the solid phase is given by:

$$\frac{\partial}{\partial t}\left(\alpha_s \rho_s \vec{v}_s\right) + \nabla \cdot \left(\alpha_s \rho_s \vec{v}_s \vec{v}_s\right) = -\alpha_s \nabla p_l + \nabla \cdot \bar{\bar{\tau}}_s + \alpha_s \rho_s \vec{g} + \beta\left(\vec{v}_l - \vec{v}_s\right) \tag{4}$$

where τ_l and τ_s is the shear stress tensor of the fluid phase or solid phase, Pa; β is the momentum exchange coefficient between two phases, dimensionless; and g is the acceleration of gravity, m/s^2.

The granular temperature Θ_s for the solid phase represents the kinetic energy of the random motion of the solid particles [26]. The transport equation derived from the kinetic theory is taken as the following form:

$$\frac{3}{2}\rho_s\left[\frac{\partial}{\partial t}(\alpha_s \Theta_s) + \nabla \cdot \left(\alpha_s \vec{v}_s \Theta_s\right)\right] = \nabla \cdot (\kappa_s \nabla \Theta_s) + \tau_s : \nabla \vec{v}_s - J_s + \Pi_\Theta \tag{5}$$

where Θ_s is the granular temperature, m^2/s^2; κ_s is the diffusion coefficient, kg/(m·s); J_s is the rate of the dissipation granular energy by inelastic collisions per unit volume, kg/(m·s^3); Π_Θ is the rate of the dissipation granular energy by inelastic collisions per unit volume, kg/(m·s^3).

In the artificial fracture, due to the low viscosity and high pumping rate, the flow state of the liquid phase may become turbulent flow. According to the research of Pfleger et al. [29], the predictions of the liquid phase turbulence kinetic energy k and its rate of dissipation ε are obtained from the following modified liquid phase $k - \varepsilon$ turbulence model:

$$\frac{\partial}{\partial t}(\alpha_l \rho_l k) + \nabla \cdot \left(\alpha_l \rho_l \vec{v}_l k\right) = \nabla \cdot \left(\alpha_l\left(\mu_l + \frac{\mu_l^t}{\sigma_k}\right)\nabla k\right) + G_{k,l} + \Pi_k - \alpha_l \rho_l \varepsilon \tag{6}$$

$$\frac{\partial}{\partial t}(\alpha_l \rho_l \varepsilon) + \nabla \cdot \left(\alpha_l \rho_l \vec{v}_l \varepsilon\right) = \nabla \cdot \left(\alpha_l\left(\mu_l + \frac{\mu_l^t}{\sigma_\varepsilon}\right)\nabla \varepsilon\right) + \alpha_l \frac{\varepsilon}{k}(C_{1\varepsilon} G_{k,l} - C_{2\varepsilon} \rho_l \varepsilon) + \Pi_\varepsilon \tag{7}$$

where k and ε is turbulence energy and its rate of dissipation, m^2/s^2 and m^2/s^3; μ_l^t is the turbulence viscosity coefficient and μ_l is the liquid molecular viscosity, Pa·s, $G_{k,l}$ is the production term of the turbulence kinetic energy, kg/(m·s^3). According to Pfleger et al. [29], the source terms Π_k and Π_ε are not considered yet. The constant parameters used in different equations are as follows: $\sigma_k = 1.0$, $\sigma_\varepsilon = 1.3$, $C_\mu = 0.09$, $C_{1\varepsilon} = 1.44$, $C_{2\varepsilon} = 1.92$.

2.2. Constitutive Equations

In the simulation presented here, the accumulation of high-concentration proppant in the proppant dune and the low-concentration on the dune appear at the same time. So, the drag correlation of Gidaspow et al. [30] was used due to the flexibility under different solid-phase concentrations:

$$\beta_{ls} = \begin{cases} \dfrac{3}{4}C_D \dfrac{\rho_l \alpha_l \alpha_s \left|\vec{v}_l - \vec{v}_s\right|}{d_s}\alpha_l^{-2.65} & \alpha_l \geq 0.8 \\[2ex] \dfrac{150\alpha_s(1-\alpha_l)\mu_l}{\alpha_l d_s^2} + \dfrac{1.75\rho_l \alpha_s \left|\vec{v}_l - \vec{v}_s\right|}{d_s} & \alpha_l < 0.8 \end{cases} \tag{8}$$

where α_l is the liquid volume fraction and the drag coefficient between phases C_D is given by:

$$C_D = \begin{cases} \dfrac{24}{Re_s}\left[1 + 0.15(Re_s)^{0.687}\right] & Re_s < 1000 \\[2ex] 0.44 & Re_s \geq 1000 \end{cases} \tag{9}$$

where d_s is the particle diameter, m. Re_s is Reynolds number defined by the relative velocity between two phases:

$$Re_s = \frac{\rho_l d_s \alpha_l \left| \vec{v_s} - \vec{v_l} \right|}{\mu_l}. \tag{10}$$

The Reynolds stress tensor for the liquid phase is given by:

$$\tau_l = \alpha_l \mu_t \left(\nabla \vec{v_1} + \left(\nabla \vec{v_1} \right)^{\mathrm{T}} \right) \tag{11}$$

where $\mu_t = \mu_l + \mu_l^t$ is liquid phase effective viscosity, Pa·s; μ_l^t is the turbulent viscosity, Pa·s.

The stress tensor for the solid phase is given by:

$$\tau_s = \left(-(p_s + p_f) + \eta \mu_b \nabla \cdot \vec{v_s} \right) I + 2(\mu_s + \mu_f) S_s \tag{12}$$

where I is the second-order identity tensor, dimensionless; S_s is the stress strain of solid phase, s^{-1}; p_f is the frictional pressure of the solid phase, Pa.

The strain stress for the solid phase is given by:

$$S_s = \frac{1}{2}\left[\nabla \vec{v_s} + \left(\nabla \vec{v_s} \right)^T \right] - \frac{1}{3}\left(\nabla \cdot \vec{v_s} \right) I. \tag{13}$$

Solid kinetic-collisional pressure is given by:

$$p_s = \alpha_s \rho_s \Theta_s (1 + 4\eta \alpha_s g_0) \tag{14}$$

where $\eta = (1 + e)/2$ and e is the particle-wall restitution coefficient, dimensionless.

The radial distribution function at contact is given by:

$$g_0 = \frac{1 - 0.5\alpha_s}{(1 - \alpha_s)^3}. \tag{15}$$

Solid-phase shear viscosity model is given by:

$$\mu_s = \left(\frac{2 + c}{3} \right) \left[\frac{\frac{\mu^*}{g_0 \eta (2 - \eta)}\left(1 + \frac{8}{5}\eta \alpha_s g_0 \right)}{\left(1 + \frac{8}{5}\eta(3\eta - 2)\alpha_s g_0 \right) + \frac{3}{5}\eta \mu_b} \right] \tag{16}$$

$$\mu^* = \mu \left[1 + \frac{2\beta\mu}{(\alpha_s \rho_s)^2 g_0 \Theta_s} \right]^{-1} \tag{17}$$

$$\mu = \frac{5 \rho_s d_s \sqrt{\Theta_s \pi}}{96} \tag{18}$$

$$\mu_b = \frac{256}{5\pi}\mu \alpha_s^2 g_0 \tag{19}$$

where μ^* is the granular viscosity with the effect of the interstitial fluid, Pa·s; μ is the solids phase dilute granular viscosity, Pa·s; and μ_b is the bulk viscosity of the solid phase, Pa·s [22].

Granular energy conductivity coefficient is given by:

$$k_s = \left(\frac{\kappa^*}{g_0} \right) \left[\frac{\left(1 + \frac{12}{5}\eta \alpha_s g_0 \right)\left(1 + \frac{12}{5}\eta^2(4\eta - 3)\alpha_s g_0 \right)}{+ \frac{64}{25\pi}(41 - 33\eta)\eta^2 (\alpha_s g_0)^2} \right] \tag{20}$$

$$\kappa^* = \kappa \left[1 + \frac{6\beta\kappa}{5(\alpha_s \rho_s)^2 g_0 \Theta_s} \right]^{-1} \tag{21}$$

$$\kappa = \frac{75\rho_s d_s \sqrt{\pi\Theta_s}}{48\eta(41 - 33\eta)} \tag{22}$$

where κ^* is the granular conductivity with the effect of the interstitial fluid, and κ is the solid phase dilute granular conductivity, W/(m·K).

Considering the complexity between the particles in the proppant dune with high volume fraction, solid-phase frictional pressure and viscosity model [22] are given by:

$$\frac{p_f}{p_c} = \left(1 - \frac{\nabla \cdot \vec{v}_s}{n \sqrt{2}\sin(\varphi)\sqrt{S_s : S_s + \Theta_s/d_p^2}}\right)^{n-1} \tag{23}$$

$$\mu_f = \frac{p_f \sin(\varphi)}{\sqrt{2}\sqrt{S_s : S_s + \Theta_s/d_p^2}}\left\{n - (n-1)\left(\frac{p_f}{p_c}\right)^{\frac{1}{n-1}}\right\} \tag{24}$$

where

$$p_c = \begin{cases} 10^{25}(\alpha_s - \alpha_s^{max})^{10} & \alpha_s > \alpha_s^{max} \\ Fr\dfrac{(\alpha_s - \alpha_s^{min})^r}{(\alpha_l - \alpha^*)^s} & \alpha_s^{max} \geq \alpha_s > \alpha_s^{min} \\ 0 & \alpha_s \leq \alpha_s^{min} \end{cases} \tag{25}$$

$$n = \begin{cases} \frac{\sqrt{3}}{2}\sin(\delta) & \nabla \cdot \vec{v}_s \geq 0 \\ 1.03 & \nabla \cdot \vec{v}_s < 0 \end{cases} \tag{26}$$

where P_c is the solid critical pressure, Pa; P_f is the solid friction pressure, Pa. δ is the internal frictional angle of the solid phase; α_s^{max} is the maximum packing concentration of the solid phase, and α_s^{min} is the minimum frictional solid concentration, dimensionless.

The generation term of the turbulent kinetic energy is given by

$$G_{k,l} = \varepsilon_l \mu_t \left(\nabla \vec{v}_l + \nabla \vec{v}_l^t\right) : \nabla \vec{v}_l. \tag{27}$$

Collisional dissipation term of the granular energy is given by

$$J_s = \frac{48}{\sqrt{\pi}}\eta(1 - \eta)\frac{\rho_s \alpha_s^2 g_0}{d_s}\Theta_s^{3/2}. \tag{28}$$

The interaction model of turbulence [31] is given by:

$$\Pi_\Theta = -3\beta\Theta_s + 81\frac{\alpha_s \mu_l^2 \left|\vec{v}_l - \vec{v}_s\right|^2}{g_0 d_s^3 \rho_s \sqrt{\pi\Theta_s}}. \tag{29}$$

In this study, the parameters used in upper equations are set as: $e = 0.9, c = 1.6, F_r = 0.05, r = 2$ $s = 5, \alpha_s^{max} = 0.63, \alpha_s^{min} = 0.5, \delta = \pi/6$.

2.3. Boundary Conditions

The values of the liquid phase velocity $v_{l,in}$ the solid phase velocity $v_{s,in}$, and the volume concentration were given as the inlet conditions. The pressure value of 0 Pa was set as the outlet conditions.

The nonslip condition was used for the liquid phase, and Johnson-Jackson model [20] was used for the solid phase.

$$\left(\mu_s + \mu_f\right)\frac{\partial v_{sl}}{\partial x} = -\frac{\varphi\pi\rho_s \alpha_s g_0 \sqrt{\Theta_s}}{2\sqrt{3}\alpha_s^{max}}v_{sl} \tag{30}$$

$$\kappa_s \frac{\partial \Theta_s}{\partial x} = \frac{\phi \pi v_{sl} \rho_s \alpha_s g_0 \sqrt{\Theta_s}}{2\sqrt{3}\alpha_s^{max}} - \frac{\sqrt{3}\pi \rho_s \alpha_s g_0 \left(1 - e_w^2\right)\sqrt{\Theta_s}}{4\alpha_s^{max}}\Theta_s \tag{31}$$

where $v_{s,l}$ is the slip velocity between the solid particles and the wall, m/s; ϕ is the specularity coefficient at the wall, dimensionless; e_w is the particle–wall restitution coefficient, dimensionless. The parameters used in the upper equations are as follows: $\phi = 0.001$, $e_w = 0.9$.

The use of wall function allows the computation of turbulent flows without the need to resolve the turbulent boundary layer. The boundary layer for liquid phase [31] is defined as

$$\frac{\partial v_l}{\partial x} = -\frac{\rho_l \kappa_v C_\mu^{1/4}\sqrt{k}}{\left(\mu_l + \mu_l^t\right)\ln(Ex^*)}v_l \tag{32}$$

$$x^* = \frac{\rho_l C_\mu^{1/4}\sqrt{k}\Delta x/2}{\mu_l} \cdot \frac{\partial k}{\partial x} = 0 \quad \frac{\partial \varepsilon}{\partial x} = 0 \tag{33}$$

The production P_k and dissipation D_k of k as well as ε at the fluid cells adjacent to the wall, is given by:

$$P_k = \alpha_l \rho_l \sqrt{C_\mu k}\frac{v_l}{\Delta x/2\ln(Ex^*)} + \beta k_{12} \tag{34}$$

$$D_k = \alpha_l \rho_l \varepsilon \tag{35}$$

where E is a Constant in wall function formulation, 9.81; κ_v is Von Karmen constant of value, 0.42; Δx is width of computational cell next to the wall, m.

2.4. Geometric Model and Solution Strategy

As shown in Figure 2, the length, width, and height of the slot were 4 m, 0.01 m, and 0.4 m, respectively. Three types of inlet positions of the top inlet, the middle inlet, and the bottom inlet were chosen to discuss the influence of the perforation positions in vertical hydraulic fractures. Eight outlets with a height of 0.02 m and a width of 0.01 m are uniformly distributed at the right wall.

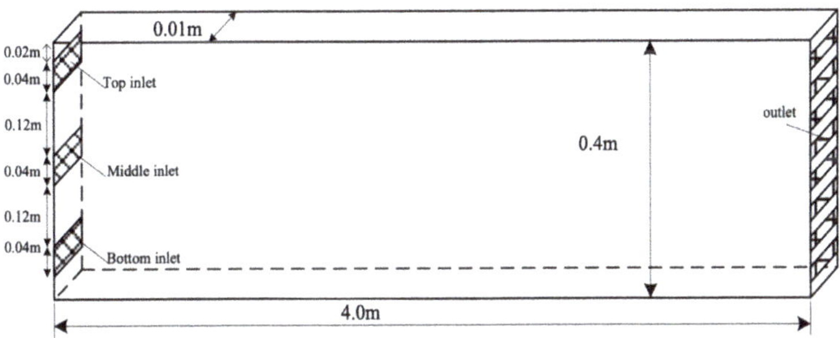

Figure 2. Schematic diagram of slot geometry.

In order to verify the independence of the mesh, Figure 3 shows the proppant placement profile with different kinds of mesh conditions at T = 15 s. Obviously, mesh sizes will affect the final calculation results. But when the grid size is changed from 200 × 40 × 4 to 200 × 40 × 6 in the fracture length, height, and width directions, the proppant placement profile will no longer change. In order to obtain more detailed flow field information, we chose 200 × 40 × 6 as our computing grid in this study.

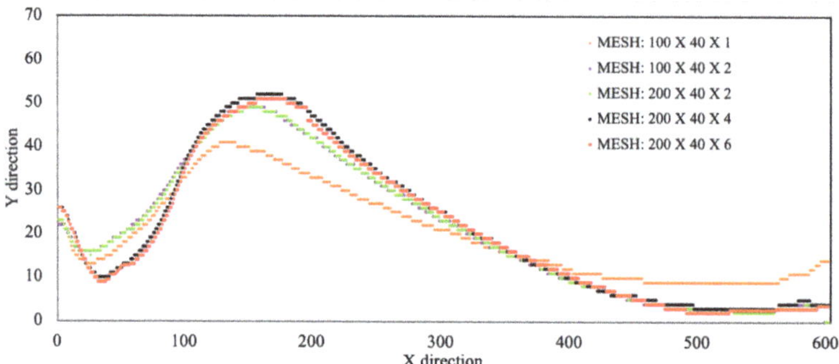

Figure 3. Numerical simulation result of proppant placement profile under different mesh conditions at T = 15 s.

Multiphase solver MFIX was used for the simulation. MFIX (multiphase flow with interphase exchanges) is an open-source code used for multiphase computational fluid dynamics simulations that were developed at National Energy Technology Laboratory (NETL, Pittsburgh, United States) [14,21,24,26]. The governing equations were discretized with the finite volume method and solved on a staggered grid. A second-order super bee discretization scheme was used for all variables. The transient numerical solution was obtained within a residual tolerance of less than 10^{-4}. It takes about 74 h to complete the calculation of a single working condition in 60 s, with i7-7700 cpu, calling 4 threads. The plane at the center along the slot width was used to display the simulation results.

3. Results and Discussions

3.1. Verification of Simulation on Proppant Transport

To verify the Eulerian two-phase simulation on proppant transport, the proppant transport experiments on laboratory scale were conducted. The experiment was conducted in a fracture with a length, width, and height of 4 × 0.006 × 0.6 m, which comprised two pieces of parallel transparent plexiglass (shown in Figure 4). The proppant-water mixture was injected into a slot from perforation to mimic an artificial fracture. The parameters for experiment and simulation were set as the same, and the average Reynolds number is 3000 for both cases. The specific details were shown in Table 1.

Figure 4. Proppant transport experimental system.

Table 1. Parameters set for experiment and simulation.

Items	Experiment	Simulation
Scale L(m)·W(mm)·H(m)	$4 \times 10 \times 0.6$	$4 \times 10 \times 0.4$
Inlet position	Central	Central
Inlet velocity (m/s)	3	3
Particle diameter (mm)	0.4–0.8	0.8
Particle Density (kg/m^3)	3200	3200
Particle volume concentration (%)	15	15
Fluid viscosity (mPa·s)	1	1
Fluid density (kg/m^3)	1000	1000
Reynolds number	3000	3000

Figure 5 shows the comparison of the proppant placement structure between experiment and simulation in the whole transportation process. In the early stage, the experiment result showed the proppant settled near the entrance and formed two dunes (Figure 5a), which was coinciding with the simulation case (Figure 5e). As proppant continued to be injected, the proppant dune first grew vertically and then grew horizontally (Figure 5b,c), and these characteristics were also can be demonstrated through numerical simulation (Figure 5f,g). Finally, the whole process had reached the equilibrium state for experiment and simulation cases. The final placement of proppant obtained by numerical simulation was basically consistent with the experimental results (Figure 5d,h).

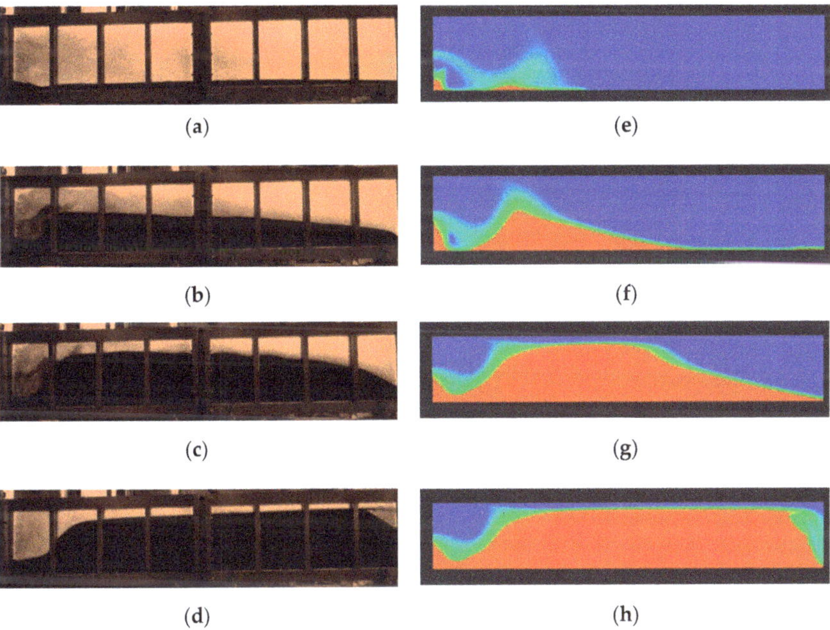

Figure 5. Experimental and numerical simulation results of proppant transport and settlement process. (**a–d**): Proppant placement in the experiment process; (**e–h**): Corresponding proppant placement in the simulation process.

Specifically, Figure 6 shows the comparison of the proppant placement pattern obtained from the experiment and the simulation at the final equilibrium state. Due to the fracture height were different, the relative height was used to compare the equilibrium height in the experiment and simulation, which was defined as the height of the proppant bank to the fracture height. The background picture was obtained from the experiment result, while the white dot line was extracted from the proppant placement profile obtained from the simulation result. The result shows that the profile of the proppant bank for the simulation case is generally consistent with the experiment result, even though some small differences exist in the inlet and outlet place. The reason for this difference may be: (1). The proppant particle size used in the experiment is nonuniform (0.4–0.8 mm), while the proppant size for simulation is set as constant at 0.8 mm; (2). the fracture outlet in the experiment is a single outlet at the top of the wellbore, while six outlets were uniformly distributed along the wellbore in numerical simulations. Hence, the placement near the outlet of the two cases was slightly different from each other. Despite these differences, the Eulerian two-phase model can effectively simulate proppant dynamic behavior and static placement pattern under HRNCs, which proves that this model can be used to simulate and predict the proppant transport and settlement in the fracture.

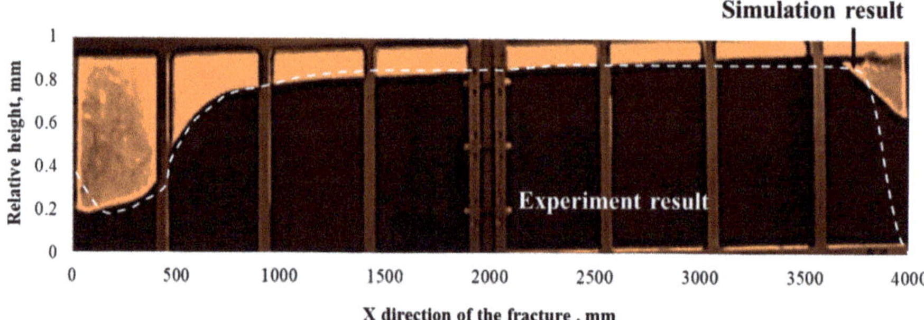

Figure 6. Comparison of proppant placement pattern obtained from experiment and simulation at the final equilibrium state.

3.2. Proppant Transport Process

In this section, by choosing the same simulation parameters in Table 1, proppant transport characteristics and mechanisms under HRNCs were comprehensively studied by analyzing the solid phase velocity field. As we discussed above, the typical proppant transportation process was divided into four typical main stages; those are: Initial settlement stage, vertically grow stage, horizontally grow stage, and equilibrium state stage.

3.2.1. Stage1: Initial Settlement Stage

In the initial settlement stage, as the water–proppant mixture was injected form the middle inlet, two opposite vortexes were generated at the upper and lower ends of the slurry jet, and the lower one was smaller because of the inhibition generated by the slurry jet, as shown in Figure 7a. The injected proppant tended to quickly settle to the bottom and formed a proppant dune with a distance from the left boundary due to the gravity effect and the energy dissipation generated by the vortex. We defined this dune as the primary dune. Meanwhile, the small part of the proppant was dragged by the lower vortex move to the opposite direction and settle near the left boundary. We defined the small dune generated here as the secondary dune. Due to the slurry flush, a none proppant placement area was created between the two dunes. In this stage, this process was governed by the slurry vortex and the settlement.

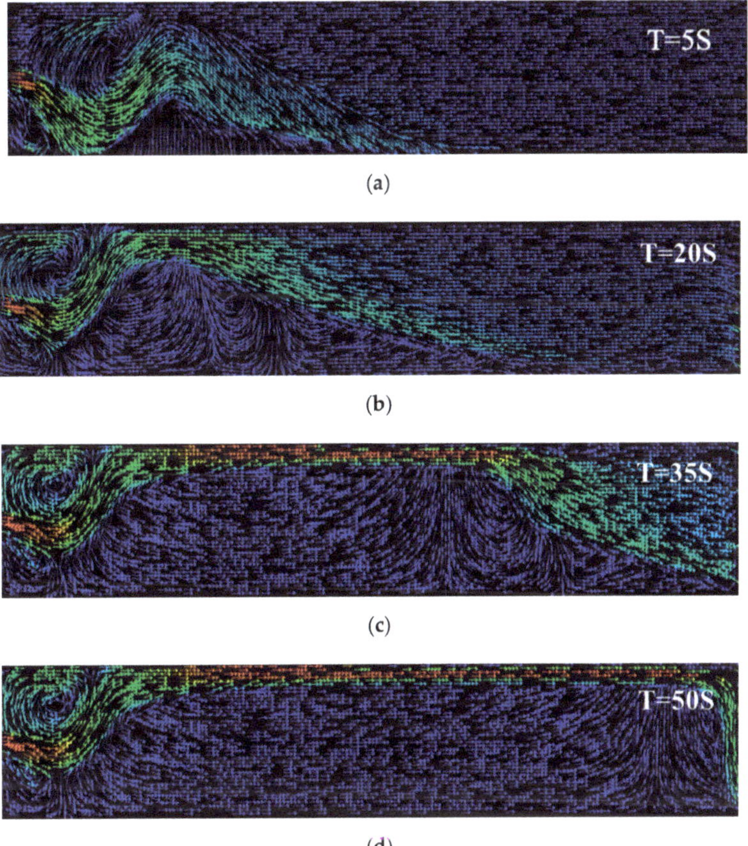

Figure 7. Four stages for proppant transport under high Reynolds number condition. (**a**): initial settlement Stage; (**b**) vertically grow Stage; (**c**) horizontally grow Stage; (**d**) equilibrium State Stage.

3.2.2. Stage2: Vertically Grow Stage

Figure 7b shows the vertically grow stage. The flow region in the fracture was divided into three main regions; those are free clean fluid flow area in the upper side of the primary dune, transition flow area on the surface of the dune, and the immobile area beneath the surface. In this stage, the liquid phase velocity near the primary dune was too small to carry proppant to a distant place. Therefore, most of the injected particles settled quickly on the proppant dune, and only a small part of them was carried to the deep part of the fracture. With a continuous injection of slurry, more proppant settled on the primary and secondary dunes, and the flow path between the dune and fracture upper surface became narrower. Therefore, the fluid velocity became larger, which increased the fluidization of the settled particles and reduce the proppant settlement. Besides, the primary dune reached a specific height at 0.28 m and stopped growing vertically, and we defined the height as the primary equilibrium height (PEH). Additionally, the secondary dune grew gradually near the entrance. The lower vortex was getting weaker in this stage due to the collective effects caused by the boundary, secondary dune, and the inject slurry flow. Consequently, the lower vortex disappeared while the secondary dune stopped growing. The valley between the two dunes was still existed due to the flushing. In this stage, the proppant transport process was governed by the fluidization, settlement, and suspension.

3.2.3. Stage3: Horizontally Grow Stage

Figure 7c shows the proppant bank development process in a horizontal direction. In this process, because of the presence of the narrow gap between the proppant bank and the upper boundary, water had enough speed to fluidize and suspend the proppant. Therefore, a great amount of proppant transported through the currently generated proppant bank rather than retarded on the surface. However, once the proppant passes through the channel between the primary dune and wall, the velocity decreased dramatically because of the increased flow cross-section. Besides, another vortex was generated just behind the dune. Due to these two reasons, water cannot carry proppant into the deep of the fracture. Proppant gradually accumulated in the front of the formed proppant dune until reaching the equilibrium height, in which the water can suspend the proppant again. The settlement-reach equilibrium height process will repeat again and again, and from a macro perspective, the proppant dune will gradually grow horizontally. In this stage, the proppant distribution is determined by the settlement, fluidization and saltation.

3.2.4. Stage4: Equilibrium State Stage

As the proppant dune grew horizontally and reached the outlet boundary, the proppant dune remained unchanged and reached a comprehensive equilibrium height (CEH), as shown in Figure 7d. The CEH was defined as the average equilibrium height of the primary dune when the proppant bank was stable during the injection process. The suspension and saltation were the main mechanisms to control the transport.

Figure 8 shows the proppant dune growth trajectory from 5 s to 65 s, in which the characteristic of the four main stages during the injection can be obviously obtained. Once the primary dune was formed, the dune will gradually grow vertically, and the front settlement angel almost remains constant at 20° until reaching the primary equilibrium height (PEH) by the end of the second stage (at around T = 20 s). The third stage began at T = 25 s, in which the proppant mainly grows horizontally, and the settlement angle of the upper side becomes more steep form 40° to nearly 90°. At T = 60 s, the proppant dune reaches the CEH.

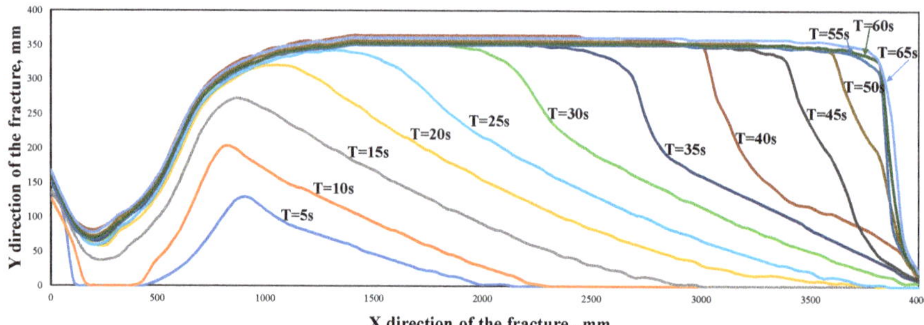

Figure 8. Proppant dune growth trajectory from for the simulation of the control sample.

3.3. Parametric Study

In this section, the parametric study for different injection velocity, inject position, particle diameter, density, and concentration were conducted for a better understanding of the effects caused by these parameters on the proppant transport process under HRNCs. The parameter sets are listed in Table 2. The bold items are set as the control sample in the parametric study. In each section, besides the investigated parameter, other parameters were set as same as the control sample.

Table 2. Parameter sets for parametric study and primary data for simulation.

Property	Symbol	Unit	Values
Injection velocity (L, s)	vin,l, vin,s	m/s	$vin,l = vin,s = 2, 3, 4, 5$
Inject position	/	/	Top, **middle**, bottom
Particle diameter	d_s	mm	0.6, 0.7, 0.8, 0.9, **1.0**
Particle density	ρ_s	kg/m^3	1800, 2200, 2500, 2800, **3200**
Particle concentration	αs	%	10, **15**, 20, 25
Fluid density	ρ_l	kg/m^3	1000
Fluid viscosity	μ_l	mPa·s	1

The bold items are the setting value for the control sample.

3.3.1. Inlet Velocity Effect

Figure 9 shows the distribution of the liquid volume concentration during the transport process for different injection velocity. In this study, the injection velocity of fluid and proppant was set to be the same, and the velocity uniformly expressed as v. The parameters except velocity were set as the same as the control sample. The average Reynolds number are 2000–5000 for v = 2–5 m/s, respectively. Compared with the control sample, the lower velocity case for $v = 2$ m/s (Figure 9A) showed a closer distance from the primary dune to the entrance and more particle placement between the dunes, which was mainly because the vortex is weaker than that of the control sample. Meanwhile, the proppant dune grew rate for v = 2 m/s is a little bit slower than that of the control sample. On the contrary, as the injection velocity became faster, the vortex generated near the entrance became larger and stronger, resulting in a larger distance for primary dunes and disappeared secondary dune. As the velocity of the slurry increases from 2 m/s to 5 m/s, the non-propped area near the inlet increase by 5.3 times.

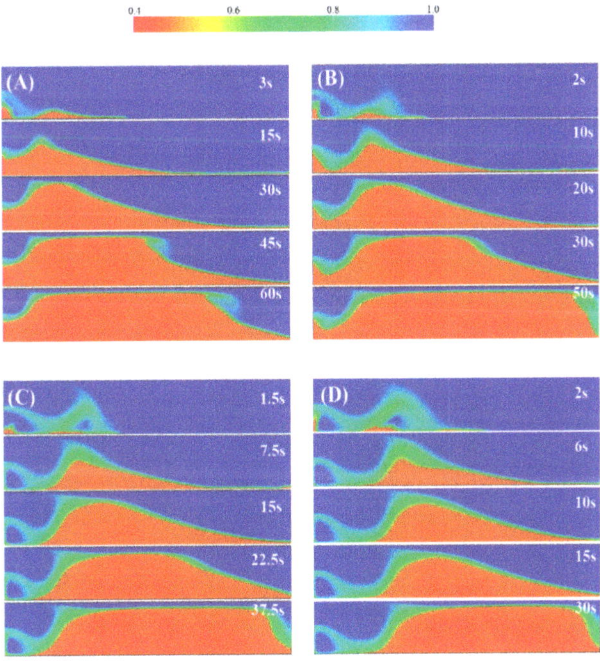

Figure 9. Volume concentration of liquid phase during the transport process for different injection velocity cases: (**A**) v = 2 m/s, (**B**) v = 3 m/s, (**C**) v = 4 m/s, and (**D**) v = 5 m/s.

The efficiency of proppant placement is also an essential factor in evaluating the effect of proppant placement. Therefore, we defined the proppant occupation as a stationary proppant bed area over the total fracture area. The proppant occupation in equilibrium stages was defined as equilibrium proppant occupation (EPO), which indicated the maximum proppant accumulation in the fracture under the specific condition. Proppant occupation, together with the EPO and CEH, were used to characterize the proppant transport and settlement for each case.

Figure 10 shows the proppant occupation of different time steps and comparison of equilibrium height for different injection velocity cases. The case for v = 5 m/s reached the equilibrium stage at T = 22 s, which was the fastest among all instances. As velocity decreased, it takes longer for the proppant dune to reach equilibrium in the fracture. However, the v = 2 m/s cases yielded the highest CEH at 35.9 cm, compared with other cases, which were at 35.2 cm, 34.6 cm, and 34.1 cm, respectively. For the low-velocity instance, the gap between the primary dune needed to be narrower to obtain enough speed to balance the amount of settled and rolled-up proppant, resulting in higher CEH. Besides, due to the smaller (none) secondary dunes and the relatively shorter primary dune, it was observed that the EPO for large velocity is lower than those of slower cases during the transportation process, which are 80.5%, 74.6%, 66.4%, and 56.5% for the velocity from 2 to 5 m/s. The small EPO, especially caused by the no proppant placement neat the entrance, will cause worse conductivity in artificial fracture, which should be avoided in the field fracturing design.

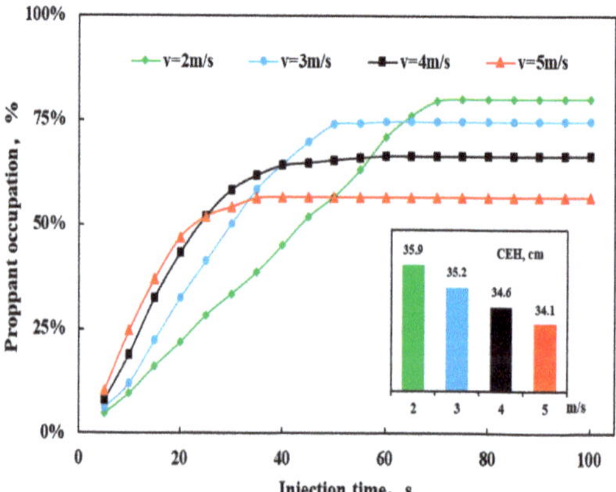

Figure 10. Proppant occupation percentage in fracture change with time a comprehensive equilibrium height (CEH) for different velocity cases.

3.3.2. Effects of Inlet Position

Figure 11 shows the distribution of the proppant placement in the transport process under different inlet positions. According to Figure 11A, when the slurry was injected from the bottom, the distance between the initial settlement location and left wall was 1.1 m, which indicated that the flow velocity at this position was not able to overcome the frictional force between the proppant particles and the bottom wall. Since no flow vortex was formed upon the bottom wall, no proppant particles were situated in the region near the left wall. As a result, only the primary dune was generated in the fracture. Proppant particles injected later would get over the dune and settle at the rear of the dune. Finally, the proppant dune gradually grew until it reached the CEH.

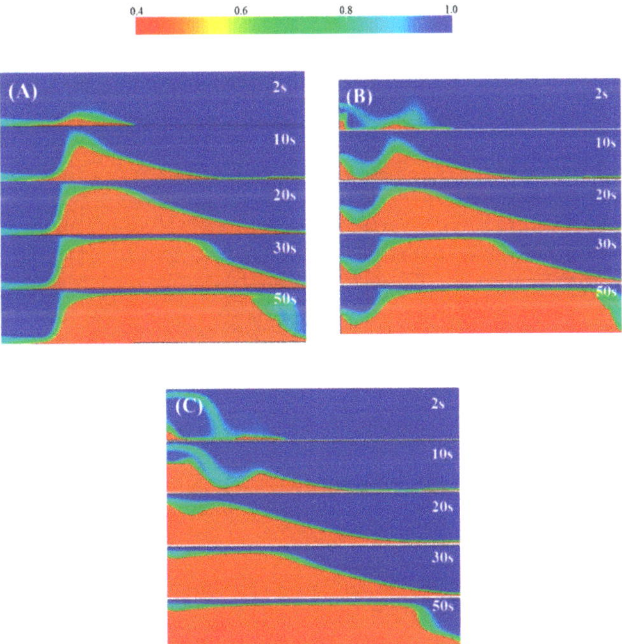

Figure 11. Proppant placement during the transport process for different inlet position cases: (**A**) Inject from the bottom inlet; (**B**) inject from the middle inlet; and (**C**) inject from the top inlet.

Figure 11C shows the proppant transport process of the top inlet condition at different time points. When the mixture was injected into the slot, the proppant quickly settled to the bottom at the position about 0.65 m from the left wall. A big clockwise vortex was generated near the entrance and brought proppant back to the inlet side, forming a sizeable secondary dune. As the proppant was injected continuously, the heights of both dunes increase, and the trough between two dunes gradually filled with the proppant. Consequently, the primary and secondary dunes gradually became an integrated dune, and the CEH was obtained almost alone the whole length of the fracture.

According to Figure 12, it took almost the same time (55 s) to reach CEH for the three cases with different inlet positions. Injecting proppant from the top could produce the highest equilibrium height at 36.1 cm, compared with that of the middle case at 35.2 cm and the bottom situation at 34.3 cm. Besides, the proppant occupation curve indicated that the proppant injected through the top yielded the highest proppant placement percentage, and the EPO reaches 82.3% compared with the middle case at 74.6% and the bottom case at 67.8%. The result identified that the vortex at the front end of the fracture is the dominant factor for proppant placement near the entrance. The large vortex could carry the proppant back to the inlet place. The large non-propped area in the fracture may cause fracture closure after pumping, resulting in the permeability significantly decreased. Therefore, choosing the relatively high inlet positions will result in better patterns of the proppant placement and improve fracture conductivity.

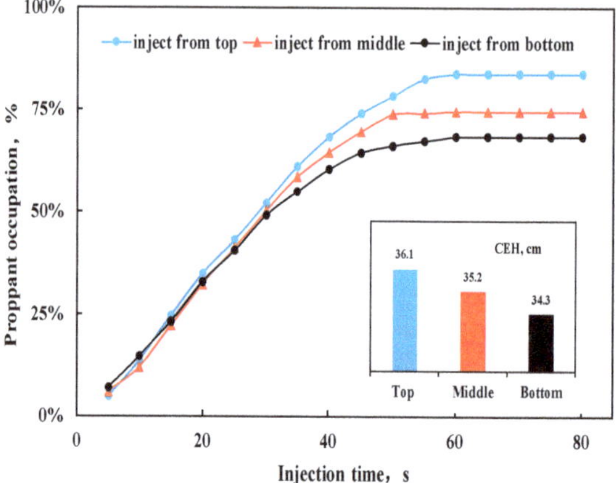

Figure 12. The curve for proppant occupation percentage in fracture change with time and CEH for different inlet position cases.

3.3.3. Effects of Particle Diameter

Figure 13 shows the effects of particle diameter on transport processes. The accumulation curve demonstrated that the larger particles settle more in the fracture at any time during the transport process because smaller particles were easier carried by the fluid. The smaller particle was carried into the far end of the fracture rather than accumulated during the transportation. For the same reason, it took a long time for smaller particles to reach CEH. However, the CEH difference is tiny among different cases, which indicated the proppant diameters had little effect on the primary dune for a relatively long time in the transport process. At last, EPO for 1 mm to 0.6 mm are 67.6%, 68.8%, 71.0%, 73.1%, and 74.6% respectively.

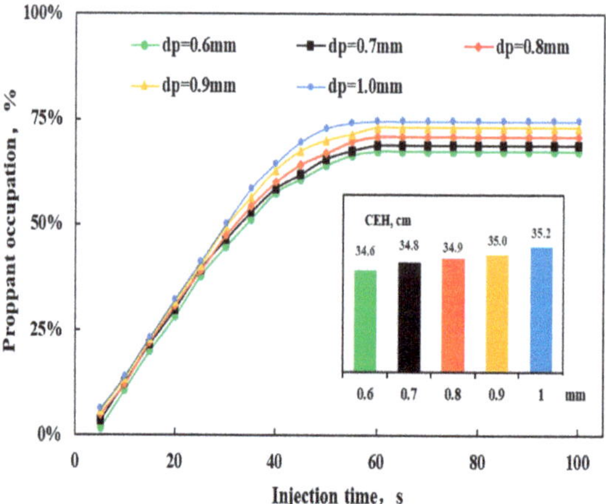

Figure 13. The curve for proppant occupation percentage in fracture change with time and comprehensive equilibrium H = height (CEH) for different particle diameter cases.

3.3.4. Effects of Particle Density

Figure 14 shows the detailed data extracted from the simulation results for different particle density. The results indicated proppant density has a significant influence on the proppant settlement in the fracture. The curve shows that the lighter particle has the least amount of settlement in the fracture during the injection process. The reason for that phenomenon was that the lighter particles were more likely to be fluidized and dragged by the fluid to the further side of the fracture. All the cases almost reached the equilibrium height at the same time, around 45 s. The EPO was 60.39% when the density was 1800 kg/m^3 is, which was 21.3% smaller than that of 3200 kg/m^3. For the CEH study, it was interesting to obtain a proppant density threshold for the particle dune growth. When the density is small, the equilibrium height increases rapidly as the proppant density increases, which changed from 32.8 cm in the 1800 kg/m^3 case to 35.8 cm in the 2500 kg/m^3 case. However, when the density was larger than 2500 kg/m^3, the CEH almost unchanged.

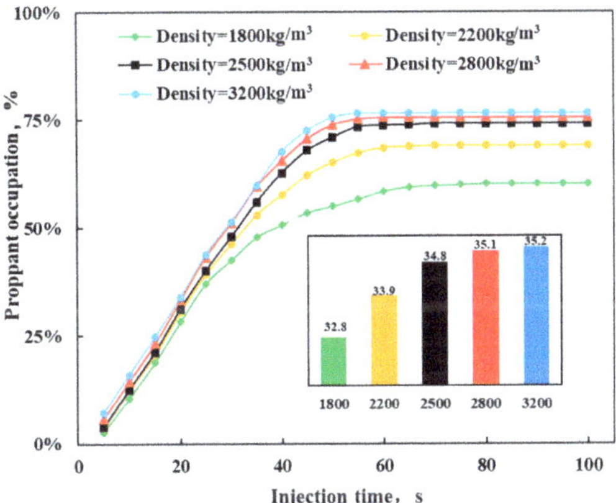

Figure 14. The curve for proppant occupation percentage in fracture change with time and CEH for different inlet particle density.

3.3.5. Effects of Solid-Phase Volume Concentration

Figure 15 shows the particle concentration effects on proppant transport. From the curves, higher proppant concentration resulted in a greater amount of settlement in the fracture at the same time. Besides, the lowest concentration (10%) case was the last one to reach the CEH at T = 76 s, which almost double that of 25% concentration This phenomenon was caused by two reasons; those are fewer injection particle amounts and easy fluidization for proppant when the concentration was low. Moreover, the CEH for lower concentration cases was also shorter, which may result from more liquid fluid flowing past over the top of the proppant bed or the higher ability of lower solid phase volume concentration to fluidize the proppant particles. In conclude, the lower concentration in the transport process lead to a more proppant place in the further part of the fracture, which was good for longer effective propped fractures. However, less proppant in the fracture may result in poor conductivity. The proppant particle concentration selection in the field should be considered comprehensively for both sides.

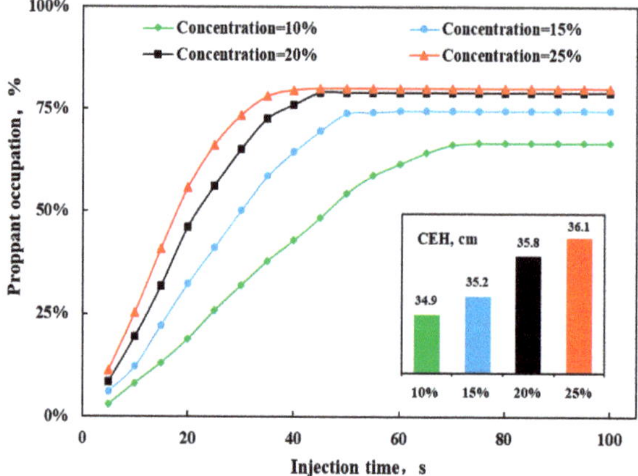

Figure 15. The curve for proppant occupation percentage in fracture change with time and CEH for different inlet position cases.

3.4. Equilibrium Height Prediction Model

The height of the equilibrium gap H, which was defined as the distance between the top of the proppant bed and the upper wall boundary, was used to represent the equilibrium height. J. Wang et al. [13] established a Bi-power law correlation for the equilibrium gap of the proppant bed, and it was given by:

$$\frac{H_1}{W} = \left[-2.3 \times 10^{-4} \ln(R_G) + 2.92 \times 10^{-3} \right] R_f^{1.2 - 1.26 \times 10^{-3} \lambda^{-0.428} \times [15.2 - \ln(R_G)]} R_p^{[-0.0172 \ln(R_G) - 0.12]}. \tag{36}$$

The gravity Reynolds number R_G is defined as:

$$R_G = \rho_l (\rho_s - \rho_l) g d_s^2 / \mu_l^2. \tag{37}$$

The gravity Reynolds number for the fluid λ is defined as:

$$\lambda = (\mu_l / \rho_l) / \left(W^{3/2} \sqrt{g} \right). \tag{38}$$

The fluid Reynolds number R_f is shown as:

$$R_f = (\rho_l \overline{v} W) / \mu_l. \tag{39}$$

The proppant Reynolds number R_p is defined as:

$$R_p = \left(\rho_p \overline{v} W \right) / \mu_l \tag{40}$$

where H_1 is the height of equilibrium gap, ml; W is the fracture width, m; H/W is gap height over fracture width, dimensionless;

Along the altitude direction of the fracture, the flow field can be divided into three layers. The bottom of the proppant bank is an immobile layer in which the velocity of the proppant particles is approaching zero. The middle layer is a mobile zone that is above the stationary bed, in which the proppant particles move by sliding and rolling or a combination of both. The top layer is a clean liquid zone. As the proppant concentration increases gradually from the fresh liquid layer to the mobile layer and then to the immobile layer, the point with the proppant concentration equal to 0.5 was chosen to be the top of the stationary bed.

Figure 16 shows the comparison of dune height between the simulation result and Wang's model under different conditions. As the parameters change, they had the same trend. However, compared with Wang's model, the equilibrium height obtained by the simulation was always higher. The mean deviation for the case of density, diameter, velocity, and solid concentration was 45.8%, 36.5%, 57.7%, and 59.4%, respectively, which indicated it is difficult to accurately predict the equilibrium height by using this model.

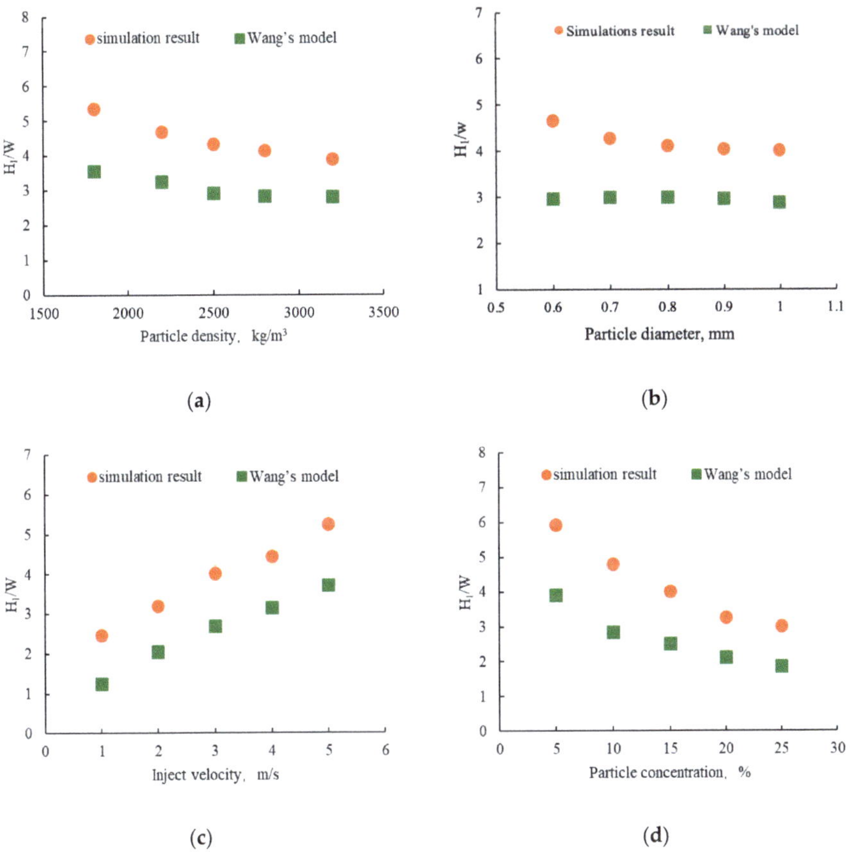

Figure 16. Comparison of the equilibrium dune height between simulation result and Wang's model. (**a**) Comparison of particle density; (**b**) comparison of particle diameters; (**c**) comparison of injection velocity; and (**d**) comparison of particle concentration.

In this paper, some modification about Wang's model was conducted based on the simulation result, to accurately predict the equilibrium height of the proppant dune in the fracture under HRNCs. The new prediction model is shown as:

$$\frac{H_1}{W} = \left[-5.76 \times 10^{-3} \ln(R_G) + 1.87 \times 10^{-3} \right] R_f^{1.2 - 1.008 \times 10^{-3} \lambda^{-0.428} \times [15.2 - \ln(R_G)]} R_p^{[-0.04 \ln(R_G) - 0.12]}. \tag{41}$$

Figure 17 shows the relationship between the new prediction model and simulation results. It can be seen from the comparison that this prediction model can fit the simulation result very well. Through calculation, the mean deviation is only 3.8%. The applicable range for this prediction model is the case where the Reynolds number is 2000–5000. This correlation can be used to quickly predict proppant placement in a fracturing simulator, which can greatly improve the simulation accuracy.

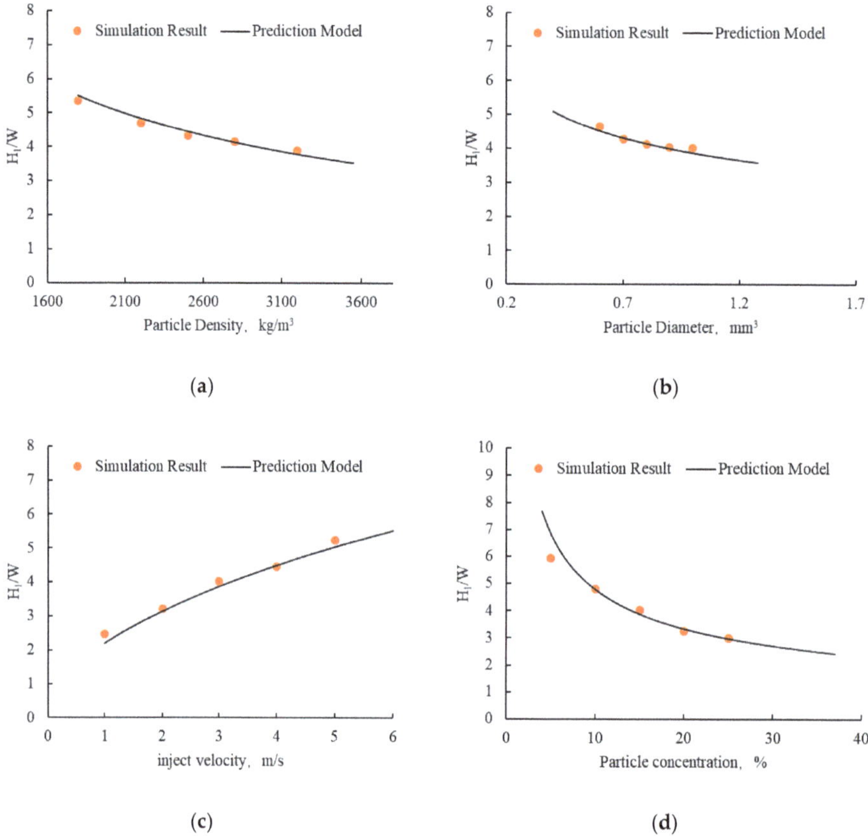

Figure 17. Comparison of the equilibrium dune height between simulation result and new prediction model. (**a**) Comparison of particle density; (**b**) comparison of particle diameters; (**c**) comparison of injection velocity; and (**d**) comparison of particle concentration.

4. Conclusions

In this work, a Eulerian two phases model is used to simulate the proppant transport process in low viscosity fluid and high-speed rate conditions (HRNCs). The turbulence effect and particle-particle/wall effect are the primary concern in this paper. According to the comparison of the results of the numerical simulations and the experimental study, the Eulerian multiphase model can capture the transport and settling behaviors of proppant in the water fracture treatments. Based on the simulation results, the following conclusions have been drawn on:

(1)	The transport process can be classified into four main stages. In addition to the mechanisms such as fluidization, suspension, and settlement presented by other studies, the vortex is also a critical mechanism in the transport process under HRNCs, especially near the entrance.

(2)   The inlet positions significantly influence the vortex and the proppant placement patterns. There is a nearly non-propped zone near the wellbore for the bottom inlet condition since no reversed flow is generated to take the proppant particles back. Higher inlet position increases the propped area near the wellbore by 7.1 times. But the blocking of the entrance may lead to high risk in pumping proppant.

(3)   Increasing the injection velocity from 2 m/s to 5 m/s, the proppant particles easier to be fluidized and increase the distance to which the proppant particles are transported in the hydraulic fracture. The non-propped area near the inlet increase by 5.3 times and the equilibrium height decrease by 5.2%.

(4)   A new prediction model is proposed to predict the equilibrium height of the proppant bank in the fracture under high Reynolds number condition. Compared with the simulated data, the mean deviation between them is only 3.8%, which indicates the model is suitable under HRNCs (2000–5000).

Author Contributions: Conceptualization, T.Z. and J.G.; Data curation, R.Y.; Formal analysis, R.Y. and J.Z.; Investigation, J.G.; Methodology, T.Z.; Project administration, J.G.; Supervision, T.Z.; Validation, T.Z.; Writing—original draft, R.Y.; Writing—review and editing, J.G. and J.Z. All authors have read and agreed to the published version of the manuscript.

Funding: This study was supported by the National Science and Technology Major Project (2016ZX05048-004-006), National Science and Technology Major Project (2016ZX05002-002), and the National Science Fund for Distinguished Young People (51525404).

Conflicts of Interest: The authors declare no conflict of interest.

References

1.   Palisch, T.T.; Vincent, M.C.; Handren, P.J. Slickwater fracturing: Food for thought. In Proceedings of the SPE Annual Technical Conference and Exhibition, Denver, CO, USA, 21–24 September 2008.

2.   King, G.E. Thirty Years of Gas Shale Fracturing: What Have We Learned? In Proceedings of the SPE Annual Technical Conference and Exhibition, Florence, Italy, 20–22 September 2010.

3.   Handren, P.; Pslish, T. Successful Hybrid Slickwater-Fracture Design Evolution: An East Texas Cotton Valley Taylor Case History. *SPE Prod. Oper.* **2009**, *24*, 415–423. [CrossRef]

4.   Guo, J.; Li, Y.; Wang, S. Adsorption damage and control measures of slick-water fracturing fluid in shale reservoirs. *Pet. Explor. Dev.* **2018**, *45*, 336–342. [CrossRef]

5.   Sahai, R. Laboratory Evaluation of Proppant Transport in Complex Fracture Systems. Ph.D. Thesis, Colorado School of Mines, Arthur Lakes Library, Golden, CO, USA, 2012.

6.   Zeng, J.; Li, H.; Zhang, D. Numerical simulation of proppant transport in hydraulic fracture with the upscaling CFD-DEM method. *J. Nat. Gas Sci. Eng.* **2016**, *33*, 264–277. [CrossRef]

7.   Britt, L.K.; Smith, M.B.; Haddad, Z.A.; Lawrence, J.P.; Chipperfield, S.T.; Hellman, T.J. Waterfracs: We Do Need Proppant After All. In Proceedings of the SPE Annual Technical Conference and Exhibition, Antonio, TX, USA, 24–27 September 2006.

8.   Kern, L.R.; Perkins, T.K.; Wyant, R.E. The mechanics of sand movement in fracturing. *J. Pet. Technol.* **1959**, *11*, 55–57. [CrossRef]

9.   Patankar, N.A.; Joseph, D.D.; Wang, J.; Barree, R.D.; Conway, M.; Asadi, M. Power law correlations for sediment transport in pressure driven channel flows. *Int. J. Multiph. Flow.* **2002**, *28*, 1269–1292. [CrossRef]

10.   Wookworth, T.R.; Miskimins, J.L. Extrapolation of laboratory proppant placement behavior to the field in slickwater fracturing application. In Proceedings of the SPE Hydraulic Fracturing Technology Conference, College Station, TX, USA, 29–31 January 2007.

11.   Liu, Y. Settling and Hydrodynamic Retardation of Proppants in Hydraulic Fractures. Ph.D. Thesis, The University of Texas, Austin, TX, USA, 2006.

12.   Palisch, T.T.; Vincent, M.; Handren, P.J. Slickwater fracturing: Food for thought. *SPE Prod. Oper.* **2010**, *25*, 327–344. [CrossRef]

13.   Wang, J.; Joseph, D.D.; Patankar, N.A.; Conway, M.; Barree, R.D. Bi-power law correlations for sediment transport in pressure driven channel flows. *Int. J. Multiph. Flow* **2003**, *29*, 475–494. [CrossRef]

14. Barree, R.D.; Conway, M.W. Experimental and numerical modeling of convective proppant transport. In Proceedings of the SPE Annual Technical Conference and Exhibition, New Orleans, LA, USA, 25–28 September 1994.

15. Warpinski, N.R.; Mayerhofer, M.J.; Vincent, M.C.; Cipolla, C.L.; Lolon, E.P. Stimulating unconventional reservoirs: Maximizing network growth while optimizing fracture conductivity. *J. Can. Pet. Technol.* **2009**, *48*, 39–51. [CrossRef]

16. Patankar, N.A.; Huang, P.Y.; Ko, T.; Joseph, D.D. Lift-off of a single particle in Newtonian and viscoelastic fluids by direct numerical simulation. *J. Fluid Mech.* **2001**, *438*, 67–100. [CrossRef]

17. Patankar, N.A.; Ko, T.; Choi, H.G.; Joseph, D.D. A correlation for the lift-off of many particles in plan Poiseuille of Newtonian fluids. *J. Fluid Mech.* **2001**, *445*, 55–76. [CrossRef]

18. Tsai, K.; Fonseca, E.; Lake, E.; Degaleesan, S. Advanced Computational Modeling of Proppant Settling in Water Fractures for Shale Gas Production. *SPE J.* **2012**, *18*, 1389–1394.

19. Zhou, Z.Y.; Pinson, D.; Zou, R.P.; Yu, A.B. Discrete particle simulation of gas fluidization of ellipsoidal particles. *Chem. Eng. Sci.* **2011**, *66*, 6128–6145. [CrossRef]

20. Benyahia, S.; Syamlal, M.; O'Brien, T.J. Evaluation of boundary conditions used to model dilute, turbulent gas/solids flows in a pipe. *Powder Technol.* **2005**, *156*, 62–72. [CrossRef]

21. Pannala, S.; Syamal, M.; O'Brien, T.J. *Computational Gas-Solids Flows and Reacting Systems: Theory, Methods and Practice*; IGI Global: Hershey, PA, USA, 2010.

22. Agrawal, K.; Loezos, P.N.; Syamlal, M. The role of meso-scale structures in rapid gas-solid flows. *J. Fluid Mech.* **2001**, *445*, 151–185. [CrossRef]

23. Srivastava, A.; Sundaresan, S. Analysis of a frictional–kinetic model for gas–particle flow. *Powder Technol.* **2003**, *129*, 72–85. [CrossRef]

24. Benyahia, S. Analysis of Model Parameters Affecting the Pressure Profile in a Circulating Fluidized Bed. *AIChE J.* **2012**, *58*, 427–439. [CrossRef]

25. Kaushal, D.R.; Thinglas, T.; Tomita, Y.; Kuchii, S.; Tsukamoto, H. CFD modeling for pipeline flow of fine particles at high concentration. *Int. J. Multiph. Flow.* **2012**, *43*, 85–100. [CrossRef]

26. Benyahia, S. On the effect of subgrid drag closures. *Ind. Eng. Chem. Res.* **2010**, *49*, 5122–5131. [CrossRef]

27. Zhang, T.; Guo, J.C.; Liu, W. CFD Modeling of Proppant Transport and Settling in Water Fracture Treatments. *J. Southwest Pet. Univ.* **2014**, *36*, 74–82.

28. Li, P.; Zhang, X.; Lu, X. Numerical simulation on solid-liquid two-phase flow in cross fractures. *Chem. Eng. Sci.* **2018**, *181*, 1–18. [CrossRef]

29. Pfleger, D.; Gomes, S.; Gilbert, N.; Wagner, H.G. Hydrodynamic simulations of laboratory scale bubble columns fundamental studies of the Eulerian-Eulerian modelling approach. *Chem. Eng. Sci.* **1999**, *54*, 5091–5099. [CrossRef]

30. Gidaspow, D.; Bezburuah, R.; Ding, J. Hydrodynamics of Circulating Fluidized Beds, Kinetic Theory Approach. In Proceedings of the 7th Engineering Foundation Conference on Fluidization, Brisbane, Australia, 3–8 May 1992; pp. 75–82.

31. Benyahia, S.; Syamlal, M.; O'Brien, T.J. Study of the Ability of Multiphase Continuum models to predict core-annulus flow. *AICHE* **2007**, *53*, 2549–2568. [CrossRef]

Publisher's Note: MDPI stays neutral with regard to jurisdictional claims in published maps and institutional affiliations.

 *energies*

Article

A Critical Review of Osmosis-Associated Imbibition in Unconventional Formations

Zhou Zhou [1], Xiaopeng Li [2,*] and Tadesse Weldu Teklu [3]

[1] State Key Laboratory of Petroleum Resources and Prospecting, China University of Petroleum, Beijing 102249, China; zhouzhou@cup.edu.cn
[2] Society of Petroleum Engineers, Aurora, CO 80015, USA
[3] Department of Petroleum Engineering, Colorado School of Mines University, Golden, CO 80401, USA; tteklu@mines.edu
* Correspondence: roy.li.inbox@gmail.com

Abstract: In petroleum engineering, imbibition is one of the most important elements for the hydraulic fracturing and water flooding processes, when extraneous fluids are introduced to the reservoir. However, in unconventional shale formations, osmosis has been often overlooked, but it can influence the imbibition process between the working fluid and the contacting formation rocks. The main objective of this study is to understand effects of fluid–rock interactions for osmosis-associated imbibition in unconventional formations. This paper summarizes previous studies on imbibition in unconventional formations, including shale, tight carbonate, and tight sandstone formations. Various key factors and their influence on the imbibition processes are discussed. Then, the causes and role of osmotic forces in fluid imbibition processes are summarized based on previous and recent field observations and laboratory measurements. Moreover, some numerical simulation approaches to model the osmosis-associated imbibition are summarized and compared. Finally, a discussion on the practical implications and field observations of osmosis-associated imbibition is included.

Keywords: imbibition; osmosis; unconventional formations; shale; EOR; hydraulic fracturing; water flooding

Citation: Zhou, Z.; Li, X.; Teklu, T.W. A Critical Review of Osmosis-Associated Imbibition in Unconventional Formations. *Energies* **2021**, *14*, 835. https://doi.org/10.3390/en14040835

Received: 3 December 2020
Accepted: 21 January 2021
Published: 5 February 2021

Publisher's Note: MDPI stays neutral with regard to jurisdictional claims in published maps and institutional affiliations.

1. Introduction

Imbibition is defined as a movement in which the wetting fluid occupies pore space through the displacement of the non-wetting fluid. Osmosis refers to a process of water molecules' spontaneous movement through a semi-permeable membrane, such as clay, from a low-salinity to high-salinity region against concentration gradient [1]. Previously, the major mechanism of imbibition was narrowed to capillary pressure; however, osmosis due to molecule diffusion and the semipermeable membrane effect has been overlooked.

In unconventional formations, such as shale and tight sandstone, the imbibition exists when the injected working fluid contacts the formation rock during hydraulic fracturing and water flooding. The capillary pressure is considered as one of the driving forces for imbibition. Since this force is in inverse proportion to the pore size, it is particularly significant in the formation of nanopores and micropores. Thus, unconventional formation with smaller pore size usually has much higher capillary pressure and larger imbibition effect. In recent years, the imbibition effect was considered as a potential explanation for the low percentage of flow back after hydraulic fracturing in the shale gas reservoir. Roychaudhuri et al. [2], Makhanov et al. [3,4], and Zhou et al. [5,6] proved that a large volume of the fracturing fluid can be imbibed by the shale samples. Imbibition has also been studied as part of the research related to water flooding to achieve a sweep area as large as possible [7].

Osmosis requires a semi-permeable membrane and concentration differences. Clay can act the membrane because of its salt-exclusionary behavior. In tight clay-rich formations, such as shale, clay distributes on the wall of the porous space so that the membrane has a high efficiency to exclude the passage of salt ions, which is called membrane efficiency. In addition, due to the salinity differences between the injected fluid and formation brine, the imbibition process in clay-rich formations is often associated with osmosis. Fakcharoenphol et al. [8] and Zhou et al. [9,10] both indicated the osmosis effect on water flooding and hydraulic fracturing.

Therefore, besides capillarity, it is necessary to study the effect of osmosis on imbibition for water flooding and hydraulic fracturing in unconventional formations. In this paper, previous studies are reviewed and summarized for the osmosis-associated imbibition. The details and conclusions are described to provide insights about osmosis-associated imbibition in petroleum engineering applications.

2. Imbibition

The main study of imbibition is through experiments. The spontaneous imbibition experimental setup is indicated in Figure 1. There is no extra pressure applied during the spontaneous imbibition. Figure 2 shows the forced imbibition experimental setup, which can apply injection pressure on a sample.

(a) (b)

Figure 1. Two typical spontaneous imbibition experiments (**a**) measuring sample weight change by hanging on the balance; (**b**) measuring fluid volume change in imbibition vessel.

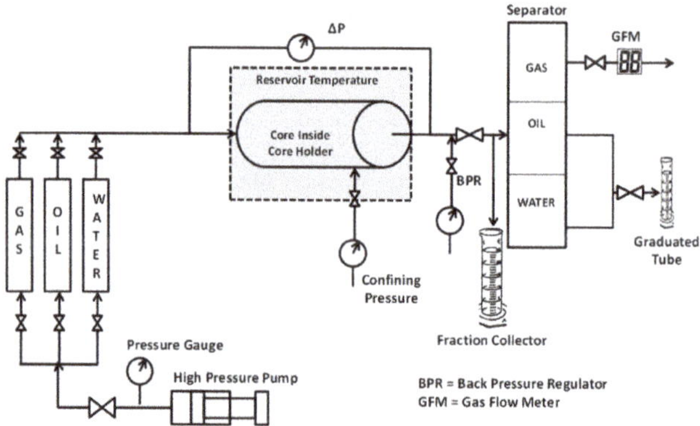

Figure 2. Typical forced imbibition experiment [11].

According to the experimental results, the imbibition rate and volume are significantly affected by various factors including wettability, initial water saturation, temperature, flow direction, fluid and rock properties, and clay content, which are described respectively in the following subsections.

Wettability. Wettability is an important factor that can impact the imbibition process. The wettability indicates the ability of a fluid to adhere to the walls of a solid. The fluid includes gas and liquid phases. When the three phases (gas, liquid, and solid) interact, the contact angles between the gas–liquid interface and the solid–liquid interface can be used to indicate wetting and non-wetting states. In the reservoir, the rock with the smaller contact angle exhibits a faster imbibition rate for the wetting phase [12]. Therefore, adjustment of the contact angle is an effective way to control imbibition and is widely investigated [13–18]. Zhou et al. [5,6] designed experiments to compare imbibition rates in shale samples with various contact angles. The results showed that the imbibition rate was strongly influenced by changing contact angles.

Wettability alteration is one of the main mechanisms to mobilize residual fluid during water flooding. The alteration is dependent on fluid composition, rock surface mineralogy, system temperature, pressure, and saturation history [19–23]. Imbibition process can also alter the wettability of formation rocks. In previous studies, there are several situations of wettability alteration due to imbibition, including the imbibition fluids being acid, water with specific ionic content, and surfactant.

Dilute acid imbibition pre, post, or during hydraulic fracturing could improve the wettability of carbonate-rich shale formations and hence improve production [24–28]. The reaction between acid and rock can alter the wettability of formations by weakening the oil–rock surface bonds on the oil-wet thin layer. The alteration is to improve the pore connectivity; thereby, the trapped oil and gas are easier to be produced.

When the ionic content in water is changed, the wettability of reservoirs can be altered. This usually happens during low-salinity or salinity-modified waterflooding. These mechanisms of the wettability alteration include fines migration and rock dissolution [29–32]; PH increases [33–35]; multi-component ion exchange [36–40]; and surface charge changes or double-layer expansion [41–51].

The purpose of surfactant in imbibition fluid is to mobilize residual saturation [52–55]. Wettability alteration toward the hydrophilic state and a decrease in interfacial tension (IFT) are caused by surfactant during imbibition [55]. Cationic surfactant adsorption in negatively charged sandstone cores can decrease the performance to lower IFT and wettability alteration [56]. Alameri et al. [57] reported that surfactants in combination with low-salinity water flooding could be applied to circumvent the high salinity challenge and improve recovery in oil–wet carbonate reservoirs. Surfactants or a hybrid of surfactant with low-salinity water were observed to improve the wettability and IFT of formations at the laboratory scale [11,18,49,58–60]. Figure 3 shows the wettability alteration (through contact angle measurements) of carbonate formation, sandstone formation, and shale (Three Forks shale formation) comparisons when the bulk fluids are seawater, seawater + CO_2, and low-salinity water + CO_2 [11].

Contact angle measurements within the pore space are not realistic; thus, the spontaneous imbibition and zeta potential measurements can provide realistic indicators of wettability alteration in porous media, especially in unconventional shale reservoirs [49,51]. Nontheless, Mahani et al. [44] measured contact angle and zeta (ζ)-potential of the carbonate–brine interface on crushed carbonate fragments. Their experiments showed that at lower brine salinities, the ζ-potential of the limestone–brine interface become more negative, which is indicative of a weaker electrostatic adhesion of the rock–brine interfaces and implies a wettability alteration to a less oil-wet condition. Alvarez and Schechter [49] performed spontaneous imbibition, contact angle, and ζ-potential measurements on siliceous unconventional liquid-rich Permian basin reservoir cores using surfactants and fracturing brine. Alvarez and Schechter [49] experiments show anionic surfactants superior wettability alteration.

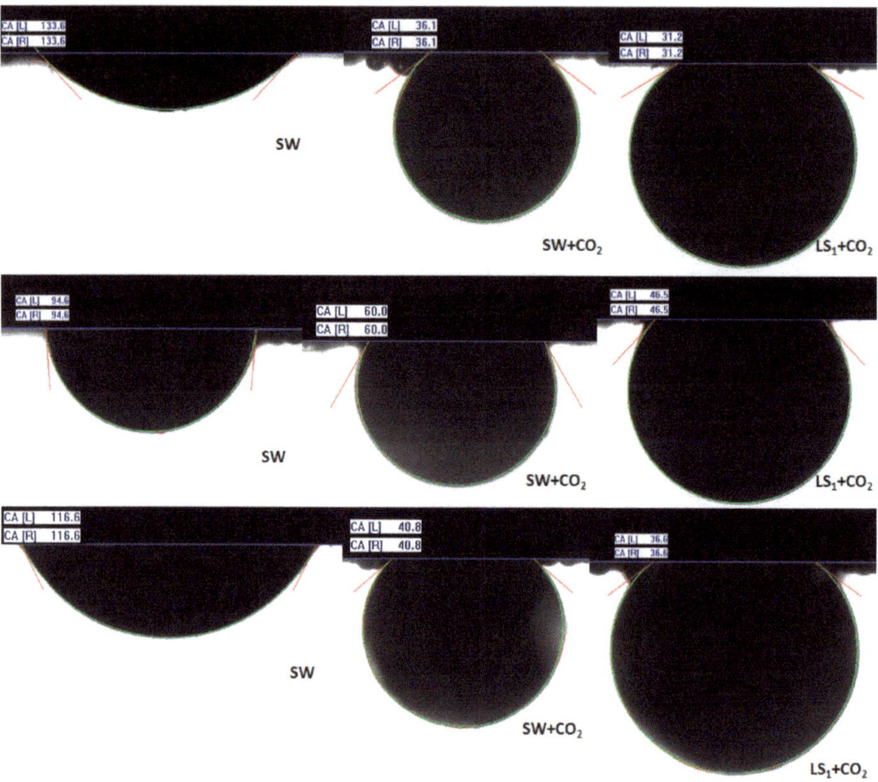

Figure 3. Contact angle of carbonate (top row), sandstone (middle row), shale (bottom row) core samples after they were soaked in seawater (SW), carbonated seawater (SW+CO$_2$), and carbonated low salinity water (LS$_1$+CO$_2$). The volume of oil droplets ranges from 4 to 15 μL [11].

Initial Water Saturation. It is difficult to determine the effect of initial water saturation on imbibition through experiments due to several reasons. When the formation has higher initial water saturation, the amount volume of imbibition can be smaller or larger. The smaller imbibition amount was indicated by Blair [61] and Li et al. [62]. The larger imbibition volume was pointed out by Cil et al. [63], Zhou et al. [64], and Morrow et al. [65]. Bennion and Thomas [66] discussed the existence of the state of noncapillary equilibrium in a low-permeability gas reservoir with abnormally low initial water saturation, and the undersaturated matrix will imbibe a significant amount of water during drilling and completion, resulting in phase trap damage to the formation. In addition, some studies indicated there was little effect from the initial water saturation on imbibition [62,67,68]. The reason for the contradictory conclusions is that the capillary pressure and effective permeability both depend on water saturation. The capillary pressure has an inverse correlation to the water saturation, while the effective permeability of water has a positive correlation to the water saturation. Thus, when the initial water saturation is high, the capillary pressure is normally low, but the effective permeability of water is high. The imbibition volume is controlled by the opposite effects from low capillary pressure but high permeability. Therefore, Morrow and Mason [69] said that the influence of initial water saturation should be investigated specific to a formation. Zhou et al. [70] found that in shale, the lower initial water saturation could cause a faster imbibition rate and higher volume of the imbibition. This is due to the very small pore size of the shale, which causes the capillary pressure to dominate the imbibition process when the initial water saturation is low.

Temperature. Temperature impacts imbibition because wettability and fluid properties change under various temperatures. Handy [71], Pooladi-Darvish, and Firoozabadi [72] indicated that higher temperature caused a faster imbibition rate. Peng and Kovscek [73] proved this conclusion through the forced imbibition experiment system with temperature control. Elevated temperature may result in formation damage due to fine migrations [74].

Flow Direction. The process of same flow direction for wetting and non-wetting phases is called co-current imbibition, which usually happens during the water flooding operation [75]. The opposite flow direction, which is regarded as counter-current imbibition, mainly occurs in unconventional formations and fractured water-wet reservoirs [75,76]. Qin [77] pointed out that the direction of the flow was determined through the wettability of the formation rock, fracture and boundary condition in the reservoir, and injection rate during operation. In the studies of Bourbiaux and Kalaydjian [78], Kantzas et al. [79], Pow et al. [80], and Li and Horne [81], it is found that co-current imbibition can cause a faster rate of recovery than counter-current imbibition.

Fluid and Rock Properties. Fluid viscosity, rock permeability, and pore size can impact the imbibition rate. When water displaces oil and gas in the reservoir, the imbibition rate increases as water viscosity increases [82].

Rock permeability is also a factor to influence the imbibition rate. The higher permeability is expected to have a higher imbibition rate [83,84]. However, Graham and Richardson [12] found that the high permeability ratio of fracture to matrix was difficult to relate to imbibition rate.

Pore size indirectly affects imbibition rate. The small size of pores can cause a high capillary pressure that is one of the driving forces of imbibition. Thus, the imbibition rate would be higher in the small size of pores. However, Egermann et al. [85] indicated that in unconventional formations, the pore size is small, while the permeability is also low. Hence, the imbibition could be still slow in unconventional formation.

Clay Content. In clay-rich formations, such as shale, imbibition is strongly affected by clay mineral [86]. Zhou et al. [6] analyzed the relationship between clay content and imbibition. In shale, the sample with higher clay content could imbibe more volume and at a faster rate than the sample with lower clay content. This was later confirmed in other experimental measurements, such as NMR [87], and the excessive imbibed water beyond capillary-driven water remains as irreducible water in the clay of shale. In addition, the imbibition of fluid with additives was also different in various clay content shale samples. In the high clay content sample, the fluid with 0.07% friction reducer has a greater imbibed volume than the fluid with KCl or KCl substitute (choline chloride, magnesium chloride, and tetramethyl ammonium chloride). However, the fluid with the 2% KCl was imbibed more than other fluids in the shale with less than 10% clay content.

3. Osmosis

Imbibition in tight formations is usually accompanied by osmosis, especially in high-salinity shale formations. Osmosis is a spontaneous net movement of solvent molecules toward a higher concentration region so as to minimize the concentration difference between two sides of a semi-permeable membrane. Solute–membrane interactions are more frequent on the higher solute concentration side than the low concentration side. Thus, more solute particles, such as salt, try to pass through the membrane, but they are excluded by the membrane due to its semi-permeable property. As those particles are pushed, a momentum is generated and pulls water molecules through the membrane from the lower concentration side [88].

Previously, osmosis was overlooked, since its effect is negligible during fluid flow in porous medium of conventional formations. However, in tight (unconventional) formations, which contain high clay mineral contents, a semi-permeable membrane can arise to generate osmosis. Neuzil and Provost [89] observed the anomalous fluid pressure in a subsurface when they performed osmosis measurements on moderately compacted high clay content Pierre shale. This can be illustrated by electric double layer (EDL) theory [90]. Clay particles are naturally and commonly negatively charged. To neutralize the surface charges, cations or counter-ions will be attracted to the clay surface and form a diffuse double layer. However, the concentration of attracted ions decrease away from the clay platelets as electrostatic force weakens. If there is enough distance between clay platelets, there may exist a charge-neutral zone in the middle. However, due to some compaction effect, the diffused layers overlap each other, where the excessive charges accumulate. This overlap region carries charges and provides exclusion forces to any of the charged particles that try to pass through it but not water molecules [91–94].

Small pore sizes, which are a distinguishing characteristic of porous medium in tight formations, also contribute to osmosis occurrence. First, small rock pores with high clay contents can increase the quantity and quality of semi-permeable membrane. Second, the disassociated ions from salt in the aqueous solution are usually hydrated and complexed by water molecules. Thus, when the pore size is small enough, water molecules are more mobile than the larger-size hydrated ions, which will experience more restriction through the rock. Hence, as being excluded by the small rock pores, those hydrated ions may acquire enough momentum to overcome the diffusive flux and pull the water out of low concentration solution [95].

Hence, osmosis in tight formations has attracted research attention. Osmosis study in drilling engineering is mainly related to wellbore stability, which is strongly affected by water-based drilling fluid. When drilling fluid is invaded into formation rocks, it can decrease rock strength and elastic modulus and increase pore pressure, which are all causes of wellbore instability [96]. In shale formations, osmosis is considered as a significant mechanism to result in fluid invasion [97,98]. However, osmosis is a particular mechanism that allows fluid to have a bi-directional flow through controlling the salinity of the drilling fluid. Abass et al. [99] indicated that a designed drilling fluid can extract formation fluid out of shale to strengthen wellbore. The designed considerations are to increase osmotic flow to the wellbore, which requires increasing the salinity and fluid viscosity as well as reducing the shale permeability. High-salinity drilling fluid can induce osmotic flow to the wellbore. High viscosity and small permeability can inhibit capillary flow into formations. Membrane efficiency is also a consideration that is hard to control but should be considered when designing. Membrane efficiency is a ratio between actual osmotic pressure and theoretical osmotic pressure. Abass et al. [99] measured membrane efficiency in shale samples from the Zuluf field of Saudi Arabia. The measurement showed that its membrane efficiency was 4.2%. This result proves that osmosis cannot be neglected in shale formations. Schlemmer et al. [100] discussed factors that can improve membrane efficiency and hence increase osmosis. These factors are clay type of high cation exchange capacity, shale pore structure with more compacted clay, formation fluid with lower salt concentration, and compositions of drilling fluid that can affect the interface of clay.

However, those studies argued about the actual role of osmosis in fluid movement because it is difficult to distinguish osmotic flow from capillary flow in the imbibition process in tight formations. Zhou et al. [9] indicated this combinational mechanism during fracturing fluid flow in shale gas formation rocks.

In Zhou et al.'s experiments, it was found that the weight of rocks increased and decreased alternately when they were immersed into high salinity fluid, which can cause osmotic extraction [9]. Weight increase indicated that fluid invaded into rocks because capillary-driven imbibition was the dominant force. With the continuous fluid invasion, capillary pressure was decreased so that fluid was imbibed into rocks less and less. When osmotic extraction was stronger than capillary imbibition, the rock weight decreased because fluid flowed out of the tight pores more than it flowed in. However, capillary imbibition became stronger again when fluid saturation was declined, so that the weight of rock increased again after a certain point of time. Hence, it is difficult to distinguish the capillary pressure and osmotic forces in the fluid movement processes, as they are dynamically changing and interacting with each other.

The osmotic effect can also contribute to the enhanced oil recovery (EOR) mechanism of unconventional shale formations. Recent studies show that low-salinity waterflooding EOR is can significantly improve oil production from shale formations, especially in high-salinity tight shale formations [8,101–104]. The concept is to enhance osmotic flow through a smaller salinity of waterflooding fluid than that of formation fluid. Figure 4 shows osmosis effect in clay-rich rocks. The application also depends on membrane efficiency. Fakcharoenphol et al. [8] and Teklu et al. [102] proved in clay-rich tight formations that osmosis can improve oil displacement under low-salinity fluid. Figure 5 shows experimental results of osmosis effect on oil displacements based on authors' work. However, it is challenging to quantify the osmosis effects on oil recovery, since fluid chemical equilibrium and rock–fluid interactions all change dynamically with salinity change in this extremely complicated process. Thus, direct measurement of osmosis during oil recovery experiments may reveal new important insights.

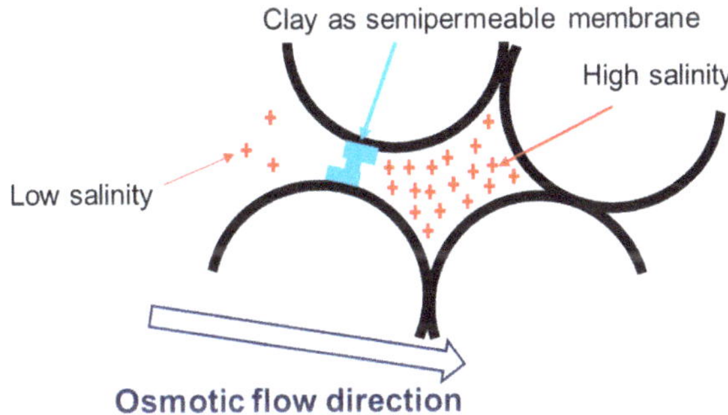

Figure 4. The schematic showing osmosis effect on fluid flow in clay-rich rocks.

Figure 5. A preserved one-inch diameter Bakken shale core sample submersed in high-salinity brine (240,000 ppm KCl) (**top**) and low-salinity brine (20,000 ppm KCl) (**bottom**) showing oil expulsion vs.

imbibition period. At imbibition Day 5 (**top**), the core was removed from a high-salinity brine beaker and produced/expelled oil was wiped and immersed in low-salinity brine from Day 5 until Day 10 (**bottom**). This shows that more oil is expelled due to osmosis during a low-salinity brine imbibition period.

A study by Padin et al. [103] performed high-pressure high-temperature chemical osmosis-driven fluid flow experiments in carbonate-rich mud rocks (shales). Their experiment showed a gradual, slow (within 120 days of experiment) increase of pressure within the samples. Based on their experiments, they concluded that chemical osmosis in organic-rich carbonate rocks could create a significant amount of driving force for oil mobilization or EOR; also, they stated that water imbibition in their experiment cannot be explained by only capillary forces.

4. Simulation for Osmosis-Associated Imbibition

The model that simulates spontaneous imbibition has been predominately attributed to capillary action [71,105–117]. Osmosis has been overlooked for a long time, as it is not as significant as other mechanisms, such as capillarity and gravity, because the membrane efficiency in a conventional reservoir is too low to make a real impact. However, shale and other unconventional formations present a significant osmosis effect due to their mineralogy and pore size structure [9]. Recently, several modeling efforts have been made to investigate the osmosis effect in unconventional reservoir development.

Fakcharoenphol et al. [118] proposed a triple-porosity fracture-matrix model and incorporated the effects of matrix wettability, capillary pressure, relative permeability, and osmotic pressure to investigate the impact of shut-in time on well productivity. In the model, the fracture forms a continuum of an interconnected network created during the hydraulic fracturing, while the organic and non-organic matrices are embedded in the fracture continuum. Fakcharoenphol et al. [8] used a numerical model to calculate osmotic pressure by tracking the salinity concentration. The simulation results indicated that osmotic pressure can be a viable mechanism by promoting water–oil counter-current flow. Wang and Rahman [119] proposed a numerical model to investigate both capillary pressure and osmosis effects on fluid leak-off during shale gas reservoir stimulation. The results showed that rock composition greatly affects the leak-off rate, and the invaded water due to capillary and osmotic pressures significantly increases the pore pressure. There is a strong non-linear relationship between imbibition volume and square root of time. Li et al. [60] developed a multi-component matrix imbibition model to investigate the effects of low-salinity water and surfactant on unconventional recovery. Simulation results matched with experimental data and revealed some important insights on the effects of water salinity and surfactant that the combination effects of contact angle and interfacial tension determine capillary pressure imbibition and that the concentration of charged ions and surfactant molecules affect the osmosis imbibition. All these processes are associated with different rock components and mineralogy, and there exists an optimum water salinity for maximum imbibition. Different from previous simulation studies, Li et al. [90] proposed a multi-mechanistic numerical shale matrix imbibition model by dividing the rock into non-membrane and membrane components. Figure 6 introduces the coupling in the model. The model considered capillary pressure and osmotic pressure as a function of water saturation, could track the dynamic water and salt movement, and was validated by matching with experimental measurements. The principles associated with the imbibition and osmosis behind each model of these reviewed works are summarized in Table 1.

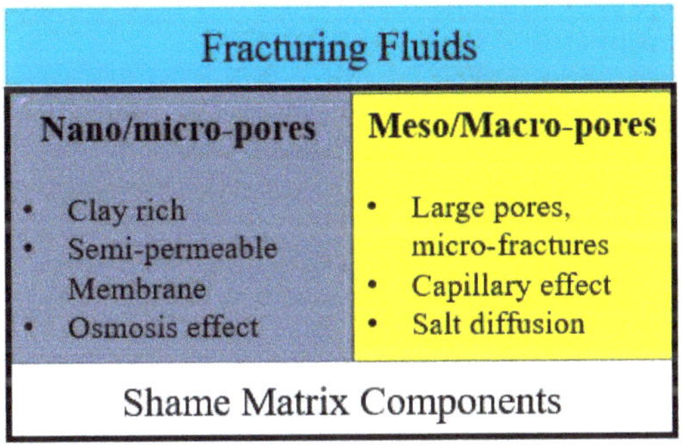

Figure 6. Numerical model with imbibition and osmosis coupling.

Table 1. Imbibition mechanisms for different models that considered imbibition and osmosis.

Model	Capillarity	Osmosis	Diffusion	Gravity
Fakcharoenphol et al. [118]	Water saturation-based capillary pressure curve for organic, non-organic, and fracture media	Osmotic pressure as a function of salt concentration and membrane efficiency	Fick's Law	Gravity effect between matrix and fracture media
Fakcharoenphol et al. [8]	Water saturation-based capillary pressure curve for matrix and fracture system	Osmotic pressure as a function of salt concentration gradient and membrane efficiency	Fick's Law	Gravity effect between matrix and fracture media
Wang and Rahman [119]	Bundle of capillary tubes associated with pore size distribution for clay, hydrophobic, hydrophilic components matrix system	Osmotic pressure as a function of salt concentration gradient and membrane efficiency	Neglected	N/A
Li et al. [90]	Water saturation-based capillary pressure curve for two-component matrix system	Osmotic pressure as a function of salt concentration gradient and membrane efficiency	Fick's Law	N/A

5. Discussion

In summary, those studies on osmosis-associated imbibition provide a good understanding of the fluid flow in unconventional formations. Osmosis is a critical mechanism of fluid movement especially in clay-rich tight formations, and it can enhance the fluid extraction process as an additional force to other forces, such as capillary pressure. However, further studies are required to investigate the impacts of osmosis on the fluid imbibition process to maximize production for field applications. Here, some discussions on the practical implications of osmosis-associated imbibition are summarized.

The common perception is that water imbibition could cause water blockage and clay swelling, which decrease the permeability of the formation. There were many experiments and studies to indicate the decrease due to water blockage [108,120–123] and clay swelling [124,125]. There was also an argument about whether the decrease is temporary or permanent. On one hand, the damage could be eliminated by increasing drawdown pressure or applying alcohol and alcohol surfactant. Therefore, the damage is temporary to the formation [15,126–130]. However, on the other hand, in some formations, additives of alcohol have little effect on gas permeability, so that water blockage can be permanent [131–134].

In recent studies, it was found that osmosis-associated imbibition could improve permeability in unconventional formations. In shale, the experiments established the relationship between the amount volume of imbibition and permeability. It was found that imbibition in shale could cause natural fractures instability and regrowth, thereby causing a significant permeability increase [25,70,135,136]. While positive capillary pressure is the main sucking force for strong imbibition in water-wet rocks, the imbibition volume and rate can be significantly boosted by osmotic effect and lead to more natural fracture reactivations and regrowth. However, that conclusion requires further investigation, because those results were taken from the experiments under non-reservoir conditions. In another word, it is not clear how many fractures can be reopened under reservoir high pressure due to osmosis-associated imbibition. Hence, more studies are necessary to determine the relationship between osmosis-associated imbibition and formation permeability under reservoir conditions for unconventional formations.

In addition, there are studies that investigate the influence of osmosis-associated imbibition on water recovery and production in shale. Those studies provided some explanations on the field observations that extended shut-in time after fracturing treatment often results in a lower percentage of water recovery and but higher production in shale [137–141]. One of the potential reasons is that the near wellbore water blockage is mitigated deeper to the reservoir due to strong imbibition enhanced by osmosis during the extended shut-in. However, more evidence is required to establish a more accurate relationship between osmosis-associated imbibition and water recovery and production.

Therefore, the findings through further investigation between osmosis-associated imbibition and permeability change, water recovery, and production will be beneficial for the hydraulic fracturing design and reservoir management, such that the injection amount and fluid types of fracturing treatment and the shut-in time and flowback rate can be specially designed to minimize the water blockage effect but maximize natural fracture reactivations and well production performance.

6. Conclusions

Osmosis-associated imbibition has strong effects on fluid flow in porous mediums in unconventional formations. Those effects are important to be related to production after hydraulic fracturing and during the water-flooding process. Wettability and clay content are the main factors in the imbibition behavior of unconventional formations. Wettability can impact capillary pressure, which is one of the driving forces of imbibition. Clay content has a strong relationship with osmotic pressure. Therefore, understanding those factors can distinguish the dominant mechanism of fluid flow in unconventional reservoirs.

Although imbibition has been extensively studied in formations, it is necessary to further investigate its impact in unconventional formations, especially after hydraulic fracturing treatment. The effect of osmosis on imbibition was neglected before; however, more evidence has shown that osmosis has a significant impact on the fluid flow in clay-rich formations, such as shale. Thus, more experimental investigations are necessary to prove its effect, such as high-pressure and high-temperature reservoir conditions with tight pore space. From these further investigations, some insights can be drawn, such as how osmosis-associated imbibition affects the production, and how to manage and control it.

In addition, the numerical modeling for a multi-mechanistic imbibition process has shed some light on the interplay of multiple forces during the imbibition process, especially the changes between capillary pressure and osmotic pressure. The simulation results matched the experimental measurements and field observations, such as the dynamic changes in salt concentration in laboratory and low flow-back recovery in the field. However, further investigations are recommended for simulations under reservoir conditions to match the actual production data with proper upscaling from laboratory-based modeling.

Author Contributions: Resources, X.L.; Supervision, Z.Z.; Writing—original draft, review, and editing, Z.Z., X.L. and T.W.T. All authors have read and agreed to the published version of the manuscript.

Funding: This research was funded by National Natural Science Foundation of China (Grant No. 51974338).

Institutional Review Board Statement: Not applicable.

Informed Consent Statement: Not applicable.

Data Availability Statement: Data sharing not applicable.

Conflicts of Interest: The authors declare no conflict of interest.

References

1. Satter, A.; Iqbal, G.M. *Reservoir Engineering*; Gulf Professional Publishing: Houston, TX, USA, 2016; pp. 155–169.
2. Roychaudhuri, B.; Tsotsis, T.; Jessen, K. An Experimental Investigation of Spontaneous Imbibition in Gas Shales. In Proceedings of the SPE Annual Technical Conference and Exhibition, Denver, CO, USA, 30 October–2 November 2011.
3. Makhanov, K.; Dehghanpour, H.; Kuru, E. An Experimental Study of Spontaneous Imbibition in Horn River Shales. In Proceedings of the SPE Canadian Unconventional Resources Conference, Calgary, AB, Canada, 30 October–1 November 2012.
4. Makhanov, K.; Dehghanpour, H.; Kuru, E. Measuring Liquid Uptake of Organic Shales: A Workflow to Estimate Water Loss during Shut-in Periods. In Proceedings of the SPE Canadian Unconventional Resources Conference, Calgary, AB, USA, 5–7 November 2013.
5. Zhou, Z.; Hoffman, T.; Bearinger, D.; Li, X. Experimental and Numerical Study on Spontaneous Imbibition of Fracturing Fluids in Shale Gas Formation. In Proceedings of the SPE/CSUR Unconventional Resources Conference—Canada, Calgary, AB, Canada, 30 September–2 October 2014.
6. Zhou, Z.; Hoffman, T.; Bearinger, D.; Li, X.; Hazim, A. Experimental and Numerical Study on Spontaneous Imbibition of Fracturing Fluids in the Horn River Shale Gas Formation. *SPE Drill. Completion* **2016**, *31*, 168–177. [CrossRef]
7. Warner, H.R. *The Reservoir Engineering Aspects of Waterflooding*, 2nd ed.; Society of Petroleum Engineers: Richardson, TX, USA, 2015; pp. 25–40.
8. Fakcharoenphol, P.; Kurtoglu, B.; Kazemi, H.; Charoenwongsa, S.; Wu, Y.S. The Effect of Osmotic Pressure on Improve Oil Recovery from Fractured Shale Formations. In Proceedings of the SPE Unconventional Resources Conference, The Woodlands, TX, USA, 1–3 April 2014.
9. Zhou, Z.; Hazim, A.; Li, X.; Bearinger, D.; Frank, W. Mechanisms of Imbibition During Hydraulic Fracturing in Shale Formations. *J. Pet. Sci. Eng.* **2016**, *141*, 125–132. [CrossRef]
10. Zhou, Z.; Teklu, T.; Li, X.; Hazim, A. Experimental study of the osmotic effect on shale matrix imbibition process in gas reservoirs. *J. Nat. Gas Sci. Eng.* **2018**, *49*, 1–7. [CrossRef]
11. Teklu, T.W. Experimental and Numerical Study of Carbon Dioxide Injection Enhanced Oil Recovery in Low-Permeability Reservoirs. Ph.D. Thesis, Colorado School of Mines, Golden, CO, USA, 2015.
12. Graham, J.W.; Richardson, J.G. Theory and Application of Imbibition Phenomena in Recovery of Oil. *J. Pet. Technol.* **1959**, *11*, 65–69. [CrossRef]
13. Wagner, O.R.; Leach, R.O. Improving Oil Displacement Efficiency by Wettability Adjustment. *Trans. AIME* **1959**, *216*, 65–72. [CrossRef]
14. Froning, H.R.; Leach, R.O. Determination of Chemical Requirements and Applicability of Wettability Alteration Flooding. *J. Pet. Technol.* **1967**, *19*, 839–843. [CrossRef]
15. Penny, G.S.; Soliman, M.Y.; Briscoe, J.E. Enhanced Load Water-Recovery Technique Improves Stimulation Results. In Proceedings of the Annual Technical Conference and Exhibition, San Francisco, CA, USA, 5–8 October 1983.
16. Wardlaw, N.C.; McKellar, M. Wettability and Connate Water Saturation in Hydrocarbon Reservoirs with Bitumen Deposits. *J. Pet. Sci. Eng.* **1998**, *20*, 141–146. [CrossRef]
17. Li, K.; Firoozabadi, A. Experimental Study of Wettability Alteration to Preferential Gas-Wetting in Porous Media and Its Effects. *SPE Reserv. Eval. Eng.* **2000**, *3*, 139–149.
18. Wang, D.; Butler, R.; Zhang, J.; Seright, R. Wettability Survey in Bakken Shale Using Surfactant Formulation Imbibition. *SPE Reserv. Eval. Eng.* **2012**, *15*, 695–705. [CrossRef]
19. Anderson, W.G. Wettability literature survey-Part 1: Rock/oil/brine interactions and the effects of core handling on wettability. *J. Pet. Technol.* **1986**, *38*, 1125–1144. [CrossRef]
20. Morrow, N.R. Wettability and Its Effect on Oil Recovery. *J. Pet. Technol.* **1990**, *42*, 1476–1484. [CrossRef]
21. Buckley, J.S.; Liu, Y.; Monsterleet, S. Mechanisms of Wetting Alteration by Crude Oils. *SPE J.* **1998**, *3*, 54–61. [CrossRef]
22. Drummond, C.; Israelachvili, J. Surface forces and wettability. *J. Pet. Sci. Eng.* **2002**, *33*, 123–133. [CrossRef]
23. Morsy, S.; Sheng, J.J.; Gomaa, A.M.; Soliman, M.Y. Potential of Improved Waterflooding in Acid-Hydraulically-Fractured Shale Formations. In Proceedings of the SPE Annual Technical Conference and Exhibition, New Orleans, LA, USA, 30 September–2 October 2013.

24. Morsy, S.; Gomaa, A.; Sheng, J.J. Imbibition Characteristics of Marcellus Shale Formation. In Proceedings of the SPE Improved Oil Recovery Symposium, Tulsa, OK, USA, 12–16 April 2014.
25. Morsy, S.; Sheng, J.J. Imbibition Characteristics of the Barnett Shale Formation. In Proceedings of the SPE Unconventional Resources Conference, The Woodlands, TX, USA, 1–3 April 2014.
26. Morsy, S.; Hetherington, C.J.; Sheng, J.J. Effect of Low-Concentration HCl on the Mineralogy, Physical and Mechanical Properties, and Recovery Factors of Some Shales. *J. Unconv. Oil Gas Resour.* **2015**, *9*, 94–102. [CrossRef]
27. Teklu, T.W.; Abass, H.; Hanashmooni, R.; Carratu, J.C.; Ermila, M. Experimental Investigation of Acid Imbibition on Matrix and Fractured Carbonate Rich Shales. *J. Nat. Gas Sci. Eng.* **2017**, *45*, 706–725. [CrossRef]
28. Teklu, T.W.; Park, D.; Jung, H.; Amini, K.; Abass, H. Effect of Dilute Acid on Hydraulic Fracturing of Carbonate-Rich Shales: Experimental Study. *SPE Prod. Oper.* **2018**, *34*, 170–184. [CrossRef]
29. Tang, G.Q.; Morrow, N. Influence of Brine Composition and Fines Migration on Crude Oil/Brine/Rock Interactions and Oil Recovery. *J. Pet. Sci. Eng.* **1999**, *24*, 99–111. [CrossRef]
30. Hiorth, A.; Cathles, L.; Madland, M. The Impact of Pore Water Chemistry on Carbonate Surface Charge and Oil Wettability. *Transp. Porous Media* **2010**, *85*, 1–21. [CrossRef]
31. Yousef, A.A.; Al-Saleh, S.H.; Al-Kaabi, A.; Al-Jawfi, M.S. Laboratory Investigation of the Impact of Injection-Water Salinity and Ionic Content on Oil Recovery from Carbonate Reservoirs. *SPE Reserv. Eval. Eng.* **2011**, *14*, 578–593. [CrossRef]
32. Yi, Z.; Sarma, H. Improving Waterflood Recovery Efficiency in Carbonate Reservoirs through Salinity Variations and Ionic Exchanges: A Promising Low-Cost "Smart-Waterflood" Approach. In Proceedings of the Abu Dhabi International Petroleum Exhibition & Conference, Abu Dhabi, UAE, 11–14 November 2012.
33. McGuire, P.L.; Chatham, J.R.; Paskvan, F.K.; Sommer, D.M.; Carini, F.H. Low Salinity Oil Recovery: An Exciting New EOR Opportunity for Alaska's North Slope. In Proceedings of the SPE Western Regional Meeting, Irvine, CA, USA, 30 March–1 April 2005.
34. Austad, T.; RezaeiDoust, A.; Puntervold, T. Chemical Mechanism of Low-Salinity Water Flooding in Sandstone Reservoirs. In Proceedings of the SPE Improved Oil Recovery Symposium, Tulsa, OK, USA, 24–28 April 2010.
35. Morrow, N.R.; Buckley, J. Improved Oil Recovery by Low-Salinity Waterflooding. *J. Pet. Technol.* **2011**, *63*, 106–112. [CrossRef]
36. Strand, S.; Hogensen, E.J.; Austad, T. Wettability alteration of Carbonates—Effects of potential determining ions (Ca^{2+} and SO_4^{2-}) and temperature. *Colloids Surf. A Physicochem. Eng. Asp.* **2006**, *275*, 1–10. [CrossRef]
37. Zhang, P.; Austad, T. Wettability and Oil Recovery from Carbonates: Effects of Temperature and Potential Determining Ions. *Colloids Surf. A Physicochem. Eng. Asp.* **2006**, *279*, 179–187. [CrossRef]
38. Lager, A.; Webb, K.J.; Black, C.J.J.; Singleton, M.; Sorbie, K.S. Low salinity Oil Recovery—An Experimental Investigation. *Petrophysics* **2008**, *49*, 28–35.
39. Austad, T.; Sgariatpanahi, S.F.; Strand, S.; Black, C.J.J.; Webb, K.J. Conditions for a Low-Salinity Enhanced Oil Recovery (EOR) Effect in Carbonate Oil Reservoirs. *Energy Fuels* **2011**, *26*, 569–575. [CrossRef]
40. Zahid, A.; Shapiro, A.; Skauge, A. Experimental Studies of Low-salinity Water Flooding Carbonate: A New Promising Approach. In Proceedings of the SPE EOR Conference at Oil and Gas West Asia, Muscat, Oman, 16–18 April 2012.
41. Jadhunandan, P.P.; Morrow, N.R. Effect of Wettability on Waterflood Recovery for Crude-Oil/Brine/Rock Systems. *SPE Reserv. Eng.* **1995**, *10*, 40–46. [CrossRef]
42. Ligthelm, D.J.; Gronsveld, J.; Hofman, J.P.; Brussee, N.J.; Marcelis, F.; van der Linde, H. Novel Waterflooding Strategy by Manipulation of Injection Brine Composition. In Proceedings of the EUROPEC/EAGE Conference and Exhibition, Amsterdam, The Netherlands, 8–11 June 2009.
43. Alameri, W.; Teklu, T.W.; Graves, R.M.; Kazemi, H.; AlSumaiti, A.M. Wettability Alteration during Low-Salinity Waterflooding in Carbonate Reservoir Cores. In Proceedings of the SPE Pacific Oil & Gas Conference and Exhibition, Adelaide, Australia, 14–16 October 2014.
44. Mahani, H.; Keya, A.L.; Berg, S.; Bartels, W.B.; Nasralla, R.; Rossen, W.R. Insights into the mechanism of wettability alteration by low-salinity flooding (LSF) in carbonates. *Energy Fuels* **2015**, *29*, 1352–1367. [CrossRef]
45. Mahani, H.; Menezes, R.; Berg, S.; Fadili, A.; Nasralla, R.; Voskov, D.; Joekar-Niasar, V. Insights into the impact of temperature on the wettability alteration by low salinity in carbonate rocks. *Energy Fuels* **2017**, *31*, 7839–7853. [CrossRef]
46. Mahani, H.; Keya, A.L.; Berg, S.; Nasralla, R. Electrokinetics of Carbonate/Brine Interface in Low-Salinity Waterflooding: Effect of Brine Salinity, Composition, Rock Type, and pH on ζ-Potential and a Surface-Complexation Model. *SPE J.* **2017**, *22*, 53–68. [CrossRef]
47. Chen, Y.; Xie, Q.; Sari, A.; Brady, P.V.; Saeedi, A. Oil/water/rock wettability: Influencing factors and implications for low salinity water flooding in carbonate reservoirs. *Fuel* **2018**, *215*, 171–177. [CrossRef]
48. Tian, H.; Wang, M. Electrokinetic mechanism of wettability alternation at oil-water-rock interface. *Surface Sci. Rep.* **2018**, *72*, 369–391. [CrossRef]
49. Alvarez, J.O.; Schechter, D.S. Wettability Alteration and Spontaneous Imbibition in Unconventional Liquid Reservoirs by Surfactant Additives. *SPE Reserv. Eval. Eng.* **2017**, *20*, 107–117. [CrossRef]
50. Ayirala, S.; Al-Saleh, S.H.; Enezi, S.; Yousef, A. Effect of Salinity and Water Ions on Electrokinetic Interactions in Carbonate Reservoir Cores at Elevated Temperatures. *SPE Reserv. Eval. Eng.* **2018**, *21*, 733–746. [CrossRef]

51. Khaleel, O.; Teklu, T.W.; Alameri, W.; Abass, H.; Kazemi, H. Wettability Alteration of Carbonate Reservoir Cores—Laboratory Evaluation Using Complementary Techniques. *SPE Reserv. Eval. Eng.* **2019**, *22*, 911–922. [CrossRef]

52. Bae, J.H. Glenn Pool Surfactant-Flood Expansion Project-A Technical Summary. *SPE Reserv. Eng.* **1995**, *10*, 123–127. [CrossRef]

53. Chen, H.L.; Lucas, L.R.; Yang, H.D.; Kenyon, D.E. Laboratory monitoring of surfactant imbibition using computerized tomography. *SPE Reserv. Eval. Eng.* **2001**, *4*, 16–25. [CrossRef]

54. Manrique, E.J.; Muci, V.E.; Gurfinkel, M.E. EOR Field Experiences in Carbonate Reservoirs in the United States. *SPE Reserv. Eval. Eng.* **2007**, *10*, 667–686. [CrossRef]

55. Hirasaki, G.; Miller, C.A.; Puerto, M. Recent Advances in Surfactant EOR. *SPE J.* **2011**, *16*, 889–907. [CrossRef]

56. Yassin, M.R.; Ayatollahi, S.; Rostami, B.; Hassani, K.; Taghikhani, V. Micro-Emulsion Phase Behavior of a Cationic Surfactant at Intermediate Interfacial Tension in Sandstone and Carbonate Rocks. *J. Energy Resour. Technol.* **2015**, *137*, 012905–012912. [CrossRef]

57. Alameri, W.; Teklu, T.W.; Graves, R.M.; Kazemi, H.; AlSumaiti, A.M. Low-Salinity Water-Alternate-Surfactant in Low-Permeability Carbonate Reservoirs. In Proceedings of the 18th European Symposium on Improved Oil Recovery conference, Dresden, Germany, 14–16 April 2015.

58. Teklu, T.W.; Alameri, W.; Kazemi, H.; Graves, R.M.; AlSumaiti, A.M. Low-Salinity-Water-Surfactant-CO_2 EOR: Theory and Experiments. In Proceedings of the 18th European Symposium on Improved Oil Recovery conference, Dresden, Germany, 14–16 April 2015.

59. Alvarez, J.O.; Neog, A.; Jais, A.; Schechter, D.S. Impact of Surfactants for Wettability Alteration in Stimulation Fluids and the Potential for Surfactant EOR in Unconventional Liquid Reservoirs. In Proceedings of the SPE Unconventional Resources Conference, The Woodlands, TX, USA, 1–3 April 2014.

60. Li, X.; Teklu, T.W.; Abass, H.; Cui, Q. The Impact of Water Salinity/Surfactant on Spontaneous Imbibition through Capillarity and Osmosis for Unconventional IOR. In Proceedings of the Unconventional Resources Technology Conference, San Antonio, TX, USA, 1–3 August 2016.

61. Blair, P.M. Calculation of Oil Displacement by Countercurrent Water Imbibition. *SPE J.* **1964**, *4*, 195–202. [CrossRef]

62. Li, K.; Chow, K.; Horne, R.N. Effect of Initial Water Saturation on Spontaneous Water Imbibition. In Proceedings of the SPE Western Regional/AAPG Pacific Section Joint Meeting, Anchorage, AK, Canada, 20–22 May 2002.

63. Cil, M.; Reis, J.C.; Miller, M.A.; Misra, D. An Examination of Countercurrent Capillary Imbibition Recovery from Single Matrix Blocks and Recovery Predictions by Analytical Matrix/Fracture Transfer Functions. In Proceedings of the SPE Annual Technical Conference and Exhibition, New Orleans, LA, USA, 27–30 September 1998.

64. Zhou, X.; Morrow, N.R.; Ma, S. Interrelationship of Wettability, Initial Water Saturation, Aging Time, and Oil Recovery by Spontaneous Imbibition and Waterflooding. *SPE J.* **2000**, *5*, 199–207. [CrossRef]

65. Morrow, N.R.; Tong, Z.; Xie, X. Scaling of Viscosity Ratio for Oil Recovery by Imbibition from Mixed-Wet Rocks. *Petrophysics* **2002**, *43*, 43–51.

66. Bennion, D.B.; Thomas, F.B. Formation Damage Issues Impacting the Productivity of Low Permeability, Low Initial Water Saturation Gas Producing Formations. *J. Energy Resour. Technol.* **2005**, *127*, 240–247. [CrossRef]

67. Viksund, B.G.; Morrow, N.R.; Ma, S.; Wang, W.; Graue, A. Initial Water Saturation and Oil Recovery from Chalk and Sandstone by Spontaneous Imbibition. In Proceedings of the International Symposium of the Society of Core Analysts, Hague, The Netherlands, 14–16 September 1998.

68. Akin, S.; Schembre, J.M.; Bhat, S.K.; Kovscek, A.R. Spontaneous Imbibition Characteristics of Diatomite. *J. Pet. Sci. Eng.* **2000**, *25*, 149–165. [CrossRef]

69. Morrow, N.R.; Mason, G. Recovery of oil by spontaneous imbibition. *Curr. Opin. Colloid Interface Sci.* **2001**, *6*, 321–337. [CrossRef]

70. Zhou, Z.; Hazim, A.; Li, X.; Teklu, T.W. Experimental Investigation of the Effect of Imbibition on Shale Permeability during Hydraulic Fracturing. *J. Nat. Gas Sci. Eng.* **2016**, *29*, 413–430. [CrossRef]

71. Handy, L.L. Determination of Effective Capillary Pressures for Porous Media from Imbibition Data. *Pet. Trans.* **1960**, *219*, 75–80. [CrossRef]

72. Pooladi-Darvish, M.; Firoozabadi, A. Experiments and Modelling of Water Injection in Water-wet Fractured Porous Media. *J. Can. Pet. Technol.* **2000**, *39*, 31–42. [CrossRef]

73. Peng, J.; Kovscek, A.R. Temperature-Induced Fracture Reconsolidation of Diatomaceous Rock during Forced Water Imbibition. In Proceedings of the SPE Western North America Regional Meeting, Anaheim, CA, USA, 27–29 May 2011.

74. Schembre, J.M.; Kovscek, A.R. Mechanism of Formation Damage at Elevated Temperature. *J. Energy Resour. Technol.* **2005**, *127*, 171–180. [CrossRef]

75. Pooladi-Darvish, M.; Firoozabadi, A. Cocurrent and Countercurrent Imbibition in a Water-Wet Matrix Block. *SPE J.* **2000**, *5*, 3–11. [CrossRef]

76. Cuiec, L.E.; Bourbiaux, B.; Kalaydjian, F. Oil Recovery by Imbibition in Low-Permeability Chalk. *SPE Form. Eval.* **1994**, *9*, 200–208. [CrossRef]

77. Qin, B. Numerical Study of Recovery Mechanisms in Tight Gas Reservoirs. Master's Thesis, University of Oklahoma, Norman, OK, USA, 2007.

78. Bourbiaux, B.J.; Kalaydjian, F.J. Experimental Study of Cocurrent and Countercurrent Flows in Natural Porous Media. *SPE Reserv. Eng.* **1990**, *5*, 361–368. [CrossRef]

79. Kantzas, A.; Pow, M.; Allsopp, K.; Marentette, D. Co-Current and Counter-Current Imbibition Analysis for Tight Fractured Carbonate Gas Reservoirs. In Proceedings of the Technical Meeting/Petroleum Conference of The South Saskatchewan Section, Regina, SK, Canada, 19–22 October 1997.

80. Pow, M.; Kantzas, A.; Allan, V.; Mallmes, R. Production of Gas from Tight Naturally Fractured Reservoirs with Active Water. *J. Can. Pet. Technol.* **1999**, *38*, 38–45. [CrossRef]

81. Li, K.; Horne, R.N. An Analytical Scaling Method for Spontaneous Imbibition in Gas/Water/Rock Systems. In Proceedings of the SPE/DOE Improved Oil Recovery Symposium, Tulsa, OK, USA, 13–17 April 2002.

82. Ma, S.; Zhang, X.; Morrow, N.R. Influence of Fluid Viscosity on Mass Transfer between Rock Matrix and Fractures. *J. Can. Pet. Technol.* **1999**, *38*, 25–30. [CrossRef]

83. Garg, A.; Zwahlen, E.; Patzek, T.W. Experimental and Numerical Studies of One-Dimensional Imbibition in Berea Sandstone. In Proceedings of the 16th Annual American Geophysical Union Hydrology Days, Fort Colllins, CO, USA, 15–18 April 1996.

84. Li, K.; Horne, R.N. Characterization of Spontaneous Water Imbibition into Gas-Saturated Rocks. In Proceedings of the SPE/AAPG Western Regional Meeting, Long Beach, CA, USA, 19–23 June 2000.

85. Egermann, P.; Laroche, C.; Manceau, E.; Delamaide, E.; Bourbiaux, B. Experimental and Numerical Study of Water/Gas Imbibition Phenomena in Vuggy Carbonates. In Proceedings of the SPE/DOE Symposium on Improved Oil Recovery, Tulsa, OK, USA, 17–21 April 2004.

86. Lan, Q.; Ghanbari, E.; Dehghanpour, H.; Hawkes, R. Water Loss versus Soaking Time: Spontaneous Imbibition in Tight Rocks. In Proceedings of the SPE/EAGE European Unconventional Resources Conference and Exhibition, Vienna, Austria, 25–27 February 2014.

87. Jiang, Y.; Fu, Y.; Lei, Z.; Gu, Y.; Qi, L.; Cao, Z. Experimental NMR Analysis of Oil and Water Imbibition during Fracturing in Longmaxi Shale, SE Sichuan Basin. *J. Jpn. Pet. Inst.* **2019**, *62*, 1–10. [CrossRef]

88. Nelson, P. *Biological Physics: Energy, Information, Life*; W. H. Freeman and Co.: New York, NY, USA, 2013.

89. Neuzil, C.E.; Provost, A.M. Recent Experimental Data May Point to a Greater Role for Osmotic Pressures in the Subsurface. *Water Resour. Res.* **2009**, *45*. [CrossRef]

90. Li, X.; Abass, H.; Teklu, T.W.; Cui, Q. A Shale Matrix Imbibition Model—Interplay between Capillary Pressure and Osmotic Pressure. In Proceedings of the SPE Annual Technical Conference and Exhibition, Dubai, UAE, 26–28 September 2016.

91. White, D.E.; Young, A.; Galley, J. Saline waters of sedimentary rocks. In *Fluids in Subsurface Environments*; American Association of Petroleum Geologists Memoir: Tulsa, OK, USA, 1965; p. 414.

92. Grim, R.E. *Clay Mineralogy*; McGraw-Hill: New York, NY, USA, 1968; p. 596.

93. Stumm, W.; Morgan, J. *Aquatic Chemistry: An Introduction Emphasizing Chemical Equilibria in Natural Waters*; Wiley Interscience: New York, NY, USA, 1970; p. 583.

94. Fritze, S.J. Ideality of Clay Membranes in Osmotic Processes: A Review. *Clays Clay Miner.* **1986**, *34*, 214–223. [CrossRef]

95. Whitworth, T.M. Hyperfiltration-Induced Isotopic Fractionation: Mechanisms and Role in the Subsurface. Ph.D. Thesis, Purdue University, West Lafayette, IN, USA, 1993.

96. Hale, A.H.; Mody, F.K.; Salisbury, D.P. The Influence of Chemical Potential on Wellbore Stability. *SPE Drill. Completion* **1993**, *8*, 207–216. [CrossRef]

97. Simpson, J.P.; Dearing, H.L. Diffusion Osmosis-An Unrecognized Cause of Shale Instability. In Proceedings of the IADC/SPE Drilling Conference, New Orleans, LA, USA, 23–25 February 2000.

98. Al-Bazali, T.M.; Zhang, J.; Chenevert, M.E.; Sharma, M.M. An Experimental Investigation on the Impact of Capillary Pressure, Diffusion Osmosis, and Chemical Osmosis on the Stability and Reservoir Hydrocarbon Capacity of Shales. In Proceedings of the SPE Offshore Europe Oil and Gas Conference and Exhibition, Aberdeen, UK, 8–11 September 2009.

99. Abass, H.; Shebatalhamd, A.; Khan, M.; Al-Shobaili, Y.; Ansari, A.; Ali, S.; Mehta, S. Wellbore Instability of Shale Formation; Zuluf Field, Saudi Arabia. In Proceedings of the SPE Technical Symposium of Saudi Arabia Section, Dhahran, Saudi Arabia, 21–23 May 2006.

100. Schlemmer, R.; Friedheim, J.E.; Growcock, F.B.; Bloys, J.B.; Headley, J.A.; Polnaszek, S.C. Chemical Osmosis, Shale, and Drilling Fluids. *SPE Drill. Complet.* **2003**, *18*, 318–331. [CrossRef]

101. Bui, B.T.; Tutuncu, A.N. Contribution of osmotic transport on oil recovery from rock matrix in unconventional reservoirs. *J. Pet. Sci. Eng.* **2017**, *157*, 392–408. [CrossRef]

102. Teklu, T.W.; Li, X.; Zhou, Z.; Alharthy, N.; Wang, L.; Abass, H. Low-Salinity water and surfactants for hydraulic fracturing and EOR of shales. *J. Pet. Sci. Eng.* **2018**, *162*, 367–377. [CrossRef]

103. Padin, A.; Torcuk, M.A.; Katsuki, D.; Kazemi, H.; Tutuncu, A.N. Experimental and Theoretical Study of Water and Solute Transport in Organic-Rich Carbonate Mudrocks. *SPE J.* **2018**, *23*, 704–718. [CrossRef]

104. Torcuk, M.A.; Uzun, O.; Padin, A.; Kazemi, H. Impact of Chemical Osmosis on Brine Imbibition and Hydrocarbon Recovery in Liquid-Rich Shale Reservoirs. In Proceedings of the SPE Annual Technical Conference and Exhibition, Calgary, AB, Canada, 30 September–2 October 2019.

105. Sherman, J.B.; Holditch, S.A. Effect of Injected Fracture Fluids and Operating Procedures on Ultimate Gas Recovery. In Proceedings of the SPE Gas Technology Symposium, Houston, TX, USA, 22–24 January 1991.

106. Babadagli, T.; Ershaghi, I. Imbibition Assisted Two-Phase Flow in Natural Fractures. In Proceedings of the SPE Western Regional Meeting, Bakersfield, CA, USA, 30 March–1 April 1992.

107. Babadagli, T. Injection Rate Controlled Capillary Imbibition Transfer in Fractured Systems. In Proceedings of the SPE Annual Technical Conference and Exhibition, New Orleans, LA, USA, 25–28 September 1994.
108. Mogensen, K.; Stenby, E. A Dynamic Pore-Scale Model of Imbibition. In Proceedings of the SPE/DOE Improved Oil Recovery Symposium, Tulsa, OK, USA, 19–22 April 1998.
109. Schembre, J.M.; Akin, S.; Castanier, L.M.; Kovscek, A.R. Spontaneous Water Imbibition into Diatomite. In Proceedings of the Western Regional Meeting, Bakersfield, CA, USA, 10–13 May 1998.
110. Civan, F.; Rasmussen, M.L. Analytical Matrix-Fracture Transfer Models for Oil Recovery by Hindered-Capillary Imbibition. In Proceedings of the SPE/DOE Improved Oil Recovery Symposium, Tulsa, OK, USA, 13–17 April 2002.
111. Ramirez, B.; Kazemi, H.; Al-Kobaisi, M.; Ozkan, E.; Atan, S. A Critical Review for Proper Use of Water/Oil/Gas Transfer Functions in Dual-Porosity Naturally Fractured Reservoirs: Part I. *SPE Reserv. Eval. Eng.* **2009**, *12*, 200–210. [CrossRef]
112. Al-Kobaisi, M.; Kazemi, H.; Ramirez, B.; Ozkan, E.; Atan, S. A Critical Review for Proper Use of Water/Oil/Gas Transfer Functions in Dual-Porosity Naturally Fractured Reservoirs: Part II. *SPE Reserv. Eval. Eng.* **2009**, *12*, 211–217. [CrossRef]
113. Teklu, T.W.; Akinboyewa, J.; Alharthy, N.; Torcuk, M.A.; AlSumaiti, A.M.; Kazemi, H.; Graves, R.M. Pressure and Rate Analysis of Fractured Low Permeability Gas Reservoirs: Numerical and Analytical Dual-Porosity Models. In Proceedings of the SPE Middle East Unconventional Gas Conference & Exhibition, Muscat, Oman, 28–30 January 2013.
114. Zhang, R.; Wu, Y.-S.; Fakcharoenphol, P. Non-Darcy Displacement in Linear Composite and Radial Aquifer during CO_2 Sequestration. *Int. J. Oil Gas Coal Technol.* **2014**, *7*, 244–262. [CrossRef]
115. Zhang, R.; Winterfeld, P.H.; Yin, X.; Xiong, Y.; Wu, Y.-S. Sequentially Coupled THMC Model for CO_2 Geological Sequestration into a 2D Heterogeneous Saline Aquifer. *J. Nat. Gas Sci. Eng.* **2015**, *27*, 579–615. [CrossRef]
116. Zhang, R.; Xiong, Y.; Winterfeld, P.H.; Yin, X.; Wu, Y.-S. A Novel Computational Framework for Thermal-Hydrological-Mechanical-Chemical Processes of CO_2 Geological Sequestration into a Layered Saline Aquifer and a Naturally Fractured Enhanced Geothermal System. *Greenh. Gases Sci. Technol.* **2016**, *6*, 370–400. [CrossRef]
117. Zhang, R.; Yin, X.; Winterfeld, P.H.; Wu, Y.-S. A Fully Coupled Thermal-Hydrological-Mechanical-Chemical Model for CO_2 Geological Sequestration. *J. Nat. Gas Sci. Eng.* **2016**, *28*, 280–304. [CrossRef]
118. Fakcharoenphol, P.; Torcuk, M.A.; Wallace, J.; Bertoncello, A.; Kazemi, H.; Wu, Y.; Honarpour, M. Managing Shut-in Time to Enhance Gas Flow Rate in Hydraulic Fractured Shale Reservoirs: A Simulation Study. In Proceedings of the SPE Annual Technical Conference and Exhibition, New Orleans, LA, USA, 30 September–2 October 2013.
119. Wang, J.; Rahman, S.S. An Investigation of Fluid Leak-off Due to Osmotic and Capillary Effects and Its Impact on Micro-Fracture Generation during Hydraulic Fracturing Stimulation of Gas Shale. In Proceedings of the EUROPEC 2015, Madrid, Spain, 1–4 June 2015.
120. Hadley, G.F.; Handy, L.L. A Theoretical and Experimental Study of the Steady State Capillary End Effect. In Proceedings of the Fall Meeting of the Petroleum Branch of AIME, Los Angeles, CA, USA, 14–17 October 1956.
121. Li, Y.; Wardlaw, N.C. The influence of wettability and critical pore-throat size ratio on snap-off. *J. Colloid Interface Sci.* **1986**, *109*, 461–472.
122. Li, Y.; Wardlaw, N.C. Mechanisms of non-wetting phase trapping during imbibition at slow rates. *J. Colloid Interface Sci.* **1986**, *109*, 473–486.
123. Zhang, D.; Kang, Y.; You, L.; Li, J. Investigation of Formation Damage Induced During Drill-In Process of Ultradeep Fractured Tight Sandstone Gas Reservoirs. *J. Energy Resour. Technol.* **2019**, *141*, 072901–072910. [CrossRef]
124. Dutta, R.; Lee, C.; Odumabo, S.; Ye, P.; Walker, S.C.; Karpyn, Z.T.; Ayala, L.F. Quantification of Fracturing Fluid Migration due to Spontaneous Imbibition in Fractured Tight Formations. In Proceedings of the SPE Americas Unconventional Resources Conference, Pittsburgh, PA, USA, 5–7 June 2012.
125. Ghanbari, E.; Xu, M.; Dehghanpour, H.; Bearinger, D. Advances in Understanding Liquid Flow in Gas Shales. In Proceedings of the SPE/CSUR Unconventional Resources Conference—Canada, Calgary, AB, Canada, 30 September–2 October 2014.
126. Holditch, S.A. Factors Affecting Water Blocking and Gas Flow from Hydraulically Fractured Gas Wells. *J. Pet. Technol.* **1979**, *31*, 1515–1524. [CrossRef]
127. Abrams, A.; Vinegar, H.J. Impairment Mechanisms in Vicksburg Tight Gas Sands. In Proceedings of the SPE/DOE Low Permeability Gas Reservoirs Symposium, Denver, CO, USA, 19–22 March 1985.
128. Mahadevan, J.; Sharma, M.M. Clean-Up of Water Blocks in Low Permeability Formations. In Proceedings of the SPE Annual Technical Conference and Exhibition, Denver, CO, USA, 5–8 October 2003.
129. Mahadevan, J.; Sharma, M.M. Factors Affecting Cleanup of Water Blocks a Laboratory Investigation. *SPE J.* **2005**, *10*, 238–246. [CrossRef]
130. Putthaworapoom, N.; Miskimins, J.L.; Kazemi, H. Sensitivity Analysis of Hydraulic Fracturing Damage Factors: Reservoir Properties and Operation Variables. In Proceedings of the SPE/EAGE European Unconventional Resources Conference and Exhibition, Vienna, Austria, 20–22 March 2012.
131. Land, C.S. Calculation of Imbibition Relative Permeability for Two- and Three-Phase Flow from Rock Properties. *SPE J.* **1968**, *8*, 149–156. [CrossRef]
132. Ehrlich, R. The Effect of Temperature on Water—Oil Imbibition Relative Permeability. In Proceedings of the Eastern Regional Meeting, Pittsburgh, PA, USA, 5–6 November 1970.

133. Soliman, M.Y.; Hunt, J.L. Effect of Fracturing Fluid and Its Cleanup on Well Performance. In Proceedings of the SPE Eastern Regional Meeting, Morgantown, WV, USA, 6–8 November 1985.
134. Ding, M. Gas Trapping and Mobilization through Water Influx in Natural Gas Reservoirs. Ph.D. Thesis, University of Calgary, Calgary, AB, Canada, 2005.
135. Yuan, B.; Wang, Y.; Shunpeng, Z. Effect of Slick Water on Permeability of Shale Gas Reservoirs. *J. Energy Resour. Technol.* **2018**, *140*, 112901–112907. [CrossRef]
136. Xu, M.; Gupta, A.; Dehghanpour, H. How significant are strain and stress induced by water imbibition in dry gas shales? *J. Pet. Sci. Eng.* **2019**, *176*, 428–443. [CrossRef]
137. Settari, A.; Sullivan, R.B.; Bachman, R.C. The Modeling of the Effect of Water Blockage and Geomechanics in Waterfracs. In Proceedings of the SPE Annual Technical Conference and Exhibition, San Antonio, TX, USA, 29 September–2 October 2002.
138. Wang, Q.; Guo, B.; Gao, D. Is Formation Damage an Issue in Shale Gas Development? In Proceedings of the SPE International Symposium and Exhibition on Formation Damage Control, Lafayette, LA, USA, 15–17 February 2012.
139. Cheng, Y. Impact of Water Dynamics in Fractures on the Performance of Hydraulically Fractured Wells in Gas-Shale Reservoirs. *J. Can. Pet. Technol.* **2012**, *51*, 143–151. [CrossRef]
140. Sharma, M.; Agrawal, S. Impact of Liquid Loading in Hydraulic Fractures on Well Productivity. In Proceedings of the SPE Hydraulic Fracturing Technology Conference, The Woodlands, TX, USA, 4–6 February 2013.
141. Tian, L.; Feng, B.; Zheng, S.; Gu, D.; Ren, X.; Yang, D. Performance Evaluation of Gas Production with Consideration of Dynamic Capillary Pressure in Tight Sandstone Reservoirs. *J. Energy Resour. Technol.* **2018**, *141*, 022902–022917. [CrossRef]

Article

Experimental Investigation and Mechanism Analysis on Rock Damage by High Voltage Spark Discharge in Water: Effect of Electrical Conductivity

Zhixiang Cai, Hui Zhang *, Kerou Liu, Yufei Chen and Qing Yu

College of Petroleum Engineering, China University of Petroleum-Beijing, 18 Fuxue Road, Changping, Beijing 102249, China; Zhixiang.cai05@gmail.com (Z.C.); liukerou1011@gmail.com (K.L.); yufeichen4323@gmail.com (Y.C.); yuqing1381@gmail.com (Q.Y.)
* Correspondence: zhanghuicup2018@163.com; Tel.: +86-132-6112-9527

Received: 23 September 2020; Accepted: 10 October 2020; Published: 18 October 2020

Abstract: High voltage spark discharge (HVSD) could generate strong pressure waves that can be combined with a rotary drill bit to improve the penetration rate in unconventional oil and gas drilling. However, there has been little investigation of the effect of electrical conductivity on rock damage and the fragmentation mechanism caused by HVSD. Therefore, we conducted experiments to destroy cement mortar, a rock-like material, in water with five conductivity levels, from 0.5 mS/cm to 20 mS/cm. We measured the discharge parameters, such as breakdown voltage, breakdown delay time, and electrical energy loss, and investigated the damage mechanism from stress waves propagation using X-ray computed tomography. Our study then analyzed the influence of conductivity on the surface damage of the sample by the pore size distribution and the cumulative pore area, as well as studied the dependence of internal damage on conductivity by through-transmission ultrasonic inspection technique. The results indicated that the increase in electrical conductivity decreased the breakdown voltage and breakdown delay time and increased the energy loss, which led to a reduction in the magnitude of the pressure wave and, ultimately, reduced the sample damage. It is worth mentioning that the relationship between the sample damage and electrical conductivity is non-linear, showing a two-stage pattern. The findings suggest that stress waves induced by the pressure waves play a significant role in sample damage where pores and two types of tensile cracks are the main failure features. Compressive stresses close horizontal cracks inside the sample and propagate vertical cracks, forming the tensile cracks-I. Tensile stresses generated at the sample–water interface due to the reflection of stress waves produce the tensile cracks-II. Our study is the first to investigate the relationship between rock damage and electrical conductivity, providing insights to guide the design of drilling tools based on HVSD.

Keywords: high voltage spark discharge; electrohydraulic effect; electrical conductivity; drilling; rock damage; pressure waves

---

1. Introduction

Improving the rate of penetration (ROP) is the focus of research in petroleum industries, because ROP is inversely proportional to drilling cost in unconventional reservoirs, which is typically 30% to 40% of the total well costs. Drilling at great depth is challenging, due to the hostile environment and the enhanced mechanical properties of the rock, making conventional rotary drilling methods inadequate. Over the past few decades, researchers and engineers have proposed some unconventional drilling techniques based on a different rock damage mechanism rather than using the drill bit's mechanical force to cut the rock. Laser drilling applies a continuous high-power laser beam to remove the rock; researchers in the Gas Technology Institute have determined its technical feasibility and investigated the effects of specific laser energy on various rock types [1–3]. Electrical plasma drilling uses high thermal loads at thousands of degrees Celsius to spall, melt, and vaporize rock, where the thermal conductivity of the rock is a critical factor in breaking the rock [4,5]. However, lasers and plasma can generate high temperatures in downhole, disabling sensors near the drill bit that measure drilling parameters and formation characteristics, which are essential for directional drilling and risk analysis [6,7].

High voltage spark discharge (HVSD) can generate strong pulse pressure waves in water, known as the electrohydraulic effect (EHD) [8]. It has been adopted in a wide range of applications, such as underwater sound source, extracorporeal shock wave lithotripsy, well cleaning [9–11], and alternative hydraulic fracturing [12–16]. We have recently proposed a new drilling technology that couples the pressure waves of HVSD with the drill bit's mechanical force, which can potentially crush hard rock and increase the ROP without generating high temperatures like other new drilling techniques [17]. We conducted a series of laboratory experiments using pressure waves to destroy shale, sandstone, and concrete, and investigated the effects of discharge voltage, discharge energy, and the number of discharges on rock damage [18].

We have conceptually designed a new drilling system inspired by conventional rotary drilling (Figure 1). In this system, the drill string is connected to a custom drill bit with several HVSD reactors integrated into the bit's nozzle, and the power required to generate the electrohydraulic effect comes from either surface power supply or a downhole electric generator [19,20]. During drilling operations, the HVSD reactors create pressure waves amplified by ellipsoidal reflectors that act upon the rock. The rock is crushed to a depth of only a few millimeters, but this dramatically reduces the mechanical properties of the surface material, making it easier to be crushed by the drill bit, and increasing the ROP in hard rock and abrasive formations.

Numerous previous experiments employing HVSD to crush rocks used tap water with very low conductivity as the reaction medium, ignoring the effect of conductivity on pressure waves. However, in drilling practice, drilling fluids are complex compositional mixtures with a wide range of conductivities. Conductivity can influence the characteristics of high-voltage spark discharges, such as breakdown voltage, energy loss, and breakdown delay time, but previous studies have been inconsistent on the effect of conductivity on pressure wave intensity [21,22]. Moreover, few studies have been done on the effect of conductivity on HVSD fragmentation and the potential mechanism of rock breaking.

This research aims to investigate the effect of electrical conductivity on the discharge characteristics of high voltage spark discharge, the magnitude of pressure waves and rock damage, and to analyze the fragmentation mechanism from stress wave propagation. Then, we performed breaking experiments on cement mortar, an analogue for natural rock, with a single pulse energy of 1444 J in water with various conductivities. Utilizing mortar rather than natural rock is because the controllability and consistency of the mortar properties facilitate quantitative analysis of the relationship between conductivity and sample breaking. We reconstructed the model of the mortar sample before and after damage using X-ray computed tomography to analyze the damage mechanism, measured the surface damage by the size distribution and cumulative area of the pores, applied through transmission technique to detect microcracks inside the sample, and quantified the internal damage by the acoustic amplitude attenuation coefficient.

The results indicate that an increase in conductivity led to a nonlinear decrease in breakdown voltage, breakdown delay time, and an increase in energy loss. Both surface and internal damage of the samples exhibited a typical two-stage pattern, with pores and two types of tensile cracks being the predominant forms of damage. Stress waves induced by pressure waves propagating inside the sample play a crucial role, and tensile waves formed by reflection at the edge of the sample are among the leading causes of sample damage. The study offers some critical insights into the effect of electrical conductivity on sample damage, which will provide significant guidance for drilling tools design, and facilitate the implementation of the drilling technology based on HVSD.

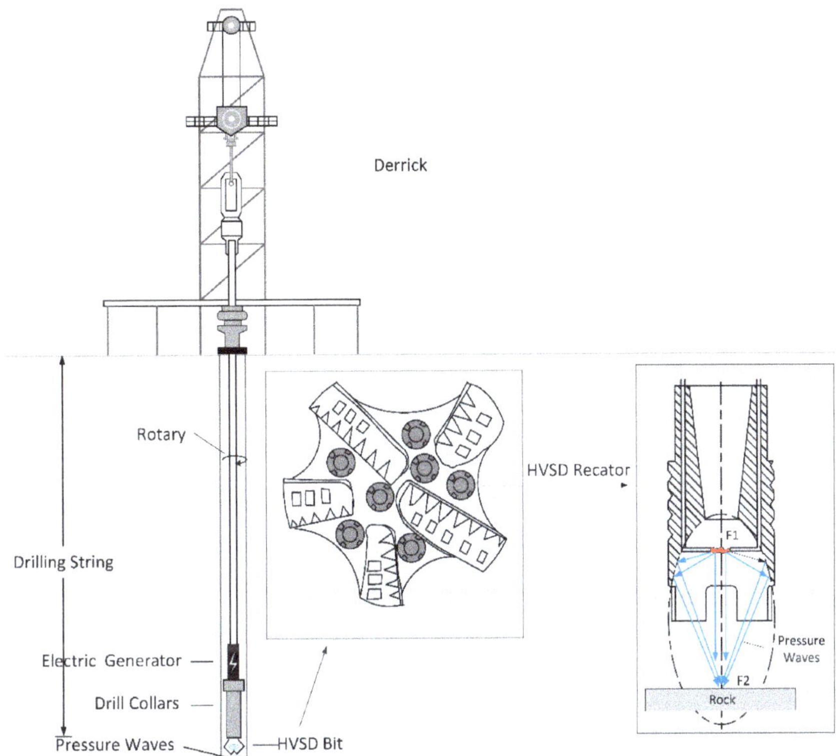

Figure 1. Conceptual design of drilling system based on the high voltage spark discharge.

2. Experiment

2.1. Sample Preparation

The absolute mass ratio of water, ordinary 42.5R Portland cement, and sand for pouring mortar was 0.5:1:1.8. The mechanical properties of the samples after 28 days of curing in water are shown in Table 1. Cubic samples of 100 mm in length, 100 mm in width, and 50 mm in height were prepared, from which some cylindrical samples of 25 mm in diameter and 30 mm in thickness were cut out (Figure 2). Mortar, like natural rocks, is a brittle material, and is often used in rock-breaking experiments, so the rock-breaking mechanism obtained from mortar is also applicable to natural rocks [13,23]. Since the mechanical properties of mortar samples from the same batch are almost identical, whereas the properties of natural rocks vary greatly, we used mortar to quantitatively study the effect of conductivity on rock damage.

Table 1. Mechanical properties of the mortar sample.

Sample	Density (g/cm^3)	Sound Velocity (m/s)	Acoustic Impedance (g/m^2s)	Compressive Strength (MPa)	Tensile Strength (MPa)	Elastic Modulus (MPa)
Mortar	2.44	4000	9.76×10^6	42	3.5	14,950

Figure 2. Two types of concrete sample.

2.2. Experimental Procedure

Experiments were conducted with high voltage discharge equipment, which consisted of a high voltage pulse power supply, a pressure wave generator, and a monitoring system. Figure 3a–c shows the diagram of the experimental equipment, the pulsed power supply components, and the pressure wave generator, respectively.

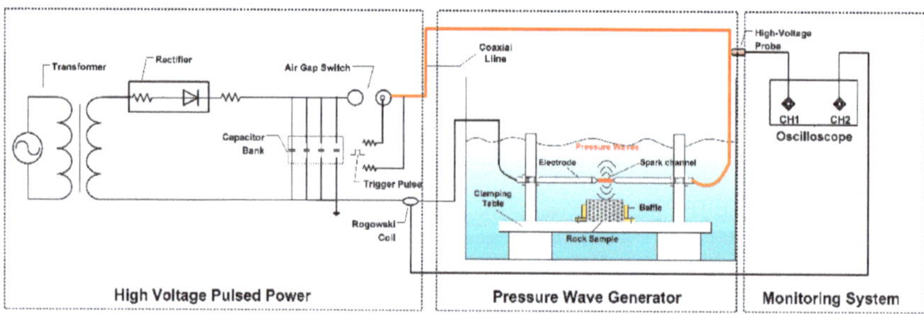

(**a**) Diagram of the experimental system

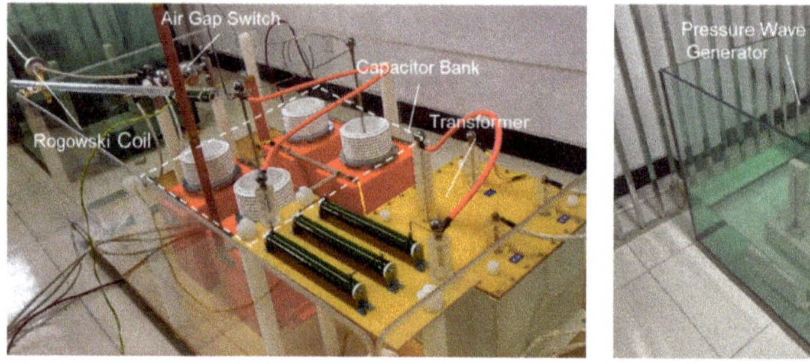

(**b**) High voltage pulsed power supply   (**c**) Pressure wave generator

Figure 3. High voltage spark discharge equipment.

The power supply mainly comprised a transformer, a rectifier, a high voltage capacitor bank, and an air-gap switch. When the power supply worked, the transformer raised the 220 V alternating current (AC) to a maximum of 70 kV, and then the rectifier converted the AC into the direct current (DC) to fully charge the capacitor bank. When the air-gap switch was closed, the energy stored on the capacitor was released into the pressure wave generator through high-voltage coaxial cables, creating strong pressure waves from the electrohydraulic effect.

The capacitor bank contained four nominally 0.5 μF capacitors in parallel, providing operating pulse energy from 50 J/pulse up to 4.9 kJ/pulse. The pressure wave generator consisted of a cubic tank with 80 cm sides filled with water of varying conductivity, a pair of electrodes, and a sample clamping table. The electrodes were made of stainless steel with a tapered tip, 17 cm long and 1.5 cm in diameter. The sample was placed on the clamping table and confined by a baffle to prevent it from being moved by pressure waves. The gap between electrodes and the distance between the sample surface and the electrodes were then adjusted to 5 mm. Water with different conductivities was filled in the pressure wave generator until the electrodes were completely submerged.

Next, the pulse power was turned on to increase the voltage across the capacitor bank to a preset value. Once the capacitor bank was fully charged, the air-gap switch was closed to provide electrical energy to the electrodes.

A strong electric field was built up between the electrodes to break down the water and form a spark channel, bridging the electrodes under the Joule heating effect (Figure 3a). The temperature and pressure in the plasma-filled channel rise dramatically in a short time (about tens of microseconds), pushing the water around the channel and radiating pressure waves. The magnitude of the pressure waves can reach several hundred MPa, and can destroy rocks with a wide range of mechanical properties.

The voltage and current waveforms were measured by a high-voltage probe (Pintech P6039A) and a Rogowski coil, respectively, and recorded with an oscilloscope (Tektronix TPS2024B, the bandwidth of 200 MHz, and sample rates of 2 GS/s).

A charging voltage of 38 kV, a capacitance of 2 μF, single 158 pulse energy of 1444 J, and a discharge number of five were set. Five solutions with different conductivities of 0.5 mS/cm (tap water), 5 mS/cm, 10 mS/cm, 15,136 mS/cm, and 20 mS/cm were prepared by mixing tap water and NaCl (99.5% purity). Three mortar samples were used for each type of water with different conductivity.

2.3. Analysis of Damage Characteristics

The X-ray computed tomography (CT, nanoVoxel3502E, and imaging resolution of ≥0.5 μm) was applied to observe the microscopic features of cylindrical sample before and after damage, to understand the mechanism of rock damage caused by pressure waves. Our study used the number of pores, and the distribution of pore sizes on the sample surface, to quantify the effect of conductivity on the surface damage. The pores were approximated as two-dimensional circles. Those with a diameter of less than 1 mm were neglected because they were mostly air bubbles caused during the sample casting process.

Apart from surface damage, it is also necessary to study the internal damage of the sample. Microcracks, the main form of the internal damage, occurring inside the sample due to the impacts of pressure waves and the interaction of the mortar particles with the stress waves, will hinder the propagation of acoustic waves, thereby attenuating the acoustic wave amplitude.

The acoustic wave amplitude was measured by an ultrasonic flaw detector (OLYMPUS 5077PR) with the through-transmission ultrasonic inspection technique (Figure 4). A transmitter (transmitting waves) and a receiver (receiving waves) were attached to opposite sides of the sample that were not directly damaged by the pressure waves. The two transducers and the sample were coupled by the ultrasonic coupling agent to ensure the transmission efficiency of acoustic energy. The attenuation of the wave amplitude, A_c, can be written in the form

$$A_c = \frac{(A_0 - A_1)}{A_0} \times 100\% \tag{1}$$

where A_0 is the ultrasonic wave amplitude measured before the sample was impacted by the pressure wave and A_1 is the wave amplitude after the impact of pressure wave.

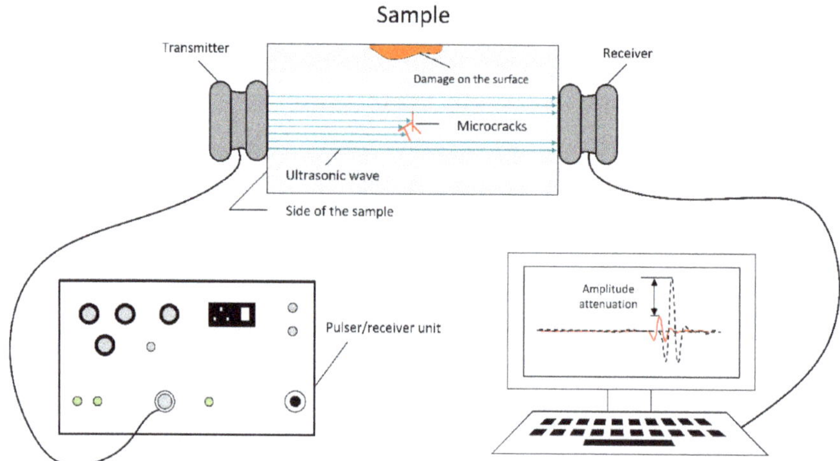

Figure 4. Schematic of ultrasonic transmission test.

3. Results and Discussion

3.1. Electrical Characteristics of Underwater Discharge at Different Conductivity

The voltage waveforms on the pair of electrodes were measured in water with different conductivity, and typical voltage waveforms are shown in Figure 5. The discharge voltage refers to the voltage on the positive electrode. The breakdown voltage is the voltage across the electrodes when the plasma channel is formed, and the time to breakdown is called breakdown delay time.

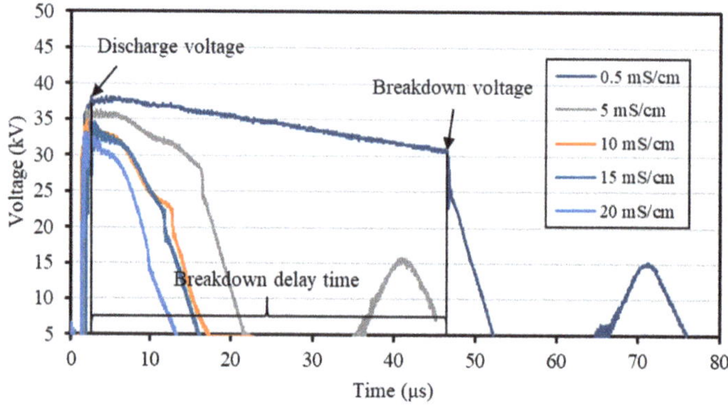

Figure 5. Voltage waveforms under the electrical conductivity from 0.5 mS/cm to 20 mS/cm.

Equation (2) expresses the efficiency of electrical energy in terms of the total energy stored in the capacitors, E_t (J), and the energy injected into the plasma channel, E_{pl} (J):

$$\eta = \frac{E_{pl}}{E_t} \times 100\% \tag{2}$$

where $E_t = \frac{1}{2} C V^2_d$, $E_{pl} = \frac{1}{2} C V^2_b$ (V_d is the discharge voltage, V_b is the breakdown voltage, and C is the capacitance of the capacitor bank).

Figure 6 shows the experimental data on the breakdown voltage under different electrical conductivity. Each value is an average of 15 replicates with a 95% confidence interval. As the electrical conductivity ranged from 0.5 mS/cm to 20 mS/cm, the breakdown voltage dropped from 31.1 kV to 14.8 kV. A similar result was also reported by Wang et al., where the breakdown voltage in 36 mS/cm KCL solution was lower than that in deionized water [24]. The mechanism of high voltage pulsed electrical breakdown in water indicates that the bubbles generated at the tip of the electrode are responsible for the breakdown of the water and the formation of the plasma channel [25–27]. In high-conductivity water, the electric field near the electrode tip produces a more intense field emission current, which will cause more water to evaporate and form more bubbles under the Joule heating effect, thus making electrical breakdown more likely to occur, which is manifested as a lower breakdown voltage on the voltage waveform.

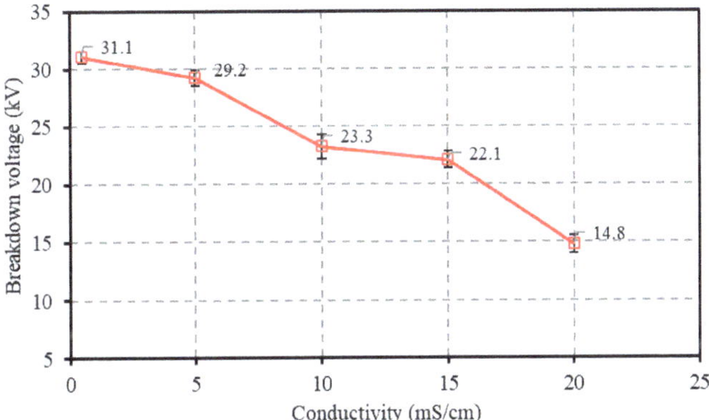

Figure 6. Breakdown voltage in water with different conductivity.

Figure 7 shows the relationship between breakdown delay time, t_b, and electrical conductivity. The most interesting aspect in this figure is that the t_b at 0.5 mS/cm is significantly larger than that of the other conductivities, while the value of t_b slightly decreases from 13 µs to 10.2 µs when the conductivity is increased from 5 mS/cm to 20 mS/cm. This result may be related to the Joule energy loss, E_{loss} (J), before water breakdown occurs, calculated by Equation (3)

$$E_{loss} = E_t - E_{pl} = \frac{1}{2}C\left(V_d^2 - V_b^2\right) \tag{3}$$

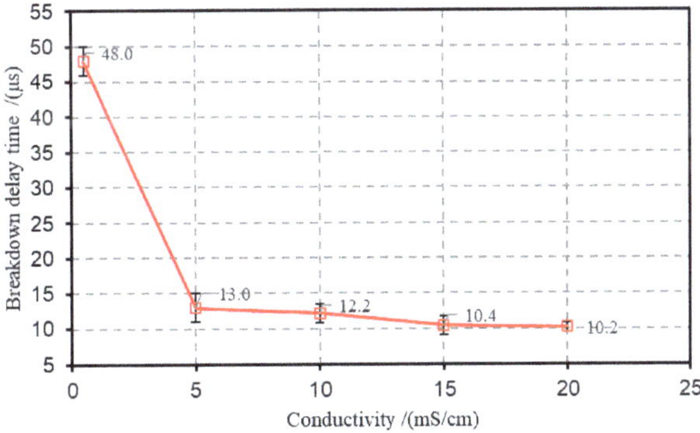

Figure 7. Breakdown delay time under different electrical conductivity.

Figure 8 shows that the electrical energy leaking into the water increases with increasing electrical conductivity. The major role of E_{loss} is to generate Joule heat in the water and form bubbles that contribute to the formation of electrical breakdown. Therefore, increasing the E_{loss} helps to reduce the breakdown delay time. When the conductivity is increased from 0.5 mS/cm to 5 mS/cm, E_{loss} increases by 71%, which explains the significant decrease in the breakdown delay time from 48 μs to 13 μs in this conductivity range (Figure 7).

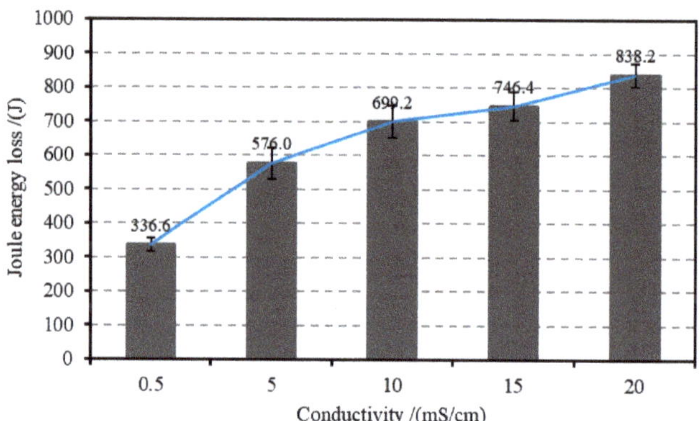

Figure 8. Joule energy loss under different electrical conductivity.

3.2. Damage of Samples under Different Electrical Conductivity

3.2.1. CT Results and Discussion

We applied the X-ray computed tomography (CT) to study the microscopic characteristics of a cylindrical sample with a diameter of 2.5 cm and a height of 3 cm. Figure 9 presents a sample model reconstructed by 3D visualization software before and after it was subjected to the pressure waves generated by underwater discharge. In Figure 9a, the pores (less than 1 mm in diameter) were air bubbles caused during the sample preparation process. As can be seen from Figure 9b, the sample was severely damaged after being hit by the pressure waves five times to a crushing depth of 6 mm, which significantly deteriorated the mechanical properties of the sample and made it easier to be crushed by the drill bit. Besides, some pits and cracks extended to the edge, causing the boundary of the sample to fall off.

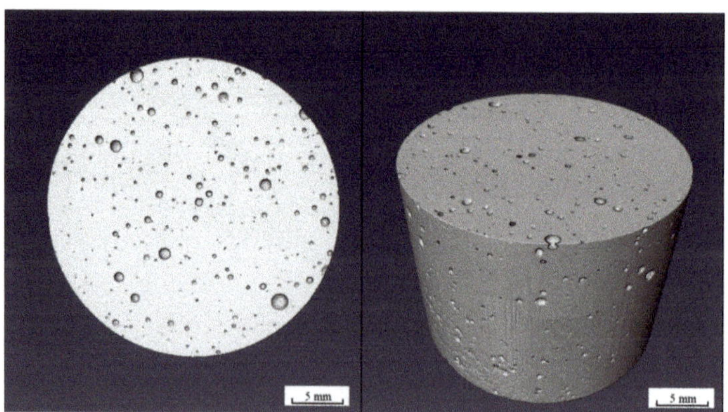

(**a**) Sample morphology before the pressure waves

Figure 9. *Cont.*

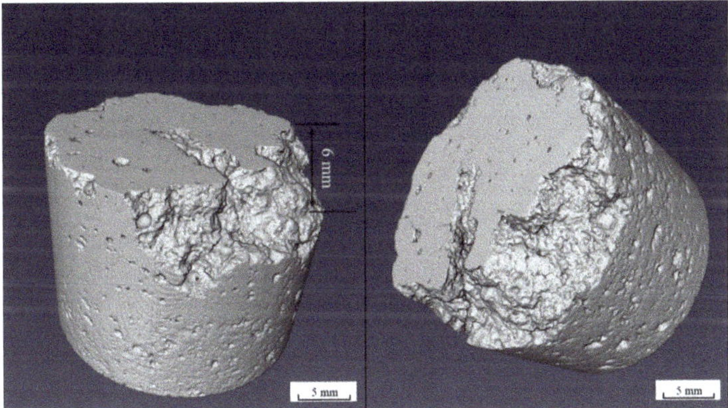

(**b**) Sample morphology after the pressure waves

Figure 9. Three-dimensional reconstructed sample.

Figure 10 illustrates the sample surface's tomographic images and the images of 6 mm depth from the surface, respectively. It reveals that cracking and erosion are the main forms of damage caused by pressure waves. These results can be explained by the interaction between pressure waves and the mortar specimen (Figure 11a). A stress wave can be decomposed into a compressive stress wave and a shear stress wave. Rock samples often have pre-existing flaws, such as vertical cracks and horizontal cracks. When the sample is subjected to compressive stresses perpendicular to the sample surface, the horizontal cracks in the sample will close, and the vertical cracks will open and extend. According to the Inglis theory [28], stresses are concentrated and amplified at the tip of a vertical crack, and when the compressive stresses are higher than the strength at the crack tip, the crack grows parallel to the direction of compressive stresses. This type of crack refers to cracks-I in Figure 11a, common in uniaxial compression testing of brittle materials, and is also referred to as longitudinal splitting [29–31].

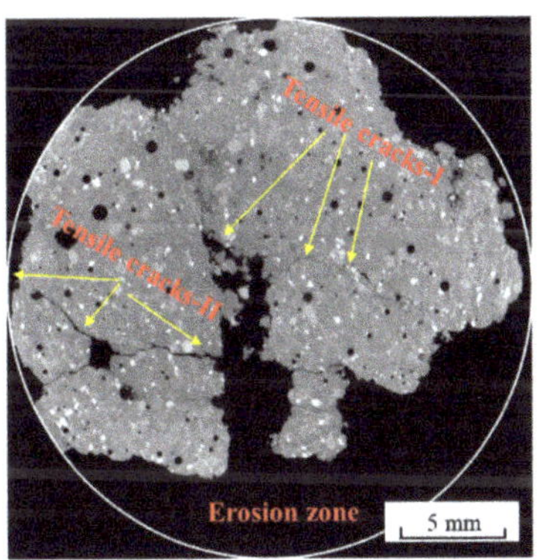

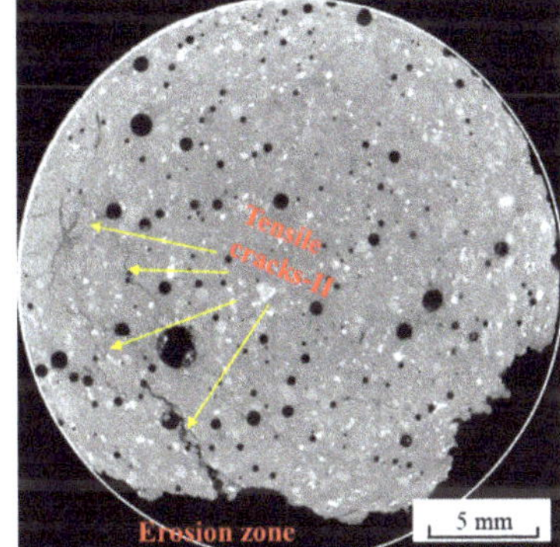

(**a**) Tomogram at the sample surface   (**b**) Tomogram at 6 mm from the surface

Figure 10. Tomographic images of the sample.

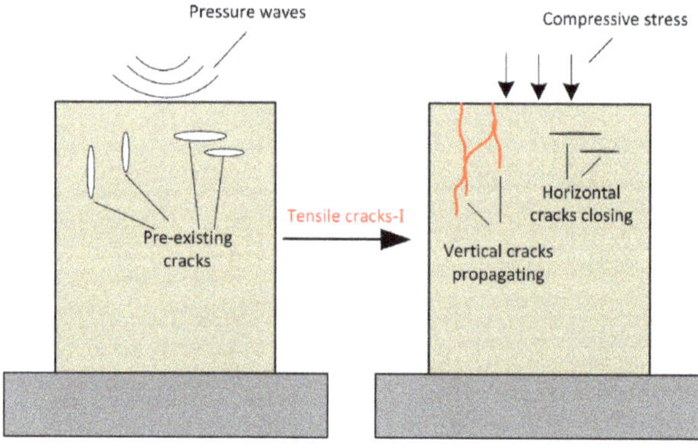

(**a**) Development of tensile cracks-I

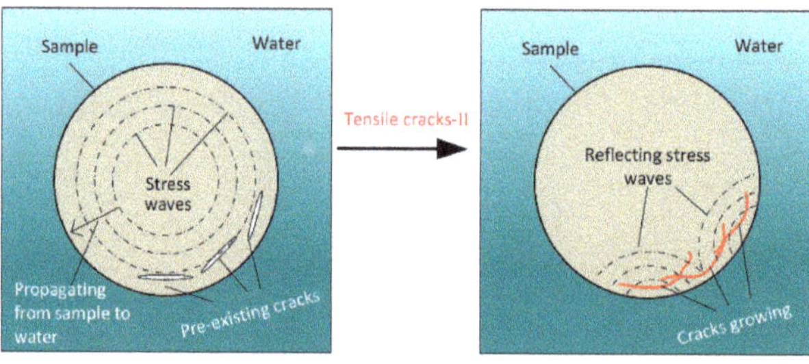

(**b**) Development of tensile cracks-II

Figure 11. Schematic for the formation mechanism of tensile cracks-I and -II.

Tensile cracks-II, located at the edge of the sample, has a different formation mechanism than cracks-I. A stress wave propagates from the interior of the sample to the sample–water interface, where it is reflected and converted into a tensile wave, since the acoustic impedance of the sample is much higher than that of the water. This tensile stress causes tensile cracks if it exceeds the material's tensile strength, known as spalling damage, and is typical of blast-induced damage and shock wave lithotripsy [32–34]. The cracks propagate further and interconnect with other cracks to form a continuous fault plane. When the stress exceeds the frictional resistance on the plane, frictional sliding and erosion will occur.

There were tensile cracks-I, tensile cracks-II, and extensive erosion of the sample surface (Figure 10a). As the depth increased, the damage degree decreased, and the damage was dominated by tensile cracks-II and erosion (Figure 10b).

3.2.2. Surface Damage Results and Discussion

Figure 12a shows the surface features of a raw sample, and Figure 12b–f show the typical surface macroscopic damage characteristics produced by pressure waves at different conductivities. The raw sample surface is smooth, while there are many cracks and pores on the surface of the samples crushed by the pressure waves. As the conductivity increased from 0.5 mS/cm to 20 mS/cm, the number of pores and the density of cracks gradually decreased.

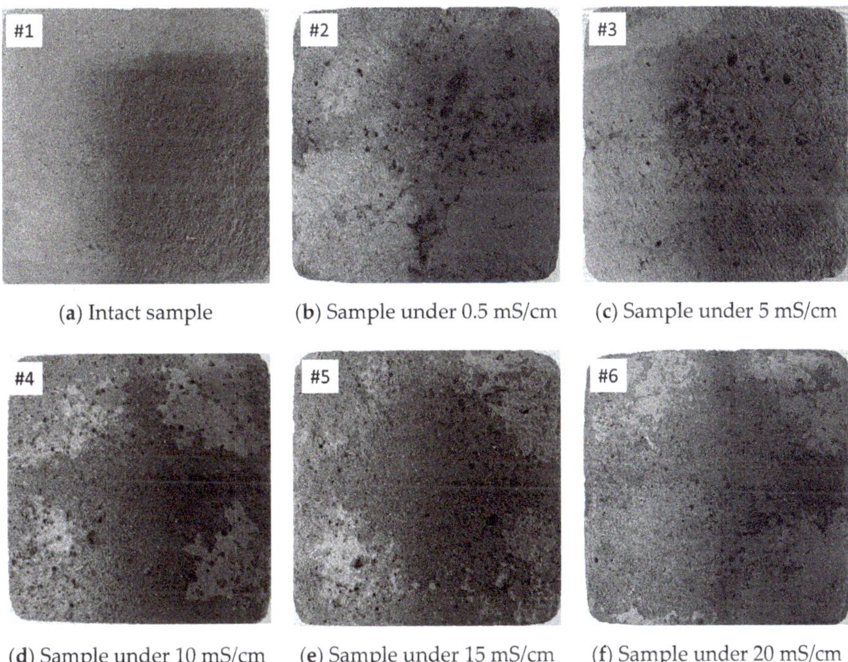

(**a**) Intact sample   (**b**) Sample under 0.5 mS/cm   (**c**) Sample under 5 mS/cm

(**d**) Sample under 10 mS/cm   (**e**) Sample under 15 mS/cm   (**f**) Sample under 20 mS/cm

Figure 12. Typical surface macro-damage features of samples under different electrical conductivity.

We used the cumulative area of the pores, S, to quantitatively assess the effect of electrical conductivity on sample damage, with a higher S value indicating more significant damage. As shown in Figure 13, the value of S decreases significantly in water with high conductivity. For example, S at 20 mS/cm conductivity is only one-fifth of the value at 5 mS/cm. This result is related to the peak pressure of the pressure waves. The increase in conductivity leads to more energy leaking in the water, and less energy to form pressure waves, decreasing the peak pressure, P, and the S value (Figure 14). The P (MPa) is given in Equation (4) [35]:

$$P = 900 \frac{\left(\frac{E_{pl}}{1000} \right)^{\alpha}}{d} \tag{4}$$

where E_{pl} (J) is the electrical energy deposited into plasma channel, d (mm) is the distance between the channel and sample surface, and α denotes a coefficient, which is 0.35 in the experiment.

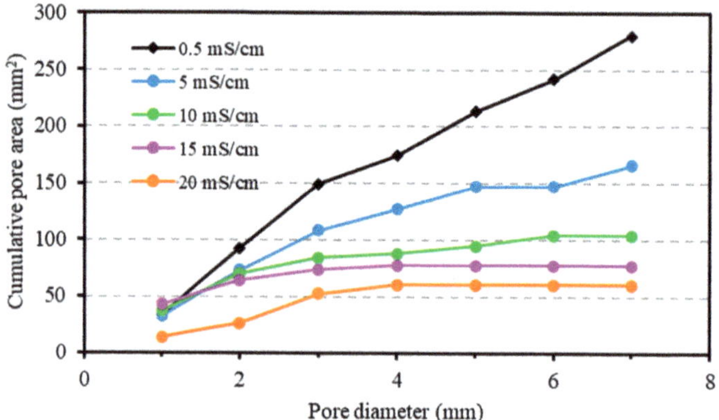

Figure 13. Cumulative pore area of samples broken by pressure waves.

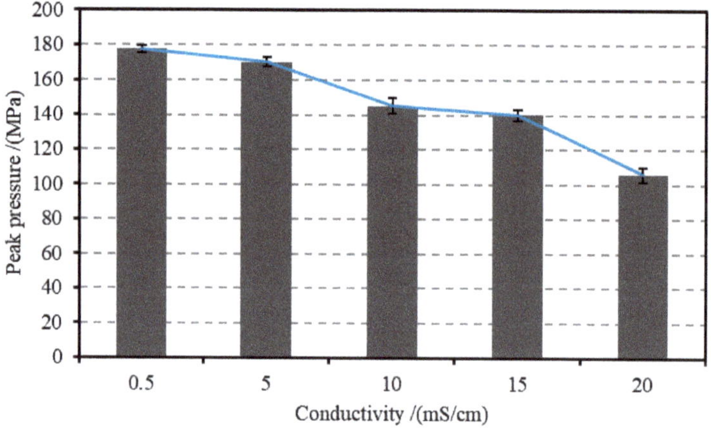

Figure 14. Peak pressure generated by electrical discharge under different conductivity.

Figure 15 illustrates the relationship between the average size distribution of pores on the sample surface and electrical conductivity. The pore size was dominated by 1 to 3 mm, and no pores larger than 6 mm in diameter were formed when the conductivity exceeded 10 mS/cm. Intuitively, the number of pores should be inversely proportional to the conductivity, but the number of pores with a 1 mm diameter increased first, and then decreased. One possible explanation is that in water with conductivities of 0.5 mS/cm and 5 mS/cm, many 1 mm pores are initiated early, but soon develop into larger dimensions because of the higher amplitude of the pressure waves (Figure 14). Therefore, in low conductivity, the number of 1 mm pores is fewer than that of relatively higher conductivity. However, when the conductivity is 20 mS/cm, the peak of pressure waves is considerably reduced by about 40%, resulting in a significant reduction in the number of pores of various sizes.

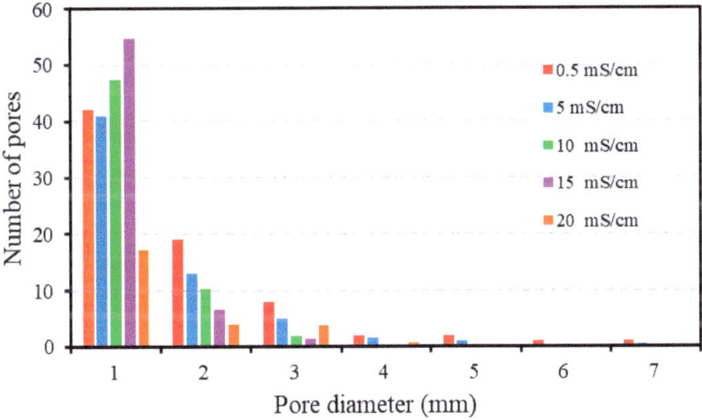

Figure 15. Number of pores on the sample surface under different electrical conductivity.

3.3. Internal Damage Results and Discussion

The through-transmission ultrasonic inspection technique allows the measurement of damage inside the sample, which is revealed in the reduced amplitude of received ultrasound waves, due to internal cracks and flaws that block the propagation of ultrasound waves.

Figure 16 shows the amplitude attenuation rate of samples, A_c, at different electrical conductivity. Each value is an average of three samples with 95% confidence interval precision. Interestingly, we found a two-stage pattern of internal damage. The A_c experienced a significant reduction from 61.1% at 0.5 mS/cm to 37.1% at 10 mS/cm, suggesting a rapid decrease in internal damage to the samples in this conductivity range. In contrast, there is little variation in A_c between 10 mS/cm and 20 mS/cm, indicating the number of flaws and microcracks within the samples are reducing, but at a slower rate. This two-stage pattern is also reflected in the sample surface damage (Figure 13), where the difference in surface damage between 0.5 mS/cm and 10 mS/cm is much more significant than the difference between 10 mS/cm and 20 mS/cm. The positive correlation between the internal and surface damages may suggest that pores and cracks on the surface grow downward under stress, creating new microcracks inside the sample, and deteriorating the mechanical properties of the sample. It seems that only when the stresses generated by the pressure wave are sufficiently higher than the compressive and tensile strengths, severe damage occurs.

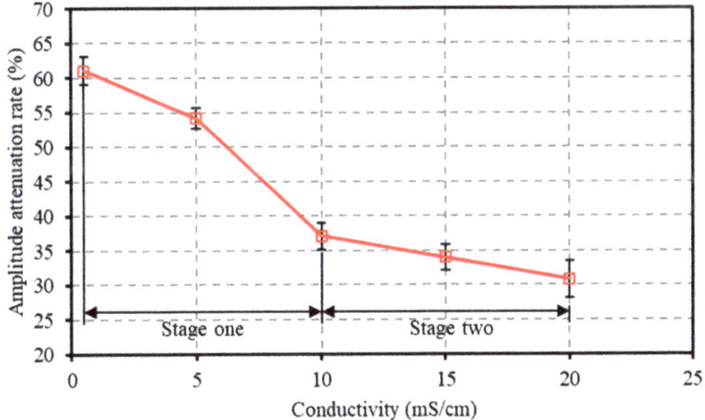

Figure 16. Amplitude attenuation rate under different conductivity.

4. Conclusions

This paper aims to investigate the characteristics of pulse discharge and the rock damage at different electrical conductivity, as well as the damage mechanism of pressure waves generated by HVSD.

Our study reveals that an increase in conductivity decreased the breakdown voltage and increased energy loss, thus decreasing the energy efficiency and the pressure wave's magnitude. However, the increase in conductivity reduces the difficulty and shortens the time required to form electrical breakdown, and accelerates the generation of pressure waves. As the conductivity increased from 0.5 mS/cm to 5 mS/cm, the breakdown delay time decreased rapidly, and remained almost constant as the conductivity increased. The sample damage exhibited a two-stage pattern in the range of 0.5 mS/cm to 20 mS/cm. Therefore, in practice, to balance energy efficiency, electrical breakdown time, and sample damage, the conductivity of water is preferably 5 mS/cm. The damage is mainly caused by pores and two types of tensile cracks, which are created by induced compressive stresses and tensile stresses induced by stress wave reflection at the mortar–water interface, respectively.

A limitation of this study is that we used the empirical formula from the literature [32] to calculate the peak pressure of the pressure wave, rather than measuring it experimentally. However, considering that our experimental system is the same as in the literature, the error in the peak pressure should be quite small. Besides, we neglected the effect of confining pressure and temperature on the damage process. In the future, we will establish an experimental system that can simulate the real stress state in the underground environment, and conduct HVSD rock-breaking tests at different temperatures and confining pressures to lay the foundation for the practical application of HVSD drilling technology in the oilfield.

Author Contributions: All the authors conceived and designed the study. Experiments: Z.C. and K.L.; data process: Z.C. and H.Z.; writing—original draft: Z.C. and H.Z.; writing—review and editing: Y.C., Q.Y., and K.L. All authors have read and agreed to the published version of the manuscript.

Funding: This research was funded by the National Natural Science Foundation of China (Grant No. 51774304, Grant No.51734010, Grant No. U1762211, Grant No. 51574262, Grant No. 51774063, Grant No. U19B6003), National Oil and Gas Major Project (Grant No. 2017ZX05009), the Foundation for Innovative Research Groups of the National Natural Science Foundation of China (Grant No. 51821092), Strategic Cooperation Technology Projects of CNPC and CUPB (Grant No. ZLZX2020-01), Sinopec Joint Fund—Topic 5 (Grant No. U19B6003-5).

Acknowledgments: Thanks to Igor Timoshkin of the University of Strathclyde for his advice and help on the mechanism of underwater electrical discharge breakdown.

References

1. Xu, Z.; Reed, C.B.; Konercki, G.; Parker, R.A.; Gahan, B.C.; Batarseh, S.; Graves, R.M.; Figueroa, H.; Skinner, N. Specific energy for pulsed laser rock drilling. *J. Laser Appl.* **2003**, *15*, 25–30. [CrossRef]
2. Sinha, P.; Gour, A. Laser Drilling Research and Application: An Update. In Proceedings of the SPE/IADC Indian Drilling Technology Conference and Exhibition, Mumbai, India, 16–18 October 2006; Society of Petroleum Engineers: Richardson, TX, USA, 2006.
3. Salehi, I.A.; Gahan, B.C.; Batarseh, S. *Laser Drilling-Drilling with the Power of Light*; Institute of Gas Technology: Des Plaines, IL, USA, 2007.
4. Gajdos, M.; Kocis, I.; Kristofic, T.; Horvath, G.; Jankovic, S. Utilization of Electrical Plasma for Hard Rock Drilling and Casing Milling. In Proceedings of the SPE/IADC Drilling Conference and Exhibition, London, UK, 17–19 March 2015.
5. Gajdos, M.; Kristofic, T.; Jankovic, S.; Horvath, G.; Kocis, I. Use of Plasma-Based Tool for Plug and Abandonment. In Proceedings of the SPE Offshore Europe Conference and Exhibition, Aberdeen, UK, 8–11 September 2015; Society of Petroleum Engineers: Richardson, TX, USA, 2015.
6. Gravley, W. Review of downhole measurement-while-drilling systems. *J. Pet. Technol.* **1983**, *35*, 1439–1445. [CrossRef]

7. Noureldin, A.; Irvine-Halliday, D.; Mintchev, M.P. Measurement-while-drilling surveying of highly inclined and horizontal well sections utilizing single-axis gyro sensing system. *Meas. Sci. Technol.* **2004**, *15*, 2426. [CrossRef]

8. Yutkin, L.A. Electrohydraulic Effect. Available online: https://scholar.google.com.hk/scholar?hl=zh-CN&as_sdt=0%2C5&q=+8.%09Yutkin%2C+L.A.+Electrohydraulic+Effect%3B+Air+Force+Systems+Command+Wright-Patterson+Afb+oh+Foreign+Technology+Division%2C+1961.&btnG= (accessed on 14 October 2020).

9. Chung, K.; Lee, S.; Hwang, Y.S.; Kim, C.Y. Modeling of pulsed spark discharge in water and its application to well cleaning. *Curr. Appl. Phys.* **2015**, *15*, 977–986. [CrossRef]

10. Bodykov, D.U.; Abdikarimov, M.S.; Seitzhanova, M.A.; Nazhipkyzy, M.; Mansurov, Z.A.; Kabdoldina, A.O.; Ualiyev, Z.R. Processing of Oil Sludge with the Use of the Electrohydraulic Effect. *J. Eng. Phys. Thermophys.* **2017**, *90*, 1096–1101. [CrossRef]

11. Xiong, L.; Liu, Y.; Yuan, W.; Huang, S.; Li, H.; Lin, F.; Pan, Y.; Ren, Y. Cyclic shock damage characteristics of electrohydraulic discharge shockwaves. *J. Phys. D* **2020**, *53*, 185502. [CrossRef]

12. Chen, W.; Maurel, O.; Borderie, C.L.; Reess, T.; Ferron, A.D.; Matallah, M.; Pijaudier-Cabot, G.; Jacques, A.; Rey-Bethbeder, F. Experimental and numerical study of shock wave propagation in water generated by pulsed arc electrohydraulic discharges. *Heat Mass Transf.* **2014**, *50*, 673–684. [CrossRef]

13. Chen, W.; La Borderie, C.; Maurel, O.; Pijaudier Cabot, G.; Rey Bethbeder, F. Simulation of damage–permeability coupling for mortar under dynamic loads. *Int. J. Numer. Anal. Methods Geomech.* **2014**, *38*, 457–474. [CrossRef]

14. Chen, W.; Maurel, O.; Reess, T.; De Ferron, A.S.; La Borderie, C.; Pijaudier-Cabot, G.; Rey-Bethbeder, F.; Jacques, A. Experimental study on an alternative oil stimulation technique for tight gas reservoirs based on dynamic shock waves generated by pulsed arc electrohydraulic discharges. *J. Pet. Sci. Eng.* **2012**, *88*, 67–74. [CrossRef]

15. Li, C.; Duan, L.; Tan, S.; Chikhotkin, V. Influences on high-voltage electro pulse boring in granite. *Energies* **2018**, *11*, 2461. [CrossRef]

16. Li, C.; Duan, L.; Tan, S.; Chikhotkin, V.; Fu, W. Damage model and numerical experiment of high-voltage electro pulse boring in granite. *Energies* **2019**, *12*, 727. [CrossRef]

17. Cai, Z.; Zhang, H.; Li, J.; Zheng, J.; Yu, Q.; Liu, K.; Liu, Y. New Technology to Assist Drilling to Improve Drilling Rate in Unconventional Gas Resources: Pulsed Arc Plasma Shockwave Technology. In Proceedings of the Abu Dhabi International Petroleum Exhibition & Conference, Abu Dhabi, UAE, 12–15 November 2018; Society of Petroleum Engineers: Richardson, TX, USA, 2018.

18. Cai, Z.; Zhang, H.; Li, J.; Yang, M.; Yu, Q.; Zheng, J.; Liu, K. An Experimental Study of Using Plasma Shock Wave for Rock Damage. In Proceedings of the Offshore Technology Conference, Houston, TX, USA, 6–9 May 2019.

19. Hui, Z.; Zhixiang, C. Well Drilling System and Method. CN108533172B, 28 May 2019.

20. Hui, Z.; Cai, Z. Plasma Generating Device, Fracturing System and Fracturing Method. CN111101915A, 5 May 2020.

21. Cathignol, D.; Mestas, J.L.; Gomez, F.; Lenz, P. Influence of water conductivity on the efficiency and the reproducibility of electrohydraulic shock wave generation. *Ultrasound Med. Biol.* **1991**, *17*, 819–828. [CrossRef]

22. Zhu, L.; He, Z.; Gao, Z.; Tan, F.; Yue, X.; Chang, J. Research on the influence of conductivity to pulsed arc electrohydraulic discharge in water. *J. Electrostat.* **2014**, *72*, 53–58. [CrossRef]

23. Yaşar, E.; Ranjith, P.G.; Viete, D.R. An experimental investigation into the drilling and physico-mechanical properties of a rock-like brittle material. *J. Pet. Sci. Eng.* **2011**, *76*, 185–193. [CrossRef]

24. Wang, H.; Wandell, R.J.; Tachibana, K.; Voráč, J.; Locke, B.R. The influence of liquid conductivity on electrical breakdown and hydrogen peroxide production in a nanosecond pulsed plasma discharge generated in a water-film plasma reactor. *J. Phys. D* **2018**, *52*, 75201. [CrossRef]

25. Marinov, I.; Guaitella, O.; Rousseau, A.; Starikovskaia, S.M. Modes of underwater discharge propagation in a series of nanosecond successive pulses. *J. Phys. D* **2013**, *46*, 464013. [CrossRef]

26. Fujita, H.; Kanazawa, S.; Ohtani, K.; Komiya, A.; Kaneko, T.; Sato, T. Initiation process and propagation mechanism of positive streamer discharge in water. *J. Appl. Phys.* **2014**, *116*, 213301. [CrossRef]

27. Timoshkin, I.V.; Fouracre, R.A.; Given, M.J.; Macgregor, S.J. Hydrodynamic modeling of transient cavities in fluids generated by high voltage spark discharges. *J. Phys. D* **2006**, *39*, 4808. [CrossRef]
28. Inglis, C.E. Stresses in a Plate Due to the Presence of Cracks and Sharp Corners. *Trans. Inst. Naval Archit* **1913**, *55*, 219–241.
29. Fishman, Y.A. Features of compressive failure of brittle materials. *Int. J. Rock Mech. Min. Sci.* **2008**, *45*, 993–998. [CrossRef]
30. Han, L.; He, Y.; Zhang, H. Study of rock splitting failure based on griffith strength theory. *Int. J. Rock Mech. Min. Sci.* **2016**, *83*, 116–121. [CrossRef]
31. Wong, R.; Lin, P.; Tang, C.A. Experimental and numerical study on splitting failure of brittle solids containing single pore under uniaxial compression. *Mech. Mater* **2006**, *38*, 142–159. [CrossRef]
32. Wang, Z.; Li, Y.; Wang, J. Numerical analysis of blast-induced wave propagation and spalling damage in a rock plate. *Int. J. Rock Mech. Min. Sci.* **2008**, *45*, 600–608. [CrossRef]
33. Cleveland, R.O.; Sapozhnikov, O.A. Modeling elastic wave propagation in kidney stones with application to shock wave lithotripsy. *J. Acoust. Soc. Am.* **2005**, *118*, 2667–2676. [CrossRef] [PubMed]
34. Cao, S.; Zhang, Y.; Liao, D.; Zhong, P.; Wang, K.G. Shock-induced damage and dynamic fracture in cylindrical bodies submerged in liquid. *Int. J. Solids Struct.* **2019**, *169*, 55–71. [CrossRef]
35. Touya, G.; Reess, T.; Pécastaing, L.; Gibert, A.; Domens, P. Development of subsonic electrical discharges in water and measurements of the associated pressure waves. *J. Phys. D* **2006**, *39*, 5236. [CrossRef]

Publisher's Note: MDPI stays neutral with regard to jurisdictional claims in published maps and institutional affiliations.

Article

Pore-Scale Lattice Boltzmann Simulation of Gas Diffusion–Adsorption Kinetics Considering Adsorption-Induced Diffusivity Change

Zhigao Peng [1,2,]*, Shenggui Liu [3], Yingjun Li [4], Zongwei Deng [1] and Haoxiong Feng [1]

[1] College of Civil Engineering, Hunan City University, Yiyang 413000, China;
dengzongwei@hncu.edu.cn (Z.D.); fenghaoxiong@hncu.edu.cn (H.F.)

[2] Key Laboratory of Key Technologies of Digital Urban-Rural Spatial Planning of Hunan Province,
Hunan City University, Yiyang 413000, China

[3] School of Mechanics & Civil Engineering, China University of Mining and Technology (Beijing),
Beijing 100083, China; liushenggui@cumtb.edu.cn

[4] State Key Laboratory for Geomechanics and Deep Underground Engineering, China University of Mining
and Technology (Beijing), Beijing 100083, China; lyj@aphy.iphy.ac.cn

* Correspondence: pengzhgg@hncu.edu.cn; Tel.: +86-181-0136-7968

Received: 13 August 2020; Accepted: 17 September 2020; Published: 20 September 2020

Abstract: The diffusion–adsorption behavior of methane in coal is an important factor that both affecting the decay rate of gas production and the total gas production capacity. In this paper, we established a pore-scale Lattice Boltzmann (LB) model coupled with fluid flow, gas diffusion, and gas adsorption–desorption in the bi-dispersed porous media of coalbed methane. The Knudsen diffusion and dynamic adsorption–desorption of gas in clusters of coal particles were considered. Firstly, the model was verified by two classical cases. Then, three dimensionless numbers, Re, Pe, and Da, were adopted to discuss the impact of fluid velocity, gas diffusivity, and adsorption/desorption rate on the gas flow–diffusion–adsorption process. The effect of the gas adsorption layer in micropores on the diffusion–adsorption–desorption process was considered, and a Langmuir isotherm adsorption theory-based method was developed to obtain the dynamic diffusion coefficient, which can capture the intermediate process during adsorption/desorption reaches equilibrium. The pore-scale bi-disperse porous media of coal matrix was generated based on the RCP algorithm, and the characteristics of gas diffusion and adsorption in the coal matrix with different Pe, Da, and pore size distribution were discussed. The conclusions were as follows: (1) the influence of fluid velocity on the diffusion–adsorption process of coalbed methane at the pore-scale is very small and can be ignored; the magnitude of the gas diffusivity in macropores affects the spread range of the global gas diffusion and the process of adsorption and determines the position where adsorption takes place preferentially. (2) A larger Fickian diffusion coefficient or greater adsorption constant can effectively enhance the adsorption rate, and the trend of gas concentration- adsorption is closer to the Langmuir isotherm adsorption curve. (3) The gas diffusion–adsorption–desorption process is affected by the adsorption properties of coal: the greater the p_L or V_m, the slower the global gas diffusivity decay. (4) The effect of the gas molecular adsorption layer has a great impact on the kinetic process of gas diffusion–adsorption–desorption. Coal is usually tight and has low permeability, so it is difficult to ensure that the gas diffusion and adsorption are sufficient, the direct use of a static isotherm adsorption equation may be incorrect.

Keywords: coalbed methane; Lattice Boltzmann method; gas diffusion; adsorption–desorption; pore-scale

---

1. Introduction

In recent years, the adsorption–desorption behavior of gases in porous media has gained wide attention, and many theories such as single-layer adsorption, multilayer adsorption, capillary condensation, and pore-filling have been successfully applied to many areas [1,2]. The adsorption effect of gas–solid in nanopores of the unconventional natural gas (UNG) reservoirs is an important factor that both affecting the decay rate of gas production and the total gas production capacity [3,4]. However, for the UNG, including shale gas, coalbed methane, due to the involved complex pore geometries and multi-scale spatial distribution, which makes it difficult to reveal the underlying characteristics of the gas adsorption and storage in the reservoir. Therefore, further research and deeper understanding in this area can better cope with the challenges in the development and production of UNG.

The coal matrix provides abundant adsorptive sites for gas molecules, which contributes greatly to methane storage. Methane in the coal reservoir is mainly present in three states: (1) dissolved in water, (2) as the free gas in the fractures or pores, and (3) adsorbed on the inner surface of micro-fractures or nanopores [2]. The amount of gas in the adsorption state accounts for more than 85% of the total gas [5]. In a typical UNG production process, free gases in fractures or large pores first diffuse out, which was related to the rapid production cycle of a gas productivity curve [6]. However, due to the limit of the gas diffusion rate and the desorption kinetics of gas molecules on the micropore surface [7], the desorbed gas will dominate the subsequent gas production process. Accordingly, understanding the gas adsorption/desorption behavior in the matrix cannot only obtain the main controlling factors on gas adsorption/desorption and provide a theoretical basis for developing new methods to enhance gas desorption but also can supply a reference for the optimization of production process design and evaluation of the total gas content in reservoirs.

The migration and storage of methane in the coal matrix are closely related to pore size [8]. The International Federation of Applied Chemistry (IUPAC) classified the pore into three types: micropore (less than 2 nm), mesopore (2–50 nm), and macropore (more than 50 nm) [9]. Typically, Coal is a typical tight porous media, with a wide distribution of pore size, which is distributed in the range of 1–100 nm and contains a great number of micropores, some mesopores, and a small number of macropores [10]. Meng [5] analyzed the pore distribution of coal samples (from the Xishan coal mining area, China) by using the static nitrogen adsorption capacity method (SY/T6154-1995) and found that the average radius of pores in coal ranged from 7.729 to 15.338 nm, with the largest proportion of micropore reaching 65.4%. Since the wide distribution of pore size, it is difficult to characterize the pore structure of the coal matrix, which makes it complex to understand the behavior of gas flow in the reservoir. At present, the most common method for simplifying the coal matrix is to treat the matrix as a bi-dispersed porous media, which with the heterogeneous geometry and the homogeneous material, and regardless of the specific structure and distribution characteristics of micropores inside it, only considering its statistical characteristics, namely the porosity and the average pore diameter [11–16]. Zhao [17] tested six groups of coal samples with middle and high rank by applying synchrotron small-angle X-ray scattering (SAXS) and found the average pore sizes were 2.9, 5.1, 6.4, 6.9, 14.3, and 23.1 nm, respectively, and obtained the logarithmic normal distribution of pore sizes based on Beaucage [18].

Gas adsorption behavior, which involved some complicated adsorption mechanism, is a key factor in the coalbed methane exploitation process. However, most of the previous studies [19,20] for investigating the gas adsorption behavior in coal were mainly based on the isothermal adsorption equation, which can give an important reference for reservoir evaluation and numerical modeling. Nevertheless, the isothermal data only provide static and macroscopic information, ignoring the dynamic processes of adsorption–desorption and the detailed information of the nanoporous surface. Do and Wang [21] observed the adsorption delay of activated carbon in the process of measuring the adsorption and desorption equilibrium data. They believed that the adsorption of the pore surface was non-uniform, which resulted in gas adsorption occurs not instantaneously and took some time to reach equilibrium. These above indicated that the time scale between the process of gas adsorption and gas diffusion was comparable [21,22]. Besides, gas flow in coal reservoirs usually involves multiple physical processes, including fluid dynamics, thermodynamics, chemical kinetics, and electrodynamics (because most natural media surfaces are charged), all of which are ultimately governed by interface phenomena at the pore-scale [23]. Since the length scale of the coal matrix is much larger than the typical pore or mineral particle sizes, the pore size inside the matrix differs greatly from the outside. Therefore, the continuity formula based on the spatial average is often employed to characterize the matrix, and the spatial heterogeneity of the smaller scale can be ignored.

The present work is based on the Lattice Boltzmann method (LBM). LBM originated from classical statistical physics and is a mesoscopic method based on simplified dynamic equations. In LBM, fluid flow is simulated by a series of virtual particles that propagate and collide on discrete lattice domains. Through rigorous mathematical analysis, mesoscopic continuity and momentum equations can be obtained from this propagation and collision dynamics. Particle properties and local dynamics provide advantages for complex boundaries and parallel computing. Besides, the dynamic nature of LBM makes new physical considerations in the LBM framework easy, which is especially useful for building multi-physical phenomena. Using the multi-scale expansion technique (Chapman–Enskog), the LB equation can be recovered to the Navier–Stokes equation under the incompressible limit, and the LB equation of mass transport can be recovered to the advection–diffusion equation [24,25]. Nowadays, the LB method has been successfully applied to a variety of fluid flow and gas transport phenomena, such as fluid flow, turbulence, multiphase multicomponent flow, particle suspension, heat transfer, and diffusion–reaction flow in porous media [26–31].

The adsorption characteristics of coalbed methane in the reservoir are affected by many factors, such as fluid flow, gas transport, gas adsorption/desorption, etc. In this paper, a multi-component gas flow–diffusion–adsorption coupled LB model was established to investigate the effects of fluid flow, gas transport, adsorption/desorption on adsorption, and adsorption-induced matrix diffusion coefficients and porosity.

2. Lattice Boltzmann Method

Coal is a low-permeability porous media with pore systems ranging from fractures (cleats) to nanopores. The present work focuses on the modeling of multi-scale systems with considering the gas viscous flow in macropores and Fick diffusion in the matrix, besides, the Knudsen diffusion and adsorption kinetics in nanopores inside the matrix is also included in the simulation. Due to the gas velocity under typical reservoir condition is very low and the Mach number is far less than 0.3, it acceptable to describe the methane flow in the coal matrix under the incompressible limit [24].

2.1. The LB Equation for Fluid Flow

The LB method for gas viscous flow at pore-scale can be expressed as follows [32]:

$$f_i(x + \mathbf{e}_i \delta_t, t + \delta_t) - f_i(x, t) = -\frac{1}{\tau}[f_i(x, t) - f_i^{eq}(x, t)] \tag{1}$$

$$f_i^{eq} = \rho \omega_i [1 + \frac{\mathbf{e}_i \cdot \mathbf{u}}{c_s^2} + \frac{\mathbf{u}\mathbf{u} : (\mathbf{e}_i \mathbf{e}_i - c_s^2 \mathbf{I})}{2c_s^4}] \tag{2}$$

where f_i is the discrete density distribution function, f_i^{eq} is the local equilibrium function, e_i is the discrete velocity of the particle, c_s is the lattice sound velocity, c is the sound velocity, τ is the relaxation time, ω_i is the weight coefficient, and δ_t is the time step.

2.2. The LB Equation for Gas Diffusion–Reaction

For the process of methane transport, it is acceptable to employ the passive scalar LB model due to the gas velocity in the reservoir is low enough and its effect on solution density and velocity can be negligible [25]. Therefore, gas transport in the reservoir with adsorption can be described as follows:

$$g_i(x + \mathbf{e}_i\delta_t, t + \delta_t) - g_i(x, t) = -\frac{1}{\tau_g}[g_i(x, t) - g_i^{eq}(x, t)] + \omega_i R_s \delta_t \tag{3}$$

$$g_i^{eq} = C_s \omega_i[1 + \frac{\mathbf{e}_i \cdot \mathbf{u}}{c_s^2}] \tag{4}$$

where g_i is a discrete concentration distribution function; g_i^{eq} is the corresponding local equilibrium distribution function; τ_g is the diffusion-related relaxation time; R_s is the source/sink term associated with gas adsorption/desorption.

For the standard LB equation, the macroscopic density and velocity can be defined as $\rho = \sum_i f_i$, $\rho\mathbf{u} = \sum_i f_i\mathbf{e}_i$, the concentration can be defined as $C_s = \sum_i g_i$. Using the Chapman–Enskog technique to expand the LB equation of viscous flow, and with $p = c_s^2\rho$, $\mu = c_s^2(\tau - 0.5)\delta_t$ the Equations (1) and (2) can recover the Navier–Stokes equation under the incompressible limit. Similarly, the LB equation of diffusion-reaction with $D_s = c_s^2(\tau_g - 0.5)\delta_t$, the Equations (3) and (4) can recover the advection–diffusion equation with a source/sink term.

The diffusion coefficient is a key parameter for the gas transport process in porous media. According to the previous study [33], diffusion coefficients can be divided into Knudsen diffusion coefficients (KDC), corrected diffusion coefficients (CDC), and Fickian diffusion coefficients (FDC). The KDC can be employed to describe the phenomenon of random walk and the CDC corresponding to the chemical potential-driven diffusion, and the FDC is usually adopted to concentration-driven diffusion. The KDC is more closely related to the microscopic features of molecules, while FDC is more directly related to the macroscopic transport of mass, which is more experimentally available.

At pore-scale, gas diffusion coefficients related to methane migration in the coal matrix includes FDC and KDC. The FDC can be obtained by experiment, and based on the parallel capillary model of porous media, the KDC can be written as [34]:

$$D_K = \frac{2r}{3}\sqrt{\frac{8RT}{\pi M}} \tag{5}$$

where R is the gas constant, T is the temperature, M is the molar mass of the gas, and r is the radius of the capillary.

When adsorption is considered, since gas molecules are adsorbed on the inner surface of the nanopore, the cross-sectional area of the free gas transfer is reduced, and the porosity associated with the pore pressure in the coal matrix is decreased. Xiong et al. [35] proposed an effective capillary radius of the adsorption layer for the single capillary of porous media:

$$r_{eff} = r - d_m\frac{p}{p + p_L} \tag{6}$$

where d_m is the diameter of the methane molecule, e.g., $d_m = 0.38$ nm [35]; p is the pressure, Pa; p_L is the Langmuir pressure, Pa; and $p/(p + p_L)$ is the adsorption saturation of the pore surface.

The model is based on the Langmuir isotherm adsorption equation, which corrects the influence of gas molecular radius and adsorption saturation on the effective radius of the pore but ignores the time effect of concentration/adsorption in the adsorption process. In the present work, the adsorption saturation is introduced by the ratio of the instantaneous adsorption amount to the saturated adsorption amount, and the effective pore radius is then considered:

$$r_{eff} = r - d_m \frac{V(x,t)}{V_m} \tag{7}$$

where V_m and V are the saturated adsorption amount and the adsorption amount at a current time, respectively.

The modified effective porosity of the capillary can be written as:

$$\varepsilon_{eff} = \varepsilon_0 \frac{r_{eff}^2}{r^2} \tag{8}$$

Due to the complex geometric composition of coal, the effects of porosity and tortuosity of the porous media also should be considered. Therefore, effective KDC can be given by:

$$D_{eff} = \frac{\varepsilon}{\tau_w} D_K = \frac{\varepsilon_0}{\tau_w} \frac{r_{eff}^2}{r^2} \frac{2r}{3} \frac{r_{eff}}{r} \sqrt{\frac{8RT}{\pi M}} = \frac{\varepsilon_0}{\tau_w} \frac{r_{eff}^3}{r^3} D_K \tag{9}$$

where ε_0 is the initial porosity of the coal matrix and τ_w is the tortuosity.

2.3. Langmuir Adsorption Kinetic Equation

The adsorption mechanism of gas in coal is very complicated, and various adsorption forms coexist [2], such as monolayer adsorption, multi-layer adsorption, capillary condensation, and capillary filling, etc. At present, many adsorption models have been established and applied to the methane–coal adsorption, among them, the Langmuir isothermal adsorption model is the most commonly used one [2,36,37]. In this paper, we employ the Langmuir adsorption kinetics equation to control the evolution of gas adsorption–desorption:

$$\frac{\partial V}{\partial t} = k_a C(V_m - V) - k_d V \tag{10}$$

where k_a and k_d are the adsorption and desorption rate constants, respectively.

The methane–coal adsorption of coal is typical physical adsorption and is a reversible process [21], which means that there are simultaneous gas adsorption and desorption in adsorption sites. At a specific time, if $\partial V/\partial t < 0$, the adsorption rate is lower than the desorption rate, the gas desorption from pores; if $\partial V/\partial t > 0$, the adsorption rate is greater than the desorption rate, the net gas adsorption amount increases; when $\partial V/\partial t = 0$, the right part of the Equation (10) can recover the classical Langmuir isotherm adsorption equation:

$$V = \frac{V_m p}{p + P_L} = \frac{V_m C}{C + C_L} \tag{11}$$

where p is the gas pressure, $p = CMc_s^2$, Pa; C_L is the concentration corresponding to the Langmuir pressure, $C_L = k_d/k_a$, mol/m^3.

It can be found that p_L is only related to the ratio of k_a and k_d and independent of their magnitude. The value of k_a, k_d, p_L can be determined based on the experiment, and anyone can be obtained from the other two of them.

To consider the gas–solid dynamic adsorption process, we developed a LB model to realize the adaptive conversion of gas in porous media between adsorption and desorption based on the model proposed by He [38]. In the model, each site in the porous media can be considered as a good adsorption position, the Langmuir rate equation can be incorporated into Equation (3) through a source/sink term:

$$R_s = \frac{\partial V}{\partial t} = k_a C (V_m - V) - k_d V \tag{12}$$

In the evolution of the LB equation, the amount of gas adsorption/desorption is updated with time, and the new amount can be obtained from a first-order difference scheme of Equation (10):

$$V_{t+1} - V = \delta_t [k_a C (V_m - V) - k_d V] \tag{13}$$

Recently, the bi-dispersed porous media has been drawn wide attention due to the similar pore size distribution and fractal characteristics with the geometric characteristics of coal. The bi-dispersed porous structure of the coal matrix, as shown in Figure 1, is composed of clusters, which are agglomerated by small particles. In the bi-dispersed porous media, there are numerous intraparticle pores within the clusters, which contain micro- and mesopores, and some interparticle pores between the clusters, mainly macropores. The previous study has been reported that the micro- and mesopores will play an important role in methane adsorption, and the macropores act primarily as transport channels [2]. The fluid flow and gas diffusion in macropores are governed by the double distribution LB equation; due to the pore size of the clusters being extremely small, the effect of Knudsen diffusion and adsorption should be considered.

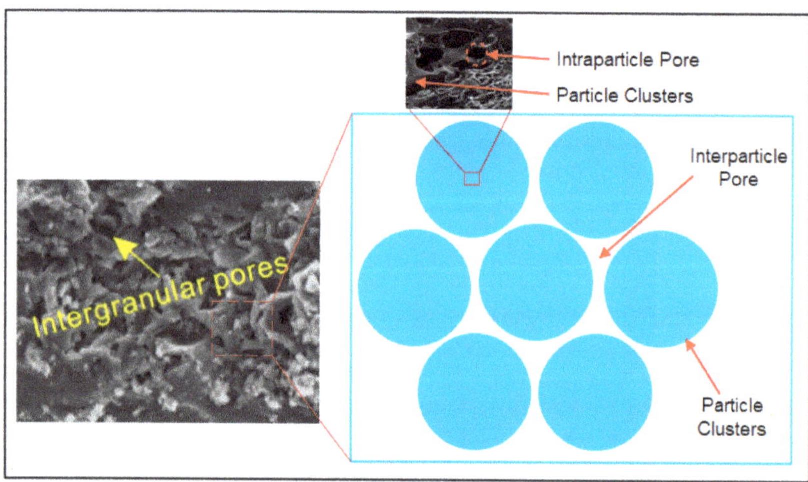

Figure 1. Schematic of the bi-dispersed porous structure of the coal matrix. Note that the scanning electron microscope (SEM) image of coal is copied from reference [39].

3. Physical Model and Verification

In this paper, the D2Q9-LB model is used to describe fluid flow and mass transfer. Firstly, the validity of the model is verified by two classic single-channel steady-state flow models, the diagram of the convection–diffusion system with surface adsorption reaction is shown in Figure 2. The size of the first model is set to 200×100, the parabolic flow rate is set at the inlet, $u_{\max} = 0.06$; the initial concentration in the field is 0, the gas with a diffusion coefficient of 1/6 spreads from the left, the inlet concentration is 1, the outlet boundary is set to $\partial C / \partial \mathbf{n} = 0$, and the Henry adsorption kinetic boundary condition is set at the bottom boundary [40,41]:

$$D_s \frac{\partial C}{\partial n} = kC \tag{14}$$

The Lévesque analytical solution after the adsorption has stabilized is:

$$\frac{L}{C_0}\frac{\partial C}{\partial n} = 0.854\left(\frac{u_{max}L^2}{xD_s}\right)^{1/3} \tag{15}$$

The comparison between the LB results and the analytical solution are shown in Figure 3a. The LB result agrees well with the analytical solution except for slight deviations at the entrance due to the singularity.

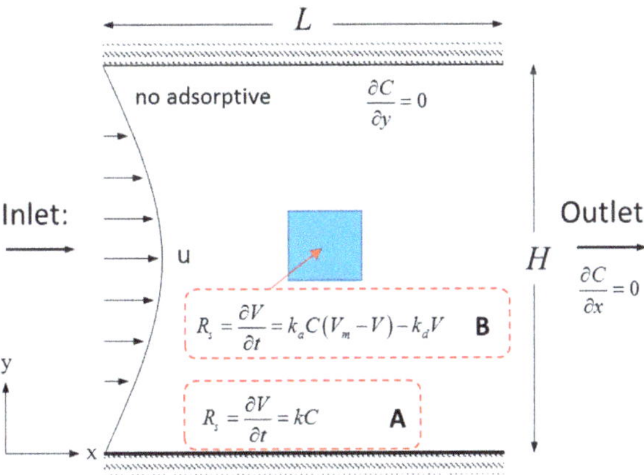

Figure 2. Diagram of the convection–diffusion system with surface adsorption reaction.

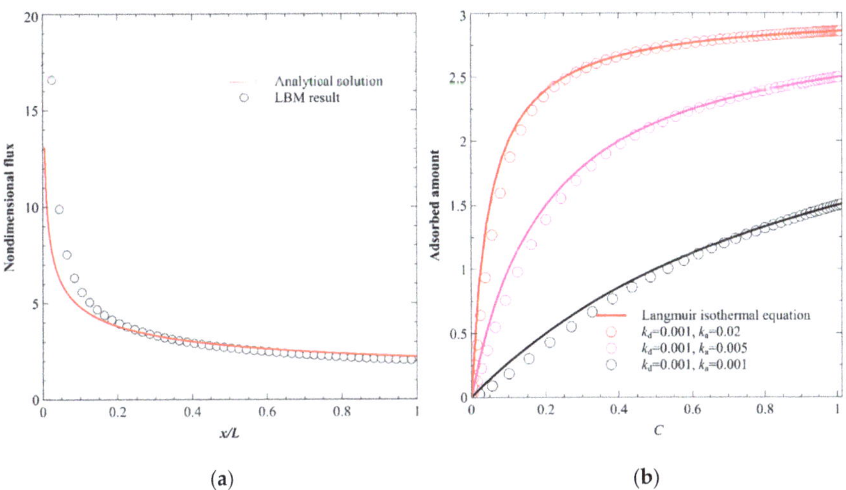

Figure 3. Comparison of the present LB results and the analytical solution. (**a**) Comparison of the LB results and the analytical solution of Henry adsorption. (**b**) Comparison of the LB results and the analytical solution of the Langmuir isotherm adsorption.

The size of the second model is set to 100×100. There is a 40×40 solid block in the center of the flow field, which represents a cluster of coal particles. Assuming that the solid block is a homogeneous material, the internal micro- and nanopores adsorb gas molecules follow the Langmuir adsorption rate equation [42]. The fluid is driven by a pressure gradient, $\nabla p = 0.01$, ignoring the fluid velocity inside the solid block ($u = 0$, $v = 0$), the four boundaries are all periodic boundaries; the initial concentration in the field is $C_0 = 0$, the gas diffuses from the left, the concentration at the inlet is unity, the diffusion coefficient inside and outside the solid is consistent, the diffusion coefficient is 1/6, and the other boundaries are set to $\partial C / \partial \mathbf{n} = 0$. Three sets of simulations with different adsorption constants were carried out, respectively. The simulation results are shown in Figure 3b. The LB results fit well with the Langmuir isotherm adsorption curves.

4. Results and Discussion

4.1. Diffusion –Adsorption of Gases in Simple Porous Media

In this paper, a simplified bi-dispersed porous media model is taken as an example to investigate the effects of fluid flow, gas diffusion (Fickian diffusion), and gas adsorption/desorption on the coupling process. As shown in Figure 4, the simulated area is 200×200, and the resolution of each grid is 10 nm. There are 16 clusters of coal particles with a length of 30×30 distributed in the field.

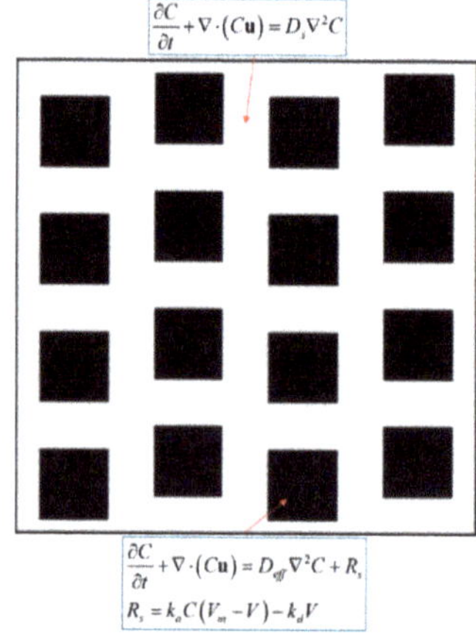

Figure 4. The simplified 2D reconstructed geometry of the coal matrix.

The fluid is driven by the pressure gradient, and the boundaries are set as the periodic boundary conditions. Besides, the left entrance of the field is set to the fixed concentration boundary ($C = C_0$), the other boundary condition is set to $\partial C / \partial \mathbf{n} = 0$, and the initial concentration in the field is 0. Since the extremely low permeability and low porosity of the coal seam, fluid velocity inside the clusters of coal particles is low enough to assume the clusters as an impermeable material ($\mathbf{u} = 0$). Gas transport inside the clusters in the form of the Knudsen diffusion and gas–solid adsorption/desorption only occurs in micropores inner the clusters. The local adsorption amount and concentration update with time. Due to the interaction between the adsorptive sites and the gas molecules that do not cause a sharp change during the period, the numerical stability in the adsorption process can be ensured. The specific parameters in the simulation are shown in Table 1.

Table 1. The input parameters in the simulation.

Simulation area $L \times L$ (um $\times$ um)	2×2
Gas density ρ_g (kg/m^3)	0.7
Gas viscosity v (Pa·s)	1.12×10^{-5}
Coal density ρ_c (kg/m^3)	1400
Pressure gradient p (MPa/m)	0.46
Input concentration C_0 (m^3/t)	1.6×10^{-2}
Gas diffusion coefficient in micro-fracture D_s (m^2/s)	7.84×10^{-5}
Langmuir volume V_m (m^3/t)	20
Gas desorption constant k_d (/s)	2.94×10^6
Langmuir pressure P_L (MPa)	2

Note: the relevant parameters in the table are taken from typical geological exploration data of coal reservoir in Qinshui Basin [43]; the adsorption and desorption constant is from reference [41,44].

For the sake of analysis, three dimensionless numbers, namely the Reynolds number (Re), Péclet number (Pe), and Damkohler number (Da), choose to quantitatively analyze the effects of fluid flow, gas diffusion, and adsorption/desorption on the coupling process. This cannot only quantify the simulation results but also extends the results to other similar situations based on the characteristics of the dimensionless number. Where $Re = UL/v$, $Pe = UL/D_s$, $Da = k_dL^2/D_s$, and U, L is the characteristic velocity and length of the system, respectively. In the simulation, the Re characterizes fluid flow, the Pe represents the relative proportion of advection and diffusion, and the Da describes the relative time scale of adsorption and gas diffusion in the system. In the present work, five kinds of combinations of Re, Pe, and Da are used to investigate the effects of the combination of fluid flow, gas diffusion, and gas–solid adsorption in the adsorption process: (i) $Re = 5.25 \times 10^{-2}$, $Pe = 7.5 \times 10^{-2}$, $Da = 1.5$, (ii) $Re = 5.25 \times 10^{-3}$, $Pe = 7.5 \times 10^{-3}$, $Da = 1.5$, (iii) $Re = 5.25 \times 10^{-3}$, $Pe = 7.5 \times 10^{-4}$, $Da = 0.15$, (iv) $Re = 5.25 \times 10^{-3}$, $Pe = 7.5 \times 10^{-5}$, $Da = 0.015$, (v) $Re = 5.25 \times 10^{-3}$, $Pe = 7.5 \times 10^{-5}$, $Da = 0.15$. The simulation results are shown in Figure 5.

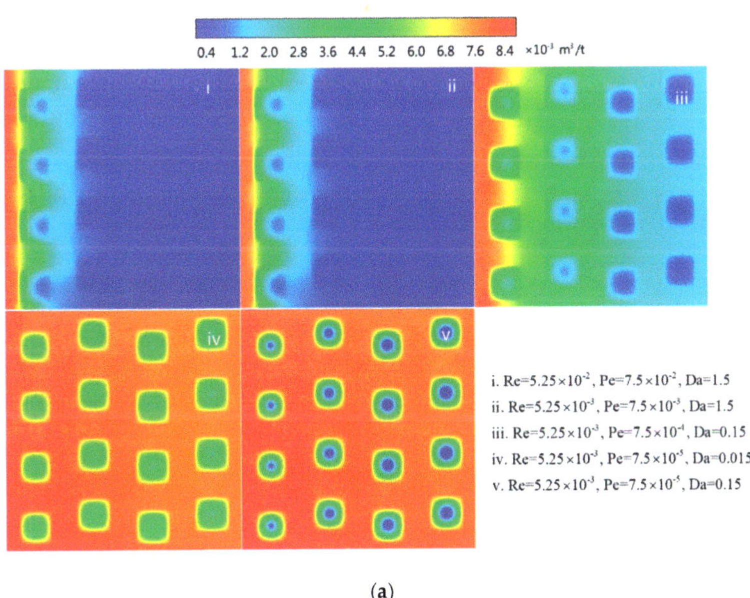

(**a**)

Figure 5. *Cont.*

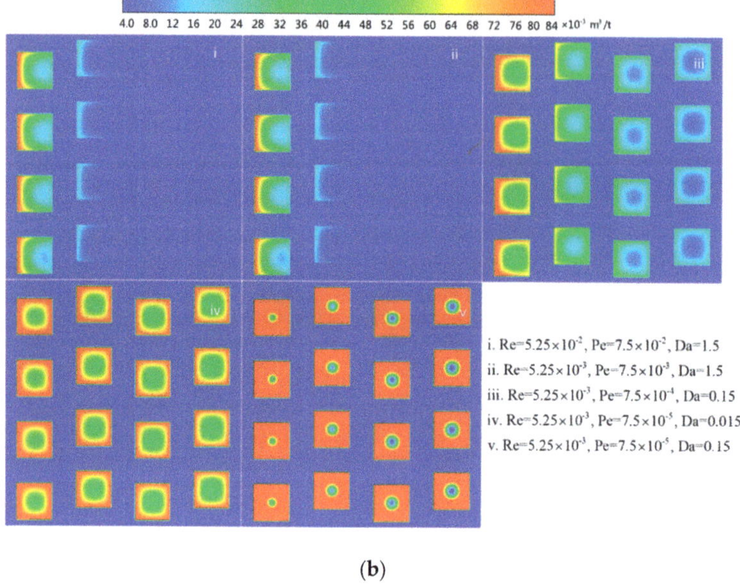

(b)

Figure 5. Concentration and adsorbed amount magnitude distribution. (**a**) Gas concentration distribution and (**b**) adsorbed amount distribution.

Figure 5 shows the gas concentration and adsorbed amount distribution of the five cases, respectively. We can observe that: (i) The influence of concentration distribution for fluid flow and gas diffusion is non-uniform in the field; a large number of gases accumulate near the inlet and the effect on the rest area is limited by the gas diffusivity. Gas adsorption mainly occurs on the first column of the solid particles, and the maximum adsorbed amount appears near the solid wall facing the inlet boundary. The adsorbed amount on the solid particles along the flow direction decreases gradually. (ii) When the fluid velocity is reduced by 10 times and the other conditions are maintained unchanged, the performance of gas diffusion–adsorption is almost identical to that of the case (i), which indicates that the fluid flow rate is not a main controlling factor. (iii) As the gas diffusion rate increases by 10 times only, the extent of gas adsorption effects is increased, and a more uniform concentration gradient is presented along the flow direction. The appearance of gas adsorption is similar to gas diffusion, which indicates that this is an adsorption-limited diffusion process. (iv) The Pe and Da number are both small (the FDC is increased by 10 times, other conditions are maintained unchanged), the gas diffusion rate is relatively large, the gas concentration distribution of the whole system is more consistent, the gas adsorption of the organic particles expands evenly from the boundary, and the adsorption amount of all particles is almost identical, which means the diffusion process is limited by the adsorption rate and the adsorption rate is low enough to keep the concentration field uniform at all times so that adsorption on all solid walls occurs uniformly. The results of cases (ii)–(iv) indicate that the magnitude of the FDC determines the distribution of the gas concentration and the location of the adsorption, which has a significant effect on the adsorption. (v) When the Pe number is small, but the Da number is large (which means the adsorption constant is increased by 10 times compared with the former case, other conditions are maintained unchanged), a higher inlet concentration and adsorption rate can accelerate the concentration update in the field, the adsorption intensity is higher, the adsorption occurs uniformly, and the adsorption amount increases remarkably.

By comparing the results of the case (i) and (ii), it can be found that the fluid velocity is not the main controlling factor in the methane migration and adsorption process at the pore-scale. The difference among the results of the cases (ii)–(iv) indicates that the magnitude of the FDC controls the extent of the gas diffusion, affecting the position and form of adsorption (whether uniform adsorption). The results of the case (iv) and case (v) show that the magnitude of the adsorption constant determines the adsorption strength/rate of the clusters of the coal particles.

4.2. Diffusion–Adsorption of Gas in 2D Reconstituted Porous Media

Coal has a complex pore structure and is difficult to characterize. In this section, the coal matrix is reconstructed based on the Random Circles Packing (RCP) algorithm. The reconstruction process is controlled by the random growth kernel and porosity, and the reconstructed image is shown in Figure 6. Due to the existence of the nanopores in the clusters of the coal particles, the effect of adsorbed gas molecules on internal pores of the clusters (adsorption layer effect) should be considered, and other conditions are set in the same way as in Section 4.1.

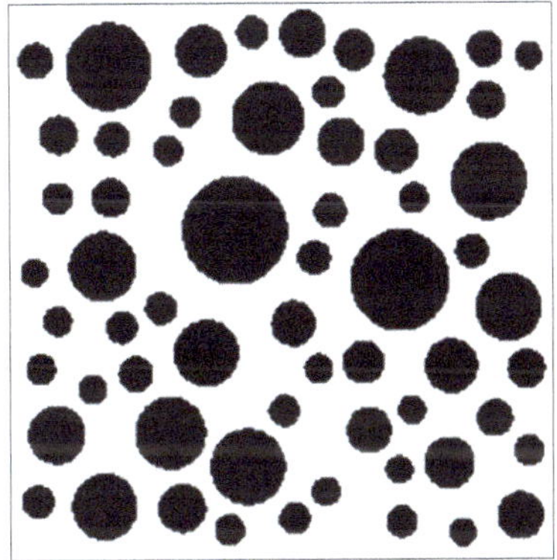

Figure 6. The reconstructed bi-dispersed porous media based on the Random Circles Packing (RCP) algorithm.

From the results of Section 4.1, it can be found that the adsorption is limited by the FDC and the adsorption constant, both of which are related to the time. This section employed the adsorption saturation (Equation (8) to update the effective pore radius and investigated the impact of the effective KDC and adsorption constant on the adsorption process and compared the corresponding results with Xiong's model [35]. The simulation results can be seen in Figures 7–9.

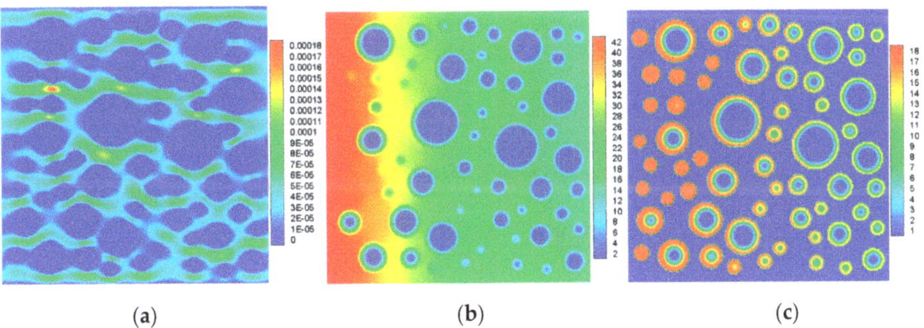

(a)   (b)   (c)

Figure 7. Simulation results at 500,000 steps. (**a**) The fluid velocity distribution, (**b**) the gas concentration distribution, and (**c**) the gas adsorption amount.

Figure 7 shows the simulation results at 500,000 steps as follows: (a) the fluid velocity distribution, (b) the gas concentration distribution, and (c) the gas adsorption amount, respectively. It illustrates that the gas diffusion behavior inside and outside of clusters of the coal particles is quite different. Outside the clusters, the diffusivity is large and gas diffusion is relatively uniform. While inside the clusters, due to the limitation of the pore size, the diffusion process is very slow, leading to the big concentration difference, and thus an amount of gas accumulated near the interface. Consequently, the gas adsorption first occurs at the periphery of the particle, and then slowly advances inward. In the simulation, the impact of the fluid flow on gas diffusion and adsorption is negligible due to its weak effect at the pore-scale, which has been proved in Section 4.1. Therefore, only the influence of the Pe and Da are discussed in the following. For the sake of comparison, we defined an average diffusion coefficient of the clusters of coal particles to describe the global gas diffusion characteristics:

$$D_{OM}(t) = \sum_x \sum_y D_{eff}(x, y) / W_{sum} \tag{16}$$

where $D_{eff}(x, y)$ is the effective KDC at a position (x, y) at a certain time and W_{sum} is the total area of clusters in the region.

Figure 8a shows the variation of the global diffusion coefficient (GDC) for nine groups of Pe and Da (at 11.2 million steps), respectively. The GDC of the organic particle is normalized by the inherent KDC, $D_{OM,0}$ (which ignores the effect of the gas adsorption layer). It can be found that with the Pe increased, the local gas diffusivity drops due to the gas adsorption was limited by diffusion, and the decay rate of the GDC becomes slow. When Pe remains unchanged, the local adsorption/desorption rate is accelerated as Da is increased, resulting in the gas molecules quickly filled the nanopores and the process of gas adsorption–desorption reaching dynamic equilibrium, and the GDC decay rate becomes faster. From the trend of groups 4–9 (red and black lines), we can find that the decay of the GDC will eventually reach a unified value. Groups 1–3 (blue lines) will also reach this unified value but will take a relatively long time.

To further explore the effect of the adsorption/desorption rate on the adsorption process, Figure 8b separately extracted three sets of data at $Pe = 7.5 \times 10^{-4}$ and compared with the results obtained by the method based on the Langmuir isotherm adsorption equation (the red line). Due to the time for the gas diffuse to the adsorptive site was ignored, the GDC decay rate of the Langmuir equation reaches the maximum value. However, sufficient time is required for gas to diffuse to an adsorptive site, even the pressure of the site reaches a specific value, the adsorption amount will not immediately reach the corresponding amount calculated by the Langmuir isothermal adsorption equation, which is just a maximum amount of gas that can be adsorbed under the pressure conditions. For a coal matrix with a fixed space size, due to the limited gas adsorption amount, the results obtained by the kinetic diffusion–adsorption–desorption model will eventually return to the analytical solution of the static isotherm adsorption equation with sufficient time; however, coal is usually tight and has low permeability, so it is difficult to ensure that the gas diffusion and adsorption are sufficient, the direct use of static isotherm adsorption equation may be incorrect.

Figure 8c,d show the effects of p_L and V_m on diffusivity, respectively. It shows that the greater the p_L or V_m, the slower the GDC decay. This is because the p_L is related to the ability of gas–solid adsorption and desorption: the greater the p_L means the faster the increment of the desorption rate than the adsorption rate, and the greater the V_m, the greater the adsorption capacity.

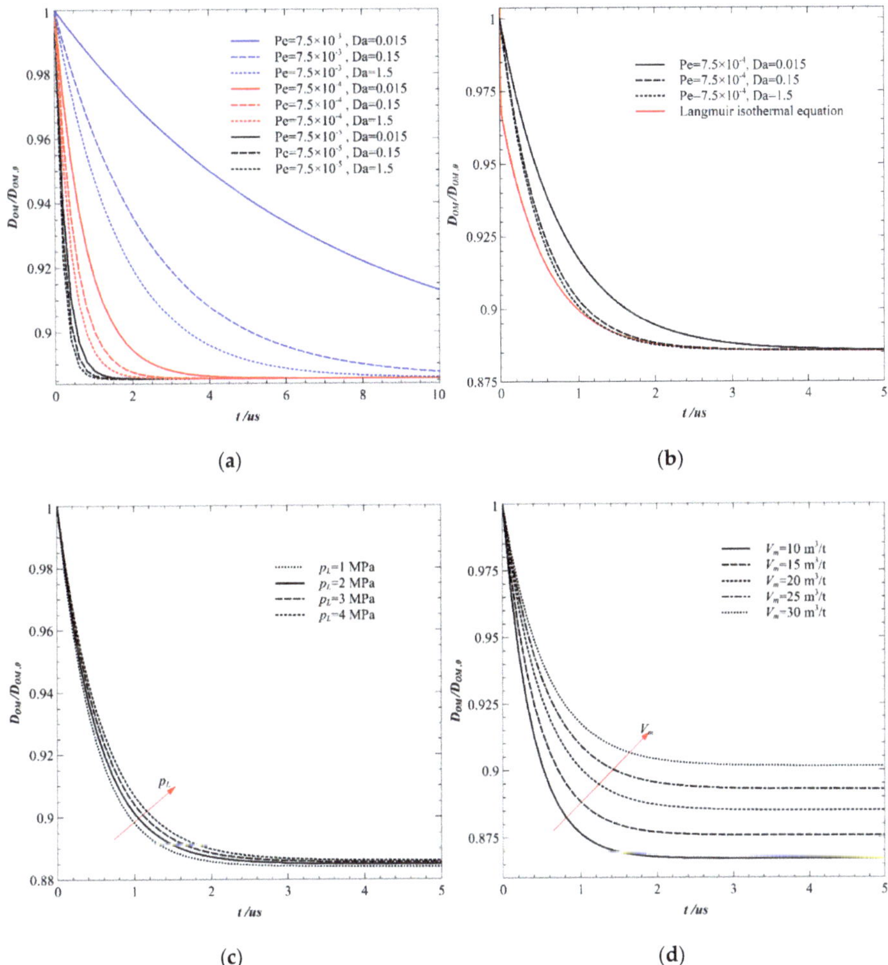

Figure 8. Changes of the global diffusion coefficient (GDC) with the adsorbed layer. (**a**) The variation of the GDC for 9 groups of *Pe* and *Da*, (**b**) the variation of the GDC at $Pe = 7.5 \times 10^{-4}$, (**c**) the effect of p_L on diffusivity, and (**d**) the effect of V_m on diffusivity.

The pore radius is a key parameter that affects the diffusion performance of gas in clusters. This section assumes that the pore size distribution inside the clusters conforms to the logarithmic normal distribution law, and we employed 6 groups of the typical pore size distribution of coal to investigate the dynamic effect of pore size on gas diffusion, as shown in Figure 9—the corresponding results can be seen in Figure 10.

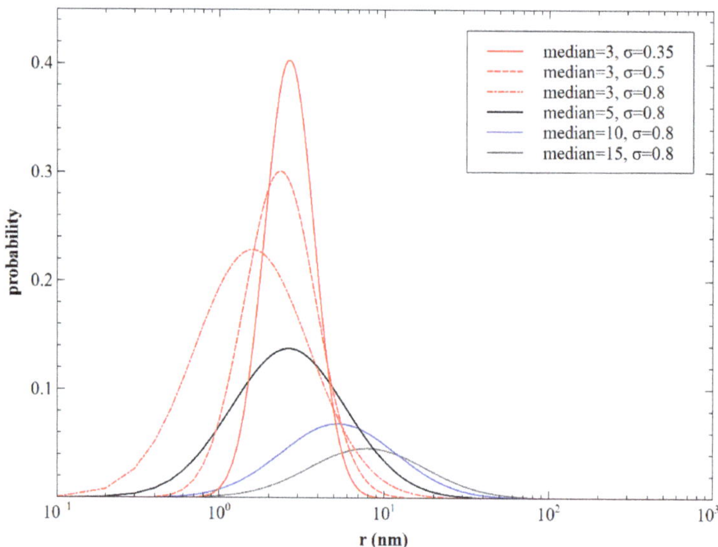

Figure 9. Pore size distribution with different median and logarithmic standard deviation.

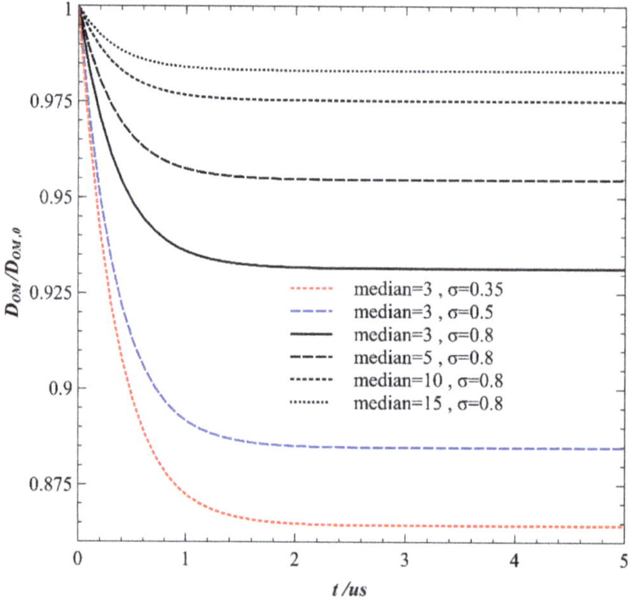

Figure 10. The variation of the GDC with the different pore size distribution.

Figure 10 shows the GDC variation of clusters with different pore size distributions. It can be found that as the median or logarithmic standard deviation σ increases, the GDC decreases more slowly, and the final value of the steady-state is smaller. The main reason for this phenomenon is that the greater of the median or σ means a wider range of pore radius distribution, which is with a higher proportion of pores with larger pore diameters in the particles. The blocking effect of the adsorbed gas molecules on the pore space will be weaker.

In Section 4.1, the characteristics of gas flow–diffusion–adsorption with different characteristic parameters *Re*, *Pe*, and *Da* were visualized through a simple bi-dispersed porous media. In Section 4.2, the influence of different parameters on the gas diffusion–adsorption process is analyzed. This paper is a continuation of our previous studies [45], which studied gas transport in a coal reservoir with dynamic adsorption. The results further explain the effects of Langmuir pressure and volume, matrix porosity on the methane diffusion–adsorption process at pore-scale. What's more, Chen [46] investigated the impact of various parameters on the production of coalbed methane at the macroscale and found that gas-production rate increases with Langmuir pressure, matrix porosity, and desorption rate, which is consistent with the results of this paper and verifies the correctness of the results.

In the research, many key parameters of coalbed methane, such as Langmuir pressure and volume, cover a relatively wide range. Moreover, due to many characteristics, including geological environment, complex components, fractures, etc., of coal are excluded at pore-scale, the results obtained can be widely applied in the coalbed methane exploitation.

5. Conclusions

In this paper, we established the pore-scale LB model coupled with fluid flow, gas diffusion, and gas adsorption–desorption in the bi-dispersed porous media of coalbed methane. The Knudsen diffusion and dynamic adsorption–desorption of gas in clusters of coal particles were considered. Firstly, the model was verified by two classical cases. Then, three dimensionless numbers, *Re*, *Pe*, and *Da*, were adopted to discuss the impact of fluid velocity, gas diffusivity, and adsorption/desorption rate on the gas flow–diffusion–adsorption process. The effect of the gas adsorption layer in micropores on the diffusion–adsorption–desorption process was considered, and a Langmuir isotherm adsorption theory-based method was developed to obtain the dynamic diffusion coefficient, which can capture the intermediate process when adsorption/desorption reaches equilibrium. The pore-scale bi-disperse porous media of coal matrix was generated based on the RCP algorithm, and the characteristics of gas diffusion and adsorption in the coal matrix with different *Pe*, *Da*, and pore size distribution were discussed. The conclusions were as follows:

(1) The influence of fluid velocity on the diffusion–adsorption process of coalbed methane at the pore-scale is very small and can be ignored; the magnitude of the FDC affects the spread range of gas diffusion and the process of adsorption and determines the position where adsorption takes place preferentially.

(2) The magnitude of the adsorption constant controls the strength/rate of gas adsorption. A larger FDC or greater adsorption constant can effectively enhance the adsorption rate, and the trend of gas concentration- adsorption is closer to the Langmuir isotherm adsorption curve.

(3) The gas diffusion–adsorption–desorption process is affected by the adsorption properties of coal. The specific performance is that the greater the p_L or V_m, the slower the GDC decay. This is because the p_L is related to the ability of gas–solid adsorption and desorption, the greater the p_L means the faster the increment of the desorption rate than the adsorption rate, and the greater the V_m, the greater the adsorption capacity.

(4) The effect of the gas molecular adsorption layer has a great impact on the kinetic process of gas diffusion–adsorption–desorption. For a coal matrix with a fixed space size, due to the limited gas adsorption amount, the results obtained by the kinetic diffusion–adsorption–desorption model will eventually return to the analytical solution of the static isotherm adsorption equation with sufficient time; however, coal is usually tight and has low permeability, so it is difficult to ensure that the gas diffusion and adsorption are sufficient, the direct use of static isotherm adsorption equation may be incorrect.

In the research, many key parameters of coalbed methane, such as Langmuir pressure and volume, cover a relatively wide range. Moreover, due to many characteristics, including geological environment, complex components, fractures, etc., of coal are excluded at pore-scale, the results obtained can be widely applied in the coalbed methane exploitation.

Author Contributions: Funding acquisition, Z.P., Z.D., and H.F.; methodology, Z.D.; supervision, Y.L.; validation, H.F.; writing—original draft, Z.P.; writing—review and editing, S.L. All authors have read and agreed to the published version of the manuscript.

Funding: This work was supported by the Project of Hunan Education Department (Grant No. 19C0370), and the General project of Hunan Natural Science Foundation (Grant No. 2020JJ4156), and the Outstanding Youth Project of Hunan Provincial Education Department (Grant No. 18B437).

Conflicts of Interest: The authors declare no conflict of interest.

Data Availability: The data used to support the findings of this study are included in the article.

References

1. Rezaee, R. *Fundamentals of Gas Shale Reservoirs*; John Wiley & Sons: Hoboken, NJ, USA, 2015.
2. Moore, T.A. Coalbed methane: A review. *Int. J. Coal Geol.* **2012**, *101*, 36–81. [CrossRef]
3. Fathi, E.; Akkutlu, I.Y. Matrix Heterogeneity Effects on Gas Transport and Adsorption in Coalbed and Shale Gas Reservoirs. *Transp. Porous Med.* **2009**, *80*, 281. [CrossRef]
4. Fu, X.; Qin, Y.; Wei, Z. *Coalbed Methane Geology*; China University of Mining and Technology Press: Xuzhou, China, 2007.
5. Meng, Z.; Tian, Y.; Li, G. *Theory and Method of Coalbed Methane Development Geology*; Science Press: Beijing, China, 2010.
6. Liu, K.; Ostadhassan, M.; Zhou, J.; Gentzis, T.; Rezaee, R. Nanoscale pore structure characterization of the Bakken shale in the USA. *Fuel* **2017**, *209*, 567–578. [CrossRef]
7. Yuan, W.; Pan, Z.; Li, X.; Yang, Y.; Zhao, C.; Connell, L.D.; Li, S.; He, J. Experimental study and modelling of methane adsorption and diffusion in shale. *Fuel* **2014**, *117*, 509–519. [CrossRef]
8. Feng, Y.-Y.; Yang, W.; Chu, W. Coalbed methane adsorption and desorption characteristics related to coal particle size. *Chinese Phys. B* **2016**, *25*, 068102. [CrossRef]
9. Sing, K.S. Reporting physisorption data for gas/solid systems with special reference to the determination of surface area and porosity (Recommendations 1984). *Pure Appl. Chem.* **1985**, *57*, 603–619. [CrossRef]
10. Zhao, Y.; Sun, Y.; Liu, S.; Chen, Z.; Yuan, L. Pore structure characterization of coal by synchrotron radiation nano-CT. *Fuel* **2018**, *215*, 102–110. [CrossRef]
11. Zhang, D.; Kang, Q. Pore Scale Simulation of Solute Transport In Fractured Porous Media. *Geophys. Res. Lett.* **2004**, *31*. [CrossRef]
12. Zhao, L.; Li, H. Stochastic Modeling of the Permeability of Randomly Generated Porous Media via the Lattice Boltzmann Method and Probabilistic Collocation Method. *Transp. Porous. Med.* **2019**, *128*, 613–631. [CrossRef]
13. Zhao, Y.; Wang, Z. Prediction of apparent permeability of porous media based on a modified lattice Boltzmann method. *J. Pet. Sci. Eng.* **2019**, *174*, 1261–1268. [CrossRef]
14. Yu, B.; Li, J. Some Fractal Characters of Porous Media. *Fractals* **2001**, *9*, 8. [CrossRef]
15. Ruckenstein, E.; Vaidyanathan, A.S.; Youngquist, G.R. Sorption by solids with bidisperse pore structures. *Chem. Eng. Sci.* **1971**, *26*, 1305–1318. [CrossRef]
16. Karbownik, M.; Krawczyk, J.; Schlieter, T. The Unipore and Bidisperse Diffusion Models for Methane in Hard Coal Solid Structures Related to the Conditions in the Upper Silesian Coal Basin. *Arch. Min. Sci.* **2020**, *65*, 605–625. [CrossRef]
17. Zhao, Y.; Liu, S.; Elsworth, D.; Jiang, Y.; Zhu, J. Pore Structure Characterization of Coal by Synchrotron Small-Angle X-ray Scattering and Transmission Electron Microscopy. *Energy Fuels* **2014**, *28*, 3704–3711. [CrossRef]
18. Beaucage, G.; Kammler, H.K.; Pratsinis, S.E. Particle size distributions from small-angle scattering using global scattering functions. *J. Appl. Cryst. J. Appl. Crystallogr.* **2004**, *37*, 523–535. [CrossRef]

19.   Wang, K.; Wang, G.; Ren, T.; Cheng, Y. Methane and CO_2 sorption hysteresis on coal: A critical review. *Int. J. Coal Geol.* **2014**, *132*, 60–80. [CrossRef]

20.   Clarkson, C.R.; Bustin, R.M. The effect of pore structure and gas pressure upon the transport properties of coal: A laboratory and modeling study. 2. Adsorption rate modeling. *Fuel* **1999**, *78*, 1345–1362. [CrossRef]

21.   Do, D.D.; Wang, K. A new model for the description of adsorption kinetics in heterogeneous activated carbon. *Carbon* **1998**, *36*, 1539–1554. [CrossRef]

22.   Do, D.D.; Wang, K. Dual diffusion and finite mass exchange model for adsorption kinetics in activated carbon. *AIChE J.* **1998**, *44*, 68–82. [CrossRef]

23.   Yang, Z.; Wang, W.; Dong, M.; Wang, J.; Li, Y.; Gong, H.; Sang, Q. A model of dynamic adsorption–diffusion for modeling gas transport and storage in shale. *Fuel* **2016**, *173*, 115–128. [CrossRef]

24.   Guo, Z.; Shu, C. *Lattice Boltzmann Method and Its Applications in Engineering*; Advances in Computational Fluid Dynamics; World Scientific: Singapore, 2013; Volume 3, ISBN 978-981-4508-29-2.

25.   He, Y.; Wang, Y.; Li, Q. *Lattice Boltzmann Method: Theory and Applications*; Science Press: Beijing, China, 2009.

26.   Kang, Q.; Wang, M.; Mukherjee, P.P.; Lichtner, P.C. Mesoscopic Modeling of Multiphysicochemical Transport Phenomena in Porous Media. *Adv. Mech. Eng.* **2010**, *2*, 142879. [CrossRef]

27.   Tian, Z.; Xing, H.; Tan, Y.; Gao, J. A coupled lattice Boltzmann model for simulating reactive transport in CO_2 injection. *Phys. A Stat. Mech. Its Appl.* **2014**, *403*, 155–164. [CrossRef]

28.   Tian, Z.; Xing, H.; Tan, Y.; Gu, S.; Golding, S.D. Reactive transport LBM model for CO_2 injection in fractured reservoirs. *Comput. Geosci.* **2016**, *86*, 15–22. [CrossRef]

29.   Kang, Q.; Lichtner, P.C.; Zhang, D. Lattice Boltzmann Pore-Scale Model For Multicomponent Reactive Transport In Porous Media. *J. Geophys. Res.* **2006**, *111*. [CrossRef]

30.   Gan, Y.; Xu, A.; Zhang, G.; Succi, S. Discrete Boltzmann modeling of multiphase flows: Hydrodynamic and thermodynamic non-equilibrium effects. *Soft Matter* **2015**, *11*, 5336–5345. [CrossRef]

31.   Peng, Z.; Liu, S.; Li, Y.; Yao, Q. Apparent Permeability Prediction of Coal Matrix with Generalized Lattice Boltzmann Model considering Non-Darcy Effect. *Geofluids* **2020**, *2020*, 8830831. [CrossRef]

32.   Chen, S.; Doolen, G.D. Lattice Boltzmann Method for Fluid Flows. *Annu. Rev. Fluid Mech.* **1998**, *30*, 329–364. [CrossRef]

33.   Zhao, W.; Cheng, Y.; Pan, Z.; Wang, K.; Liu, S. Gas diffusion in coal particles: A review of mathematical models and their applications. *Fuel* **2019**, *252*, 77–100. [CrossRef]

34.   Welty, J.; Wicks, C.E.; Rorrer, G.L.; Wilson, R.E. *Fundamentals of Momentum, Heat and Mass Transfer*, 6th ed.; Wiley: Hoboken, NJ, USA, 2014.

35.   Xiong, X.; Devegowda, D.; Michel Villazon, G.G.; Sigal, R.F.; Civan, F. A Fully-Coupled Free and Adsorptive Phase Transport Model for Shale Gas Reservoirs Including Non-Darcy Flow Effects. In Proceedings of the SPE Annual Technical Conference and Exhibition; Society of Petroleum Engineers, San Antonio, TX, USA, 8–10 October 2012.

36.   Clarkson, C.R.; Bustin, R.M. Binary gas adsorption/desorption isotherms: Effect of moisture and coal composition upon carbon dioxide selectivity over methane. *Int. J. Coal Geol.* **2000**, *42*, 241–271. [CrossRef]

37.   Bustin, R.M.; Clarkson, C.R. Geological controls on coalbed methane reservoir capacity and gas content. *Int. J. Coal Geol.* **1998**, *38*, 3–26. [CrossRef]

38.   He, X.Y.; Li, N.; Goldstein, B. Lattice Boltzmann simulation of diffusion-convection systems with surface chemical reaction. *Mol. Simulat.* **2000**, *25*, 145–156. [CrossRef]

39.   Du, Z.; Zhang, X.; Huang, Q.; Zhang, S.; Wang, C. Investigation of coal pore and fracture distributions and their contributions to coal reservoir permeability in the Changzhi block, middle-southern Qinshui Basin, North China. *Arab J. Geosci.* **2019**, *12*, 505. [CrossRef]

40.   Montessori, A.; Prestininzi, P.; La Rocca, M.; Falcucci, G.; Succi, S.; Kaxiras, E. Effects of Knudsen diffusivity on the effective reactivity of nanoporous catalyst media. *J. Comput. Sci.* **2016**, *17*, 377–383. [CrossRef]

41.   Zhou, L.; Qu, Z.G.; Chen, L.; Tao, W.Q. Lattice Boltzmann simulation of gas–solid adsorption processes at pore scale level. *J. Comput. Phys.* **2015**, *300*, 800–813. [CrossRef]

42.   Zhao, W.; Wang, K.; Liu, S.; Cheng, Y. Gas transport through coal particles: Matrix-flux controlled or fracture-flux controlled? *J. Nat. Gas Sci. Eng.* **2020**, *76*, 103216. [CrossRef]

43.   Su, X.; Lin, X.; Liu, S.; Zhao, M.; Song, Y. Geology of coalbed methane reservoirs in the Southeast Qinshui Basin of China. *Int. J. Coal Geol.* **2005**, *62*, 197–210. [CrossRef]

44. Bohn, C.D.; Scott, S.A.; Dennis, J.S.; Müller, C.R. Validation of a lattice Boltzmann model for gas–solid reactions with experiments. *J. Comput. Phys.* **2012**, *231*, 5334–5350. [CrossRef]
45. Peng, Z.; Liu, S.; Tang, S.; Zhao, Y.; Li, Y. Multicomponent Lattice Boltzmann Simulations of Gas Transport in a Coal Reservoir with Dynamic Adsorption. *Geofluids* **2018**, *2018*, 1–13. [CrossRef]
46. Chen, Z.; Liu, J.; Kabir, A.; Wang, J.; Pan, Z. Impact of Various Parameters on the Production of Coalbed Methane. *SPE J.* **2013**, *18*, 910–923. [CrossRef]

Article

Seismic Identification of Unconventional Heterogenous Reservoirs Based on Depositional History—A Case Study of the Polish Carpathian Foredeep

Anna Łaba-Biel [1], Anna Kwietniak [2] and Andrzej Urbaniec [1,*

[1] Oil and Gas Institute–National Research Institute, 25A Lubicz Str., 31-503 Krakow, Poland; laba-biel@inig.pl
[2] Faculty of Geology, Geophysics and Environmental Protection, AGH University of Science and Technology, 30 Mickiewicz Av., 30-059 Krakow, Poland; anna.kwietniak@agh.edu.pl
*   Correspondence: urbaniec@inig.pl

Received: 29 October 2020; Accepted: 17 November 2020; Published: 19 November 2020

Abstract: An integrated geological and geophysical approach is presented for the recognition of unconventional targets in the Miocene formations of the Carpathian Foredeep, southern Poland. The subject of the analysis is an unconventional reservoir built of interlayered packets of sandstone, mudstone and claystone, called a "heterogeneous sequence". This type of sequence acts as both a reservoir and as source rock for hydrocarbons and it consists of layers of insignificant thickness, below the resolution of seismic data. The interpretation of such a sequence has rarely been based on seismic stratigraphy analysis; however, such an approach is proposed here. The subject of interpretation is high-quality seismic data of high resolution that enable detailed depositional analysis. The reconstruction of the depositional history was possible due to the analysis of flattened chronostratigraphic horizons (Wheeler diagram). The identification of depositional positions in a sedimentary basin was the first step for the indication of potential target areas. These areas were also subject to seismic attribute analysis (sweetness) and spectral decomposition. The seismic attribute results positively verified the previously proposed prospects. The results obtained demonstrate that the interpretation of the Miocene sediments in the Carpathian Foredeep should take into account the depositional history reconstruction and paleogeographical analysis.

Keywords: seismic interpretation; depositional environments characteristics; Wheeler diagram; seismic attributes; heterogeneous sequence

---

1. Introduction

The terms "heterolithic bedding" or "heterogeneous sequence" are used commonly in the description of sedimentary series that are built of interlayered packets (laminae or lenses) of sandstone, mudstone and claystone. The heterogeneous sequence is not only a reservoir rock, but also a source rock that exhibits a high concentration of total organic matter [1–4]. The thicknesses of the separate intervals can be various, from millimeters to decimeters [5,6]. Considering the ratio of sandstone to mudstone, the thickness of the layers and the frequency of their appearance (i.e., the frequency of sandstone layers appearing in mudstone intervals), various types of heterogeneous sequences can be defined. The increasing interest in the heterogeneous sequence in the Carpathian Foredeep is due to the recent discoveries of gas fields [7–9]. Gas flows are present for intervals of low porosity and permeability, which, in well log analysis, are associated with low gas saturation. These intervals are challenging for well log interpretation. The low resolution of well log data and the high clay content associated with these sequences strongly influence the results. The data often suggest increased water content in the heterogeneous sequence, while high dry gas flow is measured in gas tests. For this reason, an interpretation of the heterogeneous sequence based on other geophysical data and comprehensive depositional architecture analysis is needed.

The Carpathian Foredeep is part of a huge Pannonian Basin System (PBS) that is the subject of extensive geological analysis in the aspects of conventional and unconventional reservoirs. The Carpathian Foredeep area is not as detail differentiated in terms of stratigraphy and lithology as other parts of the PBS [10–13].

In our work, we present a seismostratigraphic-based approach for indicating the possible gas-bearing intervals within the heterogeneous complex. These sequences make standard well log interpretation difficult due to the substantial horizontal and vertical variations in terms of lithology and position in a sedimentary basin. Moreover, low-resolution seismic data cannot be used for the reliable interpretation of these sequences within the Miocene sediments [14]. The seismic data presented in this study, however, have a higher than average resolution, therefore enabling a detailed depositional interpretation. In comparison to the previously published works [7,8,15] the presented data were designed to increase the resolution of a seismic image significantly. Only with such detailed seismic data is it possible to proceed with depositional analysis, which, in terms of heterogeneous sequences, yields accurate results.

Chronostratigraphic interpretations of the high-resolution seismic data and Wheeler diagrams (chronostratigraphic horizons identification and flattening) enable the definition of geometrical relations within the analyzed interval. With such an approach, it was possible to define the elements of the depositional architecture, such as slope fans, basin floor fans and barriers, then identify the depositional sequences [16,17]. Such an analysis is crucial for understanding the depositional architecture of the wider area, as well as facial change definitions that might enable us to understand the paleogeography and paleoenvironment of the Miocene succession, which could lead to hydrocarbon prospecting. The number of conventional hydrocarbon reservoirs within the Miocene formations recently started to decease, and fewer conventional prospects have been discovered in the last few years, hence the need to change the approach and begin an interpretation focused on unconventional, heterolithic targets [18]. Such a situation enforces the need for the development of an alternative methodology suited to heterogeneous sequences.

In order to indicate the possible gas-saturated intervals, we incorporated the spectral decomposition technique, which uses the concept of attenuative properties in relation to frequency modulations of gas-bearing sequences [19]. Similarly, we incorporated a sweetness attribute (based on instantaneous trace analysis), which was proven to give reliable results in clastic deposits [20].

The final set of hydrocarbon prospects was verified by the identification of the depositional architecture elements and their position within the sedimentary basin. Simultaneously, attribute analysis was performed in order to cross-check the prospected target areas. The results imply that, for the heterogeneous sequence definition, the chronostratigraphic approach plays a key role and should be routinely applied to the Miocene sediments of the Carpathian Foredeep.

2. Geological Setting and Data Description

2.1. Geological Setting, the Survey Area

The study area is located in southern Poland within the central part of the Carpathian Foredeep area (Figure 1).

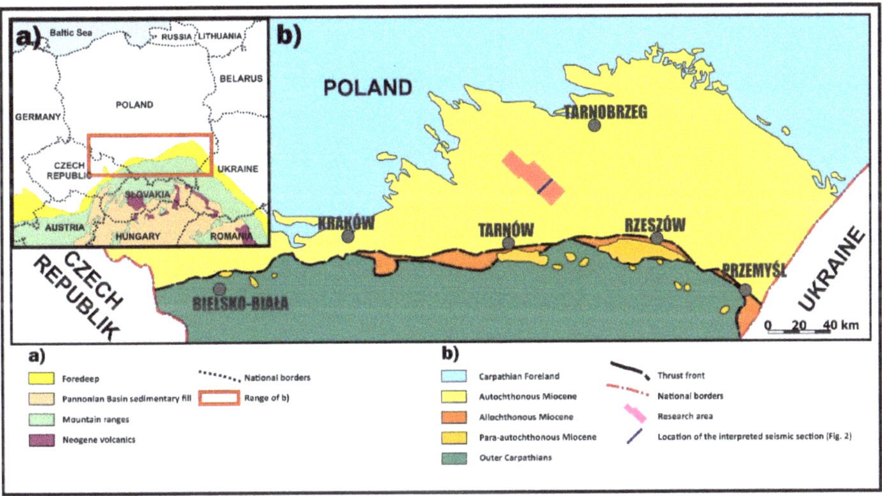

Figure 1. Geological map of the studied area: (**a**) generalized map of the Carpathian system range, after Picha [21] (modified); (**b**) location of the research area against the range of the Carpathian Foredeep in Poland; ranges of geological units according to Porębski and Warchoł [22].

The oldest structural stage in the research area is represented by a series of Neoproterozoic anchimetamorphic rocks. The Ediacaran age of that complex is confirmed by the results of micropaleontological analyses carried out on samples from the boreholes [23,24]. The middle stage is composed of Meso-Paleozoic rocks of a considerable summary thickness up to 2000 m. In the southern part of the analyzed region, directly on the Ediacaran interval, the Lower Paleozoic deposits (Ordovician and Silurian) are situated, but their ranges are not well documented by deep boreholes and insignificant thicknesses (based on well log data and seismic interpretation between 20–100 m) [25,26]. Higher up in the profile lie the carbonate series of Devonian and Carboniferous sediments [27,28]. The Mesozoic interval is represented by carbonate and clastic Triassic sediments, and above them lie the carbonate complexes of the Jurassic and Cretaceous periods [29–31].

The youngest structural stage is formed by the Miocene formations (Badenian–Sarmatian), which were initially deposited in the Carpathian Foredeep basin. The sedimentary basin of the Carpathian Foredeep was a fragment of a large foreland graben basin stretching along the whole Carpathian arc. The complex of autochthonous Miocene strata in the research area can be divided into three main units: the Lower Badenian clastic sub-evaporite series, the Upper Badenian evaporite series and the Upper Badenian–Sarmatian clastic series [32,33].

The sub-evaporite sediments, distinguished as the Skawina Formation [34], are mainly represented by a set of claystone and mudstone, with thicknesses up to a few dozen meters. The genesis of the Badenian evaporite series is connected with the Badenian Salinity Crisis (BSC), which probably commenced due to a rather sudden drop in sea level due to global cooling [35–38]. The sulphate rocks (gypsum and anhydrite) are distinguished as the Krzyżanowice Formation [34]. The siliciclastic sediments, classified as the Machów Formation [32–34], are an essential part of the autochthonous Miocene profile in the analyzed region of the Carpathian Foredeep. This formation manifests significant lithofacial diversity. The Machów Formation is assigned to NN6/NN7 calcareous nano plankton zones, although some authors qualify the uppermost part of this formation up to NN9 [39]. A detailed seismostratigraphic analysis of the middle part of the Machów Formation is the main subject of this article.

2.2. Definition of Heterogeneous Sequence

Lis and Wysocka [40], on the basis of well data analysis, show that the Carpathian Foredeep region is defined by three main types of heterogeneous sequences:

1. dominated by mudstone;
2. with an approximately equal proportion of sandstone/siltstone and mudstone;
3. dominated by sandstone and siltstone.

The specific types of bedding are characteristic of the heterogeneous sequence, predominantly flaser, wavy and lenticular laminations that are created by the interchanging deposition form traction and suspension. During the traction-based sedimentation, sand and silt fractions dominate with small-scale cross-bedding or planar bedding. The suspension-based sedimentation favors the deposition of mudstone and claystone without lamination, as well as flaser, wavy or lenticular laminations.

The heterogeneous sequence is commonly found in the Miocene sediment profile in the central part of the Carpathian Foredeep. The sequences might be present in many depositional environments and at various depths, even though they are mostly associated with shallow and tidal regions [41–45]. In the study area, the heterogeneous sequences are typical for the studied part of the Miocene complex.

In the analyzed interval, we can define the existence of the heterolithic facies with all depositional zones, from shallow to deep water environments. The interpreted interval reveals diverse and dynamic facial changes observed at the chronostratigraphic horizons and defined in the Wheeler diagram. The geometrical analysis of these chronostratigraphic horizons enables uniformly defined diagnostic elements of the depositional architecture, such as shelf, shelf margin, slope and basin floor, associated to the sedimentation environments [46,47].

2.3. Data Description

The seismic data used for the project are 3D seismic data that were acquired in 2015 and processed in 2016 by Geofizyka Kraków SA, Kraków, Poland for the Polish Oil and Gas Company. The survey covered the area of about 150 km^2 and was obtained by both dynamite and vibroseis methods. The research area covered approximately 150 km^2; the analyzed seismic section is situated in the southern part of the seismic survey (Figure 1) in a part that was obtained by the vibroseis method. The data were processed with a relative amplitude preservation scheme, and the resulting volume is a migrated seismic section in the time domain with a sampling rate of 2 ms. The volume was calibrated to the geological information with the use of 15 wells with sonic logs that exist within the survey area. The presented seismic section is situated away from the well data.

3. Methods

3.1. Depositional Analysis

The method is based on seismic sequence stratigraphy and depositional sequence characterization. A depositional sequence is understood as a lithostratigraphic unit that was deposited during a specific depositional episode in a sedimentary basin. Each seismic horizon is associated with a certain position within a depositional sequence. The chronostratigraphic horizons can be linked with a specific element of the paleoenvironment within the sedimentary basin.

The seismic volume was subject to depositional sequence interpretation. It should be noted that the sequence stratigraphy analysis based on seismic volume is less detailed than the interpretation based on well and outcrop data.

The first step of the analysis was to define a set of seismic sections to find those in directions parallel to the transportation of sediments to the basin. For the analysis, several sections were chosen. The area of analysis was limited to an exact time range by two seismic horizons. Afterwards, the apparent dip in seismic reflections was computed, and the resulting seismic volume enabled us to proceed to the next step, which was a chronostratigraphic horizon extraction. According to the information about the apparent dip and its direction, it was possible to apply a semi-automatic algorithm that tracks the seismic horizon in a given time range. The analysis was performed several times to adjust the parameters and take into account the resolution of the seismic data.

The defined chronostratigraphic horizons were transformed into the Wheeler domain by flattening [16,17] in order to analyze their relative temporal position. Wheeler diagrams were constructed for several seismic sections for the identification and visualization of the directions of deposition, hiatuses (a lack of deposition or erosion periods) and lateral continuities of depositional sequences.

With such an approach, it was possible to identify local changes in depositional directions that affected the exact shape and geometry of the chronostratigraphic horizons. The results are presented for the chosen cross-section.

3.2. Seismic Analysis

Seismic attributes were incorporated for possible gas-bearing zone interpretation. The subject of the analysis is a clastic sequence; hence, attributes that are suited to such environments give the best results. The sweetness attribute is calculated by dividing the instantaneous amplitude by the square root of the instantaneous frequency [48] and is sensitive to changes in amplitude and changes in frequency, both of which are considered direct hydrocarbon indicators (DHI) in clastic sequences [49]. The sweetness attribute might highlight thick, clean reservoirs and is sensitive to hydrocarbon content due to the incorporation of an instantaneous frequency [50]. This is a robust attribute, and its calculation is fast and efficient.

The second tool incorporated was spectral decomposition based on Fast Fourier Transform (FFT). The transform, based on the set of sine and cosine functions, decomposes the seismic trace to a time–frequency domain, in which specific frequency components hold the fraction of the amplitude of the original seismic trace. In this way, the filtering of a seismic signal is performed. The decomposition is performed in the time window; its length can be adjusted to the seismic resolution, but should not be too small in order for the decomposition to be stable. The interpretational value of the frequency volumes is that the reasoning can be limited to lower frequencies that tend to hold information about the hydrocarbon saturation.

4. Results

4.1. Depositional Environments Characterization

The first step of the interpretation was the identification of reflection discordances and lapouts such as onlaps, downlaps, offlaps, toplaps, and truncations [51]. The direction of the presented profile is SW–NE, which is parallel to the main deposition direction (see Figure 1).

A detailed interpretation of the chronostratigraphic horizons and the Wheeler diagram are presented in Figure 2. The heterogeneous sequence exhibits high diversity and complexity in its chronostratigraphic horizons. The whole analyzed interval, which spans approximately 170 ms (about 300 m), consists of 17 depositional sequences and 18 sequence boundaries (SB).

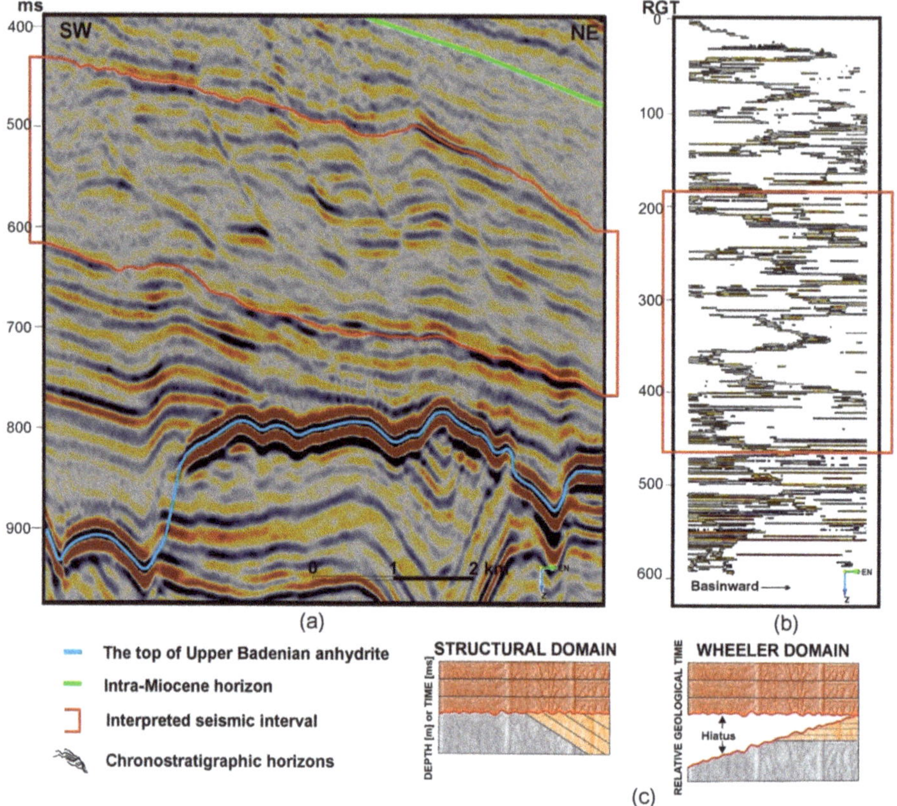

Figure 2. Detailed interpretation of (**a**) chronostratigraphic horizons in structural domain showed against the migrated seismic section (**b**) and Wheeler diagram transformed from seismic data. (**c**) Model presenting a chronostratigraphic nature of the stratigraphic units (Reproduced from Qayyum et al. [52], John Wiley & Sons Ltd, European Association of Geoscientists & Engineers and International Association of Sedimentologists, 2015, Page: 340).

Within the interpreted seismic interval (Figure 3, red frame), it was possible to indicate two kinds of depositional sequences connected to different types of sequence boundaries: the type I depositional sequence with sequence boundary no 1 (SB1) and the type II depositional sequence with sequence boundary no 2 (SB2).

Depositional sequences (type I) linked to SB1 are characterized by subaerial unconformities. These sequences change their character basinward into correlative conformities. The configuration of chronostratigraphic horizons of type I enables the identification of architectural elements that are typical of falling stage system tract (FSST) deposits that manifest hydrocarbon potential (incised valleys, slope fans and basin floor fans) [53]. These elements are also characterized by anomalous values of seismic attributes (see next subsection). The complete type I sequence consists of four depositional systems: falling stage system tract (FSST: Figure 3, brown—slope fan, yellow—basin floor fan), lowstand system tracks (LST: Figure 3, pink—lowstand wedge), transgressive system tracks (TST: Figure 3, green) and highstand system tract (HST: Figure 3, orange). In the bases of these sequences, it was possible to identify the borders of SB1, which are characterized by terrestrial and marine erosion [47].

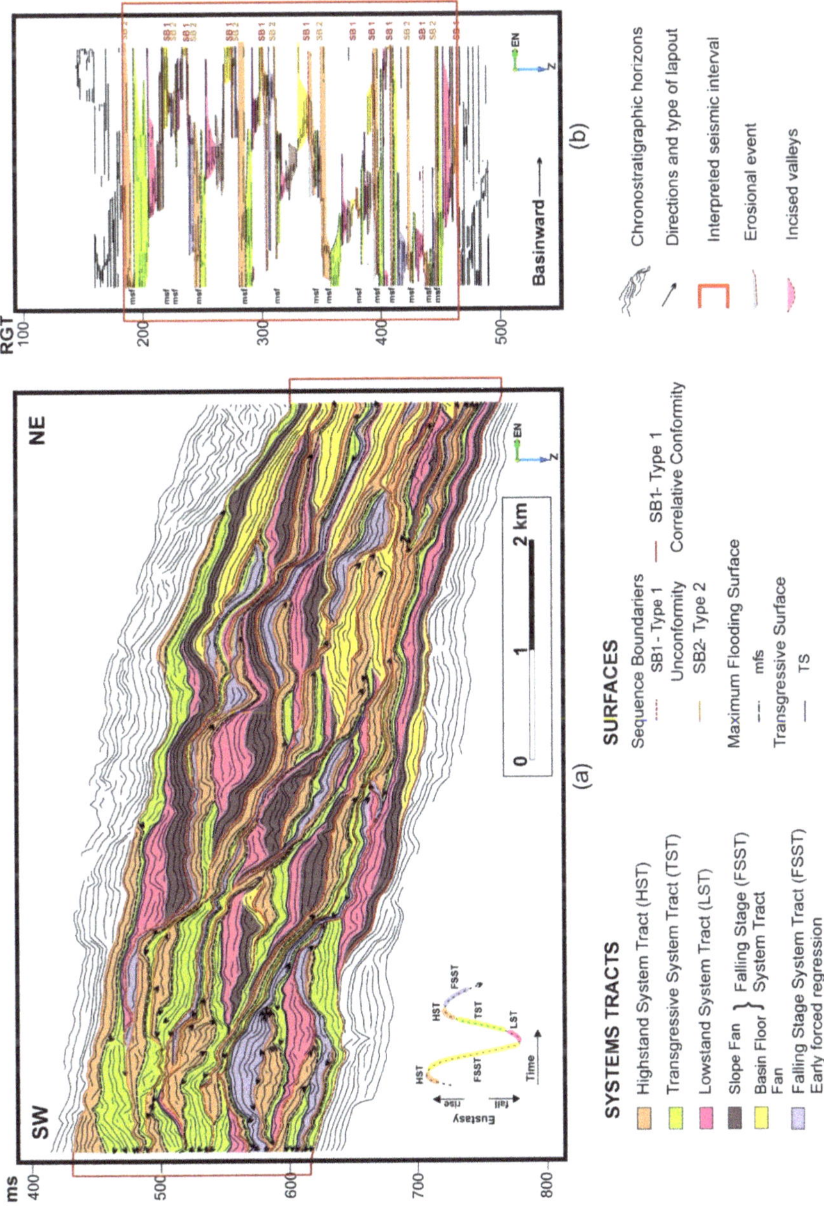

Figure 3. Detailed interpretation of (**a**) chronostratigraphic horizons in structural domain and (**b**) Wheeler diagram transformed from seismic data.

Sequence boundaries (SB2) that are associated with type II deposition manifest the aggradational deposition of the falling stage system tract (FSST). At the shelf margin, aggradational deposition can be found, but the shelf area is not intensively eroded. Additionally, at the slope and basin floor, there is a lack of fans. The influence of tectonic activity is visible. At the base of the slope, and also basinwards (e.g., slope fans—sf; basin floor fans—bff), there exist areas where there is a lack of older deposits that cannot be linked to a gap in sedimentation, but can be linked to the erosional cut (underwater erosion). Three levels/areas with high erosional relief can be defined and are generated by a cohesive debris floor (mudflow). Such processes might be linked to early forced regression [51]. The type II depositional sequence consists of an early falling stage system tract (FSST, Figure 3, purple), which corresponds to forced regression in type I sequences (or, e.g., forced regression of the shelf margin wedge), TST (Figure 3, green) and HST (Figure 3, orange). The lower boundary of these sequences (SB2) is characterized by an erosional surface in a proximal shelf zone. Basinwards, this boundary becomes correlative conformity.

Higher up in the profile, after each type I depositional sequence, a change in the depositional axis can be observed. The character of the sequence elements is also modified (type II depositional sequence). The direction of the siliciclastic material transport is SW, but the sea level drop is not rapid, and the shelf area is not substantially eroded. These circumstances do not favor fan creation within the slope and basin floor. At the shelf margin, aggradational deposition of falling stage system tracts (FSSTs) takes place.

The facies that dominate the shelf (mainly in the SW part of the cross-section) are transgressive deposits (TST, Figure 3, green), their architecture indicates shelf edge progradation. A substantial part of these deposits are highstand deltas (Figure 3, orange) that are built of clinoforms and deposits corresponding to the shallow sea level.

The heterogeneous complex has retrogradational–progradational characteristics with significant thickness changes in specific depositional elements and a diverse architecture. This is a result of few components that changed over time: sediment flux directions, subsidence and eustatic sea level changes. The primary influence on the deposition must have been related to relative sea level changes linked to subsidence rate fluctuations. The high distribution rate in the sedimentary basin seems to limit fluvial erosion and move the shelf margin forward into the area of higher subsidence (depocenter).

Some of the depositional sequences are characterized by the lack of a link between the shelf and deep basin sediments (see Figure 4a,b, around the second "TST" from the left). This is the result of an erosional cut in the older deposits caused by younger deposits. This suggests that the rebuilding of the sedimentary basin architecture took place, which is most likely linked to tectonic processes. We observe a shift in the basin axis, which is crucial information from a prospecting point of view.

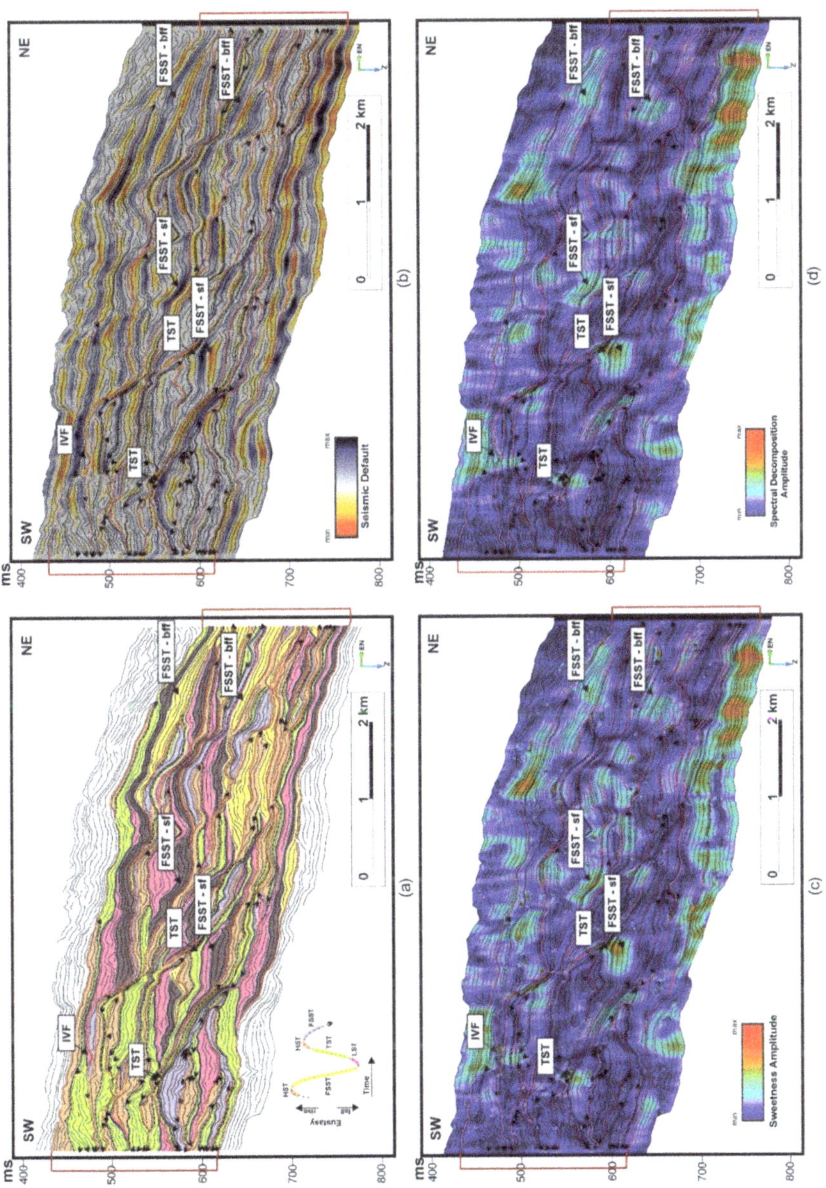

Figure 4. Depositional sequence interpretation: (**a**) interpreted sequences; (**b**) chronostratigraphic horizon interpretation; (**c**) sweetness attribute; (**d**) spectral decomposition. Depositional sequences: IVF—incised valley; FSST—failing stage system tract; sf—slope fans; bff—basin flow fans; TST—transgressive system tract.

4.2. Seismic Signatures for Unconventional Targets Prospecting

The interpretation of possible prospecting targets took into account the results of the sweetness attribute and spectral decomposition. These two attributes give similar results, but their images are slightly different. The sweetness attribute (Figure 4c) reveals more details that are associated with the definition of the attribute and instantaneous character. The value of the attribute is assigned to each sample at every time point. The amplitude image of the frequency slice (Figure 4d) is more blurred than for the sweetness attribute. This is a result of the averaging effect of the FFT computed within the specific time window. Nonetheless, for the presented low frequency volume (20 Hz), it is possible to indicate the anomalous zones. These zones are affiliated with the attenuation of high frequencies, which results in a shift in the dominant frequencies in the lower part of a spectrum. The presented image indicates such regions (Figure 4d).

The zones of the highest hydrocarbon potential are linked to the FSST, especially the slope fans and basin floor fans (Figure 4a, yellow; Figure 4c,d, bright colors—high amplitudes). Elements of the transgressive sequence are also prominent (Figure 4a, brown). The transgressive sequence is built of the fine clastic material and can play the role of a caprock (see Figure 4, "TST"—green). There also exist anomalies suggesting hydrocarbon saturation in a transgressive sequence at the shelf, but they have limited volume (Figure 4, "TST"—green, Figure 4c,d, bright colors—high amplitudes). The high-resolution seismic image enabled us to indicate the potential hydrocarbon saturation in the incised valley (see Figure 4, "IVF").

From a hydrocarbon prospecting point of view, transgressive deposits might be prominent exploration targets (TST, Figure 4a, green, Figure 4b). The main prism of these sediments is deposited in the SW part of the section, near the shelf. The interpretation of the seismic reflection pattern in the zone of maximal thickness indicates development towards the shelf. However, TST deposits manifest a retrogradational character and hence cannot be the most prominent from a hydrocarbon prospecting point of view. The most promising areas for hydrocarbon prospecting are as follows: (1) transgressive sequences that fill incised valleys; (2) onshore barriers in the central part of a profile; (3) transgressive deposits covering slope fans.

5. Discussion

Due to the high resolution of the data, it was possible to indicate incised valleys, which are rarely seen on seismic sections. Their existence proves that we provided the correct definition of the sequence boundaries. Incised valleys have very high hydrocarbon potential (both for conventional and unconventional plays), here verified by the results of the sweetness attribute and spectral decomposition. However, despite being small in size in the presented case, their existence shows the applicability of the proposed methodology. Moreover, erosional truncations that are associated with the sub-aerial erosion and diverse relief of the onshore part of the highstand delta (HST) could have been indicated by chronostratigraphic interpretation.

It should be noted that there is an ongoing discussion regarding the methodology of sequence boundary interpretation and the specialistic nomenclature [46,51,54,55]. Nevertheless, in the presented study, we decided to divide sequence boundaries into two types, type I and type II [56], even though type II discontinuities were recently considered to be an integral part of the failing stage system tract (FSST; a shift in the shoreline basinward during one sequence with a fall in the relative sea level).

The high-resolution seismic image enabled the identification of differences in the failing stage system tracts between type I and type II depositional sequences. The interpretation of a Wheeler diagram made it possible to define and differentiate between FSST type II from lower-lying highstand deposits (HST).

The heterogeneous sequence under analysis exhibits significant thickness changes and the rebuilding of architecture. This is a result of continuous fluctuations in sediment flux, the rate of subsidence and eustatic sea level. Depositional architecture is also controlled by intensive tectonics; its influence is prominent in the non-according contacts between chronostratigraphic horizons and in the erosional cutting of older deposits by younger sediments [47,55]. Depositional sequences are characterized by a geometry that ensures that the lack of a link between the shelf and basin deposits can be proposed. After each erosional episode, signatures that prove the rebuilding of the sedimentary basin can be defined; a shift in the basin axis toward the NE direction can also be indicated. The increase in the transportation rate brings fluvial erosion to a halt and shifts the shelf margin towards a higher subsidence area. This is proved by the increasing thicknesses of the sediments and the development of basin elements (slope fans, basin floor fans). After stabilization, the sediment flux rate is lower than the subsidence rate and the deposits exhibit retrogradational character until the next tectonic episode.

6. Conclusions

With the application of the seismic sequence stratigraphy method, it was possible to interpret system tracts and depositional sequences. This helped us to understand the depositional history and recovery of the depositional architecture of the sedimentary basin in the research area. Afterwards, seismic attributes were applied in order to verify the reservoir properties within the analyzed interval. The integration of these results enabled the identification of the prospective targets. Figure 5 summarizes the main findings of the analysis.

It should be stipulated that, otherwise all of the indicated prospective targets are both source and reservoir rocks, they should be considered as unconventional reservoirs. Moreover, there was no hydrocarbon migration to the Machów Formation from outside. Such an approach gives great insights into the interpretation of unconventional reservoirs.

Seismic sequence stratigraphy at the level of detail presented above is possible for high-resolution, preferably 3D, seismic surveys. A sufficient resolution, in this case, was acquired by the detailed design of the acquisition parameters, the importance of the survey design should be strongly advocated at this point. The data that were shown are unprecedented in the Carpathian Foredeep area, manifesting high vertical and horizontal resolutions. The results give weight to the argument that only high-resolution seismic data for heterogeneous sequence interpretation might reveal details that help to understand the depositional history of the Miocene formations in the Carpathian Foredeep sedimentary basin.

The detailed seismic image presented enabled accurate attribute interpretation, which was guided by sedimentary analysis. The attribute values are clearly changing, and the shapes of the presented anomalies are in agreement with the depositional interpretation. This proves that the choice of attributes and their parametrization suited the given problem.

For heterogeneous sequence interpretation, sedimentary analysis is a necessary step that can not only identify possible prospecting targets, but also define their geometry and tectonic involvement. The tectonic setting is crucial for a reliable estimation of the hydrocarbon reserves and for well planning. Such an approach is not routinely applied to the Miocene deposits of the Carpathian Foredeep. Instead, these prospects are mostly treated as convenient targets, for which structural positioning is almost exclusively applied. Based on the research results obtained, we propose a new methodological attitude utilizing both seismic sequence stratigraphy and attribute analysis, with a particular emphasis on frequency analysis.

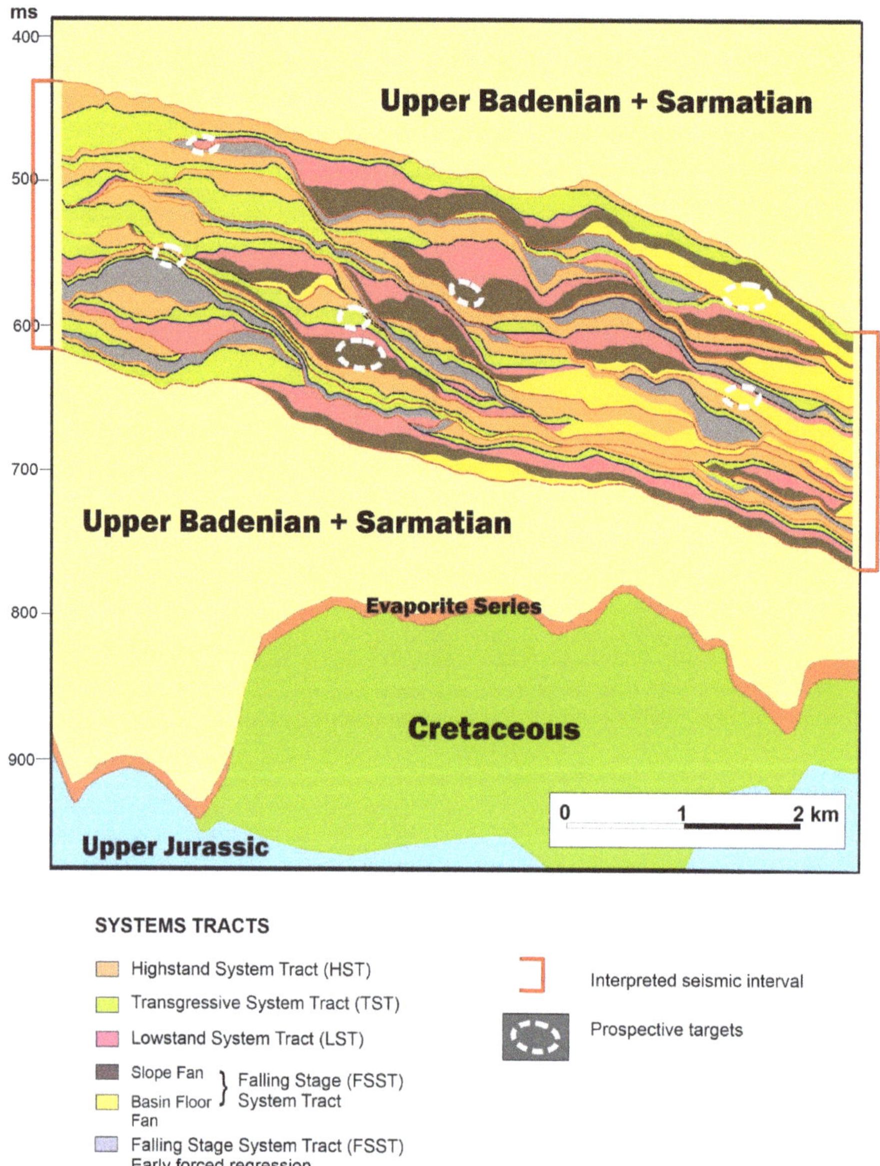

Figure 5. Geological cross-section with marked system tracts and prospecting targets within the interpreted seismic interval.

Author Contributions: Conceptualization, A.K. and A.U.; methodology, A.K.; software, A.K.; validation, A.Ł.-B., A.U. and A.K.; formal analysis, A.U.; investigation, A.Ł.-B., A.K.; writing—original draft preparation, A.K., A.U., A.Ł.-B.; writing—review and editing, A.K.; visualization, A.Ł.-B. All authors have read and agreed to the published version of the manuscript.

Funding: This research received no external funding. The work was part of the ongoing investigation being carried out by the Department of Geophysics, University of Science and Technology (11.11.140.645). This paper was written on the basis of the statutory work entitled "Seismostratigraphic analyses of the Upper Jurassic carbonate complex in the central part of the Carpathian Foreland". The work of the Oil and Gas Institute–National

Energies **2020**, *13*, 6036

Research Institute was commissioned by the Ministry of Science and Higher Education (order number: 37/SR/2020; archive number: DK-4100-25/2020).

Acknowledgments: We would like to acknowledge the Polish Oil and Gas Company for sharing their data for research purposes and for their permission to publish the results. We extend our gratitude to DGB EarthSciences for granting the University of Science and Technology AGH's OpendTect application for the HorizonCube plugin that was used for the depositional analysis. The authors would like to thank two anonymous reviewers for their suggestions and recommendations, which helped to improve the manuscript.

Conflicts of Interest: The authors declare no conflict of interest.

References

1. Kotarba, M.J.; Wilczek, T.; Kosakowski, P.; Kowalski, A.; Więcław, D. A study of organic matter and habitat of gaseous hydrocarbons in the Miocene strata of the Polish part of the Carpathian Foredeep. *Prz. Geol.* **1998**, *46*, 742–750.

2. Kotarba, M.J. Origin of natural gases in the autochthonous miocene strata of the Polish Carpathian Foredeep. *Ann. Soc. Geol. Pol.* **2011**, *81*, 409–424.

3. Matyasik, I.; Myśliwiec, M.; Leśniak, G.; Such, P. Relationship between Hydrocarbon Generation and Reservoir Development in the Carpathian Foreland (Poland). In *Proceedings of the Thrust Belts and Foreland Basins*; Springer: Berlin/Heidelberg, Germany, 2007; pp. 413–427.

4. Kotarba, M.J.; Peryt, T.M.; Koltun, Y.V. Microbial Gassystem and Prospectives of Hydrocarbon Exploration in Miocene Strata of the Polish and Ukrainian Carpathian Foredeep. *Ann. Soc. Geol. Pol.* **2011**, *81*, 523–548.

5. Thomas, R.G.; Smith, D.G.; Wood, J.M.; Visser, J.; Calverley-Range, E.A.; Koster, E.H. Inclined heterolithic stratification—Terminology, description, interpretation and significance. *Sediment. Geol.* **1987**, *53*, 123–179. [CrossRef]

6. Lettley, C.D.; Pemberton, S.G. Speciation of McMurray Formation Inclined Heterolithic Strata: Varying Depositional Character Along a Riverine Estuary System. In Proceedings of the CSPG/CSEG GeoConvention, Calgary, AB, Canada, 31 May–4 June 2004; CSPG/CSEG Datapages: Calgary, AB, Canada, 2015; Volume 51067.

7. Myśliwiec, M. Mioceńskie skały zbiornikowe zapadliska przedkarpackiego. *Prz. Geol.* **2004**, *52*, 581–592.

8. Myśliwiec, M.; Plezia, B.; Świętnicka, G. Nowe odkrycia złóz gazu ziemnego w osadach miocenu północno-wschodniej części zapadliska przedkarpackiego na podstawie interpretacji bezpośredniego wpływu nasycenia węglowodorami na zapis sejsmiczny. *Prz. Geol.* **2004**, *52*, 395–402.

9. Liszka, B.; Madej, K.; Paszkowski, M.; Porębski, S.J.; Ratusznik, Z.; Słyś, M.; Soliński, B.; Warchoł, M. Rozpoznanie heterolitowych obiektów złożowych w miocenie zapadliska przedkarpackiego pod kątem eksploatacji gazu otworami poziomymi. *Pr. Inst. Naft. Gazu* **2009**, *158*, 37–68.

10. Nagymarosy, P.; Muller, P. Some Aspects of Neogene Biostratigraphic in the Pannonian Basin. In *The Pannonian Basin: A Study in Basin Evolution Basin Evolution*; American Association of Petroleum Geologists: Tulsa, OK, USA, 1988; pp. 69–77.

11. Malvić, T. Review of Miocene shallow marine and lacustrine depositional environments in Northern Croatia. *Geol. Q.* **2012**, *56*, 493–504. [CrossRef]

12. Pavelić, D.; Kovačić, M. Sedimentology and stratigraphy of the Neogene rift-type North Croatian Basin (Pannonian Basin System, Croatia): A review. *Mar. Pet. Geol.* **2018**, *91*, 455–469. [CrossRef]

13. Sacchi, M.; Horváth, F. Towards a new time scale for the Upper Miocene continental series of the Pannonian basin (Central Paratethys). *EGU Stephan Mueller Spec. Publ. Ser.* **2002**, 79–94. [CrossRef]

14. Pietsch, K.; Porębski, S.J. The use of seismostratigraphy for exploration of Miocene gas-bearing reservoir facies in the NE part of the Carpathian Foreland Basin. *Geologia* **2010**, *36*, 173–186.

15. Kwietniak, A.; Cichostępski, K.; Kasperska, M. Spectral Decomposition Using the CEEMD Method: A Case Study from the Carpathian Foredeep. *Acta Geophys.* **2016**, *64*, 1525–1541. [CrossRef]

16. Łaba-Biel, A.; Smółka-Gnutek, P. Horizon cube i diagram Wheelera—Przykłady zastosowania w interpretacji obrazu sejsicznego. *Pr. Nauk. INiG-PIB* **2016**, *209*, 533–537.

17. Łaba-Biel, A.; Smółka-Gnutek, P. Detaliczna, geologiczna analiza obrazu sejsmicznego w oparciu o interpretację diagramu Wheelera i Horizon Cube. *Pr. Nauk. INiG-PIB* **2016**, *209*, 223–229.

18. Cichostępski, K.; Kwietniak, A.; Dec, J. Verification of bright spots in the presence of thin beds by AVO and spectral analysis in Miocene sediments of Carpathian Foredeep. *Acta Geophys.* **2019**, *67*, 1731–1745. [CrossRef]

19. Oyem, A.; Castagna, J. Sorting and visualization of Spectral-decomposition data. *Lead. Edge* **2015**, *34*, 42–47. [CrossRef]

20. Hart, B.S. Channel detection in 3-D seismic data sing sweetness. *Am. Assoc. Pet. Geol. Bull.* **2008**, *92*, 733–742. [CrossRef]

21. Picha, F.J. Exploring for hydrocarbons under thrust belts—A challenging new frontier in the Carpathians and elsewhere. *Am. Assoc. Pet. Geol. Bull.* **1996**, *80*, 1547–1564.

22. Porębski, S.J.; Warchoł, M. Hyperpycnal flows and deltaic clinoforms—Implications for sedimentological interpretations of late Middle Miocene fill in the Carpathian Foredeep Basin. *Przegląd Geol.* **2006**, *54*, 421–429.

23. Moryc, W.; Jachowicz, M. Utwory prekambryjskie w rejonie Bochnia-Tarnów-Debica. *Prz. Geol.* **2000**, *48*, 601–606.

24. Jachowicz-Zdanowska, M. Organic microfossil as sem blages from the late Ediacaran rocks of the Małopolska Block, southeast ern Poland. *Geol. Q.* **2011**, *55*, 85–94.

25. Moryc, W.; Nehring-Lefeld, M. Ordovician between Pilzno and Busko in the Carpathian Foreland (Southern Poland). *Geol. Q.* **1997**, *41*, 139–150.

26. Buła, Z.; Habryn, R. Precambrian and Palaeozoic basement of the Carpathian Foredeep and the adjacent outer Carpathians (SE Poland and Western Ukraine). *Ann. Soc. Geol. Pol.* **2011**, *81*, 221–239.

27. Zając, R. Stratygrafia i rozwój facjalny dewonu i dolnego karbonu południowej części podłoza zapadliska przedkarpackiego. *Kwart. Geol.* **1984**, *28*, 291–316.

28. Moryc, W. Budowa geologiczna podłoża miocenu w rejonie Kraków—Pilzno. Część 1. Prekambr i paleozoik (bez permu). *Nafta-Gaz* **2006**, *62*, 197–216.

29. Urbaniec, A.; Bobrek, L.; Świetlik, B. Litostratygrafia i charakterystyka mikropaleontologiczna urworów krédy dolnej w środkowej części przedgórza Karpat. *Prz. Geol.* **2010**, *58*, 1161–1175.

30. Gutowski, J.; Urbaniec, A.; Złonkiewicz, Z.; Bobrek, L.; Świetlik, B.; Gliniak, P. Upper Jurassic and Lower Cretaceous of the middle polish Carpathian foreland. *Biul. Państwowego Inst. Geol.* **2007**, *426*, 1–26.

31. Urbaniec, A.; Bartoń, R.; Bajewski, Ł.; Wilk, A. Wyniki interpretacji strukturalnej utworów triasu i paleozoiku przedgórza Karpat opartej na nowych danych sejsmicznych. *Nafta-Gaz* **2020**, *76*, 559–568. [CrossRef]

32. Jasionowski, M. Zarys litostratygrafii osadów mioceńskich wschodniej części zapadliska przedkarpackiego. *Biul. Państwowego Inst. Geol.* **1997**, *375*, 43–59.

33. Urbaniec, A.; Stadtmüller, M.; Bartoń, R. Possibility of a more detailed seismic interpretation within the Miocene formations of the Carpathian Foredeep based on the well logs interpretation. *Nafta-Gaz* **2019**, *75*, 527–544. [CrossRef]

34. Alexandrowicz, S.W.; Garlicki, A.; Rutkowski, J. Podstawowe jednostki litostratygraficzne miocenu zapadliska przedkarpackiego. *Kwart. Geol.* **1982**, *26*, 470–471.

35. Báldi, K. Paleoceanography and climate of the Badenian (Middle Miocene, 16.4-13.0 Ma) in the Central Paratethys based on foraminifera and stable isotope (δ18O and δ13C) evidence. *Int. J. Earth Sci.* **2006**, *95*, 119–142. [CrossRef]

36. De Leeuw, A.; Bukowski, K.; Krijgsman, W.; Kuiper, K.F. Age of the Badenian salinity crisis; Impact of Miocene climate variability on the circum-mediterranean region. *Geology* **2010**, *38*, 715–718. [CrossRef]

37. Karami, M.P.; de Leeuw, A.; Krijgsman, W.; Meijer, P.T.; Wortel, M.J.R. The role of gateways in the evolution of temperature and salinity of semi-enclosed basins: An oceanic box model for the Miocene Mediterranean Sea and Paratethys. *Glob. Planet. Change* **2011**, *79*, 73–88. [CrossRef]

38. Bukowski, K. *Rozprawy i Monografie*; Wydawnictwa AGH: Kraków, Poland, 2011; pp. 1–184.

39. Gaździcka, E. Nannoplankton stratigraphy of the Miocene deposits in Tarnobrzeg area (northeastern part of the Carpathian Foredeep). *Kwart. Geol.* **1994**, *38*, 553–569.

40. Lis, P.; Wysocka, A. Middle Miocene deposits in Carpathian foredeep: Facies analysis and implications for hydrocarbon reservoir prospecting. *Ann. Soc. Geol. Pol.* **2012**, *82*, 239–253.

41. Reineck, H.; Wunderlich, F. Classification and Origin of Flaser and Lenticular Bedding. *Sedimentology* **1968**, *11*, 99–104. [CrossRef]

42. Terwindt, J.H.J.; Breusers, H.N.C. Experiments on the Origin of Flaser, Lenticular and Sand-Clay Alternating Bedding. *Sedimentology* **1972**, *19*, 85–98. [CrossRef]

43. Martin, A.J. Flaser and wavy bedding in ephemeral streams: A modern and an ancient example. *Sediment. Geol.* **2000**, *136*, 1–5. [CrossRef]

44. Jackson, M.D.; Yoshida, S.; Muggeridge, A.H.; Johnson, H.D. Three-dimensional reservoir characterization and flow simulation of heterolithic tidal sandstones. *Am. Assoc. Pet. Geol. Bull.* **2005**, *89*, 507–528. [CrossRef]

45. Donselaar, C.R.; Geel, M.E. Facies architecture of heterolithic tidal deposits: The Holocene Holland Tidal Basin. *Neth. J. Geosci.* **2007**, *86*, 389–402. [CrossRef]

46. Catuneanu, O.; Galloway, W.E.; Kendall, C.G.S.C.; Miall, A.D.; Posamentier, H.W.; Strasser, A.; Tucker, M.E. Sequence Stratigraphy: Methodology and Nomenclature. *Newsl. Stratigr.* **2011**, 173–245. [CrossRef]

47. Krzywiec, P. Startygrafia sekwencji. *Przegląd Geol.* **1993**, *41*, 681–687.

48. Taner, M.T.; Koehler, F.; Sheriff, R.E. Complex Seismic Trace Analysis. *Geophysics* **1979**, *44*, 1041–1063. [CrossRef]

49. Semb, P.H. Possible seismic hydrocarbon indicators in offshore Cyprus and Lebanon. *GeoArabia* **2009**, *14*, 49–66.

50. Zelenika, K.N.; Mavar, K.N.; Brnada, S. Comparison of the sweetness seismic attribute and porosity–thickness maps, sava depression, Croatia. *Geosciences* **2018**, *8*, 426. [CrossRef]

51. Catuneanu, O. *Principles of Sequence Similarity*; Elsevier: Amsterdam, The Netherlands, 2006.

52. Qayyum, F.; Catuneanu, O.; de Groot, P. Historical developments in Wheeler diagrams and future directions. *Basin Res.* **2015**, *27*, 336–350. [CrossRef]

53. Morse, D.G. Siliciclastic resevoir rocks; the petroleum system—from source to trap. *AAPG Mem.* **1994**, *60*, 121–139.

54. Embry, A.F. Sequence boundaries and sequence hierarchies: Problems and proposals. *Stratigr. Northwest Eur. Margin Spec. Publ.* **1995**, *5*, 1–11.

55. Donaldson, W.S.; Plint, A.G.; Longstaffe, F.J. Tectonic and eustatic control on deposition and preservation of Upper Cretaceous ooidal ironstone and associated facies: Peace River Arch area, NW Alberta, Canada. *Sedimentology* **1999**, *46*, 1159–1182. [CrossRef]

56. Porębski, S.J. Podstawy stratigrafii sekwencji w sukcesjach klastycznych. *Prz. Geol.* **1996**, *44*, 995–1006.

Publisher's Note: MDPI stays neutral with regard to jurisdictional claims in published maps and institutional affiliations.

 *energies*

Article

Application of Waveform Stacking Methods for Seismic Location at Multiple Scales

Lei Li [1,2,*], Yujiang Xie [3] and Jingqiang Tan [1,2]

1   Key Laboratory of Metallogenic Prediction of Nonferrous Metals and Geological Environment Monitoring, Central South University, Changsha 410083, China; tanjingqiang@csu.edu.cn
2   School of Geosciences and Info-Physics, Central South University, Changsha 410083, China
3   Ocean and Earth Science, University of Southampton, Southampton SO14 3ZH, UK; yujiang.xie@soton.ac.uk
*   Correspondence: leili08@csu.edu.cn

Received: 17 June 2020; Accepted: 12 August 2020; Published: 11 September 2020

Abstract: Seismic source location specifies the spatial and temporal coordinates of seismic sources and lays the foundation for advanced seismic monitoring at all scales. In this work, we firstly introduce the principles of diffraction stacking (DS) and cross-correlation stacking (CCS) for seismic location. The DS method utilizes the travel time from the source to receivers, while the CCS method considers the differential travel time from pairwise receivers to the source. Then, applications with three field datasets ranging from small-scale microseismicity to regional-scale induced seismicity are presented to investigate the feasibility, imaging resolution, and location reliability of the two stacking operators. Both of the two methods can focus the source energy by stacking the waveforms of the selected events. Multiscale examples demonstrate that the imaging resolution is not only determined by the inherent property of the stacking operator but also highly dependent on the acquisition geometry. By comparing to location results from other methods, we show that the location bias is consistent with the scale size, as well as the frequency contents of the seismograms and grid spacing values.

Keywords: seismic location; microseismic events; waveform stacking; hydraulic fracturing; induced seismicity

1. Introduction

Seismicity can occur naturally in seismogenic areas or be induced by industrial operations, ranging from laboratory acoustic emission events to large-scale global earthquakes. Under controlled conditions, laboratory and small-scale experiments can reveal the mechanism of fracture initiation and propagation, quantify the changes of reservoir permeability, and evaluate the stiffness of natural fractures or faults [1,2]. At the exploration scale, passive seismic monitoring has been used to delineate fracture propagation, monitor reservoir deformation and fluid migration, and assess seismic risks associated with many subsurface operations, such as underground coal mining, hydraulic stimulation of unconventional oil and gas reservoirs, geothermal exploitation, and carbon dioxide injection and geological storage [3–9]. A complete and accurate earthquake catalogue is an important basis for subsequent data processing and even directly determines the reliability of seismic monitoring and earthquake prediction at the regional or global scale [10]. Seismic source locations are key information of earthquakes and play an important role in characterizing the geometries of multiscale fractures/faults, evaluating seismic activities, and inverting the source mechanism and in situ stress state. For instance, the spatial and temporal distribution of seismic events can help reveal the mechanism and propagation of rock fractures at the laboratory/small scale, as well as provide important information for the assessment of tectonic and volcanic seismicity at local and regional scales. Seismic location as a typical inverse problem, covering geophysical, seismological, acoustic, and engineering applications at multiple scales, has experienced significant methodological and application progress during the past century [11–15].

With the development of modern seismic instrumentation for dense acquisition and induced (micro) seismicity monitoring technology, new challenges and opportunities have emerged for noise-resistant, automatic, and real-time seismic location methods. Specifically, a new category of waveform-based location methods (e.g., waveform stacking methods and time reverse imaging) has emerged as a counterpart of conventional travel time-based inversion [13,16–18]. Waveform-based methods locate the seismic source by combing the travel time, amplitude, and phase information of seismic waveforms or wavefields to reconstruct and focus the source energy into an image profile [13]. There are three important advantages for waveform-based seismic location methods. First, they are noise-resistant, since multichannel waveforms are involved, and the coherence is enhanced to detect and locate more events. Second, the methods are basically automatic and data-driven, which enables a more efficient location process and avoids potential subjective interference, such as phase picking. Third, the source locations are resolved as images instead of simple dots, offering more insights into source processes and surrounding structures. Stacking-based methods are the most mature and successful methods considering their wide applications across scales. The basic principle of stacking-based location methods is reconstructing and focusing the radiated seismic energy from the source with a certain stacking operator, for example, the diffraction stacking (DS) operator [19–21] or the cross-correlation stacking (CCS) operator [22–24]. The origin of stacking-based seismic location methods can be traced back to the 1990s. Kiselevitch et al. (1991) proposed to utilize the maxima of the semblance over space, time, and channels to detect and locate microseismic events, and named the method "seismic emission tomography" [25]. Later on, passive seismic emission tomography was developed to locate microseismic events using recorded waveforms from surface arrays [26]. These methods share the same principle with DS in reflection seismology, that is, stacking the waveforms of individual receivers with the corresponding one-way travel time moveout to enhance the diffracted/scattered seismic energy. CCS is another well-established stacking-based method for source location. CCS exploits correlation waveforms corresponding to differential travel times at pairwise receivers from common events.

Stacking-based methods have been applied to field data at multiple scales, including experimental microseismic/acoustic emission events, mining-induced seismicity, hydraulic-fracturing-induced seismicity, volcanic–tectonic seismicity, and regular earthquakes [27–32]. However, more systematic benchmarking studies of these approaches across scales are still needed. Specifically, the imaging resolution characteristics and location reliability resulting from different frequency contents at different scales need to be studied further. Different from the comprehensive literature review of the methodological and application progress of waveform-based methods in Li et al. (2020) [13], the presented work is a case study and is the first attempt at investigating the performance of two specific stacking-based methods with common field datasets across scales. We aim to investigate the feasibility and reliability of the methods through mutually comparing the imaging results and validating the location results with reference locations, which can provide guidance for further evaluations and improvements of the methods.

2. Methodology

Stacking-based methods have been widely used to detect and locate seismic events at different scales. These methods usually consider primary phases only; thus, they can also be called partial waveform stacking [33]. This category mainly includes the diffraction stacking (DS) and cross-correlation stacking (CCS) methods. Figure 1 shows a sketch of the principles for the two methods. The generalized equation of stacking-based seismic location methods is summarized in the following Equation (1) [13]:

$$S(\mathbf{x}, t_0) = \sum_N CF(t) \cdot M(t) \tag{1}$$

where $S(\mathbf{x}, t_0)$ is the imaging value, $\mathbf{x}$ denotes the source coordinates, t_0 denotes the origin time, and in the waveform data, N is the total number of receivers, $CF(t)$ is the characteristic function (CF), t denotes the time sample, and M is the imaging or migration operators.

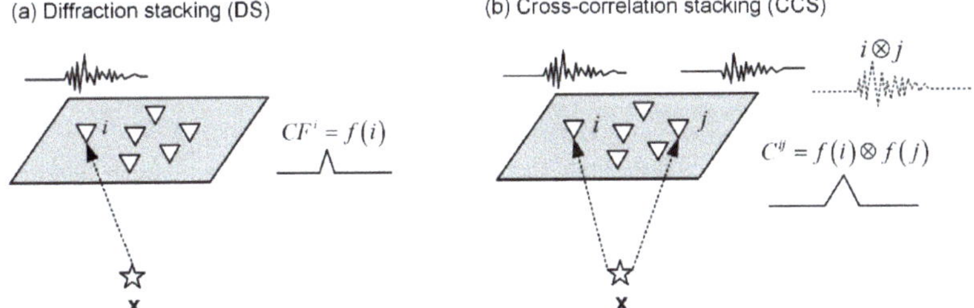

Figure 1. Sketch of the principles for the DS method (**a**) and the CCS method (**b**).

Equations (2) and (3) are the formulae of the DS and CCS methods [30]:

$$S_{DS}(\mathbf{x}, t_0) = \sum_{i=1}^{N} CF^i(t)\delta[t - (t_0 + t_{i,\mathbf{x}})] \tag{2}$$

$$S_{CCS}(\mathbf{x}) = \sum_{\substack{i = 1; \\ j = i+1}}^{N} C^{ij}(t)\delta\left[t - \left(t_{i,\mathbf{x}} - t_{j,\mathbf{x}}\right)\right] \tag{3}$$

where $S_{DS}(\mathbf{x}, t_0)$ and $S_{CCS}(\mathbf{x})$ are the stacking values of the DS and CCS methods, $CF^i(t)$ is the CF of the waveform from receiver i, $C^{ij}(t)$ is the cross-correlation waveforms of the CFs from the pair of receivers i and j, and $\delta[t - (t_0 + t_{i,\mathbf{x}})]$ and $\delta\left[t - \left(t_{i,\mathbf{x}} - t_{j,\mathbf{x}}\right)\right]$ are the imaging operators for the DS and CCS methods, where δ is the Dirac delta function and $t_{i,\mathbf{x}}$ is the theoretical travel time from receiver i to the source $\mathbf{x}$. As shown in the imaging operators, DS and CCS utilize different travel time information and involve different imaging patterns, which resemble spherical surface intersection and hyperboloid intersection in the 3D scenario, respectively [30]. Consequently, the DS method generally exhibits a higher vertical imaging resolution than CCS for surface monitoring. The basic workflow of stacking-based methods includes signal pre-processing (e.g., band-pass filtering and CF calculation), model discretization, travel time table generation, waveform stacking, source location picking, and post-processing (e.g., enhancing the imaging resolution using a probability density distribution or Gaussian filtering).

As indicated in the above equations, DS searches both the spatial location and the origin time, while CCS decouples the origin time by making use of the cross-correlograms of the waveforms from pairwise receivers. Elimination of the origin time makes the traditional 4D source imaging problem a 3D problem for seismic location in a 3D scenario and alleviates the inherent origin time-depth trade-off in seismic source location [27,34]. Generally, the origin time is a minor parameter and can be calculated posteriorly by subtracting the theoretical travel time from the observed arrival time with respect to the same event. In this application study, we estimate the performance of stacking-based methods by only evaluating the reliability of spatial source parameters. There are numerous CFs being studied and utilized, including waveform envelope, semblance, kurtosis, and short-term average to long-term average ratio (STA/LTA) [28,32,34–38]. The STA/LTA is used in this study, and different lengths of short- and long-term windows are adopted for waveforms recorded at different scales (see Table 1 for more details). The above stacking-based location function can be naturally treated as a global optimization problem, in which the spatial and temporal coordinates corresponding to the maximum imaging value, max$\{S_{DS}(\mathbf{x}, t_0)\}$ or max$\{S_{CCS}(\mathbf{x})\}$, are generally selected as the inverted source location [39]. When multicomponent data are available, one can consider incorporating information of both compressional (P-) and shear (S-) waves to enhance the constraints. However, the complex seismic phases (e.g., uneven distributions of S-wave energy due to complex source mechanisms and the acquisition geometry) and unreliable multiple velocities can potentially deteriorate the location results by introducing additional interference instead of constructive constraints [30]. Therefore, one should be careful with multiple phases, and the waveform stacking of a single phase (i.e., P-wave) is considered here.

Table 1. Basic parameters of the field datasets for stacking-based location.

Parameters	Small Scale	Exploration Scale	Regional Scale
Receiver channel	32×3	15×3	approx. 1800×1
Number of events	186	200	2
Sampling rate	1 MHz	200 Hz	500 Hz
Target volume	30 m × 35 m× 20 m	5 km × 5 km× 5 km	35.2 km × 45.2 km × 8 km
Velocity model	homogeneous isotropic	homogeneous isotropic	layered isotropic
Grid spacing	1 m	50 m	400 m
Band-pass filter	1–50 kHz	5–30 Hz	/
Length of short-/long-term window	80/160	25/50	25/50

3. Multiscale Field Data Examples

In this section, we investigate and compare the performance of the two stacking operators based on three field datasets associated with induced seismicity at different scales. Table 1 lists the basic parameters for the seismic source location of these examples. More detailed descriptions of the datasets and the associated projects can be found in the following subsections and related references.

3.1. The Small-Scale Example

Small-scale hydraulic fracturing experiments can build a bridge between experimental study and field operations, improve the understanding of the role of reservoir stimulation, and enhance the applicability of results from laboratory experiments. An in situ stimulation and circulation experiment at the Grimsel Test Site (GTS) was conducted in 2017 [40,41]. A series of small volumes of water (up to 1 m^3) were injected into pre-existing faults to induce rock fracturing, which was accompanied by abundant microseismic events. A 32-channel acquisition system was used to record microseismic signals with a 1 MHz sampling rate. Here we tested the two stacking operators on field microseismic events recorded during fracturing tests in SBH-3. The homogeneous model with a P-wave velocity of 5150 m/s and band-pass filtered (1–50 kHz) waveforms from vertical components were used to locate these events. The raw waveforms of two sample events and the imaging results are shown in Figure 2. The two-dimensional (2D) imaging results are three orthogonal slices intersecting at the location of the maximum imaging value.

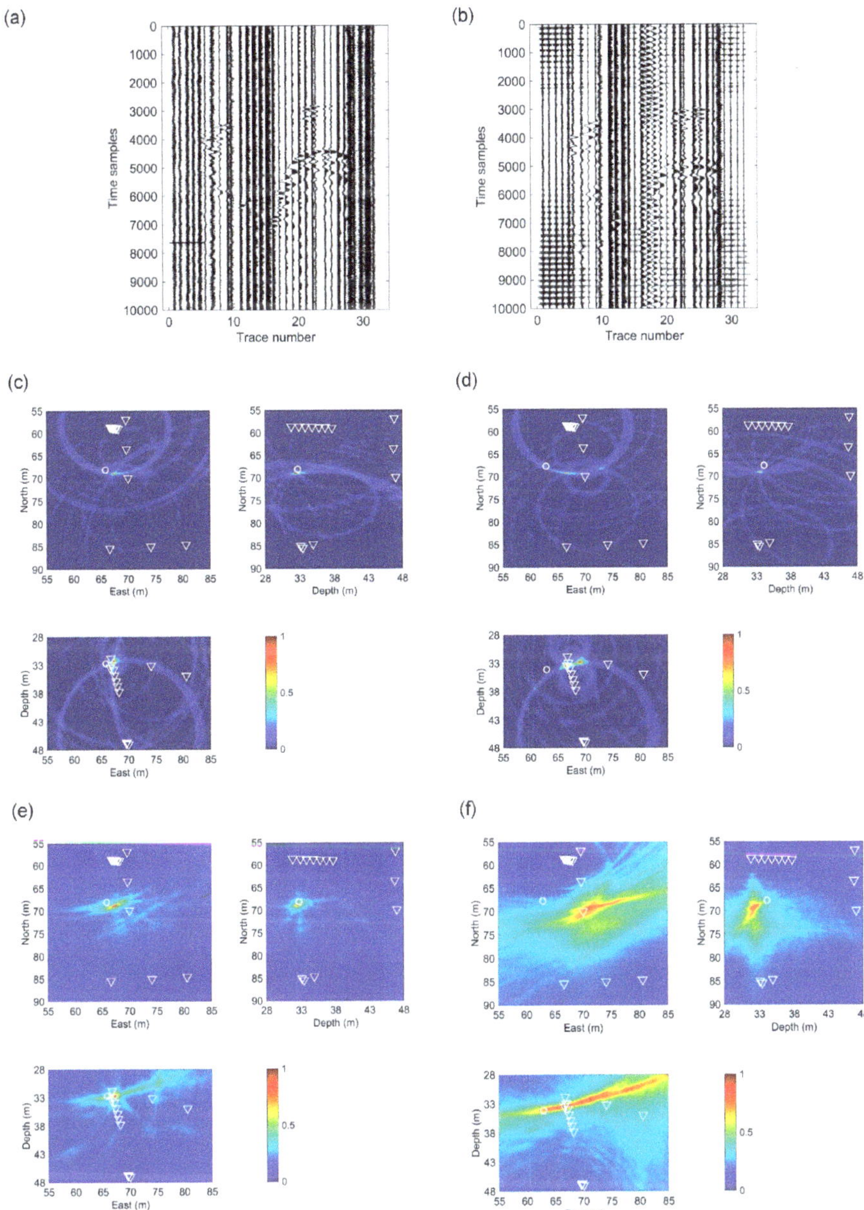

Figure 2. Waveforms and imaging results of two sample events. (**a**,**b**) Raw waveforms of the vertical components of two sample events; (**c**,**d**) 2D slices from the DS method for the events shown in (**a**,**b**); (**e**,**f**) corresponding results from the CCS method. White reversed triangles denote sensors, and white circles denote the reference locations.

Although the source energy of most events is well focused, the imaging resolutions of the two waveform stacking methods for many events are low and show obvious band-like artifacts in the east direction due to the irregular and limited coverage of the sensors. The white reversed triangles in Figures 2–4 denote the sensor locations, which mainly cover the north and depth directions and overlap in the east direction. Resultantly, the source energy is focused well in the depth–north profile, but there are strong artifacts along the east direction, indicating large uncertainty of the locations (Figure 2f) and leading to large horizontal bias of the location results when compared with those by joint hypocenter determination in a previous study [41] (Figure 3a,c and Figure 4a,c). Table 2 lists the detailed location biases of these examples. The sensor network here resembles a joint surface and downhole monitoring array, which provides good constraints for the depth and yields an average depth bias smaller than 2 m (see Figures 3b and 4b). Other factors contributing to the location uncertainty include the lack of sensor calibration and the simplified homogeneous and isotropic velocity model, which involve arrival time and travel time errors.

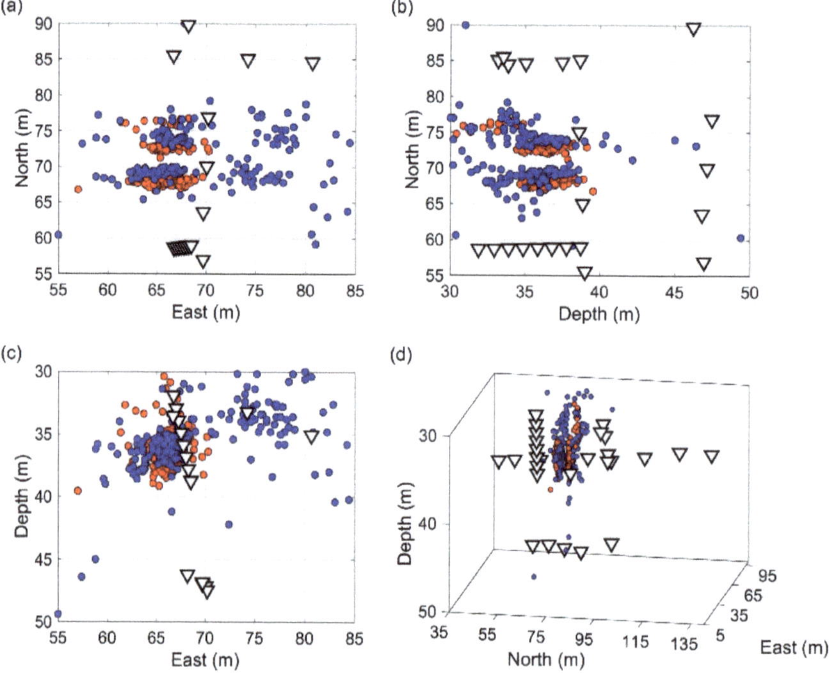

Figure 3. Location results of the DS method for the selected events. (**a–c**) 2D slices; (**d**) 3D view. Reversed triangles denote sensors, red dots are the reference locations, and blue dots are results from the DS method.

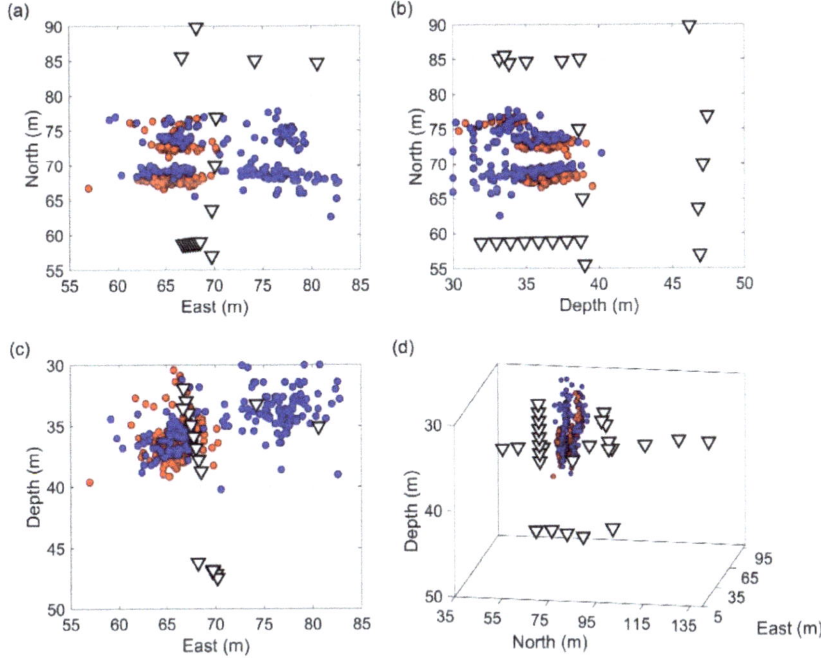

Figure 4. Location results of the CCS method for the selected events. (**a–c**) 2D slices; (**d**) 3D view. Reversed triangles denote sensors, red dots are the reference locations, and blue dots are results from the CCS method.

Table 2. Average location biases of the two stacking-based methods at multiple scales.

	Location Bias	Small Scale	Exploration Scale	Regional Scale
	E-direction (m)	5.056	94.85	729.5
	N-direction (m)	1.854	163.64	182.5
DS	Z-direction (m)	1.689	172.68	1675.5
	Overall (m)	6.156	281.59	1909.4
	E-direction (m)	6.513	72.20	156.5
	N-direction (m)	1.268	165.51	560.5
CCS	Z-direction (m)	1.799	122.11	1675.5
	Overall (m)	7.154	236.10	1782.8

3.2. The Exploration-Scale Example

Seismicity related to underground mining has been observed since the beginning of the 20th century [42]. Seismic activities are closely related to the safety of mining operations, and rock bursts are often the main cause of mining accidents [43,44]. Passive seismic/microseismic monitoring can effectively detect and evaluate the seismic activity surrounding underground mines and provide early warning of potential geological risks [3]. From June 2006 to July 2007, a temporary network (HAMNET) was set up to monitor mining-induced seismicity in the Ruhr area, Germany [45]. The network consisted of 15 three-component surface stations, covering an area of about 3 km × 2 km. The depth of the longwall was about 1 km. We selected 200 events from the dataset to test the two stacking operators, and only P-waves in the vertical components were used in the location process [30]. The sampling time was 5 ms and the raw waveforms were preprocessed by a band-pass filter (5–30 Hz). We set the target location volume as 5 km × 5 km× 5 km with grid spacing of 50 m. The imaging and location results are shown in Figures 5 and 6.

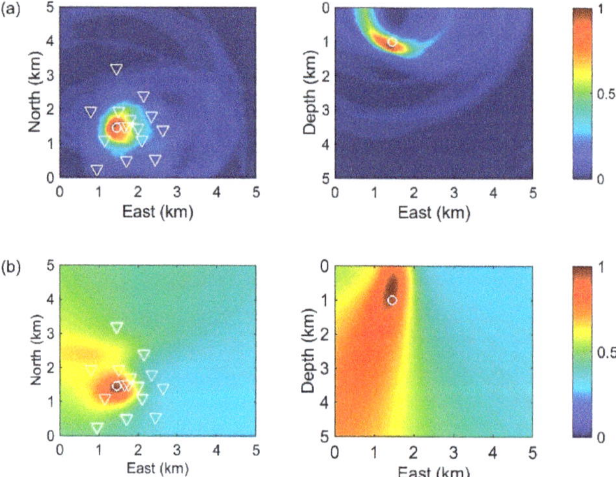

Figure 5. Imaging results of a sample event. (**a**) 2D slices from the DS method; (**b**) corresponding results of the CCS method. White reversed triangles denote sensors, and white circles denote the reference locations. The reference point (east, north) = (0, 0) corresponds to east = 411,117 and north = 5,721,611 in the Universal Transverse Mercator (UTM) coordinate system.

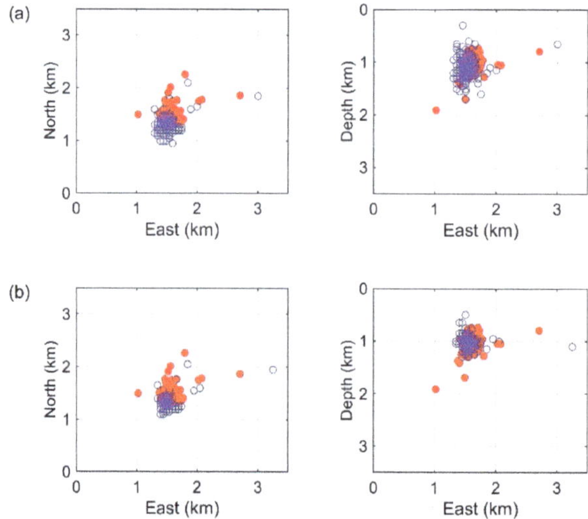

Figure 6. Location results of selected events. (**a**) Comparison between reference locations and results of the DS method; (**b**) comparison between reference locations and results of the CCS method. Red dots are the reference locations, and blue circles are results from stacking-based methods.

As shown in Figure 5, the horizontal resolutions of the imaging results for both methods are good due to the good horizontal coverage of the 15 surface stations. As mentioned before, the vertical imaging resolution of DS is higher than that of CCS in the case of surface monitoring. Figure 6 shows that the overall distributions of the selected microseismic events determined by the two methods are in good agreement with that of travel time inversion, and the CCS method has better event clustering and smaller location bias (see Table 2) than the DS method. Please note that the imaging resolution addresses the quality of the imaging profile and the location uncertainty describes the stability of the location results. The two parameters are not necessarily consistent, though they are closely related to each other. In this case, the CCS method produces more reliable and clustered location results, though it has lower imaging resolution.

3.3. The Regional-Scale Example

Induced seismicity monitoring is of great significance to both industrial production and public safety. During the past decade, several large-magnitude earthquakes related to hydraulic fracturing in unconventional reservoirs have been reported in the United States, United Kingdom, Canada, and China [7,9,46]. In particular, the number of seismic events observed in Oklahoma, the United States has increased significantly since 2008. Besides this, seismic activity related to the development of Mississippi limestone reservoirs in central and northern Oklahoma and southern Kansas has been occurring. In 2016, more than 1800 vertical component nodal seismometers, named the LArge-*n* Seismic Survey in Oklahoma (LASSO) array, were deployed in Grant County, Oklahoma, to study induced seismicity associated with production of the Mississippi limestone play [47]. The LASSO array covered an area of 25 km × 32 km. The sampling rate was 500 Hz and the grid spacing was set to 400 m (referring to the location error in the catalogue). We selected two events from the catalogue (event no. 23196, local magnitude (ML) 3 and event no. 23897, ML 1.5) and used a layered isotropic velocity model to image the sources [48]. The layout of the LASSO array and partial waveforms of the two sample events are shown in Figure 7. There were 1825 and 445 effective traces for event 23,196 (ML = 3) and event 23,897 (ML = 1.5), respectively. Our tests showed that waveforms from much fewer traces can focus the source energy. Figure 8 shows the imaging results using 10-times fewer traces.

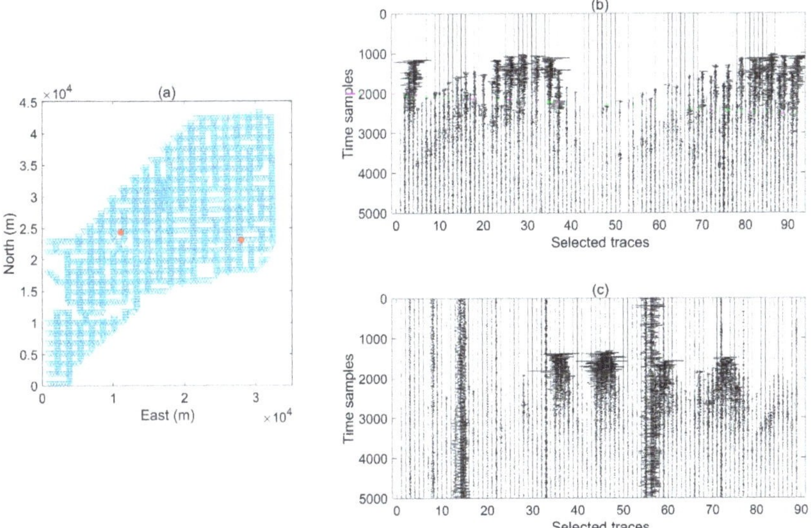

Figure 7. The LArge-*n* Seismic Survey in Oklahoma (LASSO) array (**a**) and raw waveforms of the two events (**b,c**). The blue reversed triangles are seismometers, and the red dots are the locations of the two events from the catalogue. The reference point (east, north) = (0, 0) corresponds to east = 579,000 and north = 4,051,000 in the UTM coordinate system. (**b,c**) are waveforms of event 23,196 (ML = 3) and event 23,897 (ML = 1.5).

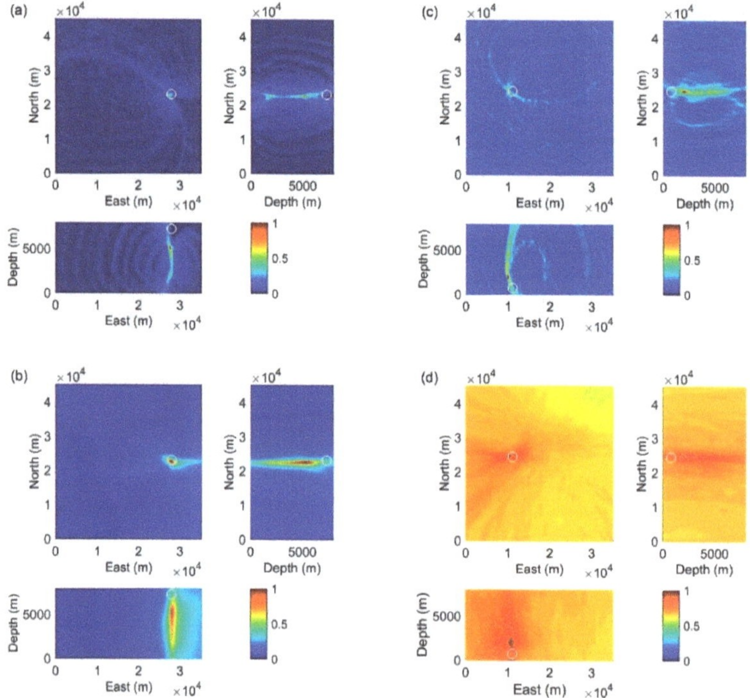

Figure 8. Imaging results of two sample events using 10-times fewer traces (183 and 45 traces in total for event 23,196 and event 23,897, respectively). (**a,c**) 2D slices from the DS method for the events shown in Figure 7b,c; (**b,d**) corresponding results from the CCS method. White circles denote the reference locations from the catalogue.

The horizontal imaging resolutions of the two methods are very high, while the vertical resolutions are much lower. The inverted depths from the two methods are quite consistent. The location bias between the stacking-based methods and the catalogue is within 2 km (see Table 2), most of which comes from uncertainty of the depth, and is acceptable considering the large scale of the monitoring array and the target volume. Although most seismic events were recorded by more than 100 or even 1000 seismometers in the LASSO array, the imaging results in Figure 8c,d indicate that several dozens of traces are sufficient to focus the source energy of these induced earthquakes, as long as a good spatial coverage of the selected traces can be ensured.

3.4. Comparison of the Results Across Scales

Compared with the DS method, the CCS method makes use of more waveform redundancy and includes more constraints for the source imaging process, which is beneficial for enhancing the signals and suppressing the artifacts. However, CCS also introduces more potential noise energy and interference information, which naturally reduces the imaging resolution (see Figures 2, 5 and 8). The results also demonstrate that the imaging resolution is highly dependent on the acquisition geometry. For the small-scale example, the sensors surround the stimulation site and ensure a relatively good imaging resolution for both the horizontal and depth directions (Figure 2c–e), though partial events have a large horizontal bias due to the limited coverage of sensors in the east direction (Figure 2f). For the exploration-scale example, the horizontal coverage of the network is two-times the target depth, which can produce reliable depth locations (Figure 6) [49]. For the regional-scale example, the vertical imaging resolutions of the two sample events are poor, due to the events being located near the margin of the array (Figures 7 and 8).

Table 2 shows the average location bias for the three examples, which was calculated by dividing the total absolute bias between the location results of stacking-based methods and the reference locations by the number of selected events. Figure 9 shows histograms of the overall location biases for all selected events in the small-scale and exploration-scale examples. The comparison of the results across the scales clearly shows that the location bias is consistent with the scale size, which naturally results from the different frequency contents of the seismic waveforms and grid spacing values (see Table 1) at different scales.

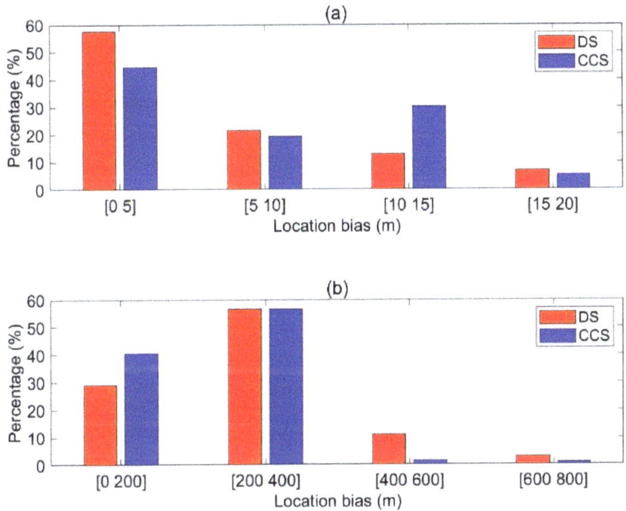

Figure 9. Histograms of the overall location biases of the DS and CCS methods for the small-scale (**a**) and exploration-scale (**b**) examples.

4. Discussion and Conclusions

In this work, we investigated the feasibility and potential of stacking-based methods using production-induced seismicity at different scales. The examples revealed the characteristics of the imaging resolution for the two stacking operators and demonstrated their feasibility in locating seismic events at multiple scales. A comparison of the location results across the scales indicated that the imaging resolution is highly dependent on the acquisition geometry, and the location bias is closely related with the scale. Although the reference locations determined by other methods or from the catalogue may not be reliable, the comparison can still indicate several potential factors that are responsible for the accuracy and reliability of the methods—that is, the layout of the monitoring network, the reliability of the velocity model, and the quality of the waveforms. The methods are noise-resistant and automatic, which means that weak events with low signal-to-noise ratio can be located and no phase picking procedure is needed.

So far, field applications of stacking-based methods have mainly focused on the exploration scale—e.g., for hydraulic fracturing monitoring with large and dense arrays in unconventional oil and gas reservoirs [27,34]. More recently, these methods have been naturally adapted to analyze low-frequency seismic events at larger scales [28,50], where a large depth uncertainty exists. There are also emerging applications for seismic events with comparably high-frequency content at smaller scales—e.g., rock burst and acoustic emission, where the issues of location uncertainty and velocity reliability are matters of considerable concern [32,51]. The events tested in this study have relatively high signal-to-noise ratio, and waveforms recorded with dozens of receivers can recover the source energy quite well. Further work will consider weaker seismic events associated with hydraulic fracturing in unconventional and geothermal reservoirs and foreshock/aftershock activity potentially preceding/following tectonic earthquakes. It is worth noting that stacking-based methods are more resistant to white noise, and one should use an advanced filtering technique to address the influence of correlated noise from, for example, vicinity to hydraulic fracturing wells [51]. For events with small magnitudes and/or low signal-to-noise ratios, wavefield reconstruction and enhancement may be required to ensure a reliable source location when using waveform stacking techniques, especially when partial traces are noisy or disabled. Another challenge for stacking-based location methods is the relatively high computation cost due to waveform storage and transmission, especially for real-time applications with large monitoring arrays. High-performance computing facilities [52,53] and advanced inversion algorithms [35,39] are promising solutions for improving the computational efficiency.

The current study is intended to be a starter for more in-depth and comprehensive investigations of stacking-based methods across scales, and the conclusions from this study still remain at a relatively qualitative level. Specifically, the stacking operators should be studied in a more systematic manner by adjusting the characteristic functions and pre-filtering techniques of the waveforms, acquisition geometries/channels, and velocity models. Although there have been several attempts at comparing the performance of different characteristic functions [17,18], field applications at multiple scales are more data- and problem-dependent. Meanwhile, novel and potential stacking operators producing better seismic energy focusing are also worth investigating. Further benchmarking studies are required to improve the applicability of stacking-based methods and build a suitable workflow for seismic location at multiple scales.

Author Contributions: Conceptualization, L.L., Y.X. and J.T.; methodology, L.L. and Y.X.; software, L.L.; validation, L.L.; writing—original draft preparation, L.L.; writing—review and editing, Y.X. and J.T.; supervision, J.T.; funding acquisition, L.L. All authors have read and agreed to the published version of the manuscript.

Funding: This research was funded by Natural Science Foundation of Hunan Province, grant number 2019JJ50762, the Open Research Fund Program of State Key Laboratory of Acoustics, Chinese Academy of Sciences, grant number SKLA201911, and China Postdoctoral Science Foundation, grant number 2019M652803. The work of Y.X. is funded by the Natural Environment Research Council, grant numbers NE/M003507/1 and NE/K010654/1, and the European Research Council, grant number GA 638665.

Acknowledgments: We acknowledge the open source datasets from the GTS experiment, the HAMNET network, and the LASSO array.

Conflicts of Interest: The authors declare no conflict of interest.

References

1. Lockner, D.; Byerlee, J.D.; Kuksenko, V.; Ponomarev, A.; Sidorin, A. Quasi-static fault growth and shear fracture energy in granite. *Nature* **1991**, *350*, 39–42. [CrossRef]
2. Rutqvist, J. Determination of hydraulic normal stiffness of fractures in hard rock from well testing. *Int. J. Rock. Mech. Min. Geomech. Abstr.* **1995**, *32*, 513–523. [CrossRef]
3. Gibowicz, S.J.; Kijko, A. *An Introduction to Mining Seismology*; Academic Press: San Diego, CA, USA, 1994; ISBN 978-0-12-282120-2.
4. Mendecki, A.J. (Ed.) *Seismic Monitoring in Mines*, 1st ed.; Chapman & Hall: London, UK, 1997; ISBN 978-0-412-75300-8.
5. Luo, X.; Hatherly, P. Application of microseismic monitoring to characterise geomechanical conditions in longwall mining. *Explor. Geophys.* **1998**, *29*, 489–493. [CrossRef]
6. Maxwell, S.C. *Microseismic Imaging of Hydraulic Fracturing*; Distinguished Instructor Series; Society of Exploration Geophysicists: Tulsa, OK, USA, 2014; ISBN 978-1-56080-315-7.
7. Grigoli, F.; Cesca, S.; Priolo, E.; Rinaldi, A.P.; Clinton, J.F.; Stabile, T.A.; Dost, B.; Fernandez, M.G.; Wiemer, S.; Dahm, T. Current challenges in monitoring, discrimination, and management of induced seismicity related to underground industrial activities: A European perspective. *Rev. Geophys.* **2017**, *55*, 310–340. [CrossRef]

8. Eaton, D.W. *Passive Seismic Monitoring of Induced Seismicity: Fundamental Principles and Application to Energy Technologies*, 1st ed.; Cambridge University Press: Cambridge, UK, 2018; ISBN 978-1-107-14525-2.

9. Li, L.; Tan, J.; Wood, D.A.; Zhao, Z.; Becker, D.; Lyu, Q.; Shu, B.; Chen, H. A review of the current status of induced seismicity monitoring for hydraulic fracturing in unconventional tight oil and gas reservoirs. *Fuel* **2019**, *242*, 195–210. [CrossRef]

10. Shearer, P.M. *Introduction to Seismology*, 2nd ed.; Cambridge University Press: Cambridge, UK, 2009; ISBN 978-0-521-70842-5.

11. Thurber, C.H.; Rabinowitz, N. *Advances in Seismic Event Location*; Springer: Dordrecht, The Netherlands, 2000; ISBN 978-90-481-5498-2.

12. Lomax, A.; Michelini, A.; Curtis, A. Earthquake Location, Direct, Global-Search Methods. In *Encyclopedia of Complexity and Systems Science*; Meyers, R.A., Ed.; Springer: New York, NY, USA, 2009; pp. 2449–2473, ISBN 978-0-387-75888-6.

13. Li, L.; Tan, J.; Schwarz, B.; Staněk, F.; Poiata, N.; Shi, P.; Diekmann, L.; Eisner, L.; Gajewski, D. Recent advances and challenges of waveform-based seismic location methods at multiple scales. *Rev. Geophys.* **2020**, *58*, e2019RG000667. [CrossRef]

14. Dong, L.; Hu, Q.; Tong, X.; Liu, Y. Velocity-Free MS/AE Source Location Method for Three-Dimensional Hole-Containing Structures. *Engineering* **2020**, in press. [CrossRef]

15. Peng, P.; Jiang, Y.; Wang, L.; He, Z. Microseismic Event Location by Considering the Influence of the Empty Area in an Excavated Tunnel. *Sensors* **2020**, *20*, 574. [CrossRef]

16. Cesca, S.; Grigoli, F. Chapter two-full waveform seismological advances for microseismic monitoring. *Adv. Geophys.* **2015**, *56*, 169–228.

17. Trojanowski, J.; Eisner, L. Comparison of migration-based location and detection methods for microseismic events. *Geophys. Prospect.* **2017**, *65*, 47–63. [CrossRef]

18. Beskardes, G.D.; Hole, J.A.; Wang, K.; Michaelides, M.; Wu, Q.; Chapman, M.C.; Davenport, K.K.; Brown, L.D.; Quiros, D.A. A comparison of earthquake backprojection imaging methods for dense local arrays. *Geophys. J. Int.* **2018**, *212*, 1986–2002. [CrossRef]

19. Kao, H.; Shan, S.J. The source-scanning algorithm: Mapping the distribution of seismic sources in time and space. *Geophys. J. Int.* **2004**, *157*, 589–594. [CrossRef]

20. Baker, T.; Granat, R.; Clayton, R.W. Real-time earthquake location using Kirchhoff reconstruction. *Bull. Seismol. Soc. Am.* **2005**, *95*, 699–707. [CrossRef]

21. Gajewski, D.; Anikiev, D.; Kashtan, B.; Tessmer, E. Localization of seismic events by diffraction stacking. In Proceedings of the SEG Technical Program Expanded Abstracts, San Antonio, TX, USA, 23–28 September 2007; Society of Exploration Geophysicists: Tulsa, OK, USA, 2007; pp. 1287–1291.

22. Schuster, G.T.; Yu, J.; Sheng, J. Interferometric/daylight seismic imaging. *Geophys. J. Int.* **2004**, *157*, 838–852. [CrossRef]

23. Xiao, X.; Luo, Y.; Fu, Q.; Jervis, M.; Dasgupta, S.; Kelamis, P. Locate microseismicity by seismic interferometry. In Proceedings of the Second EAGE Passive Seismic Workshop-Exploration and Monitoring Applications, Limassol, Cyprus, 22–25 March 2009; The European Association of Geoscientists and Engineers: Houten, The Netherlands, 2009; p. A22.

24. Li, L.; Chen, H.; Wang, X.-M. Weighted-elastic-wave interferometric imaging of microseismic source location. *Appl. Geophy.* **2015**, *12*, 221–234. [CrossRef]

25. Kiselevitch, V.L.; Nikolaev, A.V.; Troitskiy, P.A.; Shubik, B.M. Emission tomography: Main ideas, results, and prospects. In Proceedings of the SEG Technical Program Expanded Abstracts, Houston, TX, USA, 10–14 September 1991; Society of Exploration Geophysicists: Tulsa, OK, USA, 1991; p. 1602.

26. Duncan, P.M. Is there a future for passive seismic? *First Break* **2005**, *23*, 111–115.

27. Anikiev, D.; Valenta, J.; Staněk, F.; Eisner, L. Joint location and source mechanism inversion of microseismic events: Benchmarking on seismicity induced by hydraulic fracturing. *Geophys. J. Int.* **2014**, *198*, 249–258. [CrossRef]

28. Poiata, N.; Satriano, C.; Vilotte, J.-P.; Bernard, P.; Obara, K. Multiband array detection and location of seismic sources recorded by dense seismic networks. *Geophys. J. Int.* **2016**, *205*, 1548–1573. [CrossRef]

29. Grigoli, F.; Cesca, S.; Vassallo, M.; Dahm, T. Automated Seismic Event Location by Travel-Time Stacking: An Application to Mining Induced Seismicity. *Seismol. Res. Lett.* **2013**, *84*, 666–677. [CrossRef]

30. Li, L.; Becker, D.; Chen, H.; Wang, X.; Gajewski, D. A systematic analysis of correlation-based seismic location methods. *Geophys. J. Int.* **2018**, *212*, 659–678. [CrossRef]

31. Shi, P.; Nowacki, A.; Rost, S.; Angus, D. Automated seismic waveform location using Multichannel Coherency Migration (MCM)—II. Application to induced and volcano-tectonic seismicity. *Geophys. J. Int.* **2019**, *216*, 1608–1632. [CrossRef]

32. Zhang, C.; Qiao, W.; Che, X.; Lu, J.; Men, B. Automated microseismic event location by amplitude stacking and semblance. *Geophysics* **2019**, *84*, KS191–KS210. [CrossRef]

33. Pesicek, J.D.; Child, D.; Artman, B.; Cieślik, K. Picking versus stacking in a modern microearthquake location: Comparison of results from a surface passive seismic monitoring array in Oklahoma. *Geophysics* **2014**, *79*, KS61–KS68. [CrossRef]

34. Staněk, F.; Anikiev, D.; Valenta, J.; Eisner, L. Semblance for microseismic event detection. *Geophys. J. Int.* **2015**, *201*, 1362–1369. [CrossRef]

35. Gharti, H.N.; Oye, V.; Roth, M.; Kühn, D. Automated microearthquake location using envelope stacking and robust global optimization. *Geophysics* **2010**, *75*, MA27–MA46. [CrossRef]

36. Zhang, W.; Zhang, J. Microseismic migration by semblance-weighted stacking and interferometry. In Proceedings of the SEG Technical Program Expanded Abstracts, Houston, TX, USA, 22–27 September 2013; Society of Exploration Geophysicists: Tulsa, OK, USA, 2013; pp. 2045–2049.

37. Grigoli, F.; Scarabello, L.; Böse, M.; Weber, B.; Wiemer, S.; Clinton, J.F. Pick-and waveform-based techniques for real-time detection of induced seismicity. *Geophys. J. Int.* **2018**, *213*, 868–884. [CrossRef]

38. Langet, N.; Maggi, A.; Michelini, A.; Brenguier, F. Continuous Kurtosis-Based Migration for Seismic Event Detection and Location, with Application to Piton de la Fournaise Volcano, La Reunion. *Bull. Seismol. Soc. Am.* **2014**, *104*, 229–246. [CrossRef]

39. Li, L.; Tan, J.; Xie, Y.; Tan, Y.; Walda, J.; Zhao, Z.; Gajewski, D. Waveform-based microseismic location using stochastic optimization algorithms: A parameter tuning workflow. *Comput. Geosci.* **2019**, *124*, 115–127. [CrossRef]

40. Amann, F.; Gischig, V.; Evans, K.; Doetsch, J.; Jalali, R.; Valley, B.; Krietsch, H.; Dutler, N.; Villiger, L.; Brixel, B.; et al. The seismo-hydromechanical behavior during deep geothermal reservoir stimulations: Open questions tackled in a decameter-scale in situ stimulation experiment. *Solid Earth* **2018**, *9*, 115–137. [CrossRef]

41. Gischig, V.S.; Doetsch, J.; Maurer, H.; Krietsch, H.; Amann, F.; Evans, K.F.; Nejati, M.; Jalali, M.; Valley, B.; Obermann, A.C.; et al. On the link between stress field and small-scale hydraulic fracture growth in anisotropic rock derived from microseismicity. *Solid Earth* **2018**, *9*, 39–61. [CrossRef]

42. Cook, N.G.W. Seismicity associated with mining. *Eng. Geol.* **1976**, *10*, 99–122. [CrossRef]

43. Zhang, H.; Chen, L.; Chen, S.; Sun, J.; Yang, J. The Spatiotemporal Distribution Law of Microseismic Events and Rockburst Characteristics of the Deeply Buried Tunnel Group. *Energies* **2018**, *11*, 3257. [CrossRef]

44. Feng, G.; Xia, G.; Chen, B.; Xiao, Y.; Zhou, R. A Method for Rockburst Prediction in the Deep Tunnels of Hydropower Stations Based on the Monitored Microseismicity and an Optimized Probabilistic Neural Network Model. *Sustainability* **2019**, *11*, 3212. [CrossRef]

45. Bischoff, S.T.; Fischer, L.; Wehling-Benatelli, S.; Fritschen, R.; Meier, T.; Friederich, W. *Spatio-Temporal Characteristics of Mining Induced Seismicity in the Eastern Ruhr-Area*; The European Center for Geodynamics and Seismology (ECGS): Walferdange, Luxembourg, 2010.

46. Ellsworth, W.L. Injection-induced earthquakes. *Science* **2013**, *341*, 1225942. [CrossRef]

47. Dougherty, S.L.; Cochran, E.S.; Harrington, R.M. The LArge-n Seismic Survey in Oklahoma (LASSO) Experiment. *Seismol. Res. Lett.* **2019**, *90*, 2051–2057. [CrossRef]

48. Rubinstein, J.L.; Ellsworth, W.L.; Dougherty, S.L. The 2013–2016 Induced Earthquakes in Harper and Sumner Counties, Southern Kansas. *Bull. Seismol. Soc. Am.* **2018**, *108*, 674–689. [CrossRef]

49. Eisner, L.; Duncan, P.M.; Heigl, W.M.; Keller, W.R. Uncertainties in passive seismic monitoring. *Leading Edge* **2009**, *28*, 648–655. [CrossRef]

50. Ruigrok, E.; Gibbons, S.; Wapenaar, K. Cross-correlation beamforming. *J. Seismol.* **2017**, *21*, 495–508. [CrossRef]

51. López-Comino, J.A.; Cesca, S.; Heimann, S.; Grigoli, F.; Milkereit, C.; Dahm, T.; Zang, A. Characterization of hydraulic fractures growth during the Äspö Hard Rock laboratory experiment (Sweden). *Rock Mech. Rock Eng.* **2017**, *50*, 2985–3001. [CrossRef]

52.	Guidarelli, M.; Klin, P.; Priolo, E. Migration-based near real-time detection and location of microearthquakes with parallel computing. *Geophys. J. Int.* **2020**, *221*, 1941–1958. [CrossRef]
53.	Lee, E.-J.; Liao, W.-Y.; Mu, D.; Wang, W.; Chen, P. GPU-Accelerated Automatic Microseismic Monitoring Algorithm (GAMMA) and Its Application to the 2019 Ridgecrest Earthquake Sequence. *Seismol. Res. Lett.* **2020**, in press. [CrossRef]

MDPI
St. Alban-Anlage 66
4052 Basel
Switzerland
Tel. +41 61 683 77 34
Fax +41 61 302 89 18
www.mdpi.com

Energies Editorial Office
E-mail: energies@mdpi.com
www.mdpi.com/journal/energies